Xpg 228

IN CASE OF ACCIDENT[1]

In case of accident notify the laboratory instructor ***immediately.***

FIRE

Burning Clothing. Prevent the person from running and fanning the flames. Rolling the person on the floor will help extinguish the flames and prevent inhalation of the flames. If a safety shower is nearby hold the person under the shower until flames are extinguished and chemicals washed away. Do not use a fire blanket if a shower is nearby. The blanket does not cool and smoldering continues. Remove contaminated clothing. Wrap the person in a blanket to avoid shock. Get prompt medical attention. Do not, under any circumstances, use a carbon tetrachloride (toxic) fire extinguisher and be very careful using a CO_2 extinguisher (the person may smother).

Burning Reagents. Extinguish all nearby burners and remove combustible material and solvents. Small fires in flasks and beakers can be extinguished by covering the container with a fiberglass-wire gauze square, a big beaker, or a watch glass. Use a dry chemical or carbon dioxide fire extinguisher directed at the base of the flames. **Do not use water.**

Burns, Either Thermal or Chemical. Flush the burned area with cold water for at least 15 min. Resume if pain returns. Wash off chemicals with a mild detergent and water. Current practice recommends that no neutralizing chemicals, unguents, creams, lotions, or salves be applied. If chemicals are spilled on a person over a large area quickly remove the contaminated clothing while under the safety shower. Seconds count, and time should not be wasted because of modesty. Get prompt medical attention.

CHEMICALS IN THE EYE: Flush the eye with copious amounts of water for 15 min using an eyewash fountain or bottle or by placing the injured person face up on the floor and pouring water in the open eye. Hold the eye open to wash behind the eyelids. After 15 min of washing obtain prompt medical attention, regardless of the severity of the injury.

CUTS: Minor Cuts. This type of cut is most common in the organic laboratory and usually arises from broken glass. Wash the cut, remove any pieces of glass, and apply pressure to stop the bleeding. Get medical attention.

Major Cuts. If blood is spurting place a pad directly on the wound, apply firm pressure, wrap the injured to avoid shock, and get **immediate** medical attention. Never use a tourniquet.

POISONS: Call 800 information (1-800-555-1212) for the telephone number of the nearest Poison Control Center, which is usually also an 800 number.

[1]Adapted from American Chemical Society Joint Board-Council Committee on Chemical Safety. *Safety in Academic Chemistry Laboratories, Vol. 1: Accident Prevention for College and University Students, 7th ed.; American Chemical Society: Washington, DC, 2003 (0-8412-3864-2).*

Macroscale and Microscale Organic Experiments

Macroscale and Microscale Organic Experiments

SIXTH EDITION

Kenneth L. Williamson
Mount Holyoke College, Emeritus

Katherine M. Masters
Pennsylvania State University

Australia • Brazil • Japan • Korea • Mexico • Singapore • Spain • United Kingdom • United States

BROOKS/COLE
CENGAGE Learning™

***Macroscale and Microscale Organic Experiments*, Sixth Edition**
Kenneth L. Williamson, Katherine M. Masters

Publisher: Charles Hartford
Senior Development Editor: Sandra Kiselica
Editorial Assistant: Jon Olafsson
Associate Media Editor: Stephanie Van Camp
Senior Marketing Manager: Nicole Hamm
Marketing Assistant: Kevin Carroll
Senior Marketing Communications Manager: Linda Yip
Content Project Management: Pre-Press PMG
Creative Director: Rob Hugel
Art Director: John Walker
Print Buyer: Judy Inouye
Rights Acquisitions Account Manager, Text: Roberta Broyer
Rights Acquisitions Account Manager, Image: Don Schlotman
Production Service: Pre-Press PMG
Copy Editor: Denise Rubens
Cover Image: © Charles D. Winters/ Cengage Learning
Compositor: Pre-Press PMG

Library of Congress Control Number: 2010923378

ISBN-13: 978-0-538-73333-5

ISBN-10: 0-538-73333-0

Brooks/Cole
10 Davis Drive
Belmont, CA 94002-3098
USA

Cengage Learning is a leading provider of customized learning solutions with office locations around the globe, including Singapore, the United Kingdom, Australia, Mexico, Brazil, and Japan. Locate your local office at **www.cengage.com/global.**

Cengage Learning products are represented in Canada by Nelson Education, Ltd.

To learn more about Brooks/Cole, visit **www.cengage.com/brookscole**

Purchase any of our products at your local college store or at our preferred online store **www.CengageBrain.com.**

Printed in the United States of America
1 2 3 4 5 6 7 14 13 12 11 10

Contents

Preface

Innovation and exploration have always been the hallmarks of *Macroscale and Microscale Organic Experiments*, and that philosophy continues with this sixth edition. We are proud to have been part of a movement toward the increased use of microscale experiments in the undergraduate organic laboratory course.

As in previous editions, ease of use continues to be a chief attribute of the pedagogical features in this text. From the first edition onward, icons have appeared in the margin that clearly indicate whether an experiment is to be conducted on a microscale or macroscale level. Wherever possible, we have expanded the popular introductory "In This Experiment" sections, giving an overall view of the experimental work to be carried out without the detail that may obscure an understanding of how the end result is achieved. "Cleaning Up" sections at the end of almost every experiment focus students' attention on all the substances produced in a typical organic reaction, and continue to highlight current laboratory safety rules and regulations, and our emphasis on green chemistry.

In preparing the sixth edition, we have attempted to build on the strengths of previous editions while continuing to add innovative and new techniques, features, and experiments.

NEW TO THIS EDITION

- **NMR Spectra** Many NMR spectra throughout the textbook have been replaced with spectra obtained by higher field spectrometers.
- **New Transition** A new section appears before the first true synthetic reaction in Chapter 16, *The S_N2 Reaction: 1-Bromobutane*. This section helps students to preview the outcome of synthetic experiments by instructing them on how to generate chemical data tables, spectral data tables, and how to calculate theoretical yields; it also reviews percent yield calculations.
- **Green Chemistry** Katherine Masters' new role as a mom has made her even more aware of the hazards of exposure to toxic chemicals, especially teratogens. With that in mind, most of the new experiments in this edition are green.
- **Online NMR Spectroscopy** For those schools with limited or no NMR spectroscopy equipment, an online NMR experiment of an unknown compound can be accessed on the textbook's website at: **http://cengage.com/chemistry/williamson**. This experiment gives students hands-on experience with processing raw NMR data.
- **New and Changed Experiments**
  - Chapter 8, *Thin-Layer Chromatography: Analyzing Analgesics and Isolating Lycopene from Tomato Paste*, now includes a six-step outline of the process of thin-layer chromatography to help students carry out this technique in an orderly fashion.

- Chapter 23, *The Cannizzaro Reaction*, is unique in that it involves grinding a liquid aldehyde with potassium hydroxide to give two new compounds. There is no heating and no solvent involved, making it the ultimate green chemistry experiment.
- Chapter 50, *A Diels–Alder Reaction Puzzle*, is the reaction of 2,4-hexadiene with maleic anhydride to give an unexpected product, the structure of which can be deduced from spectroscopic information and a simple chemical test.
- Chapter 60, *Multicomponent Reactions: The Aqueous Passerini Reaction*, uses water as the solvent.
- Chapter 64, *Virstatin, a Possible Treatment for Cholera*, is the synthesis of a drug, virstatin ethyl ester, a multicomponent reaction in which three starting materials dissolved in water react in one flask to give one product without chemical waste.
- Chapter 66, *The Synthesis of Natural Products*, describes how to synthesize two natural products: the sex attractant of the cockroach and camphor.

Green Chemistry

Ecological wisdom is one of the tenets of the green movement. The ultimate object of green chemistry is to minimize the environmental impact of all chemical processes, including those that take place in the teaching laboratory. We achieved a major advance in this regard in 1987 with the publication of *Microscale Organic Experiments*.

Green chemistry is relative. Every microscale experiment in this text is green compared to the macroscale version that uses 100 to 1000 times as much starting material and solvent. The benzoin condensation (Chapter 53) has traditionally used potassium cyanide as the catalyst. On a microscale, this is reduced to 15 mg, one-tenth the lethal dose for the average human. Clearly, this is green compared to the macroscale experiment where a potentially lethal amount of cyanide is employed. But, almost a quarter century ago, thiamine was used as the catalyst for this reaction, and it was much safer because it is simply an edible vitamin. So this is a "greener" experiment by utilizing thiamine instead of potassium cyanide, but two drawbacks are that the reaction is much slower, and the catalyst is much more expensive.

The ultimate green experiment is one that produces no by-products and uses no solvent. Reactions of this type are common in inorganic and industrial chemistry; for example, the combination of hydrogen with oxygen to give water, or the reaction of carbon (coke) with water to give carbon monoxide and hydrogen that can be further reacted, in the presence of a catalyst, to give methane and water. We have included one experiment of this type, the Cannizzaro reaction (Chapter 23), in which an aldehyde is simply mixed with sodium hydroxide to give, through a compensated oxidation-reduction process, an alcohol and the sodium salt of a carboxylic acid—with no solvent, no by-products. The challenge is to separate the two products.

Nearly 100 procedures throughout this text are designated with the green chemistry icon. In addition to the experiments cited above, household bleach is used instead of the toxic chromium ion to oxidize cyclohexanol (Chapter 22). Air is used to oxidize fluorene (Chapter 9), and synthesized materials are reused in other steps of the polymer experiment (Chapter 67).

Online NMR Experiment

We realize that student hands-on access to an NMR spectrometer may be limited or not possible at some institutions. For this reason, an online NMR experiment of an unknown compound has been implemented and can be accessed on the textbook's website at: **http://cengage.com/chemistry/williamson**. Students are required to obtain a small amount of an unknown compound from the instructor, and take the melting point and acquire an IR spectrum (if available). From the melting point range, students will establish a list of possible compounds and then download raw NMR data from the website for study. The data can be processed with SpinWorks, a free downloadable program. Once the NMR spectrum is printed, students will be able to identify the structure of the unknown.

This experiment will allow students the opportunity to process and analyze NMR data. Students can submit a high-quality spectrum with peaks in Hertz or ppm values, calculated *J* values, integration, and expanded plots superimposed on the original spectrum.

SUPPLEMENTS

For Students

The **Premium Website** provides valuable resources intended to enhance students' organic chemistry laboratory experience, and help prepare them to work more effectively and safely in the laboratory. Students can access the Premium Website by visiting this book's website at: **www.CengageBrain.com**.

Icons throughout the text direct students to the section on the website where appropriate resources appear. These resources include photos and videos that are intended to augment the descriptions and drawings in the text. The photographs amplify the figures in the text that illustrate apparatus and techniques, and the short videos demonstrate techniques that are more easily shown in motion rather than in static drawings. The videos also include quizzes that assess understanding of laboratory techniques. Several complete experiments are presented as videos on the website. The "Synthesis of Ferrocene," for example, illustrates in 3 minutes the operations needed for this complex procedure—a procedure that requires a minimum of 90 minutes in the laboratory.

In an effort to keep up with changes in web content and stay current with relevant online resources, the "Surfing the Web" sections from previous editions have been moved to the Premium Website. Helpful supplemental materials, such as tables listing the physical properties of a number of functional group classes, are also available on the website.

For Instructors

The **Faculty Companion site** at: **www.cengage.com/chemistry/williamson** centralizes these important preparatory materials and resources in one convenient location:

- **PowerPoint Lectures** supply a flexible, customizable instructor resource that is designed to enhance pre-laboratory discussions. For the technique chapters and other selected experiments, PowerPoint slides outline the main techniques, underlying concepts, common problems, challenging manipulations, special safety precautions, and typical results. Relevant illustrations of apparatus, structures, reactions, and links to photos and videos of techniques are also provided.

- The **Instructor's Guide** is an important adjunct to this text. It contains discussions about the time needed to carry out each experiment, an assessment of the relative difficulty of each experiment, problems that might be encountered, answers to end-of-chapter questions, a list of chemicals and apparatus required for each experiment (both per student and per 24-student laboratory), sources of supply for unusual items, and a discussion of hardware and software needed for running computational chemistry experiments.

ACKNOWLEDGMENTS

We wish to express our thanks to the many people at Cengage with whom we have worked closely to make this book possible: Charlie Hartford, Publisher; Stephanie Van Camp, Associate Media Editor, and Sandra Kiselica, Senior Developmental Editor, as well as Patrick Franzen, Senior Project Manager at Pre-Press PMG. We also wish to thank the reviewers of the text: Geeta Govindarajoo, Rutgers University; William Bailey, University of Connecticut; F. J. Heldrich, College of Charleston; H. Mark Perks, University of Maryland, Baltimore County; Robert M. Carlson, University of Minnesota, Duluth; John T. Barbas, Valdosta State University; Trudy A. Dickneider, University of Scranton; and Nancy I. Totah, Syracuse University for their helpful comments. A special thanks goes to John Chisolm of Syracuse University for his suggestions and comments, and to Lynn Bradley of the College of New Jersey, who served as the accuracy reviewer, reading all the page proofs for the book.

Kenneth L. Williamson
Katherine Masters

Organic Experiments and Waste Disposal

An unusual feature of this book is the advice at the end of each experiment on how to dispose of its chemical waste. Waste disposal thus becomes part of the experiment, which is not considered finished until proper disposal of waste products has transpired. This is a valuable addition to the book for several reasons.

Although chemical waste from laboratories is less than 0.1% of that generated in the United States, its disposal is nevertheless subject to many of the same federal, state, and local regulations as is chemical waste from industry. Accordingly, there are both strong ethical and legal reasons for proper disposal of laboratory wastes. In addition, there are financial concerns because the cost of waste disposal can become a significant part of the cost of operating a laboratory.

There is yet another reason to include instructions for waste disposal in a teaching laboratory. Students will someday be among those producing large amounts of hazardous waste, regulating waste disposal operations, and voting on appropriations for them. Learning the principles and methods of sound waste disposal early in their careers will benefit them and society later.

The basics of waste disposal are easy to grasp. Innocuous water-soluble wastes are flushed down the drain with a large proportion of water. Common inorganic acids and bases are neutralized, and then flushed down the drain. Containers are provided for several classes of solvents, for example, combustible solvents and halogenated solvents. (Licensed waste handlers will subsequently remove them for suitable disposal.) Some toxic substances can be oxidized or reduced to innocuous substances that can then be flushed down the drain; for example, hydrazines, mercaptans, and inorganic cyanides can be thus oxidized by a sodium hypochlorite solution, widely available as household bleach. Dilute solutions of highly toxic cations are expensive to dispose of because of their bulk; precipitation of the cation by a suitable reagent, followed by its separation, greatly reduces its bulk and disposal cost. These and many other procedures can be found throughout this book.

One other principle of waste control lies at the heart of this book. Microscale experimentation, by minimizing the scale of chemical operations, also minimizes the volume of waste. Chromatographic procedures to separate and purify products, spectroscopic methods to identify and characterize products, and well-designed small-scale equipment enable one to conduct experiments today on a tenth to a thousandth of the scale commonly in use a generation ago.

Chemists often provide great detail in their directions for preparing chemicals so that a synthesis can be repeated, but they seldom say much about how to dispose of the hazardous byproducts. Yet the proper disposal of a chemical's byproducts is as important as its proper preparation. Dr. Williamson sets a good example by providing explicit directions for such disposal.

Blaine C. McKusick

CHAPTER 1

Introduction

When you see this icon, sign in at this book's premium website at **www.cengage.com/login** to access videos, Pre-Lab Exercises, and other online resources.

PRELAB EXERCISE: Study the glassware diagrams presented in this chapter and be prepared to identify the reaction tube, the fractionating column, the distilling head, the filter adapter, and the Hirsch funnel.

Welcome to the organic chemistry laboratory! Here, the reactions that you learned in your organic lectures and studied in your textbook will come to life. The main goal of the laboratory course is for you to learn and carry out techniques for the synthesis, isolation, purification, and analysis of organic compounds, thus experiencing the experimental nature of organic chemistry. We want you to enjoy your laboratory experience and ask you to remember that safety always comes first.

EXPERIMENTAL ORGANIC CHEMISTRY

You are probably not a chemistry major. The vast majority of students in this laboratory course are majoring in the life sciences. Although you may never use the exact same techniques taught in this course, you will undoubtedly apply the skills taught here to whatever problem or question your ultimate career may present. Application of the scientific method involves the following steps:

1. Designing an experiment, therapy, or approach to solve a problem.
2. Executing the plan or experiment.
3. Observing the outcome to verify that you obtained the desired results.
4. Recording the findings to communicate them both orally and in writing.

The teaching lab is more controlled than the real world. In this laboratory environment, you will be guided more than you would be on the job. Nevertheless, the experiments in this text are designed to be sufficiently challenging to give you a taste of experimental problem-solving methods practiced by professional scientists. We earnestly hope that you will find the techniques, the apparatus, and the experiments to be of just the right complexity, not too easy but not too hard, so that you can learn at a satisfying pace.

Macroscale and Microscale Experiments

This laboratory text presents a unique approach for carrying out organic experiments; they can be conducted on either a *macroscale* or a *microscale*. Macroscale was the traditional way of teaching the principles of experimental organic chemistry and is the basis for all the experiments in this book, a book that traces its history to

1934 when the late Louis Fieser, an outstanding organic chemist and professor at Harvard University, was its author. Macroscale experiments typically involve the use of a few grams of *starting material*, the chief reagent used in the reaction. Most teaching institutions are equipped to carry out traditional macroscale experiments. Instructors are familiar with these techniques and experiments, and much research in industry and academe is carried out on this scale. For these reasons, this book has macroscale versions of most experiments.

For reasons primarily related to safety and cost, there is a growing trend toward carrying out microscale laboratory work, on a scale one-tenth to one-thousandth of that previously used. Using smaller quantities of chemicals exposes the laboratory worker to smaller amounts of toxic, flammable, explosive, carcinogenic, and teratogenic material. Microscale experiments can be carried out more rapidly than macroscale experiments because of rapid heat transfer, filtration, and drying. Because the apparatus advocated by the authors is inexpensive, more than one reaction may be set up at once. The cost of chemicals is, of course, greatly reduced. A principal advantage of microscale experimentation is that the quantity of waste is one-tenth to one-thousandth of that formerly produced. To allow maximum flexibility in the conduct of organic experiments, this book presents both macroscale and microscale procedures for the vast majority of the experiments. As will be seen, some of the equipment and techniques differ. A careful reading of both the microscale and macroscale procedures will reveal which changes and precautions must be employed in going from one scale to the other.

Synthesis and Analysis

The typical sequence of activity in synthetic organic chemistry involves the following steps:

1. Designing the experiment based on knowledge of chemical reactivity, the equipment and techniques available, and full awareness of all safety issues.
2. Setting up and running the reaction.
3. Isolating the reaction product.
4. Purifying the crude product, if necessary.
5. Analyzing the product using chromatography or spectroscopy to verify purity and structure.
6. Disposing of unwanted chemicals in a safe manner.

1. Designing the Experiment

Because the first step of experimental design often requires considerable experience, this part has already been done for you for most of the experiments in this introductory level book. Synthetic experimental design becomes increasingly important in an advanced course and in graduate research programs. Safety is paramount, and therefore it is important to be aware of all possible personal and environmental hazards before running any reaction.

2. Running the Reaction

The rational synthesis of an organic compound, whether it involves the transformation of one functional group into another or a carbon-carbon bond-forming reaction, starts with a *reaction*. Organic reactions usually take place in the liquid phase and are *homogeneous*—the reactants are entirely in one phase. The reactants can be solids and/or liquids dissolved in an appropriate solvent to mediate the reaction. Some reactions are *heterogeneous*—that is, one of the reactants is a solid and requires stirring or shaking to bring it in contact with another reactant. A few heterogeneous reactions involve the reaction of a gas, such as oxygen, carbon dioxide, or hydrogen, with material in solution.

An *exothermic* reaction evolves heat. If it is highly exothermic with a low activation energy, one reactant is added slowly to the other, and heat is removed by external cooling. Most organic reactions are, however, mildly *endothermic*, which means the reaction mixture must be heated to overcome the activation barrier and to increase the rate of the reaction. A very useful rule of thumb is that *the rate of an organic reaction doubles with a 10°C rise in temperature*. Louis Fieser introduced the idea of changing the traditional solvents of many reactions to high-boiling solvents to reduce reaction times. Throughout this book we will use solvents such as triethylene glycol, with a boiling point (bp) of 290°C, to replace ethanol (bp 78°C), and triethylene glycol dimethyl ether (bp 222°C) to replace dimethoxyethane (bp 85°C). Using these high-boiling solvents can greatly increase the rates of many reactions.

Effect of temperature

The progress of a reaction can be followed by observing: a change in color or pH, the evolution of a gas, or the separation of a solid product or a liquid layer. Quite often, the extent of the reaction can be determined by withdrawing tiny samples at certain time intervals and analyzing them by *thin-layer chromatography* or *gas chromatography* to measure the amount of starting material remaining and/or the amount of product formed.

***Chapters 8–10*: Chromatography**

The next step, product isolation, should not be carried out until one is confident that the desired amount of product has been formed.

3. Product Isolation: Workup of the Reaction

Running an organic reaction is usually the easiest part of a synthesis. The real challenge lies in isolating and purifying the product from the reaction because organic reactions seldom give quantitative yields of a single pure substance.

In some cases the solvent and concentrations of reactants are chosen so that after the reaction mixture has been cooled, the product will *crystallize* or *precipitate* if it is a solid. The product is then collected by *filtration*, and the crystals are washed with an appropriate solvent. If sufficiently pure at that point, the product is dried and collected; otherwise, it is purified by the process of *recrystallization* or, less commonly, by *sublimation*.

***Chapter 4*: Recrystallization**

More typically, the product of a reaction does not crystallize from the reaction mixture and is often isolated by the process of *liquid/liquid extraction*.

***Chapter 7*: Liquid/Liquid Extraction**

This process involves two liquids, a water-insoluble organic liquid such as dichloromethane and a neutral, acidic, or basic aqueous solution. The two liquids do not mix, but when shaken together, the organic materials and inorganic byproducts go into the liquid layer that they are the most soluble in, either organic or aqueous. After shaking, two layers again form and can be separated. Most organic products remain in the organic liquid and can be isolated by evaporation of the organic solvent.

***Chapter 5*: Distillation**

If the product is a liquid, it is isolated by *distillation*, usually after extraction. Occasionally, an extraction is not necessary and the product can be isolated by the process of *steam distillation* from the reaction mixture.

***Chapter 6*: Steam Distillation and Vacuum Distillation**

4. Purification

When an organic product is first isolated, it will often contain significant impurities. This impure or crude product will need to be further purified or cleaned up before it can be analyzed or used in other reactions. Solids may be purified by recrystallization or sublimation, and liquids by distillation or steam distillation. Small amounts of solids and liquids can also be purified by *chromatography*.

***Chapters 11–14*: Structure Analysis**

5. Analysis to Verify Purity and Structure

The purity of the product can be determined by melting point analysis for solids, boiling point analysis or, less often, refractive index for liquids, and chromatographic analysis for either solids or liquids. Once the purity of the product has been verified, structure determination can be accomplished by using one of the various spectroscopic methods, such as ^{1}H and ^{13}C nuclear magnetic resonance (NMR), infrared (IR), and ultraviolet/visible (UV/Vis) spectroscopies. Mass spectrometry (MS) is another tool that can aid in the identification of a structure.

⚠ Never smell chemicals in an attempt to identify them.

6. Chemical Waste Disposal

All waste chemicals must be disposed of in their proper waste containers. Instructions on chemical disposal will appear at the end of each experiment. It is recommended that nothing be disposed of until you are sure of your product identity and purity; you do not want to accidentally throw out your product before the analysis is complete. Proper disposal of chemicals is essential for protecting the environment in accordance with local, state, and federal regulations.

EQUIPMENT FOR EXPERIMENTAL ORGANIC CHEMISTRY

A. Equipment for Running Reactions

Organic reactions are usually carried out by dissolving the reactants in a solvent and then heating the mixture to its boiling point, thus maintaining the reaction at that elevated temperature for as long as is necessary to complete the reaction. To keep the solvent from boiling away, the vapor is condensed to a liquid, which is allowed to run back into the boiling solvent.

Microscale reactions with volumes up to 4 mL can be carried out in a *reaction tube* (Fig. 1.1a). The mass of the reaction tube is so small and heat transfer is so rapid that 1 mL of nitrobenzene (bp 210°C) will boil in 10 seconds, and 1 mL of benzene (mp 5°C) will crystallize in the same period of time. Cooling is effected by simply agitating the tube in a small beaker of ice water, and heating is effected by immersing the reaction tube to an appropriate depth in an electrically heated sand bath. This sand bath usually consists of an electric 100-mL flask heater or heating mantle half filled with sand. The temperature is controlled by the setting on a variable voltage controller, but it heats slowly and changes temperature slowly.

Turn on the sand bath about 20 minutes before you intend to use it. The sand heats slowly and changes temperature slowly.

The air above the heater is not hot. It is possible to hold a reaction tube containing refluxing solvents between the thumb and forefinger without the need for forceps or other protective devices. Because sand is a fairly poor conductor of heat, there can be a very large variation in temperature in the sand bath depending on its depth. The temperature of a 5-mL flask can be regulated by using a spatula to pile up or remove sand from near the flask's base. The heater is easily capable of producing temperatures in excess of 300°C; therefore, never leave the controller at its maximum setting. Ordinarily, it is set at 20%–40% of maximum.

⚠ Never put a mercury thermometer in a sand bath! It will break, releasing highly toxic mercury vapor.

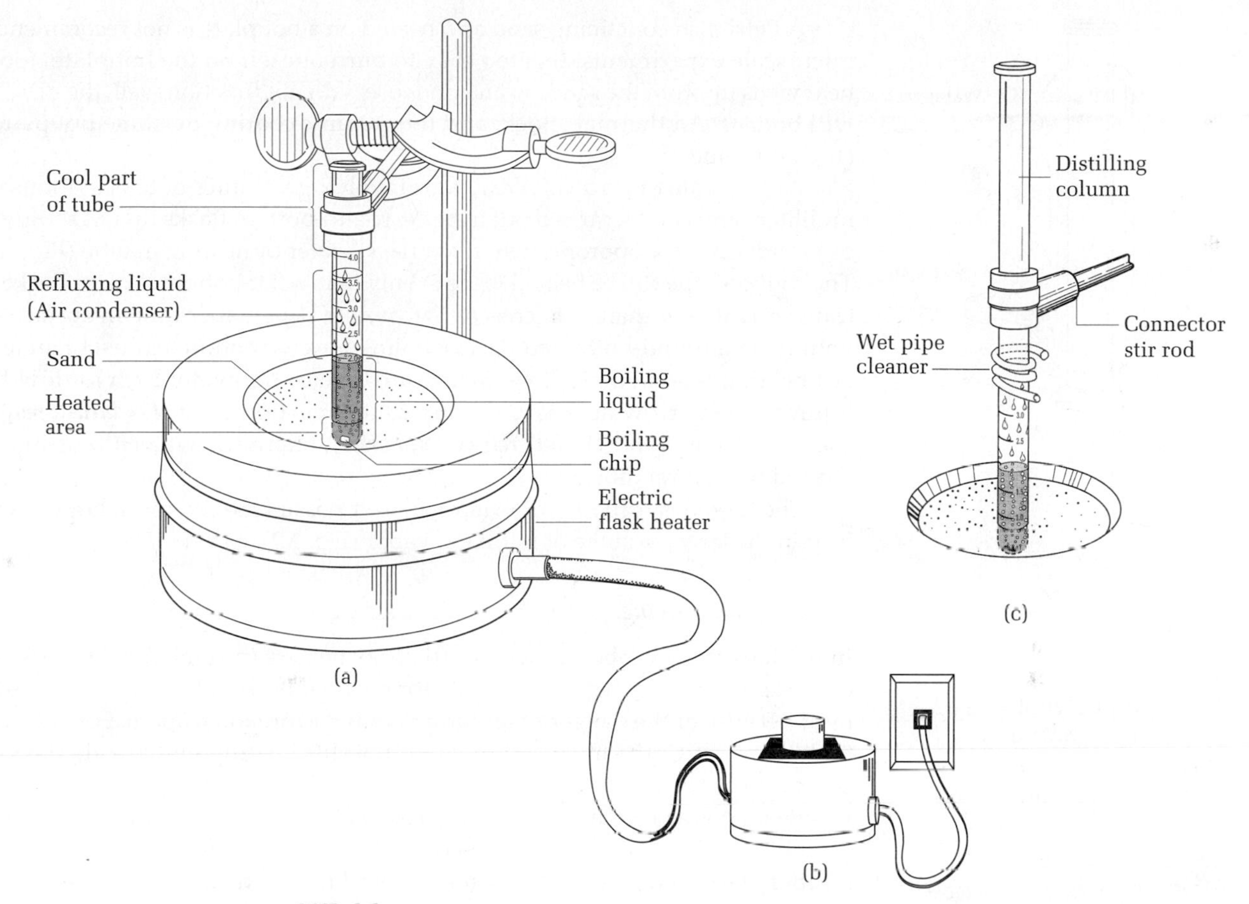

FIG. 1.1
(a) A reaction tube being heated on a hot sand bath in a flask heater. The area of the tube exposed to the heat is small. The liquid boils and condenses on the cool upper portion of the tube, which functions as an air condenser. (b) A variable voltage controller used to control the temperature of the sand bath. (c) The condensing area can be increased by adding a distilling column as an air condenser.

Photos: *Williamson Microscale Kit, Refluxing a Liquid in a Reaction Tube on a Sand Bath*; Video: *The Reaction Tube in Use*

Because the area of the tube exposed to heat is fairly small, it is difficult to transfer enough heat to the contents of the tube to cause the solvents to boil away. The reaction tube is 100 mm long, so the upper part of the tube can function as an efficient *air condenser* (Fig. 1.1a) because the area of glass is large and the volume of vapor is comparatively small. The air condenser can be made even longer by attaching the empty *distilling column* (Fig. 1.1c and 1.13o) to the reaction tube using the *connector with support rod* (Fig. 1.1c and Fig. 1.13m). The black connector is made of Viton, which is resistant to high-boiling aromatic solvents. The cream-colored connector is made of Santoprene, which is resistant to all but high-boiling aromatic solvents. As solvents such as water and ethanol boil, the hot vapor ascends to the upper part of the tube. These condense and run back down the tube. This process is called *refluxing* and is the most common method for conducting a reaction at a constant temperature, the boiling point of the solvent. For very low-boiling solvents such as diethyl ether (bp 35°C), a pipe cleaner dampened with water makes an efficient cooling device. A water-cooled condenser is also available (Fig. 1.2) but is seldom needed for microscale experiments.

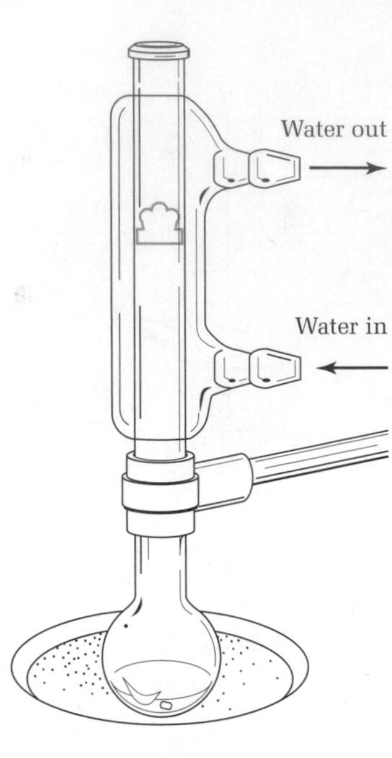

FIG. 1.2
Refluxing solvent in a 5-mL round-bottomed flask fitted with a water-cooled condenser.

Video: *How to Assemble Apparatus*

Organic reactions should be conducted in a fume hood with the sash lowered.

A Petri dish containing sand and heated on a hot plate is not recommended for microscale experiments. It is too easy to burn oneself on the hot plate; too much heat wells up from the sand, so air condensers do not function well; the glass dishes will break from thermal shock; and the ceramic coating on some hot plates will chip and come off.

Larger scale (macroscale) reactions involving volumes of tens to thousands of milliliters are usually carried out in large, round-bottom flasks that fit snugly (without sand!) into the appropriately sized flask heater or heating mantle (Figure 16.2). The round shape can be heated more evenly than a flat-bottom flask or beaker. Heat transfer is slower than in microscale because of the smaller ratio of surface area to volume in a round-bottomed flask. Cooling is again conducted using an ice bath, but heating is sometimes done in a steam bath or hot water bath for low-boiling liquids. The narrow neck is necessary for connection via a *standard-taper ground glass joint* to a water-cooled *reflux condenser*, where the water flows in a jacket around the central tube.

The high heat capacity of water makes it possible to remove a large amount of heat in the larger volume of refluxing vapor (Fig. 1.2).

Heating and Stirring

In modern organic laboratories, electric flask heaters (heating mantles), used alone or as sand baths, are used exclusively for heating. Bunsen burners are almost never used because of the danger of igniting flammable organic vapors. For solvents that boil below 90°C, the most common method for heating macroscale flasks is the *steam bath* or *hot water bath*.

Reactions are often stirred using a *magnetic stirrer* to help mix reagents and to promote smooth boiling. A Teflon-coated bar magnet (*stirring bar*) is placed in the reaction flask, and a magnetic stirrer is placed under the flask and flask heater. The stirrer contains a large, horizontally rotating bar magnet just underneath its metal surface that attracts the Teflon-coated stirring bar magnet through the glass of the flask and causes it to turn. The speed of stirring can be adjusted on the front of the magnetic stirrer.

B. Equipment for the Isolation of Products

Filtration

If the product of a reaction crystallizes from the reaction mixture on cooling, the solid crystals are isolated by *filtration*. This can be done in several ways when using microscale techniques. If the crystals are large enough and in a reaction tube, expel the air from a *Pasteur pipette*, insert it to the bottom of the tube and withdraw the solvent (Fig. 1.3). Highly effective filtration occurs between the square, flat tip of the pipette and the bottom of the tube. This method of filtration has several advantages over the alternatives. The mixture of crystals and solvent can be kept on ice during the entire process. This minimizes the solubility of the crystals in the solvent. There are no transfer losses of material because an external filtration device is not used. This technique allows several recrystallizations to be carried out in the same tube with final drying of the product under vacuum. If you know the *tare* (the weight of the empty tube), the weight of the product can be determined without removing it from the tube. In this manner a compound can be synthesized, purified by crystallization, and dried all in the same reaction tube. After removal of material for analysis, the compound in the tube can then be used for the next

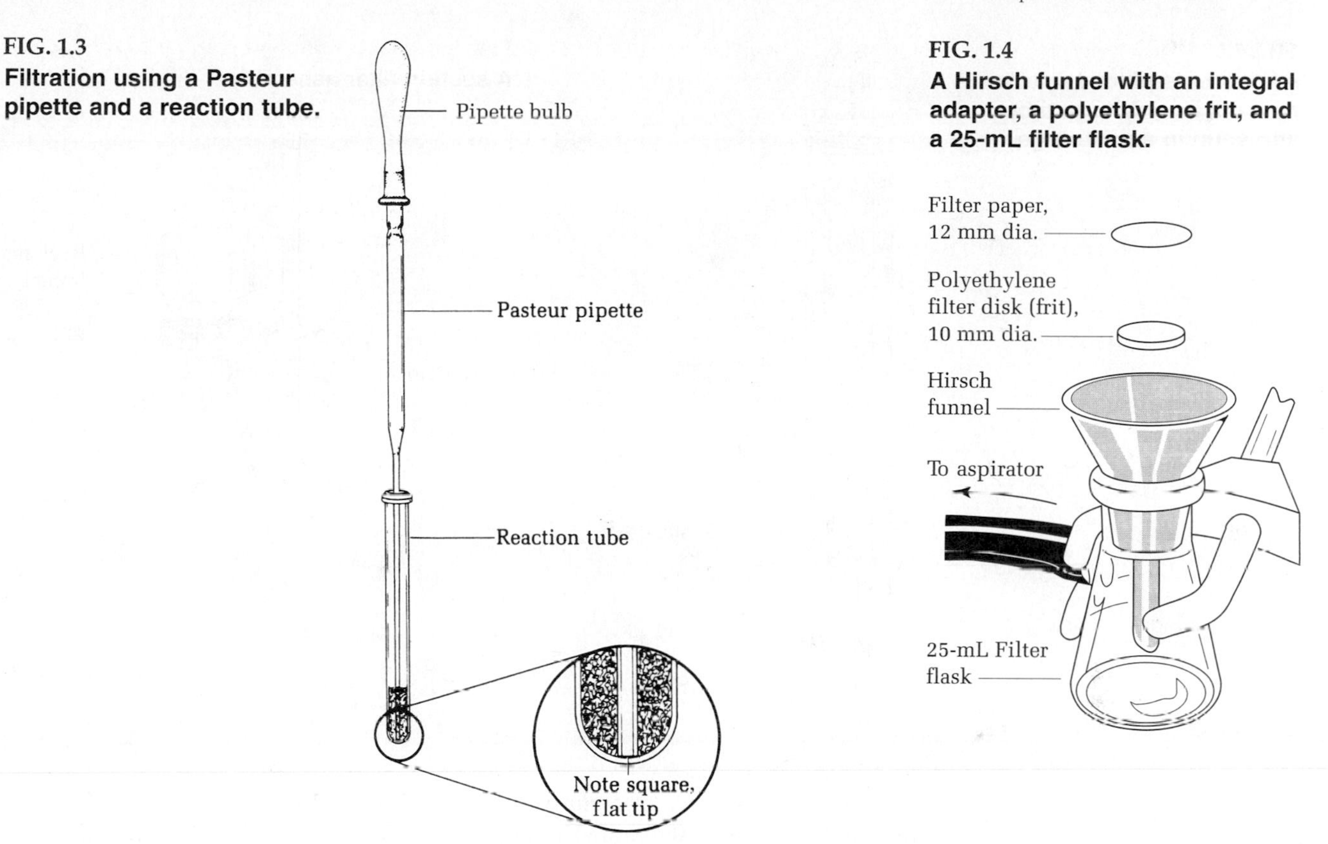

FIG. 1.3
Filtration using a Pasteur pipette and a reaction tube.

FIG. 1.4
A Hirsch funnel with an integral adapter, a polyethylene frit, and a 25-mL filter flask.

reaction. This technique is used in many of this book's microscale experiments. When the crystals are dry, they are easily removed from the reaction tube. When they are wet, it is difficult to scrape them out. If the crystals are in more than about 2 mL of solvent, they can be isolated by filtration with a *Hirsch funnel*. The one that is in the microscale apparatus kit is particularly easy to use because the funnel fits into the *filter flask* with no adapter and is equipped with a *polyethylene frit* for the capture of the filtered crystals (Fig. 1.4). The Wilfilter is especially good for collecting small quantities of crystals (Fig. 1.5).

Videos: *The Reaction Tube in Use; Filtration of Crystals Using the Pasteur Pipette*

Macroscale quantities of material can be recrystallized in conical *Erlenmeyer flasks* of the appropriate size. The crystals are collected in porcelain or plastic *Büchner funnels* fit with pieces of filter paper covering the holes in the bottom of the funnel (Fig. 1.6). A rubber *filter adapter* (*Filtervac*) is used to form a vacuum tight seal between the flask and the funnel.

Video: *Macroscale Crystallization*

Extraction

Chapter 7: **Extraction**

The product of a reaction will often not crystallize. It may be a liquid or a viscous oil, it may be a mixture of compounds, or it may be too soluble in the reaction solvent being used. In this case, an immiscible solvent is added, the two layers are shaken to effect *extraction*, and after the layers separate, one layer is removed. On a microscale, this can be done with a Pasteur pipette. The extraction process is repeated if necessary. A tall, thin column of liquid, such as that produced in a reaction tube, makes it easy to selectively remove one layer by pipette. This is more difficult to do in the usual test tube because the height/diameter ratio is small.

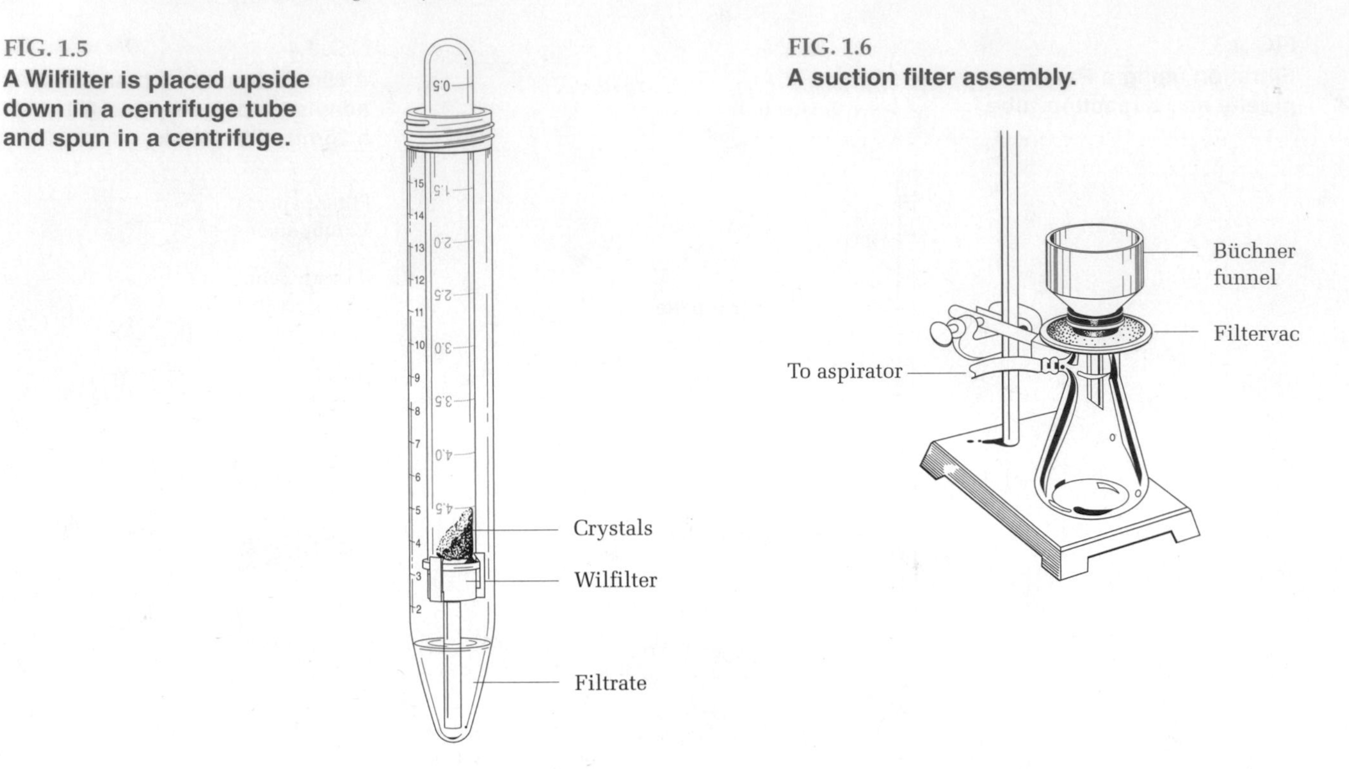

FIG. 1.5
A Wilfilter is placed upside down in a centrifuge tube and spun in a centrifuge.

FIG. 1.6
A suction filter assembly.

Photos: *Extraction with Ether and Extraction with Dichloromethane*; Videos: *Extraction with Ether, Extraction with Dichloromethane*

On a larger scale, a *separatory funnel* is used for extraction (Fig. 1.7a). The mixture can be shaken in the funnel and then the lower layer removed through the stopcock after the stopper is removed. These funnels are available in sizes from 10 mL to 5000 mL. The chromatography column in the apparatus kit is also a *micro separatory funnel* (Fig. 1.7b). Remember to remove the frit at the column base of the micro Büchner funnel and to close the valve before adding liquid.

C. Equipment for Purification

Many solids can be purified by the process of *sublimation*. The solid is heated, and the vapor of the solid condenses on a cold surface to form crystals in an apparatus constructed from a *centrifuge tube* fitted with a rubber adapter and pushed into a *filter flask* (Fig. 1.8). Caffeine can be purified in this manner. This is primarily a microscale technique, although sublimers holding several grams of solid are available.

Photos: *Sublimation Apparatus*

Mixtures of solids and, occasionally, of liquids can be separated and purified by *column chromatography*. The *chromatography column* for both microscale and macroscale work is very similar (Fig. 1.9).

Photo: *Column Chromatography*; Videos: *Extraction with Ether; Extraction with Dichloromethane*

***Chapter 9*: Column Chromatography**

Some of the compounds to be synthesized in these experiments are liquids. On a very small scale, the best way to separate and purify a mixture of liquids is by *gas chromatography*, but this technique is limited to less than 100 mg of material for the usual gas chromatograph. For larger quantities of material, *distillation* is used. For this purpose, small distilling flasks are used. These flasks have a large surface area to allow sufficient heat input to cause the liquid to vaporize rapidly so that it can be distilled and then condensed for collection in a *receiver*. The complete apparatus (Fig. 1.10) consists of a *distilling flask*, a *distilling adapter* (which also functions as an air condenser on a microscale), a *thermometer adapter*, and a *thermometer*; for macroscale, a water-cooled *condenser* and *distilling adapter* are added to the

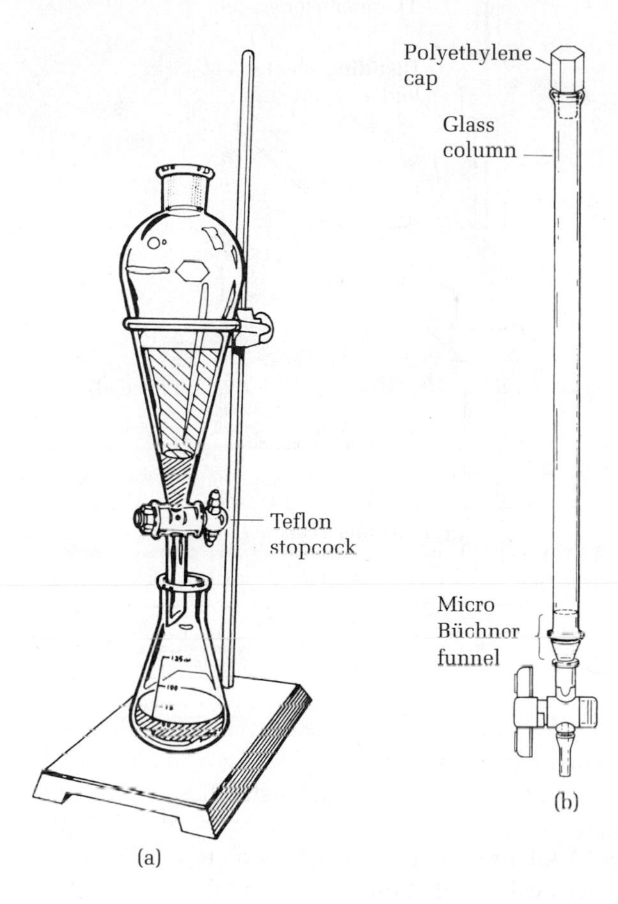

FIG. 1.7
(a) A separatory funnel with a Teflon stopcock. (b) A microscale separatory funnel. Remove the polyethylene frit from the micro Büchner funnel before using.

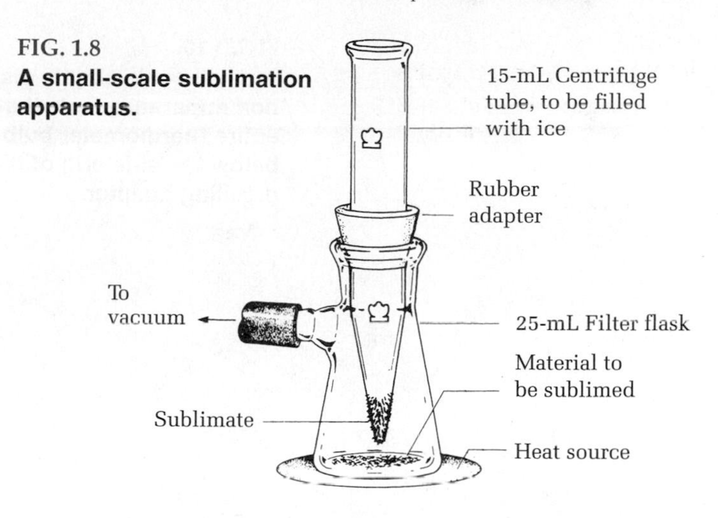

FIG. 1.8
A small-scale sublimation apparatus.

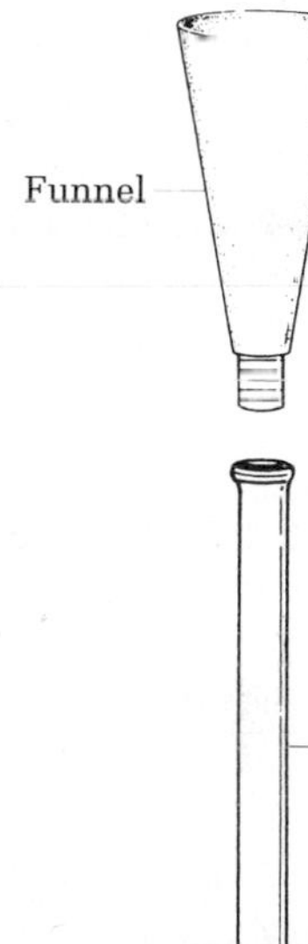

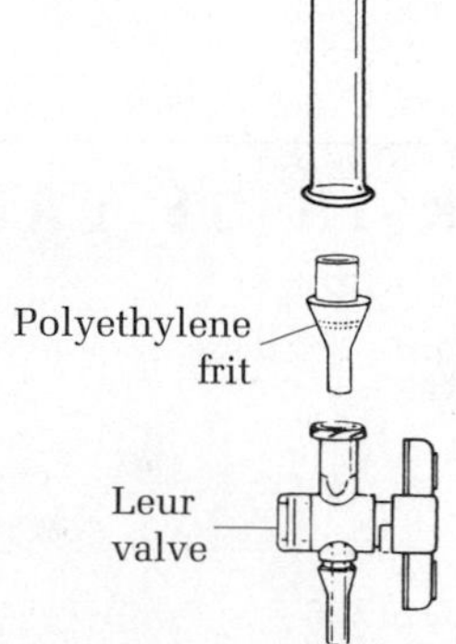

FIG. 1.9
A chromatography column consisting of a funnel, a tube, a base fitted with a polyethylene frit, and a Leur valve.

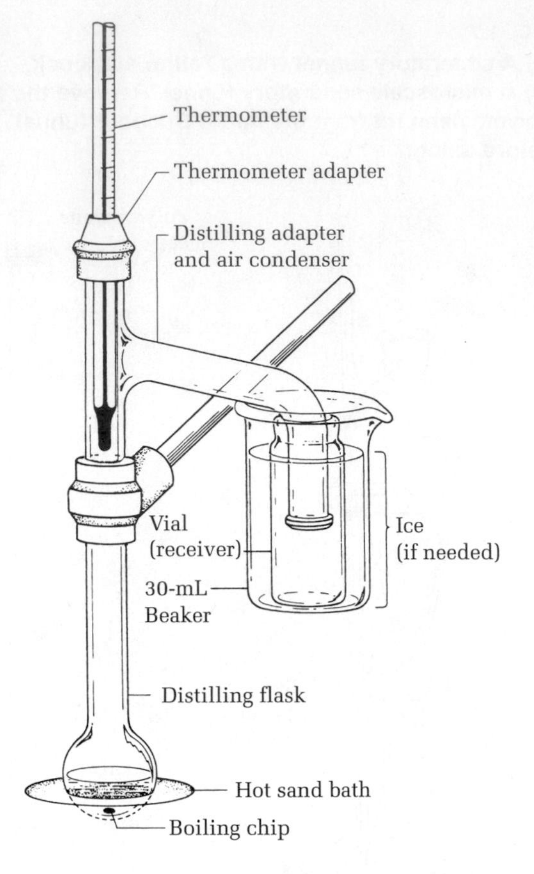

FIG. 1.10
A small-scale simple distillation apparatus. Note that the entire thermometer bulb is below the side arm of the distilling adapter.

Use mercury-free thermometers whenever possible.

Photos: *Simple and Fractional Distillation Apparatus*; Video: *How to Assemble the Apparatus*

apparatus (Fig. 1.11). *Fractional distillation* is carried out using a small, packed *fractionating column* (Fig. 1.12). The apparatus for fractional distillation is very similar for both microscale and macroscale. On a microscale, 2 mL to 4 mL of a liquid can be fractionally distilled, and 1 mL or more can be simply distilled. The usual scale in these experiments for macroscale distillation is about 25 mL.

Some liquids with a relatively high vapor pressure can be isolated and purified by *steam distillation*, a process in which the organic compound codistills with water at the boiling point of water and is then further purified and concentrated. The microscale and macroscale apparatus for this process are shown in Chapter 6.

The collection of typical equipment used for microscale experimentation is shown in Figure 1.13 and for macroscale experimentation in Figure 1.14. Other equipment commonly used in the organic laboratory is shown in Figure 1.15.

CHECK-IN OF LAB EQUIPMENT

Your first duty will be to check in to your assigned lab desk. The identity of much of the apparatus should already be apparent from the preceding outline of the experimental processes used in the organic laboratory.

Check to see that your thermometer reads about 22–25°C (71.6–77°F), which is normal room temperature. Examine the fluid column to see that it is unbroken and continuous from the bulb up. Replace any flasks that have star-shaped cracks.

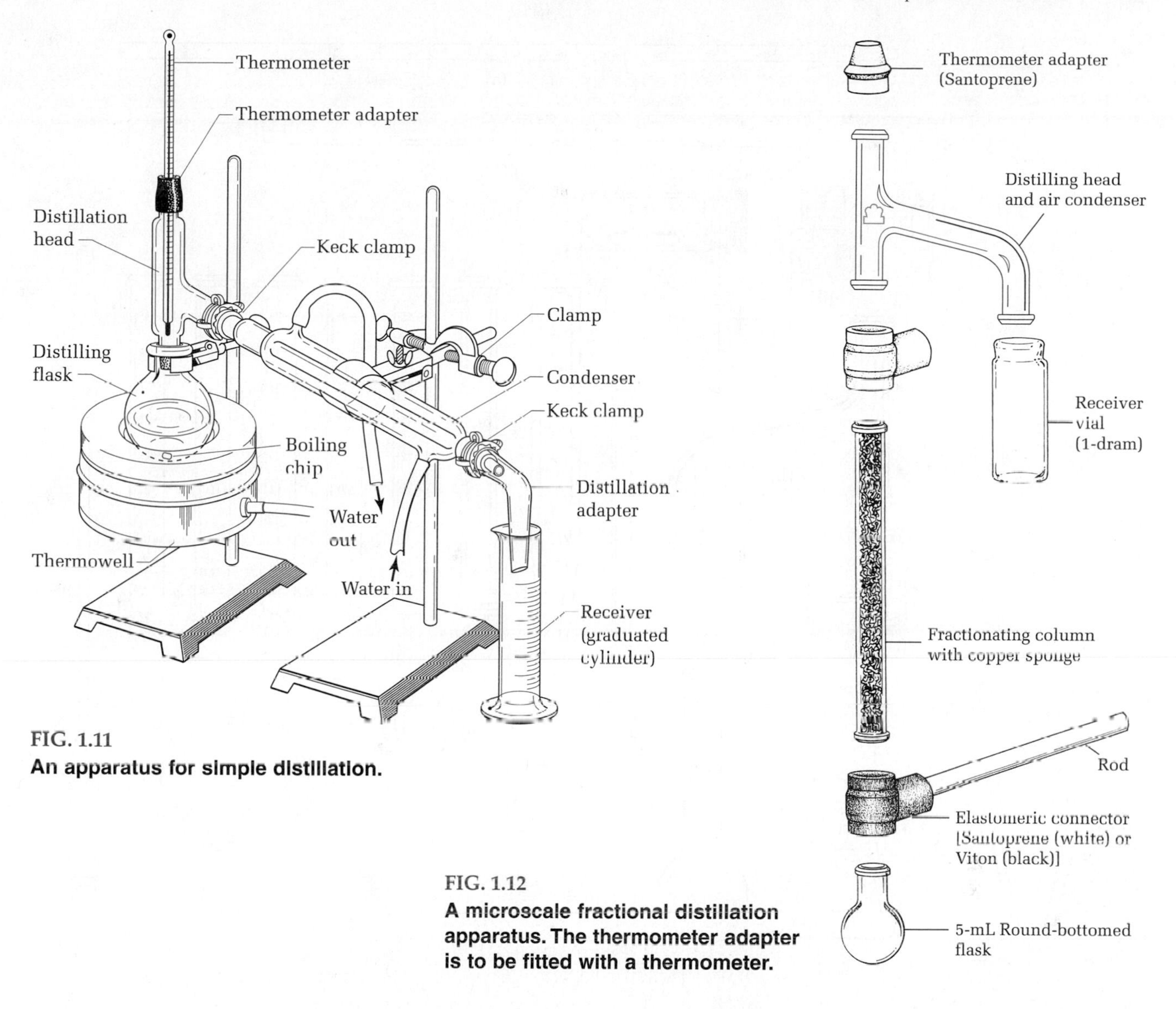

FIG. 1.11
An apparatus for simple distillation.

FIG. 1.12
A microscale fractional distillation apparatus. The thermometer adapter is to be fitted with a thermometer.

Remember that apparatus with graduations, stopcocks, or ground glass joints and anything porcelain are expensive. Erlenmeyer flasks, beakers, and test tubes are, by comparison, fairly cheap.

TRANSFER OF LIQUIDS AND SOLIDS

Borosilicate Pasteur pipettes (Fig. 1.16) are very useful for transferring small quantities of liquid, adding reagents dropwise, and carrying out recrystallizations. Discard used Pasteur pipettes in the special disposal container for waste glass. Surprisingly, the acetone used to wash out a dirty Pasteur pipette usually costs more than the pipette itself.

A plastic funnel that fits on the top of the reaction tube is very convenient for the transfer of solids to reaction tubes or small Erlenmeyer flasks for

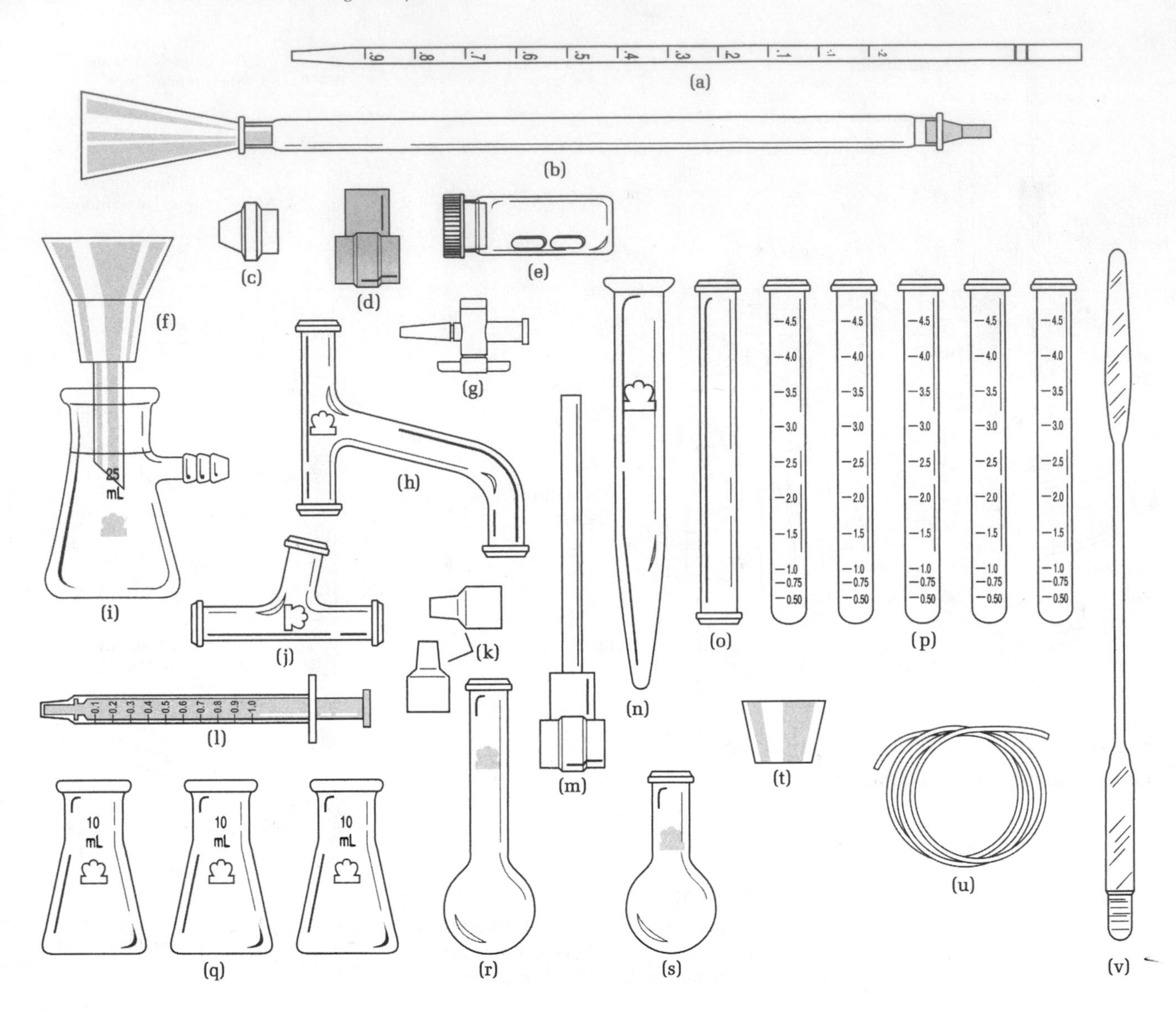

FIG. 1.13

Microscale apparatus kit.

(a) Pipette (1 mL), graduated in 1/1000ths.
(b) Chromatography column (glass) with a polypropylene funnel and 20-μm polyethylene frit in the base, which doubles as a micro Büchner funnel. The column, base, and stopcock can also be used as a separatory funnel.
(c) Thermometer adapter.
(d) Connector only (Viton).
(e) Magnetic stirring bars (4 × 12 mm) in a distillation receiver vial.
(f) Hirsch funnel (polypropylene) with a 20-μm fritted polyethylene disk.
(g) Stopcock for a chromatography column or separatory funnel.
(h) Claisen adapter/distillation head with an air condenser.
(i) Filter flask, 25 mL.
(j) Distillation head with a 105° connecting adapter.
(k) Rubber septa/sleeve stoppers, 8 mm.
(l) Syringe (polypropylene).
(m) Connector with a support rod.
(n) Centrifuge tube (15 mL)/sublimation receiver, with cap.
(o) Distillation column/air condenser.
(p) Reaction tube, calibrated, 10 × 100 mm.
(q) Erlenmeyer flasks, 10 mL.
(r) Long-necked flask, 5 mL.
(s) Short-necked flask, 5 mL.
(t) Rubber adapter for sublimation apparatus.
(u) Tubing (polyethylene), 1/16-in. diameter.
(v) Spatula (stainless steel) with scoop end.

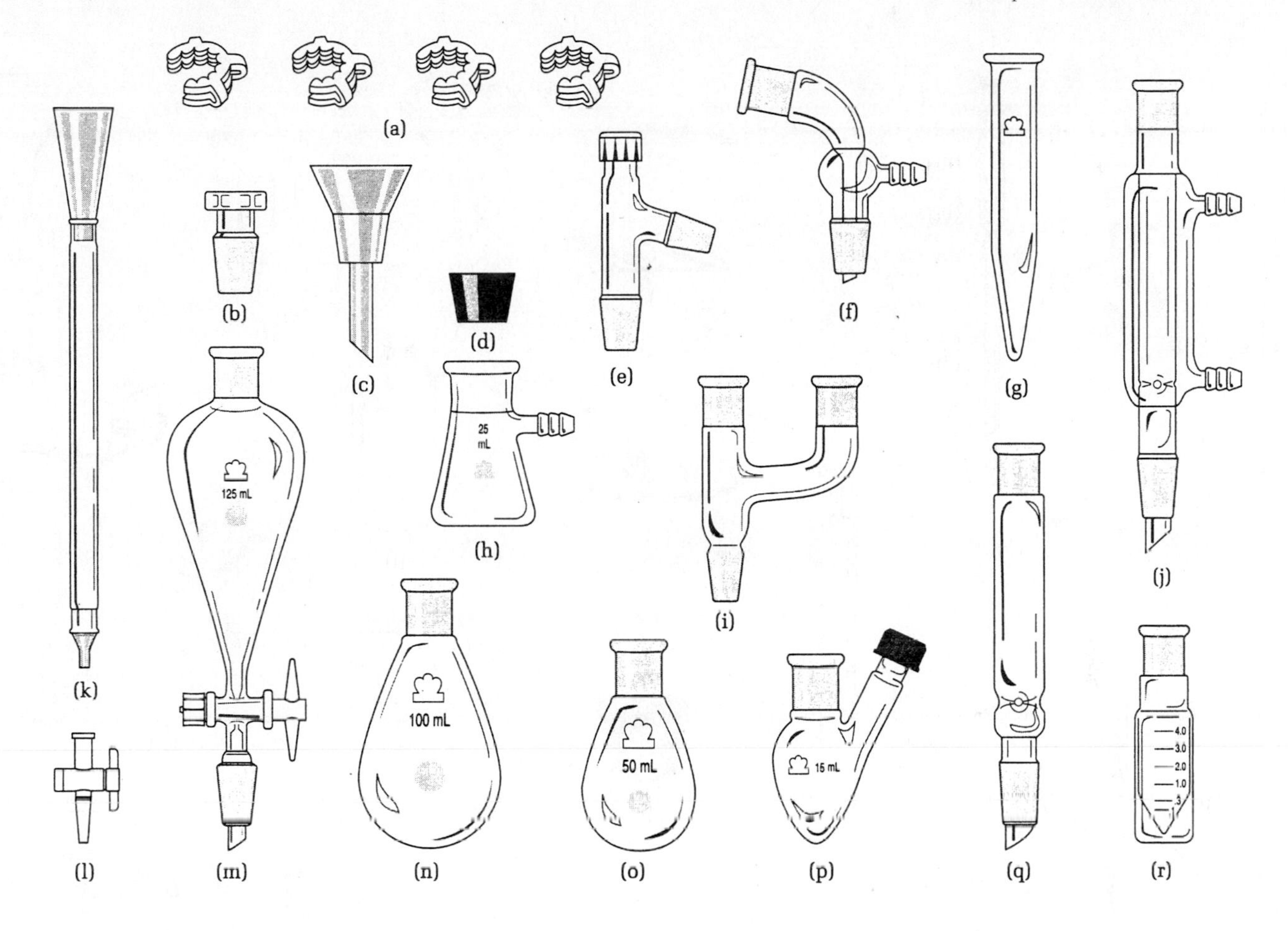

FIG. 1.14

Macroscale apparatus kit with 14/20 standard-taper ground-glass joints.

(a) Polyacetal Keck clamps, size 14.
(b) Hex-head glass stopper, 14/20 standard taper.
(c) Hirsch funnel (polypropylene) with a 20-μm fritted polyethylene disk.
(d) Filter adapter for use with sublimation apparatus.
(e) Distilling head with O-ring thermometer adapter.
(f) Vacuum adapter.
(g) Centrifuge tube (15 mL)/sublimation receiver.
(h) Filter flask, 25 mL.
(i) Claisen adapter.
(j) Water-jacketed condenser.
(k) Chromatography column (glass) with a polypropylene funnel and 20-μm polyethylene frit in the base, which doubles as a micro Büchner funnel.
(l) Stopcock for a chromatography column.
(m) Separatory funnel, 125 mL.
(n) Pear-shaped flask, 100 mL.
(o) Pear-shaped flask, 50 mL.
(p) Conical flask (15 mL) with a side arm for an inlet tube.
(q) Distilling column/air condenser.
(r) Conical reaction vial (5 mL)/distillation receiver.

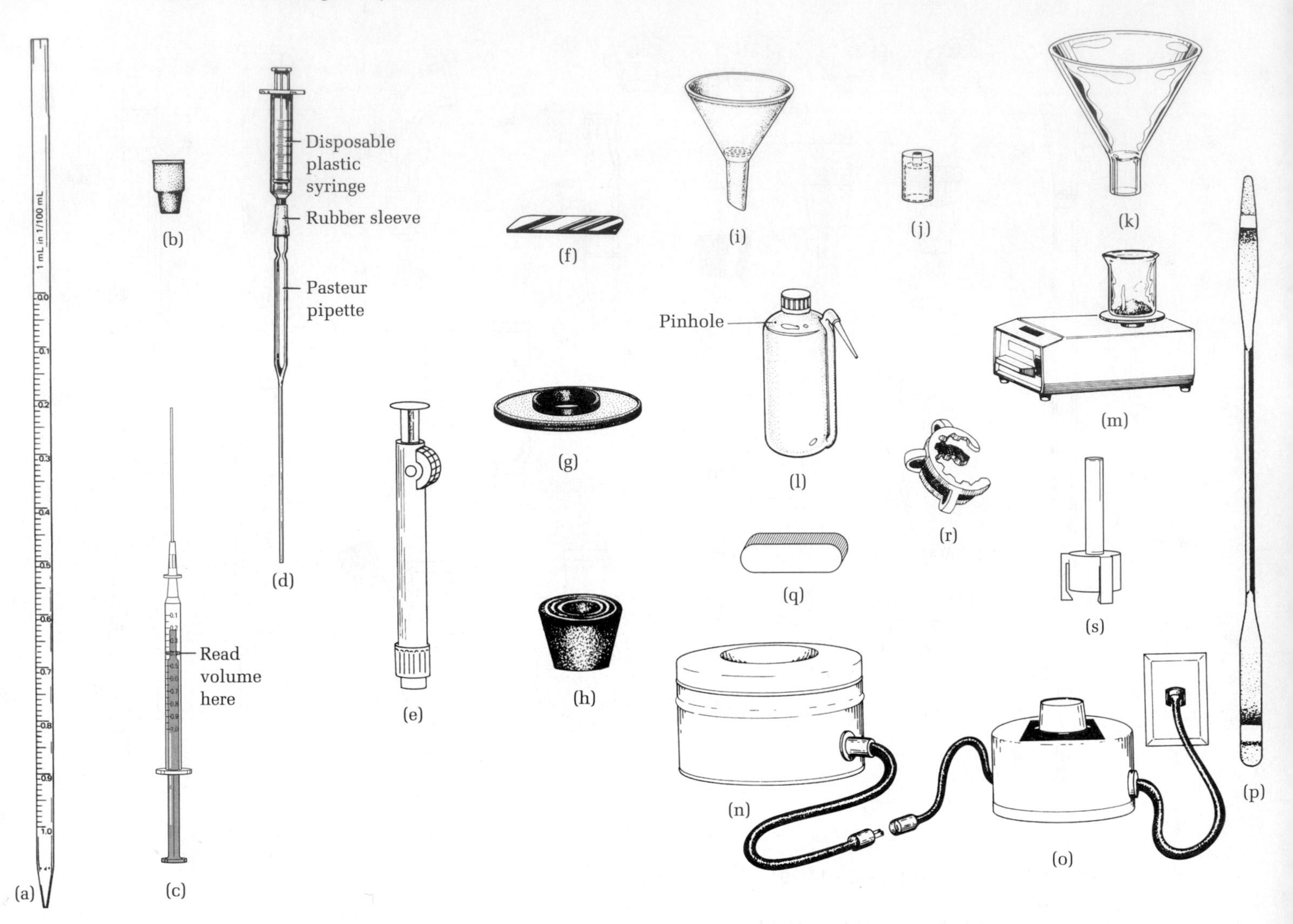

FIG. 1.15

Miscellaneous apparatus.

(a) 1.0 ± 0.01 mL graduated pipette.
(b) Septum.
(c) 1.0-mL syringe with a blunt needle.
(d) Calibrated Pasteur pipette.
(e) Pipette pump.
(f) Glass scorer.
(g) Filtervac.
(h) Set of neoprene filter adapters.
(i) Hirsch funnel with a perforated plate in place.
(j) Rubber thermometer adapter.
(k) Powder funnel.
(l) Polyethylene wash bottle.
(m) Single-pan electronic balance with automatic zeroing and 0.001 g digital readout; 100 g capacity.
(n) Electric flask heater.
(o) Solid-state control for electric flask heater.
(p) Stainless steel spatula.
(q) Stirring bar.
(r) Keck clamp.
(s) Wilfilter.

microscale experiments (Fig. 1.17). It can also function as the top of a chromatography column (Fig. 1.9). A special spatula with a scoop end (Fig. 1.13v) is used to remove solid material from the reaction tube. On a large scale, a powder funnel is useful for adding solids to a flask (Fig. 1.15k). A funnel can also be fashioned from a sheet of weighing paper for transferring lightweight solids.

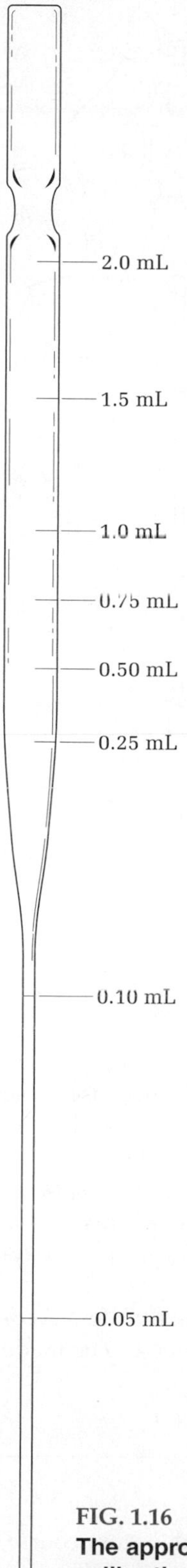

FIG. 1.16
The approximate calibration of a Kimble 9" Pasteur pipette.

WEIGHING AND MEASURING

The single-pan electronic balance (Fig. 1.15m), which is capable of weighing to ±0.001 g and having a capacity of at least 100 g, is the single most important instrument that makes microscale organic experiments possible. Most of the weighing measurements made in microscale experiments will use this type of balance. Weighing is fast and accurate with these balances as compared to mechanical balances. There should be one electronic balance for every 12 students. For macroscale experiments, a balance of such high accuracy is not necessary. Here, a balance with ±0.01 g accuracy would be satisfactory.

A container such as a reaction tube standing in a beaker or flask is placed on the balance pan. Set the digital readout to register zero and then add the desired quantity of the reagent to the reaction tube as the weight is measured periodically to the nearest milligram. Even liquids are weighed when accuracy is needed. It is much easier to weigh a liquid to 0.001 g than it is to measure it volumetrically to 0.001 mL.

It is often convenient to weigh reagents on glossy weighing paper and then transfer the chemical to the reaction container. The success of an experiment often depends on using just the right amount of starting materials and reagents. Inexperienced workers might think that if 1 mL of a reagent will do the job, then 2 mL will do the job twice as well. Such assumptions are usually erroneous.

Liquids can be measured by either volume or weight according to the following relationship:

$$\text{Volume (mL)} = \frac{\text{Weight (g)}}{\text{Density (g/mL)}}$$

Modern Erlenmeyer flasks and beakers have approximate volume calibrations fused into the glass, but these are *very* approximate. Better graduations are found on the microscale *reaction tube*. Somewhat more accurate volumetric measurements are made in 10-mL graduated cylinders. For volumes less than 4 mL, use a graduated pipette. **Never** apply suction to a pipette by mouth. Use a rubber bulb, a pipette pump, or fit the pipette with a small plastic syringe using appropriately sized rubber tubing. A Pasteur pipette can be converted into a calibrated pipette with the addition of a plastic syringe (Fig. 1.15d). Figure 1.16 also shows the calibration marks for a 9-in. Pasteur pipette. You will find among your equipment a 1-mL pipette, calibrated in hundredths of a milliliter (Fig. 1.15a). Determine whether it is designed to *deliver* 1 mL or *contain* 1 mL between the top and bottom calibration marks. For our purposes, the latter is the better pipette.

Because the viscosity, surface tension, vapor pressure, and wetting characteristics of organic liquids are different from those of water, the so-called automatic pipette (designed for aqueous solutions) gives poor accuracy in measuring organic

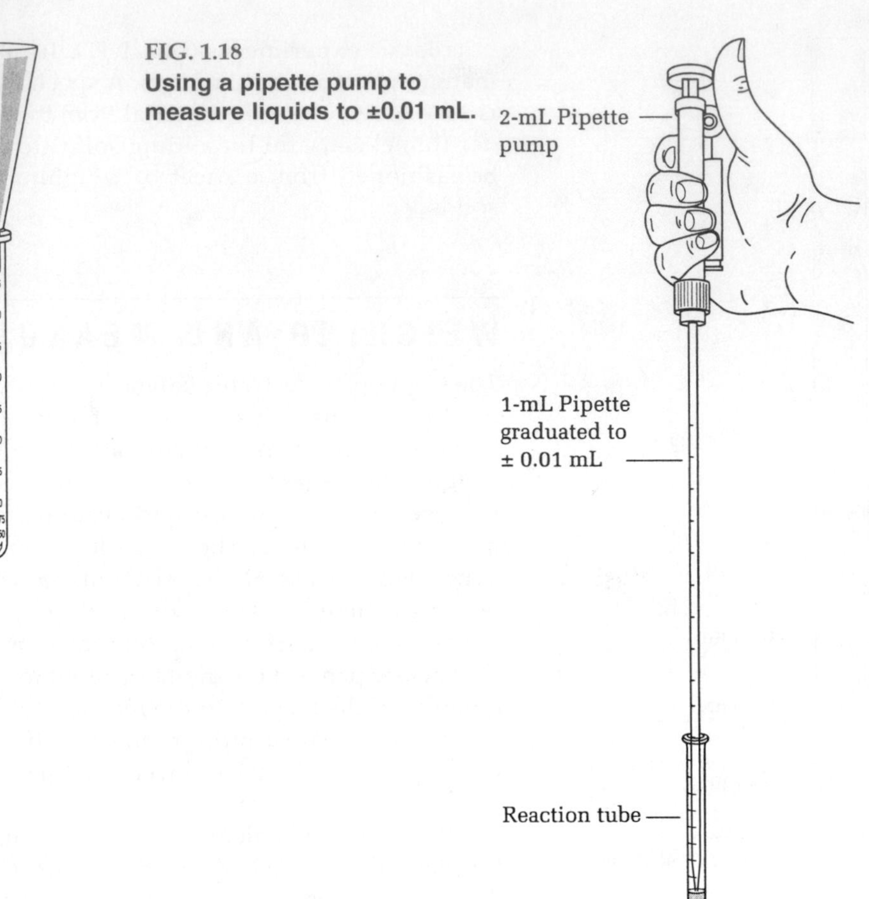

FIG. 1.17
A funnel for adding solids and liquids to a reaction tube.

FIG. 1.18
Using a pipette pump to measure liquids to ±0.01 mL.

Never pipette by mouth!

liquids. Syringes (Fig. 1.15c and Fig. 1.15d) and pipette pumps (Fig. 1.18), on the other hand, are quite useful, and these will be used frequently. Do not use a syringe that is equipped with a metal needle to measure corrosive reagents because these reagents will dissolve the metal in the needle. Because many reactions are "killed" by traces of moisture, many students' experiments are ruined by damp or wet apparatus. Several reactions that require especially dry or oxygen-free atmospheres will be run in systems sealed with a rubber septum (Fig. 1.15b). Reagents can be added to the system via syringe through this septum to minimize exposure to oxygen or atmospheric moisture.

Careful measurements of weights and volumes take more time than less accurate measurements. Think carefully about which measurements need to be made with accuracy and which do not.

TARES

Tare = weight of empty container

The tare of a container is its weight when empty. Throughout this laboratory course, it will be necessary to know the tares of containers so that the weights of the compounds within can be calculated. If identifying marks can be placed on the containers (e.g., with a diamond stylus for glassware), you may want to record tares for frequently used containers in your laboratory notebook.

To be strictly correct, we should use the word *mass* instead of *weight* because gravitational acceleration is not constant at all places on earth. But electronic balances record weights, unlike two-pan or triple-beam balances, which record masses.

WASHING AND DRYING LABORATORY EQUIPMENT

Washing

Clean apparatus immediately.

Considerable time may be saved by cleaning each piece of equipment soon after use, for you will know at that point which contaminant is present and be able to select the proper method for its removal. A residue is easier to remove before it has dried and hardened. A small amount of organic residue can usually be dissolved with a few milliliters of an appropriate organic solvent. Acetone (bp 56.1°C) has great solvent power and is often effective, but it is extremely flammable and somewhat expensive. Because it is miscible with water and vaporizes readily, it is easy to remove. Detergent and water can also be used to clean dirty glassware if an appropriate solvent cannot be found. Cleaning after an operation may often be carried out while another experiment is in process.

⚠ Both ethanol and acetone are very flammable.

Wash acetone is disposed in an organic solvents waste container; halogenated solvents go in the halogenated solvents waste container.

A *polyethylene bottle* (Fig. 1.15l) is a convenient wash bottle for acetone. Be careful not to store solvent bottles in the vicinity of a reaction where they can provide additional fuel for an accidental fire. The name, symbol, and formula of a solvent should be written on a bottle with a marker or a wax pencil. For macroscale crystallizations, extractions, and quick cleaning of apparatus, it is convenient to have a bottle of each frequently used solvent—95% ethanol, ligroin or hexanes, dichloromethane, ether, and ethyl acetate. A pinhole opposite the spout, which is covered with the finger when in use, will prevent the spout from dribbling the solvent. For microscale work, these solvents are best dispensed from 25-mL or 50-mL bottles with an attached test tube containing a graduated (1-mL) polypropylene pipette (Fig. 1.19). Be aware of any potential hazards stemming from the reactivity of these wash solvents with chemical residues in flasks. Also, be sure to dispose of wash solvents in the proper container. Acetone and most other organic solvents do not contain halogens and can therefore go in the regular organic solvents waste container. However, if dichloromethane or another halogen-containing solvent is used, it must be disposed of in the halogenated solvents waste container.

Sometimes a flask will not be clean after a washing with detergent and acetone. At that point, try an abrasive household cleaner. If still no success, try adding dilute acid or base to the dirty glassware, let it soak for a few minutes, and rinse with plenty of water and acetone.

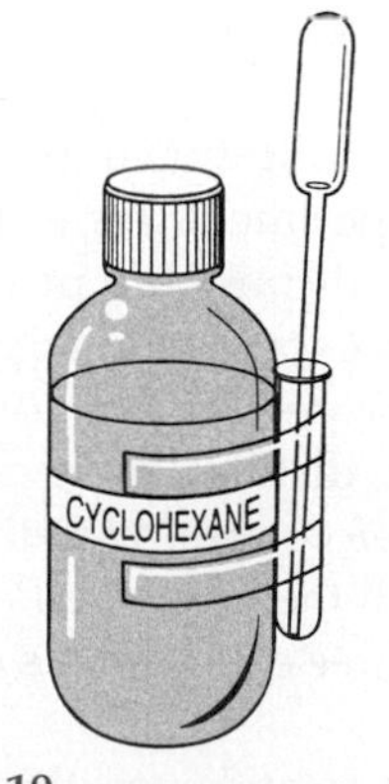

FIG. 1.19
A recrystallization solvent bottle and dispenser.

Drying

To dry a piece of apparatus rapidly, rinse with a few milliliters of acetone and invert over a beaker to drain. **Do not use compressed air**, which contains droplets of oil, water, and particles of rust. Instead, draw a slow stream of air through the apparatus using the suction of your water aspirator or house vacuum line.

MISCELLANEOUS CLEANUP

If a glass tube or thermometer becomes stuck to a rubber connector, it can be removed by painting on glycerol and forcing the pointed tip of a small spatula between the rubber and the glass. Another method is to select a cork borer that fits snugly over the glass tube, moisten it with glycerol, and slowly work it through the connector. If the stuck object is valuable, such as a thermometer, the best policy is to cut the rubber with a sharp knife. Care should be taken to avoid force that could potentially cause a thermometer to break, causing injury and the release of mercury.

THE LABORATORY NOTEBOOK

The Laboratory Notebook
What you did.
How you did it.
What you observed.
Your conclusions.

A complete, accurate record is an essential part of laboratory work. Failure to keep such a record means laboratory labor lost. An adequate record includes the procedure (what was done), observations (what happened), and conclusions (what the results mean).

Never record anything on scraps of paper. Use a lined, 8.5″ × 11″ paperbound notebook, and record all data in ink. Allow space at the front for a table of contents, number the pages throughout, and date each day's work. Reserve the left-hand page for calculations and numerical data, and use the right-hand page for notes. Never record **anything** on scraps of paper to be recorded later in the notebook. Do not erase, remove, or obliterate notes; simply draw a single line through incorrect entries.

The notebook should contain a statement or title for each experiment and its purpose, followed by balanced equations for all principal and side reactions and, where relevant, mechanisms of the reactions. Consult your textbook for supplementary information on the class of compounds or type of reaction involved. Give a reference to the procedure used; do not copy verbatim the procedure in the laboratory manual. Make particular note of safety precautions and the procedures for cleaning up at the end of the experiment.

Before coming to the lab to do preparative experiments, prepare a table (in your notebook) of reagents to be used and the products expected, with their physical properties. Use the molar ratios of reactions from your table to determine the limiting reagent, and then calculate the theoretical yield (in grams) of the desired product. Begin each experiment on a new page.

Include an outline of the procedure and the method of purification of the product in a flow sheet if this is the best way to organize the experiment (for example, an extraction; *see* Chapter 8). The flow sheet should list all possible products, by-products, unused reagents, solvents, and so on that are expected to appear in the crude reaction mixture. On the flow sheet diagram indicate how each of these is removed (e.g., by extraction, various washing procedures, distillation, or crystallization). With this information entered in your notebook before coming to the laboratory, you will be ready to carry out the experiments with the utmost efficiency. Plan your time before the laboratory period. Often two or three experiments can be run simultaneously.

When working in the laboratory, record everything you do and everything you observe **as it happens.** The recorded observations constitute the most important part of the laboratory record, since they form the basis for the conclusions you will draw at the end of each experiment. One way to do this is in a narrative form.

Alternatively, the procedure can be written in outline form on the left-hand side of the page and the observations recorded on the right-hand side.

In some colleges and universities, you will be expected to have all the relevant information about the running of an experiment entered in your notebook *before coming to the laboratory* so that your textbook will not be needed when you are conducting experiments. In industrial laboratories, your notebook may be designed so that carbon copies of all entries are kept. These are signed and dated by your supervisor and removed from your notebook each day. Your notebook becomes a legal document in case you make a discovery worth hundreds of millions of dollars!

Record the physical properties of the product from your experiment, the yield in grams, and the percent yield. Analyze your results. When things did not turn out as expected, explain why. When your record of an experiment is complete, another chemist should be able to understand your account and determine what you did, how you did it, and what conclusions you reached. That is, from the information in your notebook, a chemist should be able to repeat your work.

Preparing a Laboratory Record

Use the following steps to prepare your laboratory record. The letters correspond to the completed laboratory records that appear at the end of this chapter. Because your laboratory notebook is so important, two examples, written in alternative forms, are presented.

A. Number each page. Allow space at the front of the notebook for a table of contents. Use a hardbound, lined notebook, and keep all notes in ink.
B. Date each entry.
C. Give a short title to the experiment, and enter it in the table of contents.
D. State the purpose of the experiment.
E. Write balanced equation(s) for the reaction(s).
F. Give a reference to the source of the experimental procedure.
G. Prepare a table of quantities and physical constants. Look up the needed data in the *Handbook of Chemistry and Physics*.
H. Write equation(s) for the principal side reaction(s).
I. Write out the procedure with just enough information so that you can follow it easily. Do not merely copy the procedure from the text. Note any hazards and safety precautions. A highly experienced chemist might write a procedure in a formal report as follows: "Dibenzalacetone was prepared by condensing at room temperature 1 mmol acetone with 2 mmol benzaldehyde in 1.6 mL of 95% ethanol to which was added 2 mL of aqueous 3 *M* sodium hydroxide solution. After 30 min the product was collected and crystallized from 70% ethanol to give 0.17 g (73%) of flat yellow plates of dibenzalacetone, mp 110.5–111.5°C." Note that in this formal report, no jargon (e.g., EtOH for 95% ethanol, FCHO for benzaldehyde) is used and that the names of reagents are written out (sodium hydroxide, not NaOH). In this report the details of measuring, washing, drying, crystallizing, collecting the product, and so on are assumed to be understood by the reader.
J. Don't forget to note how to dispose of the byproducts from the experiment using the "Cleaning Up" section of the experimental procedure.
K. Record what you do as you do it. These observations are the most important part of the experiment. Note that conclusions do not appear among these observations.
L. Calculate the theoretical yield in grams. The experiment calls for exactly 2 mmol fluorobenzaldehyde and 1 mmol acetone, which will produce 1 mmol

of product. The equation for the experiment also indicates that the product will be formed from exactly a 2:1 ratio of the reactants. Experiments are often designed to have one reactant in great excess. In this experiment, a very slight excess of fluorobenzaldehyde was used inadvertently, so acetone becomes the limiting reagent.

M. Once the product is obtained, dried, and weighed, calculate the ratio of product actually isolated to the amount theoretically possible. Express this ratio as the percent yield.

N. Write out the mechanism of the reaction. If it is not given in the text of the experiment, look it up in your lecture text.

O. Draw conclusions from the observations. Write this part of the report in narrative form in complete English sentences. This part of the report can, of course, be written after leaving the laboratory.

P. Analyze TLC, IR, and NMR spectra if they are a part of the experiments. Rationalize observed versus reported melting points.

Q. Answer assigned questions from the end of the experiment.

R. This page presents an alternative method for entering the experimental procedure and observations in the notebook. Before coming to the laboratory, enter the procedure in outline form on the left side of the page. Then enter observations in brief form on the right side of the page as the experiment is carried out. Draw a single line through any words incorrectly entered. Do not erase or obliterate entries in the notebook, and never remove pages from the notebook.

Two samples of completed laboratory records follow. An alternative method for recording procedure, cleaning up, and observations is given on p. 25. The other parts of the report are the same.

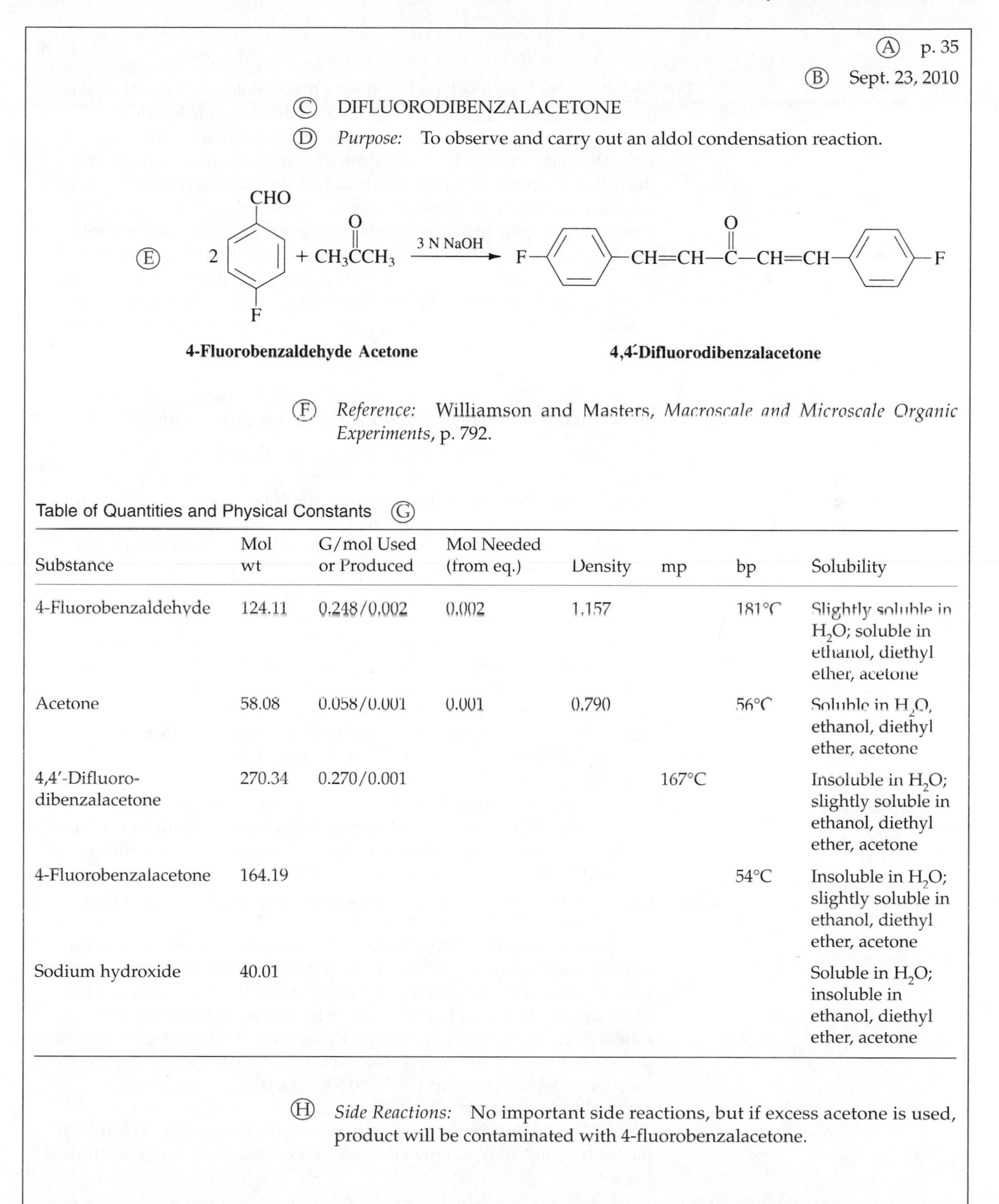

Ⓐ p. 35

Ⓑ Sept. 23, 2010

Ⓒ DIFLUORODIBENZALACETONE

Ⓓ *Purpose:* To observe and carry out an aldol condensation reaction.

Ⓔ 2 (4-F-C_6H_4-CHO) + $CH_3\overset{O}{\overset{\|}{C}}CH_3$ $\xrightarrow{\text{3 N NaOH}}$ F-C_6H_4-CH=CH-$\overset{O}{\overset{\|}{C}}$-CH=CH-$C_6H_4$-F

4-Fluorobenzaldehyde Acetone **4,4′-Difluorodibenzalacetone**

Ⓕ *Reference:* Williamson and Masters, *Macroscale and Microscale Organic Experiments*, p. 792.

Table of Quantities and Physical Constants Ⓖ

Substance	Mol wt	G/mol Used or Produced	Mol Needed (from eq.)	Density	mp	bp	Solubility
4-Fluorobenzaldehyde	124.11	0.248/0.002	0.002	1.157		181°C	Slightly soluble in H_2O; soluble in ethanol, diethyl ether, acetone
Acetone	58.08	0.058/0.001	0.001	0.790		56°C	Soluble in H_2O, ethanol, diethyl ether, acetone
4,4′-Difluoro-dibenzalacetone	270.34	0.270/0.001			167°C		Insoluble in H_2O; slightly soluble in ethanol, diethyl ether, acetone
4-Fluorobenzalacetone	164.19					54°C	Insoluble in H_2O; slightly soluble in ethanol, diethyl ether, acetone
Sodium hydroxide	40.01						Soluble in H_2O; insoluble in ethanol, diethyl ether, acetone

Ⓗ *Side Reactions:* No important side reactions, but if excess acetone is used, product will be contaminated with 4-fluorobenzalacetone.

Ⓘ *Procedure:* To a reaction tube, add 0.248 g (0.002 mole) of 4-fluorobenzaldehyde and then 1.6 mL of an ethanol solution that contains 58 mg of acetone. Mix the solution and observe it for 1 min. Remove sample for TLC. Then add 2 mL of 3 *M* NaOH(*aq*). Cap the tube with a septum and shake it at intervals over a 30-min period. Collect product on Hirsch funnel, wash crystals thoroughly with water, press as dry as possible, determine crude weight, and save sample for mp. Recrystallize product from 70% ethanol in a 10-mL Erlenmeyer flask. Cool solution slowly, scratch or add seed crystal, cool in ice, and collect product on Hirsch. Wash crystals with a few drops of ice-cold ethanol. Run TLC on silica gel plates using hexane to elute, develop in iodine chamber. Run NMR spectrum in deuterochloroform and IR spectrum as a mull.

Ⓙ *Cleaning Up:* Neutralize filtrate with HCl and flush solution down the drain. Put recrystallization filtrate in organic solvents container.

Ⓚ *Observations:* Into a reaction tube, 0.250 g of 4-fluorobenzaldehyde was weighed (0.248 called for). Using the 1.00-mL graduated pipette and pipette pump, 1.61 mL of the stock ethanol solution containing 58 mg of acetone was added (1.6 mL called for). The contents of the tube were mixed by flicking the tube, and then a drop of the reaction mixture was removed and diluted with 1 mL of hexane for later TLC. The water-clear solution did not change in appearance during 1 min.

Then about 2 mL of 3 *M* NaOH(*aq*) was added using graduations on side of reaction tube. The tube was capped with a septum and shaken. The clear solution changed to a light yellow color immediately and got slightly warm. Then after about 50 sec, the entire solution became opaque. Then yellow oily drops collected on sides and bottom of tube. Shaken every 5 min for 1/2 hr. The oily drops crystallized, and more crystals formed. At the end of 30 min, the opaque solution became clear, and yellow crystals had formed. Tube filled with crystals.

Product was collected on 12-mm filter paper on a Hirsch funnel; transfer of material to funnel was completed using the filtrate to wash out the tube. Crystals were washed with about 15 mL of water. The filtrate was slightly cloudy and yellow. It was poured into a beaker for later disposal. The crude product was pressed as dry as possible with a cork. Wt (damp) 270 mg. Sample saved for mp.

Recrystallized from about 50 mL of 70% ethanol on sand bath. Almost forgot to add boiling stick! Clear, yellow solution cooled slowly by wrapping tube in cotton. A seed crystal from Julie added to start crystallization. After about 25 min, flask was placed in ice bath for 15 min, then product was collected on Hirsch funnel (filter paper), washed with a few drops of ice-cold solvent, pressed dry and then spread out on paper to dry. Wt 190 mg. Very nice flat, yellow crystals, mp 179.5–180.5°C. The crude material before recrystallization had mp 177.5–179°C.

TLC on silica gel using hexane to elute gave only one spot, R_f 0.23, for the starting material and only one spot for the product, R_f 0.66. See attached plate.

p. 37

The NMR spectrum that was run on the Bruker in deuterochloroform containing TMS showed a complex group of peaks centered at about 7 ppm with an integrated value of 17.19 and a sharp, clear quartet of peaks centered at 6.5 ppm with an integral of 8.62. The coupling constant of the AB quartet was 7.2 Hz. See attached spectrum.

The IR spectrum, run as a Nujol mull on the Mattson FT IR, showed intense peaks at 1590 and 1655 cm^{-1} as well as small peaks at 3050, 3060, and 3072 cm^{-1}. See attached spectrum.

Cleaning Up: Recrystallization filtrate put in organic solvents container. First filtrate neutralized with IICl and poured down the drain.

Ⓛ *Theoretical Yield Calculations:*

$$0.250 \text{ g 4-fluorobenzaldehyde used } \frac{0.250 \text{ g}}{124.11 \text{ g/mole}} = 0.00201 \text{ mole}$$

$$0.058 \text{ g of acetone used } \frac{0.058 \text{ g}}{58.08 \text{ g/mole}} = 0.001 \text{ mole}$$

The moles needed (from equation) are 2 mmol of the 4-fluorobenzaldehyde and 1 mmol of acetone. Therefore, acetone is the limiting reagent, and 1.00 mmol of product should result. The MW of the product is 270.34 g/mol.

$$0.001 \text{ mol} \times 270.34 \text{ g/mol} = 0.270 \text{ g, theoretical yield of difluorodibenzalacetone}$$

Ⓜ *Percent Yield of Product:*

$$\frac{0.19 \text{ g}}{0.270 \text{ g}} \times 100 = 70\%$$

Ⓝ *The Mechanism of the Reaction:*

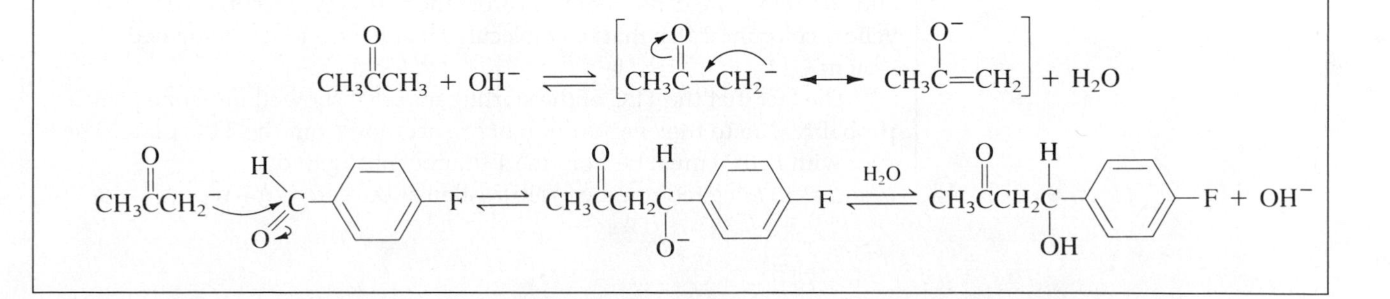

p. 38

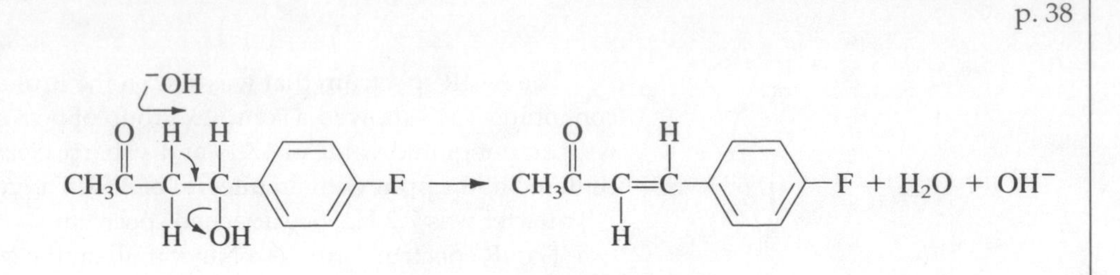

Ⓞ *Discussion and Conclusions:* Mixing the 4-fluorobenzaldehyde, acetone, and ethanol gave no apparent reaction because the solution did not change in appearance, but less than a minute after adding the NaOH, a yellow oil appeared and the reaction mixture got slightly warm, indicating that a reaction was taking place. The mechanism given in the text (above) indicates that hydroxide ion is a catalyst for the reaction, confirmed by this observation. The reaction takes place spontaneously at room temperature. The reaction was judged to be complete when the liquid surrounding the crystals became clear (it was opaque) and the amount of crystals in the tube did not appear to change. This took 30 min. The crude product was collected and washed with water. Forgot to cool it in ice before filtering it off. The crude product weighed more than the theoretical because it must still have been damp. However, this amount of crude material indicates that the reaction probably went pretty well to completion. The crude product was recrystallized from 5 mL of 70% ethanol. This was probably too much, since it dissolved very rapidly in that amount. This and the fact that the reaction mixture was not cooled in ice probably accounts for the relatively low 70% yield. Probably could have obtained second crop of crystals by concentrating filtrate. It turned very cloudy when water was added to it.

Theoretically, it is possible for this reaction to give three different products (*cis, cis-; cis, trans-;* and *trans, trans*-isomers). Because the mp of the crude and recrystallized products were close to each other and rather sharp, and because the TLC of the product gave only one spot, it is presumed that the product is just one of these three possible isomers. The NMR spectrum shows a sharp quartet of peaks from the vinyl protons, which also indicates only one isomer. Since the coupling constant observed is 7.2 Hz, the protons must be *cis* to each other. Therefore, the product is the *cis, cis*-isomer.

Ⓟ The infrared spectrum shows two strong peaks at 1590 and 1655 cm^{-1}, indicative of an α-β-unsaturated ketone. The small peaks at 3050, 3060, and 3072 cm^{-1} are consistent with aromatic and vinyl protons. The yellow color indicates that the molecule must have a long conjugated system.

The fact that the TLC of the starting material showed only one peak is probably due to the evaporation of the acetone from the TLC plate. The spot with R_f 0.23 must be from the 4-fluorobenzaldehyde.

Ⓠ Answers to assigned questions are written at the end of the report.

p. 39

Ⓡ Sept. 23, 2010 Procedure	Sept. 24, 2010 Observations
Weigh 0.248 g of 4-fluorobenzaldehyde into reaction tube.	Actually used 0.250 g.
Add 1.6 mL of ethanol stock soln that contains 0.058 g of acetone.	Actually used 1.61 mL.
Mix, observe for 1 min.	Nothing seems to happen. Water clear soln.
Add 2 mL of 3 *M* NaOH(*aq*).	Clear, then faintly yellow but clear, then after about 50 sec, the entire soln very suddenly turned opaque. Got slightly warm. Light yellow color.
Cap tube with septum; shake at 5-min intervals for 30 min.	Oily drops separate on sides and bottom of tube. These crystallized; more crystals formed. Tube became filled with crystals. Liquid around crystals became clear.
Filter on Hirsch funnel.	Filtered (filter paper) on Hirsch.
Wash crystals with much water.	Washed with about 150 mL of H_2O. Transfer of crystals done using filtrate.
Press dry.	Crystals pressed dry with cork.
Weigh crude.	Crude (damp) wt 0.27 g.
Save sample for mp.	Crude mp 177.5–179°C.
Recrystallize from 70% ethanol; . wash with a few drops of ice-cold solvent.	Used about 50 mL of 70% EtOH. Too much. Dissolved very rapidly.
Dry product.	Dried on paper for 30 min.
Weigh.	Wt 0.19 g.
Take mp.	Mp 179.5–180.5°C.
Run IR as mull.	Done.
Run NMR in $CDCl_3$.	Done.
Neutralize first filtrate with HCl, pour down drain.	Done.
Org. filtrate in waste organic bottle.	Done.

2 CHAPTER

Laboratory Safety, Courtesy, and Waste Disposal

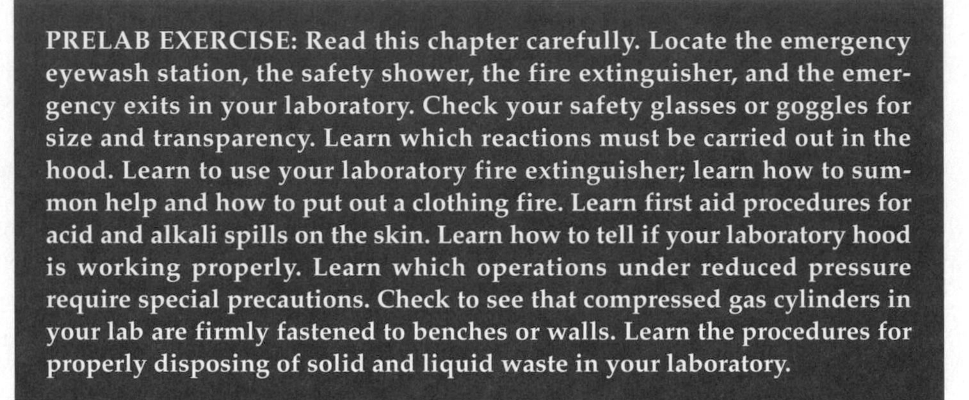

PRELAB EXERCISE: Read this chapter carefully. Locate the emergency eyewash station, the safety shower, the fire extinguisher, and the emergency exits in your laboratory. Check your safety glasses or goggles for size and transparency. Learn which reactions must be carried out in the hood. Learn to use your laboratory fire extinguisher; learn how to summon help and how to put out a clothing fire. Learn first aid procedures for acid and alkali spills on the skin. Learn how to tell if your laboratory hood is working properly. Learn which operations under reduced pressure require special precautions. Check to see that compressed gas cylinders in your lab are firmly fastened to benches or walls. Learn the procedures for properly disposing of solid and liquid waste in your laboratory.

Small-scale (microscale) organic experiments are much safer to conduct than their macroscale counterparts that are run on a scale up to 100 times larger. However, for either microscale or macroscale experiments, the organic chemistry laboratory is an excellent place to learn and practice safety. The commonsense procedures practiced here also apply to other laboratories as well as to the shop, kitchen, and studio.

General laboratory safety information—particularly applicable to this organic chemistry laboratory course—is presented in this chapter. But it is not comprehensive. Throughout this text you will find specific cautions and safety information presented as margin notes printed in red. For a brief and thorough discussion of the topics in this chapter, you should read *Safety in Academic Chemistry-Laboratories*.[1] There are also some specific admonitions regarding contact lenses (see "Eye Safety").

IMPORTANT GENERAL RULES

- Know the safety rules of your particular laboratory.
- Know the locations of emergency eyewashes, fire extinguishers, safety showers, and emergency exits.
- Never eat, drink, smoke, or apply cosmetics while in the laboratory.

[1]American Chemical Society Joint Board-Council Committee on Chemical Safety. *Safety in Academic Chemistry Laboratories, Vol. 1: Accident Prevention for College and University Students*, 7th ed.; American Chemical Society: Washington, DC, 2003 (0-8412-3864-2).

- Wear gloves and aprons when handling corrosive materials.
- Never work alone.
- Perform no unauthorized experiments and do not distract your fellow workers; horseplay has no place in the laboratory.

Dress sensibly.

- Dress properly for lab work. Do not wear open-toed shoes; your feet must be completely covered; wear shoes that have rubber soles and no heels or sneakers.
- Confine long hair and loose clothes. Do not wear shorts.
- Immediately report any accident to your instructor.
- If the fire extinguisher is used, report this to your instructor.
- Never use mouth suction to fill a pipette.
- Always wash your hands before leaving the laboratory.
- Do not use a solvent to remove a chemical from your skin. This will only hasten the absorption of the chemical through the skin.
- Do not use cell phones or tape, CD, MP3, iPods, or similar music players while working in the laboratory.
- Refer to the chemical supplier's hazard warning information or Material Safety Data Sheet (MSDS) when handling a new chemical for the first time.

Eye protection

Information on Eye Protection

Eye protection is extremely important. The requirement that you wear approved eye protection in the laboratory is mandatory and is most effective in preventing the largest number of potential accidents of serious consequence. Safety glasses or goggles of some type must be worn at all times that you are in the laboratory, whether engaged in experimental work or not. Ordinary prescription eyeglasses do not offer adequate protection.

There are three types of eye protection acceptable for use in the organic chemistry laboratory; they are described below. Laboratory safety glasses should be constructed of plastic or tempered glass. If you do not have such glasses, wear goggles that afford protection from splashes and objects coming from the side as well as from the front. Chemical splash goggles are the preferred eye protection. One of the most important features of safety glasses/goggles is the brow bar. It is critical to have proper eye protection above the eyes; a brow bar satisfies this requirement for adequate splash protection. If plastic safety glasses are permitted in your laboratory, they should have side shields (Fig. 2.1). Eye safety cannot be overemphasized in the chemistry laboratory.

1. **Goggles** are pliable, form a complete cup around the eyes, and are held in place by a strap that wraps around the head. Goggles offer the highest level of splash protection compared to other types of protective eyewear. The disadvantages are that they may fog up and limit peripheral vision. Most goggles will fit over prescription eyeglasses.
2. **Safety Glasses** look similar to prescription glasses but have side shields and a browbar to provide extra splash protection. They are generally considered to be more comfortable and to offer better peripheral vision compared to goggles. The disadvantages are that they do not fit well, if at all, over prescription eyeglasses and they offer less protection than goggles since they do not completely cover the eye.

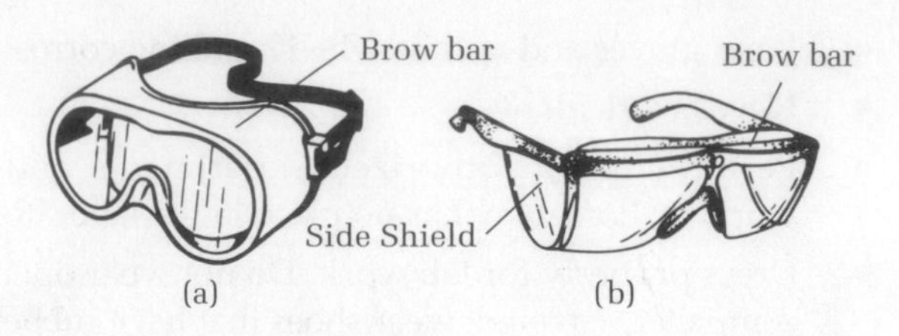

FIG. 2.1
(a) Chemical splash goggles
(b) Safety glasses

3. **Visor Goggles** are a type of "half goggle"; the upper half is similar to a goggle and the lower half is similar to safety glasses. As such, they have some of the advantages and disadvantages of both safety glasses and goggles. They are reasonably comfortable, afford good splash protection, have better peripheral vision, and show less tendency to fog than goggles. They fit over many types of prescription eyeglasses.

Contact Lens Users

In 1994, the Occupational Safety and Health Administration (OSHA) concluded that "contact lenses do not pose additional hazards to the wearer, and has determined that additional regulation addressing the use of contact lenses is unnecessary."[2] Other reports support this position.[3,4] It has been determined "that contact lenses can be worn in most work environments provided the same approved eye protection is worn as required of other workers in the area. Approved eye protection refers to safety glasses or goggles."[5]

LABORATORY COURTESY

Please show up on time and be prepared for the day's work. Clean up and leave promptly at the end of the lab period. Clean your desktop and sink before you leave the lab. Be certain that no items such as litmus paper, used filter papers, used cotton, or stir bars collect in the sink. Dispose of all trash properly. Please keep the balances clean. Always replace the caps on reagent bottles after use.

WORKING WITH FLAMMABLE SUBSTANCES

Relative flammability of organic solvents.

Flammable substances are the most common hazard in the organic laboratory. Two factors can make today's organic laboratory much safer than its predecessor: (1) making the scale of the experiments as small as possible and (2) not using flames. Diethyl ether (bp 35°C), the most flammable substance you will usually work with in this course, has an ignition temperature of 160°C, which means that a hot plate at that temperature will set it afire. For comparison, n-hexane (bp 69°C), a constituent of gasoline, has an ignition temperature of 225°C. The flash points of these organic liquids—that is, the temperatures at which they will catch fire if exposed to a flame or spark—are below 220°C. These are very flammable liquids; however, if you are

[2]OSHA Personal Protective Equipment for General Industry Standard, 29 CFR 1910; final rule, April 6, 1994, p. 16343.
[3]*Chemical Health and Safety* **1995**, 16.
[4]*Chemical Health and Safety* **1997**, 33.
[5]Ramsey, H.; and Breazeale, W. H. J. Jr. *Chem. Eng. News* **1998**, *76* (22), 6.

careful to eliminate all possible sources of ignition, they are not difficult to work with. Except for water, almost all of the liquids you will use in the laboratory are flammable.

Bulk solvents should be stored in and dispensed from appropriate safety containers (Fig. 2.2). These and other liquids will burn in the presence of the proper amount of their flammable vapors, oxygen, and a source of ignition (most commonly a flame or spark). It is usually difficult to remove oxygen from a fire, although it is possible to put out a fire in a beaker or a flask by simply covering the vessel with a flat object, thus cutting off the supply of air. Your lab notebook might do in an emergency. The best prevention is to pay close attention to sources of ignition—open flames, sparks, and hot surfaces. Remember, the vapors of flammable liquids are **always** heavier than air and thus will travel along bench tops, down drain troughs, and remain in sinks. For this reason all flames within the vicinity of a flammable liquid must be extinguished. Adequate ventilation is one of the best ways to prevent flammable vapors from accumulating. Work in an exhaust hood when manipulating large quantities of flammable liquids.

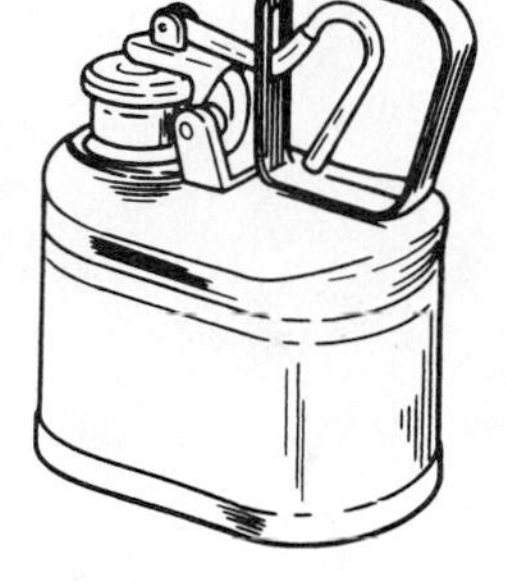

FIG. 2.2
Solvent safety can.

If a person's clothing catches fire and there is a safety shower close at hand, then shove the person under it and turn the shower on. Otherwise, push the person down and roll him or her over to extinguish the flames (stop, drop, and roll!). It is extremely important to prevent the victim from running or standing because the greatest harm comes from breathing the hot vapors that rise past the mouth. The safety shower might then be used to extinguish glowing cloth that is no longer aflame. A so-called fire blanket should not be used because it tends to funnel flames past the victim's mouth, and clothing continues to char beneath it. It is, however, useful for retaining warmth to ward off shock after the flames are extinguished.

⚠ Flammable vapors travel along bench tops.

⚠ Keep ignition sources away from flammable liquids.

An organic chemistry laboratory should be equipped with a carbon dioxide or dry chemical (monoammonium phosphate) *fire extinguisher* (Fig. 2.3). To use this type of extinguisher, lift it from its support, pull the ring to break the seal, raise the horn, aim it at the base of the fire, and squeeze the handle. Do not hold on to the horn because it will become extremely cold. Do not replace the extinguisher; report the incident so the extinguisher can be refilled.

When disposing of certain chemicals, be alert for the possibility of *spontaneous combustion*. This may occur in oily rags; organic materials exposed to strong oxidizing agents such as nitric acid, permanganate ions, and peroxides; alkali metals such as sodium; or very finely divided metals such as zinc dust and platinum catalysts. Fires sometimes start when these chemicals are left in contact with filter paper.

FIG. 2.3
Carbon dioxide fire extinguisher.

WORKING WITH HAZARDOUS CHEMICALS

If you do not know the properties of a chemical you will be working with, it is wise to regard the chemical as hazardous. The flammability of organic substances poses the most serious hazard in the organic laboratory. There is a possibility that storage containers in the laboratory may contribute to a fire. Large quantities of organic solvents should not be stored in glass bottles; they should be stored in solvent safety cans. Do not store chemicals on the floor.

A flammable liquid can often be vaporized to form, with air, a mixture that is explosive in a confined space. The beginning chemist is sometimes surprised to learn that diethyl ether is more likely to cause a laboratory fire or explosion than a worker's accidental anesthesia. The chances of being confined in a laboratory with a concentration of ether high enough to cause a loss of consciousness are extremely small, but a spark in such a room would probably destroy the building.

⚠ Flammable vapors plus air in a confined space are explosive.

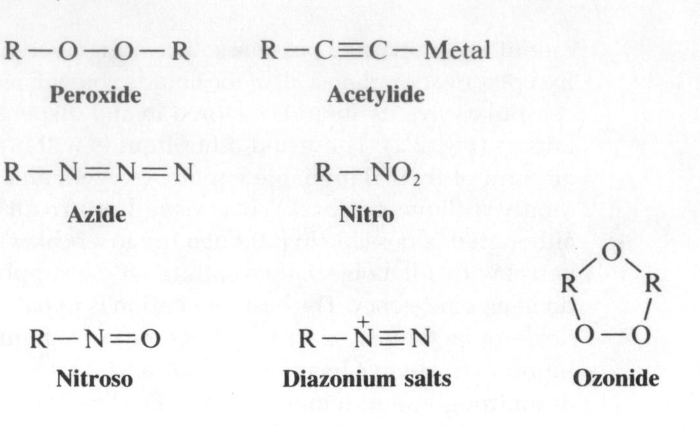

FIG. 2.4
Functional groups that can be explosive in some compounds.

The probability of forming an explosive mixture of volatile organic liquids with air is far greater than that of producing an explosive solid or liquid. The chief functional groups that render compounds explosive are the *peroxide, acetylide, azide, diazonium, nitroso, nitro, and ozonide groups* (Fig. 2.4). Not all members of these groups are equally sensitive to shock or heat. You would find it difficult to detonate trinitrotoluene (TNT) in the laboratory, but nitroglycerine is treacherously explosive. Peroxides present special problems that are discussed in the next section.

⚠ Safety glasses or goggles must be worn at all times.

You will need to contend with the corrosiveness of many of the reagents you will handle. The principal danger here is to the eyes. Proper eye protection is mandatory, and even small-scale experiments can be hazardous to the eyes. It takes only a single drop of a corrosive reagent to do permanent damage. Handling concentrated acids and alkalis, dehydrating agents, and oxidizing agents calls for commonsense care to avoid spills and splashes and to avoid breathing the often corrosive vapors.

Certain organic chemicals present acute toxicity problems from short-duration exposure and chronic toxicity problems from long-term or repeated exposure. Exposure can result from ingestion, contact with the skin, or, most commonly, inhalation. Currently, great attention is being focused on chemicals that are teratogens (chemicals that often have no effect on a pregnant woman but cause abnormalities in a fetus), mutagens (chemicals causing changes in the structure of the DNA, which can lead to mutations in offspring), and carcinogens (cancer-causing chemicals). Small-scale experiments significantly reduce these hazards but do not completely eliminate them.

WORKING WITH EXPLOSIVE HAZARDS

1. Peroxides

⚠ Ethers form explosive peroxides.

Certain functional groups can make an organic molecule become sensitive to heat and shock, such that it will explode. Chemists work with these functional groups only when there are no good alternatives. One of these functional groups, the peroxide group (R—O—O—R), is particularly insidious because it can form spontaneously when oxygen and light are present (Fig. 2.5). Ethers, especially cyclic ethers and those made from primary or secondary alcohols (such as tetrahydrofuran, diethyl ether, and diisopropyl ether), form peroxides. Other compounds that form peroxides are aldehydes, alkenes that have allylic hydrogen atoms (such as

FIG. 2.5
Some compounds that form peroxides.

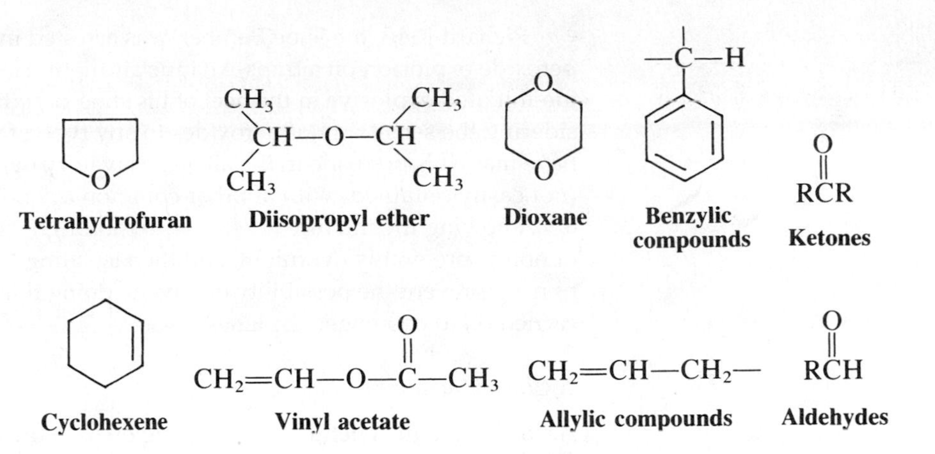

cyclohexene), compounds having benzylic hydrogens on a tertiary carbon atom (such as isopropyl benzene), and vinyl compounds (such as vinyl acetate). Peroxides are low-power explosives but are extremely sensitive to shock, sparks, light, heat, friction, and impact. The greatest danger from peroxide impurities comes when the peroxide-forming compound is distilled. The peroxide has a higher boiling point than the parent compound and remains in the distilling flask as a residue that can become overheated and explode. For this reason, one should never distill a liquid for too long a period of time so that the distilling flask completely dries out, and the distillation of a peroxide-containing liquid should be run in a hood with the sash down to help contain a possible explosion.

Never distill to dryness.

The Detection of Peroxides	The Removal of Peroxides
To a solution of 0.01 g of sodium iodide in 0.1 mL of glacial acetic acid, add 0.1 mL of the liquid suspected of containing a peroxide. If the mixture turns brown, a high concentration of peroxide is present; if it turns yellow, a low concentration of peroxide is present.	Pouring the solvent through a column of activated alumina will remove peroxides and simultaneously dry the solvent. Do not allow the column to dry out while in use. When the alumina column is no longer effective, wash the column with 5% aqueous ferrous sulfate and discard the column as nonhazardous waste.

Problems with peroxide formation are especially critical for ethers. Ethers (R-O-R) form peroxides readily. Because ethers are frequently used as solvents, they are often used in quantity and then removed to leave reaction products. Cans of diethyl ether should be dated when opened. If opened cans are not used within one month, they should be treated for peroxides and disposed of.

t-Butyl methyl ether, $(CH_3)_3C\text{-}O\text{-}CH_3$, with a primary carbon on one side of the oxygen and a tertiary carbon on the other, does not form peroxides easily. It is highly desirable to use this in place of diethyl ether for extraction. Refer to the discussion in Chapter 7.

You may have occasion to use 30% *hydrogen peroxide*. This material causes severe burns if it contacts the skin and decomposes violently if contaminated with metals or their salts. Be particularly careful not to contaminate the reagent bottle.

Richard Reid, the Shoe Bomber, was arrested in 2001 for attempting to set off a peroxide explosion on a trans-Atlantic air flight. He had packed 240 g of the peroxide-initiated explosive in the heel of his shoe, a highly hazardous undertaking considering the sensitivity of peroxides to any type of friction. Theoretically, he could have made his peroxide in the airplane lavatory by mixing 30% hydrogen peroxide (not easily obtained) with another common organic liquid and an acid catalyst at 0°C, allowing the mixture to react with thorough stirring and cooling for a number of hours, preferably overnight, and then isolating the crystalline peroxide by filtration. To prevent the possibility of anyone doing this in the future, no liquids can be carried on to passenger airplanes.

2. Closed Systems

A closed system is defined as not being open to the atmosphere. Any sealed system is a closed system. If a closed system is not properly prepared, an explosion may result from the system being under pressure, caused from gas or heat evolution from the reaction or from applied heat to the system. One way to prevent an explosion of a closed system is to use glassware that can withstand the pressure, and to evacuate the system under vacuum before it is closed to the atmosphere.

Most reactions are run in open systems; that is, they are run in apparatus that are open to atmosphere, either directly or through a nitrogen line hooked up to a bubbler, which is open to atmosphere.

WORKING WITH CORROSIVE SUBSTANCES

Handle strong acids, alkalis, dehydrating agents, and oxidizing agents carefully so as to avoid contact with the skin and eyes, and to avoid breathing the corrosive vapors that attack the respiratory tract. All strong, concentrated acids attack the skin and eyes. Concentrated *sulfuric acid* is both a dehydrating agent and a strong acid and will cause very severe burns. *Nitric acid and chromic acid* (used in cleaning solutions) also cause bad burns. *Hydrofluoric* acid is especially harmful and causes deep, painful, and slow-healing wounds. It should be used only after thorough instruction. Do not add water to concentrated sulphuric acid. The heat of solution is so large that the acid may boil and spatter. It can best be diluted by pouring the acid into water, with stirring. You should wear approved safety glasses or goggles, protective gloves, and an apron when handling these materials.

Sodium, potassium, and ammonium *hydroxides* are common bases that you will encounter. Sodium and potassium hydroxides are extremely damaging to the eyes, and ammonium hydroxide is a severe bronchial irritant. Like sulfuric acid, sodium hydroxide, *phosphorous pentoxide*, and *calcium oxide* are powerful dehydrating agents. Their great affinity for water will cause burns to the skin. Because they release a great deal of heat when they react with water, to avoid spattering they should always be added to water rather than water being added to them. That is, the heavier substance should always be added to the lighter one so that layers don't form where one liquid floats on another; the desired result is a rapid mixing that occurs as a consequence of the law of gravity.

⚠ Add H_2SO_4, P_2O_5, CaO, and NaOH to water, not the reverse.

You will receive special instructions when it comes time to handle metallic sodium, lithium aluminum hydride, and sodium hydride, three substances that can react explosively with water.

Wipe up spilled hydroxide pellets *rapidly*.

Among the strong oxidizing agents, *perchloric acid* ($HClO_4$) is probably the most hazardous. It can form heavy metal and organic *perchlorates* that are *explosive*, and it can react explosively if it comes in contact with organic compounds.

If one of these substances gets on the skin or in the eyes, wash the affected area with very large quantities of water, using the safety shower and/or eyewash station (Fig. 2.6) until medical assistance arrives. Do not attempt to neutralize the reagent chemically. Remove contaminated clothing so that thorough washing can take place. Take care to wash the reagent from under the fingernails.

Take care not to let the reagents, such as sulfuric acid, run down the outside of a bottle or flask and come in contact with your fingers. Wipe up spills immediately with a very damp sponge, especially in the area around the balances. Pellets of sodium and potassium hydroxide are very hygroscopic and will dissolve in the water they pick up from the air; they should therefore be wiped up very quickly. When handling large quantities of corrosive chemicals, wear protective gloves, a face mask, and a neoprene apron. The corrosive vapors can be avoided by carrying out work in a good exhaust hood.

Do not use a plastic syringe with a metal needle to dispense corrosive inorganic reagents, such as concentrated acids or bases.

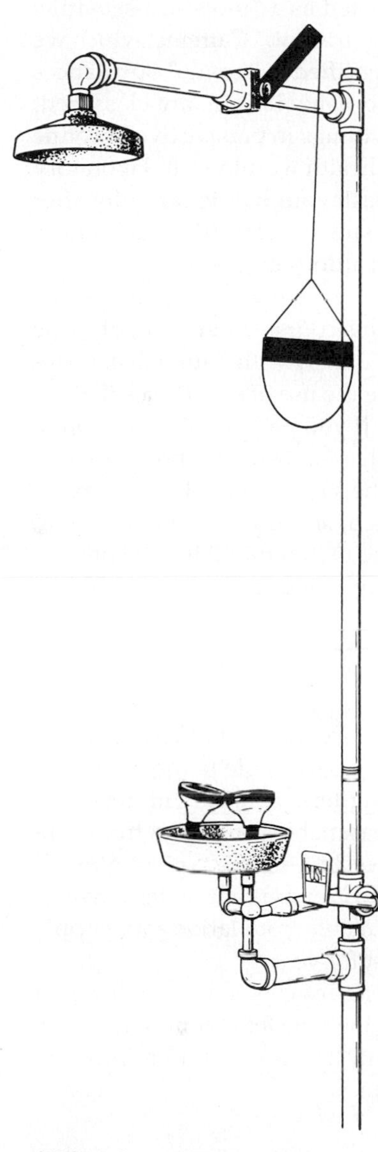

FIG. 2.6
Emergency shower and eyewash station.

WORKING WITH TOXIC SUBSTANCES

Many chemicals have very specific toxic effects. They interfere with the body's metabolism in a known way. For example, the cyanide ion combines irreversibly with hemoglobin to form cyanometmyoglobin, which can no longer carry oxygen. Aniline acts in the same way. Carbon tetrachloride and other halogenated compounds can cause liver and kidney failure. Carcinogenic and mutagenic substances deserve special attention because of their long-term insidious effects. The ability of certain carcinogens to cause cancer is very great; for example, special precautions are needed when handling aflatoxin B_1. In other cases, such as with dioxane, the hazard is so low that no special precautions are needed beyond reasonable, normal care in the laboratory.

Women of childbearing age should be careful when handling any substance of unknown properties. Certain substances are highly suspected as teratogens and will cause abnormalities in an embryo or fetus. Among these are benzene, toluene, xylene, aniline, nitrobenzene, phenol, formaldehyde, dimethylformamide (DMF), dimethyl sulfoxide (DMSO), polychlorinated biphenyls (PCBs), estradiol, hydrogen sulfide, carbon disulfide, carbon monoxide, nitrites, nitrous oxide, organolead and mercury compounds, and the notorious sedative thalidomide. Some of these substances will be used in subsequent experiments. Use care when working with these (and all) substances. One of the leading known causes of embryotoxic effects is ethyl alcohol in the form of maternal alcoholism, but the amount of ethanol vapor inhaled in the laboratory or absorbed through the skin is so minute that it is unlikely to have morbid effects.

It is impossible to avoid handling every known or suspected toxic substance, so it is wise to know what measures should be taken. Because the eating of food or the consumption of beverages in the laboratory is strictly forbidden and because one should never taste material in the laboratory, the possibility of poisoning by mouth is remote. Be more careful than your predecessors—the hallucinogenic properties of LSD and **all** artificial sweeteners were discovered by accident. The two most important measures to be taken, then, are (1) avoiding skin contact by wearing the *proper* type of protective gloves (*see* "Gloves") and (2) avoiding inhalation by working in a good exhaust hood.

Many of the chemicals used in this course will be unfamiliar to you. Their properties can be looked up in reference books, a very useful one being the *Aldrich Handbook of Fine Chemicals*.[6] Note that 1,4-dichlorobenzene is listed as a "toxic irritant" and naphthalene is listed as an "irritant." Both are used as mothballs. Camphor, used in vaporizers, is classified as a "flammable solid irritant." Salicylic acid, which we will use to synthesize aspirin (Chapter 41), is listed as a "moisture-sensitive toxic." Aspirin (acetylsalicylic acid) is classified as an "irritant." Caffeine, which we will isolate from tea or cola syrup (Chapter 7), is classified as "toxic." Substances not so familiar to you, for example, 1-naphthol and benzoic acid, are classified, respectively, as "toxic irritant" and "irritant." To put things in perspective, nicotine is classified as "highly toxic." Pay attention to these health warnings. In laboratory quantities, common chemicals can be hazardous. Wash your hands carefully after coming in contact with laboratory chemicals. Consult the *Hazardous Laboratory Chemicals Disposal Guide*[7] for information on truly hazardous chemicals.

Because you have not had previous experience working with organic chemicals, most of the experiments you will carry out in this course will not involve the use of known carcinogens, although you will work routinely with flammable, corrosive, and toxic substances. A few experiments involve the use of substances that are suspected of being carcinogenic, such as hydrazine. If you pay proper attention to the rules of safety, you should find working with these substances no more hazardous than working with ammonia or nitric acid. The single, short-duration exposure you might receive from a suspected carcinogen, should an accident occur, would probably have no long-term consequences. The reason for taking the precautions noted in each experiment is to learn, from the beginning, good safety habits.

GLOVES

Be aware that protective gloves in the organic laboratory may not offer much protection. Polyethylene and latex rubber gloves are very permeable to many organic liquids. An undetected pinhole may bring with it long-term contact with reagents. Disposable polyvinyl chloride (PVC) gloves offer reasonable protection from contact with aqueous solutions of acids, bases, and dyes, but no one type of glove is useful as a protection against all reagents. It is for this reason that no less than 25 different types of chemically resistant gloves are available from laboratory supply houses. Some gloves are quite expensive and will last for years.

If disposable gloves are available, fresh nitrile gloves can be worn whenever handling a corrosive substance and disposed of once the transfer is complete. When not wearing gloves, it is advised that you wash your hands every 15 minutes to remove any traces of chemicals that might be on them.

USING THE LABORATORY HOOD

Modern practice dictates that in laboratories where workers spend most of their time working with chemicals, there should be one exhaust hood for every two people. However, this precaution is often not possible in the beginning organic chemistry laboratory. In this course you will find that for some experiments, the hood

[6]Free copies of this catalog can be obtained from http://www.sigmaaldrich.com/Brands/Aldrich.html.
[7]Armour, M-A. *Hazardous Laboratory Chemicals Disposal Guide*, 3rd ed.; CRC Press LLC: Boca Raton, FL, 2003. This extremely useful book is now available without charge on Google Books.

must be used and for others it is advisable; in these instances, it may be necessary to schedule experimental work around access to the hoods. Many experiments formerly carried out in the hood can now be carried out at the lab desk because the concentration of vapors is significantly minimized when working at a microscale.

Keep the hood sash closed.

The hood offers a number of advantages when working with toxic and flammable substances. Not only does it draw off the toxic and flammable fumes, but it also affords an excellent physical barrier on all four sides of a reacting system when the sash is pulled down. If a chemical spill occurs, it may be contained within the hood.

It is your responsibility each time you use a hood to see that it is working properly. You should find some type of indicating device that will give you this information on the hood itself. A simple propeller on a cork works well. Note that the hood is a backup device. Never use it alone to dispose of chemicals by evaporation; use an aspirator tube or carry out a distillation. Toxic and flammable fumes should be trapped or condensed in some way and disposed of in the prescribed manner. The sash should be pulled down unless you are actually carrying out manipulations with the experimental apparatus. The water, gas, and electrical controls should be on the outside of the hood, so it is not necessary to open the hood to make adjustments. The ability of the hood to remove vapors is greatly enhanced if the apparatus is kept as close to the back of the hood as possible, where the air movement is the strongest. Everything should be at least 15 cm back from the hood sash. Chemicals should not be permanently stored in the hood, but should be removed to ventilated storage areas. If the hood is cluttered with chemicals, you will not achieve a good, smooth airflow or have adequate room for experiments.

WORKING AT REDUCED PRESSURE

Implosion

Whenever a vessel or system is evacuated, an implosion could result from atmospheric pressure on the empty vessel. It makes little difference whether it's a total vacuum or just 10 mm Hg; the pressure difference is almost the same (760 versus 750 mm Hg). An implosion may occur if there is a star crack in the flask or if the flask is scratched or etched. Only with heavy-walled flasks specifically designed for vacuum filtration is the use of a safety shield (Fig. 2.7) ordinarily unnecessary. Although caution should still be observed, the chances of implosion of the apparatus used for microscale experiments are remote.

Dewar flasks (thermos bottles) are often found in the laboratory without shielding. These should be wrapped with friction tape or covered with a plastic net to prevent the glass from flying about in case of an implosion (Fig. 2.8). Similarly, vacuum desiccators should be wrapped with tape before being evacuated.

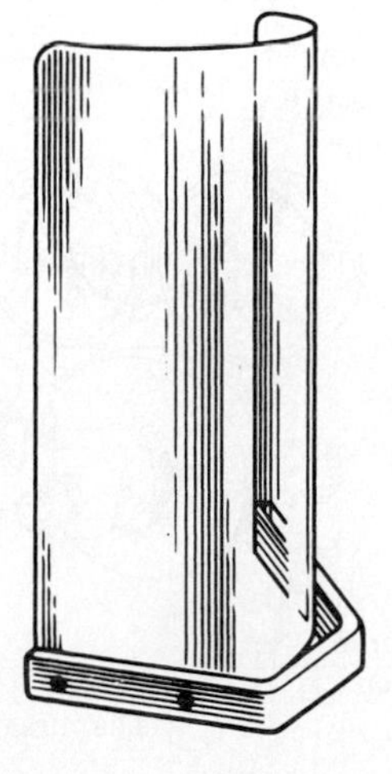

FIG. 2.7
Safety shield.

FIG. 2.8
Dewar flask with safety net in place.

WORKING WITH COMPRESSED GAS CYLINDERS

Many reactions are carried out under an inert atmosphere so that the reactants and/or products will not react with oxygen or moisture in the air. Nitrogen and argon are the inert gases most frequently used. Oxygen is widely used both as a reactant and to provide a hot flame for glassblowing and welding. It is used in the oxidative coupling of alkynes (Chapter 24). Helium is the carrier gas used in gas chromatography. Other gases commonly used in the laboratory are ammonia, which is often used as a solvent; chlorine, used for chlorination reactions; acetylene, used in combination with oxygen for welding; and hydrogen, used for high- and low-pressure hydrogenation reactions.

⚠ Always clamp gas cylinders

The following rule applies to all compressed gases: Compressed gas cylinders should be firmly secured at all times. For temporary use, a clamp that attaches to the laboratory bench top and has a belt for the cylinder will suffice (Fig. 2.9). Eyebolts and chains should be used to secure cylinders in permanent installations. Flammable gases should be stored 20 feet from oxidizing gases.

A variety of outlet threads are used on gas cylinders to prevent incompatible gases from being mixed because of an interchange of connections. Both right- and left-handed external and internal threads are used. Left-handed nuts are notched to differentiate them from right-handed nuts. Right-handed threads are used on nonfuel and oxidizing gases, and left-handed threads are used on fuel gases, such as hydrogen. Never grease the threads on tank or regulator valves because there is the possibility that the grease could ignite under certain conditions.

Cylinders come equipped with caps that should be left in place during storage and transportation. These caps can be removed by hand. Under these caps is a cylinder valve. It can be opened by turning the valve counterclockwise. However, because most compressed gases in full cylinders are under very high pressure (commonly up to 3000 lb/in.^{2}), a pressure regulator must be attached to the cylinder. This pressure regulator is almost always of the diaphragm type and has two gauges—one indicating the pressure in the cylinder, the other the outlet pressure (Fig. 2.10). On the outlet, low-pressure side of the regulator is a small needle valve and then the outlet connector. After connecting the regulator to the cylinder, unscrew the diaphragm valve (turn it counterclockwise) before

***Clockwise* movement of the diaphragm valve handle *increases* pressure.**

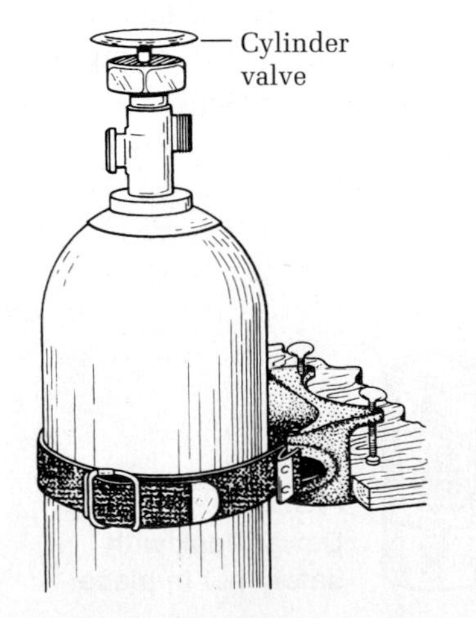

FIG. 2.9
Gas cylinder clamp.

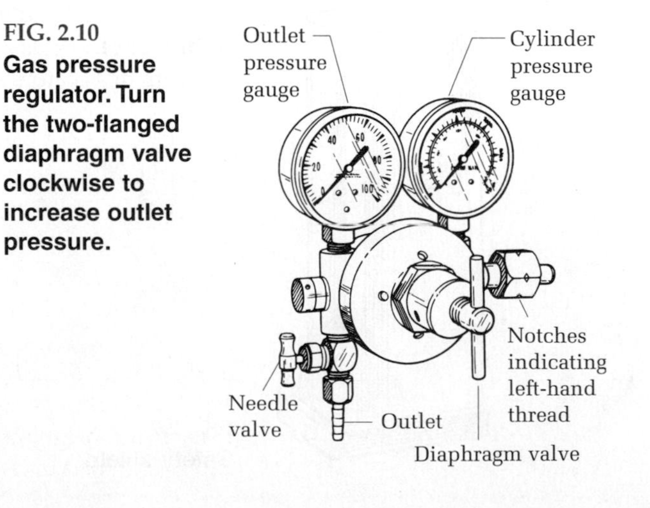

FIG. 2.10
Gas pressure regulator. Turn the two-flanged diaphragm valve clockwise to increase outlet pressure.

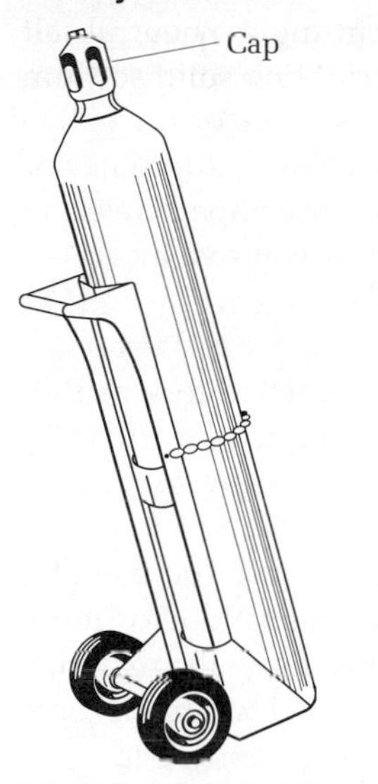

FIG. 2.11
Gas cylinder cart.

opening the cylinder valve on the top of the cylinder. This valve should be opened only as far as necessary. For most gas flow rates in the laboratory, this will be a very small amount. The gas flow or pressure is increased by turning the two-flanged diaphragm valve **clockwise**. When the gas is not being used, turn off the cylinder valve (clockwise) on the top of the cylinder (Fig. 2.9). Before removing the regulator from the cylinder, reduce the flow or pressure to zero. Cylinders should never be emptied to zero pressure and left with the valve open because the residual contents will become contaminated with air. Empty cylinders should be labeled "empty." Their valves should be closed and capped, and the cylinders should be returned to the storage area and separated from full cylinders. Gas cylinders should never be dragged or rolled from place to place, but should be fastened onto and moved in a cart designed for that purpose (Fig. 2.11). The cap should be in place whenever the cylinder is moved. If you detect even a small leak from any valve or connection, immediately seek the help of an instructor to remedy the problem. If there is a major leak of a corrosive or flammable gas, notify those around you to leave the area and seek help immediately.

ODORIFEROUS CHEMICALS

⚠ Never attempt to identify an unknown organic compound by smelling it.

Some organic chemicals just smell bad. Among these are the thiols (organic derivatives of hydrogen sulfide, also called mercaptans), isonitriles, many amines (e.g., cadaverine), and butyric acid. Washing apparatus and, if necessary, hands in a solution of a quaternary ammonium salt may solve the problem. These compounds apparently complex with many odoriferous substances, allowing them to be rinsed away. Commercial products (e.g., Zephiran, Roccal, San-O-Fec, and others) are available at pet and farm supply stores.

Clean up spills rapidly.

WASTE DISPOSAL—CLEANING UP

Spilled solids should simply be swept up and placed in the appropriate solid waste container. This should be done promptly because many solids are hygroscopic and become difficult if not impossible to sweep up in a short time. This is particularly true of sodium hydroxide and potassium hydroxide; these strong bases should be dissolved in water and neutralized with sodium bisulfate before disposal.

The method used to clean up spills depends on the type and amount of chemical spilled. If more than 1 or 2 g or mL of any chemical, particularly a corrosive or volatile one, is spilled, you should consult your instructor for the best way to clean up the spill. If a large amount of volatile or noxious liquid is spilled as might happen if a reagent bottle is dropped and broken, advise those in the area to leave the laboratory and contact your instructor immediately. If a spill involves a large amount of flammable liquid, be aware of any potential ignition sources and try to eliminate them. Large amounts of spilled acid can be neutralized with granular limestone or cement; large amounts of bases can be neutralized with solid sodium bisulfate, $NaHSO_4$. Large amounts of volatile liquids can be absorbed into materials such as vermiculite, diatomaceous earth, dry sand, kitty litter, or paper towels and these materials swept up and placed in a separate disposal container.

⚠ Mercury requires special measures—see your instructor

For spills of amounts less than 2 g of chemical, proceed as follows. Acid spills should be neutralized by dropping solid sodium carbonate onto them, testing the pH, wiping up the neutralized material with a sponge, and rinsing the neutral salt solution down the drain. Bases should be neutralized by sprinkling solid sodium bisulfate onto them, checking the pH, and wiping up with a sponge or towel. Do not use paper towels to wipe up spills of strong oxidizers such as dichromates or nitrates; the towels can ignite. Bits of sodium metal will also cause paper towels to ignite. Sodium metal is best destroyed with n-butyl alcohol. Always wear gloves when cleaning up a spill.

Cleaning Up. In the not-too-distant past, it was common practice to wash all unwanted liquids from the organic laboratory down the drain and to place all solid waste in the trash basket. For environmental reasons, this is never a wise practice and is no longer allowed by law.

Organic reactions usually employ a solvent and often involve the use of a strong acid, a strong base, an oxidant, a reductant, or a catalyst. None of these should be washed down the drain or placed in the wastebasket. Place the material, classified as waste, in containers labeled for nonhazardous solid waste, organic solvents, halogenated organic solvents, or hazardous wastes of various types.

Waste containers:
Nonhazardous solid waste
Organic solvents
Halogenated organic solvents
Hazardous waste (various types)

Nonhazardous waste encompasses such solids as paper, corks, TLC plates, solid chromatographic absorbents such as alumina or silica that are dry and free of residual *organic solvents*, and solid drying agents such as calcium chloride or sodium sulfate that are also dry and free of residual organic solvents. These will ultimately end up in a sanitary landfill (the local dump). Any chemicals that are leached by rainwater from this landfill must not be harmful to the environment. In the organic solvents container are placed the solvents that are used for recrystallization and for running reactions, cleaning apparatus, and so forth. These solvents can contain dissolved, solid, nonhazardous organic solids. This solution will go to an incinerator where it will be burned. If the solvent is halogenated (e.g., dichloromethane) or contains halogenated material, it must go in the *halogenated organic solvents* container. Ultimately, this will go to a special incinerator equipped with a scrubber to remove HCl from the combustion gases. The organic laboratory should also have several other waste disposal containers for special hazardous, reactive, and noncombustible wastes that would be incompatible with waste organic solvents and other materials. For example, it would be dangerous to place oxidants in lysts with many organics. In particular, separate waste containers should be provided for toxic heavy metal wastes containing mercury, chromium, or lead salts, and so forth. The cleaning up sections throughout this text will call your attention to these special hazards.

Hazardous wastes such as sodium hydrosulfite (a reducing agent), platinum catalysts, and Cr^{6+}an oxidizing agent) cannot be burned and must be shipped to a secure landfill. To dispose of small quantities of a hazardous waste (e.g., solid mercury hydroxide), the material must be carefully packed in bottles and placed in a 55-gal (≈-208-L) drum called a lab pack, to which an inert material has been added. The lab pack is carefully documented and hauled off to a site where such waste is disposed of by a bonded, licensed, and heavily regulated waste disposal company. Formerly, many hazardous wastes were disposed of by burial in a secure landfill. The kinds of hazardous waste that can be thus disposed of have become extremely limited in recent years, and much of the waste undergoes various kinds of treatment at the disposal site (e.g., neutralization, incineration, or reduction) to put it in a form that can be safely buried in a secure landfill or flushed to a sewer. There are relatively few places for approved disposal of hazardous waste. For example, there are none in New England, so most hazardous waste from this area is trucked to South Carolina. The charge to small generators of waste is usually based on the volume of waste. So, 1000 mL of a 2% cyanide solution would cost far more to dispose of than 20 g of solid cyanide, even though the total amount of this poisonous substance is the same. It now costs far more to dispose of most hazardous chemicals than it does to purchase them new.

Waste disposal is very expensive.

American law states that a material is not a waste until the laboratory worker declares it a waste. So—for pedagogical and practical reasons—we want you to regard the chemical treatment of the byproducts of each reaction in this text as a part of the experiment.

The law: A waste is not a waste until the laboratory worker declares it a waste.

In the section titled "Cleaning Up" at the end of each experiment, the goal is to reduce the volume of hazardous waste, to convert hazardous waste to less hazardous waste, or to convert it to nonhazardous waste. The simplest example is concentrated sulfuric acid. As a byproduct from a reaction, it is obviously hazardous. But after careful dilution with water and neutralization with sodium carbonate, the sulfuric acid becomes a dilute solution of sodium sulfate, which in almost every locale can be flushed down the drain with a large quantity of water. Anything flushed down the drain must be accompanied by a large excess of water. Similarly, concentrated bases can be neutralized, oxidants such as Cr^{6+} can be reduced, and reductants such as hydrosulfite can be oxidized (by hypochlorite or household bleach). Dilute solutions of heavy metal ions can be precipitated as their insoluble sulfides or hydroxides. The precipitate may still be a hazardous waste, but it will have a much smaller volume.

Cleaning up: reducing the volume of hazardous waste or converting hazardous waste to less hazardous or nonhazardous waste.

One type of hazardous waste is unique: a harmless solid that is damp with an organic solvent. Alumina from a chromatography column and calcium chloride used to dry an ether solution are examples. Being solids, they obviously cannot go in the organic solvents container, and being flammable they cannot go in the nonhazardous waste container. A solution to this problem is to spread the solid out in the hood to let the solvent evaporate. You can then place the solid in the nonhazardous waste container. The savings in waste disposal costs by this operation are enormous. However, be aware of the regulations in your area, as they may not allow evaporation of small amounts of organic solvents in a hood. If this is the case, special containers should be available for disposal of these wet solids.

Disposing of solids wet with organic solvents: alumina and anhydrous calcium chloride pellets.

Our goal in "Cleaning Up" is to make you more aware of all aspects of an experiment. Waste disposal is now an extremely important aspect. Check to be sure the procedure you use is permitted by law in your location. Three sources of information have been used as the basis of the procedures at the end of each experiment: the *Aldrich Catalog Handbook of Fine Chemicals*,[8] which gives brief disposal procedures for every chemical in their catalog; *Prudent Practices in the Laboratory: Handling and Disposal of Chemicals*[9]; and the *Hazardous Laboratory Chemicals Disposal Guide*.[10] The last title listed here should be on the bookshelf of every laboratory. This 464-page book gives detailed information about hundreds of hazardous substances, including their physical properties, hazardous reactions, physiological properties, health hazards, spillage disposal, and waste disposal. Many of the treatment procedures in "Cleaning Up" are adaptations of these procedures. *Destruction of Hazardous Chemicals in the Laboratory*[11] complements this book.

[8]See footnote 6 on page 34.
[9]National Research Council. *Prudent Practices in the Laboratory: Handling and Disposal of Chemicals* National Academy Press: Washington, DC, 1995.
[10]See footnote 7 on page 34.
[11]Lunn, G.; Sansone, E. B. *Destruction of Hazardous Chemicals in the Laboratory*; Wiley: New York, 1994.

The area of waste disposal is changing rapidly. Many levels of laws apply—local, state, and federal. What may be permissible to wash down the drain or evaporate in the hood in one jurisdiction may be illegal in another, so before carrying out any waste disposal, check with your college or university waste disposal officer.

BIOHAZARDS

The use of microbial growth bioassays is becoming common in chemistry laboratories. The use of infectious materials presents new hazards that must be recognized and addressed. The first step in reducing hazards when using these materials is to select infectious materials that are known not to cause illness in humans and are of minimal hazard to the environment. A number of procedures should be followed to ensure safety when these materials are used: Individuals need to wash their hands after they handle these materials and before they leave the laboratory; work surfaces need to be decontaminated at the end of each use; and all infectious materials need to be decontaminated before disposal.

QUESTIONS

1. Write a balanced equation for the reaction between the iodide ion, a peroxide, and the hydrogen ion. What causes the orange or brown color?
2. Why does the horn of the carbon dioxide fire extinguisher become cold when the extinguisher is used?
3. Why is water not used to extinguish most fires in an organic laboratory? Consider the following scenario. An organic chemistry student is setting up an experiment on the benchtop, just outside the hood. He is wearing gloves as he dispenses diethyl ether, a volatile, flammable solvent. By accident he spills some ether on his hand but continues on with the experiment. The experiment is complete, and he cleans up. He puts the ether down the drain and washes the glassware. He then takes off his gloves and washes his hands. List all of the safety issues (if any) with this scenario.

CHAPTER 3

Melting Points and Boiling Points

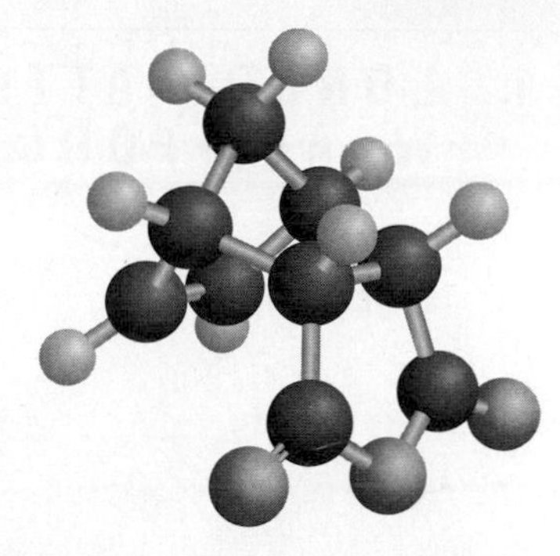

When you see this icon, sign in at this book's premium website at **www.cengage.com/login** to access videos, Pre-Lab Exercises and other online resources.

PRELAB EXERCISE: Draw the organic compounds, identify the intermolecular attractive forces for each, and list them in order of increasing boiling point as predicted by the relative strength of those intermolecular forces: (a) butane, (b) *i*-butanol, (c) potassium acetate, (d) acetone.

PART 1: FIVE CONCEPTS FOR PREDICTING PHYSICAL PROPERTIES

In organic chemistry, structure is everything. A molecule's structure determines both its physical properties and its reactivity. Since the dawn of modern chemistry 200 years ago, over 20 million substances, most of them organic compounds, have been isolated, and their properties and reactions have been studied. It became apparent from these studies that certain structural features in organic molecules would affect the observed properties in a predictable way and that these millions of organic compounds could be organized into classes based on molecular size, composition, and the pattern of bonds between atoms. Chemists also saw trends in certain properties based on systematic changes in these structural features. This organized knowledge allows us to look at a compound's structure and to predict the physical properties of that compound.

Physical properties, such as melting point, boiling point, and solubility, are largely determined by *intermolecular attractive forces.* You learned about these properties in previous chemistry courses. Because a solid understanding of these concepts is critical to understanding organic chemistry, we will review the different types of forces in the context of structural organic chemistry. Using five simple concepts, you should be able to look at the structures of a group of different organic molecules and predict which might be liquids, gases, or solids and which might be soluble in water. You can often predict the boiling point, melting point, or solubility of one structure relative to other structures. In fact, as your knowledge grows, you may be able to predict a compound's approximate melting or boiling temperature based on its structure. Your understanding of intermolecular attractive forces will be very useful in this chapter's experiments on melting and boiling points, and those in Chapters 5 and 6 that involve distillation and boiling points.

1. LONDON ATTRACTIVE FORCES (OFTEN CALLED VAN DER WAALS FORCES)

Organic molecules that contain only carbon and hydrogen (hydrocarbons) are weakly attracted to each other by London forces. Though weak, these attractive forces increase as molecular size increases. Thus, the larger the molecule, the greater the attractive force for neighboring molecules and the greater the energy required to get two molecules to move apart. This trend can be seen if we compare the melting points and boiling points of three hydrocarbons of different size: methane, hexane, and tetracosane.

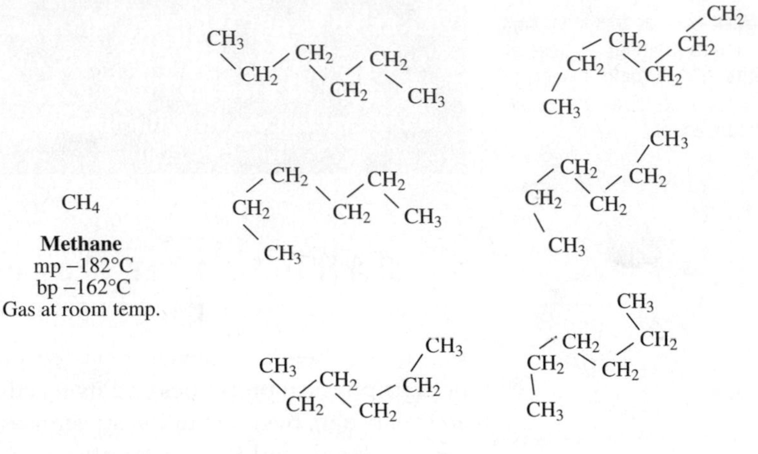

Hexane, C_6H_{14}
mp −95°C
bp +69°C
Liquid at room temp.

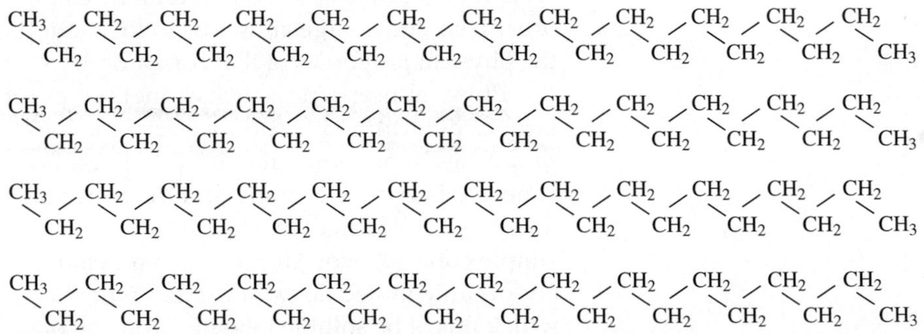

CH_3 CH_2 CH_2 CH_2 CH_2 CH_2 CH_2 CH_2 CH_2 CH_2 CH_2 CH_2
CH_2 CH_2 CH_2 CH_2 CH_2 CH_2 CH_2 CH_2 CH_2 CH_2 CH_2 CH_3

CH_3 CH_2 CH_2 CH_2 CH_2 CH_2 CH_2 CH_2 CH_2 CH_2 CH_2 CH_2
CH_2 CH_2 CH_2 CH_2 CH_2 CH_2 CH_2 CH_2 CH_2 CH_2 CH_2 CH_3

CH_3 CH_2 CH_2 CH_2 CH_2 CH_2 CH_2 CH_2 CH_2 CH_2 CH_2 CH_2
CH_2 CH_2 CH_2 CH_2 CH_2 CH_2 CH_2 CH_2 CH_2 CH_2 CH_2 CH_3

CH_3 CH_2 CH_2 CH_2 CH_2 CH_2 CH_2 CH_2 CH_2 CH_2 CH_2 CH_2
CH_2 CH_2 CH_2 CH_2 CH_2 CH_2 CH_2 CH_2 CH_2 CH_2 CH_2 CH_3

Tetracosane, $C_{24}H_{50}$
mp +51°C
bp +391°C
Solid at room temp.

We know that methane is called natural gas because methane's physical state at room temperature (20°C = 68°F) is a gas. Its London forces are so weak that methane must be cooled to −162°C at 1 atm of pressure before the molecules will stick together enough to form a liquid. Hexane is a very common liquid solvent found on most organic laboratory shelves. The intermolecular forces between its molecules are strong enough to keep them from flying apart, but the molecules are still able to flex and slide by each other to form a fluid. Hexane must be heated above 69°C, which is 231°C hotter than methane, to convert all its molecules to a gas. Tetracosane, a C_{24} solid hydrocarbon, is four times larger than hexane, and its London forces are strong enough to hold the molecules rigidly in place at room temperature. Tetracosane is one of the many long-chain hydrocarbons found in candle wax, which must be heated in order to disrupt the intermolecular forces and melt the wax into a liquid. A lot of energy is required to convert liquid tetracosane to a gas, as evidenced by its extremely high boiling point (391°C).

2. DIPOLE-DIPOLE ATTRACTIVE FORCES

The attractive forces between molecules increases when functional groups containing electronegative atoms such as chlorine, oxygen, and nitrogen are present because these atoms are more electronegative than carbon. These atoms pull electrons toward themselves, making their end of the bond slightly negatively charged (δ^-) and leaving the carbon slightly positively charged (δ^+), as shown for isopropyl chloride and acetone.

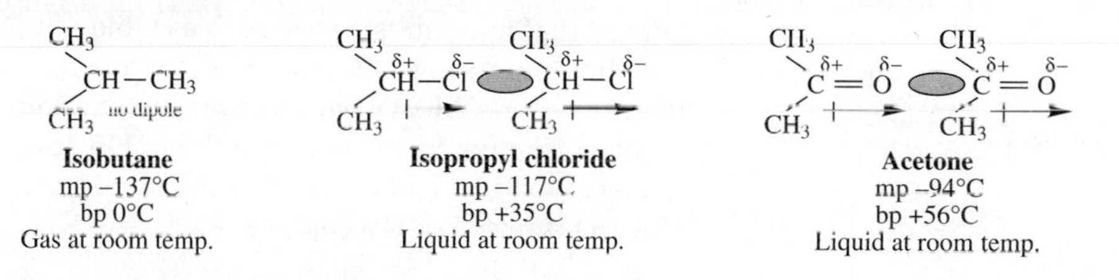

A bond with a slight charge separation is termed a *polar* bond, and polar atomic bonds often give a molecule a *dipole*: slightly positive and negative ends symbolized by an arrow in the direction of the negative charge (+→). Attraction of the positive end of one molecule's dipole to the negative end of another's dipole occurs between polar molecules, which increases the intermolecular attractive force. Dipole-dipole attractive forces are stronger than London forces, as demonstrated in the previous examples that show an increase in melting point and boiling point when a methyl group of isobutane is replaced by chlorine or oxygen.

3. HYDROGEN BONDING

Hydrogen bonding is an even stronger intermolecular attractive force, as evidenced by the large increase in the melting point and boiling points of the alcohol methanol (MW = 32) relative to those of the hydrocarbon ethane (MW = 30), both of comparable molecular weight. Hydrogen bonding occurs with organic molecules containing O—H groups (for example, alcohols and carboxylic acids) or N—H groups (for example, amines or amides). The hydrogen in these groups is attracted to the unshared pair of electrons on the O or N of another molecule, forming a hydrogen bond, often symbolized by a dashed line, which is shown for methanol.

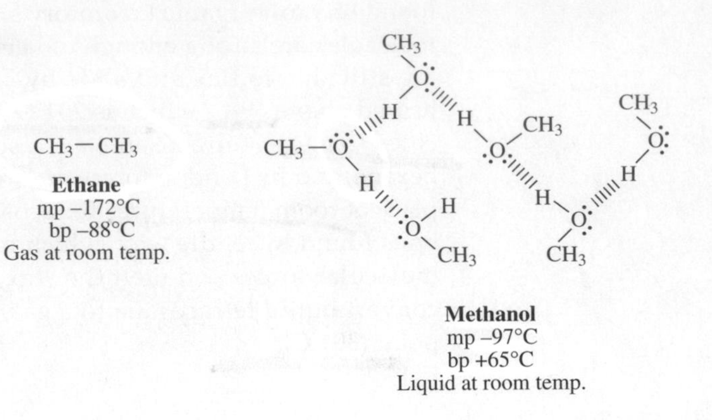

As this example indicates, the hydrogen bonds extend throughout the liquid. One can think of these hydrogen bonds as molecular Velcro that can be pulled apart if there is sufficient energy. Hydrogen bonding plays a major role in the special physical behavior of water and is a major determinant of the chemistry of proteins and nucleic acids in living systems.

4. IONIC ATTRACTIVE FORCES

Recall that ionic substances, such as table salt (NaCl), are usually solids with high melting points (>300°C) due to the strong attractive forces between positive and negative ions. Most organic molecules contain only covalent bonds and have no ionic attractive forces between them. However, there are three important exceptions involving acidic or basic functional groups that can form ionic structures as the pH is raised or lowered.

1. The hydrogen on the —OH of the carboxyl group in carboxylic acids, such as acetic acid, is acidic (H^+ donating) and reacts with bases such as potassium hydroxide (KOH) and sodium bicarbonate ($NaHCO_3$) to form salts. The process is reversed by adding an acid to lower the pH.

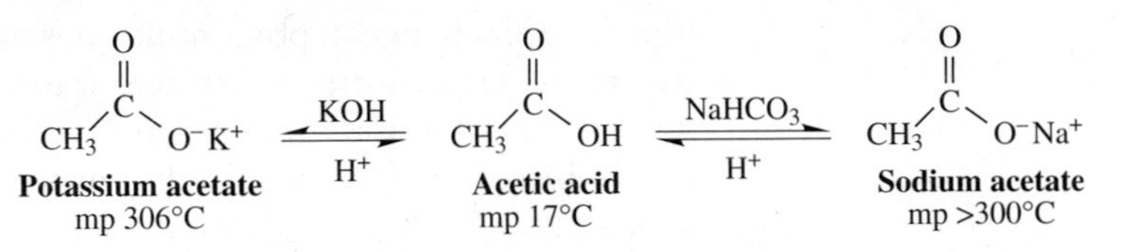

The dry salts are ionic and have very high melting points, which is expected for ionic substances. Note that this acidity is *not* observed for alcohols where the —OH group is attached to a singly bonded (sp^3 or saturated) carbon.

2. The hydrogen on an —OH group that is attached to an aromatic ring is weakly acidic and reacts with strong bases such as sodium hydroxide (NaOH) to form high melting ionic salts, as evidenced by the reaction of phenol to sodium phenolate. Again, the reaction is reversed by the addition of an acid.

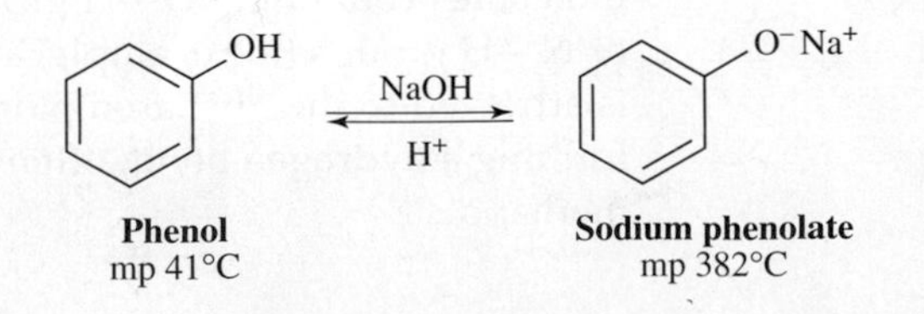

Note again that this acidity is *not observed* for alcohols where the —OH group is attached to a singly bonded (sp^3 or saturated) carbon.

3. Amines (but not amides) are basic (H^+ accepting) and will react with acid to form ionic amine salts with elevated melting points, as shown, for example, for isopropyl amine.

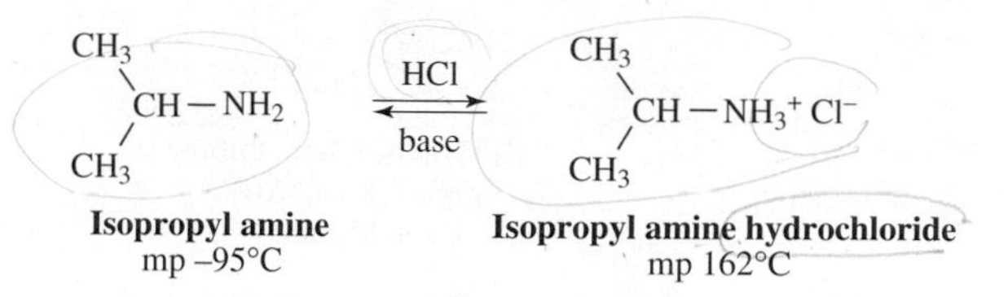

Amine salts can be converted back to amines by adding a base to raise the pH.

5. COMPETING INTERMOLECULAR FORCES AND SOLUBILITY

For pure compounds containing identical molecules, the total attractive force between molecules is the sum of all the attractive forces listed previously, both weak and strong. These forces tend to work together to raise melting and boiling points as the size of the molecule's hydrocarbon skeleton increases and as polar, hydrogen bonded, or ionic functional groups are incorporated into the molecule. However, solubility involves the interaction of two different molecules, which may have different types of attractive forces. When we try to dissolve one substance in another, we have to disrupt the attractive forces in both substances to get the molecules of the two substances to intermingle. For example, to get water (polar with hydrogen bonding) and the hydrocarbons (nonpolar and no hydrogen bonding) in motor oil to mix and dissolve in one another, we would have to disrupt the London attractive forces between the oil molecules and the hydrogen bonds between the water molecules. Because London forces are weak, separating the oil molecules does not require much energy. However, breaking apart the much stronger hydrogen bonds by inserting oil molecules between the water molecules requires considerable energy and is unfavorable. Therefore, oil or even the simplest hydrocarbon, methane, is insoluble in water. This is the molecular basis of the old adage "Oil and water don't mix."

In addition to carbon and hydrogen, the majority of organic molecules contain other elements, such as nitrogen and oxygen, in functional groups that can be polar, exhibit hydrogen bonding, show ionic tendencies, or have any combination thereof. Can we predict the water solubility of these organic substances? Let's look at some examples and see.

Figure 3.1 shows a collection of small organic molecules of about the same size and molecular weight, listed in order of increasing boiling point or melting point, which is consistent with the types of intermolecular forces discussed. With the exception of the hydrocarbon butane, all of these substances are very soluble in water—at least 100 g will dissolve in 1 L of water. It appears that the intermolecular attractive forces between these polar, hydrogen bonded, or ionic molecules and water compensates for the disruption of hydrogen bonding between water molecules so the organic molecules can move into and intermingle with the water molecules—in other words, dissolve.

Figure 3.2 shows a collection of larger organic molecules than those in Figure 3.1, again listed in order of increasing melting and boiling point, which is consistent with our knowledge of the strength of intermolecular attractive forces. The important difference for this group is that the hydrocarbon portion of each molecule is

FIG. 3.1
Some small organic molecules containing 4 atoms (carbon, oxygen, and nitrogen) in order of increasing melting or boiling points.

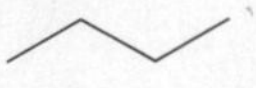

C_4 Hydrocarbon, butane
mp –138°C; bp 0°C
Insoluble in H_2O

C_3 Amine
mp –43°C; bp 48°C
Soluble in H_2O; Soluble in organic solvents

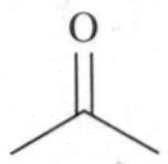

C_3 Ketone, acetone
mp –94°C; bp 56°C
Soluble in H_2O; Soluble in organic solvents

C_3 Alcohol
mp –127°C; bp 97°C
Soluble in H_2O; Soluble in organic solvents

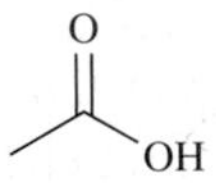

C_2 Carboxylic acid. acetic acid
mp 17°C; bp 117°C
Soluble in H_2O; Soluble in organic solvents

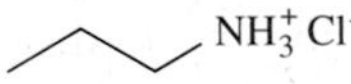

C_3 Amine hydrochloride
mp 161°C
Soluble in H_2O; Insoluble in organic solvents

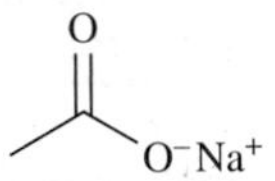

Sodium salt of C_2 carboxylic acid
mp 324°C
Soluble in H_2O; Insoluble in organic solvents

FIG. 3.2
Some organic molecules containing 8 atoms (carbon, oxygen, and nitrogen) in order of increasing melting or boiling points.

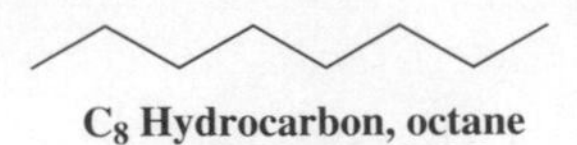

C_8 Hydrocarbon, octane
mp –57°C; bp 126°C
Insoluble in H_2O

C_7 Amine
mp –18°C; bp 154°C
Insoluble in H_2O; Soluble in organic solvents

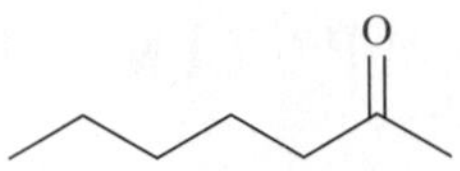

C_7 Ketone
mp –35°C; bp 150°C
Insoluble in H_2O; Soluble in organic solvents

C_7 Alcohol
mp –34°C; bp 175°C
Insoluble in H_2O; Soluble in organic solvents

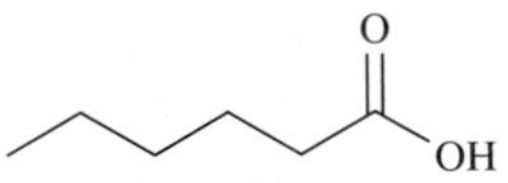

C_6 Carboxylic acid
mp –2°C; bp 200°C
Insoluble in H_2O; Soluble in organic solvents

C_7 Amine hydrochloride
mp 242°C
Soluble in H_2O; Insoluble in organic solvents

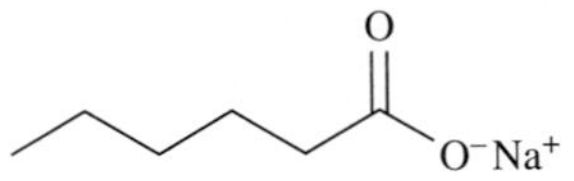

Sodium salt of C_6 carboxylic acid
mp 245°C
Soluble in H_2O; Insoluble in organic solvents

four carbons larger than for those in Figure 3.1. We might predict that the larger hydrocarbon portion makes them behave more like the water-insoluble hydrocarbon octane. Indeed, the larger hydrocarbon portion of these molecules greatly reduces their solubility in water to less than 5 g/L of water except for the two ionic compounds. These two, the amine hydrochloride and the sodium carboxylate compounds, have higher solubility—tens of grams per liter of water—proof that ionic charges can interact strongly with water molecules.

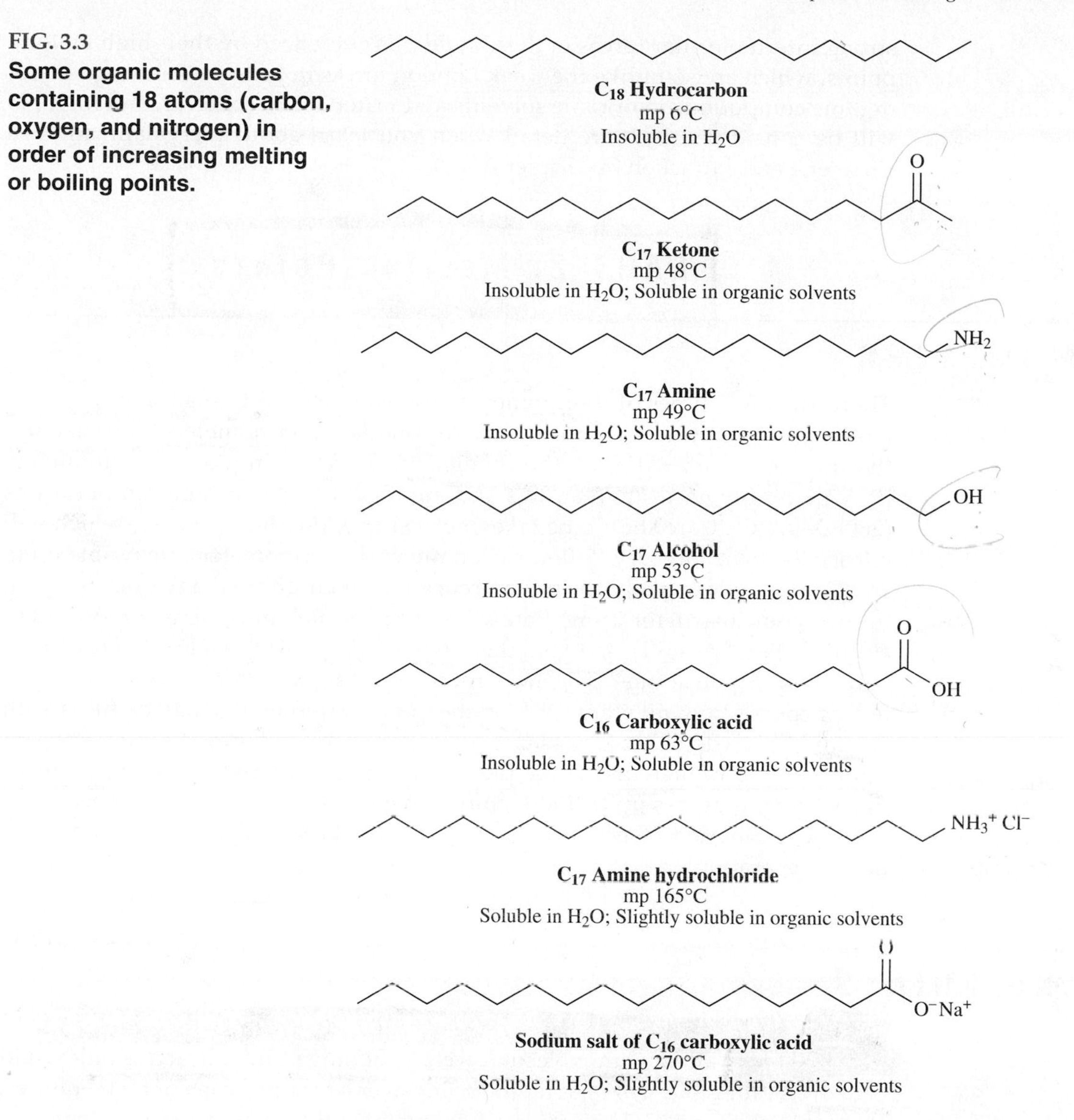

FIG. 3.3
Some organic molecules containing 18 atoms (carbon, oxygen, and nitrogen) in order of increasing melting or boiling points.

This trend in water solubility based on the size of the hydrocarbon portion continues for the set of even larger organic molecules shown in Figure 3.3 containing C_{16} to C_{18} hydrocarbon chains. Most are virtually insoluble in water, and even the ionic forms have solubilities of less than 1 g/L of water.

In addition to water, many other liquid solvents are used in the organic laboratory to dissolve substances, including acetone, dichloromethane (CH_2Cl_2), ethanol (CH_3CH_2OH), diethyl ether ($CH_3CH_2OCH_2CH_3$), hexane, methanol, and toluene ($C_6H_5CH_3$). Predicting the solubility of different organic compounds, such as those shown in Figures 3.1, 3.2, and 3.3, in these solvents can be done using the intermolecular attractive force concepts discussed here. For example, because the molecules in Figure 3.3 have long hydrocarbon skeletons, you would predict that these would probably be soluble in the hydrocarbon hexane. This predictive rule can be summed up as "Like dissolves like." You might also predict that the two ionic materials would be the least soluble in hexane because of the strong intermolecular forces in these solids, as evidenced by their high melting points, which are so unlike the weak London forces in hexane. The solubility of organic compounds in organic solvents and water at low, neutral, and high pH will be considered in more detail when you learn about recrystallization in Chapter 4 and extraction in Chapter 7.

PART 2: MELTING POINTS

A. THERMOMETERS

There are a few types of thermometers that can be used to read melting point (and boiling point) temperatures: mercury-in-glass thermometers, non-mercury thermometers, and digital thermometers. Mercury-in-glass thermometers provide highly accurate readings and are ideal for use at high temperatures (260°C–400°C). Care should be taken not to break the thermometer, which will release the toxic mercury. Teflon-coated mercury thermometers are usable up to 260°C and are less likely to spill mercury if broken. If breakage does happen, inform your instructor immediately because special equipment is required to clean up mercury spills. A digital thermometer (Fig. 3.4) has a low heat capacity and a fast response time. It is more robust than a glass thermometer and does not, of course, contain mercury. Non-mercury thermometers may be filled with isoamyl benzoate (a biodegradable liquid) or with a custom organic red-spirit liquid instead of mercury. These thermometers give reasonably accurate readings at temperatures up to 150°C, but above this temperature they need to be carefully calibrated. These thermometers should be stored vertically to prevent thread separation.

Mercury is toxic. Immediately report any broken thermometers to your instructor so that proper clean-up with a mercury disposal kit can occur.

B. MELTING POINTS

The melting point of a pure solid organic compound is one of its characteristic physical properties, along with molecular weight, boiling point, refractive index, and density. A pure solid will melt reproducibly over a narrow range of temperatures, typically less than 1°C. The process of determining this melting point is done on a truly micro scale using less than 1 mg of material. The apparatus is simple, consisting of a thermometer, a capillary tube to hold the sample, and a heating bath.

Melting points—a micro technique

Melting points are determined for three reasons: (1) If the compound is a known one, the melting point will help to characterize the sample; (2) If the compound is new, then the melting point is recorded to allow future characterization by others; (3) The range of the melting point is indicative of the purity of the compound – an impure compound will melt over a wide range of temperatures. Recrystallization of the compound will purify it and decrease the melting point range. In addition, the entire range will be displaced upward. For example, an impure sample might melt from 120°C to 124°C, and after recrystallization will melt at 125°C–125.5°C. A solid is considered pure if the melting point does not rise after recrystallization.

Characterization

An indication of purity

A crystal is an orderly arrangement of molecules in a solid. As heat is added to the solid, the molecules will vibrate and perhaps rotate, but still remain a solid. At a characteristic temperature, the crystal will suddenly acquire the necessary

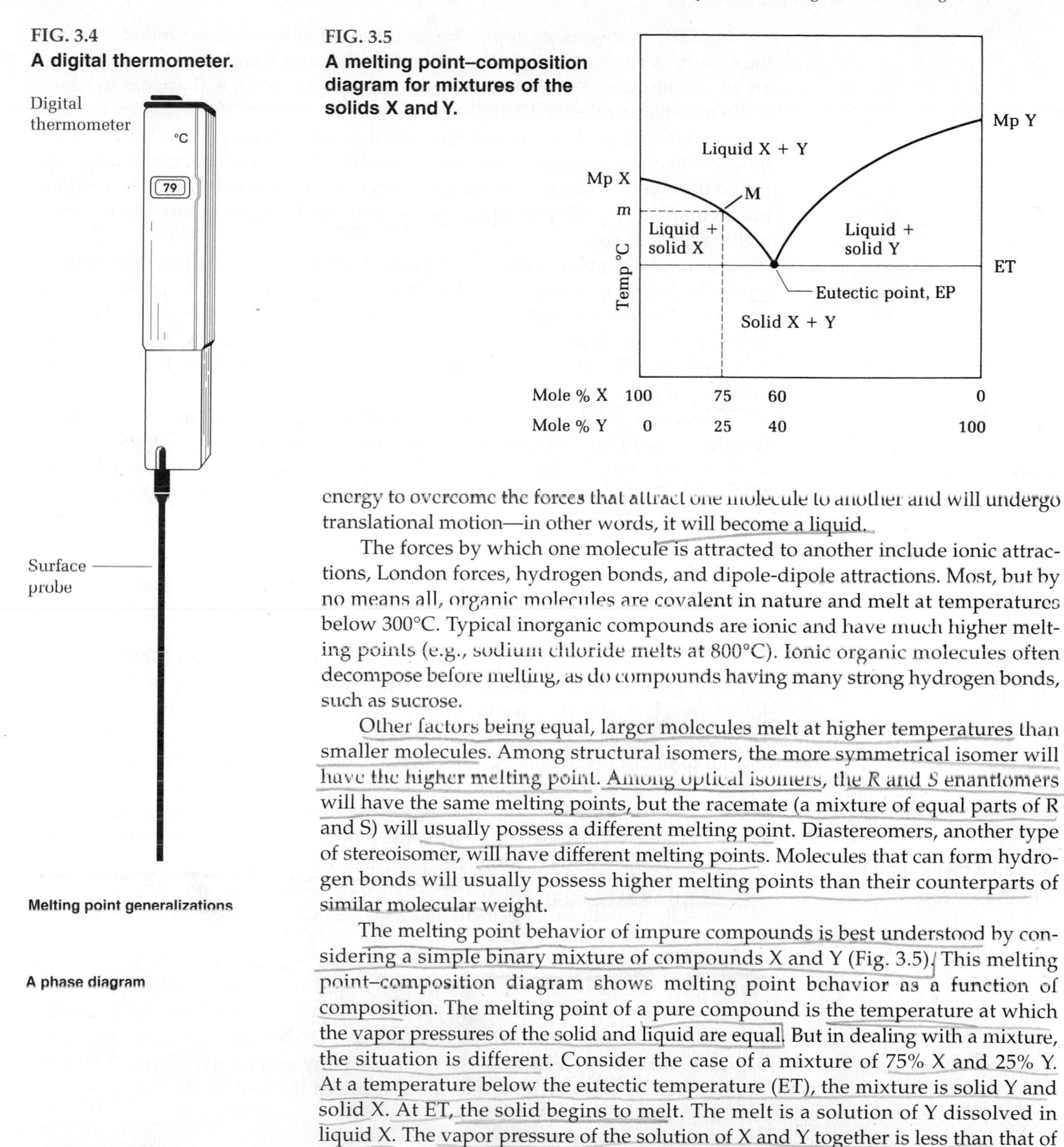

FIG. 3.4
A digital thermometer.

FIG. 3.5
A melting point–composition diagram for mixtures of the solids X and Y.

energy to overcome the forces that attract one molecule to another and will undergo translational motion—in other words, it will become a liquid.

The forces by which one molecule is attracted to another include ionic attractions, London forces, hydrogen bonds, and dipole-dipole attractions. Most, but by no means all, organic molecules are covalent in nature and melt at temperatures below 300°C. Typical inorganic compounds are ionic and have much higher melting points (e.g., sodium chloride melts at 800°C). Ionic organic molecules often decompose before melting, as do compounds having many strong hydrogen bonds, such as sucrose.

Melting point generalizations

Other factors being equal, larger molecules melt at higher temperatures than smaller molecules. Among structural isomers, the more symmetrical isomer will have the higher melting point. Among optical isomers, the *R* and *S* enantiomers will have the same melting points, but the racemate (a mixture of equal parts of R and S) will usually possess a different melting point. Diastereomers, another type of stereoisomer, will have different melting points. Molecules that can form hydrogen bonds will usually possess higher melting points than their counterparts of similar molecular weight.

A phase diagram

The melting point behavior of impure compounds is best understood by considering a simple binary mixture of compounds X and Y (Fig. 3.5). This melting point–composition diagram shows melting point behavior as a function of composition. The melting point of a pure compound is the temperature at which the vapor pressures of the solid and liquid are equal. But in dealing with a mixture, the situation is different. Consider the case of a mixture of 75% X and 25% Y. At a temperature below the eutectic temperature (ET), the mixture is solid Y and solid X. At ET, the solid begins to melt. The melt is a solution of Y dissolved in liquid X. The vapor pressure of the solution of X and Y together is less than that of pure X at the melting point; therefore, the temperature at which X will melt is lower when mixed with Y. This is an application of Raoult's law (*see* Chapter 5). As the temperature is increased, more and more of solid X melts until it is all gone at point **M** (temperature m). The melting point range is thus from ET to *m*. In practice, it is very difficult to detect the ET when a melting point is determined in a capillary because it represents the point at which an infinitesimal amount of the solid mixture has begun to melt.

Melting point depression

In this hypothetical example, the liquid solution becomes saturated with Y at the eutectic point (EP). This is the point at which X and Y and their liquid solutions are in equilibrium. A mixture of X and Y containing 60% X will appear to have a sharp melting point at the ET.

The eutectic point

The melting point range of a mixture of compounds is generally broad, and the breadth of the range is an indication of purity. The chances of accidentally coming on the eutectic composition are small. Recrystallization will enrich the predominant compound while excluding the impurity and will, therefore, decrease the melting point range.

It should be apparent that the impurity must be soluble in the compound, so an insoluble impurity such as sand or charcoal will not depress the melting point. The impurity does not need to be a solid. It can be a liquid such as water (if it is soluble) or an organic solvent, such as the one used to recrystallize the compound; this advocates the necessity for drying the compound before determining its melting point.

Advantage is taken of the depression of melting points of mixtures to prove whether or not two compounds having the same melting points are identical. If X and Y are identical, then a mixture of the two will have the same melting point; if X and Y are not identical, then a small amount of X in Y or of Y in X will reduce the melting point.

Mixed melting points

Apparatus

Melting Point Capillaries

Before using a melting point apparatus, the sample needs to be prepared for analysis. Most melting point determinations require that the sample be placed in a capillary tube. The experiments in this book require capillary tubes for sample preparation. Capillaries may be obtained commercially or may be produced by drawing out 12-mm soft-glass tubing. The tubing is rotated in the hottest part of a Bunsen burner flame until it is very soft and begins to sag. It should not be drawn out during heating; it is removed from the flame and after a slight hesitation drawn steadily and not too rapidly to arm's length. With some practice it is possible to produce 10–15 good tubes in a single drawing. The long capillary tube can be cut into 100-mm lengths with a glass scorer. Each tube is sealed by rotating one open end in the edge of a small flame until the glass melts and seals the bottom, as seen in Figure 3.6.

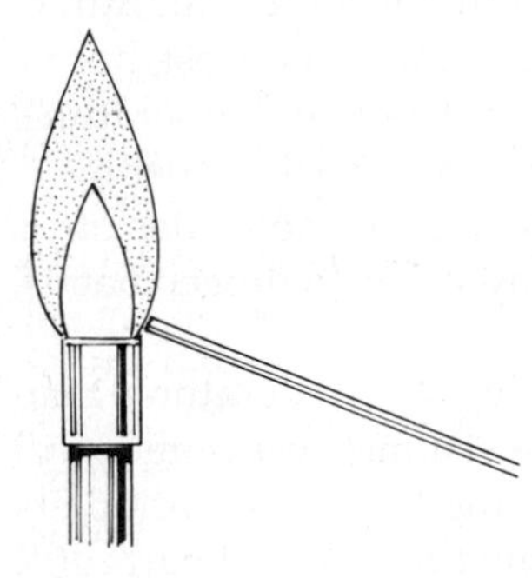

FIG. 3.6
Sealing a melting point capillary tube.

Filling Melting Point Capillaries. The dry sample is ground to a fine powder on a watch glass or a piece of weighing paper on a hard surface using the flat portion of a spatula. It is formed into a small pile, and the open end of the melting point capillary is forced down into the pile. The sample is shaken into the closed end of the capillary by rapping sharply on a hard surface or by dropping it down a 2-ft length of glass tubing onto a hard surface. The height of the sample should be no more than 2–3 mm.

Sealed Capillaries. Some samples sublime (go from a solid state directly to the vapor phase without appearing to melt), or undergo rapid air oxidation and decompose at the melting point. These samples should be sealed under vacuum. This can be accomplished by forcing a capillary through a hole previously made in a rubber septum and evacuating the capillary using a water aspirator or a mechanical vacuum pump (Fig. 3.7). Using the flame from a small micro burner, the tube is gently heated about 15 mm above the tightly packed sample. This will cause any

Samples that sublime

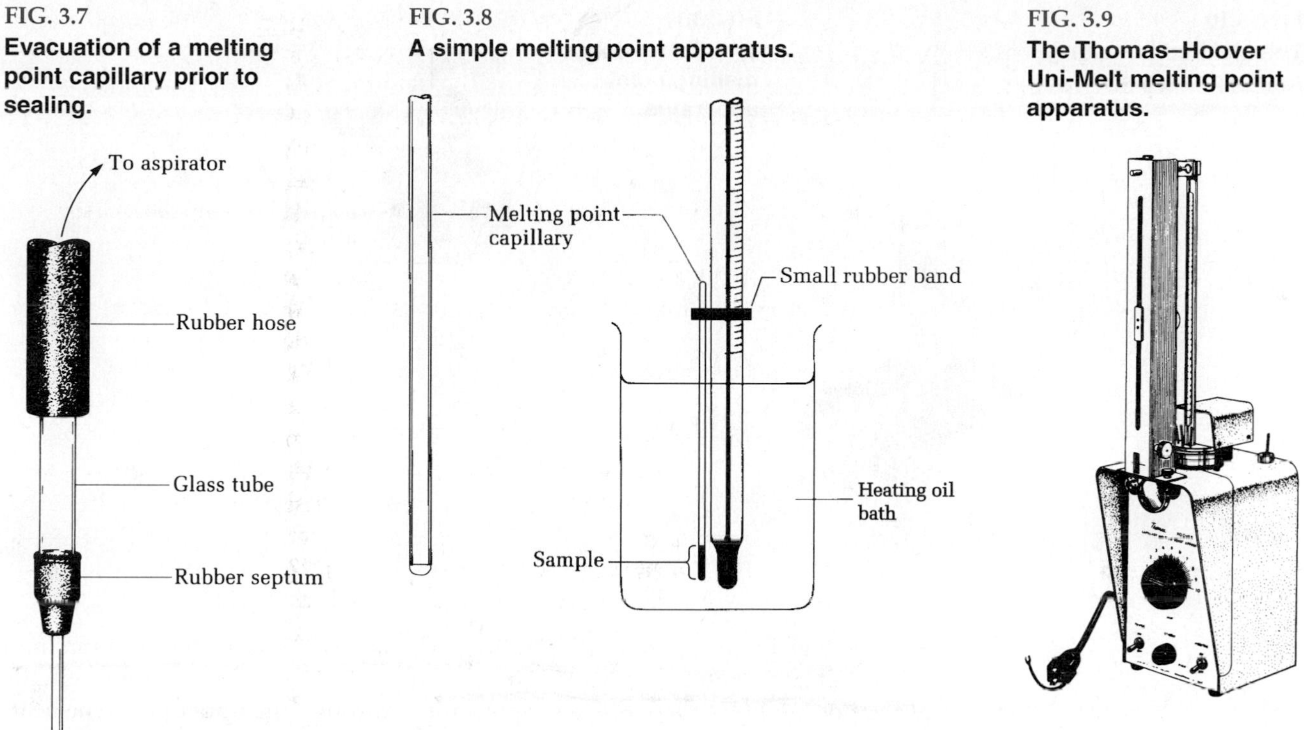

FIG. 3.8
A simple melting point apparatus.

FIG. 3.9
The Thomas–Hoover Uni-Melt melting point apparatus.

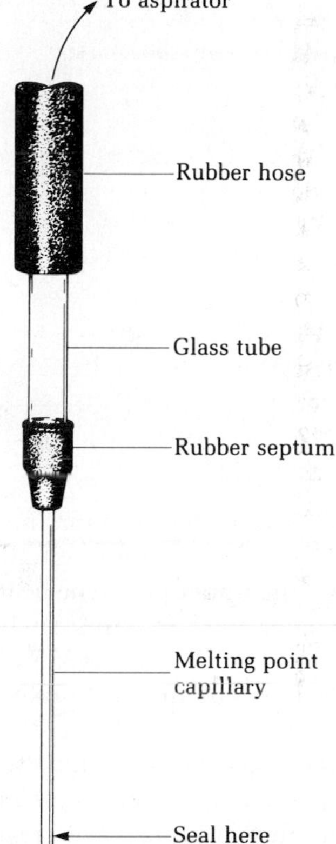

FIG. 3.7
Evacuation of a melting point capillary prior to sealing.

material in this region to sublime away. It is then heated more strongly in the same place to collapse and seal the tube, taking care that the tube is straight when it cools. It is also possible to seal the end of a long Pasteur pipette, add the sample, pack it down, and seal off a sample under vacuum in the same way.

Melting Point Devices. The apparatus required for determining an accurate melting point need not be elaborate. The same results are obtained on the simplest as on the most complex devices. The simplest setup involves attaching the sample-filled, melting point capillary to a thermometer using a rubber band and immersing them into a silicone oil bath (Fig. 3.8). This rubber band must be above the level of the oil bath; otherwise, it will break in the hot oil. The sample should be close to and on a level with the center of the thermometer bulb. This method can analyze compounds whose melting points go up to ~350°C. If determinations are to be done on two or three samples that differ in melting point by as much as 10°C, two or three capillaries can be secured to the thermometer together; the melting points can be observed in succession without removing the thermometer from the bath. As a precaution against the interchange of tubes while they are being attached, use some system of identification, such as one, two, and three dots made with a marking pencil.

A small rubber band can be made by cutting off a very short piece of 1/4" gum rubber tubing.

More sophisticated melting point devices, some of which can attain temperatures of 500°C, will now be described.

Thomas–Hoover Uni-Melt

The Thomas–Hoover Uni-Melt apparatus (Fig. 3.9) will accommodate up to seven capillaries in a small, magnified, lighted beaker of high-boiling silicone oil that is stirred and heated electrically. The heating rate is controlled with a variable transformer that is part of the apparatus. The rising mercury column of the thermometer can be observed with an optional traveling periscope device so the eye need not move away from the capillary. For industrial, analytical, and control work, the Mettler apparatus determines the melting point automatically and displays the result in digital form.

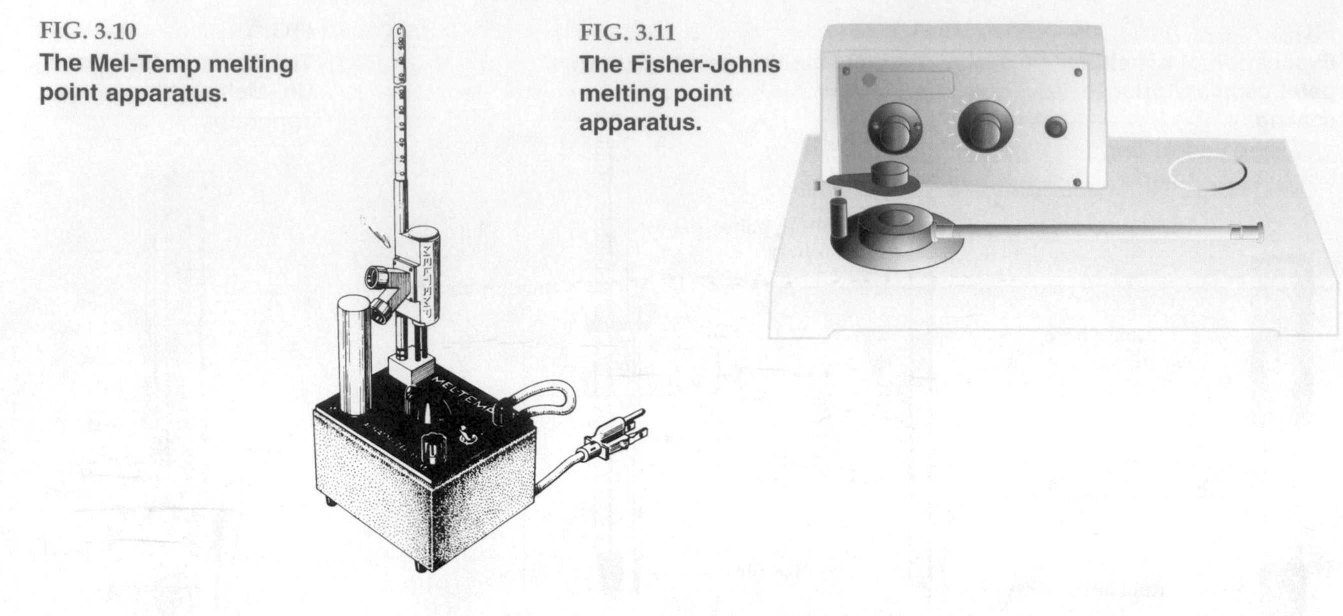

FIG. 3.10
The Mel-Temp melting point apparatus.

FIG. 3.11
The Fisher-Johns melting point apparatus.

The Mel-Temp apparatus (Fig. 3.10) consists of an electrically heated aluminum block that accommodates three capillaries. The sample is illuminated through the lower port and observed with a six-power lens through the upper port. The heating rate can be controlled and, with a special thermometer, the apparatus can be used up to 500°C—far above the useful limit of silicone oil (which is about 350°C). For this melting point apparatus, it is advisable to use a digital thermometer rather than a mercury-in-glass thermometer.

The Fisher-Johns melting point apparatus (Fig. 3.11) is used to determine the melting point of a single sample. Instead of a capillary tube, the sample is placed between two thin glass disks that are placed on an aluminum heating stage. Heating is controlled by a variable transformer, and melting is observed through a magnifier; the melting temperature is read from a mercury-in-glass thermometer. This apparatus can be used for compounds that melt between 20°C and 300°C.

Determining the Melting Point

⚠ The rate of heating is the most important factor in obtaining accurate melting points. Heat no faster than 1°C per minute.

The accuracy of the melting point depends on the accuracy of the thermometer, so the first exercise in the following experiments will be to calibrate the thermometer. Melting points of pure, known compounds will be determined and deviations recorded so a correction can be applied to future melting points. Be forewarned, however, that thermometers are usually fairly accurate.

The most critical factor in determining an accurate melting point is the rate of heating. At the melting point, the temperature increase should not be greater than 1°C per minute. This may seem extraordinarily slow, but it is necessary in order for heat from the oil bath or the heating block to be transferred equally to the sample and to the glass and mercury of the thermometer.

From your own experience, you know the rate at which ice melts. Consider performing a melting point experiment on an ice cube. Because water melts at 0°C, you would need to have a melting point bath a few degrees below zero. To observe the true melting point of the ice cube, you would need to raise the temperature extraordinarily slowly. The ice cube would appear to begin to melt at 0°C and, if you waited for temperature equilibrium to be established, it would all be melted at

0.5°C. If you were impatient and raised the temperature too rapidly, the ice might appear to melt over a range of 0°C to 20°C. Similarly, melting points determined in capillaries will not be accurate if the rate of heating is too fast.

EXPERIMENTAL PROCEDURES

1. CALIBRATION OF THE THERMOMETER

Determine the melting point of standard substances (Table 3.1) over the temperature range of interest. The difference between the values found and those expected constitutes the correction that must be applied to future temperature readings. If the thermometer has been calibrated previously, then determine one or more melting points of known substances to familiarize yourself with the technique. If the determinations do not agree within 1°C, then repeat the process. Both mercury-in-glass and digital thermometers will need to be calibrated; non-mercury thermometers are not typically used for melting point determination.

TABLE 3.1 *Melting Point Standards*

Compound		*Structure*	*Melting Point (°C)*
Naphthalene	(a)		80–82
Urea	(b)	O ‖ H_2NCNH_2	132.5–133
Sulfanilamide	(c)	NH_2 / SO_2NH_2	164–165
4-Toluic acid	(d)	COOH / CH_3	180–182
Anthracene	(e)		214–217
Caffeine (evacuated capillary)	(f)	O, CH_3, H_3C, N, N, N, N, O, CH_3	234–236.5

2. MELTING POINTS OF PURE UREA AND CINNAMIC ACID

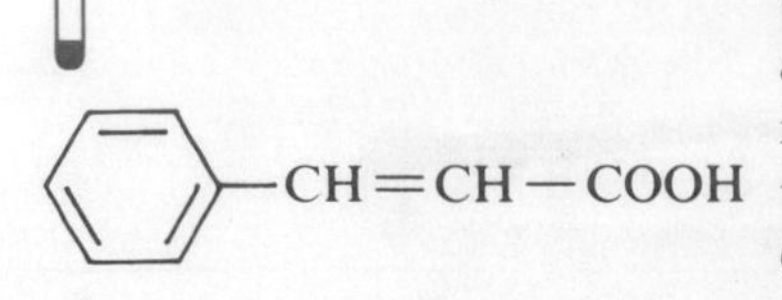

Cinnamic acid

Heat rapidly to within 20°C of the melting point.

Heat slowly (<1°C/min) near the melting temperature.

Using a metal spatula, crush the sample to a fine powder on a hard surface such as a watch glass. Push the open end of a melting point capillary into the powder and force the powder down in the capillary by tapping the capillary or by dropping it through a long glass tube held vertically and resting on a hard surface. The column of solid should be no more than 2–3 mm in height and should be tightly packed.

If the approximate melting temperature is known, the bath can be heated rapidly until the temperature is about 20°C below this point; the heating during the last 15°C–20°C should slow down considerably *so the rate of heating at the melting point is no more than 1°C per minute* while the sample is melting. As the melting point is approached, the sample may shrink because of crystalline structure changes. However, the melting process begins when the first drops of liquid are seen in the capillary and ends when the last trace of solid disappears. For a pure compound this whole process may occur over a range of only 0.5°C; hence, it is necessary to slowly increase the temperature during the determination.

Determine the melting point (mp) of either urea (mp 132.5°C–133°C) or cinnamic acid (mp 132.5°C–133°C). Repeat the determination; if the two determinations do not check within 1°C, repeat a third time.

3. MELTING POINTS OF UREA-CINNAMIC ACID MIXTURES

IN THIS EXPERIMENT you will see dramatic evidence of the phenomenon of melting point depression, which will allow you to prepare a phase diagram like that shown in Figure 3.5 (on page 49).

Make mixtures of urea and cinnamic acid in the approximate proportions 1:4, 1:1, and 4:1 by putting side by side the correct number of equal-sized small piles of the two substances and then mixing them. Grind the mixture thoroughly for at least a minute on a watch glass using a metal spatula. Note the ranges of melting of the three mixtures and use the temperatures of complete liquefaction to construct a rough diagram of melting point versus composition.

4. UNKNOWNS

Determine the melting point of one or more of the unknowns selected by your instructor and identify the substance based on its melting point (Table 3.2). Prepare two capillaries of each unknown. Run a very fast determination on the first sample to ascertain the approximate melting point. Cool the melting point bath to just below the melting point and make a slow, careful determination using the other capillary.

5. AN INVESTIGATION: DETERMINATION OF MOLECULAR WEIGHT USING MELTING POINT DEPRESSION

Before the mass spectrometer came into common usage, the molal freezing point depression of camphor was used to determine molecular weights. Whereas a 1% solid solution of urea in cinnamic acid will cause a relatively small melting

TABLE 3.2 *Melting Point Unknowns*

Compound	*Melting Point (°C)*
Benzophenone	49–51
Maleic anhydride	52–54
4-Nitrotoluene	54–56
Naphthalene	80–82
Acetanilide	113.5–114
Benzoic acid	121.5–122
Urea	132.5–133
Salicylic acid	158.5–159
Sulfanilamide	165–166
Succinic acid	184.5–185
3,5-Dinitrobenzoic acid	205–207
p-Terphenyl	210–211

point depression, a 1%-by-weight solid solution in camphor of any organic compound with a molecular weight of 100 will cause a 4.0°C depression in the melting point of the camphor. Quantitative use of this relationship can be used to determine the molecular weight of an unknown. You can learn more about the details of this technique in very old editions of *Organic Experiments*[1] or by searching the Web for "colligative properties, molal freezing point depression." Visit this book's website for more information.

PART 3: BOILING POINTS

The boiling point of a pure organic liquid is one of its characteristic physical properties, just like its density, molecular weight, and refractive index, and the melting point of its solid state. The boiling point is used to characterize a new organic liquid, and knowledge of the boiling point helps to compare one organic liquid to another, as in the process of identifying an unknown organic substance.

A comparison of boiling points with melting points is instructive. The process of determining the boiling point is more complex than that for the melting point: It requires more material, and because it is less affected by impurities, it is not as good an indication of purity. Boiling points can be determined on a few microliters of a liquid, but on a small scale it is difficult to determine the boiling point range. This requires enough material to distill—about 1 to 2 mL. Like the melting point, the boiling point of a liquid is affected by the forces that attract one molecule to another—ionic attraction, London forces, dipole-dipole interactions, and hydrogen bonding, as discussed in Part 1 of this chapter.

STRUCTURE AND BOILING POINT

In a homologous series of molecules, the boiling point increases in a perfectly regular manner. The normal saturated hydrocarbons have boiling points ranging from −161°C for methane to 330°C for $n\text{-}C_{19}H_{40}$, an increase of about 27°C for each

[1]Fieser L. F. *Organic Experiments*, 2nd ed.; D.C. Heath: Lexington, MA, 1968; 38–42.

CH_2 group. It is convenient to remember that *n*-heptane with a molecular weight of 100 has a boiling point near 100°C (98.4°C). A spherical molecule such as 2,2-dimethylpropane has a lower boiling point than *n*-pentane because it does not have as many points of attraction to adjacent molecules. For molecules of the same molecular weight, those with dipoles, such as carbonyl groups, will have higher boiling points than those without, and molecules that can form hydrogen bonds will boil even higher. The boiling point of such molecules depends on the number of hydrogen bonds that can be formed. An alcohol with one hydroxyl group will boil at a lower temperature than an alcohol with two hydroxyl groups if they both have the same molecular weight. A number of other generalizations can be made about boiling point behavior as a function of structure; you will learn about these throughout your study of organic chemistry.

BOILING POINT AS A FUNCTION OF PRESSURE

Boiling points decrease about 0.5°C for each 10-mm decrease in atmospheric pressure.

Because the boiling point of a pure liquid is defined as the temperature at which the vapor pressure of the liquid exactly equals the pressure exerted on it, the boiling point will be a function of atmospheric pressure. At an altitude of 14,000 ft, the boiling point of water is 81°C. At pressures near that of the atmosphere at sea level (760 mm), the boiling point of most liquids decreases about 0.5°C for each 10-mm decrease in atmospheric pressure. This generalization does not hold for greatly reduced pressures because the boiling point decreases as a nonlinear function of pressure (see Fig. 5.2 on page 89). Under these conditions, a nomograph relating the observed boiling point, the boiling point at 760 mm, and the pressure in millimeters should be consulted (see Fig. 6.19 on page 124). This nomograph is not highly accurate because the change in boiling point as a function of pressure also depends on the type of compound (polar, nonpolar, hydrogen bonding, etc.). Consult the *CRC Handbook of Chemistry and Physics*[2] for the correction of boiling points to standard pressure.

⚠ Mercury is toxic. Immediately report any broken thermometers to your instructor.

Most mercury-in-glass laboratory thermometers have a mark around the stem that is 3 in. (76 mm) from the bottom of the bulb. This is the immersion line; the thermometer will record accurate temperatures if immersed to this line. If you break a mercury thermometer, immediately inform your instructor, who will use special apparatus to clean up the mercury. Mercury vapor is very toxic and can be fatal if the fumes are produced by the heating of liquid mercury.

CALIBRATING THE THERMOMETER

If you have not previously carried out a calibration, test the 0°C point of your thermometer with a well-stirred mixture of crushed ice and distilled water. To check the 100°C point, put 2 mL of water in a test tube with a boiling chip to prevent bumping, and boil the water gently over a hot sand bath with the thermometer in the vapor from the boiling water. Make sure that the thermometer does not touch the side of the test tube. Then immerse the bulb of the thermometer into the liquid and see if you can observe superheating. Check the atmospheric pressure to determine the true boiling point of the water.

[2]Lide, D. R., ed. *CRC Handbook of Chemistry and Physics*, 86th ed.; CRC Press: Boca Raton, FL, 2005.

DISTILLATION CONSIDERATIONS

Prevention of Superheating—Boiling Sticks and Boiling Stones

Superheating occurs when a very clean liquid in a very clean vessel is heated to a temperature above its boiling point without ever actually boiling. That is, if a thermometer is placed in the superheated liquid, the thermometer will register a temperature higher than the boiling point of the liquid. If boiling does occur under these conditions, it occurs with explosive violence. To avoid this problem, boiling stones or boiling sticks are always added to liquids before heating them to boiling—whether to determine a boiling point or to carry out a reaction or distillation. These boiling stones or sticks provide the nuclei on which the bubbles of vapor indicative of a boiling liquid can form; be careful not to confuse the bubbling for boiling. Some boiling stones, also called boiling chips, are composed of porous unglazed porcelain. When dry, this material is filled with air in numerous fine capillaries. With heating, this air expands to form the fine bubbles upon which even boiling can take place. Once the liquid cools, it will fill these capillaries and the boiling chip will become ineffectual, so another chip must be added each time the liquid is reheated to boiling. Wooden boiling sticks about 1.5 mm in diameter—often called applicator sticks—also promote even boiling and, unlike stones, are removed easily from the solution. Neither boiling sticks nor stones work well for vacuum distillation (*see* Chapter 6).

Closed Systems

Distillations that are run at atmospheric pressure need to be open to the atmosphere to avoid pressure buildup, which could lead to an explosion. Therefore, always make sure that a distillation setup is not a closed system—unless, of course, you are running a vacuum distillation.

BOILING POINT DETERMINATION: APPARATUS AND TECHNIQUE

Boiling Point Determination by Distillation

All procedures involving volatile and/or flammable solvents should be conducted in a fume hood.

When enough material is available (at least 3 mL), the best method for determining the boiling point of a liquid is to distill it (*see* Chapter 5). Distillation allows the boiling range to be determined and thus gives an indication of purity. Bear in mind, however, that a constant boiling point is not a guarantee of homogeneity and thus purity. Constant-boiling azeotropes such as 95% ethanol are common.

Boiling Point Determination Using a Digital Thermometer and a Reaction Tube

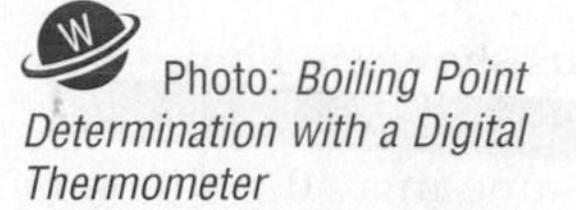

Photo: *Boiling Point Determination with a Digital Thermometer*

Boiling points can be measured rapidly and accurately using an electronic digital thermometer, as depicted in Figure 3.12. Although digital thermometers are currently too expensive for each student to have, several of these in the laboratory can greatly speed up the determination of boiling points. Also, digital thermometers are much safer to use because there is no danger from toxic mercury vapor if the thermometer is accidentally dropped.

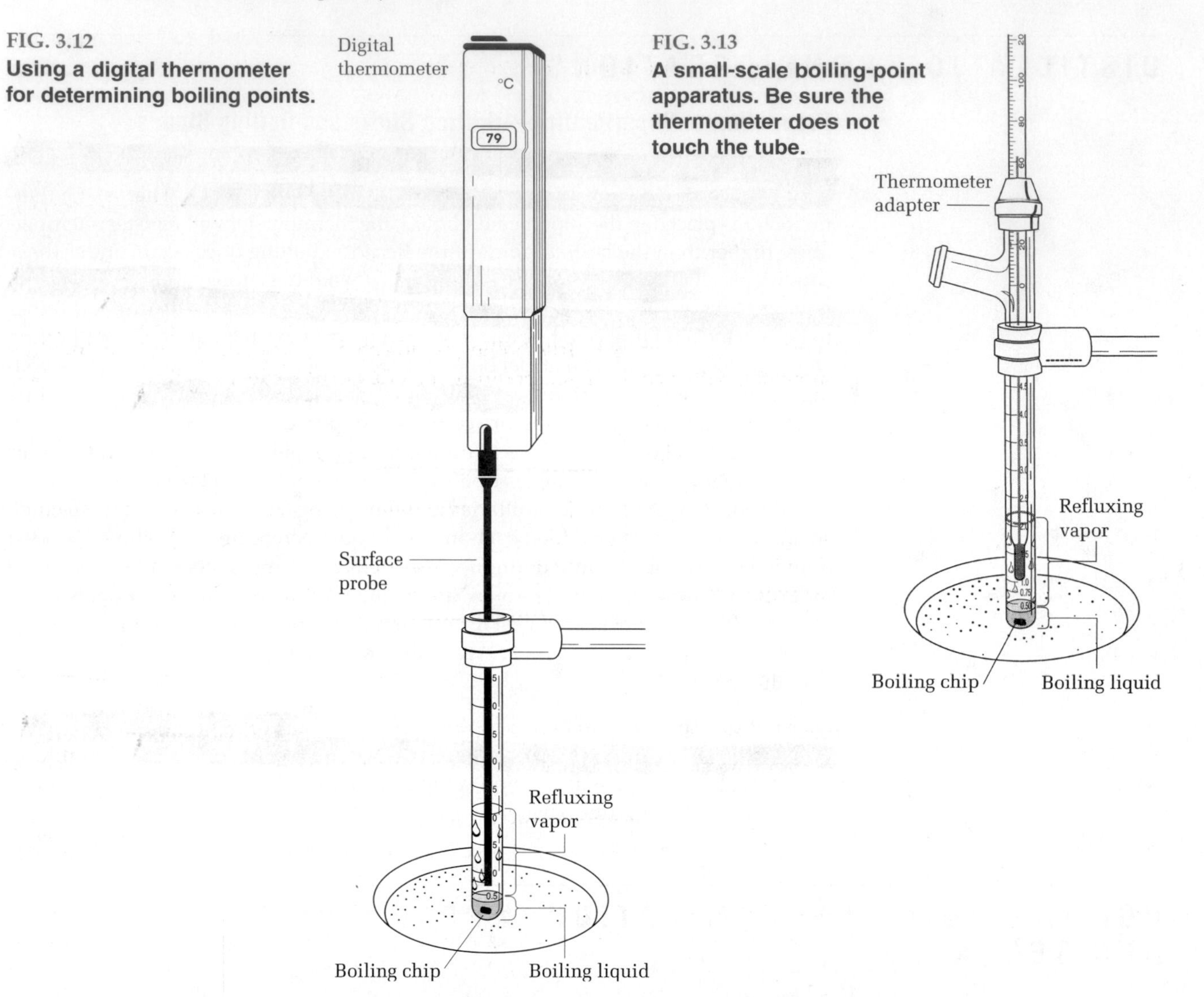

FIG. 3.12
Using a digital thermometer for determining boiling points.

FIG. 3.13
A small-scale boiling-point apparatus. Be sure the thermometer does not touch the tube.

The surface probe of the digital thermometer is the active element. Unlike the bulb of mercury at the end of a thermometer, this element has a very low heat capacity and a very fast response time, so boiling points can be determined very quickly with this apparatus. About 0.2 mL to 0.3 mL of the liquid and a boiling chip are heated on a sand bath until the liquid refluxes about 3 cm up the tube. The thermometer probe should not touch the side of the reaction tube, but should be placed approximately 5 mm above the liquid. The boiling point is the highest temperature recorded by the thermometer and is maintained for about 1 min. The application of heat will drive tiny bubbles of air from the boiling chip; do not mistake these tiny bubbles for true boiling; this mistake can readily happen if the unknown has a very high boiling point.

Boiling Point Determination in a Reaction Tube

If a digital thermometer is not available, use the apparatus shown in Figure 3.13. Using a distilling adapter at the top of a reaction tube allows access to the atmosphere. Place 0.3 mL of the liquid along with a boiling stone in a 10 × 100-mm reaction tube, clamp a thermometer so that the bulb is just above the level of the liquid, and then heat the liquid with a sand bath. It is *very important* that no part of the thermometer touch the reaction tube. Heating is regulated so that the boiling liquid refluxes about 3 cm up the thermometer, but does not boil out of the apparatus. If you cannot see the refluxing liquid, carefully run your finger down the side of the reaction tube until you feel heat. This indicates where the liquid is refluxing. Droplets of liquid must drip from the thermometer bulb to thoroughly heat the mercury. The boiling point is the highest temperature recorded by the thermometer and is maintained over about a 1-minute time interval.

The application of heat will drive out tiny air bubbles from the boiling chip. Do not mistake these tiny bubbles for true boiling. This can occur if the unknown has a very high boiling point. It may take several minutes to heat up the mercury in the thermometer bulb. True boiling is indicated by drops dripping from the thermometer, with a constant temperature recorded on the thermometer. If the temperature is not constant, then you are probably not observing true boiling.

Boiling Point Determination Using a 3-mm to 5-mm Tube

Smaller-scale boiling point apparatus

For smaller quantities, a 3-mm to 5-mm diameter tube is attached to the side of a thermometer by a rubber band (Fig. 3.14) and heated in a liquid bath. The tube, which can be made from tubing 3 mm to 5 mm in diameter, contains a small inverted capillary. This is made by cutting a 6-mm piece from the sealed end of a melting point capillary, inverting it, and sealing the closed end of the capillary. A centimeter ruler is printed on the inside back cover of this book.

When the sample is heated in this device, the air in the inverted capillary will expand, and an occasional bubble will escape. At the true boiling point, a continuous and rapid stream of bubbles will emerge from the inverted capillary. When this occurs, the heating is stopped, and the bath is allowed to cool. A time will come when bubbling ceases and the liquid just begins to rise in the inverted capillary. The temperature at which this happens is recorded. The liquid is allowed to partially fill the small capillary, and then heat is applied carefully until the first bubble emerges from the capillary. The temperature is recorded at that point. The two temperatures approximate the boiling point range for the liquid.

As the liquid was being heated, the air expanded in the inverted capillary and was replaced by vapor of the liquid. The liquid was actually slightly superheated when rapid bubbles emerged from the capillary, but on cooling, a point was

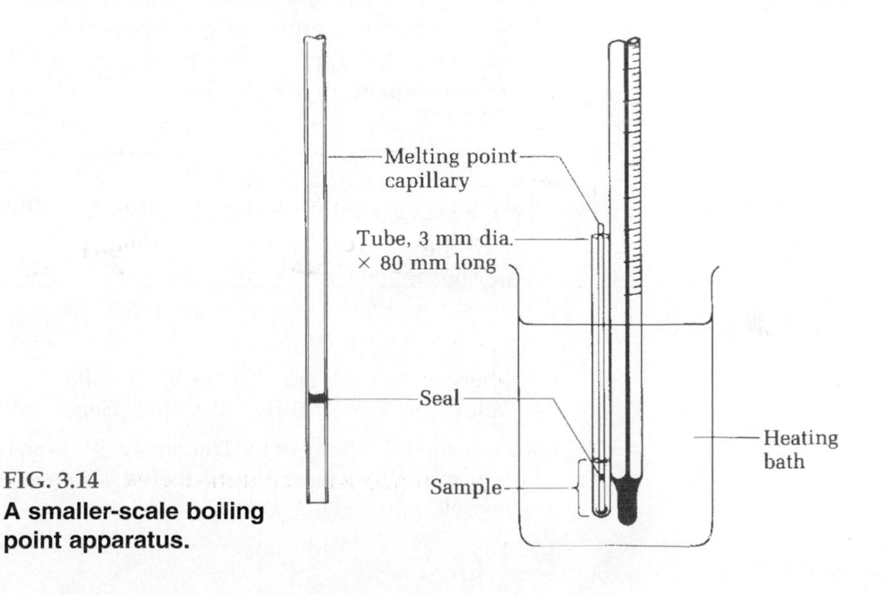

FIG. 3.14
A smaller-scale boiling point apparatus.

reached at which the pressure on the inside of the capillary matched the outside (atmospheric) pressure. This is, by definition, the boiling point.

Cleaning Up. Place the boiling point sample in either the halogenated or nonhalogenated waste container. Do not pour it down the sink.

QUESTIONS

1. What effect would poor circulation of the melting point bath liquid have on the observed melting point?
2. What is the effect of an insoluble impurity, such as sodium sulfate, on the observed melting point of a compound?
3. Three test tubes, labeled A, B, and C, contain substances with approximately the same melting points. How could you prove that the test tubes contain three different chemical compounds?
4. One of the most common causes of inaccurate melting points is too rapid heating of the melting point bath. Under these circumstances, how will the observed melting point compare to the true melting point?
5. Strictly speaking, why is it incorrect to speak of a melting *point?*
6. What effect would the incomplete drying of a sample (for example, the incomplete removal of a recrystallization solvent) have on the melting point?
7. Why should the melting point sample be finely powdered?
8. You suspect that an unknown is acetanilide (mp 113.5°C–114°C). Give a qualitative estimation of the melting point when the acetanilide is mixed with 10% by weight of naphthalene.
9. You have an unknown with an observed melting point of 90°C–93°C. Is your unknown compound A with a reported melting point of 95.5°C–96°C, or compound B with a reported melting point of 90.5°C–91°C? Explain.
10. Why is it important to heat the melting point bath or block slowly and steadily when the temperature gets close to the melting point?
11. Why is it important to pack the sample tightly in the melting point capillary?
12. An unknown compound is suspected to be acetanilide (mp 113.5°C–114°C). What would happen to the melting point if this unknown were mixed with (a) an equal quantity of pure acetanilide? (b) an equal quantity of benzoic acid?
13. Which would be expected to have the higher boiling point, *t*-butyl alcohol (2-methyl-2-propanol) or *n*-butyl alcohol (1-butanol)? Explain.
14. What is the purpose of the side arm of the thermometer adapter in Figure 3.13?

See *Web Links*

REFERENCES

Weissberger, Arnold, and Bryant W. Rossiter (eds.). *Physical Methods of Chemistry*, Vol. 1, Part V. New York: Wiley-Interscience, 1971.

J. Kofron and J.K. Hardy of the University of Akron have an excellent website with photographs illustrating mixed melting points: http://ull.chemistry.uakron.edu/organic_lab/melting_point.

CHAPTER 4

Recrystallization

PRELAB EXERCISE: Write an expanded outline for the seven-step process of recrystallization.

Recrystallization: the most important purification method for solids, especially for small-scale experiments.

Recrystallization is the most important method for purifying solid organic compounds. It is also a very powerful, convenient, and efficient method of purification, and it is an important industrial technique that is still relevant in the chemical world today. For instance, the commercial purification of sugar is done by recrystallization on an enormous scale.

A pure, crystalline organic substance is composed of a three-dimensional array of molecules held together primarily by London forces. These attractive forces are fairly weak; most organic solids melt in the range between 22°C and 250°C. An impure organic solid will not have a well-defined crystal lattice because impurities do not allow the crystalline structure to form. The goal of recrystallization is to remove impurities from a solid to allow a perfect crystal lattice to grow.

There are four important concepts to consider when discussing the process of recrystallization:

1. Solubility
2. Saturation level
3. Exclusion
4. Nucleation

Recrystallization involves dissolving the material to be purified (the *solute*) in an appropriate hot *solvent* to yield a solution (*solubility*). As the solvent cools, the solution becomes saturated with respect to the solute (*saturation level*), which then recrystallizes. As the perfectly regular array of a crystal is formed, impurities are excluded (*exclusion*), and the crystal is thus a single pure substance. Soluble impurities remain in solution because they are not concentrated enough to saturate the solution. Recrystallization of a solute is initiated at a point of *nucleation*, which can be a seed crystal, a speck of dust, or a scratch on the wall of the test tube.

In this chapter, you will carry out the recrystallization process, one of the most important laboratory operations of organic chemistry, by following its seven steps. Then you will perform several actual recrystallization experiments.

THE SEVEN STEPS OF RECRYSTALLIZATION

The process of recrystallization can be broken into seven discrete steps: (1) choosing the solvent and solvent pairs; (2) dissolving the solute; (3) decolorizing the solution with pelletized Norit; (4) filtering suspended solids; (5) recrystallizing the solute; (6) collecting and washing the crystals; and (7) drying the crystals. A detailed description of each of these steps is given in the following sections.

STEP 1. CHOOSING THE SOLVENT AND SOLVENT PAIRS

Similia similibus solvuntur.

In choosing the solvent, the chemist is guided by the dictum "like dissolves like." Even the nonchemist knows that oil and water do not mix, and that sugar and salt dissolve in water but not in oil. Hydrocarbon solvents such as hexane will dissolve hydrocarbons and other nonpolar compounds, and hydroxylic solvents such as water and ethanol will dissolve polar compounds. It is often difficult to decide, simply by looking at the structure of a molecule, just how polar or nonpolar it is and which solvent would be best. Therefore, the solvent is often chosen by experimentation. If an appropriate single solvent cannot be found for a given substance, a solvent pair system may be used. The requirement for this solvent pair is miscibility; both solvents should dissolve in each other for use as a recrystallization solvent system.

Video: *The Reaction Tube in Use*

The ideal solvent

The best recrystallization solvent (and none is ideal) will dissolve the solute when the solution is hot but not when the solution is cold; it will either not dissolve the impurities at all or it will dissolve them very well (so they do not recrystallize out along with the solute); it will not react with the solute; and it will be nonflammable, nontoxic, inexpensive, and very volatile (so it can be removed from the crystals).

Miscible: capable of being mixed

Some common solvents and their properties are presented in Table 4.1 in order of decreasing polarity of the solvent. Solvents adjacent to each other in the list will dissolve in each other; that is, they are miscible with each other, and each solvent will, in general, dissolve substances that are similar to it in chemical structure. These solvents are used both for recrystallization and as solvents in which reactions are carried out.

Procedure

Choosing a Solvent

Video: *Picking a Solvent*

To choose a solvent for recrystallization, place a few crystals of the impure solute in a small test tube or centrifuge tube and add a very small drop of the solvent. Allow the drop to flow down the side of the tube and onto the crystals. If the crystals dissolve instantly at about 22°C, that solvent cannot be used for recrystallization because too much of the solute will remain in solution at low temperatures. If the crystals do not dissolve at about 22°C, warm the tube on a hot sand bath and observe the crystals. If they do not go into solution, add 1 more drop of the solvent. If the crystals go into solution at the boiling point of the solvent and then recrystallize when the tube is cooled, you have found a good recrystallization solvent. If not, remove the solvent by evaporation and try a different solvent. In this trial-and-error process, it is easiest to try low-boiling solvents first because they can be easily removed. Occasionally, no single satisfactory solvent can be found, so mixed solvents, or *solvent pairs*, are used.

TABLE 4.1 *Recrystallization Solvents*

Solvent	*Boiling Point (°C)*	*Density (g/mL)*	*Remarks*
Water (H_2O)	100	1.000	It is the solvent of choice because it is cheap, nonflammable, nontoxic, and will dissolve a large variety of polar organic molecules. Its high boiling point and high heat of vaporization make it difficult to remove from crystals.
Acetic acid (CH_3COOH)	118	1.049	It will react with alcohols and amines, and it is difficult to remove from crystals. It is not a common solvent for recrystallizations, although it is used as a solvent when carrying out oxidation reactions.
Dimethyl sulfoxide (DMSO; CH_3SOCH_3)	189	1.100	It is not a commonly used solvent for recrystallization, but it is used for reactions.
Methanol (CH_3OH)	64	0.791	It is a very good solvent that is often used for recrystallization. It will dissolve molecules of higher polarity than other alcohols.
95% Ethanol (CH_3CH_2OH)	78	0.789	It is one of the most commonly used recrystallization solvents. Its high boiling point makes it a better solvent for less polar molecules than methanol. It evaporates readily from crystals. Esters may undergo an interchange of alcohol groups on recrystallization.
Acetone (CH_3COCH_3)	56	0.791	It is an excellent solvent, but its low boiling point means there is not much difference in the solubility of a compound at its boiling point compared to about 22°C.
2-Butanone; also known as methyl ethyl ketone (MEK; $CH_3COCH_2CH_3$)	80	0.805	It is an excellent solvent that has many of the most desirable properties of a good recrystallization solvent.
Ethyl acetate ($CH_3COOC_2H_5$)	78	0.902	It is an excellent solvent that has about the right combination of moderately high boiling point and the volatility needed to remove it from crystals.
Dichloromethane; also known as methylene chloride (CH_2Cl_2)	40	1.325	Although a common extraction solvent, dichloromethane has too low a boiling point to make it a good recrystallization solvent. It is useful in a solvent pair with ligroin.
Diethyl ether; also known as ether ($CH_3CH_2OCH_2CH_3$)	35	0.706	Its boiling point is too low for recrystallization, although it is an extremely good solvent and fairly inert. It is used in a solvent pair with hexanes.
Methyl *t*-butyl ether ($CH_3OC(CH_3)_3$)	52	0.741	It is a relatively new and inexpensive solvent. It does not easily form peroxides; it is less volatile than diethyl ether, but it has the same solvent characteristics. (*See also* Chapter 7.)
1,4-Dioxane ($C_4H_8O_2$)	101	1.034	It is a very good solvent that is not too difficult to remove from crystals; it is a mild carcinogen, and it forms peroxides.
Toluene ($C_6H_5CH_3$)	111	0.865	It is an excellent solvent that has replaced the formerly widely used benzene (a weak carcinogen) for the recrystallization of aryl compounds. Because of its boiling point, it is not easily removed from crystals.

(continued)

TABLE 4.1 *(continued)*

Solvent	*Boiling Point (°C)*	*Density (g/mL)*	*Remarks*
Pentane (C_5H_{12})	36	0.626	It is a widely used solvent for nonpolar substances. It is not often used alone for recrystallization, but it is good in combination with several other solvents as part of a solvent pair.
Hexane (C_6H_{14})	69	0.659	It is frequently used to recrystallize nonpolar substances. It is inert and has the correct balance between boiling point and volatility. It is often used as part of a solvent pair.
Cyclohexane (C_6H_{12})	81	0.779	It is similar in all respects to hexane.

Note: The solvents in this table are listed in decreasing order of polarity. Adjacent solvents in the list are, in general, miscible with each other.

TABLE 4.2 *Solvent Pairs*

Acetic acid–water	Ethyl acetate–cyclohexane
Ethanol-water	Acetone-hexanes
Acetone-water	Ethyl acetate–hexanes
Dioxane-water	*t*-Butyl methyl ether–hexanes
Acetone-ethanol	Dichloromethane-hexanes
Ethanol–*t*-butyl methyl ether	Toluene-hexanes

Solvent Pairs

To use a mixed solvent pair, dissolve the crystals in the better solvent (more solubilizing) and add the poorer solvent (less solubilizing) to the hot solution until it becomes cloudy, and the solution is saturated with the solute. The two solvents must, of course, be miscible with each other. Some useful solvent pairs are given in Table 4.2.

STEP 2. DISSOLVING THE SOLUTE

Microscale Procedure

Do not use too much solvent.

Once a recrystallization solvent has been found, the impure crystals are placed in a reaction tube, the solvent is added dropwise, the crystals are stirred with a microspatula or a small glass rod, and the tube is warmed on a steam bath or a sand bath until the crystals dissolve. Care must be exercised to use the minimum amount of solvent at or near the boiling point. Observe the mixture carefully as solvent is being added. Allow sufficient time for the boiling solvent to dissolve the solute and note the rate at which most of the material dissolves. To hasten the solution process, crush large crystals with a stirring rod, taking care not to break the reaction tube. When you believe most of the material has been dissolved, stop adding solvent. There is a possibility that your sample is contaminated with a small quantity of an insoluble impurity that never will dissolve.

Video: *Recrystallization*

If the solution contains no undissolved impurities and is not colored from impurities, you can simply let it cool, allowing the solute to recrystallize (step 5), and then collect the crystals (step 6). On the other hand, if the solution is colored, it must be treated with activated (decolorizing) charcoal and then filtered before recrystallization (step 3). If it contains solid impurities, it must be filtered before recrystallization takes place (step 4).

On a microscale, there is a tendency to use too much solvent so that on cooling the hot solution, little or no material recrystallizes. This is not a hopeless situation. The remedy is to evaporate some of the solvent (by careful boiling) and repeat the cooling process. Inspect the hot solution to see if crystals form, and if not, continue to evaporate solvent.

Prevention of bumping

Do not use wood applicator sticks (boiling sticks) in place of boiling chips in a reaction. Use them only for recrystallization.

A solution (solute dissolved in solvent) can become *superheated*; that is, heated above its boiling point without actually boiling. When boiling does suddenly occur, it can happen with almost explosive violence, a process called *bumping*. To prevent this from happening, a *wood applicator stick* can be added to the solution (Fig. 4.1). Air trapped in the wood comes out of the stick and forms the nuclei on which even boiling can occur. Porous porcelain *boiling chips* work in the same way; only a single chip is required to prevent bumping. Never add a boiling chip or a boiling stick to a hot solution because the hot solution may be superheated and boil over or bump.

Macroscale Procedure

Place the substance to be recrystallized in an Erlenmeyer flask (never use a beaker), add enough solvent to cover the crystals, and then heat the flask on a steam bath (if the solvent boils below 90°C) or a hot plate until the solvent boils. (Note: Adding a boiling stick or a boiling chip to the solution will promote even boiling. It is easy to superheat the solution; that is, heat it above the boiling point with no boiling taking place. Once the solution does boil, it does so with explosive violence, it bumps.) Never add a boiling chip or boiling stick to a hot solution because this might cause the solution to boil over.

Video: *Macroscale Crystallization*

All procedures involving volatile and/or flammable solvents should be conducted in a fume hood.

Stir the mixture or, better, swirl it (Fig. 4.2) to promote dissolution. Add solvent gradually, keeping it at the boiling point, until all of the solute dissolves. A glass rod with a flattened end can sometimes be useful in crushing large particles of solute to speed up the dissolving process. Be sure no flames are nearby when working with flammable solvents.

Be careful not to add too much solvent. Note how rapidly most of the material dissolves; stop adding solvent when you suspect that almost all of the desired material has dissolved. It is best to err on the side of too little solvent rather than too much. Undissolved material noted at this point could be an insoluble impurity

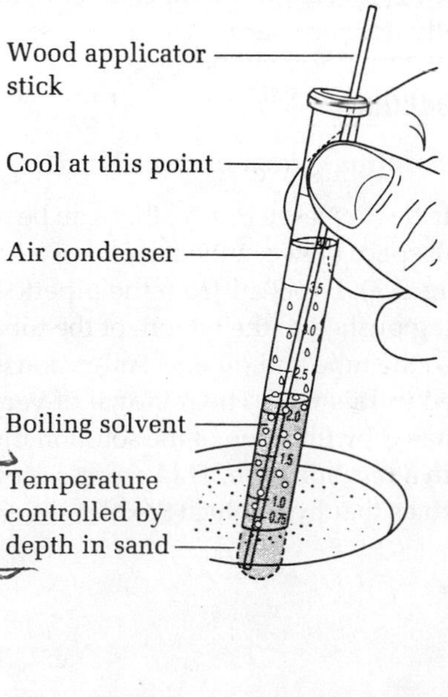

Wood applicator stick
Cool at this point
Air condenser
Boiling solvent
Temperature controlled by depth in sand

FIG. 4.1
A reaction tube being used for recrystallization. The wood applicator stick ("boiling stick") promotes even boiling and is easier to remove than a boiling chip. The Thermowell sand is cool on top and hotter deeper down, so it provides a range of temperatures. The reaction tube is long and narrow; it can be held in the hand while the solvent refluxes. Do not use a boiling stick in place of a boiling chip in a reaction.

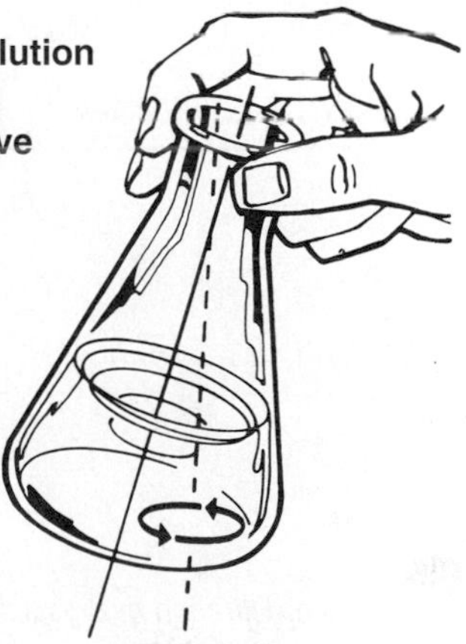

FIG. 4.2
Swirling of a solution to mix contents and help dissolve material to be recrystallized.

that never will dissolve. Allow the solvent to boil, and if no further material dissolves, proceed to step 4 to remove suspended solids from the solution by filtration, or if the solution is colored, go to step 3 to carry out the decolorization process. If the solution is clear, proceed to step 5.

STEP 3. DECOLORIZING THE SOLUTION WITH PELLETIZED NORIT

Video: *Decolorization of a Solution with Norit*

The vast majority of pure organic chemicals are colorless or a light shade of yellow; consequently, this step is not usually required. Occasionally, a chemical reaction will produce high molecular weight byproducts that are highly colored. The impurities can be adsorbed onto the surface of activated charcoal by simply boiling the solution with charcoal. Activated charcoal is made by the pyrolysis of carbonaceous material such as coconut shells, wood and lignite and activated with steam. It has an extremely large surface area per gram (several hundred square meters) and can bind a large number of molecules to this surface. On a commercial scale, the impurities in brown sugar are adsorbed onto charcoal in the process of refining sugar.

Activated charcoal = decolorizing carbon = Norit

Add a small amount (0.1% of the solute weight is sufficient) of pelletized Norit to the colored solution and then boil the solution for a few minutes. Be careful not to add the charcoal pieces to a superheated solution; the charcoal will function like hundreds of boiling chips and will cause the solution to boil over. Remove the Norit by filtration as described in step 4.

STEP 4. FILTERING SUSPENDED SOLIDS

The filtration of a hot, saturated solution to remove solid impurities or charcoal can be performed in a number of ways. These processes include gravity filtration, pressure filtration, decantation, or removing the solvent with a Pasteur pipette. Vacuum filtration is not used because the hot solvent will cool during the process, and the product will recrystallize in the filter. Filtration can be one of the most vexing operations in the laboratory if the desired compound recrystallizes during filtration. Test the solution or a small portion of it before filtration to ensure that no crystals form at about 22°C. Like decolorization with charcoal, the removal of solid impurities by filtration is usually not necessary.

Microscale Procedure

(A) Removal of Solution with a Pasteur Pipette

If the solid impurities are large in size, they can be removed by filtration of the liquid through the small space between the flat end of a Pasteur pipette and the bottom of a reaction tube (Fig. 4.3). Expel air from the pipette by squeezing the pipette bulb as the pipette is being pushed to the bottom of the tube. Use a small additional quantity of solvent to rinse the tube and pipette. Anhydrous calcium chloride, a drying agent, is easily removed in this way. The removal of very fine material, such as traces of charcoal, is facilitated by filtration of the solution through a small piece of filter paper (3 mm^2) placed in a reaction tube. This process is even easier if the filter paper is the thick variety, such as that from which Soxhlet extraction thimbles are made.[1]

Video: *Filtration of Crystals Using the Pasteur Pipette*

[1]Belletire, J. L.; Mahmoodi, N. O. *J. Chem. Educ.* **1989**, 66, 964.

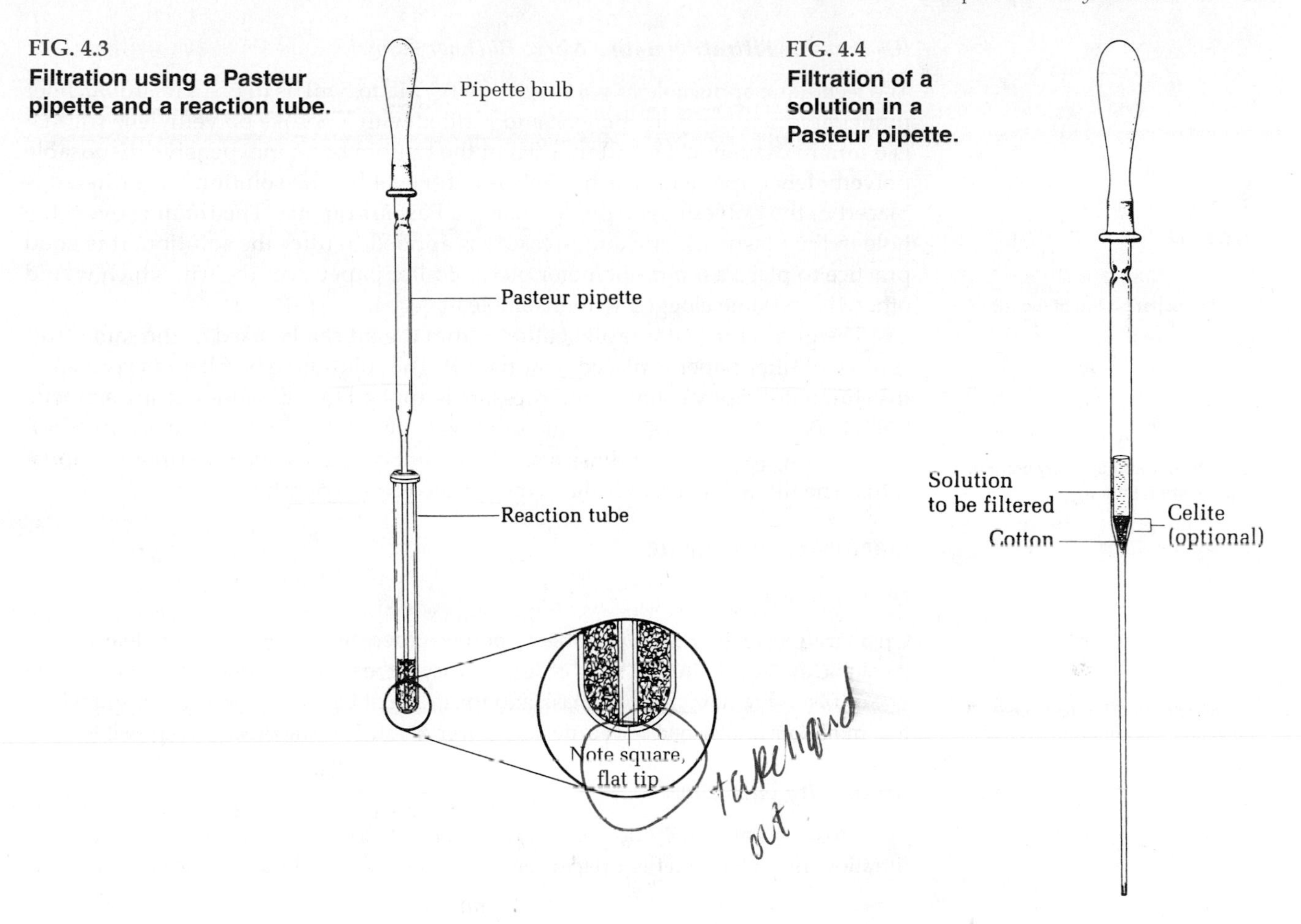

FIG. 4.3
Filtration using a Pasteur pipette and a reaction tube.

FIG. 4.4
Filtration of a solution in a Pasteur pipette.

(B) Filtration in a Pasteur Pipette

To filter 0.1 mL to 2 mL of a solution, dilute the solution with enough solvent so that the solute will not recrystallize at about 22°C. Prepare a filter pipette by pushing a tiny bit of cotton into a Pasteur pipette, put the solution to be filtered into this filter pipette using another Pasteur pipette, and then force the liquid through the filter using air pressure from a pipette bulb (Fig. 4.4). Fresh solvent should be added to rinse the pipette and cotton. The filtered solution is then concentrated by evaporation. One problem encountered with this method is using too much cotton packed too tightly in the pipette so that the solution cannot be forced through it. To remove very fine impurities such as traces of decolorizing charcoal, a 3-mm to 4-mm layer of Celite filter aid can be added to the top of the cotton.

Photo: *Preparation of a Filter Pipette*

(C) Removal of Fine Impurities by Centrifugation

To remove fine solid impurities from up to 4 mL of solution, dilute the solution with enough solvent so that the solute will not recrystallize at about 22°C. Counterbalance the reaction tube and centrifuge for about 2 minutes at high speed in a laboratory centrifuge. The clear supernatant can be decanted (poured off) from the solid on the bottom of the tube. Alternatively, with care, the solution can be removed with a Pasteur pipette, leaving the solid behind.

(D) Pressure Filtration with a Micro Büchner Funnel

The technique applicable to volumes from 0.1 mL to 5 mL is to use a micro Büchner funnel. It is made of polyethylene and is fitted with a porous polyethylene frit that is 6 mm in diameter. This funnel fits in the bottom of an inexpensive disposable polyethylene pipette in which a hole is cut (Fig. 4.5). The solution to be filtered is placed in the polyethylene pipette using a Pasteur pipette. The thumb covers the hole in the plastic pipette and pressure is applied to filter the solution. It is good practice to place a 6-mm-diameter piece of filter paper over the frit, which would otherwise become clogged with insoluble material.

Use filter paper on top of the frit.

The glass chromatography column from the kit can be used in the same way. A piece of filter paper is placed over the frit. The solution to be filtered is placed in the chromatography column, and pressure is applied to the solution using a pipette bulb. In both procedures, dilute the solution to be filtered so that it does not recrystallize in the apparatus, and use a small amount of clean solvent to rinse the apparatus. The filtered solution is then concentrated by evaporation.

Using the chromatography column for pressure filtration.

Macroscale Procedure

(A) Decantation

On a large scale, it is often possible to pour off (decant) the hot solution, leaving the insoluble material behind. This is especially easy if the solid is granular like sodium sulfate. The solid remaining in the flask and the inside of the flask should be rinsed with a few milliliters of the solvent in order to recover as much of the product as possible.

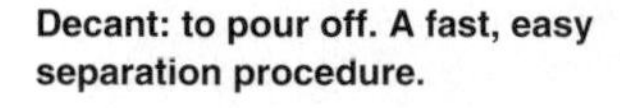
Decant: to pour off. A fast, easy separation procedure.

(B) Gravity Filtration

The most common method for the removal of insoluble solid material is gravity filtration through fluted filter paper (Fig. 4.6). This is the method of choice for removing

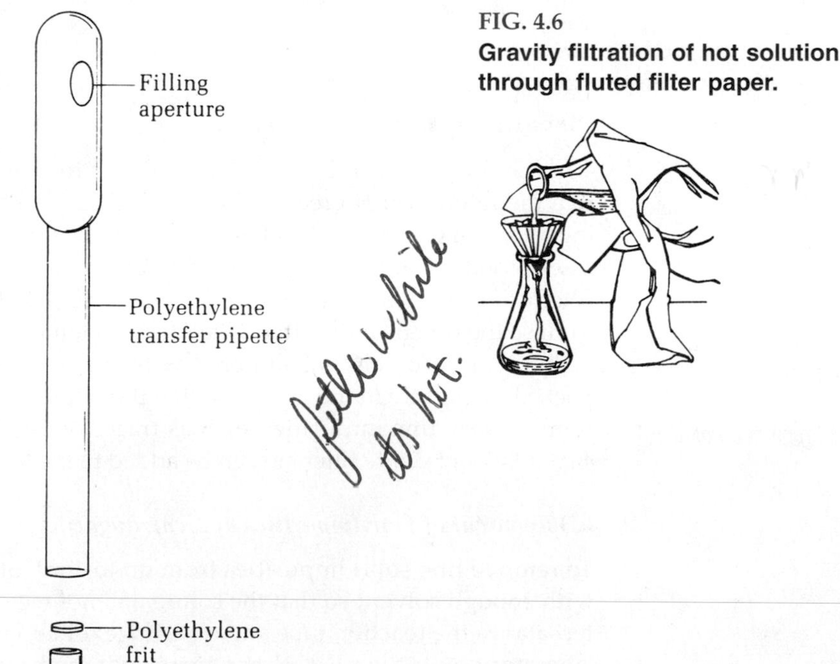

FIG. 4.5
A pressure filtration apparatus. The solution to be filtered is added through the aperture, which is closed by a finger as pressure is applied.

FIG. 4.6
Gravity filtration of hot solution through fluted filter paper.

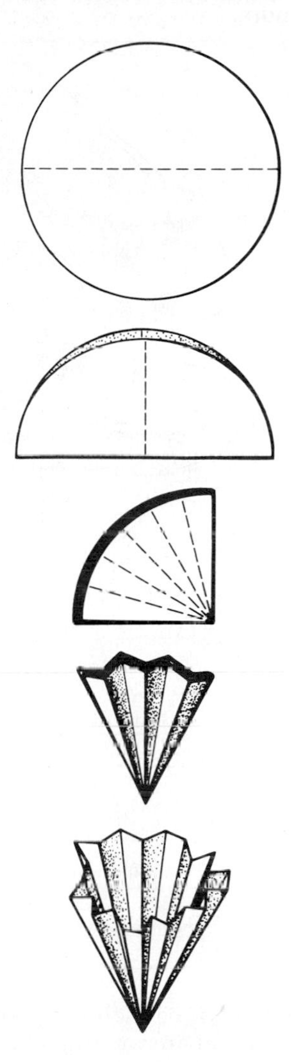

FIG. 4.7
Fluting a piece of filter paper.

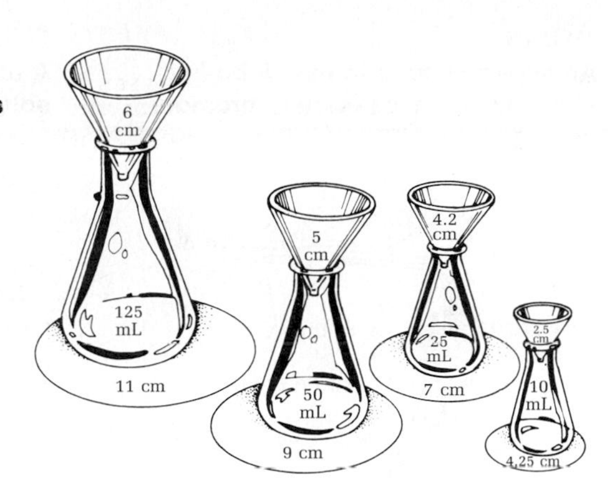

FIG. 4.8
Assemblies for gravity filtration. Stemless funnels have diameters of 2.5, 4.2, 5.0, and 6.0 cm.

⚠ Be aware that the vapors of low-boiling solvents can ignite on an electric hot plate.

finely divided charcoal, dust, lint, and so on. The following equipment is needed for this process: three labeled Erlenmeyer flasks on a steam bath or a hot plate (flask A contains the solution to be filtered, flask B contains a few milliliters of solvent and a stemless funnel, and flask C contains several milliliters of the crystallizing solvent to be used for rinsing purposes), a fluted piece of filter paper, a towel for holding the hot flask and drying out the stemless funnel, and boiling chips for all solutions.

A piece of filter paper is fluted as shown in Figure 4.7 and is then placed in a stemless funnel. Appropriate sizes of Erlenmeyer flasks, stemless funnels, and filter paper are shown in Figure 4.8. The funnel is stemless so that the saturated solution being filtered will not have a chance to cool and clog the stem with crystals. The filter paper should fit entirely inside the rim of the funnel; it is fluted to allow rapid filtration. Test to see that the funnel is stable in the neck of flask B. If not, support it with a ring attached to a ring stand. A few milliliters of solvent and a boiling chip should be placed in flask B into which the solution is to be filtered. This solvent is brought to a boil on a steam bath or hot plate, along with the solution to be filtered.

The solution to be filtered (in flask A) should be saturated with the solute at the boiling point. Note the volume and then add 10% more solvent (from flask C). The resulting slightly dilute solution is not as likely to recrystallize in the funnel during filtration. Bring the solution to be filtered to a boil, grasp flask A with a towel, and pour the solution into the filter paper in the stemless funnel equipped in flask B (Fig. 4.6). The funnel should be warm to prevent recrystallization from occurring in the funnel. This can be accomplished in two ways: (1) Invert the funnel over a steam bath for a few seconds, pick up the funnel with a towel, wipe it perfectly dry, place it on top of flask B, and then add the fluted filter paper; or (2) place the stemless funnel in the neck of flask B and allow the solvent to reflux into the funnel, thereby warming it.

Pour the solution to be filtered (in flask A) at a steady rate onto the fluted filter paper (equipped in flask B). Check to see whether recrystallization is occurring in the filter. If it does, add boiling solvent (from flask C heated on a steam bath or a hot plate) until the crystals dissolve, dilute the solution being filtered, and carry on. Rinse flask A with a few milliliters of boiling solvent (from flask C) and rinse the fluted filter paper with this same solvent.

Because the filtrate has been diluted to prevent it from recrystallizing during the filtration process, the excess solvent must now be removed by boiling the solution. This process can be sped up somewhat by blowing a slow current of air into the flask in the hood or using an aspirator tube to pull vapors into the aspirator (Fig. 4.9 and Fig. 4.10). However, the fastest method is to heat the solvent in the

FIG. 4.9
An aspirator tube in use. A boiling stick may be necessary to promote boiling.

FIG. 4.10
A tube being used to remove solvent vapors.

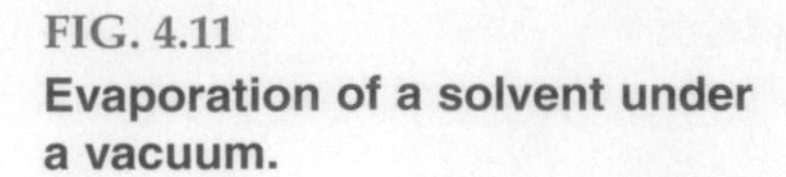

FIG. 4.11
Evaporation of a solvent under a vacuum.

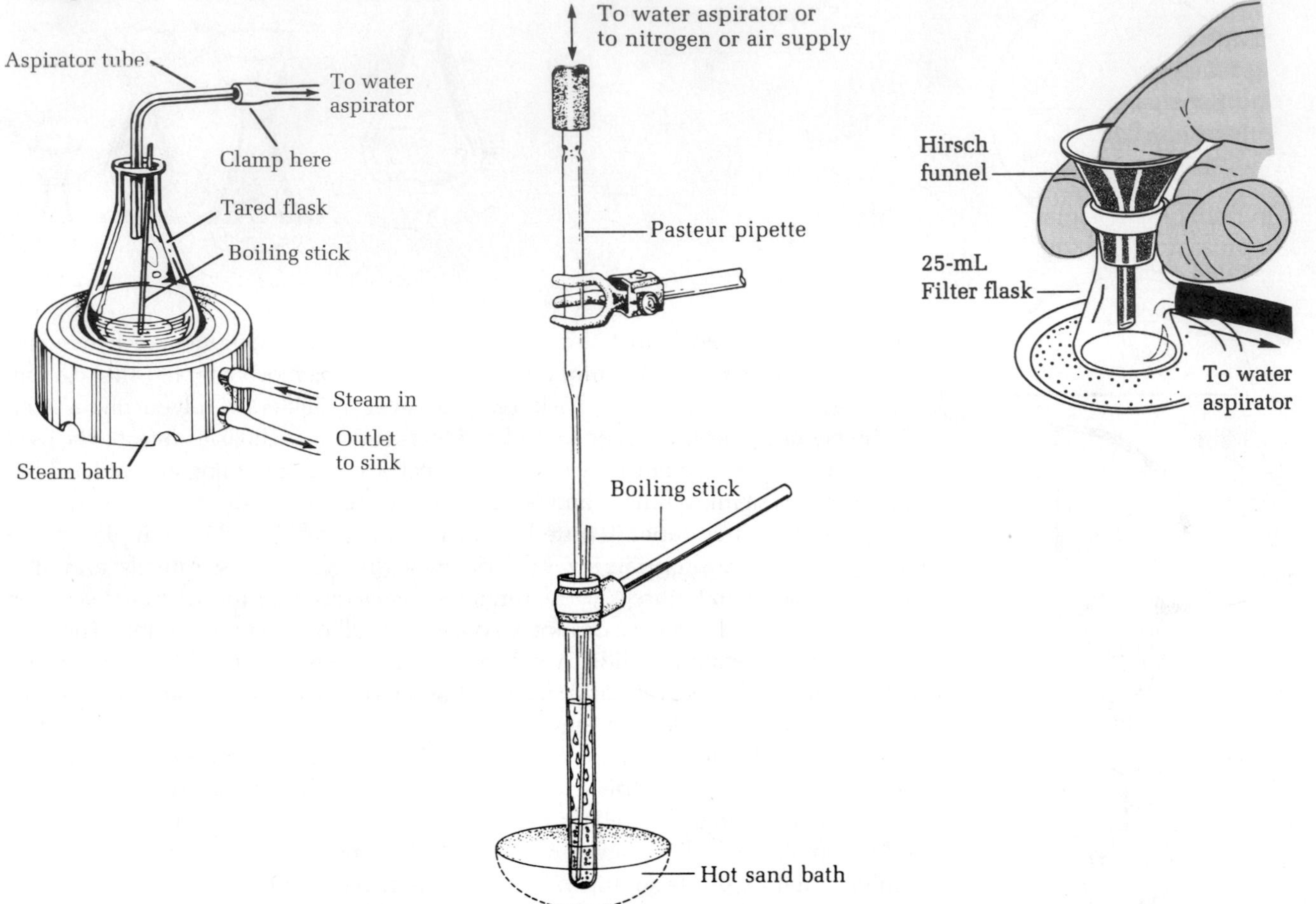

filter flask on a sand bath while the flask is connected to a water aspirator. The vacuum is controlled with the thumb (Fig. 4.11).[2] Be sure that you are wearing gloves when doing this step! If your thumb is not large enough, put a one-holed rubber stopper into the Hirsch funnel or the filter flask and again control the vacuum with your thumb. If the vacuum is not controlled, the solution may boil over and go out the vacuum hose.

⚠ Be sure that you are wearing gloves when doing this step!

STEP 5. RECRYSTALLIZING THE SOLUTE

On both a macroscale and a microscale, the recrystallization process should normally start from a solution that is saturated with the solute at the boiling point. If it has been necessary to remove impurities or charcoal by filtration, the solution has been diluted. To concentrate the solution, simply boil off the solvent under an aspirator tube as shown in Figure 4.9 (macroscale) or blow off solvent using a gentle stream of air or, better, nitrogen in the hood as shown in Figure 4.10 (microscale). Be sure to have a boiling chip (macroscale) or a boiling stick (microscale) in the

[2]See also Mayo, D. W.; Pike, R. M.; Butcher, S. M. *Microscale Organic Laboratory;* Wiley: New York, 1986; 97.

solution during this process, but make sure you remove it before initiating recrystallization.

A saturated solution.

Slow cooling is important.

Once it has been ascertained that the hot solution is saturated with the compound just below the boiling point of the solvent, allow it to cool slowly to about 22°C. Slow cooling is a critical step in recrystallization. If the solution is not allowed to cool slowly, precipitation will occur, resulting in impurities "crashing out" of solution along with the desired solute; thus, no exclusion will occur. On a microscale, it is best to allow the reaction tube to cool in a beaker filled with cotton or paper towels which act as insulation, so cooling takes place slowly. Even insulated in this manner, the small reaction tube will cool to about 22°C within a few minutes. Slow cooling will guarantee the formation of large crystals, which are easily separated by filtration and easily washed free of adhering impure solvent. On a small scale, it is difficult to obtain crystals that are too large and occlude impurities. Once the tube has cooled to about 22°C without disturbance, it can be cooled in ice to maximize the amount of product that comes out of solution. On a macroscale, the Erlenmeyer flask is set atop a cork ring or other insulator and allowed to cool gradually to about 22°C. If the flask is moved during recrystallization, many nuclei will form, and the crystals will be small and have a large surface area. They will not be easy to filter and wash clean of the mother liquor. Once recrystallization ceases at about 22°C, the flask should be placed in ice to cool further. Make sure to clamp the flask in the ice bath so that it does not tip over.

Videos: *Recrystallization, Microscale Crystallization*

Add a seed crystal or scratch the tube.

With slow cooling, recrystallization should begin immediately. If not, add a seed crystal or scratch the inside of the tube with a glass rod at the liquid-air interface. Recrystallization must start on some nucleation center. A minute crystal of the desired compound saved from the crude material will suffice. If a seed crystal is not available, recrystallization can be started on the rough surface of a fresh scratch on the inside of the container. use glass stirring rod

STEP 6. COLLECTING AND WASHING THE CRYSTALS

Once recrystallization is complete, the crystals must be separated from the ice-cold mother liquor, washed with ice-cold solvent, and dried. * ethanol

Microscale Procedure

(A) Filtration Using a Pasteur Pipette

Videos: *Filtration of Crystals Using the Pasteur Pipette*

The most important filtration technique used in microscale organic experiments employs a Pasteur pipette (Fig. 4.12). About 70% of the crystalline products from the experiments in this text can be isolated in this way. The others will be isolated by filtration on a Hirsch funnel.

The ice-cold crystalline mixture is stirred with a Pasteur pipette and, while air is being expelled from the pipette, forced to the bottom of the reaction tube. The bulb is released, and the solvent is drawn into the pipette through the very small space between the flat tip of the pipette and the curved bottom of the reaction tube. When all the solvent has been withdrawn, it is expelled into another reaction tube. It is sometimes useful to rap the tube containing the wet crystals against a hard surface to pack them so that more solvent can be removed. The tube is returned to the ice bath, and a few drops of cold solvent are added to the crystals. The mixture is stirred to wash the crystals, and the solvent is again removed. This process can be repeated as many times as necessary. Volatile solvents can be removed from the

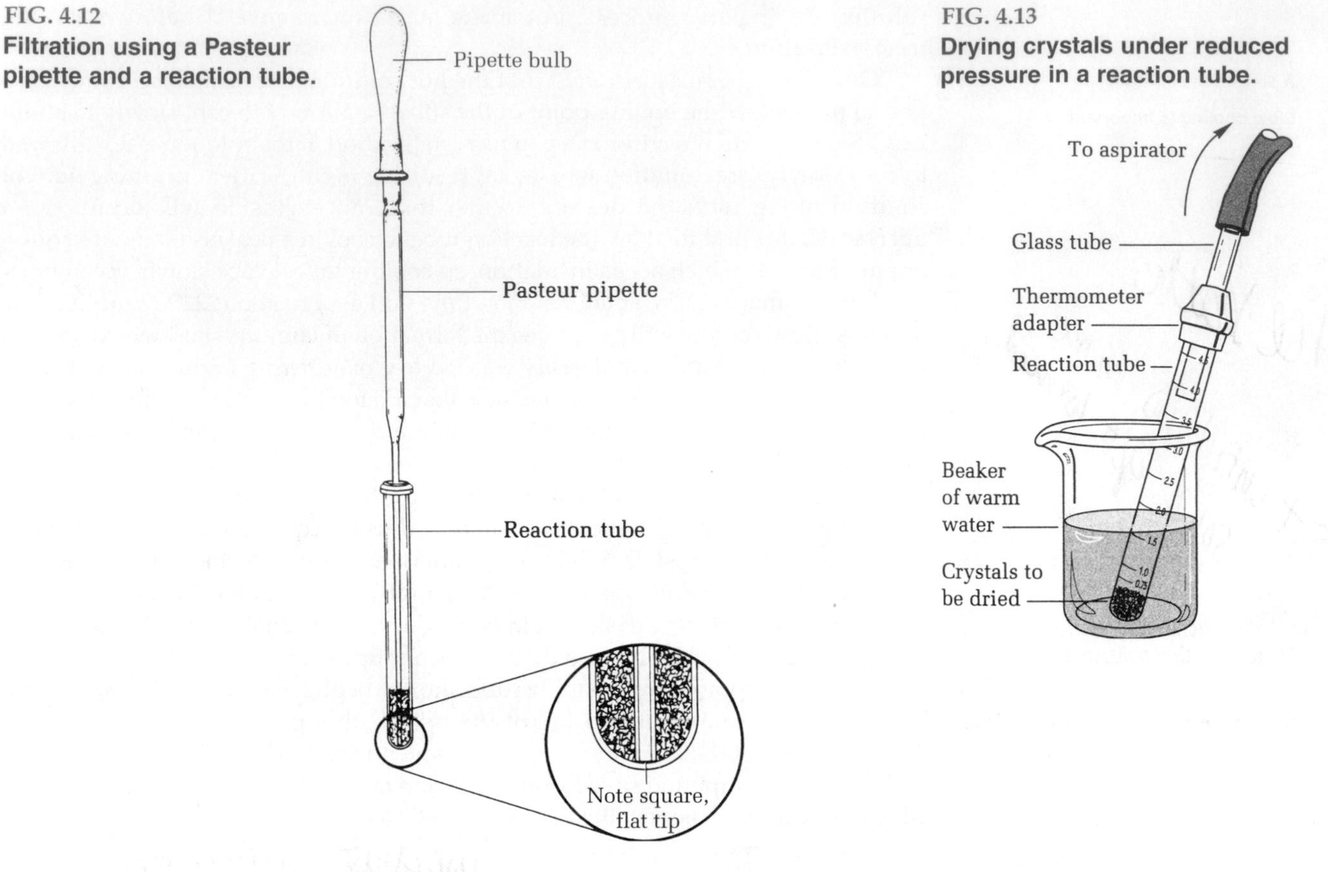

FIG. 4.12
Filtration using a Pasteur pipette and a reaction tube.

FIG. 4.13
Drying crystals under reduced pressure in a reaction tube.

damp crystals under vacuum (Fig. 4.13). Alternatively, the last traces of solvent can be removed by centrifugation using a Wilfilter (Fig. 4.14).

(B) Filtration Using a Hirsch Funnel

Video: *Microscale Filtration on the Hirsch Funnel*

When the volume of material to be filtered is greater than 1.5 mL, collect the material on a Hirsch funnel. The Hirsch funnel in the Williamson/Kontes kit[3] is unique. It is composed of polypropylene and has an integral molded stopper that fits a 25-mL filter flask. It comes fitted with a 20-μm polyethylene fritted disk, which is not meant to be disposable, although it costs only about twice as much as an 11-cm piece of filter paper (Fig. 4.15). Although products can be collected directly on this disk, it is good practice to place an 11- or 12-mm diameter piece of No. 1 filter paper on the disk. In this way, the frit will not become clogged with insoluble impurities. The disk of filter paper can be cut to size with a cork borer or leather punch. A piece of filter paper *must* be used on the old-style porcelain Hirsch funnels.

Clamp the clean, dry 25-mL filter flask in an ice bath to prevent it from falling over and place the Hirsch funnel with filter paper in the flask. The reason for cooling the filter flask is to keep the mother liquor cold so it will not dissolve the crystals on the Hirsch funnel when fresh cold solvent is used to wash crystals from the container onto the funnel. In a separate flask, cool ~10 mL of solvent in an ice bath; this solvent is used for washing the recrystallization flask and for washing the crystals. Wet the filter paper with the solvent used in the recrystallization, turn on the water aspirator (see "The Water Aspirator and the Trap"), and ascertain that

[3]The microscale kit is available through Kontes (www.kontes.com).

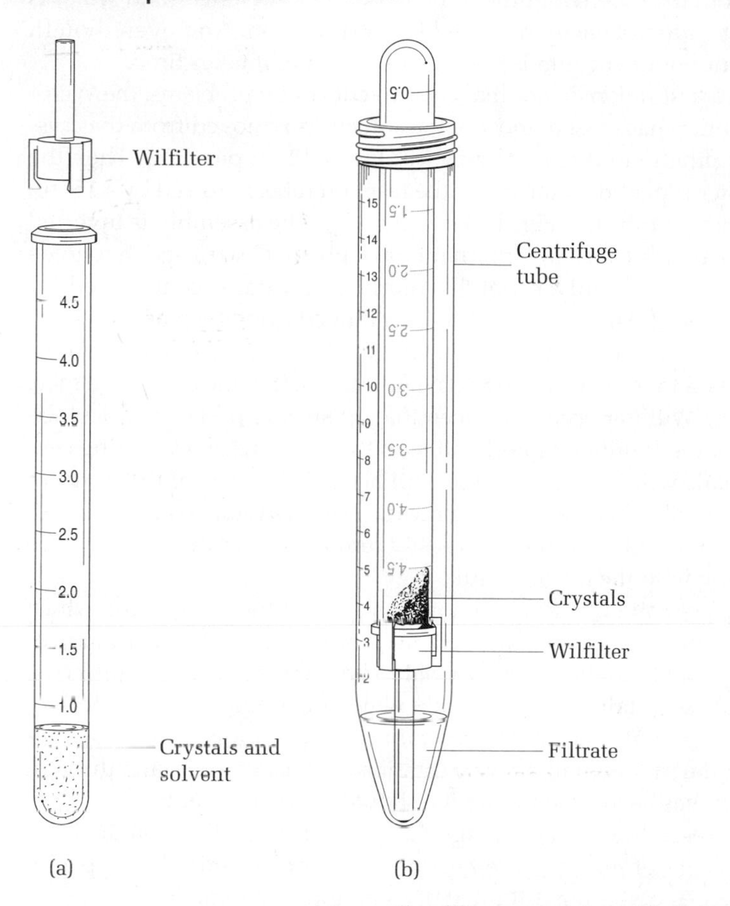

FIG. 4.14
The Wilfilter filtration apparatus. Filtration occurs between the flat face of the polypropylene Wilfilter and the top of the reaction tube.

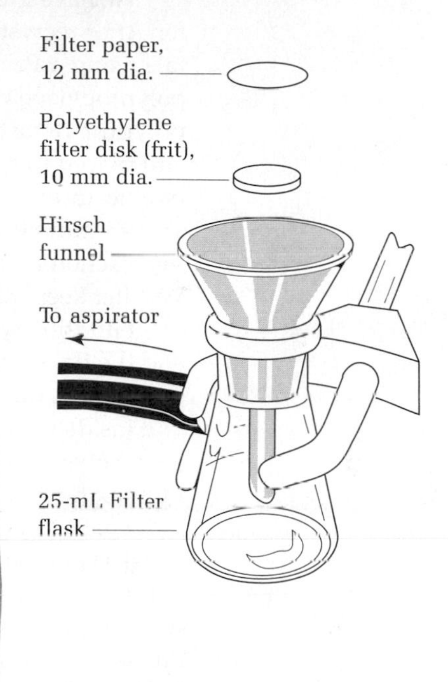

FIG. 4.15
The Hirsch funnel being used for vacuum filtration. This unique design has a removable and replaceable 20-μm polyethylene frit. No adapter is needed because there is a vacuum-tight fit to the filter flask. Always use a piece of filter paper with this funnel.

Suction filtration

the filter paper is pulled down onto the frit. Pour and scrape the crystals and mother liquor onto the Hirsch funnel and, as soon as the liquid is gone from the crystals, break the vacuum at the filter flask by removing the rubber hose.

Break the vacuum, add a very small quantity of ice-cold wash solvent, and reapply the vacuum.

Rinse the recrystallization flask with ice-cold fresh solvent. Pour this rinse through the Hirsch funnel and reapply vacuum to the filter flask. As soon as all the liquid has disappeared from the crystals, wash the crystals with a few drops of ice-cold solvent. Repeat this washing process as many times as necessary to remove colored material or other impurities from the crystals. In some cases, only one very small wash will be needed. After the crystals have been washed with ice-cold solvent, the vacuum can be left on to dry the crystals. A cork can be used to press the solvent from the crystals, if necessary.

(C) Filtration with a Wilfilter (Replacing a Craig Tube)

The isolation of less than 100 mg of recrystallized material from a reaction tube (or any other container) is not easy. If the amount of solvent is large enough (1 mL or more), the material can be recovered by filtration with a Hirsch funnel. But when the volume of liquid is less than 1 mL, much product is left in the tube during the transfer to a Hirsch funnel. The solvent can be removed with a Pasteur pipette pressed against the bottom of the tube, a very effective filtration technique, but scraping the damp crystals from the reaction tube results in major losses. If the solvent has a relatively low boiling point, it can be evaporated by connecting the tube to a water aspirator (*see* Fig. 4.13 on page 72). Once the crystals are dry, they are easily scraped from the tube with little or no loss. Some solvents—and water is the principal culprit—are not easily removed by evaporation. And even though removal of the solvent under vacuum is not terribly difficult, it takes time.

We have invented a filtration device that circumvents these problems: the Wilfilter. After recrystallization has ceased, most of the solvent is removed from the crystals using a Pasteur pipette in the usual way (*see* Fig. 4.12 on page 72). Then the polypropylene Wilfilter is placed on the top of the reaction tube, followed by a 15-mL polypropylene centrifuge tube (*see* Fig. 4.14 on page 73). The assembly is inverted and placed in a centrifuge such as the International Clinical Centrifuge that holds twelve 15-mL tubes. The assembly, properly counterbalanced, is centrifuged for about 1 minute at top speed. The centrifuge tube is removed from the centrifuge, and the reaction tube is then removed from the centrifuge tube. The three fingers on the Wilfilter keep it attached to the reaction tube. The filtrate is left in the centrifuge tube.

Filtration with the Wilfilter occurs between the top surface of the reaction tube and the flat surface of the Wilfilter. Liquid will pass through that space during centrifugation, but crystals will not. The crystals will be found on top of the Wilfilter and inside the reaction tube. The very large centrifugal forces remove all the liquid, so the crystals will be virtually dry and thus easily removed from the reaction tube by shaking or scraping with the metal spatula.

The Wilfilter replaces an older device known as a Craig tube (Fig. 4.16), which consists of an outer tube of 1-, 2-, or 3-mL capacity with an inner plunger made of Teflon (expensive) or glass (fragile). The material to be recrystallized is transferred to the outer tube and recrystallized in the usual way. The inner plunger is added, and a wire hanger is fashioned so that the assembly can be removed from the centrifuge tube without the plunger falling off. Filtration in this device occurs through the rough surface that has been ground into the shoulder of the outer tube.

The Wilfilter possesses several advantages: a special recrystallization device is not needed, no transfers of material are needed, it is not as limited in capacity (which is 4.5 mL), and its cost is one-fifth that of a Craig tube assembly.

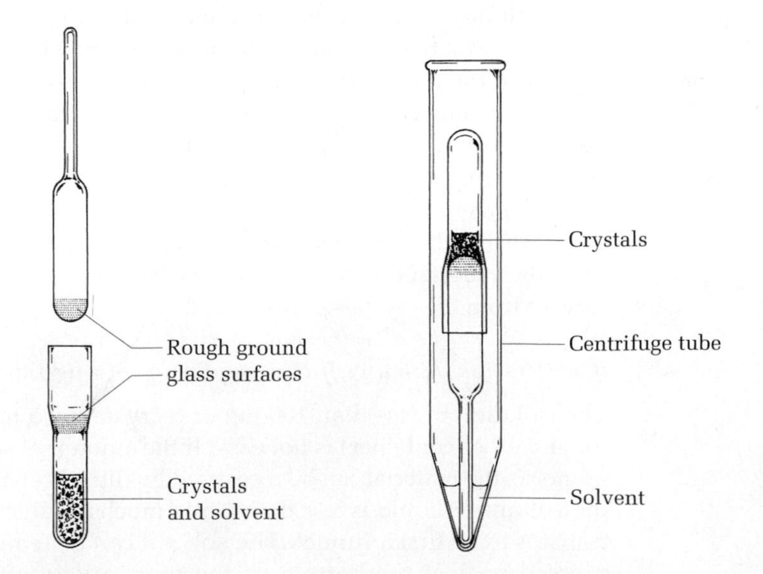

FIG. 4.16
The Craig tube filtration apparatus. Filtration occurs between the rough ground glass surfaces when the apparatus is centrifuged.

(D) Filtration into a Reaction Tube on a Hirsch Funnel

Photo: *Vacuum Filtration into a Reaction Tube through a Hirsch Funnel*

If it is desired to have the filtrate in a reaction tube instead of spread on the bottom of a 25-mL filter flask, then the process described in the previous section can be carried out in the apparatus shown in Figure 4.17. The vacuum hose is connected to the side arm of the flask using a thermometer adapter and a short length of glass tubing. Evaporate the filtrate in the reaction tube to collect a second crop of crystals.

(E) Filtration into a Reaction Tube with a Micro Büchner Funnel

Video: *Microscale Crystallization*

If the quantity of material being collected is very small, use the bottom of the chromatography column as a micro Büchner funnel, which can be fitted into the top of the thermometer adapter, as shown in Figure 4.18. Again, it is good practice to cover the frit with a piece of 6-mm filter paper (cut with a cork borer).

(F) The Micro Büchner Funnel in an Enclosed Filtration Apparatus

In the apparatus shown in Figure 4.19, recrystallization is carried out in the upper reaction tube in the normal way. The apparatus is then turned upside down, the crystals are shaken down onto a micro Büchner funnel, and a vacuum is applied through the side arm. In this apparatus, crystals can be collected in an oxygen-free atmosphere (Schlenk conditions).

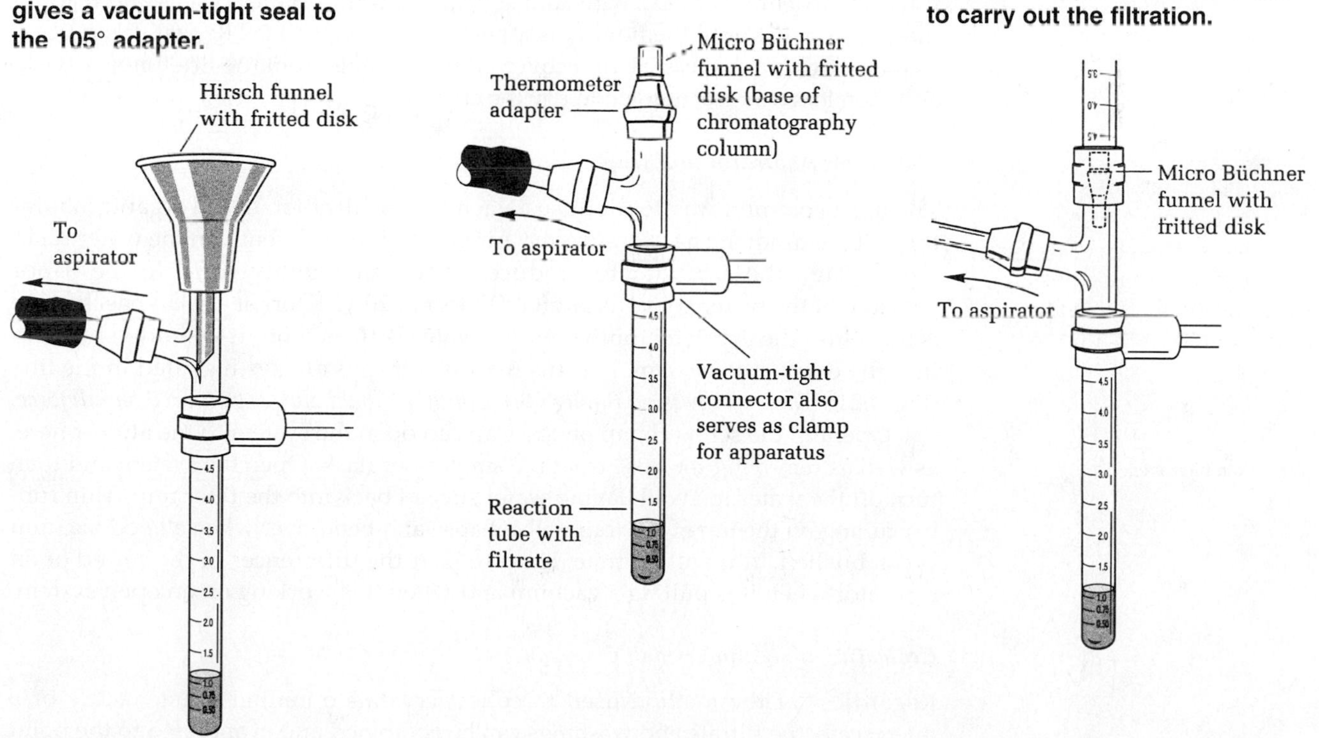

FIG. 4.17
A microscale Hirsch filtration assembly. The Hirsch funnel gives a vacuum-tight seal to the 105° adapter.

FIG. 4.18
Filtration using a microscale Büchner funnel.

FIG. 4.19
The Schlenk-type filtration apparatus. The apparatus is inverted to carry out the filtration.

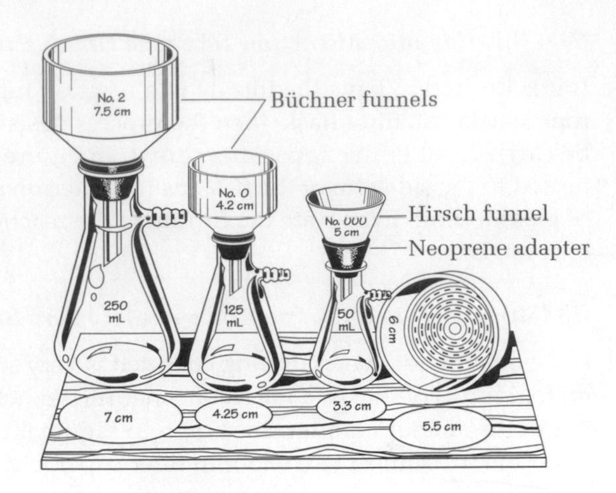

FIG. 4.20
Matching filter assemblies. The 6.0-cm polypropylene Büchner funnel (right) resists breakage and can be disassembled for cleaning.

Macroscale Apparatus

Filtration on a Hirsch Funnel and a Büchner Funnel

If the quantity of material is small (<2 g), a Hirsch funnel can be used in exactly the way that was described in a previous section. For larger quantities, a Büchner funnel is used. Properly matched Büchner funnels, filter paper, and flasks are shown in Figure 4.20. The Hirsch funnel shown in the figure has a 5-cm bottom plate to accept 3.3-cm round filter paper.

Video: *Microscale Filtration with the Hirsch Funnel*

Place a piece of filter paper in the bottom of the Büchner funnel. Wet it with solvent and be sure it lies flat so that crystals cannot escape around the edge and go under the filter paper. Then, with the vacuum off, pour the cold slurry of crystals into the center of the filter paper. Apply the vacuum; as soon as the liquid disappears from the crystals, break the vacuum to the flask by disconnecting the hose. Rinse the Erlenmeyer flask with cold solvent. Add this to the crystals and reapply the vacuum just until the liquid disappears from the crystals. Repeat this process as many times as necessary to recover all the crystals from the Erlenmeyer flask, and then leave the vacuum on to dry the crystals.

The Water Aspirator and Trap

The most common way to produce a vacuum for filtration in the organic laboratory is by employing a *water aspirator*. Air is entrained efficiently in the water rushing through the aspirator to produce a vacuum roughly equal to the vapor pressure of the water going through it (17 torr at 20°C, 5 torr at 4°C). A check valve is built into the aspirator, but when the water is turned off, it will often back up into the evacuated system. For this reason, a trap is always installed in the line (Fig. 4.21). *The water passing through the aspirator should always be turned on full force.*

Clamp the filter flask

Opening the screw clamp on the trap can open the system to the atmosphere, as well as removing the hose from the small filter flask. Open the system and then turn off the water to avoid having water sucked back into the filter trap. Thin rubber tubing on the top of the trap will collapse and bend over when a good vacuum is established. You will, in time, learn to hear the differences in the sound of an aspirator when it is pulling a vacuum and when it is working on an open system.

Collecting a Second Crop of Crystals

Regardless of the method used to collect crystals on either a macroscale or a microscale, the filtrate and washings can be combined and evaporated to the point of saturation to obtain a second crop of crystals—hence this advocates having a

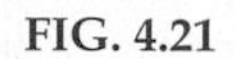

FIG. 4.21
An aspirator, a filter trap, and a funnel. Clamp the small filter flask to prevent it from tipping over.

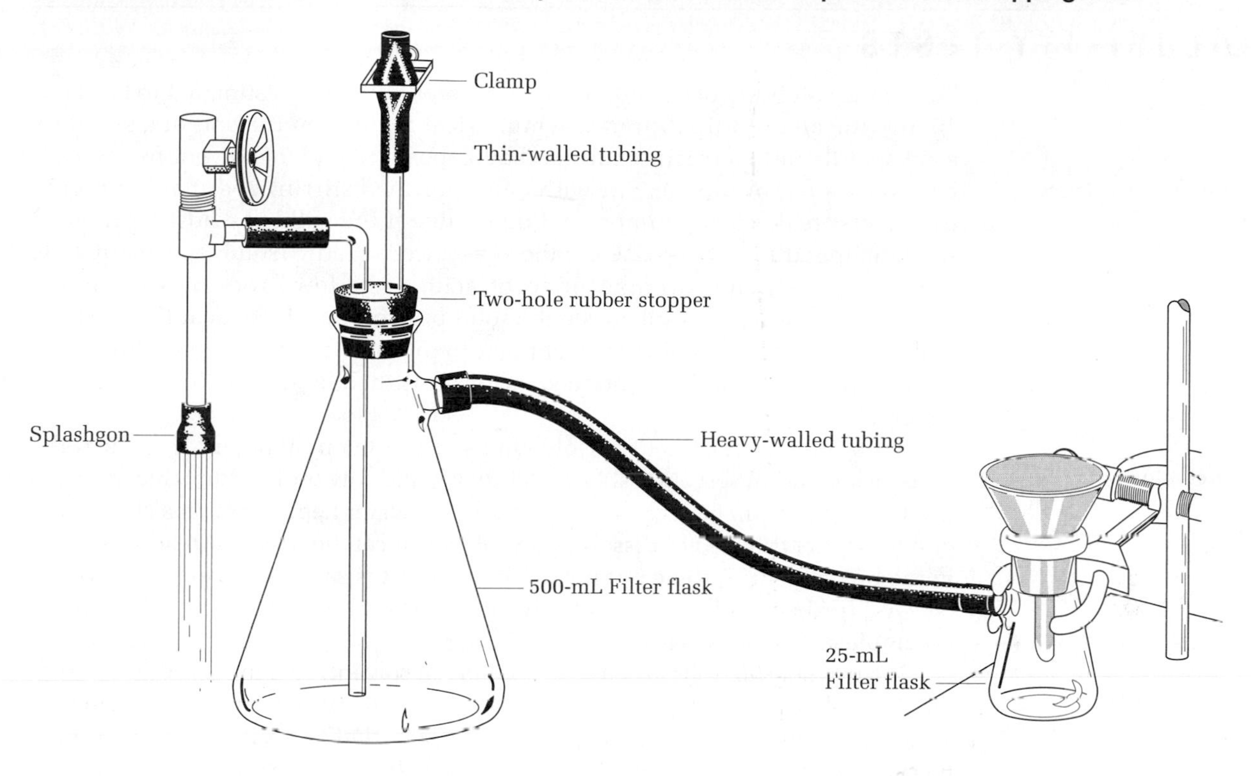

clean receptacle for the filtrate. This second crop will increase the overall yield, but the crystals will not usually be as pure as the first crop.

STEP 7. DRYING THE PRODUCT

Microscale Procedure

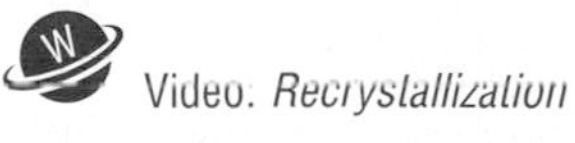

If possible, dry the product in the reaction tube after removing the solvent with a Pasteur pipette. Simply connecting the tube to the water aspirator can do this. If the tube is clamped in a beaker of hot water, the solvent will evaporate more rapidly under vacuum, but make sure not to melt the product (*see* Fig. 4.13 on page 72). Water, which has a high heat of vaporization, is difficult to remove in this way. Scrape the product onto a watch glass and allow it to dry to constant weight, which will indicate that all the solvent has been removed. If the product is collected on a Hirsch funnel or a Wilfilter, the last traces of solvent can be removed by squeezing the crystals between sheets of filter paper before drying them on the watch glass.

FIG. 4.22
Drying a solid by reduced air pressure.

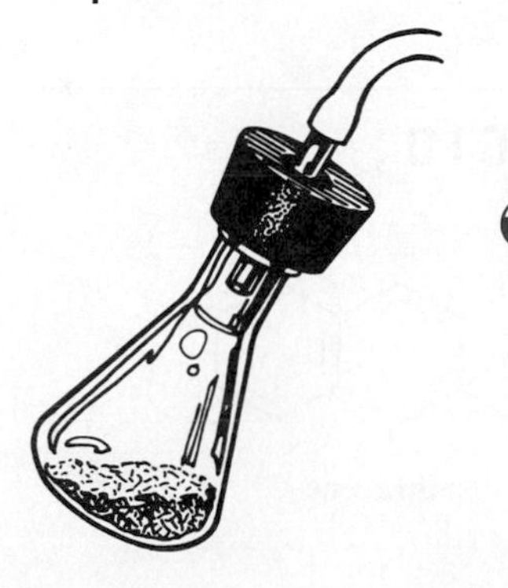

Macroscale Procedure

Once the crystals have been washed on a Hirsch funnel or a Büchner funnel, press them down with a clean cork or other flat object and allow air to pass through them until they are substantially dry. Final drying can be done under reduced pressure (Fig. 4.22). The crystals can then be turned out of the funnel and squeezed between sheets of filter paper to remove the last traces of solvent before the final drying on a watch glass.

EXPERIMENTS

1. SOLUBILITY TESTS

Video: *Picking a Solvent*

Test Compounds:

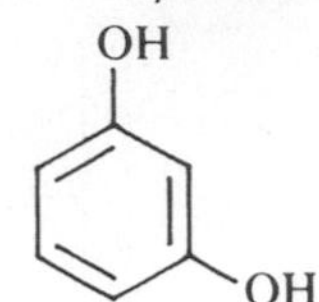

Resorcinol

Anthracene

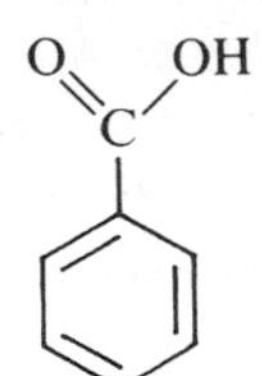

Benzoic acid

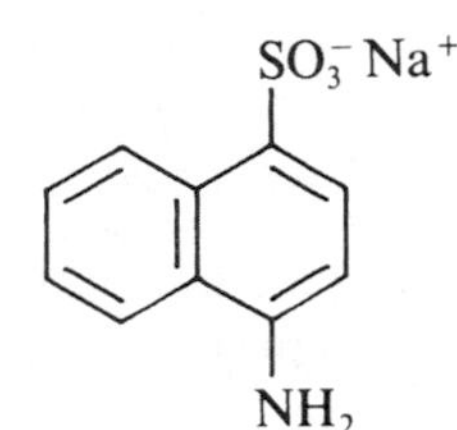

4-Amino-1-naphthalenesulfonic acid, sodium salt

To test the solubility of a solid, transfer an amount roughly estimated to be about 10 mg (the amount that forms a symmetrical mound on the end of a stainless steel spatula) into a reaction tube and add about 0.25 mL of solvent from a calibrated dropper or pipette. Stir with a fire-polished stirring rod that is 4 mm in diameter, break up any lumps, and determine if the solid is readily soluble at room temperature (about 22°C). If the substance is readily soluble at about 22°C in methanol, ethanol, acetone, or acetic acid, add a few drops of water to the solution from a wash bottle to see if a solid precipitates. If so, heat the mixture, adjust the composition of the solvent pair to produce a hot solution saturated at the boiling point, let the solution stand undisturbed, and note the character of the crystals that form.

If the substance fails to dissolve in a given solvent at about 22°C, heat the suspension and observe if a solution occurs. If the solvent is flammable, heat the test tube on a steam bath or in a small beaker of water kept warm on a steam bath or a hot plate. If the solid dissolves completely, it can be declared readily soluble in the hot solvent; if some but not all dissolves, it is said to be moderately soluble, and further small amounts of solvent should then be added until dissolution is complete.

When a substance has been dissolved in hot solvent, cool the solution by holding the flask under the tap and, if necessary, induce recrystallization by rubbing the walls of the tube with a stirring rod to make sure that the concentration permits recrystallization. Then reheat to dissolve the solid, let the solution stand undisturbed, and inspect the character of the ultimate crystals.

Perform solubility tests on the test compounds that are shown in the margin, in each of the following solvents: water (hydroxylic and ionic), toluene (an aromatic hydrocarbon), and hexanes (a mixture of isomers of hexane). Note the degree of solubility in the solvents—cold and hot—and suggest suitable solvents, solvent pairs, or other expedients for the recrystallization of each substance. Record the crystal form, at least to the extent of distinguishing between needles (pointed crystals), plates (flat and thin), and prisms. How do your observations conform to the generalization that like dissolves like?

Cleaning Up. Place organic solvents and solutions of the compounds in the organic solvents waste container. Dilute the aqueous solutions with water and flush down the drain. (For this and all other "Cleaning Up" sections, refer to the complete discussion of waste disposal procedures in Chapter 2.)

2. RECRYSTALLIZATION OF PURE PHTHALIC ACID, NAPHTHALENE, AND ANTHRACENE

COOH
COOH

Phthalic acid **Naphthalene** **Anthracene**

The process of recrystallization can be readily observed using phthalic acid. In the *CRC Handbook of Chemistry and Physics*, in the table "Physical Constants of Organic Compounds," the entry for phthalic acid gives the following solubility data (in grams of solute per 100 mL of solvent). The superscripts refer to temperature in degrees Celsius:

Water	Alcohol	Ether, etc.
0.54^{14} 18^{99}	11.71^{18}	0.69^{15} eth., i. chl.

The large difference in solubility in water as a function of temperature suggests that water is the solvent of choice. The solubility in alcohol is high at about 22°C. Ether is difficult to use because it is so volatile; the compound is insoluble in chloroform (i. chl.).

Microscale Procedure for Phthalic Acid

Video: *Recrystallization*

Recrystallize 60 mg (0.060 g) of phthalic acid from the minimum volume of water, using the previous data to calculate the required volume. First, turn on an electrically heated sand bath. Add the solid to a 10 × 100 mm reaction tube and add water dropwise with a Pasteur pipette. Use the calibration marks found in Chapter 1 (*see* Fig. 1.18 on page 19) to measure the volume of water in the pipette and the reaction tube. Add a boiling stick (a wood applicator stick) to facilitate even boiling and prevent bumping. After a portion of the water has been added, gently heat the solution to boiling on a hot sand bath in the electric heater. The deeper the tube is placed in the sand, the hotter it will be. As soon as boiling begins, continue to add water dropwise until the entire solid just dissolves. Remove the boiling stick. Cork the tube, clamp it as it cools, and observe the phenomenon of recrystallization.

Set the heater control to about 20% of the maximum.

After the tube reaches about 22°C, cool it in ice, stir the crystals that have formed with a Pasteur pipette, and expel the air from the pipette as the tip is pushed to the bottom of the tube. When the tip is firmly and squarely seated in the bottom of the tube, release the bulb and withdraw the water. Rap the tube sharply on a wood surface to compress the crystals and remove as much of the water as possible with the pipette. Then cool the tube in ice and add a few drops of ice-cold ethanol to the tube to remove water from the crystals. Connect the tube to a water aspirator and warm it in a beaker of hot water (*see* Fig. 4.13 on page 72). Once all the solvent is removed, use a stainless steel spatula to scrape the crystals onto a piece of filter paper, fold the paper over the crystals, and squeeze out excess water before allowing the crystals to dry to constant weight. Weigh the dry crystals and calculate the percent recovery of—the purified compound.

Alternate procedure: Dry the crystals under vacuum in a steam bath in the reaction tube.

Microscale Procedure for Naphthalene and Anthracene

Following the previous procedure, recrystallize 40 mg of naphthalene from 80% aqueous methanol or 10 mg of anthracene from ethanol. You may find it more convenient to use a hot water bath to heat these low-boiling alcohols. These are more typical of compounds to be recrystallized in later experiments because they are soluble in organic solvents. It will be much easier to remove these solvents from the crystals under vacuum than it is to remove water from phthalic acid. You will seldom encounter the need to recrystallize less than 30 mg of a solid in these experiments.

These compounds can also be isolated using a Wilfilter.

Cleaning Up. Dilute the aqueous filtrate with water and flush the solution down the drain. Phthalic acid is not considered toxic to the environment and if desired, can be recycled for future recrystallization experiments. Methanol and ethanol filtrates go into the organic solvents waste container.

Macroscale Procedure

Using the solubility data for phthalic acid to calculate the required volume, recrystallize 1.0 g of phthalic acid from the minimum volume of water. Add the solid to the smallest practical Erlenmeyer flask and then, using a Pasteur pipette, add water dropwise from a full 10-mL graduated cylinder. A boiling stick (a wood applicator stick) facilitates even boiling and will prevent bumping. After a portion of the water has been added, gently heat the solution to boiling on a hot plate. As soon as boiling begins, continue to add water dropwise until the entire solid dissolves. Remove the boiling stick. Place the flask on a cork ring or other insulator and allow it to cool undisturbed to about 22°C, during which time the recrystallization process can be observed. Slow cooling favors large crystals. Then cool the flask in an ice bath, decant (pour off) the mother liquor (the liquid remaining with the crystals), and remove the last traces of liquid with a Pasteur pipette. Scrape the crystals onto a filter paper using a stainless steel spatula, squeeze the crystals between sheets of filter paper to remove traces of moisture, and allow the crystals to dry. Alternatively, the crystals can be collected on a Hirsch funnel. Compare the calculated volume of water to the volume of water actually used to dissolve the acid. Calculate the percent recovery of dry, recrystallized phthalic acid.

Video: *Macroscale Crystallization*

Cleaning Up. Dilute the filtrate with water and flush the solution down the drain. Phthalic acid is not considered toxic to the environment and if desired, can be recycled for future recrystallization experiments.

3. DECOLORIZING A SOLUTION WITH DECOLORIZING CHARCOAL

Into a reaction tube, place 1.0 mL of a solution of methylene blue dye that has a concentration of 10 mg per 100 mL of water. Add to the tube about 10 or 12 pieces of decolorizing charcoal, shake, and observe the color over a period of 1–2 minutes. Heat the contents of the tube to boiling (reflux) and observe the color by holding the tube in front of a piece of white paper from time to time. How rapidly is the color removed? If the color is not removed in a minute or so, add more charcoal pellets.

Video: *Decolorization of a Solution with Norit, Decolorizing using pelletized Norit*

Cleaning Up. Place the Norit charcoal pellets in the nonhazardous solid waste container.

4. DECOLORIZATION OF BROWN SUGAR (SUCROSE, $C_{12}H_{22}O_{11}$)

Raw sugar is refined commercially with the aid of decolorizing charcoal. The clarified solution is seeded generously with small sugar crystals, and excess water is removed under vacuum to facilitate recrystallization. The pure white crystalline product is collected by centrifugation. Brown sugar is partially refined sugar and can be easily decolorized using charcoal.

In a 50-mL Erlenmeyer flask, dissolve 15 g of dark brown sugar in 30 mL of water by heating and stirring. Pour half the solution into another 50-mL flask. Heat one of the solutions nearly to its boiling point, allow it to cool slightly, and add 250 mg (0.25 g) of decolorizing charcoal (Norit pellets) to it. Bring the solution back to near the boiling point for 2 minutes; then filter the hot solution into an Erlenmeyer flask through a fluted filter paper held in a previously heated funnel. Treat the other half of the sugar solution in exactly the same way but use only 50 mg of decolorizing charcoal. Collaborate with a fellow student who will heat the solution for only 15 seconds after adding the charcoal. Compare your results.

Cleaning Up. Decant (pour off) the aqueous layer. Place the Norit in the nonhazardous solid waste container. The sugar solution can be flushed down the drain.

5. RECRYSTALLIZATION OF BENZOIC ACID FROM WATER AND A SOLVENT PAIR

Benzoic acid

Recrystallize 50 mg of benzoic acid from water in the same way that phthalic acid was recrystallized. Then in a dry reaction tube, dissolve another 50-mg sample of benzoic acid in the minimum volume of hot methanol and add water to the hot solution dropwise. When the hot solution becomes cloudy and recrystallization has begun, allow the tube to cool slowly to about 22°C; then cool it in ice and collect the crystals. Compare recrystallization in water to that in the solvent pair.

Cleaning Up. The methanol-water solution can be disposed in the organic solvents waste container or, if regulations permit, diluted with water and flushed down the drain.

6. RECRYSTALLIZATION OF NAPHTHALENE FROM A MIXED SOLVENT

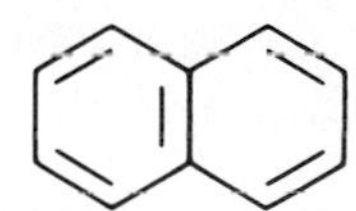

Naphthalene

Add 2.0 g of impure naphthalene (a mixture of 100 g of naphthalene, 0.3 g of a dye such as Congo Red, and another substance such as magnesium sulfate, or dust) to a 50-mL Erlenmeyer flask along with 3 mL of methanol and a boiling stick to promote even boiling. Heat the mixture to boiling over a steam bath or a hot plate, and then add methanol dropwise until the naphthalene just dissolves when the solvent is boiling. The total volume of methanol should be 4 mL. Remove the flask from the heat and cool it rapidly in an ice bath. Note that the contents of the flask set to a solid mass, which would be impossible to handle. Add enough methanol to bring the total volume to 25 mL, heat the solution to its boiling point, remove the flask from the heat, allow it to cool slightly, and add 30 mg of decolorizing charcoal pellets to remove the colored impurity in the solution. Heat the solution to its boiling point for 2 minutes; if the color is not gone, add more Norit and boil again, and then filter through a fluted filter paper in a previously warmed stemless funnel into a 50-mL Erlenmeyer flask. Sometimes filtration is slow because the funnel fits so snugly into the mouth of the flask that a back pressure develops. If you note that raising the funnel increases the flow of filtrate, fold a small strip of paper two or three times and insert it between the funnel and flask. Wash the used flask with 2 mL of hot methanol and use this liquid to wash the filter paper, transferring the

Do not try to grasp Erlenmeyer flasks with a test tube holder.

Support the funnel in a ring stand.

solvent with a Pasteur pipette in a succession of drops around the upper rim of the filter paper. When the filtration is complete, the volume of methanol should be 15 mL. If it is not, evaporate the excess methanol.

Because the filtrate is far from being saturated with naphthalene at this point, it will not yield crystals on cooling. However, the solubility of naphthalene in methanol can be greatly reduced by the addition of water. Heat the solution to its boiling point and add water dropwise from a 10-mL graduated cylinder using a Pasteur pipette (or a precalibrated pipette). After each drop of water, the solution will become cloudy for an instant. Swirl the contents of the flask and heat to redissolve any precipitated naphthalene. After the addition of 3.5 mL of water, the solution will almost be saturated with naphthalene at the boiling point of the solvent. Remove the flask from the heat and place it on a cork ring or other insulating surface to cool, without being disturbed, to about 22°C.

Immerse the flask in an ice bath along with another flask containing a 30:7 mixture of methanol and water. This cold solvent will be used for washing the crystals. The crystals will be separated from the cold recrystallization mixture using vacuum filtration on a small 50-mm Büchner funnel (Fig. 4.23). The water flowing through the aspirator should always be turned on full force. In collecting the product by suction filtration, use a spatula to dislodge crystals and ease them out of the flask. If crystals still remain in the flask, some filtrate can be poured back into the recrystallization flask as a rinse for washing as often as desired because it is saturated with solute. To further purify the crystals, break the suction, pour a few milliliters of the fresh cold solvent mixture into the Büchner funnel, and immediately reapply suction. Repeat this process until the crystals and the filtrate are free of color. Press the crystals with a clean cork to eliminate excess solvent, pull air through the filter cake for a few minutes, and then put the large, flat, platelike crystals out on a filter paper to dry. The yield of pure white crystalline naphthalene should be about 1.6 g. The mother liquor contains about 0.25 g, and about 0.15 g is retained in the charcoal and on the filter paper.

Cleaning Up. Place the Norit in the nonhazardous solid waste container. The methanol filtrate and washings are placed in the organic solvents waste container.

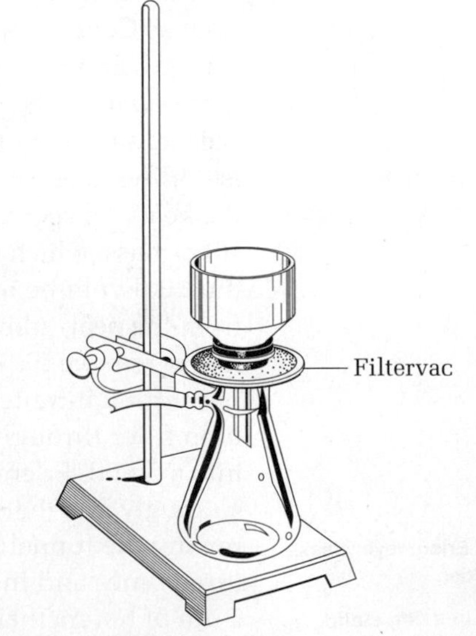

FIG. 4.23
A suction filter assembly clamped to provide firm support. The funnel must be pressed down on the Filtervac to establish reduced pressure in the flask.

7. PURIFICATION OF AN UNKNOWN

Recall the seven-step recrystallization procedure:

1. Choose the solvent.
2. Dissolve the solute.
3. Decolorize the solution (if necessary).
4. Filter suspended solids (if necessary).
5. Recrystallize the solute.
6. Collect and wash the crystals.
7. Dry the product.

You will purify an unknown provided by your instructor, 2.0 g if working on a macroscale and 100 mg if working on a microscale. Conduct tests for solubility and the ability to recrystallize in several organic solvents, solvent pairs, and water. Conserve your unknown by using very small quantities for the solubility tests. If only a drop or two of solvent is used, heating the test tube on a steam bath or a sand bath can evaporate the solvent, and the residue can be used for another test. If decolorization is necessary, dilute the solution before filtration. It is very difficult to filter a hot, saturated solution from decolorizing carbon without recrystallization occurring in the filtration apparatus. Evaporate the decolorized solution to the point of saturation and proceed with the recrystallization. Submit as much pure product as possible with evidence of its purity (i.e., the melting point). From the posted list, identify the unknown. If an authentic sample is available, your identification can be verified by a mixed melting point determination (see Chapter 3).

Cleaning Up. Place decolorizing charcoal, if used, and filter paper in the nonhazardous solid waste container. Put organic solvents in the organic solvents waste container and flush aqueous solutions down the drain.

Recrystallization Problems and Their Solutions

INDUCTION OF CRYSTALLIZATION

Seeding

Occasionally, a sample will not crystallize from solution on cooling, even though the solution is saturated with the solute at an elevated temperature. The easiest method for inducing crystallization is to add to the supersaturated solution a seed crystal that has been saved from the crude material (if it was crystalline before crystallization was attempted). In a probably apocryphal tale, the great sugar chemist Emil Fischer merely had to wave his beard over a recalcitrant solution, and the appropriate seed crystals would drop out, causing recrystallization to occur. In the absence of seed crystals, scratching the inside of the flask with a stirring rod at the liquid-air interface can often induce recrystallization. One theory holds that part of the freshly scratched glass surface has angles and planes corresponding to the crystal structure, and crystals start growing on these spots. Recrystallization is often very slow to begin. Placing the sample in a refrigerator overnight will bring success. Other expedients are to change the solvent (usually to a less soluble one) and to place the sample in an open container where slow evaporation and dust from the air may help induce recrystallization.

Scratching

OILS AND "OILING OUT"

Video: *Formation of an Oil Instead of Crystals*

When cooled, some saturated solutions—especially those containing water—deposit not crystals but small droplets referred to as oils. "Oiling out" occurs when the temperature of the solution is above the melting point of the crystals. If these droplets solidify and are collected, they will be found to be quite impure. Similarly, the melting point of the desired compound may be depressed to a point such that a low-melting eutectic mixture of the solute and the solvent comes out of solution. The simplest remedy for this problem is to lower the temperature at which the solution becomes saturated with the solute by simply adding more room-temperature solvent. In extreme cases, it may be necessary to lower this temperature well below 22°C by cooling the solution with dry ice. If oiling out persists, use another solvent.

Crystallize at a lower temperature

RECRYSTALLIZATION SUMMARY

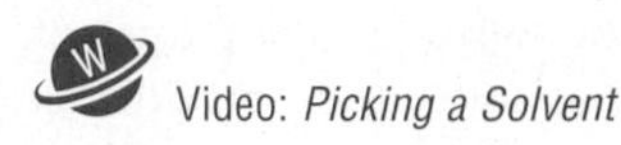

Video: *Picking a Solvent*

1. **Choosing the solvent.** "Like dissolves like." Some common solvents are water, methanol, ethanol, hexanes, and toluene. When you use a solvent pair, dissolve the solute in the better solvent and add the poorer solvent to the hot solution until saturation occurs. Some common solvent pairs are ethanol-water, ether-hexanes, and toluene—hexanes.

Video: *Recrystallization*

2. **Dissolving the solute.** In an Erlenmeyer flask or reaction tube, add solvent to the crushed or ground solute and heat the mixture to boiling. Add more solvent as necessary to obtain a hot, saturated solution.

Video: *Decolorization of a Solution with Norit*

3. **Decolorizing the solution.** If it is necessary to remove colored impurities, cool the solution to about 22°C and add more solvent to prevent recrystallization from occurring. Add decolorizing charcoal in the form of pelletized Norit to the cooled solution and then heat it to boiling for a few minutes, making sure to swirl the solution to prevent bumping. Remove the Norit by filtration and then concentrate the filtrate.

Photo: *Preparation of a Filter Pipette*; Video: *Microscale Crystallization*

4. **Filtering suspended solids.** If it is necessary to remove suspended solids, dilute the hot solution slightly to prevent recrystallization from occurring during filtration. Filter the hot solution. Add solvent if recrystallization begins in the funnel. Concentrate the filtrate to obtain a saturated solution.

Photo: *Recrystallization*; Video: *Recrystallization*

5. **Recrystallizing the solute.** Let the hot saturated solution cool to about 22°C spontaneously. Do not disturb the solution. Then cool it in ice. If recrystallization does not occur, scratch the inside of the container or add seed crystals.

Photos: *Use of the Wilfilter, Filtration Using a Pasteur Pipette*; Videos: *Microscale Filtration on the Hirsch Funnel, Filtration of Crystals Using the Pasteur Pipette*

6. **Collecting and washing the crystals.** Collect the crystals using the Pasteur pipette method, the Wilfilter, or by vacuum filtration on a Hirsch funnel or a Büchner funnel. If the latter technique is employed, wet the filter paper with solvent, apply vacuum to secure the paper, break vacuum, add crystals and liquid, apply vacuum until solvent just disappears, break vacuum, add cold wash solvent, apply vacuum, and repeat until crystals are clean and filtrate comes through clear.

Photo: *Drying Crystals Under Vacuum*; Video: *Recrystallization*

7. **Drying the product.** Press the wet product on the filter to remove solvent. Then remove it from the filter, squeeze it between sheets of filter paper to remove more solvent, and spread it on a watch glass to dry.

QUESTIONS

1. A sample of naphthalene, which should be pure white, was found to have a grayish color after the usual purification procedure. The melting point was correct, and the melting point range was small. Explain the gray color.
2. How many milliliters of boiling water are required to dissolve 25 g of phthalic acid? If the solution were cooled to 14°C, how many grams of phthalic acid would recrystallize out?
3. Why should activated carbon be used during a recrystallization?
4. If a little activated charcoal does a good job removing impurities in a recrystallization, why not use a larger quantity?
5. Under which circumstances is it wise to use a mixture of solvents to carry out a recrystallization?
6. Why is gravity filtration rather than suction filtration used to remove suspended impurities and charcoal from a hot solution?
7. Why is a fluted filter paper used in gravity filtration?
8. Why are stemless funnels used instead of long-stem funnels to filter hot solutions through fluted filter paper?
9. Why is the final product from the recrystallization process isolated by vacuum filtration rather than gravity filtration?
10. Why should wood applicator sticks not be used when carrying out a chemical reaction?
11. Why should you never use a beaker for recrystallization?

5 CHAPTER

Distillation

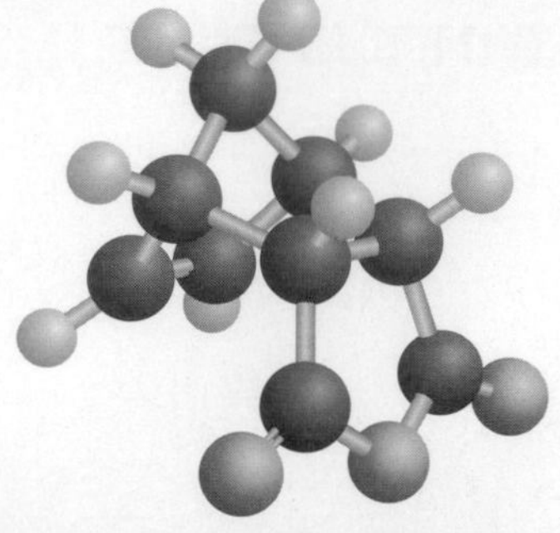

PRELAB EXERCISE: Predict what a plot of temperature versus the volume of distillate will look like for the simple distillation and the fractional distillation of (a) a cyclohexane-toluene mixture and (b) an ethanol-water mixture.

Distillation is a common method for purifying liquids and can be used to determine their boiling points.

The origins of distillation are lost in antiquity, when humans in their thirst for more potent beverages found that dilute solutions of fermented alcohol could be separated into alcohol-rich and water-rich portions by heating the solution to boiling and condensing the vapors above the boiling liquid—the process of distillation.

Because ethyl alcohol (ethanol) boils at 78°C and water boils at 100°C, one might naïvely assume that heating a 50:50 mixture of ethanol and water to 78°C would cause the ethanol molecules to leave the solution as a vapor that could be condensed as pure ethanol. However, in such a mixture of ethanol and water, the water boils at about 87°C, and the vapor above the mixture is not 100% ethanol.

A liquid contains closely packed but mobile molecules of varying energy. When a molecule of the liquid approaches the vapor-liquid boundary and possesses sufficient energy, it may pass from the liquid phase into the gas phase. Some of the molecules present in the vapor phase above the liquid may, as they approach the surface of the liquid, reenter the liquid phase and thus become part of the condensed phase. In so doing, the molecules relinquish some of their kinetic energy (i.e., their motion is slowed). Heating the liquid causes more molecules to enter the vapor phase; cooling the vapor reverses this process.

When a closed system is at equilibrium, many molecules are escaping into the vapor phase from the liquid phase, and an equal number are returning from the vapor phase to the liquid phase. The extent of this equilibrium is measured as the vapor pressure. Even when energy is increased and more molecules in the liquid phase have sufficient energy to escape into the vapor phase, equilibrium is maintained because the number moving from the vapor phase into the liquid phase also increases. However, the number of molecules in the vapor phase increases, which increases the vapor pressure. The number of molecules in the vapor phase depends primarily on the volume of the system, the temperature, the combined pressure of all the gaseous components, and the strength of the intermolecular forces exerted in the liquid phase. Review the introduction to Chapter 3 about the types of intermolecular forces.

SIMPLE DISTILLATION

Simple distillation involves boiling a liquid in a vessel (a distilling flask) and directing the resulting vapors through a condenser, in which the vapors are cooled and converted to a liquid that flows down into a collection vessel (a receiving flask). (*See* Fig. 5.5 on page 93.) Simple distillation is used to purify liquid mixtures by separating one liquid component either from nonvolatile substances or from another liquid that differs in boiling point by at least 75°C. The initial condensate will have essentially the same mole ratio of liquids as the vapor just above the boiling liquid. The closer the boiling points of the components of a liquid mixture, the more difficult they are to completely separate by simple distillation.

FRACTIONAL DISTILLATION

Fractional distillation differs from simple distillation in that a fractionating column is placed between the distilling flask and the condenser. This fractionating column allows for successive condensations and distillations and produces a much better separation between liquids with boiling points closer than 75°C. The column is packed with material that provides a large surface area for heat exchange between the ascending vapor and the descending liquid. As a result, multiple condensations and vaporizations occur as the vapors ascend the column. Condensing of the higher-boiling vapor releases heat, which causes vaporization of the lower boiling liquid on the packing so that the lower-boiling component moves up while the higher-boiling component moves down. Some of the lower-boiling component will run back into the distilling flask. Each successive condensation-vaporization cycle, also called a *theoretical plate*, produces a vapor that is richer in the more volatile fraction. As the temperature of the liquid mixture is increased, the lower-boiling fractions become enriched in the vapor.

Heat exchange between ascending vapor and descending liquid

A large surface area for the packing material is desirable, but the packing cannot be so dense that pressure changes take place within the column to cause nonequilibrium conditions. Also, if the column packing has a very large surface area, it will absorb (hold up) much of the material being distilled. Several different packings for distilling columns have been tried, including glass beads, glass helices, and carborundum chips. One of the best packings in our experience is a copper or steel sponge (brand name–Chore Boy). It is easy to insert into the column; it does not come out of the column as beads do; and it has a large surface area, good heat transfer characteristics, and low holdup. It can be used in both microscale and macroscale apparatus.

Column packing

Holdup (noun): unrecoverable distillate that wets the column packing

The ability of different column packings to separate two materials of differing boiling points is evaluated by calculating the number of theoretical plates. Each theoretical plate corresponds to one condensation-vaporization cycle. Other things being equal, the number of theoretical plates is proportional to the height of the column, so various packings are evaluated according to the *height equivalent to a theoretical plate (HETP)*; the smaller the HETP, the more plates the column will have and the more efficient it will be. The calculation is made by analyzing the proportion of lower- to higher-boiling material at the top of the column and in the distillation pot.[1]

[1]Weissberger, A. ed. *Techniques of Organic Chemistry*, Vol. IV; Wiley-Interscience: New York, 1951.

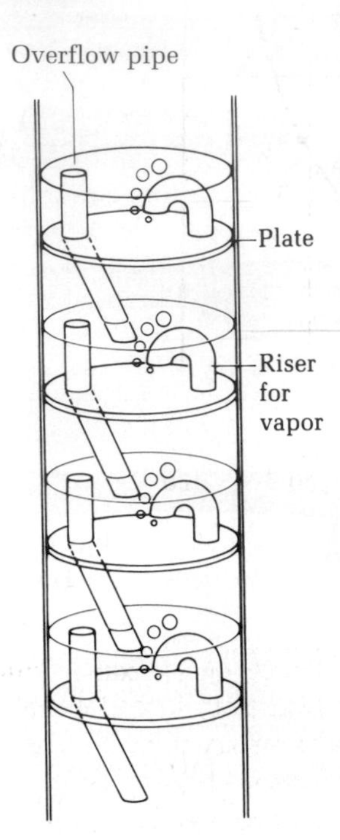

FIG. 5.1
A bubble plate distilling column.

Although not obvious, the most important variable that contributes to good fractional distillation is the rate at which the distillation is carried out. When a series of simple distillations take place within a fractionating column, it is important that complete equilibrium be attained between the ascending vapors and the descending liquid. This process is not instantaneous. It should be an adiabatic process; that is, heat should be transferred from the ascending vapor to the descending liquid with no gain of heat or net heat loss to the surroundings. Advanced distillation systems use thermally insulated, vacuum-jacketed fractionating columns. They also allow the adjustment of the ratio between the amount of condensate that is directed to the receiving flask and the amount returned to the distillation column. A reflux ratio of 30:1 or 50:1 is not uncommon for a 40-plate column. Although a distillation of this type takes several hours, this is far less time than if one to had to do 40 distillations, one after the other, and yields much better separated compounds.

Height equivalent to a theoretical plate (HETP)

Equilibration is slow.

Good fractional distillation takes a long time.

Perhaps it is easiest to understand the series of redistillations that occur in fractional distillation by examining the bubble plate column used to fractionally distill crude oil (Fig. 5.1). These columns dominate the skyline at oil refineries, with some being 150 ft high and capable of distilling 200,000 barrels of crude oil per day. The crude oil enters the column as a hot vapor. Some of this vapor with high-boiling components condenses on one of the plates. The more volatile substances travel through the bubble cap to the next higher plate, where some of the less-volatile components condense. As high-boiling liquid material accumulates on a plate, it descends through the overflow pipe to the next lower plate, and vapor rises through the bubble cap to the next higher plate. The temperature of the vapor that is rising through a cap is above the boiling point of the liquid on that plate. As bubbling takes place, heat is exchanged, and the less volatile components on that plate vaporize and go on to the next plate. The composition of the liquid on a plate is the same as that of the vapor coming from the plate below. So, on each plate, a simple distillation takes place. At equilibrium, vapor containing low-boiling material is ascending through the column, and high-boiling liquid is descending.

As a purification method, distillation, particularly fractional distillation, requires larger amounts of material than recrystallization, liquid/liquid extraction, or chromatography. Performing a fractional distillation on less than 1 g of material is virtually impossible. Fractional distillation can be carried out on a scale of about 3–4 g. As will be seen in Chapters 8, 9, and 10, various types of chromatography are employed for separations of milligram quantities of liquids.

LIQUID MIXTURES

If two different liquid compounds are mixed, the vapor above the mixture will contain some molecules of each component. Let us consider a mixture of cyclohexane and toluene. The vapor pressures, as a function of temperature, are plotted in Figure 5.2. When the vapor pressure of the liquid equals the applied pressure, the liquid boils. Figure 5.2 shows that at 760 mm Hg (standard atmospheric pressure), these pure liquids boil at about 81°C and 111°C, respectively. If one of these pure liquids were to be distilled, we would find that the boiling point of the liquid would equal the temperature of the vapor, and that the temperature of the vapor would remain constant throughout the distillation.

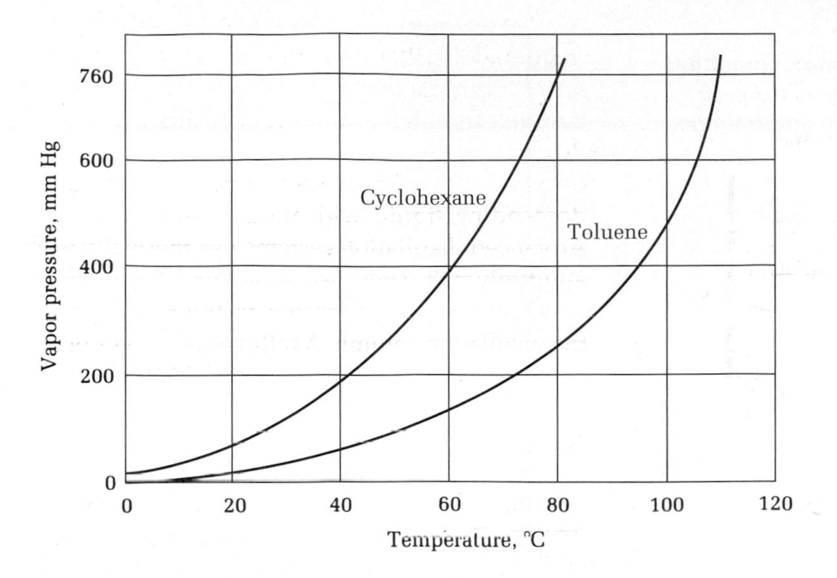

FIG. 5.2
Vapor pressure versus temperature plots for cyclohexane and toluene.

Figure 5.3 is a boiling point composition diagram for the cyclohexane-toluene system. If a mixture of 75 mole percent toluene and 25 mole percent cyclohexane is heated, we find that it boils at 100°C (point A). Above a binary mixture of cyclohexane and toluene, the vapor pressure has contributions from each component. Raoult's law states that the vapor pressure of the cyclohexane is equal to the product of the vapor pressure of pure cyclohexane and the mole fraction of cyclohexane in the liquid mixture:

Raoult's law of partial pressures

$$P_c = P_c^{\circ} N_c$$

where P_c is the partial pressure of cyclohexane, P_c° is the vapor pressure of pure cyclohexane at the given temperature, and Nc is the mole fraction of cyclohexane in the mixture. Similarly, for toluene,

The mole fraction of cyclohexane is equal to the moles of cyclohexane in the mixture divided by the total number of moles (cyclohexane plus toluene) in the mixture.

$$P_t = P_t^{\circ} N_t$$

The total vapor pressure above the solution (P_{Tot}) is given by the sum of the partial pressures due to cyclohexane and toluene:

$$P_{Tot} = P_c + P_t$$

Dalton's law states that the mole fraction of cyclohexane (X_c) in the vapor at a given temperature is equal to the partial pressure of the cyclohexane at that temperature divided by the total pressure:

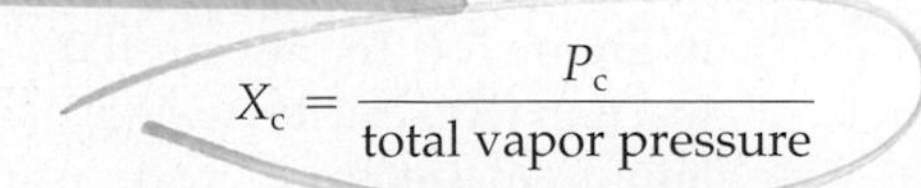

$$X_c = \frac{P_c}{\text{total vapor pressure}}$$

At 100°C, cyclohexane has a partial pressure of 433 mm Hg, and toluene has a partial pressure of 327 mm Hg; the sum of the partial pressures is 760 mm Hg, so the liquid boils. If some of the liquid in equilibrium with this boiling mixture were condensed and analyzed, it would be found to be 433/760, or 57 mole percent cyclohexane (point B, Fig. 5.3). This is the best separation that can be achieved with a simple distillation of this mixture. As the simple distillation proceeds, the boiling point of the mixture moves toward 111°C along the line from point A, and the vapor

FIG. 5.3
Boiling point-composition curves for a mixture of cyclohexane and toluene.

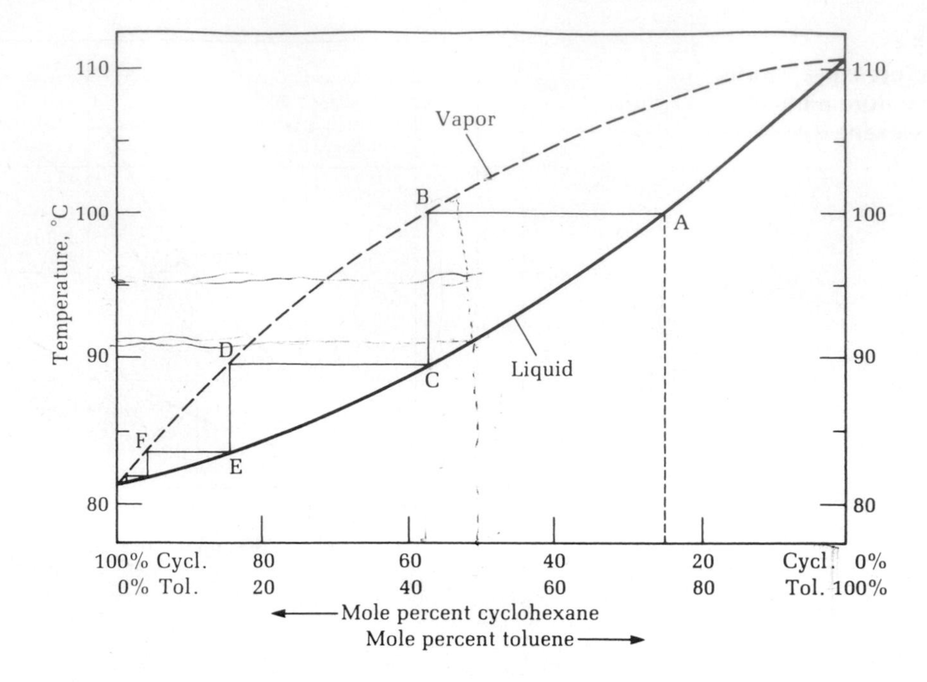

composition becomes richer in toluene as it moves from point B to 110°C. To obtain pure cyclohexane, it would be necessary to condense the liquid at point B and redistill it. When this is done, it is found that the liquid boils at 90°C (point C), and the vapor equilibrium with this liquid is about 85 mole percent cyclohexane (point D). Therefore, to separate a mixture of cyclohexane and toluene, a series of fractions would be collected, and each of these would be partially redistilled. If this fractional distillation were done enough times, the two components could be completely separated.

AZEOTROPES

Not all liquids form ideal solutions and conform to Raoult's law. Ethanol and water are two such liquids. Because of molecular interaction, a mixture of 95.5% (by weight) of ethanol and 4.5% of water boils *below* the boiling point of pure ethanol (78.15°C versus 78.3°C). Thus, no matter how efficient the distilling apparatus, 100% ethanol cannot be obtained by distillation of a mixture of, say, 75% water and 25% ethanol. A mixture of liquids of a certain definite composition that distills at a constant temperature without a change in composition is called an *azeotrope;* 95% ethanol is such an azeotrope. The boiling point-composition curve for the ethanol-water mixture is seen in Figure 5.4. To prepare 100% ethanol, the water can be removed chemically (by reaction with calcium oxide) or it can be removed as an azeotrope with still another liquid. An azeotropic mixture of 32.4% ethanol and 67.6% benzene (bp 80.1°C) boils at 68.2°C. A ternary azeotrope containing 74.1% benzene, 18.5% ethanol, and 7.4% water boils at 64.9°C. Absolute alcohol (100% ethanol) is made by adding benzene to 95% ethanol. The water is removed by distilling the ternary azeotrope of benzene, ethanol, and water, bp 64.9°C, followed by the distillation of 100% ethanol, bp 78.4°C

The ethanol-water azeotrope.

Ethanol and water form a minimum boiling azeotrope. Substances such as formic acid (bp 100.7°C) and water (bp 100°C) form maximum boiling azeotropes. The boiling point of a formic acid-water azeotrope is 107.3°C.

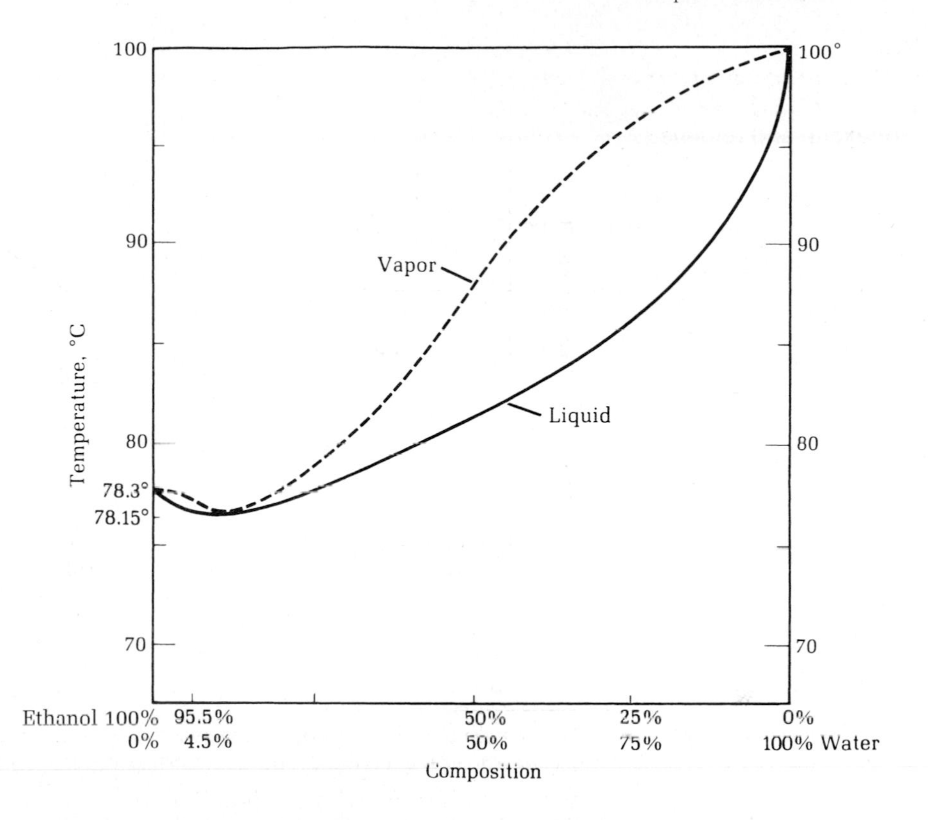

FIG. 5.4
Boiling point–composition curves for a mixture of ethanol and water.

BOILING POINTS AND DISTILLATION

A constant boiling point on distillation does not guarantee that the distillate is a single pure compound.

A pure liquid has a constant boiling point. A change in boiling point during distillation is an indication of impurity. The converse proposition, however, is not always true; that is, constancy of a boiling point does not necessarily mean that the liquid consists of only one compound. For instance, two miscible liquids of similar chemical structure that boil at the same temperature individually will have nearly the same boiling point as a mixture. And, as noted previously, azeotropes have constant boiling points that can be either above or below the boiling points of the individual components.

Distilling a mixture of sugar and water.

When a solution of sugar in water is distilled, the boiling point recorded on a thermometer located in the vapor phase is 100°C (at 760 torr) throughout the distillation, whereas the temperature of the boiling sugar solution itself is initially somewhat above 100°C and continues to rise as the concentration of sugar in the remaining solution increases. The vapor pressure of the solution is dependent on the number of water molecules present in a given volume; hence, with increasing concentration of nonvolatile sugar molecules and decreasing concentration of water, the vapor pressure at a given temperature decreases, and a higher temperature is required for boiling. However, sugar molecules do not leave the solution, and the drop clinging to the thermometer is pure water in equilibrium with pure water vapor.

Boiling point changes with pressure.

When a distillation is carried out in a system open to the air (the boiling point is thus dependent on existing air pressure), the prevailing barometric pressure should be noted and allowance made for appreciable deviations from the accepted boiling point temperature (Table 5.1). Distillation can also be done at the lower pressures that can be achieved using an oil pump or an aspirator, which produces a substantial reduction in the boiling point.

TABLE 5.1 *Variation in Boiling Point with Pressure*

Pressure (mm Hg)	*Boiling Point*	
	Water (°C)	*Benzene (°C)*
780	100.7	81.2
770	100.4	80.8
760	100.0	80.3
750	99.6	79.9
740	99.2	79.5
584*	92.8	71.2

*Instituto de Quimica, Mexico City, altitude 7700 ft (2310 m).

EXPERIMENTS

Before beginning any distillation, calibrate the thermometer to ensure accurate readings are made; refer to Part 3 of Chapter 3 for calibration instructions.

1. SIMPLE DISTILLATION

Apparatus for simple distillation.

IN THIS EXPERIMENT, the two liquids to be separated are placed in a 5-mL round-bottomed, long-necked flask that is fitted to a distilling head (Fig. 5.5). The flask has a larger surface area exposed to heat than does the reaction tube, so the necessary thermal energy can be put into the system to cause the materials to distill. The hot vapor rises and completely envelops the bulb of the thermometer before passing over it and down toward the receiver. The downward-sloping portion of the distilling head functions as an air condenser. The objective is to observe how the boiling point of the mixture changes during the course of its distillation. (Another simple distillation apparatus is shown in Figure 5.6. Here, the long air condenser will condense even low-boiling liquids, and the receiver is far from the heat.)

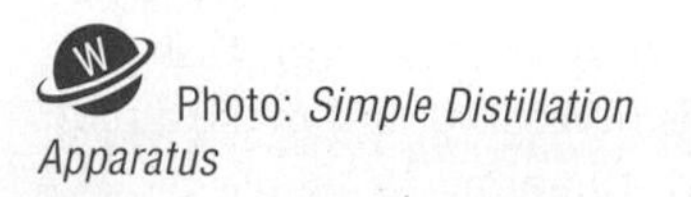
Photo: *Simple Distillation Apparatus*

The rate of distillation is determined by the heat input to the apparatus. This is most easily and effectively controlled by using a spatula to pile up or scrape away hot sand from around the flask.

(A) Simple Distillation of a Cyclohexane-Toluene Mixture

To a 5-mL long-necked, round-bottomed flask, add 2.0 mL of dry cyclohexane, 2.0 mL of dry toluene, and a boiling chip (*see* Fig. 5.5). This flask is joined by means of a Viton (black) connector to a distilling head fitted with a thermometer using a rubber connector. The thermometer bulb should be completely below the side arm of the Claisen head so that the mercury reaches the same temperature as the vapor that distills. The end of the distilling head dips well down into a receiving vial, which rests on the bottom of a 30-mL beaker filled with ice. The distillation is started by piling up hot sand to heat the flask. As soon as boiling begins, the vapors can be seen to rise up the

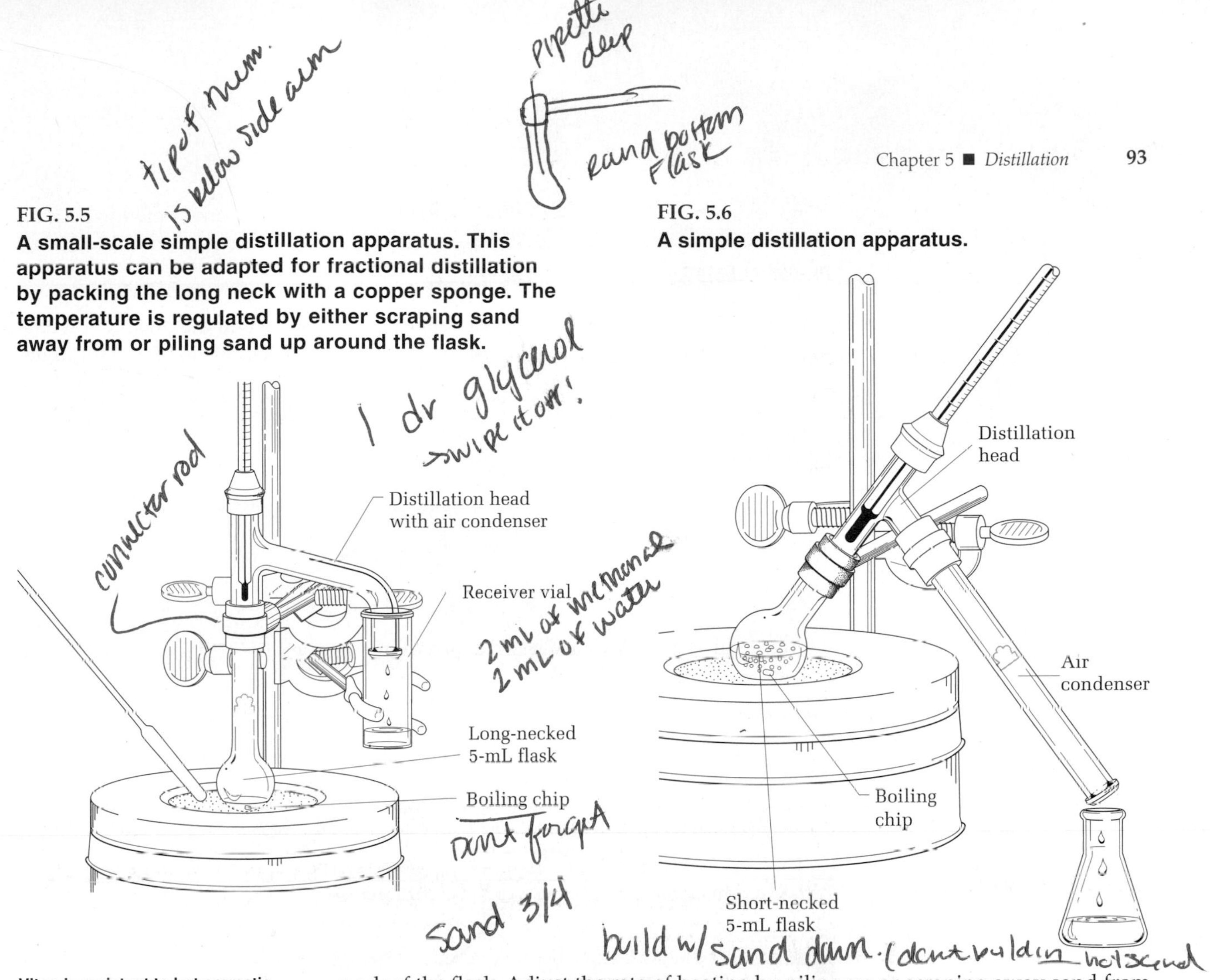

FIG. 5.5
A small-scale simple distillation apparatus. This apparatus can be adapted for fractional distillation by packing the long neck with a copper sponge. The temperature is regulated by either scraping sand away from or piling sand up around the flask.

FIG. 5.6
A simple distillation apparatus.

Viton is resistant to hot aromatic vapors.

The thermometer bulb must be *completely below* the side arm.

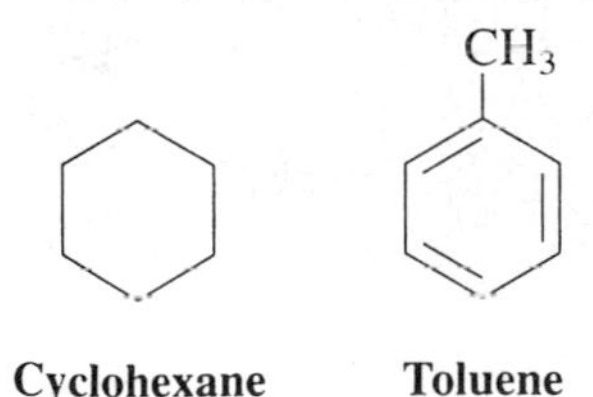

Cyclohexane
bp 81°C
MW 84.16
n_D^{20} 1.4260

Toluene
bp 111°C
MW 92.14
n_D^{20} 1.4960

Throughout this text, information regarding the physical properties of substances has been placed with each structure in the margin. MW is molecular weight, bp is boiling point, den. is density in g/mL, and nD 20 is the refractive index.

There are 21 ± 3 drops per milliliter.

neck of the flask. Adjust the rate of heating by piling up or scraping away sand from the flask so that it takes *several minutes* for the vapor to rise to the thermometer. **The rate of distillation should be no faster than 2 drops per minute.**

Record the temperature versus the number of drops during the entire distillation process. If the rate of distillation is as slow as it should be, there will be sufficient time between drops to read and record the temperature. Continue the distillation until only about 0.4 mL remains in the distilling flask. **Never distill to dryness.** On a larger scale, explosive peroxides can sometimes accumulate. At the end of the distillation, measure as accurately as possible, perhaps with a syringe, the volume of the distillate and, after it cools, the volume left in the pot (round-bottomed flask); the difference is the holdup of the column if none has been lost by evaporation. Note the barometric pressure, make any thermometer corrections necessary, and make a plot of milliliters (drop number) versus temperature for the distillation.

Cleaning Up. The pot residue should be placed in the organic solvents container. The distillate can also be placed there or recycled.

(B) Simple Distillation of an Ethanol-Water Mixture

In a 5-mL round-bottomed, long-necked flask, place 4 mL of a 10% to 20% ethanol-water mixture. Assemble the apparatus as described previously and carry out the distillation until you believe a representative sample of ethanol has collected in the receiver. In the hood, place 3 drops of this sample on a Pyrex watch glass and try to ignite it with the blue cone of a microburner flame. Does it burn? Is any unburned residue observed? There was a time when alcohol-water mixtures were mixed with gunpowder and ignited to give proof that the alcohol had not been diluted. One hundred proof alcohol is 50% ethanol by volume.

Cleaning Up. The distillate and pot residue can be disposed in the organic solvents waste container or, if regulations permit, diluted with water and flushed down the drain.

2. FRACTIONAL DISTILLATION

Photos: *Column Packing with Chore Boy for Fractional Distillation, Fractional Distillation Apparatus*

IN THIS EXPERIMENT, just as in the last one, you will distill a mixture of two liquids and again record the boiling points as a function of the volume of distillate (drops). The necessity for a very slow rate of distillation cannot be overemphasized.

Apparatus

Assemble the apparatus shown in Figure 5.7. The 10-cm column is packed with 1.5 g of copper sponge and connected to the 5-mL short-necked flask using a black (Viton) connector. The column should be vertical and care should be taken to ensure that the bulb of the thermometer does not touch the side of the distilling head. The column, but not the distilling head, will be insulated with glass wool or cotton at the appropriate time to ensure that the process is adiabatic. Alternatively,

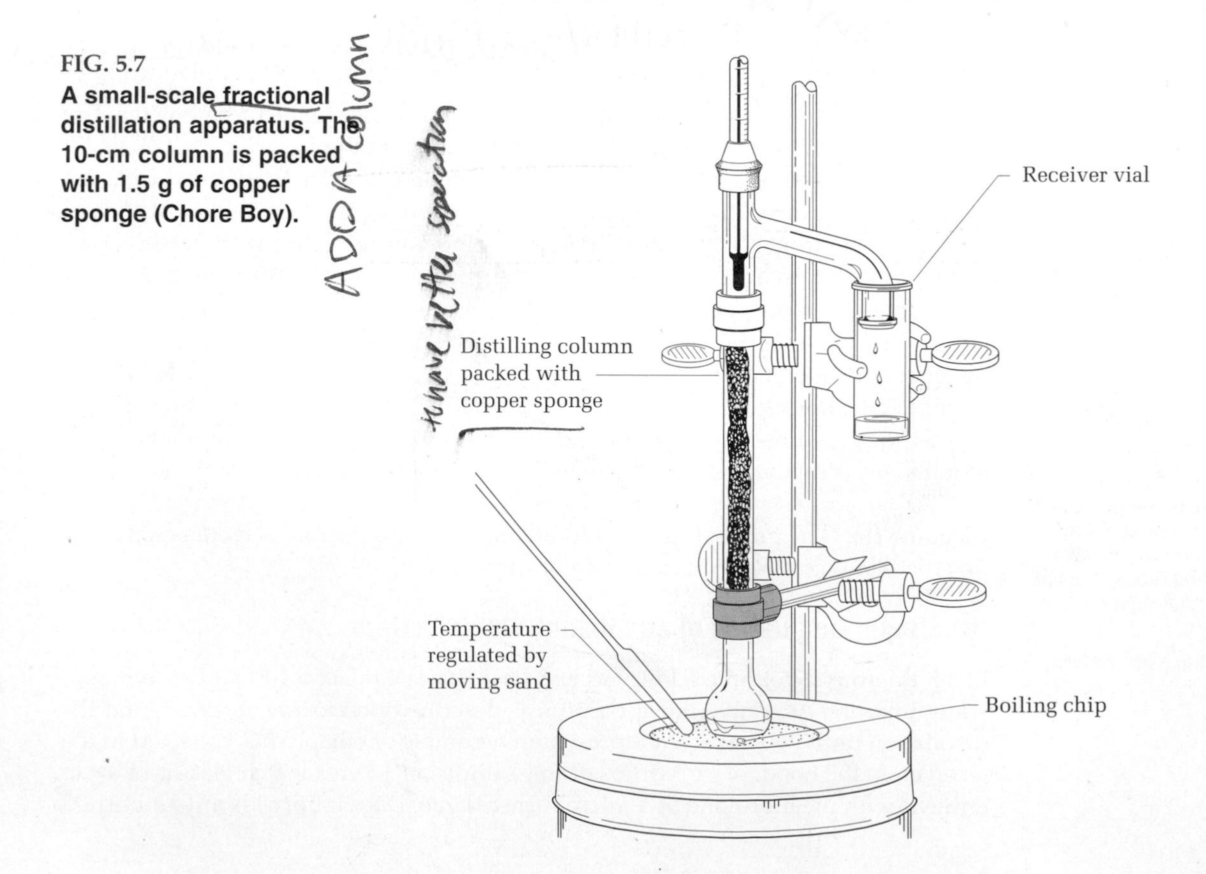

FIG. 5.7
A small-scale fractional distillation apparatus. The 10-cm column is packed with 1.5 g of copper sponge (Chore Boy).

the column can be insulated with a 15-mL polyethylene centrifuge tube that has the conical bottom cut off.

(A) Fractional Distillation of a Cyclohexane-Toluene Mixture

Never distill in an airtight system.

Adjust the heat input to the flask by piling up or scraping away sand around the flask.

To a short-necked flask, add 2.0 mL of cyclohexane, 2.0 mL of toluene, and a boiling chip. The distilling column is packed with 1.5 g of copper sponge (Fig. 5.7). The mixture is brought to a boil over a hot sand bath. Observe the ring of condensate that should rise slowly through the column; if you cannot at first see this ring, locate it by touching the column with your fingers. It will be cool above the ring and hot below. Reduce the heat by scraping sand away from the flask and wrap the column, but not the distilling head, with glass wool or cotton if it is not already insulated.

Insulate the distilling column but not the Claisen head.

21 ± 3 drops = 1 mL

The distilling head and the thermometer function as a small reflux condenser. Again, apply the heat, and as soon as the vapor reaches the thermometer bulb, reduce the heat by scraping away sand. **Distill the mixture at a rate no faster than 2 drops per minute** and record the temperature as a function of the number of drops. If the heat input has been *very* carefully adjusted, the distillation will cease, and the temperature reading will drop after the cyclohexane has distilled. Increase the heat input by piling up the sand around the flask to cause the toluene to distill. Stop the distillation when only about 0.4 mL remains in the flask, and measure the volume of distillate and the pot residue as before. Make a plot of the boiling point versus the milliliters of distillate (drops) and compare it to the simple distillation carried out in the same apparatus. Compare your results with those in Figure 5.8.

Cleaning Up. The pot residue should be placed in the organic solvents container. The distillate can also be placed there or recycled.

(B) Fractional Distillation of an Ethanol-Water Mixture

Distill 4 mL of the same ethanol-water mixture used in the simple distillation experiment, following the procedure used for the cyclohexane-toluene mixture with either the short or the long distilling column. Remove what you regard to be the ethanol fraction and repeat the ignition test. Is any difference noted?

Cleaning Up. The pot residue and distillate can be disposed in the organic solvents waste container or, if regulations permit, diluted with water and flushed down the drain.

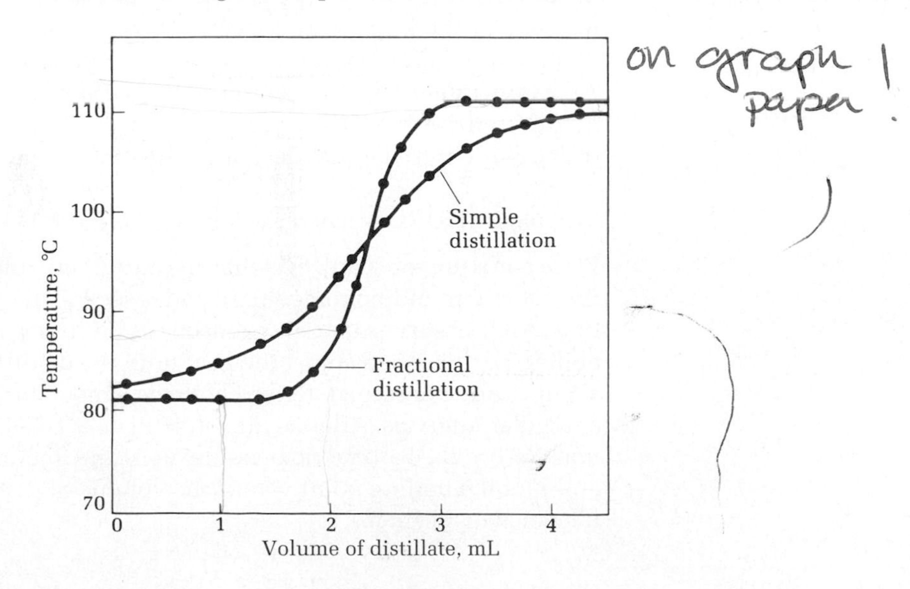

FIG. 5.8
Simple and fractional distillation curves for cyclohexane and toluene.

3. INSTANT MICROSCALE DISTILLATION

Video: *Instant Microscale Distillation*

Frequently, a very small quantity of freshly distilled material is needed in an experiment. For example, two compounds that need to be distilled freshly are aniline, which turns black because of the formation of oxidation products, and benzaldehyde, a liquid that easily oxides to solid benzoic acid. The impurities that arise in both of these compounds have much higher boiling points than the pure compounds, so a very simple distillation suffices to separate them. This can be accomplished as follows.

Place a few drops of the impure liquid in a reaction tube along with a boiling chip. Clamp the tube in a hot sand bath and adjust the heat so that the liquid refluxes about halfway up the tube. Expel the air from a Pasteur pipette, thrust it down into the hot vapor, and then pull the hot vapor into the cold upper portion of the pipette. The vapor will immediately condense and can then be expelled into another reaction tube that is held adjacent to the hot one (Fig. 5.9). In this way enough pure material can be distilled to determine a boiling point, run a spectrum, make a derivative, or carry out a reaction. Sometimes the first drop or two will be cloudy, which indicates the presence of water. This fraction should be discarded in order to obtain pure dry material.

4. SIMPLE DISTILLATION APPARATUS

In any distillation, the flask should be no more than two-thirds full at the start. Great care should be taken not to distill to dryness because, in some cases, high-boiling explosive peroxides can become concentrated.

Assemble the apparatus for macroscale simple distillation, as shown in Figure 5.10, starting with the support ring followed by the electric flask heater and then the flask. One or two boiling stones are put in the flask to promote even boiling. Each ground joint is greased by putting three or four stripes of grease lengthwise around the male joint and pressing the joint firmly into the other without twisting. The air is thus eliminated, and the joint will appear almost transparent. (Do not use excess grease because it will contaminate the product.) Water enters the condenser at the tubulature nearest the receiver. Because of the large heat capacity of water, only a very small stream (3 mm diameter) is needed; too much water pressure will cause the tubing to pop off. A heavy rubber band, or better, a Keck clamp, can be used to hold the condenser to the distillation head. Note that the bulb of the thermometer is below the opening into the side arm of the distillation head.

(A) Simple Distillation of a Cyclohexane-Toluene Mixture

Place a mixture of 30 mL cyclohexane and 30 mL toluene, and a boiling chip in a dry 100-mL round-bottomed flask and assemble the apparatus for simple distillation. After assuring that all connections are tight, heat the flask strongly until boiling begins. Then adjust the heat until the distillate drops at a regular rate of about 1 drop per second. Record both the temperature and the volume of distillate at regular intervals. After 50 mL of distillate is collected, discontinue the distillation. Record the barometric pressure, make any thermometer correction necessary, and plot the boiling point versus the volume of distillate. Save the distillate for fractional distillation.

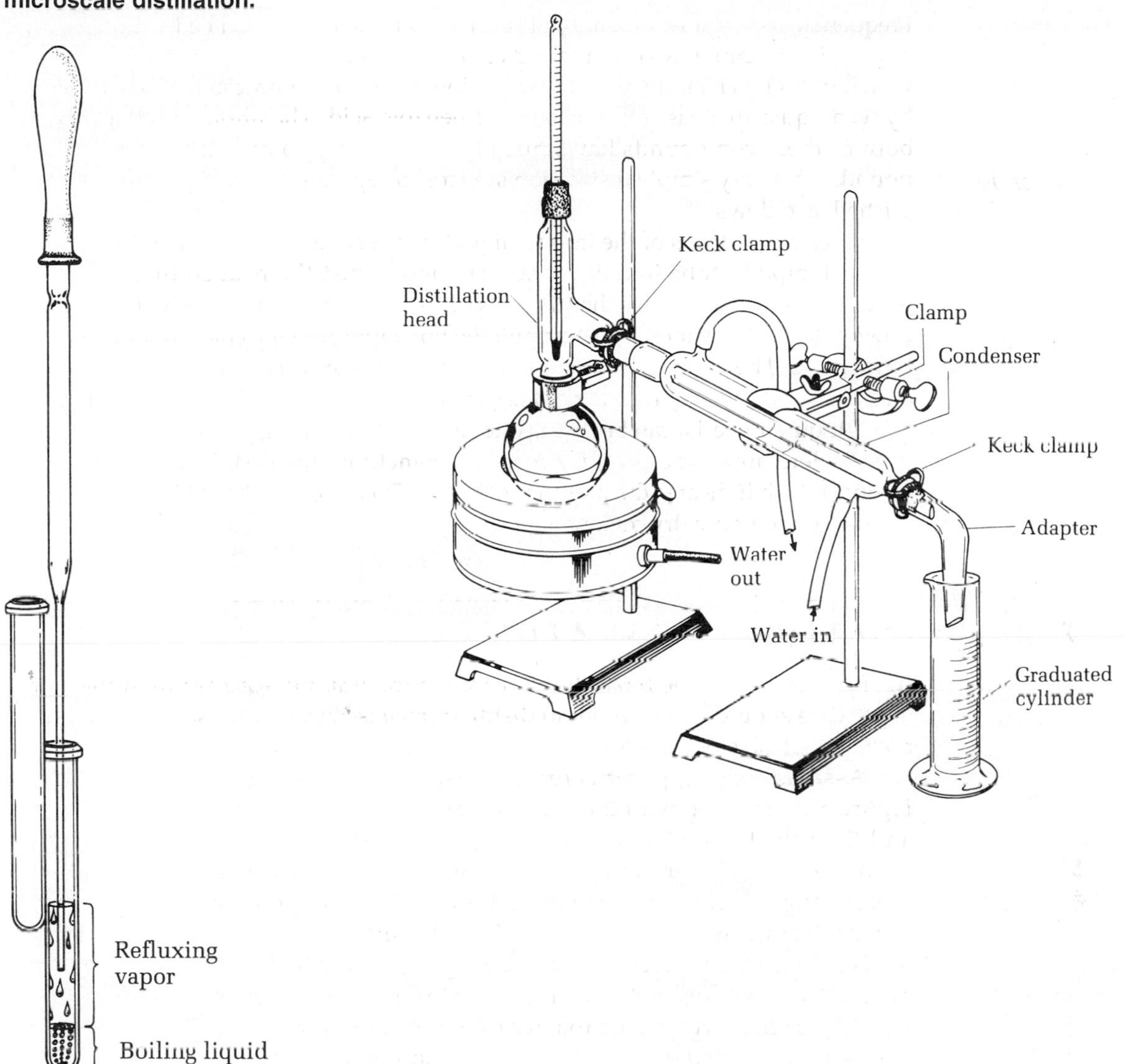

FIG. 5.9
An apparatus for instant microscale distillation.

FIG. 5.10
An apparatus for macroscale simple distillation.

⚠ Cyclohexane and toluene are flammable; make sure the distilling apparatus is tight.

Do not add a boiling chip to a hot liquid. It may boil over.

Dispose of cyclohexane and toluene in the waste container provided. Do not pour them down the drain.

Cleaning Up. The pot residue should be placed in the organic solvents container. The distillate can also be placed there or recycled.

(B) Simple Distillation of an Ethanol-Water Mixture

In a 500-mL round-bottomed flask, place 200 mL of a 20% aqueous solution of ethanol. Follow the previous procedure for the distillation of a cyclohexane-toluene mixture. Discontinue the distillation after 50 mL of distillate has been collected. Working in the hood, place 3 drops of distillate on a Pyrex watch glass and try to ignite it with the blue cone of a microburner flame. Does it burn? Is any unburned residue observed?

Cleaning Up. The pot residue and distillate can be disposed in the organic solvents waste container or, if regulations permit, diluted with water and flushed down the drain.

5. FRACTIONAL DISTILLATION

Apparatus

Assemble the apparatus shown in Figures 5.11 and 5.12. The fractionating column is packed with one-fourth to one-third of a metal sponge. The column should be perfectly vertical and be insulated with glass wool covered with aluminum foil (shiny side in). However, insulation is omitted for this experiment so that you can observe what is taking place in the column.

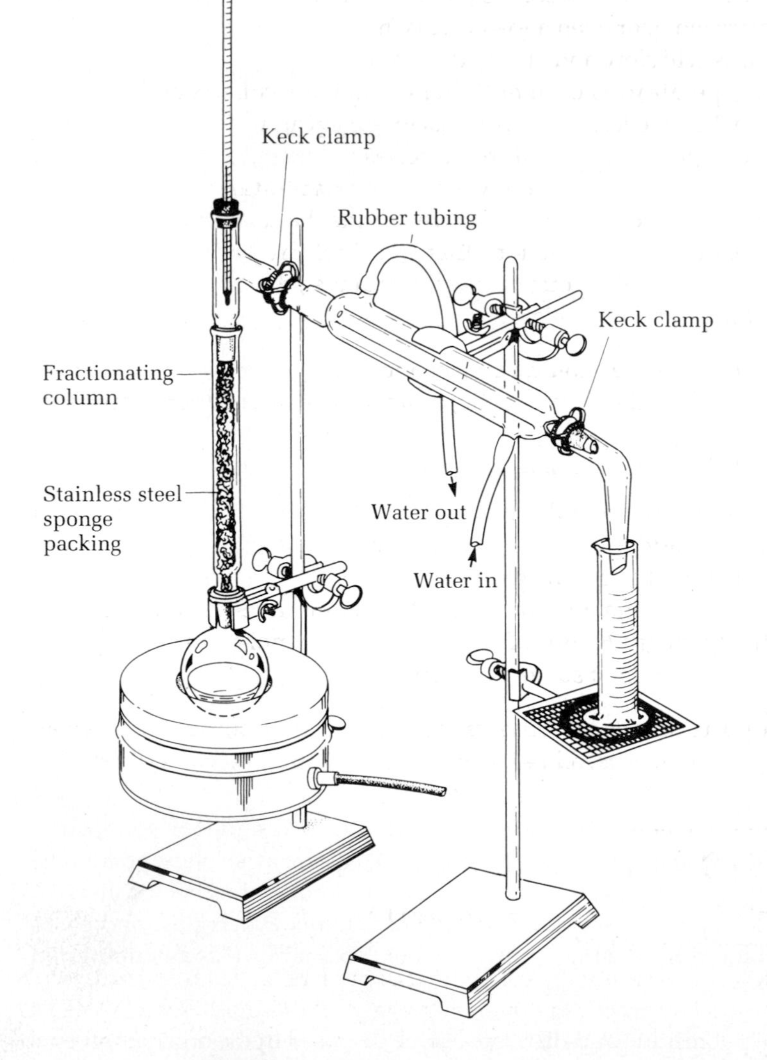

FIG. 5.11
An apparatus for macroscale fractional distillation. The position of the thermometer bulb is critical.

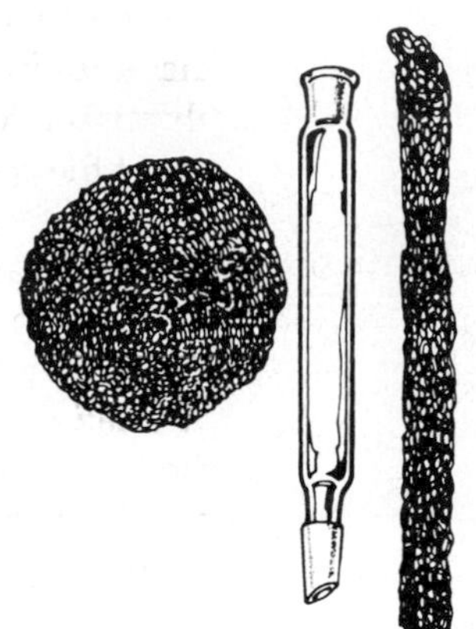

FIG. 5.12
A fractionating column and its packing. Use one-third of a copper sponge (Chore Boy).

(A) Fractional Distillation of a Cyclohexane-Toluene Mixture

After the flask from the simple macroscale distillation experiment has cooled, pour the 50 mL of distillate back into the distilling flask, add one or two new boiling chips, and assemble the apparatus for fractional distillation. The stillhead delivers into a short condenser fitted with a bent adapter leading into a 10-mL graduated cylinder. Gradually turn up the heat to the electric flask heater until the mixture of cyclohexane and toluene just begins to boil. As soon as boiling starts, turn down the power. Heat slowly at first. A ring of condensate will rise slowly through the column; if you cannot at first see this ring, locate it by cautiously touching the column with your fingers. The rise in temperature should be very gradual so that the column can acquire a uniform temperature gradient. Do not apply more heat until you are sure that the ring of condensate has stopped rising; then increase the heat gradually. In a properly conducted operation, the vapor-condensate mixture reaches the top of the column only after several minutes. Once distillation has commenced, it should continue steadily without any drop in temperature at a rate no greater than 1 mL in 1.5–2 min. Observe the flow and keep it steady by slight increases in heat as required. Protect the column from drafts by wrapping it with aluminum foil, glass wool, or even a towel. This insulation will help prevent flooding of the column, as will slow and steady distillation.

Record the temperature as each milliliter of distillate collects and take more frequent readings when the temperature starts to rise abruptly. Each time the graduated cylinder fills, quickly empty it into a series of labeled 25-mL Erlenmeyer flasks. Stop the distillation when a second constant temperature is reached. Plot a distillation curve and record what you observed inside the column in the course of the fractionation. Combine the fractions that you think are pure, and turn in the product in a bottle labeled with your name, desk number, the name of the product, the boiling point range, and the weight.

Cleaning Up. The pot residue should be placed in the organic solvents waste container. The cyclohexane and toluene fractions can also be placed there or recycled.

(B) Fractional Distillation of Ethanol-Water Mixture

Place the 50 mL of distillate from the simple distillation experiment into a 100-mL round-bottomed flask, add one or two boiling chips, and assemble the apparatus for fractional distillation. Follow the previous procedure for the fractional distillation of a cyclohexane-toluene mixture. Repeat the ignition test. Is any difference noted? Alternatively, distill 60 mL of the 10%–20% ethanol-water mixture that results from the fermentation of sucrose (*see* Chapter 64).

Cleaning Up. The pot residue and distillate can be disposed in the organic solvents waste container or, if regulations permit, diluted with water and flushed down the drain.

6. FRACTIONAL DISTILLATION OF UNKNOWNS

You will be supplied with an unknown, prepared by your instructor, that is a mixture of two solvents listed in Table 5.2, only two of which form azeotropes. The solvents in the mixture will be mutually soluble and differ in boiling point by more than 20°C. The composition of the mixture (in percentages of the two components) will be either 20:80, 30:70, 40:60, 50:50, 60:40, 70:30, or 80:20. Identify the two compounds and determine the percent composition of each. Perform a fractional distillation on

TABLE 5.2 *Some Properties of Common Solvents*

Solvent	*Boiling Point (°C)*
Acetone	56.5
Methanol	64.7
Hexane	68.8
1-Butanol	117.2
2-Methyl-2-propanol	82.2
Water	100.0
Toluene*	110.6

*Methanol and toluene form an azeotrope with a boiling point of 63.8°C (69% methanol).

4 mL of the unknown for microscale; use at least 50 mL of unknown for macroscale. Fractionate the unknown and identify the components from the boiling points. Prepare a distillation curve. You may be directed to analyze your distillate by gas chromatography (*see* Chapter 10) or refractive index (*see* Chapter 14).

Cleaning Up. Organic material goes in the organic solvents waste container. Water and aqueous solutions can be flushed down the drain.

QUESTIONS

1. In either of the simple distillation experiments, can you account for the boiling point of your product in terms of the known boiling points of the pure components of your mixture? If so, how? If not, why not?
2. From the plots of the boiling point versus the volume of distillate in the simple distillation experiments, what can you conclude about the purity of your product?
3. From the plots of the boiling point versus the volume of distillate in either of the fractional distillations of the cyclohexane-toluene mixture, what conclusion can you draw about the homogeneity of the distillate?
4. From the plots of the boiling point versus the volume of distillate in either of the fractional distillations of the ethanol-water mixture, what conclusion can you draw about the homogeneity of the distillate? Does it have a constant boiling point? If constant, is it a pure substance?
5. What is the effect on the boiling point of a solution (e.g., water) produced by a soluble nonvolatile substance (e.g., sodium chloride)? What is the effect of an insoluble substance such as sand or charcoal? What is the temperature of the vapor above these two boiling solutions?
6. In the distillation of a pure substance (e.g., water), why does all of the water not vaporize at once when the boiling point is reached?
7. In fractional distillation, liquid can be seen running from the bottom of the distillation column back into the distilling flask. What effect does this returning condensate have on the fractional distillation?
8. Why is it extremely dangerous to attempt to carry out a distillation in a completely closed apparatus (one with no vent to the atmosphere)?

9. Why is better separation of two liquids achieved by slow rather than fast distillation?

10. Explain why a packed fractionating column is more efficient than an unpacked one.

11. In the distillation of the cyclohexane-toluene mixture, the first few drops of distillate may be cloudy. Explain this occurrence.

12. What effect does the reduction of atmospheric pressure have on the boiling point? Can cyclohexane and toluene be separated if the external pressure is 350 mm Hg instead of 760 mm Hg?

See *Web Links*

13. When water-cooled condensers are used for distillation or for refluxing a liquid, the water enters the condenser at the lowest point and leaves at the highest. Why?

REFERENCE

An excellent review of distillation at the level of this textbook is *Distillation, an Introduction* by M.T. Tham, http://lorien.ncl.ac.uk/ming/distil/distilop.htm

6 CHAPTER

Steam Distillation, Vacuum Distillation, and Sublimation

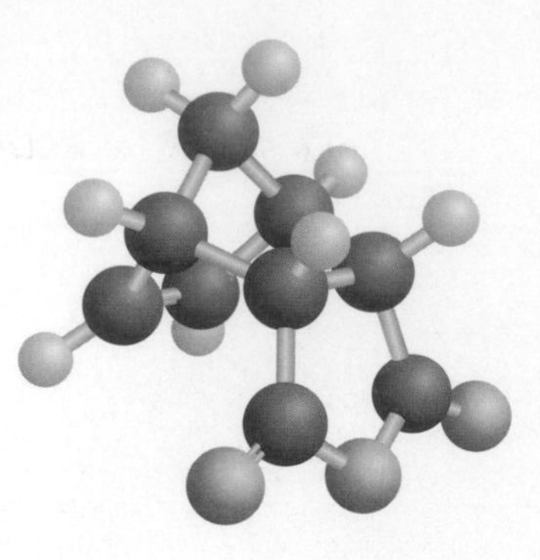

PRELAB EXERCISE: If a mixture of toluene and water is distilled at 97°C, what weight of water would be necessary to carry over 1 g of toluene? Compare the operation of a barometer, which is used to measure atmospheric pressure, with that of a mercury manometer, which is used to measure the pressure in an evacuated system.

PART 1: STEAM DISTILLATION

The boiling point of a pair of immiscible liquids is below the boiling point of either pure liquid.

Each liquid exerts its own vapor pressure.

When a mixture of cyclohexane and toluene is distilled (see Chapter 5), the boiling point of these two miscible liquids is between the boiling points of each of the pure components. By contrast, if a mixture of benzene and water (immiscible liquids) is distilled, the boiling point of the mixture will be below the boiling point of each pure component. Because benzene and water are essentially insoluble in each other, the molecules in a droplet of benzene are not diluted by water molecules from nearby water droplets; thus, the vapor pressure exerted by benzene in the mixture is the same as that of benzene alone at the existing temperature. The same is true for water. These two immiscible liquids exert pressures against the common external pressure independently, and when the sum of the two partial pressures equals the external pressure, boiling occurs. Benzene has a vapor pressure of 760 torr at 80.1°C, and if it is mixed with water, the combined vapor pressure must equal 760 torr at some temperature below 80.1°C. This temperature, the boiling point of the mixture, can be calculated from known values of the vapor pressures of the separate liquids at that temperature. Vapor pressures for water and benzene in the range 50–80°C are plotted in Figure 6.1. The dashed line cuts the two curves at points where the sum of the vapor pressures is 760 torr; hence the boiling point of the mixture is 69.3°C.

Practical use can sometimes be made of the fact that many water-insoluble liquids and solids behave as benzene does when mixed with water, volatilizing at temperatures below their boiling points. Naphthalene, a solid, boils at 218°C but distills with water at a temperature below 100°C. Because naphthalene is not very

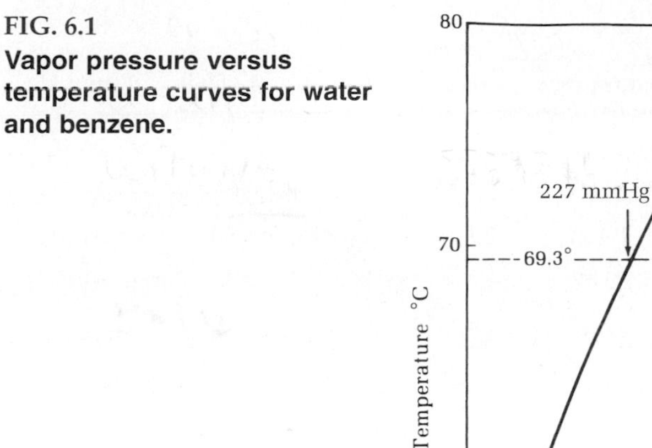

FIG. 6.1
Vapor pressure versus temperature curves for water and benzene.

volatile, considerable water is required to entrain it, and the conventional way of conducting the distillation is to pass steam into a boiling flask containing naphthalene and water, a process called *steam distillation*. With more volatile compounds, or with a small amount of material, the substance can be heated with water in a simple distillation flask, with the steam generated in situ.

Some high-boiling substances decompose before the boiling point is reached and, if impure, cannot be purified by ordinary distillation. However, these substances can be freed from contaminating substances by steam distillation at a lower temperature, at which they are stable. Steam distillation also offers the advantage of selectivity because some water-insoluble substances are volatile with steam while others are not; some volatilize so slowly that sharp separation is possible. This technique is useful in processing natural oils and resins, which can be separated into steam-volatile and non-steam-volatile fractions. It is useful for the recovery of a non-steam-volatile solid from its solution in a high-boiling solvent such as nitrobenzene (bp 210°C); all traces of the solvent can be eliminated, and the temperature can be kept low.

Used for the isolation of perfume and flavor oils.

The boiling point remains constant during a steam distillation as long as adequate amounts of both water and the organic component are present to saturate the vapor space. Determination of the boiling point and correction for any deviation from normal atmospheric pressure permits calculation of the amount of water required for the distillation of a given amount of organic substance. According to Dalton's law, the molecular proportion of the two components in the distillate is equal to the ratio of their vapor pressures (p) in the boiling mixture; the more volatile component contributes the greater number of molecules to the vapor phase. Thus,

Dalton's law

$$\frac{\text{moles of water}}{\text{moles of substance}} = \frac{p_{\text{water}}}{p_{\text{substance}}}$$

TABLE 6.1 *The Vapor Pressure of Water at Different Temperatures*

t (°C)	p (mm Hg)	t (°C)	p (mm Hg)	t (°C)	p (mm Hg)	t (°C)	p (mm Hg)
0	4.58	20	17.41	24	22.18	28	28.10
5	6.53	21	18.50	25	23.54	29	29.78
10	9.18	22	19.66	26	24.99	30	31.55
15	12.73	23	20.88	27	26.50	35	41.85
60	149.3	70	233.7	80	355.1	90	525.8
61	156.4	71	243.9	81	369.7	91	546.0
62	163.8	72	254.6	82	384.9	92	567.0
63	171.4	73	265.7	83	400.6	93	588.6
64	179.3	74	277.2	84	416.8	94	610.9
65	187.5	75	289.1	85	433.6	95	633.9
66	196.1	76	301.4	86	450.9	96	657.6
67	205.0	77	314.1	87	468.7	97	682.1
68	214.2	78	327.3	88	487.1	98	707.3
69	223.7	79	341.0	89	506.1	99	733.2

The vapor pressure of water (p_{water}) at the boiling temperature in question can be found by interpolating the data in Table 6.1. For the organic substance, the vapor pressure is equal to $760 - p_{water}$. Hence, the weight of water required per gram of substance is given by the expression:

$$\text{wt of water per g of substance} = \frac{18 \times p_{water}}{\text{MW of substance} \times (760 - p_{water})}$$

From the data given in Figure 6.1 for benzene and water, the fact that the mixture boils at 69.3°C, and the molecular weight of benzene is 78.11, the water required for the steam distillation of 1 g of benzene is only $227 \times 18/533 \times 1/78 = 0.10$ g. Nitrobenzene (bp 210°C, MW 123.11) steam distills at 99°C and requires 4.0 g of water per gram of nitrobenzene. The low molecular weight of water makes water a favorable liquid for the two-phase distillation of organic compounds.

Steam distillation is employed in several experiments in this text. On a small scale, steam is generated in the flask that contains the substance to be steam distilled by simply boiling a mixture of water and the immiscible substance.

ISOLATION OF CITRAL

Terpenes

Citral is an example of a very large group of natural products called terpenes. They are responsible for the characteristic odors of plants such as eucalyptus, pine, mint, peppermint, and lemon. The odors of camphor, menthol, lavender, rose, and hundreds of other fragrances are due to terpenes, which have 10 carbon atoms with double bonds along with aldehyde, ketone, or alcohol functional groups (Fig. 6.2).

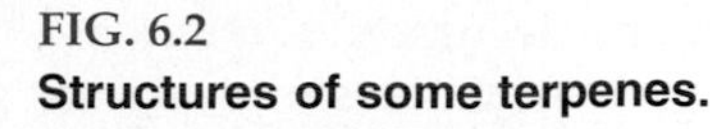

FIG. 6.2
Structures of some terpenes.

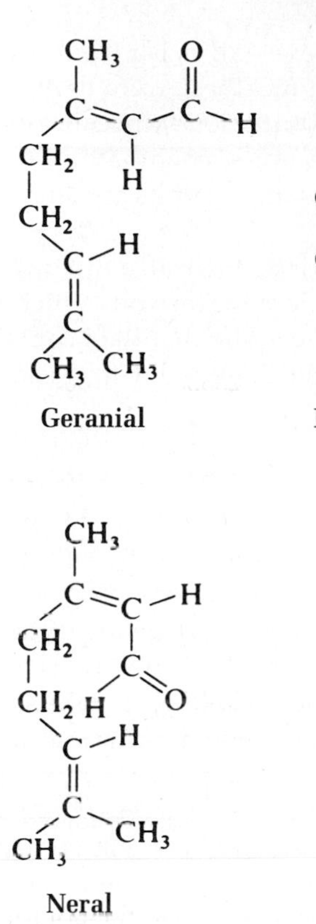

Geranial

Neral

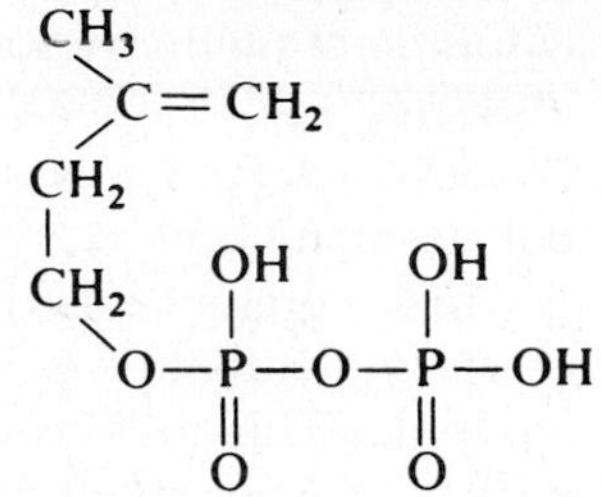

Isopentenyl pyrophosphate

CH₂ ‖ C CH₃ CH CH₂

Isoprene
(2-Methyl-1,3-butadiene)

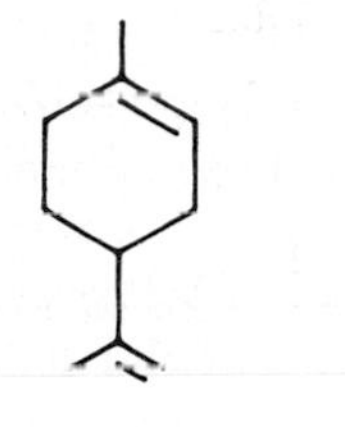

Limonene
(lemons)

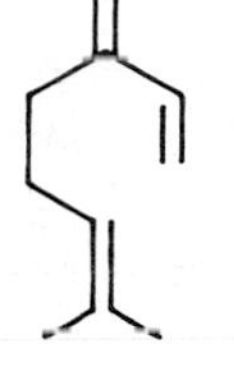

Myrcene
(bayberry)

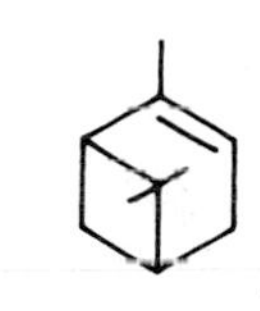

α-Pinene
(turpentine)

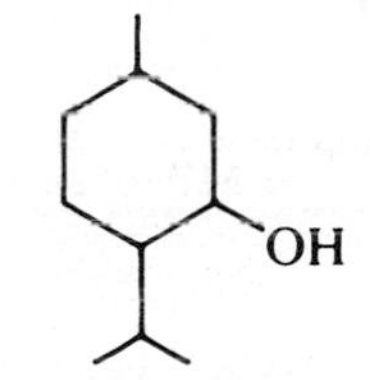

Menthol
(mint)

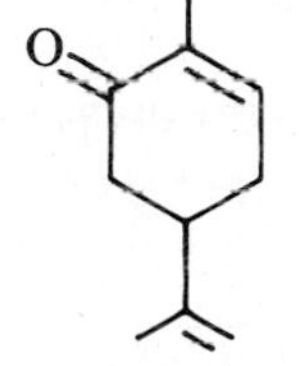

Carvone
(caraway seeds)

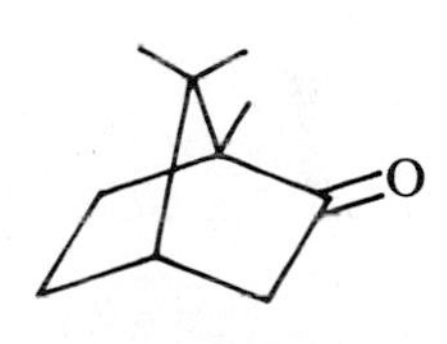

Camphor

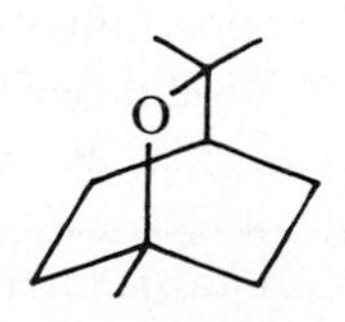

1,8-Cineole
(eucalyptus)

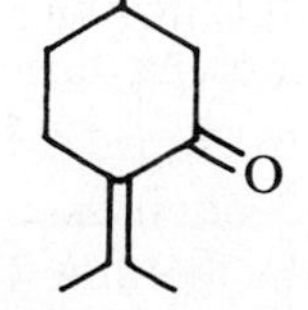

Pulegone
(pennyroyal oil)

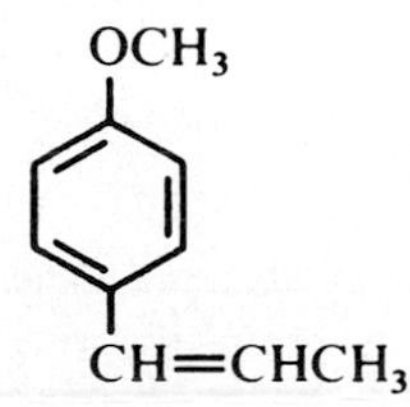

Anethole
(anise seed)

These natural terpenes all arise from a common precursor: isopentenyl pyrophosphate. At one time, these were thought to come from the simple diene, isoprene (2-methyl-1,3-butadiene), because the skeletons of terpenes can be dissected into isoprene units, each having five carbon atoms arranged as in 2-methylbutane. These isoprene units are almost always arranged in a "head-to-tail" fashion.

The isoprene rule.

In the present experiment, citral, a mixture of geranial and neral, is isolated by steam distillation of lemongrass oil, which is used to make lemongrass tea, a popular drink in Mexico. The distillate contains 90% geranial and 10% neral, with the isomer about the 2,3-bond. Citral is used in a commercial synthesis of vitamin A.

Geranial and neral are geometric isomers.

Lemongrass oil contains a number of substances; simple or fractional distillation would not be a practical method for obtaining pure citral. And because lemongrass oil boils at 229°C, it has a tendency to polymerize, oxidize, and decompose during distillation. For example, heating with potassium bisulfate, an acidic compound, converts citral to 1-methyl-4-isopropylbenzene (*p*-cymene).

Steam distillation is thus a very gentle method for isolating citral. The distillation takes place below the boiling point of water. The distillate consists of a mixture of citral (and some neral) and water. It is isolated by shaking the mixture with *t*-butyl methyl ether. The citral dissolves in the *t*-butyl methyl ether, which is immiscible with water, and the two layers are separated. The *t*-butyl methyl ether is dried (it dissolves some water) and evaporated to leave citral.

Steam distillation: the isolation of heat-sensitive compounds.

Store the citral in the smallest possible container in the dark for later characterization. The homogeneity of the substance can be investigated using thin-layer chromatography (TLC; *see* Chapter 8) or gas chromatography (*see* Chapter 10). An infrared spectrum (*see* Chapter 11) would confirm the presence of the aldehyde; an ultraviolet spectrum (*see* Chapter 14) would indicate it is a conjugated aldehyde; and a nuclear magnetic resonance (NMR) spectrum (*see* Chapter 12) would clearly show the aldehyde proton, the three methyl groups, and the two olefinic protons. Chemically, the molecule can be characterized by reaction with bromine and also permanganate, which shows it contains double bonds; the Tollens test confirms the presence of an aldehyde (*see* Chapter 36).

EXPERIMENTS

1. ISOLATION OF CITRAL

Video: *Steam Distillation Apparatus*

Into a 5-mL short-necked, round-bottomed flask place a boiling chip, 0.5 mL of lemongrass oil (*not* lemon oil)[1], and 3 mL of water. Assemble the apparatus as depicted in Figure 6.3 and distill as rapidly as possible, taking care that all the distillate condenses. Using a syringe, inject water dropwise through the septum to keep the volume in the flask constant. Distill into an ice-cooled vial, as shown, or better, into a 15-mL centrifuge tube, until no more oily drops can be detected, about 10 to 12 mL. Use this citral and water mixture for the next experiment.

Wrap the vertical part of the apparatus in cotton or glass wool to speed the distillation.

[1]Lemongrass oil can be obtained from Pfaitz and Bauer, 172 E. Aurora St., Waterbury, CT 06708.

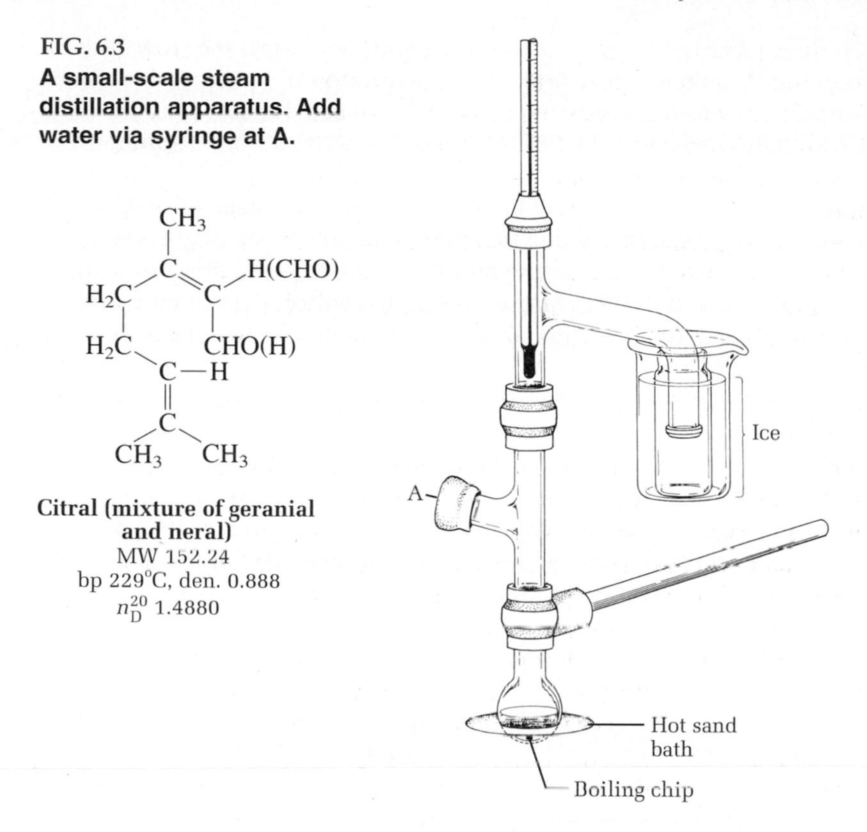

FIG. 6.3
A small-scale steam distillation apparatus. Add water via syringe at A.

Citral (mixture of geranial and neral)
MW 152.24
bp 229°C, den. 0.888
n_D^{20} 1.4880

2. EXTRACTION OF CITRAL

IN THIS EXPERIMENT, you will extract the pleasant-smelling liquid citral from water (from the previous experiment) using ether as the solvent. The ether solution is dried, and the ether is evaporated to leave pure citral. Extraction is a common isolation technique (*see* Chapter 7 for information on the theory and practice of extraction).

In a centrifuge tube, add 2.5 mL of *t*-butyl methyl ether to the cold distillate. Shake, let the layers separate, draw off the aqueous layer with a Pasteur pipette, and transfer it to another centrifuge tube. Continue this extraction process by again adding 1.5 mL of ether to the aqueous layer, shaking, and separating to extract the citral from the steam distillate. (*See* Chapter 7 for the many advantages of *t*-butyl methyl ether as an extraction solvent.)

In Sweden they say, "Add anhydrous drying agent until it begins to snow."

Place the ether in a reaction tube and then add about 2 mL of saturated sodium chloride solution to the ether. Stopper the tube, flick the tube or shake it, and then remove and discard the aqueous layer. Add anhydrous calcium chloride pellets to the ether until these no longer clump together, stopper the tube, and shake it over a period of 5–10 minutes. This removes water adhering to the reaction tube and dissolved in the ether. Force a Pasteur pipette to the bottom of the reaction tube, expel the air from the pipette, draw up the ether, and expel it into another tared reaction tube. Add fresh ether, which will serve to wash off the drying agent, and add that ether to the tared reaction tube. Add a boiling stick or boiling chip to the ether in

***See* Chapter 7 for the removal of water from ether solutions.**

the reaction tube, place it in a beaker of hot water, and evaporate the ether by boiling and drawing the ether vapors into a water aspirator or, better, by blowing a gentle stream of air or nitrogen into the tube (in the hood). The last traces of ether may be removed by connecting the reaction tube directly to a water aspirator. The residue should be a clear, fragrant oil. If it is cloudy, it is wet.

Determine the weight of the citral and calculate the percentage of citral recovered from the lemongrass oil, assuming that the density of the lemongrass oil is the same as that of citral (0.89). Transfer the product to the smallest possible airtight container and store it in the dark for later analysis to confirm the presence of double bonds (alkenes) and the aldehyde group (*see* Chapters 18 and 36). Analyze the infrared spectrum in detail.

Cleaning Up. The aqueous layer can be flushed down the drain. Any ether goes into the organic solvents waste container. Allow ether to evaporate from the calcium chloride in the hood and then place the calcium chloride in the nonhazardous solid waste container. However, be aware of the regulations in your area, as they may not allow evaporation of small amounts of organic solvents in a hood. If this is the case, special containers should be available for the disposal of these wet solids.

3. ISOLATION OF CITRAL

Citral, a fragrant terpene aldehyde made up of two isoprene units, is the main component of the steam-volatile fraction of lemongrass oil and is used in a commercial synthesis of vitamin A.

Steam Distillation Apparatus

In the assembly shown in Figure 6.4, steam is passed into a 250-mL round-bottomed flask through a section of 6-mm glass tubing fitted into a stillhead with a piece of 5-mm rubber tubing connected to a trap, which in turn is connected to a steam line.

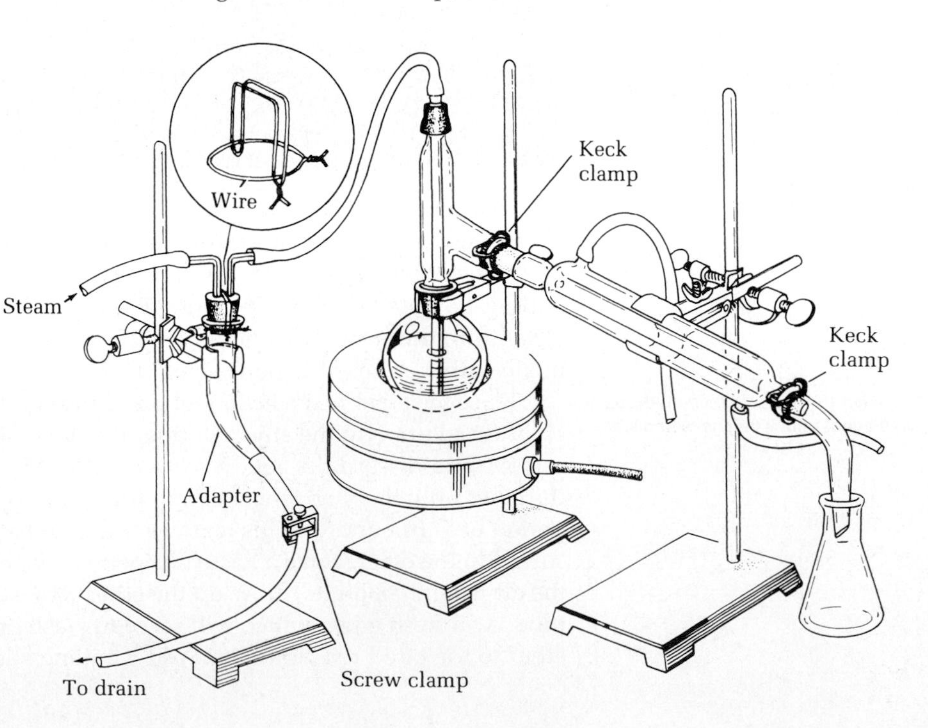

FIG. 6.4
A steam distillation apparatus.

CAUTION: Live steam causes severe burns. If the apparatus leaks steam, immediately turn off the steam valve.

Perform steam distillations within fume hoods.

The trap serves two purposes: (1) it allows water, which is in the steam line, to be removed before it reaches the round-bottomed flask; and (2) adjustment of the clamp on the hose at the bottom of the trap allows precise control of the steam flow. As shown, the stopper in the trap should be wired on as a precaution. A bent adapter attached to a long condenser delivers the condensate into a 250-mL Erlenmeyer flask.

Alternative Steam Distillation Setup

If a steam line is not available, the steam distillation setup shown in Figure 6.5 may be used. A 100-mL or 250-mL boiling flask containing the mixture to be distilled is equipped with a boiling chip or better a magnetic stirring bar and magnetic stirrer and is heated on a heating mantle. Equip the boiling flask with a distillation adapter to which is attached a thermometer adapter and a separatory funnel, which serves to add water to the boiling flask during the distillation. A condenser is attached to the thermometer adapter and is equipped with a vacuum adapter, which is attached to the receiving flask. The receiving flask can be cooled by placing it in an ice water bath. Make sure you securely clamp the distillation flask and condenser. Also make sure to use a rubber band or a Keck clip to hold the vacuum adapter on the end of the condenser.

Steam Distillation

Using a graduated cylinder, measure out 10 mL of lemongrass oil (not lemon oil) into a 250-mL boiling flask. Rinse the remaining contents of the graduated cylinder into the flask with a small amount of t-butyl methyl ether. Add 100 mL of water, make the connections as in Figure 6.4, heat the flask with a low flame, and pass in the steam.

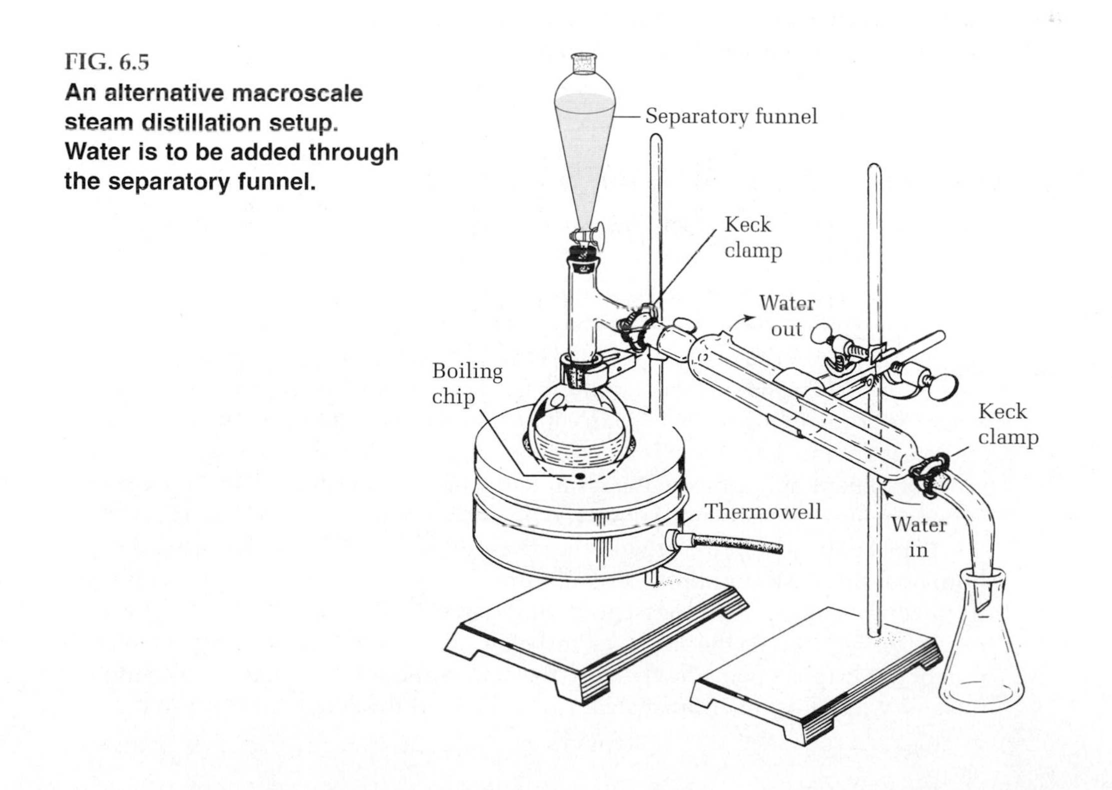

FIG. 6.5
An alternative macroscale steam distillation setup. Water is to be added through the separatory funnel.

To start the steam distillation, heat the flask containing the mixture on a Thermowell to prevent water from condensing in the flask to the point where water and product splash over into the receiver. Then turn on the steam valve, being sure that the screw clamp on the bottom of the trap is open. Slowly close the clamp and allow steam to pass into the flask. Try to adjust the rate of steam addition and the rate of heating so that the water level in the flask remains constant. Unlike ordinary distillations, steam distillations are usually run as fast as possible, with proper care to avoid splashing of material into the receiver and the escape of uncondensed steam. Distill as rapidly as the cooling facilities allow and continue until droplets of oil no longer appear at the tip of the condenser (about 250 mL of distillate).

Extraction

Pour 50 mL of t-butyl methyl ether into a 125-mL separatory funnel, cool the distillate if necessary, and pour a part of it into the funnel. Shake, let the layers separate, discard the lower layer, add another portion of distillate, and repeat. When the last portion of distillate has been added, rinse the flask with a small amount of *t*-butyl methyl ether to recover adhering citral. Use the techniques described in Chapter 7 for drying, filtering, and evaporating the t-butyl methyl ether. Take the tare of a 1-g tincture bottle, transfer the citral to it with a Pasteur pipette, and determine the weight and the yield from the lemongrass oil. Label the bottle and store (in the dark) for later testing for the presence of double bonds (alkenes) and the aldehyde group (*see* Chapters 18 and 36).

Cleaning Up. The aqueous layer can be flushed down the drain. Any *t*-butyl methyl ether goes in the organic solvents waste container. Allow *t*-butyl methyl ether to evaporate from the calcium chloride pellets in the hood and then place the calcium chloride in the nonhazardous solid waste container. However, be aware of the regulations in your area, as they may not allow evaporation of small amounts of organic solvents in a hood. If this is the case, special containers should be available for the disposal of these wet solids.

4. RECOVERY OF A DISSOLVED SUBSTANCE

Anthracene
mp 216°C

CH_3

Toluene
bp 110.8°C
den. 0.866

Measure 50 mL of a 0.2% solution of anthracene in toluene into a 250-mL round-bottomed flask and add 100 mL of water (see Fig. 6.4 on page 108). For an initial distillation to determine the boiling point and composition of the toluene-water azeotrope, fit the stillhead with a thermometer instead of a steam-inlet tube (see Fig. 5.4 on page 91). Heat the mixture with a hot plate, a sand bath, or an electric flask heater; distill about 50 mL of the azeotrope and record a value for the boiling point. After removing the heat source, pour the distillate into a graduated cylinder and measure the volumes of toluene and water. Calculate the weight of water per gram of toluene and compare the result to the theoretical value calculated from the vapor pressure of water at the observed boiling point (see Table 6.1 on page 104).

Replace the thermometer with the steam-inlet tube. To start the steam distillation, heat the flask containing the mixture with a Thermowell to prevent water from condensing, which may cause water and product to splash over into the receiver. Then turn on the steam valve, being sure that the screw clamp on the bottom of the trap is open. Slowly close the clamp and allow steam to pass into the flask. Try to adjust the rate of steam addition and the rate of heating so that the water level in the flask remains constant. Unlike ordinary distillations, steam distillations are usually run as fast as possible, with proper care to avoid splashing of material into the receiver and the escape of uncondensed steam.

Continue the distillation by passing in steam until the distillate is clear and then until fluorescence appearing in the stillhead indicates that a trace of anthracene is beginning to distill. Stop the steam distillation by first opening the clamp at the bottom of the trap and then turning off the steam valve. Grasp the round-bottomed flask with a towel when disconnecting it and, using the clamp to support it, cool it under the tap. The bulk of the anthracene can be dislodged from the flask walls and collected on a small suction filter. To recover any remaining anthracene, add a small amount of acetone to the flask, warm the flask on a steam bath to dissolve the material, add water to precipitate it, and then collect the precipitate on the same suction filter. About 80% of the hydrocarbon in the original toluene solution should be recoverable. When dry, recrystallize the material from about 1 mL of toluene and note that the crystals are more intensely fluorescent than the solution or the amorphous solid. The characteristic fluorescence is quenched by mere traces of impurities.

Cleaning Up. Place the toluene and the anthracene in the organic solvents waste container. The aqueous layer can be flushed down the drain.

5. ISOLATION AND BIOASSAY OF EUGENOL FROM CLOVES

In the 15th century, Philippus Aureolus Theophrastus Bombast von Hohenheim, otherwise known as Paracelsus, urged all chemists to make extractives for medicinal purposes rather than try to transmute base metals into gold. He believed every extractive possessed a quintessence that was the most effective part in effecting cures. There was then an upsurge in the isolation of essential oils from plant materials, which are highly concentrated extracts that may consist of one or two to hundreds of chemical compounds. Some of the essential oils that were isolated centuries ago are still used today for medicinal purposes. Among these are camphor, quinine, oil of cloves, cedarwood, turpentine, cinnamon, gum benzoin, and myrrh. Clove oil, which consists almost entirely of eugenol and its acetate, is a food flavoring agent as well as a dental anesthetic. The Food and Drug Administration has declared clove oil to be the most effective nonprescription remedy for toothache.

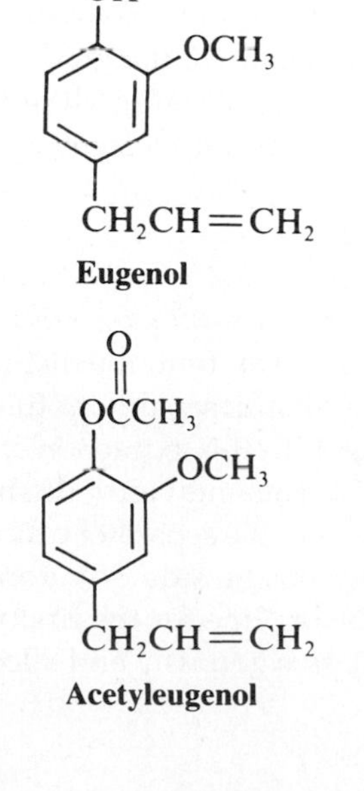

Eugenol

Acetyleugenol

The biggest market for essential oils is for perfumes, and prices for these oils reflect their rarity. Recently, worldwide production of orange oil was 1500 tons and sold for $2 per oz, whereas 400 tons of clove oil sold for $4 per oz, and 10 tons of jasmine oil sold for $150 per oz. These three oils represent the most common extractive processes. Orange oil is obtained by expression (squeezing) of the peel in presses; clove oil is obtained by steam distillation, as performed in this experiment; and jasmine oil is obtained by the extraction of the flower petals using ethanol. An even more gentle method of extraction is to use supercritical carbon dioxide. At a pressure of 73 atmospheres and temperatures above 32°C, carbon dioxide becomes a supercritical fluid. It is like a gas, but will dissolve many non-polar organic substances. It evaporates below room temperature when the pressure is released, so delicate perfumes such as rose oil can be extracted with virtually no chemical change. The resulting oil costs several hundred dollars per ounce.

Cloves are the dried flower buds from a tropical tree that can grow to 50 ft in height. The buds are handpicked in the Moluccas (the Spice Islands) of Indonesia, the East Indies, and the islands of the Indian Ocean, where a single tree can yield up to 75 lb of the sun-dried buds we know as cloves.

Cloves contain between 14% and 20% by weight of essential oil, of which only about 50% can be isolated. The principal constituent of clove oil is eugenol and its acetyl derivative. Eugenol boils at 255°C, but because it is insoluble in water, it will form an azeotrope with water and steam distill at a temperature slightly below the boiling point of water. Being a phenol, eugenol will dissolve in aqueous alkali to form the phenolate anion, which forms the basis for its separation from its acetyl derivative. The relative amounts of eugenol and acetyleugenol, and the effectiveness of the alkali extraction in separating them can be analyzed by thin-layer chromatography.

Procedure

IN THIS EXPERIMENT the essential oil—a mixture of eugenol and cetyleugenol—is steam distilled from cloves. These compounds are isolated from the aqueous distillate by extraction into dichloromethane. The dichloromethane solution is shaken with aqueous sodium hydroxide, which reacts with the acidic eugenol. The dichloromethane layer is dried and evaporated to give acetyleugenol. The basic layer is acidified, and the eugenol is extracted into dichloromethane, which is in turn dried and evaporated to give pure eugenol. The isolated crude oil, eugenol, and acetyleugenol can then be evaluated for antibacterial properties.

Steam Distillation

Place 25 g of whole cloves in a 250-mL round-bottomed flask, add 100 mL of water, and set up an apparatus for steam distillation (*see* Fig. 6.4 on page 108 or Fig. 6.5 on page 109) or simple distillation (*see* Fig. 5.10 on page 97). In the latter procedure, steam is generated in situ. Heat the flask strongly until boiling begins, and then reduce the heat just enough to prevent foam from being carried over into the receiver. Instead of using a graduated cylinder as a receiver, use an Erlenmeyer flask and transfer the distillate periodically to the graduated cylinder; then, if any material does foam over, the entire distillate will not be contaminated. Collect 60 mL of distillate, remove the flame, and add 60 mL of water to the flask. Resume the distillation and collect an additional 60 mL of distillate.

Extraction

The eugenol is sparingly soluble in water and is extracted easily from the distillate with dichloromethane. Place the 120 mL of distillate in a 250-mL separatory funnel and extract with three 15-mL portions of dichloromethane. In this extraction, very gentle shaking will fail to remove all of the product, and long and vigorous shaking will produce an emulsion of the organic layer and water. The separatory funnel will appear to have three layers in it. It is better to err on the side of vigorous shaking and draw off the clear lower layer to the emulsion line for the first two extractions. For the third extraction, shake the mixture less vigorously and allow a longer period of time for the two layers to separate.

Combine the dichloromethane extracts (the aqueous layer can be discarded) and add just enough anhydrous calcium chloride pellets so that the drying agent no longer clumps together but appears to be a dry powder as it settles in the solution. This may require as little as 2 g of drying agent. Swirl the flask for 1–2 minutes to complete the drying process and then decant the solvents into an Erlenmeyer flask. It is quite easy to decant the dichloromethane from the drying agent, so it is not necessary to set up a filtration apparatus to accomplish this separation.

Measure the volume of the dichloromethane in a dry graduated cylinder, place exactly one-fifth of the solution in a tared Erlenmeyer flask, add a wood boiling stick (easier to remove than a boiling chip), and evaporate the solution on a steam bath or a hot plate in a hood. The residue of crude clove oil will be used in a TLC (thin-layer chromatography) analysis and/or in a bioassay. From its weight, the total yield can be calculated. This residue can then be dissolved in acetone (a 5%–10% solution) for TLC analysis and/or for a bioassay.

***See* Chapter 7 for the theory of acid/base extractions.**

To separate eugenol from acetyleugenol, extract the remaining four-fifths of the dichloromethane solution (about 30 mL) with 5% aqueous sodium hydroxide solution. Carry out this extraction three times, using 10-mL portions of sodium hydroxide each time. Dry the dichloromethane layer over anhydrous calcium chloride pellets, decant the solution into a tared Erlenmeyer flask, and evaporate the solvent. The residue should consist of acetyleugenol and other steam-volatile neutral compounds from cloves.

Acidify the combined aqueous extracts to pH 1 with concentrated hydrochloric acid (use Congo Red or litmus paper to test) and then extract the liberated eugenol with three 8-mL portions of dichloromethane. Dry the combined extracts, decant into a tared Erlenmeyer flask, and evaporate the solution on a steam bath or a hot plate. Calculate the weight percent yields of eugenol and acetyleugenol on the basis of the weight of cloves used.

***Chapter 11*: Infrared Spectroscopy**

Analyze your products using infrared spectroscopy and thin-layer chromatography. Obtain an infrared spectrum of eugenol or acetyleugenol using the thin-film method. Compare your spectrum to the spectrum of a fellow student who has examined the other compound.

***Chapter 8*: Thin-Layer Chromatography**

Use plastic sheets precoated with silica gel for thin-layer chromatography. One piece the size of a microscope slide should suffice. Spot crude clove oil, eugenol, and acetyleugenol (1% solutions) side by side about 5 mm from one end of the plate. The spots should be very small. Immerse the end of the slide in a dichloromethane–hexane mixture (1:2 or 2:1) about 3 mm deep in a covered beaker. After running the chromatogram, observe the spots under an ultraviolet lamp or by developing them in an iodine chamber.

Antibiotic Behavior of Spice Essential Oils

Wash hands after handling bacteria.

You will bioassay your isolated clove oil with Bacillus cereus to test for its antibiotic behavior. Many bacterial cultures are available commercially, but Bacillus cereus is preferred because it is nonpathogenic, requires no special medium for growth, and grows at room temperature in a short time. Cultures of E. coli have to be handled with care because of their potential pathogenicity, depending on the strain. Their optimal growth is at 37°C, which means that an incubator is needed to work with these cultures.

Agar culture plates are prepared in sterile Petri dishes or can be purchased.[2] These plates are stored in the refrigerator. Whenever you handle the sterile agar

[2]Agar plates and bacteria can be purchased from Carolina Biological Supply Co. www.carolina.com. Nutrient Agar Plates: part number CE-82–1862 (packages of 10); *Bacillus cereus*: part number CE-15–4869.

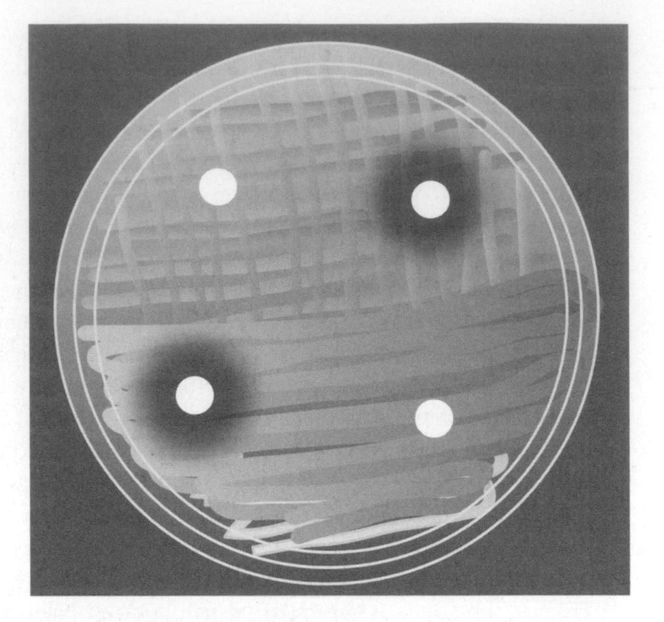

FIG. 6.6
Two sets of test disks (clockwise from top left) showing the control, 2 mg of clove oil, control, 2 mg of clove oil. The clove oil disks show inhibition of bacterial growth; there is no bacterial growth in the immediate area of the disks.

Petri dishes (bottom dish and top cover), try to minimize any exposure to random forms of bacteria or dirt in the laboratory. Always wear gloves when handling the plates and the *Bacillus cereus* because your hands can easily contaminate the agar. Open the cover of the dish a few inches so that you have just enough room to streak the plate with the bacteria or insert a testing disk that is impregnated with the essential oil. Immediately cover the agar again with the Petri dish lid to reduce the risk of outside contamination. You will use a pair of sterile tweezers to handle the test disk, and sterile swabs to streak the plates with the *Bacillus cereus*. After preparing your assay, you will observe the growth of bacteria on the plate after 1–2 days of growth. If the bacteria do not grow in the area surrounding the disk treated with your essential oil, then one or more of the compounds in this isolated mixture is exhibiting antibiotic behavior.

Bioassay Using Agar Medium Bacterial Growth Plates

Two students can use one agar plate (typically a plastic disposable Petri plate). Draw a line down the middle of the plate to separate each student's disks from the other's. Label the bottom of the culture plate with your name and each disk's identity so that you will know which disk is which. See Figure 6.6 for an example of an agar plate.

Procedure for the transfer of sterile cultures:
(All students must be wearing gloves during this entire procedure.)

1. Obtain a prepared, sterile agar plate.
2. Label the nutrient agar plate with your last name. The plate is to be divided into 4 equal quadrants. Each student/group will have two of these quadrants, one for the control and one for the clove oil. Use a marking pen to delineate these quadrants and use this area as a guide when preparing the plates. **Be very careful not to remove the cover of the plate during this step.**
3. Next, the bacteria will be transferred to the plate:

a. Insert a sterile swab into the culture tube containing the bacteria.
b. Remove the swab from the culture tube and then tilt the cover of the agar plate at an angle to keep exposure to unwanted bacteria in the air to a minimum. Do not lay the top of the plate on the benchtop and do not allow anyone to breathe on or come in close contact with the plate.
c. The moist swab is brought in contact with the agar using a rubbing motion. Start at the center of the plate and very gently rub back and forth, moving toward the edge of the plate. Turn the plate 90 degrees and repeat the rubbing of the swab. Turn the plate 135 degrees and rub the remaining section of the plate so that the entire surface area of the plate has been covered. The plate should now be evenly coated with the bacterial culture.
d. Replace the cover on the culture dish.
e. Dispose of the used sterile swab in the appropriate biohazard waste container.

4. Next, the isolated clove oil will be placed in contact with the inoculated culture.
a. Add the required volume of acetone to your oil sample to make a 5 to 10% (wt/vol) solution of your isolated clove oil in acetone.
b. Dip the tweezers into alcohol to sterilize.
c. Remove tweezers from the alcohol and fan back and forth to evaporate the alcohol.

5. Pick up one sterile sample disk using the sterile tweezers and hold it in contact with the solution of your isolated clove oil so that it is absorbed into the disk. Allow the solvent to evaporate and then transfer the disk to the culture dish by lifting the lid slightly and placing it in the center of one of the 4 quadrants.

6. Next, transfer a control disk to the culture dish. The control disk is made sterile by dipping it in ethanol and then allowing the ethanol to evaporate.

7. Wrap the edges of the dish with Parafilm to keep the agar from drying out or to prevent contamination.

8. Allow the dish to sit on the benchtop for about 10 minutes to seat the disks in the agar.

9. Invert the culture dish so that the bottom is facing up.

10. Place the dish either at room temperature in a secure location or in an incubator oven. The results of the assay can be assessed after 24 hours or longer.

Bioassays of the isolated eugenol and acetyleugenol may also be performed and compared.

Cleaning Up. Combine all of the aqueous layers, neutralize with sodium carbonate, dilute with water, and flush down the drain. Any solutions containing dichloromethane should be placed in the halogenated organic solvents waste container. Allow the solvent to evaporate from the calcium chloride pellets in the hood and then place the drying agent in the nonhazardous solid waste container. If local regulations do not allow for the evaporation of solvent from the pellets in a hood, dispose of the wet pellets in a special container. The acetone solution of oil can be placed in the nonhalogenated solvent waste container. The plastic, disposable agar plate can be taped shut with masking tape and placed into a biohazard waste bin. If an autoclave is not available, soak the agar plates and contaminated materials in bleach for several days. The bleach solution can then be flushed down the drain with lots of water.

PART 2: VACUUM DISTILLATION

Many substances cannot be satisfactorily distilled in the ordinary way, either because they boil at such high temperatures that decomposition occurs or because they are sensitive to oxidation. Steam distillation will work only if the compound is volatile. When neither ordinary distillation nor steam distillation is practical, purification can be accomplished by distillation at reduced pressure.

VACUUM DISTILLATION ASSEMBLIES

Macroscale Vacuum Distillation

Carry out vacuum distillation behind a safety shield. Before using, check the apparatus for scratches and cracks.

A typical macroscale vacuum distillation apparatus is illustrated in Figure 6.7. It is constructed of a round-bottomed flask (often called the pot) containing the material to be distilled, a Claisen distilling head fitted with a hair-fine capillary mounted through a rubber tubing sleeve, and a thermometer with the bulb extending below the side-arm opening. The condenser fits into a vacuum adapter that is connected to the receiver and, via heavy-walled rubber tubing, to a mercury manometer and then to the trap and water aspirator.

Liquids usually bump vigorously when boiled at reduced pressure, and most boiling stones lose their activity in an evacuated system; it is therefore essential to make special provision to control bumping. This is done by allowing a fine stream of air bubbles to be drawn into the boiling liquid through a glass tube drawn to a

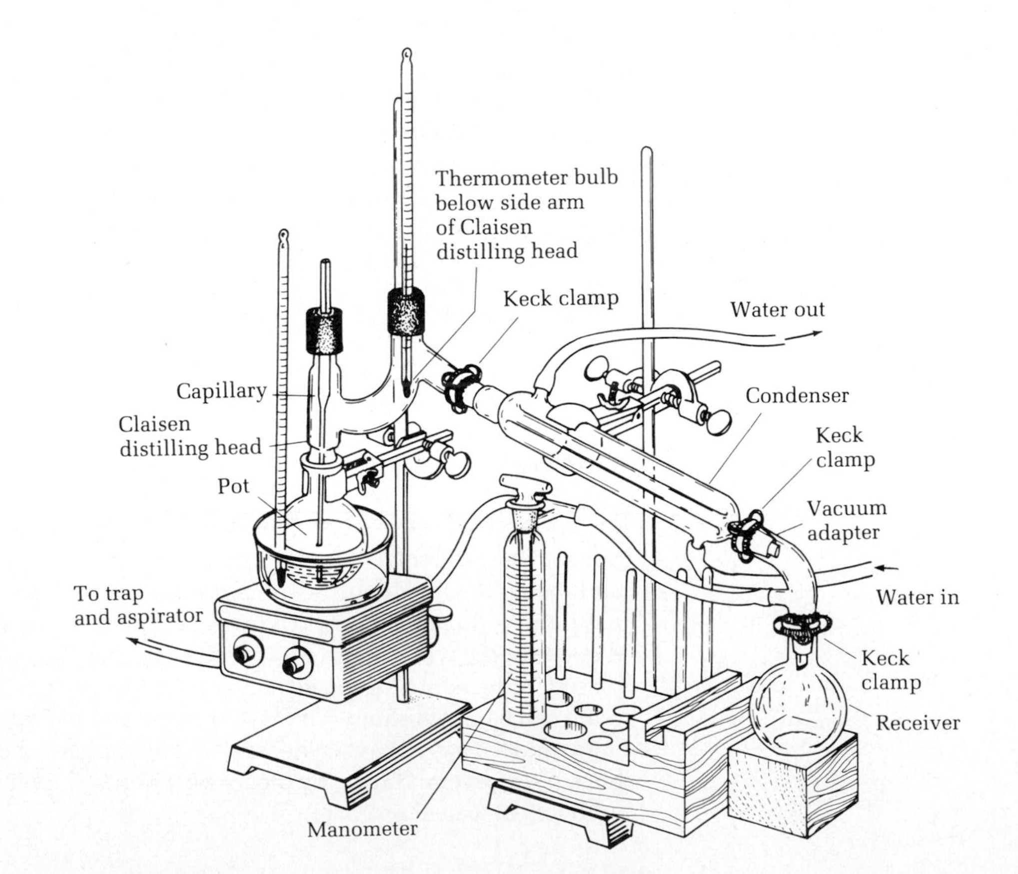

FIG. 6.7
A macroscale vacuum distillation apparatus.

hair-fine capillary. The capillary should be so fine that even under vacuum only a few bubbles of air are drawn in each second; smooth boiling will be promoted, and the pressure will remain low. The capillary should extend to the very bottom of the flask and should be slender and flexible so that it will whip back and forth in the boiling liquid. Another method used to prevent bumping when small quantities of material are being distilled is to introduce sufficient glass wool into the flask to fill a part of the space above the liquid. Rapid magnetic stirring will also prevent bumping.

Prevention of bumping.

Making a hair-fine capillary.

A hair-fine capillary is made as follows. A 6-in. length of 6-mm glass tubing is rotated and heated in a small area over a *very hot* flame to collapse the glass and thicken the side walls, as shown in Figure 6.8a. The tube is removed from the flame, allowed to cool slightly, and then drawn into a thick-walled, coarse-diameter capillary (Fig. 6.8b). This coarse capillary is heated at point X over the wing top of a Bunsen burner turned 90°. When the glass is very soft, but not soft enough to collapse the tube entirely, the coarse capillary is lifted from the flame and without hesitation drawn smoothly and rapidly into a hair-fine capillary by stretching the hands as far as they will reach (about 2 m; Fig. 6.8c). The two capillaries so produced can be snapped off to the desired length. To ascertain that there is indeed a continuous opening inside the hair-fine capillary, place the fine end into a low-viscosity liquid such as acetone or *t*-butyl methyl ether and blow in the large end. A stream of very small bubbles should be seen. If the stream of bubbles is not extremely small, make a new capillary. The flow of air through the capillary *cannot* be controlled by attaching a rubber tube and clamp to the top of the capillary tube. Should the right-hand capillary of Figure 6.8c break when in use, it can be fused to a scrap of glass (for use as a handle) and heated again at point Y (Fig. 6.8c). In this way a capillary can be redrawn many times.

Handle flames in the organic laboratory with great care.

Heating baths

The pot is heated via a heating bath rather than a Thermowell to promote even boiling and make accurate determination of the boiling point possible. The bath is filled with a suitable heat-transfer liquid (water, cottonseed oil, silicone oil, or molten metal) and heated to a temperature about 20°C higher than that at which the substance in the flask distills. The bath temperature is kept constant throughout the distillation. The surface of the liquid in the flask should be below that of the heating medium so as to lessen the tendency to bump. Heating of the flask is begun only

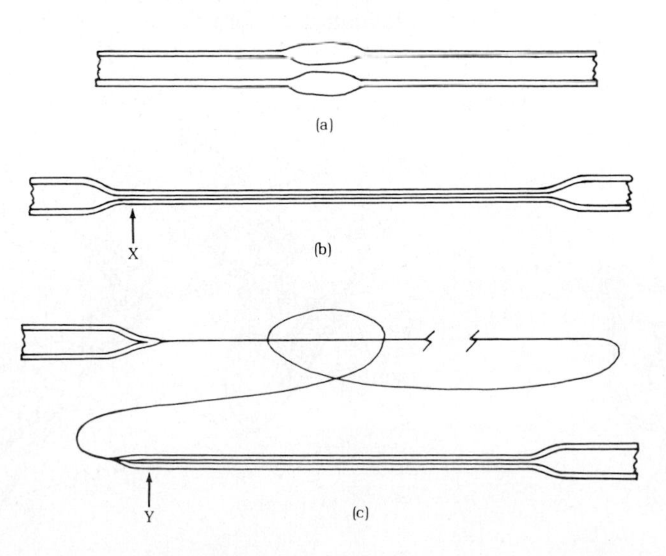

FIG. 6.8
A capillary for vacuum distillation.

after the system has been evacuated to the desired pressure; otherwise, the liquid might boil too suddenly on reduction of the pressure.

Changing fractions

To change fractions, the following must be done in sequence: Remove the source of heat, release the vacuum, change the receiver, restore the vacuum to its former pressure, and resume heating.

A fractional vacuum distillation apparatus is shown in Figure 6.9. The distillation neck of the Claisen adapter is longer than that of other adapters, and it has a series of indentations intruding from four directions such that the points nearly meet in the center. These indentations increase the surface area over which rising vapor can come to equilibrium with descending liquid, and the distillation neck then serves as a fractionating column (a Vigreux column). A column packed with a metal sponge has a high tendency to become filled with liquid (flood) at reduced pressure. The apparatus illustrated in Figure 6.9 also has a fraction collector, which allows the removal of a fraction without disturbing the vacuum in the system. While the receiver is being changed, the distillate collects in small reservoir A. The clean receiver is evacuated by another aspirator at tube B before being reconnected to the system.

Semi-microscale and Microscale Vacuum Distillation

If only a few milliliters of a liquid are to be distilled, the apparatus shown in Figure 6.10 has the advantage of low holdup—that is, not much liquid is lost to wetting the surface area of the apparatus. The fraction collector illustrated is known as a cow. Rotation of the cow about the lightly greased standard taper joint will allow four fractions to be collected without interrupting the vacuum.

A distillation head of the type shown in Figure 6.11 allows fractions to be removed without disturbing the vacuum and control of the reflux ratio (*see* Chapter 5) by manipulation of the condenser and stopcock A. These can be adjusted to remove all the material that condenses or only a small fraction, with

FIG. 6.9
A vacuum distillation apparatus with a Vigreux column and fraction collector.

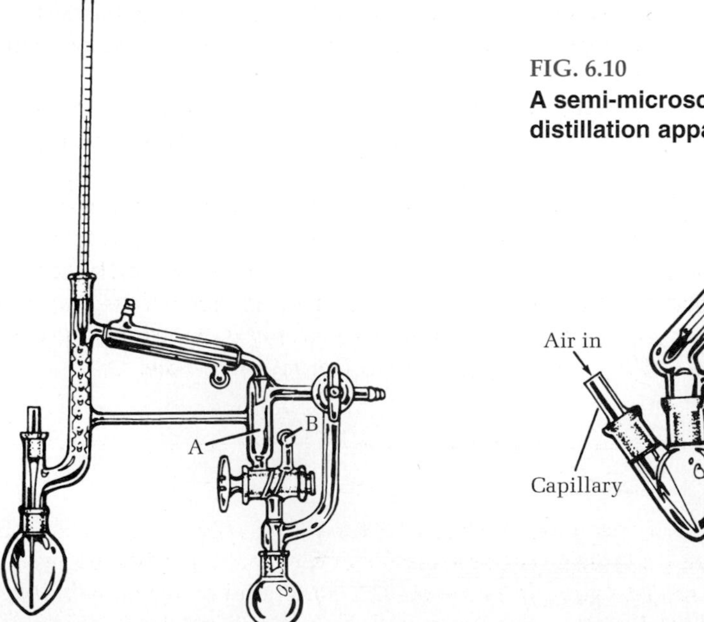

FIG. 6.10
A semi-microscale vacuum distillation apparatus.

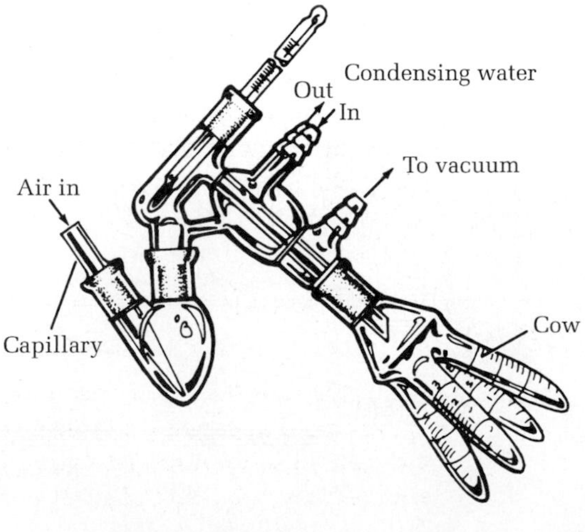

Condensing water
Out
In
To vacuum
Air in
Capillary
Cow

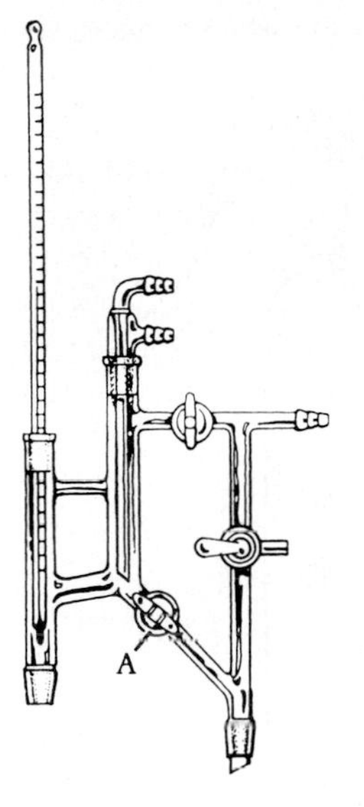

FIG. 6.11
A vacuum distillation head.

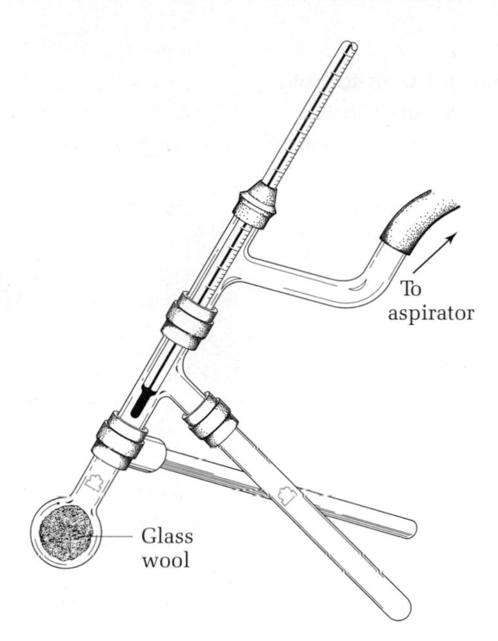

FIG. 6.12
A microscale apparatus for simple vacuum distillation.

the bulk of the liquid being returned to the distilling column to establish equilibrium between the descending liquid and the ascending vapor. Thus liquids with small boiling-point differences can be separated.

On a truly micro scale (<10 mg), simple distillation is not practical because of mechanical losses. Fractional distillation of small quantities of liquids is also difficult. The necessity for maintaining an equilibrium between ascending vapors and descending liquid means that there will inevitably be large mechanical losses. The best apparatus for the fractional distillation of 0.5 mL to 20 mL of liquid under reduced pressure is the *micro spinning band column*, which has a holdup of about 0.1 mL. In research, preparative-scale gas chromatography (*see* Chapter 10) and high-performance liquid chromatography (HPLC) have supplanted vacuum distillation of very small quantities of material.

It is possible, however, to assemble several pieces of apparatus that can be used for simple vacuum distillation using components from the microscale kit that accompanies this text. One such assembly is shown in Figure 6.12. The material to be distilled is placed in a 5-mL short-necked flask packed loosely with glass wool, which will help prevent bumping. The thermometer bulb extends entirely below the side arm, and the distillate is collected in a reaction tube, which can be cooled in a beaker of ice. The vacuum source is connected to a syringe needle; alternatively, the rubber tubing from the aspirator, if of the proper diameter, can be connected directly to the adapter, dispensing with the septum and needle.

THE KUGELROHR

The Kugelrohr (German, meaning "bulb-tube") is a widely used piece of research apparatus that consists of a series of bulbs that can be rotated and heated under vacuum. Bulb-to-bulb distillation frees the desired compound of very low-boiling and very high-boiling or nonvolatile impurities. The crude mixture is placed in

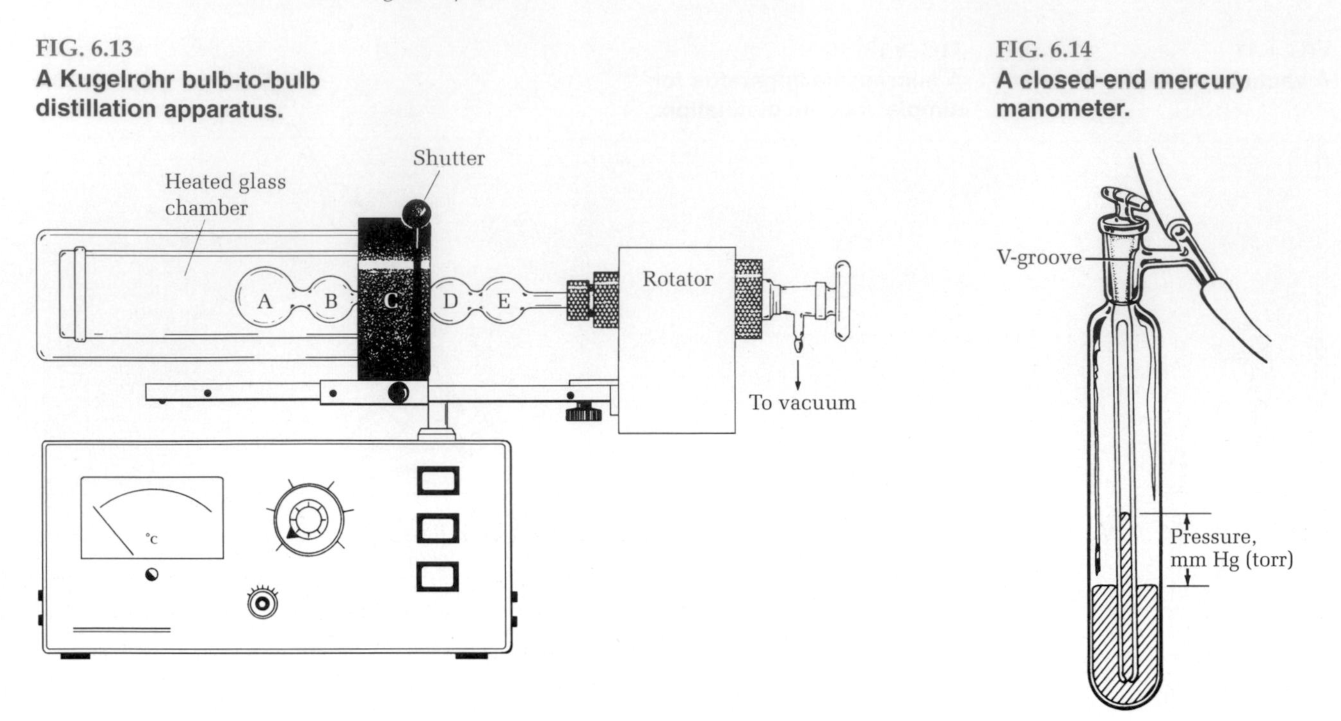

FIG. 6.13
A Kugelrohr bulb-to-bulb distillation apparatus.

FIG. 6.14
A closed-end mercury manometer.

bulb A (Fig. 6.13) in the heated glass chamber. At bulb C is a shutter mechanism that holds in the heat but allows the bulbs to be moved out of the heated chamber one by one as distillation proceeds. The lowest boiling material collects in bulb D; then bulb C is moved out of the heated chamber, the temperature increased, and the next fraction is collected in bulb C. Finally, this process is repeated for bulb B. The bulbs rotate under vacuum via a mechanism similar to the one used on rotary evaporators. The same apparatus is used for bulb-to-bulb fractional sublimation (see Part 3 for sublimation).

MANOMETERS

The pressure of the system is measured with a closed-end mercury manometer. The manometer (Fig. 6.14) is connected to the system by turning the stopcock until its V-groove is aligned with the side arm. To avoid contamination of the manometer, it should be connected to the system only when a reading is being made. The pressure (in mm Hg) is given by the height (in mm) of the central mercury column above the reservoir of mercury, and represents the difference in pressure between the nearly perfect vacuum in the center tube (closed at the top, open at the bottom) and the large volume of the manometer, which is at the pressure of the system.

Another common type of closed-end mercury manometer is shown in Figure 6.15. At atmospheric pressure the manometer is completely full of mercury, as shown in Figure 6.15a. When the open end is connected to a vacuum, the pressure is measured as the distance in millimeters between the mercury in the right and left legs. Note that the pressure in the left leg is essentially zero (mercury has a low vapor pressure). If the applied vacuum were perfect (0 mm Hg), then the two columns would have the same height.

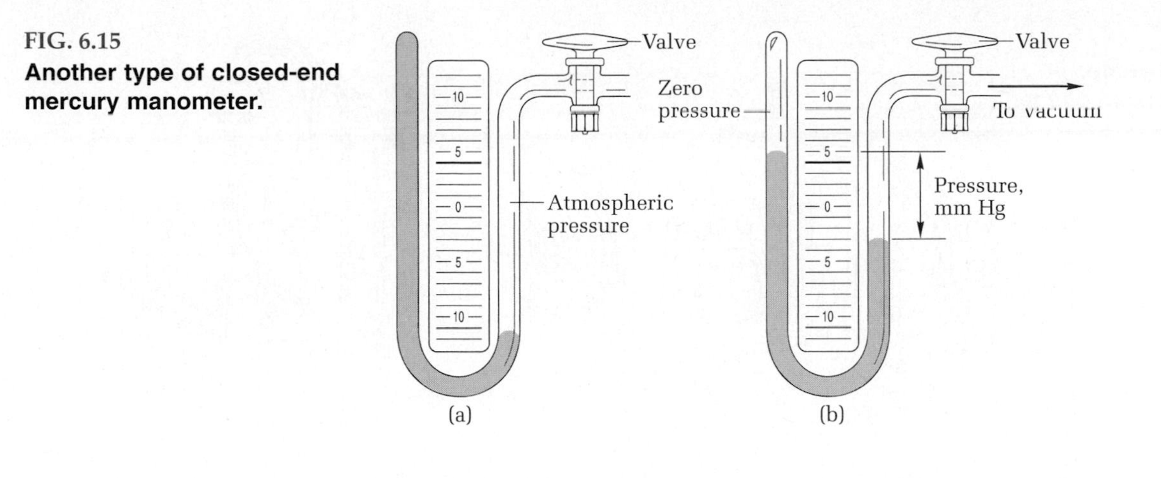

FIG. 6.15
Another type of closed-end mercury manometer.

THE WATER ASPIRATOR IN VACUUM DISTILLATION

A water aspirator in good order produces a vacuum nearly corresponding to the vapor pressure of the water flowing through it. Polypropylene aspirators perform well and are not subject to corrosion, which does affect brass aspirators. If a manometer is not available, the assembly is free of leaks, and the trap and lines are clean and dry, an approximate estimate of the pressure can be made by measuring the water temperature and reading the pressure from Table 6.1 (on page 104).

THE ROTARY OIL PUMP

To obtain pressures below 10 mm Hg, a mechanical vacuum pump of the type illustrated in Figure 6.16a is used. A pump of this type in good condition can attain pressures as low as 0.1 mm Hg. These low pressures are measured with a tilting-type McLeod gauge (Fig. 6.16b). When a reading is being made, the gauge is tilted to the vertical position shown, and the pressure is read as the difference between the heights of the two columns of mercury. Between readings, the gauge is rotated clockwise 90°.

⚠
The Dewar flask should be wrapped with tape or otherwise protected from implosion.

Never use a mechanical vacuum pump before placing a mixture of dry ice and isopropyl alcohol in a Dewar flask (Fig. 6.16c) around the trap, and never pump corrosive vapors (e.g., hydrogen chloride gas) into the pump. Should this happen, change the pump oil immediately. With care, the pump will give many years of good service. The dry-ice trap condenses organic vapors and water vapor, both of which would otherwise contaminate the vacuum pump oil and exert enough vapor pressure to destroy a good vacuum.

For an exceedingly high vacuum (5×10^{-8} mm Hg), a high-speed, three-stage mercury diffusion pump is used (Fig. 6.17).

RELATIONSHIP BETWEEN BOILING POINT AND PRESSURE

It is not possible to calculate the boiling point of a substance at some reduced pressure from a knowledge of the boiling temperature at 760 mm Hg because the relationship between boiling point and pressure varies from compound to compound

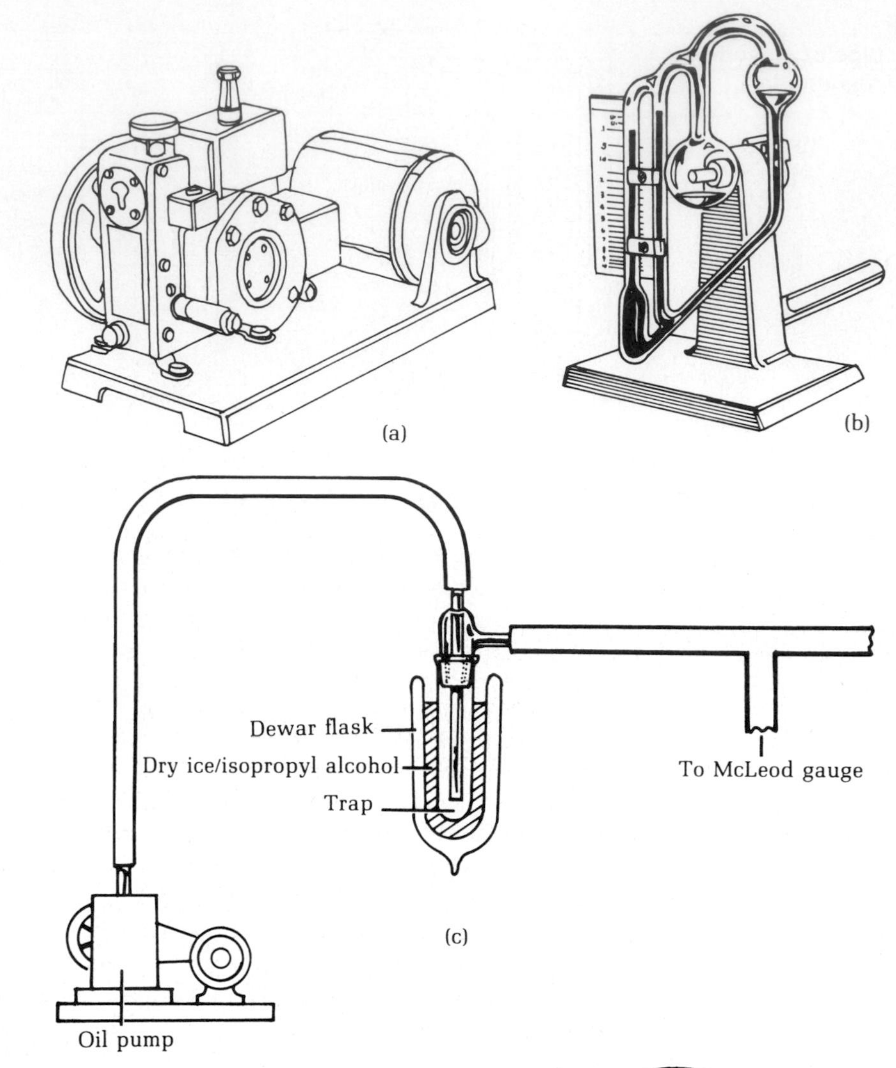

FIG. 6.16
(a) A rotary oil pump.
(b) A tilting McLeod gauge.
(c) A vacuum system.

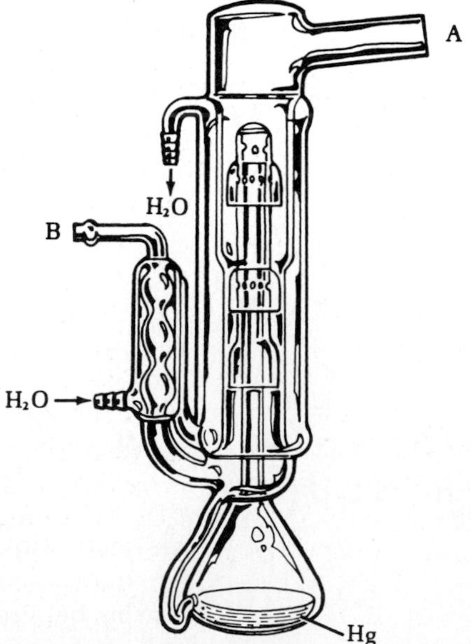

FIG. 6.17
A high-speed, three-stage mercury diffusion pump that is capable of producing a vacuum of 5×10^{-8} mm Hg. Mercury is boiled in the flask; the vapor rises in the center tube, is deflected downward in the inverted cups and entrains gas molecules, which diffuse in from the space to be evacuated, A. The mercury condenses to a liquid and is returned to the flask; the gas molecules are removed at B by an ordinary rotary vacuum pump.

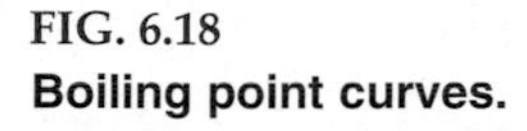

FIG. 6.18
Boiling point curves.

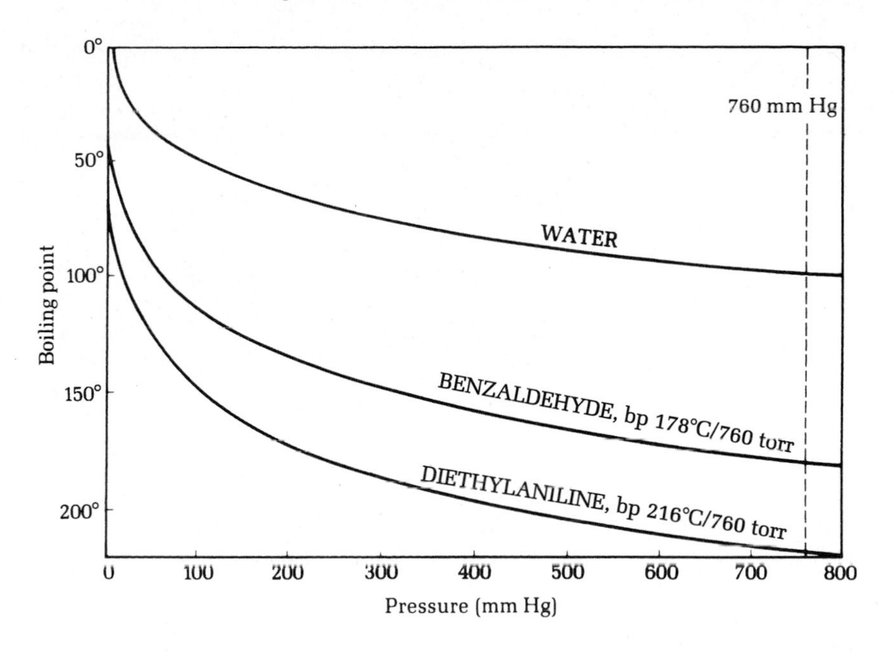

and is somewhat unpredictable. It is true, however, that boiling point curves for organic substances have much the same general disposition, as illustrated by the two lower curves in Figure 6.18. These are similar and do not differ greatly from the curve for water. For substances boiling in the range of 150°C to 250°C at 760 mm Hg, the boiling point at 20 mm Hg is 100–120°C lower than that at 760 mm Hg. Benzaldehyde, which is very sensitive to air oxidation at the normal boiling point of 178°C, distills at 76°C at 20 mm Hg, and the concentration of oxygen in the rarefied atmosphere is just 3% of what it would be in an ordinary distillation.

The curves all show a sharp upward inclination in the region of very low pressure. The lowering of the boiling point that accompanies a reduction in pressure is far more pronounced at low pressures rather than at high pressures. A drop in atmospheric pressure of 10 mm Hg lowers the normal boiling point of an ordinary liquid by less than 1°C, but a reduction of pressure from 20 mm Hg to 10 mm Hg causes a drop of about 15°C in the boiling point. The effect at pressures below 1 mm is still more striking, and with the development of highly efficient oil vapor or mercury vapor diffusion pumps, distillation at a pressure of a few thousandths or ten-thousandths of a millimeter has become a standard operation in many research laboratories. High-vacuum distillation—that is, at a pressure below 1 mm Hg—affords a useful means of purifying extremely sensitive or very slightly volatile substances. Table 6.2 indicates the order of magnitude of the reduction in boiling point attainable by operating at various pressures with different methods, and illustrates the importance of keeping vacuum pumps in good repair.

The boiling point of a substance at varying pressures can be estimated from a pressure-temperature nomograph, such as the one shown in Figure 6.19. If the boiling point of a nonpolar substance at 760 mm Hg is known, for example, 145°C (column B), and the new pressure measured, for example, 28 mm Hg (column C), then a straight line connecting these values on columns B and C when extended intersects column A to give the observed boiling point of 50°C. Conversely, a substance observed to boil at 50°C (column A) at 1.0 mm Hg (column C) will boil at approximately 212°C at atmospheric pressure (column B). If the material being

TABLE 6.2 *Distillation of a (Hypothetical) Substance at Various Pressures*

Method		*Pressure (mm Hg)*	*bp (°C)*
Ordinary distillation		760	250
Aspirator	summer	25	144
	winter	15	134
	Poor condition	10	124
Rotary oil pump	Good condition	3	99
	excellent condition	1	89
Mercury vapor pump (high vacuum distillation)		0.01	30

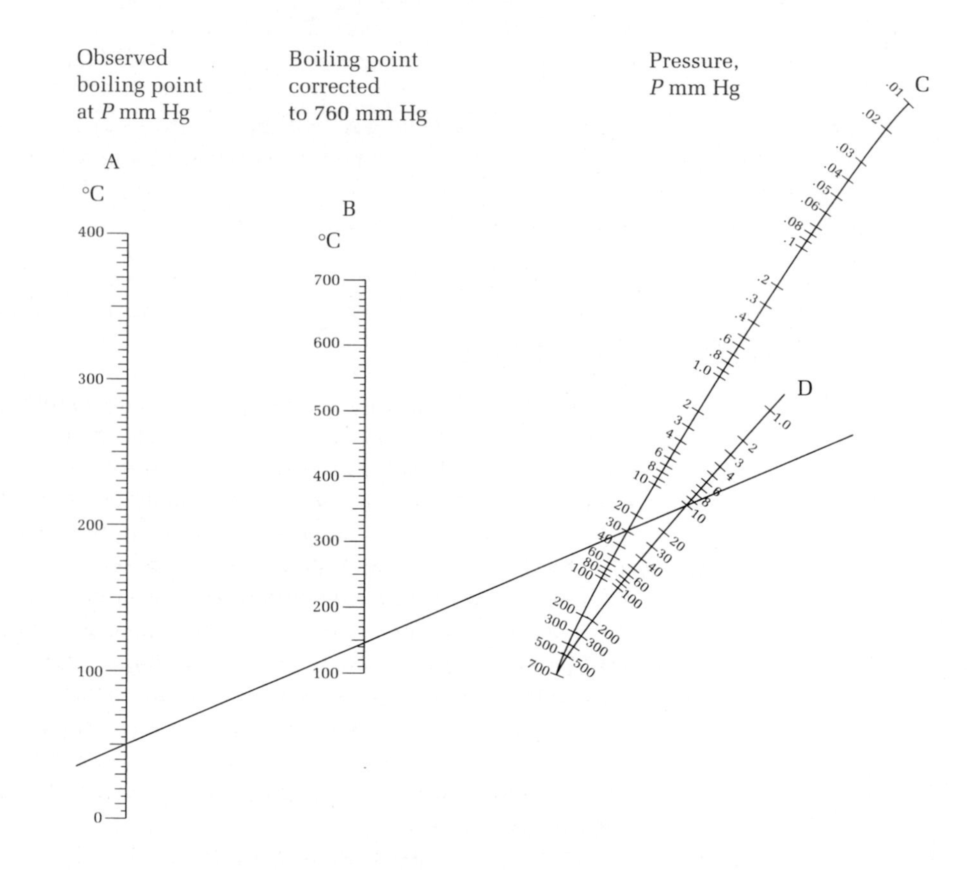

FIG. 6.19
A pressure-temperature nomograph.

distilled is polar, such as a carboxylic acid, column D is used. Thus, a substance that boils at 145°C at atmospheric pressure (column B) will boil at 28 mm Hg (column C) if it is nonpolar, and at a pressure of 10 mm Hg (column D) if it is polar. The boiling point will be 50°C (column A).

Vacuum distillation is not confined to the purification of substances that are liquid at ordinary temperatures; it can often be used to advantage for solid substances. This operation is conducted for a different purpose and by a different technique. A solid is seldom distilled in order to separate constituents of differing degrees of volatility; distillation is intended to purify the solid. Often, one vacuum distillation can remove foreign coloring matter and tar without any appreciable loss of product, whereas several wasteful recrystallizations might be required to

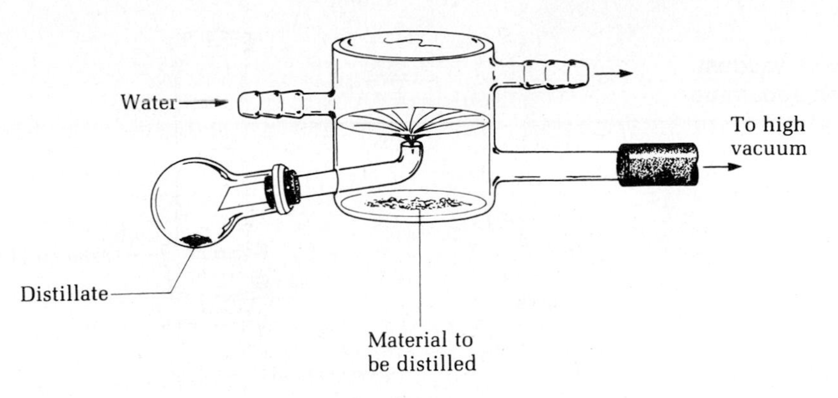

FIG. 6.20
A molecular distillation apparatus.

attain the same purity. It is often good practice to distill a crude product and then to recrystallize it. Time is saved in the recrystallization process because the hot solution usually requires neither filtration nor clarification. The solid must be dry, and a test should be made to determine if it will distill without decomposition at the pressure of the available pump. A compound lacking the required stability at high temperatures is sometimes indicated by its structure, but a high melting point should not be taken as an indication that distillation will fail. Substances melting as high as 300°C have been distilled with success at the pressure of an ordinary rotary vacuum pump.

MOLECULAR DISTILLATION

At 1×10^{-3} mm Hg, the mean free path of nitrogen molecules is 56 mm at 25°C. This means that an average N_2 molecule can travel 56 mm before bumping into another N_2 molecule. Specially designed apparatus makes it possible to distill almost any volatile molecule by operating at a very low pressure and having the condensing surface close to the material being distilled. The thermally energized molecule escapes from the liquid, moves about 10 mm without encountering any other molecules, and condenses on a cold surface. A simple apparatus for molecular distillation is illustrated in Figure 6.20. The bottom of the apparatus is placed on a hot surface, and the distillate moves a very short distance before condensing. If it is not too viscous, it will drip from the point on the glass condensing surface and run down to the receiver.

The efficiency of this molecular distillation apparatus is less than one theoretical plate (*see* Chapter 5). It is used to remove very volatile substances such as solvents and to separate volatile, high-boiling substances at low temperatures from nonvolatile impurities.

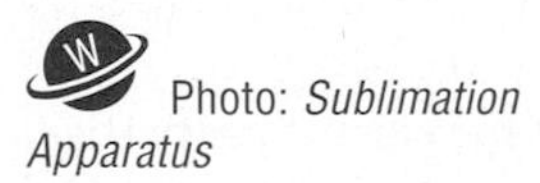
Photo: *Sublimation Apparatus*

In more sophisticated apparatus of quite different design, the liquid to be distilled falls down on a heated surface, which is again close to the condenser. This movement of a film prevents a buildup of nonvolatile material on the surface of the material to be distilled, which would cause the distillation to cease. Vitamin A is distilled commercially in this manner.

PART 3: SUBLIMATION

Sublimation is the process whereby a solid evaporates from a warm surface and condenses on a cold surface, again as a solid (Fig. 6.21 and Fig. 6.22). This technique is particularly useful for the small-scale purification of solids because there is

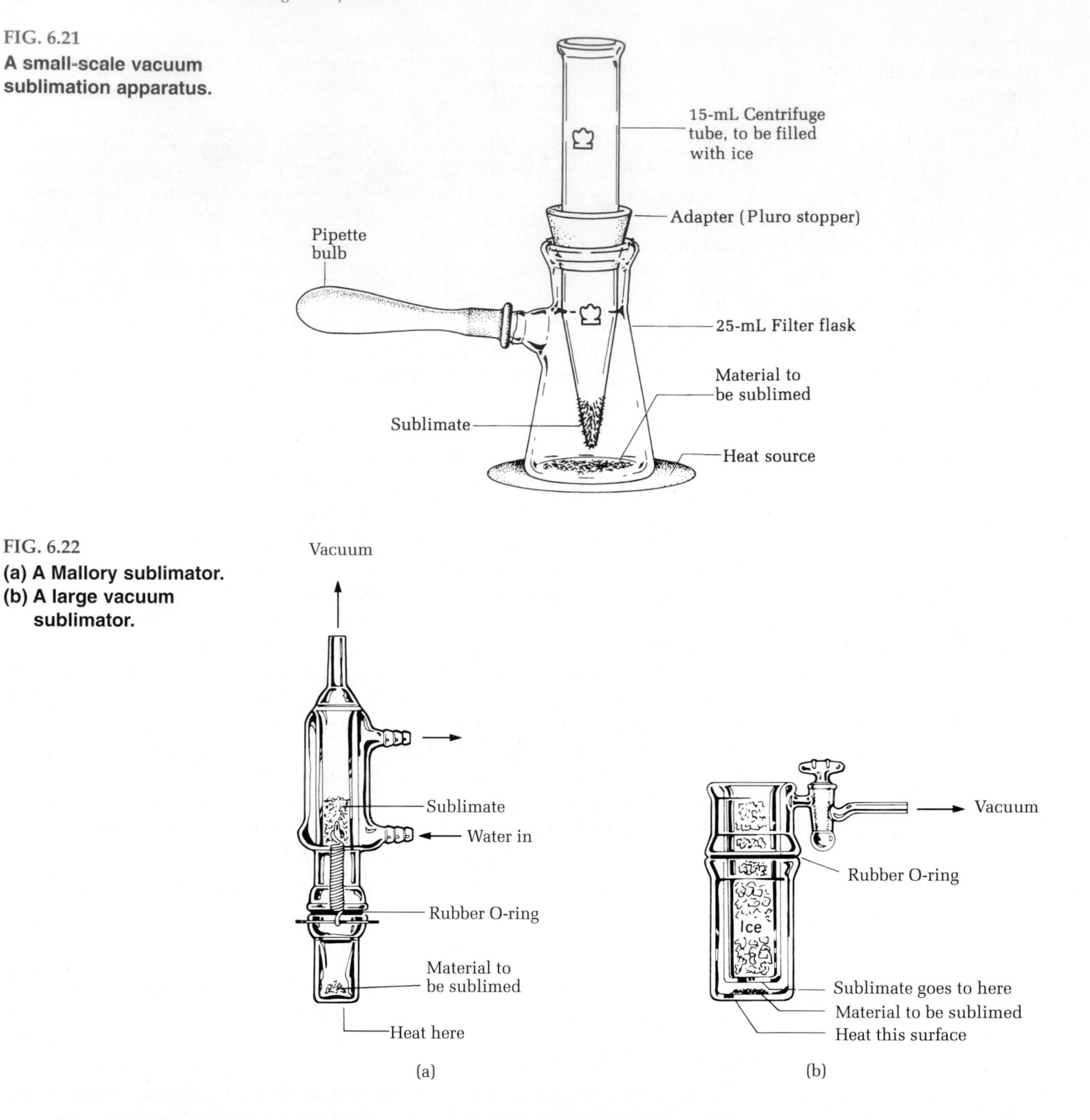

FIG. 6.21
A small-scale vacuum sublimation apparatus.

FIG. 6.22
(a) A Mallory sublimator.
(b) A large vacuum sublimator.

so little loss of material in transfer. If the substance possesses the correct properties, sublimation is preferred over recrystallization when the amount of material to be purified weighs less than 100 mg.

As demonstrated in the experiment that follows, sublimation can occur readily at atmospheric pressure. For substances with lower vapor pressures, vacuum sublimation is used. At very low pressures, sublimation becomes very similar to molecular distillation, where the molecule leaves the warm solid and passes unobstructed to a cold condensing surface and condenses in the form of a solid.

Because sublimation occurs from the surface of the warm solid, impurities can accumulate and slow down or even stop the sublimation, in which case it is necessary to open the apparatus, grind the impure solid to a fine powder, and restart the sublimation.

Sublimation is much easier to carry out on a small scale than on a large one. It would be unusual for a research chemist to sublime more than about 10 g of material because the sublimate tends to fall back to the bottom of the apparatus. In an apparatus such as that illustrated in Figure 6.22a, the solid to be sublimed is placed in the lower flask and connected via a lubricant-free rubber O-ring to the condenser, which in turn is connected to a vacuum pump. The lower flask is immersed in an oil bath at the appropriate temperature, and the product is sublimed and condensed onto the cool walls of the condenser. The parts of the apparatus are gently separated, and the condenser is inverted; the vacuum connection serves as a convenient funnel for product removal. For large-scale work, the sublimator in Figure 6.22b is used. The inner well is filled with a coolant (ice or dry ice). The sublimate clings to this cool surface, from which it can be removed by scraping and dissolving in an appropriate solvent.

A very simple sublimation apparatus: an inverted stemless funnel on a watch glass.

LYOPHILIZATION

Lyophilization, also called freeze-drying, is the process of subliming a solvent, usually water, with the object of recovering the solid that remains after the solvent is removed. This technique is used extensively to recover heat- and oxygen-sensitive substances of natural origin, such as proteins, vitamins, nucleic acids, and other biochemicals, from a dilute aqueous solution. The aqueous solution of the substance to be lyophilized is usually frozen to prevent bumping and then subjected to a vacuum of about 1×10^{-3} mm Hg. The water sublimes and condenses as ice on the surface of a large and very cold condenser. The sample remains frozen during this entire process, without any external cooling being supplied, because of the very high heat of vaporization of water; thus the temperature of the sample never exceeds 0°C. Freeze-drying on a large scale is employed to make "instant" coffee, tea, soup, and rice and all sorts of dehydrated foods.

EXPERIMENT

SUBLIMATION OF AN UNKNOWN SUBSTANCE

Apparatus and Technique

The apparatus consists of a 15-mL centrifuge tube thrust through an adapter (use a drop of glycerol to lubricate the adapter) fitted in a 25-mL filter flask. Once the adapter is properly positioned on the centrifuge tube, it should remain there permanently. This tube is used only for sublimation.

The sample is either ground to a fine powder and uniformly distributed on the bottom of the flask or introduced into the flask as a solution, and then the solvent evaporated to deposit the substance on the bottom of the filter flask. If the compound sublimes easily, care must be exercised using the latter technique to ensure that the sample does not evaporate as the last of the solvent is being removed.

TABLE 6.3 *Sublimation Unknowns*

Substance	*mp (°C)*	*Substance*	*mp (°C)*
1,4-Dichlorobenzene	55	Benzoic acid	122
Naphthalene	82	Salicylic acid	159
1-Naphthol	96	Camphor	177
Acetanilide	114	Caffeine	235

The 15-mL centrifuge tube and adapter are placed in the flask so that the tip of the centrifuge tube is about 3–8 mm above the bottom of the flask (*see* Fig. 6.21). The flask is clamped with a three-prong clamp. Many substances sublime at atmospheric pressure, including the unknowns listed in Table 6.3. For vacuum sublimation, the side arm of the flask leads to an aspirator or vacuum pump for reduced pressure sublimation. The centrifuge tube is filled with ice and water. The ice is not added before closing the apparatus, so atmospheric moisture will not condense on the tube.

Sublimation: Add sample to flask, close flask openings, add ice, sublime sample, remove ice, add room-temperature water, open apparatus, and scrape off product.

The filter flask is warmed cautiously on a hot sand bath until the product just begins to sublime. The heat is maintained at that temperature until sublimation is complete. Because some product will collect on the cool upper parts of the flask, it should be fitted with a loose cone of aluminum foil to direct heat there and cause the material to collect on the surface of the centrifuge tube. Alternatively, tilt the flask and rotate it in the hot sand, or remove the flask from the sand bath, and then use a heat gun to warm the flask.

Roll the filter flask in the sand bath or heat it with a heat gun to drive material from the upper walls of the flask.

Once sublimation is judged complete, the ice water is removed from the centrifuge tube with a pipette and replaced with water at room temperature. This will prevent moisture from condensing on the product once the vacuum is turned off and the tube is removed from the flask. The product is scraped from the centrifuge tube with a metal spatula onto a piece of glazed paper. It is much easier to scrape the product from a centrifuge tube than from a round-bottomed test tube. The last traces can be removed by washing off the tube with a few drops of an appropriate solvent, if that is deemed necessary.

Procedure

Into the bottom of a 25-mL filter flask, place 50 mg of an impure unknown taken from the list in Table 6.3. These substances can be sublimed at atmospheric pressure, although some will sublime more rapidly at reduced pressure. Slip a rubber pipette bulb over the flask's side arm to close off the flask, and then place ice water in the centrifuge tube. Cautiously warm the flask until sublimation starts, and then maintain that temperature throughout the sublimation. Fit the flask with a loose cone of aluminum foil to direct heat up the sides of the flask, which causes the product to collect on the tube. Once sublimation is complete, remove the ice water from the centrifuge tube and replace it with water at room temperature. Collect the product, determine its weight and the percent recovery, and from the melting point identify the unknown. Hand in the product in a neatly labeled vial.

Cleaning Up. Wash material from the apparatus with a minimum quantity of acetone, which, except for the chloro compound, can be placed in the organic solvents waste container. The 1,4-dichlorobenzene solution must be placed in the halogenated organic waste container.

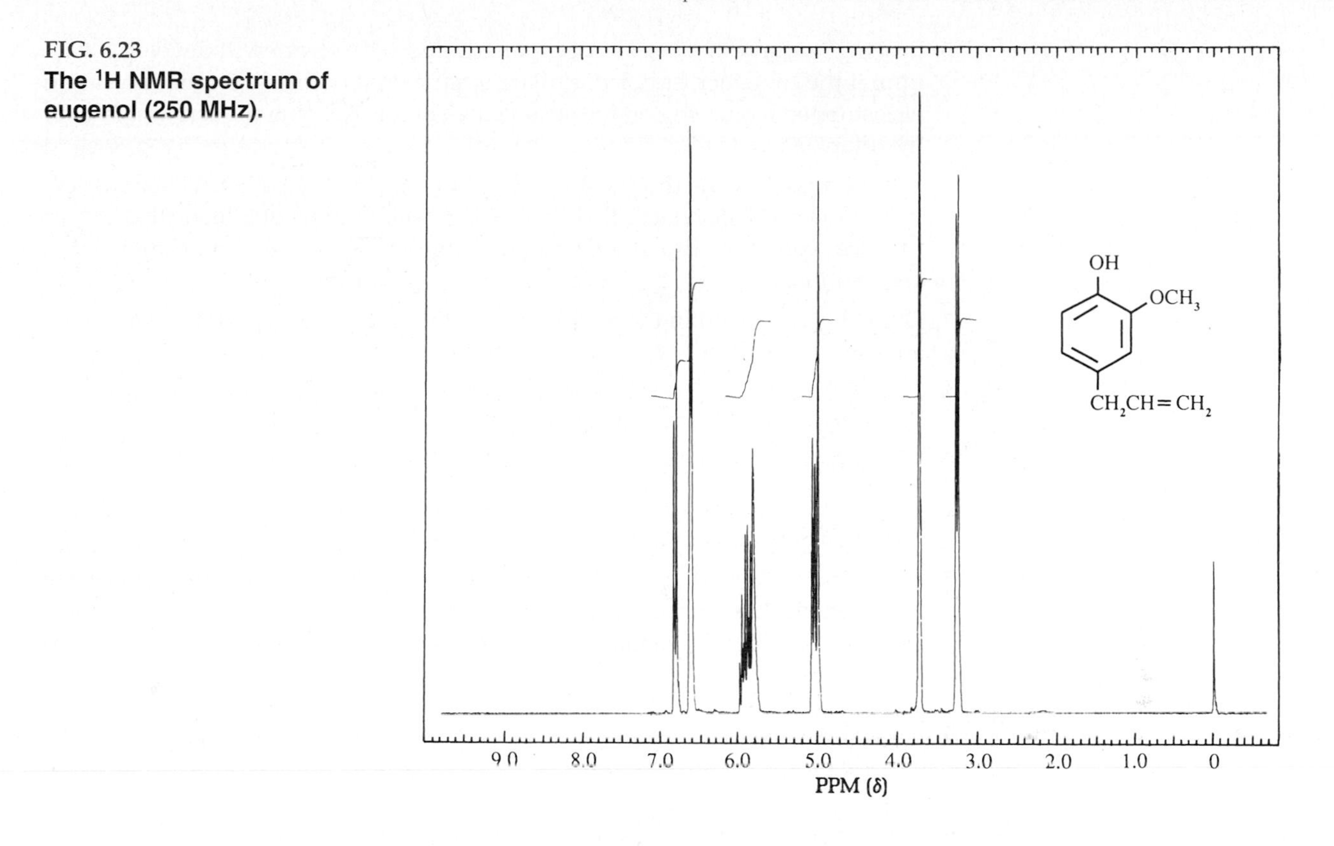

FIG. 6.23
The ^{1}H NMR spectrum of eugenol (250 MHz).

FOR FURTHER INVESTIGATION

Room deodorants, for example, the Renuzit Long Last adjustable air freshener, appear to sublime. Place about 150 mg, carefully weighed, of the green solid in the filter flask and sublime it with very little heat applied to the flask. Identify the substance that sublimes. Weigh the residue. As an advanced experiment, run the IR spectrum (see Chapter 11) of the residue (not of the substance that sublimes). Comment on the economics of air fresheners.

QUESTIONS

1. Assign the peaks in the ^{1}H NMR spectrum of eugenol (Fig. 6.23) to specific protons in the molecule. The OH peak is at 5.1 ppm.
2. A mixture of ethyl iodide (C_2H_5I, bp 72.3°C) and water boils at 63.7°C. What weight of ethyl iodide would be carried over by 1 g of steam during steam distillation?
3. Iodobenzene (C_6H_5I, bp 188°C) steam distills at 98.2°C. How many molecules of water are required to carry over 1 molecule of iodobenzene? How many grams of water per gram of iodobenzene?
4. The condensate from a steam distillation contains 8 g of an unknown compound and 18 g of water. The mixture steam distilled at 98°C. What is the molecular weight of the unknown?

5. Bearing in mind that a sealed container filled with steam will develop a vacuum if the container is cooled, explain what would occur if a steam distillation was stopped by turning off the steam valve before opening the screw clamp on the adapter trap.
6. A mixture of toluene (bp 110.8°C) and water is steam distilled. Visual inspection of the distillate reveals that there is a greater volume of toluene than water present, yet water (bp 100°C) has the higher vapor pressure. Explain this observation.
7. Did sublimation aid in the purification of the unknown in part 3? Justify your answer.
8. From your experience and knowledge of the process of sublimation, cite some examples of substances that sublime.
9. How might freeze-dried foods be prepared?
10. Describe the mercury level in a manometer like that shown in Figure 6.15a if the left arm is 1000 cm tall.
11. What criteria would you use to decide whether distillation is an appropriate technique for the purification of a reaction product? What type of distillation method—simple, fractional, vacuum, or steam—should be used?

See *Web Links*

CHAPTER 7

Extraction

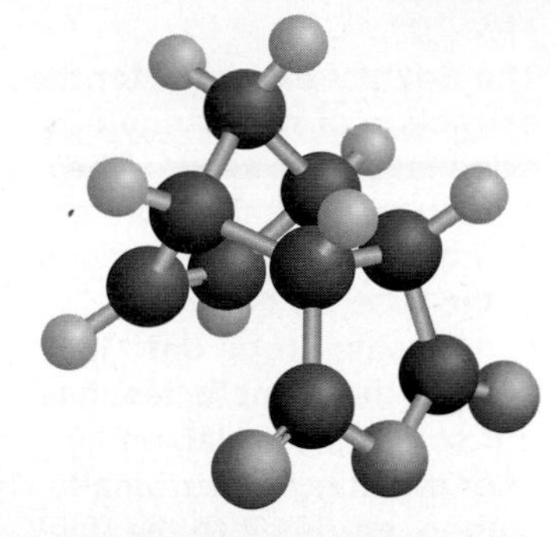

PRELAB EXERCISE: Describe how to separate a mixture of 3-toluic acid and 4′-aminoacetophenone using acid-base liquid/liquid extraction. What species will end up in the aqueous layer if you mix a solution of benzoic acid and aniline in ether with a solution of $NaHCO_3$ (aq)? Draw the structure of this species.

Extraction is one of the oldest chemical operations known to humankind. The preparation of a cup of coffee or tea involves the extraction of flavor and odor components from dried vegetable matter with hot water. Aqueous extracts of bay leaves, stick cinnamon, peppercorns, and cloves, along with alcoholic extracts of vanilla and almond, are used as food flavorings. For the past 150 years or so, organic chemists have extracted, isolated, purified, and characterized the myriad compounds produced by plants that for centuries have been used as drugs and perfumes—substances such as quinine from cinchona bark, morphine from the opium poppy, cocaine from coca leaves, and menthol from peppermint oil. The extraction of compounds from these natural products is an example of solid/liquid extraction—the solid being the natural product and the liquid being the solvent into which the compounds are extracted. In research, a Soxhlet extractor (Fig. 7.1) is often used for solid/liquid extraction.

Although solid/liquid extraction is the most common technique for brewing beverages and isolating compounds from natural products, liquid/liquid extraction is a very common method used in the organic laboratory, specifically when isolating reaction products. Reactions are typically homogeneous liquid mixtures and can therefore be extracted with either an organic or aqueous solvent. Organic reactions often yield a number of byproducts—some inorganic and some organic. Also, because some organic reactions do not go to 100% completion, a small amount of starting material is present at the end of the reaction. When a reaction is complete, it is necessary to do a *workup*, that is, separate and purify the desired product from the mixture of byproducts and residual starting material. Liquid/liquid extraction is a common separation step in this workup, which is then followed by purification of the product. There are two types of liquid/liquid extraction: neutral and acid/base. The experiments in this chapter demonstrate solid/liquid extraction and the two types of liquid/liquid extraction.

FIG. 7.1
The Soxhlet extractor for the extraction of solids such as dried leaves or seeds. The solid is put in a filter paper thimble. Solvent vapor rises in the tube on the right; condensate drops onto the solid in the thimble, leaches out soluble material, and, after initiating an automatic siphon, carries it to the flask where nonvolatile extracted material accumulates. Substances of low solubility can be extracted by prolonged operation.

Condenser
Paper thimble
Soxhlet

Organic products are often separated from inorganic substances in a reaction mixture by liquid/liquid extraction with an organic solvent. For example, in the synthesis of 1-bromobutane (*see* Chapter 16), 1-butanol, also a liquid, is heated with an aqueous solution of sodium bromide and sulfuric acid to produce the product and sodium sulfate.

$$2\ CH_3CH_2CH_2CH_2OH + 2\ NaBr + H_2SO_4 \rightarrow 2\ CH_3CH_2CH_2CH_2Br + 2\ H_2O + Na_2SO_4$$

The 1-bromobutane is isolated from the reaction mixture by extraction with *t*-butyl methyl ether, an organic solvent in which 1-bromobutane is soluble and in which water and sodium sulfate are insoluble. The extraction is accomplished by simply adding *t*-butyl methyl ether to the aqueous mixture and shaking it. Two layers will result: an organic layer and an aqueous layer. The *t*-butyl methyl ether is less dense than water and floats on top; it is easily removed/drained away from the water layer and evaporated to leave the bromo product free of inorganic substances, which reside in the aqueous layer.

PARTITION COEFFICIENT

The extraction of a compound such as 1-butanol, which is slightly soluble in water as well as very soluble in ether, is an equilibrium process governed by the solubilities of the alcohol in the two solvents. The ratio of the solubilities is known as the *distribution coefficient*, also called the *partition coefficient* (*k*), and is an equilibrium constant with a certain value for a given substance, pair of solvents, and temperature.

The *concentration* of the solute in each solvent can be well correlated with the *solubility* of the solute in the pure solvent, a figure that is readily found in solubility tables in reference books. For substance C:

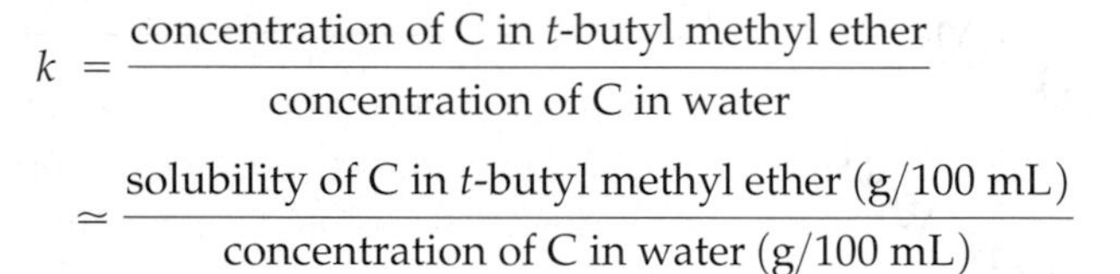

$$k = \frac{\text{concentration of C in } t\text{-butyl methyl ether}}{\text{concentration of C in water}}$$

$$\simeq \frac{\text{solubility of C in } t\text{-butyl methyl ether (g/100 mL)}}{\text{concentration of C in water (g/100 mL)}}$$

Consider compound A that dissolves in *t*-butyl methyl ether to the extent of 12 g/100 mL and dissolves in water to the extent of 6 g/100 mL.

$$k = \frac{\text{12g/100mL } t\text{-butyl methyl ether (g/100 mL)}}{\text{6g/100mL water}} = 2$$

If a solution of 6 g of A in 100 mL of water is shaken with 100 mL of *t*-butyl methyl ether, then:

$$k = \frac{x\text{g of A/100mL } t\text{-butyl methyl ether}}{6 - x\text{g of A/100mL water}}$$

from which

$$x = 4.0 \text{ g of A in the ether layer}$$

$$6 - x = 2.0 \text{ g of A left in the water layer.}$$

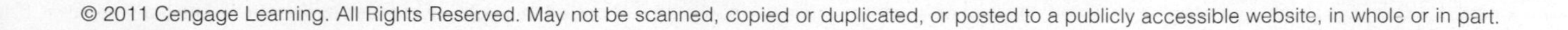

It is, however, more efficient to extract the 100 mL of aqueous solution twice with 50-mL portions of *t*-butyl methyl ether rather than once with a 100-mL portion.

$$k = \frac{x\text{g of A}/50\text{mL}}{6 - x\text{g of A}/100\text{mL}} = 2$$

from which

$$x = 3.0 \text{ g of A in the } t\text{-butyl methyl ether layer}$$
$$6 - x = 3.0 \text{ g of A in the water layer.}$$

If this 3.0 g/100 mL of water is extracted again with 50 mL of *t*-butyl methyl ether, we can calculate that 1.5 g of A will be in the ether layer, leaving 1.5 g in the water layer. So two extractions with 50-mL portions of ether will extract 3.0 g + 1.5 g = 4.5 g of A, whereas one extraction with a 100-mL portion of *t*-butyl methyl ether removes only 4.0 g of A. Three extractions with 33-mL portions of *t*-butyl methyl ether would extract 4.7 g. Obviously, there is a point at which the increased amount of A extracted does not repay the effort of multiple extractions, but remember that several small-scale extractions are more effective than one large-scale extraction.

PROPERTIES OF EXTRACTION SOLVENTS

Liquid/liquid extraction involves two layers: the organic layer and the aqueous layer. The solvent used for extraction should possess many properties, including the following:

- It should readily dissolve the substance to be extracted at room temperature.
- It should have a low boiling point so that it can be removed readily.
- It should not react with the solute or the other solvent.
- It should not be highly flammable or toxic.
- It should be relatively inexpensive.

In addition, it should not be miscible with water (the usual second phase). No solvent meets every criterion, but several come close. Some common liquid/liquid extraction solvent pairs are water-ether, water-dichloromethane, and water-hexane. Notice that each combination includes water because most organic compounds are immiscible in water, and therefore can be separated from inorganic compounds. Organic solvents such as methanol and ethanol are not good extraction solvents because they are soluble in water.

Identifying the Layers

One common mistake when performing an extraction is to misidentify the layers and discard the wrong one. It is good practice to save all layers until the desired product is in hand. The densities of the solvents will predict the identities of the top and bottom layers. In general, the densities of nonhalogenated organic solvents are less than 1.0 g/mL and those of halogenated solvents are greater than 1.0 g/mL. Table 7.1 lists the densities of common solvents used in extraction.

Although density is the physical property that determines which layer is on top or on bottom, a very concentrated amount of solute dissolved in either layer can reverse the order. The best method to avoid a misidentification is to perform a drop test. Add a few drops of water to the layer in question and watch the drop

TABLE 7.1 *Common Solvents Listed by Density*

Solvent	*Density (g/mL)*
Hexane	0.695
Diethyl ether	0.708
t-Butyl methyl ether	0.740
Toluene	0.867
Water	1.000
Dichloromethane	1.325
Chloroform	1.492

Identify layers by a drop test.

carefully. If the layer is water, then the drop will mix with the solution. If the solvent is the organic layer, then the water drop will create a second layer.

Ethereal Extraction Solvents

$CH_3CH_2—O—CH_2CH_3$

Diethyl ether
"Ether"
MW 74.12, den. 0.708
bp 34.6°C, n_D^{20} 1.3530

In the past, diethyl ether was the most common solvent for extraction in the laboratory. It has high solvent power for hydrocarbons and oxygen-containing compounds. It is highly volatile (bp 34.6°C) and is therefore easily removed from an extract. However, diethyl ether has two big disadvantages: it is highly flammable and poses a great fire threat, and it easily forms peroxides. The reaction of diethyl ether with air is catalyzed by light. The resulting peroxides are higher boiling than the ether and are left as a residue when the ether evaporates. If the residue is heated, it will explode because ether peroxides are volatile and treacherously explosive. In recent years, a new solvent has come on the scene—*tert*-butyl methyl ether.

$H_3C—C(CH_3)_2—O—CH_3$

***tert*-Butyl methyl ether**
MW 88.14, den. 0.741
bp 55.2°C, n_D^{20} 1.369

tert-Butyl methyl ether, called methyl *tert*-butyl ether (MTBE) in industry, has many advantages over diethyl ether as an extraction solvent. Most important, it does not easily form peroxides, so it can be stored for much longer periods than diethyl ether. And, in the United States, it is less than half the price of diethyl ether. It is slightly less volatile (bp 55°C), so it does not pose the same fire threat as diethyl ether, although one must be as careful in handling this solvent as in handling any other highly volatile, flammable substance. The explosion limits for *t*-butyl methyl ether mixed with air are much narrower than for diethyl ether, the toxicity is less, the solvent power is the same, and the ignition temperature is higher (224°C versus 180°C).

The weight percent solubility of diethyl ether dissolved in water is 7.2%, whereas that of *t*-butyl methyl ether is 4.8%. The solubility of water in diethyl ether is 1.2%, while in *t*-butyl methyl ether it is 1.5%. Unlike diethyl ether, *t*-butyl methyl ether forms an azeotrope with water (4% water) that boils at 52.6°C. This means that evaporation of any *t*-butyl methyl ether solution that is saturated with water should leave no water residue, unlike diethyl ether.

The low price and ready availability of *t*-butyl methyl ether came about because it replaced tetraethyl lead as the antiknock additive for high-octane gasoline and as a fuel oxygenate, which helps reduce air pollution, but its water solubility has allowed it to contaminate drinking water supplies in states where leaking underground fuel storage tanks are not well regulated. Consequently, it has been replaced with the much more expensive ethanol. In this text, *t*-butyl methyl ether is *strongly* suggested wherever diethyl ether formerly would have been used in an

FIG. 7.2
A separatory funnel with a Teflon stopcock.

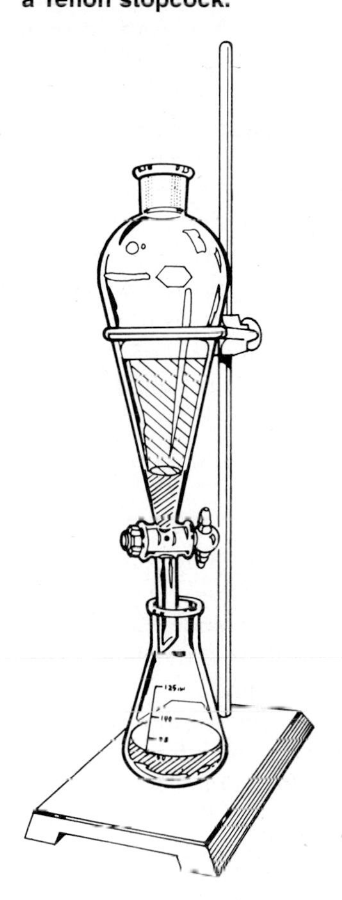

extraction. It will not, however, work as the only solvent in the Grignard reaction, probably because of steric hindrance. So whenever the word *ether* appears in this text as an extraction solvent, it is suggested that *t*-butyl methyl ether be used and not diethyl ether.

MIXING AND SEPARATING THE LAYERS

For microscale separations, mixing and separating the layers with a pipette normally incurs very little product loss. Because the two solvents are typically in a reaction tube for microscale extraction, the two layers can be mixed by drawing up and rapidly expelling them with a pipette. Then the layers are allowed to separate, and the bottom layer is separated by drawing it up into a pipette and transferring it to a different container.

For macroscale separations, a separatory funnel (Fig. 7.2) is used to mix and separate the organic and aqueous layers. In macroscale experiments, a frequently used method of working up a reaction mixture is to dilute the mixture with water and extract it with an organic solvent, such as ether, in a separatory funnel. When the stoppered funnel is shaken to distribute the components between the immiscible solvents *t*-butyl methyl ether and water, pressure always develops through volatilization of ether from the heat of the hands, and liberation of a gas (CO_2) (in acid/base extractions) can increase the pressure. Consequently, the funnel is grasped so that the stopper is held in place by one hand and the stopcock by the other, as illustrated in Figure 7.3. After a brief shake or two, the funnel is held in the inverted position shown, and the stopcock is opened cautiously (with the funnel stem pointed away from nearby persons) to release pressure. The mixture can then be shaken more vigorously, with pressure released as necessary. When equilibration is judged to be complete, the slight, constant terminal pressure due to ether is released, the stopper is rinsed with a few drops of ether delivered by a Pasteur pipette, and the layers are allowed to separate. The organic reaction product is distributed wholly or largely into the upper ether layer, whereas inorganic salts, acids or bases pass into the water layer, which can be drawn off. If the reaction was conducted in alcohol or some other water-soluble solvent, the bulk of the solvent is removed in the water layer, and the remainder can be eliminated in two or three washings with 1–2 volumes of water conducted with the techniques used in the first equilibration. The separatory funnel should be supported in a ring stand, as shown in Figure 7.2.

FIG. 7.3
The correct position for holding a separatory funnel when shaking. Point outlet away from yourself and your neighbors.

Before adding a liquid to the separatory funnel, check the stopcock. If it is glass, see that it is properly greased, bearing in mind that too much grease will clog the hole in the stopcock and also contaminate the extract. If the stopcock is Teflon, see that it is adjusted to a tight fit in the bore. Store the separatory funnel with the Teflon stopcock loosened to prevent sticking. Because Teflon has a much larger temperature coefficient of expansion than glass, a stuck stopcock can be loosened by cooling the stopcock in ice or dry ice. Do not store liquids in the separatory funnel; they often leak or cause the stopper or stopcock to freeze. To have sufficient room for mixing the layers, fill the separatory funnel no more than three-fourths full. Withdraw the lower layer from the separatory funnel through the stopcock, and pour the upper layer out through the neck.

All too often, the inexperienced chemist discards the wrong layer when using a separatory funnel. Through incomplete neutralization, a desired component may still remain in the aqueous layer, or the densities of the layers may change. Cautious workers save all layers until the desired product has been isolated.

The organic layer is not always the top layer. If in doubt, perform a drop test by adding a few drops of each to water in a test tube.

Practical Considerations When Mixing Layers

Pressure Buildup

The heat of one's hand or heat from acid/base reactions will cause pressure buildup in an extraction mixture that contains a very volatile solvent such as dichloromethane. The extraction container—whether a test tube or a separatory funnel—must be opened carefully to vent this pressure.

Sodium bicarbonate solution is often used to neutralize acids when carrying out acid/base extractions. The result is the formation of carbon dioxide, which can cause foaming and high pressure buildup. Whenever bicarbonate is used, add it very gradually with thorough mixing and frequent venting of the extraction device. If a large amount of acid is to be neutralized with bicarbonate, the process should be carried out in a beaker.

Emulsions

Imagine trying to extract a soap solution (e.g., a nonfoaming dishwashing detergent) into an organic solvent. After a few shakes with an organic solvent, you would have an absolutely intractable emulsion. An emulsion is a suspension of one liquid as droplets in another. Detergents stabilize emulsions, and so any time a detergent-like molecule is in the material being extracted, there is the danger that emulsions will form. Substances of this type are commonly found in nature, so one must be particularly wary of emulsion formation when creating organic extracts of aqueous plant material, such as caffeine from tea. Emulsions, once formed, can be quite stable. You would be quite surprised to open your refrigerator one morning and see a layer of clarified butter floating on the top of a perfectly clear aqueous solution that had once been milk, but milk is the classic example of an emulsion.

Prevention is the best cure for emulsions. This means shaking the solution to be extracted *very gently* until you see that the two layers will separate readily. If a bit of emulsion forms, it may break simply on standing for a sufficient length of time. Making the aqueous layer highly ionic will help. Add as much sodium chloride as will dissolve and shake the mixture gently. Vacuum filtration sometimes works and, when the organic layer is the lower layer, filtration through silicone-impregnated filter paper is helpful. Centrifugation works very well for breaking emulsions. This is easy on a small scale, but often the equipment is not available for large-scale centrifugation of organic liquids.

Shake gently to avoid emulsions.

DRYING AGENTS

The organic solvents used for extraction dissolve not only the compound being extracted but also water. Evaporation of the solvent then leaves the desired compound contaminated with water. At room temperature, water dissolves 4.8% of *t*-butyl methyl ether by weight, and the ether dissolves 1.5% of water. But ether is virtually insoluble in water saturated with sodium chloride (36.7 g/100 mL). If ether that contains dissolved water is shaken with a saturated aqueous solution of sodium chloride, water will be transferred from the *t*-butyl methyl ether to the aqueous layer. So, strange as it may seem, ethereal extracts are routinely dried by shaking them with an aqueous saturated sodium chloride solution.

Solvents such as dichloromethane do not dissolve nearly as much water and are therefore dried over a chemical drying agent. Many choices of chemical drying agents are available for this purpose, and the choice of which one to use is governed by four factors: (1) the possibility of reaction with the substance being extracted, (2) the speed with which it removes water from the solvent, (3) the efficiency of the process, and (4) the ease of recovery from the drying agent.

Some very good but specialized and reactive drying agents are potassium hydroxide, anhydrous potassium carbonate, sodium metal, calcium hydride, lithium aluminum hydride, and phosphorus pentoxide. Substances that are essentially neutral and unreactive and are widely used as drying agents include anhydrous calcium sulfate (Drierite), magnesium sulfate, molecular sieves, calcium chloride, and sodium sulfate.

Drierite, $CaSO_4$

Drierite, a specially prepared form of calcium sulfate, is a fast and effective drying agent. However, it is difficult to ascertain whether enough has been used. An indicating type of Drierite is impregnated with cobalt chloride, which turns from blue to red when it is saturated with water. This works well when gases are being dried, but it should not be used for liquid extractions because the cobalt chloride dissolves in many protic solvents.

Magnesium sulfate, $MgSO_4$

Magnesium sulfate is also a fast and fairly effective drying agent, but it is so finely powdered that it always requires careful filtration for removal.

Molecular sieves, zeolites

Molecular sieves are sodium alumino-silicates (zeolites) that have well-defined pore sizes. The 4 Å size adsorbs water to the exclusion of almost all organic substances, making them a fast and effective drying agent. Like Drierite, however, it is impossible to ascertain by appearance whether enough has been used. Molecular sieves in the form of 1/16-in. pellets are often used to dry solvents by simply adding them to the container.

Calcium chloride ($CaCl_2$) pellets are the drying agent of choice for small-scale experiments.

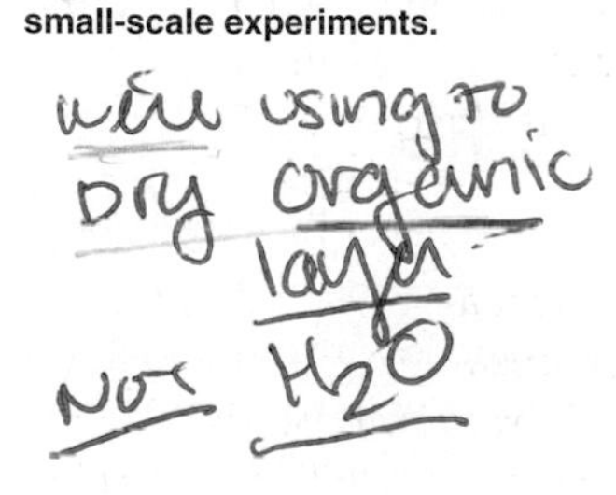

Calcium chloride, recently available in the preferred form of pellets (4 to 80 mesh[1]), is a very fast and effective drying agent. It has the advantage that it clumps together when excess water is present, which makes it possible to know how much to add by observing its behavior. Unlike the older granular form, the pellets do not disintegrate into a fine powder. These pellets are admirably suited to microscale experiments where the solvent is removed from the drying agent with a Pasteur pipette. Calcium chloride is much faster and far more effective than anhydrous sodium sulfate; after much experimentation, we have decided that this is the agent of choice, particularly for microscale experiments. These pellets are used for most of the drying operations in this text. Note, however, that calcium chloride reacts with some alcohols, phenols, amides, and some carbonyl-containing compounds. Advantage is sometimes taken of this property to remove not only water from a solvent but also, for example, a contaminating alcohol (*see* Chapter 16—the synthesis of 1-bromobutane from 1-butanol). Because *t*-butyl methyl ether forms an azeotrope with water, its solutions should, theoretically, not need to be dried because evaporation carries away the water. Drying these ether solutions with calcium chloride pellets removes water droplets that get carried into the ether solution.

Sodium sulfate, Na_2SO_4

Sodium sulfate is a very poor drying agent. It has a very high capacity for water but is slow and not very efficient in the removal of water. Like calcium chloride pellets, it clumps together when wet, and solutions are easily removed from it using a Pasteur pipette. Sodium sulfate has been used extensively in the past and should still be used for compounds that react with calcium chloride.

[1]These drying pellets are available from Fisher Scientific, Cat. No. C614–3.

PART 1: THE TECHNIQUE OF NEUTRAL LIQUID/LIQUID EXTRACTION

The workup technique of liquid/liquid extraction has four steps: (1) mixing the layers, (2) separating the layers, (3) drying the organic layer, and (4) removing the solvent. The microscale neutral liquid/liquid extraction technique is described in the following sections.

STEP 1. MIXING THE LAYERS

FIG. 7.4
Mixing the contents of a reaction tube by flicking it. Grasp the tube firmly at the very top and flick it vigorously at the bottom. The contents will mix without coming out of the tube.

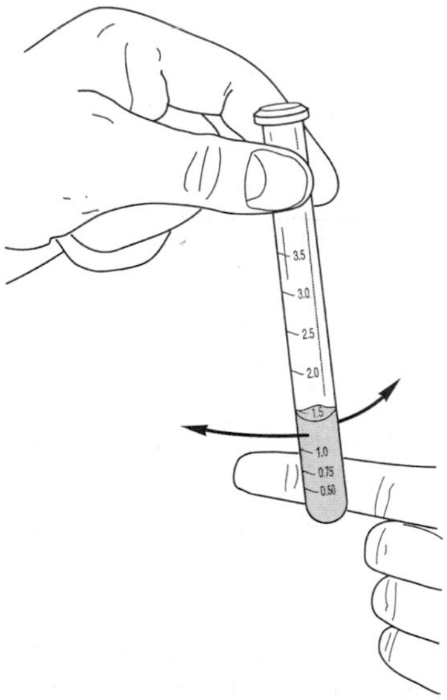

Once the organic and aqueous layers are in contact with one another, mixing is required to ensure that the desired compound(s) get extracted into the desired layer. First, place 1–2 mL of an aqueous solution of the compound to be extracted in a reaction tube. Add about 1 mL of extraction solvent, for example, dichloromethane. Note, as you add the dichloromethane, whether it is the top or the bottom layer. (Since dichloromethane is more dense than water, predict what layer it will be.) An effective way to mix the two layers is to flick the tube with a finger. Grasp the tube firmly at the very top between the thumb and forefinger and flick it vigorously at the bottom (Fig. 7.4). You will find that this violent motion mixes the two layers well, but nothing will spill out the top. Another good mixing technique is to pull the contents of the reaction tube into a Pasteur pipette and then expel the mixture back into the tube with force. Doing this several times will effect good mixing of the two layers. A stopper can be placed in the top of the tube, and the contents can be mixed by shaking the tube, but the problem with this technique is that the high vapor pressure of the solvent will often force liquid out around the cork or stopper.

STEP 2. SEPARATING THE LAYERS

Always draw out the lower layer and place it in another container.

After thoroughly mixing the two layers, allow them to separate. Tap the tube if droplets of one layer are in the other layer, or on the side of the tube. After the layers separate completely, draw up the lower dichloromethane layer into a Pasteur pipette. Leave behind any middle emulsion layer. The easiest way to do this is to attach the pipette to a pipette pump (Fig. 7.5). This allows very precise control of the liquid being removed. It takes more skill and practice to remove the lower layer cleanly with a 2-mL rubber bulb attached to a pipette because the high vapor pressure of the solvent tends to make it dribble out. To avoid losing any of the solution, it is best to hold a clean, dry, empty tube in the same hand as the full tube to receive the organic layer (Fig. 7.6).

From the discussion of the partition coefficient, you know that several small extractions are better than one large one, so repeat the extraction process with two further 1-mL portions of dichloromethane. An experienced chemist might summarize all the preceding with the following notebook entry, "Aqueous layer extracted 3 × 1-mL portions CH_2Cl_2," and in a formal report would write, "The aqueous layer was extracted three times with 1-mL portions of dichloromethane."

If you are working on a larger microscale, a microscale separatory funnel (Fig. 7.7) should be used. A separatory funnel, regardless of size, should be filled to only about two-thirds of its capacity so the layers can be mixed by shaking. The microscale separatory funnel has a capacity of 8.5 mL when full, so it is useful for an extraction with a total volume of about 6 mL.

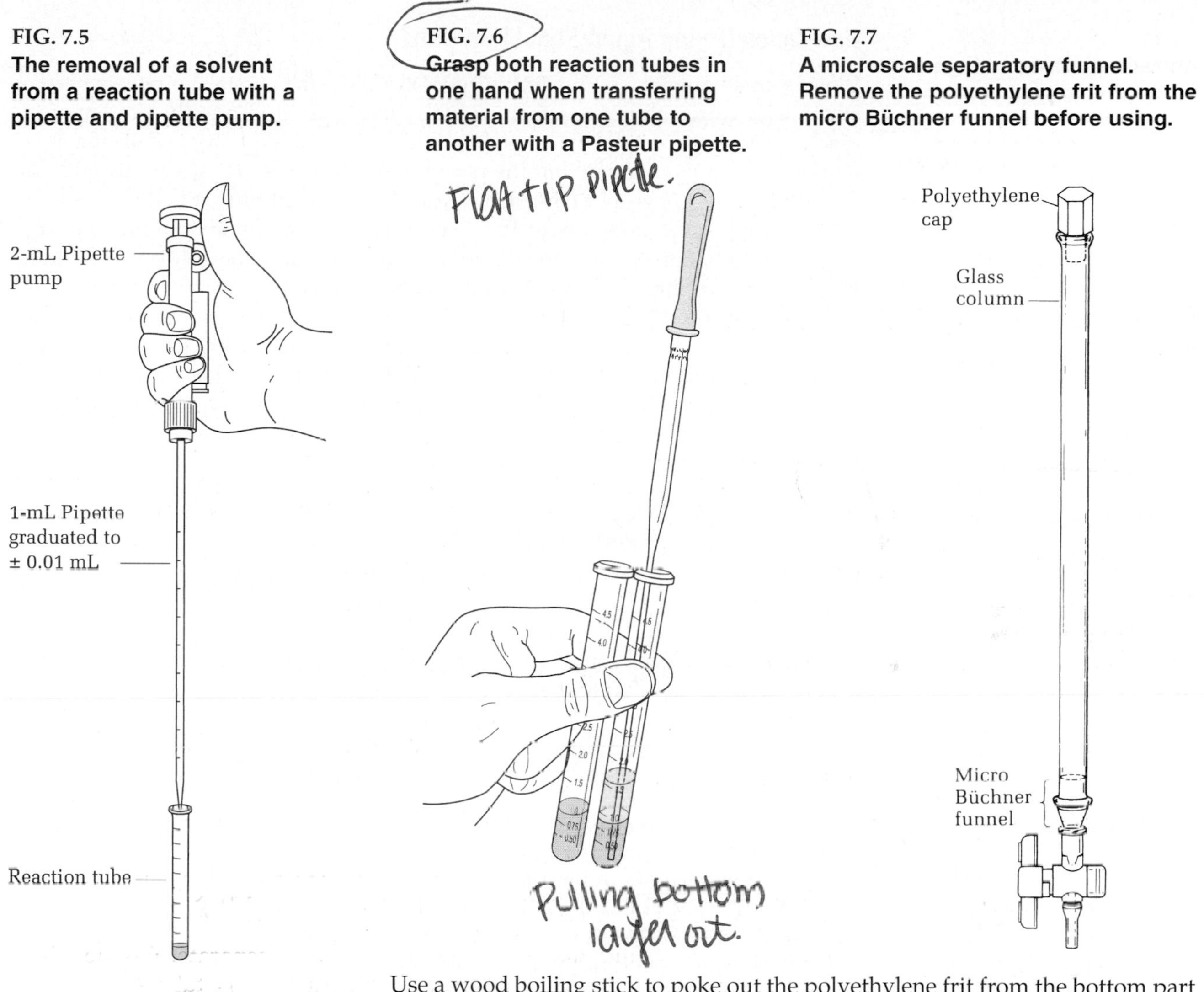

FIG. 7.5
The removal of a solvent from a reaction tube with a pipette and pipette pump.

FIG. 7.6
Grasp both reaction tubes in one hand when transferring material from one tube to another with a Pasteur pipette.

FIG. 7.7
A microscale separatory funnel. Remove the polyethylene frit from the micro Büchner funnel before using.

Use a wood boiling stick to poke out the polyethylene frit from the bottom part of the separatory funnel. Store it for later replacement. Close the valve, add up to 5 mL of the solution to be extracted to the separatory funnel, then add the extraction solvent so that the total volume does not exceed 6 mL.

Cap the separatory funnel and mix the contents by inverting the funnel several times. If the two layers separate fairly easily, then the contents can be shaken more thoroughly. If the layers do not separate easily, be careful not to shake the funnel too vigorously because intractable emulsions could form.

Remove the stopper from the funnel, clamp it, and then, grasping the valve with two hands, empty the bottom layer into an Erlenmeyer flask or other container. If the top layer is desired, pour it out through the top of the separatory funnel—don't drain it through the valve, which may have a drop of the lower layer remaining in it.

STEP 3. DRYING THE ORGANIC LAYER

Dichloromethane dissolves a very small quantity of water, and microscopic droplets of water are suspended in the organic layer, often making it cloudy. To remove the water, a drying agent, for example, anhydrous calcium chloride pellets, is added to the dichloromethane solution.

FIG. 7.8
An aspirator tube being used to remove solvent vapors.

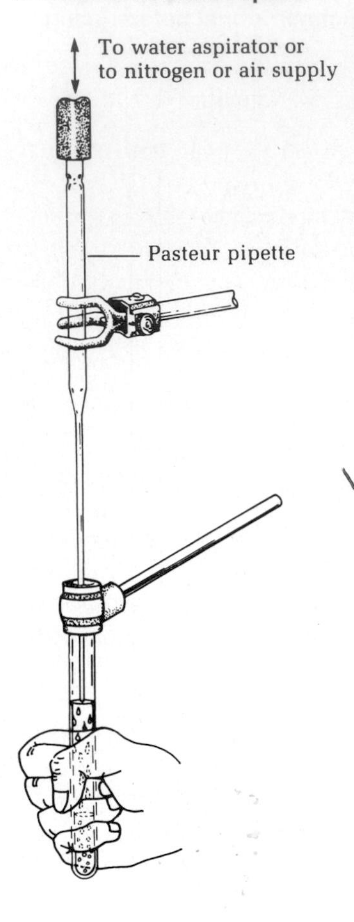

How Much Drying Agent Should Be Used?

When a small quantity of the drying agent is added, the crystals or pellets become sticky with water, clump together, and fall rapidly as a lump to the bottom of the reaction tube. There will come a point when a new small quantity of drying agent no longer clumps together, but the individual particles settle slowly throughout the solution. As they say in Sweden, "Add drying agent until it begins to snow." The drying process takes about 10–15 minutes, during which time the tube contents should be mixed occasionally by flicking the tube. The solution should no longer be cloudy, but clear (although it may be colored).

Once drying is judged complete, the solvent is removed by forcing a Pasteur pipette to the bottom of the reaction tube and pulling the solvent in. Air is expelled from the pipette as it is being pushed through the crystals or pellets so that no drying agent will enter the pipette. It is very important to wash the drying agent left in the reaction tube with several small quantities of pure solvent to transfer all the extract.

STEP 4. REMOVING THE SOLVENT

If the quantity of extract is relatively small, say 3 mL or less, then the easiest way to remove the solvent is to blow a stream of air (or nitrogen) onto the surface of the solution from a Pasteur pipette (Fig. 7.8). Be sure that the stream of air is very gentle before inserting it into the reaction tube. The heat of vaporization of the solvent will cause the tube to become rather cold during the evaporation and, of course, slow down the process. The easiest way to add heat is to hold the tube in your hand.

Another way to remove the solvent is to attach the Pasteur pipette to an aspirator and pull air over the surface of the liquid. This is not quite as fast as blowing air onto the surface of the liquid, and runs the danger of sucking up the liquid into the aspirator.

If the volume of liquid is more than about 3 mL, put it into a 25-mL filter flask, put a plastic Hirsch funnel in place, and attach the flask to the aspirator. By placing your thumb in the Hirsch funnel, the vacuum can be controlled, and heat can be applied by holding the flask in the other hand while swirling the contents (Fig. 7.9).

Record the tare of the final container.

The reaction tube or filter flask in which the solvent is evaporated should be tared (weighed empty), and this weight recorded in your notebook. In this way, the weight of material extracted can be determined by again weighing the container that contains the extract.

EXPERIMENT

PARTITION COEFFICIENT OF BENZOIC ACID

IN THIS EXPERIMENT, you will shake a solution of benzoic acid in water with the immiscible solvent dichloromethane. The benzoic acid will distribute (partition) itself between the two layers. By removing the organic layer, drying, and evaporating it, the weight of benzoic acid in the dichloromethane can be determined and thus the ratio in the two layers. This ratio is a constant known as the partition coefficient.

FIG. 7.9
The apparatus for removing a solvent under vacuum.

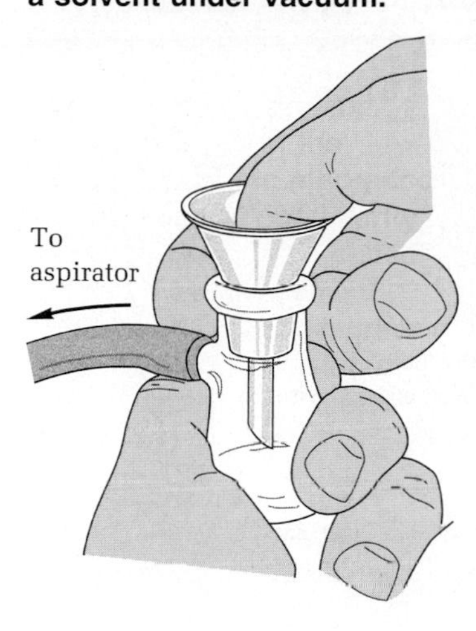

In a reaction tube, place about 100 mg of benzoic acid (weighed to the nearest milligram) and add exactly equal volumes of water followed by dichloromethane (about 1.6 mL each). While making this addition, note which layer is organic and which is aqueous. Put a septum on the tube and shake the contents vigorously for at least 2 minutes. Allow the tube to stand undisturbed until the layers separate and then carefully draw off, using a Pasteur pipette, *all* of the aqueous layer without removing any of the organic layer. It may be helpful to draw out the tip of the pipette to a finer point in a flame and, using this, to tilt the reaction tube at a 45° angle to make this separation as clean as possible.

Add anhydrous calcium chloride pellets to the dichloromethane in very small quantities until they no longer clump together. Mix the contents of the tube by flicking it, and allow it to stand for about 5 minutes to complete the drying process. Using a dry Pasteur pipette, transfer the dichloromethane to a tared dry reaction tube or a 10-mL Erlenmeyer flask containing a boiling chip. Complete the transfer by washing the drying agent with two more portions of solvent that are added to the original solution, and then evaporate the solvent. This can be done by boiling off the solvent while removing solvent vapors with an aspirator tube, or by blowing a stream of air or nitrogen into the container while warming it in one's hand (*see* Fig. 7.8). This operation should be performed in a hood.

From the weight of the benzoic acid in the dichloromethane layer, the weight in the water layer can be obtained by difference. The ratio of the weight in dichloromethane to the weight in water is the distribution coefficient because the volumes of the two solvents were equal. Report the value of the distribution coefficient in your notebook.

Cleaning Up. The aqueous layer can be flushed down the drain. Dichloromethane goes into the halogenated organic solvents waste container. After allowing the solvent to evaporate from the sodium sulfate in the hood, place the sodium sulfate in the non-hazardous solid waste container. If local regulations do not allow for the evaporation of solvents in a hood, dispose of the wet sodium sulfate in a special waste container.

PART 2: ACID/BASE LIQUID/LIQUID EXTRACTION

Acid/base liquid/liquid extraction involves carrying out simple acid/base reactions to separate strong organic acids, weak organic acids, neutral organic compounds, and basic organic substances. The chemistry involved is given in the following equations, using benzoic acid, phenol, naphthalene, and aniline as examples of the four types of compounds.

Here is the strategy (refer to the flow sheet in Fig. 7.10): The four organic compounds are dissolved in *t*-butyl methyl ether. The ether solution is shaken with a saturated aqueous solution of sodium bicarbonate, a weak base. This will react only with the strong acid, benzoic acid (**1**), to form the ionic salt, sodium benzoate (**5**), which dissolves in the aqueous layer and is removed. The ether solution now contains just phenol (**2**), naphthalene (**4**), and aniline (**3**). A 3 *M* aqueous solution of sodium hydroxide is added, and the mixture is shaken. The hydroxide, a strong base, will react only with the phenol (**2**), a weak acid, to form sodium phenoxide (**6**), an ionic compound that dissolves in the aqueous layer and is removed. The ether now contains only naphthalene (**4**) and aniline (**3**). Shaking it with dilute

FIG. 7.10

A flow sheet for the separation of a strong acid, a weak acid, a neutral compound, and a base: benzoic acid, phenol, naphthalene, and aniline (this page). Acid/base reactions of the acidic and basic compounds (opposite page).

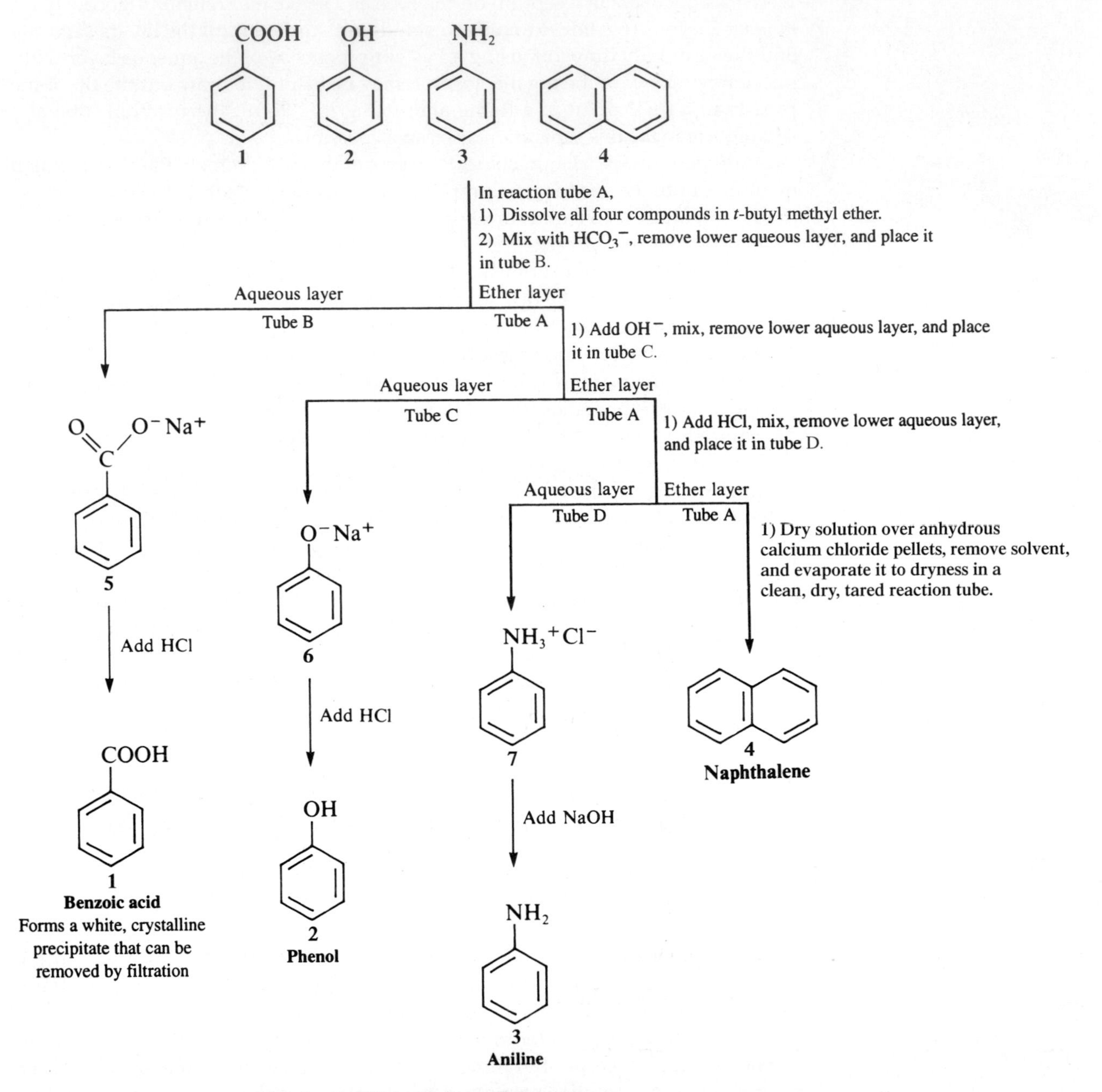

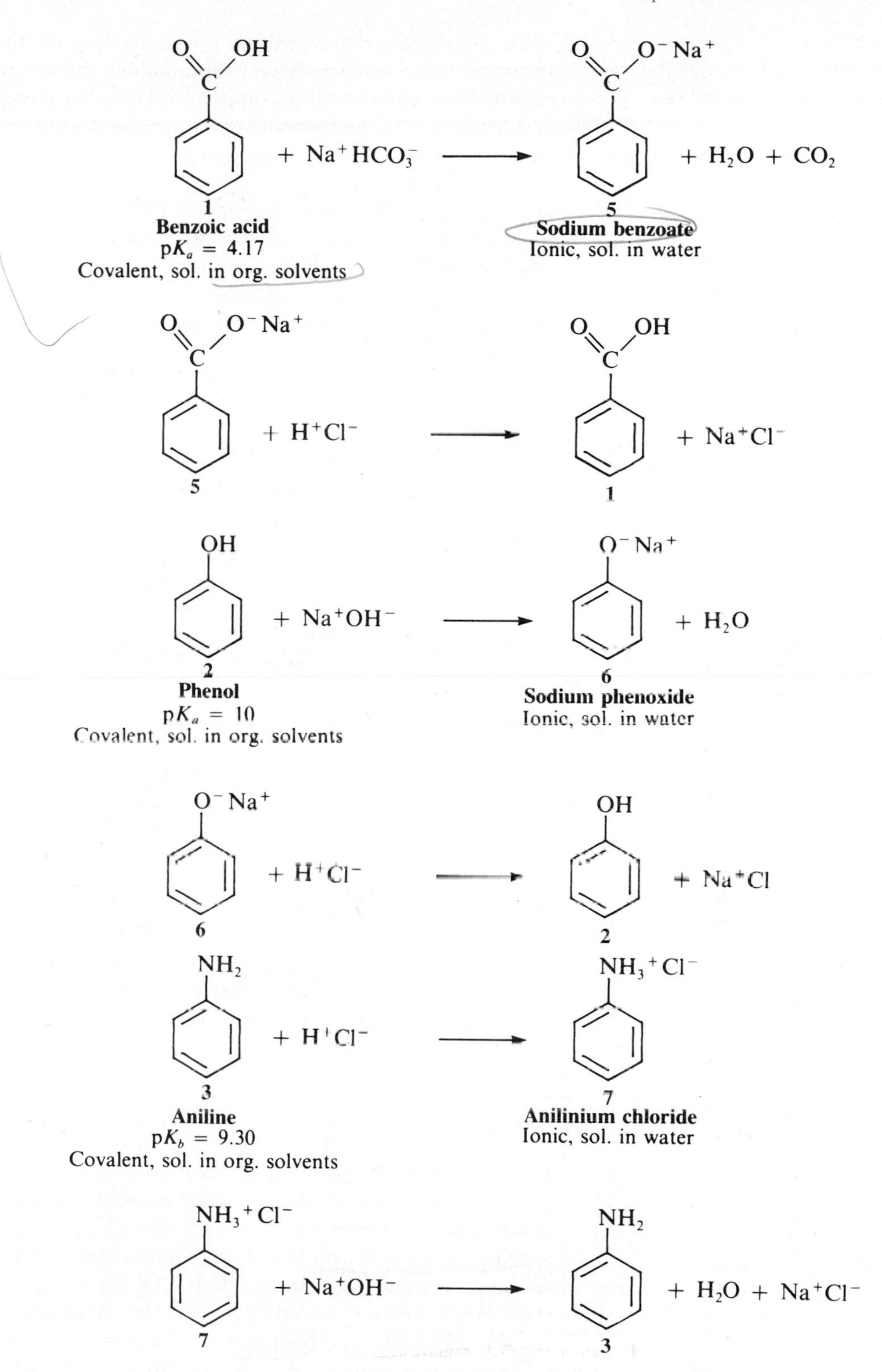

O OH
C
+ Na⁺HCO₃⁻
O O⁻Na⁺
C
+ H₂O + CO₂
1
Benzoic acid
pKₐ = 4.17
Covalent, sol. in org. solvents
5
Sodium benzoate
Ionic, sol. in water
O O⁻Na⁺
C
+ H⁺Cl⁻
5
O OH
C
+ Na⁺Cl⁻
1
OH
+ Na⁺OH⁻
2
Phenol
pKₐ = 10
Covalent, sol. in org. solvents
O⁻Na⁺
+ H₂O
6
Sodium phenoxide
Ionic, sol. in water
O⁻Na⁺
+ H⁺Cl⁻
6
OH
+ Na⁺Cl
2
NH₂
+ H⁺Cl⁻
3
Aniline
pK_b = 9.30
Covalent, sol. in org. solvents
NH₃⁺Cl⁻
7
Anilinium chloride
Ionic, sol. in water
NH₃⁺Cl⁻
+ Na⁺OH⁻
7
NH₂
+ H₂O + Na⁺Cl⁻
3

hydrochloric acid removes the aniline, a base, as the ionic anilinium chloride (**7**). The aqueous layer is removed. Evaporation of the *t*-butyl methyl ether now leaves naphthalene (**4**), the neutral compound. The other three compounds are recovered by adding acid to the sodium benzoate (**5**) and sodium phenoxide (**6**) and base to the anilinium chloride (**7**) to regenerate the covalent compounds benzoic acid (**1**), phenol (**2**), and aniline (**3**).

The pK_a of carbonic acid, H_2CO_3, is 6.35.

The ability to separate strong acids from weak acids depends on the acidity constants of the acids and the basicity constants of the bases. In the first equation, consider the ionization of benzoic acid, which has an equilibrium constant (K_a) of 6.8×10^2. The conversion of benzoic acid to the benzoate anion in equation 4 is governed by the equilibrium constant, K (equation 5), obtained by combining equations 3 and 4.

$$C_6H_5COOH + H_2O \rightleftharpoons C_6H_5COO^- + H_3O^+ \quad (1)$$

$$K_a = \frac{[C_6H_5COO^-][H_3O^+]}{[C_6H_5COOH]} = 6.8 \times 10^{-5}, pK_a = 4.17 \quad (2)$$

$$K_w = [H_3O^+][OH^-] = 10^{-14} \quad (3)$$

$$C_6H_5COOH + OH^- \rightleftharpoons C_6H_5COO^- + H_2O \quad (4)$$

$$K = \frac{[C_6H_5COO^-]}{[C_6H_5COOH][OH^-]} = \frac{K_a}{K_w} = \frac{6.8 \times 10^{-5}}{10^{-14}} = 3.2 \times 10^8 \quad (5)$$

If 99% of the benzoic acid is converted to $C_6H_5COO^-$,

$$\frac{[C_6H_5COO^-]}{[C_6H_5COOH]} = \frac{99}{1} \quad (6)$$

then from equation 5 the hydroxide ion concentration would need to be 6.8×10^{-7} *M*. Because saturated $NaHCO_3$ has $[OH^-] = 3 \times 10^{-4}$ *M*, the hydroxide ion concentration is high enough to convert benzoic acid completely to sodium benzoate.

For phenol, with a K_a of 10^{-10}, the minimum hydroxide ion concentration that will produce the phenoxide anion in 99% conversion is 10^{-2} *M*. The concentration of hydroxide in 10% sodium hydroxide solution is 10^{-1} *M*, and so phenol in a strong base is entirely converted to the water-soluble salt.

GENERAL CONSIDERATIONS

If acetic acid was used as the reaction solvent, it would also be distributed largely into the aqueous phase. If the reaction product is a neutral substance, however, the residual acetic acid in the ether can be removed by one washing with excess 5% sodium bicarbonate solution. If the reaction product is a higher molecular weight acid, for example, benzoic acid (C_6H_5COOH), it will stay in the ether layer, while acetic acid is being removed by repeated washing with water. The benzoic acid can then be separated from neutral byproducts by extraction with sodium bicarbonate or sodium hydroxide solution and acidification of the extract. Acids of high molecular weight are extracted only slowly by sodium bicarbonate, so sodium carbonate is used in its place; however, carbonate is more prone than bicarbonate to produce emulsions. Sometimes an emulsion in the lower layer can be settled by twirling the separatory funnel by its stem. An emulsion in the upper layer can be broken by grasping the funnel at the neck and swirling it. Because the tendency to emulsify increases with the removal of electrolytes and solvents, a little sodium chloride solution is added with each portion of wash water. If the layers are largely clear but an emulsion persists at the interface, the clear part of the water layer can be drawn off, and the emulsion run into a second funnel and shaken with fresh ether.

Liquid/liquid extraction and acid/base extraction are employed in the majority of organic reactions because it is unusual to have the product crystallize from the reaction mixture or to be able to distill the reaction product directly from the reaction mixture. In the research literature, one will often see the statement "the reaction mixture was worked up in the usual way," which implies an extraction process of the type described here. Good laboratory practice dictates, however, that the details of the process be written out.

EXPERIMENTS

1. SEPARATION OF A CARBOXYLIC ACID, A PHENOL, AND A NEUTRAL SUBSTANCE

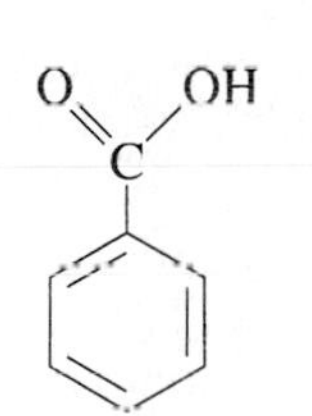

Benzoic acid
mp 123°C, pK_a 4.17

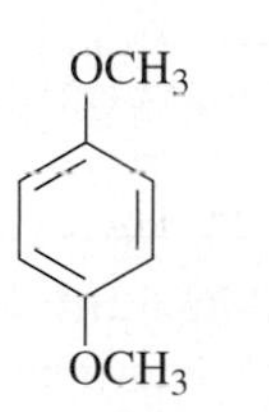

1,4-Dimethoxybenzene (Hydroquinone dimethyl ether)
mp 57°C

A mixture of equal parts of a carboxylic acid, a phenol, and a neutral substance is to be separated by extraction from an ether solvent. Note carefully the procedure for this extraction. In the next experiment, you are to work out your own extraction procedure. Your unknown will consist of either benzoic acid or 2-chlorobenzoic acid (the carboxylic acid), 4-*t*-butyl phenol or 4-bromophenol, and biphenyl or 1,4-dimethoxybenzene (the neutral substance). The object of this experiment is to identify the three substances in the mixture and to determine the percent recovery of each from the mixture.

Procedure

IN THIS EXPERIMENT, three organic solids are separated by reaction with base followed by extraction. Bicarbonate converts carboxylic acids (but not phenols) to ions. Hydroxide ion converts phenols (as well as carboxylic acids) to ions. Ionic substances are soluble in water. The addition of hydrochloric acid to the aqueous ionic solutions regenerates nonionic substances. At each step you should ask yourself, "Have I converted a nonionic substance to an ionic one?" (or vice versa). The ionic substances will be in the aqueous layer; the nonionic ones will be in the organic layer.

Photos: *Extraction with Ether*
Video: *Extraction with Ether*

Dissolve about 0.18 g of the mixture (record the exact weight) in 2 mL of *t*-butyl methyl ether or diethyl ether in a reaction tube (tube 1). Then add 1 mL of a saturated aqueous solution of sodium bicarbonate to the tube. Use the graduations on the side of the tube to measure the amounts, since they do not need to be exact. Mix the contents of the tube thoroughly by pulling the two layers into a Pasteur pipette and expelling them forcefully into the reaction tube. Do this for about 3 minutes. Allow the layers to separate completely and then draw off the lower layer into another reaction tube (tube 2). Add another 0.15 mL of sodium bicarbonate solution to the tube, mix the contents as before, and add the lower layer to tube 2.

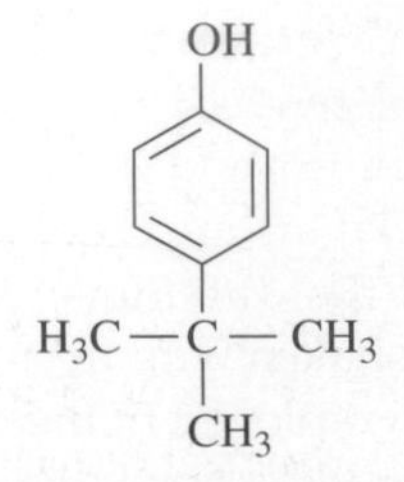

4-*tert*-Butylphenol
mp 101°C, pK_a 10.17

Add HCl with care; CO_2 is released.

Videos: *Filtration of Crystals Using the Pasteur Pipette, Microscale Filtration on the Hirsch Funnel*

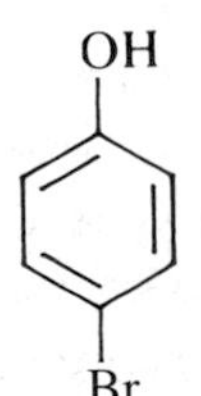

4-Bromophenol
mp 66°C, pK_a 10.2

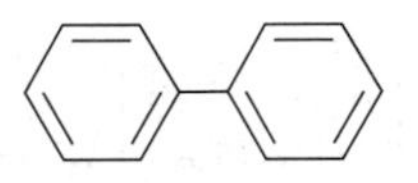

Biphenyl
mp 71°C

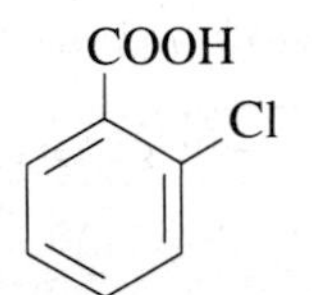

2-Chlorobenzoic acid
mp 141°C, pK_a 2.92

The best way to remove the solvent: under a gentle stream of air or nitrogen.

Exactly which chemical species is in tube 2? Add 0.2 mL of ether to tube 2, mix it thoroughly, remove the ether layer, and discard it. This is called *back-washing* and serves to remove any organic material that might contaminate the contents of tube 2.

Add 1.0 mL of 3 *M* aqueous sodium hydroxide to tube 1, shake the mixture thoroughly, allow the layers to separate, draw off the lower layer using a clean Pasteur pipette, and place it in tube 3. Extract tube 1 with two 0.15-mL portions of water, and add these to tube 3. Backwash the contents of tube 3 with 0.15 mL of ether and discard the ether wash just as was done for tube 2. Exactly which chemical species is in tube 3?

To tube 1, add saturated sodium chloride solution, mix, remove the aqueous layer, and then add to the ether anhydrous calcium chloride pellets until the drying agent no longer clumps together. Wash it off with ether after the drying process is finished. Allow 5–10 minutes for the drying of the ether solution.

Using the concentration information given in the inside back cover of this book, calculate exactly how much concentrated hydrochloric acid is needed to neutralize the contents of tube 2. Then, by dropwise addition of concentrated hydrochloric acid, carry out this neutralization while testing the solution with litmus paper. An excess of hydrochloric acid does no harm. This reaction must be carried out with *extreme care* because much carbon dioxide is released in the neutralization. Add a boiling stick to the tube and very cautiously heat the tube to bring most of the solid carboxylic acid into solution. Allow the tube to cool slowly to room temperature, and then cool it in ice. Remove the solvent with a Pasteur pipette and recrystallize the residue from boiling water. Again, allow the tube to cool slowly to room temperature and then cool it in ice. At the appropriate time, stir the crystals and collect them on a Hirsch funnel using the procedures detailed in Chapter 4. The crystals can be transferred and washed on the funnel using a small quantity of ice water. The solubility of benzoic acid in water is 1.9 g/L at 0°C and 68 g/L at 95°C. The solubility of chlorobenzoic acid is similar. Spread the crystals out onto a tared piece of paper, allow them to dry thoroughly, and determine the percent recovery of the acid. Assess the purity of the product by checking its melting point.

In exactly the same way, neutralize the contents of tube 3 with concentrated hydrochloric acid. This time, of course, there will be no carbon dioxide evolution. Again, heat the tube to bring most of the material into solution, allow it to cool slowly, remove the solvent, and recrystallize the phenol from boiling water. At the appropriate time, after the product has cooled slowly to room temperature and then in ice, it is also collected on a Hirsch funnel, washed with a very small quantity of ice water, and allowed to dry. The percent recovery and melting point are determined.

The neutral compound is recovered using a Pasteur pipette to remove the ether from the drying agent and to transfer it to a tared reaction tube. The drying agent is washed two or three times with additional ether to ensure complete transfer of the product.

Evaporate the solvent by placing the tube in a warm water bath and directing a stream of nitrogen or air onto the surface of the ether in the hood (*see* Fig. 7.8 on page 140). An aspirator tube can also be used for this purpose. Determine the weight of the crude product and then recrystallize it from methanol-water if it is the low-melting compound. Reread Chapter 4 for detailed instructions on carrying out the process of recrystallization from a mixed solvent. The product is dissolved in about 0.5–1 mL of methanol, and water is added until the solution gets cloudy, which indicates that the solution is saturated. This process is best carried out while heating the tube in a hot water bath at 50°C. Allow the tube to cool slowly to room temperature, and then cool it thoroughly in ice. If you have the high-melting compound, recrystallize it from ethanol (8 mL/g).

The products are best isolated by collection on a Hirsch funnel using an ice-cold alcohol-water mixture to transfer and wash the compounds. Determine the percent recovery and the melting point. Turn in the products in neatly labeled $1\frac{1}{2}$-in. × $1\frac{1}{2}$-in. (4 cm × 4 cm) ziplock plastic bags attached to the laboratory report. If the yield on recrystallization is low, concentrate the filtrate (the mother liquor) and obtain a second crop of crystals.

2. SEPARATION OF NEUTRAL AND BASIC SUBSTANCES

IN THIS EXPERIMENT, remember that hydrochloric acid will convert a nonionic amine to an ionic substance and that base will regenerate the nonionic substance from the ionic form. The nonionic and neutral substances are ether soluble, and the ionic substances will be found in the aqueous layer.

A mixture of equal parts of a neutral substance (naphthalene or 1,4-dichlorobenzene) and a basic substance (4-chloroaniline or ethyl 4-aminobenzoate) is to be separated by extraction from an ether solution. Naphthalene and 1,4-dichlorobenzene are completely insoluble in water. The bases will dissolve in hydrochloric acid, while the neutral compounds will remain in the ether solution. The bases are insoluble in cold water, but will dissolve to some extent in hot water and are soluble in ethanol. Naphthalene and 1,4-dichlorobenzene can be purified as described in Chapter 4. They also sublime very easily. Keep the samples covered.

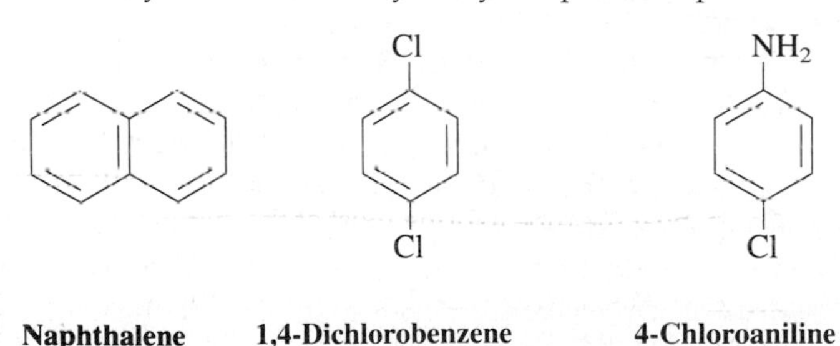

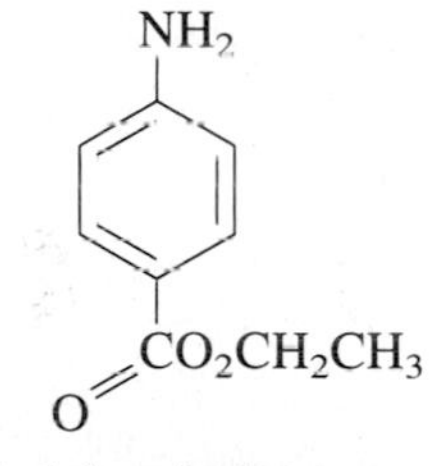

Naphthalene mp 82°C

1,4-Dichlorobenzene mp 56°C

4-Chloroaniline mp 68–71°C, pK_b 4.15

Ethyl 4-aminobenzoate mp 90°C, pK_b 4.92

Plan a step-by-step procedure for separating 200 mg of the mixture into its components and have the plan checked by your instructor before proceeding. A flow sheet is a convenient way to present the plan. Select the correct solvent or mixture of solvents for the recrystallization of the bases on the basis of solubility tests. Determine the weights and melting points of the isolated and purified products and calculate the percent recovery of each. Turn in the products in neatly labeled vials or 1-1/2-in. × 1-1/2-in. ziplock plastic bags attached to the report.

⚠ Ether vapors are heavier than air and can travel along bench tops, run down drain troughs, and collect in sinks. Be extremely careful to avoid flames when working with volatile ethers.

Cleaning Up. Combine all aqueous filtrates and solutions, neutralize them, and flush the resulting solution down the drain. Used ether should be placed in the organic solvents waste container, and the drying agent, once the solvent has evaporated from it, can be placed in the nonhazardous solid waste container. If local regulations do not allow for the evaporation of solvents in a hood, dispose of the wet sodium sulfate in a special waste container. Any 4-chloroaniline or 1,4-dichlorobenzene should be placed in the halogenated waste container.

3. SEPARATION OF ACIDIC AND NEUTRAL SUBSTANCES

A mixture of equal proportions of benzoic acid, 4-*t*-butylphenol, and 1,4-dimethoxybenzene is to be separated by extraction from *t*-butyl methyl ether. Note the detailed directions for extraction carefully. Prepare a flow sheet (*see* Fig. 7.10 on page 142) for this sequence of operations. In the next experiment you will work out your own extraction procedure.

pH = −log [H⁺]
pK_a = acidity constant
pK_b = basicity constant

Procedure

Extinguish all flames when working with *t*-butyl methyl ether! The best method for removing the ether is by simple distillation. Dispose of waste ether in the container provided.

Dissolve 3 g of the mixture in 30 mL of *t*-butyl methyl ether and transfer the mixture to a 125-mL separatory funnel (*see* Fig. 7.2 on page 135) using a little *t*-butyl methyl ether to complete the transfer. Add 10 mL of water and take note of which layer is organic and which is aqueous. Add 10 mL of a saturated aqueous solution of sodium bicarbonate to the funnel. Swirl or stir the mixture to allow carbon dioxide to escape. Stopper the funnel and cautiously mix the contents. Vent the liberated carbon dioxide and then shake the mixture thoroughly with frequent venting of the funnel. Repeat the process with another 10 mL of bicarbonate solution. Allow the layers to separate completely and then draw off the lower layer into a 50-mL Erlenmeyer flask (labeled flask 1). What does this layer contain?

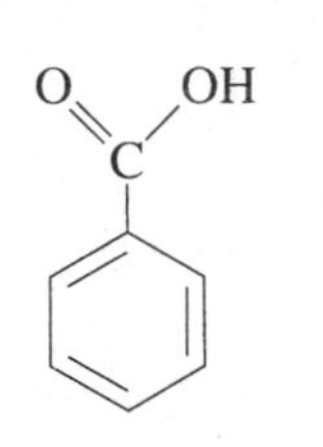

Benzoic acid
mp 123°C, pK_a 4.17

Add 10 mL of a 1.5 *M* aqueous sodium hydroxide solution to the separatory funnel, shake the mixture thoroughly, allow the layers to separate, and draw off the lower layer into a 50-mL Erlenmeyer flask (labeled flask 2). Repeat the process with another 10 mL of base. Then add an additional 5 mL of water to the separatory funnel, shake the mixture as before, and add this to flask 2. What does flask 2 contain?

Add 15 mL of a saturated aqueous solution of sodium chloride to the separatory funnel, shake the mixture thoroughly, allow the layers to separate, and draw off the lower layer, which can be discarded. What is the purpose of adding a saturated sodium chloride solution? Carefully pour the ether layer into a 50-mL Erlenmeyer flask (labeled flask 3) from the top of the separatory funnel, taking great care not to allow any water droplets to be transferred. Add about 4 g of anhydrous calcium chloride pellets to the ether extract and set it aside.

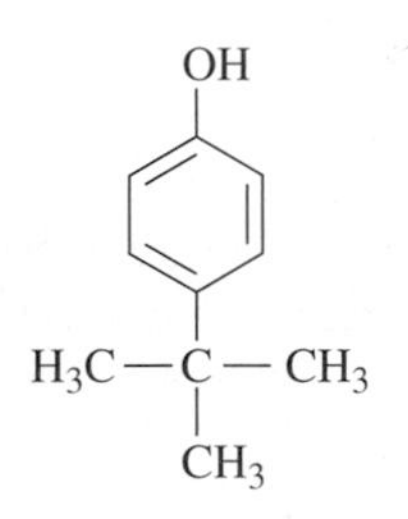

4-*tert*-Butylphenol
mp 101°C, pK_a 10.17

Acidify the contents of flask 2 by dropwise addition of concentrated hydrochloric acid while testing with litmus paper. Cool the flask in an ice bath.

Cautiously add concentrated hydrochloric acid dropwise to flask 1 until the contents are acidic to litmus, and then cool the flask in ice.

Decant (pour off) the ether from flask 3 into a tared flask, making sure to leave all the drying agent behind. Wash the drying agent with additional ether to ensure complete transfer of the product. If decantation is difficult, then remove the drying agent by gravity filtration (*see* Fig. 4.6 on page 68). Put a boiling stick in the flask and evaporate the ether in the hood. An aspirator tube can be used for this purpose (Fig. 7.11). Determine the weight of the crude *p*-dimethoxybenzene and then recrystallize it from methanol. See Chapter 4 for detailed instructions on how to carry out recrystallization.

Video: *Macroscale Crystallization*

Isolate the *t*-butylphenol from flask 2, employing vacuum filtration on a Hirsch funnel, and wash it on the filter with a small quantity of ice water. Determine the weight of the crude product and then recrystallize it from ethanol. Similarly isolate, weigh, and recrystallize from boiling water the benzoic acid in flask 1. The solubility of benzoic acid in water is 1.9 g/L at 0°C and 68 g/L at 95°C.

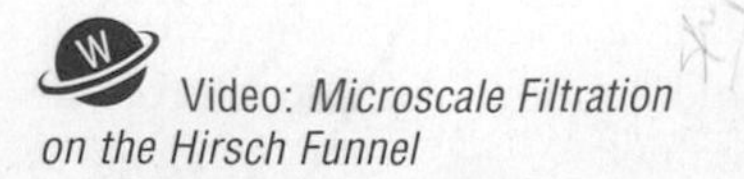
Video: *Microscale Filtration on the Hirsch Funnel*

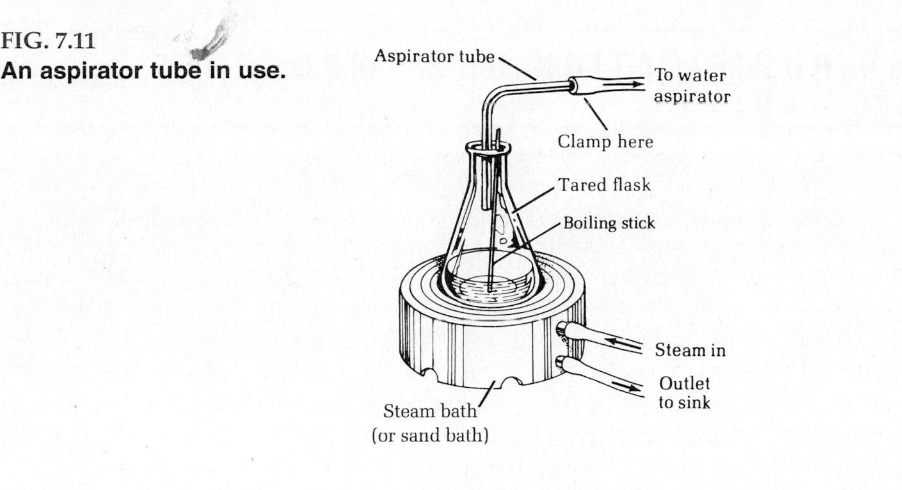

FIG. 7.11
An aspirator tube in use.

Dry the purified products, determine their melting points and weights, and calculate the percent recovery of each substance, bearing in mind that the original mixture contained 1 g of each compound. Hand in the three products in neatly labeled vials.

Cleaning Up. Combine all aqueous layers, washes, and filtrates. Dilute with water, neutralize using either sodium carbonate or dilute hydrochloric acid. This material can then be flushed down the drain with excess water. Methanol filtrate and any ether go in the organic solvents waste container. Allow ether to evaporate from the calcium chloride in the hood. Then place the calcium chloride in the nonhazardous solid waste container. If local regulations do not allow for the evaporation of solvents in a hood, dispose of the wet sodium sulfate in a special waste container.

OCH_3

OCH_3

1,4-Dimethoxybenzene (Hydroquinone dimethyl ether)
mp 57°C

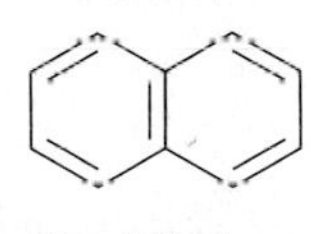

Naphthalene
mp 82°C

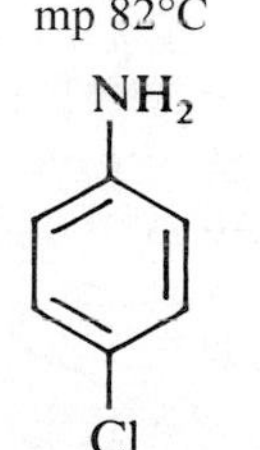

4-Chloroaniline
mp 68–71°C, pK_b 10.0

4. SEPARATION OF NEUTRAL AND BASIC SUBSTANCES

A mixture of equal parts of a neutral substance (naphthalene) and a basic substance (4-chloroaniline) is to be separated by extraction from *t*-butyl methyl ether solution. The base will dissolve in hydrochloric acid, whereas the neutral naphthalene will remain in the *t*-butyl methyl ether solution. 4-Chloroaniline is insoluble in cold water, but will dissolve to some extent in hot water and is soluble in ethanol. Naphthalene can be purified as described in Chapter 4.

Plan a procedure for separating 2.0 g of the mixture into its components and have the plan checked by your instructor before proceeding. A flow sheet is a convenient way to present the plan. Using solubility tests, select the correct solvent or mixture of solvents to recrystallize 4-chloroaniline. Determine the weights and melting points of the isolated and purified products, and calculate the percent recovery of each. Turn in the products in neatly labeled vials.

Handle aromatic amines with care. Most are toxic, and some are carcinogenic. Avoid breathing the dust and vapor from the solid and keep the compounds off the skin, which is best done by wearing nitrile gloves.

Cleaning Up. Combine all aqueous filtrates and solutions, neutralize them, and flush the resulting solution down the drain with a large excess of water. Used *t*-butyl methyl ether should be placed in the organic solvents waste container, and the drying agent, once the solvent has evaporated from it, can be placed in the nonhazardous solid waste container. If local regulations do not allow for the evaporation of solvents in a hood, dispose of the wet sodium sulfate in a special waste container. Any 4-chloroaniline should be placed in the chlorinated organic compounds waste container.

5. EXTRACTION AND PURIFICATION OF COMPONENTS IN AN ANALGESIC TABLET

IN THIS EXPERIMENT, a powdered analgesic tablet, Excedrin, is boiled with dichloromethane and filtered. The solid on the filter is boiled with ethanol, which dissolves everything but the binder. The ethanol is evaporated and from the hot solution, acetaminophen recrystallizes. The dichloromethane solution is shaken with base that converts aspirin to the water-soluble carboxylate anion. The dichloromethane is then evaporated to give caffeine that is purified by sublimation. The aqueous carboxylate anion solution is made acidic, which frees aspirin; warming the mixture and allowing it to cool allows aspirin to recrystallize. It is isolated by filtration.

Excedrin contains aspirin, caffeine, and acetaminophen as determined by thin-layer chromatography (TLC; *see* Chapter 8) or high performance liquid chromatography. A tablet is held together with a binder to prevent the components from crumbling when stored or while being swallowed. A close reading of the contents on the package will disclose the nature of the binder. Starch is commonly used, as is microcrystalline cellulose or silica gel. All of these have one property in common: They are insoluble in water and common organic solvents.

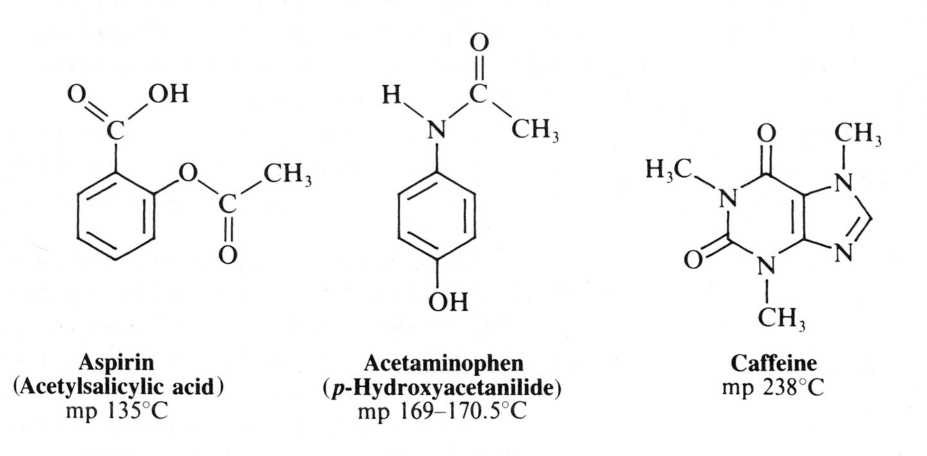

Aspirin **(Acetylsalicylic acid)** mp 135°C

Acetaminophen **(*p*-Hydroxyacetanilide)** mp 169–170.5°C

Caffeine mp 238°C

Inspection of the structures of caffeine, acetylsalicylic acid, and acetaminophen reveals that one is a base, one is a strong organic acid, and one is a weak organic acid. It might be tempting to separate this mixture using exactly the same procedure employed in separating benzoic acid, 4-*t*-butylphenol, and 1,4-dimethoxybenzene (experiment 3)—that is, dissolve the mixture in dichloromethane; separate the strongly acidic component by reaction with bicarbonate ion, a weak base; then remove the weakly acidic component by reaction with hydroxide, a strong base. This process would leave the neutral compound in the dichloromethane solution.

In the present experiment, the solubility data (see Table 7.2) reveal that the weak acid, acetaminophen, is not soluble in ether, chloroform, or dichloromethane, so it cannot be extracted by a strong base. We can take advantage of this lack of solubility by dissolving the other two components, caffeine and aspirin, in dichloromethane

TABLE 7.2 *Solubilities*

	Water	*Ethanol*	*Chloroform*	*Diethyl ether*	*Hexanes*
Aspirin	0.33 g/100 mL at 25°C; 1 g/100 mL at 37°C	1 g/5 mL	1 g/17 mL	1 g/13 mL	
Acetaminophen	v. sl. sol. cold; sol. hot	sol.	ins.	sl. sol.	ins.
Caffeine	1 g/46 mL at 25°C; 1 g/5.5 mL at 80°C; 1 g/1.5 mL at 100°C	1 g/66 mL at 25°C; 1 g/22 mL at 60°C	1 g/5.5 mL	1 g/530 mL	sl. sol.

and removing the acetaminophen by filtration. The binder is also insoluble in dichloromethane, so the solid mixture can be treated with ethanol to dissolve the acetaminophen and not the binder. These can then be separated by filtration, with the acetaminophen isolated by evaporation of the ethanol.

This experiment is a test of technique. It is not easy to separate and recrystallize a few milligrams of a compound that occurs in a mixture.

Microscale Procedure

Handle dichloromethane in the hood. It is a suspected carcinogen.

Photos: *Micro Büchner Funnel, Vacuum Filtration into Reaction Tube through Hirsch Funnel, Use of the Wilfilter*

In a mortar, grind an Extra Strength Excedrin tablet to a very fine powder. The label states that this analgesic contains 250 mg of aspirin, 250 mg of acetaminophen, and 65 mg of caffeine per tablet. Place 300 mg of this powder in a reaction tube and add 2 mL of dichloromethane. Warm the mixture briefly and note that a large part of the material does not dissolve. Filter the mixture on a microscale Büchner funnel (the base of the chromatography column in the kit; Fig. 7.12) into another reaction tube. This is done by transferring the slurry to the funnel with a Pasteur pipette, and completing the transfer with a small portion of dichloromethane which will also wash the material on the funnel. This filtrate is solution 1. A Hirsch funnel (Fig. 7.13) or a Wilfilter (Fig. 7.14) can also be used for this procedure. Pressure filtration is another alternative.

The binder can be starch, microcrystalline cellulose, or silica gel.

Transfer the powder on the filter to a reaction tube, add 1 mL of ethanol, and heat the mixture to boiling on a sand bath (with a boiling stick). Not all the material will go into solution. That which does not is the binder. Filter the mixture on the same previously used microscale Büchner funnel into a tared reaction tube, and complete the transfer and washing using a few drops of hot ethanol.

Acetaminophen

Photos: *Vacuum Filtration into a Reaction Tube through a Hirsch Funnel, Micro Büchner Funnel*; Video: *Filtration of Crystals Using the Pasteur Pipette*

Evaporate about two-thirds of the filtrate by boiling off the ethanol or, better, by warming the solution and blowing a stream of air into the reaction tube. Heat the residue to boiling (add a boiling stick to prevent bumping) and, if necessary, add more ethanol to bring the solid into solution. Allow the saturated solution to cool slowly to room temperature to deposit crystals of acetaminophen, which are reported to melt at 169°C–170.5°C. After the mixture has cooled to room temperature, cool it in ice for several minutes, remove the solvent with a Pasteur pipette, wash the crystals once with 2 drops of ice-cold ethanol, remove the ethanol, and dry the crystals under aspirator vacuum while heating the tube on a steam or sand bath.

Check product purity by TLC (Chapter 8) using 25:1 ethyl acetate–acetic acid to elute the silica gel plates.

Alternatively, the original ethanol solution can be evaporated to dryness, and the residue recrystallized from boiling water. The crystals can be collected on a Hirsch funnel (Fig. 7.13) or by use of a Wilfilter (Fig. 7.14). Once the crystals are dry, determine their weight and melting point. TLC analysis (see Chapter 8) and a determination of the melting points of these crystals and the two other components of this mixture will indicate their purity.

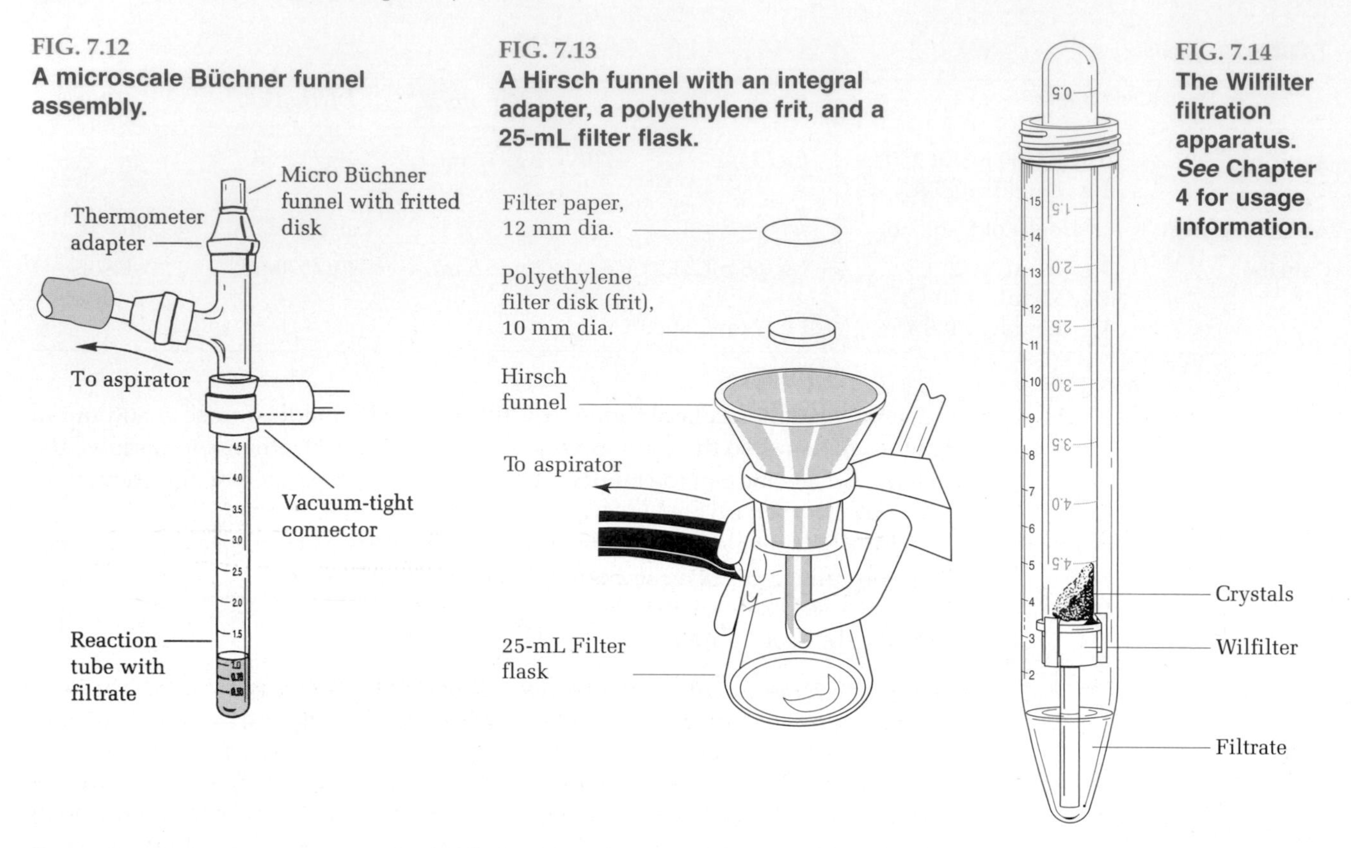

FIG. 7.12
A microscale Büchner funnel assembly.

FIG. 7.13
A Hirsch funnel with an integral adapter, a polyethylene frit, and a 25-mL filter flask.

FIG. 7.14
The Wilfilter filtration apparatus. *See* Chapter 4 for usage information.

The dichloromethane filtered from the binder and acetaminophen mixture (solution 1) should contain caffeine and aspirin. These can be separated by extraction either with acid (which will remove the caffeine as a water-soluble salt) or with base (which will remove the aspirin as a water-soluble salt). We shall use the latter procedure.

Videos: *Recrystallization, Extraction with Dichloromethane*; Photos: *Vacuum Filtration into Reaction Tube through Hirsch Funnel, Use of the Wilfilter*

To the dichloromethane solution in a reaction tube, add 1 mL of 3 *M* sodium hydroxide solution and shake the mixture thoroughly. Remove the aqueous layer, add 0.2 mL more water, shake the mixture thoroughly, and again remove the aqueous layer, which is combined with the first aqueous extract.

Photos: *Sublimation Apparatus, Filtration Using a Pasteur Pipette*; Videos: *Recrystallization, Filtration Using a Pasteur Pipette*

To the dichloromethane, add calcium chloride pellets until the drying agent no longer clumps together. Shake the mixture over a 5-minute to 10-minute period to complete the drying process. Then remove the solvent, wash the drying agent with more solvent, and evaporate the combined extracts to dryness under a stream of air to leave crude caffeine.

Caffeine
See Figure 4.13 on page 72 for drying crystals under vacuum.

The caffeine can be purified by sublimation (Fig. 7.15) or by recrystallization. Recrystallize the caffeine by dissolving it in the minimum quantity of 30% ethanol in tetrahydrofuran. It also can be recrystallized by dissolving the product in a minimum quantity of hot toluene or acetone, and adding to this solution hexanes until the solution is cloudy while at the boiling point. In any case, allow the solution to cool slowly to room temperature; then cool the mixture in ice and remove the solvent from the crystals with a Pasteur pipette. Remove the remainder of the solvent under aspirator vacuum and determine the weight of the caffeine and its melting point.

The aqueous hydroxide extract contains aspirin as the sodium salt of the carboxylic acid. To the aqueous solution, add 3 *M* hydrochloric acid dropwise until the solution tests strongly acid with indicator paper; then add 2 more drops of acid. This will give a suspension of white acetylsalicylic acid in the aqueous solution. It

FIG. 7.15
A sublimation apparatus.

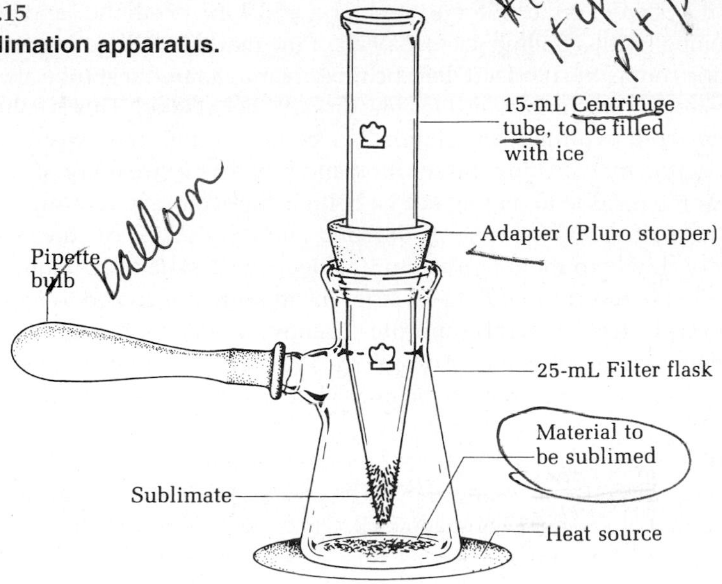

15-mL Centrifuge tube, to be filled with ice
Adapter (Pluro stopper)
Pipette bulb
25-mL Filter flask
Material to be sublimed
Sublimate
Heat source

FIG. 7.16
Recrystallization in a reaction tube.

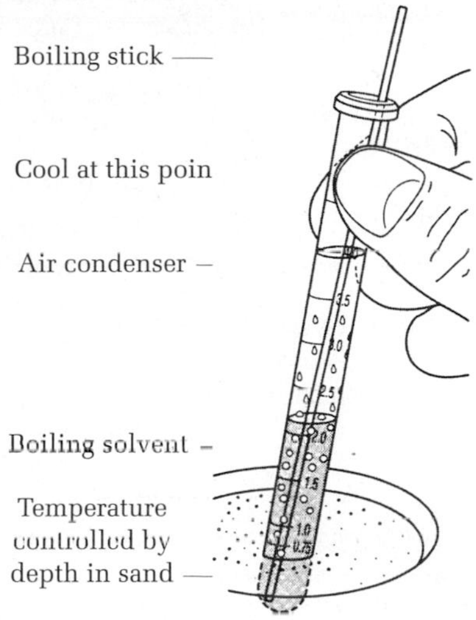

Boiling stick
Cool at this poin
Air condenser
Boiling solvent
Temperature controlled by depth in sand

could be filtered off and recrystallized from boiling water, but this would cause transfer losses. An easier procedure is to heat the aqueous solution that contains the precipitated aspirin.

Photos: *Filtration Using a Pasteur Pipette, Use of the Wilfilter;* Video: *Recrystallization, Filtration of Crystals Using the Pasteur Pipette*

Add a boiling stick and heat the mixture to boiling (Fig. 7.16), at which time the aspirin should dissolve completely. If it does not, add more water. Long boiling will hydrolyze the aspirin to salicylic acid (mp 157–159°C). Once completely dissolved, the aspirin should be allowed to recrystallize slowly as the solution cools to room temperature in an insulated container. Once the tube has reached room temperature, it should be cooled in ice for several minutes, and then the solvent is removed with a Pasteur pipette. Wash the crystals with a few drops of ice-cold water and isolate them with a Wilfilter or scrape them out onto a piece of filter paper. Squeezing the crystals between sheets of filter paper will hasten drying. Once these crystals are completely dry, determine the weight of the acetylsalicylic acid and its melting point.

Aspirin

Alternative procedure: Use a Hirsch funnel or a Wilfilter to isolate the aspirin.

Cleaning Up. Place any dichloromethane-containing solutions in the halogenated organic waste container and the other organic liquids in the organic solvents waste container. The aqueous layers should be diluted and neutralized with sodium carbonate before being flushed down the drain. After it is free of solvent, the calcium chloride can be placed in the nonhazardous solid waste container. If local regulations do not allow for the evaporation of solvents in a hood, dispose of the wet sodium sulfate in a special container.

Macroscale Procedure

In a mortar, grind two Extra Strength Excedrin tablets to a very fine powder. The label states that this analgesic contains 250 mg of aspirin, 250 mg of acetaminophen, and 65 mg of caffeine per tablet. Place this powder in a test tube and add 7.5 mL of dichloromethane. Warm the mixture briefly and note that a large part of the material does not dissolve. Filter the mixture into another test tube by using a Hirsch funnel equipped with a piece of filter paper. Use a Pasteur pipette and complete the transfer with a small portion of dichloromethane. This filtrate is solution 1.

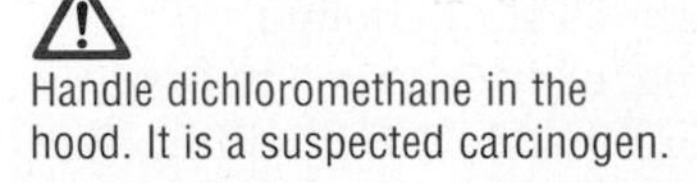

Handle dichloromethane in the hood. It is a suspected carcinogen.

The binder can be starch, microcrystalline cellulose, or silica gel.

Transfer the powder on the filter to a test tube, add 4 mL of ethanol, and heat the mixture to boiling (with a boiling stick). Not all of the material will go into solution. That which does not is the binder. Filter the mixture into a tared test tube and complete the transfer and washing by using a few drops of hot ethanol. This is solution 2.

Acetaminophen

Evaporate about two-thirds of solution 2 by boiling off the ethanol (with a boiling stick) or, better, by warming the solution and blowing a stream of air into the test tube. Heat the residue to boiling (add a boiling stick to prevent bumping) and add more ethanol, if necessary, to bring the solid into solution. Allow the saturated solution to cool slowly to room temperature to deposit crystals of acetaminophen, which is reported to melt at 169–170.5°C. After the mixture has cooled to room temperature, cool it in ice for several minutes, remove the solvent with a Pasteur pipette, wash the crystals once with 2 drops of ice-cold ethanol, remove the ethanol, and dry the crystals under aspirator vacuum while heating the tube on a steam or sand bath.

Photo: *Filtration Using a Pasteur Pipette*; Video: *Filtration of Crystals Using the Pasteur Pipette*

Alternatively, the original ethanol solution is evaporated to dryness, and the residue is recrystallized from boiling water. The crystals are best collected and dried on a Hirsch funnel (*see* Fig. 7.13 on page 152). Once the crystals are dry, determine their weight and melting point. TLC analysis (*see* Chapter 8) and a determination of the melting points of these crystals and the two other components of this mixture will indicate their purity.

Check product purity by TLC (*Chapter 8*) using 25:1 ethyl acetate–acetic acid to elute the silica gel plates.

Video: *Extraction with Dichloromethane*

The dichloromethane filtered from the binder and acetaminophen mixture (solution 1) should contain caffeine and aspirin. These can be separated by extraction either with acid (which will remove the caffeine as a water-soluble salt) or with base (which will remove the aspirin as a water-soluble salt). We shall use the latter procedure.

An alternative to shaking is pipette mixing or using a vortex stirrer, if available.

To the dichloromethane solution in a test tube, add 4 mL of 3 *M* sodium hydroxide solution and shake the mixture thoroughly. Remove the aqueous layer, add 1 mL more water, shake the mixture thoroughly, and again remove the aqueous layer, which is combined with the first aqueous extract.

To the dichloromethane, add anhydrous calcium chloride pellets until the drying agent no longer clumps together. Shake the mixture over a 5–10-minute period to complete the drying process, then remove the solvent, wash the drying agent with more solvent, and evaporate the combined extracts to dryness under a stream of air to leave crude caffeine.

Photo: *Sublimation Apparatus*

Caffeine

The caffeine can be purified by sublimation or by recrystallization. Recrystallize the caffeine by dissolving it in the minimum quantity of 30% ethanol in tetrahydrofuran. It can also be recrystallized by dissolving the product in a minimum quantity of hot toluene or acetone, and adding hexanes to this solution until the solution is cloudy while at the boiling point. In any case, allow the solution to cool slowly to room temperature; then cool the mixture in ice and remove the solvent from the crystals with a Pasteur pipette. Remove the remainder of the solvent under aspirator vacuum and determine the weight of the caffeine and its melting point.

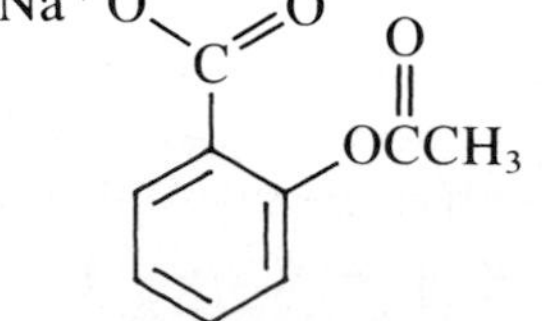

Sodium acetylsalicylate
(soluble in water)

The aqueous hydroxide extract contains aspirin as the sodium salt of the carboxylic acid. To the aqueous solution, add 3 *M* hydrochloric acid dropwise until the solution tests strongly acid with indicator paper; then add 2 more drops of acid. This will give a suspension of white acetylsalicylic acid in the aqueous solution. It could be filtered off and recrystallized from boiling water, but this would cause transfer losses. An easier procedure is to simply heat the aqueous solution that contains the precipitated aspirin and allow it to recrystallize on slow cooling.

Aspirin

Alternative procedure: Use a Hirsch funnel (*see* Fig. 4.15 on page 73) to isolate the aspirin.

Add a boiling stick and heat the mixture to boiling, at which time the aspirin should dissolve completely. If it does not, add more water. Long boiling will

Photo: *Filtration Using a Pasteur Pipette*; Video: *Filtration of Crystals Using the Pasteur Pipette*

hydrolyze the aspirin to salicylic acid (mp 157–159°C). Once completely dissolved, the aspirin should be allowed to recrystallize slowly as the solution cools to room temperature in an insulated container. Once the tube has reached room temperature, it should be cooled in ice for several minutes, and then the solvent is removed with a Pasteur pipette. The crystals are to be washed with a few drops of ice-cold water and then scraped out onto a piece of filter paper. Squeezing the crystals between sheets of the filter paper will hasten drying. Once these crystals are completely dry, determine the weight of the acetylsalicylic acid and its melting point.

Cleaning Up. Place any dichloromethane-containing solutions in the halogenated organic waste container and the other organic liquids in the organic solvents waste container. The aqueous layers should be diluted and neutralized with sodium carbonate before being flushed down the drain. After it is free of solvent, the calcium chloride can be placed in the nonhazardous solid waste container. If local regulations do not allow for the evaporation of solvents in a hood, dispose of the wet sodium sulfate in a special waste container.

EXTRACTIONS FROM COMMON ITEMS

6. Extraction of Caffeine from Tea

Tea and coffee have been popular beverages for centuries, primarily because they contain caffeine, a stimulant. Caffeine stimulates respiration, the heart, and the central nervous system, and it is a diuretic (i.e., it promotes urination). It can cause nervousness and insomnia and, like many drugs, can be addictive, making it difficult to reduce the daily dose. A regular coffee drinker who consumes just 4 cups per day can experience headache, insomnia, and even nausea upon withdrawal from the drug. On the other hand, it helps people to pay attention and can sharpen moderately complex mental skills as well as prolong the ability to exercise.

Caffeine may be the most widely abused drug in the United States. During the course of a day, an average person may unwittingly consume up to 1 g of caffeine. The caffeine content of some common foods and drugs is given in Table 7.3.

Caffeine belongs to a large class of compounds known as alkaloids. These are of plant origin, contain basic nitrogen, often have a bitter taste and a complex structure, and usually have physiological activity. Their names usually end in *-ine*; many are quite familiar by name if not chemical structure—for example, nicotine, cocaine, morphine, and strychnine.

Tea leaves contain tannins, which are acidic, as well as a number of colored compounds and a small amount of undecomposed chlorophyll (soluble in dichloromethane). To ensure that the acidic substances remain water soluble and that the caffeine will be present as the free base, sodium carbonate is added to the extraction medium.

The solubility of caffeine in water is 2.2 mg/mL at 25°C, 180 mg/mL at 80°C, and 670 mg/mL at 100°C. It is quite soluble in dichloromethane, the solvent used in this experiment to extract the caffeine from water.

Caffeine can be easily extracted from tea bags. The procedure one would use to make a cup of tea—simply "steeping" the tea with very hot water for about 7 minutes—extracts most of the caffeine. There is no advantage to boiling the tea leaves with water for 20 minutes. Because caffeine is a white, slightly bitter, odorless, crystalline solid, it is obvious that water extracts more than just

TABLE 7.3 *Caffeine Content of Common Foods and Drugs*

Espresso	120 mg per 2 oz
Coffee, regular, brewed	103 mg per cup
Instant coffee	57 mg per cup
Coffee, decaffeinated	2–4 mg per cup
Tea, black (fermented)	40–75 mg per cup
Tea, green	15–30 mg per cup
Tea, white	10–15 mg per cup
Cocoa	5–40 mg per cup
Milk chocolate	6 mg per oz
Baking chocolate	35 mg per oz
Coca-Cola, Classic	46 mg per 12 oz
Anacin, Bromo-Seltzer, Midol	32 mg per pill
Excedrin, Extra Strength	65 mg per pill
Dexatrim, Dietac, Vivarin	200 mg per pill
Dristan	16 mg per pill
No-Doz	100 mg per pill

caffeine. When the brown aqueous solution is subsequently extracted with dichloromethane, caffeine primarily dissolves in the organic solvent. Evaporation of the solvent leaves crude caffeine, which on sublimation yields a relatively pure product. When the concentrated tea solution is extracted with dichloromethane, emulsions can form very easily. There are substances in tea that cause small droplets of the organic layer to remain suspended in the aqueous layer. This emulsion formation results from vigorous shaking. To avoid this problem, it might seem that one could boil the tea leaves with dichloromethane first and then extract the caffeine from the dichloromethane solution with water. In fact, this does not work. Boiling 25 g of tea leaves with 50 mL of dichloromethane gives only 0.05 g of residue after evaporation of the solvent. Subsequent extractions yield even less material. Hot water causes the tea leaves to swell and is obviously a far more efficient extraction solvent. An attempt to sublime caffeine directly from tea leaves is also unsuccessful.

Microscale Procedure

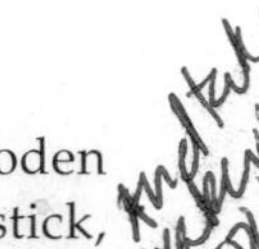

In a 30-mL beaker, place 15 mL of water, 2 g of sodium carbonate, and a wooden boiling stick. Bring the water to a boil on the sand bath, remove the boiling stick, and brew a very concentrated tea solution by immersing a tea bag (2.4 g tea) in the very hot water for 5 minutes. After the tea bag cools enough to handle, and being careful not to break the bag, squeeze as much water from the bag as possible. Again, bring the water to a boil and add a new tea bag to the hot solution. After 5 minutes, remove the tea bag and squeeze out as much water as possible. This can be done easily on a Hirsch funnel. Rinse the bag with a few mL of very hot water, but be sure the total volume of aqueous extract does not exceed 12 mL. Pour the extract into a 15-mL centrifuge tube and cool the solution in ice to below 40°C (the boiling point of dichloromethane).

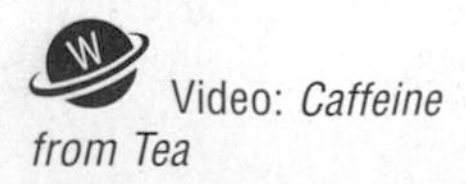
Video: *Caffeine from Tea*

Using three 2-mL portions of dichloromethane, extract the caffeine from the tea. Cork the tube and use a gentle rocking motion to carry out the extraction. Vigorous shaking will produce an intractable emulsion, whereas extremely gentle mixing will fail to extract the caffeine. If you have ready access to a centrifuge, the shaking can be very vigorous because any emulsions formed can be broken fairly well by centrifugation for about 90 seconds. After each extraction, remove the lower organic layer into a reaction tube, leaving any emulsion layer behind. Dry the combined extracts over anhydrous calcium chloride pellets for 5–10 minutes in an Erlenmeyer flask. Add the drying agent in portions with shaking until it no longer clumps together. Transfer the dry solution to a tared 25-mL filter flask, wash the drying agent twice with 2-mL portions of dichloromethane, and evaporate the contents of the flask to dryness (*see* Fig. 7.9 on page 140). The residue will be crude caffeine (determine its weight), which is to be purified by sublimation.

CAUTION: Do not breathe the vapors of dichloromethane and, if possible, work with this solvent in the hood.

Balance the centrifuge tubes.

Fit the filter flask with a Pluro stopper or No. 2 neoprene adapter through which is thrust a 15-mL centrifuge tube. Put a pipette bulb on the side arm. Clamp the flask with a large three-prong clamp, fill the centrifuge tube with ice and water, and heat the flask on a hot sand bath (*see* Fig. 7.15 on page 153). Caffeine is reported to sublime at about 170°C. Tilt the filter flask and rotate it in a hot sand bath to drive more caffeine onto the centrifuge tube. Use a heat gun to heat the upper walls of the filter flask. When sublimation ceases, remove the ice water from the centrifuge tube and allow the flask to cool somewhat before removing the centrifuge tube. Scrape the caffeine onto a tared weighing paper, weigh and, using a plastic funnel, transfer it to a small vial or a plastic bag. At the discretion of your instructor, determine the melting point with a sealed capillary. The melting point of caffeine is 238°C. Using the centrifugation technique to separate the extracts, about 30 mg of crude caffeine can be obtained. This will give you 10–15 mg of sublimed material, depending on the caffeine content of the particular tea being used. The isolated caffeine can be used to prepare caffeine salicylate (experiment 9).

Cleaning Up. Discard the tea bags in the nonhazardous solid waste container. Allow the solvent to evaporate from the drying agent and discard in the same container. If local regulations do not allow for the evaporation of solvents in a hood, dispose of the wet sodium sulfate in a special waste container. Place any unused and unrecovered dichloromethane in the chlorinated organic compounds waste container. The apparatus can be cleaned with soap and hot water. Caffeine can be flushed down the drain because it is biodegradable.

Macroscale Procedure

To an Erlenmeyer flask containing 25 g of tea leaves (or 10 tea bags) and 20 g of sodium carbonate, add 225 mL of vigorously boiling water. Allow the mixture to stand for 7 minutes and then decant into another Erlenmeyer flask. To the hot tea leaves, add another 50 mL of hot water and then immediately decant and combine with the first extract. Very little, if any, additional caffeine is extracted by boiling the tea leaves for any length of time, so it is not necessary to do this. Decantation works nearly as well as vacuum filtration and is much faster.

CAUTION: Carry out work with dichloromethane in the hood.

Cool the aqueous solution to near room temperature and extract it twice with 30-mL portions of dichloromethane. Do not shake the separatory funnel so vigorously as to cause emulsion formation, bearing in mind that if it is not shaken vigorously enough the caffeine will not be extracted into the organic layer. Use a gentle rocking motion of the separatory funnel. Drain off the dichloromethane layer on

Rock the separatory funnel very gently to avoid emulsions.

Dispose of used dichloromethane in the waste container provided.

the first extraction; include the emulsion layer on the second extraction. Dry the combined dichloromethane solutions and any emulsion layer with anhydrous calcium chloride pellets. Add sufficient drying agent until it no longer clumps together on the bottom of the flask. Carefully decant or filter the dichloromethane solution into a tared Erlenmeyer or distilling flask. Silicone-impregnated filter paper allows dichloromethane to pass but retains water. Wash the drying agent with a further portion of solvent and evaporate or distill the solvent. A wood applicator stick is better than a boiling chip to promote smooth boiling because it is easily removed once the solvent is gone. The residue of greenish-white crystalline caffeine should weigh about 0.25 g.

Recrystallization of Caffeine

To recrystallize the caffeine, dissolve it in 5 mL of hot acetone, transfer it with a Pasteur pipette to a small Erlenmeyer flask and, while it is hot, add hexanes to the solution until a faint cloudiness appears. Set the flask aside and allow it to cool slowly to room temperature. This mixed-solvent method of recrystallization depends on the fact that caffeine is far more soluble in acetone than hexanes, so a combination of the two solvents can be found where the solution is saturated in caffeine and will appear cloudy as the caffeine starts to precipitate out of the solution. Cool the solution containing the crystals and remove them by vacuum filtration, employing a Hirsch funnel or a very small Büchner funnel. Use a few drops of hexanes to transfer and wash the crystals. If you wish to obtain a second crop of crystals, collect the filtrate in a test tube, concentrate it to the cloud point using an aspirator tube (*see* Fig. 7.11 on page 149), and repeat the recrystallization process.

Cleaning Up. The filtrate can be diluted with water and washed down the drain. Any dichloromethane collected goes into the halogenated organic waste container. After the solvent is allowed to evaporate from the drying agent in the hood, the drying agent can be placed in the nonhazardous solid waste container; otherwise it goes in the hazardous waste container. If local regulations do not allow for the evaporation of solvents in a hood, dispose of the wet sodium sulfate in a special waste container. The tea leaves go in the nonhazardous solid waste container.

7. Extraction of Caffeine from Cola Syrup

Coca-Cola was originally flavored with extracts from the leaves of the coca plant and the kola nut. Coca is grown in northern South America; the Indians of Peru and Bolivia have for centuries chewed the leaves to relieve the pangs of hunger and sensitivity to high mountain cold. The cocaine from the leaves causes local anesthesia of the stomach. It has limited use as a local anesthetic for surgery on the eye, nose, and throat. Unfortunately, it is now a widely abused and illicit drug. Kola nuts contain about 3% caffeine as well as a number of other alkaloids. The kola tree is in the same family as the cacao tree from which cocoa and chocolate are obtained. Modern cola drinks do not contain cocaine; however, Coca-Cola contains 46 mg of caffeine per 12-oz serving. The acidic taste of many soft drinks comes from citric, tartaric, phosphoric, and benzoic acids.

Automatic soft drink dispensing machines mix a syrup with carbonated water. In the following experiment, caffeine is extracted from concentrated cola syrup.

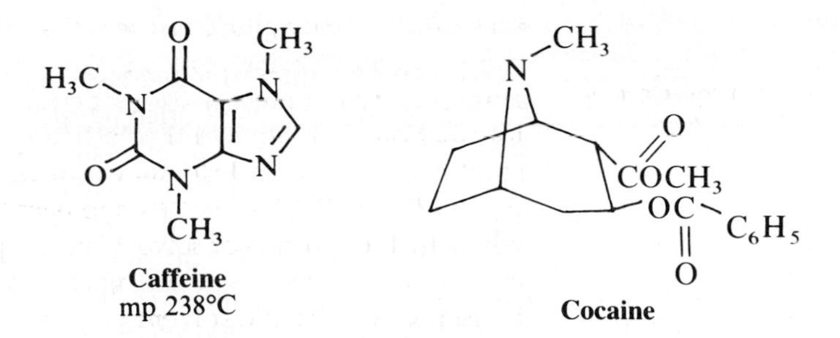

Caffeine
mp 238°C

Cocaine

Microscale Procedure

Add 1 mL of concentrated ammonium hydroxide to a mixture of 5 mL of commercial cola syrup and 5 mL of water in a 15-mL centrifuge tube. Add 1 mL of dichloromethane and tip the tube gently back and forth for 5 minutes. Do not shake the mixture as in a normal extraction because an emulsion will form, and the layers will not separate. After the layers have separated as much as possible, remove the clear lower layer, leaving the emulsion behind. Using 1.5 mL of dichloromethane, repeat the extraction in the same way two more times. At the final separation, include the emulsion layer with the dichloromethane. If a centrifuge is available, the mixture can be shaken vigorously, and the emulsion broken by centrifugation for 90 seconds. Combine the extracts in a reaction tube and dry the solution with anhydrous calcium chloride pellets. Add the drying agent with shaking until it no longer clumps together. After 5–10 minutes, remove the solution with a Pasteur pipette and place it in a tared filter flask. Wash off the drying agent with more dichloromethane and evaporate the mixture to dryness. Determine the crude weight of caffeine; then sublime it as described in the preceding experiment.

CAUTION: Do not breathe the vapor of dichloromethane. Work with this solvent in the hood.

Macroscale Procedure

Add 10 mL of concentrated ammonium hydroxide to a mixture of 50 mL of commercial cola syrup and 50 mL of water. Place the mixture in a separatory funnel, add 50 mL of dichloromethane, and swirl the mixture and tip the funnel back and forth for at least 5 minutes. Do not shake the solutions together as in a normal extraction because an emulsion will form, and the layers will not separate. An emulsion is made up of droplets of one phase suspended in the other. (Milk is an emulsion.) Separate the layers. Repeat the extraction with a second 50-mL portion of dichloromethane. From your knowledge of the density of dichloromethane and water, you should be able to predict which is the top layer and which is the bottom layer. If in doubt, add a few drops of each layer to water. The aqueous layer will be soluble in the water; the organic layer will not.

Combine the dichloromethane extracts and any emulsion that has formed in a 125-mL Erlenmeyer flask; then add anhydrous calcium chloride pellets to remove water from the solution. Add the drying agent until it no longer clumps together at the bottom of the flask, but swirls freely in solution. Swirl the flask with the drying agent from time to time over a 10-minute period. Carefully decant (pour off) the dichloromethane or remove it by filtration through a fluted filter paper, add about 5 mL more solvent to the drying agent to wash it, and decant this also. Combine the dried dichloromethane solutions in a tared flask and remove the dichloromethane by distillation or evaporation on a steam or sand bath. Remember to add a wood applicator stick to the solution to promote even boiling. Determine the weight of the crude product.

Chlorinated solvents are toxic, insoluble in water, and expensive and should never be poured down the drain.

Recrystallization of Caffeine

To recrystallize the caffeine, dissolve it in 5 mL of hot acetone, transfer it with a Pasteur pipette to a small Erlenmeyer flask, and, while it is hot, add hexanes to the solution until a faint cloudiness appears. Set the flask aside and allow it to cool slowly to room temperature. This mixed-solvent method of recrystallization depends on the fact that caffeine is far more soluble in acetone than hexanes, so a combination of the two solvents can be found where the solution is saturated in caffeine (the cloud point). Cool the solution containing the crystals and remove them by vacuum filtration, employing a Hirsch funnel or a very small Büchner funnel. Use a few drops of hexanes to transfer and wash the crystals. If you wish to obtain a second crop of crystals, collect the filtrate in a test tube, concentrate it to the cloud point using an aspirator tube (*see* Fig. 7.11 on page 149), and repeat the recrystallization process.

Cleaning Up. Combine all aqueous filtrates and solutions, neutralize them, and flush the resulting solution down the drain. Used dichloromethane should be placed in the halogenated waste container, and the drying agent, once the solvent has evaporated from it, can be placed in the nonhazardous solid waste container. If local regulations do not allow for the evaporation of solvents in a hood, dispose of the wet sodium sulfate in a special waste container. The hexanes-acetone filtrates should be placed in the organic solvents waste container.

Sublimation of Caffeine. Sublimation is a fast and easy way to purify caffeine. Using the apparatus depicted in Figure 7.15 (on page 153), sublime the crude caffeine at atmospheric pressure following the procedure in part 3 of Chapter 6.

8. Isolation of Caffeine from Instant Coffee

Photo: *Extraction Procedure*

Instant coffee, according to manufacturers, contains between 55 mg and 62 mg of caffeine per 6-oz cup, and a cup is presumably made from a teaspoon of the powder, which weighs 1.3 g; so 2 g of the powder should contain 85–95 mg of caffeine. Unlike tea, however, coffee contains other compounds that are soluble in dichloromethane, so obtaining pure caffeine from coffee is not easy. The objective of this experiment is to extract instant coffee with dichloromethane (the easy part), and then to try to devise a procedure for obtaining pure caffeine from the extract.

From TLC analysis (*see* Chapter 8), you may deduce that certain impurities have a high R_f value in hydrocarbons (in which caffeine is insoluble). Consult reference books (see especially the *Merck Index*[2]) to determine the solubility (and lack of solubility) of caffeine in various solvents. You might try trituration (grinding the crude solid with a solvent) to dissolve impurities preferentially. Column chromatography is another possible means of purifying the product. Or you might convert all of it to the salicylate and then regenerate the caffeine from the salicylate. Experiment! Or you can simply use the following procedure.

[2]O'Neill, M. J.; Smith, A.; Heckelman, P. E.; Budavari, S., eds. *The Merck Index*, 13th ed.; Merck and Co., Inc.: Rahway, NJ, 2001.

Procedure

IN THIS EXPERIMENT, a very concentrated aqueous solution of coffee is prepared and shaken vigorously with an organic solvent to make an intractable emulsion that can be broken (separated into two layers) by centrifugation. Caffeine is isolated by recrystallization.

In a 10-mL Erlenmeyer flask, place 2 g of sodium carbonate and 2 g of instant coffee powder. Add 9 mL of boiling water, stir the mixture well, bring it to a boil again with stirring, cool it to room temperature, and then pour it into a 15-mL plastic centrifuge tube fitted with a screw cap. Add 2 mL of dichloromethane, cap the tube, shake it vigorously for 60 seconds; then centrifuge it at high speed for 90 seconds. Remove the clear yellow dichloromethane layer and place it in a 10-mL Erlenmeyer flask. Repeat this process twice more. To the combined extracts, add anhydrous calcium chloride pellets until they no longer clump together and allow the solution to dry for a few minutes; then transfer it to a tared 25-mL filter flask and wash the drying agent with more solvent. Remove the solvent as was done in the tea extraction experiment and determine the weight of the crude caffeine. You should obtain about 60 mg of crude product. Sublimation of this orange powder gives an impure orange sublimate that smells strongly of coffee, so sublimation is not a good way to purify this material.

Caffeine has no odor.

Dissolve a very small quantity of the product in a drop of dichloromethane and perform a TLC analysis of the crude material. Dissolve the remainder of the material in 1 mL of boiling 95% ethanol; then dilute the mixture with 1 mL of *t*-butyl methyl ether, heat to boiling, and allow to cool slowly to room temperature. Long, needlelike crystals should form in the orange solution. Alternatively, recrystallize the product from a 1:1 mixture of hexanes and 2-propanol, using about 2 mL. Cool the mixture in ice for at least 10 minutes and then collect the product on a Hirsch funnel. Complete the transfer with the filtrate and then wash the crystals twice with cold 50/50 ethanol/*t*-butyl methyl ether. The yield of white fluffy needles of caffeine should be more than 30 mg.

Video: *Recrystallization*

Cleaning Up. Allow the solvent to evaporate from the drying agent and discard it in the nonhazardous waste container. If local regulations do not allow for the evaporation of solvents in a hood, dispose of the wet sodium sulfate in a special waste container. Place any unused and unrecovered dichloromethane in the chlorinated organic solvents waste container.

9. Caffeine Salicylate

One way to confirm the identity of an organic compound is to prepare a derivative of it. Caffeine melts and sublimes at 238°C. It is an organic base and can therefore accept a proton from an acid to form a salt. The salt formed when caffeine combines with hydrochloric acid, like many amine salts, does not have a sharp melting point; it merely decomposes when heated. But the salt formed from salicylic acid, even though ionic, has a sharp melting point and can thus be used to help characterize caffeine. Figure 7.17 is the ^1H NMR spectrum of caffeine.

Preparation of a derivative of caffeine.

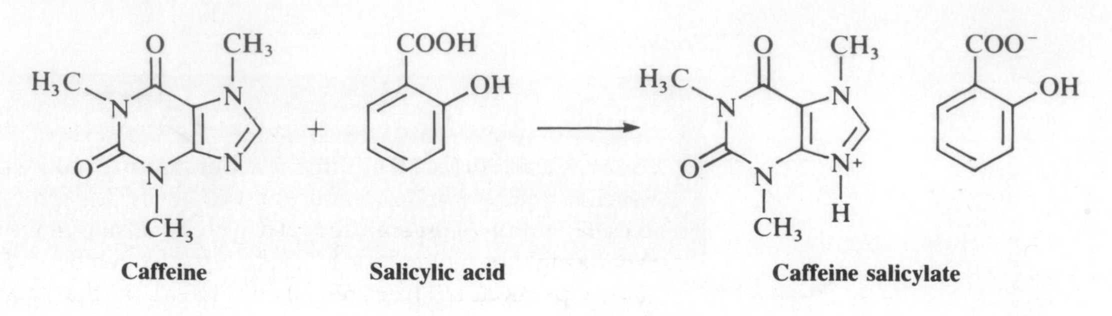

Procedure

CAUTION: Petroleum ether is very flammable. Extinguish all flames.

Recrystallization from mixed solvents.

Video: *Filtration of Crystals Using the Pasteur Pipette*; Photo: *Drying Crystals Under Vacuum*

The quantities given can be multiplied by 5 or 10, if necessary. To 10 mg of sublimed caffeine in a tared reaction tube, add 7.5 mg of salicylic acid and 0.5 mL of dichloromethane. Heat the mixture to boiling and add petroleum ether (a poor solvent for the product) dropwise until the mixture just turns cloudy, indicating that the solution is saturated. If too much petroleum ether is added, then clarify it by adding a very small quantity of dichloromethane. Insulate the tube to allow it to cool slowly to room temperature; then cool it in ice. The needlelike crystals are isolated by removing the solvent with a Pasteur pipette while the reaction tube is in the ice bath. Evaporate the last traces of solvent under vacuum (Fig. 7.18) and determine the weight of the derivative and its melting point. Caffeine salicylate is reported to melt at 137°C.

Cleaning Up. Place the filtrate in the halogenated organic solvents container.

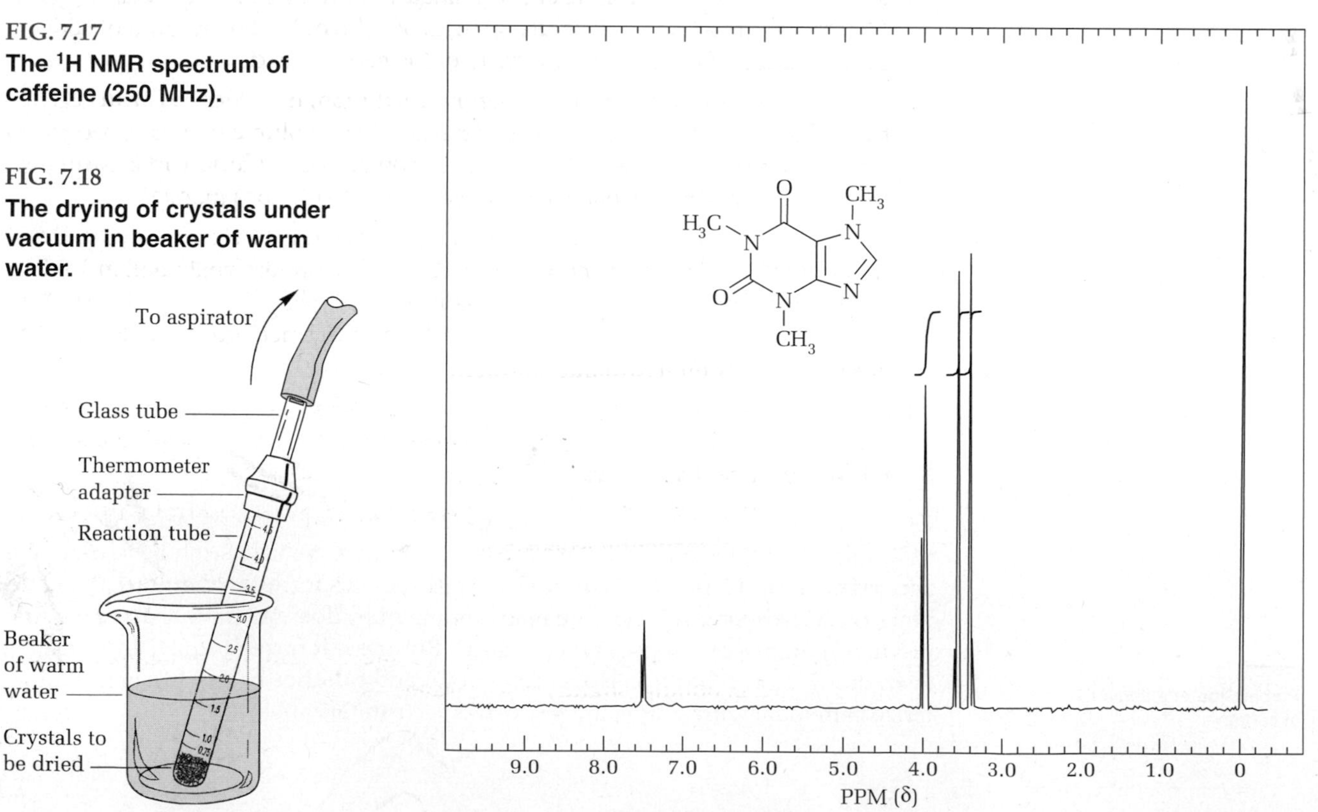

FIG. 7.17
The ^1H NMR spectrum of caffeine (250 MHz).

FIG. 7.18
The drying of crystals under vacuum in beaker of warm water.

QUESTIONS

1. Suppose a reaction mixture, when diluted with water, afforded 300 mL of an aqueous solution of 30 g of the reaction product malononitrile $[CH_2(CN)_2]$, which is to be isolated by extraction with ether. The solubility of malononitrile in ether at room temperature is 20.0 g/100 mL, and in water is 13.3 g/100 mL. What weight of malononitrile would be recovered by extraction with (a) three 100-mL portions of ether and (b) one 300-mL portion of ether? *Suggestion:* For each extraction, let x equal the weight extracted into the ether layer. In part (a), the concentration in the ether layer is $x/100$ and in the water layer is $(30 - x)/300$; the ratio of these quantities is equal to $k = 20/13.3$.
2. Why is it necessary to remove the stopper from a separatory funnel when liquid is being drained from it through the stopcock?
3. The pK_a of *p*-nitrophenol is 7.15. Would you expect this to dissolve in a sodium bicarbonate solution? The pK_a of 2,5-dinitrophenol is 5.15. Will it dissolve in bicarbonate solution?
4. The distribution coefficient, k = concentration in hexanes ÷ concentration in water, between hexanes and water for solute A is 7.5. What weight of A would be removed from a solution of 10 g of A in 100 mL of water by a single extraction with 100 mL of hexanes? What weight of A would be removed by four successive extractions with 25-mL portions of hexanes? How much hexanes would be required to remove 98.5% of A in a single extraction?
5. In experiment 1, how many moles of benzoic acid are present? How many moles of sodium bicarbonate are contained in 1 mL of a 10% aqueous solution? (A 10% solution has 1 g of solute in 9 mL of solvent.) Is the amount of sodium bicarbonate sufficient to react with all of the benzoic acid?
6. To isolate benzoic acid from a bicarbonate solution, it is acidified with concentrated hydrochloric acid, as in experiment 1. What volume of acid is needed to neutralize the bicarbonate? The concentration of hydrochloric acid is expressed in various ways on the inside back cover of this laboratory manual.
7. How many moles of 4-*t*-butylphenol are in the mixture to be separated in experiment 1? How many moles of sodium hydroxide are contained in 1 mL of 5% sodium hydroxide solution? (Assume the density of the solution is 1.0.) What volume of concentrated hydrochloric acid is needed to neutralize this amount of sodium hydroxide solution?
8. Draw a flow sheet to show how you would separate the components of a mixture containing an acid substance, toluic acid, a basic substance, *p*-bromoaniline, and anthracene, a neutral substance.
9. Write equations showing how caffeine could be extracted from an organic solvent and subsequently isolated.
10. Write equations showing how acetaminophen might be extracted from an organic solvent such as an ether, if it were soluble.
11. Write detailed equations showing the mechanism by which aspirin is hydrolyzed in boiling, slightly acidic water.

See *Web Links*

8 CHAPTER

Thin-Layer Chromatography: Analyzing Analgesics and Isolating Lycopene from Tomato Paste

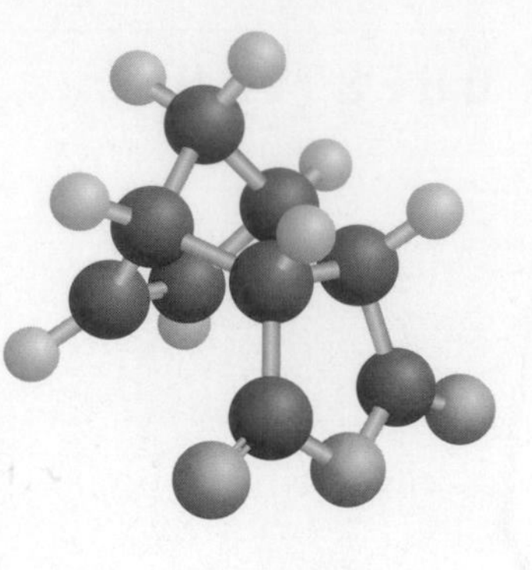

PRELAB EXERCISE: Based on the number and polarity of the functional groups in aspirin, acetaminophen, ibuprofen, and caffeine, whose structures are shown on page 184, predict which of these four compounds has the highest R_f value and which has the lowest.

Chromatography is the separation of two or more compounds or ions caused by their molecular interactions with two phases—one moving and one stationary. These two phases can be a solid and a liquid, a liquid and a liquid, a gas and a solid, or a gas and a liquid. You very likely have seen chromatography carried out on paper towels or coffee filters, which showed various colors denoting the separation of inks and food dyes. In the laboratory, cellulose paper is the stationary or solid phase, and a propanol-water mixture is the mobile or liquid phase. The samples are spotted near one edge of the paper, and this edge is dipped into the liquid phase. The solvent is drawn through the paper by capillary action, and the molecules are separated based on how they interact with the paper. Although there are several different forms of chromatography, the principles are essentially the same.

Thin-layer chromatography (TLC) is a sensitive, fast, simple, and inexpensive analytical technique that you will use repeatedly while carrying out organic experiments. It is a micro technique; as little as 10^{-9} g of material can be detected, although the usual sample size is from 1×10^{-6} g to 1×10^{-8} g. The stationary phase is normally a polar solid adsorbent, and the mobile phase can be a single solvent or a combination of solvents.

TLC requires micrograms of material.

USES OF THIN-LAYER CHROMATOGRAPHY

1. **To determine the number of components in a mixture.** TLC affords a quick and easy method for analyzing such things as a crude reaction mixture, an extract from a plant substance, or the ingredients in a pill. Knowing the

number and relative amounts of the components aids in planning further analytical and separation steps.

2. **To determine the identity of two substances.** If two substances spotted on the same TLC plate give spots in identical locations, they *may* be identical. If the spot positions are not the same, the substances cannot be the same. It is possible for two or more closely related but not identical compounds to have the same positions on a TLC plate. Changing the stationary or mobile phase will usually effect their separation.
3. **To monitor the progress of a reaction.** By sampling a reaction at regular intervals, it is possible to watch the reactants disappear and the products appear using TLC. Thus, the optimum time to halt the reaction can be determined, and the effect of changing such variables as temperature, concentrations, and solvents can be followed without having to isolate the product.
4. **To determine the effectiveness of a purification.** The effectiveness of distillation, crystallization, extraction, and other separation and purification methods can be monitored using TLC, with the caveat that a single spot does not guarantee a single substance.
5. **To determine the appropriate conditions for a column chromatographic separation.** In general, TLC is not satisfactory for purifying and isolating macroscopic quantities of material; however, the adsorbents most commonly used for TLC—silica gel and alumina—are also used for column chromatography, which is discussed in Chapter 9. Column chromatography is used to separate and purify up to 1 g of a solid mixture. The correct adsorbent and solvent to use for column chromatography can be rapidly determined by TLC.
6. **To monitor column chromatography.** As column chromatography is carried out, the solvent is collected in a number of small flasks. Unless the desired compound is colored, the various fractions must be analyzed in some way to determine which ones have the desired components of the mixture. TLC is a fast and effective method for doing this.

THE PRINCIPLES OF CHROMATOGRAPHY

To thoroughly understand the process of TLC (and other types of chromatography), we must examine the process at the molecular level. All forms of chromatography involve a dynamic and rapid equilibrium of molecules between the liquid and the stationary phases. For the chromatographic separation of molecules A and B shown in Figure 8.1, there are two states:

1. **Free**—dissolved in the liquid or gaseous mobile phase.
2. **Adsorbed**—sticking to the surface of the solid stationary phase.

Molecules A and B are continuously moving back and forth between the dissolved (free) and adsorbed states, with billions of molecules adsorbing and billions of other molecules desorbing from the solid stationary phase each second. The equilibrium between the free and adsorbed states depends on the relative strength of the attraction of A and B to the liquid phase molecules *versus* the

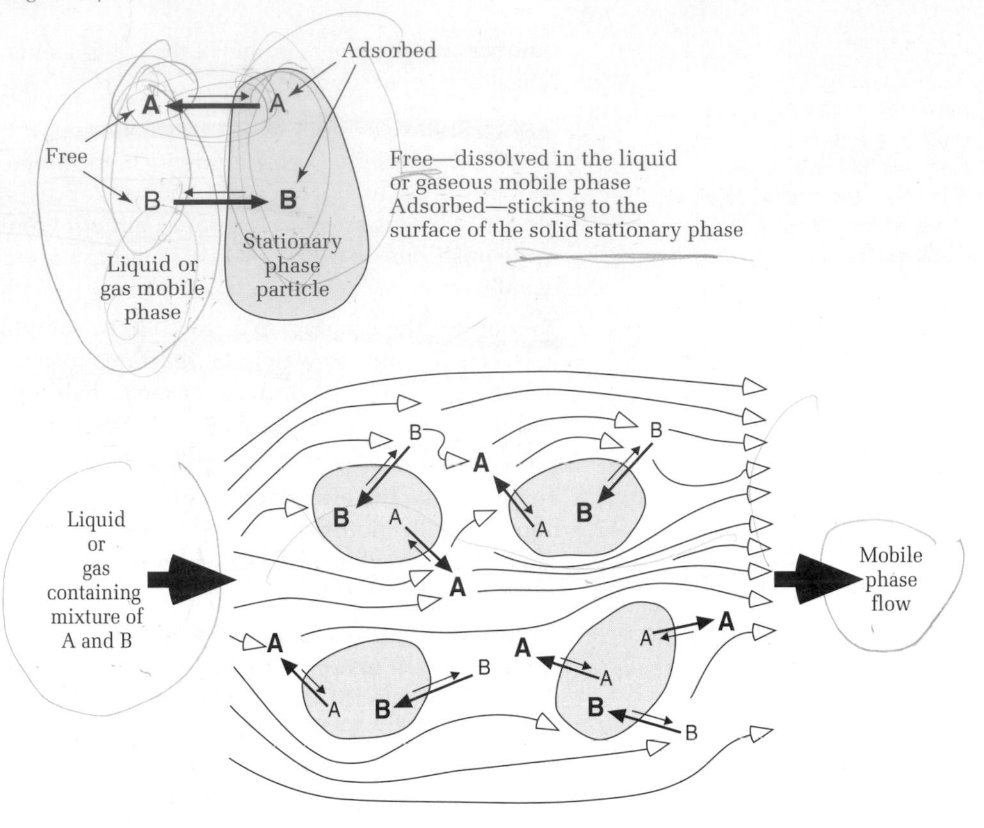

FIG. 8.1
The mixture of molecules A and B is in a dynamic equilibrium between the free and adsorbed states.

FIG. 8.2
The mixture of molecules A and B is in a dynamic equilibrium between the stationary adsorbent and a *flowing* mobile phase.

strength of attraction of A and B to the stationary phase structure. As discussed in the introduction to Chapter 3, the strength of these attractive forces depends on the following factors:

- Size, polarity, and hydrogen bonding ability of molecules A and B
- Polarity and hydrogen bonding ability of the stationary phase
- Polarity and hydrogen bonding ability of the mobile phase solvent

Molecules distribute themselves, or *partition*, between the mobile and stationary phases depending on these attractive forces. As implied by the equilibrium arrows in Figure 8.1, the A molecules are less polar and are only weakly attracted to a polar stationary phase, spending most of their time in the mobile phase. In contrast, equilibrium for the more polar B molecules lies in the direction of being adsorbed onto the polar stationary phase. The equilibrium constant k (also called the *partition coefficient*) is a measure of the distribution of molecules between the mobile phase and the stationary phase, and is similar to the distribution coefficient for liquid/liquid extraction. This constant changes with structure.

Simply adding a mixture to a combination of a liquid phase and a stationary phase will not separate it into its pure components. For separation to happen, the liquid phase must be mobile and be flowing past the stationary phase, as depicted in Figure 8.2. Because the A molecules spend more time in the mobile phase, they will be carried through the stationary phase and be eluted faster and move farther in a given amount of time. Because the B molecules are adsorbed

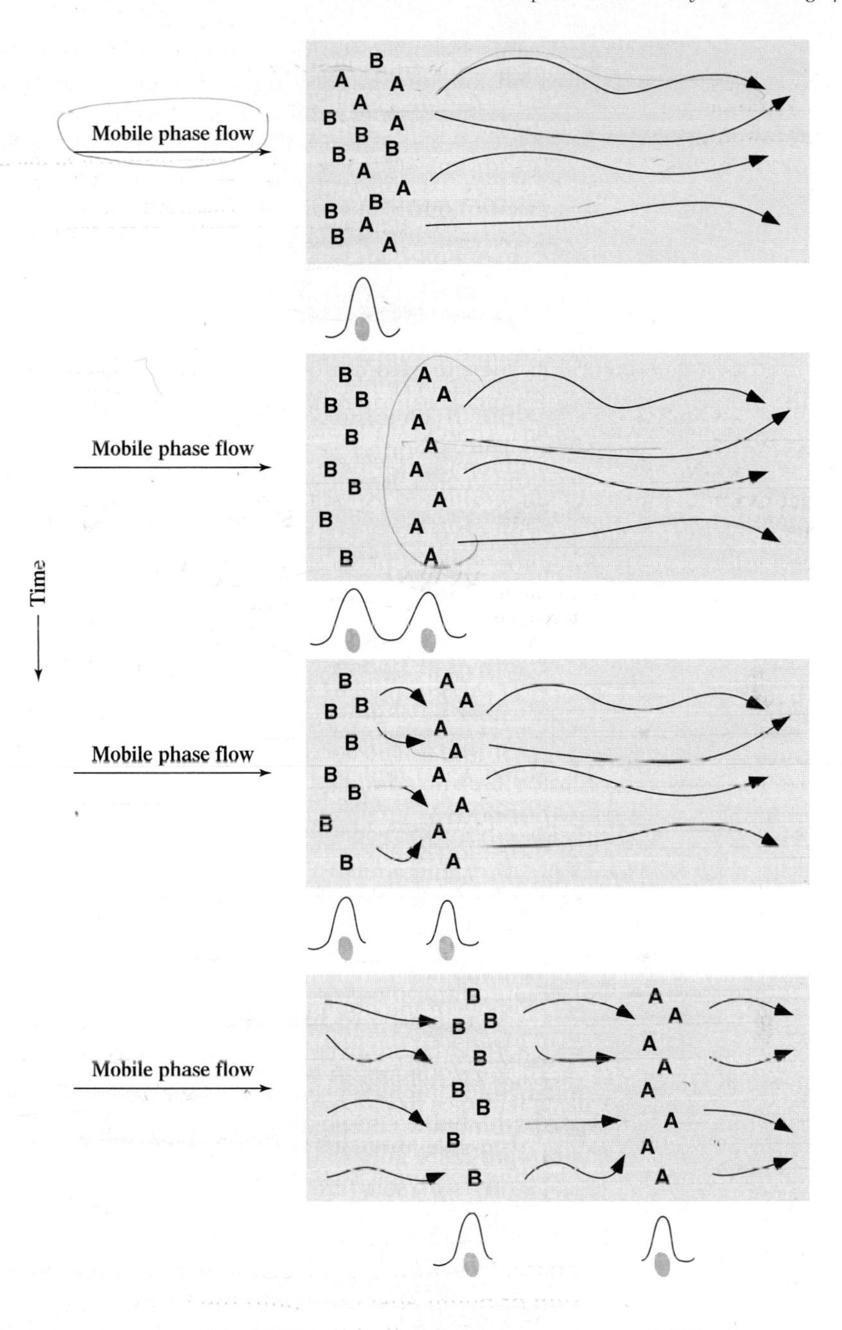

FIG. 8.3
A chromatographic separation. Over time, the mobile phase carries the less weakly adsorbed A molecules ahead of the more strongly adsorbed B molecules.

on the stationary phase more than A molecules, the B molecules spend less time in the mobile phase and therefore migrate through the stationary phase more slowly and are eluted later. The B molecules do not migrate as far in the same amount of time. The consequence of this difference is that A is gradually separated from B by moving ahead in the flowing mobile phase as time passes, as shown in Figure 8.3.

A simple analogy may help to illustrate these concepts. Imagine a group of hungry and not-so-hungry people riding a moving sidewalk (the mobile phase in this analogy) that moves beside a long buffet table covered with all sorts of delicious food (the stationary phase). Hungry people, attracted to the food, will step off and on the moving belt many times in order to fill their plates. The not-so-hungry people will step off and on to get food far less often. Consequently, the more strongly attracted hungry people will lag behind, while the not-so-hungry ones will move ahead on the belt. The two types of people are thus separated based on the strength of their attraction for the food.

Stationary Phase Adsorbents

In TLC, the stationary phase is a polar adsorbent, usually finely ground alumina $[(Al_2O_3)_x]$ or silica $[(SiO_2)_x]$ particles, coated as a thin layer on a glass slide or plastic sheet. Silica, commonly called *silica gel* in the laboratory, is simply very pure white sand. The extended covalent network of these adsorbents creates a very polar surface. Partial structures of silica and alumina are shown below. The silicon or aluminum atoms are the smaller, darker spheres:

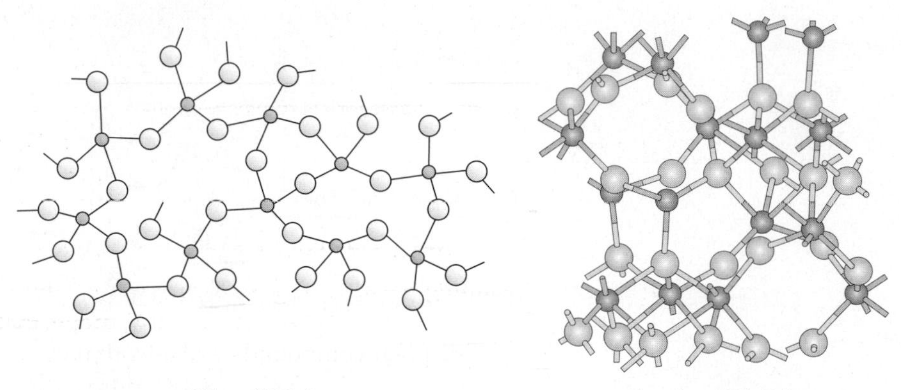

Silica, $(SiO_2)_x$ Alumina, $(Al_2O_3)_x$

FIG. 8.4
A partial silica structure showing polar Si—O bonds.

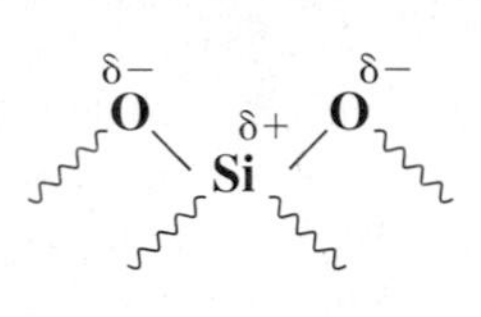

The electropositive character of the aluminum or silicon and the electronegativity of oxygen create a very polar stationary phase (Fig. 8.4). Therefore, the more polar the molecules to be separated, the stronger the attraction to the stationary phase. Nonpolar molecules will tend to stay in the mobile phase. In general, the more polar the functional group, the stronger the adsorption on the stationary phase and the more slowly the molecules will move. In an extreme situation, the molecules will not move at all. This problem can be overcome by increasing the polarity of the mobile phase so that the equilibrium between the free and adsorbed state is shifted toward the free state.

Although silica is the most common stationary phase used for TLC, many other types are used, ranging from paper to charcoal, nonpolar to polar, and reverse phase to normal phase. Several different types of stationary phases are listed according to polarity in Table 8.1.

Silica gel and alumina are commonly used in column chromatography for the purification of macroscopic quantities of material (*see* Chapter 9). Of the two, alumina, when anhydrous, is the more active; that is, it will adsorb substances more strongly. It is thus the adsorbent of choice for the separation of relatively nonpolar substrates, such as hydrocarbons, alkyl halides, ethers, aldehydes, and ketones. To separate more polar substrates, such as alcohols, carboxylic acids, and amines, the less active adsorbent, silica gel, is often used.

TABLE 8.1 *Common Stationary Phases Listed by Increasing Polarity*

Increasing polarity ↓
Polydimethyl siloxane*
Methyl- or Phenylsiloxane*
Cyanopropyl siloxane*
Carbowax [poly(ethyleneglycol)]*
Reverse phase (hydrocarbon-coated silica, e.g., C_{18})
Paper
Cellulose
Starch
Calcium sulfate
Silica (silica gel)
Florisil (magnesium silicate)
Magnesium oxide
Alumina (aluminum oxide; acidic, basic, or neutral)
Activated carbon (charcoal or Norit pellets)

*Stationary phase for gas chromatography

Molecular Polarity and Elution Sequence

Elution sequence is the order in which the components of a mixture move during chromatography.

Assuming we are using a polar adsorbent, how can we determine how rapidly the compounds in our particular mixture move, that is, their elution sequence? Because the more polar compounds will adsorb more strongly to the polar stationary phase they will move the slowest and the shortest distance on a TLC plate. Non-polar compounds will move rapidly, and will elute first or move the greatest distance on the TLC plate. Table 8.2 lists several common compound classes according to how they move or elute on silica or alumina.

Ethyl acetoacetate

Ethyl pentanoate

You should be able to look at a molecular structure, identify its functional group(s), and easily determine whether it is more or less polar than another structure with different functional groups. Note that the polarity of a molecule increases as the number of functional groups in that molecule increases. Thus, ethyl acetoacetate, with both ketone and ester groups, is more polar than ethyl pentanoate, which has only an ester group. However, it should be noted that chromatography is not an exact science. The rules discussed here can be used to help predict the order of elution; however, only performing an experiment will give definitive answers.

Mobile Phase Solvent Polarity

The key to a successful chromatographic separation is the mobile phase. You normally use silica gel or alumina as the stationary phase. In extreme situations, very polar substances chromatographed on alumina will not migrate very far from the starting point (i.e., give low R_f values), and nonpolar compounds chromatographed on silica gel will travel with the solvent front (i.e., give high R_f values). These extremes of behavior are markedly affected, however, by the solvents used to carry out the chromatography. A polar solvent will carry along with it polar substrates,

TABLE 8.2 *Elution Order for Some Common Functional Groups with a Silica or Alumina Stationary Phase*

Increasing polarity of functional group ↓
Highest/fastest (elute with nonpolar mobile phase)
Alkane hydrocarbons
Alkyl halides (halocarbons)
Alkenes (olefins)
Dienes
Aromatic hydrocarbons
Ethers
Esters
Ketones
Aldehydes
Amines
Alcohols
Phenols
Carboxylic acids
Sulfonic acids
Lowest/slowest (need polar mobile phase to elute)

and nonpolar solvents will do the same with nonpolar compounds—another example of the generalization "like dissolves like." You cannot change the polarities of the compounds in your mixture, but by using different solvents, either alone or as mixtures, you can adjust the polarity of the mobile phase and affect the equilibria between the free and adsorbed states. Changing the polarity of the mobile phase can optimize the chromatographic separation of mixtures of compounds with a wide variety of polarities.

Avoid using benzene, carbon tetrachloride, and chloroform. Benzene is known to be a carcinogen when exposure is prolonged; the others are suspected carcinogens.

Table 8.3 lists, according to increasing polarity, some solvents that are commonly used for both TLC and column chromatography. Because the polarities of benzene, carbon tetrachloride, or chloroform can be matched by other, less toxic solvents, these three solvents are seldom used. Carbon tetrachloride and chloroform are suspected carcinogens and benzene is known to be a carcinogen when exposure is prolonged. In general, the solvents for TLC and column chromatography are characterized by having low boiling points that allow them to be easily evaporated and low viscosities that allow them to migrate rapidly. A solvent more polar than methanol is seldom needed. Often, two solvents are used in a mixture of varying proportions; the polarity of the mixture is a weighted average of the two. Hexane and ether mixtures are often employed.

Finding a good solvent system is usually the most critical aspect of TLC. If the mobile phase has not been previously determined, start with a nonpolar solvent such as hexanes and observe the separation. If the mixture's components do not move very far, try adding a polar solvent such as ether or ethyl acetate to the hexanes. Compare the separation to the previous plate. In most cases, a combination of two solvents is the best choice. If the spots stay at the bottom of the plate, add more of the polar solvent. If they run with the solvent front (move to the top), increase the proportion of the nonpolar solvent. Unfortunately, some trial and error is

TABLE 8.3 *Common Mobile Phases Listed by Increasing Polarity*

Increasing polarity ↓
Helium
Nitrogen
Pentanes (petroleum ether)
Hexanes (ligroin)
Cyclohexane
Carbon tetrachloride*
Toluene
Chloroform*
Dichloromethane (methylene chloride)
t-Butyl methyl ether
Diethyl ether
Ethyl acetate
Acetone
2-Propanol
Pyridine
Ethanol
Methanol
Water
Acetic acid

*Suspected carcinogens

usually involved in determining which solvent system is the best. There is a large amount of literature on the solvents and adsorbents used in the separation of a wide variety of substances.

THE SIX STEPS OF THIN-LAYER CHROMATOGRAPHY

The process of thin-layer chromatography (TLC) can be broken down into six main steps: (1) preparing the sample; (2) spotting the TLC plate; (3) picking a solvent; (4) developing the TLC plate; (5) visualizing the TLC plate; (6) calculating R_f values. A detailed description of each of these steps is given in the following sections.

Step 1. Preparing the Sample

Too much sample is a frequent problem. Use a 1% solution of the mixture. Apply very small spots.

You need to dissolve only a few milligrams of material (a 1% solution) because one can detect a few micrograms of compound on a TLC plate. Choose a volatile solvent such as diethyl ether. Even if the material is only partially soluble, you will normally be able to observe the compound because only low concentrations are needed. If the prepared sample is too concentrated, streaking can occur on the plate. Streaking may lead to the overlap of two or more compounds, thus skewing results. See Figure 8.5 for an example of a streaking spot.

FIG. 8.5
A streaking spot.

FIG. 8.6
A marked and spotted TLC plate.

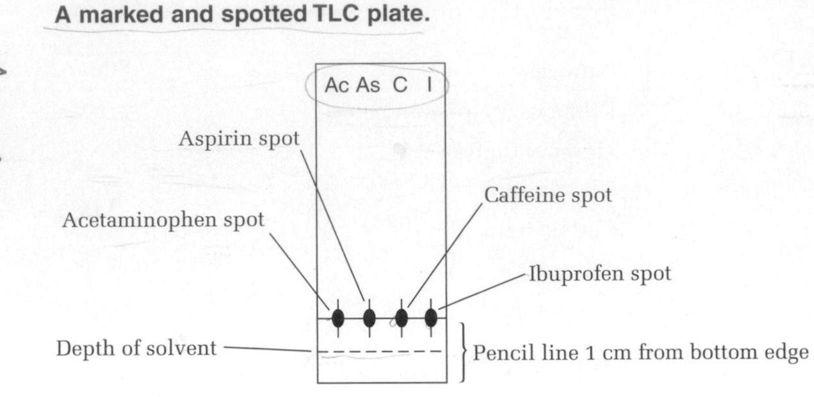

FIG. 8.7
(a) A spotting capillary. (b) Soften the glass by heating at the base of the flame as shown and then stretch to a fine capillary.

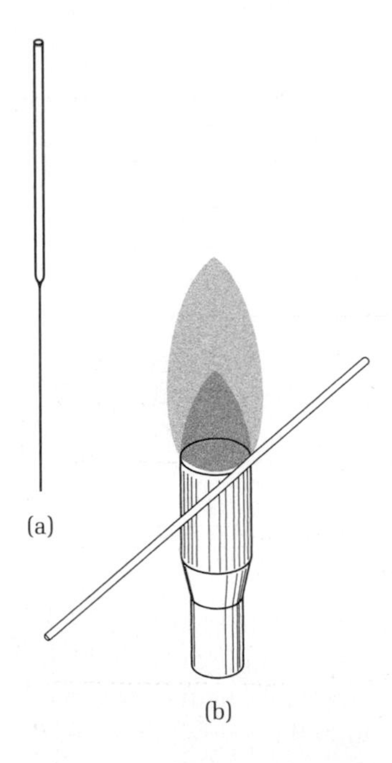

CAUTION: Bunsen burners should be used only in lab areas that are far from flammable organic solvents.

Step 2. Spotting the TLC Plate

The use of commercially available TLC plates, poly(ethylene terephthalate) (Mylar) sheets coated with silica gel using polyacrylic acid as a binder, is highly recommended; these fluoresce under ultraviolet (UV) light.[1] The TLC plates must be handled gently and by the edge, or the 100-mm-thick coating of silica gel can be easily scratched off or contamination of the surface can occur. With a pencil, lightly draw a faint line 1 cm from the end and then three or four short hash marks to guide spotting. Lightly write identifying letters at the top of the plate to keep track of the placement of the compound spots (Fig. 8.6). Note that a pencil is always used to mark TLC plates because the graphite (carbon) is inert. If ink is used to mark the plate, it will chromatograph just as any other organic compound, interfering with the samples and giving flawed results.

Once the sample is prepared, a spotting capillary must be used to add the sample to the plate. Spotting capillaries can be made by drawing out open-end melting point tubes or Pasteur pipette stems in a burner flame (Fig. 8.7).[2] The bore of these capillaries should be so small that once a liquid is drawn into them, it will not flow out to form a drop. Practice spotting just pure solvent onto an unmarked TLC plate. Dip the capillary into the solvent and let a 2–3 cm column of solvent flow into it by capillary action. Do this by holding the capillary vertically over the *coated* side of the plate, and lower the pipette until the tip just touches the adsorbent. Only then will liquid flow onto the plate; quickly withdraw the capillary when the spot is about 1 mm in diameter. The center of the letter *o* on this page is more than 1 mm in diameter. The solvent will evaporate quickly, leaving your mixture behind on the plate. You may have to spot the plate a couple of times in the same place to ensure that sufficient material is present; do not spot too much sample because this will lead to a poor separation. It is extremely important that the spots be as small as possible. If the spot is large, then two or more spots of a sample may overlap on the TLC plate, thus causing erroneous conclusions about the separation and/or the sample's purity or content. Practice placing spots a number of times until you develop good spotting technique. You are now ready to spot the solutions of mixtures as described in Experiments 1 and 2.

[1]Whatman flexible plates for TLC, cat. no. 4410 222 (Fisher cat. no. 05-713-162); cut with scissors to 1" × 3". Unlike student-prepared plates, these coated sheets give very consistent results. A supply of these plates makes it a simple matter to examine most of the reactions in this book for completeness of reaction, purity of product, and side reactions.
[2]Three-inch pieces of old and unusable gas chromatography capillary columns are also effective spotting capillaries.

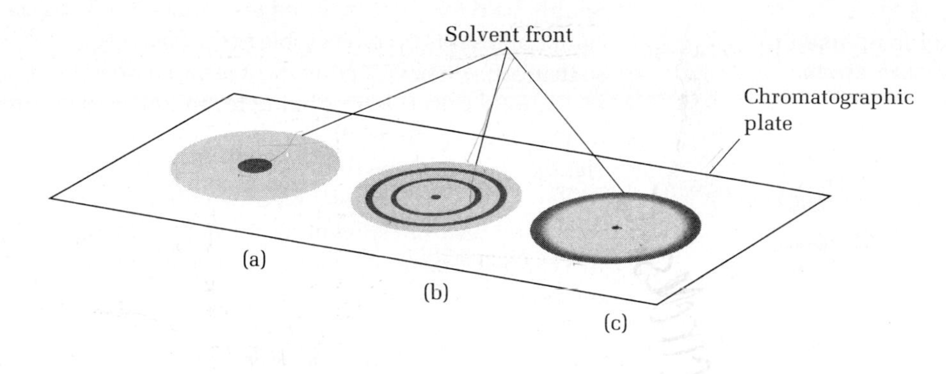

FIG. 8.8
A fast method for determining the correct solvent for TLC. See the text for the procedure.

Step 3. Picking a Solvent

One of the most crucial parts of a successful TLC separation is choosing the appropriate mobile phase; see the section "Mobile Phase Polarity" above. Make some spots of the mixture to be separated on a test TLC plate, either a new or a used one. Touch each spot with a different solvent held in a capillary. Referring to Fig. 8.8, note that the mixture did not travel away from the point of origin in Fig. 8.8a, and in Fig. 8.8b, two concentric rings are seen between the origin and the solvent front. This is how a good solvent behaves. In Fig. 8.12c, the mixture traveled with the solvent front.

FIG. 8.9
Using a wide-mouth bottle to develop a TLC plate.

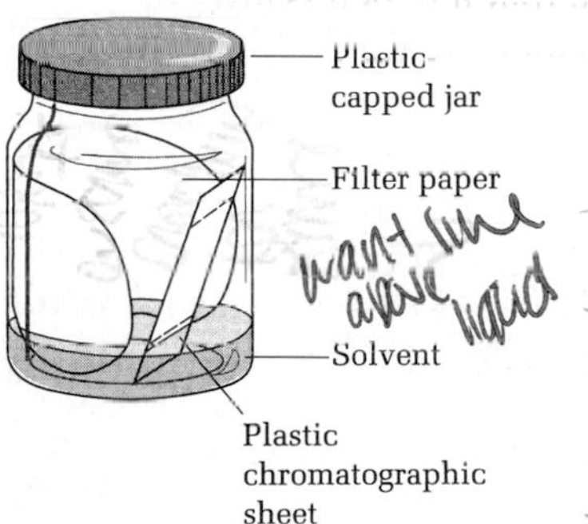

Step 4. Developing the Plate

Once the dilute solution of the mixture has been spotted on the plate, the next step is the actual chromatographic separation, called *plate development*.

Once a mobile phase has been chosen, the marked and spotted TLC plate is inserted into a 4-oz wide-mouth bottle (Fig. 8.9) or beaker (Fig. 8.10) containing 4 mL of the mobile phase, either a solvent or a solvent mixture. The bottle is lined with filter paper that is wet with solvent to saturate the atmosphere within the container. Use tweezers to place the plate in the development chamber; oils from your fingers can sometimes smear or ruin a TLC plate. Also make sure that the origin spots are not below the solvent level in the chamber. If the spots are submerged in the solvent, they are washed off the plate and lost. The top of the container is put in place and the time noted. (If a beaker is used, the beaker is to be covered with aluminum foil.) The solvent travels up the thin layer by capillary action. If the substance is a pure colored compound, one soon sees a spot traveling either along with the solvent front or, more commonly, at some distance behind the solvent front. Once the solvent has run within a centimeter of the top of the plate, remove the plate with tweezers. Immediately, before the solvent evaporates, use a pencil to draw a line across the plate where the solvent front can be seen. The proper location of this solvent front line is needed for R_f calculations.

FIG. 8.10
Using a foil-covered beaker to develop a TLC plate.

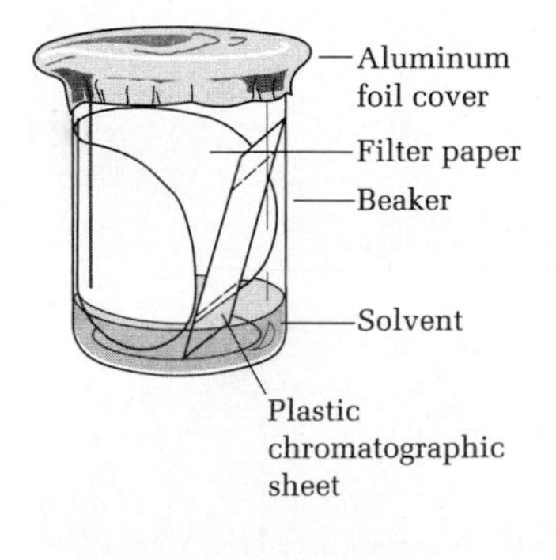

Step 5. Visualizing the Results

If you are fortunate enough to be separating organic molecules that are colored, such as dyes, inks, or indicators, then visualizing the separated spots is easy. However, because most organic compounds are colorless, this is rarely the case.

For most compounds, a UV light works well for observing the separated spots. TLC plates normally contain a fluorescent indicator that makes them glow green under UV light of wavelength 254 nm. Compounds that adsorb UV light at this wavelength will quench the green fluorescence, yielding dark purple or bluish spots on the plate.

FIG. 8.11
A UV lamp used to visualize spots.

⚠ Never look into a UV lamp.

Simply hold the plate by its edges under a UV lamp as shown in Figure 8.11, and the compound spots become visible to the naked eye. Lightly circle the spots with a pencil so that you will have a permanent record of their location for later calculations.

Another useful visualizing technique is to use an iodine (I_2) chamber. Certain compounds, such as alkanes, alcohols, and ethers, do not absorb UV light sufficiently to quench the fluorescence of the TLC plate and therefore will not show up under a UV lamp. However, they will adsorb iodine vapors and can be detected (after any residual solvent has evaporated) by placing the plate for a few minutes in a capped 4-oz bottle containing some crystals of iodine. Iodine vapor is adsorbed by the organic compound to form brown spots. These brown spots should be outlined with a pencil immediately after removing the plate from the iodine bottle because they will soon disappear as the iodine sublimes away; a brief return to the iodine chamber will regenerate the spots. Using both the UV lamp and iodine vapor visualization methods will ensure the location of all spots on the TLC plate.

Many specialized spray reagents, also known as TLC stains, have also been developed to give specific colors for certain types of compounds. A detailed listing of different TLC stains that help visualize certain compounds with specific functional groups can be found in *The Chemist's Companion: A Handbook of Practical Data, Techniques, and References*,[3] a valuable classic reference book that contains copious amounts of helpful information.

Step 6. Calculating R_f Values

R_f is the ratio of the distance the spot travels from the origin to the distance the solvent travels.

In addition to qualitative results, TLC can also provide a chromatographic parameter known as an R_f value. The R_f value is the retention factor expressed as a decimal fraction. The R_f value is the ratio of the distance the spot travels from the point of origin to the distance the solvent travels. The R_f value can be calculated as follows:

$$R_f = \frac{\text{distance spot travels}}{\text{distance solvent travels}}$$

This number should be calculated for each spot observed on a TLC plate. Figure 8.12 shows a diagram of a typical TLC plate and how the distances are

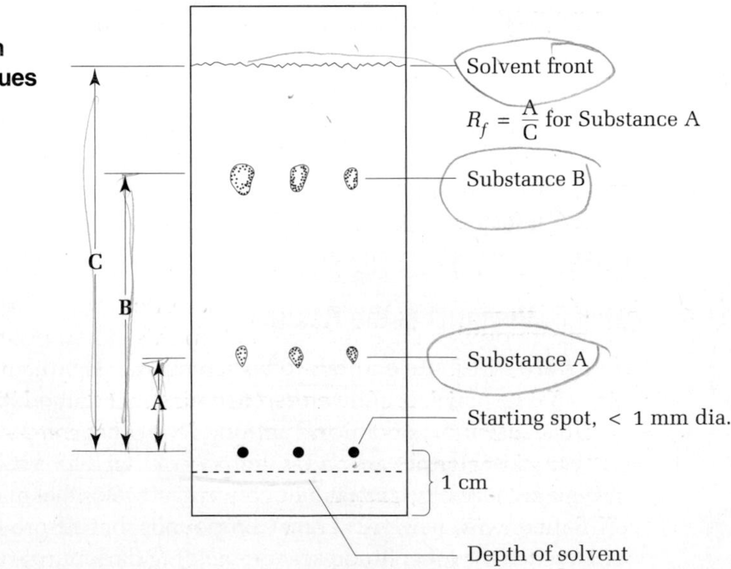

FIG. 8.12
A developed TLC plate with spots visualized and R_f values determined.

[3]Gordon, A. J., Richard A. Ford, The *Chemist's Companion: A Handbook of Practical Data, Techniques, and References.* John Wiley & Sons, Inc.: New York, 1972.

TABLE 8.4 *Chromatography Terms and Their Definitions with Examples*

Chromatography Term	*Definition*	*Examples*
Mixture	A collection of different compounds	Aspirin, ibuprofen, caffeine, and fluorene/fluorenone
Stationary phase	A fixed material that can adsorb compounds	Alumina, silica gel, and silicone gum, et al.
Mobile phase	A moving liquid or gas that dissolves compounds and carries them along	Hexane, CH_2Cl_2, and ethyl acetate (TLC and column chromatography) et al.; helium gas (gas chromatography)
Adsorption	The strength of attraction between the compounds and the stationary phase	London forces, hydrogen bonds, and dipole-dipole attractive forces
Separation	A measure of the elution or migration rate of compounds	R_f (TLC); elution volume (column chromatography); retention time (gas chromatography)

measured to calculate the R_f value. The best separations are usually achieved when the R_f values fall between 0.3 and 0.7.

If two spots travel the same distance or have the same R_f value, then it might be concluded that the two components are the same molecule. Just as many organic molecules have the same melting point and color, many can have the same R_f value, so identical R_f values do not necessarily mean identical compounds. For comparisons of R_f values to be valid, TLC plates must be run under the exact same conditions for the stationary phase, mobile phase, and temperature. Even then, additional information such as a mixed melting point or an IR spectrum should be obtained before concluding that two substances are identical.

COMPARISON OF DIFFERENT TYPES OF CHROMATOGRAPHY

Table 8.4 summarizes the terminology used in chromatography and how these apply to different types of chromatography. All of the chromatographic types involve the same principles but vary in the nature of the stationary phase and the mobile phase, and the measure of separation.

EXPERIMENTS

1. ANALGESICS

Analgesics are substances that relieve pain. The most common of these is aspirin, a component of more than 100 nonprescription drugs. In Chapter 41, the history of this most popular drug is discussed. In this experiment, analgesic tablets will be analyzed by TLC to determine which analgesics they contain and whether they contain caffeine, which is often added to counteract the sedative effects of the analgesic.

In addition to aspirin and caffeine, the most common components of the currently used analgesics are acetaminophen and ibuprofen. In addition to one or more of these substances, each tablet contains a binder—often starch, microcrystalline cellulose, or silica gel. And to counteract the acidic properties of aspirin, an inorganic buffering agent is added to some analgesics. An inspection of analgesic labels will reveal that most cold remedies and decongestants contain both aspirin and caffeine in addition to the primary ingredient.

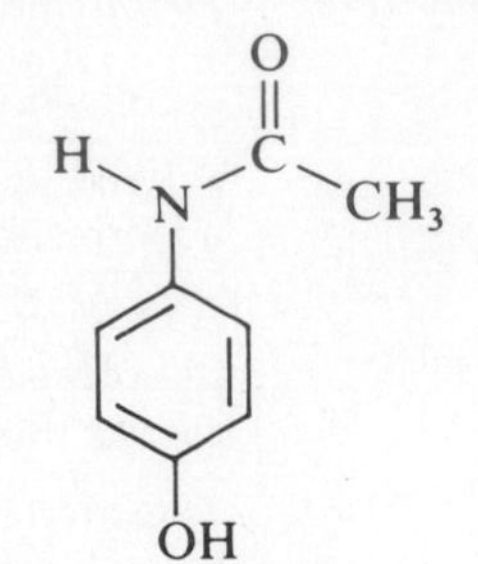

Aspirin
Acetylsalicylic acid

Acetaminophen
4-Acetamidophenol

Ibuprofen
2-(4-Isobutylphenyl)propionic acid

Caffeine

To identify an unknown by TLC, the usual strategy is to run chromatograms of known substances (the standards) and the unknown at the same time. If the unknown has one or more spots that correspond to spots with the same R_f values as the standards, then those substances are probably present.

Proprietary drugs that contain one or more of the common analgesics and sometimes caffeine are sold under some of the following brand names: Bayer Aspirin, Anacin, Datril, Advil, Excedrin, Extra Strength Excedrin, Tylenol, and Vanquish. Note that ibuprofen has a chiral carbon atom. The *S*-(+)-enantiomer is more effective than the other.

Procedure

Before proceeding, practice the TLC spotting technique described earlier. Following that procedure, draw a light pencil line about 1 cm from the end of a chromatographic plate. On this line, spot aspirin, acetaminophen, ibuprofen, and caffeine, which are available as reference standards. Use a separate capillary for each standard (or rinse the capillary carefully before reusing). Make each spot as small as possible, preferably less than 0.5 mm in diameter. Examine the plate under UV light to see that enough of each compound has been applied; if not, add more. On a separate plate, run the unknown and one or more of the standards.

Because of the insoluble binder, not all of the unknown will dissolve.

The unknown sample is prepared by crushing a part of a tablet, adding this powder to a test tube or small vial along with an appropriate amount of ethanol, and then mixing the suspension. Not all of the crushed tablet will dissolve, but enough will go into solution to spot the plate. The binder—starch or silica—will not dissolve. Weigh out only part of the tablet to try to prepare a 1% solution of the unknown. Typically, ibuprofen tablets contain 200 mg of the active ingredient, aspirin tablets contain 325 mg, and acetaminophen tablets contain 500 mg.

To the developing jar or beaker (*see* Fig. 8.9 or Fig. 8.10 on page 173), add 4 mL of the mobile phase, a mixture of 95% ethyl acetate and 5% acetic acid. Insert the spotted TLC plates with tweezers. After the solvent has risen nearly to the top of the plate, remove the plate from the developing chamber, mark the solvent front with a pencil, and allow the solvent to dry. Examine the plate under UV light to see the components as dark spots against a bright green-blue background. Outline the spots with a pencil. The spots can also be visualized by putting the plate in an iodine chamber made by placing a few crystals of iodine in the bottom of a capped 4-oz jar. Calculate the R_f values for the spots and identify the components in the unknown.

Cleaning Up. Solvents should be placed in the organic solvents waste container; dry, used chromatographic plates can be discarded in the nonhazardous solid waste container.

2. PLANT PIGMENTS

The botanist Michael Tswett discovered the technique of chromatography and applied it, as the name implies, to colored plant pigments. The leaves of plants contain, in addition to chlorophyll-a and chlorophyll-b, other pigments that are revealed in the fall when the leaves die and the chlorophyll rapidly decomposes. Among the most abundant of the other pigments are the carotenoids, which include the carotenes and their oxygenated homologs, the xanthophylls. The bright orange beta-carotene is the most important of these because it is transformed in the liver to vitamin A, which is required for night vision.

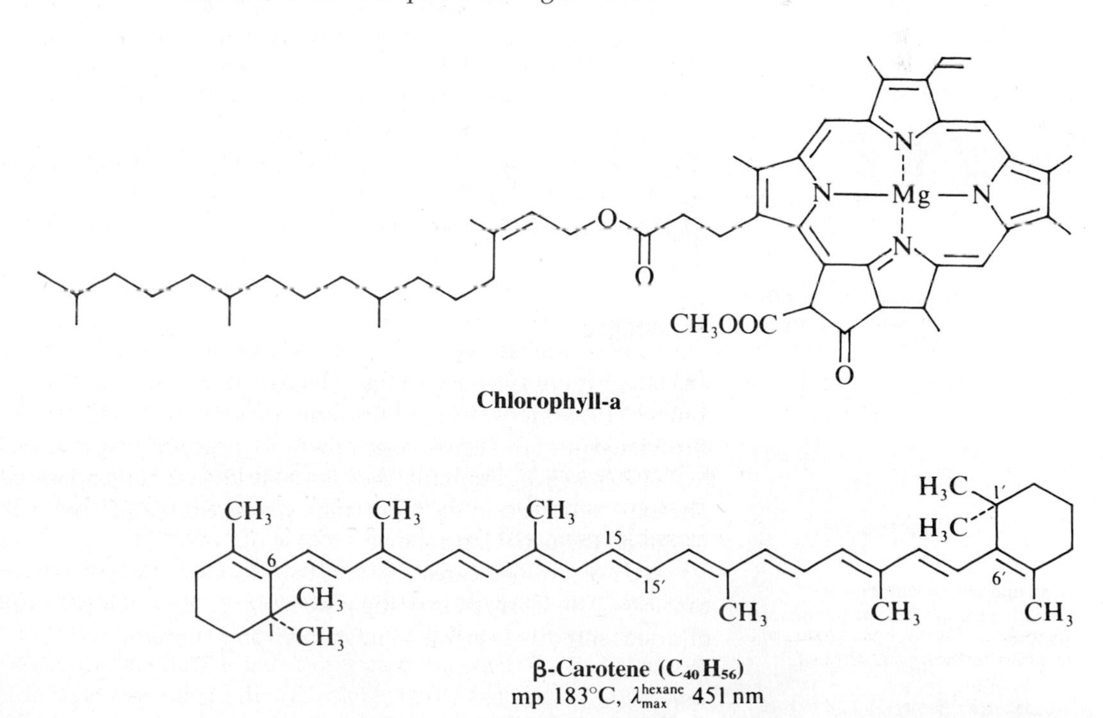

Chlorophyll-a

β-Carotene ($C_{40}H_{56}$)
mp 183°C, λ_{max}^{hexane} 451 nm

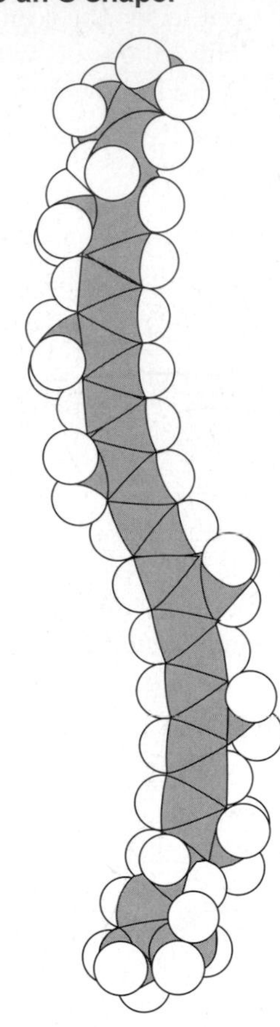

FIG. 8.13 An energy-minimized, space-filling model of lycopene. The molecule is flat, but steric hindrance of the methyl groups causes the molecule to bend into an S shape.

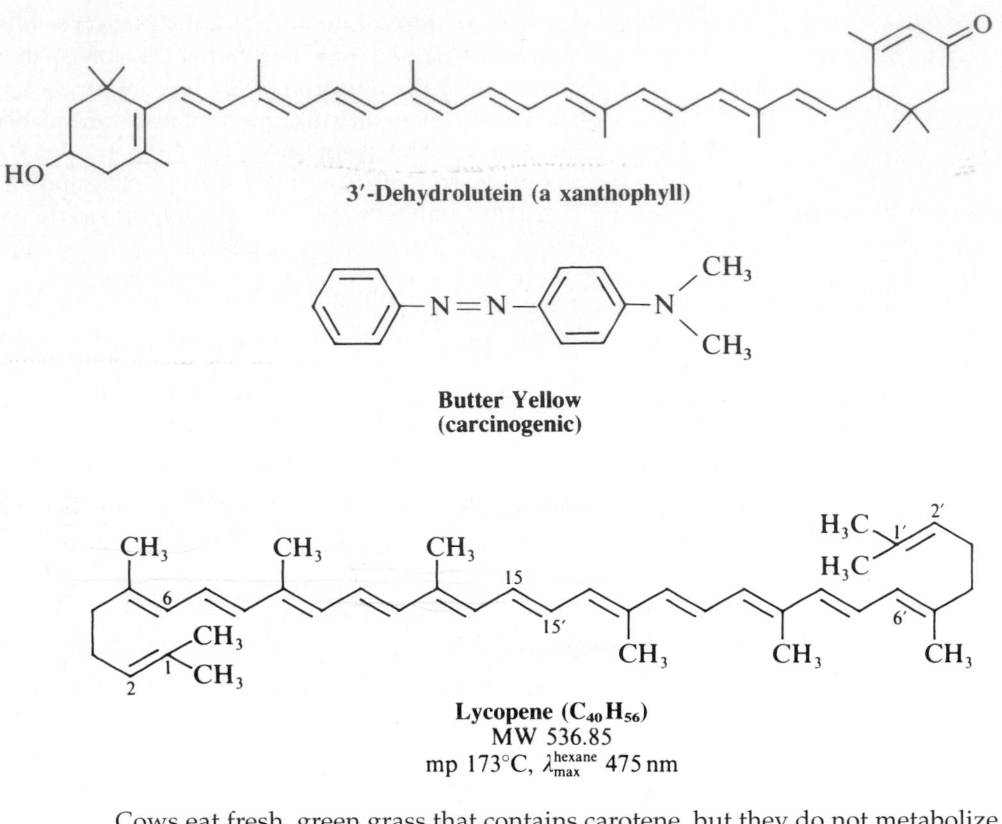

3′-Dehydrolutein (a xanthophyll)

Butter Yellow (carcinogenic)

Lycopene ($C_{40}H_{56}$)
MW 536.85
mp 173°C, λ_{max}^{hexane} 475 nm

Cows eat fresh, green grass that contains carotene, but they do not metabolize the carotene entirely, so it ends up in their milk. Butter made from this milk is therefore yellow. In the winter, the silage that cows eat does not contain carotene because that compound readily undergoes air oxidation, and the butter made at that time is white. For some time, an azo dye called Butter Yellow was added to winter butter to give it the accustomed color, but the dye was found to be a carcinogen. Now winter butter is colored with synthetic carotene, as is all margarine.

Lycopene from tomato paste and Gr beta-carotene from strained carrots.

Lycopene (Fig. 8.13), the red pigment of the tomato, is a C_{40}-carotenoid made up of eight isoprene units. (Gr beta-Carotene, the yellow pigment of the carrot, is an isomer of lycopene in which the double bonds at C_1—C_2 and C'_1—C'_2 are replaced by bonds extending from C_1 to C_6 and from C'_1 to C'_6 to form rings. The chromophore in each case is a system of 11 *all-trans* conjugated double bonds; the closing of the two rings causes Gr beta-carotene to absorb at shorter wavelengths than lycopene does, shifting its color from red to yellow.

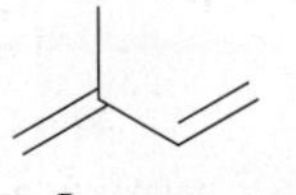

Isoprene

In 1911, Richard Willstätter and Heinrich R. Escher isolated 20 mg of lycopene per kilogram of fresh tomatoes, which contain about 96% water.[4] They then found a more convenient source in commercial tomato paste: the seeds and skin were eliminated, and the water content was reduced by evaporation in vacuum to a content of 26% solids. From this they isolated 150 mg of lycopene per kilogram of paste. The expected yield in the following experiment is 0.075 mg, which is not enough to weigh on the usual lab balance.

As an interesting variation, try extracting lycopene from commercial ketchup.

A jar of strained carrots sold as baby food serves as a convenient source of beta-carotene. The German investigators isolated 1 g of Gr beta-carotene per kilogram of dried, shredded carrots of unstated water content.

[4]Willstätter, R.; Escher, H. R. *Z Physiol. Chem.* **1911**, *64*, 47–61.

The following procedure calls for the dehydration of tomato or carrot paste with ethanol and extraction with dichloromethane, an efficient solvent for lipids.

Experimental Considerations

⚠ **CAUTION:** Do not breathe the vapors of dichloromethane. Carry out the extraction in the hood.

Carotenoids are very sensitive to light-catalyzed air oxidation. Perform this experiment as rapidly as possible; keep the solutions as cool and dark as possible. This extraction produces a mixture of products that can be analyzed by both TLC and column chromatography (*see* Chapter 9). If enough material for TLC only is desired, use one-tenth the quantities of starting material and solvents employed in the following procedure. This extraction can also be carried out with hexane if the ventilation is not adequate enough to use dichloromethane. However, hexane is more prone to form emulsions than the chlorinated solvent.

Procedure

IN THIS EXPERIMENT, some tomato or carrot paste is treated with acetone, which will remove water and lipids but not the highly colored carotenoid hydrocarbons. The carotenoids are extracted by dichloromethane and analyzed by TLC.

A 5-g sample of fresh tomato or carrot paste (baby food) is transferred to the bottom of a 25 × 150 mm test tube, followed by 10 mL of acetone. The mixture is stirred and shaken before being filtered on a Hirsch funnel. Scrape as much of the material from the tube as possible and press it dry on the funnel. Let the tube drain thoroughly. Place the filtrate in a 125-mL Erlenmeyer flask.

Return the solid residue to the test tube, shake it with a 10-mL portion of dichloromethane, and again filter the material on a Hirsch funnel. Add the filtrate to the 125-mL flask. Repeat this process two more times and then pour the combined filtrates into a separatory funnel. Add water and a sodium chloride solution (which aids in the breaking of emulsions) and shake the funnel gently. This aqueous extraction will remove the acetone and any water-soluble components from the mixture, leaving the hydrocarbon carotenoids in the dichloromethane. Dry the colored organic layer over anhydrous calcium chloride and filter the solution into a dry flask. Remove about 0.5 mL of this solution and store it under nitrogen in the dark until it can be analyzed by TLC (the carotenoids are very susceptible to air oxidation). Evaporate the remainder of the dichloromethane solution to dryness under a stream of nitrogen or under vacuum on a rotary evaporator. This material can be used for column chromatography (*see* Chapter 9). If it is to be stored, fill the flask with nitrogen and store it in a dark place.

Air can be used for the evaporation, but nitrogen is better because these hydrocarbons air oxidize with great rapidity.

Thin-Layer Chromatography

Spot the mixture on a TLC plate about 1 cm from the bottom and 8 mm from the edge. Make one spot concentrated by repeatedly touching the plate in the exact same location, but ensure that the spot is as small as possible, certainly less than 1 mm in diameter. The other spot can be of lower concentration. Develop the plate with an 80:20 hexane-acetone mixture. With other plates, you could try cyclohexane and toluene as eluents and also hexane-ethanol mixtures of various compositions.

Many spots may be seen. There are two common carotene and chlorophyll isomers and four xanthophyll isomers.

The container in which the chromatography is carried out should be lined with filter paper that is wet with the solvent so that the atmosphere in the container will be saturated with the solvent vapor. After elution is completed, remove the TLC plate and mark the solvent front with a pencil and outline the colored spots. Examine the plate under UV light. Also place the plate in an iodine chamber to visualize the spots.

Cleaning Up. The aqueous saline filtrate containing acetone can be flushed down the drain. Recovered and unused dichloromethane should be placed in the halogenated organic waste container; the solvents used for TLC should be placed in the organic solvents waste container. If local regulations allow, evaporate any residual solvent from the drying agents in the hood and place the dried solid in the nonhazardous waste container. Otherwise, place the wet drying agent in a waste container designated for this purpose. Used plant material and dry TLC plates can be discarded in the nonhazardous waste container.

Procedure

In a small mortar, grind 2 g of green or brightly colored fall leaves (do not use ivy or waxy leaves) with 10 mL of ethanol, pour off the ethanol (which serves to break up and dehydrate the plant cells), and grind the leaves successively with three 1-mL portions of dichloromethane that are decanted or withdrawn with a Pasteur pipette and placed in a test tube. The pigments of interest are extracted by the dichloromethane. Alternatively, place 0.5 g of carrot paste (baby food) or tomato paste in a test tube, stir and shake the paste with 3 mL of ethanol until the paste has a somewhat dry or fluffy appearance, remove the ethanol, and extract the dehydrated paste with three 1-mL portions of dichloromethane. Stir and shake the plant material with the solvent to extract as much of the pigments as possible.

Fill the tube containing the dichloromethane extract from leaves or vegetable paste with a saturated sodium chloride solution and shake the mixture. Remove the aqueous layer; add anhydrous calcium chloride pellets to the dichloromethane solution until the drying agent no longer clumps together. Shake the mixture with the drying agent for about 5 minutes and then withdraw the solvent with a Pasteur pipette and place it in a test tube. Add to the solvent a few pieces of Dri-Rite to complete the drying process. Gently stir the mixture for about 5 minutes, transfer the solvent to a test tube, wash off the drying agent with more solvent, and then evaporate the combined dichloromethane solutions under a stream of nitrogen while warming the tube in your hand or in a beaker of warm water. Carry out this evaporation in the hood.

These hydrocarbons air oxidize with great rapidity.

Immediately cork the tube filled with nitrogen and then add 1 or 2 drops of dichloromethane to dissolve the pigments for TLC analysis. Carry out the analysis without delay by spotting the mixture on a TLC plate about 1 cm from the bottom and 8 mm from the edge. Make one spot concentrated by repeatedly touching the plate, but ensure that the spot is as small as possible—less than 1.0 mm in diameter. The other spot can be of lower concentration. Develop the plate with a 70:30 hexane-acetone mixture. With other plates, try cyclohexane and toluene as eluents and also hexane-ethanol mixtures of various compositions. The container in which the chromatography is carried out should be lined with filter paper that is wet with the solvent so that the atmosphere in the container will be saturated with solvent vapor. After elution is completed, mark the solvent front with a pencil and outline the colored spots. Examine the plate under the UV light. Are any new spots seen? Report colors and R_f values for all of your spots and identify each as lycopene, carotene, chlorophyll, or xanthophyll.

Cleaning Up. The ethanol used for the dehydration of the plant material can be flushed down the drain along with the saturated sodium chloride solution. Recovered and unused dichloromethane should be placed in the halogenated organic waste container. The solvents used for TLC should be placed in the organic solvents waste container. If local regulations allow, evaporate any residual solvent from the drying agents in the hood and place the dried solid in the nonhazardous waste container. Otherwise, place the wet drying agent in a waste container designated for this purpose. Used plant material and dry TLC plates can be discarded in the nonhazardous waste container.

3. FOR FURTHER INVESTIGATION

Many of the pigments in plants are made up of compounds called *anthocyanins*. Grind about 4 g of colored plant tissue (flower petals, blueberries, strawberries, cranberries, apple skins, red cabbage, red or purple grapes, etc.) and a small amount of alumina or fine sand with about 4 mL of a mixture of 99% methanol and 1% hydrochloric acid. Spot the extract on cellulose TLC plates and elute with a solvent mixture of 20% concentrated hydrochloric acid, 40% water, and 4% formic acid. Note the number and color of the spots. Look up the structures of the possible anthocyanins, of which many are glycosides of the aglycones delphinidin, peonidin, malvidin, and cyanidin, among others.

4. COLORLESS COMPOUNDS

You will now apply the thin-layer technique to a group of colorless compounds. The spots can be visualized under UV light if the plates have been coated with a fluorescent indicator; chromatograms can also be developed in a 4-oz bottle containing crystals of iodine. During development, spots will appear rapidly but remember that they also disappear rapidly. Therefore, outline each spot with a pencil immediately on withdrawal of the plate from the iodine chamber. Some suggested solvents are pure cyclohexane, pure toluene, toluene (3 mL) plus dichloromethane (1 mL), or toluene (4.5 mL) plus methanol (1/2 mL).

The compounds for trial are to be selected from the following list (all 1% solutions in toluene; those compounds with an asterisk are fluorescent under UV light):

1. Anthracene*
2. Cholesterol
3. 2,7-Dimethyl-3,5-octadiyne-2,7-diol
4. Diphenylacetylene
5. *trans,trans*-1,4-Diphenyl-1,3-butadiene*
6. *p*-Di-*t*-butylbenzene
7. 1,4-Di-*t*-butyl-2,5-dimethoxybenzene
8. *trans*-Stilbene
9. 1,2,3,4-Tetraphenylnaphthalene*
10. Tetraphenylthiophene
11. *p*-Terphenyl*
12. Triphenylmethanol
13. Triptycene

Except for tetraphenylthiophene, the structures for all of these molecules will be found in this book.

Make your own selections.

It is up to you to make selections and to plan your own experiments. Do as many as time permits. One plan would be to select a pair of compounds that are estimated to be separable and that have R_f values determinable with the same solvent. One can assume that a hydroxyl compound will travel less rapidly with a hydrocarbon solvent than a hydroxyl-free compound; you will therefore know what to expect if the solvent contains a hydroxylic component. An aliphatic solvent should carry along an aromatic compound with aliphatic substituents better than one without such groups. Follow the procedure in Step 3 for picking a solvent.

Preliminary trials on used plates.

Once a solvent is chosen, run a complete chromatogram on the two compounds on a fresh plate. If separation of the two compounds seems feasible, put two spots of one compound on a plate, let the solvent evaporate, and put spots of the second compound over the first ones. Run a chromatogram and see if you can detect two spots in either lane (with colorless compounds, it is advisable not to attempt a three-lane chromatogram until you have acquired considerable practice and skill).

Cleaning Up. Solvents should be placed in the organic solvents waste container, and dry, used chromatographic plates can be discarded in the nonhazardous solid waste container.

DISCUSSION

If you have investigated hydroxylated compounds, you doubtless have found that it is reasonably easy to separate a hydroxylated from a nonhydroxylated compound or a diol from a mono-ol. How, by a simple reaction followed by a thin-layer chromatogram, could you separate cholesterol from triphenylmethanol? Heating a sample of each with acetic anhydride and a trace of pyridine catalyst for 5 minutes on a steam bath, followed by chromatography, should do it.

A first trial of a new reaction leaves questions about what has happened and how much, if any, starting material is present. A comparative chromatogram of the reaction mixture with starting material may tell the story. How crude is a crude reaction product? How many components are present? The thin-layer technique may give the answers to these questions and suggest how best to process the product. A preparative column chromatogram may afford a large number of fractions of eluent (say, 1 to 30). Some fractions probably contain nothing and should be discarded, while others should be combined for evaporation and workup. How can you identify the good and the useless fractions? Take a few used plates and put numbered circles on clean places of each; spot samples of each of the fractions and, without any chromatography, develop the plates with iodine. Negative fractions for discard will be obvious, and the pattern alone of positive fractions may allow you to infer which fractions can be combined. Thin-layer chromatograms of the first and last fractions of each suspected group would then show whether or not your inferences are correct.

FLUORESCENCE

Four of the compounds listed in Experiment 4 are fluorescent under UV light. These compounds give colorless spots that can be picked up on a chromatogram by fluorescence (after removal from the UV-absorbing glass developing bottle). If a UV light source is available, spot the four compounds on a used plate and observe the fluorescence.

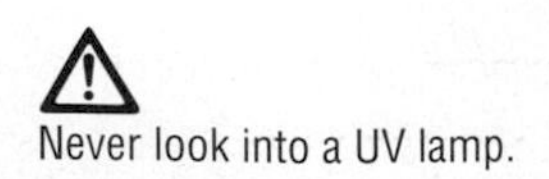
Never look into a UV lamp.

See *Web Links*

Take this opportunity to examine a white shirt or handkerchief under UV light to see if it contains a brightener, that is, a fluorescent white dye or optical bleach. These substances are added to counteract the yellow color that repeated washing gives to cloth. Brighteners of the type of Calcofluor White MR, a sulfonated *trans*-stilbene derivative, are commonly used in detergent formulations for cotton. A substituted coumarin derivative is typical of brighteners used for nylon, acetate, and wool. Detergents normally contain 0.1%–0.2% of optical bleach. The amount of dye on a freshly laundered shirt is approximately 0.01% of the weight of the fabric.

THIN-LAYER CHROMATOGRAPHY SUMMARY

1. **Preparing the sample**. Make a 1% solution of the sample in a low-boiling solvent. It is extremely important to make the solution at this concentration. If more sample is needed, it can be applied repeatedly at the same spot.
2. **Spotting the TLC Plate**. Prepare a very fine capillary from a melting point capillary. With a pencil, mark a line 1 cm from the bottom of the plate. Apply the sample on the line and make the spot no more than 1 mm in diameter.
3. **Picking a Solvent**. Apply test spots to a new or used TLC plate. Add a drop of a possible solvent to each spot. Pick the best solvent (Refer to Fig. 8.8).
4. **Developing the Plate**. Place the plate in a covered jar or beaker lined with filter paper. The depth of solvent should be about 0.5 cm. When the solvent rises to within about a cm of the top of the plate, remove it and mark the solvent front with a pencil.
5. **Visualizing the Results**. After the solvent evaporates, observe the plate under a UV light and/or place the plate in a sealed container containing a few crystals of iodine. Circle the developed spots while they are visible.
6. **Calculating the R_f value**. Measure the distance the center of the spot has traveled above the starting line and divide this distance by the distance the solvent front traveled. The result is the R_f value.

QUESTIONS

1. Why might it be very difficult to visualize the separation of *cis*- and *trans*-2-butene by TLC?
2. What error is introduced into the determination of an R_f value if the top is left off the developing chamber?
3. What problem will ensue if the level of the developing liquid is higher than the applied spot in a TLC analysis?
4. In what order (from top to bottom) would you expect to find naphthalene, butyric acid, and phenyl acetate on a silica gel TLC plate developed with dichloromethane?
5. In carrying out an analysis of a mixture, what do you expect to see when the TLC plate has been allowed to remain in the developing chamber too long, so that the solvent front has reached the top of the plate?

6. Arrange the following in order of increasing R_f with TLC: acetic acid, acetaldehyde, 2-octanone, decane, and 1-butanol.
7. What will be the result of applying too much compound to a TLC plate?
8. Why is it necessary to run TLC in a closed container and to have the interior vapor saturated with the solvent?
9. What will be the appearance of a TLC plate if a solvent of too low polarity is used for the development? A solvent of too high polarity?
10. A TLC plate showed two spots with R_f values of 0.25 and 0.26. The plate was removed from the developing chamber, the residual solvent was allowed to evaporate from the plate, and then the plate was returned to the developing chamber. What would you expect to see after the second development was complete?
11. One of the analgesics has a chiral center. Which compound is it? One of the two enantiomers is far more effective at reducing pain than the other.
12. Using a ruler to measure distances, calculate the R_f value for substance B in Figure 8.10.

CHAPTER 9

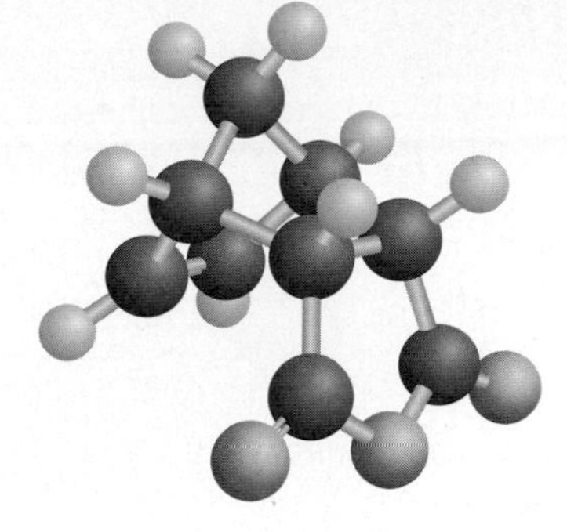

Column Chromatography: Fluorenone, Cholesteryl Acetate, Acetylferrocene, and Plant Pigments

PRELAB EXERCISE: Compare column chromatography and thin-layer chromatography (TLC) with regard to the (1) quantity of material that can be separated, (2) time needed for the analysis, (3) solvent systems used, and (4) ability to separate compounds.

Column chromatography is one of the most useful methods for the separation and purification of both solids and liquids when carrying out microscale experiments. It becomes expensive and time consuming, however, when more than about 10 g of material must be purified. Column chromatography involves the same chromatographic principles as detailed for TLC in Chapter 8, so be sure that you understand those before doing the experiments in this chapter.

As discussed in Chapter 1, organic chemists obtain new compounds by synthesizing or isolating natural products that have been biosynthesized by microbes, plants, or animals. In most cases, initial reaction products or cell extracts are complex mixtures containing many substances. As you have seen, recrystallization, distillation, liquid/liquid extraction, and sublimation can be used to separate and purify a desired compound from these mixtures. However, these techniques are frequently not adequate for removing impurities that are closely related in structure. In these cases, column chromatography is often used. The broad applicability of this technique becomes obvious if you visit any organic chemistry research lab, where chromatography columns are commonplace.

Three of the five experiments in this chapter involve synthesis and may be your first experience in running an organic reaction. Experiments 1 and 2 involve the synthesis of a ketone. In Experiment 3, an ester of cholesterol is prepared. Experiment 4 demonstrates the separation of colored compounds. Experiment 5 involves the isolation and separation of natural products (plant pigments), which is analogous to Experiment 2 in Chapter 8 but on a larger scale.

The most common adsorbents for column chromatography—silica gel and alumina—are the same stationary phases as used in TLC. The sample is dissolved in a small quantity of solvent (the eluent) and applied to the top of the column. The eluent, instead of rising by capillary action up a thin layer, flows down through the column filled with the adsorbent. Just as in TLC, there is an equilibrium established between the solute adsorbed on the silica gel or alumina and the eluting solvent flowing down through the column, with the less strongly absorbed solutes moving ahead and eluting earlier.

Three mutual interactions must be considered in column chromatography: the activity of the stationary adsorbent phase, the polarity of the eluting mobile solvent phase, and the polarity of the compounds in the mixture being chromatographed.

ADDITIONAL PRINCIPLES OF COLUMN CHROMATOGRAPHY

Adsorbents

A large number of adsorbents have been used for column chromatography, including cellulose, sugar, starch, and inorganic carbonates; but most separations employ alumina $[(Al_2O_3)_x]$ or silica gel $[(SiO_2)_x]$. Alumina comes in three forms: acidic, neutral, and basic. The neutral form of Brockmann activity grade II or III, 150 mesh, is most commonly employed. The surface area of this alumina is about 150 m^2/g. Alumina as purchased will usually be activity grade I, meaning that it will strongly adsorb solutes. It must be deactivated by adding water, shaking, and allowing the mixture to reach equilibrium over an hour or so. The amount of water needed to achieve certain activities is given in Table 9.1. The activity of the alumina on TLC plates is usually about III. Silica gel for column chromatography, 70–230 mesh, has a surface area of about 500 m^2/g and comes in only one activity.

TABLE 9.1 *Alumina Activity*

Brockmann activity grade	I	II	III	IV	V
Percent by weight of water	0	3	6	10	15

TABLE 9.2 *Elutropic Series for Solvents*

n-Pentane (least polar)
Petroleum ether
Cyclohexane
Hexanes
Carbon disulfide
t-Butyl methyl ether
Dichloromethane
Tetrahydrofuran
Dioxane
Ethyl acetate
2-Propanol
Ethanol
Methanol
Acetic acid (most polar)

Solvents

Solvent systems for use as mobile phases in column chromatography can be determined from TLC, the scientific literature, or experimentally. Normally, a separation will begin with a nonpolar or low-polarity solvent, allowing the compounds to adsorb to the stationary phase; then the polarity of the solvent is *slowly* increased to desorb the compounds and allow them to move with the mobile phase. The polarity of the solvent should be changed gradually. A sudden change in solvent polarity will cause heat evolution as the alumina or silica gel adsorbs the new solvent. This will vaporize the solvent, causing channels to form in the column that severely reduce its separating power.

Several solvents are listed in Table 9.2, arranged in order of increasing polarity (elutropic series), with *n*-pentane being the least polar. The order shown in the table reflects the ability of these solvents to dislodge a polar substance adsorbed onto either silica gel or alumina, with *n*-pentane having the lowest solvent power.

As a practical matter, the following sequence of solvents is recommended in an investigation of unknown mixtures: elute first with petroleum ether (pentanes); then hexanes; followed by hexanes containing 1%, 2%, 5%, 10%, 25%, and 50% ether; pure ether; ether and dichloromethane mixtures; followed by dichloromethane and methanol mixtures. Either diethyl ether or *t*-butyl methyl ether can be used, but *t*-butyl methyl ether is recommended. Solvents such as methanol and water are normally not used because they can destroy the integrity of the stationary phase by dissolving some of the silica gel. Some typical solvent combinations are hexanes-dichloromethane, hexanes-ethyl acetate, and hexanes-toluene. An experimentally determined ratio of these solvents can sufficiently separate most compounds.

Petroleum ether: mostly isometric pentanes.

TABLE 9.3 *Elution Order for Solutes*

Alkanes (first)
Alkenes
Dienes
Aromatic hydrocarbons
Ethers
Esters
Ketones
Aldehydes
Amines
Alcohols
Phenols
Acids (last)

Compound Mobility

The ease with which different classes of compounds elute from a column is indicated in Table 9.3. Molecules with nonpolar functional groups are least adsorbed and elute first, while more polar or hydrogen-bonding molecules are more strongly adsorbed and elute later. The order is similar to that of the eluting solvents—another application of "like dissolves like."

Sample and Column Size

Chromatography columns can be as thin as a pencil for milligram quantities to as big as a barrel for the industrial-scale separation of kilogram quantities. A microscale column for the chromatography of about 50 mg of material is shown in Figure 9.1; columns with larger diameters, as shown in Figures 9.2 and 9.3, are used for macroscale procedures. The amount of alumina or silica gel used should generally weigh at least 30 times as much as the sample, and the column, when packed, should have a height at least 10 times the diameter. The density of silica gel is 0.4 g/mL, and the density of alumina is 0.9 g/mL, so the optimum size for any column can be calculated.

Packing the Column

Microscale Procedure

Before you pack the column, tare several Erlenmeyer flasks, small beakers, or 20-mL vials to use as receivers. Weigh each one carefully and mark it with a number on the etched circle.

Uniform packing of the chromatography column is critical to the success of this technique. Two acceptable methods for packing a column are dry packing and slurry packing, which normally achieve the best results. Assemble the column as depicted in Figure 9.1. To measure the amount of adsorbent, fill the column one-half to two-thirds full; then pour the powder out into a small beaker or flask. Clamp the column in a vertical position and close the valve. Always grasp the valve with one hand while turning it with the other. Fill the column with a non-polar solvent such as hexanes almost to the top.

Photo: *Column Chromatography*, Video: *Column Chromatography*

Dry Packing Method. This is the simplest method for preparing a microscale column. Slowly add the powdered alumina or silica gel through the funnel while gently tapping the side of the column with a pencil. The solid should "float" to the bottom of the column. Try to pack the column as evenly as possible; cracks, air bubbles, and channels in the powder will lead to a poor separation.

Slurry Packing Method. To slurry pack a column, add about 8 mL of hexanes to the adsorbent in a flask or beaker, stir the mixture to eliminate air bubbles, and

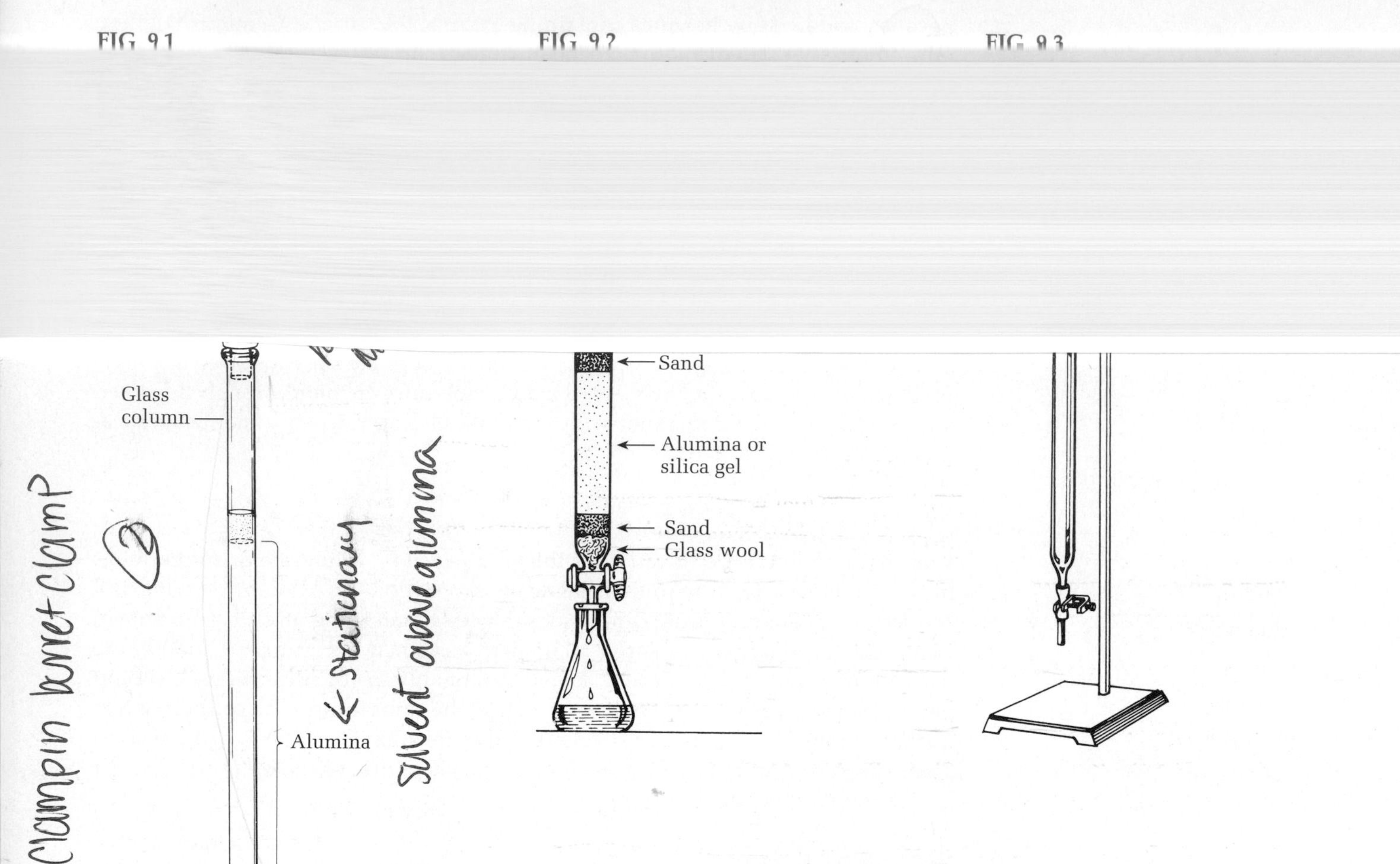

then (this is the hard part) swirl the mixture to get the adsorbent suspended in the solvent and immediately pour the entire slurry into the funnel. Open the valve, drain some solvent into the flask that had the adsorbent in it and finish transferring the slurry to the column. Place an empty flask under the column and allow the solvent to drain to about 5 mm above the top surface of the adsorbent. Tap the column with a pencil until the packing settles to a minimum height. Try to pack the column as evenly as possible; cracks, air bubbles, and channels in the packed column will lead to a poor separation.

The slurry method normally gives the best column packing, but it is also the more difficult technique to master. Whether the dry packing or slurry packing method is chosen, the most important aspect of packing the column is creating an evenly distributed and packed stationary phase. The slurry method is often used for macroscale separations.

Once the column is loaded with solvent and adsorbent, place a flask under it, open the stopcock (use two hands for the microscale column), and allow the solvent level to drop to the *top* of the packing. Avoid allowing the solvent level to go below the stationary phase (known as letting the column "run dry") because this allows air bubbles and channel formation to occur, which leads to a poor separation.

Macroscale Procedure

Extinguish all flames; work in the laboratory hood.

Before you pack the column, prepare several small Erlenmeyer flasks to use as receivers by taring (weighing) each one carefully, recording the weight in your laboratory notebook, and marking each with a number on the etched circle.

Dry Packing Method

The column can be prepared using a 50-mL burette such as the one shown in Figure 9.2 or using the less expensive and equally satisfactory chromatographic tube shown in Figure 9.3, in which the flow of solvent is controlled by a screw pinchclamp. Weigh the required amount of silica gel (12.5 g in the first experiment), close the pinchclamp on the tube, and fill about half full with a 90:10 mixture of hexanes and ether. With a wooden dowel or glass rod, push a small plug of glass wool through the liquid to the bottom of the tube, dust in through a funnel enough sand to form a 1-cm layer over the glass wool, and level the surface by tapping the tube. Unclamp the tube. With your right hand, grasp both the top of the tube and the funnel so that the whole assembly can be shaken to dislodge silica gel that may stick to the walls; with your left hand pour in the silica gel slowly (Fig. 9.4) while tapping the column with a rubber stopper fitted on the end of a pencil. If necessary, use a Pasteur pipette full of a 90:10 mixture of hexanes and ether to wash down any silica gel that adheres to the walls of the column above the liquid. When the silica gel has settled, add a little sand to provide a protective layer at the top. Open the pinchclamp, let the solvent level fall until it is just slightly above the upper layer of sand, and then stop the flow.

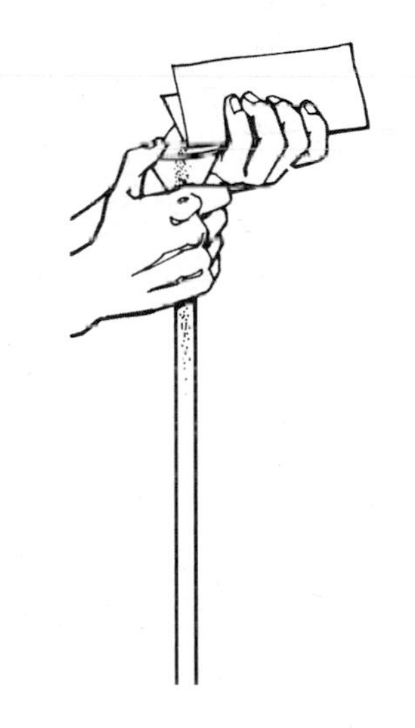

FIG. 9.4
A useful technique for filling a chromatographic tube with silica gel.

Slurry Packing Method

Alternatively, the silica gel can be added to the column (half-filled with hexanes) by slurrying the silica gel with a 90:10 mixture of hexanes and ether in a beaker. The powder is stirred to suspend it in the solvent and it is immediately poured through a wide-mouth funnel into the chromatographic tube. Rap the column with a rubber stopper to cause the silica gel to settle and to remove bubbles. Add a protective layer of sand to the top. The column is now ready for use.

Cleaning Up. After use, the tube is conveniently emptied by pointing the open end into a beaker, opening the pinchclamp, and applying gentle air pressure to the tip. If the plug of glass wool remains in the tube after the alumina leaves, wet it with acetone and reapply air pressure. Allow the adsorbent to dry in the hood and then dispose of it in the nonhazardous waste container.

Adding the Sample

Dissolve the sample completely in a very minimum volume of dichloromethane (just a few drops) in a small flask or vial. Add to this solution 300 mg of the adsorbent, stir, and evaporate the solvent completely by heating the slurry *very gently* with *constant* stirring to avoid bumping. Remember that dichloromethane boils at 41°C. Pour this dry powder into the funnel of the chromatography column, wash it down onto the column with a few drops of hexane, and then tap the column to remove air bubbles from the layer of adsorbent-solute mixture just added. Open the valve and carefully add new solvent in such a manner that the top surface of the column is not disturbed. A thin layer of fine sand can be added to the column after the sample to avoid disturbance of the column surface when the solvent is being added. Run the solvent down near to the surface several times to apply the sample as a narrow band at the top of the column.

Isolating the Separated Compounds

If the mixture to be separated contains colored compounds, then monitoring the column is very simple. The colored bands will move down the column along with the solvent, and as they approach the end of the column, you can collect the separated colors in individual containers. However, most organic molecules are colorless. In this case, the separation must be monitored by TLC. Spot each fraction on a TLC plate (Fig. 9.5). Four or five fractions can be spotted on a single plate. Before you develop the plate, do a quick examination under UV light to see if there is any compound where you spotted. If not, you can spot the next fraction in that location. Note which fraction is in which lane. Develop the plate and use the observed spot(s) to determine which compound is in each of the collected fractions. Spotting some of the starting material or the product (if available) on the TLC plate as a standard will help in the identification.

FIG. 9.5
Spot each fraction on a TLC plate. Examine under UV light to see which fractions contain the compound.

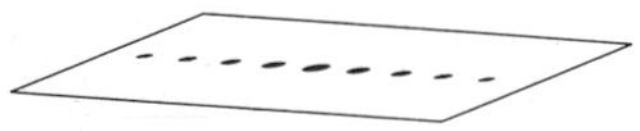

The colors of the fractions or the results from analyzing the fractions by TLC will indicate which fractions contain the compound(s) you are interested in isolating. Combine fractions containing the same compound and evaporate the solvent. Recrystallization may be used to further purify a solid product. However, on a milligram scale, there is usually not enough material to do this.

OTHER TYPES OF CHROMATOGRAPHY

Flash Chromatography

Relying only on gravity, liquid flow through a column can be quite slow, especially if the column is tightly packed. One method to speed up the process is *flash chromatography*. This method uses a pressure of about 10 psi of air or nitrogen on top of the column to force the mobile phase through the column. Normally, doing this would give a poorer separation. However, it has been found that with a finer mesh of alumina or silica gel, flash chromatography can increase the speed without lowering the quality of the separation. Go to this book's web site for an illustrated set of instructions for packing and using a flash chromatography column.

High-performance liquid chromatography.

High-performance liquid chromatography (HPLC) is a high-tech version of column chromatography, which is capable of separating complex mixtures with dozens of components. A high-pressure pump forces solvent at pressures up to 10,000 psi through a stainless steel tube packed tightly with extremely small adsorbent particles. The eluent flows from the column to a detector, such as a tiny UV absorbance cell or a mass spectrometer that is able to detect extremely small amounts of separated components, as little as a picogram (10^{-12} g).

Reverse-phase chromatography.

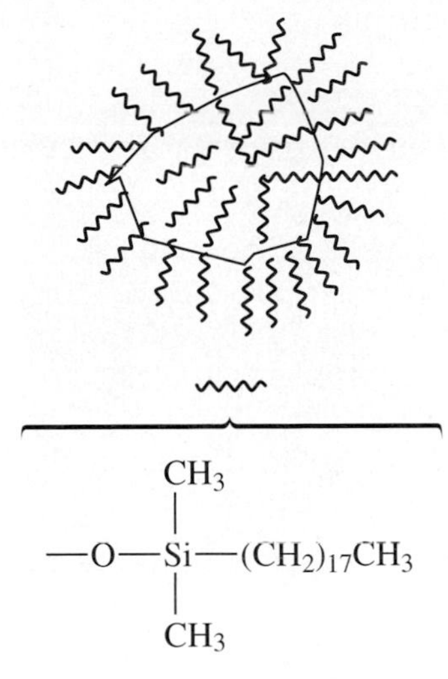

C_{18} reverse-phase packing

HPLC is used extensively in biochemistry to separate cellular components such as proteins, lipids, and nucleic acids. Mixtures of these types of compounds can be dissolved only in a predominantly aqueous mobile phase such as methanol-water or acetonitrile-water, and normal silica gel or alumina stationary phases do not work well with high concentrations of water. Rather than polar stationary phases, highly nonpolar ones called *reverse-phase packings* are used. These are manufactured by bonding lots of hydrocarbon molecules to the surfaces of silica gel particles, which convert the particles into highly nonpolar grease balls. With this packing, the order of elution is the reverse of that observed for a normal silica gel phase. On a reverse-phase column, the more nonpolar compounds will adhere to the nonpolar stationary phase more strongly, and the polar compounds will elute first.

Chiral chromatography

Is it possible to separate two enantiomers (optical isomers), both of which have the same intermolecular attractive forces? Chiral stationary phases can be used to separate enantiomers. Giving the stationary phase an asymmetry or handedness allows one enantiomer to be specifically retained on the column. Such columns are quite expensive and are limited to a particular type of separation, but they have led to great achievements in separation science. This separation technique is of great importance in the pharmaceutical industry because the U.S. Food and Drug Administration (FDA) specifies the amounts of impurities, including enantiomers, that can be found in drugs. For example, thalidomide, a drug prescribed as a sedative and an antidepressant in the 1960s, was found to be a potent teratogen that caused birth defects when pregnant women took the drug. It was quickly pulled from the market. Thalidomide has two enantiomers, and further research demonstrated that only one of the enantiomers caused the birth defects.

EXPERIMENTS

1. AIR OXIDATION OF FLUORENE TO FLUORENONE

The 9-position of fluorene is unusually reactive for a hydrocarbon. The protons on this carbon atom are acidic by virtue of being doubly benzylic, and, consequently, this carbon can be oxidized by several reagents, including oxygen from the air. Here the oxidation is carried out using a phase-transfer catalyst called methyltricapryl ammonium chloride, commonly known as Stark's catalyst or Aliquat 336. In the presence of this catalyst, the hydroxide is carried into the organic layer, where it can remove one of the acidic fluorene protons, creating a carbanion that can react with oxygen in the air. An intermediate hydroperoxide is formed and loses water to give the ketone.

Sufficiently acidic to be removed by strong base

H H

Fluorene
mp 114°C
MW 166.22

$+ O_2$

CH_3 Cl^-
N^+

Methyltricapryl ammonium chloride
(Aliquat 336, Stark's catalyst)

10 *M* NaOH
Toluene
10–60 min

O

Fluorenone
mp 83°C
MW 180.21

$+ H_2O$

Microscale Procedure

⚠ Sodium hydroxide is a strong base and is very corrosive. Wash your hands immediately if contact occurs.

This experiment requires a 50-mL to 125-mL separatory funnel. To a 25-mL Erlenmeyer flask clamped to a ring stand above a magnetic stirrer, add 5 mL of 10 *M* NaOH and 70 mg of fluorene while stirring with a $\frac{1}{2}$-inch magnetic stirring bar. Add 5 mL of toluene and stir until all of the solid has dissolved. (Observe the color and identify which layer is organic and which is aqueous.) Add approximately 3 drops of Stark's catalyst (Aliquat 336) to the solution. Stir vigorously but without splashing the solution. The reaction can take anywhere from 10 minutes to 30 minutes. Follow the rate of the reaction by TLC. Develop the TLC plate by using 20% dichloromethane in hexanes, and use a UV lamp to visualize the products. Also spot the plate with a 1% fluorene standard. When approximately half of the fluorene has been converted to fluorenone (as evidenced by the fact that the product spot is about the same size and intensity), pour the reaction mixture into a separatory funnel, rinsing the beaker with an additional 3 mL of toluene, which is also added to the separatory funnel.

Separate the organic layer from the aqueous layer. Wash the organic layer in the separatory funnel with 1.5 *M* hydrochloric acid (three separate times with 5 mL each time) and then saturated sodium chloride (three separate times with 5 mL each time). After each washing, drain the aqueous layer from the separatory funnel into a waste beaker. Dry the remaining toluene layer over anhydrous calcium chloride pellets in a 125-mL Erlenmeyer flask. Add the pellets until they no longer clump together (3–5 scoops). Allow the product to dry for 5–10 minutes before decanting the toluene from the solid calcium chloride and transferring the toluene to a 100-mL tared beaker. Wash the solid calcium chloride with 3 mL of toluene, adding this to the main portion of toluene to complete the transfer of product. The crude mixture of fluorene and fluorenone will be separated by column chromatography in the next lab period. The beaker containing the toluene extract can be allowed to stand in your hood (labeled properly) until the next lab period. The toluene will evaporate in the interim. Alternatively, if you have time, reduce the volume of solvent (~1 mL) by heating gently on a sand bath. Insert a boiling stick if you do this.

Cleaning Up. Carefully dilute the strongly basic aqueous reaction layer 10- to 20-fold with water in a large beaker, add the hydrochloric acid washes, and then neutralize by slowly adding additional hydrochloric acid. Flush this and the sodium chloride washes down the drain with lots of water. If local regulations allow, evaporate any residual solvent from the drying agents in the hood and place the dried solid in the nonhazardous waste container. Otherwise, place the wet drying agent in a waste container designated for this purpose.

Column Chromatography of the Fluorene-Fluorenone Mixture

Figure 9.1 on page 188 shows the typical setup you will use for the chromatographic separation of fluorene and fluorenone, and a general outline of the procedure is given on page 196. It is essential to have at least 10 clean 10-mL Erlenmeyer flasks, reaction tubes, small beakers, test tubes, or vials available to collect the chromatography fractions as they elute. At least two of these should be weighed, with their tare weight recorded. Once you have this done and the column is assembled, you can pack the column with alumina according to the following instructions.

Set up a sand bath at a heating setting of 50 before you do anything else.

Packing the Column

Before you assemble the column, check the small plug that fits into the bottom of the column to make sure that it has a small fritted disk inside. Next, make sure that the plug fits snugly into the glass column and is not easy to pull out. If it is loose, get a new bottom plug from the stockroom. Finish assembling the chromatography column as depicted in Figure 9.1. Be sure to clamp the column securely and vertically.

Grasp the valve with one hand and turn it with the other. Close the valve and fill the column with hexanes to the bottom of the plastic funnel. Weigh out approximately 4.5 g of activity grade III alumina in a small beaker and slowly sprinkle the dry alumina into the hexanes in the column while you tap the column with a pen or pencil. It may be necessary to drain off some of the solvent to keep it from flowing over the top. This amount of alumina should fill the column to a height of about 10 cm. It is extremely important *not* to let the column run dry at any time. This will allow air to enter the column, which will result in uneven bands and poor separation.

After all of the alumina has been added to the column, open the stopcock and continue to tap the column as you allow the solvent to drain slowly until the solvent just barely covers the surface of the alumina, collecting the solvent in an Erlenmeyer flask. Alternatively the slurry packing method (p. 198) can be used.

Adding the Sample

It is important to use a minimum amount of solvent when dissolving the sample. If too much is used, poor separation will result.

The solvent is drained just to the surface of the alumina, which should be perfectly flat. Dissolve the crude mixture of fluorene and fluorenone in 10 drops of dichloromethane and 10 drops of toluene, and add this with a pipette to the surface of the alumina. Be sure to add the sample as a solution; should any sample crystallize, add 1 more drop of dichloromethane. (This is done so that the sample to be added to the column is in the most concentrated solution possible.) Drain some liquid from the column until the dichloromethane-toluene solution just barely covers the surface of the alumina. Then add a few drops of hexanes and drain out some solvent until the liquid just covers the alumina. Repeat until the sample is seen as a narrow band at the top of the column. Carefully add a 4–5 mm layer of sand and then fill the column with hexanes. Alternatively, the sample can be added as a solid adsorbed on 300 mg of alumina (p. 199).

Collect 3-mL fractions in a combination of small vials, 10-mL Erlenmeyer flasks, 13 × 100 mm test tubes, vials, and small beakers. You will probably collect close to ten 3-mL fractions. While the chromatography is running, you will be determining the amount of fluorene or fluorenone in each fraction by TLC. Once you determine which fractions contain which compound, you will combine the "like" fractions and evaporate the solvent.

boiling stick broken in half. Tilt the flasks and vials on their side as much as possible to allow the heavy vapors to escape. As soon as all the liquid seems to have boiled off, set the flask *on its side* on the bench top in the hood to allow the last traces of heavy solvent vapors to escape. After the flask has cooled to room temperature, if that fraction contains any material, crystals may appear. If an oily, gooey residue is present, you may have to scratch it with a glass stirring rod to induce crystallization. With good organizational effort, you can do the TLC analysis and evaporate off the solvent at about the same rate at which you collect fractions; thus you can follow the progress of the chromatography simply by noting the amount of material in each flask, vial, or test tube. If, after solvent removal and cooling, the flasks are perfectly clean on careful inspection, they can be used to collect subsequent fractions.

If the yellow band has not moved one third of the way down the column after two fractions have been collected, you can speed up the elution by replacing the hexanes solvent at the top of the column with 20% dichloromethane in hexanes. Once the first component has completely eluted, you can speed up the elution of the second component by using 50% dichloromethane in hexanes. Decide when the product has been completely eluted from the column by using visual cues and TLC. Using a few drops of dichloromethane, wash all the fractions that contain fluorene as determined by TLC analysis into a tared container. Do the same for the fluorenone fractions. Evaporate the dichloromethane and obtain dry weights for the product and the recovered fluorene.

Mark all compound spots on all TLC plates with a pencil. Tape your developed TLC plates in your notebook with wide transparent tape. Calculate the theoretical yield and the percent yield of your pure fluorenone both with and without taking into account the amount of fluorene starting material recovered. Calculate the percent recovery of fluorene.

Cleaning Up. When you are done with the column, pour out the excess solvent into the proper waste container, pull out the bottom, and leave the wet column propped in a beaker in your desk hood. The column will dry out by the next lab, and the dry, used alumina can then be easily emptied out into a waste bin.

2. CHROMIUM(VI) OXIDATION OF FLUORENE TO FLUORENONE

The 9-position of fluorene is unusually reactive for a hydrocarbon. The protons on this carbon atom are acidic by virtue of being doubly benzylic, and, consequently, this carbon can be oxidized by several reagents, including elemental oxygen. In this

Chromium(VI) oxidations are less favored today because of environmental concerns based on chromium's toxicity.

experiment, the very powerful and versatile oxidizing agent chromium(VI), in the form of chromium trioxide, is used to carry out the oxidation. Chromium(VI) in a variety of other forms is used for about a dozen oxidation reactions in this text. The dust of chromium(VI) salts is reported to be a carcinogen, so avoid breathing it.

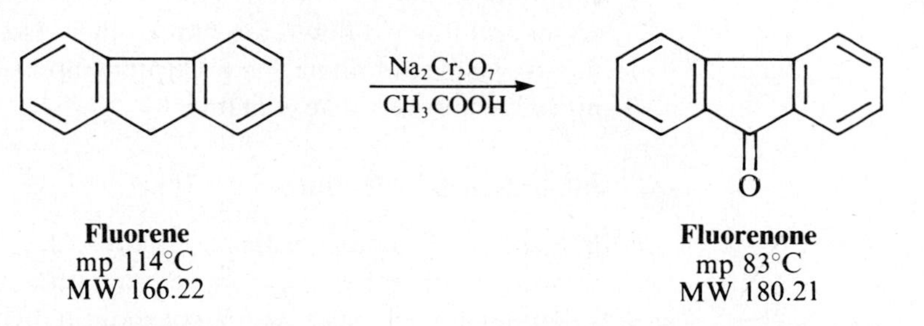

Fluorene
mp 114°C
MW 166.22

Fluorenone
mp 83°C
MW 180.21

Microscale Procedure

IN THIS EXPERIMENT, the hydrocarbon fluorene is oxidized to the ketone fluorenone by sodium dichromate in acetic acid with heating. The mixture is diluted with water, and the crude product is isolated by filtration. then it is dissolved in ether. The ether is dried and evaporated to give a mixture of fluorene and fluorenone, which is separated by column chromatography.

Sodium dichromate is toxic. The dust is corrosive to nasal passages and skin, and is a suspected carcinogen. Hot acetic acid is very corrosive to skin. Handle sodium dichromate and acetic acid in the hood; always wear gloves.

In a reaction tube, dissolve 50 mg of fluorene in 0.25 mL of acetic acid by heating, and add this hot solution to a solution of 0.15 g of sodium dichromate dihydrate in 0.5 mL of acetic acid. Heat the reaction mixture to 80°C for 15 minutes in a hot water bath; then cool it and add 1.5 mL of water. Stir the mixture for 2 minutes; then filter it on a Hirsch funnel. Wash the product well with water and press out as much water as possible. Return the product to the reaction tube, add 2 mL of ether, and add anhydrous calcium chloride pellets until they no longer clump together. Cork and shake the tube, and allow the product to dry for 5–10 minutes before evaporating the ether in another tared reaction tube. Use ether to wash off the drying agent and to complete the transfer of product. Use this ether solution to spot a TLC plate. This crude mixture of fluorene and fluorenone will be separated by column chromatography.

Photo: *Column Chromatography*; Video: *Column Chromatography*

Column Chromatography of the Fluorene-Fluorenone Mixture

Prepare a microscale chromatographic column exactly as described at the beginning of this chapter (*see* Fig. 9.1 on page 188). Use alumina as the adsorbent. Dissolve the crude mixture of fluorene and fluorenone in a mixture of 10 drops of dichloromethane and 10 drops of toluene, and add this to the surface of the alumina. Be sure to add the sample as a solution; should any sample crystallize, add 1 more drop of dichloromethane. Run the hexanes down to the surface of the alumina, add a few drops more of hexanes, and repeat the process until the sample is seen as a narrow band at the top of the column. Carefully add a 3-mm layer of sand, fill the column with hexanes, and collect 5-mL fractions in tared 10-mL Erlenmeyer flasks. Sample each flask for TLC (see Fig. 9.5 on page 190) and evaporate each to dryness. Final drying can be done under vacuum using the technique shown in Figure 9.6. You are to decide when all of the product has been eluted from the column. The TLC plates can be developed using 20% dichloromethane in hexanes. Combine fractions that are identical and determine the melting points of the two substances.

The sample can also be applied to the column using the technique described on page 196.

Cleaning Up. The filtrate probably contains unreacted dichromate. To destroy it, add 3 *M* sulfuric acid until the pH is 1; then complete the reduction by adding solid sodium thiosulfate until the solution becomes cloudy and blue colored. Neutralize

acid by swirling and heating on a hot plate. Adjust the temperature of the dichromate solution to 80°C, transfer the thermometer, and adjust the fluorene–acetic acid solution to 80°C; then, *in the hood*, pour in the dichromate solution. Note the time and the temperature of the solution and heat on a steam bath for 30 minutes. Observe the maximum and final temperature; then cool the solution and add 150 mL of water. Swirl the mixture for a full 2 minutes to coagulate the product and promote rapid filtration. Collect the yellow solid in an 8.5-cm Büchner funnel using vacuum filtration (if filtration is slow, empty the contents of the funnel and flask into a beaker and stir vigorously for a few minutes). Wash the filter cake well with water and then suck the filter cake as dry as possible. Either let the product dry overnight or dry it quickly as follows: Put the moist solid into a 50-mL Erlenmeyer flask, add ether (20 mL) and swirl to dissolve, and add anhydrous calcium chloride (10 g) to scavenge the water. Decant the ethereal solution through a cone of anhydrous calcium chloride in a funnel into a 125-mL Erlenmeyer flask; then rinse the flask and funnel with ether. Evaporate on a steam bath under an aspirator, heat until all the ether is removed, and pour the hot oil into a 50-mL beaker to cool and solidify. Scrape out the yellow solid. The yield should be about 4.0 g.

⚠ Handle dichromate and acetic acid in the laboratory hood; always wear gloves.

Cleaning Up. Follow the procedure in the immediately preceding microscale experiment for destroying excess dichromate and disposing of reagents and solvents.

One-half hour unattended heating.

Separation of Fluorene and Fluorenone

Prepare a column of 12.5 g of alumina, run out excess solvent, and pour onto the column a solution of 0.5 g of the fluorene-fluorenone mixture. Elute at first with hexanes and use tared 50-mL flasks as receivers. The yellow color of fluorenone provides one index of the course of the fractionation, and the appearance of solid around the delivery tip provides another. Frequently wash the solid on the tip into the receiver with ether. When you believe that one component has been eluted completely, change to another receiver until you judge that the second component is beginning to appear. Then, when you are sure the second component is being eluted, change to a 1:1 hexanes and ether mixture and continue until the column is exhausted. It is possible to collect practically all of the two components in the two receiving flasks, with only a negligible intermediate fraction. After evaporation of the solvent, evacuate each flask under vacuum (Fig. 9.7) and determine the weight and melting point of the products. A convenient method for evaporating fractions is to use a rotary evaporator (Fig. 9.8).

FIG. 9.6
Drying a solid by reduced air pressure.

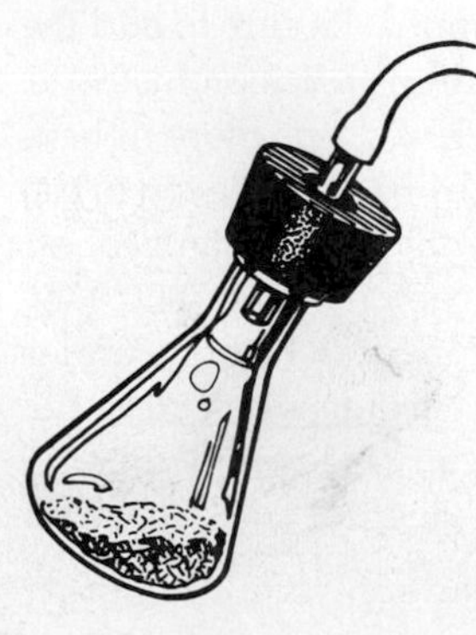

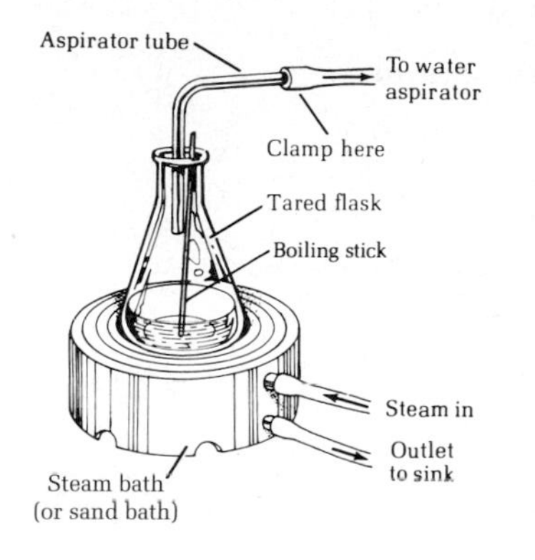

FIG. 9.7
An aspirator tube in use. A boiling stick may be necessary to promote even boiling.

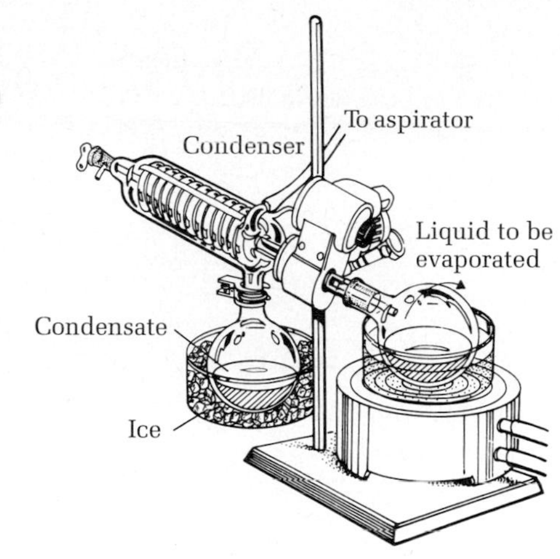

FIG. 9.8
A rotary evaporator. The rate of evaporation with this apparatus is very fast due to the thin film of liquid spread over the entire inner surface of the rotating flask, which is heated under vacuum. Foaming and bumping are also greatly reduced.

Cleaning Up. All organic material from this experiment can go in the organic solvents waste container. If local regulations allow, evaporate any residual solvent from the alumina in the hood and place the dried solid in the nonhazardous waste container. Otherwise, place the wet drying agent in a waste container designated for this purpose.

3. ACETYLATION OF CHOLESTEROL

Cholesterol is a solid alcohol; the average human body contains about 200 g distributed in the brain, spinal cord, and nerve tissue and, it occasionally clogs the arteries and the gallbladder.

In the following experiment, cholesterol is dissolved in acetic acid and allowed to react with acetic anhydride to form the ester cholesteryl acetate. The reaction does not take place rapidly, and consequently does not go to completion under the conditions of this experiment. Thus, when the reaction is over, both unreacted cholesterol and the product, cholesteryl acetate, are present. Separating these by fractional crystallization would be extremely difficult, but because they differ in polarity (the hydroxyl group of cholesterol is the more strongly adsorbed on alumina), they are easily separated by column chromatography. Both molecules are colorless and hence cannot be detected visually. Each fraction should be sampled by TLC. In that way, not only the presence but also the purity of each fraction can be assessed. It is also possible to put 1 drop of each fraction on a watch glass and evaporate it to see if the fraction contains product. Solid will also appear on the tip of the column while a compound is being eluted.

Microscale Procedure

IN THIS EXPERIMENT, cholesterol is refluxed with acetic acid and acetic anhydride. In the standard procedure, the mixture is diluted with water and extracted with ether, and the ether is dried and evaporated. The resulting mixture of cholesterol and cholesteryl acetate is separated by column chromatography on silica gel, eluting with hexanes and then hexanes and ether mixtures. Cholesteryl acetate comes off first, followed by cholesterol.

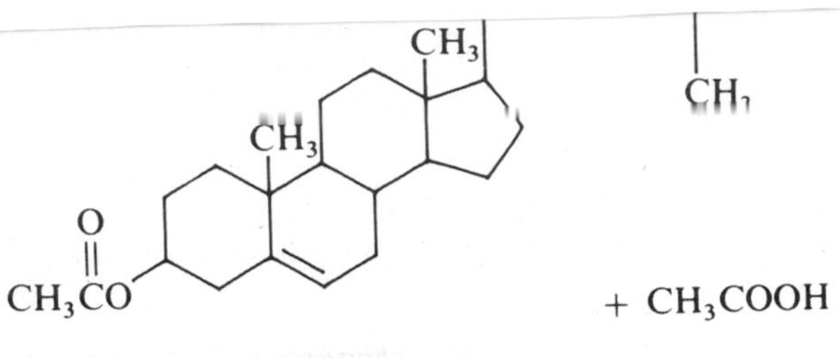

$+ CH_3COOH$

Cholesteryl acetate

In a reaction tube, add 0.5 mL of acetic acid to 50 mg of cholesterol. The initial thin slurry may set into a stiff paste of the molecular complex consisting of one molecule of cholesterol and one molecule of acetic acid. Add 0.10 mL of acetic anhydride and a boiling chip, and gently reflux the reaction mixture on a hot sand bath for no more than 30 minutes (*see* Fig. 1.1 on page 5).

While the reaction is taking place, prepare the microscale chromatography column as described previously, using silica gel as the adsorbent. Refer to Figure 9.1 on page 188 and the associated procedure earlier in the chapter. Cool the mixture, add 2 mL of water, and extract the product with three 2-mL portions of ether that are placed in a 15-mL centrifuge tube. Wash the ether extracts in the tube with two 2-mL portions of water and one 2.5-mL portion of 3 *M* sodium hydroxide (these three washes remove the acetic acid) and dry the ether by shaking it with 2.5 mL of saturated sodium chloride solution. Then complete the drying by adding enough anhydrous calcium chloride pellets to the solution so that the drying agent does not clump together.

Photo: *Column Chromatography*; Video: *Column Chromatography*

Shake the ether solution with the drying agent for 10 minutes and then transfer it in portions to a tared reaction tube and evaporate to dryness. Use 1 drop of this ether solution to spot a TLC plate for later analysis. If the crude material weighs more than the theoretical weight, you will know that it is not dry or that it contains acetic acid, which can be detected by its odor. Dissolve this crude cholesteryl acetate in a minimum quantity of ether and apply it to the top of the chromatography column.

To prevent the solution from dribbling from the pipette, use a pipette pump to make the transfer. It also could be applied as a dry powder adsorbed on silica gel, as in Experiment 1. Elute the column with hexanes, collecting two 5-mL fractions in tared 10-mL Erlenmeyer flasks or other suitable containers. Add a boiling stick to each flask and evaporate the solvent under an aspirator tube on a steam bath or a sand bath (*see* Fig. 9.7). If the flask appears empty, it can be used to collect later fractions. Lower the solvent layer to the top of the sand and elute with 15 mL of a 70:30 mixture of hexanes and ether, collecting five 3-mL fractions. Evaporate the solvent under an aspirator tube (Fig. 9.7). The last traces of solvent can be removed using reduced pressure, as shown in Figure 9.6 on page 196. Follow the 70:30 mixture with 10 mL of a 50:50 mixture of hexanes and ether, collecting four 2-mL fractions. Save any flask that has any visible residue. Analyze the original mixture and each fraction by TLC on silica gel plates using a 1:1 mixture of hexanes and ether to develop the plates, and either UV light or iodine vapor to visualize the spots.

Cholesteryl acetate (mp 115°C) and cholesterol (mp 149°C) should appear, respectively, in early and late fractions with a few empty fractions (no residue) in between. If so, combine consecutive fractions of early and late material and determine the weights and melting points. Calculate the percentage of the acetylated material compared to the total recovered and calculate the percentage yield from cholesterol.

Cleaning Up. After neutralization, acetic acid, the aqueous layers, and the saturated sodium chloride layers from the extraction can be flushed down the drain with water. Ether, hexanes, and TLC solvents should be placed in the organic solvents waste container. If local regulations allow, evaporate any residual solvent from the drying agents and the chromatography packing in the hood and place the dried solid in the nonhazardous waste container. Otherwise, place the wet materials in waste containers designated for this purpose.

Macroscale Procedure

Cover 0.5 g of cholesterol with 5 mL of acetic acid in a small Erlenmeyer flask; swirl and note that the initially thin slurry soon sets to a stiff paste of the molecular compound $C_{27}H_{45}OH \cdot CH_3CO_2H$. Add 1 mL of acetic anhydride and heat the mixture on a steam bath for any convenient period of time from 15 minutes to 1 hour; record the actual heating period. While the reaction takes place, prepare the chromatographic column. Cool the reaction mixture, add 20 mL of water, and extract with two 25-mL portions of ether. Wash the combined ethereal extracts twice with 15-mL portions of water and once with 25 mL of 3 M sodium hydroxide and dry by shaking the ether extracts with 25 mL of saturated sodium chloride solution. Then dry the ether over anhydrous calcium chloride pellets for 10 minutes in an Erlenmeyer flask, filter, and evaporate the ether. Save a few crystals of this material for TLC analysis. Dissolve the crystals in 3–4 mL of ether, transfer the solution with a Pasteur pipette onto a column of 12.5 g of silica gel, and rinse the flask with another small portion of ether.[1] In order to apply the ether solution to the top of the sand and avoid having it coat the interior of the column, pipette the solution down a 6-mm-diameter glass tube that is resting on the top of the sand. Label a series of 50-mL Erlenmeyer flasks as fractions 1 to 10. Open the pinchclamp, run the eluant solution into a 50-mL Erlenmeyer flask, and as soon as the solvent in the column has fallen to the level of the upper layer of sand, fill the column with a part of a measured 125 mL of a 70:30 mixture of hexanes and ether. When about 25 mL of eluant has collected in the flask (fraction 1), change to a fresh flask; add a boiling stone to the first flask and evaporate the solution to dryness on a steam bath under an aspirator tube (*see* Fig. 9.7 on page 197). Evacuation using an aspirator helps to remove last traces of ligroin (*see* Figure 9.6 on page 196). If fraction 1 is negative (no residue), use the flask for collecting further fractions. Continue adding the hexanes and ether mixture until the 125-mL portion is exhausted; then use 100 mL of a 1:1 hexanes and ether mixture. A convenient bubbler (Fig. 9.9) made from a 125-mL Erlenmeyer flask, a short piece of 10-mm-diameter glass tubing, and a cork will

FIG. 9.9
A bubbler for adding solvent automatically.

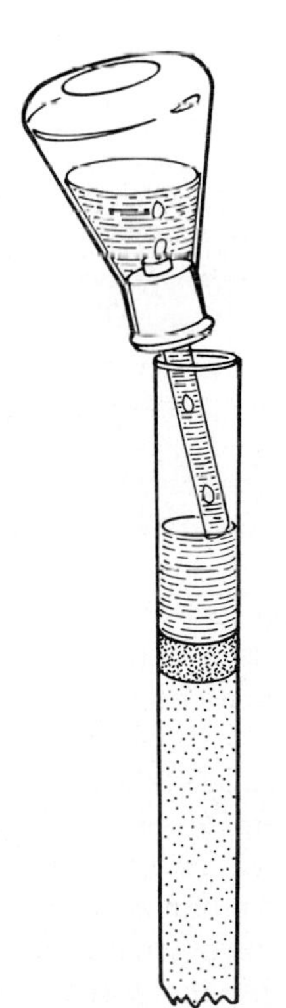

[1]Ideally, the material to be adsorbed is dissolved in hexanes (ligroin), the solvent of lowest elutant power. The present mixture is not soluble enough in hexanes, and so ether is used, but the volume is kept to a minimum.

automatically add solvent. A separatory funnel with a stopper and a partially open stopcock serves the same purpose. Collect and evaporate successive 25-mL fractions of eluant. Save any flask that has any visible solid residue. The ideal method for the removal of solvents involves the use of a rotary evaporator (*see* Fig. 9.8 on page 197). Analyze the original mixture and each fraction by TLC on silica gel plates using a 1:1 mixture of hexanes-ether to develop the plates, and either UV light or iodine vapor to visualize the spots.

Carry out the procedure in a laboratory hood.

Cholesteryl acetate (mp 115°C) and cholesterol (mp 149°C) should appear, respectively, in early and late fractions with a few empty fractions (no residue) in between. If so, combine consecutive fractions of early and of late material, and determine the weights and melting points. Calculate the percentage of the acetylated material compared to the total recovered and compare your result with those of others in your class employing different reaction periods.

Cleaning Up. After neutralization, acetic acid, the aqueous layers, and the saturated sodium chloride layers from the extraction can be flushed down the drain with water. Ether, hexanes, and TLC solvents should be placed in the organic solvents waste container. Hexanes and other solvents from the chromatography go into the organic solvents container. If local regulations allow, evaporate any residual solvent from the drying agents and silica gel in the hood and place the dried solid in the non-hazardous waste container. Otherwise, place the wet drying agent and silica gel in waste containers designated for this purpose.

4. CHROMATOGRAPHY OF A MIXTURE OF FERROCENE AND ACETYLFERROCENE

IN THIS EXPERIMENT, a mixture of two compounds is separated by chromatography on alumina. The first compound to come down the column is ferrocene (a yellow band). The solvent polarity is changed so that acetylferrocene is eluted as an orange band. The solvents are evaporated from the collection flasks and the compounds recrystallized. Both of these compounds are colored (see Chapter 49 for their preparation), so it is easy to follow the progress of the chromatographic separation.

Photo: *Column Chromatography*; Video: *Column Chromatography*

Acetylferrocene is toxic.

Prepare the microscale alumina column exactly as described at the beginning of the chapter. Then add a dry slurry of 90 mg of a 50:50 mixture of acetylferrocene and ferrocene that has been adsorbed onto 300 mg of alumina, following the procedure for preparing and adding the sample given at the beginning of the chapter.

Carefully add hexanes to the column, open the valve (use both hands), and elute the two compounds. The first to be eluted, ferrocene, will be seen as a yellow band. Collect this in a 10-mL flask. Any crystalline material seen at the tip of the valve should be washed into the flask with a drop or two of ether. Without allowing the column to run dry, add a 50:50 mixture of hexanes and ether, and elute the acetylferrocene, which will be seen as an orange band. Collect it in a 10-mL flask. Spot a silica gel TLC plate with these two solutions. Evaporate the solvents from the two flasks and determine the weights of the residues. An easy way to evaporate the solvent is to place it in a tared 25-mL filter flask and heat the flask in your hand under vacuum while swirling the contents (Fig. 9.10).

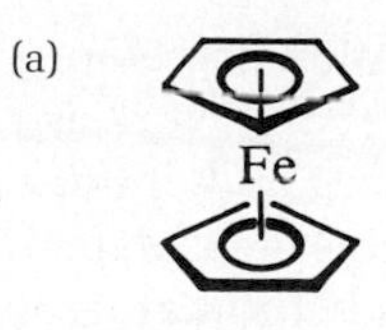

Ferrocene
MW 186.04
mp 172–174°C

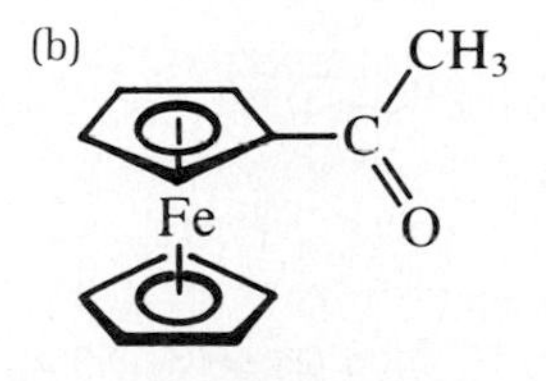

Acetylferrocene
MW 228.08
mp 85–86°C

Recrystallize the products from the minimum quantities of hot hexanes. Isolate the crystals, dry them, and determine their weights and melting points. Calculate the percent recovery of the crude and recrystallized products based on the 45 mg quantity of each in the original mixture.

The TLC plate is eluted with a 30:1 mixture of toluene and absolute ethanol. Do you detect any contamination of one compound by the other?

Cleaning Up. If regulations allow, empty the chromatography column onto a piece of aluminum foil in the hood. After the solvent has evaporated, place the alumina and sand in the nonhazardous waste container. Otherwise, place the wet alumina and sand in a designated waste container. Evaporate the crystallization mother liquor to dryness and place the residue in the hazardous waste container.

5. ISOLATION OF LYCOPENE AND β-CAROTENE

Lycopene, the red pigment in tomatoes, is a C_{40}-carotenoid made up of eight five-carbon isoprene units. β-Carotene, the yellow pigment of the carrot, is an isomer of lycopene in which the double bonds at C_1—C_2 and C'_1—C'_2 are replaced by bonds extending from C_1 to C_6 and from C'_1 to C'_6 to form rings. The chromophore in each case is a system of 11 all-trans conjugated double bonds; the closing of the two rings renders β-carotene less highly pigmented than lycopene.

These colored hydrocarbons have been encountered in the TLC experiment (*see* Chapter 8). The isolation procedure described here affords sufficient carotene and lycopene to carry out analytical spectroscopy and some isomerization reactions. It might be of interest if some students isolate carotene from strained carrot baby food while others isolate lycopene from tomato paste.

Lycopene is responsible not only for the red color of tomatoes but also of red grapefruit, watermelon,, and flamingos. If flamingos do not include foods containing lycopene in their diet, they will be white.

Lycopene is the predominant carotenoid in blood plasma. It is not converted into vitamin A as carotene is, but it is a powerful antioxidant and is an efficient scavenger of singlet oxygen.

FIG. 9.10
Evaporation of a low-boiling liquid under vacuum. Heat is supplied by the hand, the contents of the flask are swirled, and the vacuum is controlled with the thumb.

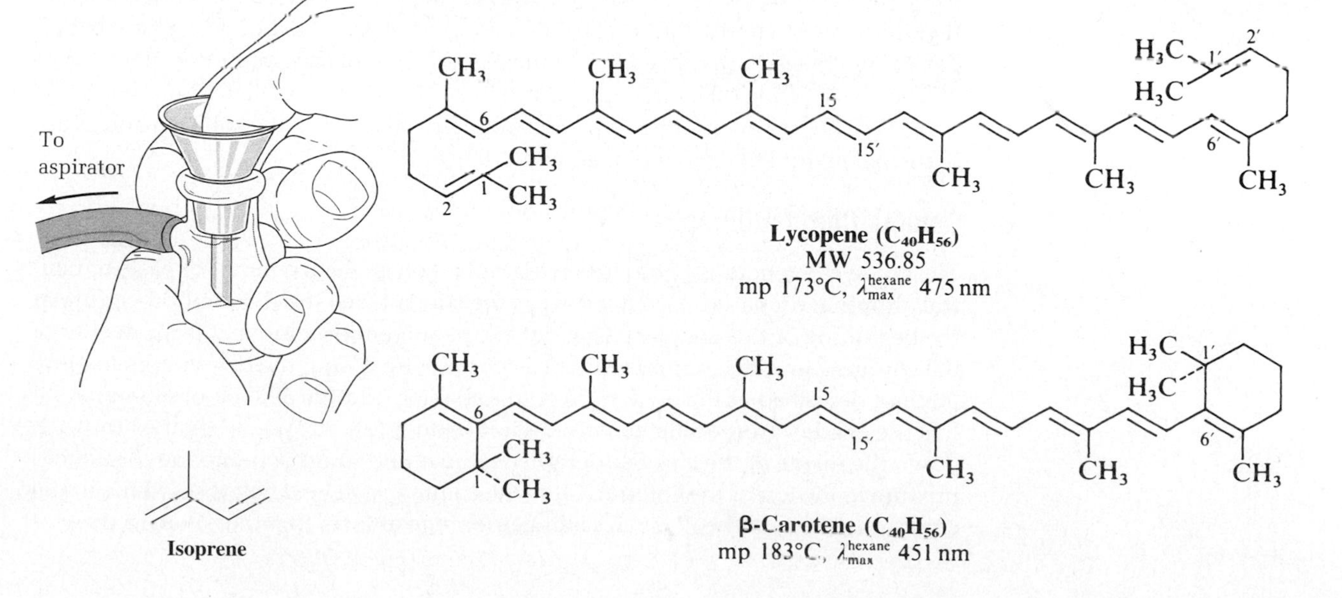

Carotenoids are highly sensitive to photochemical air oxidation; therefore, protect solutions and solids from undue exposure to light and heat and work as rapidly as possible. Do not heat solutions when evaporating solvents and, if possible, flush apparatus with nitrogen to exclude oxygen. Research workers isolate these compounds in dimly lit rooms and/or wrap all containers and chromatographic columns in aluminum foil, and carry out extractions and crystallizations using solvents that have been deoxygenated.

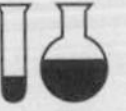

Dehydration and Extraction of Tomato or Carrot Paste

IN THIS EXPERIMENT, a vegetable paste is stirred with acetone to remove water, but not the coloring matter, from the paste. The mixture is filtered, the yellow filtrate discarded, and the material on the filter squeezed as dry as possible. This solid is then extracted three times with dichloromethane. The solution is dried over calcium chloride and evaporated at room temperature under vacuum to leave the crude carotenoids.

Add 5 g of tomato or carrot paste to a 15-mL centrifuge tube or 25 × 150-mm test tube; then add about 7 mL of acetone and stir the paste for several minutes until it is no longer gummy. This acetone treatment removes most of the water from the cellular mixture. Filter the mixture on a small Büchner funnel. Scrape out the tube with a spatula, let it drain thoroughly, and squeeze out as much liquid as possible from the solid residue in the funnel with a spatula. Discard the yellow filtrate. Then return the solid residue to the centrifuge tube and add 5 mL of dichloromethane to effect extraction. Cap the tube and shake the mixture vigorously. Filter the mixture on a Büchner funnel once more, repeat the extraction and filtration with two or three further 5-mL portions of dichloromethane, clean the tube thoroughly, and place the filtrates in it. Dry the solution over anhydrous calcium chloride pellets, filter the solution into a small flask, and evaporate the solution to dryness with a stream of nitrogen or under vacuum using a rotary evaporator (Fig. 9.8 on page 197) or the apparatus shown in Figure 9.10 on page 201, never heating the sample above 50°C. Determine the weight of the crude material. It will be very small. If the residue is dry, as it should be, add just enough dichloromethane to dissolve the residue. Save 1 drop of this solution to carry out a TLC analysis (using dichloromethane as the eluent on silica gel plates; *see* Chapter 8). Then add 200 mg of alumina to the remaining dichloromethane solution and evaporate the mixture to dryness, again without heat.

Video: *Column Chromatography*, Photo: *Column Chromatography*

Column Chromatography

The crude carotenoid is to be chromatographed on an 8-cm column of basic or neutral alumina, prepared with hexanes as the solvent (see the detailed procedure at the beginning of this chapter). Run out excess solvent or remove it from the top of the chromatography column with a Pasteur pipette. Using the dry sample loading method described at the beginning of this chapter, add the 300 mg of alumina that has the crude carotenoids absorbed on it. Add a few drops of hexanes to wash down the inside of the chromatography column and to consolidate the carotenoid mixture at the top of the column. Elute the column with hexanes, discard the initial colorless eluate, and collect all yellow or orange eluates together. Place a drop of solution on a microscope slide and evaporate the remainder to dryness using a stream of nitrogen or a rotary evaporator (*see* Fig. 9.8 on page 197). Examination of the material spotted on the slide may reveal crystallinity. If you are using tomato paste, a small amount of yellow Gr beta-carotene will come off first, followed by lycopene. Collect the red lycopene separately by eluting with a mixture of 10% acetone in hexane and also evaporate that solution to dryness.

Finally, dissolve the samples obtained by evaporating the solvent in the least possible amount of dichloromethane and carry out TLC of the two products in order to ascertain their purity (*see* Chapter 8, Experiment 2). You may want to combine your purified products with those of several other students, evaporate the solution to dryness, dissolve the residue in deuterochloroform, $CDCl_3$, and determine the ^1H NMR spectrum (*see* Chapter 12). Also obtain an infrared spectrum and a visible spectrum (in hexane).

Note that Gr beta-carotene is in demand as a source of vitamin A and is manufactured by an efficient synthesis. Until very recently no use for lycopene had been found.

Cleaning Up. Place recovered and unused dichloromethane in the halogenated organic waste container; the solvents used for TLC in the organic solvents waste container. If local regulations allow, evaporate any residual solvent from the drying agents in the hood and place the dried solid in the nonhazardous waste container. Otherwise, place the wet drying agent in a waste container designated for this purpose. Used plant material and dry TLC plates can be discarded in the nonhazardous waste container.

For Further Investigation

The carotenoids of any leaf can be isolated in the manner described in this experiment. Grind the leaf material (about 10 g) in a mortar with some sand; then follow the above procedure. Waxy leaves do not work well. The carotenoids are present in the leaf during its entire life span, so a green leaf from a maple tree or euonymus shrub, also known as burning bush, known to turn bright red in the fall, will show lycopene even when the leaf is green. In the fall, the chlorophyll decomposes before the carotenoids, so the leaves appear in a variety of orange and red hues.

It is of interest to investigate the carotenoids of the tomato, of which there are some 80 varieties. The orange-colored tangerine tomato contains an isomer of lycopene. If a hexane solution of the prolycopene from this tomato is treated with a drop of a very dilute solution of iodine in hexane and then exposed to bright light, the solution will turn deep-orange in color, indicating that a *cis*-double bond has isomerized to the *trans* form. The product is, however, still not identical to natural lycopene.

Isomerization

Prepare a hexane solution of either carotene or lycopene, and save a drop for TLC. Treat the solution with a very dilute solution of iodine in hexane, expose the resulting mixture to strong light for a few minutes, and then carry out TLC on the resulting solution. Also compare the visible spectrum before and after isomerization.

Iodine serves as a catalyst for the light-catalyzed isomerization of some of the *trans*-double bonds to an equilibrium mixture containing *cis*-isomers.

QUESTIONS

1. Predict the order of elution of a mixture of triphenylmethanol, biphenyl, benzoic acid, and methyl benzoate from an alumina column.
2. What would be the effect of collecting larger fractions when carrying out the experiments in this chapter?
3. What would have been the result if a large quantity of petroleum ether alone were used as the eluent in either experiments 1 or 2?
4. Once the chromatographic column has been prepared, why is it important to allow the level of the liquid in the column to drop to the level of the alumina before applying the solution of the compound to be separated?
5. A chemist started to carry out column chromatography on a Friday afternoon, reached the point at which the two compounds being separated were about three-fourths of the way down the column, and then returned on Monday to find that the compounds came off the column as a mixture. Speculate the reason for this. The column had not run dry over the weekend.
6. Write a detailed mechanism for the formation of fluorenone from fluorene. Explain the purpose of the phase-transfer catalyst.
7. The primary cause of low yields in the isolation of lycopene is oxidation of the product during the procedure. Once crystalline, it is reasonably stable. Speculate on the primary products formed by photochemical air oxidation of lycopene.

See *Web Links*

CHAPTER 10

Gas Chromatography: Analyzing Alkene Isomers

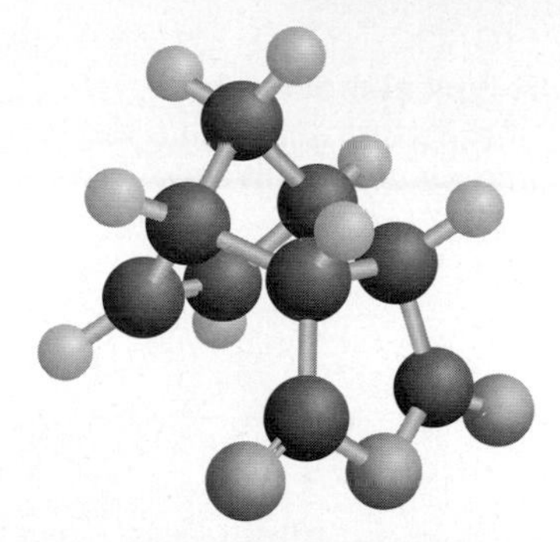

PRELAB EXERCISE: If the dehydration of 2-methyl-2-butanol occurred on a purely statistical basis, what would be the relative proportions of 2-methyl-1-butene and 2-methyl-2-butene?

Gas chromatography (GC) is a rapid and sensitive method of separating and analyzing mixtures of gaseous or liquid compounds. The information it provides can tell you:

1. if you have successfully synthesized your product;
2. whether your product contains unreacted starting material or other impurities;
3. whether your product is a mixture of isomers; and
4. the relative amounts of different materials or isomers in a mixture.

Gas chromatography is one of several *instrumental* analysis methods used by organic chemists. Unlike the apparatus for thin-layer (TLC) or column chromatography, the apparatus used for gas chromatography—a modern gas chromatograph like that shown in Figure 10.1—costs thousands of dollars and is therefore shared among many users. There are two major reasons for spending this much money for this instrument. First, very complex mixtures containing hundreds of components can be separated. Second, separated components can be detected in very small amounts, 10^{-6} to 10^{-15} g. The chromatogram in Figure 10.2 is a good example of the power of gas chromatography; it shows the separation of 1 mg of crude oil into the hundreds of compounds that are present: alkanes, alkenes, and aromatics, including many isomers. The exceptional sensitivity of GC instruments is the major reason they are often used in environmental and forensic chemistry labs, where the detection of trace amounts is necessary.

Gas chromatography involves the same principles that apply to all forms of chromatography, which were covered in Chapter 8. The mobile phase is a gas, usually helium. The stationary phase is often a fairly nonpolar polymer such as polydimethylsiloxane or poly(ethylene glycol) (Carbowax) that is stable at temperatures as high as 350°C. The most commonly used phase is polydimethyl-siloxane in which 5% of the methyl groups have been replaced by phenyl groups.

There are two types of GC columns:

1. *Packed columns* in which the stationary phase consists of solid particles similar to the alumina or silica gel used in column chromatography, but coated with a nonpolar polymer (e.g., polydimethylsiloxane or Carbowax) and packed into 3–6 mm diameter metal or glass tubes that are 1–3 m long and rolled into a compact coil.

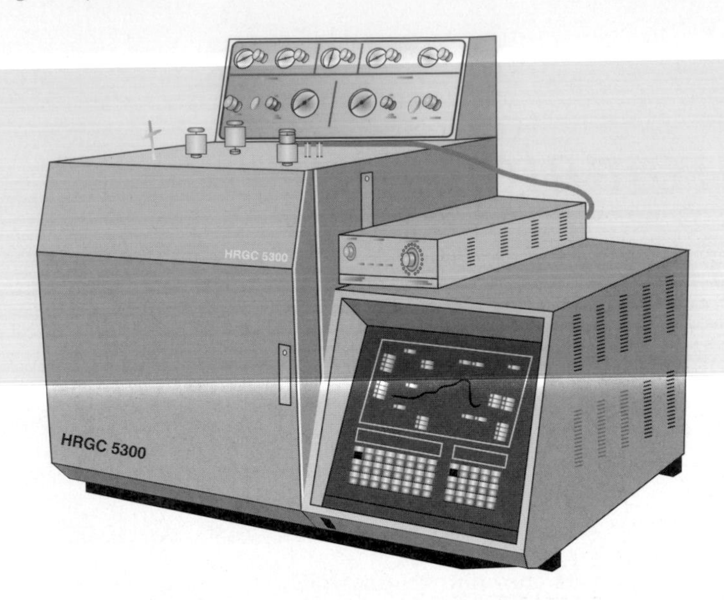

FIG. 10.1
A typical modern gas chromatograph with a data acquisition analysis computer.

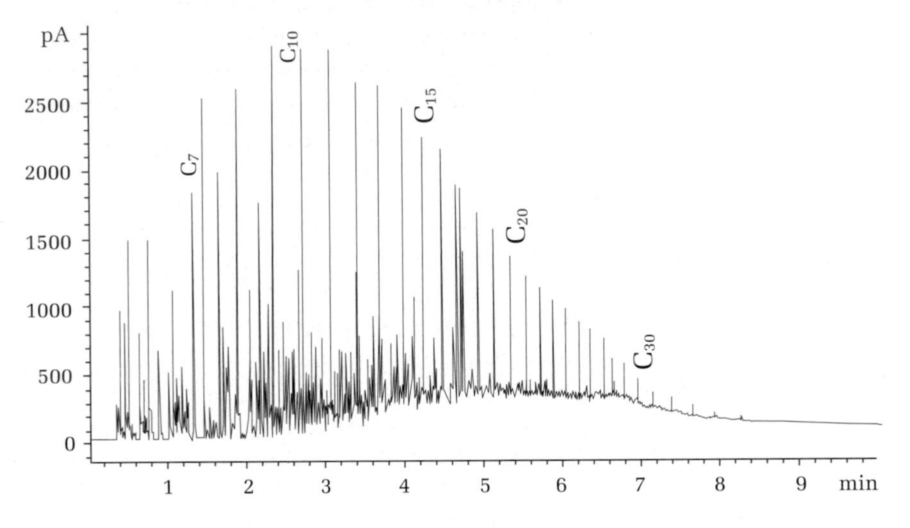

FIG. 10.2
A chromatogram from the GC analysis of 1 mg of crude petroleum. The larger peaks represent the series of *n*-alkanes from C_5 to $>C_{30}$. The column was temperature programmed from 40°C to 300°C at 20°C/minutes.

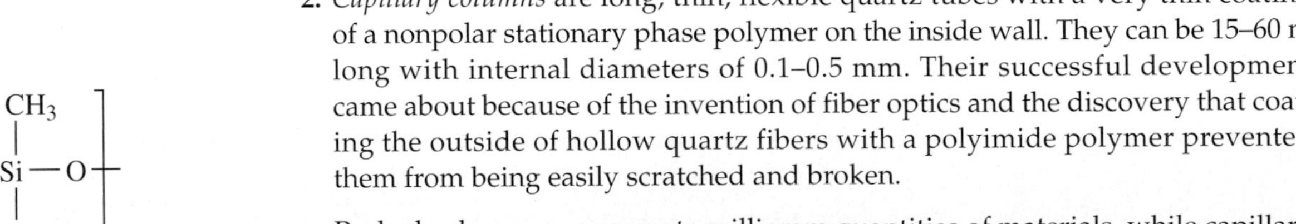

2. *Capillary columns* are long, thin, flexible quartz tubes with a very thin coating of a nonpolar stationary phase polymer on the inside wall. They can be 15–60 m long with internal diameters of 0.1–0.5 mm. Their successful development came about because of the invention of fiber optics and the discovery that coating the outside of hollow quartz fibers with a polyimide polymer prevented them from being easily scratched and broken.

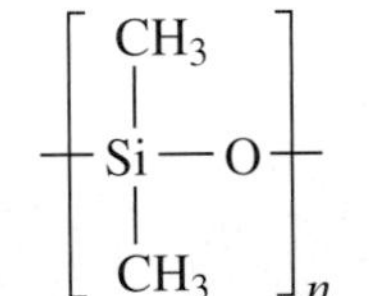

Polydimethylsiloxane

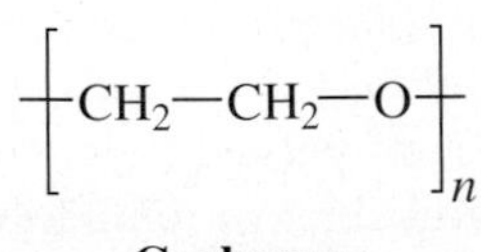

Carbowax

Packed columns can separate milligram quantities of materials, while capillary columns work best with microgram quantities or less. On the other hand, capillary columns have much better separating power than packed columns.

The GC instrument has the following components (Fig. 10.3). Helium at a pressure of 10–30 psi from a compressed gas cylinder flows through the heated injector port, the column (located in a temperature-controlled oven), and the heated detector at a flow rate of 10–60 mL/minutes for a packed column and about 1 mL/minutes for a capillary column. A microliter syringe is used to inject 1–25 μL of sample

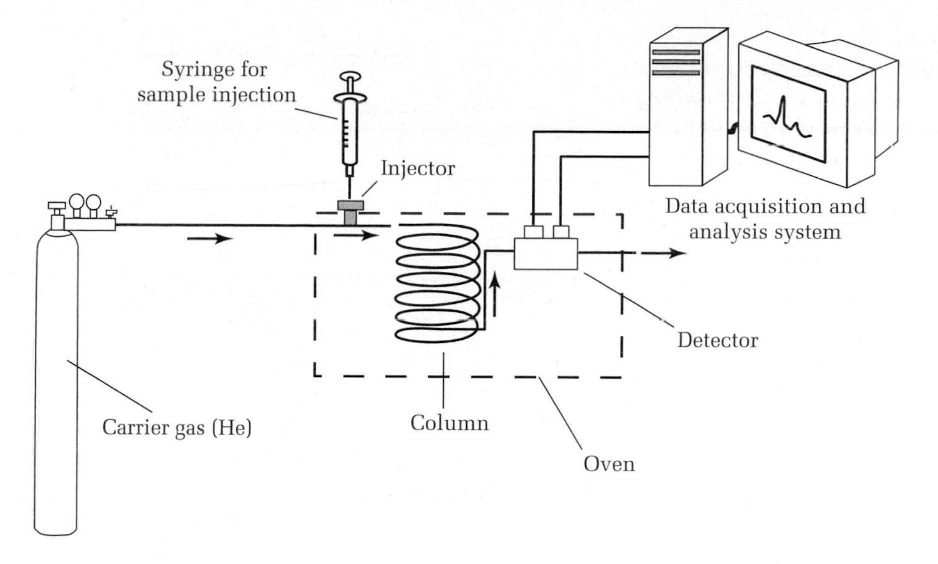

FIG. 10.3
A diagram of a gas chromatograph.

through a rubber septum into the hot (~250°C) injector, where any liquid compounds are instantly vaporized to the gaseous state. As the vaporized molecules are swept through the column, they interact with the stationary phase and are adsorbed and desorbed many times each second in a dynamic equilibrium. As discussed in Chapter 8, the equilibria between adsorbed and free states depends on each molecule's size (London forces), polarity, and ability to hydrogen bond. Molecules spending more time absorbed to the stationary phase will not be carried through the column as rapidly as those that are weakly bound and free. A molecular view of this differential adsorption and flow is depicted in Figure 10.4.

The separated components pass one after the other into the detector. In column chromatography (*see* Chapter 9), human eyes are the detectors, either seeing colored compounds flow out of the column or visualizing the presence of colorless compounds on a fluorescent TLC plate. Gas chromatographs have electronic detectors that produce a signal voltage proportional to the number of molecules passing through them at any instant in time. A record of this voltage versus time is a gas chromatogram, with peak areas representing the amount of each individual component passing through the detector. Figure 10.5 is the chromatogram produced by the injection of a mixture of two compounds that were separated on the column and detected to produce peaks A and B. The amount of each is proportional to the area under its peak, so we can say that there appears to be about twice as much B as A. If an integrating recorder or computer data acquisition system is used, the exact areas under all peaks are automatically determined and can be used to quantify the amount of each component in a mixture.

Most GC instruments use one of the following types of detectors:

- The *thermal conductivity detector* (TCD) consists of an electrically heated wire or thermistor. The temperature of the sensing element depends on the thermal conductivity of the gas flowing around it. Changes in thermal conductivity, such as when organic molecules displace some of the carrier gas, cause a temperature rise in the element that is sensed as a change in resistance. The TCD is the least sensitive (detecting micrograms per second, 10^{-6} g/s, of material) of the four detectors described here, but it is quite rugged. It will detect all types of molecules, not just those containing C—H bonds, and compounds passing through it are unchanged and can be collected.

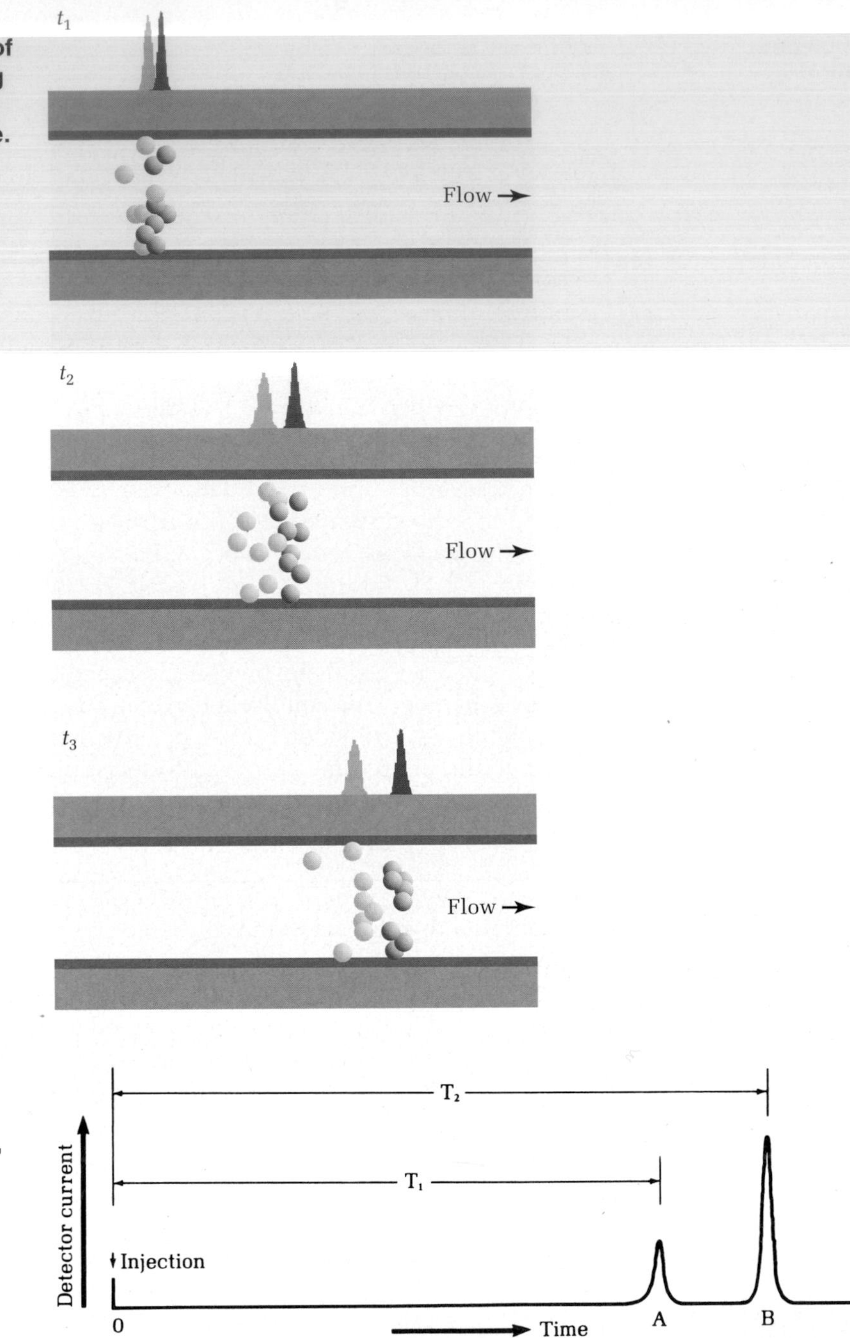

FIG. 10.4
The differential adsorption of two compounds in a flowing gas leads to the separation of the compounds over time.

FIG. 10.5
A gas chromatogram of a mixture of two compounds, A and B.

- The *flame ionization detector* (FID) is at least a thousand times more sensitive than the TCD, easily detecting nanograms per second (10^{-9} g/s) of material. The molecules, however, must contain C—H bonds because the sample is actually burned to form ions. These ions carry a tiny current between two electrodes, which is greatly amplified to produce the signal output.
- The *electron capture detector* (ECD) contains a tiny amount of a radioactive substance, such as ^{63}Ni, that emits high-energy electrons (β^- particles).

The sample molecules can capture these electrons, become charged, and carry a current between two electrodes, as in the FID detector. Halocarbons have exceptionally good cross sections for electron capture because of the electronegativity of the halogen atoms, and these types of compounds can be detected at picogram per second (10^{-12} g/s) levels. For this reason, ECD is used to measure levels of chlorocarbon pollutants, such as chloroform, DDT, and dioxin in the environment.

- A *mass spectrometer detector* is the only one of these four detectors that provides information about the structure of the molecules that pass into it. For this reason, combination gas chromatography-mass spectrometry (GC–MS) systems are often used in organic research, forensic science, and environmental analysis in spite of their high cost ($40,000 to $250,000). Sensitivity is at the nanogram to picogram per second level. Chapter 13 discusses mass spectrometry in detail.

The instant you inject your sample into the GC instrument, you normally press a button that starts a clock in the data recorder, either an electronic integrator or a computer data system. Then, as each eluting component produces a peak in the detector, the top of the peak is detected, and the elapsed time since injection, called the *retention time*, is recorded and stored. In Figure 10.5, the retention time of compound A is T_1 and of B is T_2. Which compound is more strongly absorbed by the stationary phase?

One powerful variable in gas chromatography that is not available in TLC or column chromatography is temperature. Because the GC column is in an oven whose temperature is programmable—that is, the temperature can be raised from near room temperature to as high as 350°C at a constant and reproducible rate—we are able to separate much more complex mixtures than if the oven could be set to only one temperature. At room temperature, a mixture containing lower-boiling, weakly adsorbed components (such as ether and dichloromethane) and higher-boiling, strongly adsorbed components (such as butanol and toluene) might take hours to separate because the latter would move significantly more slowly through the column at low temperature. If the column were kept at a high temperature, the higher-boiling compounds would come out in a shorter time, but the lower-boiling ones would not be retained at all and therefore would not be separated. By programming the oven temperature, we can inject such a mixture at a low initial temperature so that weakly adsorbed components are separated. Then, as the temperature slowly rises, the more strongly retained components will move through the column faster and also have reasonable retention times. Temperature programming allows the separation of very complex mixtures containing both low-boiling small molecules and high-boiling large ones, as demonstrated by the GC analysis of crude petroleum in Figure 10.2 on page 206.

Temperature programming

THE GENERAL GC ANALYSIS PROCEDURE

Most GC separations are of gaseous and liquid mixtures. Although some solids will pass through a gas chromatograph at higher temperatures, it is best if all of the samples are distilled or vacuum transferred before injection into the GC instrument to ensure that only volatile compounds are present. If samples contain materials that cannot be vaporized and swept through the GC column, even at high temperatures, these will remain on the column and can ultimately ruin its performance. GC columns are expensive and can cost up to $400 to replace.

The guidelines given here are for GC analysis using a capillary column and an FID detector, which require very dilute solutions to avoid overwhelming the column, even if 90%–95% of the sample is split away so that it does not enter the column, a common practice in capillary chromatography. Samples analyzed with a packed column and a TCD detector can be much more concentrated or even undiluted (neat).

Sample Preparation

There are different ways of preparing a GC sample, depending on whether your product is a solid, a liquid (usually obtained by distillation), or a gaseous mixture; on the type of column (capillary or packed); and on the type of detector (FID, ECD, or TCD) being used.

Solid Samples

Low-melting, sublimable solids can be analyzed at higher column temperatures. Run a GC analysis on solids only if your experimental procedure explicitly says that you can do so. To prepare a solid sample for capillary column analysis, put a *small* crystal of the solid into a small vial and dissolve it in 1 mL of dichloromethane.

Liquid Samples

Dilute solutions of liquid organic mixtures in dichloromethane are the most common form of GC samples. Place 1 or 2 drops of your distilled product into a small vial and add 1 mL of dichloromethane. Because GC columns can be damaged by moisture, it is suggested that enough *anhydrous* sodium sulfate be added to cover the bottom of the vial to a depth of 1–2 mm to ensure that all traces of water are removed. The sodium sulfate can be left in the sample because it usually will not go into the fine needle of the microliter syringe used to inject the sample. Injection of 1 μL of a sample prepared in this manner usually provides strong signals in an FID detector. In special cases, undiluted neat or pure liquids can be injected directly without dilution as noted; for example, the methylbutenes in Experiment 1. Also, if preparative gas chromatography is used to separate and collect the components of a mixture, neat liquids are usually injected.

Gaseous Samples

Gases are normally collected over water, as in Experiment 2. Do not remove the septum-capped collection tube from the beaker of water, but bring it and the beaker of water together to the gas chromatograph. A gas syringe is inserted through the rubber septum and 10–500 μL (depending on the type of the gas chromatograph) of the gas sample is withdrawn and injected into the unit.

Sample Analysis

Make sure that the required gases are flowing and that the instrument has been on at least 1 hour so that all zones, the injector, the oven, and the detector have come to the required temperatures. Enter a temperature program: initial temperature, heating rate (in degrees per minutes), and final temperature, as required by the particular experiment. Fill the microliter syringe with the sample solution, normally 1 μL. A 10-μL capacity syringe with a plunger guide is recommended. Handle this expensive syringe carefully; it has a sharp needle, and the glass barrel is easily dropped and broken. Make sure that there are no bubbles in the syringe barrel when you draw up the sample. If bubbles are present, fill the syringe about halfway and depress the plunger quickly. Draw up the sample again slowly. Repeat as necessary to remove all bubbles. To obtain a quality chromatogram, the needle should be inserted rapidly, the sample should be injected rapidly and the syringe should be removed as soon as the plunger has been depressed and the sample injected. Then quickly press the start analysis button on the gas chromatograph, computer, or integrator to start data acquisition or mark the paper on the pen recorder.

⚠ Do not touch the metal injector cap. It is very HOT.

Once all the compounds have eluted and the oven temperature program has finished, the run should be stopped. Plot your chromatogram using the computer or remove the chromatogram chart from the integrator or flat bed recorder. Record the GC analysis parameters: column diameter and length, column packing type, carrier gas and its flow rate, column temperature (initial, final, and heating rate if programmed), detector and injector temperatures, sample injection amount and solvent (if used), signal attenuation, and chart speed; keep these with your gas chromatogram. Write the structures of all identifiable peaks on the chromatogram. If you used dichloromethane to dilute your sample, the largest peak will be due to this substance and will have a short retention time.

Remember that retention times are not constants. If the programmed oven temperature rate is higher or lower than for a prior analysis, then the retention times for all of the peaks will be consistently shorter or consistently longer. Changes in helium flow and the aging of the column also affect retention times. Chemists primarily look for similarities in the pattern of eluting peaks, not for perfect matches in retention times. In general, for simple mixtures chromatographed on nonpolar stationary phases, the retention time order is the same as the order of increasing boiling points. For example, you would expect cyclohexene (bp 83°C) to have a shorter retention time than toluene (bp 110°C). You can often ascertain the identity of a certain peak by adding a small amount of a known standard to the sample and rerunning the GC analysis. The peak corresponding to the standard will increase significantly and can be identified. Often the chromatogram will show a number of small peaks that cannot be assigned to the product, the starting material, or the solvents; these are likely due to byproducts. As long as these impurities are minor, say, less than 10% of the total area (not including the solvent), they can probably be ignored. The ultimate method of identifying every peak is to analyze the sample on a GC–MS instrument (Fig. 10.6).

Cleaning Up. Rinse out the syringe by filling it with dichloromethane and squirting it onto a tissue or paper towel at least four times. Clean and return any sample collection tubes, if used.

Collecting a Sample for an Infrared Spectrum

The small amount of sample analyzed in gas chromatography is an advantage in many cases, but it precludes isolating the separated components. Some specialized chromatographs can separate samples as large as 0.5 mL per injection and automatically collect each fraction in a separate container. At the other extreme, gas chromatographs equipped with FIDs can detect micrograms of substances, such as traces of pesticides in food, or drugs in blood and urine. Clearly, a gas chromatogram gives little information about the chemical nature of the sample being detected. Gas chromatography cannot positively identify most samples, nor can it always detect all substances in a sample. All that a gas chromatogram can truly tell you is that the detector was sensitive to a compound, and it can give you the relative time that a component eluted from a column. However, certain

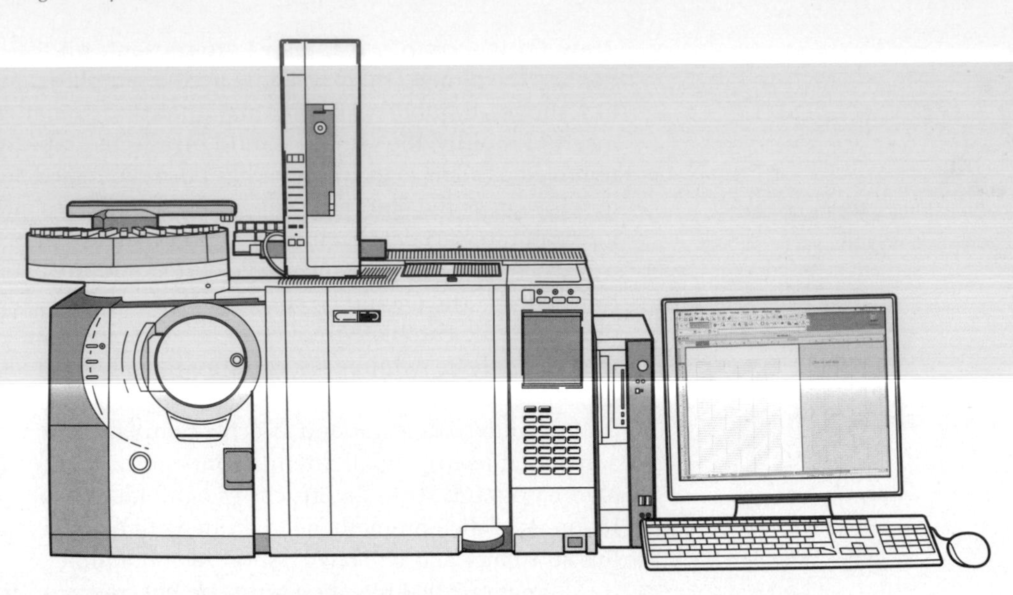

FIG. 10.6
A combined gas chromatograph/mass spectrometer, fitted with an automatic sample changer.

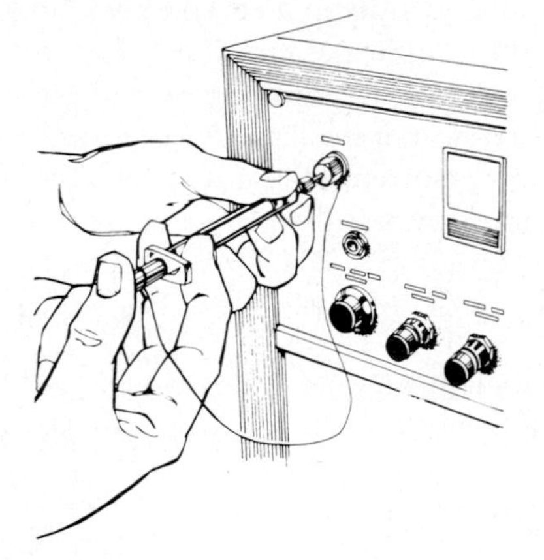

FIG. 10.7
Some gas chromatographs with packed columns allow the collection of small amounts of separated compounds.

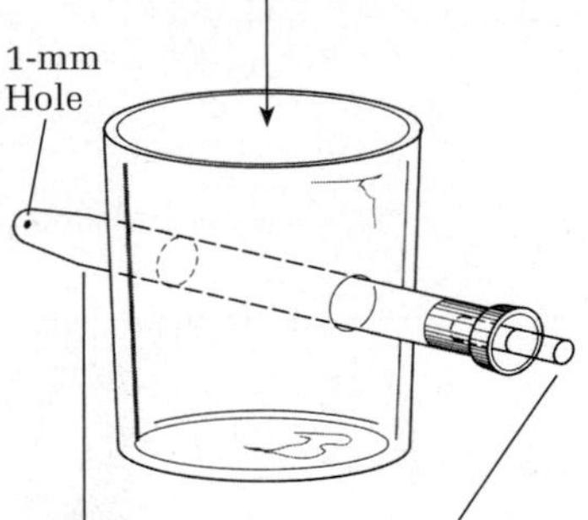

FIG. 10.8
A gas chromatographic collection device. Fill the container with ice or a dry ice–acetone mixture and attach to the outlet port of the gas chromatograph.[1]

preparative chromatographs, like that shown in Figure 10.7, allow the collection of enough sample at the exit port to obtain an infrared spectrum. About 10–15 μL of a mixture (not diluted in solvent) is injected, and as the peak for the compound of interest appears, a 2-mm-diameter glass tube, 3 in. long and packed with glass wool, is inserted into the rubber septum at the exit port. The sample, if it is not too volatile, will condense in the cold glass tube. Subsequently, the sample is washed out with 1 or 2 drops of solvent, and an infrared spectrum is obtained. This process can be repeated to collect enough sample for obtaining an NMR spectrum (*see* Chapter 12), using a few drops of deuterochloroform ($CDCl_3$) to wash out the tube each time. See Figure 10.8 for another collection device.

[1]This apparatus is available from Kimble Kontes (Vineland, NJ).

EXPERIMENTS

1. 2-METHYL-1-BUTENE AND 2-METHYL-2-BUTENE SYNTHESIS BY DEHYDRATION OF AN ALCOHOL[2]

2-Methylcyclohexanol can be used as a substrate if the volatility of the butenes is a problem.

The dilute sulfuric acid–catalyzed dehydration of 2-methyl-2-butanol (*t*-amyl alcohol) proceeds readily to give a mixture of alkenes that can be analyzed by gas chromatography. The mechanism of this reaction involves the intermediate formation of the relatively stable tertiary carbocation, followed by the loss of a proton either from a primary carbon atom to give the terminal olefin, 2-methyl-1-butene, or from a secondary carbon to give 2-methyl-2-butene.

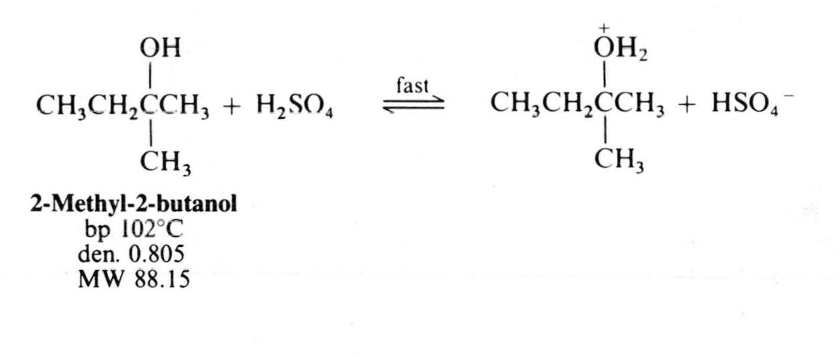

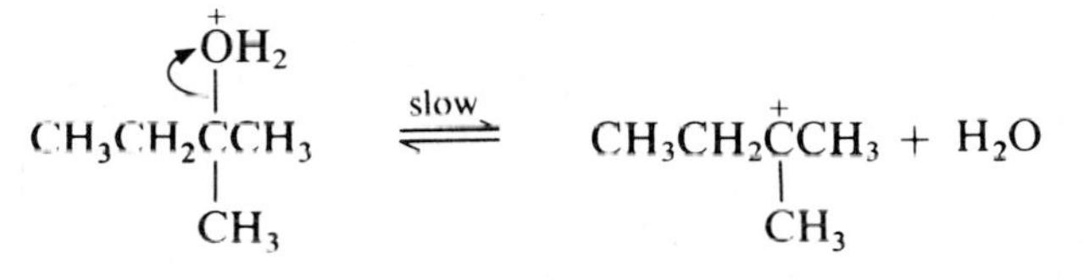

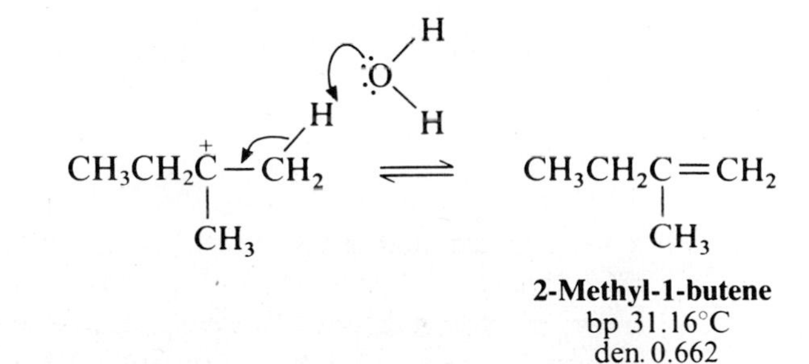

[2]Schimelpfenig, C. W. *J. Chem. Educ.* 1962, *39*, 310.

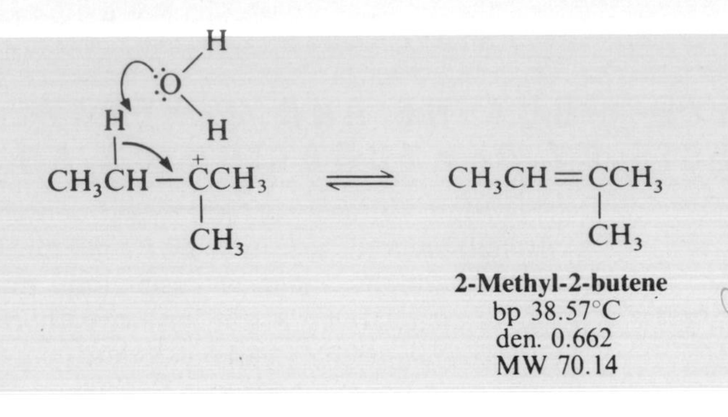

2-Methyl-2-butanol can also be dehydrated in high yield using iodine as a catalyst:

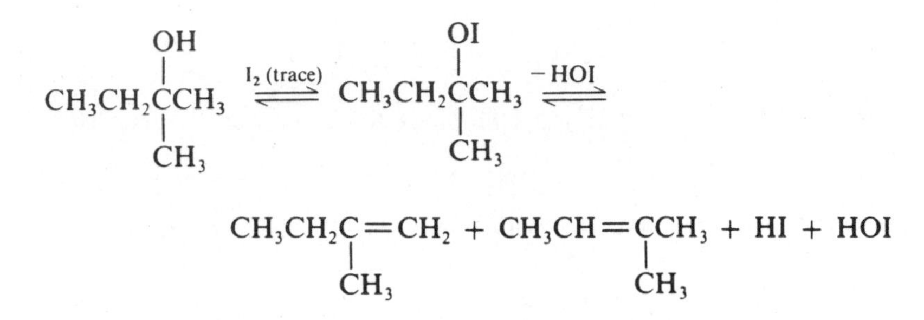

Each step of this E_1 elimination reaction is reversible, and thus the reaction is driven to completion by removing one of the products, the alkene. Often these reactions produce several alkene isomers. The Saytzeff rule states that the more substituted alkene is the more stable, and thus the one formed in larger amount. The *trans*-isomer is more stable than the *cis*-isomer. With this information, it should be possible to deduce which peak on the gas chromatogram corresponds to a given alkene and to predict the ratios of the products.

In analyzing your results from this experiment, consider the fact that the carbocation can lose any of six primary hydrogen atoms but only two secondary hydrogen atoms to give the product olefins.

Microscale Procedure

IN THIS EXPERIMENT, a tertiary alcohol is dehydrated to a mixture of alkenes in a reaction catalyzed by sulfuric acid. The products are collected by simple distillation into an ice-cold vial. Cold aqueous base is added to remove the acid, and the organic layer is dried with calcium chloride. The products are again distilled and then analyzed by gas chromatography.

Into a 5-mL long-necked, round-bottomed flask, place 1 mL of water and then add dropwise with thorough mixing 0.5 mL of concentrated sulfuric acid. Cool the hot solution in an ice bath; then add to the cold solution 1.0 mL (0.80 g) of 2-methyl-2-butanol. Mix the reactants thoroughly, add a boiling stone, and set up the apparatus for simple distillation, as shown in Figure 10.9. Hold the flask in a

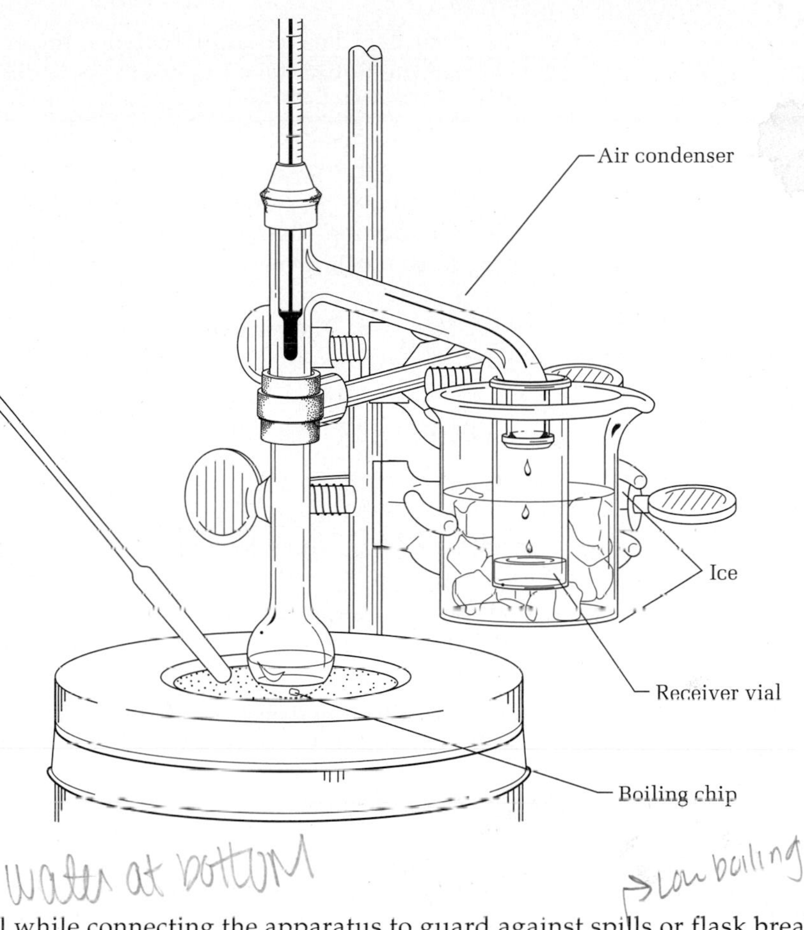

FIG. 10.9
An apparatus for the dehydration of 2-methyl-2-butanol. The beaker can be clamped with a large three-prong clamp.

towel while connecting the apparatus to guard against spills or flask breakage. Cool the receiver (a vial) in ice to avoid losing the highly volatile products. Warm the flask on a sand bath to start the reaction and distill the products over the temperature range 30°C–45°C. After all the products have distilled, the rate of distillation will decrease markedly. Cease distillation at this point. After this, the temperature registered on the thermometer will rise rapidly as water and sulfuric acid begin to distill.

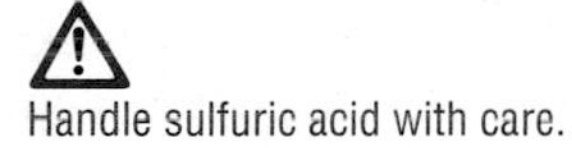
Handle sulfuric acid with care.

Keep the side arm well down inside the cold vial to avoid losing product.

To the distillate, add 0.3 mL *cold* 3 *M* sodium hydroxide solution to neutralize sulfurous acid, mix well, draw off the aqueous layer, and dry the product over anhydrous calcium chloride pellets, adding the drying agent in small quantities until it no longer clumps together. Keep the vial cold. While the product is drying, rinse out the distillation apparatus with water, ethanol, and then a small amount of acetone and dry it thoroughly by drawing air through it using an aspirator. If this is not done carefully, the products will become contaminated with acetone. Carefully transfer the dry mixture of butenes to the distilling apparatus using a Pasteur pipette, add a boiling chip to the now dry products, and distill into a tared, cold receiver. Collect the portion boiling up to 43°C. Weigh the product and calculate the yield, which is normally around 50% on this small scale compared to 84% on a macroscale. Inject a few microliters of product into a gas chromatograph maintained at room temperature (packed column suggestion: 6-mm-diameter × 3-m column packed with 10% SE-30 silicone rubber on Chromosorb-W).

Products are easily lost by evaporation. Store the mixture in a spark-proof refrigerator if GC analysis must be postponed.

For capillary column analysis, inject 1–10 μL of just the gaseous *vapors* above the liquid. Oven temperatures should be as low as room temperature (oven door open) to obtain baseline separation of the methylbutene isomers. In a few minutes, two peaks should appear. Measure the relative areas under the two peaks. From your knowledge of the mechanisms of the dehydration of secondary alcohols, which olefin should predominate? Does this agree with the boiling points? (In general, the compound with the shorter retention time has the lower boiling point.)

Cleaning Up. The residue in the reaction tube should be diluted with water, neutralized with sodium carbonate, and then flushed down the drain.

Macroscale Procedure

Cool a 100 mL flask containing 18 mL of water in an ice-water bath while slowly pouring in 9 mL of concentrated sulfuric acid. Cool this 1:2 acid mixture further with swirling while slowly pouring in 18 mL (15 g) of 2-methyl-2-butanol. Shake the mixture thoroughly and then mount the flask for fractional distillation over a flask heater, with the arrangement for ice cooling of the distillate seen in Figure 19.2 (on page 337). Cooling is needed because the olefin is volatile. Use a long condenser and an ample stream of cooling water. Heat the flask slowly with an electric flask heater until distillation of the hydrocarbon is complete. If the ice-cooled test tube will not hold 15 mL, be prepared to collect half of the distillate in another ice-cooled test tube. Transfer the distillate to a cool separatory funnel and shake it with about 5 mL of a cold 3 *M* sodium hydroxide solution to remove any traces of sulfurous acid. The aqueous solution sinks to the bottom and is drawn off. Dry the hydrocarbon layer by adding sufficient anhydrous calcium chloride pellets until the drying agent no longer clumps together. After about 5 minutes, remove the drying agent by gravity filtration or careful decantation into a dry 25-mL round-bottomed flask, and distill the dried product through a fractionating column (*see* Fig. 19.2 on page 337), taking the same precautions as before to avoid evaporation losses. Rinse the fractionating column with acetone and dry it with a stream of air to remove water from the first distillation. Collect in a tared bottle the portion boiling at 30°C–43°C. The yield reported in the literature is 84%; the average student yield is about 50%. Analyze the sample by injecting a few microliters of product into a gas chromatograph maintained at room temperature and equipped with a 6-mm-diameter × 3-m column packed with 10% SE-30 silicone rubber on Chromosorb-W or a similar inert packing. In a few minutes, two peaks should appear. From your knowledge of the mechanisms of dehydration of secondary alcohols, which olefin should predominate? Does this agree with the boiling points? (In general, the compound with the shorter retention time has the lower boiling point.) Measure the relative areas under the two peaks.

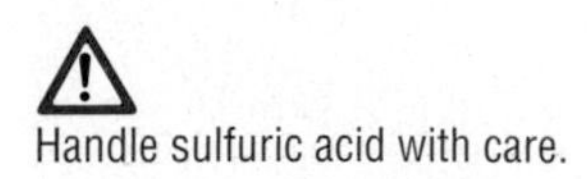
Handle sulfuric acid with care.

Cleaning Up. The pot residue from the reaction is combined with the sodium hydroxide wash and neutralized with sodium carbonate. The pot residue from the distillation of the product is combined with the acetone that was used to wash out the apparatus and placed in the organic solvents waste container. If the calcium chloride pellets are dry, they can be placed in the nonhazardous solid waste container. If they are wet with organic solvents, they must be placed in the hazardous solid waste container for solvent-contaminated drying agents.

2. COMPUTATIONAL CHEMISTRY

The steric energies of the two isomeric butenes produced in Experiment 1 can be calculated using a molecular mechanics program, but the results are not valid because the bonding type is not the same. It is possible to compare the steric energies of *cis*- and *trans*-isomers of an alkene, but not a 1,1-disubstituted alkene with a 1,2-disubstituted alkene.

Compare your yields of the two isomers to their heats of formation, calculated using AM1 or a similar semiempirical calculation as described in Chapter 15. Do the calculated heats of formation correlate with the relative percentages of the isomers? What does this correlation or lack thereof tell you about the mechanism of the reaction?

3. 1-BUTENE AND *cis*- AND *trans*-2-BUTENE SYNTHESIS BY DEHYDRATION OF AN ALCOHOL[3]

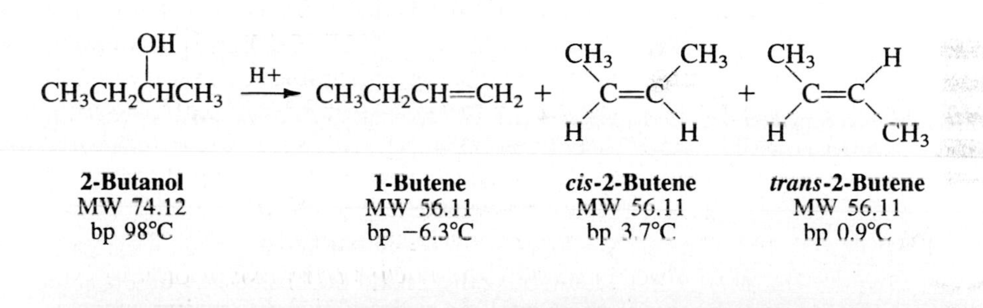

IN THIS EXPERIMENT, 2-butanol is dehydrated with sulfuric acid to give a mixture of gaseous alkenes that are collected by the downward displacement of water. The gaseous mixture is analyzed by gas chromatography.

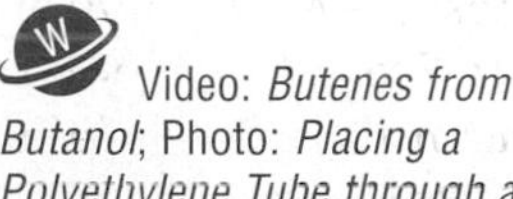

Video: *Butenes from Butanol*; Photo: *Placing a Polyethylene Tube through a Septum*

Bend the polyethylene tubing to shape in hot sand or steam.

In a 10 × 100 mm reaction tube, place 0.10 mL (81 mg) of 2-butanol and 0.05 mL (1 drop) of concentrated sulfuric acid. It is not necessary to measure the quantities of alcohol and acid exactly. Mix the reactants well and add a boiling chip. Stirring the reaction mixture is not necessary because it is homogeneous. Insert a polyethylene tube through a septum (Fig. 10.10), place this septum on the reaction tube, and lead the tubing into the mouth of the distilling column/collection tube, which is also capped with a septum. The distilling column/collection tube is filled with water, the open end capped with a finger, and inverted in a beaker of water and clamped for the collection of butene by the downward displacement of water (Fig. 10.11).

Lower the reaction tube into a hot (~100°C) sand bath and increase the heat slowly to complete the reaction. Collect a few milliliters of the butene mixture and *remove the polyethylene tube from the water bath.* If the polyethylene tube is not removed from the water bath, water will be sucked back into the reaction tube

[3]Helmkamp, G. K.; Johnson, H. W. Jr. *Selected Experiments in Organic Chemistry*, 3rd ed., Freeman: New York, 1983; 99.

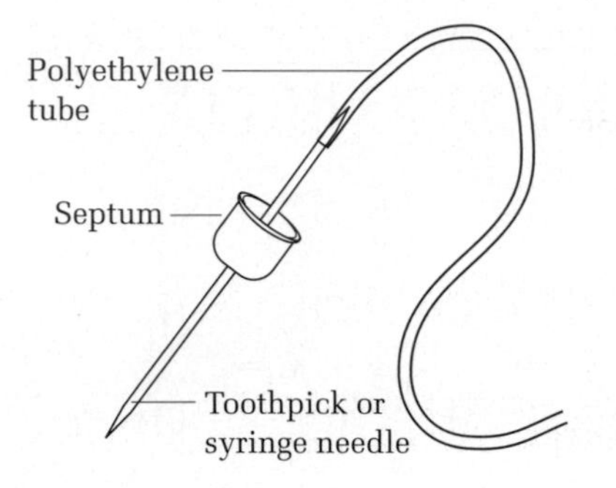

FIG. 10.10
To thread a polyethylene tube through a septum, make a hole in the septum with a needle and then push a toothpick through the hole. Push the polyethylene tube firmly onto the toothpick; then pull and push on the toothpick to thread the tubing through the septum. Finally, pull the tube from the toothpick after it has come out the other side of the septum. A blunt syringe needle can also be used instead of a toothpick.

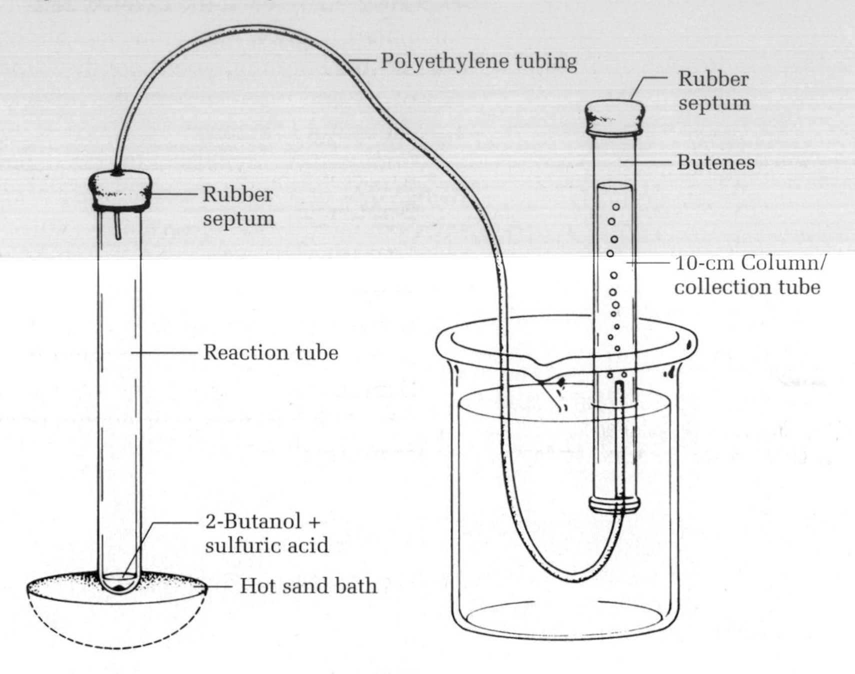

FIG. 10.11
An apparatus for the dehydration of 2-butanol. Butenes are collected by the downward displacement of water.

when it cools. Do not remove the septum-capped collection tube from the beaker of water, but bring it and the beaker of water together to the gas chromatograph. Use about 0.5 mL of the gaseous butene mixture to carry out the GC analysis with a TCD detector, or a smaller amount if an FID detector is used.

Cleaning Up. The residue in the reaction tube may be diluted with water and flushed down the drain.

QUESTIONS

1. Write the structures of the three olefins produced by the dehydration of 3-methyl-3-pentanol.
2. When 2-methylpropene is bubbled into dilute sulfuric acid at room temperature, it appears to dissolve. What new substance has been formed?
3. A student wished to prepare ethylene gas by the dehydration of ethanol at 140°C, using sulfuric acid as the dehydrating agent. A low-boiling liquid was obtained instead of ethylene. What was the liquid, and how might the reaction conditions be changed to give ethylene?
4. What would be the effect on retention time of increasing the carrier gas flow rate?
5. What would be the effect on retention time of raising the column temperature?
6. What would be the effect of raising the temperature or increasing the carrier gas flow rate on the ability to resolve two closely spaced peaks?

7. If you were to use a column one-half the length of the one you actually used, how do you think the retention times of the butenes would be affected? How do you think the separation of the peaks would be affected? How do you think the width of each peak would be affected?
8. From your knowledge of the dehydration of tertiary alcohols, which olefin should predominate in the product of the dehydration of 2-methyl-2-butanol? Why?
9. What is the maximum volume (at standard temperature and pressure, STP) of the butene mixture that could be obtained by the dehydration of 81 mg of 2-butanol?
10. What other gases are in the collection tube besides the three butenes at the end of the reaction in Experiment 3?
11. Using a computational chemistry program (*see* Chapter 15), calculate the steric energies or heats of formation of the three isomeric butenes produced in Experiment 3. Do these energies correlate with the product distributions found? What does a correlation or lack thereof tell you about the mechanism of the reaction?

For Additional Experiments, sign in at this book's premium website at **www.cengage.com/login.**

11 CHAPTER

Infrared Spectroscopy

PRELAB EXERCISE: When an infrared (IR) spectrum is run, it is possible that the chart paper is not properly placed or that the spectrometer is not mechanically adjusted. Describe how you could calibrate an IR spectrum.

The types and molecular environment of functional groups in organic molecules can be identified by infrared (IR) spectroscopy. Like nuclear magnetic resonance (NMR) and ultraviolet (UV) spectroscopy, IR spectroscopy is nondestructive. Moreover, the small quantity of sample needed, the speed with which a spectrum can be obtained, the relatively low cost of the spectrometer, and the wide applicability of the method combine to make IR spectroscopy one of the most common structural elucidation tools used by organic chemists.

IR radiation consists of wavelengths that are longer than those of visible light. It is detected not with the eyes, but by a feeling of warmth on the skin. When absorbed by molecules, radiation of these wavelengths (typically 2.5–5 μm) increases the amplitude of vibrations of the chemical bonds joining atoms.

2.5–25 μm equals 4000–400 cm^{-1}. Wavenumber(cm^{-1}) is proportional to frequency (c is the speed of light).

$$\bar{\nu}\,(\text{cm}^{-1}) = \frac{\nu}{c}$$

IR spectra are measured in units of frequency or wavelength. The wavelength is measured in micrometers[1] or microns, μ (1 μm = 1 × 10^{-6} m). The positions of absorption bands are measured in frequency units called wavenumbers ν, which are expressed in reciprocal centimeters, cm^{-1}, corresponding to the number of cycles of the wave in each centimeter.

$$\text{cm}^{-1} = \frac{10{,}000}{\mu\text{m}}$$

Examine the scale carefully.

Unlike UV and NMR spectra, IR spectra are inverted, with the strongest absorptions at the bottom (called "peaks" although they look like valleys), and are not always presented on the same scale. Some spectrometers record the spectra on an ordinate linear in microns, but this compresses the low-wavelength region. Other spectrometers present the spectra on a scale linear in reciprocal centimeters, but linear on two different scales: one between 4000 and 2000 cm^{-1}, which spreads out the low-wavelength region; and the other a smaller one between 2000 and 667 cm^{-1}. Consequently, spectra of the same compound run on two different spectrometers will not always look the same.

[1] Although micrometers are known as microns, the micron is not the official SI (International System of Units) unit.

IR spectroscopy easily detects:

Hydroxyl groups	**—OH**
Amines	**—NH_2**
Nitriles	**—C≡N**
Nitro groups	**—NO_2**

IR spectroscopy is especially useful for detecting and distinguishing among all carbonyl-containing compounds:

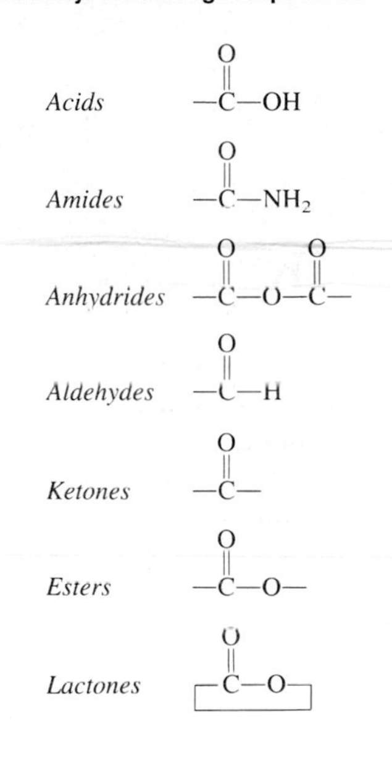

To picture the molecular vibrations that interact with IR light, imagine a molecule as being made up of balls (atoms) connected by springs (bonds). The vibration can be described by Hooke's law from classical mechanics, which says that the frequency of a stretching vibration is directly proportional to the strength of the spring (bond) and inversely proportional to the masses connected by the spring. Thus we find that C—H, N—H, and O—H bond-stretching vibrations are high frequency (short wavelength) compared to those of C—C and C—O because of the low mass of hydrogen compared to that of carbon or oxygen. The bonds connecting carbon to bromine and iodine, atoms of large mass, vibrate so slowly that they are beyond the range of most common IR spectrometers. A double bond can be regarded as a stiffer, stronger spring, so we find C=C and C=O vibrations at higher frequencies than C—C and C—O stretching vibrations. And C≡C and C≡N stretch at even higher frequencies than C=C and C=O (but at lower frequencies than C—H, N—H, and O—H). These frequencies are in keeping with the bond strengths of single (~100 kcal/mol), double (~160 kcal/mol), and triple bonds (~220 kcal/mol).

These stretching vibrations are intense and particularly easy to analyze. A nonlinear molecule of *n* atoms can undergo $3n - 6$ possible modes of vibration, which means cyclohexane with 18 atoms can undergo 48 possible modes of vibration. Each vibrational mode produces a peak in the spectrum because it corresponds to the absorption of energy at a discrete frequency. These many modes of vibration create a complex spectrum that defies simple analysis, but even in very complex molecules, certain functional groups have characteristic frequencies that can easily be recognized. Within these functional groups are the above-mentioned atoms and bonds, C—H, N—H, O—H, C=C, C=O, C≡C, and C≡N. Their absorption frequencies are given in Table 11.1.

When the frequency of IR light is the same as the natural vibrational frequency of an interatomic bond, light will be absorbed by the molecule, and the amplitude of the bond vibration will increase. The intensity of IR absorption bands is proportional to the change in dipole moment that a bond undergoes when it stretches. Thus, the most intense bands (peaks) in an IR spectrum are often from C=O and C—O stretching vibrations, whereas the C=C stretching band for a symmetrical acetylene is almost nonexistent because the molecule undergoes no net change of dipole moment when it stretches:

TABLE 11.1 *Characteristic IR Absorption Wavenumbers*

Functional Group	*Wavenumber (cm^{-1})*
O—H	3600–3400
N—H	3400–3200
C—H	3080–2760
C≡N	2260–2215
C≡C	2150–2100
C=O	1815–1650
C=C	1660–1600
C—O	1200–1050

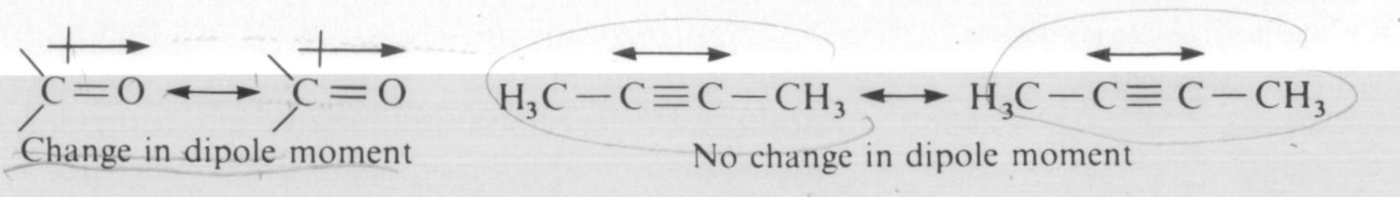

The intensity of absorption is proportional to the change in dipole moment.

Unlike proton NMR spectroscopy, where the area of the peaks is strictly proportional to the number of hydrogen atoms causing the peaks, the intensities of IR peaks are not proportional to the numbers of atoms causing them. Given the chemical shifts and coupling constants, it is not too difficult to calculate a theoretical NMR spectrum that is an exact match to the experimental one. For larger molecules, the calculation of all possible stretching and bending frequencies (the IR spectrum) requires large amounts of time on a fast computer. Every peak or group of peaks in an NMR spectrum can be assigned to specific hydrogens in a molecule, but the assignment of the majority of peaks in an IR spectrum with absolute certainty is usually not possible. Peaks to the right (longer wavelength) of 1250 cm^{-1} are the result of combinations of vibrations that are characteristic not of individual functional groups but of the molecule as a whole. This part of the spectrum is often referred to as the *fingerprint region* because it is uniquely characteristic of each molecule. Although two organic compounds can have the same melting points or boiling points and can have identical UV and NMR spectra, they cannot have identical IR spectra (except, as usual, for enantiomers). IR spectroscopy is thus the final arbiter in deciding whether two compounds are identical.

ANALYSIS OF IR SPECTRA

Only a few simple rules or equations govern IR spectroscopy. Because it is not practical to calculate theoretical spectra, the analysis is done almost entirely by *correlation* with other spectra. In printed form, these comparisons take the form of lengthy discussions, so detailed analysis of a spectrum is best done with a good reference book at hand.

In a modern analytical or research laboratory, a collection of many thousands of spectra is maintained on a computer. When the spectrum of an unknown compound is run, the analyst picks out five or six of the strongest peaks and asks the computer to list all the known compounds that have peaks within a few reciprocal centimeters of the experimental peaks. From the printout of a dozen or so compounds, it is often possible to pinpoint all the functional groups in the molecule being analyzed. There may be a perfect match of all peaks, in which case the unknown will have been identified.

For relatively simple molecules, a computer search is hardly necessary. Much information can be gained about the functional groups in a molecule from relatively few correlations.

To carry out an analysis, (1) pay most attention to the strongest absorptions; (2) pay more attention to peaks to the left (shorter wavelength) of 1250 cm^{-1}; and (3) pay as much attention to the absence of certain peaks as to the presence of others. The absence of characteristic peaks will definitely exclude certain functional groups. Be wary of weak O—H peaks because water is a common contaminant of many samples. Because potassium bromide is hygroscopic, water is often found in the spectra of samples prepared as KBr pellets.

The Step-by-Step Analysis of IR Spectra

IR spectra are analyzed as follows:

1. Is there a peak between 1820 cm^{-1} and 1625 cm^{-1}? If not, go to Step 2.
   (a) Is there a strong, wide O—H peak between 3200 cm^{-1} and 2500 cm^{-1}? If so, the compound is a carboxylic acid (Fig. 11.1, oleic acid). If not . . .
   (b) Is there a medium-to-weak N—H band between 3520 cm^{-1} and 3070 cm^{-1}? If there are two peaks in this region, the compound is a primary amide; if not, it is a secondary amide. If there is no peak in this region . . .
   (c) Are there two strong peaks, one in the region 1870 cm^{-1} to 1800 cm^{-1} and the other in the region 1800 cm^{-1} to 1740 cm^{-1}? If so, an acid anhydride is present. If not . . .
   (d) Is there a peak in the region of 2720 cm^{-1}? If so, is the carbonyl peak in the region 1715 cm^{-1} to 1680 cm^{-1}? If so, the compound is a conjugated aldehyde; if not, it is an isolated aldehyde (Fig. 11.2, benzaldehyde). However, if there is no peak near 2720 cm^{-1} . . .
   (e) Does the strong carbonyl peak fall in the region 1815 cm^{-1} to 1770 cm^{-1} and the compound give a positive Beilstein test? If so, it is an acid halide. If not . . .
   (f) Does the strong carbonyl peak fall in the region 1690 cm^{-1} to 1675 cm^{-1}? If so, the compound is a conjugated ketone. If not . . .

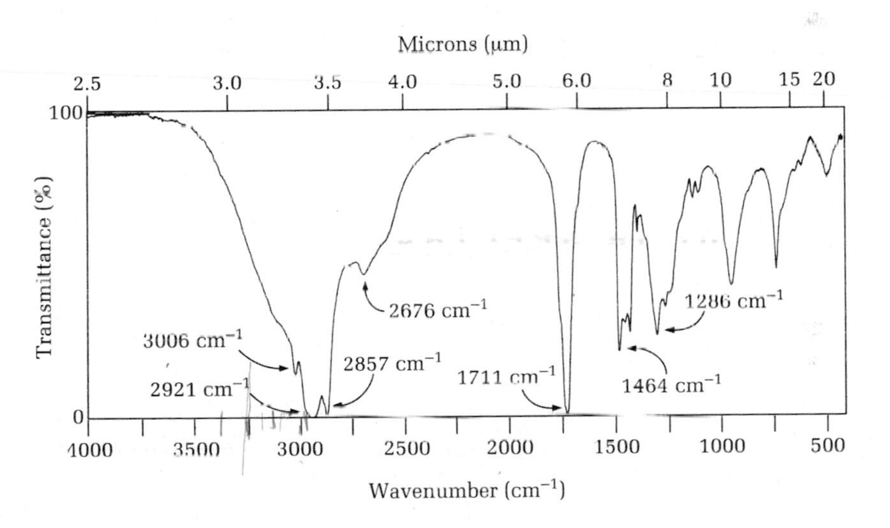

FIG. 11.1
The IR spectrum of oleic acid (thin film).

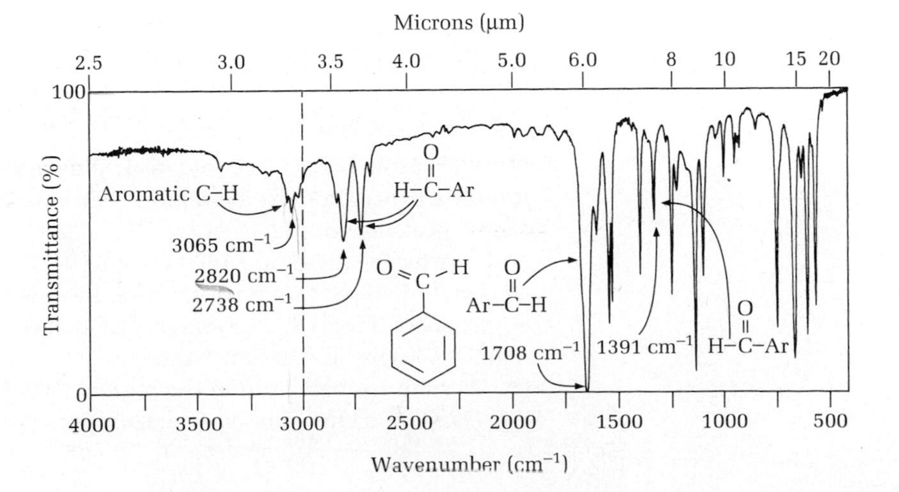

FIG. 11.2
The IR spectrum of benzaldehyde (thin film).

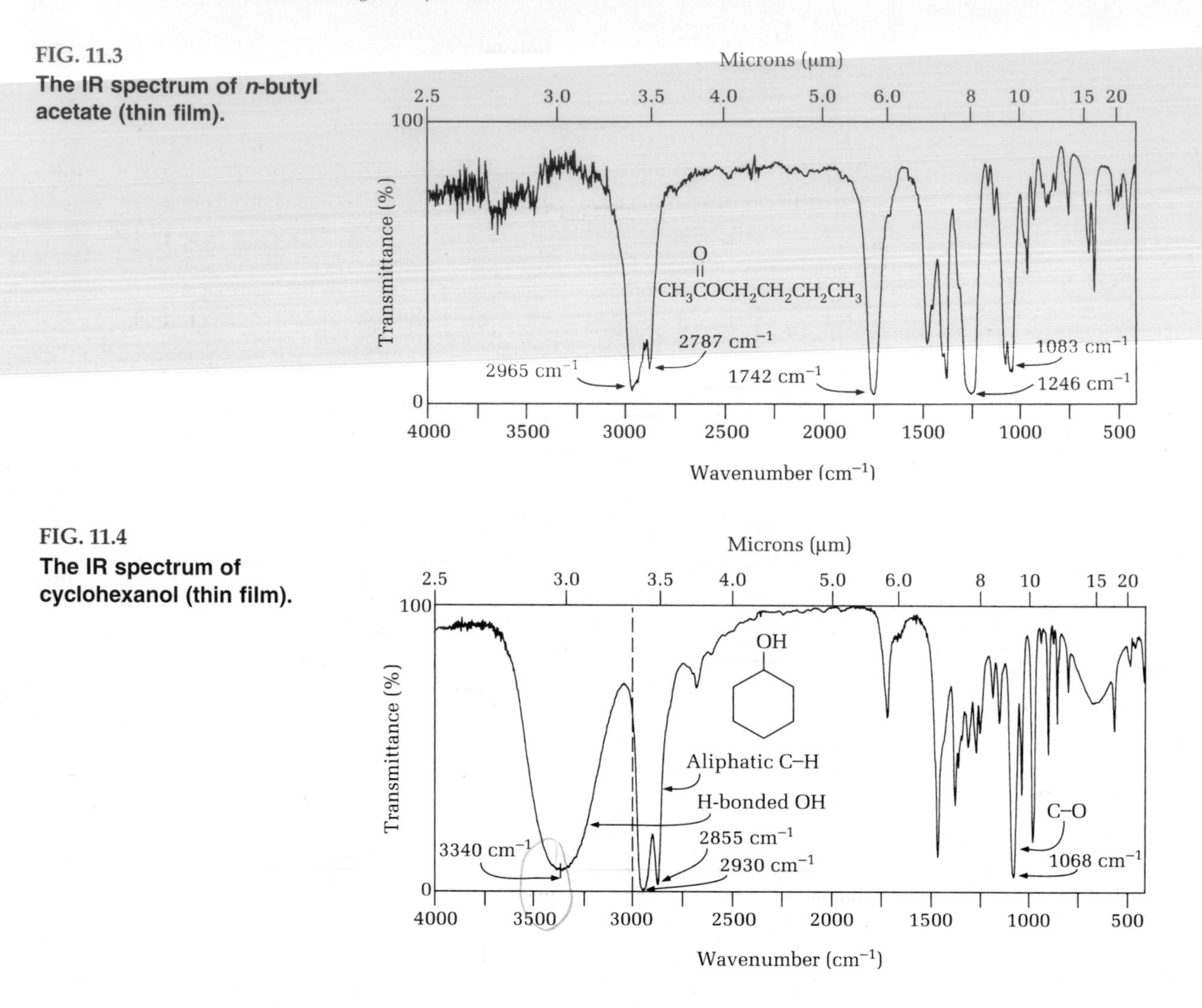

FIG. 11.3 **The IR spectrum of *n*-butyl acetate (thin film).**

FIG. 11.4 **The IR spectrum of cyclohexanol (thin film).**

(g) Does the strong carbonyl peak fall in the region 1670 cm^{-1} to 1630 cm^{-1}? If so, the compound is a tertiary amide. If not . . .

(h) Does the spectrum have a strong, wide peak in the region 1310 cm^{-1} to 1100 cm^{-1}? If so, does the carbonyl peak fall in the region 1730 cm^{-1} to 1715 cm^{-1}? If so, the compound is a conjugated ester; if not, the ester is not conjugated (Fig. 11.3, *n*-butyl acetate). If there is no strong, wide peak in the region 1310 to 1100 cm^{-1}, then . . .

(i) The compound is an ordinary nonconjugated ketone (see for example Fig. 22.8, the IR spectrum of cyclohexanone).

2. If the spectrum lacks a carbonyl peak in the region 1820 cm^{-1} to 1625 cm^{-1}, does it have a broad band in the region 3650 cm^{-1} to 3200 cm^{-1}? If so, does it also have a peak at about 1200 cm^{-1}, a C—H stretching peak to the left of 3000 cm^{-1}, and a peak in the region 1600 cm^{-1} to 1470 cm^{-1}? If so, the compound is a phenol. If the spectrum does not meet these latter three criteria, the compound is an alcohol (Fig. 11.4, cyclohexanol). However, if there is no broad band in the region 3650 cm^{-1} to 3200 cm^{-1}, then . . .

(a) Is there a broad band in the region 3500 cm^{-1} to 3300 cm^{-1}, and does the compound smell like an amine, or does it contain nitrogen? If so, are there

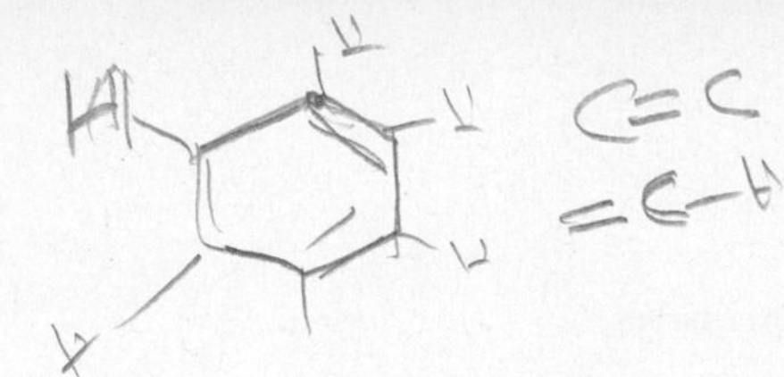

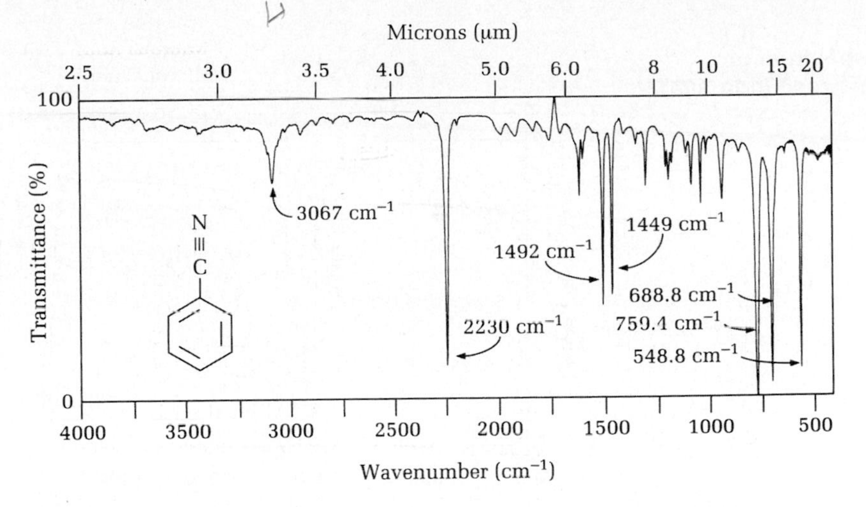

FIG. 11.5
The IR spectrum of benzonitrile (thin film).

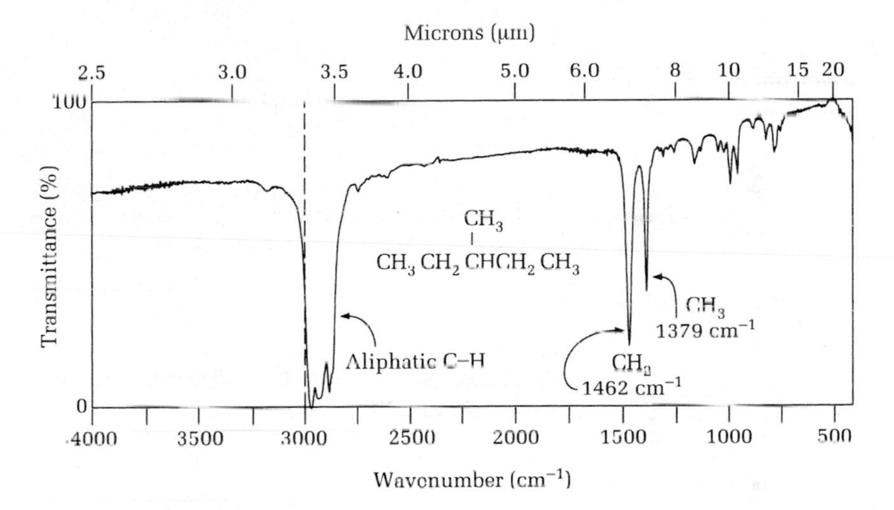

FIG. 11.6
The IR spectrum of 3-methylpentane (thin film).

two peaks in this region? If so, the compound is a primary amine; if not, it is a secondary amine. However, if there is no broad band in the region 3500 cm^{-1} to 3300 cm^{-1}, then . . .

(b) Is there a sharp peak of medium-to-weak intensity at 2260 cm^{-1} to 2100 cm^{-1}? If so, is there also a peak at 3320 cm^{-1} to 3310 cm^{-1}? If so, then the compound is a terminal acetylene. If not, the compound is most likely a nitrile (Fig. 11.5, benzonitrile), although it might be an asymmetrically substituted acetylene. If there is no sharp peak of medium-to-weak intensity at 2260 cm^{-1} to 2100 cm^{-1}, then . . .

(c) Are there strong peaks in the region 1600 cm^{-1} to 1540 cm^{-1} and 1380 cm^{-1} to 1300 cm^{-1}? If so, the molecule contains a nitro group. If not . . .

(d) Is there a strong peak in the region 1270 cm^{-1} to 1060 cm^{-1}? If so, the compound is an ether. If not . . .

(e) The compound is either a tertiary amine (odor?), a halogenated hydrocarbon (Beilstein test?), or just an ordinary hydrocarbon (Fig. 11.6, 3-methylpentane, and Fig. 11.7, *t*-butylbenzene).

Many comments can be added to this bare outline. For example, dilute solutions of alcohols will show a sharp peak at about 3600 cm^{-1} for a nonhydrogen-bonded O—H in addition to the usual broad hydrogen-bonded O—H peak.

FIG. 11.7
The IR spectrum of *t*-butylbenzene (thin film).

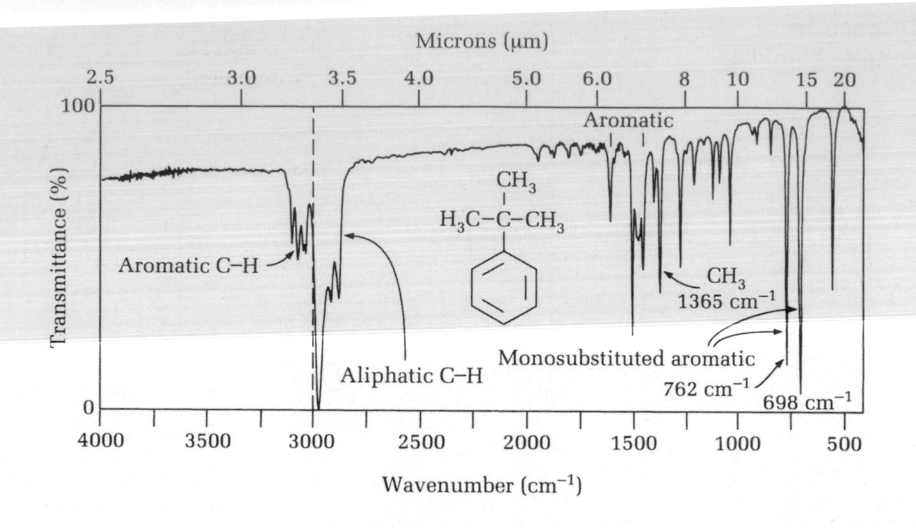

The effect of ring size on the carbonyl frequencies of lactones and esters:

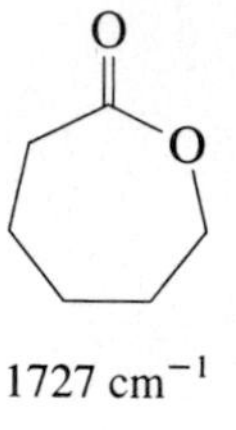
1727 cm^{-1}

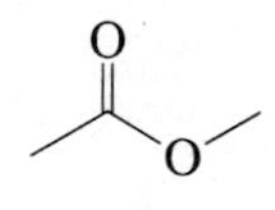
1745 cm^{-1}

1740 cm^{-1}

1775 cm^{-1}

1832 cm^{-1}

Aromatic hydrogens give peaks just to the left of 3000 cm^{-1}, whereas aliphatic hydrogens appear just to the right of 3000 cm^{-1}. However, NMR spectroscopy is the best method for identifying aromatic hydrogens.

The carbonyl frequencies listed earlier refer to an open chain or an unstrained functional group in a nonconjugated system. If the carbonyl group is conjugated with a double bond or an aromatic ring, the peak will be displaced to the right by 30 cm^{-1}. When the carbonyl group is in a ring smaller than six members or if there is oxygen substitution on the carbon adjacent to an aldehyde or ketone carbonyl, the peak will be moved to the left (refer to the margin notes about the effect of ring size and Table 11.2).

TABLE 11.2 *Characteristic IR Carbonyl Stretching Peaks (Chloroform Solutions)*

	Carbonyl-Containing Compounds	*Wavenumber (cm^{-1})*
RC(=O)R	Aliphatic ketones	1725–1705
RC(=O)Cl	Acid chlorides	1815–1785
R–C=C–C(=O)–R	α,β-Unsaturated ketones	1685–1666
ArC(=O)R	Aryl ketones	1700–1680
(cyclohexanone)=O	Cyclohexanones	1725–1705
–C(=O)–CH_2–C(=O)–	β-Diketones	1640–1540

The effect of ring size on carbonyl frequency of ketones:

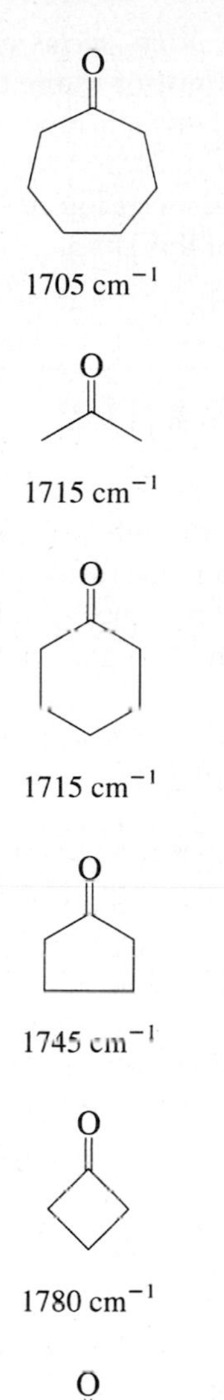

TABLE 11.2 (*continued*)

	Carbonyl-Containing Compounds	*Wavenumber (cm^{-1})*
RC(=O)H	Aliphatic aldehydes	1740–1720
R–C=C–C(=O)H	α,β-Unsaturated aldehydes	1705–1685
Ar C(=O)H	Aryl aldehydes	1715–1695
RCOOH	Aliphatic acids	1725–1700
R–C=C–COOH	α,β-Unsatur.ated acids	1700–1680
ArCOOH	Aryl acids	1700–1680
RC(=O)OR′	Aliphatic esters	1740
R–C=C–C(=O)OR′	α,β-Unsaturated esters	1730–1715
Ar C(=O)OR	Aryl esters	1730–1715
HC(=O)OR	Formate esters	1730–1715
CH_2=CHOC(=O)CH_3; C_6H_5OC(=O)CH_3	Vinyl and phenyl acetate	1776
R–C(=O)–O–C(=O)–R	Acyclic anhydrides (two peaks) 1780–1740	1840–1800
RC(=O)NH_2	Primary amides	1694–1650
RC(=O)NHR′	Secondary amides	1700–1670
RC(=O)NR'_2	Tertiary amides	1670–1630

Methyl groups often give a peak near 1375 cm^{-1}, but NMR is a better method for detecting this group.

The pattern of substitution on an aromatic ring (mono-, *ortho-*, *meta-*, and *para*, di-, tri-, tetra-, and penta-) can be determined from C—H out-of-plane bending vibrations in the region 670–900 cm^{-1}. Much weaker peaks between 1650 cm^{-1} and 2000 cm^{-1} are illustrated in Figure 11.8.

Extensive correlation tables and discussions of characteristic group frequencies can be found in the specialized references listed at the end of this chapter.

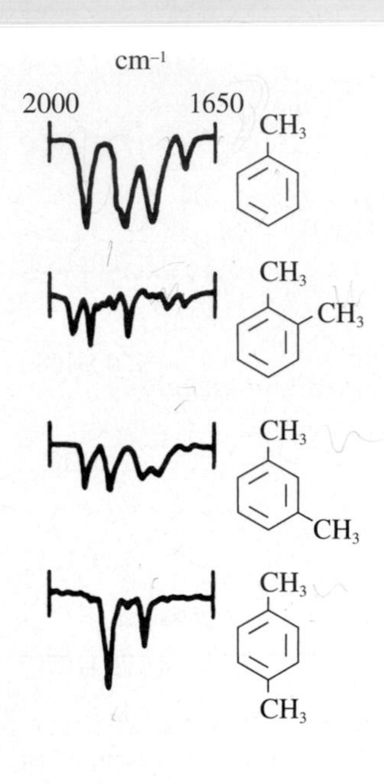

FIG. 11.8
The band patterns of toluene and *o*-, *m*-, and *p*-xylene. These peaks are *very weak.* They are characteristic of aromatic substitution patterns in general, not just for these four molecules. Try to find these patterns in other spectra throughout this text.

THE FOURIER TRANSFORM INFRARED (FTIR) SPECTROMETER

Like their NMR counterparts, Fourier transform IR instruments are computer based, can sum a number of scans to increase the signal-to-noise ratio, and allow enormous flexibility in the ways spectra can be analyzed and displayed because they are digitally oriented. In addition, they are faster and more accurate than double-beam analog spectrometers.

A schematic diagram of a single-beam FTIR spectrometer is given in Figure 11.9. IR radiation goes to a Michelson interferometer in which half the light beam passes through a partially coated mirror to a fixed mirror, and half goes to a moving mirror. The combined beams pass through the sample and are then focused on the detector. The motion of the mirror gives a signal that varies sinusoidally. Depending on the mirror position, the different frequencies of light either reinforce or cancel each other, resulting in an interferogram. The mirror, the only moving part of the spectrometer, thus affects the light frequencies that encode the very high optical frequencies, transforming them to low-frequency signals that change as the mirror moves back and forth. The Fourier transform converts the digitized signal into a signal that is a function of frequency—creating a spectrum.

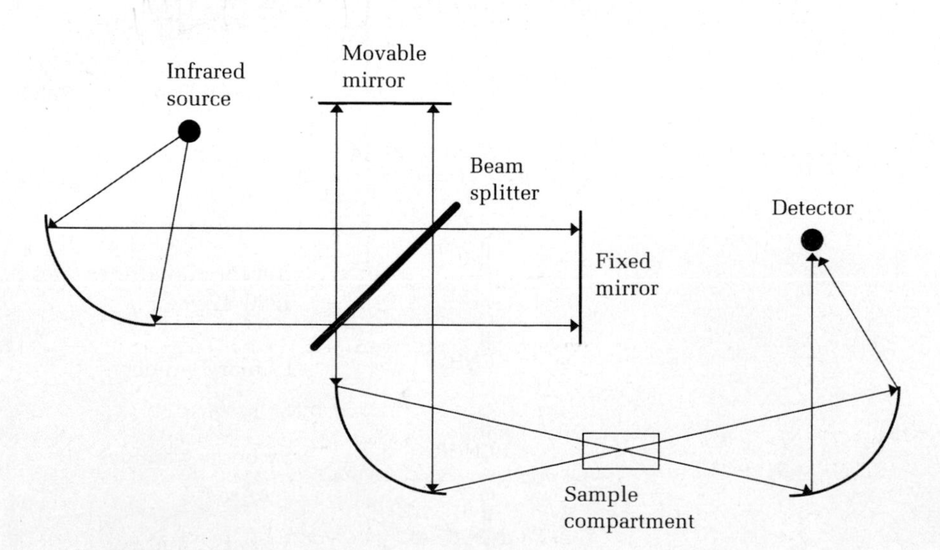

FIG. 11.9
The optical path diagram of an FTIR spectrometer.

Like Fourier transform NMR spectroscopy, all IR frequencies are sampled simultaneously instead of being scanned successively, so a spectrum is obtained in seconds. Because there is no slit to sort out the wavelengths, high resolution is possible without losing signal strength. Extremely small samples or very dilute solutions can be examined because it is easy to sum hundreds of scans in the computer. It is even possible to take a spectrum of an object seen under a microscope. The spectrum of one compound can be subtracted from, say, a binary mixture spectrum to reveal the nature of the other component. Data can be smoothed, added, resized, and so on because of the digital nature of the output from the spectrometer.

EXPERIMENTAL CONSIDERATIONS

The sample, solvents, and equipment must be dry.

Glass, quartz, and plastics are opaque to IR radiation and absorb it. Therefore, sample holders must be made of other transparent materials. Metal halide salts such as sodium or silver chloride, calcium or barium fluoride, and potassium bromide are transparent to IR radiation; these are prepared as large polished crystals for use in cell holders, with NaCl and KBr being the most common. They are fragile and are easily attacked by moisture, so they must be handled gently by the edges and kept away from moisture and aqueous solutions. IR spectra can be determined on neat (undiluted) liquids, on solutions with an appropriate solvent, and on solids using mulls (a finely ground solid suspended in oil) and KBr pellets, and by diffuse reflectance of the solid mixed with KBr. All IR cells or KBr powder must be stored in a desiccator when not in use.

THE SPECTRA OF NEAT LIQUIDS

To run a spectrum of a neat (free of water!) liquid, remove a demountable cell (Fig. 11.10) from the desiccator, place a drop of the liquid between the salt plates, press the plates together gently to remove any air bubbles, and add the top rubber gasket and metal top plate. Next, put on all four of the nuts and *gently* tighten them to apply an even pressure to the top plate. Place the cell in the sample compartment (nearest the front of the spectrometer) and run the spectrum.

Although running a spectrum on a neat liquid is convenient and results in no extraneous bands to interpret, it is not possible to control the path length of the light through the liquid in a demountable cell. A low-viscosity liquid when squeezed between the salt plates may be so thin that the short path length gives peaks that are too weak. A viscous liquid, on the other hand, may give peaks that are too intense. A properly run spectrum will have the most intense peak with a transmittance of about 10%.

A fast, simple alternative: Place a drop of the compound on a round salt plate, add a top plate, and mount the two on the holder pictured in Figure 11.11.

Another demountable cell is pictured in Figure 11.11. The plates are thin wafers of silver chloride, which is transparent to IR radiation. This cell has advantages over the salt cell in that the silver chloride disks are more resistant to breakage than NaCl plates, less expensive, and not affected by water. Because silver chloride is photosensitive, the wafers must be stored in a dark place to prevent them from turning black. Because one side of each wafer is recessed, the thickness of the sample can be varied according to the manner in which the cell is assembled. In general, the spectra of pure liquids are run as the thinnest possible films. Most of the spectra in this chapter have been obtained in this way. The disks are cleaned by

FIG. 11.10

An exploded view of a demountable salt cell for analyzing the IR spectra of neat liquids. The salt plates are fragile and expensive. Do not touch the surfaces. Use only dry solvents and samples.

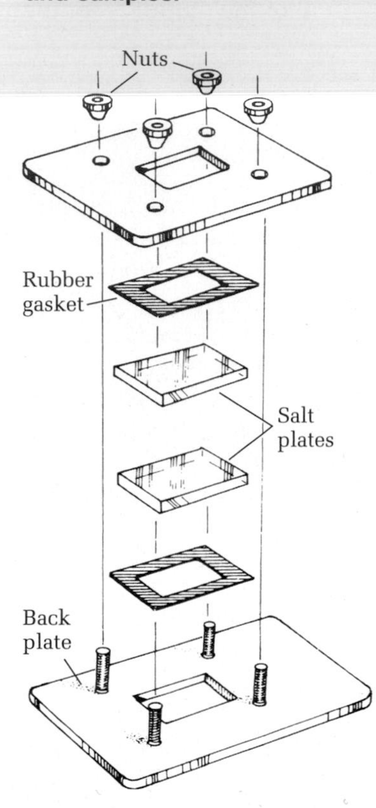

FIG. 11.11

A demountable silver chloride or sodium chloride cell.

rinsing them with an organic solvent such as acetone or ethanol, and wiping dry with an absorbent paper towel.

A very simple cell consists of two circular NaCl disks (1-in. diameter × 3/16-in. thick, 25 mm × 4 mm). The sample, 1 drop of a pure liquid or solution, is applied to the center of a disk with a polyethylene pipette (to avoid scratching the NaCl disk). The other disk is added, the sample is squeezed to a thin film, and the two disks are placed on the V-shaped holder shown in Figure 11.11.

Solvents: $CHCl_3$, CS_2, CCl_4

THE SPECTRA OF SOLUTIONS

The most widely applicable method of running the spectra of solutions involves dissolving an amount of the liquid or solid sample in an appropriate solvent to give a 10% solution. Just as in NMR spectroscopy, the best solvents to use are carbon disulfide and carbon tetrachloride. Because these compounds are not polar enough to dissolve many substances, chloroform is used as a compromise when the sample refuses to dissolve. Unlike NMR solvents, no solvent suitable in IR spectroscopy is entirely free of absorption bands in the frequency range of interest (Figs. 11.12, 11.13, and 11.14). In chloroform, for instance, no light passes through the cell between 650 cm^{-1} and 800 cm^{-1}. As can be seen from the figures, spectra obtained using carbon disulfide and chloroform cover the entire IR frequency range. Carbon tetrachloride would appear to be a good choice because it has few interfering peaks, but it is a poor solvent for polar compounds. In practice, a background spectrum is run with the same solvent in the same cell used to run the sample solution, and the data system subtracts the solvent peaks in background spectrum from the solution spectrum to yield an IR spectrum of the sample only.

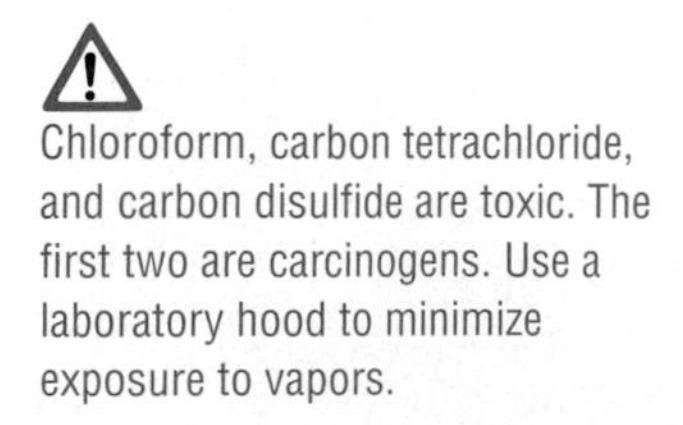

Chloroform, carbon tetrachloride, and carbon disulfide are toxic. The first two are carcinogens. Use a laboratory hood to minimize exposure to vapors.

Three large drops of solution will fill the usual sealed IR cell (Fig. 11.15). A 10% solution of a liquid sample can be approximated by diluting 1 drop of the liquid sample with 9 drops of the solvent. Because weights are more difficult to estimate, solid samples should be weighed to obtain a 10% solution.

Three drops are needed to fill the cell.

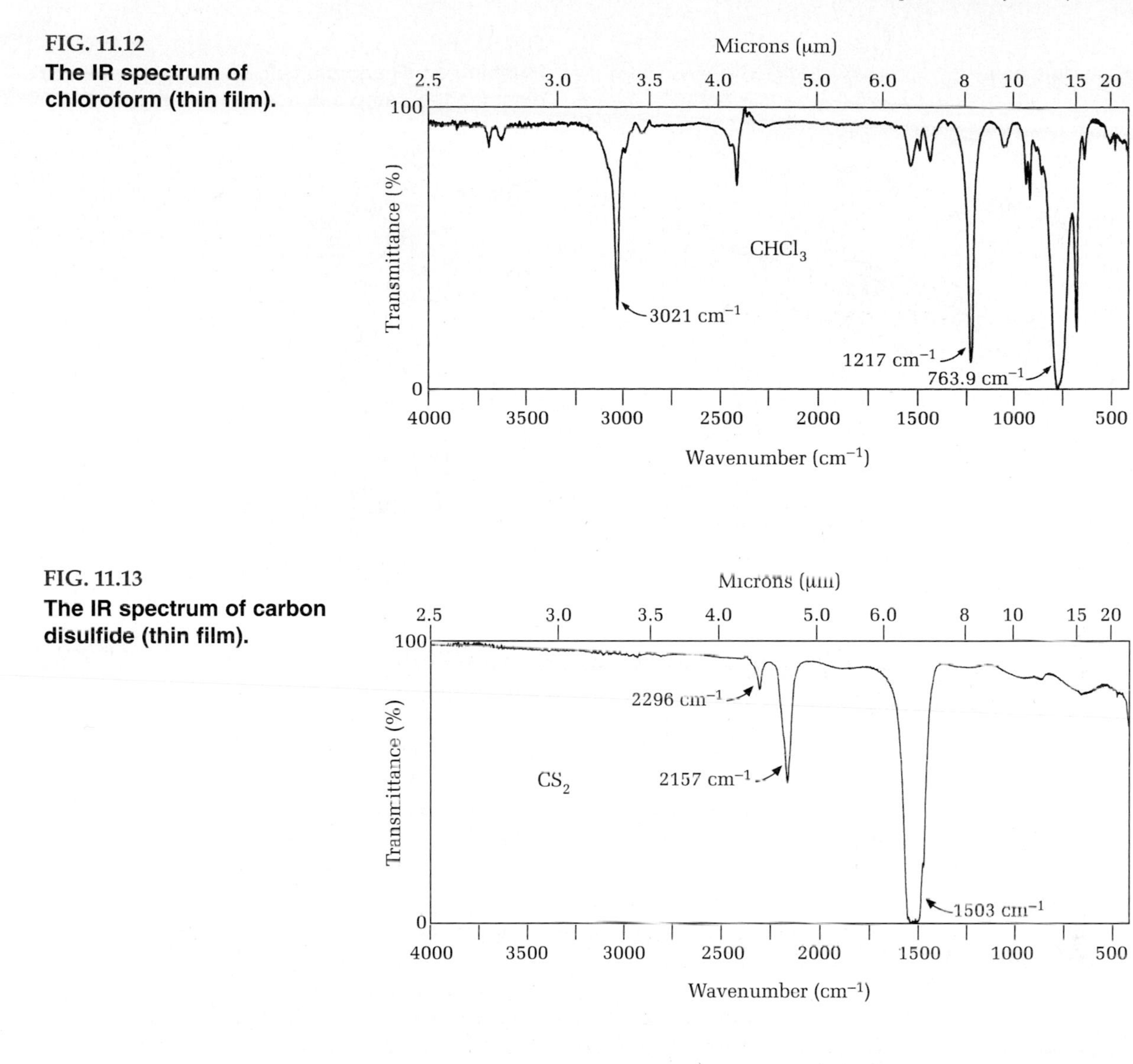

FIG. 11.12
The IR spectrum of chloroform (thin film).

FIG. 11.13
The IR spectrum of carbon disulfide (thin film).

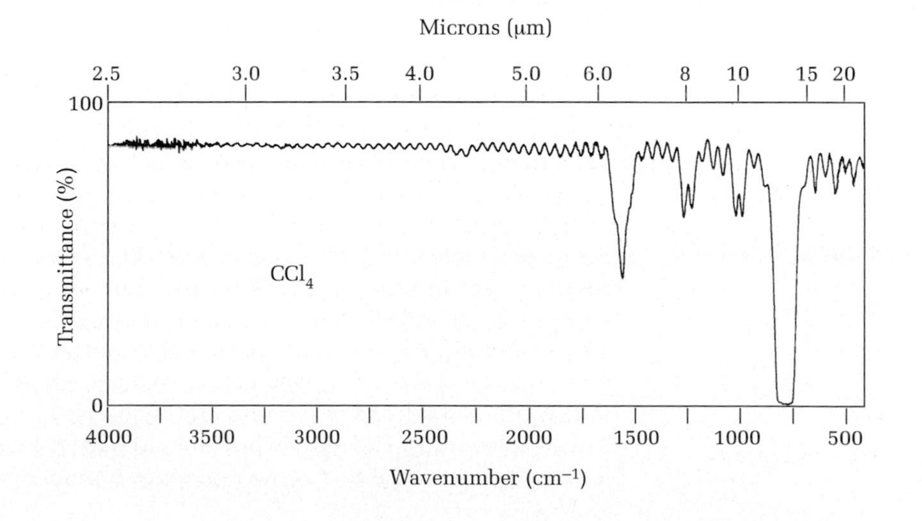

FIG. 11.14
The IR spectrum of carbon tetrachloride (thin film).

FIG. 11.15
A sealed IR sample cell.

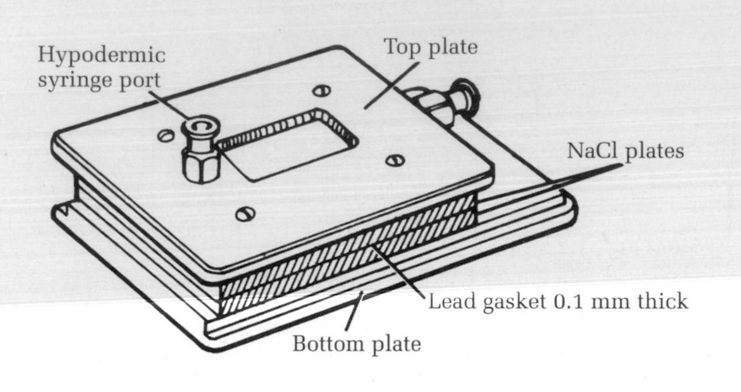

FIG. 11.16
Flushing an IR sample cell. The solvent used to dissolve the sample is used in this process.

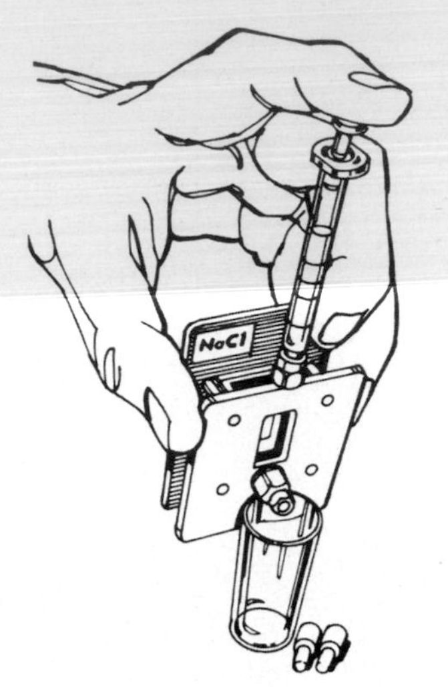

The solvent and the sample must be dry. Do not touch or breathe on NaCl plates. Cells are very expensive.

Dispose of waste solvents in the container provided.

First, obtain a background spectrum of the solvent. The IR cell is filled by inclining it slightly and placing about 3 drops of the solvent in the lower hypodermic port with a capillary dropper. The cell must be completely dry because the new sample will not enter if the cell contains solvent. The liquid can be seen rising between the salt plates by looking into the cell's window. In the most common sealed cell, the salt plates are spaced 0.1 mm apart. Be sure that the cell is filled past the top of the window and that no air bubbles are present. Then place a Teflon stopper lightly but firmly in the hypodermic port. Be particularly careful not to spill any of the sample on the outside of the cell windows. Place the cell in the instrument cell holder and acquire a background spectrum. Blow dry nitrogen or draw dry air through the cell to remove the solvent and dry the cell, fill with the sample solution prepared above, and acquire the spectrum.

After running the sample, force clean solvent through the cell using a syringe attached to the top port of the cell (Fig. 11.16). Dry using dry nitrogen or air, and store in a desiccator.

Cleaning Up. Discard halogenated liquids in the halogenated organic waste container. Other solutions should be placed in the organic solvents waste container.

MULLS

The sample must be finely ground.

Solids insoluble in the usual solvents can be run as mulls. In preparing a mull, the sample is ground to a particle size less than that of the wavelength of light going through the sample (2.5 μm) to avoid scattering light. About 15–20 mg of the sample is ground for 3–10 minutes in an agate mortar until it is spread over the entire inner surface of the mortar and has a caked and glassy appearance. Then to make a mull, add 1 or 2 drops of paraffin oil (Nujol; Fig. 11.17) and grind the sample for 2–5 more minutes. The mull, which should have the consistency of thin margarine, is transferred to the bottom salt plate of a demountable cell (*see* Fig. 11.10 on

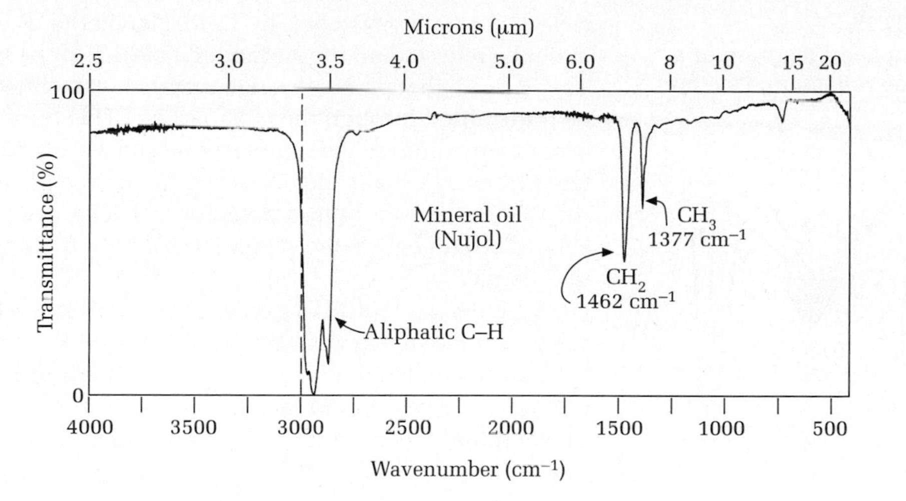

FIG. 11.17
The IR spectrum of Nujol (paraffin oil).

page 230) using a rubber policeman, the top plate is added and rotated back and forth to distribute the sample evenly to eliminate all air pockets, and the spectrum is run. Because the bands from Nujol obscure certain frequency regions, running another mull using Fluorolube as the mulling agent will allow the entire IR spectral region to be covered. If the sample has not been ground sufficiently fine, there will be marked loss of transmittance at the short-wavelength end of the spectrum. After running the spectrum, the salt plates are wiped clean with a cloth saturated with an appropriate solvent.

A fast, simple alternative: Grind a few milligrams of the solid with 1 drop of tetrachloroethylene between two round salt plates. Mount on the holder as shown in Figure 11.11 on page 230.

POTASSIUM BROMIDE DISK

The spectrum of a solid sample can also be run by incorporating the sample in a potassium bromide (KBr) disk. This procedure needs only one disk to cover the entire spectral range because KBr is completely transparent to IR radiation. Although very little sample is required, making the disk calls for special equipment and time to prepare it. Because KBr is hygroscopic, water is a problem.

Into a stainless steel capsule containing a ball bearing are weighed 1.5–2 mg of the compound and 200 mg of spectroscopic-grade KBr (previously dried and stored in a desiccator). The capsule is shaken for 2 minutes on a Wig-L-Bug (the device used by dentists to mix silver amalgam). The sample is evenly distributed over the face of a 13-mm stainless steel die and subjected to a pressure of 14,000–16,000 psi for 3–6 minutes while under vacuum in a hydraulic press. A transparent disk is produced, which is removed from the die with tweezers and placed in a holder like that shown in Figure 11.11 on page 230 prior to running the spectrum.

DIFFUSE REFLECTANCE

Reasonable IR spectra of solids can be obtained without pressing a KBr pellet if the IR instrument is equipped with diffuse reflectance optics. A mixture of the sample and KBr is put in a small sample cup, and this is placed in the diffuse

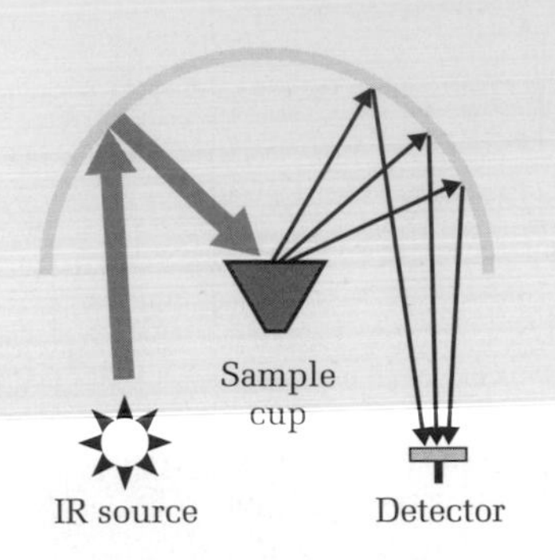

FIG. 11.18
A simplified diagram of a diffuse reflectance system.

reflectance optics system (Fig. 11.18). When the IR beam enters the sample, it can either be reflected off the surface of a particle or be transmitted through a particle. The IR beam that passes through a particle can either reflect off the next particle or be transmitted through the next particle. This transmission-reflectance event can occur many times in the sample, which increases the path length. Finally, any scattered IR energy that is not absorbed by the sample is collected by a spherical mirror that is focused onto a detector. A background spectrum of only pure KBr should be run before the sample to subtract out any absorbance that is not from the sample.

Weigh out 0.160–0.170 g of KBr quickly (KBr readily absorbs atmospheric moisture) and place it in a vial. Weigh out between 4 mg and 7 mg of the solid sample to be analyzed and add this to the vial. Mix and transfer the mixture into a small, dry mortar and pestle. Grind for 1–2 minutes until a fine powder is obtained. Transfer to weighing paper and then to the cup of the diffuse reflectance sample holder. Place the holder in the reflectance attachment and acquire the spectrum.

GAS PHASE IR SPECTROSCOPY: THE WILLIAMSON GAS PHASE IR CELL

In the past, it was difficult to obtain the IR spectra of gases. A commercial gas cell costs hundreds of dollars; the cell must be evacuated, and the dry gas carefully introduced—not a routine process. But we have found that a uniquely simple gas phase IR cell can be assembled at virtually no cost from a 105° microscale connecting adapter.[2]

To make the Williamson gas phase IR cell, attach a thin piece of polyethylene film from, for example, a sandwich bag to each end of the adapter with a small rubber band, fine wire, or thread. Place a loose wad of cotton in the side arm and mount the IR cell in the beam of the spectrometer (Fig. 11.19 and Fig. 11.20).

The adapter can be supported on a V-block as shown in Figure 11.20 or, temporarily, on a block of modeling clay. Be sure the beam, usually indicated by a red laser, strikes both ends of the cell. Run a background spectrum on the cell. The spectrum of polyethylene (Fig. 11.21) is very simple, so interfering bands will be few. This spectrum is automatically subtracted when you put a sample in the cell and run a spectrum.

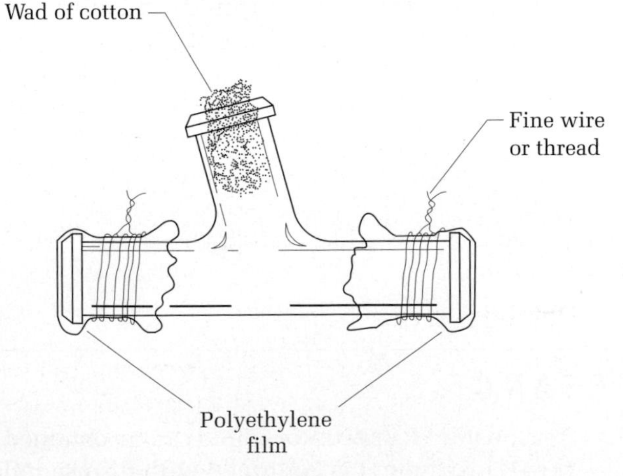

FIG. 11.19
The Williamson Gas Phase IR Cell utilizing the 105° connecting adapter from the microscale kit that accompanies this text.

[2]The microscale connecting adapter is a available from Kimble Kontes (catalog no. 748001-1000).

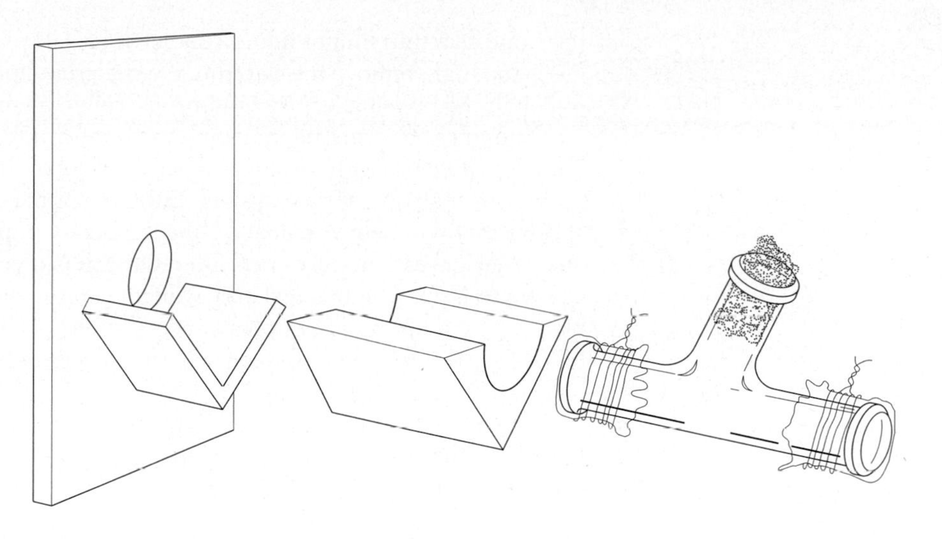

FIG. 11.20
The Williamson Gas Phase IR Cell mounted in a cell holder.

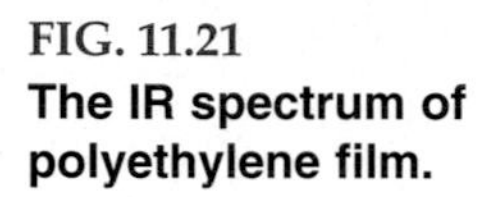

FIG. 11.21
The IR spectrum of polyethylene film.

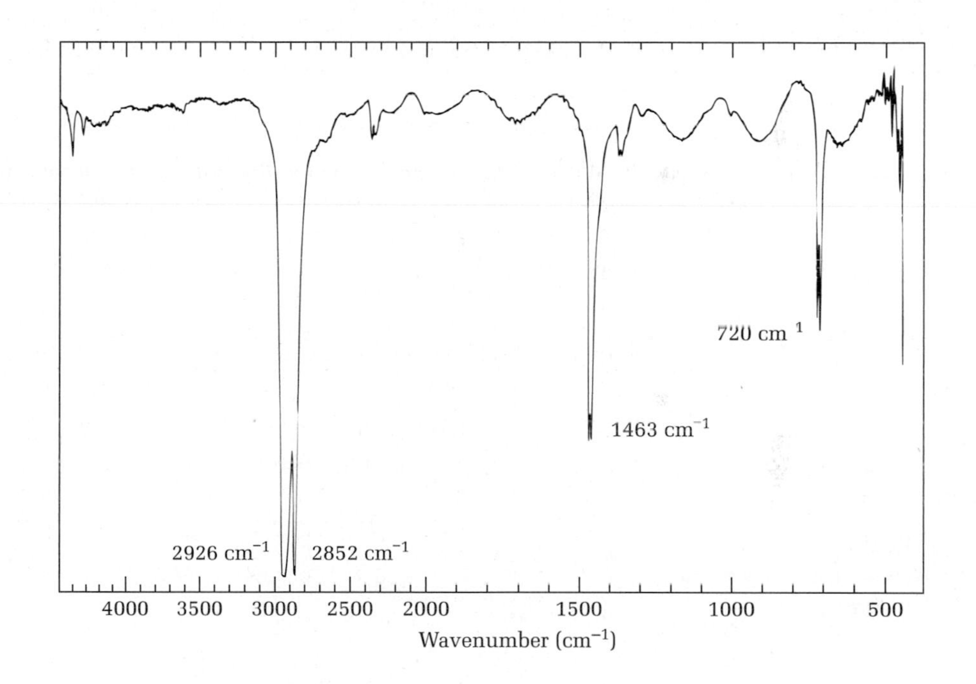

A more permanent cell can be constructed by attaching two thin plates of silver chloride to the ends of the adapter with epoxy glue. These silver chloride plates can be made in the press used for making KBr discs, or they can be purchased.[3]

Obtaining a Spectrum

To run a gas phase spectrum, remove the cell from the spectrometer and spray a very short burst of a gas from an aerosol can onto the cotton. The nonvolatile components of the spray will stick to the cotton, and the propellant gas will diffuse into the cell. Run the spectrum in the usual way. The concentration of the gas is about

[3]The silver chloride disks can be purchased from Fisher Scientific (catalog no. 14-385-860).

the same as a thin film of liquid run between salt plates. If the concentration of the gas is too high, remove the cell from the spectrometer, remove the cotton, and "pour out" half the invisible gas by holding the cell in a vertical position.

You may be fortunate in having in your school's library a copy of Volume 3 of *The Aldrich Library of FT-IR Spectra: Vapor Phase* or an online database of IR spectra in your IR instrument data system. Either resource contains hundreds of reference spectra that will help you analyze the IR spectra of propellants in aerosol cans, and any other gases you may encounter. Otherwise, you may want to compare your spectra to the few in this text and available on the web site for this text, and assemble your own library of gas phase spectra. It is possible to make approximate calculations of IR vibrational frequencies using a semi-empirical calculation at the AM1 level (*see* Chapter 15).

Throughout the remainder of this book, representative IR spectra of starting materials and products are presented, and the important bands in each spectrum are identified.

FOR FURTHER INVESTIGATION

The Gas Phase IR Spectra of Aerosol Propellants

Acquire gas phase spectra from a number of substances that come from aerosol cans. Some cans will be labeled with the name of the propellant gas; others will not. Analyze the spectra, bearing in mind that the gas will be a small molecule such as carbon dioxide (Fig. 11.22), methane (Fig. 11.23), nitrous oxide (Fig. 11.24), propane, butane, isobutane, or a fluorocarbon such as 1,1,1,2-tetrafluoroethane. A very old aerosol can may contain an illegal chlorofluorocarbon, now outlawed because these substances destroy the ozone layer. Some examples of aerosol cans include hairsprays, lubricants (WD-40, Teflon), Freeze-It, inhalants for asthma, cigarette lighters, butane torches, and natural gas from the lab.

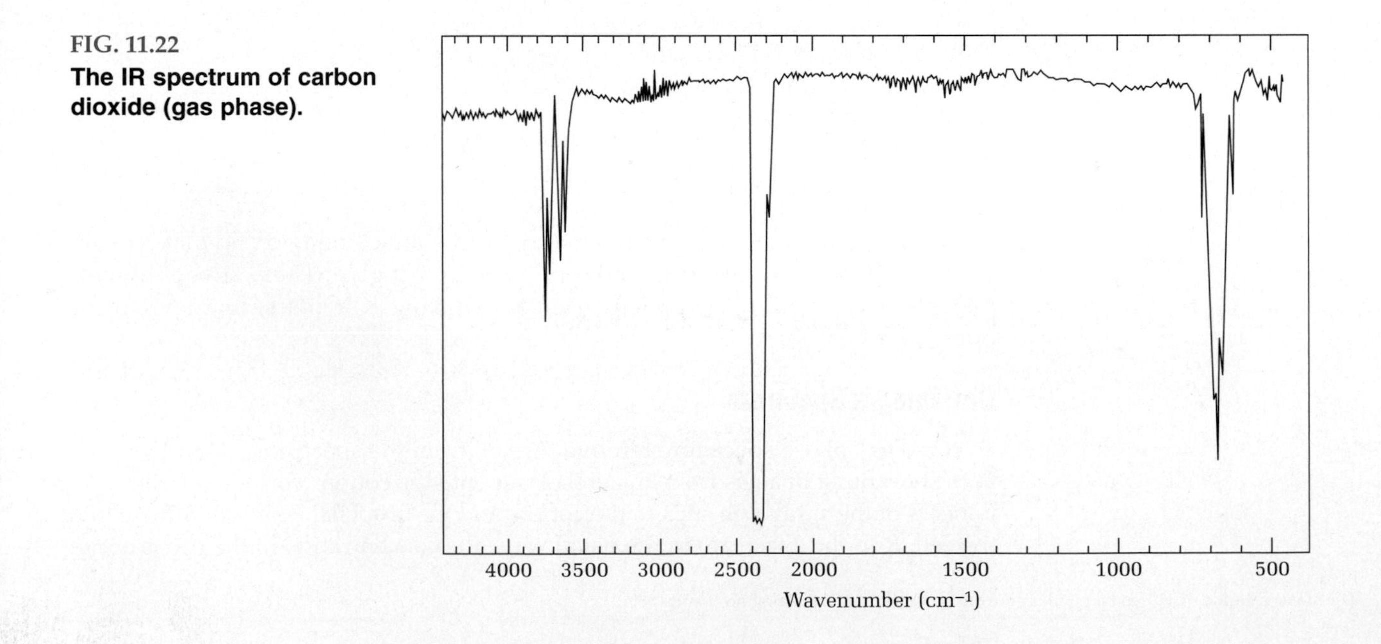

FIG. 11.22
The IR spectrum of carbon dioxide (gas phase).

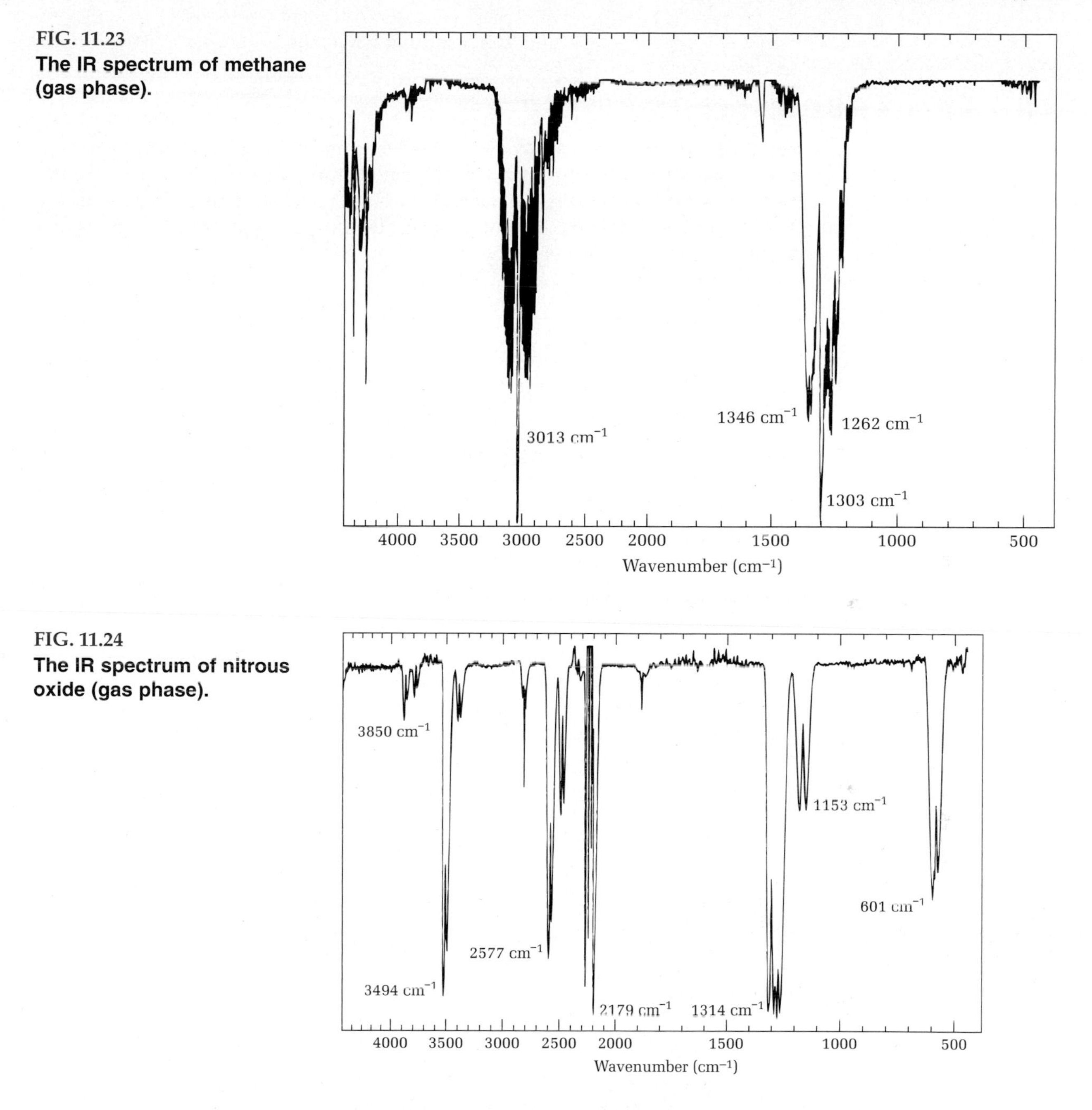

FIG. 11.23
The IR spectrum of methane (gas phase).

FIG. 11.24
The IR spectrum of nitrous oxide (gas phase).

Other Gas Phase Spectral Research

You can deliberately generate hydrogen chloride (or deuterium chloride) and sulfur dioxide, and then analyze the gas that is evolved from a thionyl chloride reaction. Can you detect carbon monoxide in automobile exhaust? How about marsh gas that you see bubbling out of ponds? With perhaps dozens of scans, can you record the spectra of vanillin being exuded over vanilla beans, eugenol vapor over cloves, anisaldehyde over anise seeds? What is the nature of the volatile components of chili peppers, orange peel, nail polish, and nail polish remover? Could you use this method to determine if someone has been drinking alcohol?

EXPERIMENT

UNKNOWN CARBONYL COMPOUND

The sample, solvents, and apparatus must be dry.

Run the IR spectrum of an unknown carbonyl compound obtained from your laboratory instructor. Be particularly careful that all apparatus and solvents are completely free of water, which will damage the NaCl plates. Determine the frequency of the carbonyl peak in the unknown *and* list the possible types of compounds that could correspond to this frequency (Table 11.2).

REFERENCES

Colthrup, Norman B., Lawrence H. Daly, and Stephen E. Wiberley, *Introduction to Infrared and Raman Spectroscopy*, 3rd ed. New York: Academic Press, 1990.

Cooper, James W., *Spectroscopic Techniques for Organic Chemists*. New York: John Wiley, 1980.

Lin-Vein, Daimay, Norman B. Colthrup, William G. Fateley, and Jeannette G. Grasselli, *The Handbook of Infrared and Raman Characteristic Frequencies of Organic Molecules*. Boston: Academic Press, 1991.

Pouchert, Charles J. *The Aldrich Library of FT-IR Spectra*, 2nd ed. Milwaukee: Aldrich Chemical Co., 1997.

Roeges, Noel P. G, *A Guide to the Complete Interpretation of Infrared Spectra of Organic Structures*. New York: John Wiley, 1994.

Silverstein, Robert M., Francis X. Webster, and David Kiemle, *Spectrometric Identification of Organic Compounds*, 7th ed.[4] New York: Wiley, 2005.

Pretsch, Ernö, Philippe Bühlmann, and Martin Badertscher, *Structure Determination of Organic Compounds: Tables of Spectral Data*, 4th ed. Berlin: Springer, 2009.

[4]This book includes IR, UV, and NMR spectra.

CHAPTER 12

Nuclear Magnetic Resonance Spectroscopy

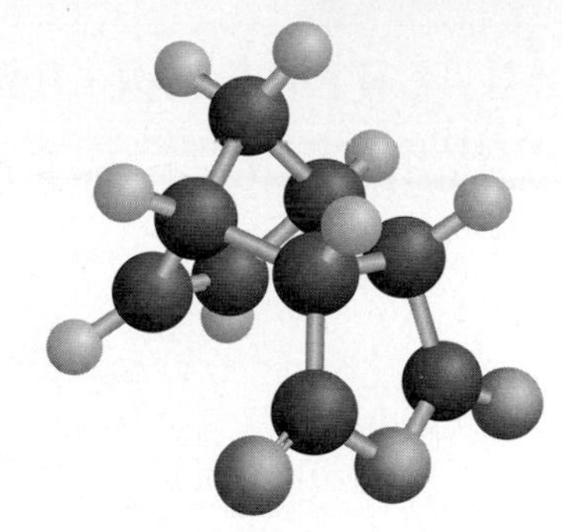

PRELAB EXERCISE: Outline the preliminary solubility experiments you would carry out using inexpensive solvents before preparing a solution of an unknown compound for nuclear magnetic resonance (NMR) spectroscopy using expensive deuterated solvents.

Most organic chemists would agree that the most powerful instrumental method for revealing the structure of organic molecules is nuclear magnetic resonance (NMR) spectroscopy. The 2002 Nobel Prize in Chemistry was awarded, in part, to Kurt Wüthrich for advances in NMR spectroscopy that allowed the determination of the three-dimensional structure of biological macromolecules in solution. Because of such capabilities, introductory organic chemistry courses devote considerable time to the study of NMR. This concise chapter assumes that you have had some prior exposure to the concepts discussed; it focuses on the practical aspects of sample preparation, data acquisition, and interpretation of NMR spectra to elucidate and confirm organic structures.

^{1}H NMR: Determination of the number, kind, and relative locations of hydrogen atoms (protons) in a molecule.

Let us briefly review the theory of NMR. Certain nuclei such as ^{1}H, ^{13}C, ^{15}N, ^{18}O, ^{19}F, and ^{31}P are said to have a spin, S, of ½ and behave like tiny magnets that can assume two energy states when placed in a magnetic field: aligned (lower energy) and opposed (higher energy). Like the energy states in electronic spectra, the difference in energy between these states is quantized, and energy is absorbed and emitted only at certain radio frequencies. These frequencies depend on the type of nucleus and the magnetic field strength, which is constant for a given instrument. More importantly, however, is the fact that slight differences in the electronegativity and bonding state (sp, sp^2, and sp^3) of surrounding atoms cause small—parts per million (ppm)— variations in the magnetic field felt by each nucleus, causing them to have absorption signals at slightly different frequencies. These small variations, called *chemical shifts*, are plotted versus signal intensity to produce the NMR spectrum. The interpretation of these signals and other spectral features such as splitting patterns and peak areas, as described in the following sections, facilitates organic structure elucidation.

Chemical shift, δ (ppm)

INTERPRETATION OF ¹H NMR SPECTRA

There are two approaches to interpreting proton NMR spectra:

1. The *structure from the spectrum* approach is the strategy of using the information in the NMR spectrum to draw the structure of the molecule based on reference tables and rules. This approach is used if the compound's structure is unknown. In this case, the NMR spectrum alone is often insufficient to "solve" the complete structure, and must be combined with knowledge about the compound's source (synthetic reaction or natural product) and complementary spectral data (infrared, ultraviolet, and/or mass spectrometric).

2. The *spectrum from the structure* approach might be thought of as the reverse of the first approach and is commonly used when verifying products of known reactions using known starting materials. A hypothetical NMR spectrum is created, based on the known compound's structure (or related structural possibilities), using the same reference tables and rules as in the first approach. Computer programs are available that allow the calculation of a hypothetical NMR spectrum for any molecule simply by entering its structure. (Refer to the supplemental information for this chapter on this book's web site.) Even better, there may be a published spectrum of the compound available in a collection of reference spectra. Either way, the sample spectrum should be compared to the hypothetical or reference spectrum, and if the two are closely matched, one can be confident that the correct compound has been obtained.

H_3C I
CH_2
Ethyl iodide

NMR spectra contain considerably more structural information than do infrared spectra, and this additional information should be used in interpreting them. The following four most informative features are the principal ones to look for when interpreting spectra. We will use the example of ethyl iodide as a specific illustration of each feature.

1. The Number of Signals Due to Equivalent Hydrogen Nuclei or Protons

The three hydrogen atoms of the methyl group are said to be chemically and magnetically equivalent. Replacing any of these hydrogens with another atom of hydrogen will give the same compound. Similarly, the two methylene protons are magnetically and chemically equivalent. This is because there is very fast rotation about the carbon-carbon bond. Individual hydrogens or each set of *equivalent* hydrogens will experience slightly different magnetic fields depending on their chemical environment and, therefore, will produce a signal at different places in the spectrum. Thus, we would predict that ethyl iodide would produce two signals, one for the three equivalent CH_3 protons, and the other from the equivalent protons on the CH_2 bonded to the electronegative iodine. The NMR spectra of ethyl iodide, regardless of whether it is obtained at low field strength (Fig. 12.1) or high field strength (Fig. 12.2), shows three signals, or peaks, at 0.00 ppm, 1.83 ppm, and 3.20 ppm, with the latter two split into patterns of lines. The peak at 0.00 ppm is due to the addition of a reference compound, tetramethylsilane (TMS), used as the zero reference point to assure that the instrument frequency assignments are accurate. Thus, there are only two sets of signal peaks from ethyl iodide as predicted.

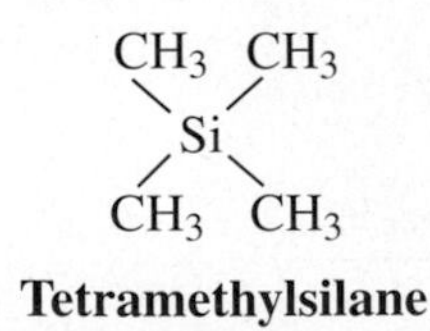

Tetramethylsilane

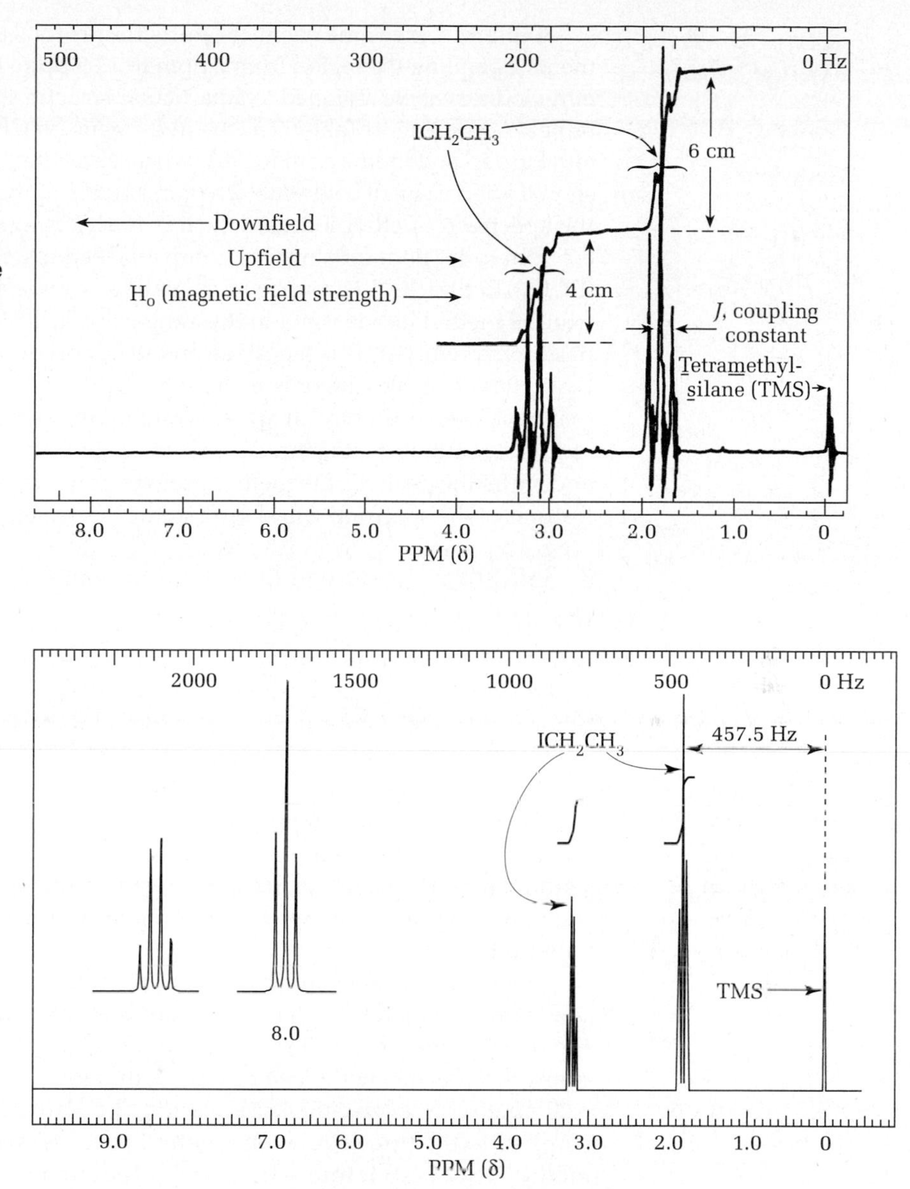

FIG. 12.1
The ^{1}H NMR spectrum of ethyl iodide (60 MHz). The stair-step-like line is the integral. In the integral mode of operation, the recorder pen moves from left to right, and moves vertically a distance proportional to the areas of the peaks over which it passes. Hence, the relative areas of the quartet of peaks at 3.20 ppm and the triplet of peaks at 1.83 ppm are given by the relative heights of the integrals (4 cm is to 6 cm as 2 is to 3). The relative numbers of hydrogen atoms are proportional to the peak areas (2 H and 3 H).

FIG. 12.2
The ^{1}H NMR spectrum of ethyl iodide (250 MHz). Compare this spectrum to Figure 12.1, run at 60 MHz. The peaks at 1.83 ppm and 3.20 ppm have been expanded and plotted on the left side of the spectrum.

2. The Position of Each Signal on the Horizontal Axis

Each signal's position on the horizontal axis is called the chemical shift, and it indicates the effective magnetic field around a single proton or group of equivalent protons as affected by nearby bonds and atoms. This *chemical environment* depends on structure, that is, the bond types (sp^2, sp^3, sp, or aromatic) and the electronegativities of the atoms one, two, and three bonds away from the particular protons. Chemical shifts, symbolized by δ, are measured in dimensionless parts per million (ppm) to the left of, or *downfield* from, the reference TMS peak, according to the equation:

$$\delta\ (\text{ppm}) = \frac{\text{shift of peak downfield TMS (in Hz)}}{\text{spectrometer frequence (in MHz)}}$$

CH_3—
Methyl

—CH_2—
Methylene

—CH—
|
Methine

Table 12.1 gives the chemical shifts for protons in different chemical environments, spanning the region from 0 ppm to 12.5 ppm. Each signal in the spectrum of ethyl iodide can be assigned to a particular structural feature using this and similar tables. According to Table 12.1, the hydrogens on CH_2 groups attached to both an alkyl group and iodine as in RCH_2I (where R can be methyl, methylene, or methine) should yield a signal between +2.3 ppm and 3.2 ppm. We see a signal at 3.20 ppm in the spectrum of ethyl iodide, which is in this range. The table shows a range of 0.7 ppm to 1.1 ppm for a methyl group attached to a methylene or methine R group (RCH_3); R is the -CH_2I group for ethyl iodide. In actuality, the other signal in the spectrum of methyl iodide is not in this range; it is at 1.83 ppm, which is further downfield than predicted. This type of inconsistency occurs occasionally and demonstrates that reference tables are only generally representative of the majority of molecules that have been examined. It appears that in this particular case, the electronegative iodine, even though it is three bonds away, still strongly affects the magnetic field and shifts the peak for the methyl protons downfield. An examination of the other features of the spectrum will show that this peak assignment is correct.

3. Splitting Patterns and Coupling Constants

Pascal's triangle

1	singlet, s
1 1	doublet, d
1 2 1	triplet, t
1 3 3 1	quartet, q
1 4 6 4 1	pentet
1 5 10 10 5 1	sextet

Patterns and coupling constants in the spectrum help define which groups are next to each other in a molecule's carbon skeleton. In an open-chain molecule with free rotation about the bonds and no chiral centers, protons couple with each other over three chemical bonds to give characteristic patterns of lines. If one or more equivalent protons couple to *one* adjacent proton, then the coupling hydrogen(s) appears as a *doublet* of equal intensity lines separated by the coupling *constant* (J) measured in Hertz (Hz). If the coupling proton or protons couple equally to two protons three chemical bonds away, they appear as a *triplet* of lines with relative intensities of 1:2:1, again separated by J. Finally, a *quartet* of lines in the ratio of 1:3:3:1 arises when one or more protons couple to the *three* protons on a methyl group.

In general, chemically equivalent protons give a pattern of lines containing one more line than the number of protons being coupled to, and the intensities of the peaks follow the binomial expansion, conveniently represented by Pascal's triangle. Thus the methylene group of ethyl iodide appears as a quartet of lines at 3.20 ppm with relative intensities of 1:3:3:1 because the two methylene protons are coupled to the three equivalent methyl protons (see Fig. 12.1 and Fig. 12.2). The methyl peak is split into a 1:2:1 triplet centered at 1.83 ppm because the methyl protons are coupled to the two adjacent methylene protons. This quartet-triplet combination is a strong indication of an ethyl group in any molecule. The J indicated in Figure 12.1 is the same 7.6 Hz between all lines in the quartet and triplet because the two groups are adjacent; therefore, the protons are coupled. In the 250 MHz spectrum, Figure 12.2, the splitting patterns are compressed and harder to distinguish because there are more Hz per ppm, but the expansion of these peaks on the left of the spectrum clearly shows a quartet and triplet, and the separation between each peak is still 7.6 Hz.

Other characteristic proton coupling constants are given in Table 12.2. In alkenes, *trans* coupling is larger than cis coupling, and both are much larger than geminal coupling (coupling that takes place between two groups on the same carbon). *Ortho, meta,* and *para* couplings in aromatic rings range from 0 Hz to 9 Hz. The couplings in a rigid system of saturated bonds are strongly dependent on the dihedral angle between the coupling protons, as seen in cyclohexane.

TABLE 12.1 *Proton Chemical Shifts*

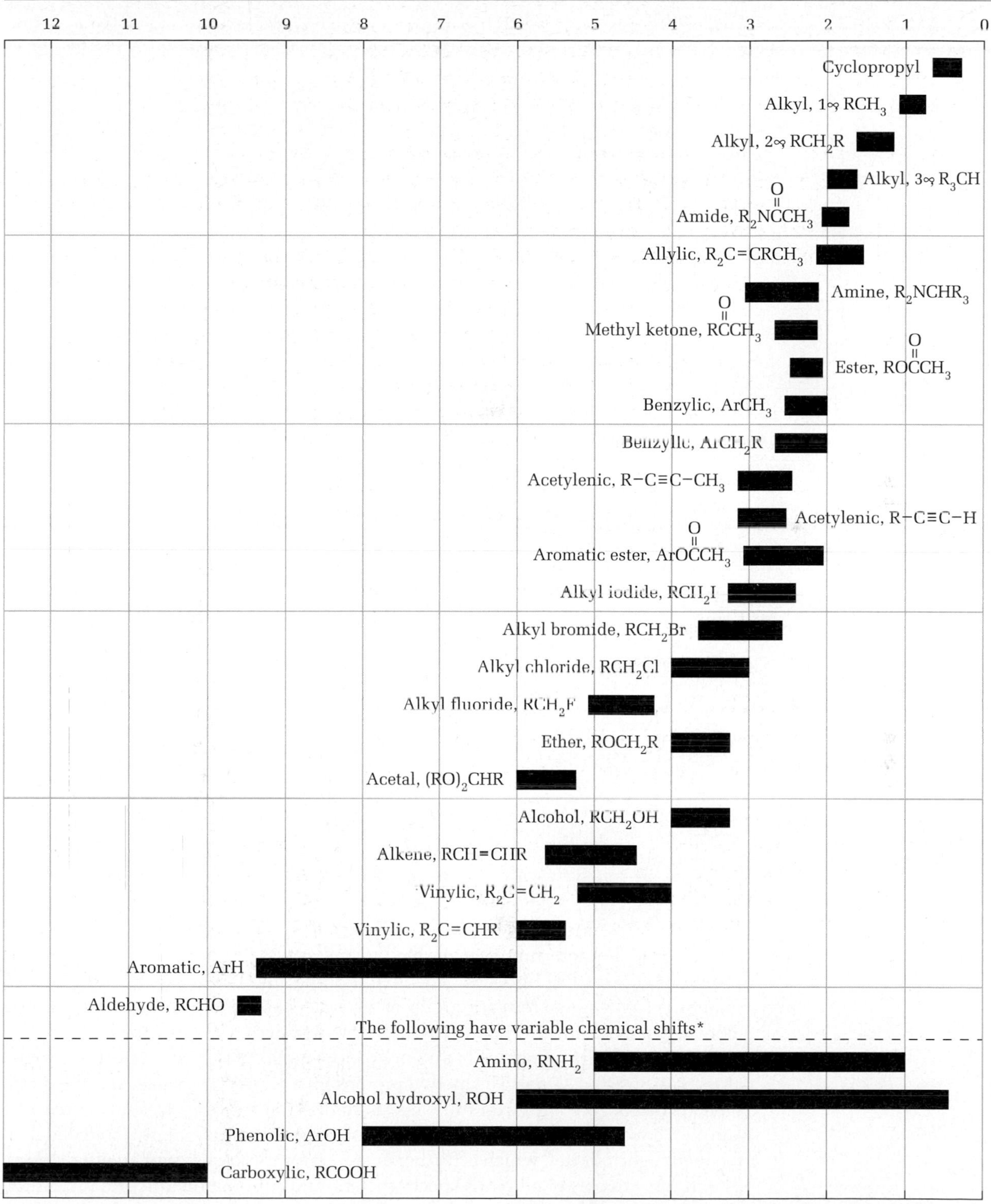

TABLE 12.2 *Spin-Spin Coupling Constants for Various Geometries*

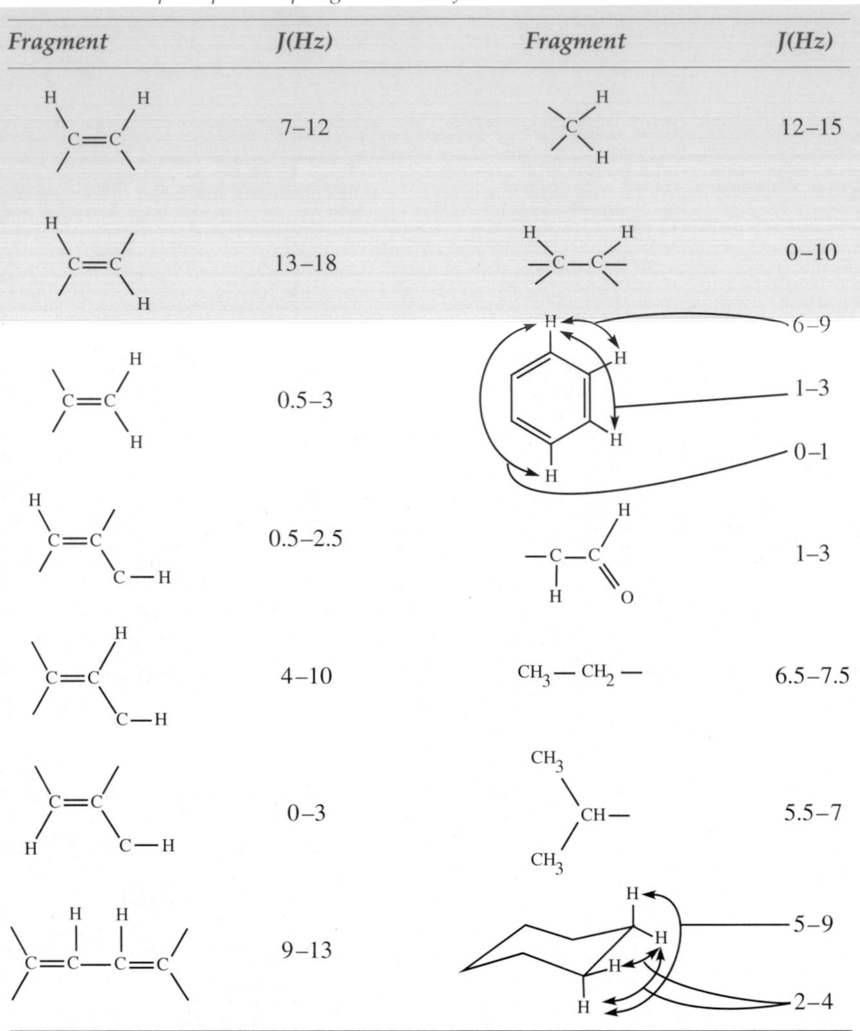

Fragment	*J(Hz)*	*Fragment*	*J(Hz)*
	7–12		12–15
	13–18		0–10
	0.5–3		6–9, 1–3, 0–1
	0.5–2.5		1–3
	4–10	CH_3— CH_2 —	6.5–7.5
	0–3	(CH_3)$_2$CH—	5.5–7
	9–13		5–9, 2–4

4. The Integral

The relative numbers of distinctive hydrogen atoms (protons) in the molecule of ethyl iodide are determined from the *integral*, the stair-step line over the peaks shown in Figures 12.1 and 12.2. The height of the step is proportional to the area under the NMR signal for each group of equivalent protons. In NMR spectroscopy (contrasted with infrared spectroscopy, for instance), the area for each signal, including all splitting peaks, is directly proportional to the number of hydrogen atoms causing that signal. The methyl protons give a triplet of peaks, and the methylene protons give a quartet of peaks. Both ethyl iodide spectra in Figures 12.1 and 12.2 have a ratio of 3 to 2 for the area of the methyl triplet relative to the area of the methylene quartet, which is further evidence for the presence of an ethyl group. Integrators are part of all NMR spectrometers, and running the integral takes little more time than running the spectrum. Most spectrometers print out a numerical value for the integral (with many more significant figures than are justified!).

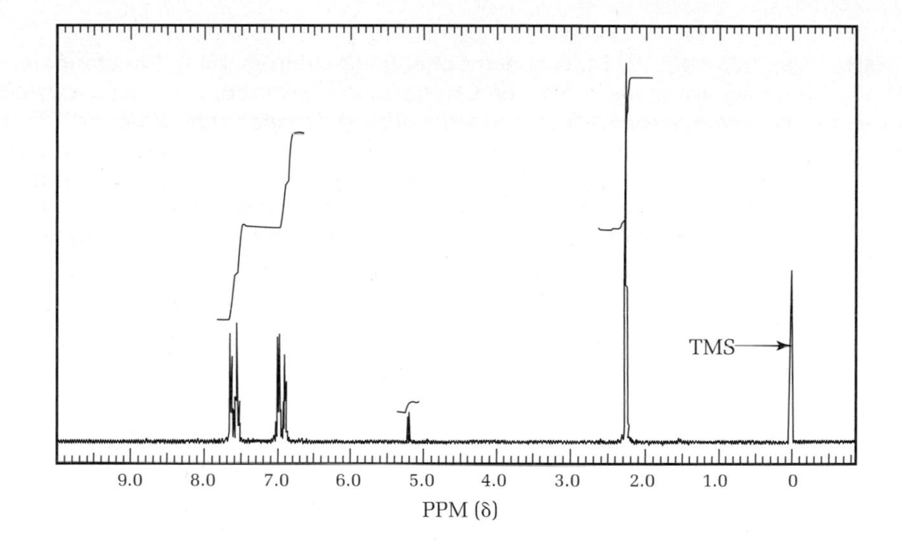

FIG. 12.3
The ^{1}H NMR spectrum of 4-iodotoluene (90 MHz in $CDCl_3$).

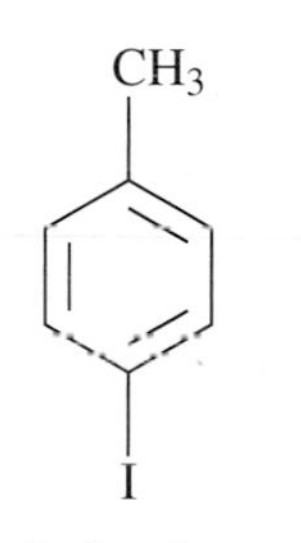

4-Iodotoluene

In actuality, when chemists interpret NMR spectra, they use a combination of the *structure from the spectrum* and the *spectrum from the structure* approaches. A typical mental analysis of the NMR spectrum of 4-iodotoluene (Fig. 12.3) might go like this:

> This molecule should produce three signals: a singlet for the methyl protons, a doublet for the two equivalent aromatic protons ortho to the methyl group, and a doublet for the other two equivalent aromatic protons ortho to the iodo group. The methyl group is attached to an aromatic ring, and according to the table of proton chemical shifts, this should be detected as a singlet peak between 1.8 ppm and 2.8 ppm. The NMR spectrum of my product shows a singlet at 2.28 ppm, which is consistent with this. The aromatic protons show up as doublets in the right range, 6 ppm to 9 ppm according to the table. The spectrum has doublets at 6.95 ppm and 7.60 ppm. It also shows a singlet at 5.2 ppm and, according to the table, this implies that my product contains hydrogen on an alkene or in an acetal. Bad news! From other information I know, my product has neither of these functional groups. But wait! If I look at spectra of common sample impurities, I see that dichloromethane, the solvent I extracted the product with, appears at 5.25 ppm as a singlet. So this peak is not from my product and can be ignored.

This back-and-forth analysis correlating structure and NMR data, including coupling constants and integrals, is continued until all NMR peaks, splitting patterns, and integrals can be accounted for, and the identity of the product is verified.

When NMR data are reported in the literature, it is usually in a concise numerical form. For example, the NMR data for ethyl iodide derived from its spectrum (Fig. 12.2) would be reported as ^{1}H NMR ($CDCl_3$): 1.83 (3H, t, J = 7.6 Hz), 3.20 (2H, q, J = 7.6 Hz), where $CDCl_3$ (deuterochloroform) is the solvent, 1.83 and 3.20 are the chemical shifts in ppm, 3H and 2H are integrals, and t = triplet, q = quartet, and J is the coupling constant in Hz.

Some NMR spectra are not as easily analyzed as the spectrum for ethyl iodide. Consider the spectra shown in Figure 12.4. The proton spectrum of this unsaturated chloroester has been run at 500 MHz. Each chemically and magnetically nonequivalent

FIG. 12.4

The 500-MHz ^1H and 75-MHz ^{13}C NMR spectra of ethyl (3-chloromethyl)-4-pentenoate.[1] *I. ^{13}C DEPT[2] spectrum.* The CH_3 and CH peaks are upright, and the CH_2 peaks are inverted. The quaternary carbon (the carbonyl carbon) does not appear. *II. The normal 75 MHz noise-decoupled ^{13}C spectrum.* Note the small size of the carbonyl peak. *III. Expansions of each group of proton NMR peaks.* Protons E and F as well as protons H and I are not equivalent to each other. These pairs of protons are diastereotopic because they are on a carbon adjacent to a chiral carbon atom. The frequencies of all peaks are found on this book's web site. *IV. The integral.* The height of the integral is proportional to the number of protons under it. *V. The 500-MHz ^1H spectrum.*

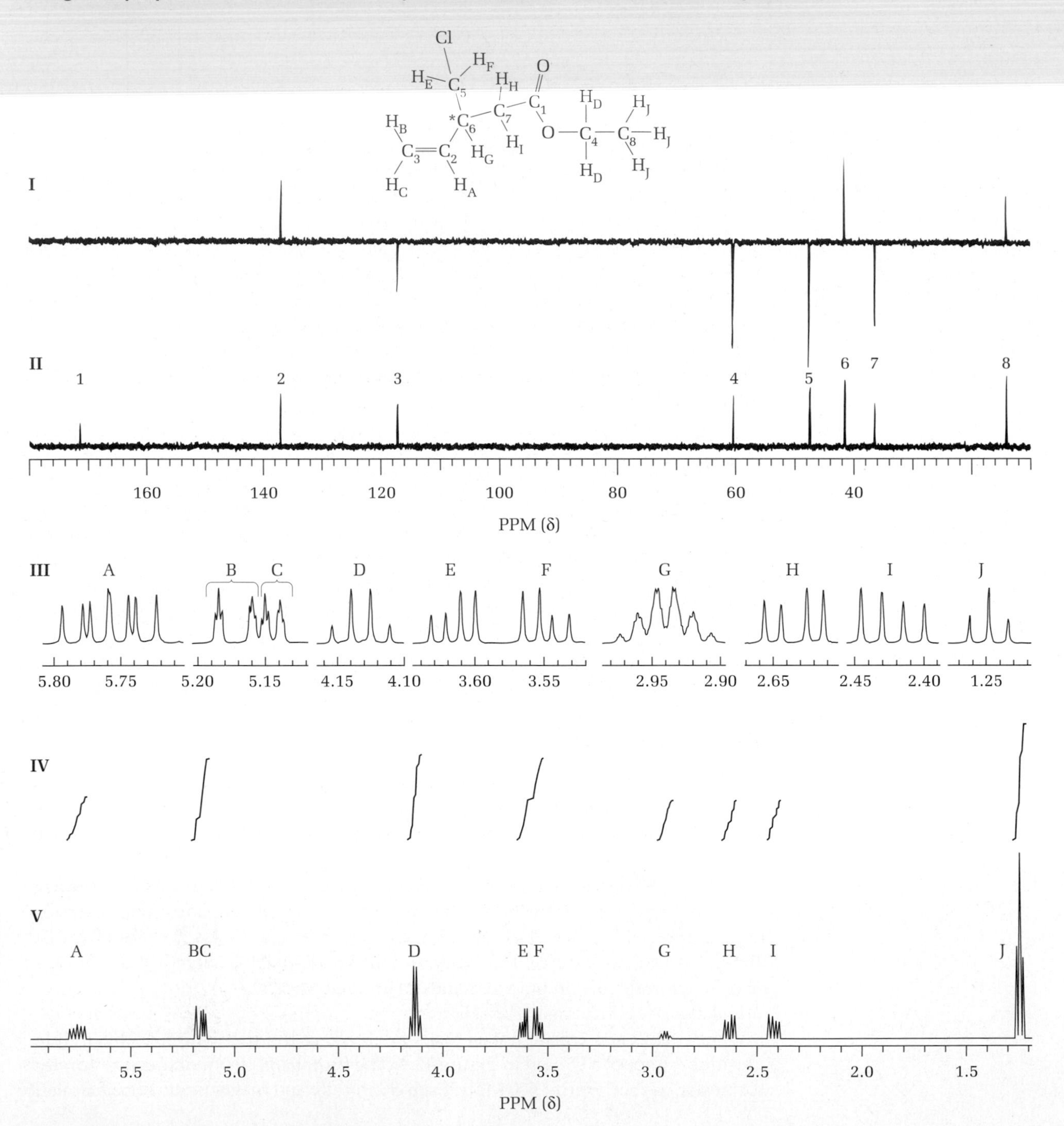

[1]Spectra courtesy of Professor Scott Virgil.

[2]DEPT: **d**istortionless **e**nhancement by **p**olarization **t**ransfer.

proton is well resolved so that all of the couplings can be seen. Only the quartet of peaks at 4.1 ppm (relative area 2) and the triplet at 1.2 ppm (relative area 3) follow the simple first-order coupling rules outlined earlier. This quartet/triplet pattern is very characteristic of the commonly encountered ethyl group.

A more complex spectrum.

Because of the chiral carbon (marked with an asterisk), the protons on carbons 5 and 7 are diastereotopic, have different chemical shifts, and couple with each other and with adjacent protons to give the patterns seen in the spectrum. Many of the peaks can be assigned to specific hydrogens based simply on their chemical shifts. The coupling patterns then confirm these assignments.

CARBON-13 SPECTROSCOPY

The element carbon consists of 98.9% carbon atoms with mass 12 (^{12}C) and spin 0 (NMR inactive) and only 1.1% carbon atoms with mass 13 (^{13}C) and spin 1/2 (NMR active). Carbon, with such a low concentration of spin 1/2 nuclei, gives a very small signal when run under the same conditions as used for a proton spectrum. Carbon resonates at 75 MHz in a spectrometer where protons resonate at 300 MHz. Because only 1 in 100 carbon atoms has mass 13, the chances of a molecule having two ^{13}C atoms adjacent to one another are small. Consequently, coupling of one ^{13}C with another is not observed.

^{13}C spectra: Broadband noise decoupling gives a single line for each carbon.

Each ^{13}C atom couples to hydrogen atoms over one, two, and three bonds. Because the coupling constants are large, there is a high probability of peak overlap. To simplify the spectra as well as to increase the signal-to-noise ratio, a special technique is routinely used in obtaining ^{13}C spectra: *broadband noise decoupling*. Decoupling has the effect of collapsing all multiplets (quartets, triplets, etc.) into a single peak. Further, the energy put into decoupling the protons will appear in the carbon spectrum in the form of an enhanced peak. This Nuclear Overhauser Enhancement (NOE) effect makes the peak appear three times larger than it would otherwise be. The result of decoupling is that every chemically and magnetically distinct carbon atom will appear as a single sharp line in the spectrum. Because the NOE effect is somewhat variable and does not affect carbons bearing no protons, one cannot do carbon counting from peak integrals in the same way that hydrogen counting is done from proton NMR spectra.

Carbon Chemical Shifts

The range of carbon chemical shifts is 200 ppm compared to the 10-ppm range for protons. It is not common to have an accidental overlap of carbon peaks. In the spectrum for the unsaturated chloroester in Figure 12.4 (spectrum II), there are eight sharp peaks corresponding to the eight carbon atoms in the molecule.

The generalities governing carbon chemical shifts are very similar to those governing proton shifts, as seen by comparing Table 12.1 to Table 12.3. Most of the downfield peaks are due to those carbon atoms near electron-withdrawing groups. In Fig. 12.4 (spectrum II), the furthest downfield peak is that from the carbonyl carbon of the ester. The attached electronegative oxygens make the peak appear at about 172 ppm. The peak is smaller than any other in the spectrum because it does not have an attached proton, and thus does not benefit from the NOE effect.

The ^{13}C spectrum of sucrose in Figure 12.5 displays a single line for each carbon atom; in Figure 20.3 (on page 345), a single line is seen for each of the 27 carbon atoms in cholesterol.

TABLE 12.3 *Carbon Chemical Shifts*

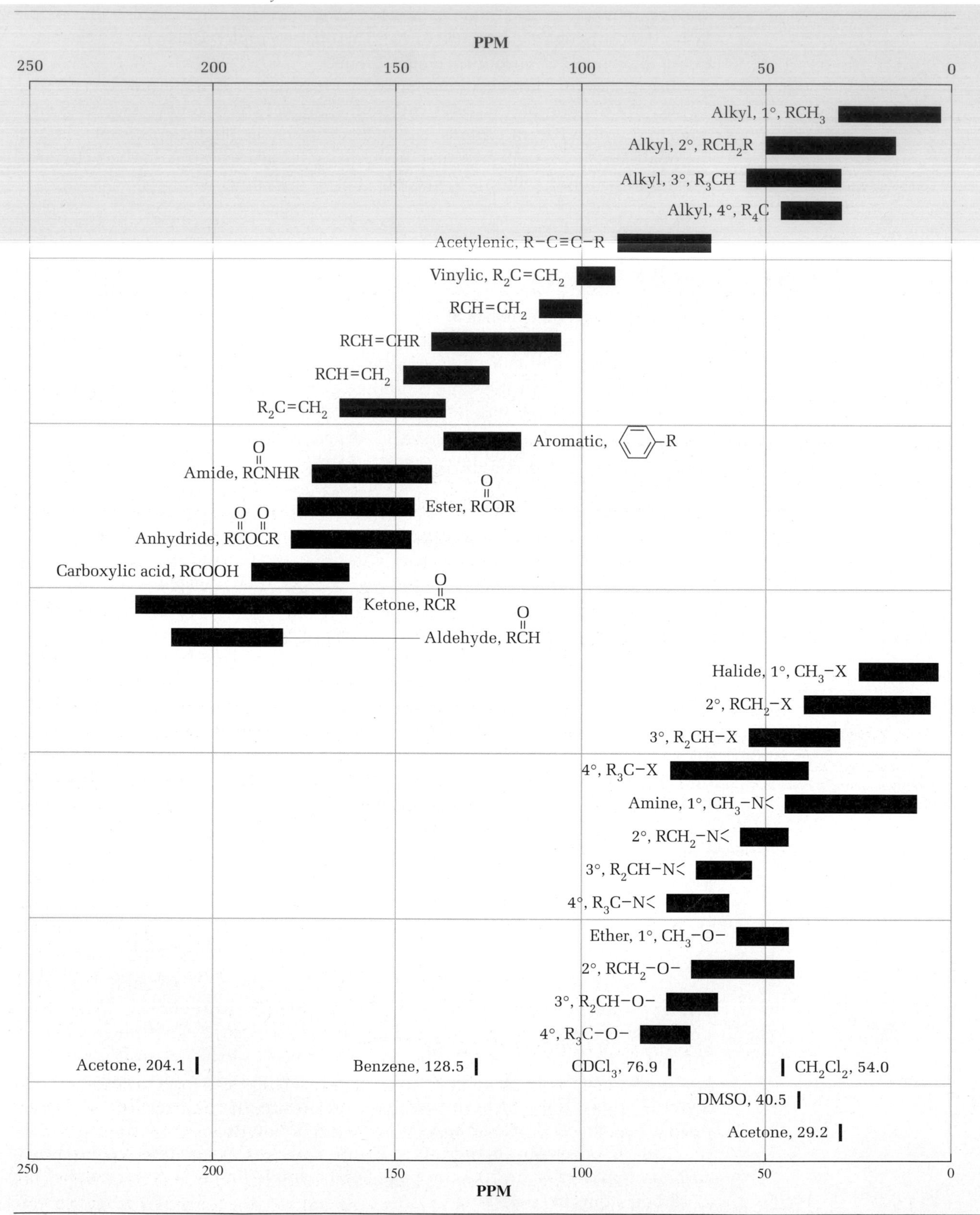

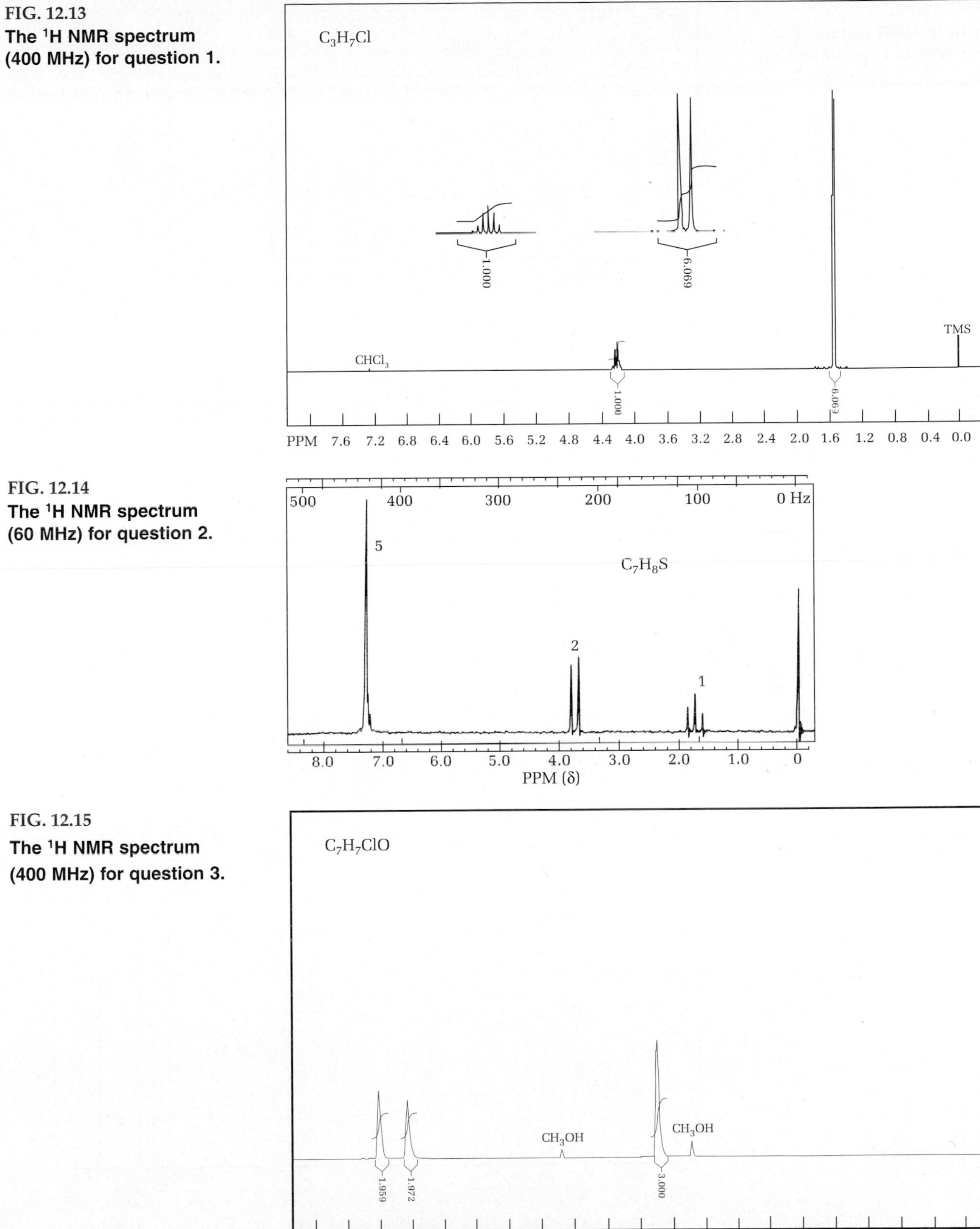

FIG. 12.13
The ^{1}H NMR spectrum (400 MHz) for question 1.

FIG. 12.14
The ^{1}H NMR spectrum (60 MHz) for question 2.

FIG. 12.15
The ^{1}H NMR spectrum (400 MHz) for question 3.

FIG. 12.16
The ^1H NMR spectrum (60 MHz) of compound (a), $C_4H_{10}O$, for question 4.

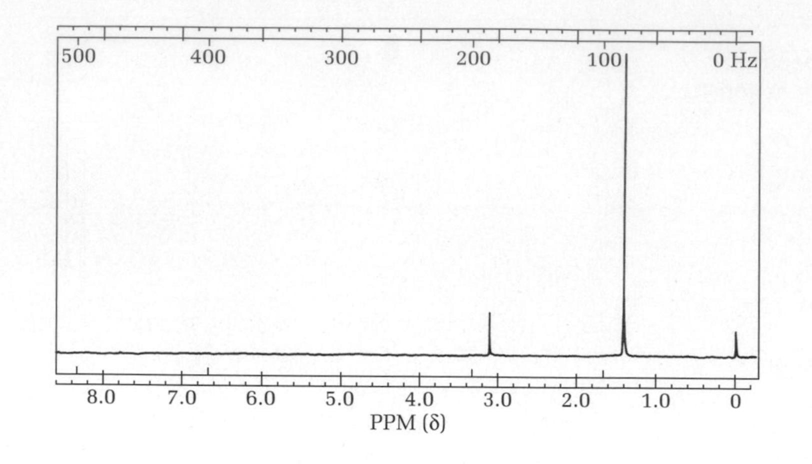

FIG. 12.17
The ^{13}C NMR spectrum (22.6 MHz) of compound (b), $C_4H_{10}O$, for question 4.

FIG. 12.18
The ^{13}C NMR spectrum (22.6 MHz) of compound (c), $C_4H_{10}O$, for question 4.

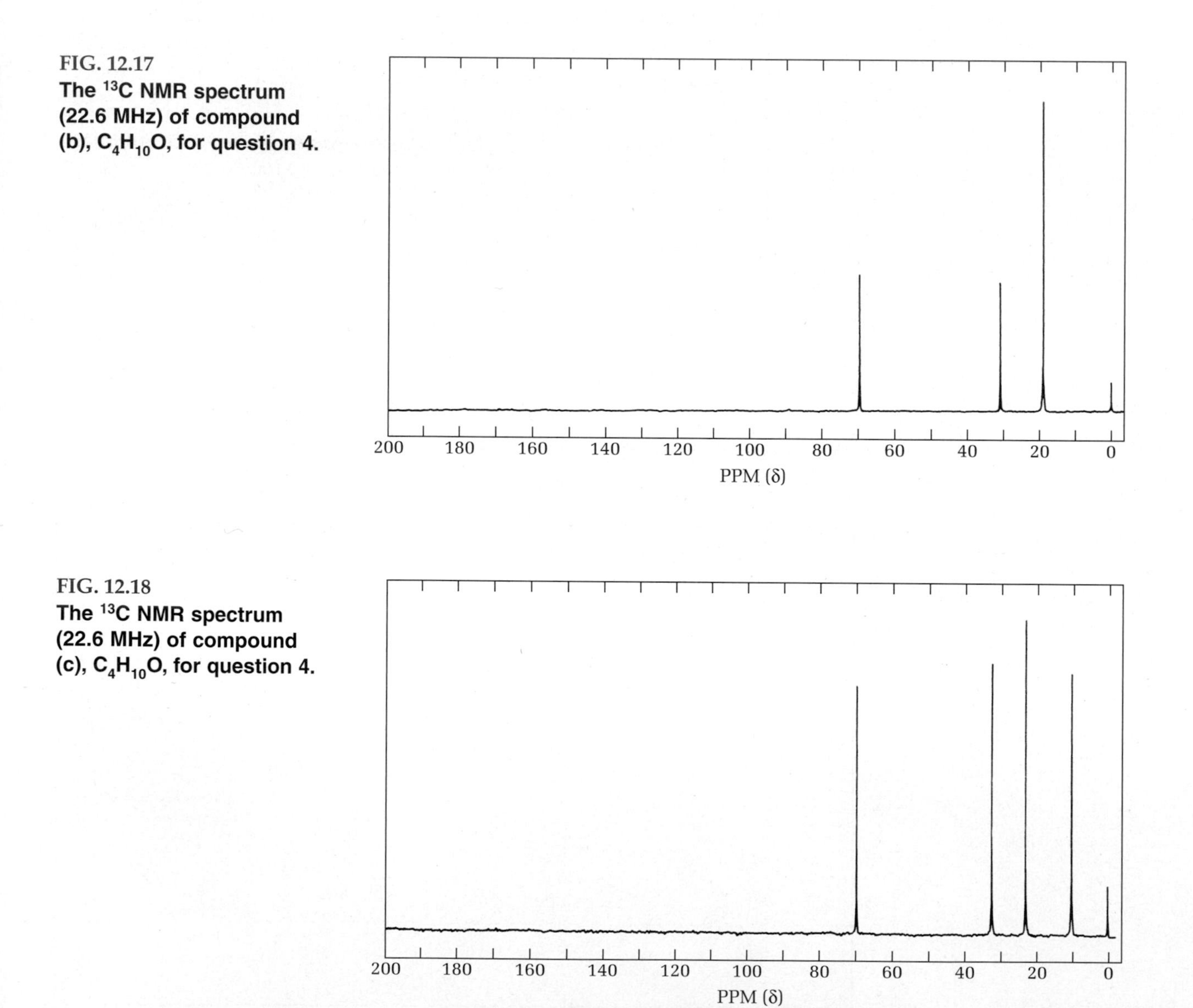

REFERENCES

Abraham, Raymond J., J. Fisher, and P. Loftus, *Introduction to NMR Spectroscopy.* Chichester, UK: John Wiley & Sons, 1988.

Croasmun, William R., and Robert M. K. Carlson, *Two-Dimensional NMR Spectroscopy: Applications for Chemists and Biochemists*, 2nd ed. New York: VCH Publishers, 1994.

Friebolin, Horst, *Basic One- and Two-Dimensional NMR Spectroscopy*, 4th ed. New York: Wiley-VCH, 2005.

Gadian, David G., *NMR and Its Applications to Living Systems*, Oxford, UK: Oxford University Press, 1996.

Pavia, Donald L., Gary M. Lampman, and George S. Kriz, *Introduction to Spectroscopy*, 4th ed.[4] Brooks/Cole Thomson Learning, 2008.

Sanders, Jeremy K. M., and Brian K. Hunter, *Modern NMR Spectroscopy: A Guide for Chemists*, 2d ed. Oxford, UK: Oxford University Press, 1993.

Silverstein, Robert M., Francis X. Webster, and David Kiemle, *Spectrometric Identification of Organic Compounds*, 7th ed.[4] New York: Wiley, 2005.

Young, Ian R., Dave M. Grant, and Ryan K. Harris, eds., *Methods in Biomedical Magnetic Resonance Imaging and Spectroscopy*, Vol 1 & 2. New York: John Wiley & Sons, 2000.

See *Web Links*

[4]This book includes IR, UV, and NMR spectra.

13 CHAPTER

Mass Spectrometry

PRELAB EXERCISE: Calculate the molecular weight of bromobutane (in grams per mole) and the exact masses of its molecular ions (in daltons) as measured by mass spectrometry.

Although its origins date back to the early part of the 20th century, mass spectrometry (MS) has experienced a renaissance in the past two decades due to major improvements in ion formation and analysis. Because mass spectrometry is capable of providing composition and structure information for a broad range of compounds at sensitivities much greater than those of other techniques, more money is invested worldwide in mass spectrometers than in any other type of instrumentation.

In the chemical and life sciences, mass spectrometry is used to identify and quantify compounds present in complex organic mixtures; to identify structures of biomolecules such as carbohydrates, nucleic acids, and steroids; to sequence proteins and oligosaccharides; to determine how drugs are used by the body; and to check fermentation processes for the biotechnology industry. In environmental science, mass spectrometry is used to detect environmental pollutants such as dioxins and mercury in fish and humans, and to determine gene damage from environmental causes. In forensic science, mass spectrometry is used to confirm and quantify drugs of abuse and steroid use by athletes, and to identify accelerants used in arson. Mass spectrometry is used to determine the age and origins of specimens in geochemistry and archaeology; to locate oil deposits by measuring petroleum precursors in rock; to report the composition of molecular species found in space; to perform ultrasensitive, multielement inorganic analyses; and to establish the elemental composition of semiconductor materials.

Mass spectrometry changes at a fast pace, primarily in the discovery of new methods for introducing ionized molecules in the gas phase into the spectrometer. Not many years ago, the idea of analyzing proteins by mass spectrometry would have been thought impossible.

Notice that the infrared (IR), nuclear magnetic resonance (NMR), and ultraviolet (UV) methods are called *spectroscopy*, but MS is mass *spectrometry*. This difference helps remind us that mass spectrometry does not involve the absorption of specific electromagnetic energies resulting in spectra of frequency versus intensity. Rather, mass spectrometry involves the production and detection of different masses to yield spectra recorded as ion mass versus ion abundance.

THE MASS SPECTROMETER

Mass spectrometers have the following components, connected as shown in Figure 13.1.

1. Ion Source

Neutral molecules in the gas state move randomly, and it is impossible to move them in any particular direction without using a pump to create a pressure gradient. However, if molecules can be positively or negatively charged to form ions, these ions can be attracted and repelled by charged metal surfaces so that they can be moved or accelerated in any chosen direction. The ionization of molecules is done in numerous ways, but the most common way is shown in Figure 13.2. Sample molecules in the gaseous state are introduced into an *ion source*, a small metal enclosure containing a hot filament, like that in a light bulb, on one side. Electrons are emitted from the filament and attracted to a small metal collector electrode on the other side by a slight positive voltage, usually +70 eV, relative to the filament. These energetic electrons can collide with the gaseous sample molecules and ionize them by knocking electrons out of their electron clouds, producing *molecular ions*, most having a single positive charge, but a few with two positive charges. Some negative ions are formed by the capture of an electron, but these are much less abundant. Therefore, most electron ionization spectra are of positive ions.

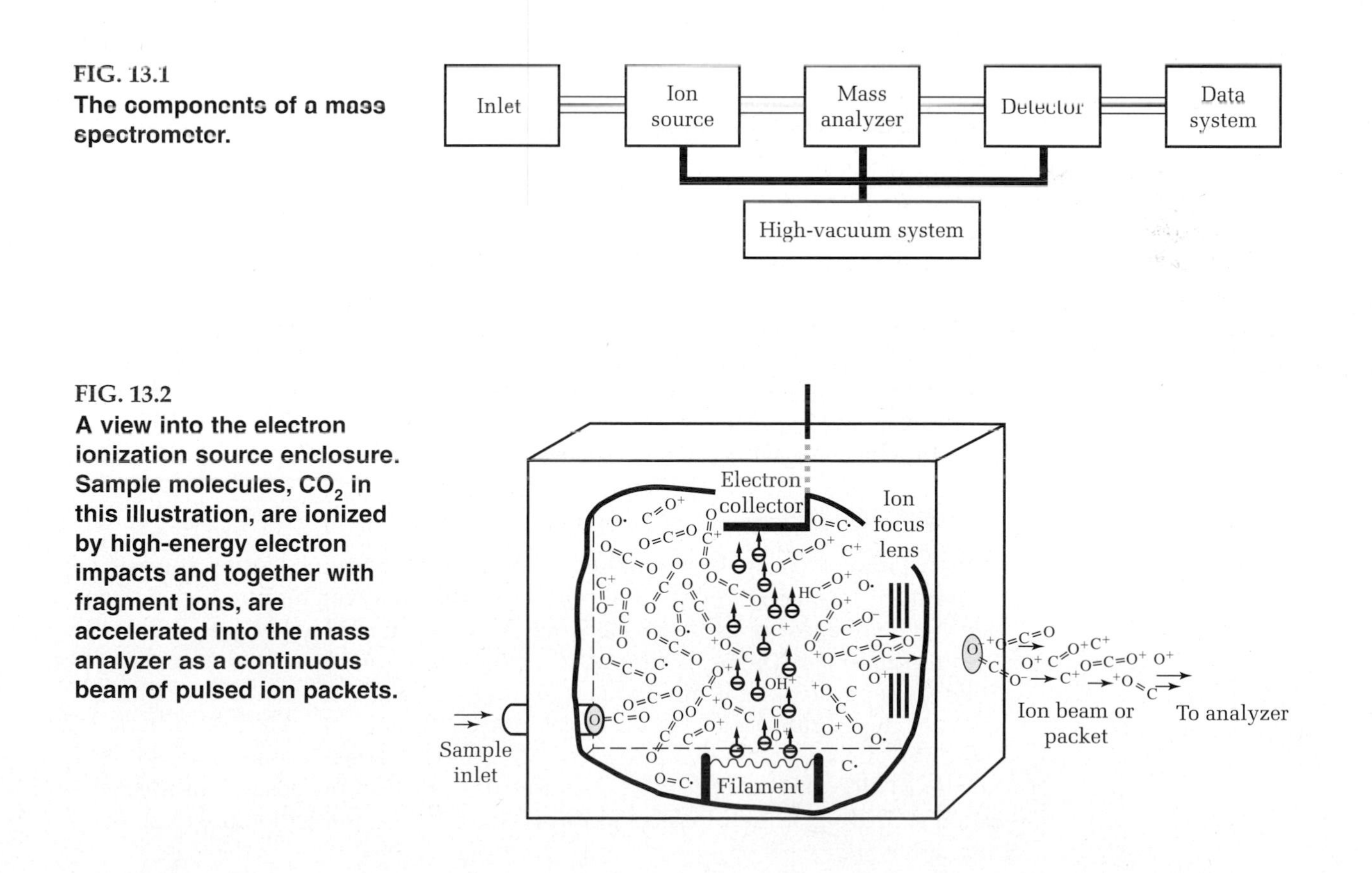

FIG. 13.1
The components of a mass spectrometer.

FIG. 13.2
A view into the electron ionization source enclosure. Sample molecules, CO_2 in this illustration, are ionized by high-energy electron impacts and together with fragment ions, are accelerated into the mass analyzer as a continuous beam of pulsed ion packets.

The ionization process is illustrated in Figure 13.2 and in the following text using a sample of carbon dioxide. Electron ionization produces $O{=}C{=}O^{+\cdot}$ molecular ions written using the cation radical symbol $^{+\cdot}$ because they are odd electron species.

Ionization:

$$O{=}C{=}O \;+\; e^{\ominus} \longrightarrow O{=}C{=}O^{+\cdot} \;+\; 2e^{\ominus}$$

Fragmentation:

$$O{=}C{=}O^{+\cdot} \longrightarrow O\cdot \;+\; C{=}O^{+} \;\text{ or }\; O^{+} \;+\; C{=}O\cdot$$

$$C{=}O^{+} \longrightarrow C^{+\cdot} \;+\; O\cdot \;\text{ or }\; C\cdot \;+\; O^{+\cdot}$$

The energy of the bombarding electrons (about 70 eV) is generally much greater than that of the bonds holding the molecule together. Thus, when high-energy electrons interact with the molecule, not only does ionization occur, but bonds are broken and *fragmentation* occurs, giving rise to *fragment ions* and *fragment neutrals*. The C=O bonds of CO_2 can be broken to produce either a $C{=}O^{+}$ ion and a neutral O radical, or an O^{+} ion and a neutral C=O radical. If both C=O bonds are broken, a C^{+} ion can be produced.

The electron ionization source, as well as the other parts of a mass spectrometer, must be maintained at very low pressure (high vacuum; ~10^{-8} atm), and only very small amounts of sample can be introduced. If the pressure is too high, the ions would collide with unionized molecules or other ions instead of being accelerated from the ion source and into the ion mass analyzer. Also, the electron filament would burn out if an atmosphere of air were present. Therefore, sample inlets are typically a fine capillary or a tiny pinhole to restrict sample flow into the source chamber. Solids are introduced on a metal rod that slides through a sealed vacuum lock. Molecules that are not ionized as well as neutral fragments are pumped away.

There are several other ionization methods that have the advantage of producing intense molecular ions or protonated molecular ions, which are just as informative. For samples that are gaseous or that can be vaporized in a vacuum, chemical ionization and field ionization are often used. Atmospheric pressure chemical ionization and liquid injection field desorption are other methods for producing ions. These will not be discussed here, but the references at the end of this chapter provide details. High molecular weight nonvolatile compounds can be ionized by electrospray ionization (ESI) or matrix-assisted laser desorption ionization (MALDI), both of which are discussed in a later section on bioanalytical mass spectrometry.

2. Mass Analyzer

Positive ions are propelled into the analyzer by maintaining the ion source at a more positive electrical potential relative to the analyzer, and by focusing the ions with voltages applied to an electronic lens system located between the source and the analyzer. The beam of ions passes through the *mass analyzer* where the ions are separated according to mass. Actually, ions are separated based on the mass to charge ratio (m/z). For example, for CO_2^{+} ions $z = 1$, so $m/z = 44/1$, or 44; for CO_2^{++} ions $z = 2$, so $m/z = 44/2$, or 22. Doubly charged ions are rare, so $z = 1$ in almost all cases, and $m/1 = m$. Therefore, for practical purposes, ions are separated according to their mass, and the ensuing discussion will mainly use mass instead of m/z.

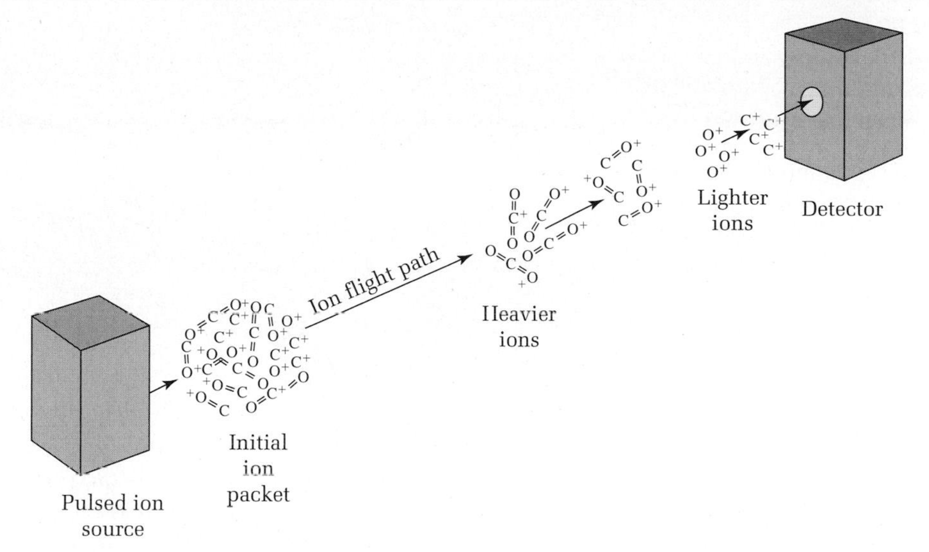

FIG. 13.3
A time-of-flight mass spectrometer. A packet of molecular and fragment ions is pulsed out of the ion source. Lower mass ions arrive at the detector first, with higher mass ions following.

In the earliest mass spectrometers, the ions of different masses were separated by passing the ion beam through a magnetic field. Today, however, two types of mass analyzers are commonly used. The simplest in theory is the *time-of-flight (TOF) analyzer* (Fig. 13.3). The ion beam is pulsed through an electronic gate in little packets, and the start time of each packet of ions is recorded. The packet of ions travels down a long tube, typically 1–2 m in length. Lower mass, lighter ions travel faster and arrive at the detector to produce signals with short times, while heavier ions produce signals with longer times. The development of high-speed electronics now allows the detection of ions that differ in mass by only 1 part in 10,000, which corresponds to a difference in flight times of a few microseconds (10^{-6} s) or less.

Quadrupole mass analyzers are commonly used in GC–MS systems (discussed in a later section of this chapter). The continuous beam of ions from the source passes between four parallel metal rods or poles (Fig. 13.4). Depending on the magnitude and frequency of the direct current (DC) and alternating current (AC) voltages on the rods, only ions of a certain *resonant* mass will pass between the rods and into the detector. Ions of other masses are nonresonant and will be deflected to the side and miss the detector. The mass range is scanned by increasing the rod voltage and AC frequency so that ions of increasing masses hit the detector.

3. Detector

After separation by mass, the ions pass into the *detector* where they impact a metal electrode called a dynode that has a high negative voltage. This releases a number of electrons (~5–10) that are drawn toward a slightly more positive dynode. Each of these electrons releases a number of electrons, which accelerate toward a third dynode, releasing even more electrons (Fig. 13.5). These *electron multipliers* have enough dynode stages that a single positive ion impact produces a cascade of a million or more electrons at the final electrode. Single ions can be detected, thus making mass spectrometry among the most sensitive methods of chemical analysis.

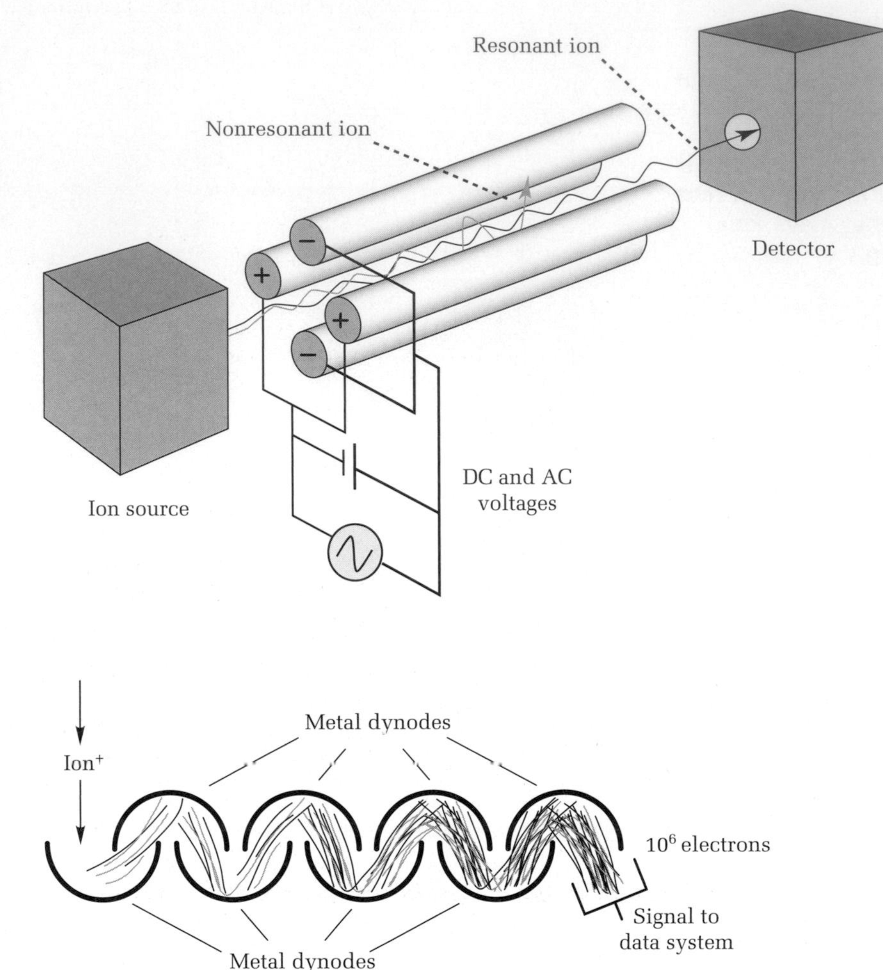

FIG. 13.4
A quadrupole mass spectrometer. Ions oscillate as they move through the electrical fields on the four rods. Changing these fields allows a specific resonant ion to pass to the detector. Ions of other masses are deflected.

FIG. 13.5
A diagram of an electron multiplier detector.

The signal output voltage is proportional to the number of ions hitting the detector at a given time (TOF analysis) or AC and DC voltages (quadrupole analysis), and is continuously digitized and stored in a computer. The computer has files of mass versus calibrated TOFs or quadrupole voltages, and can therefore plot the mass spectrum as the *percent relative abundance* versus mass or more correctly *m/z* (see earlier discussion). All signal peaks are normalized with the most abundant ion as 100%.

A dalton (Da) is a mass unit used in mass spectrometry and is defined as 1/12 of the mass of ^{12}C, which has the atomic weight 12.00000.

The electron ionization spectrum for a sample of CO2 is shown in Figure 13.6. The ionized CO_2 molecule (or molecular ion) appears at mass 44. The ion is singly charged, and the *nominal ion mass* is 44 atomic mass units or daltons (Da): carbon = 12 Da and oxygen = 16 Da (in calculating nominal ion mass, atomic masses are rounded to the nearest integer). Because the ionization process breaks up or fragments some of the CO_2 molecules, a fraction of the ions appear in the spectrum at mass values less than the molecular ion. Cleavage of a carbon-oxygen bond in the molecular ion, which produces ionized carbon monoxide (CO^+) and a neutral oxygen radical (O·), or ionized atomic oxygen (an O^+ ion) and a neutral carbon monoxide radical (C=O·), results in the fragment ions at masses 28 and 16. Loss of two neutral oxygen atoms results in an additional fragment at mass 12 for carbon (a C^+ ion). Note that only charged fragments, not neutral fragments, appear in the spectrum.

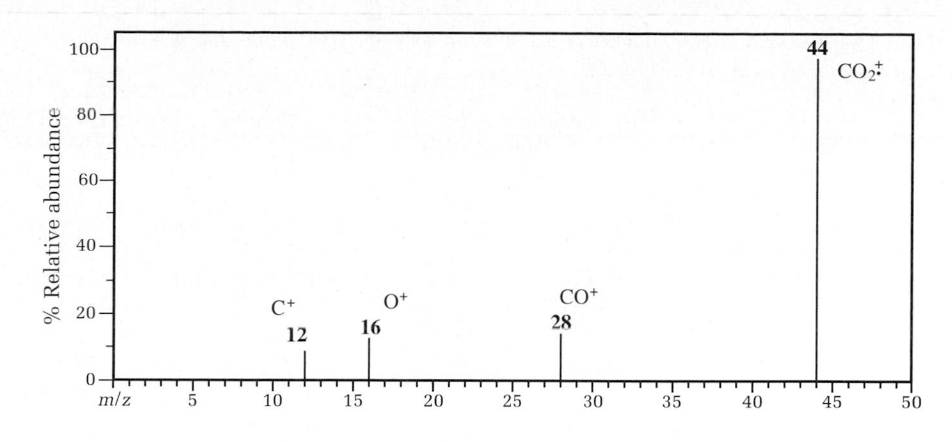

FIG. 13.6
The mass spectrum is a histogram of the relative abundances of molecular and fragment ions versus mass, illustrated here for the analysis of CO_2.

Most mass spectrometers are set up so that they do not scan below $m/z = 35$ because there are very large background signals at masses 32 (O_2), 28 (N_2), and 18 (H_2O) due to residual air and moisture in the instrument, in spite of the high vacuum used. Therefore, unlike Figure 13.6, the remaining spectra in this chapter start at mass 35. Weak background signals at masses 40 and 44 from atmospheric argon and carbon dioxide are also seen in many spectra but are usually ignored.

INTERPRETATION OF ELECTRON IONIZATION SPECTRA

Molecular Ions

The mass of a molecular ion $M^{+\cdot}$, as measured by a mass spectrometer, is not the same as the molecular weight conventionally used to weigh out mole quantities on a balance. The analyzer of a mass spectrometer separates individual molecular ions based on the masses of the isotopes in those ions. The important isotopes for organic mass spectrometry are listed in Table 13.1. Every organic compound contains both ^{12}C isotopes with mass 12.0000 Da and ^{13}C isotopes with mass 13.0034 Da, with a 1.1% probability of a given carbon being a ^{13}C. (Radioactive ^{14}C is at a much lower natural abundance.) About 89% of molecules in a sample of decane contain only ^{12}C and have the composition $^{12}C_{10}H_{22}$, and 11% of the molecules have one ^{13}C (10 carbons times the 1.1% probability of each one being a ^{13}C) and the composition $^{12}C_9{}^{13}C_1H_{22}$. When 1 mol of decane is weighed out, huge numbers of carbon atoms (~60×10^{23}) are present, and therefore it is reasonable to use an average mass of the two isotopes, 12.0107 g/mol, to calculate the molecular weight of decane as 142.285 g/mol. But in a mass spectrometer, $^{12}C_{10}H_{22}{}^+$ ions are separated from $^{12}C_9{}^{13}C_1H_{22}{}^+$ ions, as seen in the mass spectrum of decane shown in Figure 13.7.

To calculate the exact masses of these ions, isotopic masses for ^{12}C (12.0000 Da) and ^{13}C (13.0034 Da) must be used. The $^{12}C_{10}H_{22}{}^+$ molecular ion has a mass of 142.1716 Da and that of $^{12}C_9{}^{13}C_1H_{22}{}^+$ has a mass of 143.1750 Da. The heavier isotopes of hydrogen (^2H and ^3H) and oxygen (^{17}O and ^{18}O) have abundances that are so low that they are difficult to detect and are usually ignored. For decane, the *nominal* mass of the most abundant molecular ion containing just ^{12}C, obtained by rounding the exact mass to the closest integer, is 142 Da, the same as the integer molecular weight of decane, 142 g/mol. For other isotopes, this may not be true.

TABLE 13.1 *The Natural Abundance of Isotopes (atoms per 10,000 atoms and percent abundance) and the Atomic Mass of Each Isotope*

	Abundance and Mass of the Most Abundant Isotope (given 10,000 atoms)				*Abundance and Mass of Other Isotopes (given 10,000 atoms)*				
	Isotope	*No. of Atoms*	*Mass (Da)*	*% Abundance*	*Isotope*	*No. of Atoms*	*Mass (Da)*	*% Abundance*	*Average Mass (g/mol)*
C	^{12}C	9890	12.0000	98.9	^{13}C	110	13.0034	1.1	12.0107
H	^{1}H	9999	1.0078	99.99	^{2}H	1	2.0141	0.01	1.00794
O	^{16}O	9985	15.9949	99.76	^{17}O ^{18}O	4 20	16.9991 17.9992	0.04 0.2	15.9994
N	^{14}N	9963	14.0031	99.63	^{15}N	37	15.0001	0.37	14.0067
Cl	^{35}Cl	7577	34.9689	75.8	^{37}Cl	2423	36.9659	24.2	35.453
Br	^{79}Br	5069	78.9183	50.7	^{81}Br	4931	80.9163	49.3	79.904
S	^{32}S	9493	31.9721	94.93	^{33}S ^{34}S	76 429	32.9715 33.9679	0.76 4.29	32.065

The elements ^{19}F, ^{31}P, and ^{127}I are monoisotopic.

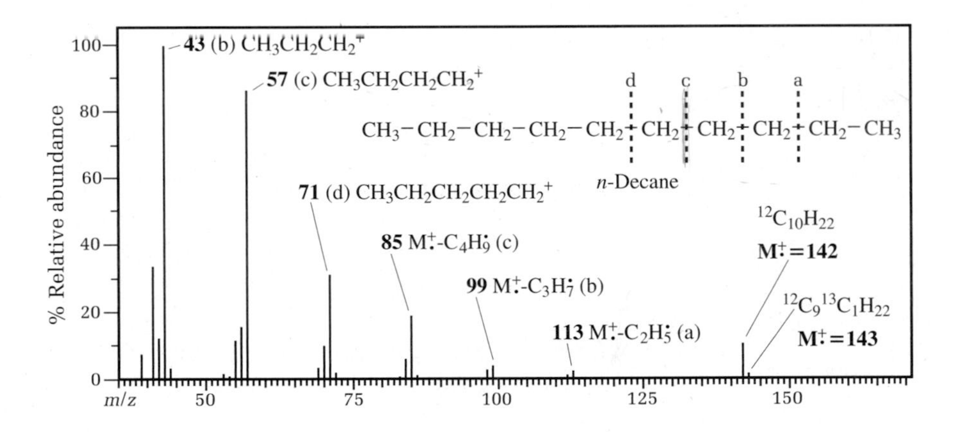

FIG. 13.7
The mass spectrum of decane showing isotopic molecular ions, fragment ions, and neutral losses due to bond cleavages at a, b, c, and d.

In the mass spectrum of 4-chloro-3-methylphenol (Fig. 13.8), there are four molecular ions with isotopic compositions shown due to the presence of not only ^{13}C, but also the two isotopes of chlorine, ^{35}Cl and ^{37}Cl. The exact mass of the largest molecular ion, $^{12}C_7H_7O^{35}Cl$ is 142.0184 Da. This is quite different than the molecular weight of this compound, 142.5774 g/mol. After rounding, the nominal mass of the ion is 142 Da, but the integer molecular weight is 143 g/mol. This difference is the primary reason that daltons are used in mass spectrometry—to help differentiate exact masses measured by mass spectrometry from the conventional molecular weight in grams/mole used in lab weighing.

Both chlorine and bromine have isotopes separated by 2 Da, creating unique and easily recognized patterns in a mass spectrum. The ratio of peaks depends on the number of chlorines or bromines in the molecule, as shown in Figure 13.9. The Cl_1 pattern is seen in the mass spectrum of the chlorophenol in Figure 13.8.

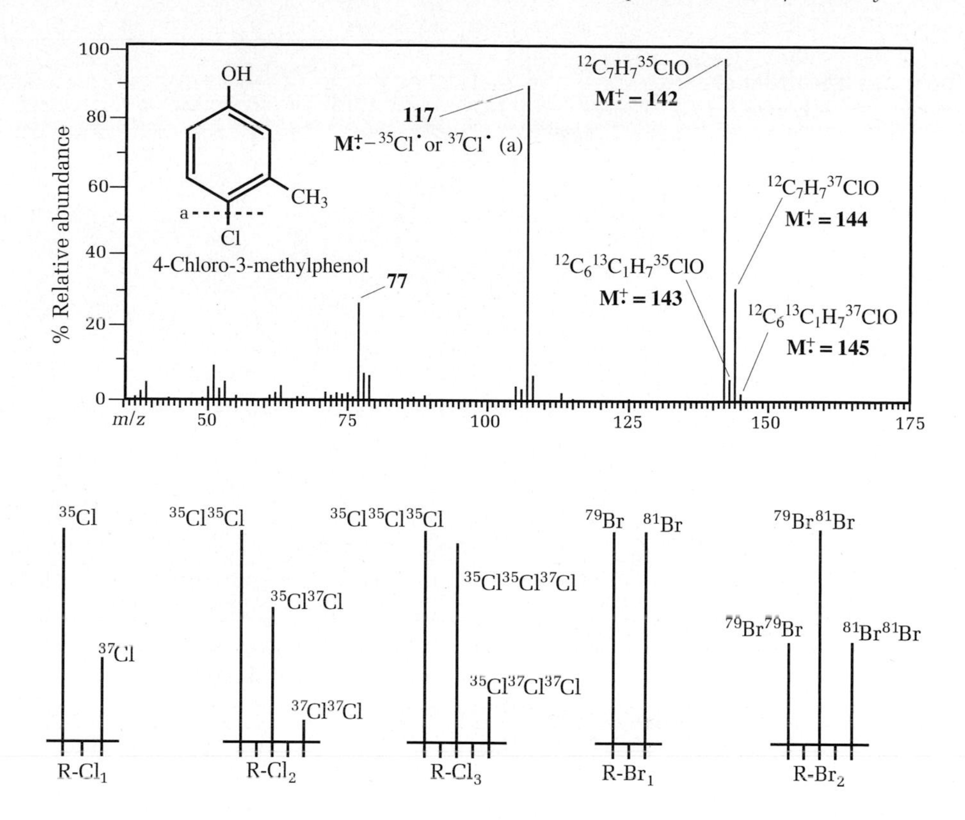

FIG. 13.8
The mass spectrum of 4-chloro-3-methyl phenol.

FIG. 13.9
Chlorine and bromine isotope patterns.

TABLE 13.2 *Six Compounds with Molecular Ions of Nominal Mass 142 That Can Be Differentiated by Accurate Mass Measurement*

Compound	*Molecular Formula*	*Exact Mass of Molecular Ion*
3-Fluoro-4-methoxyphenol	$C_7H_7FO_2$	142.0430
Octenoic acid	$C_8H_{14}O_2$	142.0993
Cyclohexylurea	$C_7H_{14}N_2O$	142.1106
3-Nonanone	$C_9H_{18}O$	142.1357
1,2-Diaminocyclooctane	$C_8H_{18}N_2$	142.1469
Decane	$C_{10}H_{22}$	142.1721

Many modern instruments are able to measure ion masses to within a few parts per million of accuracy. For example, even though all of the compounds listed in Table 13.2 (as well as hundreds of other possible compounds) have molecular ions at nominal mass 142, exact mass measurement would easily differentiate among them.

Conversely, given an exact mass measured by mass spectrometry, a computer can compare this with all possible exact masses calculated for all reasonable combinations of ^{12}C, ^{1}H, ^{14}N, ^{16}O, ^{35}Cl, and so forth and produce a list of compositions that come within a few parts per million of the measured mass. Compositions determined in this manner are required whenever a new compound is reported in

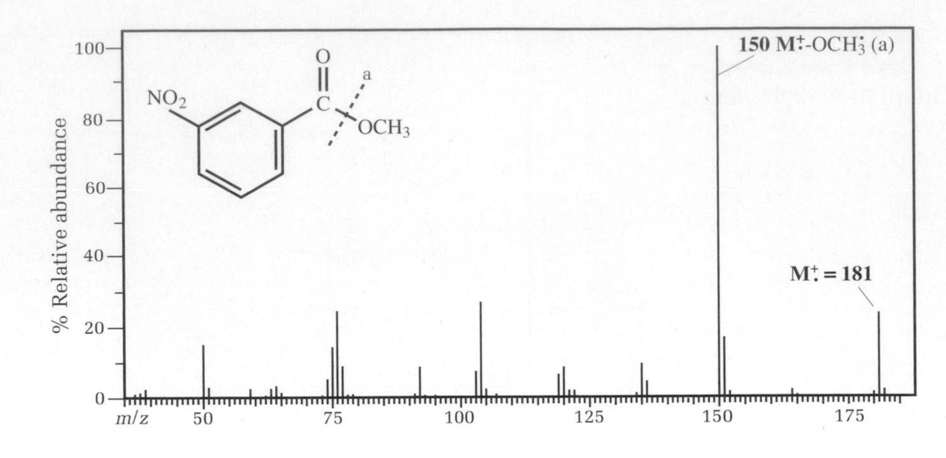

FIG. 13.10
The mass spectrum of methyl 3-nitrobenzoate.

a publication. (Note, however, that the composition $C_9H_2O_2$, for example, is not reasonable and is incorrect even though it has a mass of 142.)

Compounds containing C, H, O, F, Cl, Br, I, and S or an even number of nitrogens will have an even mass molecular ion. *The nitrogen rule* states that all molecules with an odd number of nitrogen atoms will have an odd mass molecular ion. Compare the molecular ions in the mass spectrum of a nitrogen-containing compound (Fig. 13.10) with the other spectra in this chapter. If we look at the simplest compounds of C, O, and N, the reason becomes apparent. C and O have even masses and form an even number of bonds, so H_2O has a molecular mass of 18 and CH_4 has a molecular mass of 16, both even. Nitrogen (^{14}N) has an even mass but forms an odd number of bonds, so NH_3 has an odd mass, 17. Organic compounds are simply derivatives of these simple molecules. The halogens form only single bonds (odd) but have odd masses, so replacement of a hydrogen by a halogen does not change the molecular mass from even to odd.

Fragmentation

With electron ionization, the percent relative abundance, or intensity, of the molecular ion varies dramatically with compound structure. Carbon dioxide (*see* Fig. 13.6 on page 265) and aromatic compounds (*see* Fig. 13.8) contain multiple double bonds that are stronger than single bonds, and the π bonding systems can help stabilize the positive charge so that less bond fragmentation occurs and the molecular ion abundance is high. Decane has all single bonds and fragments more readily. The most abundant ions are fragments, which exhibit a weak molecular ion signal. Although the mass of the molecular ion is considered the most important piece of information to be derived from a mass spectrum, the sad truth is that many compounds, like hydrocarbons, alcohols, and alkyl halides, do not show any molecular ion signal. For example, the mass spectrum of 4,4-dimethyloctane (Fig. 13.11) shows no signal at mass 142, which indicates that it fragments even more readily than decane. Nevertheless, the masses and mass differences of abundant fragment ions can provide useful structural information.

We have already discussed the fragment ions (detected) and fragment neutrals (not detected) generated by bond cleavage in CO_2. It is clear that these peak masses are calculated by the simple arithmetic of adding up the atomic isotopic masses of the atoms in the fragments formed by the cleavage of bonds in the molecule. Table 13.3 gives the masses and possible structures for several commonly

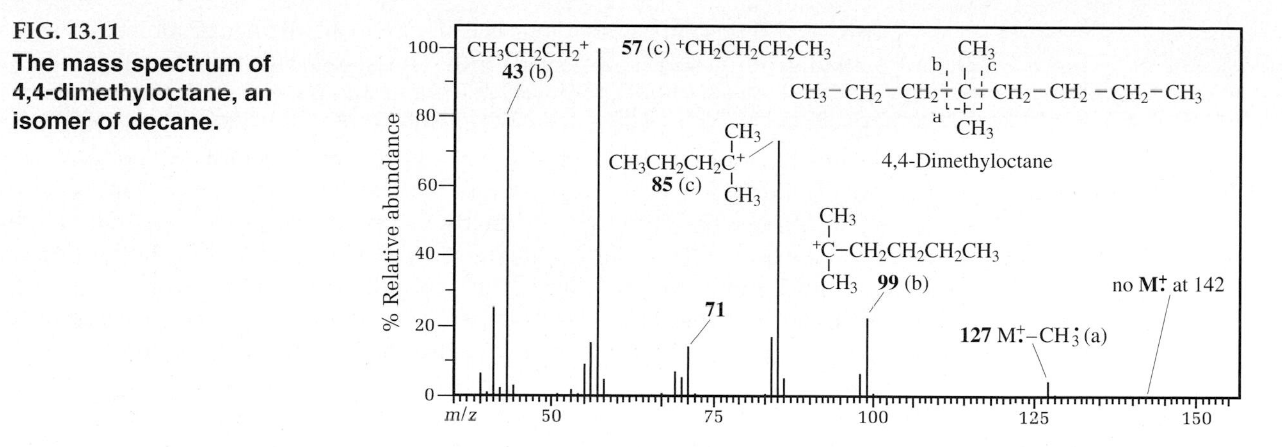

FIG. 13.11
The mass spectrum of 4,4-dimethyloctane, an isomer of decane.

TABLE 13.3 *Common Fragment Ion Masses and Structures*

Fragment Ion Mass (Da)		*Structure(s)*
. . . 113, 99, 85, 71, 57, 43 Alkyl series		. . . $C_8H_{17}^+$, $C_7H_{15}^+$, $C_6H_{13}^+$, $C_5H_{11}^+$, $C_4H_9^+$, $C_3H_7^+$ Differ by CH_2 or 14 Da
. . . 111, 97, 83, 69, 55, 41 Alkenyl series		. . . $C_8H_{15}^+$, $C_7H_{13}^+$, $C_6H_{11}^+$, $C_5H_7^+$, $C_4H_7^+$, $C_3H_5^+$ Differ by CH_2 or 14 Da
. . . 113, 99, 85, 71, 57, 43 Alkanoyl series	(a)	$C_6H_{13}C(=O)^+$, $C_5H_{11}C(=O)^+$, $C_4H_9C(=O)^+$, $C_3H_7C(=O)^+$, $C_2H_5C(=O)^+$ Differ by CH2 or 14 Da
105 plus mass of any group on ring	(b)	$C_6H_5C(=O)^+$ (benzoyl)
91 plus mass of any group on ring	(c)	$C_6H_5CH_2^+$ (benzyl)
77 plus mass of any group on ring	(d)	$C_6H_5^+$ (phenyl)

observed fragment ions. The abundance of a particular fragment ion depends on two factors: the ease with which bonds to it can be broken, and the stability of the resulting ion. In the mass spectrum of decane (*see* Fig. 13.7 on page 266), we can see a series of fragment ions at masses 41, 57, 71, 85, 99, and 113, differing by increments of 14 Da, which is the mass of a CH_2 unit. Table 13.3 lists this alkyl series of ions. These arise from the cleavage of bonds at a, b, c, and d, as marked on the decane molecular ion, to form a charged fragment and an uncharged or neutral fragment. From the ion abundances, it is clear that the positive charge prefers to reside on fragments containing three or four carbons, with masses 43 and 57, so the neutral fragments formed by bond cleavage contain the other seven or six carbons.

This alkyl series of fragment ions is common in all alkanes or molecules with large alkyl groups, and is observed in the mass spectrum 4,4-dimethyloctane (Fig. 13.11). However, there are several important differences in the percentage ion abundances in its spectrum when compared to that of decane (Fig. 13.7). These can be explained in terms of the increasing stability of carbocations as we go from primary to secondary to tertiary carbocations, the same stability observed in solution organic reactions. The structure of dimethyloctane promotes fragmentation at b and c to give the stable tertiary cations shown at masses 85 and 99. Both of these ions are much more abundant in this spectrum than in the spectrum of decane. Notice also the decreased intensity at mass 71, for the C_6H_{11} alkyl ion. Because there are no C_6 alkyl groups in 4,4-dimethyloctane, the formation of an ion of mass 71 requires the breaking of two C—C bonds and a hydrogen transfer, a much less favorable process. Finally, unlike decane, no molecular ion signal is detected, which is more evidence of the facile fragmentation of this molecule.

Strong peak intensities at masses 77, 91, and 105 are uniquely diagnostic for the presence of an aromatic ring (*see* Table 13.3). These ions are increased in mass by any groups attached to the aromatic ring. Chapter 28 describes the synthesis of methyl 3-nitrobenzoate. The mass spectrum of this product is shown in Figure 13.10. The most abundant ion at mass 150 corresponds to the benzoyl ion of mass 105 shown in Table 13.3, plus the weight of a nitro group, NO_2, mass 46, minus 1, the mass of a hydrogen that the NO_2 replaces.

Notice that the fragment ion masses of the alkanoyl series are the same as that of the alkyl series; therefore, it might seem impossible to distinguish carbonyl compounds like ketones, acids, and esters from alkanes. However, the presence of a C=O leads to some diagnostic fragmentations. For example, the mass spectrum of 3-nonanone in Figure 13.12 shows a very diagnostic fragment ion at mass 72, which is even. This ion arises by a fragmentation of the molecular ion involving a rearrangement called the McLafferty rearrangement:

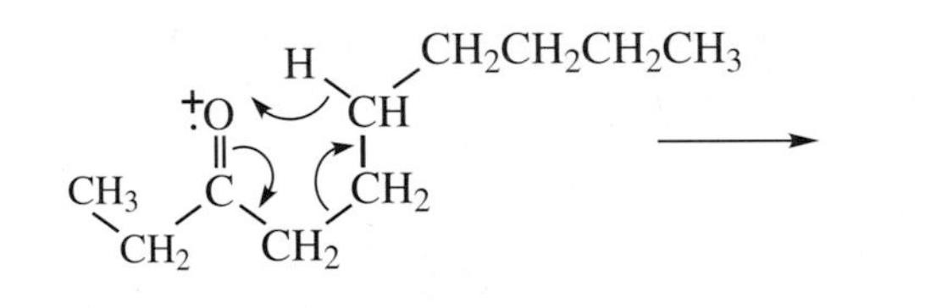

$M^{+\cdot}$ = 142

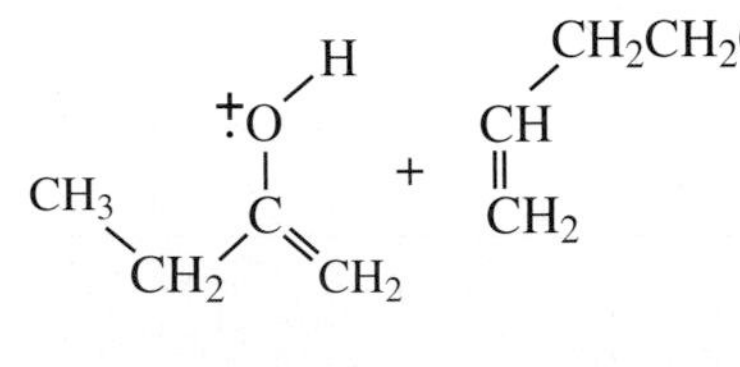

Mass 72 fragment ion **Neutral loss fragment**

Electron pair bonds shift in the six-member intermediate, and a hydrogen is transferred from carbon 6 of the ketone to the carbonyl oxygen. The bond rearrangement releases the neutral alkene, and the positive charge remains on the oxygen. Fragmentation with a hydrogen rearrangement is designated with ~H on the spectrum. Such fragmentations are observed quite frequently.

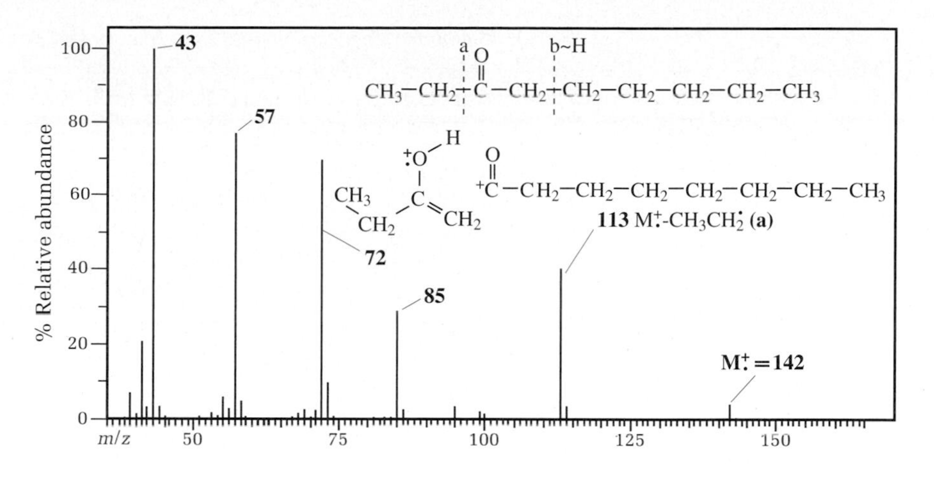

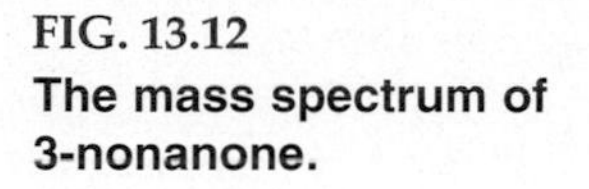
FIG. 13.12
The mass spectrum of 3-nonanone.

Neutral Loss Masses

When a molecule breaks into two fragments, only one of them carries the positive charge of the parent molecular ion and can be detected. The uncharged piece, normally a radical or in certain cases a small molecule, is not detected. However, its mass can be easily calculated by subtracting the mass of the charged fragment from the molecular ion mass. Like fragment ion masses, the neutral fragment difference masses provide structural information. And they are often easier to understand. Common neutral loss masses are shown in Table 13.4.

Fragment ions can be specified in either of two ways: the fragment structure with a positive charge attached or the $M^{+\cdot}$ minus the neutral fragment. Both ways are used in the spectra discussed here. The latter method is much more concise and readily understood for the higher mass fragments close to the $M^{+\cdot}$. For example, the fragment at mass 113 in the decane spectrum (Fig. 13.7) is much easier to designate as $M^{+\cdot}$ minus ethyl ($M^{+\cdot} - C_2H_5$) than as $CH_3CH_2CH_2CH_2CH_2CH_2CH_2CH_2^{+}$. The same is true for fragments at masses 99 and 85. In the spectrum of the chlorophenol in Figure 13.8, the difference in mass between the largest $M^{+\cdot}$ at mass 142 and the major fragment ion at mass 117 is 35. From Table 13.4, we see this can be attributed to a neutral loss of chlorine and the fragment ion is specified as $M^{+\cdot} - Cl$, where Cl is either ^{35}Cl or ^{37}Cl depending on the isotopes in the fragmenting molecular ion.

One of the major differences between the mass spectra of decane and 4,4-dimethyloctane (Fig. 13.7 and Fig. 13.11) lies in the neutral loss of methyl. This fragmentation, specified here as $M^{+\cdot} - CH_3$, is seen for the dimethyloctane because it leads to a stable tertiary carbocation fragment. This type of fragment is not possible with the linear alkane decane, and a neutral loss of methyl is not observed. The major fragmentation of the ester shown in Figure 13.10 involves a neutral loss with a mass difference of 31, which is attributable to cleavage of the neutral OCH_3 radical. In Figure 13.12, the ion at mass 113 is specified both as its fragment ion structure and neutral loss designation.

Many neutral losses involve the elimination of an intact neutral molecule such as water, carbon monoxide, carbon dioxide, or acetic acid. Figure 13.13 shows the mass spectrum of the alcohol 2-nonanol, and, as is typical of alcohols, no molecular ion is seen because of the facile neutral loss of H_2O. The mass of the molecular ion can be deduced, however, by examining other ions observed at

TABLE 13.4 *Some Common Neutral Loss Masses and Possible Fragment Structures*

Mass Difference (Da)	*Neutral Radical or Molecule Lost*
1	·H (hydrogen)
15	$\cdot CH_3$ (methyl)
17	·OH (hydroxyl)
18	H_2O (water)
28	$C{\equiv}O$ (carbon monoxide) or $CH_2{=}CH_2$ (ethene)
29	$\cdot CH_2CH_3$ (ethyl) or $\cdot C(=O){-}H$ (formyl)
31	$\cdot OCH_3$ (methoxyl)
35 or 37	$^{35}Cl\cdot$ or $^{37}Cl\cdot$ (chloro)
42	$CH_2{=}C{=}O$ (from acetyl group)
43	$CH_3{-}C(=O)\cdot$ (acetyl) or $\cdot C_3H_7$ (propyl)
44	CO_2 (carbon dioxide)
45	$\cdot OCH_2CH_3$ (ethoxyl) or $\cdot C(=O)OH$ (carboxyl)
60	$CH_3C(=O)OH$ (acetic acid)
79 or 81	$^{79}Br\cdot$ or $^{81}Br\cdot$ (bromo)

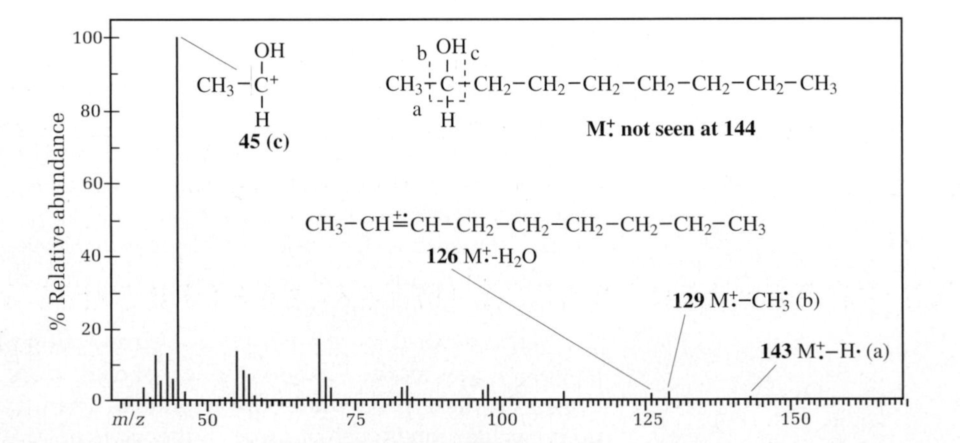

FIG. 13.13
The mass spectrum of 2-nonanol.

higher molecular weight. In particular, the presence of two ions 3 Da apart (masses 126 and 129 in Fig. 13.13) indicates neutral losses of water and a methyl group from a molecular ion. For 2-nonanol, adding 15 to mass 129 for a methyl radical equals 144, and adding 18 to mass 126 for a water molecule also equals mass 144. This is strong evidence that the molecular ion mass is indeed 144. The fact that the alkene fragment ion formed by the loss of water has an even mass (except for a compound containing nitrogen) is another indicator of the loss of a molecule rather than a radical.

Each of the bond fragmentations designated a, b, and c in Figure 13.13 is called an *alpha cleavage* because each involves the cleavage of the bond adjacent, or alpha, to the C—OH bond. The electron movement for each of the fragmentations to give ions at masses 143, 129 and 45 are shown below (R = $CH_2CH_2CH_2CH_2CH_2CH_2CH_3$):

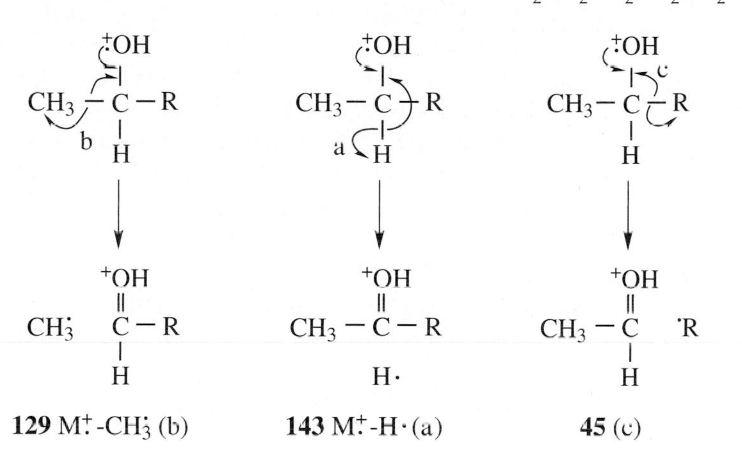

Computer-Aided Mass Spectral Identification

Today, there are computer searchable databases of digitized mass spectra of over 200,000 different organic compounds. With less than 1 mg of sample, a mass spectrum can be acquired by a mass spectrometer and stored in a computer. This digitized spectrum can be compared with all the spectra in these collections, and those reference spectra that have ion masses and intensities similar to the sample spectrum can be displayed. If the sample and a reference library spectrum match closely enough, the chemist can feel confident that the sample has been correctly identified. This sort of analytical power is critical to the success of forensic and environmental chemistry laboratories. Be aware, however, that mass spectra of many structural isomers, for example, the three possible dimethylbenzenes, will be so similar that a secure identification is not possible. In this case, other spectral data such as IR and NMR must be acquired or known standards must be run using a GC–MS system to compare retention times as well as mass spectra.

GAS CHROMATOGRAPHY–MASS SPECTROMETRY (GC–MS)

The development of capillary column gas chromatography greatly simplified the direct coupling of a gas chromatograph with a mass spectrometer. The combination is one of the most powerful organic analysis techniques available, and thousands of GC–MS instruments are found in research, analytical, and teaching labs around the world. As described in Chapter 10, gas chromatography has the ability to separate complex mixtures containing hundreds of compounds (*see* Fig. 10.2 on page 206). However, many detectors, such as the flame ionization detector or the electron capture detector, while being very sensitive, do not provide any structural information. Replacing these with a mass spectrometer maintains the sensitivity but provides the rich structural information available from this instrument.

The typical mass analyzer is of the quadrupole type (*see* Fig. 13.4 on page 264), but TOF analyzers and other types are also used. The advantages of the TOF analyzer are rapid scanning (10 ms per spectrum or less) and accurate mass measurement. Quadrupole instruments acquire spectra at the rate of about one spectrum per second for a mass range of 35–600 Da. A typical GC–MS analysis using temperature programming of the GC oven to elute all the compounds in a mixture requires about 30 minutes. Every spectrum acquired by the mass spectrometer is digitized and stored by the data system; almost 2000 digitized spectra are acquired and stored in 30 minutes. A typical capillary gas chromatography peak is 3–10 seconds wide, so from three to six spectra are recorded as each compound enters the mass spectrometer ion source from the end of the GC column. The computer constructs a gas chromatogram of the separation by plotting the sum of all the ion signal intensities in each scan, called the total ion current (TIC), versus time. The instrument operator can click on any point on the chromatogram, typically at the top of each peak, to view the mass spectrum at that time in a separate window. These can be compared to a digital library of reference spectra as discussed in the previous section and identified if there is a good match. In this way, complex mixtures of hundreds of compounds can be separated and structurally characterized.

BIOANALYTICAL MASS SPECTROMETRY

In 2002, John Fenn of the United States and Koichi Tanaka of Japan shared the Nobel Prize in Chemistry for their development of *soft ionization* methods for the analysis and structural identification of biological macromolecules such as proteins. Because the electron ionization method requires that the sample be in the gaseous vapor state in order to be ionized, most high molecular weight polar materials, such as polysaccharides, nucleic acids, and proteins, would decompose long before they could be heated to a high enough temperature to produce appreciable gaseous molecules. Therefore, for most of its history, mass spectrometry was limited to molecules of molecular weight 2000 Da or less, mostly less than 1000 Da. Then in the 1990s, two new methods were developed that allowed "soft" ionization, without heating, and thus the mass analysis of materials with molecular weights of over 1 million Da. Both of these new techniques—MALDI and ESI—have the advantage that fragmentation is minimal and primarily molecular ions are observed. Actually, the ions formed are *pseudomolecular ions* produced by the attachment of one or more protons or other positive ions, such as sodium or potassium, to the neutral molecule.

MALDI (*matrix-assisted laser desorption ionization*) uses laser pulses to sputter ions from a mixture of a high molecular weight sample embedded in a crystalline matrix of a low molecular weight compound, such as a cyanobenzoic acid (Fig. 13.14). The matrix facilitates the lifting of the large macromolecules from

FIG. 13.14
The MALDI ionization process.

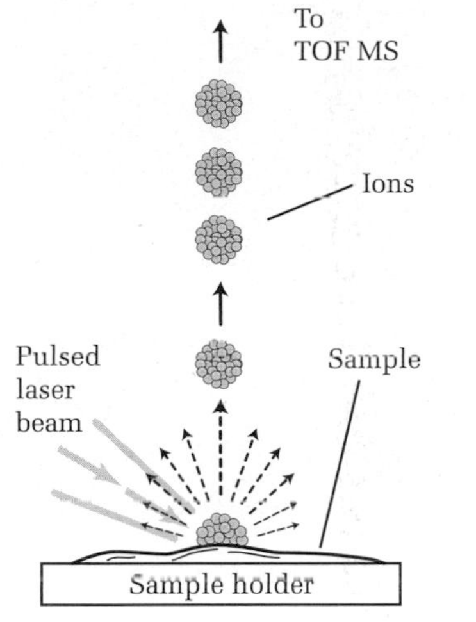

FIG. 13.15

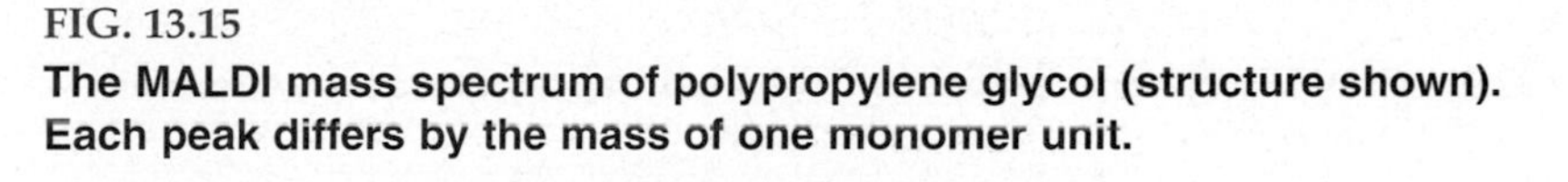

The MALDI mass spectrum of polypropylene glycol (structure shown). Each peak differs by the mass of one monomer unit.

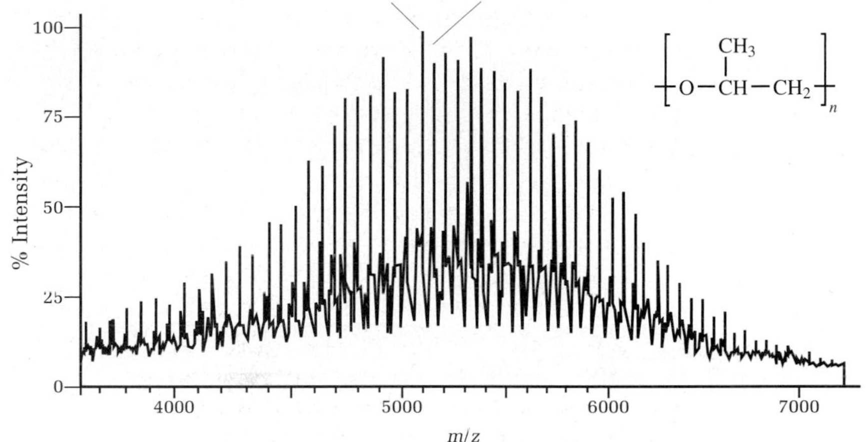

the surface into the gas phase, where they are mass analyzed, typically using TOF mass spectrometry. A MALDI mass spectrum of polypropylene glycol is shown in Figure 13.15.

ESI (electrospray ionization) is illustrated in Figures 13.16a and 13.16b. When a liquid solution of sample molecules flows from the tip of a fine metal capillary maintained at a high voltage toward a flat metal plate with a potential of thousands of volts between them (Fig. 13.16a), a fine spray of charged droplets forms. Flowing nitrogen gas aids the evaporation of the solvent, and the charged droplets shrink (Fig. 13.16b). At some point, the charge becomes too concentrated, and repulsive forces break the droplet into finer droplets. The evaporation, shrinking, and droplet fracture continues until only charged sample molecules remain. Macromolecules typically pick up many charges; the larger the molecule, the more the charges it can hold. A pin-size hole in the plate leads into the high vacuum of a mass analyzer of the quadrupole, TOF, or some other type. The fact that molecules have multiple charges complicates the mass spectrum because each different charge state appears at a different *m/z*. The electrospray spectrum of the hen egg lysozyme is shown in Figure 13.17. Ions with seven different charge states are detected. Computer-aided calculations allow the determination of these charge states, and from this a molecular weight of 14,305.67 Da is calculated.

As in the case of GC–MS, the development of ESI has allowed the coupling of high performance liquid chromatography (HPLC) with mass spectrometry to give HPLC–MS. This has become a powerful tool for the analysis of complex biological mixtures, such as protein digests.

Fast atom bombardment (FAB) is another method for introducing molecules like peptides and proteins into the spectrometer. The sample is embedded in a matrix such as glycerol and bombarded with atoms of argon or xenon at up to 10,000 electron volts. This produces mostly intact protonated or deprotonated molecules.

FIG. 13.16

(a) The electrospray ionization source. (b) The electrospray ionization process.

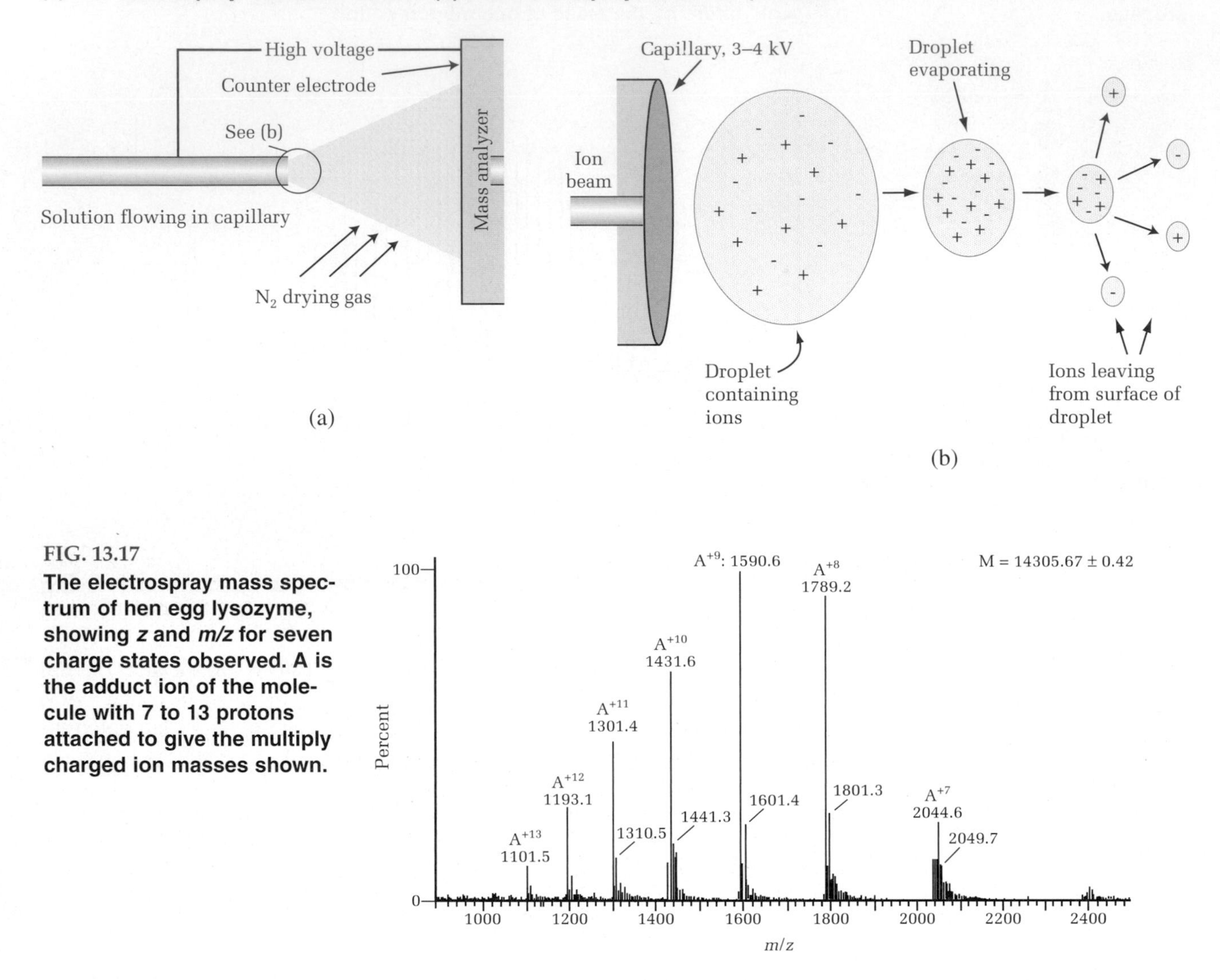

FIG. 13.17

The electrospray mass spectrum of hen egg lysozyme, showing *z* and *m/z* for seven charge states observed. A is the adduct ion of the molecule with 7 to 13 protons attached to give the multiply charged ion masses shown.

QUESTIONS

1. Show by calculation that if you weighed 10,000 molecules of decane consisting of the expected percentages of $^{12}C_{10}H_{22}$ and $^{12}C_{9}{}^{13}C_{1}H_{22}$ molecules, you would obtain an average molecular weight that is the same as that listed in reference books.

2. Give both the isotopic composition formula and the relative intensity for the two peaks marked with a question mark in the mass spectra in Figures 13.18 and 13.19.

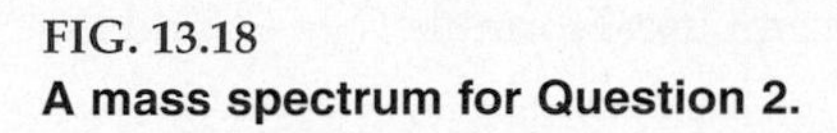

FIG. 13.18
A mass spectrum for Question 2.

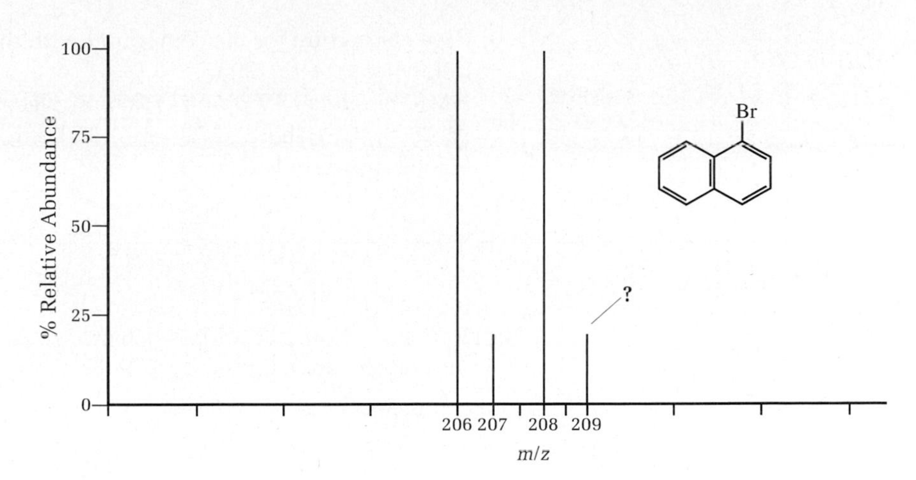

FIG. 13.19
A mass spectrum for Question 2.

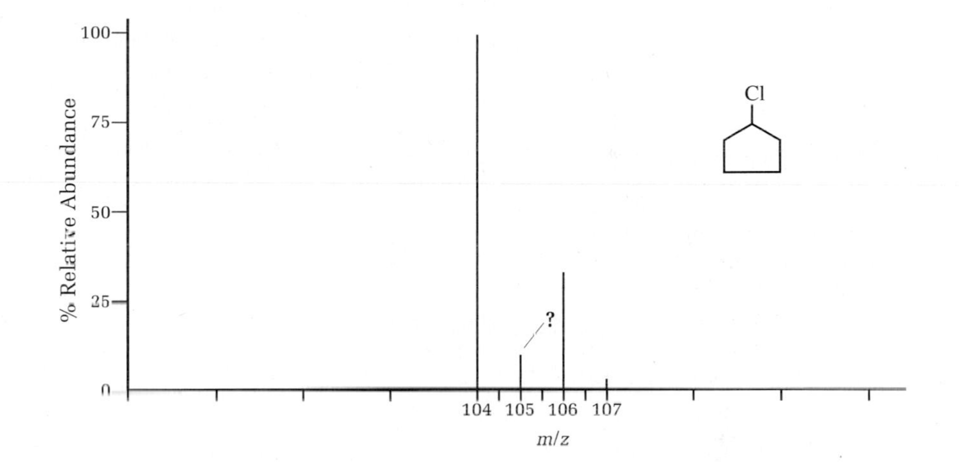

FIG. 13.20
A mass spectrum for Question 3.

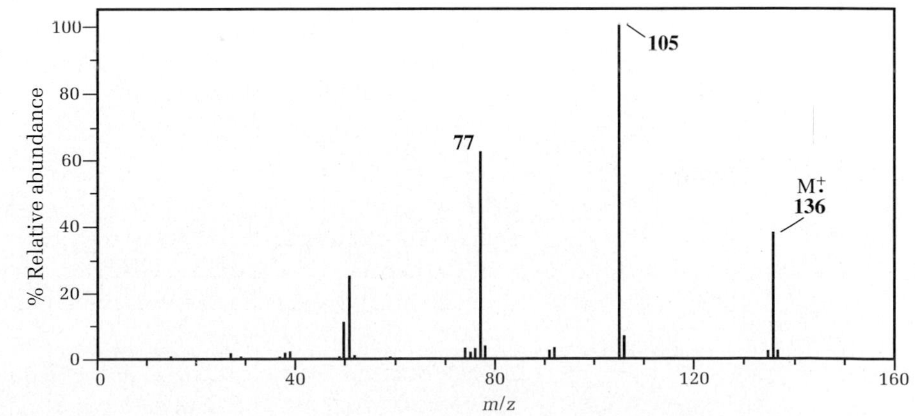

3. Give a structure for the compound with the empirical formula $C_8H_8O_2$ and the following spectral data:
NMR: Singlet at 3.7 ppm (3 H); multiplet at 7.2 ppm (5 H)
IR: 2850 cm^{-1} (sharp), 1720 cm^{-1}, 1600 cm^{-1}, 1503 cm^{-1}
MS: as shown in Figure 13.20

REFERENCES

de Hoffmann, Edmond and Stroobant, Vincent, *Mass Spectrometry: Principles and Applications*, 3rd ed., New York, Wiley, 2007.

Gross, Jurgen H., *Mass Spectrometry, A Textbook*, Springer Verlag, New York, 2006.

Lee, Terrence A., *A Beginner's Guide to Mass Spectral Interpretation*. Wiley, 1998.

McLafferty, Fred W., and František Tureček, *Interpretation of Mass Spectra*, 4th ed. Sausalito, CA: University Science Books, 1993.

McMaster, Marvin, and Christopher McMaster, *GC/MS: A Practical User's Guide*. New York: Wiley-VCH, 1998.

Silverstein, Robert M., Francis X. Webster, and David Kiemle, *Spectrometric Identification of Organic Compounds*, 7th ed.[1] New York: Wiley, 2005.

See *Web Links*

Sparkman, O. David, *Mass Spectrometry Desk Reference*. Pittsburgh: Global View Publishing, 2006.

[1]This book includes IR, UV, and NMR spectra.

CHAPTER 14

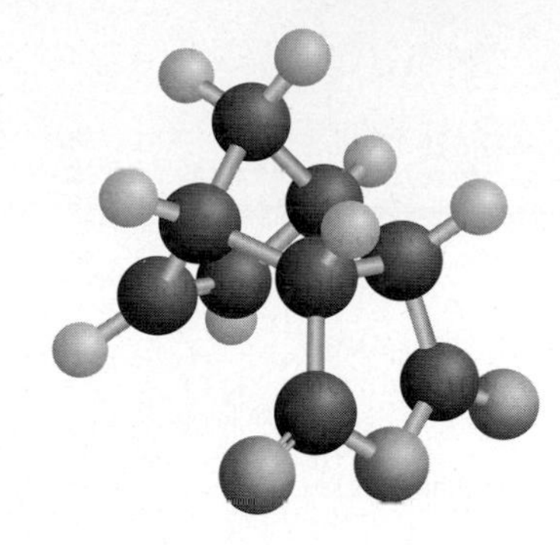

Ultraviolet Spectroscopy, Refractive Indices, and Qualitative Instrumental Organic Analysis

PRELAB EXERCISE: In the identification of an unknown organic compound, certain procedures are more valuable than others. For example, far more information is obtained from an IR or NMR spectrum than from a UV spectrum or a refractive index measurement. Outline, in the order of priority, the steps you will employ in identifying an organic unknown.

ULTRAVIOLET SPECTROSCOPY

UV: Electronic transitions within molecules

Ultraviolet (UV) spectroscopy produces data about electronic transitions within molecules. Whereas absorption of low-energy infrared (IR) radiation causes bonds in a molecule to stretch and bend, the absorption of short-wavelength, high-energy ultraviolet (UV) radiation causes electrons to move from one energy level to another with energies that are capable of breaking chemical bonds. Many molecules produce only a major absorption signal (*see* Figs. 14.2 to 14.7 on pages 280–284) due to a single, extended structural feature involving π bonded systems: carbon-carbon single bonds, carbon-oxygen or carbon-nitrogen double bonds, triple bonds, and aromatic rings. Because many different combinations of these π systems can absorb at nearly the same wavelength, the structure is usually elucidated by other techniques, and its UV spectrum is interpreted based on this structure.

We are most concerned with transitions of π electrons in conjugated and aromatic ring systems. These transitions occur in the wavelength region of 200–800 nm (nanometers, 10^{-9} meters). Most common UV spectrometers cover the region of 200–400 nm, as well as the visible spectral region of 400–800 nm. Below 200 nm, air (oxygen) absorbs UV radiation; spectra in that region must therefore be obtained in a vacuum or in an atmosphere of pure nitrogen.

Consider ethylene, even though it absorbs UV radiation in the normally inaccessible region at 163 nm. The double bond in ethylene has two *s* electrons in a π molecular orbital and two, less tightly held, *p* electrons in a π molecular orbital. Two unoccupied, high-energy-level, antibonding orbitals are associated with these orbitals. When ethylene absorbs UV radiation, one electron moves up from the bonding π molecular orbital to the antibonding π^* molecular orbital (Fig. 14.1). As Figure 14.1 indicates, this change requires less energy than the excitation of an electron from the σ to the σ^* molecular orbital.

By comparison with IR spectra and nuclear magnetic resonance (NMR) spectra, UV spectra are mostly featureless (Fig. 14.2). This condition results as molecules in a number of different vibrational states undergo the same electronic transition to produce a band spectrum instead of a line spectrum.

Band spectra

Unlike IR spectroscopy, UV spectroscopy lends itself to precise quantitative analysis of substances. The intensity of an absorption band is usually given by the molar extinction coefficient, ε, which, according to the Beer-Lambert law, is equal to the absorbance (A) divided by the product of the molar concentration (c) and the path length (l) in centimeters.

Beer-Lambert law

$$\varepsilon = \frac{A}{cl}$$

λ_{max} = wavelength of maximum absorption

The wavelength of maximum absorption (the tip of the peak) is given by λ_{max}. Because UV spectra are so featureless, it is common practice to describe a spectrum like that of cholesta-3,5-diene (Fig. 14.2) as λ_{max} = 234 nm (ε = 20,000) and not bother to reproduce the actual spectrum.

ε, extinction coefficient

The extinction coefficients of conjugated dienes and enones are in the range of 10,000–20,000, so only very dilute solutions are needed for spectra. In the example in Figure 14.2, the absorbance at the tip of the peak is 1.2, and the path length is the usual 1 cm; so the molar concentration, c, needed for this spectrum is 6×10^{-5} mol/L, which is 0.221 mg/10 mL of solvent. The usual laboratory balance cannot accurately weigh such small quantities; therefore sample preparation usually requires the quantitative serial dilution of more concentrated solutions.

The path length (*l*) is the distance (in centimeters) that light travels through a sample.

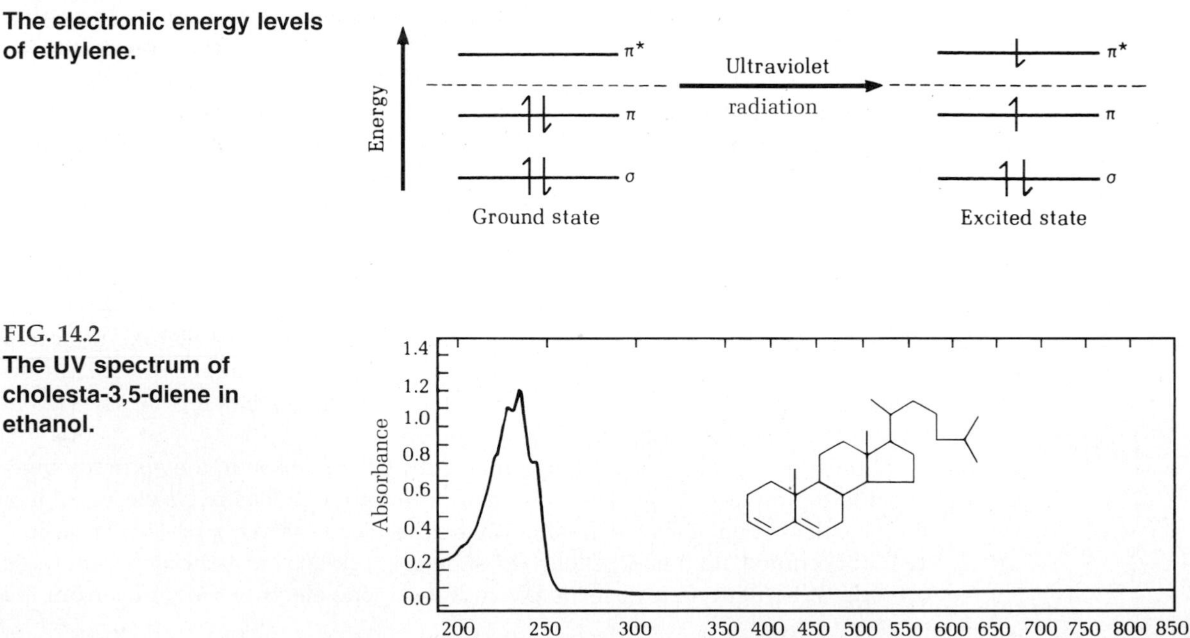

FIG. 14.1
The electronic energy levels of ethylene.

FIG. 14.2
The UV spectrum of cholesta-3,5-diene in ethanol.

***Spectro* grade solvents**

The usual solvents for UV spectroscopy are 95% ethanol, methanol, water, and also saturated hydrocarbons such as hexane, trimethylpentane, and isooctane. The three hydrocarbons are often especially purified to remove impurities that absorb in the UV region. Any transparent solvent can be used for spectra in the visible region.

UV quartz cells are expensive; handle with care.

Sample cells for spectra in the visible region are made of glass or clear plastic, but UV cells must be composed of the more expensive fused quartz because glass absorbs UV radiation. The cells and solvents must be clean and pure because very little of a substance produces a UV spectrum. A single fingerprint will give a spectrum!

Ethylene has $\lambda_{max} = 163$ nm ($\varepsilon = 15{,}000$), and butadiene has $\lambda_{max} = 217$ nm ($\varepsilon = 20{,}900$). As the conjugated system is extended, the wavelength of maximum absorption moves to longer wavelengths (toward the visible region). For example, lycopene, with 11 conjugated double bonds, has $\lambda_{max} = 470$ nm ($\varepsilon = 185{,}000$; Fig. 14.3). Because lycopene absorbs blue visible light at 470 nm, the substance appears bright red. It is responsible for the color of tomatoes; its isolation and analysis are described in Chapters 8 and 9.

Woodward and Fieser rules for dienes and dienones.

The wavelengths of maximum absorption of conjugated dienes and polyenes and conjugated enones and dienones are given by the Woodward and Fieser rules (Tables 14.1 and 14.2). The application of the rules is demonstrated by the spectra of pulegone (1) and carvone (2) in Figure 14.4. The solvent correction is given in Table 14.3. The calculations are given in Tables 14.4 and 14.5.

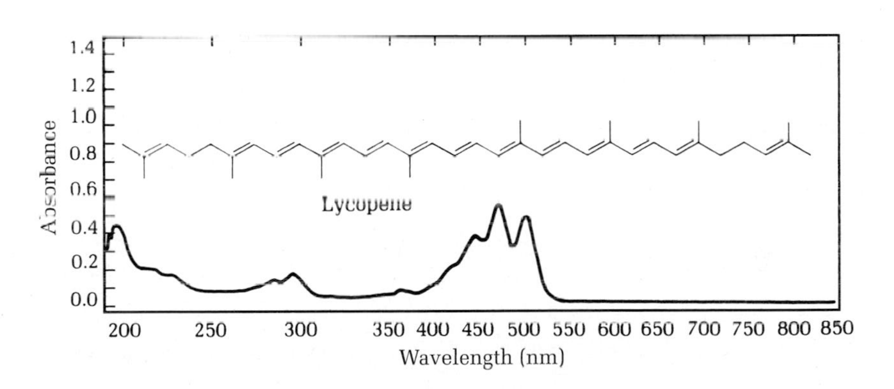

FIG. 14.3
The UV/Vis spectrum of lycopene in isooctane.

TABLE 14.1 *Rules for Predicting the λ_{max} for Conjugated Dienes and Polyenes*

	Increment (nm)
Parent acyclic diene (butadiene)	217
Parent heteroannular diene	214
Double bond extending the conjugation	30
Alkyl substituent or ring residue	5
Exocyclic location of double bond to any ring	5
Groups: OAc, OR	0
Solvent correction (see Table 14.3)	( )
$\lambda_{max}^{EtOH} =$	Total

TABLE 14.2 *Rules for Predicting λ_{max} for Conjugated Enones and Dienones:*

$$\beta\text{—}\overset{\beta}{\text{C}}\text{=}\overset{\alpha}{\text{C}}\text{—}\overset{R}{\text{C}}\text{=O} \quad \text{and} \quad \delta\text{—}\overset{\delta}{\text{C}}\text{=}\overset{\gamma}{\gamma}\text{—}\overset{\beta}{\text{C}}\text{=}\overset{\alpha}{\text{C}}\text{—}\overset{R}{\text{C}}\text{=O}$$

	Increment (nm)
Parent α,β-unsaturated system	215
Double bond extending the conjugation	30
R (alkyl or ring residue), OR, $OCOCH_3$	
α	10
β	12
γ, δ and higher	18
α-Hydroxyl, enolic	35
α-Cl	15
α-Br	23
Exocyclic location of double bond to any ring	5
Homoannular diene component	39
Solvent correction (see Table 14.3)	( )
λ_{max}^{EtOH} =	Total

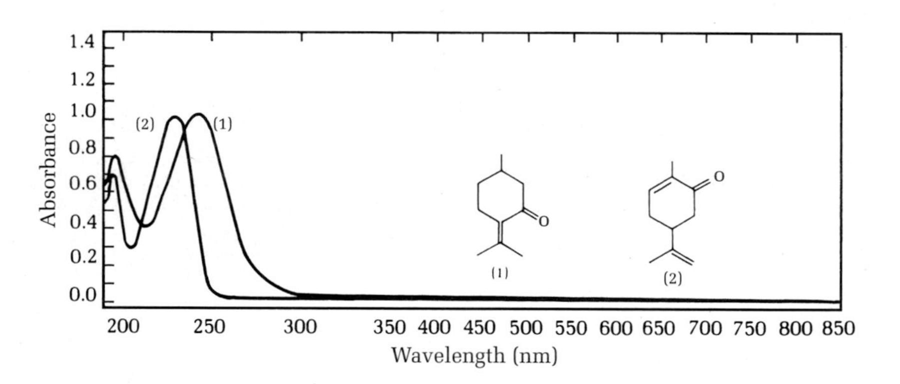

FIG. 14.4
The UV spectra of (1) pulegone and (2) carvone in hexane.

TABLE 14.3 *Solvent Correction*

Solvent	*Factor for Correction to Ethanol*
Hexane	+11
Ether	+7
Dioxane	+5
Chloroform	+1
Methanol	0
Ethanol	0
Water	–8

TABLE 14.4 *Calculation of λ_{max} for Pulegone (See Fig. 14.4)*

Parent α,β-unsaturated system	215 nm
α-Ring residue, R	10
β-Alkyl group (two methyls)	24
Exocyclic double bond	5
Solvent correction (hexane)	–11
Calculated λ_{max} =	243 nm; found = 244 nm

TABLE 14.5 *Calculation of λ_{max} for Carvone (See Fig. 14.4)*

Parent α,β-unsaturated system	215 nm
α-Alkyl group (two methyls)	10
β-Ring residue	12
Solvent correction (hexane)	–11
Calculated λ_{max} =	226 nm; found = 229 nm

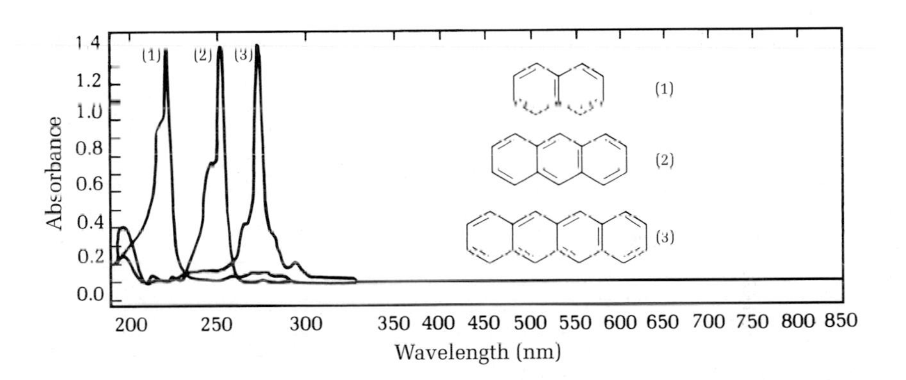

FIG. 14.5
The UV spectra of (1) naphthalene, (2) anthracene, and (3) tetracene.

No simple rules exist for the calculation of aromatic ring spectra, but several generalizations can be made. From Figure 14.5, it is obvious that as polynuclear aromatic rings are extended linearly, λ_{max} shifts to longer wavelengths.

Effect of acid and base on λ_{max}.

As alkyl groups are added to benzene, λ_{max} shifts from 255 nm for benzene to 261 nm for toluene to 272 nm for hexamethylbenzene. Substituents bearing non-bonding electrons also cause shifts of λ_{max} to longer wavelengths—for example, from 255 nm for benzene to 257 nm for chlorobenzene, 270 nm for phenol, and 280 nm for aniline (ε = 6200–8600). That these effects are the result of the interaction of the π-electron system with nonbonded electrons is seen dramatically in the spectra of vanillin and the anion derived by deprotonation of its phenolic OH (Fig. 14.6). The two additional nonbonding electrons in the anion cause λ_{max} to shift from 279 nm to 351 nm and ε to increase. Protonation of the non-bonding electrons on the nitrogen of aniline to give the anilinium cation causes λ_{max} to decrease from 280 nm to 254 nm (Fig. 14.7). These changes of λ_{max} as a function of pH have obvious analytical applications.

FIG. 14.6
The UV spectra of (1) neutral vanillin and (2) the anion of vanillin.

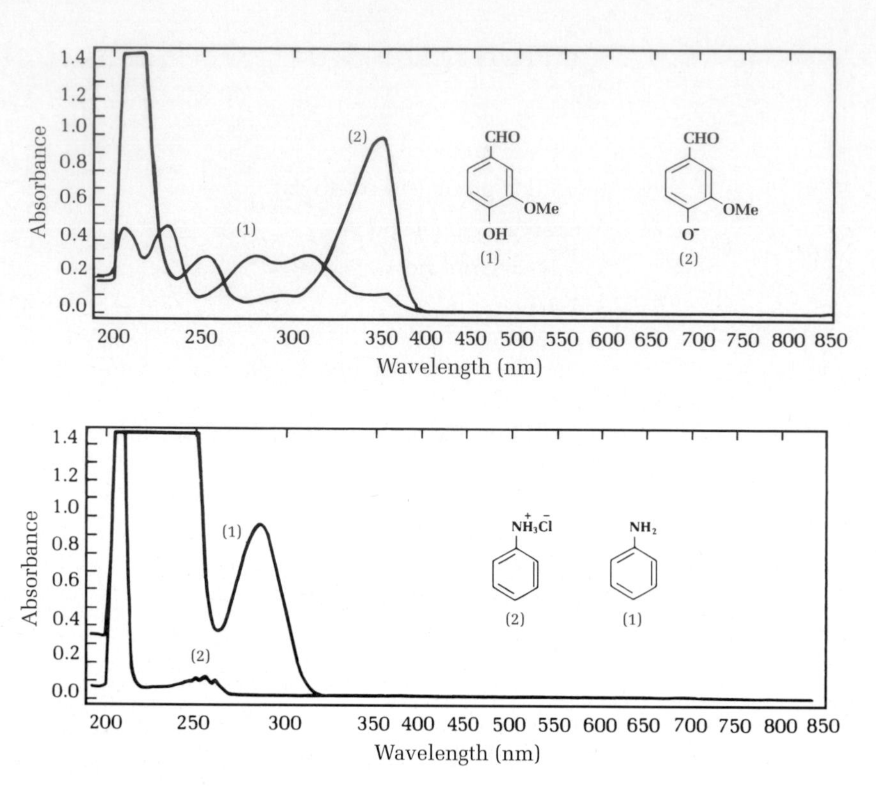

FIG. 14.7

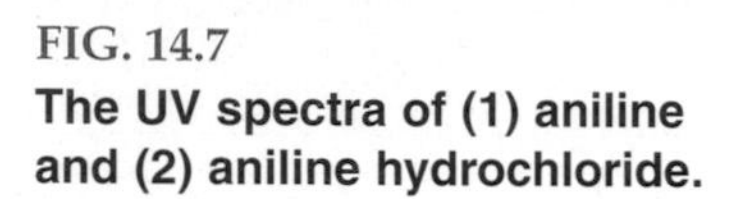
The UV spectra of (1) aniline and (2) aniline hydrochloride.

Intense bands result from π-π conjugation of double bonds and carbonyl groups with the aromatic ring. Styrene, for example, has λ_{max} = 244 nm (ε = 12,000), and benzaldehyde has λ_{max} = 244 nm (ε = 15,000).

EXPERIMENT

UV SPECTRUM OF AN UNKNOWN ACID, BASE, OR NEUTRAL COMPOUND

Determine whether an unknown compound obtained from your instructor is acidic, basic, or neutral from UV spectra in a neutral solvent such as pure ethanol or methanol, as well as under acidic conditions (add 1 drop of 5% HCl to the solution in the cuvette and mix) and basic conditions (add 1 drop of 1 M NaOH to the acidic solution in the cuvette, mix, and check that the pH is basic).

Cleaning Up. Because UV samples are extremely dilute solutions in ethanol or methanol, they can normally be flushed down the drain.

REFRACTIVE INDICES

The *refractive index*, symbolized by n, is a physical constant that, like the boiling point, can be used to characterize liquids. It is the ratio of the velocity of light traveling in air to the velocity of light moving in the liquid (Fig. 14.8). It is also equal to the ratio of the sine of the angle of incidence ϕ to the sine of the angle of refraction ϕ':

$$n = \frac{\text{velocity of light in air}}{\text{velocity of light in liquid}} = \frac{\sin \phi}{\sin \phi'}$$

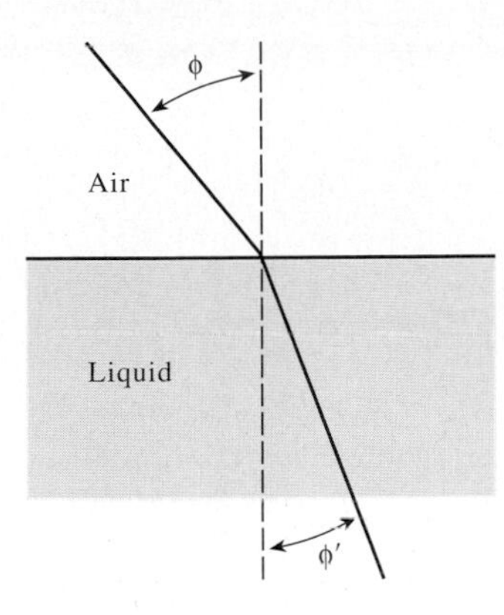

FIG. 14.8
The refraction of light.

The angle of refraction is also a function of temperature and the wavelength of light (consider the dispersion of white light by a prism). Because the velocity of light in air (strictly speaking, a vacuum) is always greater than that through a liquid, the refractive index is a number greater than 1, for example, hexane, $n_D^{20} = 1.3751$; diiodobenzene, $n_D^{20} = 1.7179$. The superscript 20 indicates that the refractive index was measured at 20°C, and the subscript D refers to the yellow D-line from a sodium vapor lamp, which produces light with a wavelength of 589 nm.

The measurement is made on a refractometer using a few drops of liquid. Compensation is made within the instrument for the fact that white light, not sodium vapor light, is used, and a temperature correction must be applied to the observed reading by using the following equation which automatically compensates for temperatures higher or lower than 20°C:

$$n_D^{20} = n_D^t + 0.00045(t - 20°C)$$

The refractive index can be determined to 1 part in 10,000, but because the value is quite sensitive to impurities, full agreement in the literature regarding the last figure does not always exist. For this reason, the refractive indices in this book have been rounded to the nearest 1 part in 1000, as have the refractive indices reported in the Aldrich catalog of chemicals. To master the technique of using the refractometer, measure the refractive indices of several known, pure liquids before measuring an unknown.

Specialized hand-held refractometers that read over a narrow range are used to determine the concentration of sugar, salt, or alcohol in water.

USING A REFRACTOMETER

Refractometers come in many designs. In the most common, the Abbé design (Fig. 14.9), two or three drops of the sample are placed on the measuring prism using a polyethylene Beral pipette (to avoid scratching the prism face). The illuminating prism is closed, and the lamp is turned on and positioned for maximum brightness as seen through the eyepiece. If the refractometer is set to a nearly correct value, then a partially gray image will be seen, as shown in Figure 14.10a. Turn the index knob so that the line separating the dark and light areas is at the crosshairs, as shown in Figure 14.10b. Sometimes the line separating the dark and light areas is fuzzy and colored (Fig. 14.10c). Turn the chromatic adjustment until the demarcation line is sharp and colorless. Then read the refractive index. On a newer instrument, press a button or hold down the on/off switch to light up the scale in the field of vision or activate the digital readout. On older models, read the refractive index through a separate eyepiece. Read the temperature on the thermometer attached to the refractometer, and make the appropriate temperature correction to the observed index of refraction. When the measurement is completed, open the prism and wipe off the sample with lens paper, using ethanol, acetone, or hexane only as necessary.

For most organic liquids, the index of refraction decreases approximately 0.00095 ± 0.0001 for every °C increase in temperature.

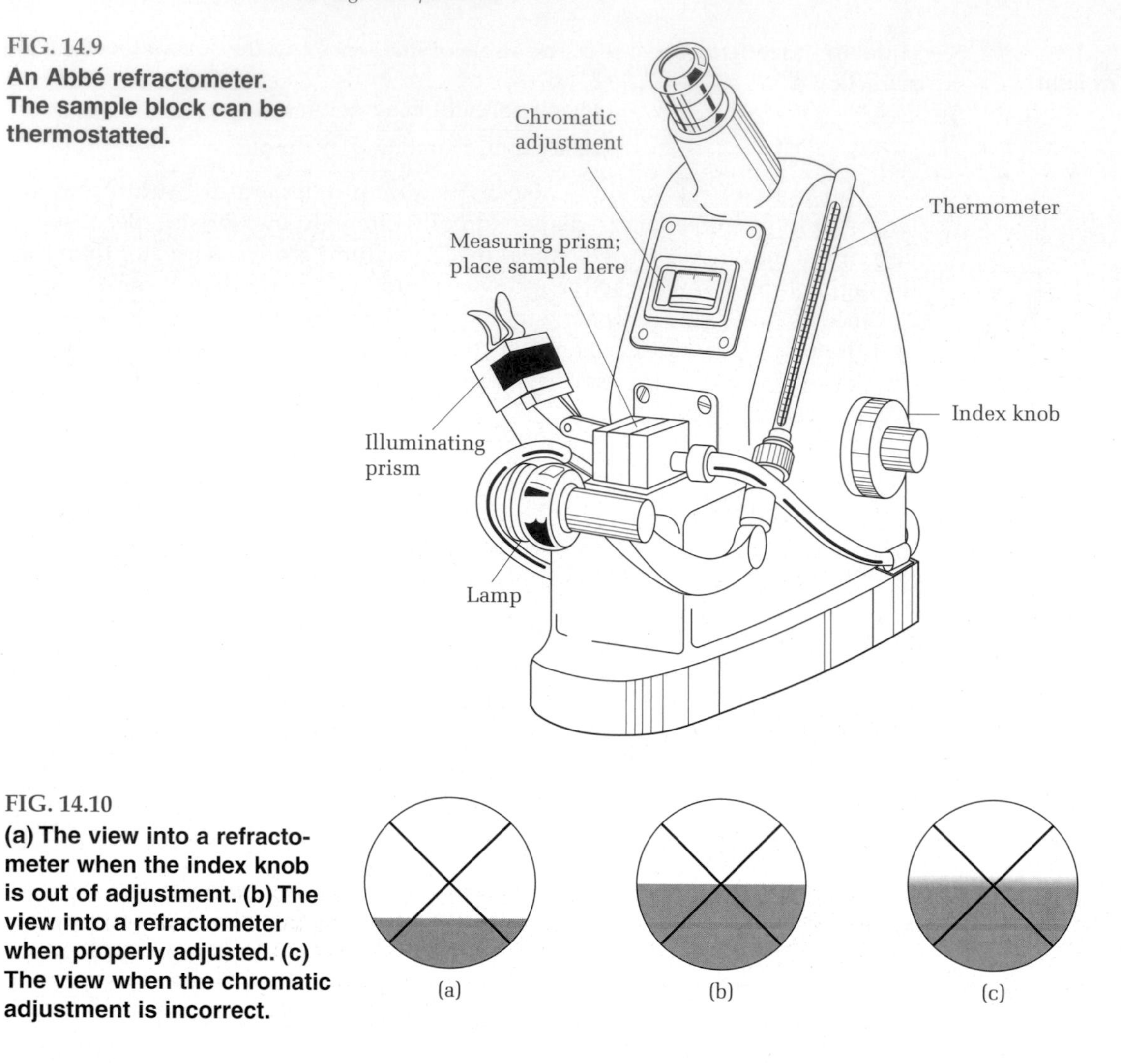

FIG. 14.9
An Abbé refractometer. The sample block can be thermostatted.

FIG. 14.10
(a) The view into a refractometer when the index knob is out of adjustment. (b) The view into a refractometer when properly adjusted. (c) The view when the chromatic adjustment is incorrect.

QUALITATIVE INSTRUMENTAL ORGANIC ANALYSIS

As indicated in many of the previous chapters, physical characterization and structural elucidation are major activities in organic chemistry. In many areas, such as drug metabolism studies and forensic or environmental chemistry, only milligram or microgram quantities of organic compounds are available. Fortunately, today's instrumental methods have the requisite sensitivity to meet this challenge. It may be possible to establish the structure of a compound on the basis of spectra alone (IR, NMR, MS [mass spectrometry], and/or UV), but often these spectra must be supplemented with other information about the unknown: physical state, solubility, and confirmatory tests for functional groups. Before spectra are run, other information about the unknown must be obtained. Is it pure or a mixture (test by thin-layer [TLC], gas, or liquid chromatography)? Once a substance is known to be homogeneous, it can often be identified by spectroscopy alone. What are its physical properties: melting point, boiling point, and color and solubility in various solvents, such as those commonly used in NMR? A mass spectrum can determine a compound's molecular weight and, if measured with sufficient accuracy, identify the elements present and the molecular formula.

For the millions of organic compounds that have been synthesized or isolated from nature, spectral data are included when they are reported in the chemical literature. Thousands of the more common chemicals that can be purchased or easily synthesized have had their IR, NMR, MS, and UV spectra printed in multivolume collections that can be searched by compound class, formula, or spectral features. Today, these collections have been converted into digital form that can be searched by computer and compared with the spectra obtained for an unknown, yielding a list of all closely matching compounds. If two substances have identical IR spectra, they can be regarded as identical. Such is not always true of other spectra. Many substances can have identical or nearly identical UV spectra, and it is possible for the MS or NMR spectra for two different substances to be almost identical. When new substances are encountered in research laboratories, their spectra can be compared with those in commercial databases. Even though the particular new compound is not represented in those databases, a list of very similar substances can be generated, which will guide the determination of the structure of the new substance.

EXPERIMENT

IDENTIFYING AN UNKNOWN COMPOUND

Most students consider the identification of an unknown organic compound to be one of the most enjoyable and challenging organic lab activities. In this laboratory course you may receive an "unknown" substance, which, of necessity, is usually a commercially available compound. At least initially, you may not be given access to a commercial spectral database for searching and comparison, but you will use the skills you have learned to interpret the spectral data you acquire. After you have done the best you can with your IR, NMR, and UV spectra and have arrived at a short list of possible compounds, you will be provided with the mass spectrum of your unknown. Interpretation of this spectrum usually eliminates most of the candidate structures and should allow you to complete the identification of the unknown.

Physical State

Check for Sample Purity

Distill or recrystallize as necessary. A constant boiling point and sharp melting point are indicators of purity, but beware of azeotropes and eutectic substances. Check homogeneity by TLC, GC, or HPLC.

Note the Color

Common colored compounds include nitro and nitroso compounds (yellow), diketones (yellow), quinones (yellow to red), azo compounds (yellow to red), and polyconjugated olefins and ketones (yellow to red). Phenols and amines are often brown to dark purple because of traces of air oxidation products.

Note the Odor

Some liquid and solid amines are recognizable by their fishy odors; esters are often pleasantly fragrant. Alcohols, ketones, aromatic hydrocarbons, and aliphatic olefins have characteristic odors. On the unpleasant side in terms of odor are thiols, isonitriles, and low-molecular-weight carboxylic acids.

Ignition Test

Heat a small sample on a spatula; first hold the sample near the side of a microburner to see if it melts normally and then burns. Then heat directly in the flame. If a large ashy residue is left after ignition, the unknown is probably a metal salt. Aromatic compounds often burn with a smoky flame.

Beilstein Test for Halogens

Although the presence of a halogen can usually be determined by mass spectrometry, this test is so simple that it can easily be run to confirm the MS data. Heat the tip of a copper wire in a burner flame until no further coloration of the flame is noticed. Allow the wire to cool slightly, then dip it into the unknown (solid or liquid), and again heat it in the flame. A green flash is indicative of chlorine, blue-green of bromine, and blue of iodine; fluorine is not detected because copper fluoride is not volatile. The Beilstein test is very sensitive; halogen-containing impurities may give misleading results. Run the test on a compound known to contain a halogen for comparison to your unknown.

Spectra

Obtain IR and NMR spectra following the procedures in Chapters 11 and 12. If these spectra indicate the possible presence of conjugated double bonds, aromatic rings, or conjugated carbonyl compounds, obtain the UV spectrum following the procedures in this chapter. Interpret the spectra as fully as possible by investigating the reference sources cited at the end of the spectroscopy chapters. Once you have interpreted these spectra to the best of your ability, you may obtain or be provided with a mass spectrum (Chapter 13), which should make a secure identification of the unknown possible.

Solubility Tests

There is a logical sequence for determining the solubility of an organic compound in order, for example, to dissolve it for spectroscopic analysis. In the process of determining the solubility of the unknown, much information about the nature of the compound can be obtained.

Like dissolves like.

Like dissolves like; a substance is most soluble in that solvent to which it is most closely related in structure. This statement serves as a useful classification scheme for all organic molecules. The solubility measurements are done at room temperature with 1 drop of a liquid or 5 mg of a solid (finely crushed) and 0.2 mL of solvent. The mixture should be rubbed with a rounded stirring rod and agitated vigorously. Lower members of a homologous series are easily classified; higher members become more like the hydrocarbons from which they are derived. If a very small amount of the sample fails to dissolve when added to some of the solvent, it can be considered insoluble. Conversely, if several portions dissolve readily in a small amount of the solvent, the substance is obviously soluble. If an unknown seems to be more soluble in dilute acid or base than in water, the observation can be

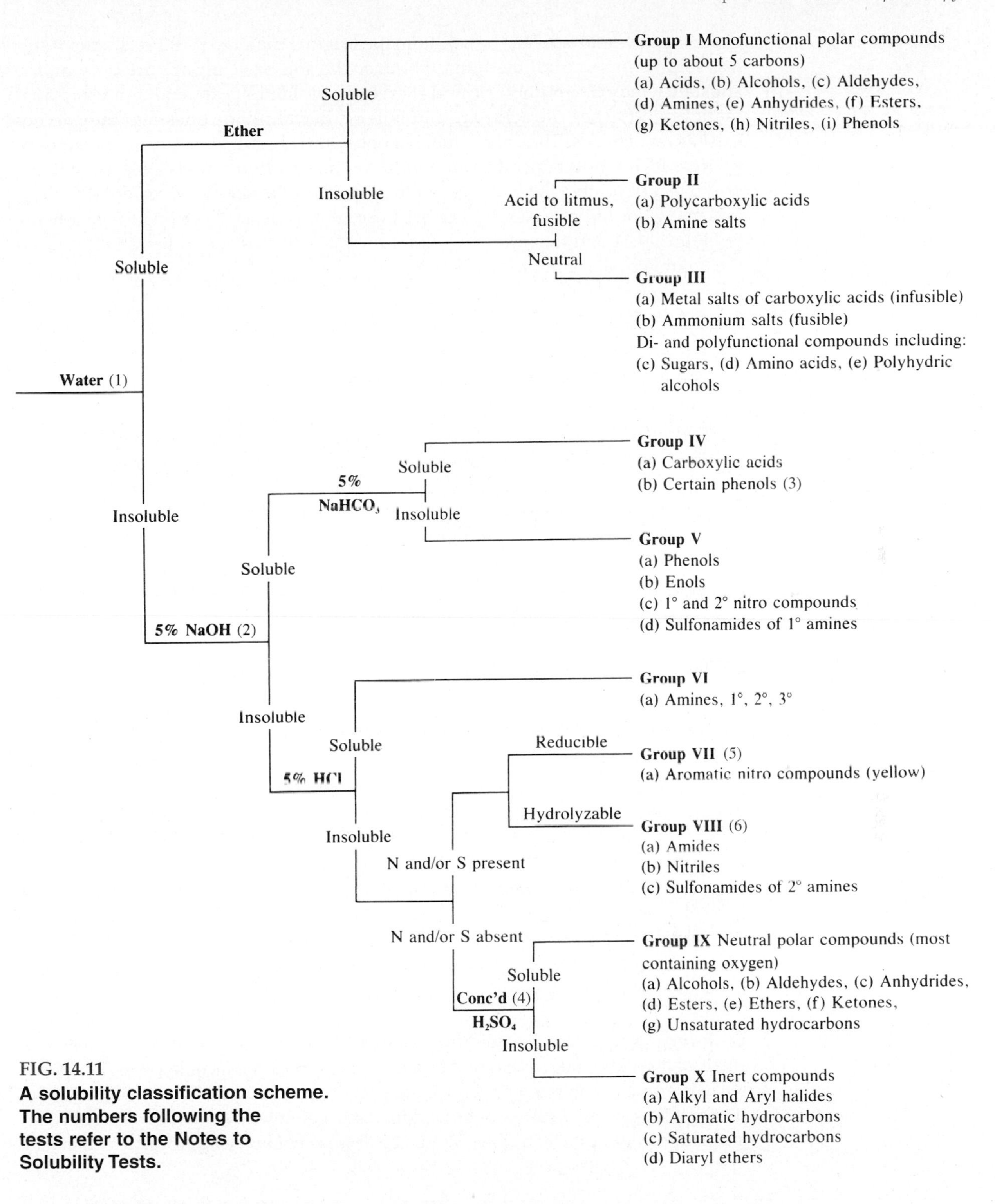

FIG. 14.11
A solubility classification scheme. The numbers following the tests refer to the Notes to Solubility Tests.

confirmed by neutralizing the solution; the original material will precipitate if it is less soluble in a neutral medium. If both acidic and basic groups are present, the substance may be amphoteric and therefore soluble in both acid and base. Aromatic aminocarboxylic acids are amphoteric, like aliphatic ones, but they do not exist as zwitterions. They are soluble in both dilute hydrochloric acid and sodium hydroxide, but not in bicarbonate solution. Aminosulfonic acids exist as zwitterions; they are soluble in alkali but not in acid. The solubility tests are not infallible, and many borderline cases are known. Carry out the tests according to the scheme in Figure 14.11 and the following Notes to Solubility Tests, and tentatively assign the unknown to one of the groups I–X.

Notes to Solubility Tests

1. Groups I, II, III (soluble in water). Test the solution with pH paper. If the compound is not easily soluble in cold water, treat it as water insoluble but test with indicator paper.
2. If the substance is insoluble in water but dissolves partially in 5% sodium hydroxide, add more water; the sodium salts of some phenols are less soluble in alkali than in water. If the unknown is colored, be careful to distinguish between the *dissolving and the reacting* of the sample. Some quinones (colored) *react* with alkali and give highly colored solutions. Some phenols (colorless) *dissolve and then* become oxidized to give colored solutions. Some compounds (e.g., benzamide) are hydrolyzed with such ease that careful observation is required to distinguish them from acidic substances.
3. Nitrophenols (yellow), aldehydophenols, and polyhalophenols are sufficiently strongly acidic to react with sodium bicarbonate.
4. Oxygen- and nitrogen-containing compounds form oxonium and ammonium ions in concentrated sulfuric acid and dissolve.
5. On reduction in the presence of hydrochloric acid, nitro compounds form water-soluble amine hydrochlorides. Dissolve 250 mg of tin(II) chloride in 0.5 mL of concentrated hydrochloric acid, add 50 mg of the unknown, and warm. The material should dissolve with the disappearance of the color and give a clear solution when diluted with water.
6. Most amides can be hydrolyzed by short boiling with a 10% sodium hydroxide solution; the acid dissolves with evolution of ammonia. Reflux 100 mg of the sample and a 10% sodium hydroxide solution for 15–20 minutes. Test for the evolution of ammonia, which confirms the elementary analysis for nitrogen and establishes the presence of a nitrile or amide.

Cleaning Up. Because the quantities of material used in these tests are extremely small, and because no hazardous substances are handed out as unknowns, it is possible to dilute the material with a large quantity of water and flush it down the drain, unless it is very smelly or is dissolved in water-insoluble nonhalogenated or halogenated solvents, in which case it should be disposed of in the appropriate containers.

QUESTIONS

1. Calculate the UV absorption maximum for 2-cyclohexene-1-one.
2. Calculate the UV absorption maximum for 3,4,4-trimethyl-2-cyclohexene-1-one.

3. Calculate the UV absorption maximum for

4. What concentration, in g/mL, of a substance with MW 200 should be prepared to give an absorbance value equal to 0.8 if the substance has $\varepsilon = 16{,}000$ and a cell with a path length of 1 cm is employed?

5. When borosilicate glass (Kimax, Pyrex), n_D^{20} 1.474, is immersed in a solution having the same refractive index, it is almost invisible. Soft glass with n^{20} 1.52 is quite visible. This is an easy way to distinguish between the two types of glass. Calculate the mole percents of toluene and heptane that will have a refractive index of 1.474, assuming a linear relationship between the refractive indices of the two.

REFERENCES

See *Web Links*

Gillam, Albert. E., Edward S. Stern, and Christopher J. Timmons, *Gillam and Stern's Introduction to Electronic Absorption Spectroscopy in Organic Chemistry*, 3rd. ed. London: Edward Arnold, 1970.

Jaffe, Hans H., and Milton Orchin, *Theory and Applications of Ultraviolet Spectroscopy*. New York: John Wiley, 1962.

Lambert, Joseph B., Herbert F. Shurvell, David A. Lightner, and Robert G. Cooks, *Organic Structural Spectroscopy*. Upper Saddle River, NJ: Prentice-Hall, 1998.

Rao, Chintamani N. R., *Ultraviolet and Visible Spectroscopy: Chemical Applications*, 3rd ed. London: Butterworths, 1975.

Silverstein, Robert M., Francis X. Webster, and David Kiemle, *Spectrometric Identification of Organic Compounds*, 7th ed.[1] New York: Wiley, 2005.

Williams, Dudley H., and Ian Fleming, *Spectroscopic Methods in Organic Chemistry*, 5th ed. New York: McGraw-Hill, 1995.

[1]This book includes IR, UV, and NMR spectra.

15 CHAPTER

Computational Chemistry

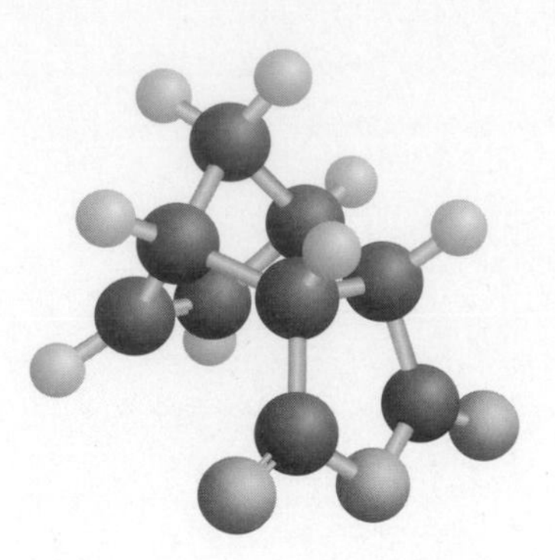

The computer has, as in so many other aspects of modern life, changed the way we do and visualize organic chemistry. The ability to predict the outcome, in a quantitative sense, of a chemical reaction and to visualize organic molecules in three dimensions, once the province of specialists, can now be carried out on a desktop computer. This computational approach to chemistry is being used extensively in drug design, protein mutagenesis, biomimetics, catalysis, studies of DNA-protein interactions, and the determination of structures of molecules using nuclear magnetic resonance (NMR) and infrared (IR) spectroscopy, to name just a few applications.

GRAPHICAL AND MATHEMATICAL MODELS

Physical properties and chemical reactivity are a direct consequence of molecular structure, commonly represented by graphical models showing atomic geometry and electron distribution. Models can be simple or complex; although all models are simplified representations of reality, it is not absolutely necessary that a particular model correspond exactly to reality. The level of complexity depends on how much information is needed to understand and predict chemical behavior. For example, all of the models for methanol shown in Figure 15.1 are correct, but certain models provide more information than others. What is essential is that models allow chemists to predict some measurable quantity with sufficient accuracy. For example, the simplest model of methanol in Figure 15.1, the formula CH_4O, provides all the information necessary to calculate the amount of CO_2 that will be released if one burns a certain quantity of methanol. However, this simple model is not adequate for predicting the hydrogen bonding that occurs between methanol molecules. To do this, a Lewis structure that shows the nonbonding pairs of electrons on the oxygen is a more appropriate model. Chemists often use Lewis structures and think about these representations in terms of the valence bond theory. A Lewis structure shows a molecule as a skeleton of atoms with their accompanying *valence shell electrons*. Valence shell electrons are shown as either electron pair bonds connecting atoms or as nonbonded pairs localized on atoms.

Lewis structures are simple to draw, and the information conveyed is usually sufficient, within the rules of *valence shell electron pair repulsion (VSEPR) theory*, to obtain an approximate prediction of the three-dimensional shape of a molecule. Lewis structures are also usually adequate to give a general picture of the electronic changes that occur during the course of a reaction, and are used to write mechanisms involving electron "pushing," which is represented by arrows. Simple hand-drawn structures provide enough information and insight that most of the

FIG. 15.1
Some different models for visualizing the structure and understanding the physical and chemical behavior of methanol. The method for producing the models is also given in most cases.

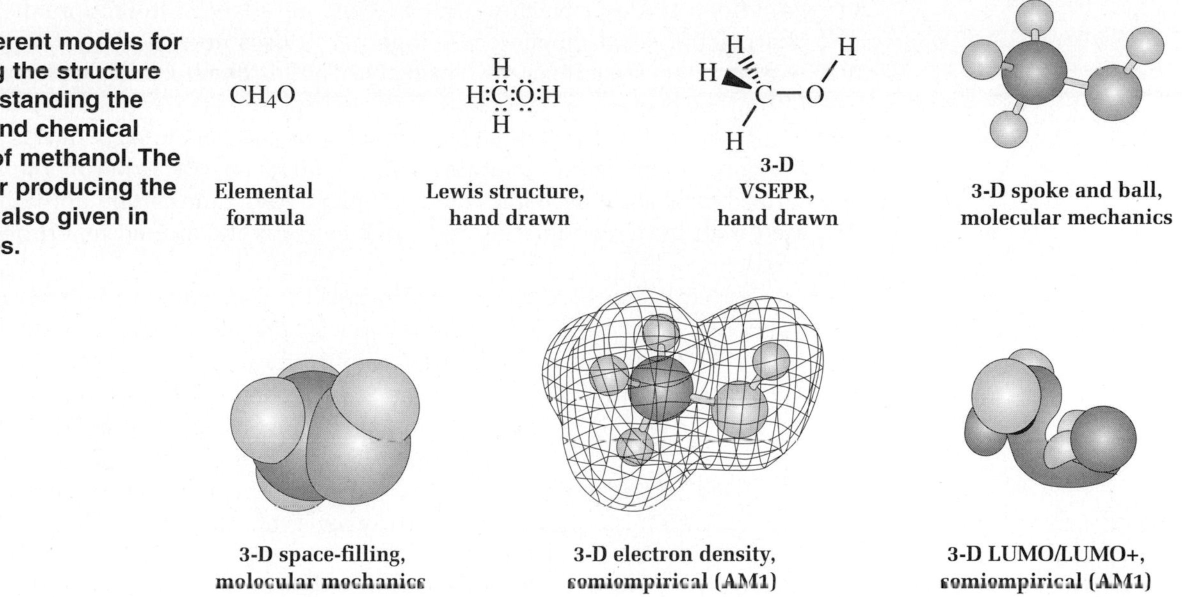

organic structure and reactivity of a molecule can be understood using just these. However, hand-drawn structures do not provide detailed quantitative information about bond lengths, angles, and strengths; ionization potentials; electron affinities; or dipoles and molecular energies. There are many times when such information is useful in solving a problem.

By way of analogy, if only the skeletal structure of an animal were presented to an untrained person, that person would probably make mistakes when trying to predict the animal's complete shape and behavior. A trained anatomist, however, will more likely be able to make reasonable predictions based on past experience. Hand-drawn organic chemical structures present a similar situation. Novices may have more difficulty making correct predictions, whereas a trained chemist is more likely to derive useful information about the structure and properties of a molecule based on its Lewis structure. When drawing accurate structures and predicting physical properties and chemical behavior as determined by molecular energies, however, both the expert and the novice alike encounter problems. Clearly, the development of more numerically accurate, mathematical models would help. They yield tables of numbers such as bond lengths, molecular energies, and orbital information that can be used directly or converted to graphical representations. These more accurate and detailed graphical structures can be rotated in any direction on a computer screen and would be hard to draw by hand. They help to visualize complex numerical data, providing much greater insight into molecular and reaction chemistry.

There are two major types of mathematical models. In the *molecular mechanics* model, molecules are made up of atoms and bonds (as opposed to nuclei and electrons), and the model uses the ball and spring equations developed in physics for classical mechanics. Atom positions are adjusted to best match known empirical parameters, such as bond lengths and angles derived from experimental measurement. In the *quantum chemical* model, molecules are considered to be positively charged atomic nuclei in a cloud of negative electrons. This model relies on the pure theory of quantum physics and mathematical equations to describe the attractive and repulsive interactions between the charges that hold the molecule together. It uses no experimental parameters, working only from theoretical principles, and is therefore also called the *ab initio* ("from the beginning") model.

There is also a third model, called the *semiempirical* model, which is a hybrid of the pure theory-based quantum chemical model and the experimentally parameterized molecular mechanics model. It helps overcome certain limitations associated with these two models and can be effectively applied to a broad range of problems.

You do not need to be a highly trained theoretician to run these calculations, but you should be critical and even skeptical of results produced by computational chemistry. One of the major limitations of all these computations is that they usually assume noninteracting molecules, in other words, gaseous molecules in a vacuum. Obviously, reactions run in solvents involve solvent-molecule interactions that can significantly change the experimentally measured energetics from those predicted by calculation. (There are some models that can estimate aqueous solvent effects.) It is quite possible to run a series of calculations that produce results that are absolutely meaningless. Therefore, it is imperative that you use your knowledge of organic chemistry and repeatedly ask, "Are these results consistent with what is observed experimentally? Do they make sense using the traditional hand-drawn, arrow-pushing chemistry taught in textbooks?" If they don't, it probably means that you have chosen the wrong computational model to do the calculation. Computers just generate numbers; they cannot tell you whether the chosen method is valid for the problem being studied. People have to make these decisions to correctly solve real problems.

The biggest challenge in computational chemistry lies in understanding the capabilities and limitations of each model. These will be emphasized in the following sections. One also has to choose which computational method will provide useful information in a minimum amount of time. Ab initio calculations on larger molecules can take hours and even days to complete. If the same information can be derived from a semiempirical calculation in one-tenth the time, the fancier method is unnecessary and wasteful of computer resources.

MOLECULAR MECHANICS MODELS

The simplest, fastest, and easiest-to-understand calculations involve molecular mechanics, and the results from this approach are usually adequate for many of the modeling exercises in this text. Molecular mechanics is used to calculate the structures of molecules based not on a complete solution of the Schrödinger equation but on a mechanical model for molecules. This model regards molecules as being masses (atoms) connected by forces somewhat like springs (bonds) at certain preferred lengths and angles.

Every organic chemist routinely uses a set of models to examine the three-dimensional structures of molecules because it is difficult to visualize these from a two-dimensional drawing. Molecular mechanics programs allow one to determine quickly the best (lowest-energy) structure for a molecule. Experienced chemists have been doing this intuitively for years, but simple plastic or metal models can lead to incorrect structures in many cases, and, of course, one obtains no quantitative data regarding the stability of one structure compared to another.

Modern computational chemistry software makes it quite easy to draw an approximate structure of a molecule on the computer screen. After the rough structure is drawn and converted to a digital format, a molecular mechanics program such as SYBYL or MMFF automatically steps through a series of iterative calculations. Such programs adjust every bond angle, bond length, dihedral angle, and van der Waals interaction to produce a new structure, often called the *equilibrium geometry*, having the lowest possible *strain energy*. The absolute value of the strain energy often differs from one program to another, and so it can be used only in a relative sense to determine the difference in energies between two possible structural geometries or conformations. Molecular mechanics assumes that all bond properties are the same in all molecules. For exceptional molecules such as cyclopropane and cyclobutane, it merely substitutes new parameters for these unusual rings. The experimental data on which many of the parameters for molecular mechanics rest come from X-ray diffraction studies of crystals, electron diffraction studies of gases, and microwave spectroscopy. Even if an ab initio quantum chemical calculation is ultimately planned, the starting structure should always be of a minimal energy that is obtained by using the faster molecular mechanics program before going to the next level.

Molecular mechanics calculations do *not* provide any information about the locations of electrons and other electronic properties such as ionization potentials, and they *cannot* be used to explore processes that involve bond breaking or bond making. This computational method cannot evaluate the energy of a transition state, and thus the activation energy from which reaction rates can be derived, nor can it calculate the heat of formation or of combustion. These sorts of calculations are the province of the quantum chemical computational models; unlike them, molecular mechanics calculations are empirical in nature and are not derived from first principles.

If a molecule has a single minimum-energy conformation and that conformation has been found, then the shape of the final molecule can be used to predict or confirm approximate vicinal (three-bond) proton-proton NMR coupling constants, which are dependent on the dihedral angle between the protons. (Couplings are also dependent to a smaller degree on the electronegativity of substituents, bond angles, etc.)

In a typical empirical force field used in molecular mechanics, the steric energy of a molecule can be represented as the sum of five energies given by the following equation:

$$E_{steric} = E_{str} + E_{bnd} + E_{tor} + E_{oop} + E_{vdW}$$

where E_{str} is the energy needed to stretch or compress a bond. Plastic student molecular models are realistic in that there is no way to adjust this value; in molecular mechanics calculations, one finds that there is a big force constant (more correctly, potential constant) for this parameter. Most sets of molecular models are constructed so that modest bending of bond angles is possible, and similarly, one finds that the force constant for angle bending (E_{bnd}) has an intermediate value. The force constant for torsional motion (E_{tor}) is very small, as seen in the free rotation about bonds in most mechanical models. The out-of-plane bending term (E_{oop}) arises in molecules such as methylene cyclobutene where, without this term, the methylene group would want to be bent out of the plane to relieve angle strain. The energy due to van der Waals interactions (E_{vdW}) is very important. It takes two forms, either attractive or repulsive, depending on internuclear distance.

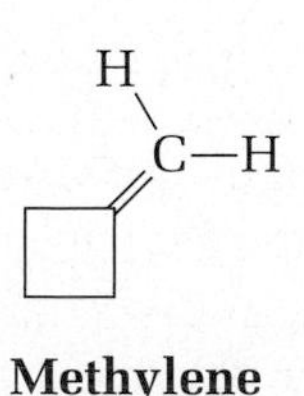

Methylene cyclobutene

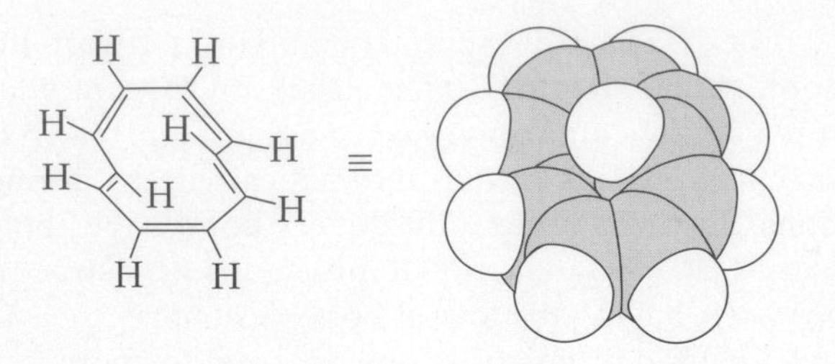

FIG. 15.2
Cyclodecapentene. From this energy-minimized structure, it is easy to see that the molecule cannot be flat, and thus cannot benefit from aromatic stabilization.

Molecular mechanics allows the calculation of the equilibrium geometry of a molecule (Fig. 15.2). The individual mathematical equations describing the mechanics of a molecule are not terribly complex. For a typical organic molecule containing 30–40 atoms, only a few seconds at most are needed to carry out the millions of calculations and summations for the pair-wise interactions between atoms described by these equations.

All of these energies are empirical in nature, not calculated from first principles.

Let us examine in detail the five most important factors entering into molecular mechanics calculations. The bond between two atoms can be regarded just like a spring. It can be stretched, compressed, or bent and, provided these distortions are small, will follow Hooke's law from classical physics.

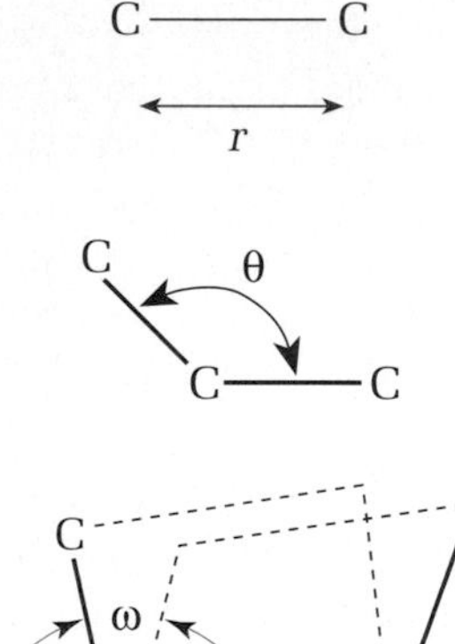

For bond stretching, the energy is given by:

$$E_{str} = (1/2)k_r(r - r_0)^2$$

where E_{str} is the bond stretching energy in kJ/mol, k_r is the bond stretching force constant in kJ/mol · Å^2, r is the bond length, and r_0 is the equilibrium bond length.

Similarly, the energy for bond angle bending (in-plane bending) is given by:

$$E_{bnd} = (1/2)k_\theta(\theta - \theta_0)^2$$

where E_{bnd} is the bond bending energy in kJ/mol, k_θ is the angle-bending force constant in kJ/mol · deg^2, θ is the angle between two adjacent bonds in degrees, and θ_0 is the equilibrium value for the angle between the two bonds in degrees.

The energy for twisting a bond, E_{tor}, is given by:

$$E_{tor} = (1/2)k_\omega[1 - \cos(j\omega)]$$

where E_{tor} is the torsional energy in kJ/mol, k_ω is the torsional force constant in kJ/mol · deg, ω is the torsion angle in degrees, and j is usually 2 or 3 depending on the symmetry of the bond (e.g., twofold in ethylene or threefold in ethane).

The energy term for out-of-plane bending (E_{oop}) is not used very often, but it is a necessary term. It arises in cases such as cyclobutanone, where the C—C—O bond angle is not the sp^2 bond angle of 120° but 133°. To relieve this bond angle strain, the oxygen could tip up out of the plane, but this would require a large amount of energy because of the distortion of the π bonding of the carbonyl. So the force constant for out-of-plane bending is different from that for in-plane bending (bond angle bending; E_{bnd}) discussed earlier.

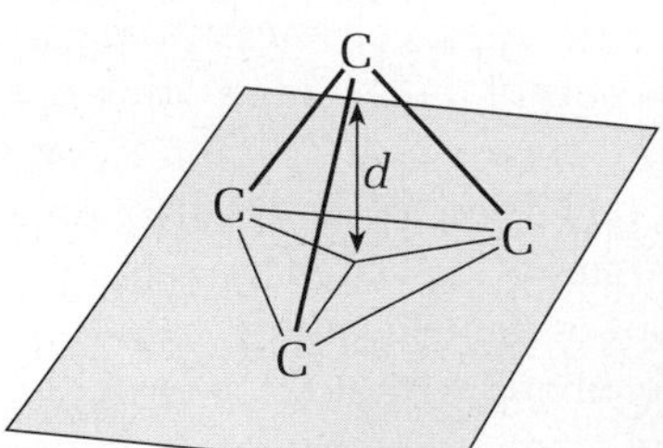

$$E_{oop} = (1/2)k_\delta(d^2)$$

where E_{oop} is the out-of-plane bending energy in kJ/mol, k_δ is the out-of-plane bending constant in kj/mol · Å^2, and d is the height of the central atom in angstroms above the plane of its substituents.

Very important is the van der Waals energy term (E_{vdW}) given by:

$$E_{vdw} = \varepsilon[(r_0/r)^{12} - 2(r_0/r)^6]$$

FIG. 15.3
The van der Waals energy function.

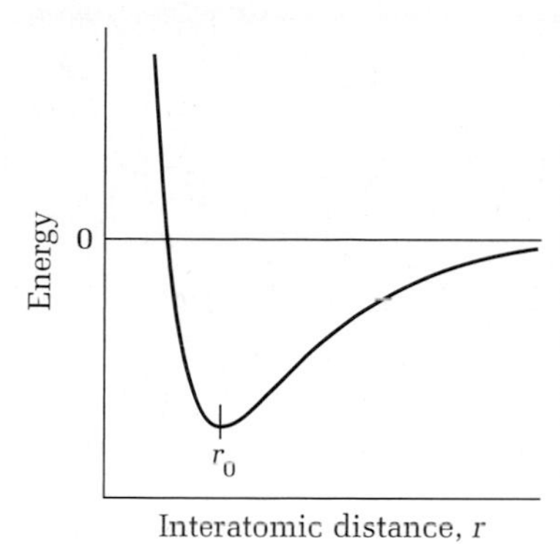

where E_{vdW} is the van der Waals energy in kJ/mol, ε is the van der Waals potential of the two interacting atoms in kJ/mol, r_0 is the equilibrium distance between the two atoms, and r is the interatomic distance in the molecule in angstroms.

This important function is best understood by referring to Figure 15.3, which shows a curve that describes the energy of two atoms as a function of their distance (r) from each other. At very short distances, the two atoms repel each other very strongly. This interaction raises the energy of the system as the 12th power of the interatomic distance. At a slightly longer distance, a weak attraction is described by an r^6 term, and then at large separation, the interaction energy goes asymptotically to zero. The value of ε determines the depth of the potential well. At distance r_0, when the attractive forces overwhelm the repulsive forces, the system has minimum energy. It is r_0 that we call the *bond length.* The molecular mechanics van der Waals term is just an approximation of the ideal curve. It comes close enough and is easy to compute.

Carrying out molecular mechanics calculations on a personal computer is not difficult. One first draws the structure of the molecule on a computer screen in the same way or even using the same program, such as ChemDraw that is used to draw organic structures for publication in reports and papers. This results in a wire model such as the hexagon we associate with cyclohexane with the proper x, y, z coordinates for each carbon. The program can be directed to add hydrogens at all the unfilled valences. Then the molecular mechanics program is told to minimize the structure, and in a few seconds, a minimum energy structure is calculated.

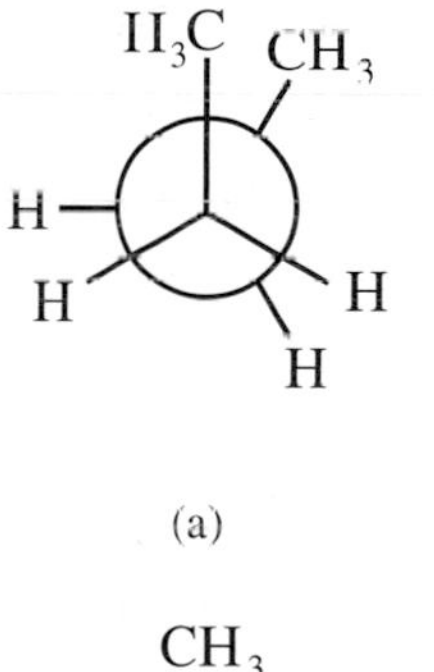

(a)

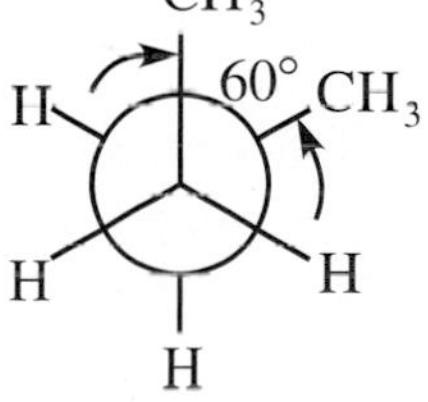

(b)

(c)

Consider, for example, the molecule butane. When you initially draw the molecule, you will undoubtedly not draw a perfectly correct structure in terms of bond lengths, bond angles, and dihedral angles. The computer will calculate the energy of the structure you draw, and then make a slight change in the coordinates of the atoms and recalculate the energy of the molecule. If the newly calculated energy is lower than the previously calculated one, it will continue to change the coordinates until a minimum energy is reached. This iterative process is continued until it is found that no change in any bond angle, bond length, torsion angle, or van der Waals interaction can be made without raising the energy of the molecule. At this point, the calculation stops. You can then query the results and ask for specific bond lengths, dihedral angles, etc.

But there is one possible pitfall in this process: The computer program usually has no way of knowing whether it has calculated the structure with the lowest possible energy for the molecule (the *global minimum*) or just a *local minimum*. A geographical analogy may make this statement more clear. It is as if you were instructing the computer to roll a ball to the lowest place in California (Death Valley). The computer might find the lowest place in the Sacramento Valley, but the algorithm on which it operates has no way of knowing whether that is the global minimum (Death Valley) or if there is any way to climb the mountains that separate the local minimum (the Sacramento Valley) from the global minimum.

Now consider once more the molecule butane. If the initial structure put into the computer has a conformation with a torsion angle (dihedral angle) of between 0° and just a bit less than 120° (margin figure a), the molecular mechanics program will calculate the *gauche* conformation (margin figure b) with a 60° torsion angle as the lowest-energy form. In fact, the lowest-energy form (margin figure c) is *anti* and has a torsion angle of 180°. The global minimum is the *anti* conformer, but the

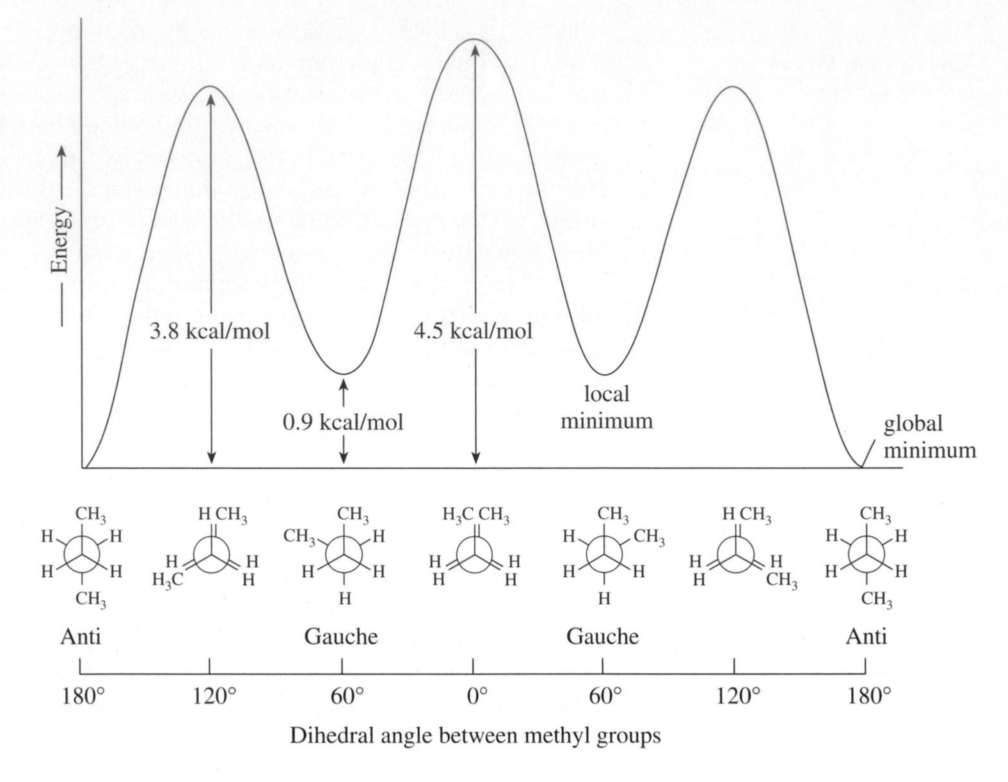

FIG. 15.4
The potential energy of butane as a function of the dihedral angle between the two central carbons.

calculation can easily fall into one of the two local minima, the *gauche* conformations, which are not the lowest-energy conformations. The full potential energy diagram for all rotational conformers of butane generated by molecular mechanics, shown in Figure 15.4, illustrates this concept clearly.

There are several approaches to the solution for this problem, but no one approach is completely satisfactory. If one has enough experience and intuition, it is often possible to spot the fact that the minimum-energy conformation calculated by the computer is not the global minimum. A better solution, however, is to calculate energies for an arbitrary set of dihedral angles about one of the bonds. In butane, one could ask for the energies to be calculated at dihedral angles of 0°, 30°, 60°, 90°, 120°, 150°, and 180°. In this way you would be sure to catch the lowest-energy form.

Now consider a larger molecule. If one were to search for the best conformation at angles of 0°, 30°, 60°, 90°, 120°, 150°, and 180° around each carbon-carbon bond, it would be necessary to make 7^n calculations, where n is the number of carbon-carbon bonds. In 2-bromononane, for example, this would be $7^8 = 5,764,801$ calculations. The problem gets out of hand very rapidly.

It is not possible to use a dihedral driver within a ring, so the lowest-energy form is found by other means (ring flip and flap, molecular dynamics, and Monte Carlo methods).

Molecular mechanics can be used to calculate IR spectra.

An elaboration of this computational method, called *molecular dynamics,* is often used to study the motions of large molecules such as proteins and other large polymers. Molecular dynamics is basically a series of molecular mechanics calculations, with each conformation being a frame in a movie animation of the movements of a molecule in the process of folding into a preferred shape.

QUANTUM CHEMICAL OR AB INITIO MODELS

Hartree-Fock [HF] approximation; ab initio models are often called HF models.

In the 1920s, a more sophisticated theory of atomic and molecular structure was developed that improved on Lewis structures and yielded a more accurate representation of molecular properties. This new theory relied on quantum mechanics and the Schrödinger equation for predicting the behavior of an electron in the presence of a positively charged nucleus. However, the mathematics for the exact solution of the equations involved was so complex that a full calculation could only be performed on the one-electron hydrogen atom. However, by assuming that the nuclei don't move (Born-Oppenheimer approximation) and that electrons move independently of one another, the Schrödinger equation could be simplified so that it was practical to perform useful calculations on multi-atom systems. Many scientists have worked on further refinements of the equations and computer programs that have helped to gradually improve the quality of the results. Because of the intensive nature of the calculations, the availability of fast desktop computers and efficient computational programs has also been very important. Today, calculations involving more than 100 nonhydrogen atoms are feasible. Two commonly used HF model programs are 3–21G, which deals with s and p electrons and 6–31G*, which deals with d electrons, in addition to s and p electrons.[1] These methods can also calculate NMR and UV spectra, as well as calculate and plot COSY and NOSY NMR spectra. If further accuracy is required, one can enhance the HF models by using perturbation theory to compensate for electron-electron repulsion within a molecular orbital. Approximations that add perturbation theory were developed by Möller and Plesset, and a commonly used MP-level model is MP2/6–31G*. This model, unlike the simple HF model, allows the calculation of thermochemical properties when bonds are broken and new ones made as it all happens in a chemical reaction. Note that different approximations often lead to different results. The mathematical complexity of quantum calculations puts a full explanation of the principles beyond the scope of this text.

Unlike molecular mechanics, ab initio models use no empirical data. The calculation is truly from first principles, using approximate Schrödinger equations to calculate the locations of the atomic nuclei and *all* electrons: valence, unpaired, and inner shell. The equilibrium (minimum-energy) geometries are usually improved relative to those from molecular mechanics, better reflecting the true structure. Many properties that are defined by electron distribution, such as dipole moments, ionization potentials, and the shape and sign of molecular orbitals, can be calculated.

These capabilities are best demonstrated by the following example. In the presence of a free radical initiator (AIBN, azoisobutylnitrile; *see* Chapter 18), tributyltin

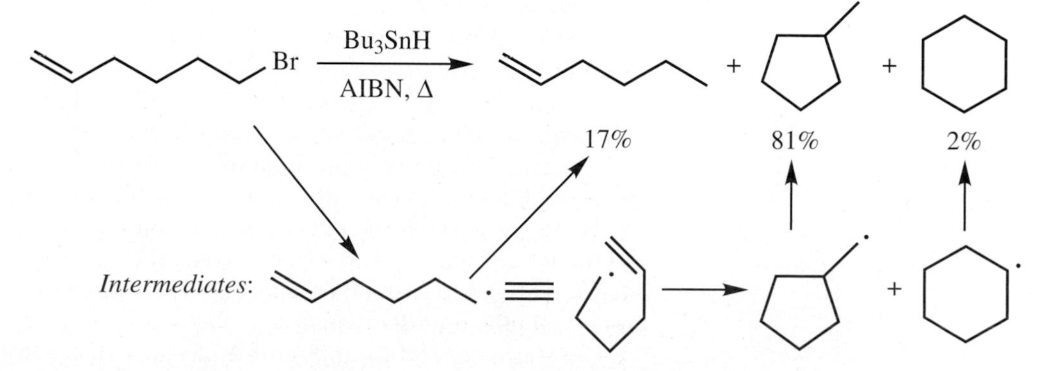

[1]The asterisk means that the program has been improved.

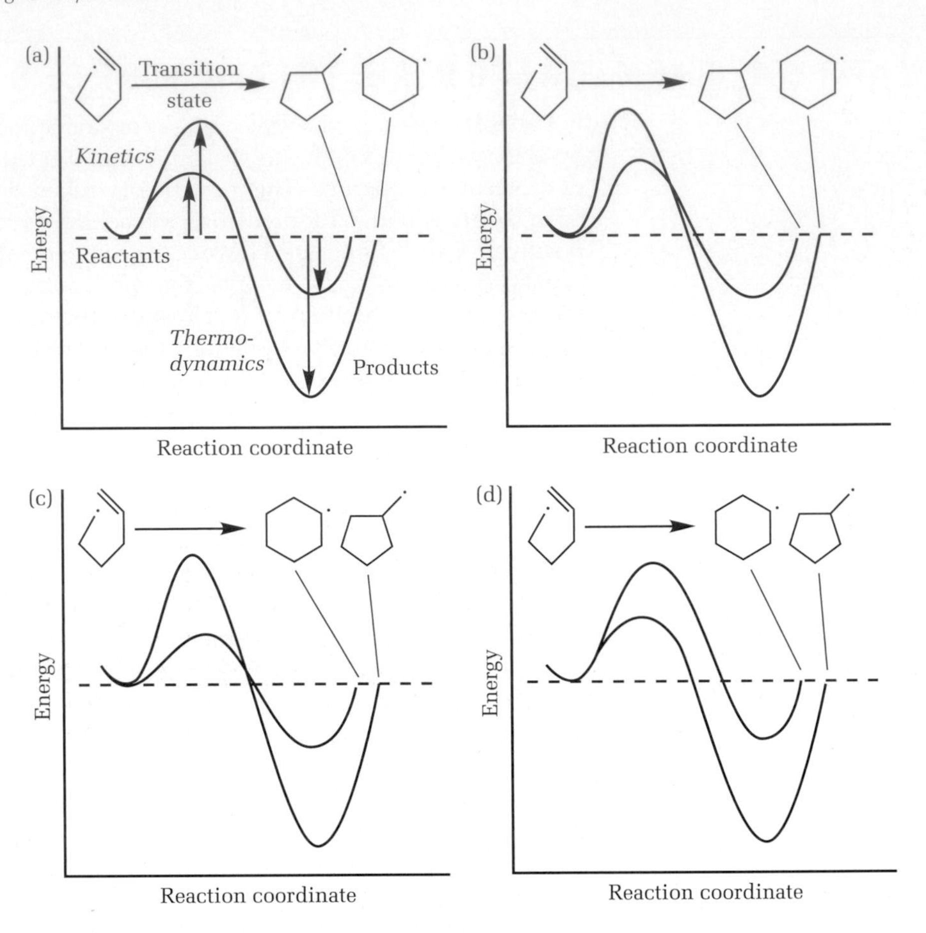

FIG. 15.5
The four possible energy–reaction coordinate diagrams for the intramolecular cyclization of the hexenyl radical to either the cyclopentylmethyl radical or the cyclohexyl radical.

hydride cleaves carbon-bromine bonds to give free radicals. In the case of 6-bromo-1-hexene, the initially formed radical can either abstract a hydrogen from surrounding molecules to form 1-hexene, or undergo intramolecular radical cyclization to form either a cyclopentylmethyl or cyclohexyl radical, either of which also picks up hydrogen to give the products shown.

Organic chemistry fundamentals would predict that cyclohexane should be the major cyclic product because the formation of six-membered rings and secondary radicals in the intermediate cyclohexyl radical should be favored. Contrary to this prediction, the major product (81%) is derived from the five-membered cyclopentylmethyl radical, with the less stable primary radical. Let us see whether an analysis of the energetics of this reaction using ab initio methods provides insight into this apparent contradiction.

As in the energy analysis of the conformers of butane (Fig. 15.4), it is useful to use energy-reaction coordinate diagrams to understand the possible reaction pathways. Figure 15.5 shows four diagrams depicting the energy needed to cause the radical and π bond to react (the transition state), and the energy released as a new bond is formed. The former energy determines the kinetics or rate of the reaction, and the latter determines the thermodynamic stability of the products relative to the reactant and to each other. There are four different energy pathways possible, as shown in the four diagrams.

Diagrams (a) and (b) in Figure 15.5 show the cyclohexyl radical to be more thermodynamically stable than the cyclopentylmethyl radical, consistent with the

TABLE 15.1 *The Relationship of Product Ratio to Energy Differences*

Energy Difference (kcal/mol)	*Energy Difference (kJ/mol)*	*Product Ratio, Major:Minor*
0.5	2	80:20
1	4	90:10
2	8	95:5
3	12	99:1

chemical fundamentals discussed previously. Diagrams (c) and (d) make the cyclopentylmethyl radical the more stable one, which would be inconsistent. When the energy minima of these two radicals are calculated using a 6–31G* ab initio computation, the cyclohexyl radical is roughly 5 kcal/mol more stable than the cyclopentylmethyl radical, consistent with our chemical knowledge. Energy diagrams (c) and (d) are therefore incorrect depictions of the reaction.

If the cyclohexyl radical is the more stable, as calculated and depicted in diagrams (a) and (b), there must be another explanation of why the cyclopentyl-methyl product is formed 40 times more readily. A possible explanation is that the transition state energy necessary to get the initial hexenyl radical to react at carbon 2, forming a five-membered ring, is less than the energy needed to react at carbon 6. Because ab initio methods also allow calculation of the energy maxima of transition states, this question can also be examined computationally. The calculated difference in transition state energies is 2.7 kcal/mol lower in favor of ring closure to the cyclopentylmethyl radical, making diagram (b) the only correct representation of the energetics. Computational chemistry shows that the final product ratio is determined by kinetics, with the rate of cyclopentylmethyl closure being faster than the rate of cyclohexyl closure due to the lower transition state energy.

Note that the addition of the initial hexenyl radical to the double bond is *irreversible* for this case because the reverse transition state energy is quite high. If the energy difference was low enough, a dynamic equilibrium between product and reactant radicals might eventually lead to the formation of the more thermodynamically stable product (cyclohexyl radical), given sufficient time and energy. Most *reversible* reaction product ratios are determined by thermodynamics, not kinetics.

Product ratios and reactant:product ratios, whether determined by thermodynamics or transition state energy differences, are roughly proportional to their energy differences, as shown in Table 15.1. Notice that even small energy differences lead to large differences in these ratios.

SEMIEMPIRICAL MODELS

Most molecular computations done by organic chemists, especially those examining minimum energy geometries, are done using a semiempirical model because such a model provides the best compromise between speed and accuracy. This kind of model can be thought of as a hybrid of molecular mechanics models based on experimentally measured *empirical* data and pure quantum chemical theory or ab initio models, thus the name *semiempirical*. Semiempirical models use Schrödinger equation approximations like those described previously but, in order to make the calculations less time-consuming, *only* the locations of *valence electrons* are calculated, *not all electrons*. For the inner shell electrons, empirical data from typical organic molecules is used to estimate their locations. This and other approximations allow semiempirical models to be used for molecules containing up to several hundred or more nonhydrogen atoms, much larger than those that can be modeled by the more rigorous ab initio models. However, semiempirical models are often less accurate, especially for calculations on novel molecules with unusual bonding or reactivity. Commonly used programs for organic molecules are AM1 and RM1.

Most computational chemistry texts not only provide explanations of the different mathematical model equations, but also do a large number of comparisons of calculated properties (bond lengths, angles, dipoles, various energies, etc.) from the different mathematical models versus actual properties measured experimentally in the lab. The data can be quite sobering, with differences between calculated and experimental values averaging 7%–10% and with many examples having much larger errors. As seen from Table 15.1, these types of errors in energy calculations would lead to even larger discrepancies in calculated and experimental product ratios. As stated earlier, every computational result has to be examined critically to make sure it fits with both chemical logic and experimental results. Nevertheless, being able to model transient intermediates such as free radicals or transition states, species that are virtually impossible to isolate and study in the lab, provides some exciting ways to extend our understanding of chemistry.

TYPES OF CALCULATIONS

The model used to perform the calculation of any particular property or graphical representation determines the amount and accuracy of information generated in the output. The first two types of calculations, geometry optimization and single point calculations, can be performed with any of the three computational models described previously. The last three types of calculations, involving electron locations and energies, are not solvable by molecular mechanics.

1. Geometry Optimization

Geometry optimization is a standard computational chemistry calculation to find the lowest-energy, or most relaxed, conformation for a molecule. The approach is the same for all levels of calculation, involving an iterative jiggling process like that described for molecular mechanics. At each step, the molecular geometry is modified slightly, and the energy of the molecule is compared with the last cycle.

The computer moves the atoms in the molecule a little, calculates the energy, moves them a little more, and keeps going until it finds the lowest energy. This is the energy minimum of the molecule and is obtained at the optimized geometry. Recall that the energies from molecular mechanics can only be used in a relative sense, while those from quantum electronic structure models can be compared in an absolute sense, like heats of formation.

Be wary of structures that may be stuck in a local, not global, energy minimum. As you may know, the chair conformation of cyclohexane is lower in energy than the boat conformation (Fig. 15.6). If you entered the boat structure, geometry optimization should find the chair as the structure with the lowest energy. Remember that sometimes, however, the computer gets "stuck" trying to find a minimum due

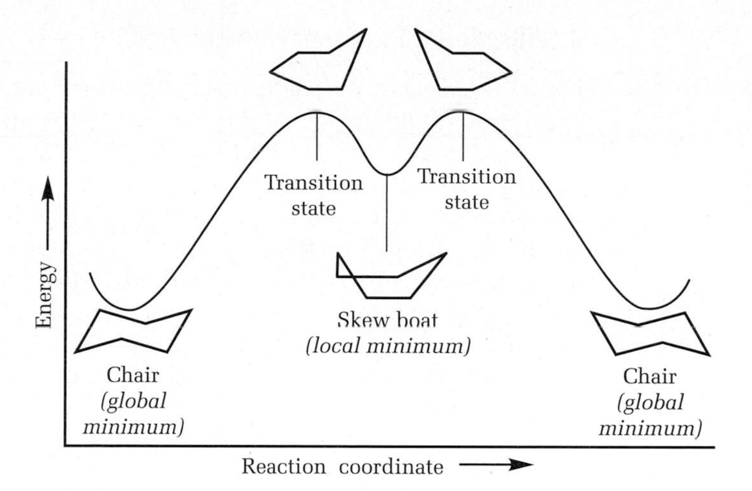

FIG. 15.6
A reaction coordinate diagram for cyclohexane, showing local skew boat and global chair minimum-energy geometries.

to the presence of a local minimum. For example, if you start with a boat structure of cyclohexane, the computer may get to a "skew-boat" local energy minimum but may not flip the molecule completely into the chair conformation that corresponds to the *global minimum*. This can happen when the energy jump from skew boat to chair is too large. A knowledge of fundamental organic chemistry as presented in introductory textbooks helps to recognize these types of problems.

2. Single Point Calculations

Single point calculations are often used in combination with geometry optimization to investigate steric hindrance. In this case, the method only performs one computational cycle to calculate the energy of a particular fixed geometry. In a thermodynamically controlled reaction, the energetic difference between two conformations is often due to steric hindrance. If the product molecule optimizes in one conformation, you can use single point calculations to determine how much more energy is needed to form the nonpreferred conformation. The structure drawing and manipulation part of the software will allow you to move only that part of the molecule that changes in the higher energy form, leaving the rest of the molecule optimized. The single point calculation performed on this modified molecule will give an energy that you can directly compare with the optimized energy to find the energy difference between the lower and higher energy conformers. For example, the energetic difference between having a constituent in the axial or equatorial position on a cyclohexane ring can be determined.

3. Transition State Calculations

Transition state calculations can be thought of as the reverse of geometry optimizations. In this case, the method searches for a structure of *maximum* energy, a transient intermediate that cannot be isolated experimentally. For example, this type of calculation allows one to examine transition state energies and geometries of intermediates involved in carbocation rearrangements. The literature contains standard models that should be used as the starting point for these calculations. It takes an appreciable amount of effort and experience to properly analyze transition state structures and energies.

4. Vibrational Frequency Calculations

Vibrational frequency calculations allow the calculation of IR stretching and bending absorption frequencies, and it is a lot of fun to view animations of these types of motions in molecules. The vibrational frequency of a two-atom system is proportional to the square root of the force constant (the second derivative of the energy with respect to the interatomic distance) divided by the reduced mass of the system (which depends on the masses of the two atoms). In fact, the analysis of transition states (as well as other essential quantities such as entropy) really starts by exploring slight movements of atoms that can be considered as vibrations.

The frequencies calculated using semiempirical models are usually higher than the frequencies measured on an IR instrument by about 10%; the accuracy depends on the computational model used. The reason for this discrepancy is that the calculations do not take solvent effects into account, but all peaks are displaced by about the same amount. Simply offsetting the calculated spectrum will allow comparison of the calculated and experimental spectra. The best way to check the calculated values is to see whether the vibration animations make sense. For example, if an animation shows a C=O stretching vibration, the calculated frequency can be matched with the measured frequency of that type, which is usually easy to assign in the acquired IR spectrum.

5. Electron Density and Spin Calculations, Graphical Models, and Property Maps

Electron density and spin calculations, graphical models, and property maps allow the visualization of electronic properties such as electron densities, electrostatic potentials, spin densities, and the shapes and signs of molecular orbitals. The values for a particular property at each point in the three-dimensional space around a molecule are displayed on a two-dimensional computer screen as a surface of constant numerical value, often called an *isosurface*, which can be rotated on the screen to study it. Alternately, numerical variations of a given property (such as electron density) at a defined distance from the molecule can be displayed as a property map using color as a key. Carrying out surface calculations and viewing their graphical representations are major activities in computational chemistry, and can provide useful insight into the mechanisms of organic reactions. A few examples of isosurfaces, orbitals, and property maps are shown in Figure 15.1 (on page 293). More examples and explanations, including color images, are available on the web site for this book.

See Molecular Property Maps

EXAMPLES OF CALCULATIONS

Try to reduce computational time by understanding the various computational methods and answering the question, "What kind of information do I need?" If only geometry or ring strain information is needed, using the more intensive ab initio calculation to produce high-precision molecular orbitals is not necessary. Even if you need the high accuracy of an ab initio calculation, do not start there. First, approximate the molecular geometry using a molecular mechanics calculation. Then use this structure to run an RM1 semiempirical calculation for a good approximation of the valence electron properties. Finally, if even more accuracy is required, use the RM1 output for the 3–21G or 6–31G* input to see how the inner shell electrons affect the electronic properties.

An instructive exercise is to calculate the energy of butane at dihedral angles of 0°, 60°, 120°, and 180°, using single point energy calculations for each. You should be able to reproduce the *form* of the curve shown in Figure 15.4 (on page 298), but you usually will not obtain the experimentally derived energies shown.

Remember that the calculated steric energy has no physical significance. Differences in steric energies can be useful, however, if the molecules being compared have the same bonding pattern. The molecule with the lower energy will be the more stable isomer (geometric, conformational, or stereo). Other types of comparisons cannot be made. A molecular mechanics calculation cannot be used to compare the stabilities of unrelated molecules.

If the program you are using breaks down the steric energy into the six terms E_{steric}, E_{str}, E_{bnd}, E_{tor}, E_{oop}, and E_{vdW}, do not give the individual terms much credence. Differing parameter sets in other programs may weight these differently. You should pay the most attention to the total steric energy.

Molecular mechanics calculations, like IR and NMR spectroscopy, are a tool available to the chemist. As such, they are presented in this text as adjuncts to experimental work in the laboratory, not as an end in themselves. In most of the computational problems presented in the following chapters, you are asked to carry out computations of the steric energies or the heats of formations of isomers.

One must be careful when using molecular mechanics computations to draw conclusions. The computation can be made on any molecule, but the resulting steric energy cannot always be compared to the steric energy of another molecule, even though they are isomers. For example, *cis*- and *trans*-butene and other similar *cis*- and *trans*-isomers can be compared to each other because they are both 1,2-disubstituted alkenes, but a valid comparison cannot be made between 1,2-and 1,1-disubstituted alkenes.

Although not intending to mislead you deliberately, we have included computations on reactions and products that for one reason or another are not expected to give valid comparisons or answers. It is for you to decide whether the computations are in agreement with your experiment.

The increase in time needed for a calculation on a molecule having *n* atoms:

- **molecular mechanics, n^2**
- **semiempirical, n^3**
- **ab initio, $>n^4$**

Higher levels of computation, the semiempirical molecular orbital methods and the even higher ab initio methods, will give heats of formation that can be compared to one another. The problem with these higher levels of computation is that they require more time and more computer power. But often an RM1 semiempirical molecular orbital calculation on a relatively small molecule can be done in a few seconds. The cost (in computer time) of ab initio calculations goes up more than the fourth power of the number of atoms, whereas molecular mechanics calculations go up as the square of the number of atoms in the molecule.

The radical chlorination of 1-chlorobutane gives four isomeric dichlorobutanes. In Chapter 18 these chlorobutanes are synthesized, and their relative amounts are measured by gas chromatography with the object of determining the relative reactivity of the various hydrogens in the starting material. Using a semiempirical program, you can calculate the energies of the four products to see whether or not these energies correlate with the product distributions. A correlation, or lack thereof, may give information regarding the mechanism of the reaction.

Similarly, the dehydration of 2-butanol (*see* Chapter 10) gives three isomeric butenes that are separated and analyzed by gas chromatography. Calculation of their heats of formation or steric energies may or may not correlate with the percentage of each isomer formed in the reaction. Again, a correlation or lack thereof may give information about the mechanism of the reaction or about the applicability of computational chemistry to problems of this type.

Treatment of *cis*-norbomene-5,6-*endo*-dicarboxylic acid with concentrated sulfuric acid gives the isomeric compound X (Chapter 48). The energies of any proposed structures for X can be calculated and compared with that of the starting material. If the reaction conditions (hot concentrated sulfuric acid) are regarded as favoring equilibrium between the product and the starting material, then X would be expected to have the lower energy if the correct structure is proposed.

Hexaphenylethane is a molecule that does not exist, but this does not preclude calculation of its steric energy and bond lengths (Chapter 31). One can calculate the energy of the isomeric dimer that does form. This calculation gives a clear picture of why the triphenylmethyl radical does not simply dimerize, even though the steric energy may not be correct.

The oxidation of citronellol can give four possible isomeric isopulegols (*see* Chapter 23). Molecular mechanics calculations can indicate which of these is the most stable. Similarly, oxidation of these alcohols can give two possible isopulegones that theoretically can equilibrate. Molecular mechanics calculations can again indicate which is the more stable. These rearrange to pulegone. Again, calculation may disclose whether this product is more stable than the starting material, throwing light on the mechanism of the isomerization reaction.

The Wittig reaction usually gives a mixture of *cis*- and *trans*-isomers. In Chapter 39, two Wittig reactions are carried out, in which one isomer predominates in each reaction. A molecular mechanics calculation discloses not only what these isomers look like, but also their energies.

A very easy reaction to carry out is the aldol condensation of benzaldehyde with acetone to give high-yield dibenzalacetone (mp 110°C–111°C). It is interesting to explore the products of this reaction with molecular mechanics because three geometric isomers can be formed, and seven single-bond *cis*- or *trans*-isomers of the geometric isomers lie at the energy minima. The question is, which of these 10 isomers is formed in the reaction? Computational aspects of this question are dealt with at length in Chapter 37.

In all of these calculations, not only will you calculate the energies of the molecules, you will also be able to see what the lowest-energy conformation of the molecule looks like. In this regard, it is interesting to look at the calculated low-energy conformation of benzophenone (Chapter 38), tetraphenylcyclopentadienone (Chapter 51), and pseudopellitierene (Chapter 66). *Z*- and *E*-stilbene are synthesized in Chapters 58 and 60, and in Chapter 59, the Perkin reaction is used to make a mixture of *Z*- and *E*-phenylcinnamic acids. Energy calculations, as well as a picture of the molecular conformations, help rationalize the relative amounts of the isomers formed in these synthetic reactions as well as their ultraviolet (UV) spectra.

For Additional Experiments, sign in at this book's premium web site at **www.cengage.com/login.**

Other examples of synthetic experiments with computational chemistry supplements are available on this book's web site.

REFERENCES

Burkert, Ulrich, and Norman L. Allinger, *Molecular Mechanics* (ACSMonograph 177). Washington, DC: American Chemical Society, 1982.

Clark, Tim, *A Handbook of Computational Chemistry: A Practical Guide to Chemical Structure and Energy Calculations*. New York: John Wiley & Sons, 1985.

Earl, Boyd L., David W. Emerson, Brian J. Johnson, and Richard L. Titus, "Teaching Practical Computer Skills to Chemistry Majors," *J. Chem. Educ.* 71 (1994): 1065.

Freeman, Fillmore, Zufan M.Tsegai, K. Marc Lasner, Warren J. Hehre, "A Comparison of the ab Initio Calculated and Experimental Conformational Energies of Alkylcyclohexanes," *J. Chem. Educ.* 77 (2000): 661.

Grant, Guy H., and W. Graham Richards, *Computational Chemistry*. Oxford, UK: Oxford University Press, 1995.

Hehre, Warren J., *A Guide to Molecular Mechanics and Quantum Chemical Calculations*. Irvine, CA: Wavefunction Press, 2003.

Hehre, Warren J., *A Laboratory Book of Computational Organic Chemistry*. Irvine, CA: Wavefunction Press, 1998.

Hehre, Warren J., Lonnie D. Burke, Alan J. Shusterman, and William J. Pietro, *Experiments in Computational Organic Chemistry*. Irvine, CA: Wavefunction Press, 1993.

Hehre, Warren J., *The Molecular Modeling Workbook for Organic Chemistry*, 2nd ed. Irvine, CA: Wavefunction Press, 2005.

Jarret, Ronald M., and Ny Sin, "Molecular Mechanics as an Organic Chemistry Laboratory Exercise," *J. Chem. Educ.* 67 (1990): 153–155.

Jensen, Frank, *Introduction to Computational Chemistry*. New York: John Wiley & Sons, 1999.

Lipkowitz, Kenny B., Raima Larter, and Tom Cundari, eds., *Reviews in Computational Chemistry*, Vol. 20. New York: John Wiley & Sons, 2004.

Lipkowitz, Kenny B., and Daniel Robertson, "Conformer Hunting: An Open-Ended Computational Chemistry Exercise That Expresses Real-World Complexity and Student Forethought," *J. Chem. Educ.* 77 (2000): 206.

The *Journal of Chemical Education*. Since 1989, many articles on molecular mechanics calculations and computational chemistry in general have been published. Use electronic journal searches to find these articles.

The Synthetic Experiments

INTRODUCTION

At this point, you have practiced the important basic techniques of recrystallization, distillation, extraction, and chromatography. You will now carry out the synthesis of various compounds; each of these synthetic experiments will require one or more of the basic techniques. Before starting any synthetic procedure, you should prepare for the work by generating a chemical data table and, if applicable, a spectral data table. Once you're finished with the experiment, you will usually be required to calculate the percent yield of the product. The purpose of this section is to introduce you to the preparation of these tables and to review how to determine percent yield.

CHEMICAL DATA TABLE

A synthetic procedure specifies a certain proportion of all reagents. The prescription of quantities is based on considerations of stoichiometry. Before undertaking a preparative experiment, you should analyze the procedure and calculate the molecular proportions of the reagents. A chemical data table of the properties of the starting material, reagents, products, and by-products provides guidance in regulating temperature and in separating and purifying the product, and should be entered in your laboratory notebook. An example is given below in Table 16.1; the reagents and amounts correspond to those used in Chapter 16, Experiment 2.

The first step in generating a chemical data table is to look up the properties of the reagents. These properties include molecular weight (MW), density (in g/mL), melting point (Mp °C), and boiling point (Bp °C). The molecular weight will be used to calculate the number of moles of each reagent, given the gram amount or milliliter amount (using density to convert to grams) of each reagent used in the procedure. The mole amount will be used to determine the number of equivalents of each reagent, as well as the limiting reagent and those reagents used in excess.

The second step in generating a chemical data table is to consider the balanced reaction equation of the synthetic step in order to determine the mole ratio of starting reagents. Chapter 16, Experiment 2 is the synthesis of 1-bromobutane, which is made from 1-butanol, sodium bromide, and sulfuric acid:

$$CH_3CH_2CH_2CH_2OH \xrightarrow{NaBr,\ H_2SO_4} CH_3CH_2CH_2CH_2Br + NaHSO_4 + H_2O$$

One mole of 1-butanol theoretically requires 1 mole each of sodium bromide and sulfuric acid to produce hydrobromic acid, HBr, the reactive bromine species in this reaction:

$$NaBr + H_2SO_4 \rightleftharpoons HBr + NaHSO_4$$

Thus, in theory, each reagent is in 1:1:1 mole ratio. The procedure of Experiment 2 in Chapter 16 calls for 13.3 g of sodium bromide, 10 mL of 1-butanol, and 11.5 mL of concentrated sulfuric acid. The molecular weight of sodium bromide is 102.91 g/mol, which can be used to determine the mole amount of sodium bromide:

$$\frac{13.3 \text{ g NaBr} \times 1 \text{ mole}}{102.91 \text{ g/mole}} = 0.11 \text{ moles of NaBr}$$

Similar calculations can be used to determine the mole amounts of 1-butanol and of sulfuric acid (be sure to convert milliliters to grams using density). You will see that this procedure calls for using a slight excess of sodium bromide (0.13 moles; 1.2 equivalents) and nearly twice the theoretical amount of sulfuric acid (0.20 moles; 1.8 equivalents). Excess sulfuric acid is used to shift the equilibrium in favor of a high concentration of hydrobromic acid. It is recommended that you check your calculations before weighing out chemicals to ensure the proper amounts for the procedure.

Theoretical Yield

The importance of calculating the mole amounts and equivalents of each reagent is revealed when determining the theoretical yield of the product. The first step is to establish the limiting reagent, which is defined as the reagent used in the lowest mole or equivalent amount. In this experiment, the limiting reagent is 1-butanol (see Table 16.1). The balanced equation of the reaction (shown above) reveals that the mole ratio of 1-butanol to 1-bromobutane is 1:1. Thus, the theoretical yield of product is 0.11 mole. The maximal weight of 1-bromobutane is then calculated and inserted into Table 16.1:

$$0.11 \text{ (mole of alcohol)} \times 137.03 \text{ (MW of product)}$$
$$= 14.8 \text{ g 1-bromobutane (theoretical yield of product)}$$

Percent Yield

Rarely do organic reactions give 100% yields of one pure product; an important objective of every experiment is to obtain the highest yield of the desired product. As determined above, the theoretical yield of 1-bromobutane is 14.8 g; it could be obtained if the reaction proceeded perfectly. If the actual yield is only 10.5 g, then the reaction would be said to give a 71% yield

$$\frac{10.8 \text{ g}}{14.8 \text{ g}} \times 100 = 71\%$$

Typical student yields are included throughout this text. These are not theoretical yields; they suggest what an average or above-average student can expect to obtain for the experiment.

Most synthetic experiments require the evaluation of the product by one or more spectroscopic methods to aid in confirming formation of the product. ^{13}C and ^{1}H NMR spectroscopic analyses are suggested for 1-bromobutane. A spectral data table can be created in order to compare diagnostic information of the starting material and of the product (and by-products, if applicable). Once the spectrum is collected, it can be analyzed with the spectral data table to confirm quickly the presence or absence of product (and by-product). Table 16.2 gives diagnostic information that would be obtained by ^{1}H NMR spectroscopy for 1-butanol and 1-bromobutane, as well as 1-butene and di-*n*-butyl ether (possible by-products).

Spectra of starting materials and/or products are included throughout this text to serve as verification of your work.

TABLE 16.1 *Chemical Data Table*

Reagents							
					Moles		
Reagent	*MW*	*Den.*	*Bp (°C)*	*Wt used (g)*	*Theory*	*Used*	*Equivalents*
n-C_4H_9OH	74.12	0.810	118	8.0	0.11	0.11	1.0
NaBr	102.91	—	—	13.3	0.11	0.13	1.2
H_2SO_4	98.08	1.84	—	20.0	0.11	0.20	8

Products								
			Bp (°C)		*Theoretical Yield*		*Found*	
Compound	*MW*	*Den.*	*Given*	*Found*	*Moles*	*G*	*g*	*%*
n-C_4H_9Br	137.03	1.275	101.6	____	0.11	14.8	____g	____%

TABLE 16.2 *Spectral Data Table*

Compound	*Diagnostic* ^{1}*H NMR Information*
1-butanol	5 signals; 1 broad singlet at 4.2 ppm
1-bromobutane	4 signals
1-butene	7 signals; alkene protons at ~5 ppm with complex splitting
di-*n*-butyl ether	4 signals

CHAPTER 16

The S_N2 Reaction: 1-Bromobutane

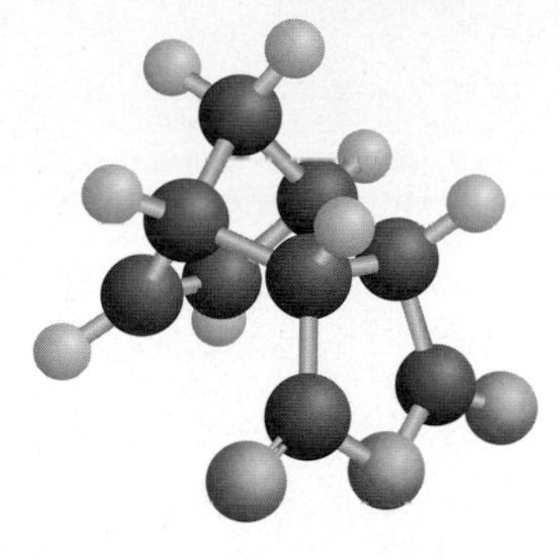

When you see this icon, sign in at this book's premium website at **www.cengage.com/login** to access videos, Pre-Lab Exercises, and other online resources.

PRELAB EXERCISE: Prepare a detailed flow sheet for the isolation and purification of *1-bromobutane*. Indicate how each reaction byproduct is removed, and which layer is expected to contain the product in each separation step.

Choice of reagents

In this experiment, 1-butanol is converted to 1-bromobutane by an S_N2 reaction. In general, a primary alkyl bromide can be prepared by heating the corresponding alcohol with (1) constant-boiling hydrobromic acid (47% HBr); (2) an aqueous solution of sodium bromide and excess sulfuric acid, which is an equilibrium mixture containing hydrobromic acid; or (3) a solution of hydrobromic acid produced by bubbling sulfur dioxide into a suspension of bromine in water. Reagents 2 and 3 contain sulfuric acid at a concentration high enough to dehydrate secondary and tertiary alcohols to undesirable byproducts (alkenes and ethers); hence the HBr method is preferred for preparing halides of the types R_2CHBr and R_3CBr. Primary alcohols are more resistant to dehydration and can be converted efficiently to the bromides by the more economical methods (2 and 3), unless these are of such high molecular weight that they lack adequate solubility in the aqueous mixtures. The $NaBr$-H_2SO_4 method is preferred to the Br_2-SO_2 method because of the unpleasant, choking property of sulfur dioxide. Here is the overall equation, along with key properties of the starting material and principal product.

$$CH_3CH_2CH_2CH_2OH \xrightarrow{NaBr,\ H_2SO_4} CH_3CH_2CH_2CH_2Br + NaHSO_4 + H_2O$$

1-Butanol	**1-Bromobutane**
bp 118°C	bp 101.6°C
den. 0.810	den. 1.275
MW 74.12	MW 137.03
n_D^{20} 1.399	n_D^{20} 1.439

Chemical Data Table Template

The procedure that follows specifies a certain proportion of 1-butanol, sodium bromide, sulfuric acid, and water; defines the reaction temperature and time; and describes operations to be performed in working up the reaction mixture. A chemical data table should be completed and entered in your laboratory notebook (see Table 16.1 in "The Synthetic Experiments" section, which uses the

The laboratory notebook

quantities in Experiment 2 of this chapter). Also, construct a flow sheet of the operations to be performed.

One mole of 1-butanol theoretically requires 1 mole each of sodium bromide and sulfuric acid, but this procedure calls for using a slight excess of bromide and twice the theoretical amount of acid. Excess acid is used to shift the equilibrium in favor of a high concentration of hydrobromic acid, the reactive bromine species. The amount of sodium bromide used is arbitrarily set at 1.2 times the theoretical amount as an insurance measure:

$$NaBr + H_2SO_4 \rightleftharpoons HBr + NaHSO_4$$

The probable byproducts are 1-butene, dibutyl ether, and the starting alcohol. The alkene is easily separated by distillation, but the other substances are in the same boiling point range as the product. However, all three possible byproducts can be eliminated by extraction with concentrated sulfuric acid.

Experiments

SYNTHESIS OF 1-BROMOBUTANE

$$CH_3CH_2CH_2CH_2OH \xrightarrow{NaBr,\ H_2SO_4} CH_3CH_2CH_2CH_2Br + NaHSO_4 + H_2O$$

1-Butanol	**1-Bromobutane**
bp 118°C	bp 101.6°C
den. 0.810	den. 1.275
MW 74.12	MW 137.03
n_D^{20} 1.399	n_D^{20} 1.439

Handle sulfuric acid with care because it is corrosive to tissue.

On a microscale, greater accuracy is achieved by weighing than by volumetric measurement. It is much easier to weigh 0.80 g of a liquid than to measure 0.80 mL with a pipette.

IN THIS EXPERIMENT, sodium bromide and 1-butanol are dissolved in water. Sulfuric acid is added cautiously, which generates hydrobromic acid, which in turn reacts with the alcohol upon heating to make 1-bromobutane. This alkyl bromide is then codistilled with water (steam distillation). The bromobutane is removed from the water and then treated with sulfuric acid to remove impurities. The sulfuric acid layer is removed, the product is treated with aqueous base, the base is removed, and the product is dried over calcium chloride. A small quantity of xylene is added to the product, and the mixture is fractionally distilled to give pure 1-bromobutane. The xylene acts as a chaser in this distillation. This is a fairly complex experiment.

In a 5-mL round-bottomed, long-necked flask, dissolve 1.33 g of sodium bromide in 1.5 mL of water and 0.80 g of 1-butanol. Cautiously, while constantly swirling, add 1.1 mL (2.0 g) of concentrated sulfuric acid dropwise to the solution. The NaBr will dissolve during heating. Fit the flask with a distillation head (Fig. 16.1) and reflux the reaction mixture on a sand bath for 45 minutes, taking care that none of the reactants distill during the reaction period. Wrap the upper end of the apparatus with a damp pipe cleaner if escaping vapor is a problem. The upper layer that soon separates in the reaction flask is the alkyl bromide because the

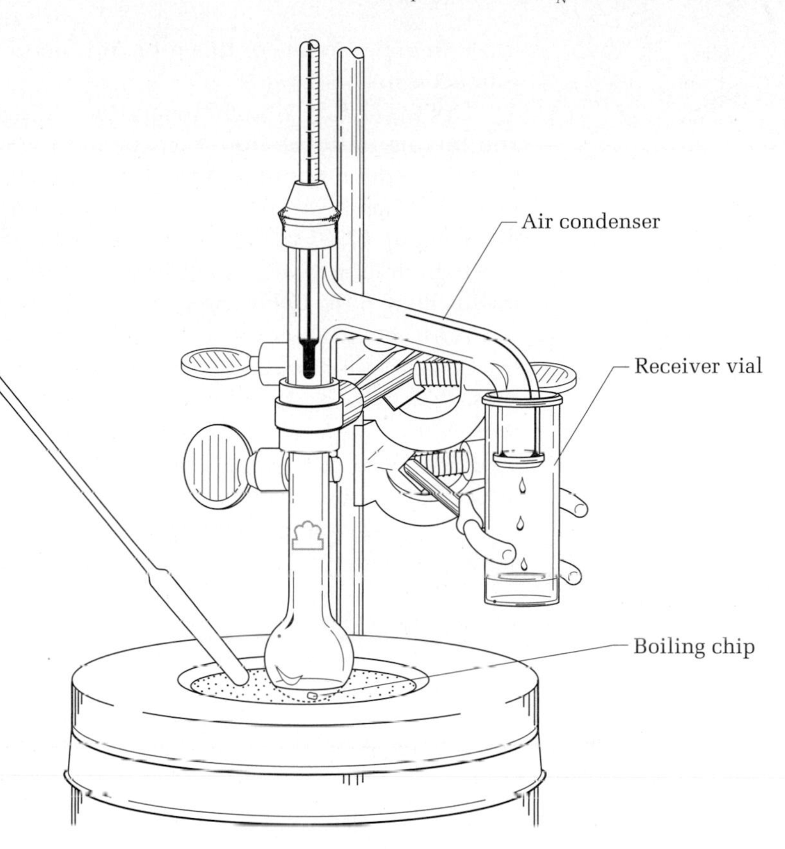

FIG. 16.1
An apparatus for refluxing and distilling a reaction mixture, which is shown in the distillation mode.

aqueous solution of inorganic salts has the greater density and sinks to the bottom. Remove the pipe cleaner, insulate the flask with cotton or glass wool, and distill the product into a collection vial until no more water-insoluble droplets come over, by which time the temperature of the distillate should have reached 115°C. If in doubt about whether the product has completely distilled, collect some of the distillate in a small tube and examine it carefully. The sample collected in the receiver is an azeotrope of 1-bromobutane and water containing some sulfuric acid, 1-butene, unreacted 1-butanol, and di-*n*-butyl ether. Rinse the distillation head with acetone so that it will be dry for use later in the experiment. Also, to prepare for the final step, take time to clean the round-bottomed flask by rinsing it with 1 mL of ethanol followed by 1 mL of acetone and then drawing air through it using a water aspirator. It is very important when drying an apparatus of this type to remove all the wash solvent, acetone in this case, otherwise, it will contaminate the final product.

Control heating by piling sand around the flask or by scraping it away.

Transfer the distillate to a reaction tube, rinsing the vial with about 1 mL of water, which is then mixed with the sample in the reaction tube. Note that the 1-bromobutane now forms the lower layer. Remove the 1-bromobutane with a Pasteur pipette and place it in a dry reaction tube. Add 1 mL of concentrated sulfuric acid and mix the contents well by flicking the tube. The acid removes any unreacted starting material as well as any alkenyl or ethereal byproducts. Allow the two layers to separate completely and then remove the sulfuric acid layer. The relative densities given previously will help identify the two layers. An empirical method of distinguishing the layers is to remove a drop of the lower layer into a test tube of water to see whether the material is soluble (H_2SO_4) or not (1-bromobutane). Separate the layers and wash the 1-bromobutane layer with 1 mL of 3 *M* sodium hydroxide solution (den. 1.11) to remove traces of acid, separate, and be careful to save the proper

Identifying the organic layer.

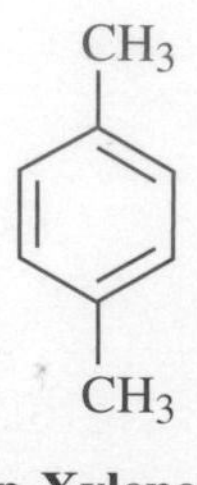

p-Xylene
MW 106.17
bp 137–138°C

layer. In experiments of this type, it is good practice to save all layers until the product is in hand.

Dry the cloudy 1-bromobutane by adding anhydrous calcium chloride pellets and mixing until the liquid clears and the calcium chloride no longer clumps together. After 5 minutes, decant the dried liquid into the 5-mL round-bottomed flask that has been dried. Rinse the drying agent with two 1-mL portions of *p*-xylene (bp 137–138°C), which is then transferred to the round-bottomed flask. The high-boiling xylene is a "chaser" solvent; it chases all the bromobutane from the distilling flask. Otherwise, about 0.3 mL would remain behind.

Using a chaser solvent.

Viton is very resistant to aromatic solvents.

Add a boiling stone, pack the neck of the flask with a stainless steel sponge for fractional distillation, and fit it with a dry distilling head and thermometer. Use a Viton (black) connector between the flask and the distilling head. Wrap the column with cotton or glass wool and distill, collecting material boiling in the range of 99–103°C. Stop collecting the moment the temperature begins to rise above 103°C. Most of the product will boil at 102°C. A typical yield is in the range of 1–1.2 g.

Put the sample in a vial of appropriate size, and make a neatly printed label stating the name and formula of the product and your name.

Cleaning Up. Carefully dilute all nonorganic material with water (the pot residue, the sulfuric acid wash, the sodium hydroxide wash) and combine and neutralize with sodium carbonate before flushing down the drain. Because all the organic material has been contaminated with small quantities of product, it must be placed in the halogenated organic waste container. If local regulations allow, evaporate any residual solvent from the drying agents in the hood and place the dried solid in the nonhazardous waste container. Otherwise, place the wet drying agent in a waste container designated for this purpose.

SYNTHESIS OF 1-BROMOBUTANE

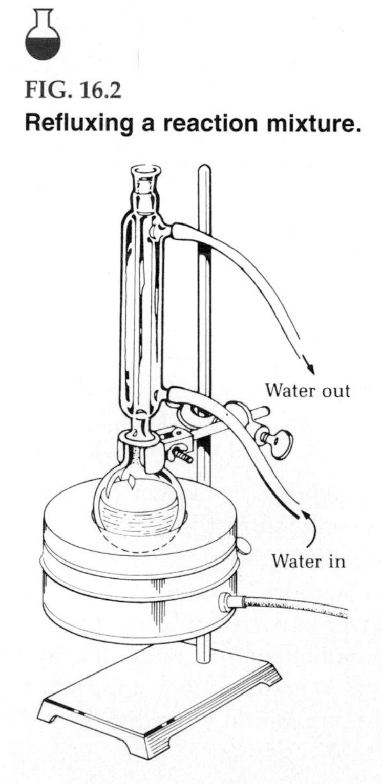

FIG. 16.2
Refluxing a reaction mixture.

Put 13.3 g of sodium bromide, 15 mL of water, and 10 mL of *n*-butyl alcohol in a 100-mL round-bottomed flask, cool the mixture in an ice-water bath, and slowly add 11.5 mL of concentrated sulfuric acid while swirling and cooling. Place the flask in an electric flask heater, clamp it securely, and fit it with a short condenser for reflux condensation (Fig. 16.2). Heat to the boiling point, note the time, and adjust the heat for brisk and steady refluxing. The upper layer that soon separates is the alkyl bromide because the aqueous solution of inorganic salts has a greater density and sinks. Reflux for 45 minutes, remove the heat, and let the condenser drain for a few minutes (extension of the reaction period to 1 hour increases the yield by only 1–2%). Remove the condenser, mount a stillhead in the flask, and set up the condenser for simple distillation (*see* Fig. 5.10 on page 97) through a bent or vacuum adapter into a 50-mL Erlenmeyer flask. Distill the mixture, make frequent readings of the temperature, and continue to distill until no more water-insoluble droplets come over, by which time the temperature should have reached 115°C (collect a few drops of distillate in a test tube and see if they are water soluble). The increasing boiling point is due to azeotropic distillation of *n*-butyl bromide with water containing increasing amounts of sulfuric acid, which raises the boiling point.

Pour the distillate into a separatory funnel, shake with about 10 mL of water, and note that *n*-butyl bromide (1-bromobutane) now forms the lower layer. A pink coloration in this layer due to a trace of bromine can be discharged by adding a pinch of sodium bisulfite and shaking again. Drain the lower layer of 1-bromobutane into a clean flask, clean and dry the separatory funnel, and return the 1-bromobutane to it. Then cool 10 mL of concentrated sulfuric acid thoroughly in an ice bath, add the acid to the funnel, shake well, and allow 5 minutes for separation of the layers. Use care in handling concentrated sulfuric acid. Check to see that the stopcock and stopper do not leak. Identify the two layers by comparing the density values; an empirical method of telling the layers apart is to draw off a few drops of the lower layer into a test tube and see whether the material is soluble in water (H_2SO_4) or insoluble in water (bromobutane). Separate the layers, allow 5 minutes for further drainage, and separate again. Then wash the 1-bromobutane with 10 mL of 3 *M* sodium hydroxide (den. 1.11) solution to remove traces of acid, separate, and be careful to save the proper layer.

Handle sulfuric acid with care because it is corrosive to tissue.

Dry the cloudy 1-bromobutane by adding 1 g of anhydrous calcium chloride pellets while swirling until the liquid clears. After 5 minutes, decant the dried liquid into a 25-mL flask or filter it through a fluted filter paper, add a boiling stone, distill, and collect material boiling in the range of 99–103°C. A typical student yield is in the range of 10–12 g. Note the approximate volumes of forerun[1] and residue.

Calcium chloride removes both water and alcohol from a solution.

Put the sample in a narrow-mouth bottle of appropriate size; make a neatly printed label stating the name and formula of the product and your name. Press the label onto the bottle using a piece of filter paper to smooth it out, and be sure that it is secure. After all the time spent on the preparation, the final product should be worthy of a carefully executed and secured label.

A proper label is important.

Check purity by infrared, refractive index, or gas chromatographic analysis.

Cleaning Up. Carefully dilute all nonorganic material with water (the reaction pot residue, the sulfuric acid wash, and the sodium hydroxide wash) and combine and neutralize with sodium carbonate before flushing down the drain with excess water. The residue from the distillation of 1-bromobutane goes in the waste container for halogenated organic solvents. If local regulations allow, evaporate any residual solvent from the drying agents in the hood and place the dried solid in the nonhazardous waste container. Otherwise, place the wet drying agent in a waste container designated for this purpose.

The ^{13}C NMR spectra of 1-bromobutane (Fig. 16.3) and 1-butanol (Fig. 16.4), as well as the ^{1}H NMR spectrum of 1-butanol (Fig. 16.5) are shown as a verification for your work.

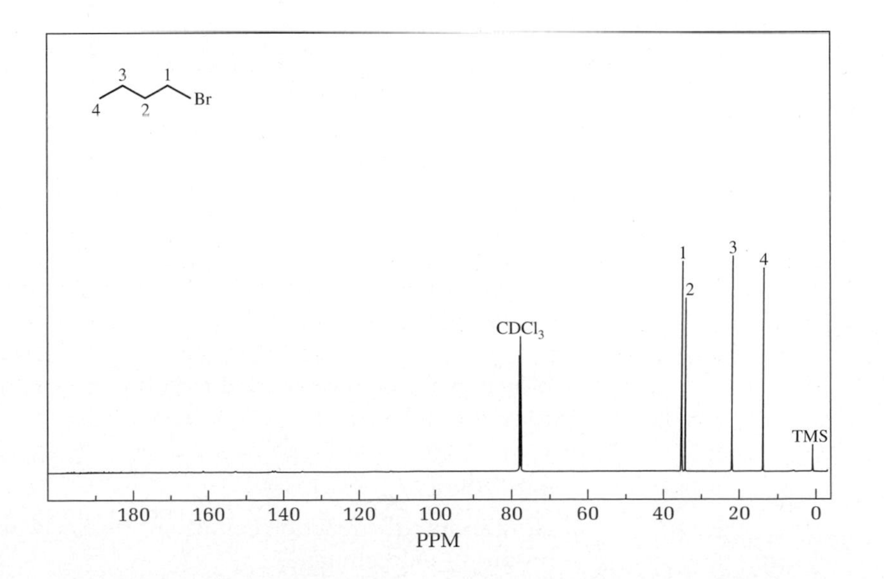

FIG. 16.3
The ^{13}C NMR spectrum of *n*-butyl bromide (100 MHz).

[1]The forerun is the material that distills before the temperature levels off between 99°C and 103°C, which is the boiling point of the product.

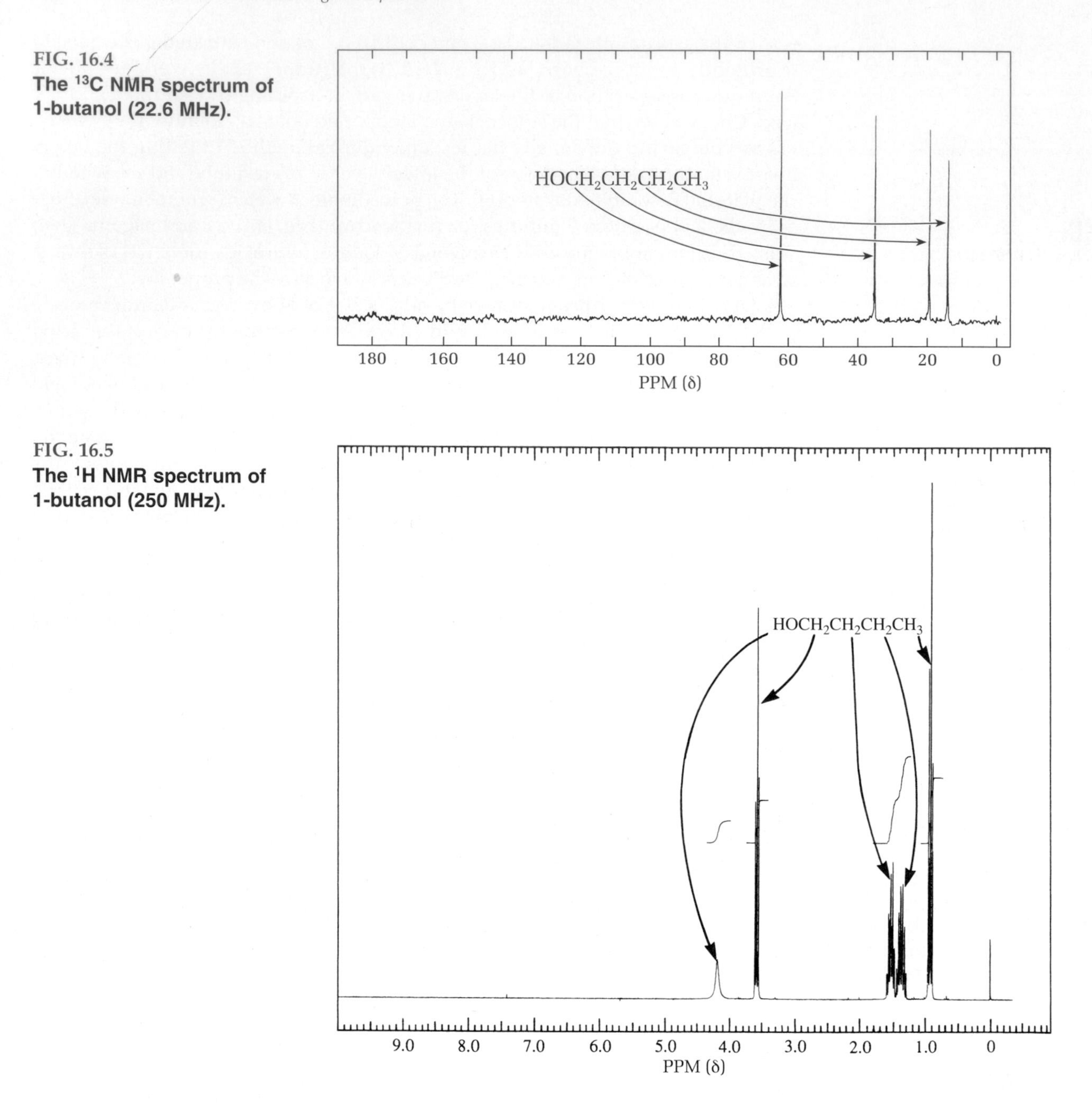

FIG. 16.4
The ^{13}C NMR spectrum of 1-butanol (22.6 MHz).

FIG. 16.5
The ^{1}H NMR spectrum of 1-butanol (250 MHz).

QUESTIONS

1. Which experimental method would you recommend for the preparation of 1-bromooctane? of *t*-butyl bromide?
2. Explain why the crude product is apt to contain certain definite organic impurities.
3. How should the reaction conditions in this experiment be changed to try to produce 1-chlorobutane?

4. Write a balanced equation for the reaction of sodium bisulfite with bromine.
5. What is the purpose of refluxing the reaction mixture for 45 minutes? Why not simply boil the mixture in an Erlenmeyer flask?
6. Why is it necessary to remove all water from 1-bromobutane before distilling it?
7. Write reaction mechanisms showing how 1-butene and di-*n*-butyl ether are formed.
8. Why is the resonance of the bromine-bearing carbon atom and the hydroxyl-bearing carbon atom the farthest downfield in Figures 16.3 and 16.4?

17 CHAPTER

Nucleophilic Substitution Reactions of Alkyl Halides

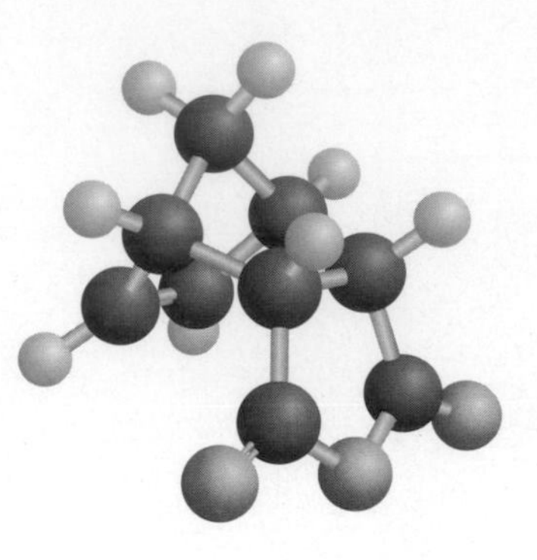

PRELAB EXERCISE: Predict the outcomes of the two sets of experiments to be carried out with the 11 halides used in this chapter.

The alkyl halides, R—X (where X = Cl, Br, I, and sometimes F), play a central role in organic synthesis. These can easily be prepared from, among others, alcohols, alkenes, and, industrially, alkanes. In turn, they are the starting materials for the synthesis of a large number of new functional groups. The syntheses are often carried out by nucleophilic substitution reactions in which the halide is replaced by another group such as cyano, hydroxyl, ether, ester, alkyl—the list is long. As a consequence of the importance of this substitution reaction, it has been studied carefully by employing reactions such as the two used in this chapter. Some of the questions that can be asked are as follows:

- How does the structure of the alkyl part of the alkyl halide affect the reaction?
- What is the effect of changing the . . .

  nature of the halide?

  nature of the solvent?

  relative concentrations of the reactants?

  temperature of the reaction?

  nature of the nucleophile?

In this chapter, we explore the answers to some of these questions.

In free radical reactions, the covalent bond undergoes homolysis when it breaks:

$$\mathrm{R{:}\ddot{\underset{\cdot\cdot}{Cl}}{:}} \longrightarrow \mathrm{R\cdot} + \mathrm{\cdot\ddot{\underset{\cdot\cdot}{Cl}}{:}}$$

A carbocation is planar and has an empty *p*-orbital.

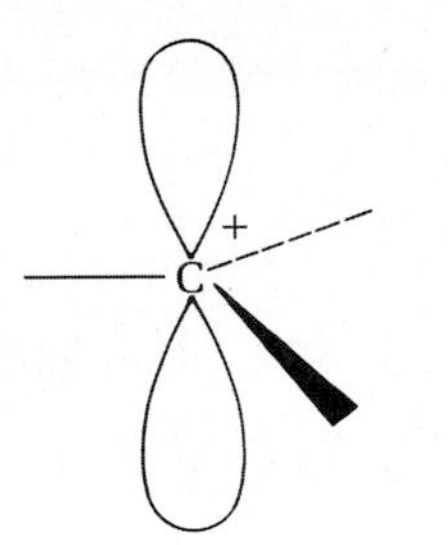

whereas in ionic reactions, it undergoes heterolysis:

$$R:\ddot{\underset{..}{Cl}}: \longrightarrow R^+ + :\ddot{\underset{..}{Cl}}:^-$$

A carbocation is often formed as a reactive intermediate in these reactions. This carbocation is sp^2 hybridized and trigonal-planar in structure with a vacant *p*-orbital. Much experimental evidence of the type obtained in this chapter indicates that the order of stability of carbocations is:

Order of carbocation stability

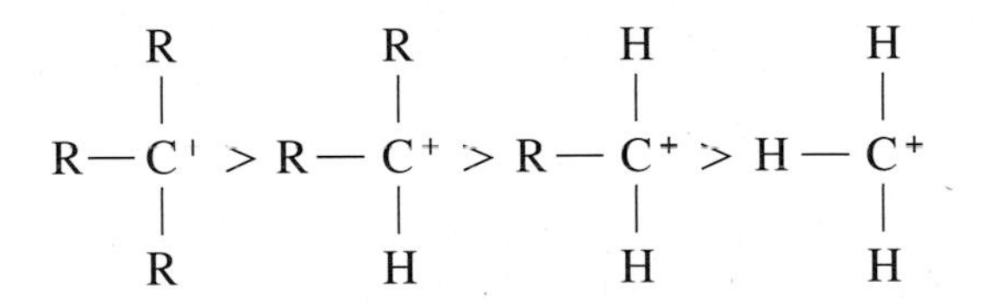

The alkyl (R) groups stabilize the positive charge of the carbocation by displacing or releasing electrons toward the positive charge. As the number of alkyl groups attached to the carbocation increases, the stabilization increases.

Nucleophilic substitution

Many organic reactions occur when a nucleophile (a species with an unshared pair of electrons) reacts with an alkyl halide to replace the halogen with the nucleophile.

$$Nu:^- + R:\ddot{\underset{..}{X}}: \longrightarrow Nu:R + :\ddot{\underset{..}{X}}:^-$$

One step

This substitution reaction can occur in one smooth step:

$$Nu:^- + R:\ddot{\underset{..}{X}}: \longrightarrow \left[\overset{\delta-}{Nu} \cdots\cdots R \cdots\cdots \overset{\delta-}{\ddot{\underset{..}{X}}:}\right] \longrightarrow Nu:R + :\ddot{\underset{..}{X}}:^-$$

Two steps

or it can occur in two discrete steps:

$$R:\ddot{\underset{..}{X}}: \longrightarrow R^+ + :\ddot{\underset{..}{X}}:^-$$

$$Nu:^- + R^+ \longrightarrow Nu:R$$

depending primarily on the structure of the R group. The nucleophile ($Nu:^-$) can be a substance with a full negative charge, such as $:\ddot{\underset{..}{I}}:^-$ or $H:\ddot{\underset{..}{O}}:^-$, or an uncharged molecule with an unshared pair of electrons such as exists on the oxygen atom in water, $H—\ddot{\underset{..}{O}}—H$. Not all of the halides, $:\ddot{\underset{..}{X}}:^-$, depart with equal ease in nucleophilic substitution reactions. In this chapter, we investigate the ease with which the different halogens leave in one of the substitution reactions.

To distinguish between the reaction that occurs as one smooth step and the reaction that occurs as two discrete steps, it is necessary to study the kinetics of the reaction. If the reaction were carried out with several different concentrations of $R:\ddot{\underset{..}{X}}:^-$ and $Nu:^-$, we could determine if the reaction is bimolecular or unimolecular.

Reaction kinetics

In the case of the smooth, one-step reaction, the nucleophile must collide with the alkyl halide. The kinetics of the reaction:

$$\mathrm{Nu{:}^- + R{:}\ddot{X}{:} \longrightarrow \left[\overset{\delta-}{Nu} \cdots\cdots R \cdots\cdots \overset{\delta-}{\ddot{X}}{:}\right] \longrightarrow Nu{:}R + {:}\ddot{X}{:}^-}$$

$$\mathrm{Rate} = k\left[\mathrm{Nu{:}^-}\right]\left[\mathrm{R{:}\ddot{X}{:}}\right]$$

are found to depend on the concentration of both the nucleophile and the halide. Such a reaction is said to be a bimolecular nucleophilic substitution reaction, S_N2. If the reaction occurs as a two-step process,

$$\mathrm{R{:}\ddot{X}{:} \xrightarrow{slow} R^+ + {:}\ddot{X}{:}^-}$$

$$\mathrm{Nu{:}^- + R^+ \xrightarrow{fast} Nu{:}R}$$

$$\mathrm{Rate} = k\left[\mathrm{R{:}\ddot{X}{:}}\right]$$

the rate of the first step, the slow step, depends only on the concentration of the halide, and it is said to be a unimolecular nucleophilic substitution reaction, S_N1.

Racemization: S_N1

The S_N1 reaction proceeds through a planar carbocation. Even if the starting material were chiral, the product would be a 50:50 mixture of enantiomers because the planar intermediate can form a bond with the nucleophile on either face.

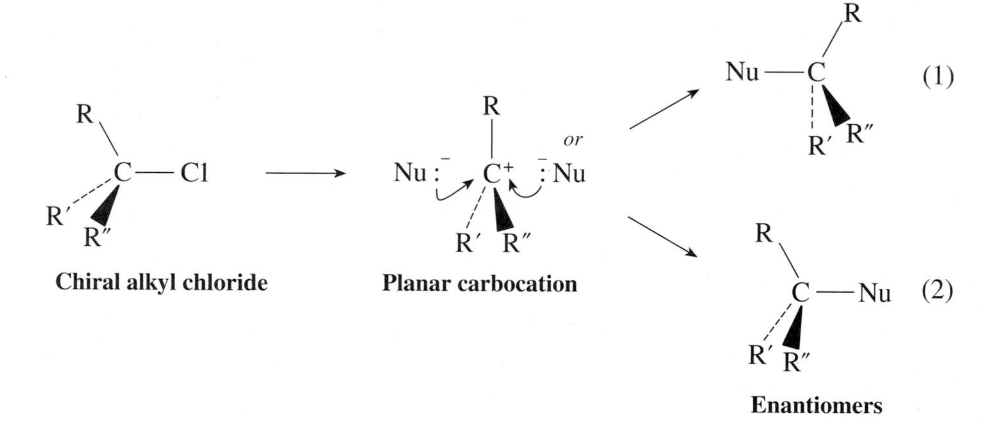

Inversion: S_N2

The S_N2 reaction occurs with inversion of configuration to give a product of the opposite chirality from the starting material.

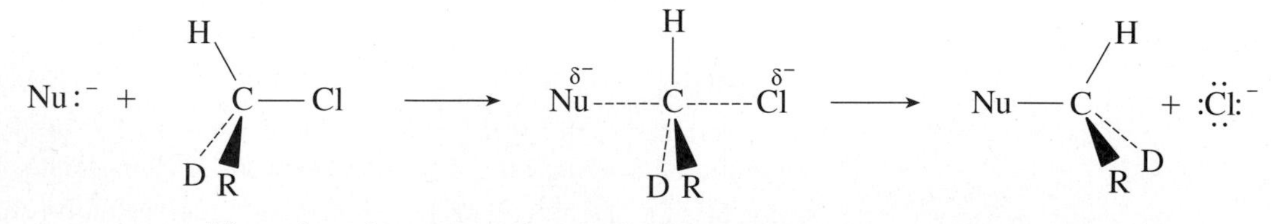

Order of S_N2 reactivity

The order of reactivity for *simple* alkyl halides in the S_N2 reaction is:

$$\mathrm{CH_3{-}X > R{-}CH_2{-}X > R{-}\underset{\large R}{\underset{|}{CH}}{-}X > R{-}\overset{R}{\overset{|}{\underset{R}{\underset{|}{C}}}}{-}X}$$

The tertiary halide is in parentheses because it usually does not react by an S_N2 mechanism. The primary factor in this order of reactivity is steric hindrance, that is, the ease with which the nucleophile can come within bonding distance of the alkyl halide; 2,2-dimethyl-1-bromopropane, even though it is a primary halide, reacts 100,000 times slower than ethyl bromide (CH_3CH_2Br) because of steric hindrance that prevents attack on the bromine atom in the dimethyl compound.

Steric hindrance toward S_N2

$$CH_3-\underset{\large CH_3}{\overset{\large CH_3}{\underset{|}{\overset{|}{C}}}}-CH_2-Br \qquad\qquad CH_3CH_2Br$$

Rate = 1 Rate = 100,000

The primary factor in S_N1 reactivity is the relative stability of the carbocation that is formed. For simple alkyl halides, this means that only tertiary halides react by this mechanism. The tertiary halide must be able to form a planar carbocation. Only slightly less reactive are the allyl carbocations, which derive their great stability from the delocalization of the charge on the carbon by resonance.

The allyl carbocation

$$CH_2{=}CH-CH_2-\ddot{\underset{..}{Br}}: \longrightarrow CH_2{=}CH-\overset{+}{C}H_2 \longleftrightarrow \overset{+}{C}H_2-CH{=}CH_2 + :\ddot{\underset{..}{Br}}:^-$$

The nature of the solvent has a large effect on the rates of S_N2 reactions. In a solvent with a hydrogen atom attached to an electronegative atom such as oxygen, the protic solvent forms hydrogen bonds to the nucleophile. These solvent molecules get in the way during an S_N2 reaction.

Solvent effects

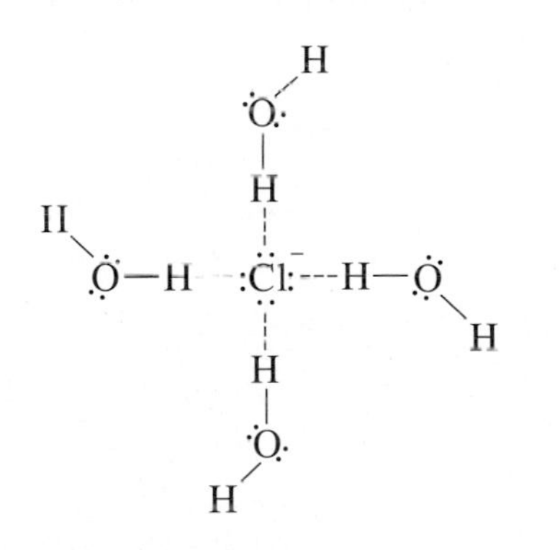

Aprotic: no ionizable protons

If the solvent is polar and aprotic, solvation of the nucleophile cannot occur, and the S_N2 reaction can occur up to a million times faster. Some common polar, aprotic solvents are N,N-dimethylformamide and dimethylsulfoxide:

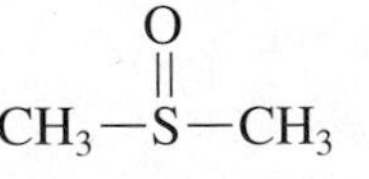

N,N-Dimethylformamide **Dimethylsulfoxide**

In the S_N1 reaction, a polar protic solvent such as water stabilizes the transition state more than it does the reactants, lowering the energy of activation for the reaction and thus increasing the rate, relative to the rate in a nonpolar solvent. Acetic acid, ethanol, and acetone are relatively nonpolar solvents and have lower dielectric constants than the polar solvents water, dimethylsulfoxide, and N,N-dimethylformamide.

The leaving group

The rate of S_N1 and S_N2 reactions depends on the nature of the leaving group, the best leaving groups being the ones that form stable ions. Among the halogens, we find that the iodide ion (I^-) is the best leaving group as well as the best nucleophile in the S_N2 reaction.

Vinylic and aryl halides

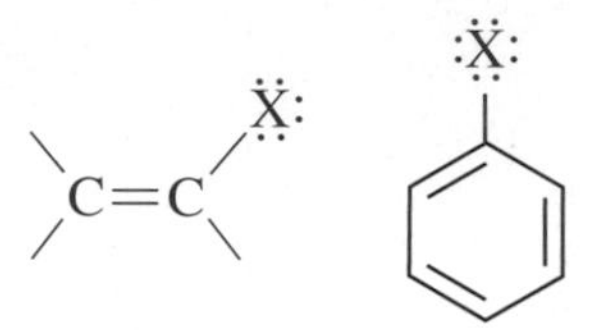

do not normally react by S_N1 reactions because the resulting carbocations

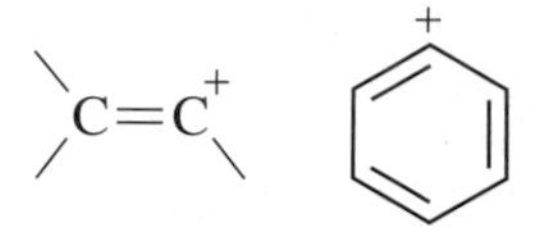

are relatively difficult to form. For S_N2 reactions, electrons in the nearby double bonds repel the nucleophile, which is either an ion or a polarized neutral species.

Temperature dependence

The rates of both S_N1 and S_N2 reactions depend on the temperature of the reaction. As the temperature increases, the kinetic energy of the molecules increases, leading to a greater rate of reaction. The rate of many organic reactions will approximately double when the temperature increases about 10°C. This information is summarized in Table 17.1.

TABLE 17.1 *Summary of S_N1 and S_N2 Reactions*

	Unimolecular Nucleophilic Substitution (S_N1)	*Second-Order Nucleophilic Substitution (S_N2)*
Kinetics	First order	Second order
Mechanism	Two steps; unimolecular in the rate-determining step *via* a carbocation	One-step, bimolecular
Stereochemistry	Racemization predominates	Inversion of configuration
Reactivity of structure	$3° > 2° > 1° > CH_3 >$ vinyl	$3° < 2° < 1° < CH_3$
Rearrangements	May occur	No rearrangements because no carbocation intermediate
Effect of leaving group	—I > —Br > —Cl>> —F	—I > —Br > —Cl >> —F
Effect of nucleophile	Not important because it is not in the rate-determining step	$I^- > Br^- > Cl^- > F^-$
Concentration of nucleophile	S_N1 is favored by low concentration	S_N2 is favored by high concentration
Solvent polarity	High, favors S_N1	Low, favors S_N2

EXPERIMENTS

In the experiments that follow, 11 representative alkyl halides are treated with sodium iodide in acetone and with an ethanolic solution of silver nitrate.

1. SODIUM IODIDE IN ACETONE

Organic halides that can react by an S_N2 mechanism give a precipitate of NaX with sodium iodide in acetone.

Acetone, with a dielectric constant of 21, is a relatively nonpolar solvent that will readily dissolve sodium iodide. The iodide ion is an excellent nucleophile, and the nonpolar solvent (acetone) favors the S_N2 reaction; it does not favor ionization of the alkyl halide. The extent of reaction can be observed because sodium bromide and sodium chloride are not soluble in acetone, and precipitate from solution if a reaction occurs.

$$Na^+I^- + R{-}Cl \longrightarrow R{-}I + NaCl\downarrow$$
$$Na^+I^- + R{-}Br \longrightarrow R{-}I + NaBr\downarrow$$

2. ETHANOLIC SILVER NITRATE SOLUTION

Organic halides that can react by an S_N1 mechanism give a precipitate of AgX with an ethanolic silver nitrate solution.

When an alkyl halide is treated with an ethanolic solution of silver nitrate, the silver ion coordinates with an electron pair of the halogen. This weakens the carbon-halogen bond because a molecule of insoluble silver halide is formed, thus promoting an S_N1 reaction of the alkyl halide. The solvent, ethanol, favors ionization of the halide, and the nitrate ion is a very poor nucleophile, so alkyl nitrates do not form by an S_N2 reaction.

$$R{-}\ddot{\underset{..}{X}}: \underset{}{\overset{Ag^+}{\rightleftharpoons}} \overset{\delta^+}{R}\ddot{\underset{..}{X}}\overset{\delta^+}{Ag} \longrightarrow R^+ + AgX\downarrow$$

On the basis of the foregoing discussion, tertiary halides would be expected to react with silver nitrate most rapidly, and primary halides least rapidly. The R^+ ion can react with the solvent to give either an alkene or an ether.

MICROSCALE PROCEDURE

Label 11 small containers (reaction tubes, 3-mL centrifuge tubes, 10 × 75-mm test tubes, or 1-mL vials), and place 0.1 mL or 100 mg of each of the following halides in the tubes.

$CH_3CH_2CH_2CH_2Cl$
1-Chlorobutane
bp 77–78°C

$CH_3CH_2CH_2CH_2Br$
1-Bromobutane
bp 100–104°C

$CH_3CH_2CH(Cl)CH_3$
2-Chlorobutane
bp 68–70°C

$(CH_3)_3C{-}Cl$
2-Chloro-2-methylpropane
bp 51–52°C

Br

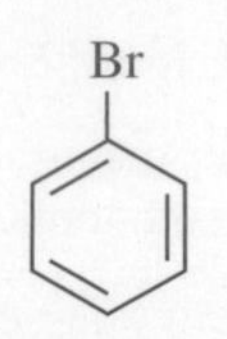

Bromobenzene
bp 156°C

$CH_3CH{=}CHCH_2Cl$

1-Chloro-2-butene mixture of *cis* and *trans* isomers
bp 63.5°C (*cis*)

bp 68°C (*trans*)

CH_3
|
CH_3CHCH_2Cl

1-Chloro-2-methylpropane
bp 68–69°C

Br
|
$CH_3CH_2CHCH_3$

2-Bromobutane
bp 91°C

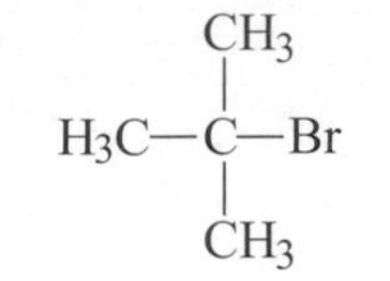

2-Bromo-2-methylpropane
bp 71–73°C

Once all of the halides have been placed into labeled tubes or vials, arrange them into the following groupings:

Group 1: 1-bromobutane, 2-bromobutane, 2-bromo-2-methylpropane, and bromobenzene

Group 2: 1-chlorobutane, 2-chloro-2-methylpropane, 1-chloro-2-butene

Group 3: 1-bromobutane, 2-bromobutane, 1-chlorobutane, 2-chlorobutane

When you perform experiments on these halides, be sure to do them in the order in which they appear in their group lists in order to observe trends and make conclusions about the reactivity of each halide within its respective group.

To each tube of each group, rapidly add 1 mL of an 18% solution of sodium iodide in acetone, stopper each tube, mix the contents thoroughly, and note the time. Note the time of first appearance of any precipitate. If no reaction occurs within about 5 minutes, place those tubes in a 50°C water bath and watch for any reaction over the next 5 or 6 minutes.

Empty the tubes, rinse them with ethanol, place the same amount of each of the alkyl halides in each tube as in the first part of the experiment and organize them by the groupings described above. Add 1 mL of 1% ethanolic silver nitrate solution to each tube, mix the contents well, and note the time of addition as well as the time of appearance of the first traces of any precipitate. If a precipitate does not appear in 5 minutes, heat the tubes containing these unreactive halides in a 50°C water bath for 5 to 6 minutes and watch for any reaction.

To test the effect of solvent on the rate of S_N1 reactivity, compare the time needed for a precipitate to appear when 2-chlorobutane is treated with a 1% ethanolic silver nitrate solution, and when treated with 1% silver nitrate in a mixture of 50% ethanol and 50% water.

In your analysis of the results from these experiments, consider the following for both S_N1 and S_N2 conditions:

- the nature of the leaving group (Cl vs. Br) in the 1-halobutanes
- the effect of structure, that is, compare simple primary, secondary, and tertiary halides, unhindered primary vs. hindered primary halides, a simple tertiary halide vs. a complex tertiary halide, and an allylic halide vs. a tertiary halide

- the effect of solvent polarity on the S_N1 reaction
- the effect of temperature on the reaction

See *Web Links*

Cleaning Up. Because all of the test solutions contain halogenated material, all test solutions and washes as well as unused starting materials should be placed in the halogenated organic waste container.

QUESTIONS

1. What would be the effect of carrying out the sodium iodide in acetone reaction with the alkyl halides using an iodide solution half as concentrated?

2. The addition of sodium or potassium iodide catalyzes many S_N2 reactions of alkyl chlorides or bromides. Explain.

3. In S_N1 reactions, the intermediate carbocations can eliminate a proton to yield alkenes or react with the solvent to yield ethers. Draw the structures of the byproducts of this type that would be derived from the reaction of the carbocation derived from 2-bromo-2-methylbutane in ethanol.

18 CHAPTER

Radical Initiated Chlorination of 1-Chlorobutane

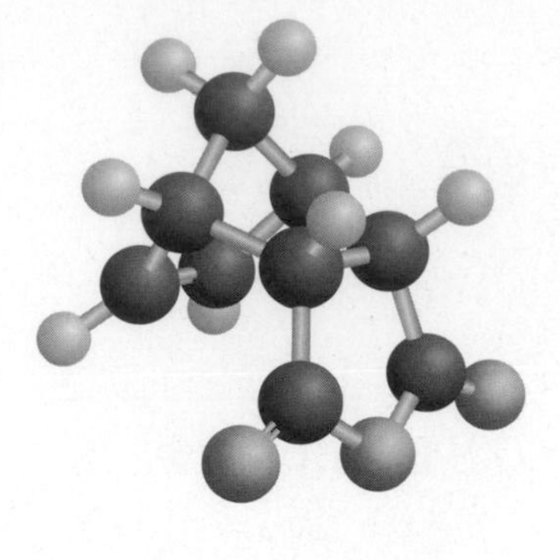

PRELAB EXERCISE: Write two chain propagation steps for the reaction of butane with sulfuryl chloride.

Most of the alkanes from petroleum are used to produce energy by combustion, but a small percentage are converted to industrially useful compounds by controlled reaction with oxygen or chlorine. The alkanes are inert to attack by most chemical reagents, but will react with oxygen and halogens under the special conditions of radical-initiated reactions.

At room temperature, an alkane such as butane will not react with chlorine. In order for the following reaction to occur,

$$CH_3CH_2CH_2CH_3 + Cl_2 \longrightarrow CH_3CH_2CH_2CH_2Cl + HCl$$

a precisely oriented four-center collision of the Cl_2 molecule with the butane molecule must occur, with two bonds broken and two bonds formed simultaneously:

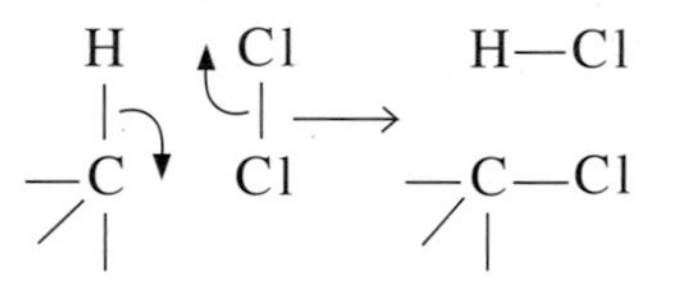

It is unlikely that the necessarily precise orientation of the two reacting molecules will be found. They must, of course, be within bonding distance of each other as well, so steric factors play a part.

If a concerted four-center reaction does not work, then an alternative possibility is a stepwise mechanism:

$$:\ddot{\underset{..}{Cl}}:\ddot{\underset{..}{Cl}}: \xrightarrow{\text{slow}} 2\ :\ddot{\underset{..}{Cl}}\cdot \quad (1)$$

$$CH_3CH_2CH_2CH_3 \xrightarrow{\text{slow}} CH_3CH_2CH_2CH_2\cdot + H\cdot \quad (2)$$

$$:\ddot{\underset{..}{Cl}}\cdot + CH_3CH_2CH_2CH_2\cdot \xrightarrow{\text{fast}} CH_3CH_2CH_2CH_2Cl \quad (3)$$

$$:\ddot{\underset{..}{Cl}}\cdot + H\cdot \xrightarrow{\text{fast}} H:\ddot{\underset{..}{Cl}}: \quad (4)$$

For this series of reactions to occur, the chlorine molecule must dissociate into two chlorine atoms. Because chlorine has a bond energy of 58 kcal/mole, we would not expect any significant numbers of molecules to dissociate at room temperature. Thermal motion at 25°C can only break bonds having energies less than 30–35 kcal/mole. Although thermal dissociation would require a high temperature, the dissociation of chlorine into atoms (chlorine radicals) can be caused by violet and ultraviolet light:

$$Cl_2 \xrightarrow{h\nu} 2\ :\ddot{\underset{..}{Cl}}\cdot$$

The photon energy of red light is 48 kcal/mole, whereas light of 300 nm (ultraviolet) has a photon energy of 96 kcal/mole.

The chlorine radical can react with butane by abstraction of a hydrogen atom:

$$CH_3CH_2CH_2CH_3 + :\ddot{\underset{..}{Cl}}\cdot \rightleftharpoons CH_3CH_2CH_2CH_2\cdot + HCl$$

a reaction that is very slightly exothermic (and thus written as a reversible reaction). The butyl radical can react with chlorine:

$$CH_3CH_2CH_2CH_2\cdot + Cl_2 \longrightarrow CH_3CH_2CH_2CH_2Cl + :\ddot{\underset{..}{Cl}}\cdot$$

a reaction that evolves 26 kcal/mole of energy. The net result of these two reactions is a reaction of Cl_2 with butane catalyzed by $:\ddot{\underset{..}{Cl}}\cdot$ because a chlorine radical is produced for every chlorine radical consumed. The whole process can be terminated by the reaction of radicals with each other:

$$CH_3CH_2CH_2CH_2\cdot + :\ddot{\underset{..}{Cl}}\cdot \longrightarrow CH_3CH_2CH_2CH_2Cl$$

$$2\ CH_3CH_2CH_2CH_2\cdot \longrightarrow CH_3(CH_2)_6CH_3$$

$$2\ :\ddot{\underset{..}{Cl}}\cdot \longrightarrow Cl_2$$

To summarize, the process of light-induced radical chlorination involves three steps: chain initiation in which chlorine radicals are produced, chain propagation that involves no net consumption of chlorine radicals, and chain termination that destroys radicals.

$$Cl_2 \longrightarrow 2\ :\ddot{\underset{..}{Cl}}\cdot \qquad \textit{Chain initiation}$$

$$\left.\begin{array}{l} CH_3CH_2CH_2CH_3 + :\ddot{\underset{..}{Cl}}\cdot \rightleftharpoons CH_3CH_2CH_2CH_2\cdot + HCl \\ CH_3CH_2CH_2CH_2\cdot + Cl_2 \longrightarrow CH_3CH_2CH_2CH_2Cl + :\ddot{\underset{..}{Cl}}\cdot \end{array}\right\} \textit{Chain propagation}$$

$$\left.\begin{array}{l} CH_3CH_2CH_2CH_2\cdot + :\ddot{\underset{..}{Cl}}\cdot \longrightarrow CH_3CH_2CH_2CH_2Cl \\ 2\ CH_3CH_2CH_2CH_2\cdot \longrightarrow CH_3(CH_2)_6CH_3 \\ 2\ :\ddot{\underset{..}{Cl}}\cdot \longrightarrow Cl_2 \end{array}\right\} \textit{Chain termination}$$

RELATIVE REACTIVITIES OF HYDROGEN

In the preceding example, the product is shown to be 1-chlorobutane, but in fact the reaction produces a mixture of 1-chlorobutane and 2-chlorobutane. If the reaction were to occur purely by chance, we would expect the ratio of products to be 6:4 because there are six primary hydrogens in the two CH_3 groups, and four secondary hydrogens in the two CH_2 groups of butane. But because a secondary C—H bond is weaker than a primary C—H bond (95 kcal/mole vs. 98 kcal/mole), we might expect more 2-chlorobutane than chance would dictate.

TABLE 18.1 *Product Distribution for the Chlorination of 2-Methylbutane*

Reactant		*Products*	*Statistical Expectation* *Ratio*	*Statistical Expectation* *Percent*	*Found (%)*	*Relative Reactivity*
$CH_3-CH(CH_3)-CH_2CH_3$	$\xrightarrow[300°]{Cl_2}$	$CH_3-CH(CH_3)-CHCl-CH_3$	2/12	17	33	3.3
		$ClCH_2-CH(CH_3)-CH_2-CH_3$	6/12	50	30	1
		$CH_3-CCl(CH_3)-CH_2-CH_3$	1/12	8	22	4.4
		$CH_3-CH(CH_3)-CH_2-CH_2Cl$	3/12	25	15	1

The chlorination of 2-methylbutane is summarized in Table 18.1. As can be seen from the table, the relatively weak tertiary C—H bond (92 kcal/mole) gives rise to 22% of product in contrast to the 8% expected on the basis of a random attack of :C̈l· on the starting material.

The relative reactivities of the various hydrogens of 2-methylbutane on a per-hydrogen basis (referred to the primary hydrogens of C-4 as 1.0) can be calculated. The molecule has nine primary hydrogens, chlorination of which accounts for 45% of the products. This means that each primary hydrogen is responsible for $45\% \div 9 = 5.0\%$ of the product. Similarly, the two secondary hydrogens account for 33% of the product, so each secondary hydrogen is responsible for 16.5% of the product. Finally, the one tertiary hydrogen is responsible for 22% of the product. These results mean that tertiary hydrogens are $22 \div 5.0 = 44$ times as reactive as primar ward chlorination, and secondary hydrogens are $16.5 + 5.0 = 3.3$ times as reactive toward chlorination as primary hydrogens. Results very similar to these have been found for a large number of hydrocarbons. The relative reactivity toward chlorination is as follows:

$$\begin{array}{ccccc} R_3CH & > & R_2CH_2 & > & RCH_3 \\ 4.4 & & 3.3 & & 1 \end{array}$$

A reaction of this type is of little use unless you happen to need the four products in the ratios found, and can manage to separate them (their boiling points are very similar). Industrially, however, the radical chlorination of methane and ethane is important, and the products can be separated easily:

$$CH_4 \xrightarrow[\Delta]{Cl_2} CH_3Cl + CH_2Cl_2 + CHCl_3 + CCl_4$$

	CH_3Cl	CH_2Cl_2	$CHCl_3$	CCl_4
	Chloromethane	**Dichloromethane**	**Trichloromethane**	**Tetrachloromethane**
Common names:	Methyl chloride	Methylene chloride	Chloroform	Carbon tetrachloride
Boiling points:	−24°C	40°C	62°C	77°C

In the first experiment, we will chlorinate 1-chlorobutane because it is easier to handle in the laboratory than gaseous butane, and we will use sulfuryl chloride as our source of chlorine radicals because it is easier to handle than gaseous chlorine. Instead of using light to initiate the reaction, we will use a chemical initiator, 2,2′-azobis-(2-methylpropionitrile). This azo compound (R—N=N—R) decomposes at moderate temperatures (80–100°C) to give two relatively stable radicals and nitrogen gas:

The common name for this catalyst is AIBN, which stands for azoisobutylnitrile.

$$CH_3-C(CN)(CH_3)-N{=}N-C(CN)(CH_3)-CH_3 \xrightarrow{80-100^\circ} 2\ CH_3-\dot{C}(CN)(CH_3) + N_2\uparrow$$

2,2′-Azobis-(2-methylpropionitrile)
MW 164.21, mp 102–103°(dec.)

$$CH_3-\dot{C}(CN)(CH_3) + Cl-SO_2-Cl \longrightarrow CH_3-C(CN)(CH_3)-Cl + \cdot SO_2-Cl$$

Sulfuryl chloride

$$\cdot SO_2-Cl \longrightarrow SO_2 + :\ddot{C}l\cdot$$

Before conducting the experiment, try to predict the ratios of products.

The radical monochlorination of 1-chlorobutane can give four products: 1,1-, 1,2-, 1,3-, and 1,4-dichlorobutane. If the reaction occurred completely at random, we would expect products in the ratios of the number of hydrogen atoms on each carbon, that is, 2:2:2:3, respectively (22%, 22%, 22%, 33%).

Will these ratios correlate with the calculated heats of formation of the products?

The object of the following experiments is to carry out the radical chlorination of 1-chlorobutane, and then to determine the ratios of products using gas chromatography. From these ratios, the relative reactivities of the hydrogens can be calculated.

$$CH_3CH_2CH_2CH_2Cl + SO_2Cl_2 \xrightarrow{R\cdot} CH_3CH_2CH_2CHCl_2 + CH_3CH_2CHClCH_2Cl + CH_3CHClCH_2CH_2Cl + CH_2ClCH_2CH_2CH_2Cl + SO_2 + HCl$$

1-Chlorobutane: MW 92.57, den. 0.886, bp 77–78°C

Sulfuryl chloride: MW 134.97, den. 1.67, bp 69°C

$CH_3CH_2CH_2CHCl_2$: bp 114°C; $CH_3CH_2CHClCH_2Cl$: bp 124°C; $CH_3CHClCH_2CH_2Cl$: bp 134°C; $CH_2ClCH_2CH_2CH_2Cl$: bp 162°C

EXPERIMENTS

1. RADICAL CHLORINATION OF 1-CHLOROBUTANE

IN THIS EXPERIMENT, chlorine radicals (from sulfuryl chloride and a catalyst) react with 1-chlorobutane to form four different dichlorobutanes. The amounts of each product depend on the number and kinds of hydrogens in the starting material. The product is analyzed by gas chromatography. Because the reaction gives off both hydrogen chloride and sulfur dioxide, these gases are trapped. The crude product is diluted with water, the organic phase is washed with dilute base and water, and then dried before carrying out the gas chromatography.

To a 10 × 100 mm reaction tube add 1-chlorobutane (0.50 mL, 0.432 g), sulfuryl chloride (0.16 mL, 0.27 g), 2,2′-azobis-(2-methylpropionitrile) (4 mg,), and a boiling chip. Use a 1-mL graduated pipette to measure the 1-chlorobutane, and use a dispenser or a 0.5-mL syringe to measure the sulfuryl chloride (in the hood). Rinse the syringe immediately after use and leave it disassembled. Weigh the azo compound on a balance. In this experiment, none of the reagents need to be measured with great care; the 1-chlorobutane is in large excess, and the azo compound is present in catalytic amounts.

Fit the reaction tube with a rubber septum and a piece of polyethylene tubing that leads down into another reaction tube, the mouth of which has a piece of damp cotton placed in it (Fig. 18.1). During this reaction, sulfur dioxide and hydrogen chloride gas are evolved, and the damp cotton will absorb the gas. The technique for threading a polyethylene tube through a septum is shown in Figure 18.2. The amount of HCl evolved in this experiment (2 mmol) is equal to 0.15 mL of concentrated hydrochloric acid. Be sure that the end of the polyethylene tube does not touch any water because it would be sucked back into the reaction tube. Clamp the reaction tube with the reactants in a beaker of hot water maintained at 80–85°C so that just the tip of the tube is immersed in the water. This will cause the contents to boil gently and the vapors to condense on the cool upper walls of the reaction tube.

Photos: *Placing a Polyethylene Tube through a Septum, Gas Trap*

At the end of the reaction period, remove the tube from the beaker of water, allow it to cool, and then carefully add 0.5 mL of water dropwise to the tube from a Pasteur pipette. Note which layer is the aqueous one. Mix the contents thoroughly and then draw off and discard the water layer. Wash the organic phase in the same way with a 0.5-mL portion of 5% sodium bicarbonate solution, and once with 0.5 mL water. Carefully remove all the water with a Pasteur pipette and then add anhydrous calcium chloride pellets to dry the product. Transfer the dry product (it should be perfectly clear, not cloudy) to a small, tared, screw-capped sample vial or corked reaction tube. Do not store the product in a polyethylene or rubber-capped container because it will be readily absorbed by these materials. It is not a good idea to store the product for a prolonged period before analysis because the composition will change, depending on which components evaporate or are absorbed by the cap on the container. Determine the weight of the product and then analyze it by gas chromatography.

FIG. 18.1

The microscale apparatus for the chlorination of 1-chlorobutane with HCl and SO_2 trap. Heat the polyethylene tubing carefully in a steam bath to bend it permanently. The tubes are to be clamped appropriately.

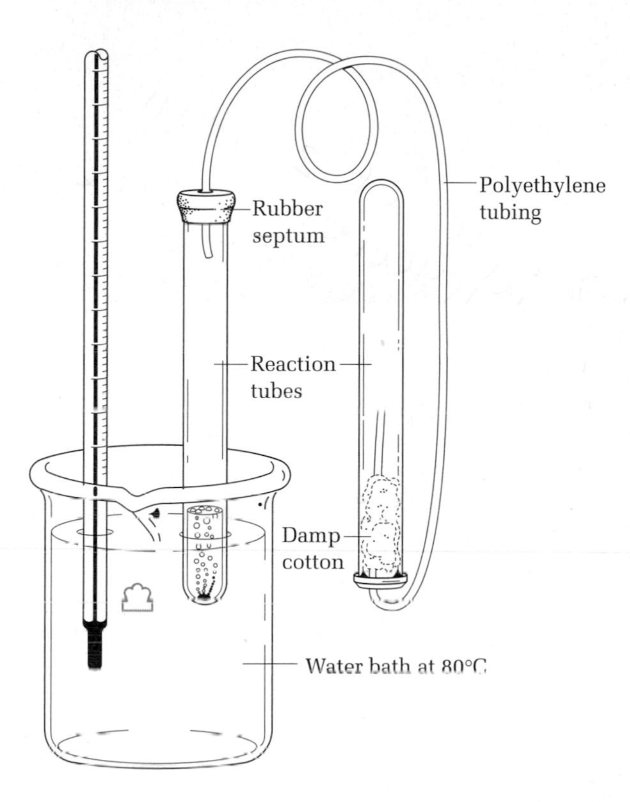

FIG. 18.2

To thread a polyethylene tube through a septum, make a hole through the septum with a needle and then push a toothpick through the hole. Push the polyethylene tube firmly onto the toothpick, then pull and push on the toothpick to thread the tubing through the septum. Finally, pull the tube from the toothpick after it has come out the other side of the septum. A blunt syringe needle can be used instead of a toothpick.

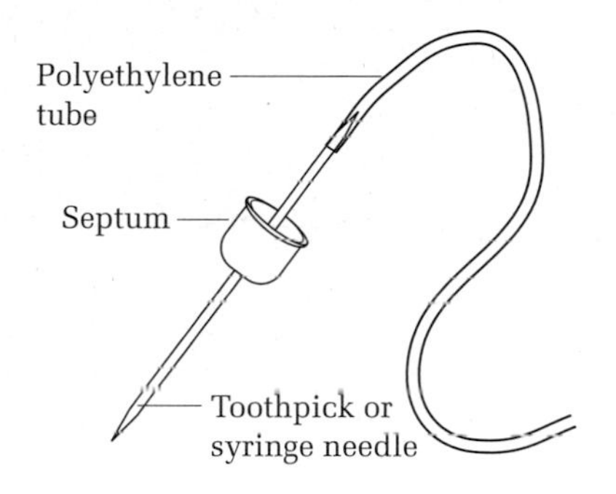

Cleaning Up. Rinse the damp cotton with water and combine the rinse with all the aqueous layers and washes. Neutralize the aqueous solution with sodium carbonate and flush the solution down the drain. The drying agent will be coated with chlorinated product and must therefore be disposed of in the hazardous waste container. Any unused starting material and product must be placed in the halogenated organic waste container.

2. FREE-RADICAL CHLORINATION OF 1-CHLOROBUTANE

Conduct this experiment in a laboratory hood.

To a 25-mL round-bottomed flask, add 1-chlorobutane (5 mL, 4.32 g), sulfuryl chloride (1.6 mL, 2.7 g), 2,2′-azobis-(2-methylpropionitrile) (0.03 g), and a boiling chip. Equip the flask with a condenser and gas trap (Fig. 18.3). Heat the mixture to gently reflux on a steam bath for 20 minutes. Remove the flask from the steam bath, allow it to cool somewhat, and quickly—to minimize the escape of sulfur dioxide and hydrochloric acid—lift the condenser from the flask and add a second 0.03-g portion of the initiator. Heat the reaction mixture for an additional 10 minutes, remove the flask and condenser from the steam bath, and cool the flask in a beaker of water. Pour the contents of the flask through a funnel into about 10 mL of water in a small separatory funnel, shake the mixture, and separate the two phases. Wash

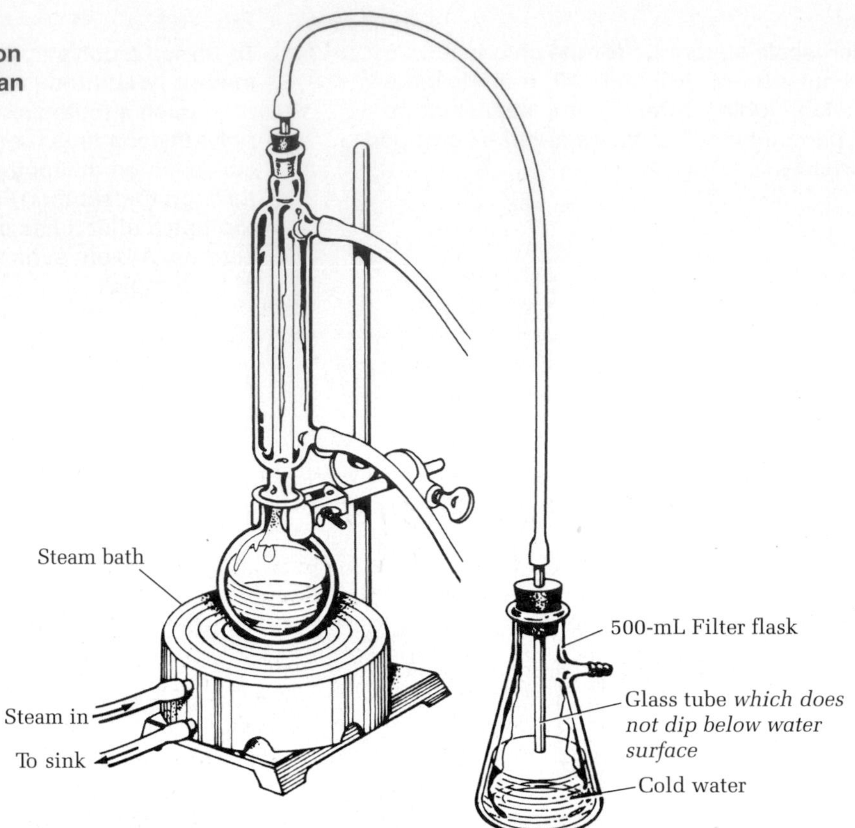

FIG. 18.3
The apparatus for the chlorination of 1-chlorobutane with SO_2 and an HCl gas trap.

the organic phase with two 4-mL portions of 0.5 *M* sodium bicarbonate solution, and once with a 4-mL portion of water; then dry the organic layer over anhydrous calcium chloride pellets (about 1 g) in a dry Erlenmeyer flask. The mixture can be analyzed by gas chromatography at this point, or the unreacted 1-chlorobutane can be removed by fractional distillation (up to bp 85°C) and the pot residue analyzed by gas chromatography.

Cleaning Up. Empty the gas trap and combine the contents with all the aqueous layers and washes. Neutralize the aqueous solution with sodium carbonate and flush the solution down the drain with a large excess of water. The drying agent will be coated with chlorinated product and must therefore be disposed of in the hazardous waste container for solvent-contaminated drying agents. Any unused starting material and product must be placed in the halogenated organic waste container.

GAS CHROMATOGRAPHY

See Chapter 10 for information about gas chromatography. A 30 m × 0.32 mm i.d. 95% methyl/5% phenyl capillary column programmed from 35° to 150°C at 6°/minute should separate all of the dichlorobutanes, as well as any trichlorobutane isomers. Alternately, a Carbowax packed column works well, although any other nonpolar phase such as methyl silicone should work as well. A typical set of packed-column operating conditions would be a column temperature of 100°C, a helium flow rate of 35 mL/min, a column size of 5-mm dia × 2 m, a sample size of 5 μL, and attenuation of 16. With a nonpolar column, the products are expected to come out in the order of their boiling points.

The molar amounts of each compound present in the reaction mixture are proportional to the areas under the peaks in the chromatogram. Because 1-chlorobutane is present in large excess, let this peak run off the paper, but be sure to keep the four product peaks on the paper. If the chromatograph is not equipped with an integrator, determine the relative peak areas by simply cutting out the peaks with a pair of scissors and weighing them. This method of peak integration works very well and depends on the uniform thickness of paper, which results in its weight being proportional to its area. Calculate the relative percent of each product molecule and the partial rate factors relative to the primary hydrogens on carbon-4. Compare your results to those for the chlorination of 2-methylbutane, the data for which were given in Table 18.1.

See *Web Links*

COMPUTATIONAL CHEMISTRY

Construct the four dichlorobutanes, and minimize their energies using a molecular mechanics program. Compare the steric energies for the different conformers to determine if the global minimum for each one has been found. Record the four steric energies and then submit each energy-minimized molecule to an AM1 semi-empirical molecular orbital calculation to calculate the heat of formation. Because one would not expect a correlation to exist between the steric energies and the heats of formation, only the heats of formation can be used to compare the stabilities of the different isomers. Do the heats of formation correlate with the product distributions of the four dichlorobutanes? Does this correlation or lack thereof give information regarding the mechanism of the reaction? Remember that the isomers would need to be in equilibrium with each other so that the most stable is also the one formed in highest yield.

QUESTIONS

1. Using the relative reactivities of the primary and secondary hydrogens in butane as given in Table 18.1, and given that the reactivity of the hydrogens of the—CH_2Cl group is only 2 times that of the primary hydrogens on the methyl group, calculate the ratio of all products based on the number of each type of hydrogen and the relative reactivity.
2. Arrange the following radicals in order of increasing stability, low to high:

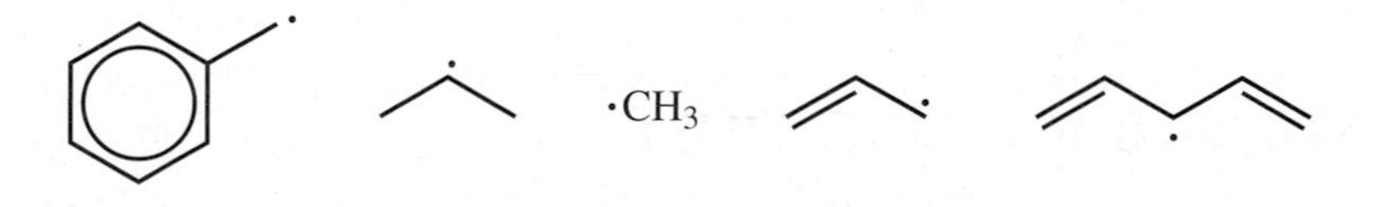

19 CHAPTER

Alkenes from Alcohols: Cyclohexene from Cyclohexanol

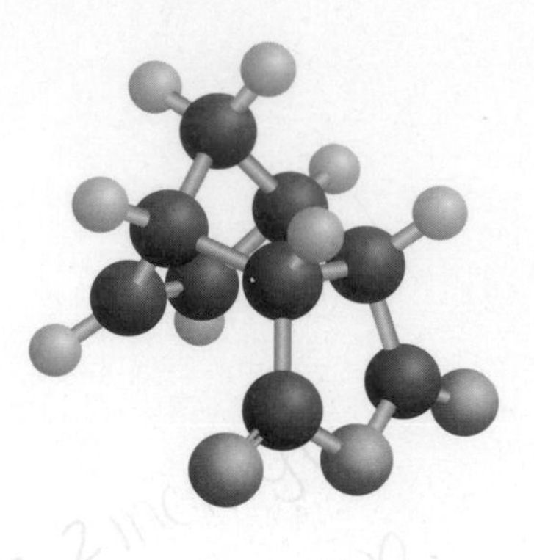

PRELAB EXERCISE: Prepare a detailed flow sheet for the preparation of cyclohexene, indicating at each step which layer contains the desired product.

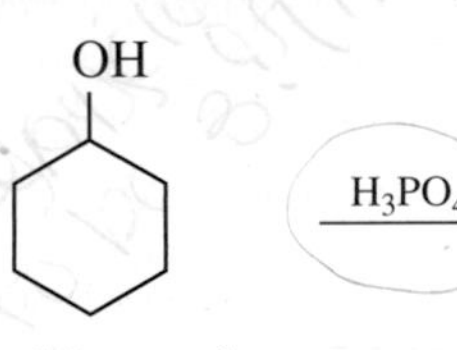

Cyclohexanol
mp 25°C, bp 161°C
den. 0.96, MW 100.16

Cyclohexene
bp 83°C
den. 0.81, MW 82.14

The dehydration of cyclohexanol to cyclohexene can be accomplished by pyrolysis of the cyclic secondary alcohol with an acid catalyst at a moderate temperature, or by distillation over alumina or silica gel. The procedure selected for these experiments involves catalysis by phosphoric acid; sulfuric acid is not more efficient, causes charring, and gives rise to sulfur dioxide. When a mixture of cyclohexanol and phosphoric acid is heated in a flask equipped with a fractionating column, the formation of water soon becomes evident. On further heating, the water and the cyclohexene that form distill together by the principle of steam distillation, and any higher-boiling cyclohexanol that may volatilize is returned to the flask. However, after dehydration is complete and the bulk of the product has distilled, the fractionating column remains saturated with a water-cyclohexene mixture that merely refluxes and does not distill. Hence, for the recovery of otherwise lost reaction product, a chaser solvent is added, and distillation is continued. A suitable chaser solvent is the water-immiscible, aromatic solvent toluene (bp 110°C); as it steam distills, it carries over the more volatile cyclohexene. When the total water-insoluble layer is separated, dried, and redistilled through the dried column, the chaser again drives the cyclohexene from the column; the difference in boiling points is such that a sharp separation is possible. The holdup in a metal sponge-packed column is so great (about 1.0 mL) that if a chaser solvent is not used in the procedure to free it from the column, the yield will be much lower.

The mechanism of this reaction involves initial rapid protonation of the hydroxyl group by the phosphoric acid:

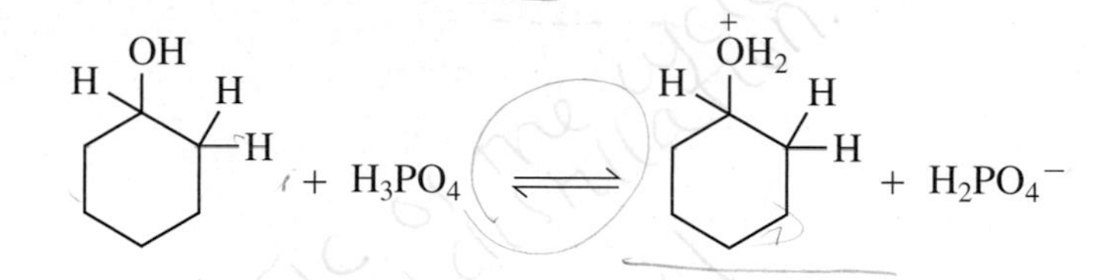

This is followed by loss of water to give the unstable secondary carbocation, which quickly loses a proton to water to give the alkene:

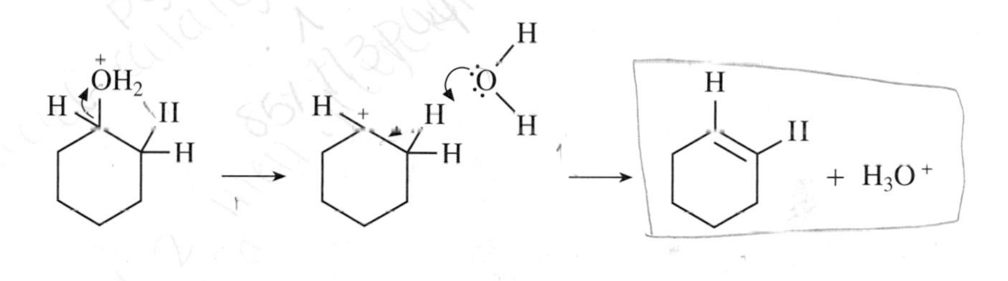

EXPERIMENTS

1. THE PREPARATION OF CYCLOHEXENE

IN THIS EXPERIMENT, cyclohexanol undergoes acid-catalyzed dehydration to give cyclohexene. The product is distilled from the reaction flask along with the water generated. Toluene is used as a chaser solvent. The distillate is washed with salt solution, dried, and fractionally distilled.

⚠ Handle phosphoric acid with care because it is corrosive to tissue.

Introduce 2.0 g of cyclohexanol followed by 0.5 mL of 85% phosphoric acid and a boiling chip into a 5-mL round-bottomed, long-necked flask. Add a copper sponge to the neck of the flask and shake to mix the layers. Heat will evolve. Use the arrangement for fractional distillation as shown in Figure 19.1. Note that the bulb of the thermometer must be completely below the side arm of the distilling head. Wrap the fractionating column and distilling head with glass wool or cotton. (Refer to the fractional distillation experiments in Chapter 5 for details of this technique.)

Use a black Viton connector.

Heat the mixture gently on a sand bath and then distill until the residue in the flask has a volume of about 0.5 mL to 1.0 mL, and very little distillate is being formed; note the temperature range. Let the assembly cool slightly after removing it from the sand, remove the thermometer briefly, and add 2 mL of toluene (the chaser solvent) into the top of the column using a Pasteur pipette. Note the amount of the upper layer in the boiling flask and distill again until the volume of

Using a chaser solvent

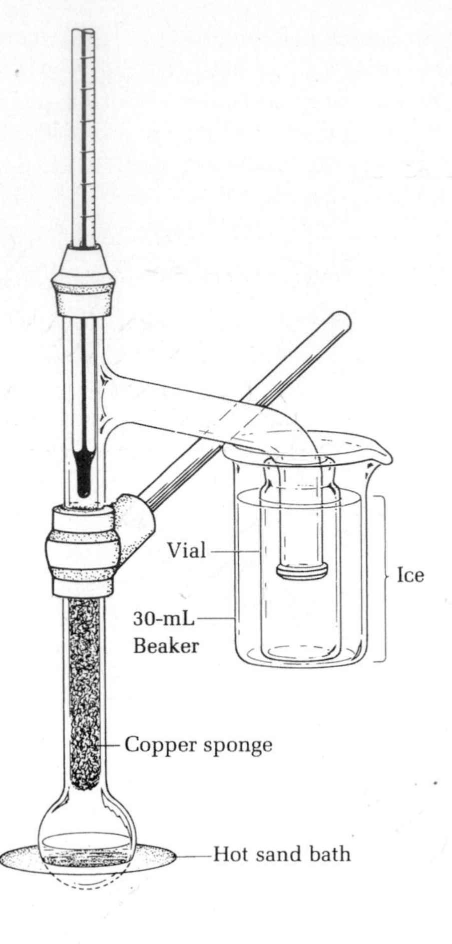

FIG. 19.1
The apparatus for synthesizing cyclohexene from cyclohexanol. The column is packed with a copper sponge.

this layer has been reduced by about half. Transfer the contents of the vial into a reaction tube and rinse out the vial with a little toluene. Use toluene for rinsing in subsequent operations. Wash the mixture with an equal volume of saturated sodium chloride solution, remove the aqueous layer, and then add sufficient anhydrous calcium chloride to the reaction tube so that it does not clump together. Shake the solution with the drying agent and let it dry for at least 5 minutes.

While this is taking place, clean the distilling apparatus first with water, then ethanol, and finally a little acetone. It is absolutely essential that the apparatus be completely dry; otherwise, the product will become contaminated with whatever solvent remains in the apparatus. Before starting the distillation, note the barometric pressure and determine the reading expected given that the boiling point is 83°C at 1 atm. Transfer the dry cyclohexene solution to the distilling flask, add a boiling chip, and distill the product. At the moment the temperature starts to rise above the plateau at which the product distills, stop the distillation to avoid contamination of the cyclohexene with toluene. A typical yield of this volatile alkene is about 1 g. Report your yield in grams and your percentage yield. Run an infrared (IR) spectrum and interpret it for purity. Look for peaks due to starting material and toluene. Gas chromatography is especially useful in the analysis of this compound because the expected impurities differ markedly in their boiling points from the product.

⚠ Cyclohexene is extremely flammable; keep away from an open flame.

Cleaning Up. The aqueous solutions (pot residues and washes) should be diluted with water and neutralized before flushing down the drain. The ethanol wash and sodium chloride solution also can be flushed down the drain, whereas the acetone wash and all toluene-containing solutions should be placed in the organic solvents waste container. Once dried and free of solvent, the calcium chloride can be placed in the nonhazardous solid waste container, if local regulations permit. Otherwise, the wet drying agent should be placed in the designated waste container.

2. THE PREPARATION OF CYCLOHEXENE

Handle phosphoric acid with care because it is corrosive to tissue.

Introduce 20.0 g of cyclohexanol (technical grade), 5 mL of 85% phosphoric acid, and a boiling stone into a 100-mL round-bottomed flask and shake to mix the layers. Note the evolution of heat. Use the typical arrangement for fractional distillation as shown in Figure 5.11 (on page 102), but modified by the use of a bent adapter delivering into an ice-cooled test tube in a 125-mL Erlenmeyer receiver, as shown in Figure 19.2.

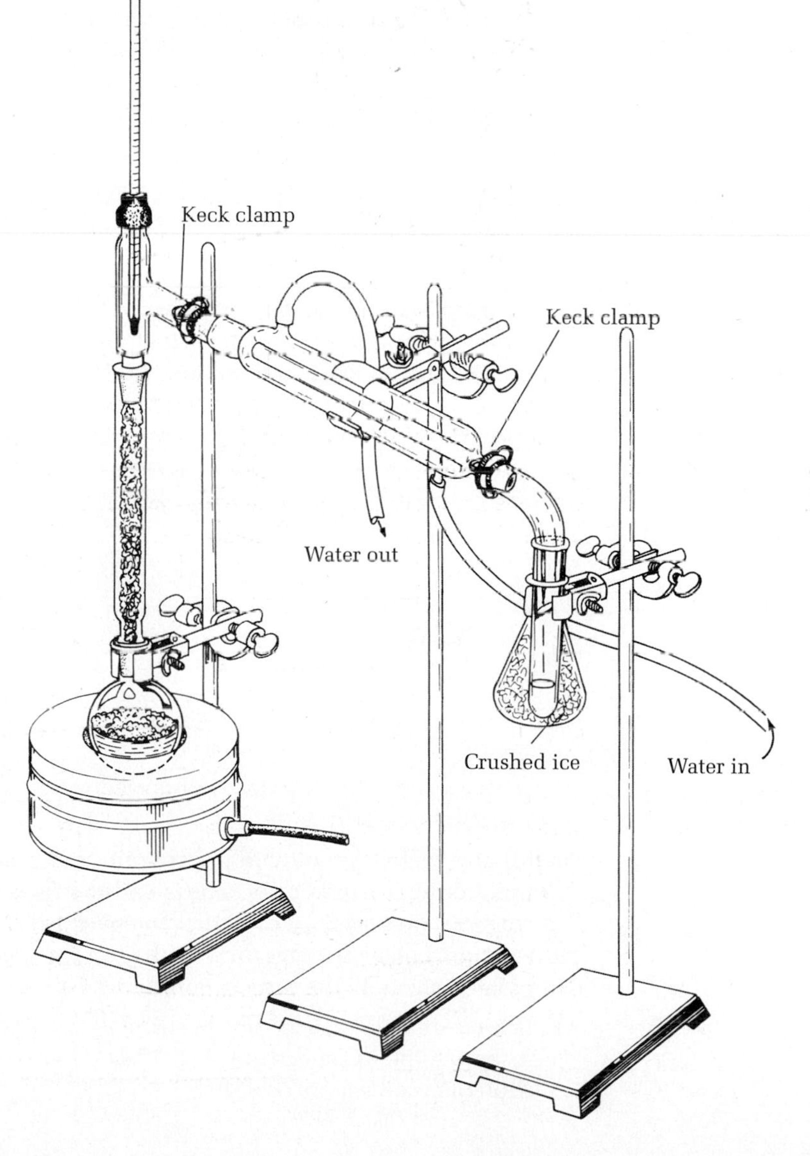

FIG. 19.2
Fractionation into an ice-cooled receiver.

Note the initial effect of heating the mixture, and then distill until the residue in the flask has a volume of 5-10 mL and very little distillate is being formed; note the temperature range. Then let the assembly cool slightly, remove the thermometer briefly, and pour 20 mL of toluene (the chaser solvent) into the top of the column through a long-stemmed funnel. Note the amount of the upper layer in the boiling flask and distill again until the volume of the layer has been reduced by about half. Pour the contents of the test tube into a small separatory funnel and rinse with a little chaser solvent; use this solvent for rinsing in subsequent operations.

Using a chaser solvent

Wash the mixture with an equal volume of saturated sodium chloride solution, separate the water layer, run the upper layer into a clean flask, and add 5 g of anhydrous calcium chloride (10 × 75-mm test tube-full) to dry it. Before the final distillation, note the barometric pressure, apply any necessary thermometer corrections, and determine the reading expected for a boiling point of 83°C. Dry the boiling flask, column, and condenser; decant the dried liquid into the flask through a stemless funnel plugged with a bit of cotton; and fractionally distill, with all precautions against evaporation losses. The ring of condensate should rise very slowly as it approaches the top of the fractionating column so that the thermometer can record the true boiling point soon after distillation begins. Record both the corrected boiling point of the bulk of the cyclohexene fraction and the temperature range, which should be no more than 2°C. If the cyclohexene has been dried thoroughly, it will be clear and colorless; if wet, it will be cloudy. A typical student yield is 13.2 g. Report your yield in grams and your percent yield.

⚠ Cyclohexene is extremely flammable; keep away from open flame.

Cleaning Up. The aqueous solutions (pot residues and washes) should be diluted with water and neutralized before flushing down the drain with a large excess of water. The ethanol wash and sodium chloride solution can also be flushed down the drain; the acetone wash and all toluene-containing solutions should be placed in the organic solvents waste container. The calcium chloride should be placed in the contaminated drying agents waste container or, if local regulations allow, once free of solvent, the calcium chloride can be placed in the nonhazardous solid waste container.

For Additional Experiments, sign in at this book's premium website at **www.cengage.com/login.**

QUESTIONS

1. Assign the peaks in the ^{1}H NMR spectrum of cyclohexene (Fig. 19.3) to specific groups of protons on the molecule.
2. Which product(s) would be obtained by the dehydration of 2-heptanol? of 2-methyl-l-cyclohexanol?
3. Mixing cyclohexanol with phosphoric acid is an exothermic process, whereas the production of cyclohexene is endothermic. Referring to the two chemical reactions on page 335, construct an energy diagram showing the course of this reaction. Label the diagram with the starting alcohol, the oxonium ion (the protonated alcohol), the carbocation, and the product.
4. Is it reasonable that the ^{1}H spectrum of cyclohexene (Fig. 19.3) should closely resemble the ^{13}C spectrum (Fig. 19.4)? Compare these spectra with the spectrum of cyclohexanol (Fig. 19.5).

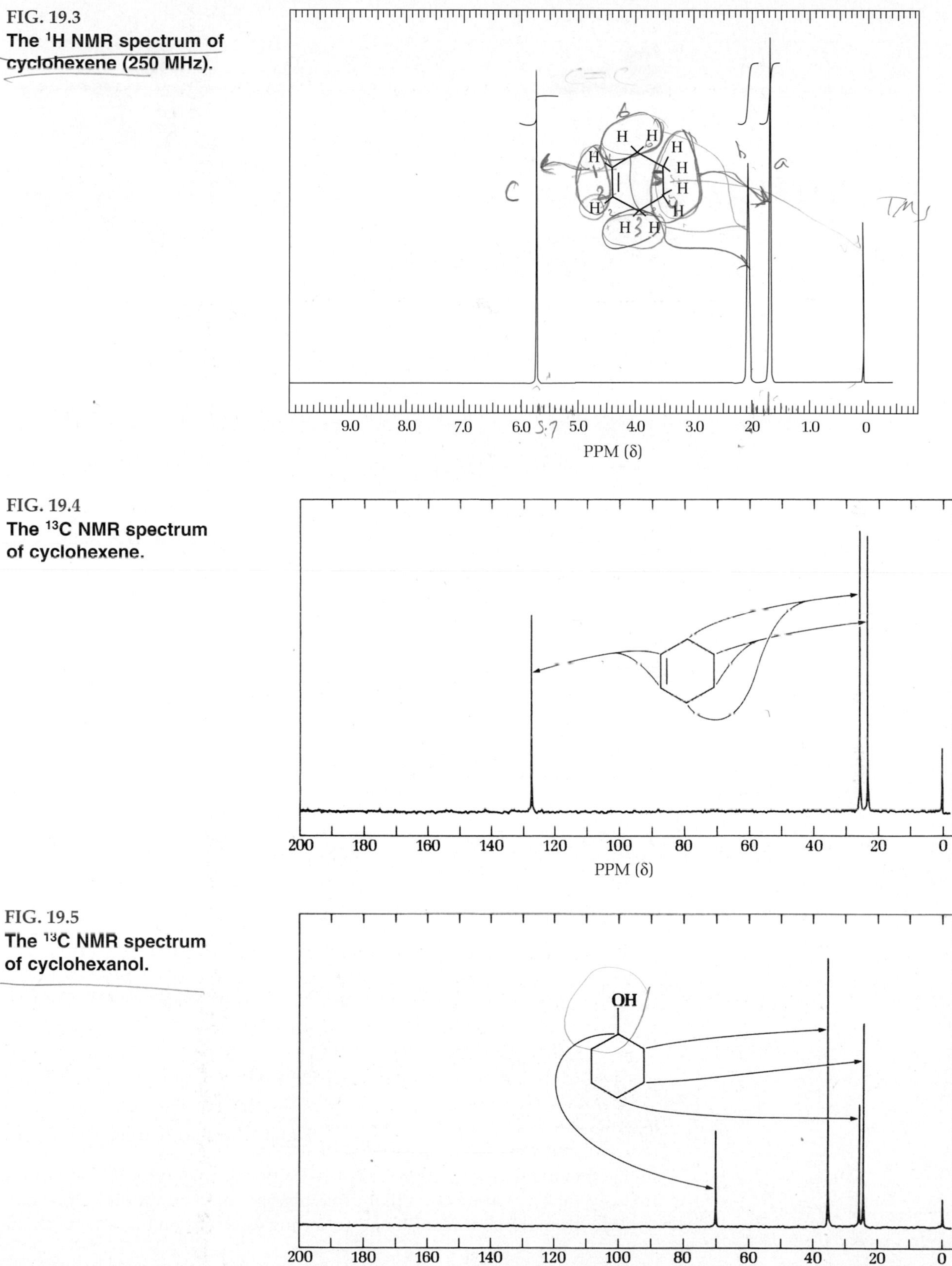

FIG. 19.3
The ^{1}H NMR spectrum of cyclohexene (250 MHz).

FIG. 19.4
The ^{13}C NMR spectrum of cyclohexene.

FIG. 19.5
The ^{13}C NMR spectrum of cyclohexanol.

20 CHAPTER

Bromination and Debromination: Purification of Cholesterol

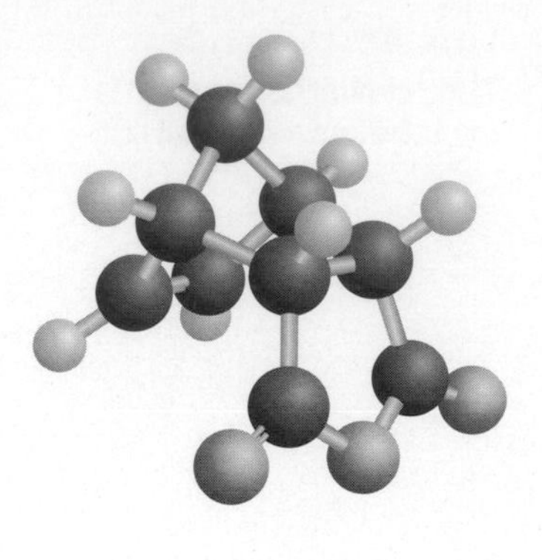

PRELAB EXERCISE: Make a molecular model of 5α, 6β-dibromocholestan-β-ol and convince yourself that the bromine atoms in this molecule are trans and diaxial.

The bromination of a double bond is an important and well-understood organic reaction. In the experiments in this chapter, it is employed for the very practical purpose of purifying crude cholesterol through the process of bromination, crystallization, and then zinc dust debromination.

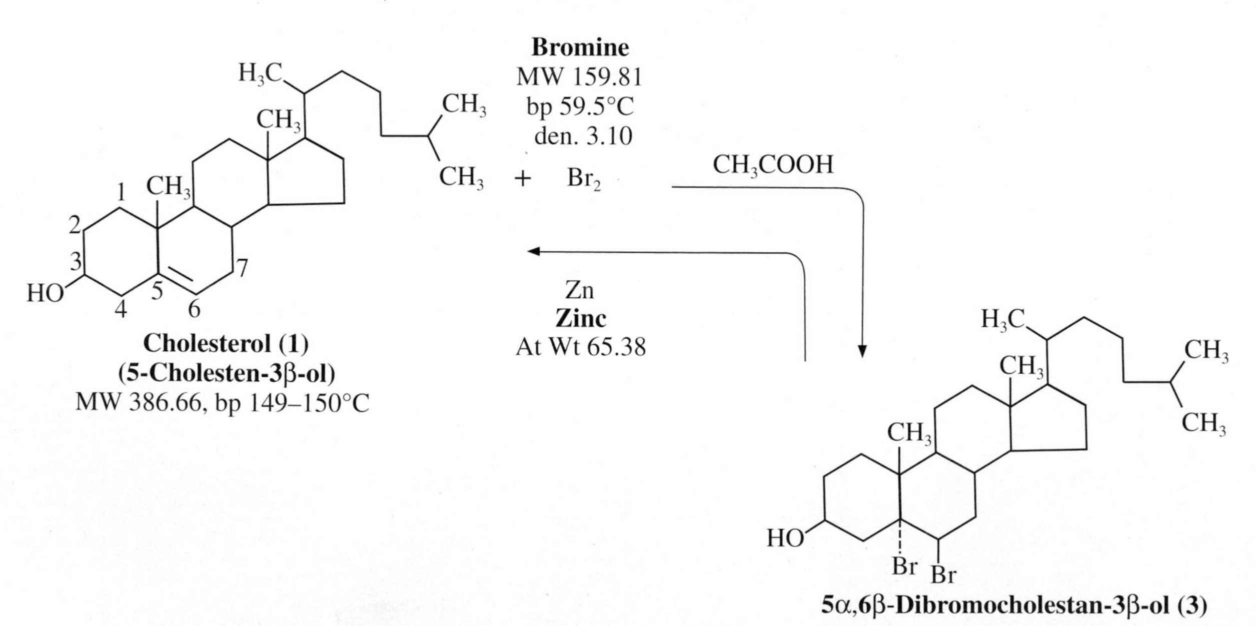

The reaction involves nucleophilic attack by the alkene on bromine with the formation of a tertiary carbocation that probably has some bromonium ion character resulting from the sharing of the nonbonding electrons on bromine with the electron-deficient C-5 carbon. This ion is attacked from the backside by the bromide ion to form dibromocholesterol, with the bromine atoms in the trans and diaxial configuration, the usual result when brominating a cyclohexene.

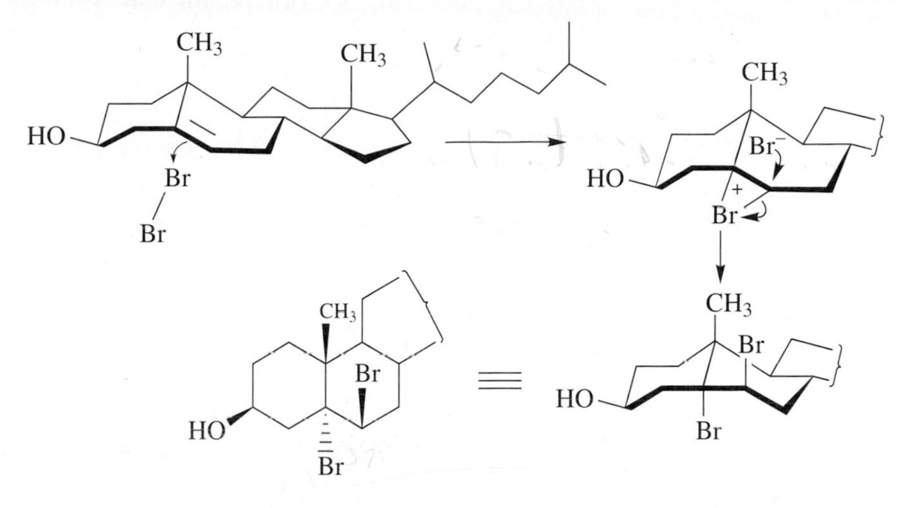

Cholesterol isolated from natural sources contains small amounts (0.1%–3%) of 3β-cholestanol, 7-cholesten-3β-ol, and 5,7-cholestadien-3β-ol.[1] These substances are so very similar to cholesterol in solubility that their removal by crystallization is not feasible. However, complete purification can be accomplished through the sparingly soluble dibromo derivative 5α,6β-dibromocholestan-3β-ol. 3β-Cholestanol is saturated and does not react with bromine; thus, it remains in the mother liquor. 7-Cholesten-3β-ol and 5,7-cholestadien-3β-ol are dehydrogenated by bromine to dienes and trienes, respectively, that likewise remain in the mother liquor and are eliminated along with colored byproducts.

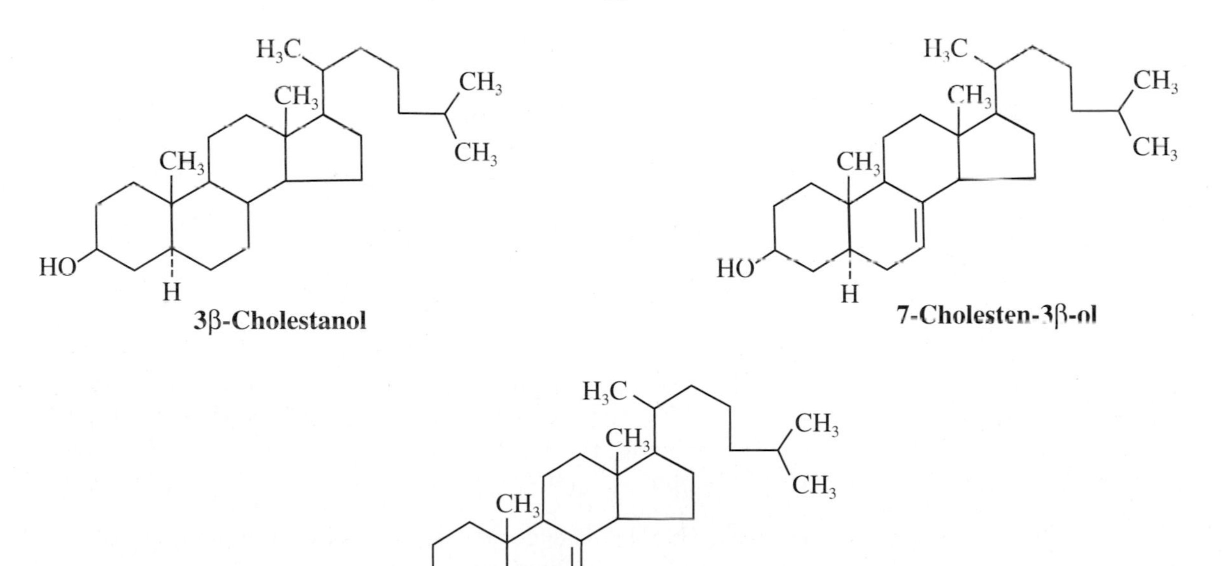

[1]A fourth companion, cerebrosterol, or 25-hydroxycholesterol, is easily eliminated by crystallization from alcohol.

The cholesterol dibromide that crystallizes from the reaction solution is collected, washed free of impurities or their dehydrogenation products, and debrominated with zinc dust, resulting in the regeneration of very pure cholesterol. Specific color tests can differentiate between pure cholesterol and tissue cholesterol purified by ordinary methods.

EXPERIMENTS

1. MICROSCALE BROMINATION OF CHOLESTEROL

Be extremely careful to avoid getting the solution of bromine and acetic acid on the skin. Carry out the reaction in the hood and wear disposable gloves for maximum safety.

Video: *Microscale Filtration on the Hirsch Funnel*

In a 10 × 75 mm test tube, dissolve 100 mg of gallstone cholesterol or commercial cholesterol (with a content of 7-cholesten-3β-ol about 0.6%) in 0.7 mL of ether by gentle warming. Then add 0.5 mL of a solution of bromine and sodium acetate in acetic acid.[2] Cholesterol dibromide begins to crystallize in 1-2 minutes. Cool in an ice bath and stir the crystalline paste with a stirring rod for about 10 minutes to ensure complete crystallization; at the same time, cool a mixture of 0.3 mL of ether and 0.7 mL of acetic acid in ice. Then collect the crystals on a Hirsch funnel and wash with the iced ether-acetic acid solution to remove the yellow mother liquor. A short test tube is used to facilitate the transfer of the crystalline paste. This material is not easy to filter on a very small scale because the crystals are tiny. Alternatively, carry out the bromination in a reaction tube and isolate the product with a Wilfilter (Fig. 20.1). Finally, wash the crystals on the filter with a few drops of ethanol, while continuing to apply suction, and transfer the white solid without drying it (dry weight 120 mg) to a reaction tube.

Cleaning Up. The filtrate from this reaction contains halogenated material and must be placed in the halogenated organic waste container.

2. MACROSCALE BROMINATION OF CHOLESTEROL

Be extremely careful to avoid getting the solution of bromine and acetic acid on the skin. Carry out the reaction in the hood and wear disposable gloves for maximum safety.

In a 25-mL Erlenmeyer flask, dissolve 1 g of commercial cholesterol (with a content of 7-cholesten-3β-ol about 0.6%) in 7 mL of ether by gentle warming and, with a pipette fitted with a pipetter or a plastic syringe, or from a burette in the hood, add 5 mL of a solution of bromine and sodium acetate in acetic acid.[3] Cholesterol dibromide begins to crystallize in 1-2 minutes. Cool in an ice bath and stir the crystalline paste with a stirring rod for about 10 minutes to ensure complete crystallization; at the same time, cool a mixture of 3 mL of ether and 7 mL of acetic acid in ice. Then collect the crystals on a Hirsch funnel and wash with the iced ether-acetic acid solution to remove the yellow mother liquor. Finally, wash with a little methanol, while continuing to apply suction, and transfer the white solid without drying it (dry weight 1.2 g) to a 50-mL Erlenmeyer flask.

Cleaning Up. The filtrate from this reaction contains halogenated material and so must be placed in the halogenated organic waste container.

[2]Weigh a 125-mL Erlenmeyer flask on a balance placed in the hood, and add 4.5 g bromine using a Pasteur pipette (avoid breathing the vapor); then add 50 mL acetic acid and 0.4 g sodium acetate (anhydrous).

[3]See footnote 2.

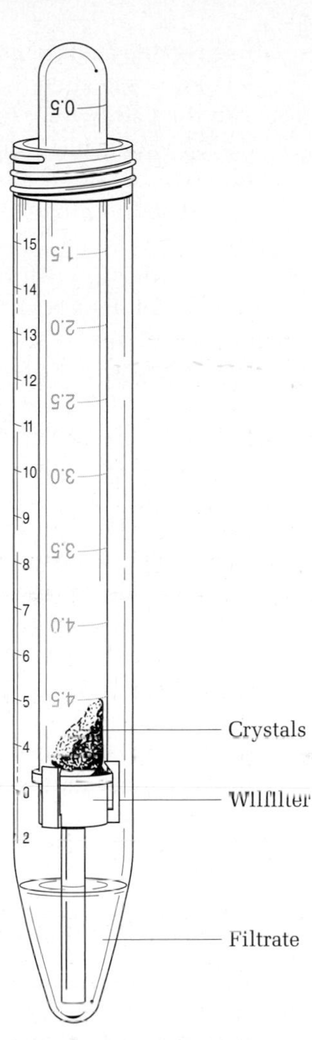

FIG. 20.1
A Wilfilter filtration apparatus. See Chapter 4 for the use of the Wilfilter.

3. MICROSCALE ZINC DUST DEBROMINATION

IN THIS EXPERIMENT, the dibromo compound from Experiment 1 is dissolved in ether and acetic acid and debrominated with zinc dust to regenerate the cholesterol. The ether solution, after separation from any zinc and zinc acetate, is washed with water, base, and salt solution. The solution is dried over calcium chloride and then separated from the drying agent. Methanol is added, and the solution is evaporated until the pure product begins to crystallize.

Add 2 mL of ether, 0.5 mL of acetic acid, and 40 mg of zinc dust[4] to the dibromocholesterol from Experiment 1 and mix the contents. In about 3 minutes, the dibromide dissolves; after 5–10 minutes of mixing (a magnetic stirrer could be

[4]If the reaction is slow, add more zinc dust or place the tube or flask in an ultrasonic cleaning bath, which will activate the zinc. The amount specified is adequate if material is taken from a freshly opened bottle, but zinc dust deteriorates on exposure to air.

Video: *Microscale Filtration on the Hirsch Funnel;* Photo: *Use of the Wilfilter*

used), zinc acetate usually separates to form a white precipitate (the dilution sometimes is such that no separation occurs). Stir for an additional 5 minutes and then add water (no more than 1 drop) until any solid present (zinc acetate) dissolves to make a clear solution. Decant the solution from the zinc into a reaction tube, wash the ethereal solution twice with 2 mL of water and then with 1 mL of 3 M sodium hydroxide and 1 mL of saturated sodium chloride solution. Then shake the ether solution with anhydrous calcium chloride pellets, pipette off the dry solution, wash the drying agent with a little ether, add 1 mL methanol (and a boiling stick), and evaporate the solution to the point where most of the ether is removed and the purified cholesterol begins to crystallize. Remove the solution from the heat; let crystallization proceed at room temperature and then in an ice bath. Collect the crystals on a Hirsch funnel or use a Wilfilter (Fig. 20.1). Wash the crystals with cool methanol; you should obtain 60 to 70 mg of cholesterol (mp 149–150°C).

Collect the product on a Wilfilter.

Cleaning Up. The unreacted zinc dust can be discarded in the nonhazardous solid waste container after it has been allowed to dry and then sit exposed to the air on a watch glass for about 30 minutes. Sometimes the zinc dust at the end of this reaction will get quite hot as it air oxidizes. The aqueous and acetic acid solutions after neutralization can be flushed down the drain. The organic filtrates containing ether and methanol should be placed in the organic solvents waste container. If local regulations allow, evaporate any residual solvent from the drying agents in the hood and place the dried solid in the nonhazardous waste container. Otherwise, place the wet drying agent in a waste container designated for this purpose.

4. MACROSCALE ZINC DUST DEBROMINATION

To the flask containing dibromocholesterol, add 20 mL of ether, 5 mL of acetic acid, and 0.2 g of zinc dust and swirl.[5] In about 3 minutes the dibromide dissolves; after swirling for 5–10 minutes, zinc acetate usually separates to form a white precipitate (the dilution sometimes is such that no separation occurs). Stir for an additional 5 minutes and then add water by drops (no more than 0.5 mL) until any solid present (zinc acetate) dissolves to make a clear solution. Decant the solution from the zinc into a separatory funnel and wash the ethereal solution twice with water and then with 10% sodium hydroxide (to remove traces of acetic acid). Then shake the ether solution with an equal volume of saturated sodium chloride solution to reduce the water content, dry the ether with anhydrous calcium chloride, remove the drying agent, add 10 mL of methanol (and a boiling stone), and evaporate the solution on a steam bath to the point where most of the ether is removed and the purified cholesterol begins to crystallize. Remove the solution from the steam bath; let crystallization proceed at room temperature and then in an ice bath, collect the crystals, and then wash them with cold methanol; you should obtain 0.6–0.7 g of cholesterol (mp 149–150°C). Figures 20.2 and 20.3 show, respectively, the proton and carbon NMR spectra of cholesterol.

Cleaning Up. The unreacted zinc dust can be discarded in the nonhazardous solid waste container after it has been allowed to dry and then sit exposed to the air on a watch glass for about 30 minutes. Sometimes the zinc dust at the end of

[5]See footnote 4.

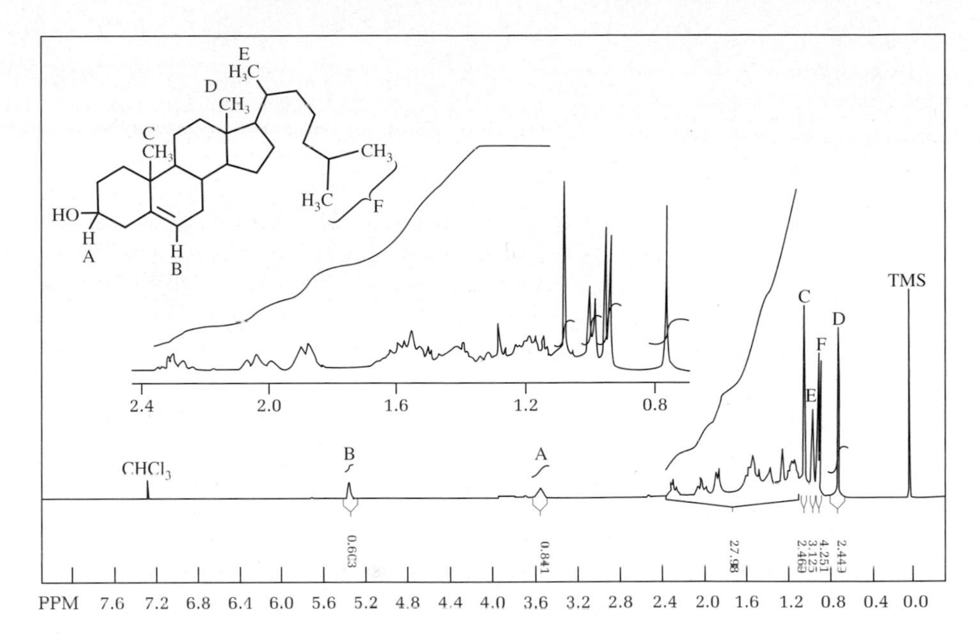

FIG. 20.2
The ^{1}H NMR spectrum of cholesterol (400 MHz).

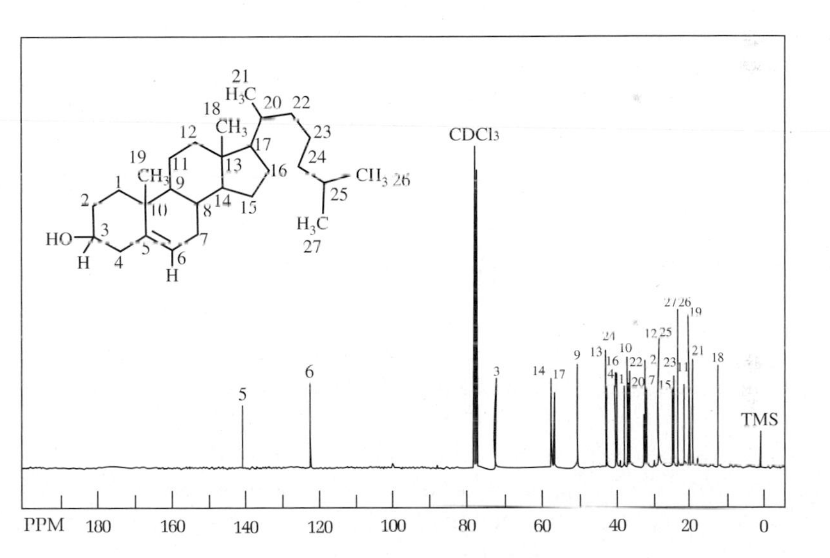

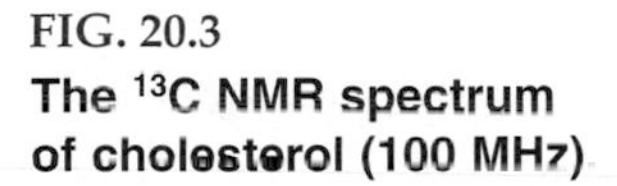

FIG. 20.3
The ^{13}C NMR spectrum of cholesterol (100 MHz)

this reaction will get quite hot as it air oxidizes. The aqueous and acetic acid solutions after neutralization can be flushed down the drain. The organic filtrates containing ether and methanol should be placed in the organic solvents waste container. If local regulations allow, evaporate any residual solvent from the drying agents in the hood and place the dried solid in the nonhazardous waste container. Otherwise, place the wet drying agent in a waste container designated for this purpose.

QUESTIONS

1. What is the purpose of the acetic acid in this reaction?
2. Why might old zinc dust not react in the debromination reaction?

FIG. 20.4
The IR spectrum of cholesterol (KBr disk).

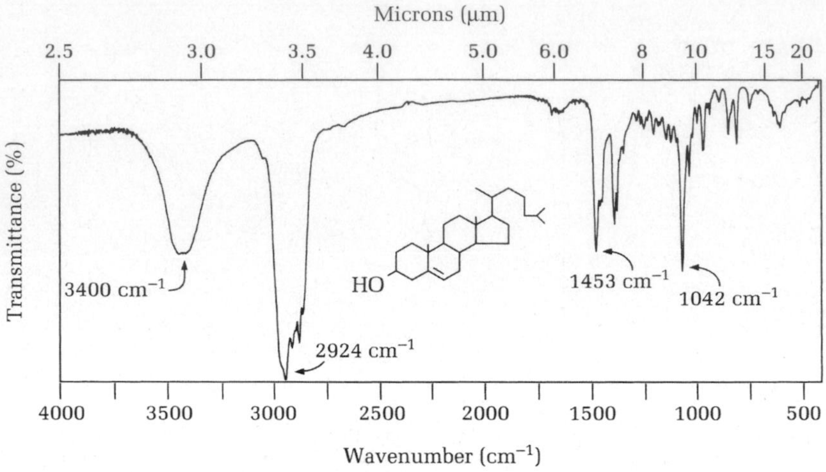

See *Web Links*

3. Why does the acetate ion not attack the intermediate bromonium ion to give the 5-acetoxy-6-bromo compound in the bromination of cholesterol?
4. Assign the 3400 cm^{-1} peak in the infrared (IR) spectrum of cholesterol (Fig. 20.4).

CHAPTER 21

Dichlorocarbene

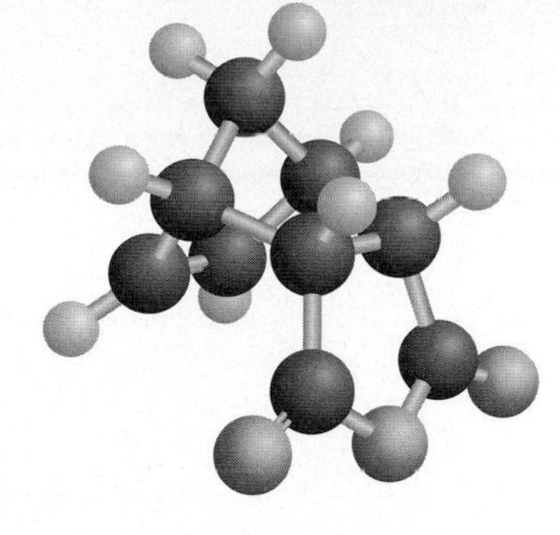

PRELAB EXERCISE: Propose a synthesis of dichlorocarbene using the principles of phase-transfer catalysis. Why is phase-transfer catalysis particularly suited to this reaction?

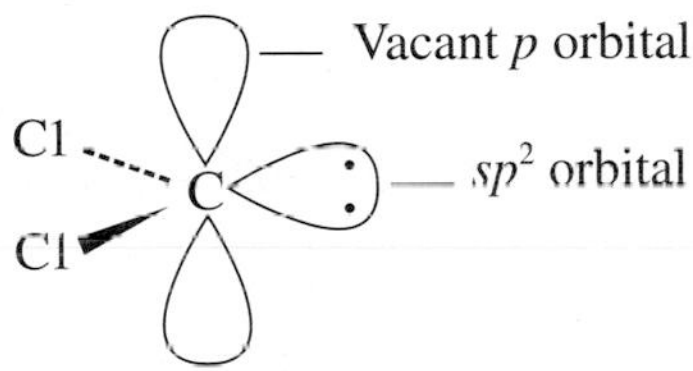

Dichlorocarbene

Like a carbanion—electrons in sp² *orbital. Like a carbocation—vacant* p *orbital*

Dichlorocarbene is a highly reactive intermediate of bivalent carbon with only six valence electrons around the carbon. It is electrically neutral and a powerful electrophile. As such, it reacts with alkenes, forming cyclopropane derivatives by cis-addition to the double bond.

There are a dozen or so ways by which dichlorocarbene can be generated. In this experiment, the thermal decomposition of anhydrous sodium trichloroacetate in an aprotic solvent (diglyme) in the presence of *cis, cis*-1,5-cyclooctadiene generates dichlorocarbene to give 5,5,10,10-tetrachlorotricyclo[7.1.0.0^{4,6}]decane (*see* page 348).

THE THERMAL DECOMPOSITION OF SODIUM TRICHLOROACETATE; REACTION OF DICHLOROCARBENE WITH 1,5-CYCLOOCTADIENE

$CH_3OCH_2CH_2OCH_2CH_2OCH_3$
2-Methoxyethyl ether
Diglyme
MW 134.17, bp 161°C
miscible with water

The thermal decomposition of sodium trichloroacetate initially gives the trichloromethyl anion, which, in the absence of a proton-donating solvent, gives dichlorocarbene by loss of a chloride ion (equation 1). In the presence of a proton-donating solvent (or moisture), this anion gives chloroform (equation 2).

The conventional method for carrying out the reaction is to add the salt portionwise, over 1–2 hours, to a magnetically stirred solution of the olefin in diethylene glycol dimethyl ether (2-methoxyethyl ether; diglyme) at a temperature maintained in a bath at 120°C. Under these conditions the reaction mixture becomes almost black, isolation of a pure product is tedious, and the yield is low.

Tetrachloroethylene, a nonflammable solvent used in the dry-cleaning industry, boils at 121°C and is relatively inert toward electrophilic dichlorocarbene. On generation of dichlorocarbene from either chloroform or sodium trichloroacetate in the presence of tetrachloroethylene, the yield of hexachlorocyclopropane

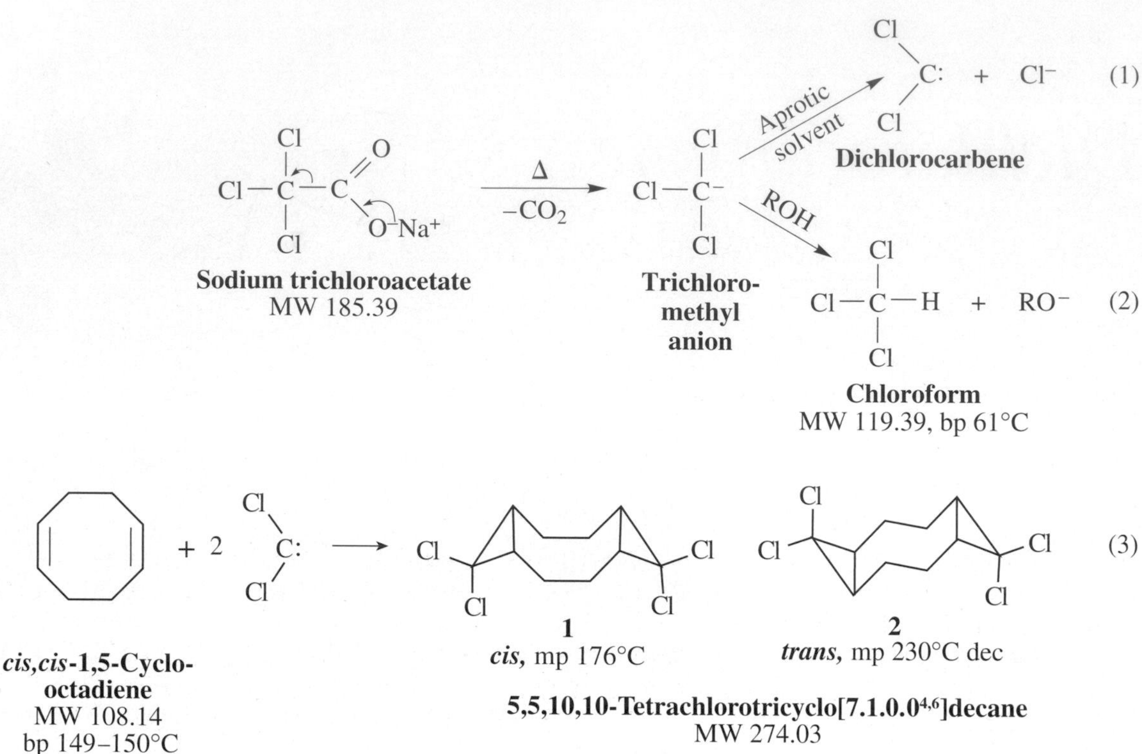

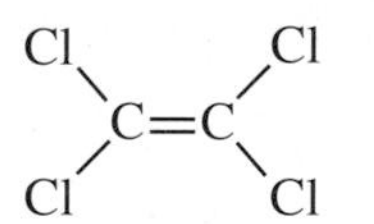

Tetrachloroethylene
bp 121°C, den. 1.623

(mp 104°C) is only 0.2%–10%. The first idea for simplifying the procedure[1] was to use tetrachloroethylene to control the temperature to the desired range, but sodium trichloroacetate is insoluble in this solvent, and no reaction occurs. Diglyme, or an equivalent, is required to provide some solubility. The reaction proceeds better in a 7:10 mixture of diglyme to tetrachloroethylene than in diglyme alone, but because the salt dissolves rapidly in this mixture, it has to be added in several small portions; the reaction mixture becomes very dark. The situation is vastly improved by the simple expedient of decreasing the amount of diglyme to a 2.5:10 ratio. The salt is so sparingly soluble in this mixture that it can be added at the start of the experiment; it will dissolve slowly as the reaction proceeds. The boiling and evolution of carbon dioxide provide adequate stirring, the mixture can be left unattended, and the little color that develops is eliminated by washing the crude product with methanol.

The main reaction product, 5,5,10,10-tetrachlorotricyclo[7.1.0.0^{4,6}]decane, crystallizes from ethyl acetate in the form of beautiful prismatic needles (mp 174–175°C), and this was the sole product encountered in runs made in the solvent mixture recommended. In earlier runs made in diglyme with the manual control of temperature, the ethyl acetate mother liquor material, on repeated crystallization from toluene, afforded small amounts of a second isomer (mp 230°C, dec). Elemental analyses indicated a pair of *cis-trans* isomers (**1** and **2** on page 348) had been formed, and both gave negative permanganate tests. To distinguish between them, the junior author of the paper undertook an X-ray analysis that showed that the lower melting isomer is *cis* and the higher melting isomer is *trans*.

X-ray analysis

[1]Fieser LF, Sachs DH. *J Org Chem.* **1964**;29:1113.

1. SODIUM TRICHLOROACETATE

IN THIS EXPERIMENT, sodium trichloroacetate is prepared by exactly neutralizing the acid with sodium hydroxide. The resulting solution is evaporated to complete dryness.

Use caution in handling the corrosive trichloroacetic acid. Carry out the procedure in the hood.

Sodium trichloroacetate will decompose completely to sodium chloride in a drying oven at 150°C.

Be sure that the salt is completely dry.

Use commercial sodium trichloroacetate (dry) or prepare it as follows. Place 6.4 g of trichloroacetic acid in a 125-mL Erlenmeyer flask, dissolve 1.6 g of sodium hydroxide pellets in 6 mL of water in a 25-mL Erlenmeyer flask, cool the solution thoroughly in an ice bath, and swirl the flask containing the acid in the ice bath while slowly dropping in about nine-tenths of the alkali solution. Then add 1 drop of 0.04% Bromocresol green solution to produce a faint yellow color, visible when the flask is dried and placed on white paper. With a Pasteur pipette, titrate the solution to an endpoint where a single drop produces a change from yellow to blue. If the endpoint is overshot, add a few crystals of acid and titrate more carefully. Close the flask with a rubber stopper, connect to an aspirator, place the flask in a boiling water bath, and wrap a towel around it for maximum heat. Turn on the water at full force for maximum vacuum. The evaporation requires no further attention and should be complete in 15–20 minutes. When you have an apparently dry white solid, scrape it out with a spatula and break up the large lumps. If you see any evidence of moisture, or if the weight exceeds the calculated value, dry it to constant weight under vacuum.

2. DICHLOROCARBENE REACTION

IN THIS EXPERIMENT, absolutely dry sodium trichloroacetate is refluxed in a special solvent mixture with a diene. The trichloroacetate decomposes to give sodium chloride (a precipitate), dichlorocarbene (which reacts with the diene), and carbon dioxide (the evolution of which is used to follow the progress of the reaction). At the end of the reaction, the solvent is removed by steam distillation, and the solid product is extracted into dichloromethane, which is dried and evaporated. Crude product is washed with methanol and recrystallized from ethyl acetate.

Tetrachloroethylene is a suspected cancer-causing agent and mutagen. Handle in the hood.

Place 0.91 g of dry sodium trichloroacetate in a 5-mL round-bottomed short-necked flask mounted over a hot sand bath. Add a magnetic stirring bar, 1 mL of tetrachloroethylene, 0.25 mL of diglyme, and 0.25 mL (220 mg) of *cis,cis*-1,5-cyclooctadiene.[2] Attach the empty distilling column as an air condenser and cap it with a septum bearing polyethylene tubing, which in turn dips below the surface of 0.5 mL of diglyme contained in a reaction tube (Fig. 21.1). This bubbler will show when the evolution of carbon dioxide ceases. The solvent should be the same as that of the reaction mixture (tetrachloroethylene), should there be a suckback. Heat the reaction mixture to boiling, note the time, and reflux gently until the

[2]Diglyme, tetrachloroethylene, and 1,5-cyclooctadiene should be stored over Linde 5A molecular sieves or anhydrous magnesium sulfate to guarantee that the reagents are dry. The reaction will not work if any reagent is wet. For safety reasons, dispense tetrachloroethylene in the hood. Do the same with the diene; it has a bad odor.

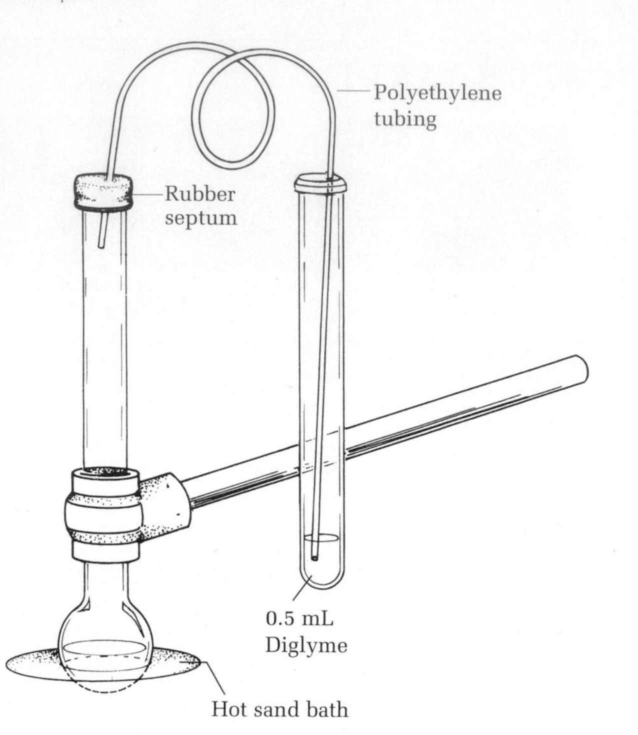

FIG. 21.1
A carbon dioxide bubbler, reflux air condenser, and reaction flask for dichlorocarbene synthesis. *See* Figure 18.2 (on page 331) for the technique of threading a polyethylene tube through a septum.

Photo: *Placing a Polyethylene Tube through a Septum*

reaction is complete. You will notice foaming, due to liberated carbon dioxide, and separation of finely divided sodium chloride. Inspection of the bottom of the flask will show lumps of undissolved sodium trichloroacetate, which gradually disappear.

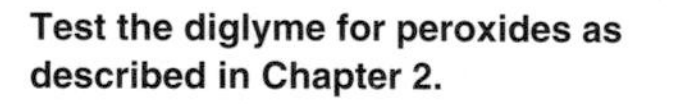
Test the diglyme for peroxides as described in Chapter 2.

When the reaction is complete, add 3.5 mL of water to the hot mixture, connect an adapter and a distilling head to the flask (Fig. 21.2), add a boiling chip, heat the mixture to boiling, and steam distill (*see* Chapter 6) until the tetrachloroethylene is eliminated—the product will separate in the flask as an oil or semisolid. If necessary, inject more water through the septum to complete the steam distillation. Cool the flask to room temperature, transfer the contents to a reaction tube using dichloromethane, and remove the aqueous layer. Dry the solution over anhydrous calcium chloride pellets, remove the solvent, wash the drying agent with more dichloromethane, and evaporate the solvent in a reaction tube. The weight of the crude product should be about 400 mg.

Cover the crude product with methanol, break up the cake with a flattened stirring rod, and crush the lumps. Cool in ice, collect the product crystals, and wash the product with methanol. The yield of colorless, or nearly colorless, 5,5,10,10-tetrachlorotricyclo[7.1.0.$0^{4,6}$]decane is about 300 mg. This material (mp 174–175°C) consists almost entirely of the *cis* isomer. Dissolve it in ethyl acetate (1.5–2 mL) and let the solution stand undisturbed for crystallization at room temperature. The pure *cis* isomer separates and forms large, prismatic needles (mp 175–176°C).

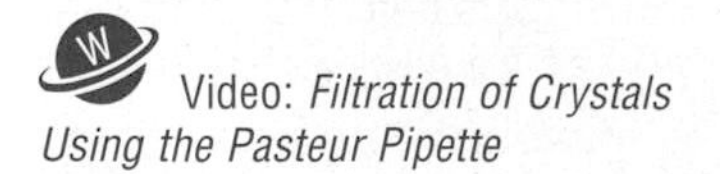
Video: *Filtration of Crystals Using the Pasteur Pipette*

Cleaning Up. The aqueous layer should be neutralized with hydrochloric acid, diluted with water, and flushed down the drain. Organic layers that contain halogenated material must be placed in the halogenated organic solvents container. If local regulations allow, evaporate any residual solvent from the drying agents in

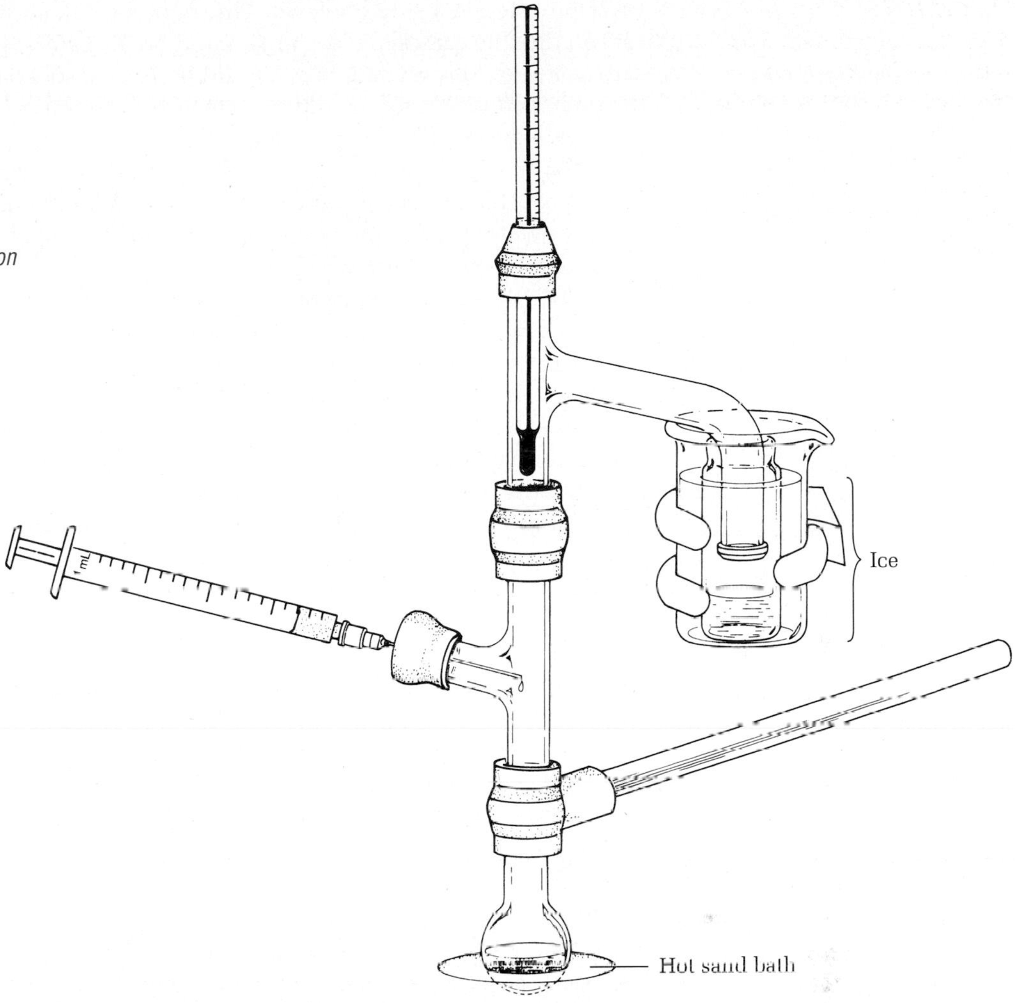

FIG. 21.2
A microscale steam distillation apparatus. Add water via a syringe.

Video: *Steam Distillation Apparatus*

Ice

Hot sand bath

the hood and place the dried solid in the nonhazardous waste container. Otherwise, place the wet drying agent in a waste container designated for this purpose. Methanol and ethyl acetate used in crystallization can be placed in the organic solvents waste container.

3. DICHLOROCARBENE REACTION

Place 7.1 g of dry sodium trichloroacetate in a 100-mL round-bottomed flask mounted over a flask heater and add 10 mL of tetrachloroethylene, 2.5 mL of diglyme, and 2.5 mL (2.2 g) of *cis,cis*-1,5-cyclooctadiene.[3] Attach a reflux condenser and, in the top opening of the condenser, insert a rubber stopper carrying a glass tube connected by rubber tubing to a short section of glass tube, which can be inserted below the surface of 2–3 mL of tetrachloroethylene in a 20 × 150-mm test

[3]Diglyme, tetrachloroethylene, and 1,5-cyclooctadiene should be stored over Linde 5A molecular sieves or anhydrous magnesium sulfate to guarantee that the reagents are dry. The reaction will not work if any reagent is wet. For safety reasons, dispense tetrachloroethylene in the hood. Do the same with the diene; it has a bad odor.

Reflux time: about 75 minutes

tube mounted at a suitable level. This bubbler will show when the evolution of carbon dioxide ceases; the solvent should be the same as that of the reaction mixture so as not to contaminate it, should there be a suckback. Heat to boiling, note the time, and reflux gently until the reaction is complete. You will notice foaming, due to liberated carbon dioxide, and the separation of finely divided sodium chloride. Inspection of the bottom of the flask will show lumps of undissolved sodium trichloroacetate, which gradually disappear. A large flask is specified because it will serve later for the removal of tetrachloroethylene by steam distillation. Make advance preparations for this operation.

Easy workup

When the reaction is complete, add 40 mL of water to the hot mixture, heat over a hot electric flask heater, and steam distill (*see* Chapter 6) until the tetrachloroethylene is eliminated—the product separates as an oil or semisolid. Cool the flask to room temperature, decant the supernatant liquid into a separatory funnel, and extract with dichloromethane. Run the extract into the reaction flask and use enough additional dichloromethane to dissolve the product; use a Pasteur pipette to rinse down material adhering to the adapter. Run the dichloromethane solution of the product into an Erlenmeyer flask through a cone of anhydrous calcium chloride pellets in a funnel and evaporate the solvent (bp 41°C). The residue is a tan or brown solid (14 g). Cover it with methanol, break up the cake with a flattened stirring rod, and crush the lumps. Cool in ice, collect the product, and wash with methanol. The yield of colorless, or nearly colorless, 5,5,10,10-tetrachlorotricyclo[7.1.0.0^{4,6}]decane is 3.3 g. This material (mp 174–175°C) consists almost entirely of the *cis* isomer. Dissolve it in ethyl acetate (15–20 mL) and let the solution stand undisturbed for crystallization at room temperature. The pure *cis* isomer separates in large, prismatic needles (mp 175–176°C). The proton NMR spectrum is shown in Figure 21.3.

Cleaning Up. The aqueous layer can be diluted with water and flushed down the drain. Organic layers, which contain halogenated material, must be placed in the halogenated organic solvents waste container. If local regulations allow, evaporate any residual solvent from the drying agents in the hood and place the dried solid in the nonhazardous waste container. Otherwise, place the wet drying agent in a waste container designated for this purpose. Methanol and ethyl acetate used in crystallization can be placed in the organic solvents waste container.

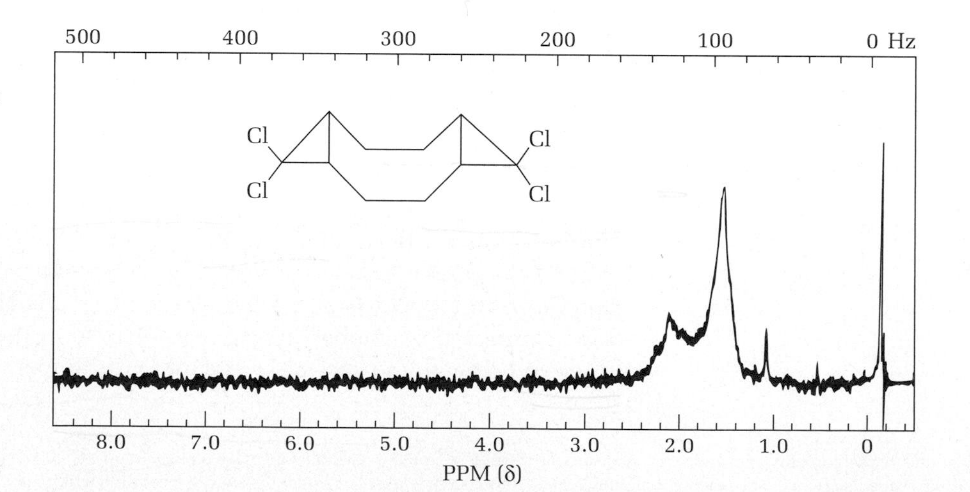

FIG. 21.3
The ^{1}H NMR spectrum of 5,5,10,10-tetrachlorotricyclo [7.1.0.0^{4,6}]decane (60 MHz).

4. PHASE-TRANSFER CATALYSIS—REACTION OF DICHLOROCARBENE WITH CYCLOHEXENE

$$CCl_3^- + H_2O \xrightarrow{} CHCl_3 + OH^-$$
(3)

$$:CCl_2 + 2\,H_2O \xrightarrow{} CH_2Cl_2 + 2\,OH^-$$
(4)

Ordinarily, water must be scrupulously excluded from carbene-generating reactions, as mentioned earlier. Both the intermediate trichloromethyl anion (**3**) and dichlorocarbene (**4**) react with water. But in the presence of a phase-transfer catalyst (a quaternary ammonium salt such as benzyltriethylammonium chloride), it is possible to carry out the reaction in aqueous medium. In the presence of a catalytic amount of the quaternary ammonium salt, chloroform, 50% aqueous sodium hydroxide, and the olefin are stirred for a few minutes to produce an emulsion. After the exothermic reaction is complete (about 30 minutes), the product is isolated.

Both benzyltriethylammonium chloride (**1**) and benzyltriethylammonium hydroxide (**2**) partition between the aqueous and organic phases. In the aqueous phase, the quaternary ammonium chloride (**1**) reacts with concentrated hydroxide to give the quaternary ammonium hydroxide (**2**). In the chloroform phase, **2** reacts with chloroform to give the trichloromethyl anion (**3**), which eliminates chloride ion to give dichlorocarbene, $:CCl_2$ (**4**), and **1**. The carbene (**4**) reacts with the alkene to give the product (**5**).

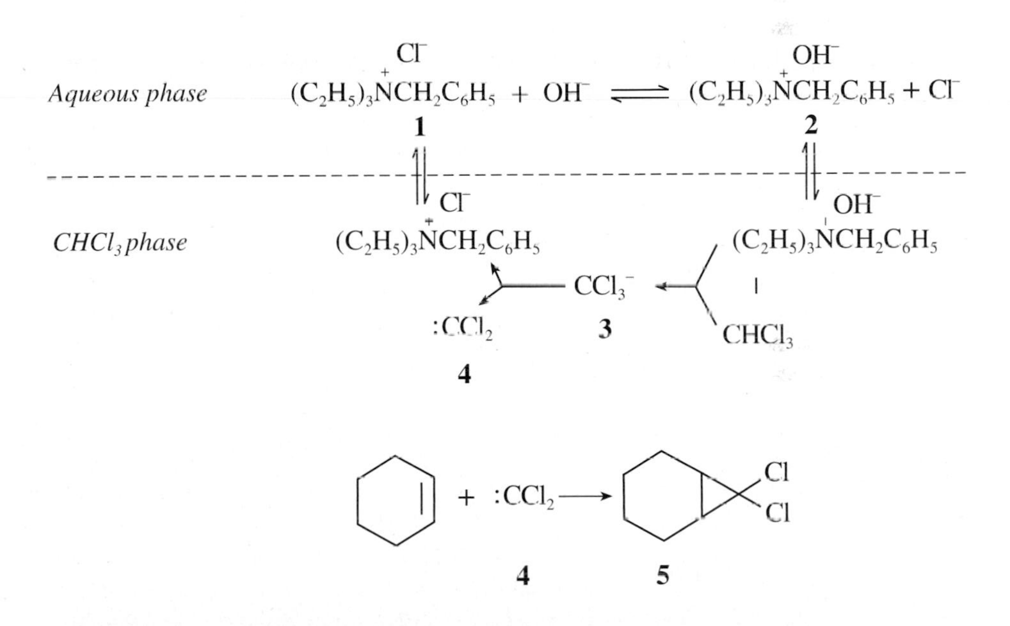

Macroscale Procedure

IN THIS EXPERIMENT, a mixture of cyclohexene, chloroform, some cyclohexanol (see "Emulsion Formation"), and a catalytic amount of a quaternary ammonium chloride ($R_4N^+Cl^-$) are stirred to produce a thick emulsion and an exothermic reaction. In a phase-transfer reaction, dichlorocarbene is formed and reacts with the alkene to form the product. The product is extracted into ether, washed, dried, and distilled to give the pure product (5).

Chloroform is a carcinogen and should be handled in the hood. Avoid all contact with the skin.

Fifty percent sodium hydroxide solution is very corrosive. Handle with care.

To a mixture of 8.2 g of cyclohexene, 12.0 g of chloroform and 20 mL of 50% aqueous sodium hydroxide in a 125-mL Erlenmeyer flask containing a thermometer, add 0.2 g of benzyltriethylammonium chloride.[4] Swirl the mixture to produce a thick emulsion. The temperature of the reaction will rise gradually at first, and then markedly accelerate. As it approaches 60°C, prepare to immerse the flask in an ice bath. With the flask alternately in and out of the ice bath, stir the thick paste and maintain the temperature between 50°C and 60°C. After the exothermic reaction is complete (about 10 minutes), allow the mixture to cool spontaneously to 35°C and then dilute with 50 mL of water. Separate the layers (test to determine which is the organic layer), extract the aqueous layer once with 10 mL of ether, and wash the combined organic layer and ether extract once with 20 mL of water. Dry the cloudy organic layer by shaking it with anhydrous calcium chloride pellets until the liquid is clear. Decant into a 50-mL round-bottomed flask and distill the mixture (with a boiling chip), first from a steam bath to remove the ether and some unreacted cyclohexene and chloroform, then over a very hot sand bath or a free flame. Collect 7,7-dichlorobicyclo[4.1.0]heptane over the range of 195–200°C. The yield is about 9 g. A purer product will result if the distillation is carried out under reduced pressure.

Emulsion Formation

The success of this reaction is critically dependent on forming a thick emulsion, for only then will the surface between the two phases be large enough for a complete reaction to occur. James Wilbur at Southwest Missouri State University found that when freshly distilled, pure reagents were used, no exothermic reaction occurred, and very little or no dichloronorcarane was formed. We have subsequently found that old, impure cyclohexene gives very good yields of product. Cyclohexene is notorious for forming peroxides at the allylic position. Under the reaction conditions, this peroxide will be converted to the ketone:

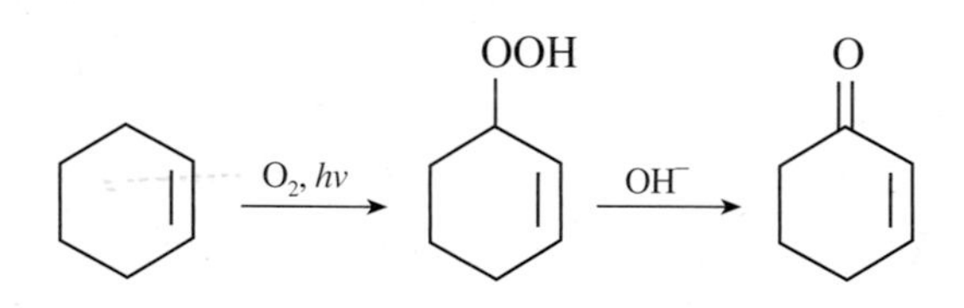

Apparently, small amounts of this ketone are responsible for the formation of stable emulsions. This appears to be one of the rare cases when impure reagents lead to successful completion of a reaction. If only pure reagents are available, then the reaction mixture must be stirred vigorously overnight using a magnetic stirrer. Alternatively, a small quantity of cyclohexene hydroperoxide can be prepared by pulling air through the alkene for a few hours, employing an aspirator, and using this impure reagent, or—as Wilbur found—by adding 1 mL of cyclohexanol to the reaction mixture. Detergents such as sodium lauryl sulfate do not increase the rate of reaction or emulsion formation.

Cleaning Up. Organic layers go in the organic solvents waste container. The aqueous layer should be diluted with water and flushed down the drain. If local regulations allow, evaporate any residual solvent from the drying agents in the hood and

[4]Many other quaternary ammonium salts, commercially available, work equally well (e.g., cetyltrimethy–l–ammonium bromide). Fabric softeners also work. See: Lee AWM, Yip WC. *J Chem Educ*. 1991;68(1):69–70.

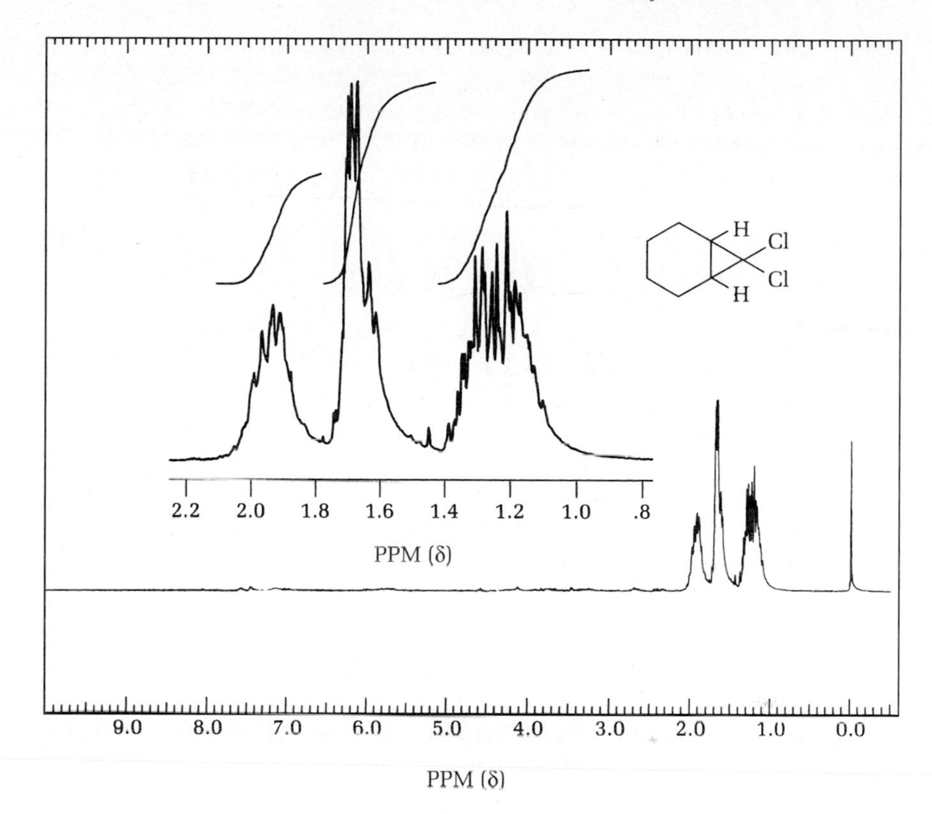

FIG. 21.4
The ^{1}H NMR spectrum of 7,7-dichlorobicyclo[4.1.0] heptane (250 MHz).

place the dried solid in the nonhazardous waste container. Otherwise, place the wet drying agent in a waste container designated for this purpose. The pot residue from the distillation goes in the halogenated organic solvents waste container.

Computational Chemistry

Construct models of the *cis*- and *trans*-dichlorocarbene adducts of cyclooctadiene, **1** and **2**, using a molecular mechanics program. Carry out energy minimization and note the steric energies of each isomer. Then carry out RM1 semiempirical molecular orbital calculations on each isomer to determine their heats of formation. Do these calculations confirm that the predominant product, the *cis*-isomer, is also the most stable? Would you have predicted that the *cis*-isomer would be the most stable by simple inspection of the energy-minimized conformers?

For Additional Experiments, sign in at this book's premium web site at **www.cengage.com/login**.

QUESTIONS

1. If water is not rigorously excluded from sodium trichloroacetate dichlorocarbene synthesis, what side reaction occurs?
2. Why is vigorous stirring or emulsion formation necessary in a phase-transfer reaction?
3. Assign the three groups of peaks in the NMR spectrum of 7,7-dichlorobicyclo [4.1.0]heptane (Fig. 21.4).

22 CHAPTER

Oxidation: Cyclohexanol to Cyclohexanone; Cyclohexanone to Adipic Acid

When you see this icon, sign in at this book's premium web site at **www.cengage.com/login** to access videos, Pre-Lab Exercises, and other online resources.

PRELAB EXERCISE: Write balanced equations for the dichromate and hypochlorite oxidations of cyclohexanol to cyclohexanone, and for the permanganate oxidation of cyclohexanone to adipic acid.

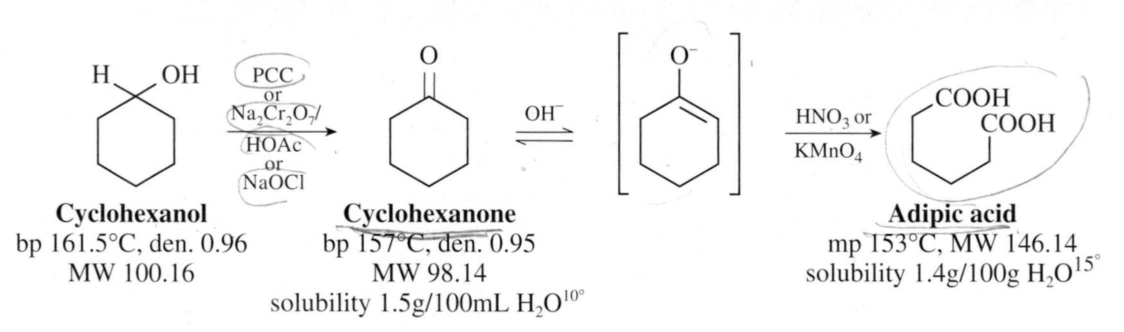

The oxidation of a secondary alcohol to a ketone is accomplished by many different oxidizing agents, including sodium dichromate, pyridinium chlorochromate, and sodium hypochlorite (household bleach). The ketone can be oxidized further to the dicarboxylic acid, producing adipic acid. Both of these oxidations can be carried out by the permanganate ion to give the diacid. Nitric acid is a powerful oxidizing agent that can oxidize cyclohexane, cyclohexene, cyclohexanol, or cyclohexanone to adipic acid.

THE OXIDATION OF AN ALCOHOL TO A KETONE

Dichromate oxidation

The dichromate mechanism of oxidizing an alcohol to a ketone appears to be the following:

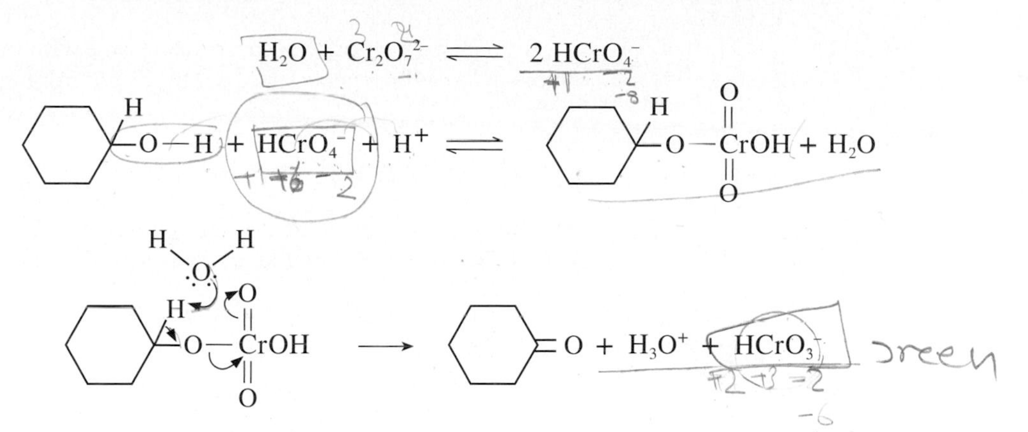

A number of intermediate valence states of chromium are involved in this reaction—the orange Cr^{6+} ion is ultimately reduced to the green Cr^{3+} ion. The course of the oxidation can be followed by these color changes.

In microscale Experiment 1, cyclohexanol is oxidized to cyclohexanone using pyridinium chlorochromate in dichloromethane. The progress of the reaction can be followed by thin-layer chromatography (TLC). In Experiment 2, the macroscale version of this reaction is carried out using sodium dichromate in acetic acid because the reagents are less expensive, the reaction is faster, and much less solvent is required.

Chromium(VI) is probably the most widely used and versatile laboratory oxidizing agent; it is used in a number of different forms to carry out selective oxidations in this text. From an environmental standpoint, however, it is far from ideal. The inhalation of the dust from insoluble Cr(VI) compounds may cause cancer of the respiratory system. The product of the reaction [Cr(III)] should not be flushed down the drain because it is toxic to aquatic life at extremely low concentrations. Therefore, as stated in the "Cleaning Up" section of Experiments 1 and 2, the Cr(III) must be precipitated as insoluble $Cr(OH)_3$, and this material is considered a hazardous waste.

Experiments 3 and 4 use an alternative oxidant for secondary alcohols that is just as efficient and much safer from an environmental standpoint: 5.25% (0.75 *M*) sodium hypochlorite solution, which is available in the grocery store as household bleach.[1] The mechanism of the reaction is not clear. It is not a free radical reaction, the reaction is much faster in acid than in base, elemental chlorine is presumably the oxidant, and hypochlorous acid must be present. It may form an intermediate alkyl hypochlorite ester, which by an E_2 elimination gives the ketone and chloride ion.

Sodium hypochlorite oxidation

$$\text{C}_6\text{H}_{11}\text{OH} + \text{HOCl} \rightleftharpoons \text{C}_6\text{H}_{11}\text{O—Cl} + \text{H}_2\text{O}$$

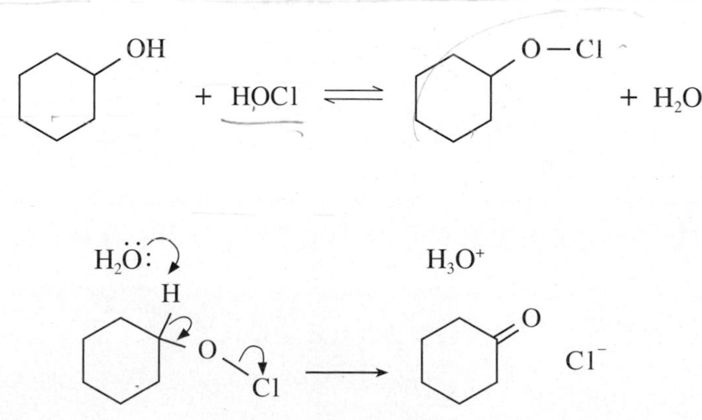

[1]Mohrig JR, Nienhuis DM, Linck CF, Van Zoeren C, Fox BG. *J Chem Educ.* **1985**;*62*:519.

Excess hypochlorite is easily destroyed with bisulfite; the final product is the chloride ion, which is far less toxic to the environment than Cr(III).

THE OXIDATION OF A KETONE TO A CARBOXYLIC ACID

Nitric acid oxidation

In microscale Experiment 5, nitric acid is the oxidant. The balanced equation for the oxidation of cyclohexanone to adipic acid is as follows:

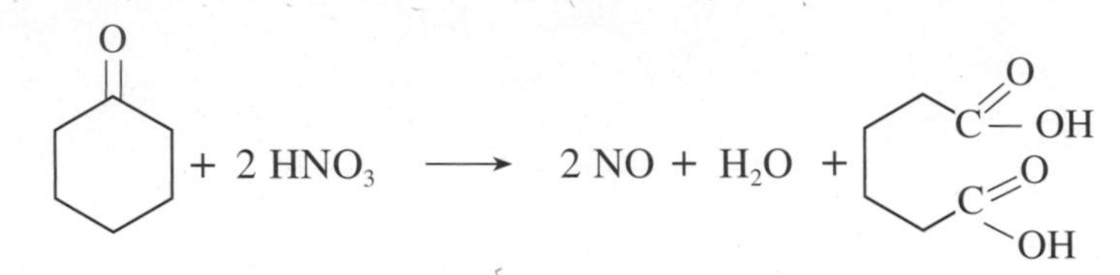

In this reaction, nitric acid is reduced to nitric oxide.

Permanganate oxidation

Experiment 6 involves the permanganate oxidation of a ketone to a dicarboxylic acid. The reaction can be followed as the bright purple permanganate solution reacts to give a brown precipitate of manganese dioxide. A possible mechanism for this oxidation starts by the reaction of MnO_4 with the enol form of the ketone and continues as shown here:

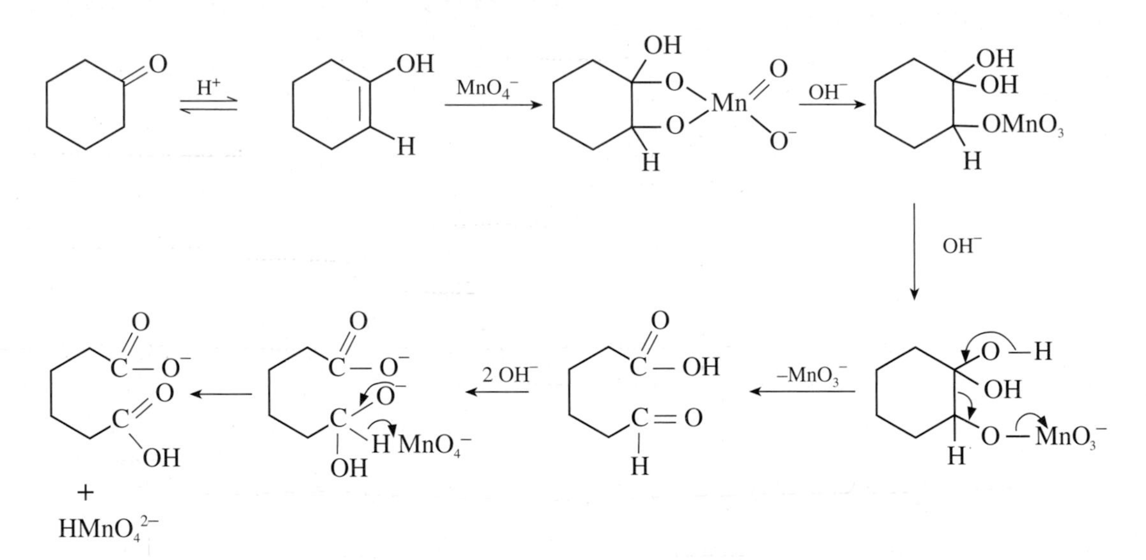

$$3\,HMnO_4^{2-} + H_2O \longrightarrow 2\,MnO_2 + MnO_4^- + 5\,OH^-$$

EXPERIMENTS

1. CYCLOHEXANONE FROM CYCLOHEXANOL

IN THIS EXPERIMENT, an alcohol is dissolved in dichloromethane and oxidized at room temperature to a ketone with a Cr^{6+} reagent: pyridinium chlorochromate (PCC). The ketone is isolated by simply evaporating the solvent from the filtered reaction mixture.

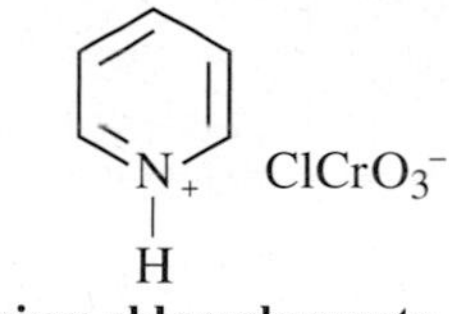

Pyridinium chlorochromate, PCC
MW 215.56

H OH
PCC or $Na_2Cr_2O_7$
O

Cyclohexanol
bp 161.5°C, den. 0.96
MW 100.16

Cyclohexanone
bp 157°C, den. 0.95
MW 98.14
solubility 1.5g/100mL $H_2O^{10°}$

Handle open containers of dichloromethane in the hood; it is a suspected cancer-causing agent.

To a 10-mL Erlenmeyer flask, add 0.62 g of finely powdered pyridinium chlorochromate (*see* Chapter 23 for its preparation), 4 mL of dichloromethane, and 250 mg of cyclohexanol. The mixture is stirred magnetically or shaken over a period of days (a week does no harm) until TLC on silica gel indicates that the reaction is complete. Elute the TLC plates with dichloromethane and visualize with iodine.

At the end of the reaction period, remove the chromium salts by filtration on a Hirsch funnel or, if the salts are coarse enough, by removing the solvent with a Pasteur pipette. Wash the chromium salts with a few drops of dichloromethane and evaporate the solvent under a gentle stream of nitrogen or air to isolate the product—cyclohexanone. Confirm the structure of this product by obtaining an infrared (IR) spectrum of the pure liquid between sodium chloride or silver chloride plates. Look for a band at 3600 to 3400 cm^{-1}, which is indicative of unoxidized alcohol (or water if you are not careful).

Videos: *Microscale Filtration on the Hirsch Funnel, Filtration of Crystals Using the Pasteur Pipette;* Photo: *Filtration Using a Pasteur Pipette*

Cleaning Up. Place used dichloromethane in the halogenated organic waste container. The chromium salts should be dissolved in hydrochloric acid and diluted to <5%. The solution, which should have a pH < 3, should then be treated with a 50% excess of sodium bisulfite to reduce the orange dichromate ion to the green chromic ion. Make the solution basic with ammonium hydroxide to precipitate chromium as the hydroxide. This precipitate is collected on filter paper, which is placed in the hazardous solid waste container for heavy metals. The filtrate from this latter treatment can be flushed down the drain.

Using Celite will speed the filtration.

2. CYCLOHEXANONE FROM CYCLOHEXANOL

In a 50-mL Erlenmeyer flask, dissolve 7.5 g of sodium dichromate dihydrate in 12.5 mL of acetic acid by swirling the mixture on a hot plate; then cool the solution with ice to 15°C. In a second Erlenmeyer flask, chill a mixture of 7.5 g of cyclohexanol and 5 mL of acetic acid in ice. After the first solution is cooled to 15°C, transfer the thermometer and adjust the temperature in the second flask to 15°C. Wipe the outside of the flask containing the dichromate solution to dry it, pour the solution into the cyclohexanol–acetic acid mixture, rinse the flask with a little acetic acid, note the time, and remove the initially light orange solution from the ice bath. Keep the ice bath ready for use when required.[2] The exothermic reaction that is soon evident can

The dust from Cr^{6+} salts is toxic and a suspected carcinogen; weigh them in the hood.

[2]If the acetic acid solutions of cyclohexanol and dichromate are mixed at 25°C rather than at 15°C, the yield of crude cyclohexanone is only 3.5 g. A clue to the evident importance of the initial temperature is suggested by an experiment in which the cyclohexanol was dissolved in 6 mL of benzene instead of 5 mL of acetic acid, and the two solutions were mixed at 15°C. Within a few minutes, orange-yellow crystals separated and soon filled the flask; the substance probably is the chromate ester $(C_6H_{11}O)_2CrO_2$. When the crystal magma was allowed to stand at room temperature, the crystals soon dissolved, exothermic oxidation proceeded, and cyclohexanone was formed in high yield. Perhaps a low initial temperature ensures complete conversion of the alcohol into the chromate ester before side reactions set in.

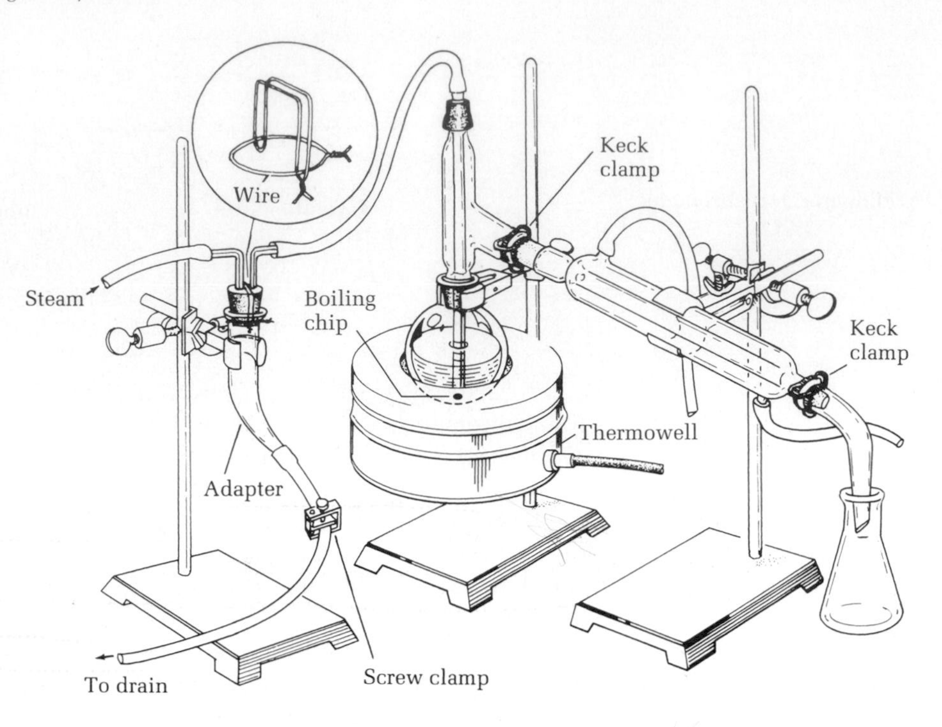

FIG. 22.1
A steam distillation apparatus.

get out of hand unless controlled. When the temperature rises to 60°C, cool the flask in ice just enough to prevent a further rise and then, by intermittent brief cooling, keep the temperature close to 60°C for 15 minutes. No further cooling is needed, but the flask should be swirled occasionally and the temperature monitored. The usual maximal temperature is 65°C (25–30 minutes). When the temperature begins to drop and the solution becomes pure green, the reaction is complete. Allow 5–10 minutes more reaction time; then pour the green solution into a 100-mL round-bottomed flask, rinse the Erlenmeyer flask with 50 mL of water, and add the solution to the flask for steam distillation (Chapter 6) of the product (Fig. 22.1). Distill as long as any oil passes over with the water and, because cyclohexanone is appreciably soluble in water, continue somewhat beyond this point (about 40 mL will be collected).

Reaction time: 45 minutes

Alternatively, instead of setting up the apparatus for steam distillation, simply add a boiling chip to the 100-mL flask and distill 20 mL of liquid, cool the flask slightly, add 20 mL of water to it, and distill 20 mL more. Note the temperature during the distillation. This is a steam distillation in which steam is generated in situ rather than from an outside source.

The Isolation of Cyclohexanone from the Steam Distillate

Cyclohexanone is fairly soluble in water. Dissolving inorganic salts such as potassium carbonate or sodium chloride in the aqueous layer will decrease the solubility of cyclohexanone such that it can be completely extracted with ether. This process is known as *salting out*.

To salt out the cyclohexanone, add to the distillate 0.2 g of sodium chloride per milliliter of water present and swirl to dissolve the salt. Next, pour the mixture into a separatory funnel, rinse the flask with ether, add more ether to a total volume of 10–15 mL, shake, and draw off the water layer. Then wash the ether layer with 10 mL of 3 *M* sodium hydroxide solution to remove the acetic acid, test a drop of the wash liquor to be sure it contains excess alkali, and draw off the aqueous layer.

To dry the ether containing dissolved water, shake the ether layer with an equal volume of saturated aqueous sodium chloride solution. Draw off the aqueous layer, pour the ether out of the neck of the separatory funnel into an Erlenmeyer flask, add about 2.5 g of anhydrous calcium chloride pellets, and complete the final drying of the ether solution by occasional swirling of the solution over a 5-minute period. Remove the drying agent by decantation or gravity filtration into a tared Erlenmeyer flask, and rinse the flask that contained the drying agent (the calcium chloride) and the funnel with ether. Add a boiling chip to the ether solution and evaporate the ether on a steam bath under an aspirator tube (*see* Figure 7.11 on page 149). Cool the contents of the flask to room temperature, evacuate the crude cyclohexanone under aspirator vacuum to remove the final traces of ether (*see* Fig. 9.7 on page 197 and Fig. 9.10 on page 201), and weigh the product. The yield is 15–16 g.

The crude cyclohexanone can be purified by simple distillation or used directly to make adipic acid (Experiment 6).

Using Celite will speed up the filtration.

Cleaning Up. To the residue from the steam distillation, add sodium bisulfite to destroy any excess dichromate ion. Collect the precipitate of chromium hydroxide that forms when the solution is made just slightly basic with ammonium hydroxide, and dispose of the filter paper and precipitate in the hazardous waste container designated for heavy metals. The filtrate can be flushed down the drain. Combine the water layers from the extraction process, neutralize with dilute hydrochloric acid, and flush the solution down the drain. Any ether should be placed in the organic solvents waste container. If local regulations allow, evaporate any residual solvent from the drying agents in the hood and place the dried solid in the nonhazardous waste container. Otherwise, place the wet drying agent in a waste container designated for this purpose.

3. CYCLOHEXANONE FROM CYCLOHEXANOL BY HYPOCHLORITE OXIDATION

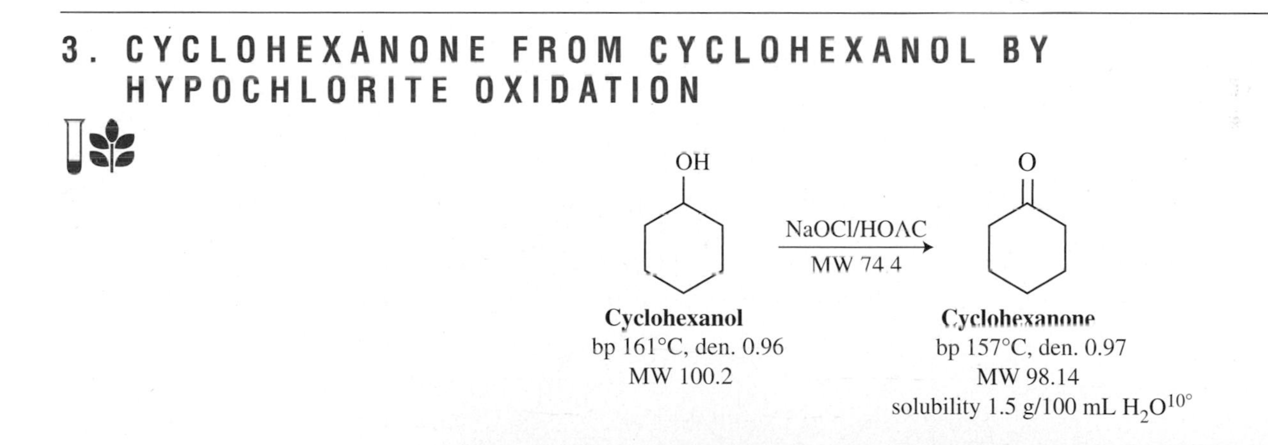

IN THIS EXPERIMENT, an alcohol is oxidized to a ketone with household bleach (sodium hypochlorite). The product is isolated by steam distillation and is extracted into the distillate with ether that is dried in a column of drying agents. Removal of the ether leaves the product—cyclohexanone. If a 6% hypochlorite solution is used, you might find it more convenient to carry out the experiment on twice the scale indicated. Using a $\frac{1}{2}$-inch stir bar with a magnetic stirrer is suggested.

To a 5-mL long-necked, round-bottomed flask fitted with a connector that has a support rod, add 150 mg of cyclohexanol followed by a mixture of 80 mg of acetic acid and 2.3 mL of a 5.25% solution of sodium hypochlorite (e.g., Clorox). Do not use scented bleach and note the concentration. Ultra Clorox is a 6% solution, so use 2 mL of it and 70 mg of acetic acid. This mixture should be acidic. Test with pH paper and if it is not acidic, add more acetic acid. Add a thermometer. The temperature will rise to 45°C. Swirl the contents of the flask; as soon as the temperature begins to drop, remove the thermometer, cap the flask with a septum, and shake it vigorously for about 3 minutes. Then place the flask in a beaker of water at 45°C for 15 minutes to complete the reaction.

Destroy excess oxidizing agent by adding saturated sodium bisulfite solution (about 1 or 2 drops). At this point, the reaction mixture no longer gives a positive test with starch-iodide paper (a positive test for an oxidant is a blue-purple color from the starch-triiodide complex).

Add a drop of thymol blue indicator to the solution, and then neutralize the solution with 0.3 or 0.4 mL 6 *M* sodium hydroxide solution. Add a boiling chip to the solution; attach to the flask a distilling head, a thermometer adapter, and a thermometer. Then distill about 0.8 mL of a mixture of water and cyclohexanone (Fig. 22.2). This is a steam distillation in which the steam is generated in situ (*see* Chapter 6).

Isolation of Cyclohexanone from the Steam Distillate

Cyclohexanone is fairly soluble in water. Dissolving inorganic salts such as potassium carbonate or sodium chloride in the aqueous layer will decrease the solubility of cyclohexanone such that it can be completely extracted with ether. This process is known as *salting out.*

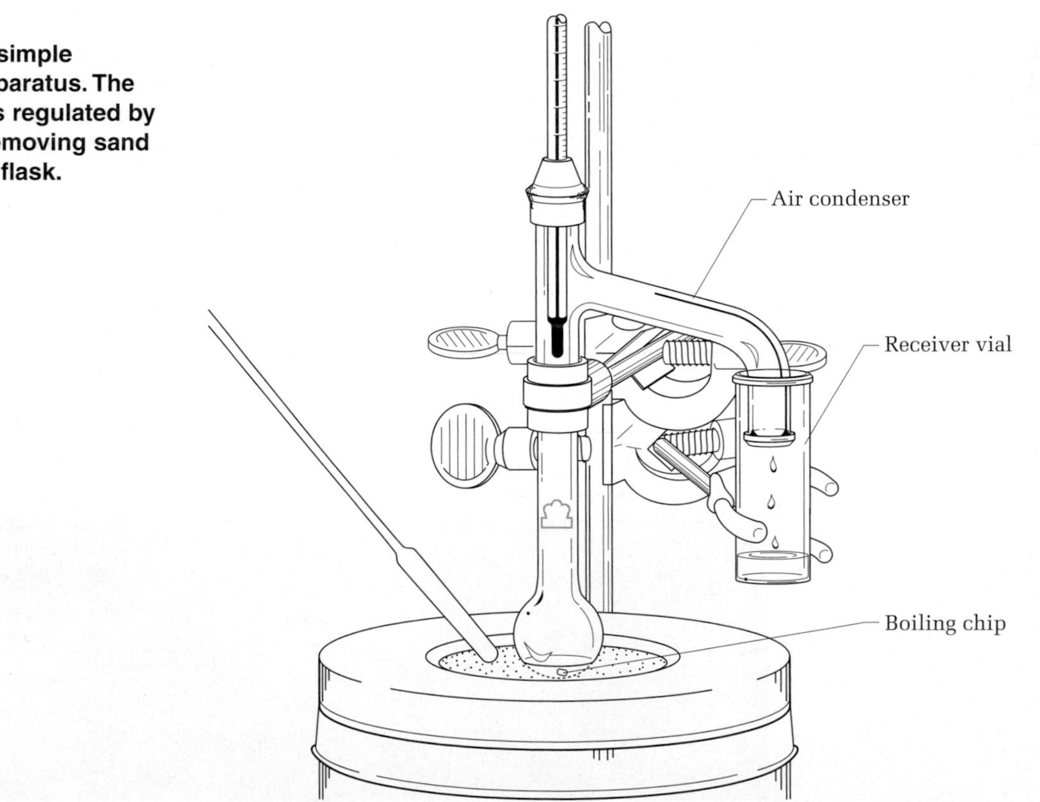

FIG. 22.2
A microscale simple distillation apparatus. The temperature is regulated by piling up or removing sand from near the flask.

To salt out the cyclohexanone, transfer the distillate to a reaction tube, add 0.1 g of sodium chloride to the distillate, stir, and shake the tube until most of the sodium chloride dissolves. Then extract the solution with three 0.4-mL portions of ether. As a reminder, this process consists of adding 0.4 mL of the ether to the reaction tube, shaking the tube, allowing the layers to separate, drawing off the top layer with a Pasteur pipette, and placing this top layer (the ether extract) in a clean, dry reaction tube.

Because ether dissolves water to a small extent, the ether extract is wet. A convenient way to dry the extract is to pass it through a short column of drying agent and alumina (a powerful drying agent). To the chromatography column, add about a 1-in.-high column of alumina and on top of this, a 1-in.-high column of anhydrous sodium sulfate. Transfer the ether solution of cyclohexanone to the column, let it sit for 1–2 minutes, and then collect the dry ether in a tared 5-mL round-bottomed flask. Wash out the column with about 2 mL of ether, which is collected in the same flask.

Add a boiling chip to the dry ether solution and fit the flask for simple distillation. Distill all the ether using the short-path microscale distillation apparatus shown in Figure 22.3. The perfectly water-clear residue is cyclohexanone, which can be analyzed by IR and nuclear magnetic resonance (NMR) spectroscopy. If it is cloudy, it is wet. Determine the weight of the product and calculate the percentage yield.

Cleaning Up. The aqueous reaction mixture after extraction of the product contains primarily sodium, chloride, and acetate ions and can therefore be flushed down the drain. Any unused ether goes in the organic solvents waste container. If local regulations allow, evaporate any residual solvent from the drying agents and alumina in

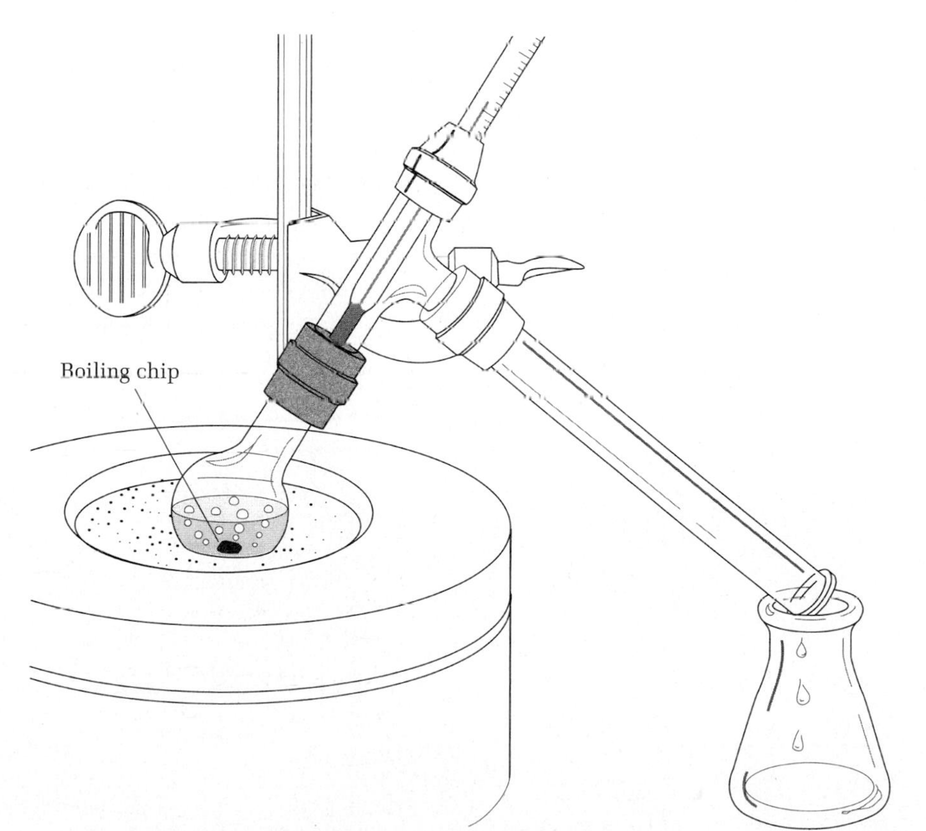

FIG. 22.3
A short-path microscale simple distillation apparatus.

the hood, and place the dried solid in the nonhazardous waste container. Otherwise, place the wet materials in a waste container designated for this purpose.

4. CYCLOHEXANONE FROM CYCLOHEXANOL BY HYPOCHLORITE OXIDATION

Into a 250-mL Erlenmeyer flask in the hood, place 8 mL (0.075 mol) of cyclohexanol. Introduce a thermometer and slowly add to the flask with swirling a mixture of 4 mL of acetic acid and 115 mL of a commercial household bleach such as Clorox (usually 5.25% by weight sodium hypochlorite, which is 0.75 molar). Do not use scented bleach and note the concentration. Ultra Clorox is a 6% solution, so use 100.6 mL of it and 3.5 mL of acetic acid. This can be added from a separatory funnel clamped to a ring stand or from another Erlenmeyer flask. Take care not to come in contact with the reagent. It should be acidic. Test it with indicator paper and, if necessary, add more acetic acid. During the addition, keep the temperature in the range of 40°C–50°C. Have an ice bath available in case the temperature goes above 50°C, but do not allow the temperature to go below 40°C because the oxidation will be incomplete. The addition should take about 15–20 minutes. Swirl the reaction mixture periodically for the next 20 minutes to complete the reaction.

The addition should take 15–20 minutes.

Because the exact concentration of the hypochlorite in the bleach depends on its age, it is necessary to analyze the reaction mixture for unreacted hypochlorite ion and to reduce any excess to chloride ion with bisulfite. Add a drop of the reaction mixture to a piece of starch-iodide paper. Any unreacted hypochlorite will cause the appearance of the blue starch-triiodide complex. Add 1-mL portions of saturated sodium bisulfite solution to the reaction mixture until the starch-iodide test is negative.

Add a few drops of thymol blue indicator solution to the mixture, and then slowly add with swirling 15–20 mL of 6 *M* sodium hydroxide solution until the mixture is neutral, which is indicated by a blue color. Transfer the reaction mixture to a 250-mL round-bottomed flask, add a boiling chip, and set up the apparatus for simple distillation. Distillation under these conditions is a steam distillation (*see* Chapter 6) in which the steam is generated in situ. Continue the distillation until no more cyclohexanone comes over with the water (40 mL of distillate should be collected).

Note the temperature of the vapor during this distillation.

Refer to the section "Isolation of Cyclohexanone from Steam Distillate" in Experiment 2 to complete the reaction sequence.

Cleaning Up. The residue from the steam distillation contains only chloride, sodium, and acetate ions and can therefore be flushed down the drain.

5. ADIPIC ACID FROM CYCLOHEXANONE

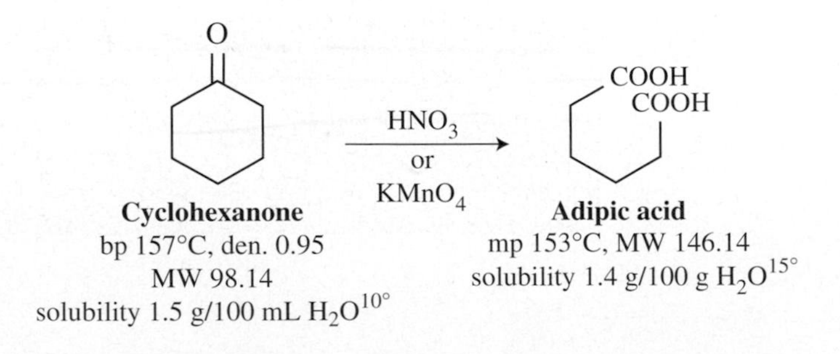

IN THIS EXPERIMENT, 150 mg of cyclohexanone (no more) is carefully oxidized with concentrated nitric acid in a highly exothermic reaction. The adipic acid crystallizes from the reaction mixture. It must be very carefully washed free of excess nitric acid because it is quite soluble in water.

Video: *Filtration of Crystals Using the Pasteur Pipette*; Photos: *Filtration using the Pasteur Pipette, Use of the Wilfilter*

Because of the highly exothermic nature of this reaction with nitric acid, it is not advisable to scale up this procedure. In a 10 × 100 mm reaction tube, place 1.0 mL of concentrated nitric acid and a boiling chip. Clamp the tube and also clamp a Pasteur pipette so that it dips into the tube and is connected to a water aspirator pulling a gentle vacuum in order to remove nitrogen oxides. Add to the tube 1 small drop of cyclohexanone from a vial containing 150 mg of the ketone. Warm the tube on a sand bath until brown oxides of nitrogen are seen emanating from the nitric acid and an exothermic reaction begins. Do not add more cyclohexanone until it is quite clear (evolution of brown oxides of nitrogen, generation of heat) that the reaction has started. Remove the tube from the heat. Add the remaining cyclohexanone to the hot nitric acid over a period of about 3 minutes. The reaction is extremely exothermic; no external heating is necessary, but add the ketone rapidly enough to keep the reaction going. After completing the cyclohexanone addition, heat the tube to boiling for 1 minute.

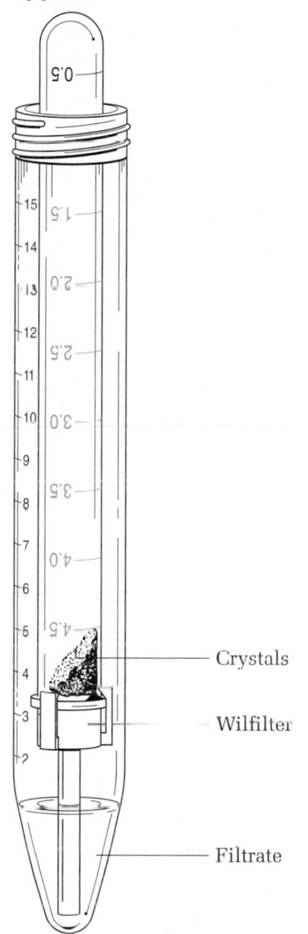

FIG. 22.4
A Wilfilter filtration apparatus.

As the tube cools to room temperature, fine crystals of adipic acid should appear. If they do not, scratch the inside of the tube at the liquid-air interface with a small glass rod to initiate crystallization. Cool the tube in a mixture of ice and water for at least 3 minutes, then stir the crystals with the tip of a glass Pasteur pipette, and remove the solvent by forcing the pipette to the bottom of the tube and pulling the liquid into the pipette. Rap the tube on a hard surface to remove more solvent from the crystals, taking care to keep the tube cold. Add 0.2 mL of ice water to the crystals, cool thoroughly in ice, and remove as much of the wash liquid as possible with the Pasteur pipette, bearing in mind that the product is very soluble in water. Collect the damp crystals on a Wilfilter by centrifugation (Fig. 22.4) or dry the crystals under vacuum in the reaction tube while heating it on a steam bath. Determine the melting point, the IR spectrum, and the purity of the product by TLC. For TLC, the eluent should be 3 parts of 95% ethanol and 1 part of ammonium hydroxide.

Cleaning Up. The nitric acid filtrate and aqueous wash should be neutralized with sodium carbonate before flushing down the drain. The TLC solvent should be neutralized with dilute hydrochloric acid and then flushed down the drain.

6. ADIPIC ACID FROM CYCLOHEXANONE

IN THIS EXPERIMENT, cyclohexanone is oxidized by potassium permanganate under basic conditions. The precipitated manganese dioxide is filtered off; the filtrate is then concentrated, acidified with hydrochloric acid, and cooled to yield crystalline adipic acid.

The reaction to prepare adipic acid is conducted with 5.0 g of cyclohexanone, 15 g of potassium permanganate, and amounts of water and alkali that can be adjusted to provide an attended exothermic reaction period of 1/2 hour or an unattended overnight reaction.

Short-Term Procedure

Choose either controlling the temperature for 30 minutes or running an unattended overnight reaction.

For the short-term reaction, mix the cyclohexanone and permanganate with 125 mL of water in a 250-mL Erlenmeyer flask, adjust the temperature to 30°C, note that there is no spontaneous temperature rise, and then add 1 mL of 3 *M* sodium hydroxide solution. A temperature rise is soon registered by the thermometer. It may be of interest to determine the temperature at which you can just detect warmth, by holding the flask in the palm of your hand or by touching the flask to your cheek. When the temperature reaches 45°C (15 minutes), slow the oxidation process by brief ice-cooling, and keep the temperature at 45°C for 20 minutes. Wait for a slight further rise (47°C) and an eventual drop in temperature (about 25 minutes later), and then heat the mixture by swirling it on a hot plate to complete the oxidation and to coagulate the precipitated manganese dioxide. Make a spot test by withdrawing some reaction mixture on the tip of a stirring rod and touching it to filter paper; permanganate, if present, will appear in a purple ring around the spot of manganese dioxide. If permanganate is still present, add small amounts of sodium bisulfite until the spot test is negative. Then filter the mixture by suction on an 11-cm Büchner funnel, wash the brown precipitate well with water, add a boiling chip, and evaporate the filtrate on a hot plate or over a flame from a large beaker to a volume of 35 mL. If the solution is not clear and colorless, clarify it with decolorizing charcoal and evaporate again to 35 mL. Acidify the hot solution with concentrated hydrochloric acid to pH 1–2, add 5 mL acid in excess, and let the solution stand to crystallize. Collect the crystals on a small Büchner funnel, wash them with a very small quantity of cold water, press the crystals between sheets of filter paper to remove excess water, and set them aside to dry. A typical yield of adipic acid (mp 152–153°C) is 3.5 g.

Video: *Microscale Filtration on the Hirsch Funnel*

Long-Term Procedure

In the alternative procedure, the weights of cyclohexanone and permanganate are the same, but the amount of water is doubled (250 mL) to moderate the reaction and make temperature control unnecessary. All the permanganate must be dissolved before the reaction is begun. Heat the flask and swirl the contents vigorously. Test for undissolved permanganate with a glass rod (as done in the short-term procedure). After adjusting the temperature of the solution to 30°C, 5 mL of 3 *M* sodium hydroxide is added, and the mixture is swirled briefly and left to stand overnight (the maximum temperature should be 45–46°C). The workup is the same as in the short-term procedure; a typical yield of adipic acid is 4.1 g.

A longer oxidizing time does no harm.

Cleaning Up. Place the manganese dioxide precipitate in the hazardous waste container for heavy metals. Neutralize the aqueous solution with sodium carbonate and flush it down the drain.

The proton and ^{13}C NMR spectra of cyclohexanone are given in Figures 22.5 and 22.6. The IR spectrum of cyclohexanol is presented in Figure 22.7, and for cyclohexanone in Figure 22.8. The ^{13}C spectrum of adipic acid is given in Figure 22.9. Compare any spectra you acquired with these reference spectra.

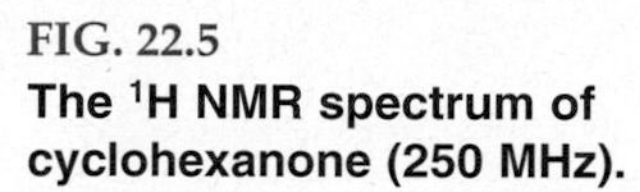

FIG. 22.5
The ^1H NMR spectrum of cyclohexanone (250 MHz).

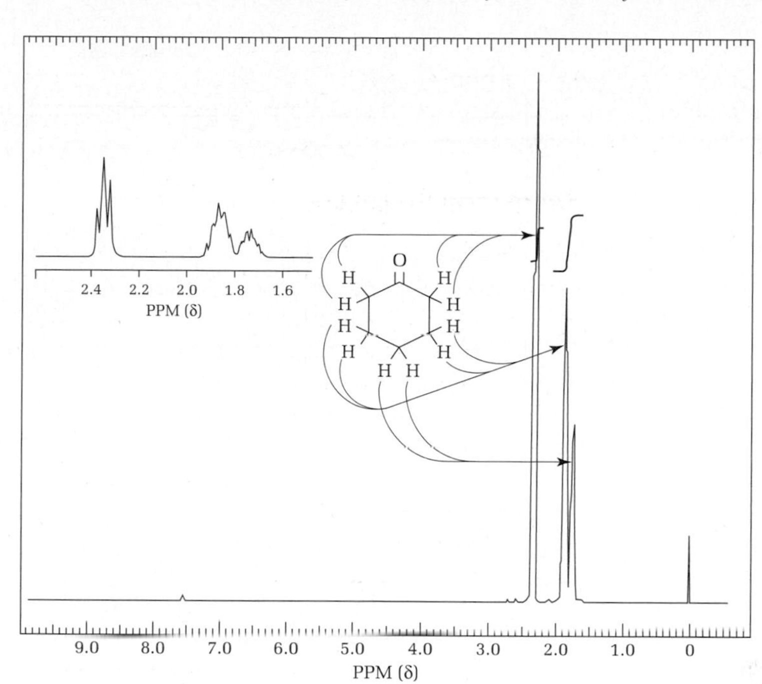

FIG. 22.6
The ^{13}C NMR spectrum of cyclohexanone (100 MHz).

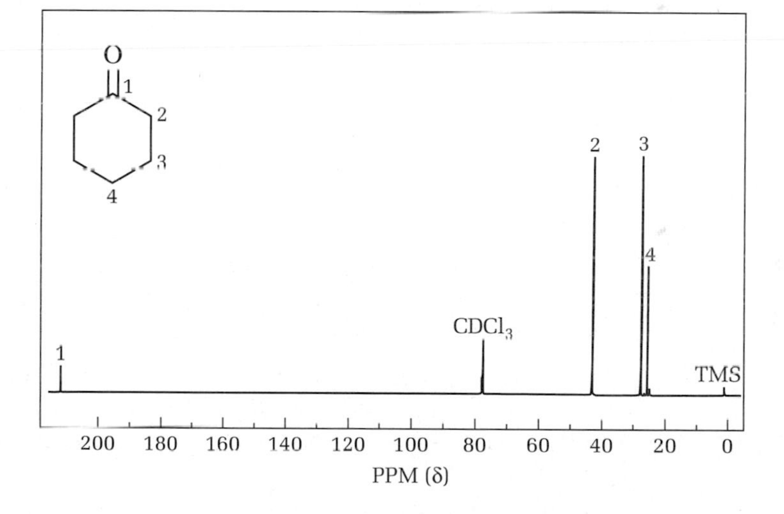

FIG. 22.7
The IR spectrum of cyclohexanol (thin film).

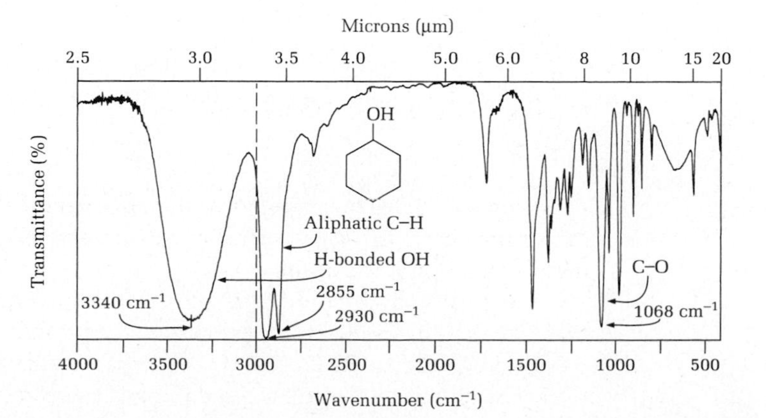

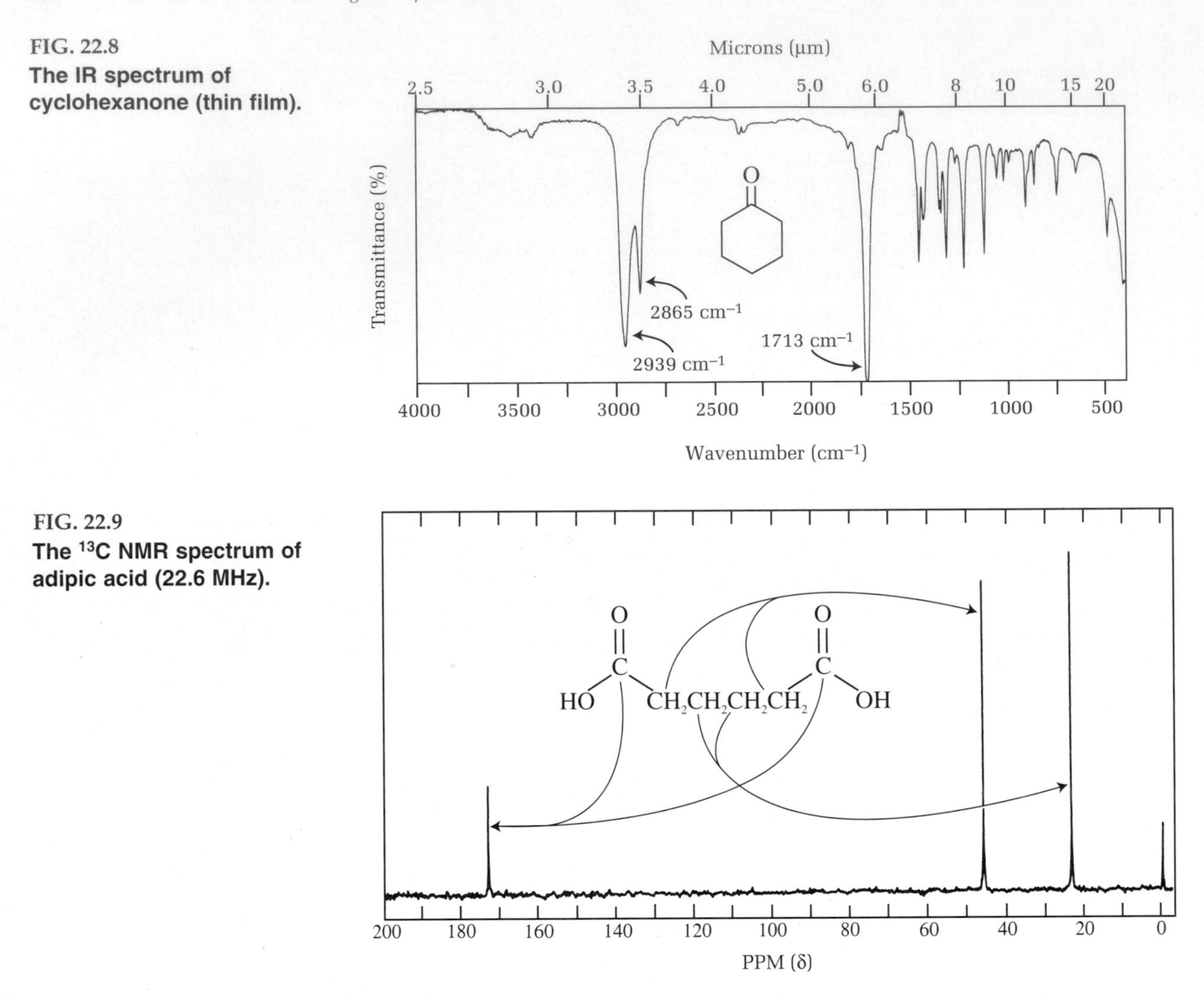

FIG. 22.8
The IR spectrum of cyclohexanone (thin film).

FIG. 22.9
The ^{13}C NMR spectrum of adipic acid (22.6 MHz).

QUESTIONS

1. In the hypochlorite oxidation of cyclohexanol to cyclohexanone, what purpose does the acetic acid serve?
2. Explain the order of the chemical shifts of the carbon atoms in the ^{13}C spectra of cyclohexanone (Fig. 22.6) and adipic acid (Fig. 22.9).

CHAPTER 23

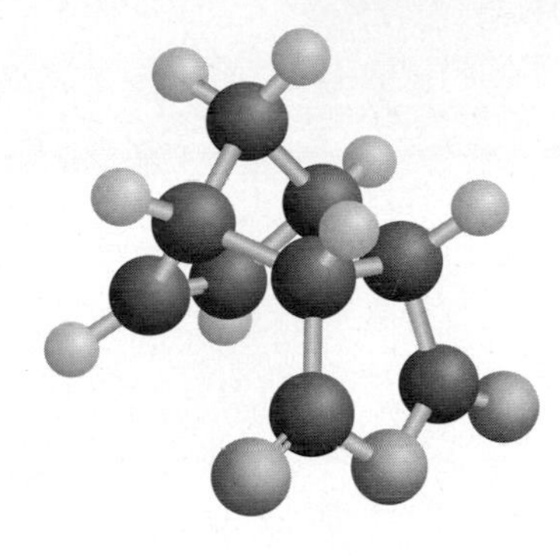

The Cannizzaro Reaction: Simultaneous Synthesis of an Alcohol and an Acid in the Absence of Solvent

PRELAB EXERCISE: Devise a procedure to isolate both products of this experiment's reaction. Hint: Consider the physical properties of both products.

Stanislao Cannizzaro discovered in 1853 that benzaldehyde on treatment with potassium carbonate gave equimolar quantities of benzoic acid and benzyl alcohol:

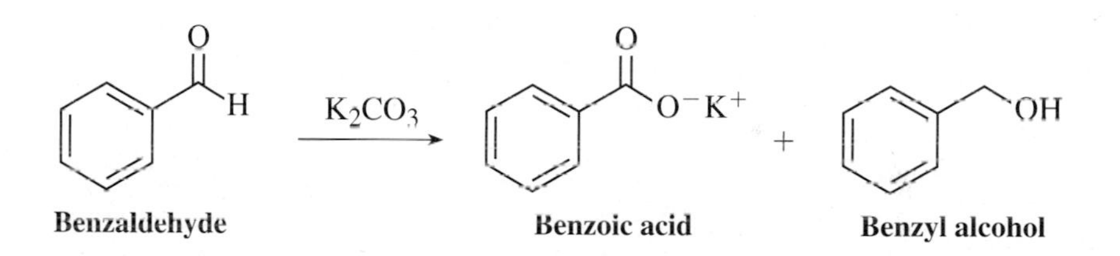

This is a disproportionation reaction that can be looked upon as an internal oxidation/reduction. In the present reaction, 2-chlorobenzaldehyde is ground with powdered potassium hydroxide to give the corresponding alcohol, and, after acidification, the carboxylic acid. The challenge is for you to devise procedures for isolating the two products.

This experiment is adapted from the work of Phonchalya, et al. who found that the procedure could be run in the absence of solvent. This, and the fact that all the organic atoms of the starting material are in the products, makes this a very green reaction.

This reaction of an aldehyde with a strong base takes place only if the aldehyde has no hydrogen atoms alpha to the carbonyl group. If the aldehyde has an alpha hydrogen, then an aldol condensation will occur (*see* Chapter 37). The mechanism of the Cannizzaro involves a hydride ion (H^-) transfer, which can occur only in the presence of strong base.

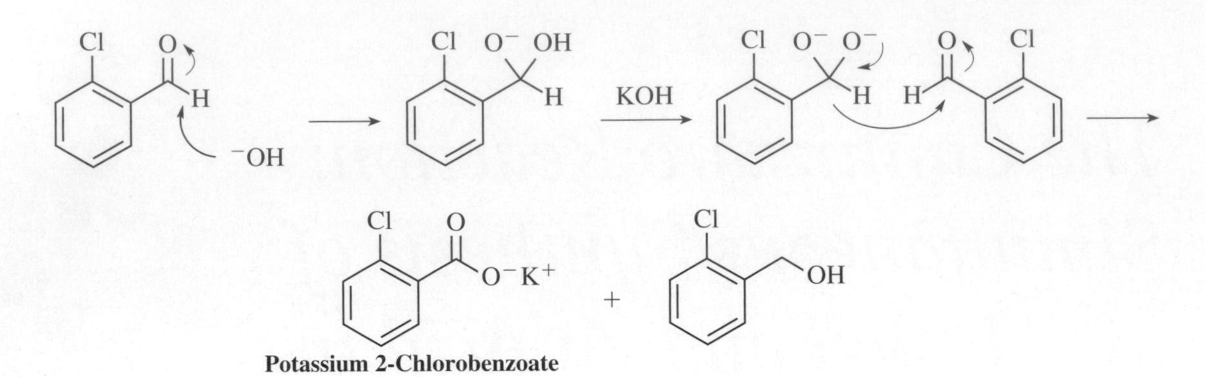

Note that at the completion of the reaction, there should be no chlorobenzaldehyde present and that the 2-chlorobenzoic acid will be in ionic form.

Using molecular modeling software, calculate the heats of formation of the products and explain why this reaction would be expected to occur.

2-Chlorobenzaldehyde	2-Chlorobenzoic acid	2-Chlorobenzyl alcohol
MW 140.569	MW 156.568	MW 142.58
bp 210–215°C	mp 138–141°C	mp 69–71°C
den. 1.248	den. 1.544	den. 1.211
ins H_2O	ins H_2O cold, sol hot	ins H_2O

EXPERIMENT

MICROSCALE CANNIZZARO REACTION OF 2-CHLOROBENZALDEHYDE

To a reaction tube, add 2 mmol of 2-chlorobenzaldehyde followed by 2.5 mmol of finely powdered potassium hydroxide.[1] In calculating the amount of potassium hydroxide to add, look up its properties in a handbook (doing so is important!). Stir the mixture with a glass stirring rod until there is no further change in its appearance, which will be about 30 minutes. Follow the course of the reaction using thin layer chromatography. Spot a plate with the reaction mixture, the reaction mixture plus 2-chlorobenzaldehyde (a co-spot) and 2-chlorobenzaldehyde alone as 1% solutions in dichloromethane. Remember that one component of the mixture will not dissolve. Develop the plates in the usual manner with 70/30 hexanes/ethyl acetate and examine the dry plates under an ultraviolet lamp.

[1]Potassium hydroxide is easily ground into a very fine powder in 25 g batches by processing the pellets in an ordinary food blender (e.g., Waring Osterizer) for 1 minute. The finely powdered base is transferred in a hood to a number of small bottles with tightly fitting caps.

At this point, isolate and purify the two products using the procedure you have devised (have it checked by your laboratory instructor).

QUESTIONS

1. There are other reaction conditions that yield 2-chlorobenzoic acid from 2-chlorobenzaldyde. Name one.
2. There are other reaction conditions that yield 2-chlorobenzyl alcohol from 2-chlorobenzaldehyde. Name one.

24 CHAPTER

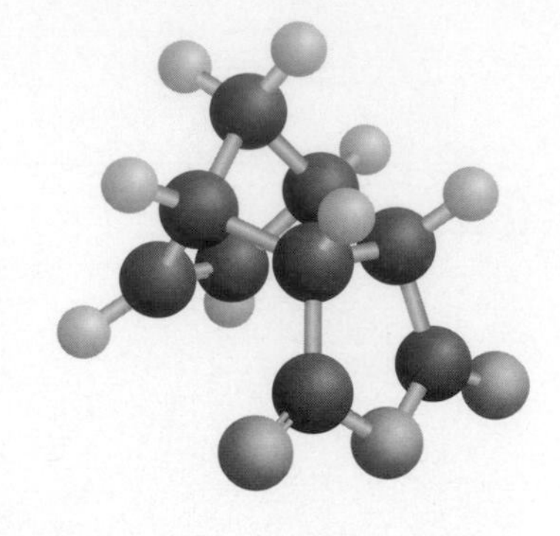

Oxidative Coupling of Alkynes: 2,7-Dimethyl-3,5-octadiyn-2,7-diol

PRELAB EXERCISE: Show the reactions for a two-step method that might be used to convert 2-methyl-3-butyn-2-ol to isoprene.

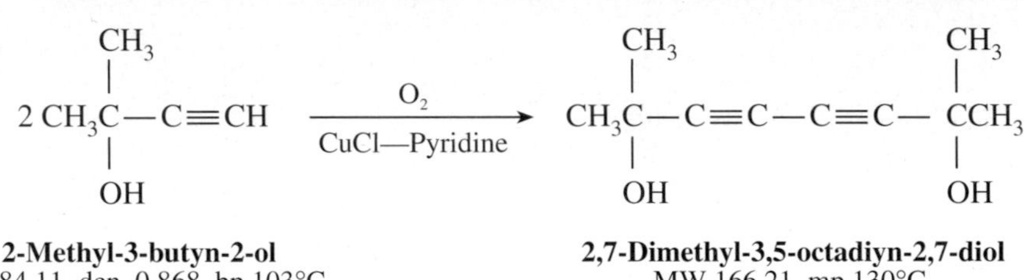

2-Methyl-3-butyn-2-ol
MW 84.11, den. 0.868, bp 103°C

2,7-Dimethyl-3,5-octadiyn-2,7-diol
MW 166.21, mp 130°C

Isoprene

The starting material, 2-methyl-3-butyn-2-ol, is made commercially from acetone and acetylene and is convertible into isoprene. This experiment illustrates the oxidative coupling of a terminal acetylene to produce a diacetylene, commonly known as the Glaser reaction.

Glaser reported the coupling of acetylenes using the cuprous ion in 1869. Using density functional theory, a mechanism was reported only in 2002 by Fomina, et al.[1] The complex mechanism involves Cu^{+}, Cu^{2+} and Cu^{3+} ions. As noted by Glaser, the cuprous acetylide is oxidized by oxygen. The new work shows that a dicopper-dioxo complex is formed in this oxidation, but Cu^{2+} is the actual oxidizing agent. The ammonium hydroxide seems to be needed to keep the acetylide in solution.

$$C_6H_5C{\equiv}CH \xrightarrow[NH_4OH]{CuCl} C_6H_5C{\equiv}CCu \xrightarrow[O_2]{air} C_6H_5C{\equiv}C{-}C{\equiv}CC_6H_5$$

The reaction is very useful in the synthesis of polyenes, vitamins, fatty acids, and the annulenes. Johann Baeyer used the reaction in his historic synthesis of

[1]Fomina L, Vazquez B, Tkatchouk E, Fomine S. *Tetrahedron*. **2002**; *58*:6641–6647.

indigo back in 1882. The reaction allowed the unequivocal establishment of the carbon skeleton in this dye:

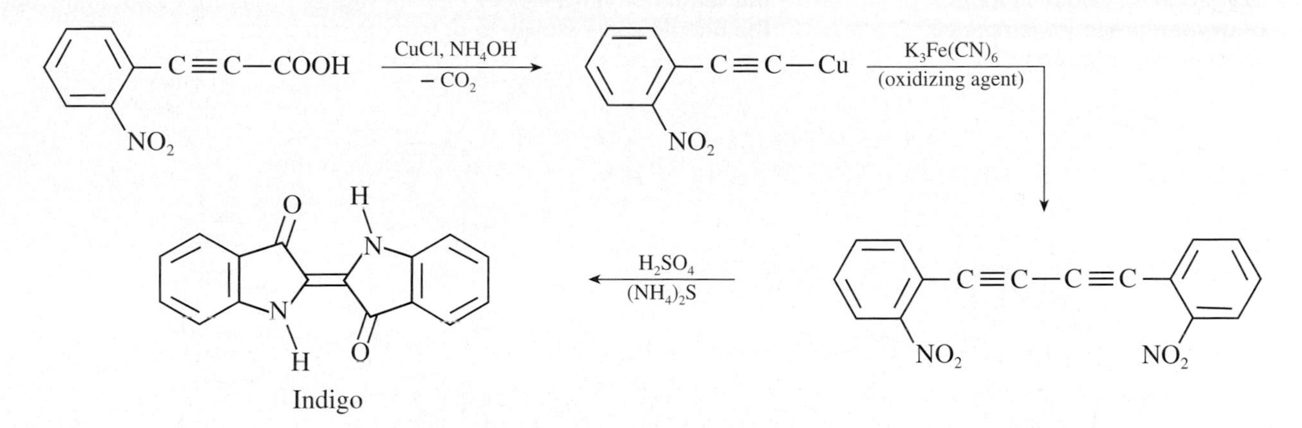

EXPERIMENT

PREPARATION OF 2,7-DIMETHYL-3,5-OCTADIYN-2,7-DIOL

Microscale Procedure

IN THIS EXPERIMENT, the reaction time for this coupling reaction is shortened using an excess of catalyst and by supplying one of the reactants, oxygen, under pressure. Pressurized oxygen is obtained by inflating a white rubber pipette bulb. The progress of the reaction is followed by observing the decrease in diameter of the balloon. Water is the other product of the reaction. Swirling or stirring the reaction mixture for about 20 minutes completes the reaction. Acid is added to neutralize the basic pyridine, and sodium chloride solution is added to precipitate the product. The crystalline product is wet and so it is dissolved in ether, which, in the usual way, is washed, dried, and evaporated. The residue is crystallized from toluene to give the diol product.

Measure pyridine in the hood.

The reaction vessel is a 25-mL filter flask equipped with a stirring bar and a rubber bulb secured to the side arm with copper wire, as shown in Figure 24.1. Add 1.0 mL of 2-methyl-3-butyn-2-ol, 1.0 mL of methanol, 0.3 mL of pyridine, and 0.1 g of copper(I) chloride. Before going to the oxygen station,[2] practice capping the flask with a rubber septum until you can do this quickly. You are to flush out the flask with oxygen and cap it before air can diffuse in; the reaction will be about twice as fast in an atmosphere of oxygen as in air. Cut the top off a 1-mL plastic syringe barrel, insert it into the oxygen tank supply hose, and put a needle on the other end of the syringe barrel. Remove the rubber septum from the flask, insert the needle into

[2]The valves of a cylinder of oxygen should be set to deliver gas at a pressure of 10 lb/in.^{2} when the needle valve is opened. The barrel of a 2.5-mL plastic syringe is cut off and thrust into the end of a $\frac{1}{4}$- × $\frac{3}{16}$-in. rubber delivery tube. Read about the handling of compressed gas cylinders in Chapter 2. Be sure the cylinder is secured to a bench or wall.

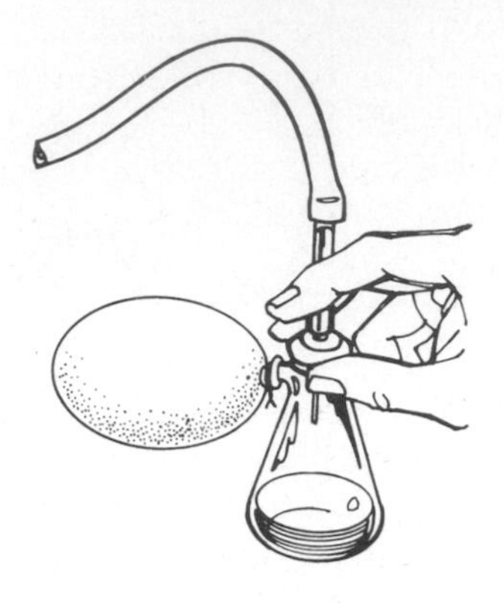

FIG. 24.1
The balloon technique of oxygenation. About 10 lb/in.² of oxygen pressure is needed to inflate the pipette bulb.

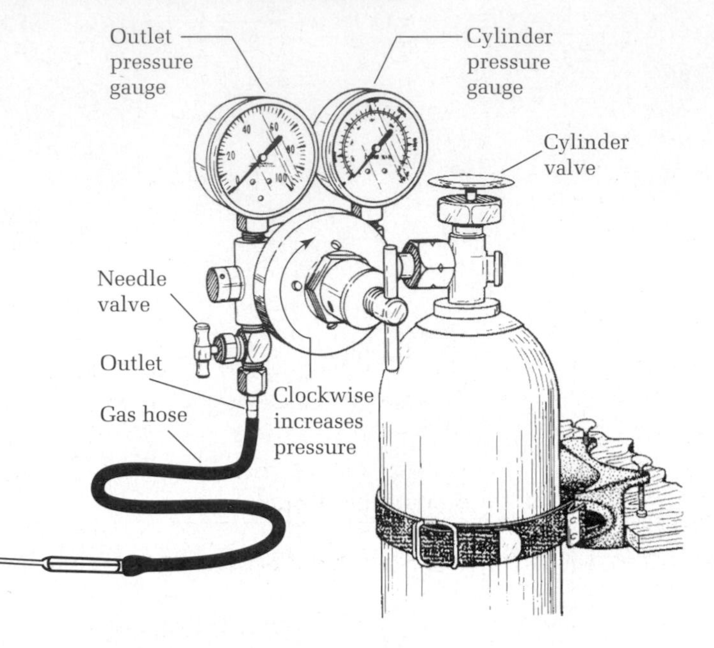

FIG. 24.2
Open the main tank valve, adjust the gas pressure to 10 lb/in.² with the regulator valve—clockwise for higher pressure—and finally open the needle valve slowly to obtain oxygen gas.

the reaction liquid, open the needle valve on the oxygen gas regulator (Fig. 24.2), and let oxygen bubble through the solution in a brisk stream for 2 minutes. Close the valve and quickly cap the flask, and wire the cap. Next, thrust the needle of the delivery syringe through the center of the rubber septum, open the valve, and run in oxygen until you have produced a sizable inflated bulb (Fig. 24.1). Close the valve and withdraw the needle (the hole is self-sealing), note the time, and start swirling the reaction mixture. The rate of oxygen uptake depends on the efficiency of mixing of the liquid and gas phases. Swirl or stir the reaction mixture continuously for 20–25 minutes.[3] The reaction mixture warms up and becomes deep green. Introduce a second charge of oxygen the same size as the first, note the time, and swirl. In 25–30 minutes, the balloon reaches a constant volume, indicating that the reaction is complete. A pair of calipers is helpful in recognizing the constant size of the balloon, and thus the end of the reaction (Fig. 24.3).

FIG. 24.3
The appearance of the pipette bulb after oxygenation.

Open the reaction flask, cool if warm, add 0.5 mL of concentrated hydrochloric acid to neutralize the pyridine and keep copper compounds in solution, and cool again; the color changes from green to yellow. Use a spatula to dislodge any copper salts adhering to the walls, leaving the salts in the flask. Then add 2.5 mL of a saturated sodium chloride solution to precipitate the diol, and stir and cool the thick paste that results. Scrape out the paste onto a Hirsch funnel with the spatula; press down the material to an even cake. Rinse out the flask with water and wash the filter cake with enough additional water to remove all the color from the solid and leave a colorless product. Because the moist product dries slowly, drying is accomplished in ether solution, and the operation is combined with the recovery of diol retained in the mother liquor. Transfer the moist product to a 10-mL Erlenmeyer

[3]A magnetic stirrer does not significantly shorten the reaction time.

Photo: *Placing a Polyethylene Tube through a Septum*

flask, extract the mother liquor in the filter flask with one 1.5-mL portion of ether, wash the extract once with water, and add the ethereal solution to the flask containing the solid product. Add enough additional ether to dissolve the product, transfer the solution to a test tube, remove the water layer, wash with a saturated sodium chloride solution, and dry over anhydrous calcium chloride. The solvent is evaporated on a steam bath, and the solid residue is heated and evacuated until the weight is constant. Recrystallization from toluene then gives colorless needles of the diol (mp 129–130°C). The yield of crude product is about 0.5 g. In the recrystallization, the recovery is practically quantitative.

Analyze the product by thin-layer chromatography (TLC) and infrared (IR) spectroscopy. Compare the diyne with the starting material to determine whether the product is pure.

Cleaning Up. Combine all aqueous and organic layers from the reaction, add sodium carbonate to make the aqueous layer neutral, and shake the mixture gently to extract pyridine into the organic layer. Place the organic layer in the organic solvents waste container. Remove any solid (copper salts) from the aqueous layer by filtration. The solid can be placed in the nonhazardous solid waste container, and the aqueous layer can be diluted with water and flushed down the drain. If local regulations allow, evaporate any residual ether from the drying agent in the hood and place the dried solid in the nonhazardous waste container. Otherwise, place the wet drying agent in a waste container designated for this purpose.

Macroscale Procedure

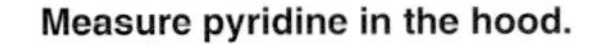

Measure pyridine in the hood.

The reaction vessel is a 125-mL or 50-mL filter flask with a rubber bulb secured to the side arm with copper wire, as shown in Figures 24.1 and 24.3. Add 5 mL of 2-methyl-3-butyn-2-ol, 5 mL of methanol, 1.5 mL of pyridine, and 0.25 g of copper(I) chloride. Before going to the oxygen station,[4] practice capping the flask with a rubber septum until you can do this quickly. You are to flush out the flask with oxygen and cap it before air can diffuse in; the reaction will be about twice as fast in an atmosphere of oxygen as in air. Insert the oxygen delivery syringe into the flask with the needle under the surface of the liquid, open the valve, and let oxygen bubble through the solution in a brisk stream for 2 minutes (Fig. 24.2). Close the valve and quickly cap the flask and wire the cap. Next, thrust the needle of the delivery syringe through the center of the rubber septum, open the valve, and run in oxygen until you have produced a sizable inflated bulb (Fig. 24.1). Close the valve and withdraw the needle (the hole is self-sealing), note the time, and start swirling the reaction mixture. The rate of oxygen uptake depends on the efficiency of mixing of the liquid and gas phases. By vigorous and continuous swirling, it is possible to effect deflation of the balloon to a constant size (*see* Fig. 24.3 on page 374) in 20–25 minutes.[5] The reaction mixture warms up and becomes deep green. Introduce a second charge of oxygen of the same size as the first, note the time, and swirl. In 25–30 minutes, the balloon reaches a constant volume (e.g., a 5-cm sphere) and the reaction is complete. A pair of calipers is helpful in recognizing the constant size of the balloon, and thus the end of the reaction.

[4]The valves of a cylinder of oxygen should be set to deliver gas at a pressure of 10 lb/in.^{2} when the needle valve is opened. The barrel of a 2.5-mL plastic syringe is cut off and thrust into the end of a $\frac{1}{4}$- × $\frac{3}{16}$-in. rubber delivery tube. Read about the handling of compressed gas cylinders in Chapter 2. Be sure the cylinder is secured to a bench or wall.

[5]A magnetic stirrer does not significantly shorten the reaction time.

Open the reaction flask, cool if warm, add 2.5 mL of concentrated hydrochloric acid to neutralize the pyridine and keep copper compounds in solution, and cool again; the color changes from green to yellow. Use a spatula to dislodge any copper salts adhering to the walls. Then add 13 mL of a saturated sodium chloride solution to dissolve the copper salts and precipitate the diol. Stir and cool the thick paste that results, and scrape it out onto a small Büchner funnel with a spatula; press down the material to an even cake. Rinse out the flask with water and wash the filter cake with enough additional water to remove all the color from the cake and leave a colorless product. Because the moist product dries slowly, drying is accomplished in ether solution, and the operation is combined with the recovery of diol retained in the mother liquor. Transfer the moist product to a 50-mL Erlenmeyer flask, extract the mother liquor in the filter flask with one portion of ether, wash the extract once with water, and run the ethereal solution into the flask containing the solid product. Add enough additional ether to dissolve the product, transfer the solution to a separatory funnel, drain off the water layer, wash with a saturated sodium chloride solution, and filter through anhydrous calcium chloride pellets into a tared flask. The solvent is evaporated on a steam bath, and the solid residue is heated and evacuated until the weight is constant. Recrystallization from toluene then gives colorless needles of the diol (mp 129–130°C). The yield of crude product is 3–4 g. In the recrystallization, the recovery is practically quantitative.

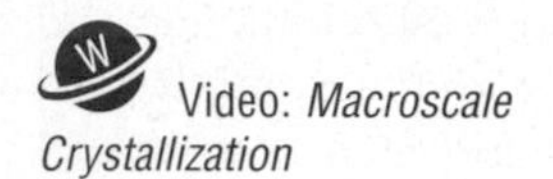

Video: *Macroscale Crystallization*

Cleaning Up. Combine all aqueous and organic layers from the reaction, add sodium carbonate to make the aqueous layer neutral, and shake the mixture gently to extract pyridine into the organic layer. Place the organic layer in the organic solvents waste container. Remove any solid (copper salts) from the aqueous layer by filtration. The solid can be placed in the nonhazardous solid waste container, and the aqueous layer can be diluted with water and flushed down the drain. If local regulations allow, evaporate any residual solvent from the drying agent in the hood and place the dried solid in the nonhazardous waste container. Otherwise, place the wet drying agent in a waste container designated for this purpose.

See *Web Links*

QUESTIONS

1. Give two oxidation states of copper in this reaction, and identify which state forms the copper acetylide. What is oxygen reduced to?
2. What volume of oxygen at standard temperature and pressure is consumed in this reaction?

CHAPTER 25

Catalytic Hydrogenation

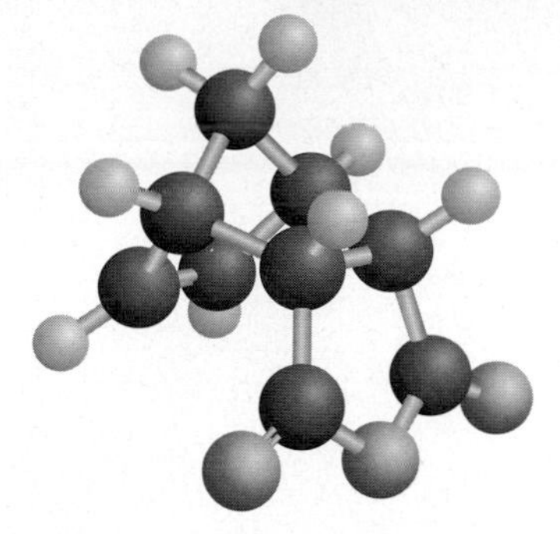

PRELAB EXERCISE: Calculate the volume of hydrogen gas generated when 3 mL of 1 *M* sodium borohydride reacts with concentrated hydrochloric acid. Write a balanced equation for the reaction of sodium borohydride with platinum chloride. Calculate the volume of hydrogen that can be liberated by reacting 1 g of zinc with acid.

Catalytic reduction is a very important and widely used industrial process; usually no harmful wastes are produced in the process. Catalytic hydrogenation and dehydrogenation are carried out on an enormous scale, for example, in the catalytic cracking and reforming of crude oil to make gasoline.

Nitrobenzene can be reduced catalytically to aniline with water as the only byproduct:

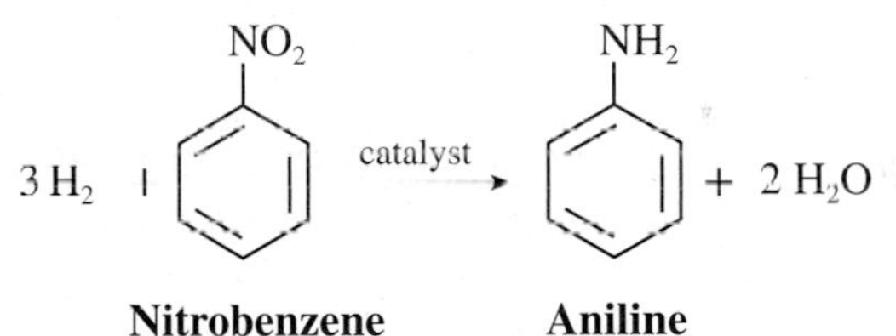

Styrene is made by the catalytic dehydrogenation of ethylbenzene at very high temperatures, but styrene can also be very easily hydrogenated back to ethylbenzene. Palladium, as a catalyst, lowers the energy barrier for the reaction in both directions.

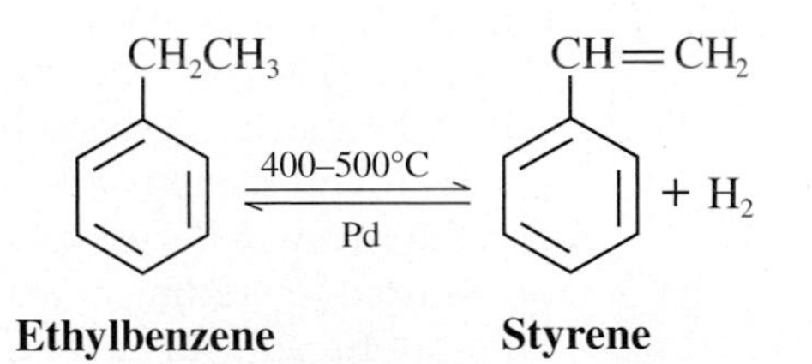

The addition of hydrogen to alkenes is one of the most common reactions. The alkene is more reactive toward this process than is the aromatic ring or functional groups such as esters or ketones.

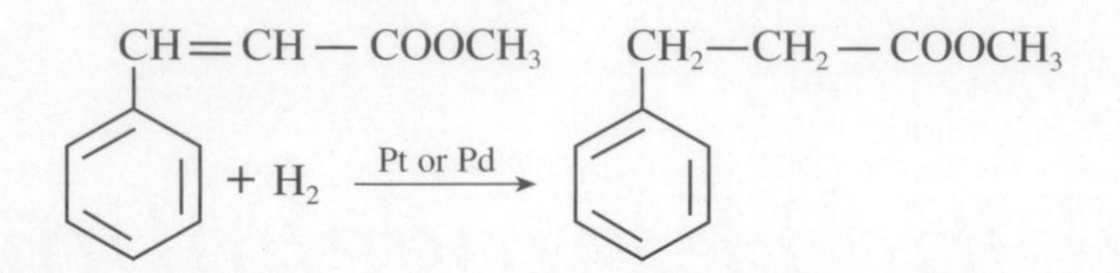

Hydrogenation is stereospecific, so alkynes are reduced to *cis*-alkenes. The metal is usually supported on a high-surface-area material such as charcoal. The alkene and the hydrogen probably are both adsorbed onto the surface of the catalyst before the transfer occurs. This heterogeneous reaction, a reaction that involves reactants in the liquid or gas phase and a catalyst in the solid phase, is difficult to study.

In the experiments in this chapter, catalytic hydrogenation is carried out in several different ways. Experiment 1 is a puzzle for you to solve. In Experiment 2, hydrogen gas from an external supply is used to hydrogenate a long-chain unsaturated alcohol to the corresponding saturated alcohol. In Experiment 3, the hydrogen is generated in situ using the Brown hydrogenation technique. Experiment 4 utilizes a process called *transfer hydrogenation* to produce a saturated fat from an unsaturated one (olive oil). In Experiment 5, olive oil is catalytically reduced by using hydrogen gas.

Experiments

1. A Small Research Project

$\xrightarrow{\text{10\% Pd/C}}$?

The purpose of this experiment is to determine whether a chemical reaction has occurred when cyclohexene is boiled for a short period of time with a small quantity of a catalyst, 10% palladium on carbon (charcoal), and, if so, to determine the structure of the product(s).

Microscale Procedure

In a reaction tube place about 50 mg of 10% palladium on carbon. To this, add about 0.8 mL of cyclohexene and a boiling chip. Reflux this mixture on a warm sand bath (low setting on controller) for about 10–15 minutes. Be careful not to boil the alkene out of the reaction tube. If necessary, place a coil of pipe cleaner moistened with water around the top of the tube. This will keep the top of the tube cool so that the cyclohexene will not escape. Further insurance against boiling away of the reactant is afforded by adding an empty distilling column to the reaction tube, although with careful adjustment of the heat input (the depth of the tube in the sand), these measures should not be necessary (Fig. 25.1).

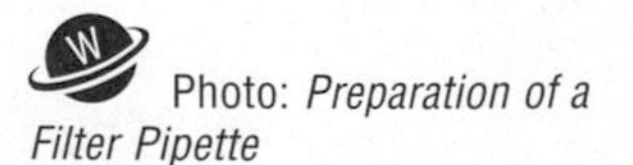
Photo: *Preparation of a Filter Pipette*

At the end of the reflux period, cool the tube to room temperature and filter the solution to remove the catalyst. This is done by stuffing a small ball of cotton into a Pasteur pipette, and using another Pasteur pipette to transfer the reaction mixture which contains charcoal. Filter the solution into a clean, dry reaction tube. If one filtration does not remove all the charcoal, recycle the filtrate through the filter (Fig. 25.2).

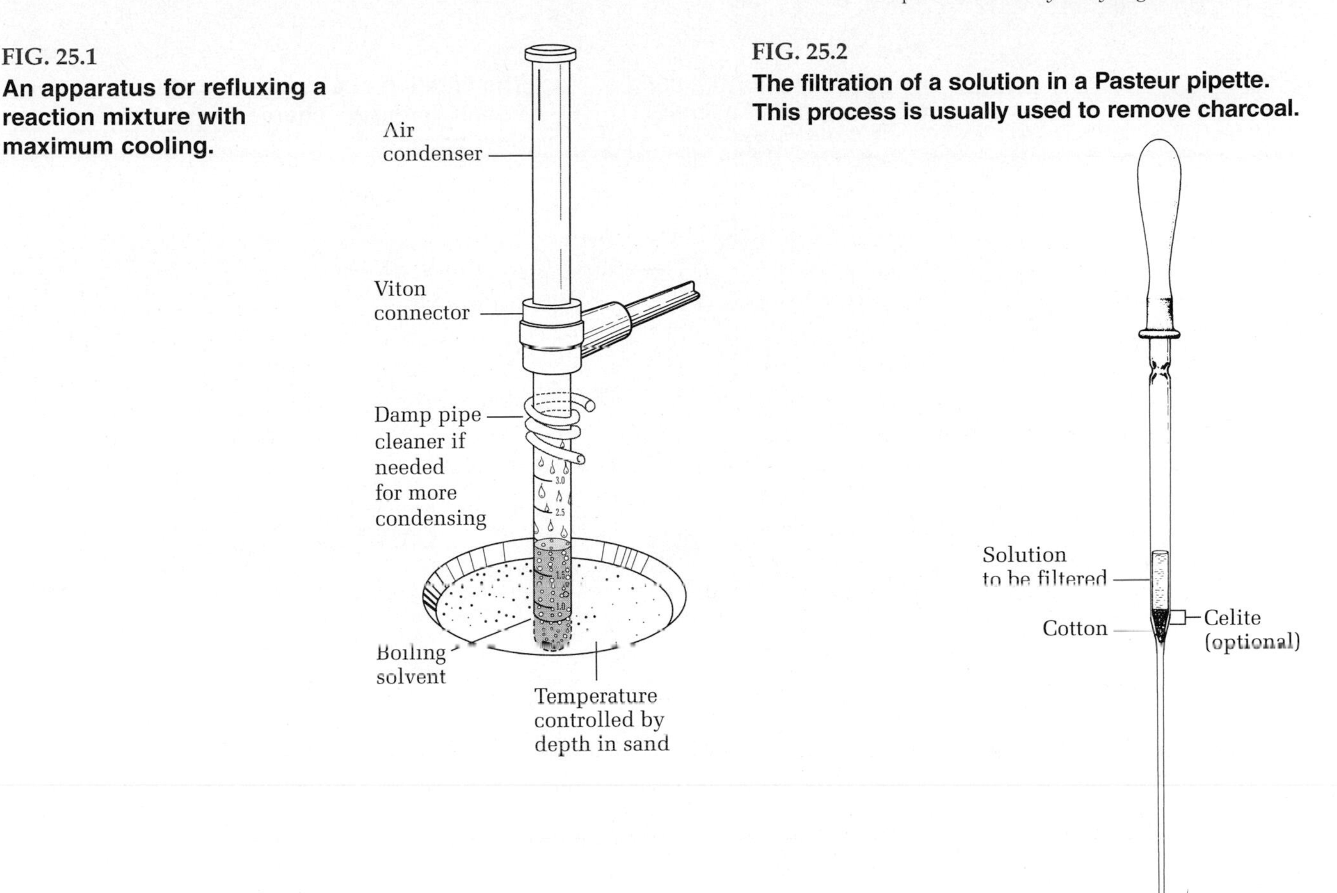

FIG. 25.1
An apparatus for refluxing a reaction mixture with maximum cooling.

FIG. 25.2
The filtration of a solution in a Pasteur pipette. This process is usually used to remove charcoal.

Macroscale Procedure

In a 15-mL tapered 14/20 standard taper flask, place about 100 mg of 10% palladium on carbon. To this, add about 2 mL of cyclohexene and a boiling chip. Reflux this mixture on a warm sand bath (low setting on controller) for at least 15 minutes (Fig. 25.3).

Photo: *Preparation of a Filter Pipette*

At the end of the reflux period, cool the flask to room temperature and filter the solution to remove the catalyst. This is done by stuffing a small ball of cotton into a Pasteur pipette and then using another Pasteur pipette to transfer the reaction mixture containing the palladium on charcoal. Filter the solution into a clean, dry reaction tube or culture tube. Alternatively, the reaction mixture can be filtered in a Hirsch funnel (Fig. 25.4). Use fine porosity filter paper in the funnel.

Analysis of the Product

Run a test for the presence of an alkene on your starting material, cyclohexene, and your product. The decolorization of bromine is probably the best test to run. Count the number of drops of a 3% solution of bromine in dichloromethane or carbon tetrachloride that are decolorized by 0.10 mL of cyclohexene and by 0.10 mL of your filtered reaction mixture.

Analyze your product by gas chromotography (or a gas chromatography–mass spectrometry system). Review Chapter 10 for the theory of gas chromatography and the procedures for analysis.

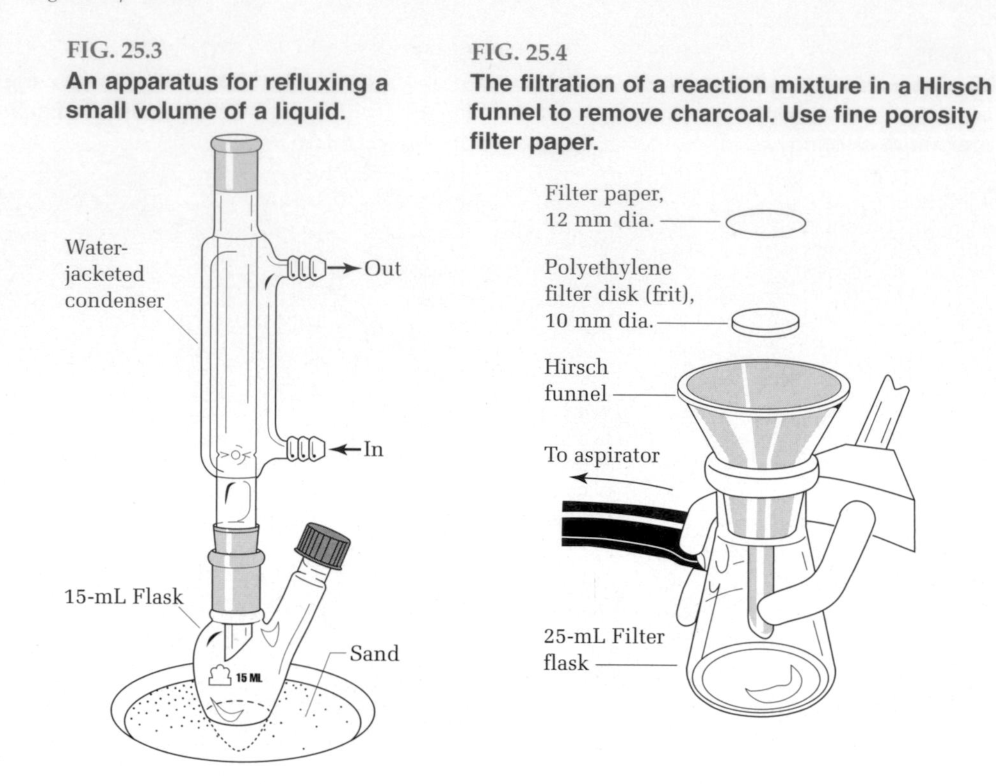

FIG. 25.3
An apparatus for refluxing a small volume of a liquid.

FIG. 25.4
The filtration of a reaction mixture in a Hirsch funnel to remove charcoal. Use fine porosity filter paper.

Run and interpret nuclear magnetic resonance (NMR), infrared (IR), and ultraviolet (UV) spectra of the starting material and product. The most useful of these is the NMR spectrum, followed by the UV spectrum. On the basis of these data, interpret the results of this experiment.

2. CATALYTIC HYDROGENATION OF OLEYL ALCOHOL USING PALLADIUM ON CHARCOAL

$$CH_3(CH_2)_7CH{=}CH(CH_2)_8OH + H_2 \xrightarrow{10\%\ Pd/C} CH_3(CH_2)_{17}OH$$

Oleyl alcohol
MW 268.49, den. 0.849
bp 207°C/13 mm
n_D^{20} 1.4600

Octadecanol
MW 270.50
mp 60°C

IN THIS EXPERIMENT, the 18-carbon unsaturated alcohol oleyl alcohol is reduced to the corresponding saturated alcohol using one of the most widely employed techniques for catalytic hydrogenation. The catalyst is 10% palladium on carbon, which currently sells for $23 per gram; because this reaction is being carried out on a microscale, we will use less than a dollar's worth.

The hydrogen could come from a compressed gas cylinder, but it is just as convenient to prepare a small quantity by the action of acid on zinc.

The apparatus for this experiment can be set up in a number of ways. If you merely want to hydrogenate a double bond and are not concerned about the quantitative aspects of the reaction, fit the apparatus with a rubber balloon that holds the hydrogen. As soon as the balloon has reached a constant size, hydrogen uptake is complete, and the product can be isolated.

In a modification of the apparatus employed in this experiment, the hydrogen uptake can be followed as water rises in the hydrogen reservoir (a graduated cylinder). With this apparatus, the volume of hydrogen absorbed as a function of time can be measured.

In research-grade apparatus, the hydrogen is stored in a gas burette over mercury. The volume of hydrogen can be determined very accurately so that quantitative estimates of the numbers of double bonds in unknowns can be made, or so that selective hydrogenation of the more reactive of several double bonds can be carried out. The reaction is stopped when the theoretical amount of hydrogen has been absorbed.

Experimental Considerations

Video: *Catalytic Hydrogenation*

The apparatus in Figure 25.5 will be assembled. The graduated cylinder is filled with hydrogen, and the reaction flask containing the solvent, the catalyst, and a

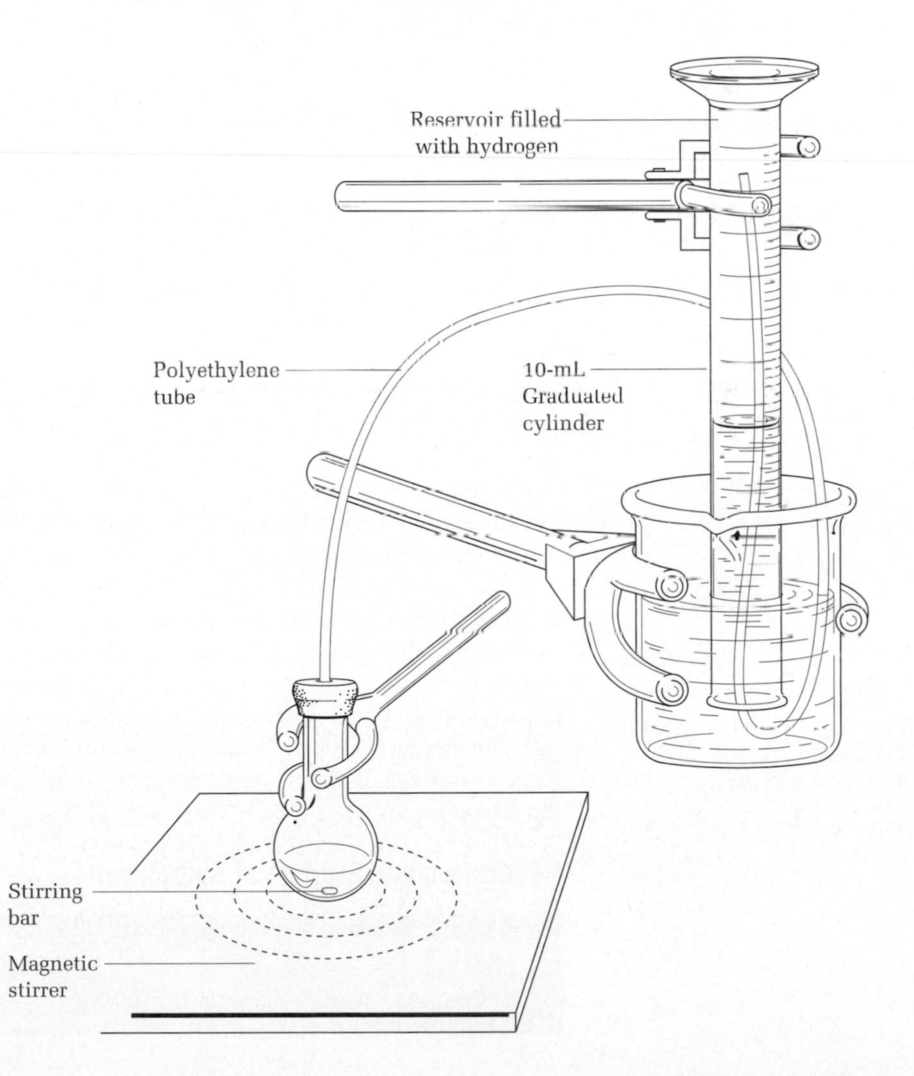

FIG. 25.5
The catalytic hydrogenation apparatus.

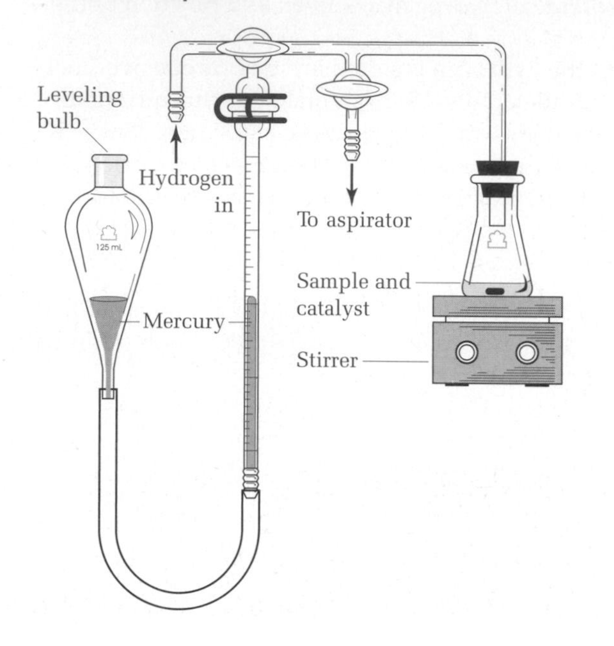

FIG. 25.6
An apparatus for quantitative hydrogenation. Air in the apparatus is removed with an aspirator, and hydrogen is admitted and brought to atmospheric pressure by raising or lowering the leveling bulb. The stirrer is started, and the amount of hydrogen taken up is measured in a gas burette after again bringing the pressure inside to that of the atmosphere.

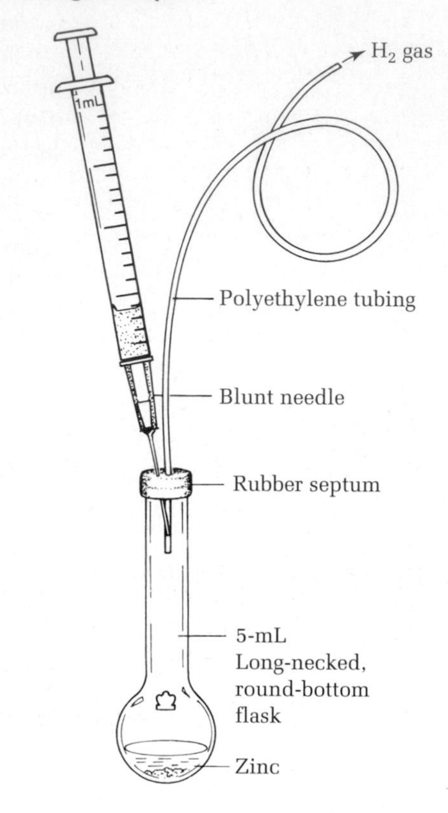

FIG. 25.7
A hydrogen generator. *See* Figure 18.2 on page 331 for threading a polyethylene tube through a septum.

magnetic stirring bar is flushed with hydrogen. The stirrer is started, and as soon as the water level in the reservoir reaches a constant level, the compound to be hydrogenated is injected through the septum of the reaction flask. It can be either a pure liquid or a solid dissolved in alcohol.

A millimole (0.001 mole) of hydrogen at standard temperature and pressure has a volume of 22.4 mL. This will hydrogenate a millimole of a compound that has one double bond. The apparatus in Figure 25.5 holds about 9 mL of gas (0.4 mmol).

For highly accurate measurement of the hydrogen uptake, we should take into account the temperature, the atmospheric pressure, and the vapor pressure of the methanol in the reaction flask. We should also be careful to equalize the internal and external pressures so that precise volume measurements can be made (Fig. 25.6). The experiments in this chapter can be run satisfactorily without this level of accuracy.

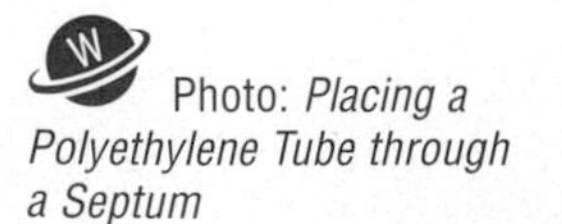
Photo: *Placing a Polyethylene Tube through a Septum*

The apparatus for hydrogen generation is in Figure 25.7. The 5-mL long-necked flask contains 1 g of zinc. As hydrogen is needed, sulfuric acid is injected through the septum.

Generation and Storage of Hydrogen

IN THIS EXPERIMENT, hydrogen is generated by the addition of acid to zinc. It is stored in a graduated cylinder by the downward displacement of water.

Assemble the apparatus shown in Figure 25.5. In a 5-mL short-necked, round-bottomed flask, place 20 mg of 10% palladium on carbon, a small magnetic stirring bar, and 1.5 mL of methanol. Fill a 250-mL or 400-mL beaker with water. Fill a 10-mL or 25-mL graduated cylinder with water, seal the top with your finger or a loose cork, and invert it in the beaker of water. Clamp it in this position in the beaker of water, removing the cork if used. In the 5-mL long-necked flask, place 1 g of granulated or mossy zinc and close it with a septum bearing a polyethylene tube, folding the rubber flange over the outside of the flask neck (Fig. 25.7). Put a kink in the tube by exposing a short section of it to a heat source. This will make insertion of the tube into the cylinder easier. Inject a few drops of 6 *M* (33%) sulfuric acid to generate hydrogen and displace the air from the 5-mL short-necked flask. Thread the polyethylene tubing into the bottom of the graduated cylinder clamped over a beaker of water (*see* Fig. 25.5). Inject more acid onto the zinc and fill the graduated cylinder reservoir with about 9 mL of hydrogen. Adjust the height of the graduated cylinder so that the water level inside is the same as that outside. Remove the septum and the tube from the gas-generating flask and connect them to the short-necked flask as in Figure 25.5.

Turn on the stirrer briefly to saturate the methanol with hydrogen. Note the level of the water in the cylinder. If it comes to the same level as the water on the outside when the cylinder is moved up and down, there is a leak in the system.

Hydrogenation

IN THIS EXPERIMENT, a liquid alkene is hydrogenated to give a solid product. Uptake of hydrogen is easily observed. The catalyst is removed by filtration, and the solvent is evaporated to give pure product.

Once the water level in the graduated cylinder has stabilized with the stirrer off, inject 0.4 mmol of a compound that has one olefinic double bond to be hydrogenated. In this particular experiment, this will be 107 mg of oleyl alcohol dissolved in 0.6 mL of methanol. If a larger reservoir of hydrogen has been used, for example, a 25-mL graduated cylinder, a proportionately larger quantity of alkene can be hydrogenated. Turn on the stirrer and note the height of the hydrogen in the reservoir as a function of time. Be sure the end of the tube is always above the water level in the reservoir. When the water level no longer changes, the reaction is complete.

Photo: *Preparation of a Filter Pipette*

Filter the reaction mixture into a tared 25-mL filter flask using a Pasteur pipette; push a piece of cotton firmly into a Pasteur pipette followed by a 5-mm layer of Celite filter aid. With a second pipette, transfer the reaction mixture to the filter pipette and allow the solution to filter through the cotton to remove the charcoal. Force the solution through the cotton with a rubber bulb (*see* Fig. 25.2 on page 379). Rinse the reaction flask, the stirring bar, and the transfer pipette with a few drops of methanol; then filter this solution as well. If the filtrate is gray, re-filter the solution in the same pipette without changing the filter.

Photo: *Evaporation of Solvent in a Filter Flask*

Remove the methanol from the filter flask under aspirator vacuum (*see* Fig. 25.9 on page 388). Heat the filter flask on a steam bath or a sand bath under vacuum to remove the last traces of methanol. The residue should be pure 1-octadecanol (mp 60°C), although the melting point will depend on the quality of the starting alcohol. Test a small portion of the product for unsaturation using a bromine solution.

Cleaning Up. Decant the sulfuric acid from the zinc, dilute the solution, neutralize with sodium carbonate, and remove zinc hydroxide by filtration. The filtrate can be flushed down the drain. Zinc and zinc hydroxide are placed in the non-hazardous solid waste container. Wash out the syringe and needle carefully.

3. BROWN HYDROGENATION[1]

COOH
COOH

$\xrightarrow[PtCl_4]{NaBH_4—HCl}$

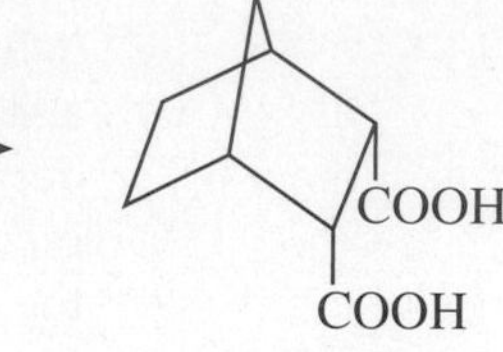

***cis*-Norbornene-5,6-*endo*-dicarboxylic acid**
mp 180–190°C, dec
MW 182.17

***cis*-Norbornane-5,6-*endo*-dicarboxylic acid**
mp 170–175°C, dec
MW 184.19

FIG. 25.8
Apparatus for macroscale Brown hydrogenation.

IN THIS EXPERIMENT, platinum chloride is reduced with borohydride to metallic platinum that is deposited on charcoal. An aqueous solution of an alkene carboxylic acid and a borohydride solution is added to the sealed flask. The hydrogen gas is rapidly absorbed while swirling the solution. The aqueous solution is extracted with ether, and in the usual way the ether is washed, dried, and evaporated. The solid product is recrystallized from aqueous hydrochloric acid.

Handle platinum chloride carefully to prevent any waste.

Ideal reaction time: about 15 minutes

See Figure 25.6 for a quantitative hydrogenation apparatus. However, in this experiment we will measure the uptake of hydrogen in a semiquantitative manner. The reaction vessel is a 125-mL filter flask with a white rubber pipette bulb wired onto the side arm (Fig. 25.8). Introduce 10 mL of water, 1 mL of platinum(IV) chloride solution,[2] and 0.5 g of decolorizing charcoal, and swirl during the addition of 3 mL of stabilized 1 *M* sodium borohydride solution.[3] While allowing 5 minutes for the formation of the catalyst, dissolve 1 g of *cis*-norbornene-5,6-*endo*-dicarboxylic acid in 10 mL of hot water. Pour 4 mL of concentrated hydrochloric acid into the reaction flask, followed by the hot solution of the unsaturated acid. Cap the flask with a large rubber septum and wire it on. Draw 1.5 mL of the stabilized sodium borohydride solution into the barrel of a plastic syringe, thrust the needle through the center of the rubber septum, and add the solution dropwise with swirling. The initial uptake of hydrogen is so rapid that the balloon may not inflate until you start injecting a second 1.5 mL aliquot of borohydride solution through the septum. When the addition is complete and the reaction appears to be reaching an endpoint (about 5 minutes), heat the flask on a steam bath while swirling and try to estimate the time at which the balloon is deflated to a constant size (about 5 minutes). When the balloon size is constant, heat and swirl for an additional 5 minutes; then release the pressure by pushing the needle of an open syringe through the rubber septum.

[1]Brown HC, Brown CA. *J Am Chem Soc.* **1962**;*84*:1495.
[2]A solution of 200 mg of $PtCl_4$ in 4 mL of water.
[3]Dissolve 1.6 g of sodium borohydride and 0.30 g of sodium hydroxide pellets (stabilizer) in 40 mL of water. When not in use, the solution should be stored in a refrigerator in a loosely stoppered container. If left for some time at room temperature in a tightly stoppered container, enough gas pressure may develop to explode the container.

Video: *Microscale Filtration on the Hirsch Funnel*; Photo: *Extraction with Ether*

Filter the hot solution by suction on a Hirsch funnel and place the catalyst in a jar marked "Catalyst Recovery." Cool the filtrate and extract it with three 15-mL portions of ether. The combined extracts are to be washed with saturated sodium chloride solution and dried over anhydrous calcium chloride pellets. Evaporation of the ether gives about 0.8–0.9 g of white solid.

Common ion effect

Acidified water appears to be the best solvent for the crystallization of the saturated *cis-diacid*. The diacid is very soluble in water and crystallizes extremely slowly with poor recovery. However, the situation is materially improved by adding a little hydrochloric acid to decrease the solubility of the diacid.

Scrape out the bulk of the solid product and transfer it to a 25-mL Erlenmeyer flask. Add 1–2 mL of water to the 125-mL filter flask, heat to boiling to dissolve residual solid, and pour the solution into the 25-mL flask. Bring the material into solution at the boiling point with a total of no more than 3 mL of water (as a guide, measure 3 mL of water into a second 25-mL Erlenmeyer). With a Pasteur pipette, add 3 drops of concentrated hydrochloric acid and let the solution stand for crystallization. Clusters of heavy prismatic needles will soon separate; the recovery is about 90%. The product should give a negative test for unsaturation with acidified permanganate solution (1% potassium permanganate in 10% sulfuric acid).

Observe what happens when a sample of the product is heated in a melting-point capillary to about 170°C. Account for the result. You may be able to confirm your inference by letting the oil bath cool until the sample solidifies, and then noting the temperature and behavior on remelting.

Cleaning Up. The catalyst removed by filtration may be pyrophoric (i.e., spontaneously flammable in air). Immediately remove it from the Hirsch funnel, wet it with water, and place it either in the hazardous waste container or in the catalyst recovery container. It should be kept wet with water at all times. The combined aqueous filtrates, after neutralization, are diluted with water and flushed down the drain. If local regulations permit, the ether can be allowed to evaporate from the calcium chloride pellets in the hood, and they can be placed in the nonhazardous solid waste container. Otherwise, place the wet pellets in the designated waste container.

4. TRANSFER HYDROGENATION OF OLIVE OIL[4,5]

IN THIS EXPERIMENT, olive oil is treated with a palladium-on-carbon catalyst and the hydrogen donor cyclohexene to produce a solid fat by transfer hydrogenation. In Experiment 5, olive oil is catalytically reduced using hydrogen gas. If both experiments are performed, the results can be compared. Analyze the two products by titration with bromine in carbon tetrachloride (or dichloromethane) and also by NMR spectroscopy. Alternatively, you can carry out just one reduction and compare your results with a classmate who used the other procedure.

The metabolism of olive oil, like other vegetable oils containing unsaturated fatty acids, results in an increased production of high-density lipoproteins (HDL) that do

[4]Discussions with Gottfried Brieger regarding transfer hydrogenation are gratefully acknowledged.

[5]Barry B. Snider of Brandeis University points out some interesting problems with this experiment. *See* the text website.

not deposit as much cholesterol in the arteries as the low-density lipoproteins (LDL); LDLs contribute to the disease arteriosclerosis (hardening of the arteries).

Olive oil is a triester consisting of a trihydric alcohol, glycerol, and three long-chain fatty (carboxylic) acids. The fatty acid in olive oil is primarily oleic acid, an 18-carbon monounsaturated compound. So olive oil can be regarded as primarily glycerol trioleate, although it contains about 15% saturated fat and an equal quantity of polyunsaturated fat. The double bond in oleic acid has the *cis* configuration; the molecule is bent in the center and does not pack well into a crystal lattice; thus, olive oil is a liquid at room temperature. If the double bonds are saturated, then the triester, glycerol tristearate, is a solid, and melts at almost 70°C. Similarly, a molecule having *trans* double bonds is a solid.

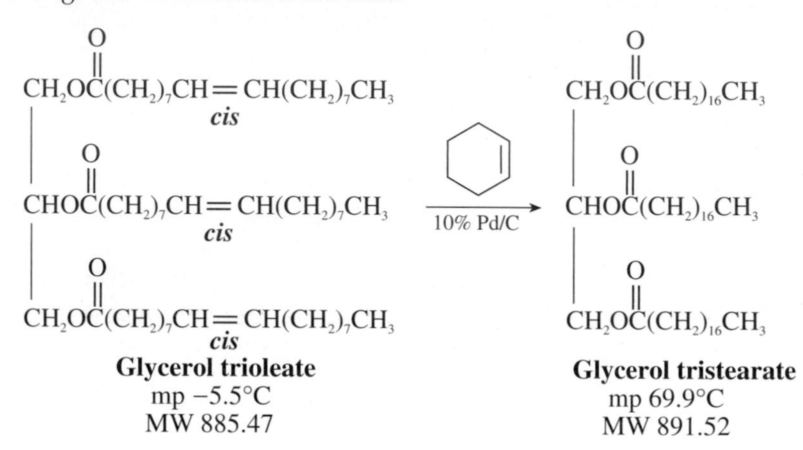

To manufacture margarine, a vegetable oil such as olive oil or, more commonly, corn or soybean oil is partially hydrogenated. This gives a mixture of liquid and solid fats that possesses the consistency of butter. In the process, some of the *cis* double bonds are also isomerized to the more stable *trans* isomers. Like saturated fatty acids, the *trans* isomers also contribute to the formation of LDL. When oils such as these are completely hydrogenated, the resulting fat is a solid. Many saturated fats are found in animals. The hard fat (tallow) on a raw steak is made up of saturated fats.

In this procedure, we use 100% extra-virgin olive oil from Italy. The label says that of each 14 g of oil, 10 are monounsaturated (mostly from oleic acid), 2 are polyunsaturated (mostly linoleic acid, an 18-carbon acid with two double bonds), and 2 are saturated (this is a mixture of predominantly 18-carbon saturated stearic acid and a smaller amount of 16-carbon palmitic acid).

From this rough analysis, it can be concluded that complete hydrogenation of all the double bonds in olive oil will give a product containing about 90%–95% stearic acid (C-18) and 5%–10% palmitic acid (C-18) esterified to glycerol. Although the product will not be pure glycerol tristearate, it should be a solid with a melting point above 50°C.

The technique of transfer hydrogenation is employed. In the presence of a catalyst, hydrogen is lost from the cyclohexene donor molecules and transferred to the double bonds in the olive oil. Although it cannot be used in all cases, this technique of hydrogenation is quite convenient because it consists of simply refluxing for a few minutes the substance to be hydrogenated with the hydrogen donor and the catalyst.

A variety of hydrogen donor molecules can be used, among which are cyclohexene, hydrazine, formic acid, cyclohexadiene, and ammonium formate. When ammonium formate decomposes in the presence of a catalyst, it produces hydrogen, ammonia, and carbon dioxide. Not only does reduction of double bonds take place, but the ammonia can react with esters to form amides. Cyclohexene and cyclohexadiene both lose hydrogen to form the same stable end product. In this experiment, cyclohexene is used as the hydrogen donor.

Microscale Procedure

In a reaction tube, place 400 mg of olive oil, 1 mL of cyclohexene, 50 mg of 10% palladium-on-carbon catalyst, and a boiling chip. Reflux the mixture for at least 30 minutes. Be careful not to boil the alkene out of the reaction tube. If necessary, place a coil of damp pipe cleaner around the top of the tube. This will keep the top of the tube cool so the cyclohexene will not escape. The empty distilling column could also be mounted on top of the reaction tube to provide even more condensing area (*see* Fig. 25.1 on page 379); however, with careful adjustment of the heat input (the depth of the tube in the sand), this should not be necessary. The isolation procedure is given in Experiment 5.

Macroscale Procedure

In a 14/20 standard taper flask equipped with a water-cooled condenser, place 0.80 g of olive oil, 2 mL of cyclohexene, 100 mg of 10% palladium-on-carbon catalyst, and a boiling chip (*see* Fig. 25.3 on page 380). Reflux the mixture for at least 15 minutes. The isolation procedure is given in Experiment 5.

5. CATALYTIC HYDROGENATION OF OLIVE OIL

Microscale Procedure

Follow the procedure of Experiment 2, using the apparatus illustrated in Figure 25.5 on page 381. If 9 mL of hydrogen is generated, this will be sufficient to completely hydrogenate 119 mg of olive oil (assuming that the olive oil is 100 percent glycerol trioleate). Demonstrate in your laboratory notebook that this calculation is correct.

Dissolve the olive oil in 0.6 mL of methanol and inject it into a flask containing 1.5 mL of methanol, 20 mg of 10 percent palladium-on-carbon catalyst, and a small magnetic stirring bar. Stir the reaction mixture until the uptake of hydrogen ceases, and then isolate the product.

Macroscale Procedure

Follow the procedure of Experiment 2. Using an apparatus similar to that in Figure 25.5 on page 381 but with a larger beaker and graduated cylinder, generate 45 mL of hydrogen and hydrogenate 590 mg of olive oil. Dissolve the olive oil in 3 mL of methanol and inject it into a 25-mL flask containing 7.5 mL of methanol, 100 mg of 10 percent palladium-on-carbon catalyst, and a small magnetic stirring bar. Stir the reaction mixture until the uptake of hydrogen ceases and then isolate the product.

Isolation of Products

Prepare a Pasteur pipette as a micro filter by forcing a small piece of cotton firmly down to the constriction (*see* Fig. 25.2 on page 379). Add the reaction mixture to the filter pipette and then force the solution through the cotton into a tared, 25-mL

Photo: *Filtration Using a Pasteur Pipette*

Vacuum evaporation while heating is necessary to ensure that the fat will solidify when cooled on ice.

filter flask. Rinse the reaction tube, filter with a few drops of hexane, and then evaporate the solution to dryness (Fig. 25.9). To remove the last traces of volatile liquid from the product, heat the filter flask in a hot sand bath under vacuum. After cooling on ice, the product should solidify to a hard, white fat. Analyze a portion of the product. The bulk of the product can be used for the synthesis of soap (*see* Experiment 6 in Chapter 40).

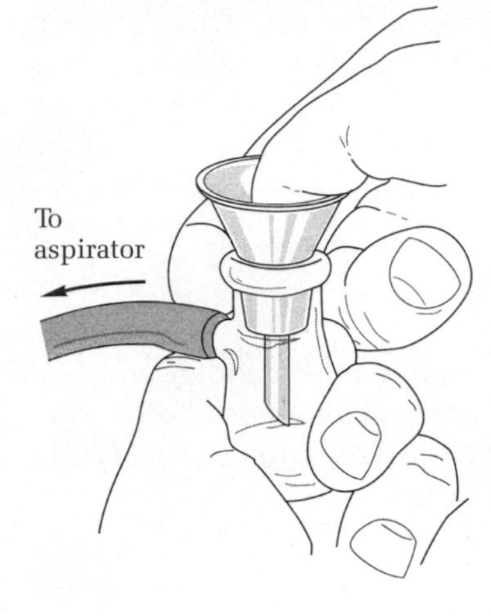

FIG. 25.9
An apparatus for removing a solvent under vacuum.

Analysis of Products

If you performed both Experiments 4 and 5, determine and compare the weights of the products and their physical properties. Count the number of drops of 3% bromine in carbon tetrachloride (or dichloromethane) that can be decolorized by equivalent quantities of each product and of olive oil.

Analyze the NMR spectra of olive oil and of the two products. In olive oil, a complex set of peaks centered at 5.38 ppm are produced by protons on a *cis* double bond. If the catalyst has isomerized the *cis* to the *trans*, double-bond peaks for the *trans* isomer will be found centered at 5.35 ppm. Completely hydrogenated olive oil will, of course, have no peaks in this region of the spectrum. A set of 5 peaks centered at 5.26 ppm will always be present and can be used to calibrate the integration. These peaks arise from the hydrogen on the central carbon of the glycerol part of the triester.

The IR spectra of olive oil (Fig. 25.10) and of the two products show very small differences. A weak peak in the 3013–3011 cm^{-1} region is characteristic of the *cis* double bond, and a weak peak in the 970–967 cm^{-1} region is characteristic of the *trans* double bond.

Computational Chemistry

Calculate, using a semiempirical molecular orbital program such as AM1, the heats of formation of cyclohexene, cyclohexane, and benzene. Use this information to

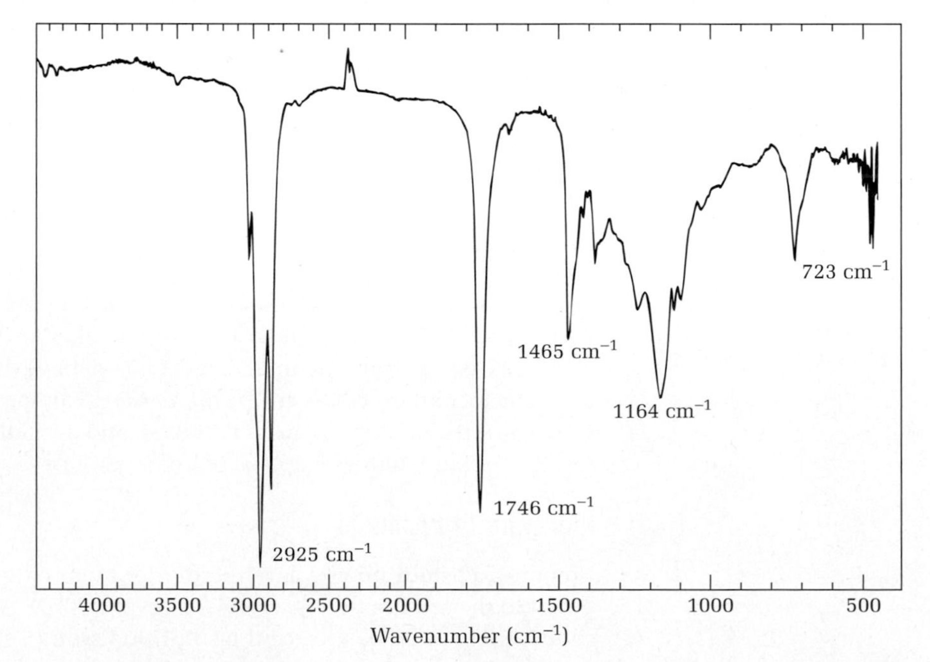

FIG. 25.10
The IR spectrum (thin film) of olive oil (Bertolli brand, extra virgin), which consists primarily of glycerol trioleate.

rationalize the results when using cyclohexene as the hydrogen donor. Molecular mechanics steric energies for these molecules do not allow valid comparisons to be made among the three.

For Further Investigation

Reflux olive oil with a 10% palladium-on-carbon catalyst in an inert solvent such as cyclohexane. Analyze the product by NMR spectroscopy. Has the catalyst alone caused the *cis* double bond to isomerize to the *trans*? Analyze samples of commercial margarine by NMR. Are *trans* double bonds present?

QUESTIONS

1. How would you convert the reduced norbornane diacid product to the corresponding bicyclic anhydride?
2. Why does the addition of hydrochloric acid in Experiment 2 cause the solubility of the bicyclic diacid in water to decrease?
3. Why is decolorizing charcoal added to the hydrogenation reaction mixture in the experiment involving Brown hydrogenation?
4. In the transfer hydrogenation of olive oil, what compound is cyclohexene ultimately converted to?
5. What is the driving force for the transfer hydrogenation of olive oil when using cyclohexene?
6. Write balanced equations for the transfer hydrogenation of pure glycerol trioleate with cyclohexene and ammonium formate as the hydrogen sources. Calculate the number of milligrams of each donor needed to hydrogenate 0.4 g of glycerol trioleate using transfer hydrogenation.

See *Web Links*

26 CHAPTER

Sodium Borohydride Reduction of 2-Methylcyclohexanone: A Problem in Conformational Analysis

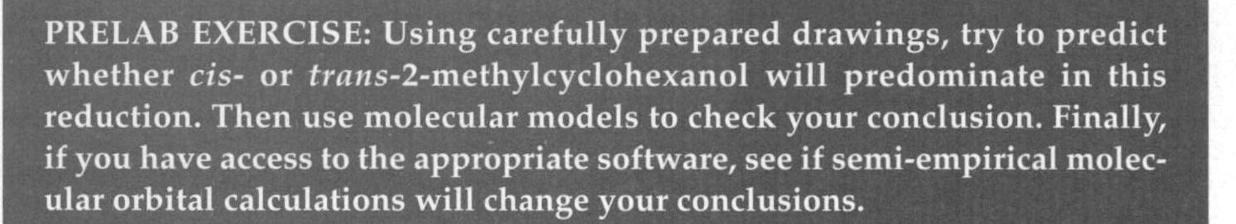

PRELAB EXERCISE: Using carefully prepared drawings, try to predict whether *cis*- or *trans*-2-methylcyclohexanol will predominate in this reduction. Then use molecular models to check your conclusion. Finally, if you have access to the appropriate software, see if semi-empirical molecular orbital calculations will change your conclusions.

The objective of this experiment is to determine the structures and relative percentage yields of the products formed when 2-methylcyclohexanone is reduced with sodium borohydride. We will also compare these results to predictions made using molecular mechanics and semiempirical molecular orbital calculations.

In the reduction of 2-methylcyclohexanone, both the *cis* and *trans* isomers of 2-methylcyclohexanol can be formed:

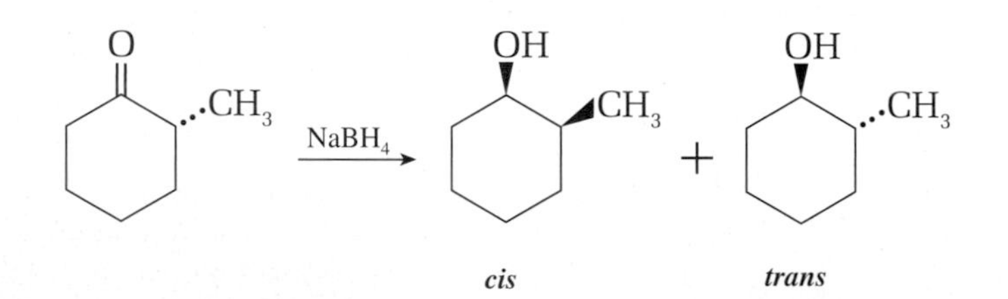

An examination of the models reveals that the two cyclohexanols can, in principle, exist in four chair conformations:

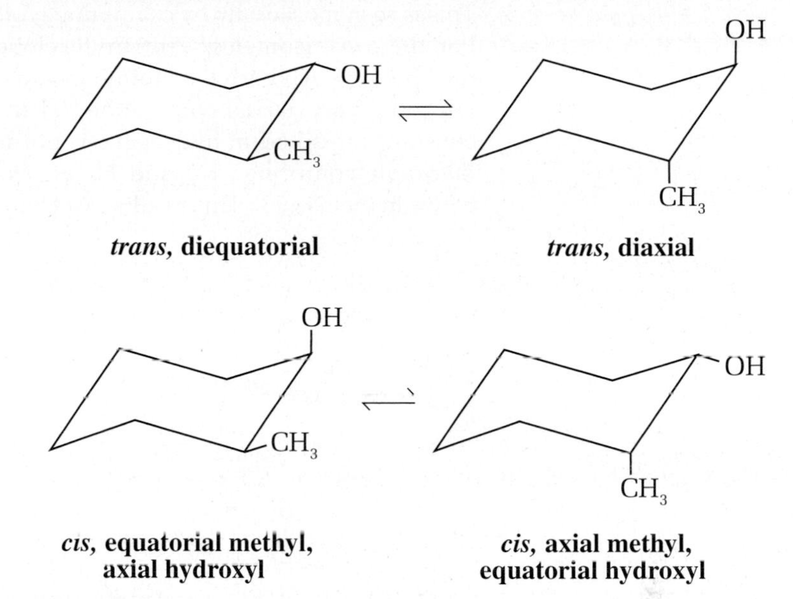

A molecular mechanics program can be used to calculate the relative energies of these four isomers and help you to predict the most stable *trans* and *cis* conformations. Even without calculation, you should be able to predict which of the two *trans* conformations is the more stable.

By carrying out semiempirical molecular orbital calculations on the starting material (*see* Chapter 15), you may be able to decide, based on the shape and location of the lowest unoccupied molecular orbital (LUMO), whether the borohydride anion will attack the carbonyl group from the top, to give predominantly the *trans* isomer, or from the bottom, to give mostly the *cis* isomer. You may also be able to make this prediction by studying a molecular model or even a drawing of 2-methylcyclohexanone:

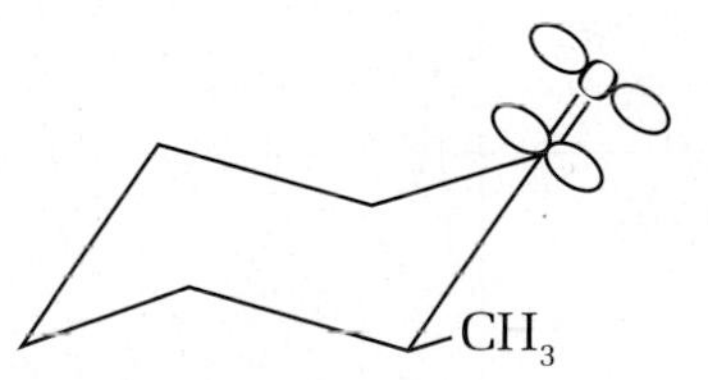

You can determine the structures and relative percentage yields of the products in this reaction using nuclear magnetic resonance (NMR) spectroscopy. The 250-MHz ^{1}H NMR spectrum of a 50:50 mixture of methylcyclohexanols is given in Figure 26.1. The two lowest field groups of peaks are from the hydrogen atom on the hydroxyl-bearing carbon atom (H_A).

The reduction of 2-methylcyclohexanone with sodium borohydride will give a mixture of products, but not necessarily a 50:50 mixture. Integration of the two low-field groups of peaks will allow determination of the actual percentages of products in this reaction. First, however, the peaks must be assigned to the *cis* and *trans* isomers. The groups of peaks have been expanded and numbered on the spectrum in Figure 26.1, and the frequencies of the 11 peaks are listed in the caption.

The NMR coupling constant of the proton on the hydroxyl-bearing carbon is a function of the dihedral angle between that proton and an adjacent vicinal proton. This is seen most easily by examining molecular models and Newman projections. For the *trans* isomer of 2-methylcyclohexanol, the predominant conformer is the one in which both the methyl group and hydroxyl group are equatorial, and H_A and H_B are diaxial with a dihedral angle of 180° between them. The coupling constant for dihedral hydrogens is normally in the 9–12 Hz range. For the less favorable conformer, H_A and H_B are both equatorial, and the dihedral angle between them is 60°. This results in a coupling constant in the 2–5 Hz range.

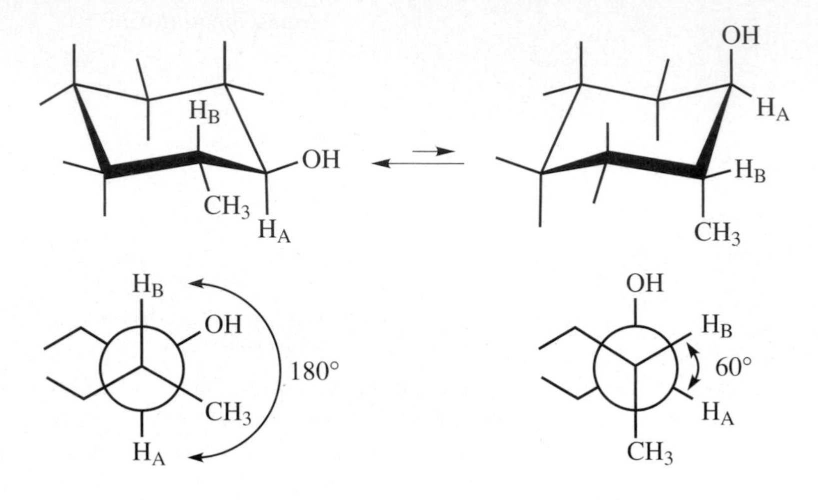

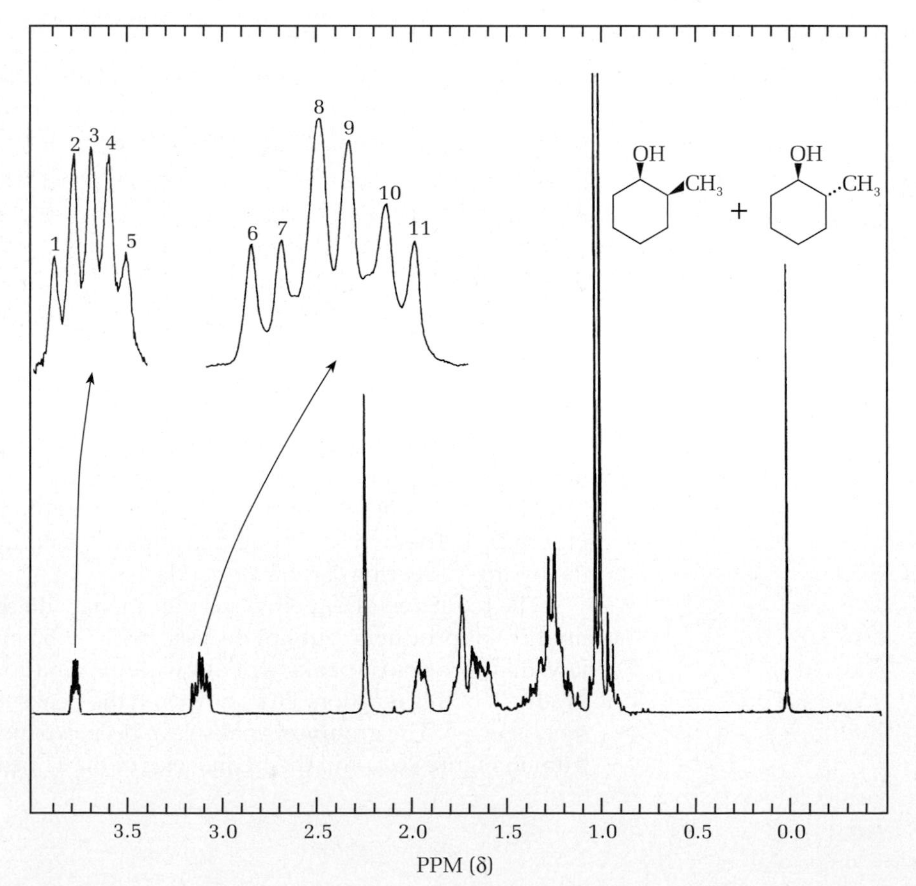

FIG. 26.1
The 250-MHz NMR spectrum of a mixture of *cis*- and *trans*-2-methylcyclohexanol. The frequencies of peaks 1 to 5 are 950.3, 947.5, 944.8, 942.2, and 939.5 Hz, respectively. The frequencies of peaks 6 to 11 are 791.2, 786.9, 781.6, 777.1, 771.6, and 767.4 Hz, respectively.

For *cis*-2-methylcyclohexanol, the two possible conformers are about equal in energy. In either conformer, H_A and H_B are in a 60° axial-equatorial stereochemical relationship, which results in a coupling constant in the 4–7 Hz range.

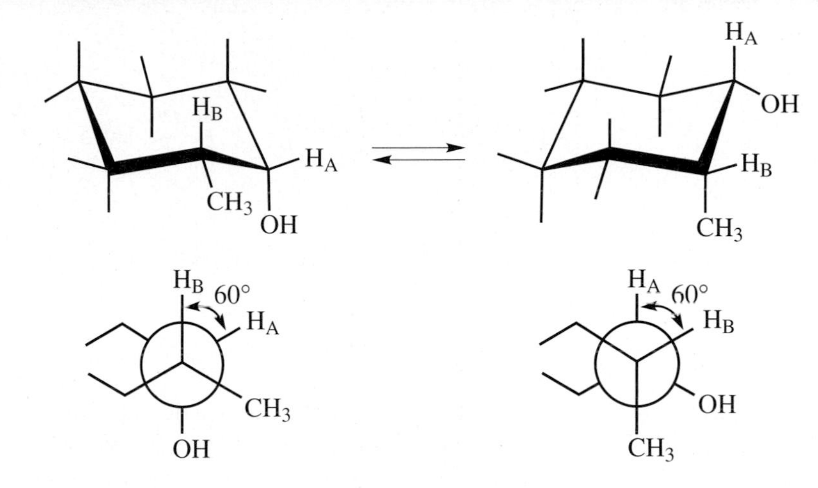

From this information, it should be possible to assign the groups of peaks at 3.1 and 3.8 ppm to either *cis*- or *trans*-2-methylcyclohexanol. From the areas of the two sets of peaks, the relative percentages of the isomers can be determined.

EXPERIMENT

BOROHYDRIDE REDUCTION OF 2-METHYLCYCLOHEXANONE

This reaction can be run on a scale four times larger in a 25-mL Erlenmeyer flask.

IN THIS EXPERIMENT a liquid ketone in methanol is reduced with solid sodium borohydride. Base is added, and the product is extracted into dichloromethane. In the usual way, this organic layer is dried and evaporated to leave the liquid alcohol. The same procedure could be used to reduce most ketones to their corresponding alcohols.

To a reaction tube, add 1.25 mL of methanol and 300 mg of 2-methyl-cyclohexanone. Cool this solution in an ice bath contained in a small beaker. While the reaction tube is in the ice bath, carefully add 50 mg of sodium borohydride to the solution. After the vigorous reaction has ceased, remove the tube from the ice bath and allow it to stand at room temperature for 10 minutes, at which time the reaction should appear to be finished. To decompose the borate ester, add 1.25 mL of 3 *M* sodium hydroxide solution. To the resulting cloudy solution, add 1 mL of water. The product will separate as a small, clear upper layer. Remove as much of this as possible, place it in a reaction tube, and then extract the remainder of the product from the reaction mixture with two 0.5-mL portions of dichloromethane. Add these dichloromethane extracts to the small product layer and dry the combined extracts over anhydrous sodium sulfate (not calcium chloride). After a few minutes, transfer the solution to a dry reaction tube containing a boiling chip. In the hood, boil off the dichloromethane (and any accompanying methanol) and use the residue to run an NMR spectrum in the usual way.

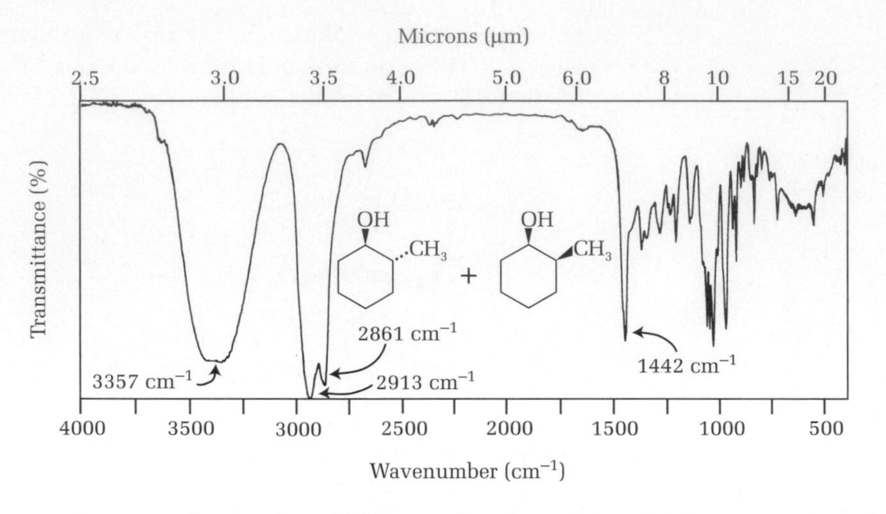

FIG. 26.2
The IR spectrum of a mixture of *cis-* and *trans*-2-methylcyclohexanol (thin film).

Integrate the two low-field sets of peaks at 3.1 and 3.8 ppm, analyze the coupling constant patterns to assign the two sets of peaks, and report the percentage distribution of *cis-* and *trans*-2-methylcyclohexanol formed in this reaction. The relative amounts of the products can, of course, also be determined by gas chromatography. How do these results compare to your predictions and calculations? Which *cis* and which *trans* conformation is the more stable?

Run an infrared (IR) spectrum of the product as a thin film to determine whether the reduction of the starting material has been completed (Fig. 26.2).

Cleaning Up. The reaction mixture is neutralized with acetic acid (to react with sodium borohydride) and flushed down the drain with water.

Computational Chemistry

For Additional Experiments, sign in at this book's premium website at **www.cengage.com/login.**

Using a molecular mechanics program, calculate the steric energies of *cis-* and *trans*-2-methylcyclohexanol. Each of these isomers has two principal conformations. Does reduction of 2-methylcyclohexanone give the more stable isomer? Explain.

QUESTIONS

1. Draw the NMR peak expected for H_A when H_A couples with H_B in the following structure.

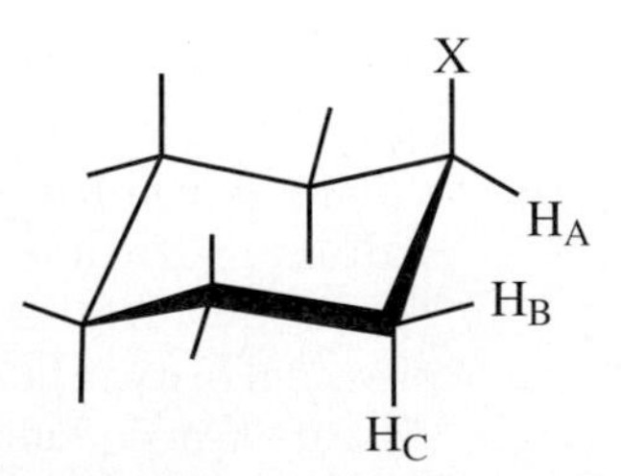

2. Draw the NMR peak expected when H_A couples with both H_B and H_C. Remember that the coupling constants are not equal.
3. In the IR spectrum of the product, what is the approximate frequency of the most important peak in the starting material that should be absent in the product?

CHAPTER 27

Epoxidation of Cholesterol

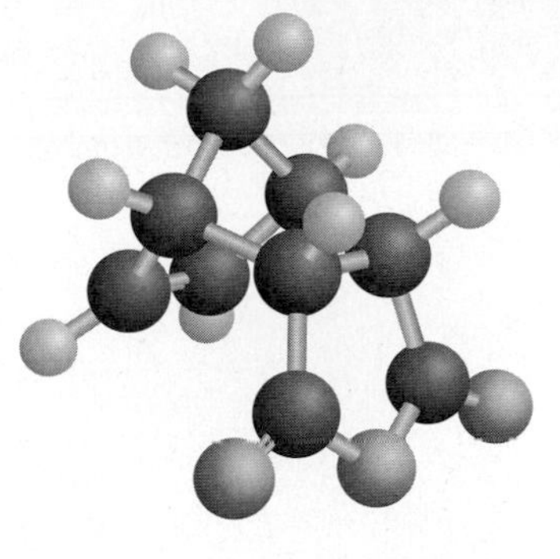

PRELAB EXERCISE: When milk is irradiated with ultraviolet (UV) light, the vitamin D content increases. What type of reaction is taking place?

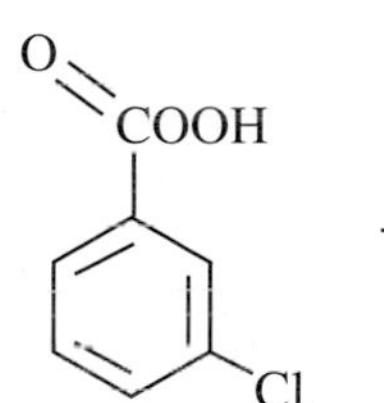

+ Cholesterol $\xrightarrow{CH_2Cl_2}$ 5α, 6α-Epoxycholestan-3β-ol + 3-Chlorobenzoic acid

3-Chloroperoxybenzoic acid	**Cholesterol**	**5α, 6α-Epoxycholestan-3β-ol**	**3-Chlorobenzoic acid**
MW 172.57	MW 386.66	MW 402.66	MW 156.57
mp 92–94°C (dec)	mp 149°C	mp 110–112°C mp 139°C	mp 157°C

In this experiment, an epoxidation reaction is carried out on cholesterol, which is representative of a very important group of molecules—the steroids. The rigid cholesterol molecule gives products of well-defined stereochemistry. The epoxidation reaction is stereospecific, and the product can be used to carry out further stereospecific reactions.

Video: *Cholesterol from Human Gallstones*

Cholesterol is the principal constituent of gallstones and can be readily isolated from them. The average person contains about 200 g of cholesterol, which is primarily found in brain and nerve tissue. The blockage of arteries by deposits of cholesterol also leads to the disease arteriosclerosis (hardening of the arteries).

Certain naturally occurring and synthetic steroids have powerful physiological effects. Progesterone and estrogen are the female sex hormones, and testosterone is the male sex hormone; these hormones are responsible for the development of secondary sex characteristics. The closely related synthetic steroid, norethisterone, is an oral contraceptive, and the addition of four hydrogen atoms (reduction of the ethynyl group to the ethyl group) and a methyl group gives an anabolic steroid, ethyltestosterone. This muscle-building steroid is now outlawed for use by Olympic athletes. RU-486, mifepristone, is an abortifacient. Fluorocortisone (structure shown on the next page) is used to treat inflammations such as arthritis, and ergosterol is converted to vitamin D_2 when it is irradiated with UV light (represented by $h\nu$).

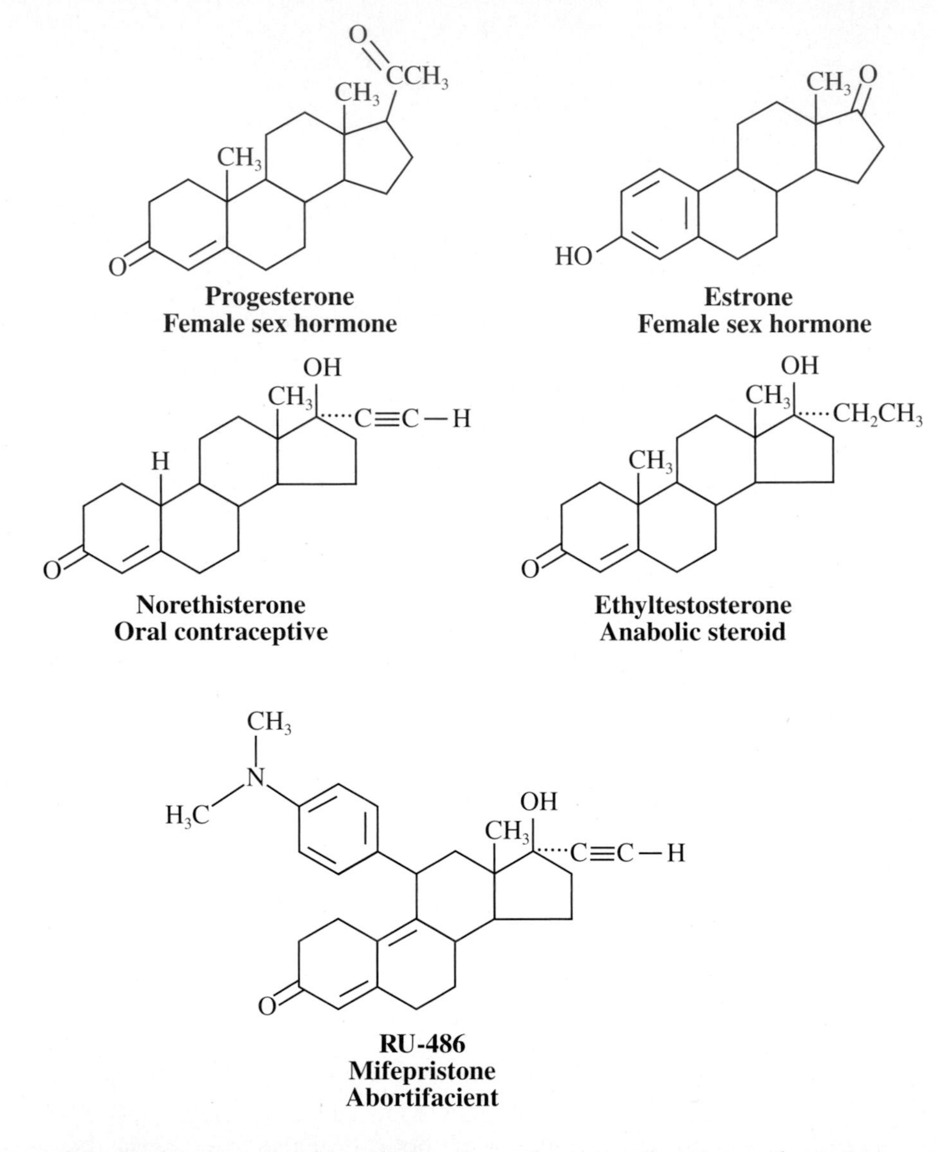

Much of our present knowledge about the stereochemistry of reactions was developed from steroid chemistry. In the epoxidation of cholesterol, the double bond of cholesterol is stereospecifically converted to the 5α, 6α epoxide. The α designation indicates that the epoxide is on the backside of the molecule, as it is usually printed. A substituent on the top side (above the plane of the paper) is designated β. A study of molecular models reveals that the angular methyl group hinders topside attack on the double bond by a perbenzoic acid; thus, the epoxide forms preferentially on the back or α side of the molecule. As evidenced by ^1H

NMR analysis, the epoxidation of cholesterol is not completely selective. A 4:1 ratio of α:β epoxides is typical.

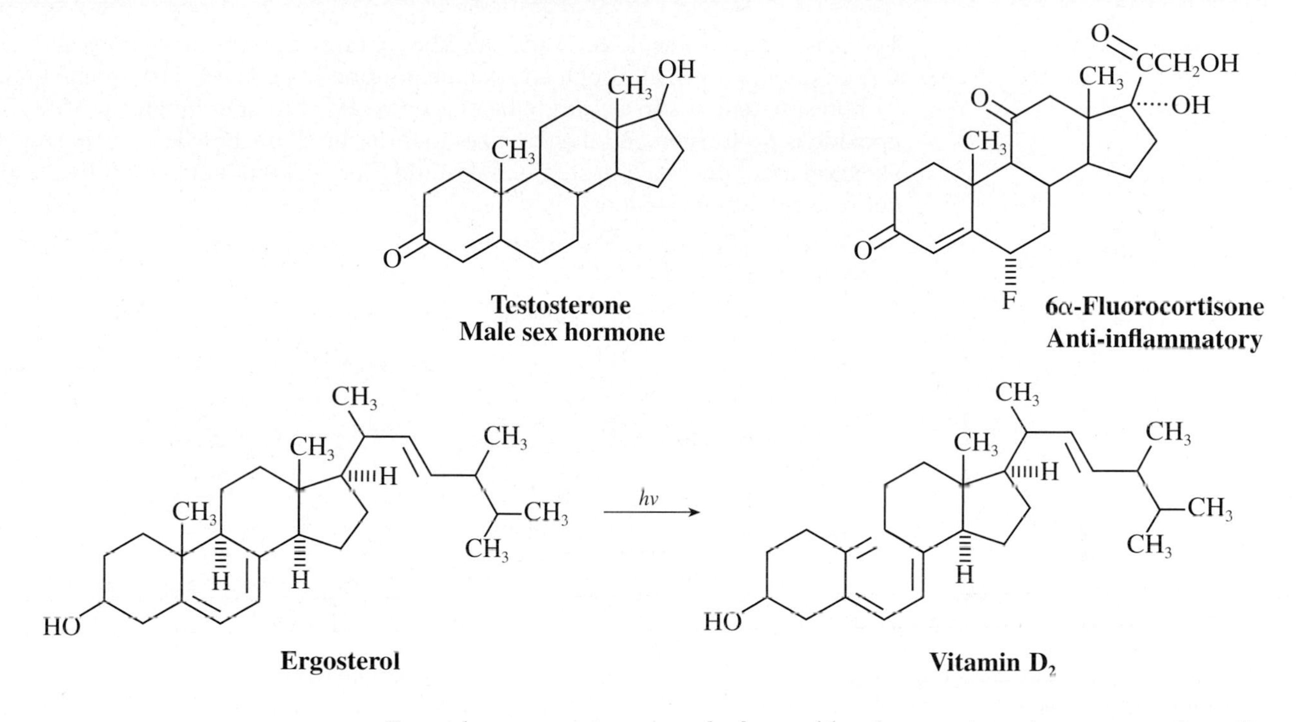

Epoxides are most commonly formed by the reaction of a peroxycarboxylic acid with an olefin at room temperature. It is a one-step cycloaddition reaction:

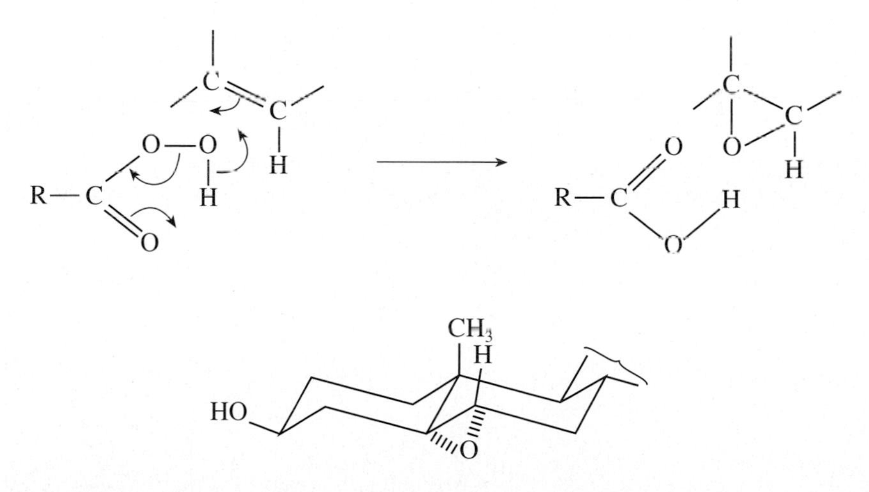

Some peroxyacids are explosive; the reagent used in Experiments 1 and 2 is a particularly stable and convenient peroxycarboxylic acid.

The reaction is carried out in an inert solvent, dichloromethane, and the product is isolated by chromatography. No great care is required in the chromatography to collect fractions because the 3-chlorobenzoic acid, being polar, is adsorbed strongly onto the silica gel, while the relatively nonpolar product is eluted easily by ether. After the removal of ether, the product is easily recrystallized from a mixture of acetone and water.

STEROID BIOSYNTHESIS THROUGH AN EPOXIDE INTERMEDIATE

The biosynthesis of cholesterol and the other naturally occurring steroids shown in this chapter proceeds through an epoxide intermediate. The C_{30} triterpene hydrocarbon squalene (Latin *squalus*, whale) is epoxidized to give squalene oxide. This epoxide is protonated and then cyclizes to form the characteristic four-ring steroid skeleton and, after several steps, cholesterol, in a series of very carefully studied enzyme-catalyzed reactions.

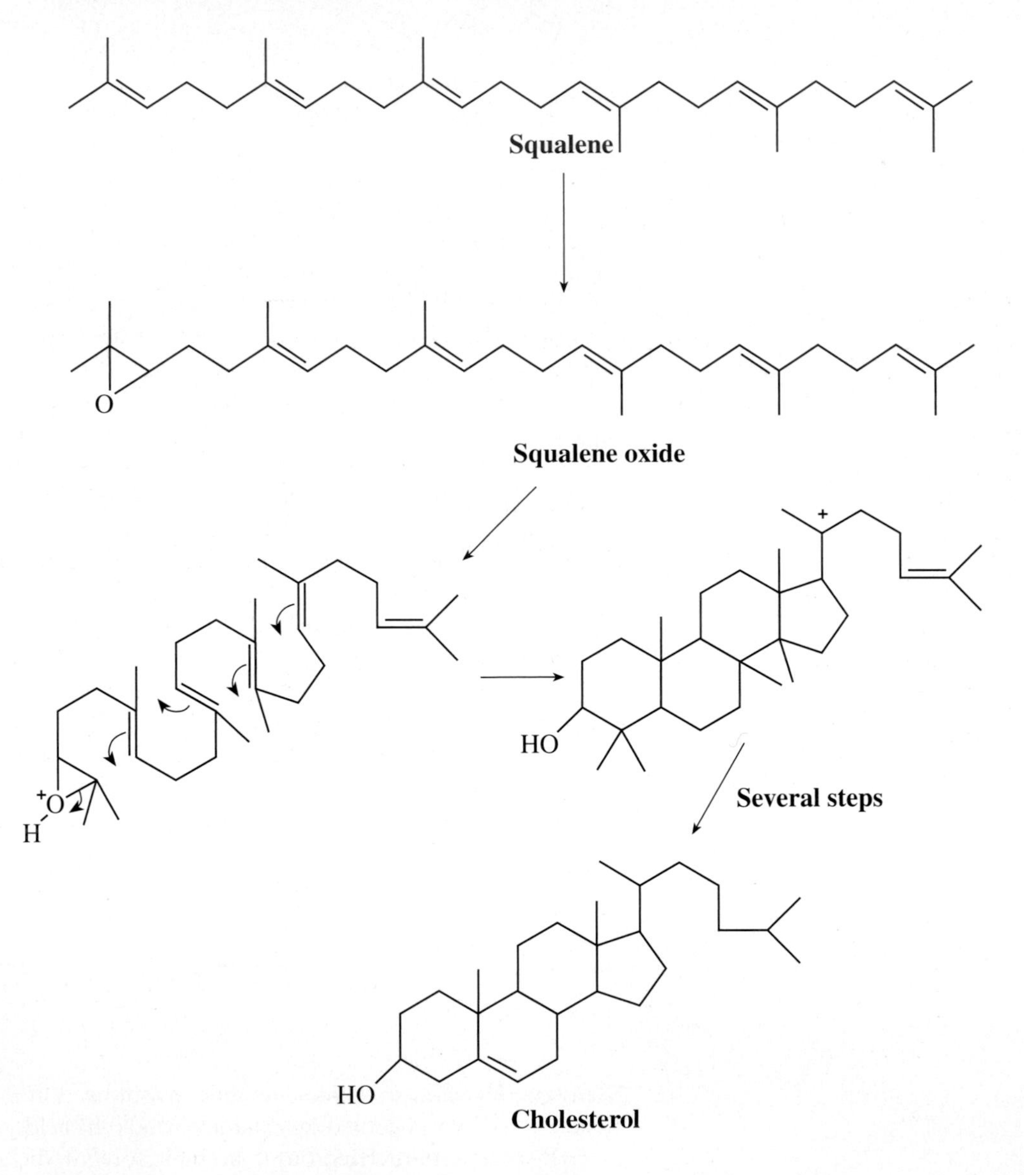

EXPERIMENTS

1. CHOLESTEROL EPOXIDE

IN THIS EXPERIMENT a steroid alkene is converted to an epoxide with meta-chloroperbenzoic acid (MCPB) in dichloromethane. The MCPB produced is removed by what amounts to a simple rapid filtration through silica gel. Evaporation of the solvent gives the epoxide, which is recrystallized from a mixture of acetone and water.

Because dichloromethane is a suspected carcinogen, handle it in a hood.

Dissolve 194 mg of cholesterol in 0.8 mL of dichloromethane by gentle warming in a 10 × 100 mm reaction tube. In another reaction tube, dissolve 117 mg of 80% 3-chloroperoxybenzoic acid (or 187 mg of 50% material) in 0.8 mL of dichloromethane by gentle warming. Cool the two solutions and then mix them together. They must be cool before mixing because the reaction is exothermic. Stopper the reaction tube and place it in a beaker of water at 40°C for 30 minutes to complete the reaction. The progress of the reaction can be followed by thin-layer chromatography (TLC) on silica gel plates, using *t*-butyl methyl ether as the eluent.

Photo and Video: *Column Chromatography*

The reaction mixture is pipetted into a chromatography column (Fig. 27.1) prepared from 3 g of silica gel. Follow the procedure described in Chapter 9, except use ether to fill the column and to prepare the silica gel slurry. The 3-chlorobenzoic acid will be strongly adsorbed by the silica gel. The product is eluted with 30 mL of ether collected in a tared 50-mL round-bottomed flask. The ether can flow through the column by gravity or can be forced out using pressure from a rubber bulb (flash chromatography).

Most of the ether is removed on a rotary evaporator; the last traces are removed using the apparatus depicted in Figure 9.6 (on page 196) for drying a solid under reduced pressure, or that depicted in Figure 9.10 (on page 201). The residue should weigh more than 150 mg. If it does not, pass more ether through the column and collect the product as before. Dissolve the product in 1.5 mL of warm acetone and, using a Pasteur pipette, transfer it to a reaction tube. Add 0.2 mL of water to the solution, warm the mixture to bring the solid into solution, and then let the tube and contents cool slowly to room temperature. Cool the mixture in ice and collect the product on a Hirsch funnel. Press the solid down on the filter to squeeze solvent from the crystals; then wash the product with 0.25 mL of ice-cold 90% acetone. Spread the product out on a watch glass to dry. Determine the weight and melting point of the product, and calculate the percentage yield.

Video: *Microscale Filtration on the Hirsch Funnel*

Cleaning Up. Place dichloromethane solutions in the halogenated organic solvents waste container, and place organic solvents in the organic solvents waste container. The silica gel should be placed in the hazardous waste container. If it should be necessary to destroy 3-chloroperoxybenzoic acid, add it to an excess of an ice-cold solution of saturated sodium bisulfite in the hood. A peracid will give a positive starch-iodide test (blue-purple color).

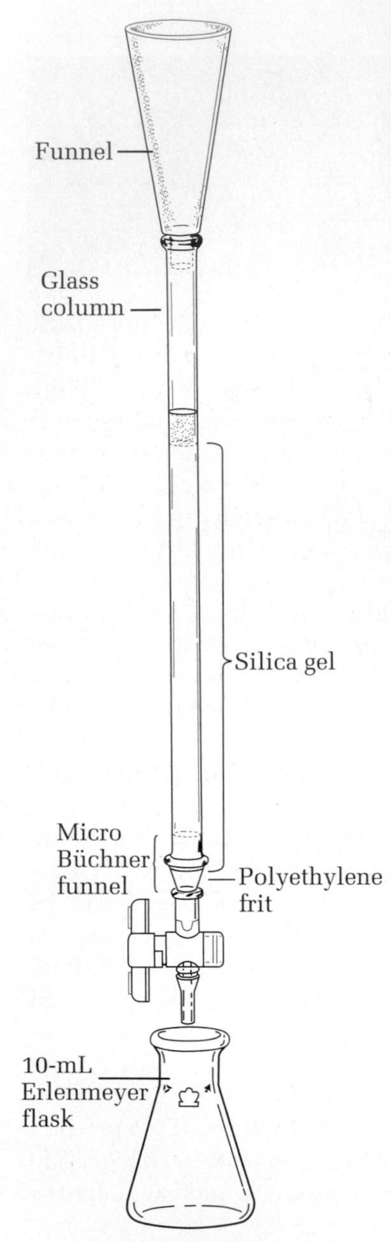

FIG. 27.1
A chromatography column.

2. CHOLESTEROL EPOXIDE

Dissolve 1 g of cholesterol in 4 mL of dichloromethane (*see* caution statement on previous page) in a 25-mL Erlenmeyer flask by gentle warming. In another 25-mL flask, dissolve 0.6 g of 80% 3-chloroperoxybenzoic acid in 4 mL of dichloromethane by gentle warming. Cool the two solutions and mix together. The two solutions must be cool before mixing because the reaction is exothermic. Clamp the flask in a beaker of water at 40°C for 30 minutes to complete the reaction. The progress of the reaction can be followed by TLC on silica gel plates using ether as the eluent.

The reaction mixture is pipetted into a chromatography column prepared from 15 g of silica gel. Follow the procedure described in Chapter 9, except use ether to fill the column and to prepare the silica gel slurry. The 3-chlorobenzoic acid will be strongly adsorbed by the silica gel. The product is eluted with 150 mL of ether collected in a tared 250-mL round-bottomed flask.

Most of the ether is removed on a rotary evaporator, and the last traces are removed using the apparatus depicted in Figure 9.6 (on page 196) for drying a solid under reduced pressure. The residue should weigh more than 0.75 g. If it does not, pass more ether through the column and collect the product as before. Dissolve the product in 7.5 mL of warm acetone and, using a Pasteur pipette, transfer it to a 25-mL Erlenmeyer flask. Add 1 mL of water to the solution, warm the mixture to bring the solid into solution, and then let the flask and contents cool slowly to room temperature. Cool the mixture in ice and collect the product on a Hirsch funnel. Press the solid down on the filter to squeeze solvent from the crystals; then wash the product with 1 mL of ice-cold 90% acetone. Spread the product out on a watch glass to dry. Determine the weight and melting point of the product and calculate the percent yield.

Cleaning Up. Place dichloromethane solutions in the halogenated organic solvents waste container and organic solvents in the organic solvents waste container. The silica gel should be placed in the hazardous waste container. If it should be necessary to destroy 3-chloroperoxybenzoic acid, add it to an excess of an ice-cold solution of saturated sodium bisulfite in the hood. A peracid will give a positive starch-iodide test (blue-purple color).

QUESTIONS

1. How many moles of each reactant are used in this experiment? Assume the 3-chloroperoxybenzoic acid is 80% pure.
2. What simple test could you perform to show that 3-chlorobenzoic acid is not eluted from the chromatography column?
3. Milk contains ergosterol. Can you imagine how "Vitamin D Enriched Milk" is made commercially? The patent on this process earned the University of Wisconsin many millions of dollars.

REFERENCE

See *Web Links*

Carl Djerassi, *This Man's Pill: Reflections on the 50th Birthday of the Pill*. Oxford, UK: Oxford University Press, 2001.

CHAPTER 28

Nitration of Methyl Benzoate

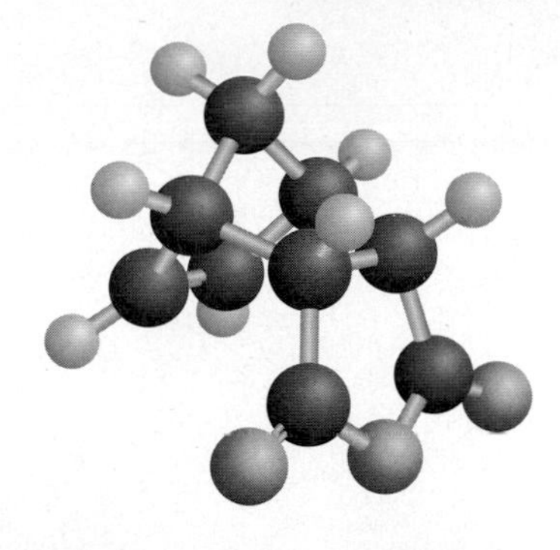

When you see this icon, sign in at this book's premium website at **www.cengage.com/login** to access videos, Pre-Lab Exercises, and other online resources.

PRELAB EXERCISE: Draw the complete mechanism for the nitration of chlorobenzene. Chlorine is an *ortho-para* director and deactivator of the benzene ring.

The nitration of methyl benzoate is a typical electrophilic aromatic substitution reaction. The electrophile is the nitronium ion generated by the interaction of concentrated nitric and sulfuric acids:

$$HNO_3 + 2H_2SO_4 \rightleftharpoons NO_2^+ + 2HSO_4^- + H_3O^+$$

Nitronium ion

Sulfuric acid protonates the methyl benzoate:

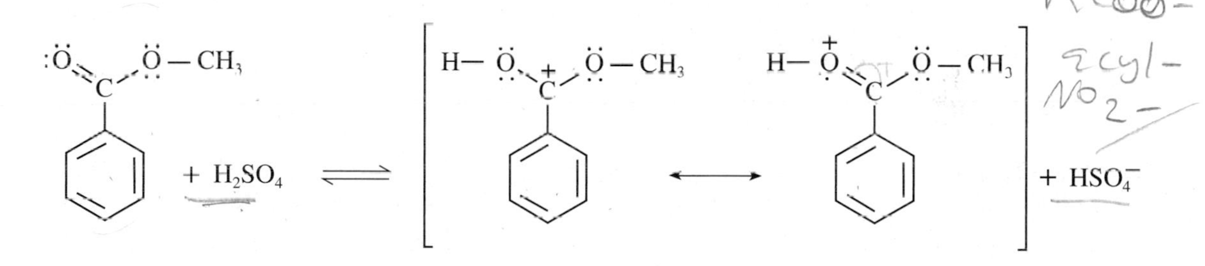

The nitronium ion then reacts with this protonated intermediate at the *meta* position, where the electron density is highest, that is, where there is no positively charged resonance form. It yields the intermediate arenium ion which has the four resonance forms shown here:

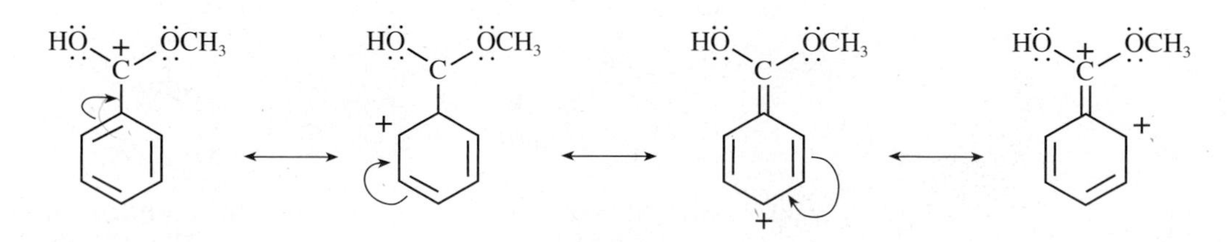

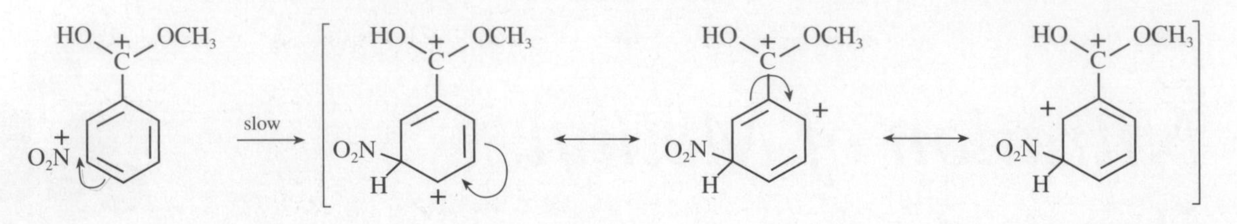

The arenium ion intermediate then transfers a proton to the basic bisulfate ion to give methyl 3-nitrobenzoate:

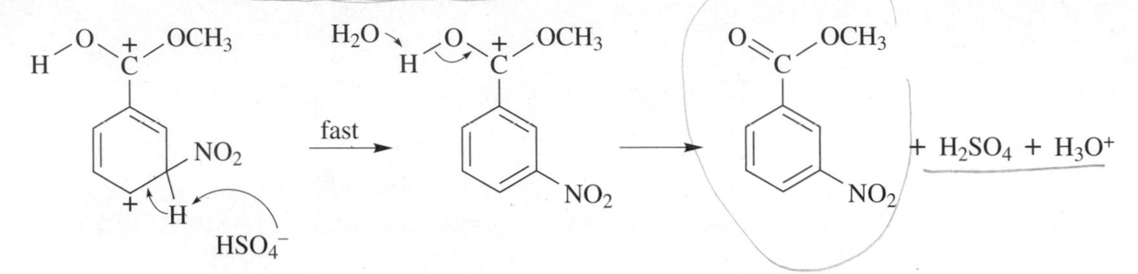

The ester group is a *meta* director and deactivator of the benzene ring. It is much easier to nitrate a molecule such as anisole, where the methoxyl group is an *ortho-para* director and an activator of the benzene ring, as the following resonance structures indicate:

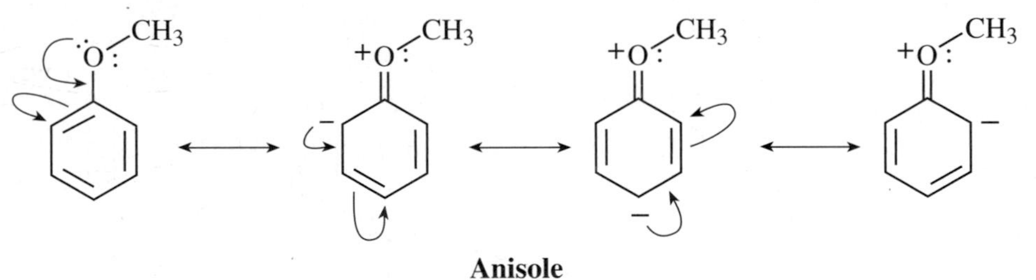

Anisole

EXPERIMENTS

1. MICROSCALE NITRATION OF METHYL BENZOATE

$$HNO_3 + 2H_2SO_4 \rightleftharpoons NO_2^+ + 2HSO_4^- + H_3O^+$$

Nitronium ion **Hydronium ion**

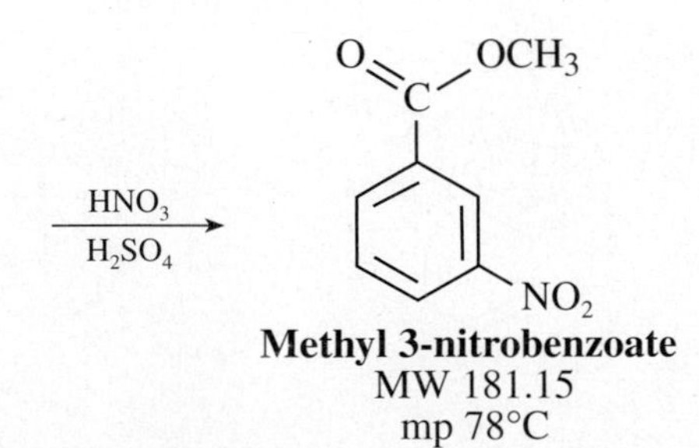

Methyl benzoate
MW 136.16
bp 199.6°C
den. 1.09
n_D^{20} 1.5170

Methyl 3-nitrobenzoate
MW 181.15
mp 78°C

IN THIS EXPERIMENT, a cold solution of an aromatic ester that has been dissolved in sulfuric acid is reacted with nitric acid. This highly exothermic reaction is kept under control by cooling; then the mixture is poured onto ice. The solid product is isolated by filtration and recrystallized from methanol, in which it is very soluble.

⚠ Do not use a plastic syringe and needle to measure either sulfuric acid or nitric acid. The acids react with the metal needle.

⚠ Use care in handling concentrated sulfuric and nitric acids.

To 0.6 mL of concentrated sulfuric acid in a 10 × 100 mm reaction tube, add 0.30 g of methyl benzoate. Flick the tube or use magnetic stirring. Cool the mixture to 0°C; add dropwise, using a Pasteur pipette, a mixture of 0.2 mL of concentrated sulfuric acid and 0.2 mL of concentrated nitric acid. Keep the reaction mixture in ice. Using a stirring rod, keep the reaction well mixed during the addition of the acids and do not allow the temperature of the mixture to rise above about 15°C, as judged by touching the reaction tube.

After all the nitric acid has been added, warm the mixture to room temperature and, after 15 minutes, pour it onto 2.5 g of ice in a small beaker. Isolate the solid product by suction filtration using a Hirsch funnel and a 25-mL filter flask. Wash the product well with water, and then with one 0.2-mL portion of ice-cold methanol. If the methanol is not ice-cold, some product can be lost in this washing step. Save a small sample for a melting-point determination and analysis by thin-layer chromatography (TLC) and infrared (IR) spectroscopy.

The remainder is weighed and recrystallized from an equal weight of methanol in a reaction tube. Alternatively, the sample can be dissolved in a slightly larger quantity of methanol, and water added dropwise to make the hot solution saturated with the product. Slow cooling should produce large crystals with a melting point of 78°C. The crude material can be obtained with about an 80% yield and a melting point of 74–76°C. If the yield is not as large as expected, concentrate the filtrate and collect a second crop of product.

Cleaning Up. Dilute the filtrate from the reaction with water, neutralize with sodium carbonate, and flush down the drain. The methanol from the crystallization should be placed in the organic solvents waste container.

2. MACROSCALE NITRATION OF METHYL BENZOATE

⚠ Use care in handling concentrated sulfuric and nitric acids.

In a 125-mL Erlenmeyer flask, cool 12 mL of concentrated sulfuric acid to 0°C and then add 6.1 g of methyl benzoate. Again, cool the mixture to 0–10°C. Now add dropwise, using a Pasteur pipette, a cooled mixture of 4 mL of concentrated sulfuric acid and 4 mL of concentrated nitric acid. During the addition of the acids, swirl the mixture frequently (or use magnetic stirring) and maintain the temperature of the reaction mixture in the range of 5–15°C.

When all of the nitric acid has been added, warm the mixture to room temperature and, after 15 minutes, pour it on 50 g of cracked ice in a 250-mL beaker. Isolate the solid product by suction filtration using a small Büchner funnel, wash well with water, and then wash with two 10-mL portions of ice-cold methanol. A small sample is saved for a melting-point determination. The remainder is weighed and crystallized from an equal weight of methanol. The crude product should be obtained with about an 80% yield and a melting point of 74–76°C. The recrystallized product should have a melting point of 78°C. For your reference, the carbon and proton nuclear magnetic resonance (NMR) spectra of the product are presented in Figures 28.1 and 28.2.

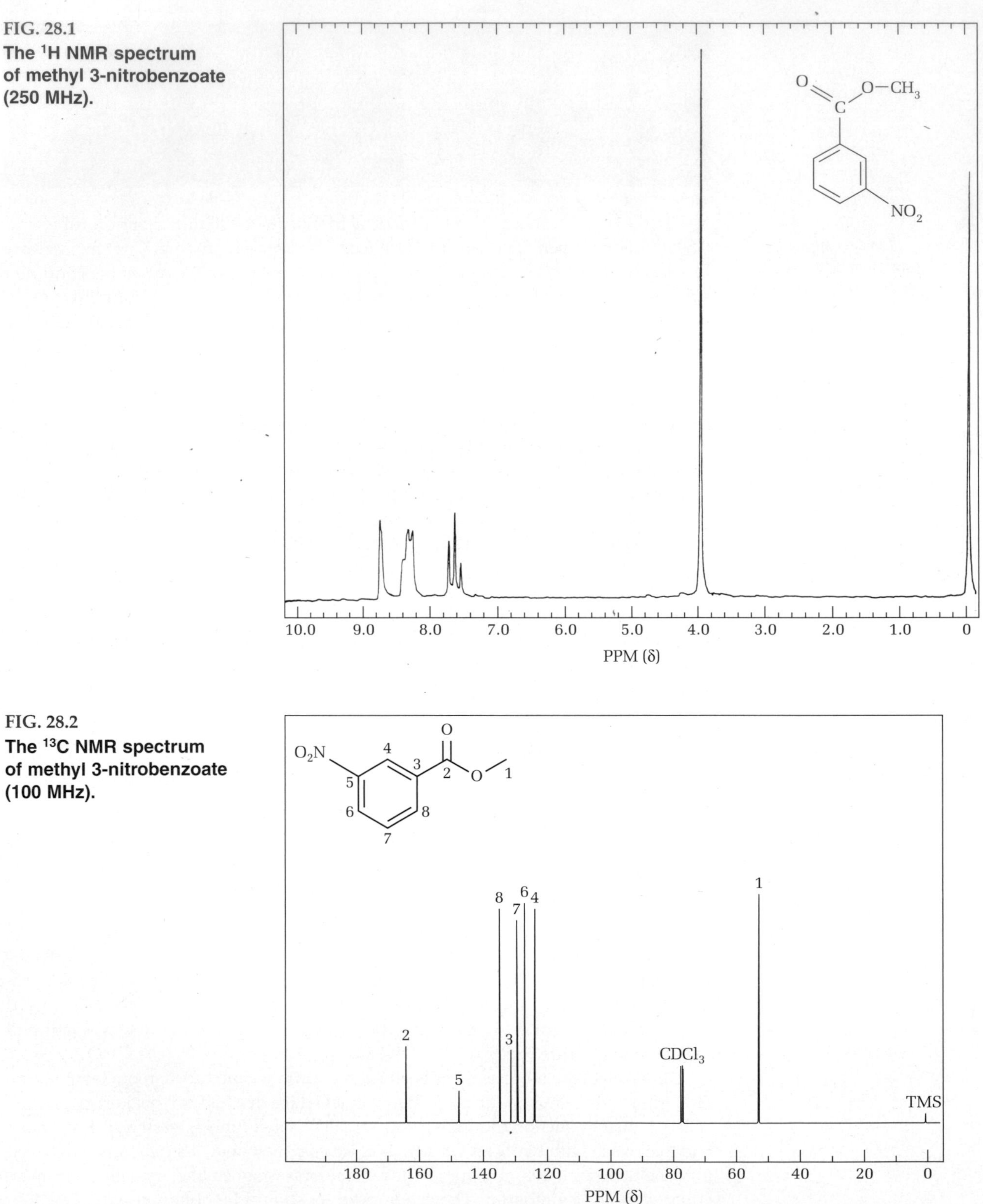

FIG. 28.1
The ^{1}H NMR spectrum of methyl 3-nitrobenzoate (250 MHz).

FIG. 28.2
The ^{13}C NMR spectrum of methyl 3-nitrobenzoate (100 MHz).

Cleaning Up. Dilute the filtrate from the reaction with water, neutralize with sodium carbonate, and flush down the drain. The methanol from the crystallization should be placed in the organic solvents waste container.

QUESTIONS

1. Hydrocarbons do not dissolve in concentrated sulfuric acid, but methyl benzoate does. Explain this difference and write an equation showing the ions that are produced.
2. What would you expect the structure of the dinitro ester to be? Consider the directing effects of the ester and the first nitro group upon the addition of the second nitro group.
3. Draw resonance structures to show in which position nitrobenzene will nitrate to form dinitrobenzene.
4. Assign the peaks at 3101 cm^{-1}, 1709 cm^{-1}, and 1390 cm^{-1} in the IR spectrum of methyl 3-nitrobenzoate (Fig. 28.3).

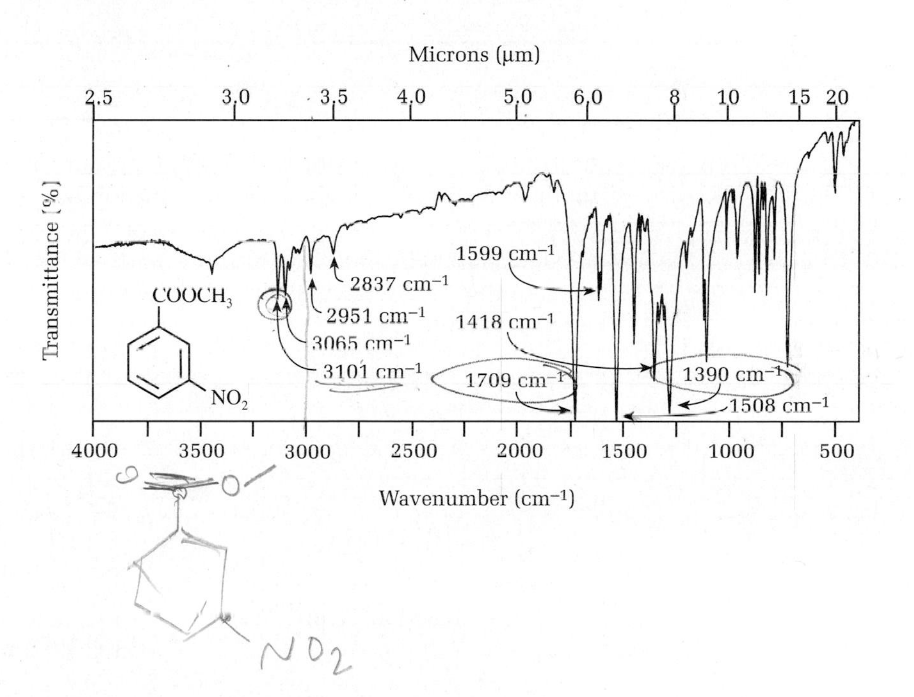

FIG. 28.3
The IR spectrum of methyl 3-nitrobenzoate. The broad peak at 3400 cm^{-1} comes from water in the KBr disk.

29 CHAPTER

Friedel–Crafts Alkylation of Benzene and Dimethoxybenzene; Host-Guest Chemistry

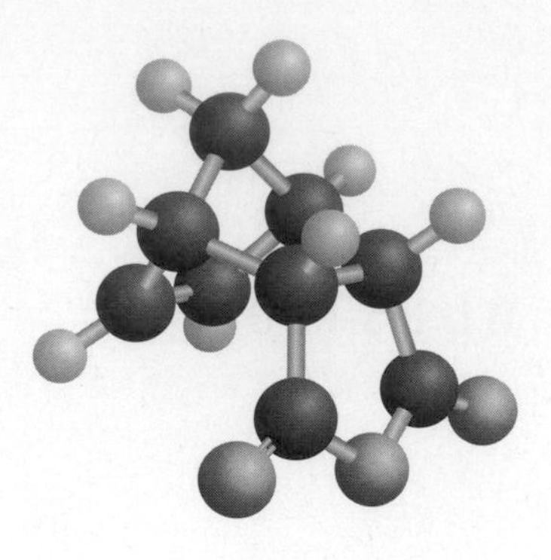

PRELAB EXERCISE: Prepare a flow sheet for the alkylation of benzene and the alkylation of dimethoxybenzene, indicating how the catalysts and unreacted starting materials are removed from the reaction mixture.

The Friedel–Crafts[1] alkylation of aromatic rings most often employs an alkyl halide and a strong Lewis acid catalyst. Some of the catalysts that can be used, in order of decreasing activity, are the halides of aluminum, antimony, iron, titanium, tin, bismuth, and zinc. Although useful, the reaction has several limitations. The aromatic ring must be unsubstituted or bear activating groups; because the product—an alkylated aromatic molecule—is more reactive than the starting material, multiple substitution usually occurs. Furthermore, primary halides will rearrange under the reaction conditions:

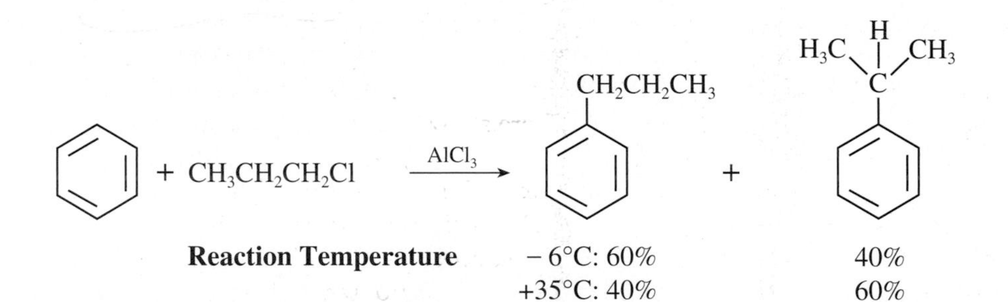

In this reaction, a tertiary halide and the most powerful Friedel–Crafts catalyst, $AlCl_3$, are allowed to react with benzene. (If you prefer not to work with benzene, you can carry out alkylations of dimethoxybenzene or *m*-xylene.) The initially formed *t*-butylbenzene is a liquid, whereas the product, 1,4-di-*t*-butylbenzene, which has a symmetrical structure, is a beautiful crystalline solid. The alkylation reaction probably proceeds through the carbocation under the conditions of the experiments in this chapter:

[1]Charles Friedel and James Crafts (who later became the president of MIT) discovered this reaction in 1879.

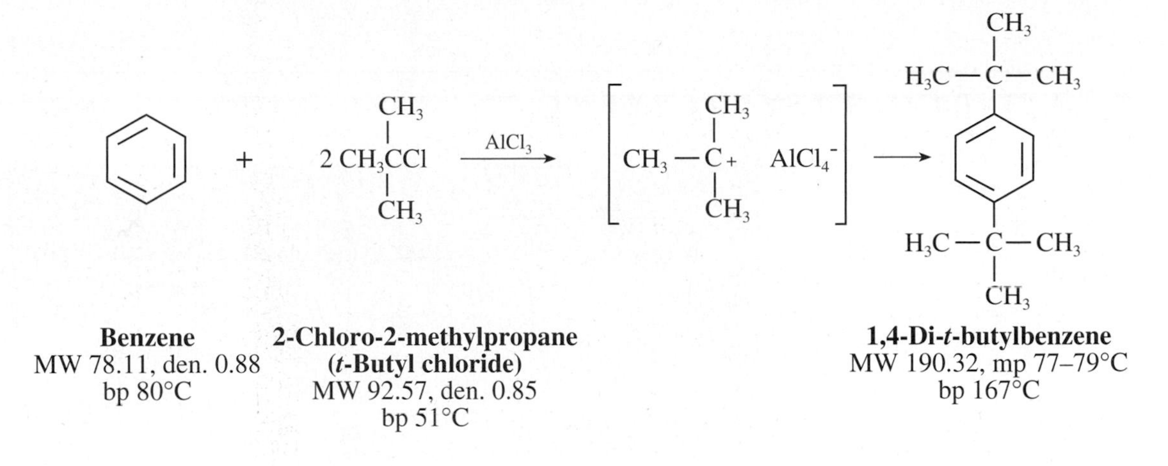

The reaction is reversible. If 1,4-di-*t*-butylbenzene is allowed to react with *t*-butyl chloride and 1.3 mol of aluminum chloride at 0°C–5°C, 1,3 di-*t*-butylbenzene, 1,3,5-tri-*t*-butylbenzene, and unchanged starting material are found in the reaction mixture. Thus, the mother liquor from crystallization of 1,4-di-*t*-butylbenzene probably contains *t*-butylbenzene, the desired 1,4-di-product, the 1,3-di-isomer, and 1,3,5-tri-*t*-butylbenzene.

INCLUSION COMPLEXES: HOST-GUEST CHEMISTRY

Although the mother liquor probably contains a mixture of several components, the 1,4-di-*t*-butylbenzene can be isolated easily as an inclusion complex. Inclusion complexes are examples of host-guest chemistry. The host molecule thiourea, NH_2CSNH_2, has the interesting property of crystallizing in a helical crystal lattice that has a cylindrical hole in it. The guest molecule can reside in this hole if it is the correct size. It is not bound to the host; nuclear magnetic resonance (NMR) studies indicate the guest molecule can rotate longitudinally within the helical crystal lattice. There are often nonintegral numbers (on the average) of host molecules per guest. The inclusion complex of thiourea and 1,4-di-*t*-butylbenzene crystallizes quite nicely from a mixture of the other hydrocarbons; thus more of the product can be obtained. Because thiourea is very soluble in water, the product is recovered from the complex by shaking it with a mixture of ether and water. The complex immediately decomposes, and the product dissolves in the ether layer, from which it can be recovered.

Compare the length of the 1,4-di-*t*-butylbenzene molecule to the length of various *n*-alkanes and predict the host-guest ratio for a given alkane. You can then check your prediction experimentally. *n*-Hexane can be isolated from the mixture of isomers sold under the name *hexanes*.

EXPERIMENTS

1. 1,4-DI-*t*-BUTYLBENZENE

CAUTION: Benzene is a mild carcinogen. Handle it in the hood; do not breathe its vapors or allow the liquid to come in contact with the skin.

IN THIS EXPERIMENT, a mixture of benzene and an alkyl chloride are treated with aluminum chloride, a Lewis acid catalyst. During the alkylation reaction, hydrogen chloride is evolved and must be trapped. The dialkyl product is isolated by adding water to the reaction mixture and extracting it with ether. In the usual way the ether solution is washed, dried, and evaporated to give the product. This crude material is recrystallized from methanol to give the pure dialkyl product.

Photos: *Removing a Reagent from a Septum-Capped Bottle, Polypropylene Syringe Containing Ether*

Aluminum chloride dust is extremely hygroscopic and irritating. It hydrolyzes to hydrogen chloride on contact with moisture. Clean up any spilled material immediately.

Photos: *Gas Trap, Placing a Polyethylene Tube through a Septum; Extraction with Ether*

Using a 1.0-mL plastic syringe, measure 0.40 mL of dry 2-chloro-2-methylpropane (*t*-butyl chloride) and 0.20 mL of dry benzene into a dry 10 × 100 mm reaction tube equipped with a septum and tubing (Fig. 29.1). The benzene and the alkyl chloride will usually be found in septum-stoppered containers. Cool the tube in ice and then add 20 mg of aluminum chloride. Weighing and transferring this small quantity is difficult because aluminum chloride rapidly reacts with moist air. Keep the reagent bottle closed as much of the time as possible while weighing the reagent into a very small, dry, capped vial. Because the aluminum chloride is a catalyst, the amount need not be exactly 20 mg.

Mix the contents of the reaction tube by flicking the tube with a finger. After an induction period of about 2 minutes, a vigorous reaction sets in, with bubbling and liberation of hydrogen chloride. The hydrogen chloride is trapped using the apparatus depicted in Figure 29.1. The wet cotton in the empty reaction tube will dissolve and trap the hydrogen chloride. Figure 29.2 illustrates how to thread a polyethylene tube through a septum. Near the end of the reaction, the product separates as a white solid. When this occurs, remove the tube from the ice and let it stand at room temperature for 5 minutes.

Add about 1.0 mL of ice water to the reaction mixture, mix the contents thoroughly, and extract the product with three 0.8-mL portions of ether. Wash the combined ether extracts with about 1.5 mL of saturated sodium chloride solution, and dry the ether over anhydrous calcium chloride pellets. Add sufficient drying agent so that it does not clump together. After 5 minutes, transfer the ether solution to a dry, tared reaction tube, using more ether to wash the drying agent, and evaporate the ether under a stream of air in the hood. Remove the last traces of ether under water aspirator vacuum (Fig. 29.3). The oily product should solidify on cooling and weigh about 300 mg.

Add anhydrous calcium chloride until it no longer clumps together.

For crystallization, dissolve the product in 0.40 mL of methanol and allow the solution to come to room temperature without disturbance. After thorough cooling at 0°C, remove the methanol with a Pasteur pipette and rinse the crystals with a drop of ice-cold methanol while keeping the reaction tube in ice. Save this methanol solution for analysis by thin-layer chromatography (**TLC**). The yield of recrystallized material after drying under aspirator vacuum should be about 160 mg. Remove a sample of crystals for analysis by infrared (**IR**) or NMR spectroscopy, TLC, and melting-point determination (Fig. 29.4 and Fig. 29.5). Using TLC, compare the pure crystalline product to the residue left after evaporation of the methanol.

Spontaneous crystallization gives beautiful needles or plates.

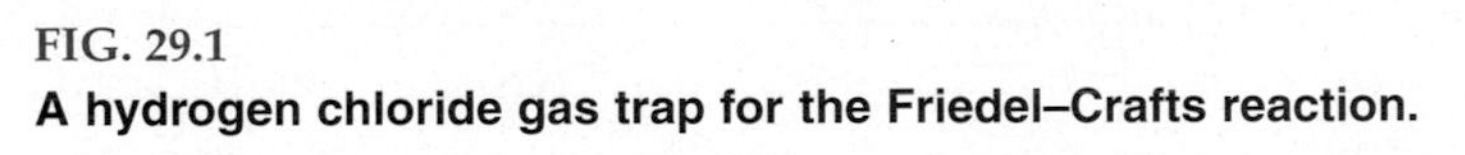

FIG. 29.1
A hydrogen chloride gas trap for the Friedel–Crafts reaction.

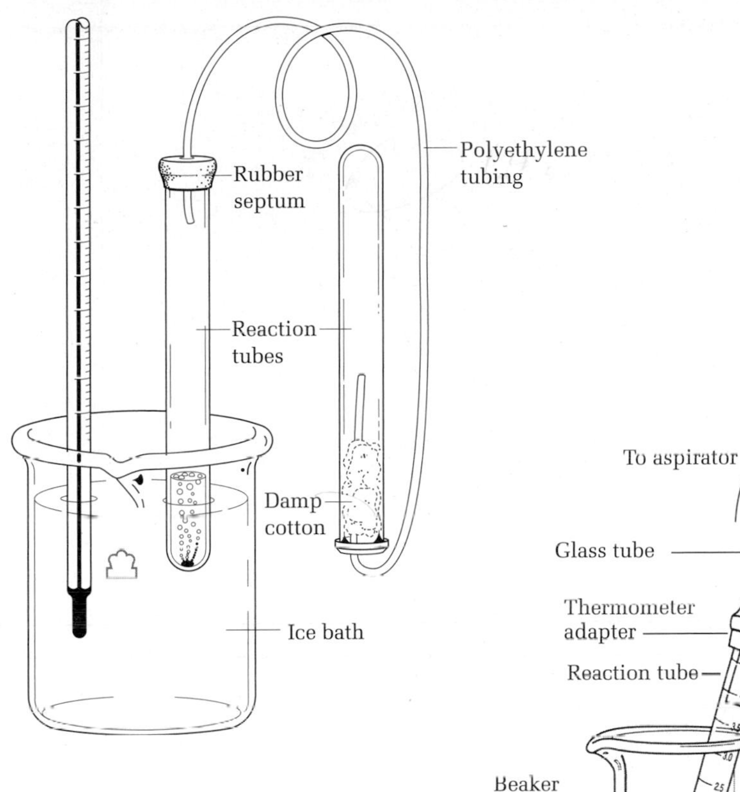

FIG. 29.2
To thread a polyethylene tube through a septum, make a hole through the septum with a needle, then push a toothpick through the hole. Push the polyethylene tube firmly onto the toothpick, then pull and push on the toothpick to thread the tubing through the septum. Finally, pull the tube from the toothpick after it has come out the other side of the septum. A blunt syringe needle can be used instead of a toothpick.

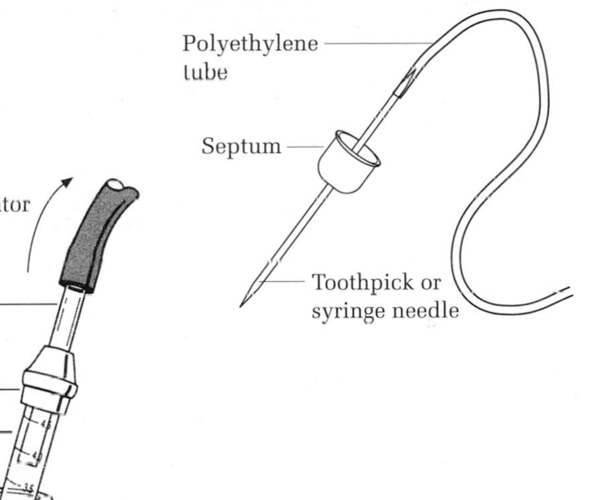

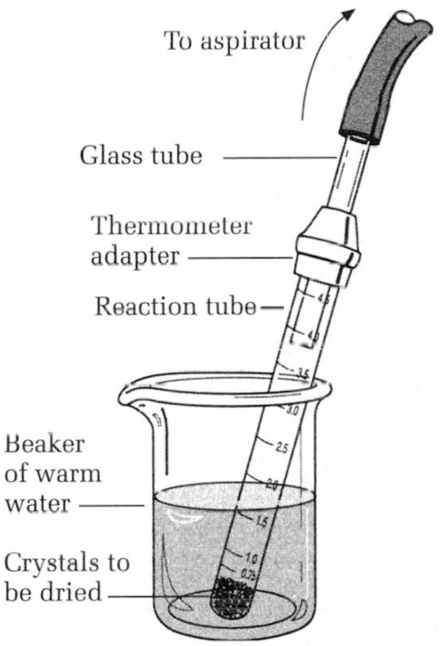

FIG. 29.3
Drying crystals under reduced pressure.

FIG. 29.4
The IR spectrum of 1,4-di-*t*-butylbenzene.

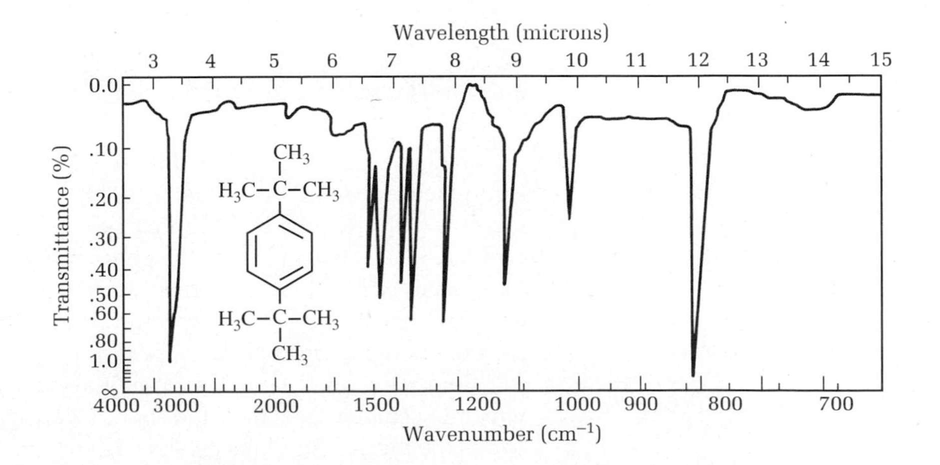

FIG. 29.5
The ^{1}H NMR spectrum of 1,4-di-*t*-butylbenzene (100 MHz).

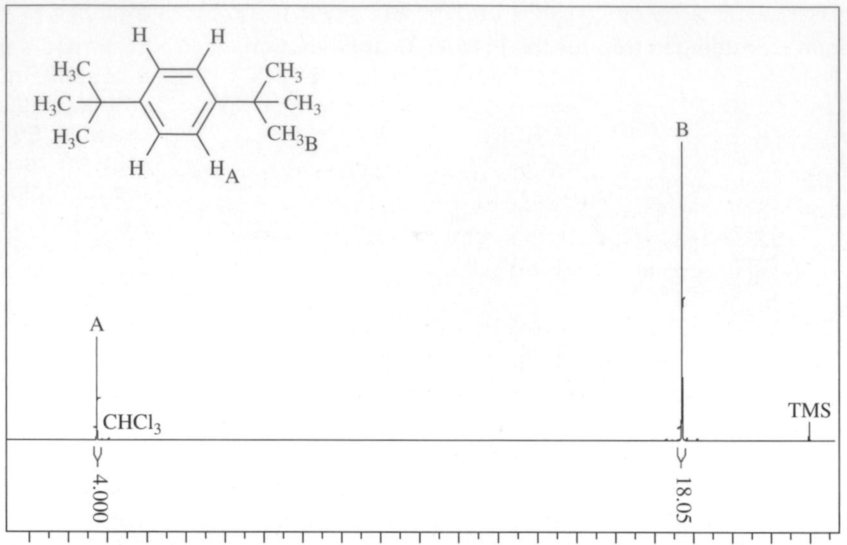

Cleaning Up. Place any unused *t*-butyl chloride in the halogenated organic waste container and any unused benzene in the hazardous waste container for benzene. Any unused aluminum chloride should be mixed thoroughly with a large excess of sodium carbonate, and the solid mixture should be added to a large volume of water before being flushed down the drain. The combined aqueous layers from the reaction should be neutralized with sodium carbonate and then flushed down the drain. Methanol from the crystallization is to be placed in the organic solvents waste container.

2. PREPARATION OF THE THIOUREA INCLUSION COMPLEX

$H_2N\overset{S}{\overset{\|}{C}}NH_2$

Thiourea
MW 76.12

IN THIS EXPERIMENT, an inclusion complex is prepared by simply cooling a solution of thiourea and di-*t*-butylbenzene. The complex is isolated, weighed, and decomposed with water; then the dialkyl benzene is extracted into ether that is dried and evaporated. The residue is weighed to calculate how many molecules of thiourea have been complexed with a molecule of the di-*t*-butylbenzene.

CAUTION: Thiourea is a mild carcinogen. Handle the solid in a hood. Do not breathe its dust.

In a tared reaction tube, dissolve 200 mg of thiourea and 120 mg of 1,4-di-*t*-butylbenzene in 2.0 mL of methanol at room temperature; then cool the mixture in ice, at which time the inclusion complex will crystallize. Using a Pasteur pipette, remove the solvent and wash the product twice with just enough methanol to cover the crystals while keeping the tube on ice. Connect the reaction tube to a water aspirator and, using the heat of your hand, evaporate the remaining methanol under reduced pressure until the weight of the tube is constant. The yield should be about 200 mg.

The inclusion complex starts to crystallize in 10 minutes.

Remove a small sample and set it aside; carefully determine the weight of the remaining complex and then add about 1.2 mL of water and 1.2 mL of ether to the tube. Shake the mixture until the crystals disappear. This causes the complex to break up, with the thiourea remaining in the aqueous layer and the 1,4-di-*t*-butylbenzene passing into the ether layer. Draw off the aqueous layer and dry the ether layer with anhydrous calcium chloride pellets. Add sufficient drying agent so that it does not clump together. More ether can be added if necessary. Transfer the ether to a tared reaction tube and wash the drying agent twice with fresh portions of ether. The objective is to make a quantitative transfer of the butylbenzene. Evaporate the ether and remove the last traces under aspirator vacuum. After the weight of the tube is constant, record the weight of the hydrocarbon. Calculate the number of molecules of thiourea per molecule of hydrocarbon (probably not an integral number).

Cleaning Up. Place any unused thiourea and the 1.2 mL of the aqueous solution containing thiourea in the hazardous waste container for thiourea. Alternatively, treat the thiourea with excess aqueous 5.25% sodium hypochlorite solution (household bleach), dilute the mixture with a large amount of water, and flush it down the drain. If local regulations allow, evaporate any residual solvent from the drying agent in the hood and place the dried solid in the nonhazardous waste container. Otherwise, place the wet drying agent in a waste container designated for this purpose.

3. 1,4-DI-*t*-BUTYLBENZENE

CAUTION: Benzene is a mild carcinogen. Handle it in the hood; do not breathe its vapors or allow the liquid to come in contact with the skin.

Aluminum chloride dust is extremely hygroscopic and irritating. It hydrolyzes to hydrogen chloride on contact with moisture. Clean up any spilled material immediately.

Reaction time: about 15 minutes

Add anhydrous calcium chloride pellets until they no longer clump together.

Spontaneous crystallization gives beautiful needles or plates.

Video: *Macroscale Crystallization*

Measure in the hood 20 mL of 2-chloro-2-methylpropane (*t*-butyl chloride) and 10 mL of benzene in a 125-mL filter flask equipped with a one-holed rubber stopper fitted with a thermometer. Place the flask in an ice-water bath to cool. Weigh 1 g of fresh aluminum chloride onto a creased paper and scrape it with a small spatula into a 10 × 75-mm test tube; close the tube at once with a cork.[2] Connect the side arm of the flask to an aspirator (preferably one made of plastic) and operate it at a rate sufficient to carry away hydrogen chloride formed in the reaction; alternatively, make a trap for the hydrogen chloride similar to that shown in Figure 18.3 (on page 332). Cool the liquid to 0–3°C, add about one-quarter of the aluminum chloride, replace the thermometer, and swirl the flask vigorously in the ice bath. After an induction period of about 2 minutes, a vigorous reaction sets in, with bubbling and liberation of hydrogen chloride. Add the remainder of the catalyst in three portions at intervals of about 2 minutes. Towards the end, the reaction product begins to separate as a white solid. When this occurs, remove the flask from the bath and let stand at room temperature for 5 minutes. Add ice and water to the reaction mixture, allow most of the ice to melt, and then add ether for extraction of the product, stirring with a rod or spatula to help bring the solid into solution. Transfer the solution to a separatory funnel and shake; draw off the lower layer and wash the upper ether layer first with water then with a saturated sodium chloride solution. Dry the ether solution over anhydrous calcium chloride pellets for 5 minutes, filter the solution to remove the drying agent, remove the ether by evaporation on a steam bath, and evacuate the flask using an aspirator to remove traces of solvent until the weight is constant; the yield of crude product should be 15 g.

The oily product should solidify on cooling. For crystallization, dissolve the product in 20 mL of hot methanol and let the solution come to room temperature without disturbance. If you are in a hurry, with minimal agitation, place it gently in an ice-water bath and observe the result. After thorough cooling at 0°C, collect the product and rinse the flask and product with a little ice-cold methanol. The yield of

[2]Alternative scheme: Put a wax pencil mark on the test tube 37 mm from the bottom and fill the tube with aluminum chloride to this mark.

1,4-di-*t*-butylbenzene from the first crop is 8.2–8.6 g of satisfactory material. Save the product for the next step as well as the mother liquor, in case you later wish to work it up for a second crop of product.

Inclusion Complex Formation

CAUTION: Thiourea is a carcinogen. Handle the solid in a hood. Do not breathe its dust.

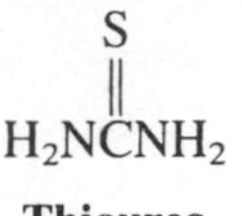

Thiourea
MW 76.12

The inclusion complex starts to crystallize in 10 minutes.

In a 25-mL Erlenmeyer flask, dissolve 5 g of thiourea and 3 g of 1,4-di-*t*-butylbenzene in 50 mL of warm methanol (break up lumps with a flattened stirring rod) and let the solution stand for crystallization of the complex, which occurs with ice cooling. Collect the crystals, rinse with a little methanol, and dry to constant weight; the yield is 5.8 g. Remove a small sample and set it aside; carefully determine the weight of the remaining complex and place the material in a separatory funnel along with about 25 mL each of water and ether. Shake until the crystals disappear, draw off the aqueous layer containing thiourea, wash the ether layer with saturated sodium chloride, and dry the ether layer over anhydrous calcium chloride pellets. Remove the drying agent by filtration and collect the filtrate in a tared 125-mL Erlenmeyer flask. Evaporate and evacuate as before, being sure the weight of hydrocarbon is constant before you record it. Calculate the number of molecules of thiourea per molecule of hydrocarbon (probably *not* an integral number).

Workup of mother liquor

To work up the mother liquor from the crystallization, first evaporate the methanol. Note that the residual oil does not solidify on ice cooling. Next, dissolve the oil, together with 5 g of thiourea, in 50 mL of methanol, collect the inclusion complex that crystallizes (about 3.2 g), and recover 1,4-di-*t*-butylbenzene from the complex as before (about 0.8 g before crystallization). The IR and ^{1}H NMR spectra of the product are seen in Figures 29.4 and 29.5.

Cleaning Up. Place any unused *t*-butyl chloride in the halogenated organic waste container and any unused benzene in the hazardous waste container for benzene. Any unused aluminum chloride should be mixed thoroughly with a large excess of sodium carbonate, and the solid mixture should be added to a large volume of water before being flushed down the drain. The combined aqueous layers from the reaction should be neutralized with sodium carbonate and then flushed down the drain. Methanol from the crystallization is to be placed in the organic solvents waste container.

4. 1,4-DI-*t*-BUTYL-2,5-DIMETHOXYBENZENE

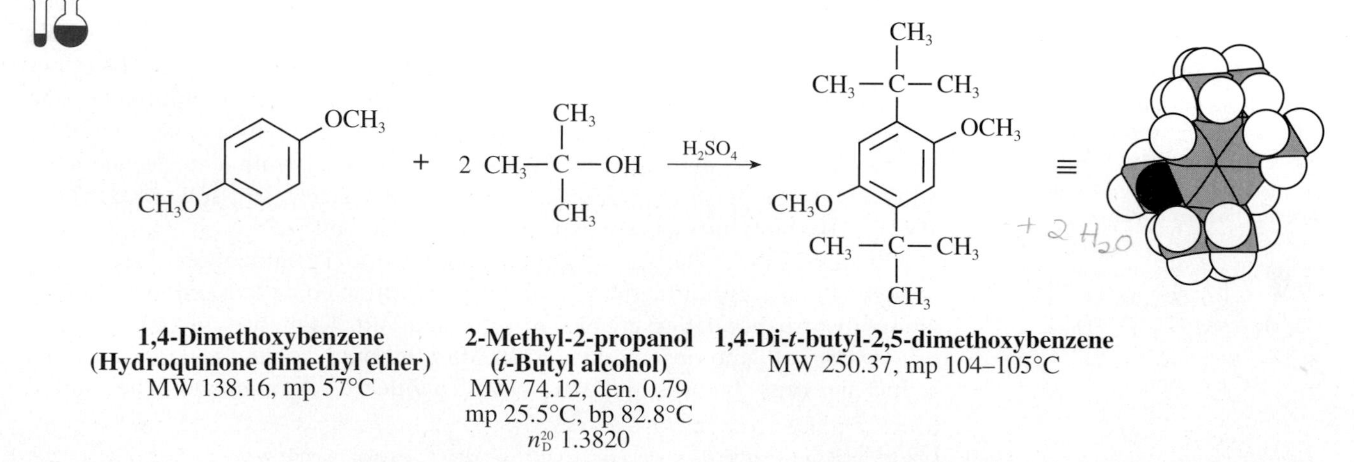

1,4-Dimethoxybenzene (Hydroquinone dimethyl ether)
MW 138.16, mp 57°C

2-Methyl-2-propanol (*t*-Butyl alcohol)
MW 74.12, den. 0.79
mp 25.5°C, bp 82.8°C
n_D^{20} 1.3820

1,4-Di-*t*-butyl-2,5-dimethoxybenzene
MW 250.37, mp 104–105°C

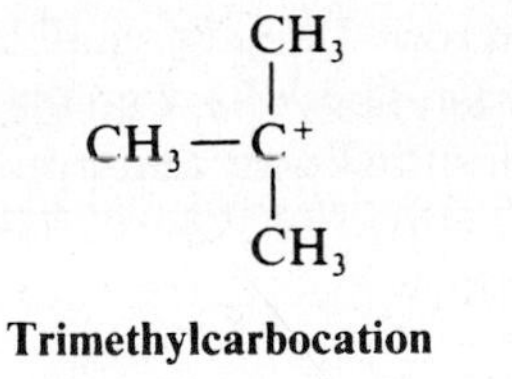

Trimethylcarbocation

This experiment illustrates the Friedel–Crafts alkylation of an activated benzene molecule with a tertiary alcohol in the presence of sulfuric acid as the Lewis acid catalyst. Like the reaction of benzene and *t*-butyl chloride, the substitution involves attack by the electrophilic trimethylcarbocation.

Microscale Procedure

IN THIS EXPERIMENT, a mixture of dimethoxybenzene and *t*-butyl alcohol is dissolved in acetic acid and treated with sulfuric acid, a Lewis acid catalyst. Water is added to the reaction mixture, and the solid product is isolated by removing the aqueous material with a Pasteur pipette. The dialkylated product is recrystallized from methanol.

CAUTION: Handle concentrated sulfuric acid with care.

In a 10 ×100 mm reaction tube, dissolve 120 mg of 1,4-dimethoxybenzene (hydroquinone dimethyl ether) in 0.4 mL of acetic acid with gentle warming and add 0.2 mL of *t*-butyl alcohol (it may be necessary to melt this alcohol). Cool the mixture in ice and then add to it 0.4 mL of concentrated sulfuric acid dropwise from a Pasteur pipette. After each drop of acid is added, mix the solution thoroughly. At the end of this addition, considerable solid reaction product should have separated. Stir the mixture thoroughly with a glass stirring rod, remove the reaction tube from the ice, and allow it to warm to room temperature and remain at 20–25°C for at least 10 minutes to complete the reaction. Next, cool the mixture in ice to cause crystallization to occur. Measure 2.5 mL of water into a container. Very carefully add 1 drop of water to the reaction mixture, stir with the glass rod, and continue to add the remainder of the water dropwise with cooling and stirring. Remove the solvent from the cold solution with a Pasteur pipette and wash the crystals thoroughly with water. Recrystallize the product from methanol. After allowing the mixture to cool to room temperature and then to 0°C in ice, remove the solvent using a Pasteur pipette. The last traces of methanol can be removed under aspirator vacuum while warming the tube in your hand or in a beaker of warm water (Fig. 29.3). The yield of large plates of 1,4-di-*t*-butyl-2,5-dimethoxybenzene should be about 80–100 mg. Analyze the product by IR spectroscopy and TLC, using ligroin as the eluent. Determine the melting point and the percentage yield.

Stir *thoroughly* after each drop of water is added.

Videos: *Filtration of Crystals Using the Pasteur Pipette; Recrystallization*; Photo: *Drying Crystals Under Vacuum*

Cleaning Up. Combine the aqueous layer, the methanol washes, and the crystallization mother liquor; dilute with water; neutralize with sodium carbonate; and flush down the drain. Any spilled sulfuric acid should be covered with a large excess of solid sodium carbonate, and the mixture should be added to water before being flushed down the drain.

Macroscale Procedure

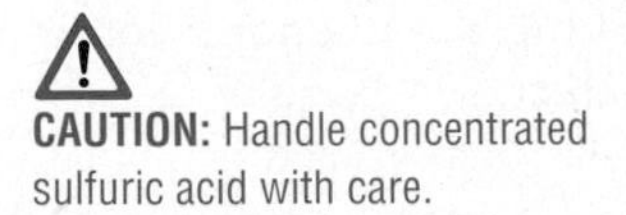

CAUTION: Handle concentrated sulfuric acid with care.

Place 3 g of 1,4-dimethoxybenzene (hydroquinone dimethyl ether) in a 125-mL Erlenmeyer flask, add 5 mL of *t*-butyl alcohol and 10 mL of acetic acid, and place the flask in an ice-water bath to cool. Measure 15 mL of concentrated sulfuric acid into a 25-mL Erlenmeyer flask and place the flask—properly supported—in an ice bath to cool. For good thermal contact, the ice bath should be an ice-water mixture. Put a thermometer in the larger flask, swirl in the ice bath until the temperature is in the range 0–3°C, and remove the thermometer (solid, if present, will dissolve later). Do not use the thermometer as a stirring rod. Clamp a small separatory funnel in a position to deliver into the 125-mL Erlenmeyer flask so that the flask can remain in the ice-water bath, wipe the smaller flask dry, and pour the chilled sulfuric acid solution into the funnel. While swirling the 125-mL flask in the ice bath, run in the chilled sulfuric acid by rapid drops during the course of 4–7 minutes.

Total reaction time: about 12 minutes

By this time, considerable solid reaction product should have separated, and insertion of a thermometer should show that the temperature is in the 15–20°C range. Swirl the mixture while maintaining the temperature at about 20–25°C for an additional 5 minutes, and then cool in ice. Add ice to the mixture to dilute the sulfuric acid; then add water to nearly fill the flask, cool, and collect the product on a Büchner funnel with suction. It is good practice to clamp the filter flask so that it does not tip over. Apply only very gentle suction at first to avoid breaking the filter paper, which is weakened by the strong sulfuric acid solution. Wash liberally with water and then turn on the suction to full force. Press down the filter cake with a spatula and let it drain well. Meanwhile, cool a 15-mL portion of methanol for washing to remove a little oil and a yellow impurity. Release the suction, cover the filter cake with one-third of the chilled methanol, and then apply suction. Repeat the washing two more times.

W Video: *Macroscale Crystallization*

Because air-drying of the crude reaction product takes time, the following short procedure is suggested: Place the moist material in a 50-mL Erlenmeyer flask, add a little dichloromethane (5–8 mL) to dissolve the organic material, and note the appearance of aqueous droplets. Add enough anhydrous calcium chloride pellets to the flask so that the drying agent no longer clumps together, let drying proceed for 10 minutes, and then remove the drying agent by gravity filtration or careful decantation into another 50-mL Erlenmeyer flask. Add 15 mL of methanol (bp 65°C) to the solution. Remove the dichloromethane (bp 41°C) using a rotary evaporator or by evaporation on a steam bath in a hood. When the volume is estimated to be about 15 mL, let the solution stand for crystallization. When crystallization is complete, cool in ice and collect.

From an environmental standpoint, it would be better to eliminate the dichloromethane and excess methanol by simple distillation using the apparatus depicted in Figure 5.10 (on page 97). Leave 15 mL in the flask and allow the mixture to cool slowly. Large crystals will form. Collect the product on a small Büchner funnel. The yield of large plates of pure 1,4-di-*t*-butyl-2,5-dimethoxybenzene is 2–2.5 g.

Antics of Growing Crystals

Robert Stolow of Tufts University reported[3] that growing crystals of the di-*t*-butyldimethoxy compound change shape in a dramatic manner: Thin plates curl and roll up and then uncurl so suddenly that they propel themselves for a distance of several centimeters. If you do not observe this phenomenon during crystallization of a small sample, you may be interested in consulting the papers cited and pooling your sample with others for trial on a large scale. The solvent mixture recommended by the Tufts workers for observing the phenomenon is 9.7 mL of acetic acid and 1.4 mL of water per gram of product.

[3]Stolow RD, Larsen JW. *Chem Ind*. **1963**;449. See also Blatchly JM, Hartshorne NH. *Trans Faraday Soc*. **1966**;62:512.

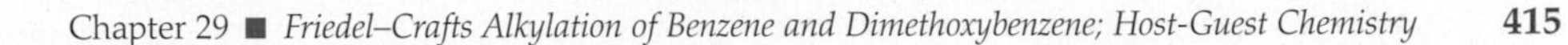

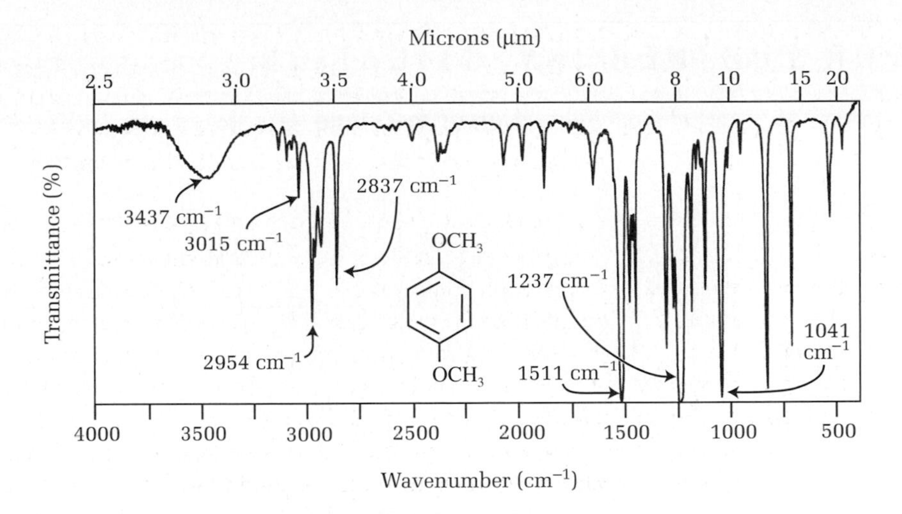

FIG. 29.6
The IR spectrum of 1,4-dimethoxybenzene (KBr disk). Note water contaminant at 3437 cm^{-1}.

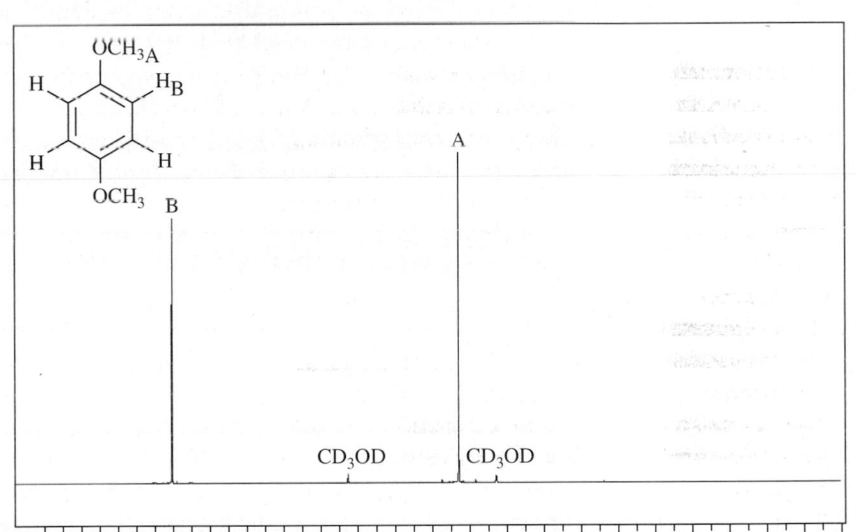

FIG. 29.7
The ^{1}H NMR spectrum of 1,4-dimethoxybenzene (400 MHz).

Figures 29.6 and 29.7 present the IR and NMR spectra of the starting hydroquinone dimethyl ether. Can you predict the appearance of the NMR spectrum of the product?

Cleaning Up. Combine the aqueous layer, the methanol washes, and the crystallization mother liquor; dilute with water; neutralize with sodium carbonate; and flush down the drain. Any spilled sulfuric acid should be covered with a large excess of solid sodium carbonate, and the mixture should be added to water before being flushed down the drain. Dichloromethane mother liquor from the crystallization is placed in the halogenated organic waste container. If local regulations allow, evaporate any residual solvent from the drying agent in the hood and place the dried solid in the nonhazardous waste container. Otherwise, place the wet drying agent in a waste container designated for this purpose.

5. FOR FURTHER INVESTIGATION

Alkylation of *m*-Xylene

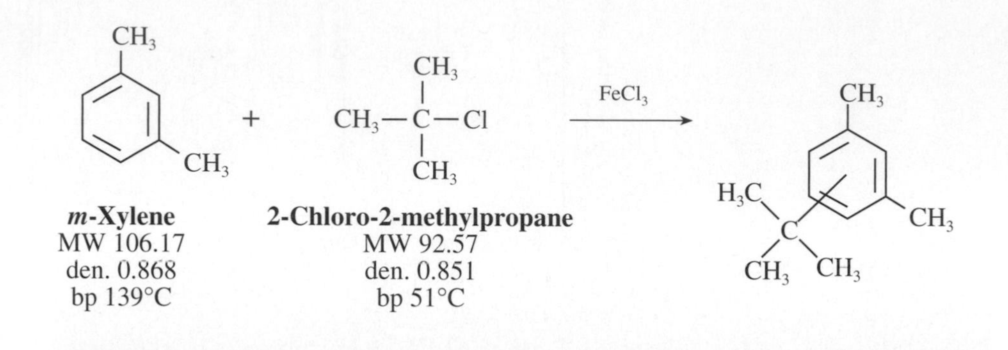

IN THIS EXPERIMENT, an excess of *m*-xylene (1,3-dimethylbenzene) and *t*-butyl chloride (2-chloro-2-methylpropane) is treated with iron(III) chloride, a Lewis acid catalyst. Hydrogen chloride gas is trapped. When the reaction is complete, water is added; the organic layer is washed, dried, and subjected to instant microscale distillation to remove excess 1,3-dimethylbenzene. The residual alkylated product is analyzed by IR spectroscopy to determine the structure of the product.

The objective of this experiment is to determine the structure of the product formed when *m*-xylene (not benzene, as in Experiment 1) is alkylated with *t*-butyl chloride. Although aluminum chloride is the catalyst most often used for the Friedel–Crafts reaction, it is difficult to store and weigh out because it reacts very rapidly with moisture in the air. Therefore, in this experiment, another Lewis acid catalyst, iron(III) chloride is used. The methyl groups in *m*-xylene are ortho-para directors and activators of the benzene ring, but other factors may intervene in this particular experiment. Read about the factors affecting alkylations in the first part of this chapter.

An excess of *m*-xylene is used in this reaction to ensure that the product will be monoalkylated by the *t*-butyl cation. Because the boiling points of the reactants and the product differ highly, a crude but very rapid and efficient distillation is done to remove unreacted xylene and *t*-butyl chloride, leaving only the product. This instant microscale distillation is carried out by boiling the mixture in a reaction tube and then pulling the hot vapors into a Pasteur pipette where they condense. After about half the material has been distilled in this way, the high-boiling residue will consist of almost pure product.

Instant microscale distillation

IR spectroscopy can be used to determine unequivocally the structure of the product. It has been found that the hydrogen atoms on a benzene ring give rise to one or two intense, characteristic peaks in the region of 730 to 885 cm^{-1} regardless of the nature of the substituents on the benzene ring. For instance, monosubstituted benzenes have two peaks: one in the range of 770 to 730 cm^{-1}, and the other in the range of 710 to 690 cm^{-1} (*see* Table 29.1). NMR spectroscopy can also be used to determine the structure of the product.

TABLE 29.1 *Infrared C—H Out-of-Plane Bending Vibrations of Substituted Benzenes*

Substituted Benzene	*Peak 1 (cm^{-1})*	*Peak 2 (cm^{-1})*
Benzene	671	—
Monosubstituted benzenes	770–730	710–690
1,2-Disubstituted	770–735	—
1,3-Disubstituted	810–750	710–690
1,4-Disubstituted	835–810	—
1,2,3-Trisubstituted	780–760	745–705
1,2,4-Trisubstituted	825–805	885–870
1,3,5-Trisubstituted	865–810	730–675
1,2,3,4-Tetrasubstituted	810–800	—
1,2,3,5-Tetrasubstituted	850–840	—
1,2,4,5-Tetrasubstituted	870 855	
Pentasubstituted	870	—

Molecular mechanics calculations on the three possible monosubstitution products from this reaction may help you to predict the outcome or help confirm conclusions drawn from experimental evidence.

Iron(III) Chloride Catalyzed Reaction

Photos: *Gas Trap, Placing a Polyethylene Tube through a Septum*

Place 0.6 mL of *m*-xylene and 0.5 mL of *t*-butyl chloride in a dry reaction tube equipped with a septum and a polyethylene tube that leads to a tube containing a small wad of damp cotton. See Figures 29.1 and 29.2 (on page 409) for threading a polyethylene tube through a septum. The cotton, dampened with 2 or 3 drops of water, serves to trap the hydrogen chloride evolved during the reaction. Cool the mixture in an ice bath (Fig. 29.1), add 30 mg of iron(III) chloride (purple, free-flowing crystals when pure and dry), and replace the septum and polyethylene tube. After a short induction period, the reaction will begin a vigorous evolution of hydrogen chloride.

When the vigorous part of this reaction is over, remove the reaction tube from the ice bath and allow the reaction mixture to warm to room temperature. When bubbles of hydrogen chloride cease to be evolved or after 15 minutes, add 1 mL of water to the reaction mixture, mix well, and then remove the water layer with a Pasteur pipette. Repeat this process using about 1 mL of saturated aqueous sodium bicarbonate solution, followed by 1 mL of saturated sodium chloride solution. Transfer the organic layer to another reaction tube and dry it with anhydrous calcium chloride pellets.

Videos: *Instant Microscale Distillation*

Transfer the organic layer once again to a dry reaction tube, add a boiling chip, and heat the mixture to boiling. Allow the refluxing vapors to rise about 3 cm in the tube, and then draw them into a Pasteur pipette. Squirt the condensate into another reaction tube held in the same hand (Fig. 29.8). Repeat this until about half the mixture has been distilled.

Analyze the residue by IR spectroscopy, which is most easily done as a thin film between sodium chloride or silver chloride plates. The IR spectrum of *m*-xylene is shown in Figure 29.9. Refer to Table 29.1 for the IR absorption frequencies

FIG. 29.8
An apparatus for instant microscale distillation.

Refluxing vapor
Boiling liquid
Boiling chip

FIG. 29.9
The IR spectrum of *m*-xylene (thin film) from 500 cm^{-1} to 1000 cm^{-1} to show C—H out-of-plane bending vibrations.

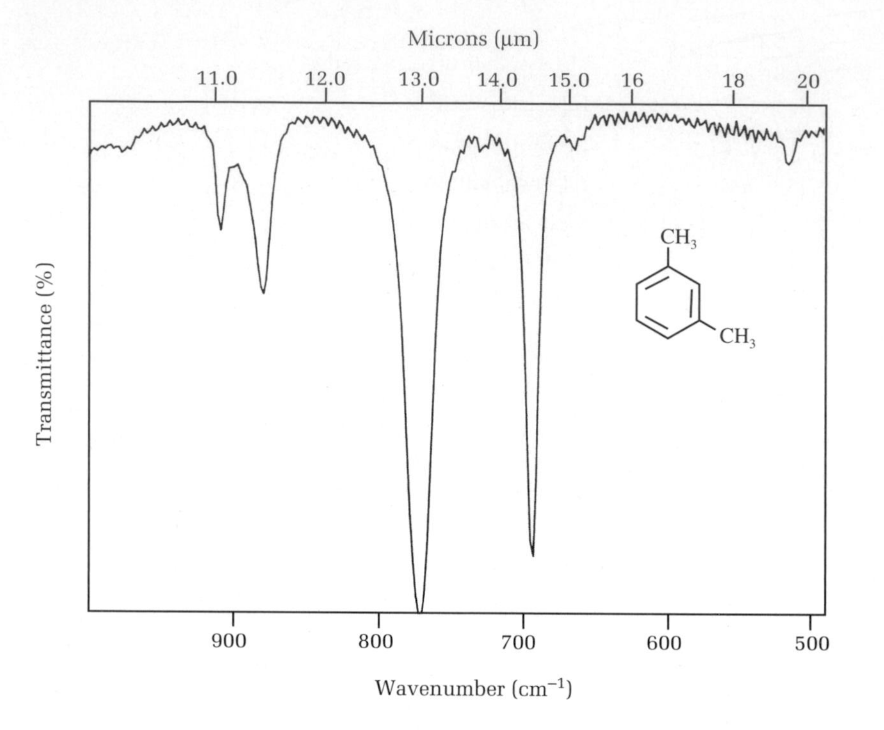

assigned to the different benzene substitution patterns. Note the two strong peaks in Figure 29.9 for *m*-xylene and the frequencies expected for 1,3-disubstitution from Table 29.1. From this correlation chart for the aromatic C—H out-of-plane bending modes, deduce the structure of your product. In your laboratory report, write a mechanism for this reaction that illustrates all the details of your conclusions. Discuss the structure of the product in terms of directive effects of substituents on the benzene ring, the steric effects of substituents, and thermodynamic versus kinetic control of the reaction.

Carry out molecular mechanics calculations on the three possible products to determine their relative steric energies or heats of formation. Do these calculations corroborate your conclusions?

Cleaning Up. Place any unused *t*-butyl chloride in the halogenated organic waste container and any unused *m*-xylene in the organic solvents waste container. Unused iron(III) chloride and all wash solutions should be combined and neutralized with sodium bicarbonate solution, diluted with water, and flushed down the drain. If local regulations allow, evaporate any residual solvent from the drying agent in the hood and place the dried solid in the nonhazardous waste container. Otherwise, place the wet drying agent in a waste container designated for this purpose.

See *Web Links*

QUESTIONS

1. Write equations to explain why the reaction of 1,4-di-*t*-butylbenzene with *t*-butyl chloride and aluminum chloride gives 1,3,5-tri-*t*-butylbenzene.
2. Why must aluminum chloride be protected from exposure to the air?
3. Would you expect to find two strong peaks at ~690 cm^{-1} and ~770 cm^{-1} (Fig. 29.9) in your product from the alkylation of *m*-xylene? Why or why not?
4. Draw a detailed mechanism for the formation of *t*-butyl-2,5-dimethoxybenzene.
5. Why is the 1,4 isomer, 1,4-di-*t*-butyl-2,5-dimethoxybenzene, the major product in the alkylation of dimethoxybenzene? Would you expect either of the following compounds to be formed as side products: 1,3-di-*t*-butyl-2,5-dimethoxybenzene or 1,4-dimethoxy-2,3-di-*t*-butylbenzene? Why or why not?
6. Suggest two other compounds that might be used in place of *t*-butyl alcohol to form 1,4-di-*t*-butyl-2,5-dimethoxybenzene.
7. Can you locate the two peaks in Figure 28.3 (on page 405) that show that methyl 3-nitrobenzoate is indeed 1,3-disubstituted? (*See* Experiment 5 and Table 29.1.)

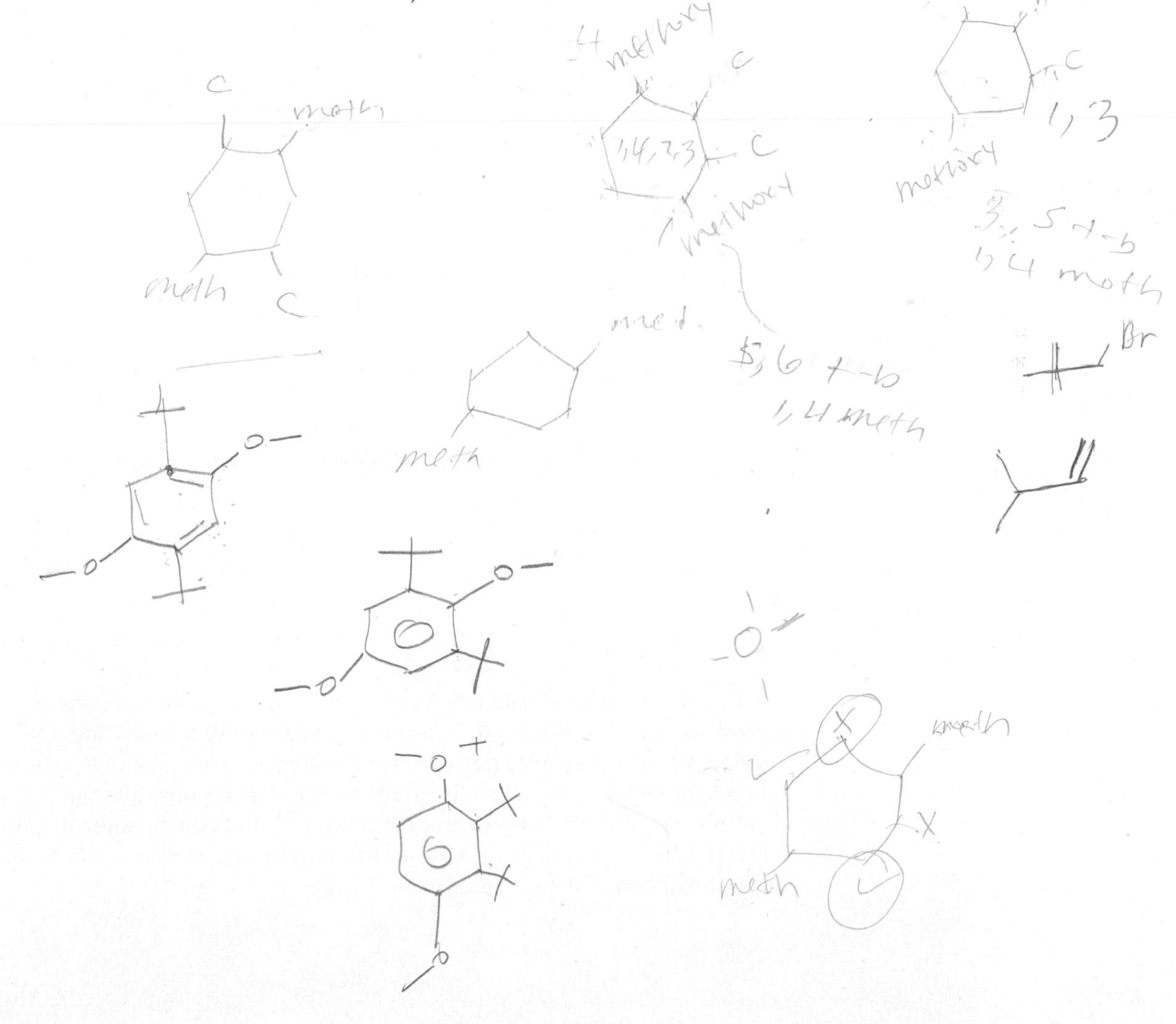

30 CHAPTER

Alkylation of Mesitylene

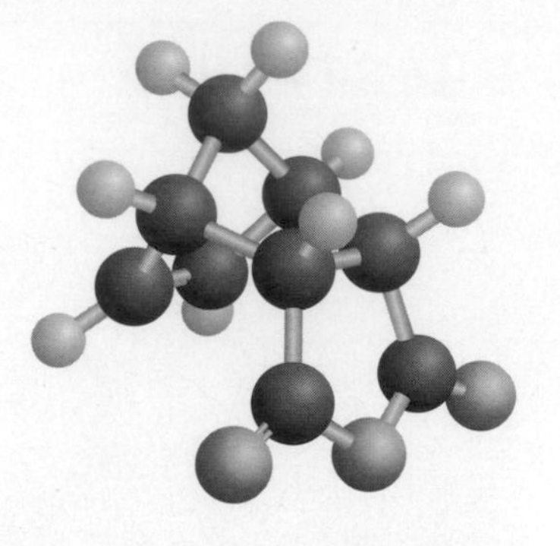

PRELAB EXERCISE: How much formic acid is consumed in this reaction?

The reaction of an alkyl halide with an aromatic molecule in the presence of aluminum chloride, the Friedel–Crafts reaction, proceeds through the formation of an intermediate carbocation:

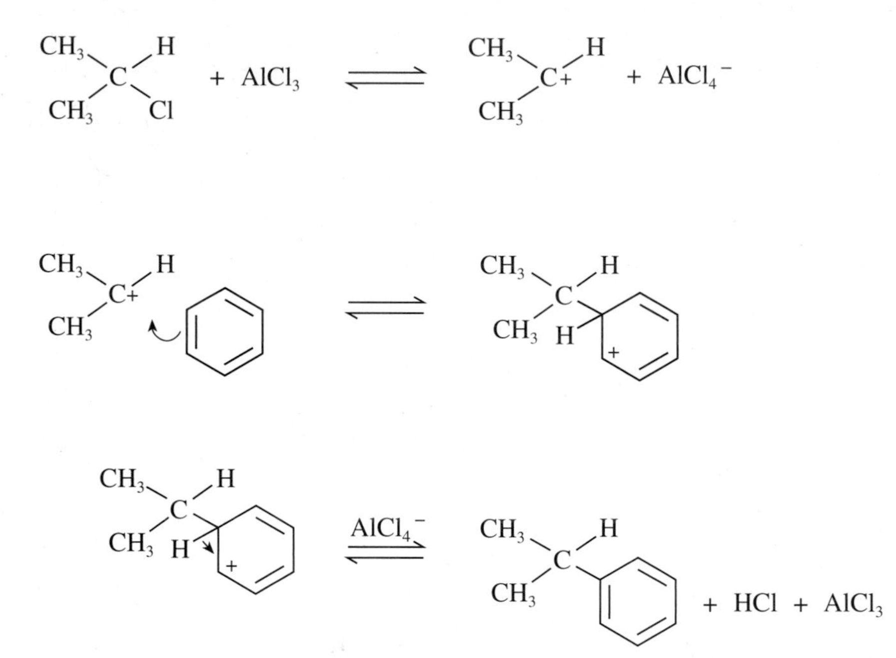

Charles Friedel and James Crafts (who later became the president of MIT) discovered this reaction in 1879, but seven years prior to this date, Johann Baeyer and his colleagues carried out very similar reactions using aldehydes as the alkylating agent and strong acids as catalysts. These reactions, like the Friedel–Crafts reaction, proceed through carbocation intermediates. Consider the synthesis of DDT (1,1,1-trichloro-2,2-di(*p*-chlorophenyl)ethane), colloquially referred to as DichloroDiphenylTrichloroethane:

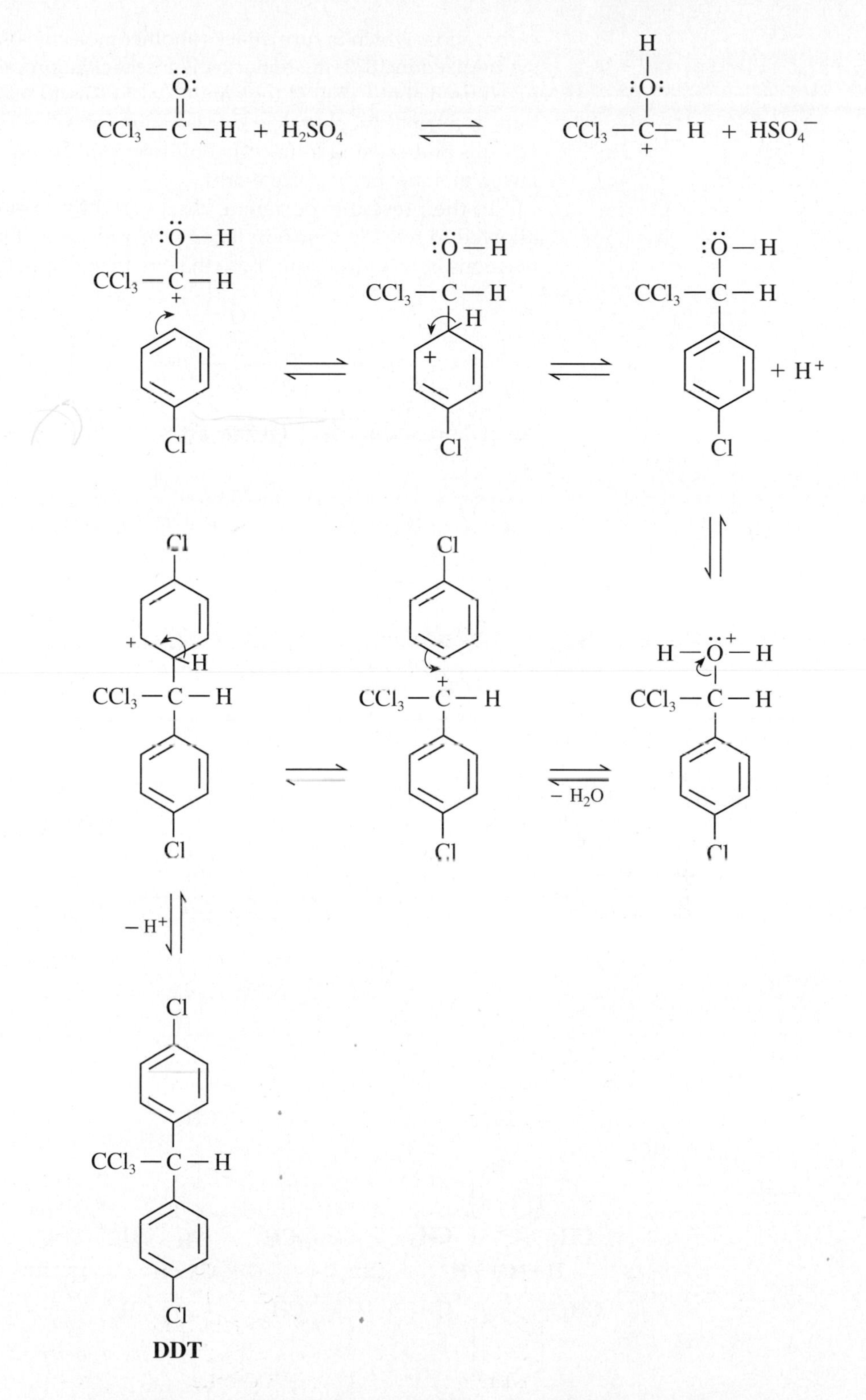

Trichloroacetaldehyde (chloral) forms a carbocation on reaction with concentrated sulfuric acid. This reacts primarily at the *para*-position of chlorobenzene; the intermediate alcohol, being benzylic, in the presence of acid readily forms a new carbocation, which in turn attacks another molecule of chlorobenzene. Even though synthesized in 1872, the remarkable insecticidal properties of DDT were not recognized until about 1940. It took another 25 years to realize that DDT, which is resistant to normal biochemical degradation, was building up in rivers, lakes, and streams and causing long-term environmental damage to wildlife. It is now outlawed in many parts of the world.

In the present experiment, discovered by Baeyer in 1872, formaldehyde is allowed to react with mesitylene in the presence of formic acid. The sequence of reactions is very similar to those that are involved in the formation of DDT:

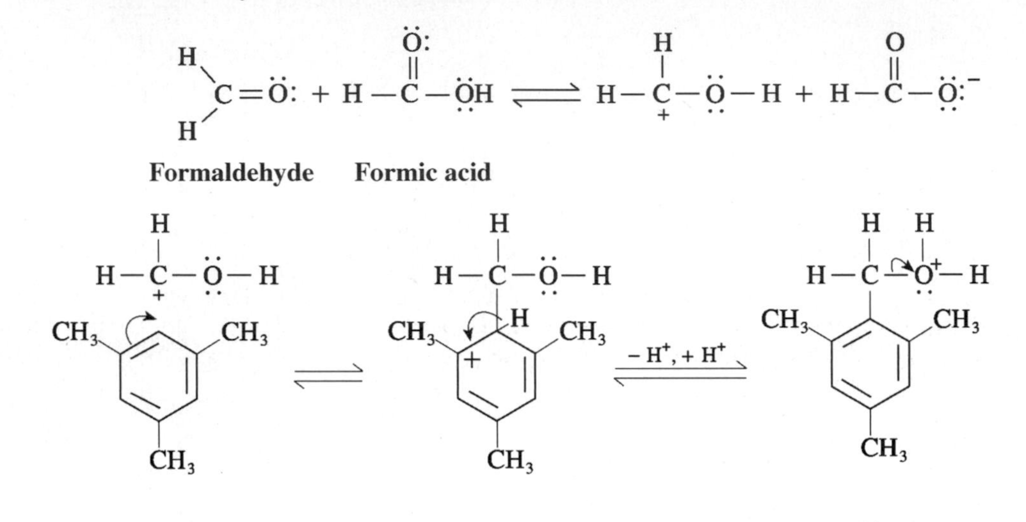

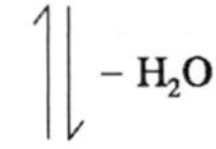

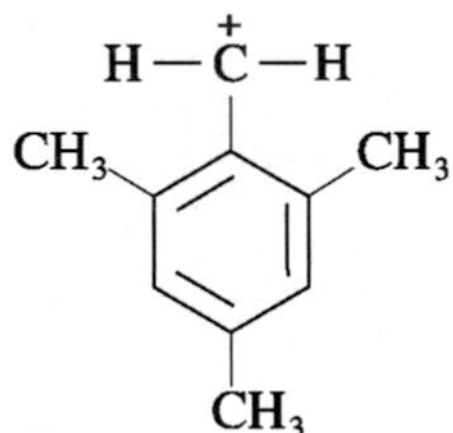

Benzylic carbocation

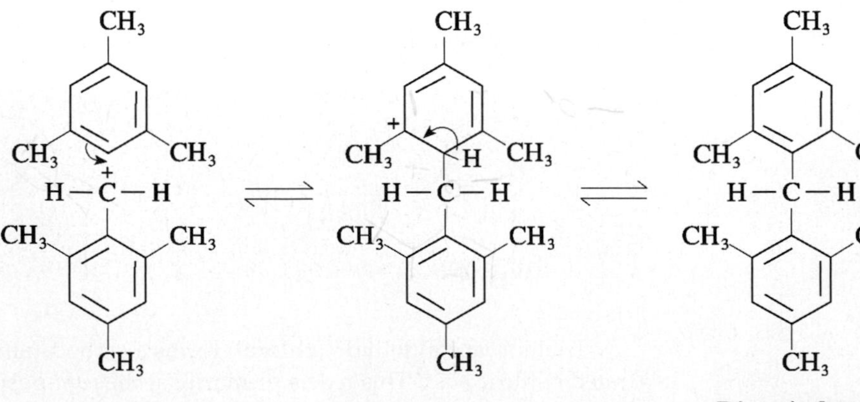

Dimesitylmethane

When aluminum chloride, a much more powerful catalyst, is used in this reaction, the methyl groups on the mesitylene rearrange and disproportionate to form a number of products, including polymeric material.

The strongly activated ring of phenol reacts with formaldehyde at the *ortho*- and *para*- positions to form a polymer called Bakelite (*see* Chapter 67).

A convenient form of formaldehyde to use in a reaction of this type is paraformaldehyde, a polymer that readily decomposes to formaldehyde:

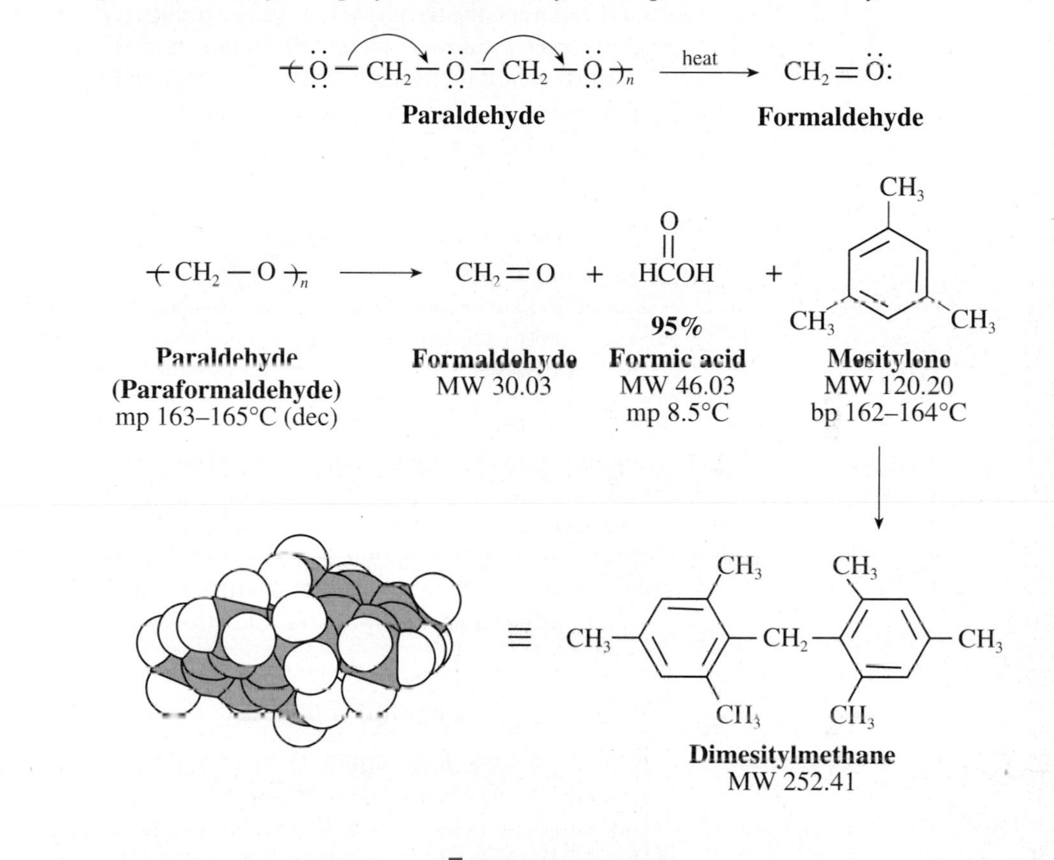

EXPERIMENT

DIMESITYLMETHANE

Microscale Procedure

CAUTION: Formic acid is corrosive to tissue. Avoid all contact with the liquid or its vapors. Should any come in contact with your skin, wash it off immediately with a large quantity of water.

IN THIS EXPERIMENT two molecules of 1,3,5-trimethylbenzene undergo a Friedel-Crafts alkylation with formaldehyde catalyzed by formic acid. On cooling, the product crystallizes from the reaction mixture. It is washed, dried, and crystallized from hexanes.

A carborundum or Teflon boiling chip is specified because most white boiling chips will react with the formic acid and stop the reaction.

To 10 mg of paraformaldehyde in a 10 × 100 mm reaction tube in the hood, add 0.06 mL of 95% formic acid and a carborundum (black) or Teflon boiling chip. Dissolve the paraformaldehyde in the formic acid by boiling on a hot sand bath;

then add 0.133 mL of mesitylene and another carborundum or Teflon boiling chip. Add the distilling column as an air condenser (Fig. 30.1) and reflux the reaction mixture for 2 hours. Cool the mixture first to room temperature and then in ice. Remove the excess formic acid using a Pasteur pipette; wash the crystals with water, aqueous sodium carbonate solution, and again with water; then scrape the crystals out onto a piece of filter paper and pick out the boiling chips. Squeeze the crystals between sheets of filter paper to dry.

Photo: *Filtration Using a Pasteur Pipette and* Videos: *Filtration of Crystals Using the Pasteur Pipette; Recrystallization*

Determine the weight of the crude product and save a few crystals for a melting-point determination. Recrystallize the product by dissolving it in a minimum quantity of boiling hexanes. Allow the solution to cool to room temperature, add a seed crystal if necessary, and then cool the mixture for at least 15 minutes in ice before removing the solvent using a Pasteur pipette. Another solvent for recrystallization is a mixture of 0.75 mL of toluene and 0.1 mL of methanol, which is adequate for 0.5 g of product. Obtain the IR (infrared) and NMR (nuclear magnetic resonance) spectra of the pure material and the melting points of the crude and recrystallized product. Turn in the pure product along with a card giving relevant data on the substance, including the number of moles of each of the starting materials. From the spectra and any tests you may wish to run, confirm the structure of the product.

Cleaning Up. At the end of this reaction, there should be no formaldehyde remaining in the reaction mixture. Should it be necessary to destroy formaldehyde, it should be diluted with water and 7 mL of household bleach added to oxidize every 100 mg of paraformaldehyde. After 20 minutes, it can be flushed down the drain. The formic acid solvent from the reaction and aqueous washings should be combined, neutralized with sodium carbonate, and flushed down the drain. Mother liquor from recrystallization is placed in the organic solvents waste container.

Macroscale Procedure

To 0.75 g of paraformaldehyde in a 50-mL round-bottomed flask in the hood, add 4.5 mL of 95% formic acid. Fit the flask with a reflux condenser (Fig. 30.2) and bring the mixture to a boil on an electrically heated sand bath. After about 5 minutes, most of the paraformaldehyde will have dissolved. Add through the condenser 10.0 mL of mesitylene. Reflux the reaction mixture for 2 hours, cool the mixture to room temperature, and then cool it in ice. Remove the product by vacuum filtration in the hood, taking great care not to come into contact with the residual formic acid. Wash the crystals with water, aqueous sodium carbonate solution, and then again with water and squeeze the crystals between sheets of filter paper to dry. Determine the weight of the crude product and save a few crystals for a melting-point determination. Recrystallize the product. A suggested solvent is a mixture of 7.5 mL of toluene and 1 mL of methanol, which is adequate for 5 g of product. Obtain the melting points of the crude and recrystallized product. Calculate the percent yield and turn in the pure product to your instructor. IR and NMR spectra of the product are found in Figures 30.3, 30.4, and 30.5.

CAUTION: Formic acid is corrosive to tissue. Avoid all contact with the liquid or its vapors. Should any come in contact with your skin, wash it off immediately with a large quantity of water. Carrying out this reaction in the hood protects not only against formic acid but also against formaldehyde, a suspected carcinogen formed transitorily in the reaction.

Cleaning Up. At the end of this reaction, there should be no formaldehyde remaining in the reaction mixture. Should it be necessary to destroy formaldehyde, it should be diluted with water and 7 mL of household bleach added to oxidize

FIG. 30.1
An apparatus for refluxing a reaction mixture with maximum cooling.

FIG. 30.2
A macroscale reflux apparatus.

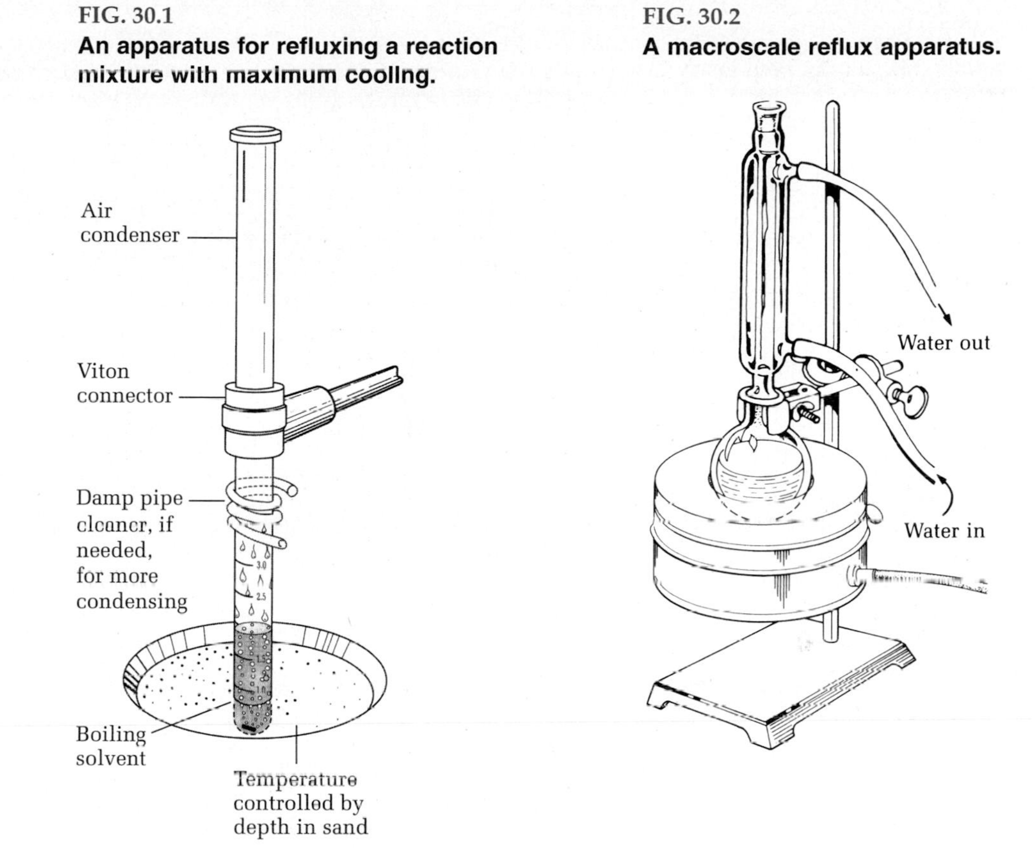

FIG. 30.3
The IR of dimesitylmethane.

Video: *Macroscale Crystallization*

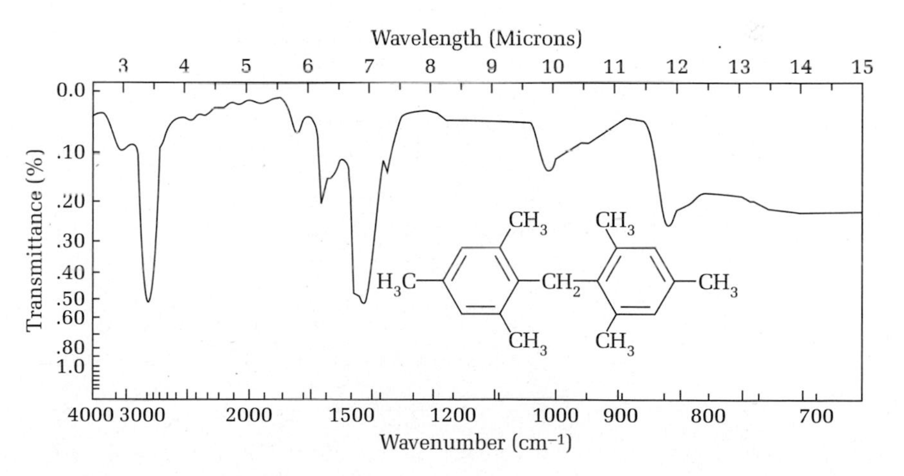

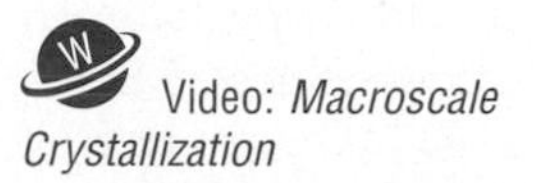

every 100 mg of paraformaldehyde. After 20 minutes, it can be flushed down the drain. The formic acid solvent from the reaction and aqueous washings should be combined, neutralized with sodium carbonate, and flushed down the drain. Mother liquor from recrystallization is placed in the organic solvents waste container.

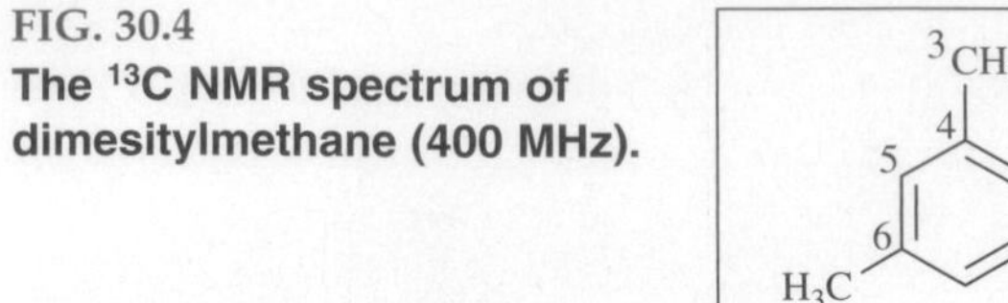

FIG. 30.4
The ^{13}C NMR spectrum of dimesitylmethane (400 MHz).

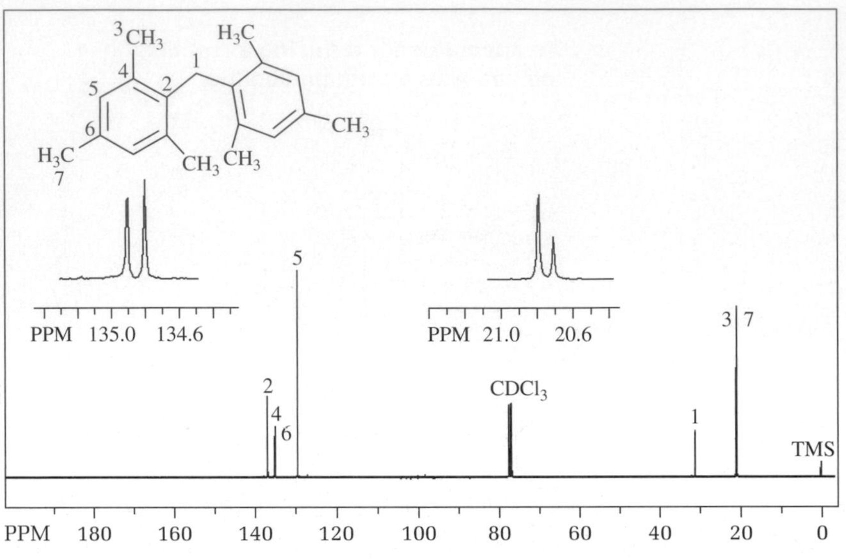

FIG. 30.5
The ^1H NMR spectrum of dimesitylmethane (100 MHz).

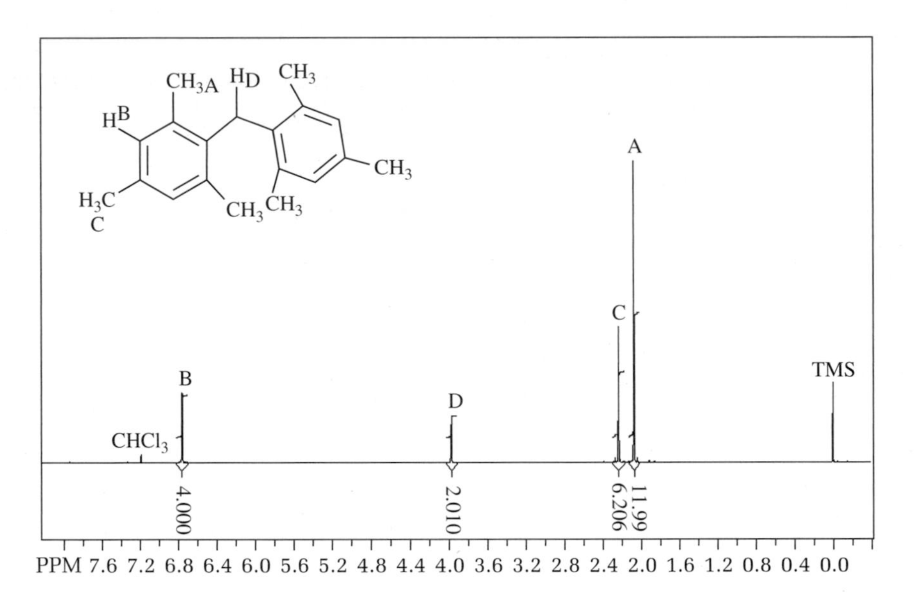

QUESTIONS

For Additional Experiments, sign in at this book's premium web site at **www.cengage.com/login.**

1. The intermediate benzylic carbocation is stabilized by resonance. Draw the contributing resonance structures.
2. Draw the structure of dimesitylmethane, designate on its NMR spectrum all the hydrogens that are equivalent, and predict the relative intensities of all the peaks corresponding to these sets of equivalent hydrogens. The two types of methyls differ from each other by 0.15 ppm.

CHAPTER 31

The Friedel–Crafts Reaction: Anthraquinone and Anthracene

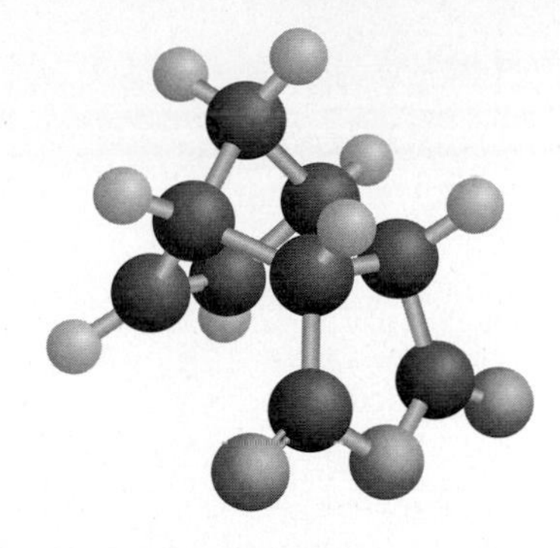

PRELAB EXERCISE: Draw the mechanism for the cyclization of 2-benzoylbenzoic acid to anthraquinone using concentrated sulfuric acid.

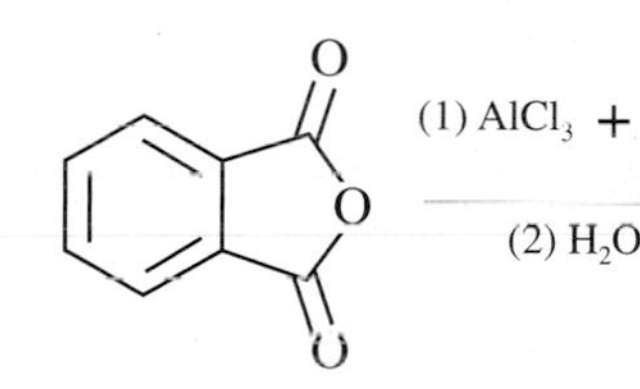

Phthalic anhydride
MW 148.11, mp 132°C

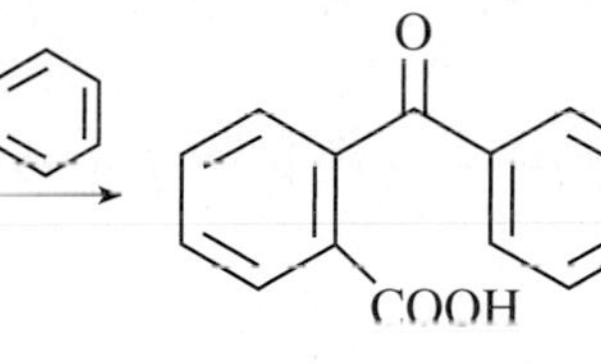

2-Benzoylbenzoic acid
MW 226.22, mp 127°C

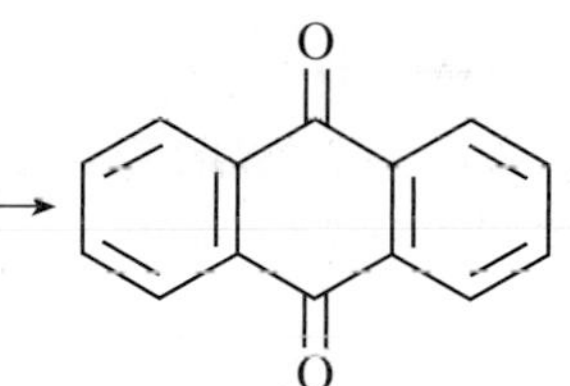

Anthraquinone
MW 208.20, mp 286°C

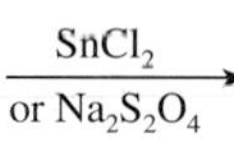

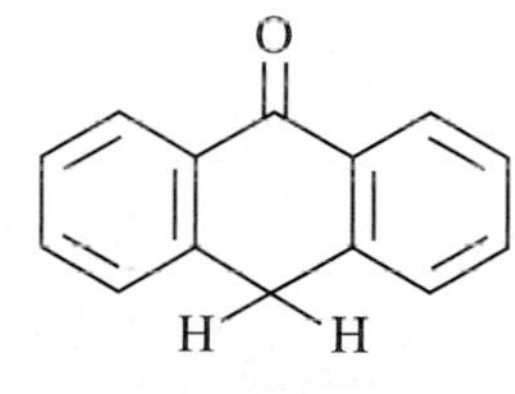

Anthrone
MW 194.22, mp 156°C

Anthracene
MW 178.22, mp 216°C

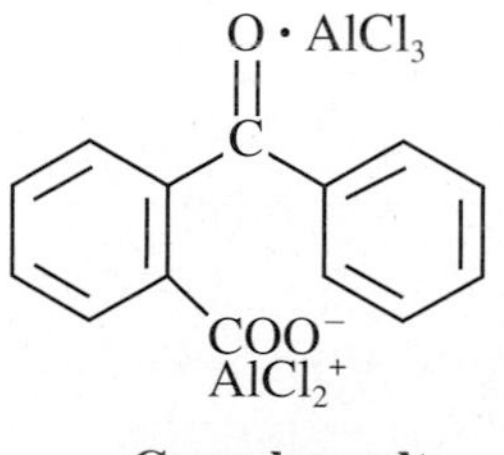

Complex salt

The Friedel-Crafts reaction of phthalic anhydride with excess benzene as solvent and two equivalents of aluminum chloride proceeds rapidly and gives a complex salt of 2-benzoylbenzoic acid in which 1 mole of aluminum chloride has reacted with the acid function to form the salt $RCO_2^-\ AlCl_2^+$; a second mole of aluminum chloride is bound to the carbonyl group. On addition of ice and hydrochloric acid, the complex is decomposed, and the basic aluminum salts are brought into solution (Experiments 1 and 2).

The treatment of 2-benzoylbenzoic acid with concentrated sulfuric acid effects cyclodehydration to anthraquinone, a pale-yellow, high-melting compound of great stability (Experiments 3 and 4). Because anthraquinone can be sulfonated (made to react with sulfuric acid) only under forcing conditions, a high temperature can be used to shorten the reaction time without loss in yield of product; the conditions are adjusted so that anthraquinone separates from the hot solution in crystalline form to favor rapid drying (Experiments 3 and 4).

Reduction of anthraquinone to anthrone (Experiments 5 and 6) can be accomplished rapidly on a small scale with tin(II) chloride in acetic acid solution. A second method, which involves refluxing anthraquinone with an aqueous solution of sodium hydroxide and sodium hydrosulfite, is interesting to observe because of the sequence of color changes: Anthraquinone is reduced first to a deep red solution containing anthrahydroquinone diradical dianion, and then the red color gives way to a yellow color characteristic of anthranol radical anion. As the alkali is neutralized by the conversion of $Na_2S_2O_4$ to 2 $NaHSO_3$, anthranol ketonizes to the more stable anthrone. The second method is preferred in industry because (1) sodium hydrosulfite costs less than half as much as tin(II) chloride, and (2) water is cheaper than acetic acid, and no solvent recovery problem is involved.

$Na_2S_2O_4$
Sodium hydrosulfite

Reduction of anthrone to anthracene (Experiments 7 and 8) is accomplished by refluxing in aqueous sodium hydroxide solution with activated zinc dust. This method has the merit of affording pure, beautifully fluorescent anthracene.

FIESER'S SOLUTION, AN OXYGEN SCAVENGER

A solution of 2 g of sodium anthraquinone-2-sulfonate and 15 g of sodium hydrosulfite in 100 mL of a 20% aqueous solution of potassium hydroxide produces a blood-red solution of the diradical dianion:

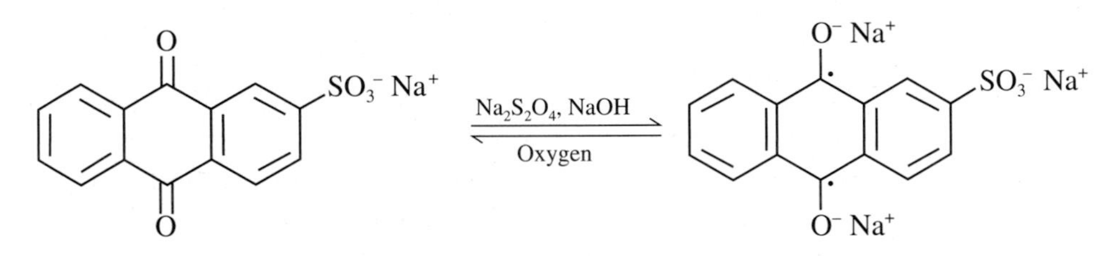

This solution has a remarkable affinity for oxygen. It is used to remove traces of oxygen from gases such as nitrogen or argon when it is desirable to render them absolutely oxygen-free. This solution has a capacity of about 800 mL of oxygen. The color fades, and the solution turns brown when it is exhausted.

EXPERIMENTS

1. 2-BENZOYLBENZOIC ACID

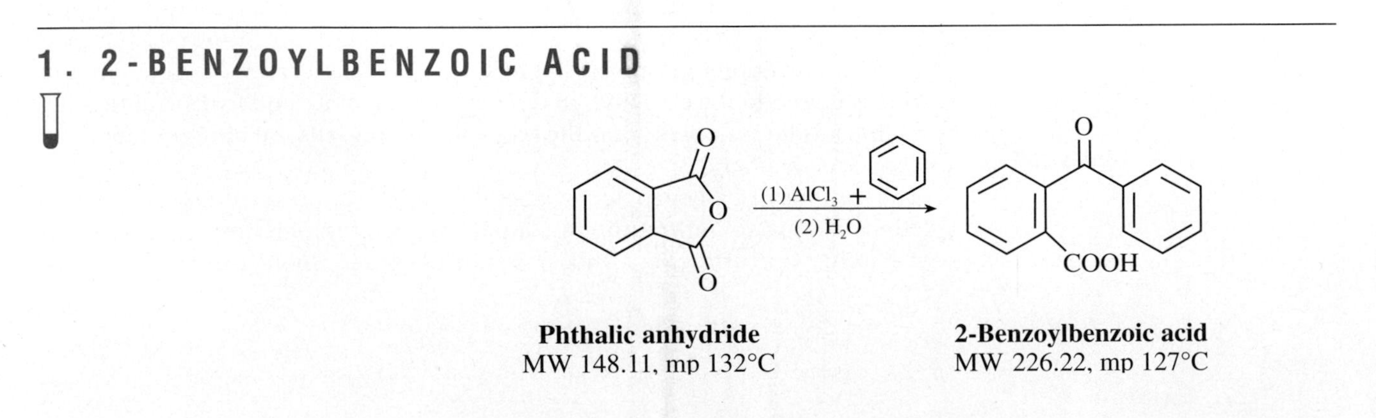

Phthalic anhydride
MW 148.11, mp 132°C

2-Benzoylbenzoic acid
MW 226.22, mp 127°C

IN THIS EXPERIMENT, an aromatic anhydride reacts with benzene and aluminum chloride to give 2-benzoylbenzoic acid and hydrogen chloride gas, which is trapped. Careful temperature control is needed for this exothermic reaction. At the end of the reaction, first ice, then water and hydrochloric acid are added, and the product is extracted into ether. The ether is dried and evaporated to a small volume (some benzene is still present), at which point hexanes is added. The product crystallizes and is isolated by filtration.

CAUTION: Benzene is a carcinogen. Carry out this experiment in a hood. Avoid skin contact with the benzene by wearing gloves. Do not carry out this reaction if adequate hood facilities are unavailable.

Photo: *Gas Trap*; Video: *The Grignard Reaction: Mixing Reaction Mixture by Flicking Reaction Tube*

Photos: *Extraction with Ether; Use of the Wilfilter*, Video: *Microscale Filtration on the Hirsch Funnel*

Into a dry 10 × 100 mm reaction tube, place 150 mg of phthalic anhydride and 0.75 mL of dry benzene. Cool the mixture in an ice bath, and then add 300 mg of anhydrous aluminum chloride.[1] Cap the tube with a septum that is connected to polyethylene tubing leading to another reaction tube, which contains a piece of damp cotton to act as a trap for the hydrogen chloride liberated in the reaction (Fig. 31.1). Figure 31.2 shows how to thread a polyethylene tube through a septum. Mix the contents of the tube thoroughly by flicking the tube. Warm the tube by the heat of your hand. If the reaction does not start, warm the tube *very gently* in a beaker of hot water, or hold it over a sand bath or a steam bath for a few seconds. At the first sign of vigorous boiling or evolution of hydrogen chloride, hold the tube over the ice bath in readiness to cool it if the reaction becomes too vigorous. Continue this gentle, cautious heating until the reaction proceeds smoothly enough to reflux it on a hot sand bath. This will take about 5 minutes.

Heat the reaction mixture on the sand bath until evolution of hydrogen chloride almost ceases; then cool it in ice and add 1 g of ice in small pieces. Allow each little piece to react before adding the next. Using a glass rod, mix the reaction mixture well during this hydrolysis. After the reaction subsides, add 0.3 mL of concentrated hydrochloric acid and 0.5 mL of water. Mix the contents of the tube thoroughly, and be sure the mixture is at room temperature. Add 0.5 mL more water, mix the solution well, ascertain that it is at room temperature, and then add 1.5 mL of ether and break up any lumps in the tube with a stirring rod. Mix thoroughly by drawing up the liquid into a Pasteur pipette and rapidly expelling it several times, or by stoppering the tube and shaking it vigorously to complete the hydrolysis and extraction. Allow the layers to separate and then remove the aqueous layer with a Pasteur pipette. Add 0.1 mL of concentrated hydrochloric acid and 0.25 mL of water. Pipette mix as above, or stopper and shake the mixture vigorously again and remove the aqueous layers. Transfer the organic layers to a small test tube containing calcium chloride pellets. Allow the solution to dry for 5 minutes; then transfer it back to the clean dry reaction tube, rinsing the drying agent with a small quantity of ether. Add a boiling chip and boil off the solvents until the volume in the tube is 0.5 mL; then add hexanes until the solution is slightly turbid, indicating that the product is beginning to crystallize. Allow the product to crystallize at room temperature and then at 0°C. Collect the crystals on a Hirsch funnel or on a Wilfilter by centrifugation. This product may be the monohydrate. Heat it at 100°C to drive off any water. Pure 2-benzoylbenzoic acid melts at 127–128°C; the yield should be about 0.2 g.

[1]Aluminum chloride should be weighed in a small, dry, stoppered vial or reaction tube. The aluminum chloride should be from a freshly opened bottle and should be weighed and transferred in the hood. It is very hygroscopic and releases hydrogen chloride upon reaction with water (including humidity in the air). Weigh the reagent rapidly to avoid exposure of the compound to air. The quality of the aluminum chloride determines the success of this experiment.

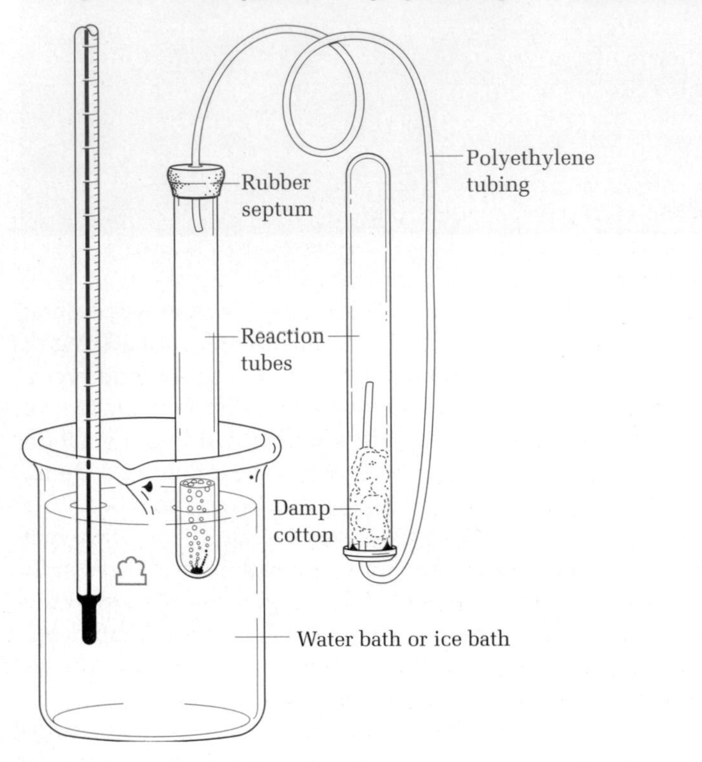

FIG. 31.1
A hydrogen chloride trap for the Friedel–Crafts reaction. The tubing can be bent permanently by heating it in a steam bath.

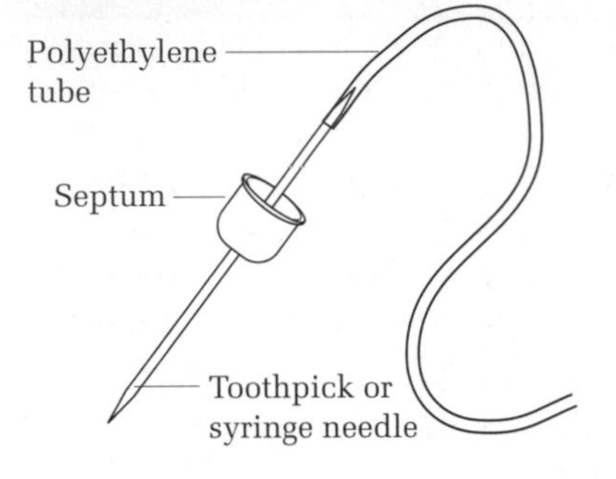

FIG. 31.2
To thread a polyethylene tube through a septum, make a hole through the septum with a needle and then push a toothpick through the hole. Push the polyethylene tube firmly onto the toothpick, and then pull and push on the toothpick to thread the tubing through the septum. Finally, pull the tube from the toothpick after it has come out the other side of the septum.

Celite may speed up the filtration of the aluminum hydroxide.

Cleaning Up. Combine all aqueous layers, neutralize with sodium carbonate, remove the aluminum hydroxide by filtration, and flush the filtrate down the drain. The aluminum hydroxide and calcium chloride pellets, after being freed of organic solvent (evaporation in the hood), can be placed in the nonhazardous solid waste container. If local regulations do not allow for the evaporation of solvents in a hood, dispose of the wet pellets in a special waste container. Benzene, ether, and hexanes go in the organic solvents waste container.

2. 2-BENZOYLBENZOIC ACID

CAUTION: Benzene is a carcinogen. Carry out this experiment in a hood. Avoid skin contact with the benzene by wearing gloves. Do not carry out this reaction if adequate hood facilities are unavailable.

This Friedel–Crafts reaction is conducted in a 100-mL round-bottomed flask equipped with a short condenser. A trap for collecting liberated hydrogen chloride is connected to the top of the condenser by rubber tubing of sufficient length to make it possible either to heat the flask on a steam bath or to plunge it into an ice bath. The trap is a suction flask half-filled with water and fitted with a delivery tube inserted to within 1 cm of the surface of the water (*see* Fig. 18.3 on page 332).

Three grams of phthalic anhydride and 15 mL of reagent grade benzene are placed in the flask, and this solution is cooled in an ice bath until the benzene begins to crystallize. Ice cooling serves to moderate the vigorous reaction, which otherwise might be difficult to control. Six grams of anhydrous aluminum chloride[2] are added, the condenser and trap are connected, and the flask is shaken well and warmed for a few minutes by the heat of your hand. If the reaction does not start, the flask is warmed *very gently* by holding it for a few seconds over a steam bath. At the first sign of vigorous boiling, or the evolution of hydrogen chloride, the flask is held over an ice bath in readiness to cool it if the reaction becomes too vigorous. This gentle, cautious heating is continued until the reaction is proceeding smoothly enough to be refluxed on a steam bath. This point is reached in about 5 minutes.

Short reaction period, high yield

Continue heating on a steam bath, swirl the mixture, and watch it carefully for sudden separation of the addition compound; the heat of crystallization is such that it may be necessary to plunge the flask into an ice bath to moderate the process. Once the addition compound has separated as a thick paste, heat the mixture for an additional 10 minutes on a steam bath, remove the condenser, and swirl the flask in an ice bath until cold. (Should no complex separate, heat for an additional 10 minutes, and then proceed as directed.) Take the flask and ice bath to the hood, weigh out 20 g of ice, add a few small pieces of ice to the mixture, swirl and cool as necessary, and wait until the ice has reacted before adding more. After the 20 g of ice have been added and the decomposition reaction has subsided, add 4 mL of concentrated hydrochloric acid and 20 mL of water, swirl vigorously, and be sure that the mixture is at room temperature. Then add 10 mL of water, swirl vigorously, and again be sure the mixture is at room temperature. Add 10 mL of ether and, with a flattened stirring rod, dislodge any solid from the neck and walls of the flask and break up lumps at the bottom. To further promote hydrolysis of the addition compound, extraction of the organic product, and solution of basic aluminum halides, stopper the flask with a cork and shake vigorously for several minutes.

When most of the solid has disappeared, pour the mixture through a funnel into a separatory funnel until the separatory funnel is nearly filled. Discard the lower aqueous layer. Pour the rest of the mixture into the separatory funnel, rinse the reaction flask with fresh ether, and again drain off the aqueous layer. To reduce the fluffy, dirty precipitate that appears at the interface, add 2 mL of concentrated hydrochloric acid and 5 mL of water, shake vigorously for 2–3 minutes, and drain off the aqueous layer. If some interfacial dirty emulsion still persists, decant the benzene-ether solution through the mouth of the separatory funnel into filter paper flute-folded for gravity filtration and use fresh ether to rinse the separatory funnel. Clean the separatory funnel and pour in the filtered benzene-ether solution. Shake the solution with a portion of dilute hydrochloric acid; then isolate the reaction product by either of the following procedures:

Alternative procedures

Discard benzene-ether filtrates in the container provided.

1. Add 10 mL of 3 *M* sodium hydroxide solution, shake thoroughly, and separate the aqueous layer.[3] Extract with a further 5-mL portion of aqueous alkali and combine the extracts. Wash with 2 mL of water and add this aqueous solution to the 15 mL of aqueous extract already collected. Discard the benzene-ether solution. Acidify the 17 mL of combined alkaline extract with concentrated hydrochloric acid to pH 1–2 and, if the *o*-benzoylbenzoic acid separates as an oil, cool in ice and rub the walls of the flask with a stirring rod to induce crystallization of the hydrate; collect the product and wash it well with water.

[2]The aluminum chloride is best weighed in a stoppered test tube. The chloride should be from a freshly opened bottle, and should be weighed and transferred in the hood. It is very hygroscopic; work rapidly to avoid exposure of the compound to air. It releases hydrogen chloride on exposure to water (including humidity in the air). The quality of the aluminum chloride determines the success of this experiment.

[3]The nature of the yellow pigment that appears in the first alkaline extract is unknown; the impurity is apparently transient because the final product dissolves in alkali to give a colorless solution.

This material is the monohydrate $C_6H_5COC_6H_4CO_2H \cdot H_2O$. To convert it into anhydrous *o*-benzoylbenzoic acid, put it in a tared, 50-mL round-bottomed flask, evacuate the flask at the full force of the aspirator, and heat it in the open rings of a steam bath or boiling water bath, covering the flask with a towel. Check the weight of the flask and contents for constancy after 45 minutes, 1 hour, and 1.25 hours. The yield is usually about 4 g (mp 126–127°C).

2. Filter the benzene-ether solution through anhydrous calcium chloride pellets for superficial drying, put it into a 50-mL round-bottomed flask, and use a rotary evaporator or distill over a steam bath or hot water bath through a condenser into an ice-cooled receiver until the volume in the distilling flask is reduced to about 11 mL. Add hexanes slowly until the solution is slightly turbid and let the product crystallize first at 25°C and then at 5°C. The yield of anhydrous, colorless, well-formed crystals (mp 127–128°C) is about 4 g.

Drying time: about 1 hour

Cleaning Up. Combine all aqueous layers, neutralize with sodium carbonate, remove the aluminum hydroxide by filtration, and flush the filtrate down the drain. The aluminum hydroxide and calcium chloride pellets, after being freed of organic solvent (evaporation in the hood), can be placed in the nonhazardous solid waste container. If local regulations do not allow for the evaporation of solvents in a hood, dispose of the wet pellets in a special waste container. Benzene, ether, and hexanes go in the organic solvents waste container.

3. ANTHRAQUINONE

Anthraquinone can be purified by sublimation.

Vapor pressure of anthraquinone

Temperature (°C)	*Pressure (mm Hg)*
200	1.8
220	4.4
240	12.6
260	33.0

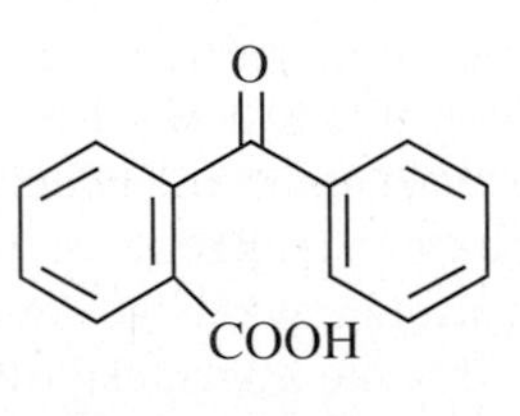

2-Benzoylbenzoic acid
MW 226.22
mp 127°C

Sulfuric acid →

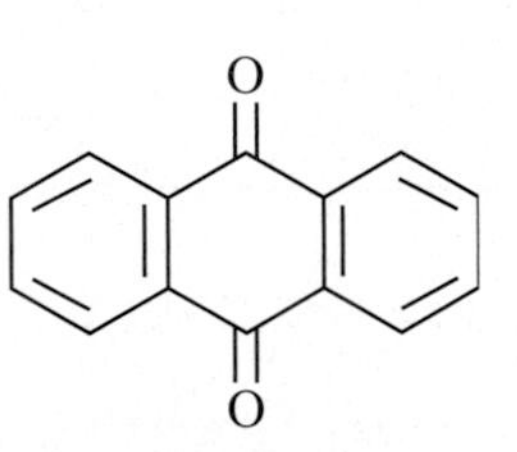

Anthraquinone
MW 208.20
mp 286°C

IN THIS EXPERIMENT, 2-benzoylbenzoic acid is dissolved in sulfuric acid and heated to cyclize the material to anthraquinone. With great caution, water is added in microdrops to the mixture, which causes the product to crystallize. It is isolated by filtration, and then heated with base to remove the unreacted carboxylic acid as a water-soluble ammonium salt. The pure material is collected by filtration, washed, and dried.

In a reaction tube, dissolve 100 mg of 2-benzoylbenzoic acid[4] in 0.5 mL of concentrated sulfuric acid by gently heating and stirring. Immerse a thermometer in the reaction mixture and heat it at 150–155°C for 45 minutes. Allow the tube to cool to

[4]Available from the Aldrich Chemical Company.

FIG. 31.3
A Wilfilter filtration apparatus. Refer to Chapter 4 for instructions on use.

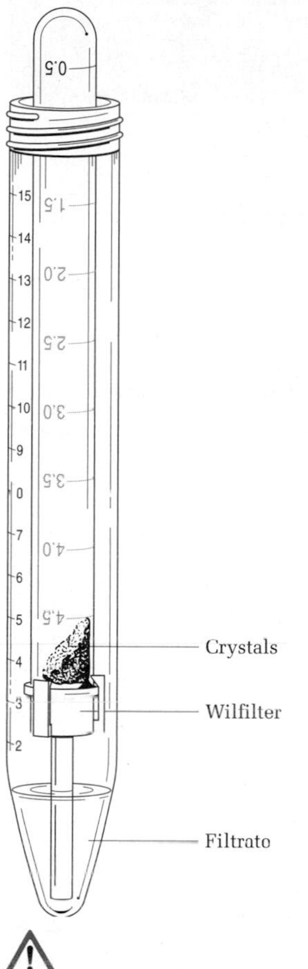

below 100°C and then, using extreme caution, add a very small drop of water to the mixture. Mix the contents of the tube and continue adding water in minute drops to the mixture. This will cause the product to crystallize and, if done slowly enough, the crystals will be large enough to collect easily by filtration. Fill the tube with water and collect the product by filtration on a Hirsch funnel or a Wilfilter (Fig. 31.3). Return the damp product to the reaction tube, and boil the product with 0.5 mL of concentrated ammonium hydroxide to remove unreacted starting material. Filter, wash the product well with water and then with acetone, dry, determine the weight, and then calculate the percent yield. Do not determine the melting point because it is so high (286°C).

Cleaning Up. The aqueous filtrate, after neutralization with sodium carbonate, is diluted with water and flushed down the drain.

4. ANTHRAQUINONE

Place 4.0 g of 2-benzoylbenzoic acid[5] (anhydrous) in a 125-mL round-bottomed flask, add 20 mL of concentrated sulfuric acid, and heat on a hot water or steam bath with swirling until the solid is dissolved. Then clamp the flask over a microburner, insert a thermometer, raise the temperature to 150°C, and heat to maintain a temperature of 150–155°C for 5 minutes. Let the solution cool to 100°C, remove the thermometer after letting it drain, and, with a Pasteur pipette, add 4 mL of water by drops with swirling to keep the precipitated material dissolved as long as possible so that it will separate as small, easily filtered crystals. Let the mixture cool further, dilute with water until the flask is full, let cool again, collect the product by suction filtration, and wash well with water. Then remove the filtrate, wash the filter flask, return the funnel to the filter flask without applying suction, and test the filter cake for unreacted starting material as follows: Dilute 8 mL of concentrated ammonia solution with 40 mL of water, pour the solution onto the filter, and loosen the cake so that it is well leached. Then apply suction, wash the cake with water, and acidify a few milliliters of the filtrate. If there is no precipitate upon acidifying, the yield of anthraquinone should be close to the theoretical because it is insoluble in water. Dry the product to constant weight, but do not determine the melting point because it is so high (mp 286°C).

Handle hot sulfuric acid with care: highly corrosive

Reaction time: 5 minutes

Video: *Microscale Filtration on the Hirsch Funnel*

Cleaning Up. The aqueous filtrate, after neutralization with sodium carbonate, is diluted with water and flushed down the drain.

5. ANTHRONE

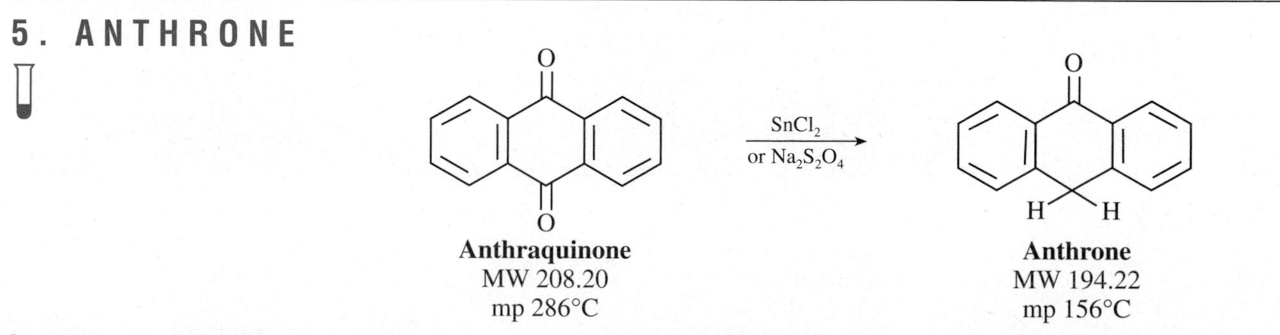

[5]Available from the Aldrich Chemical Company.

IN THIS EXPERIMENT, anthraquinone can be reduced with two different reagents. In the first reaction, acetic acid and hydrochloric acid are the solvents; in the second reaction, the solvent is water. In both reactions, the product crystallizes from the reaction mixture. Compare the two methods of reduction in the microscale experiment.

Procedure 1: Tin(II) Chloride Reduction

$SnCl_2 \cdot 2H_2O$
Tin(II) chloride dehydrate

In a reaction tube, put 50 mg of crude anthraquinone from the previous experiment, 0.40 mL of acetic acid, and a solution made by warming 0.13 g of tin(II) chloride dihydrate with 0.13 mL of concentrated hydrochloric acid. Add a boiling stone, note the time, and reflux gently until crystals of anthraquinone have completely disappeared (8–10 minutes); reflux for an additional 15 minutes and record the total time. Then add water (about 0.12 mL) dropwise until the solution is saturated. Let the solution stand for crystallization. Collect the product using a Wilfilter (Fig. 31.3), dry it, and determine the melting point. The yield of pale-yellow crystals is about 40 mg.

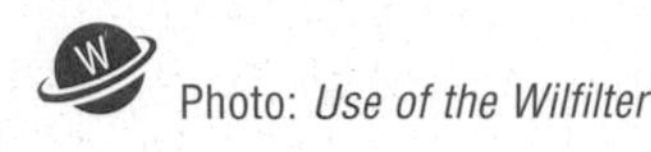
Photo: *Use of the Wilfilter*

Cleaning Up. Dilute the filtrate with a large volume of water and flush the solution down the drain.

Procedure 2: Hydrosulfite Reduction

$Na_2S_2O_4$
Sodium hydrosulfite
Decomposes on storage

In a 5-mL long-necked, round-bottomed flask, put 50 mg of anthraquinone, 60 mg of sodium hydroxide, 150 mg of fresh sodium hydrosulfite, and 1.3 mL of water. Heat over a hot sand bath and swirl for a few minutes to convert the anthraquinone to the deep-red anthrahydroquinone anion. Note that particles of different appearance begin to separate even before the anthraquinone has all dissolved. Add an empty distilling column as an air condenser to the flask, and reflux for 45 minutes. Then cool the product, and filter it using a Hirsch funnel or a Wilfilter (Fig. 31.3), wash it well with water, and let dry. Determine the weight and melting point of the crude material and crystallize it from 95% ethanol. Record the approximate volume of solvent used and, if the first crop of crystals recovered is not satisfactory, concentrate the mother liquor and secure a second crop. The usual yield is about 40 mg.

Video: *Microscale Filtration on the Hirsch Funnel*; Photo: *Use of the Wilfilter*

Cleaning Up. The aqueous filtrate is treated with household bleach (sodium hypochlorite) until the reaction is complete. Remove the solid by filtration. It may be placed in the nonhazardous solid waste container. The filtrate can be diluted with water and flushed down the drain.

6. ANTHRONE

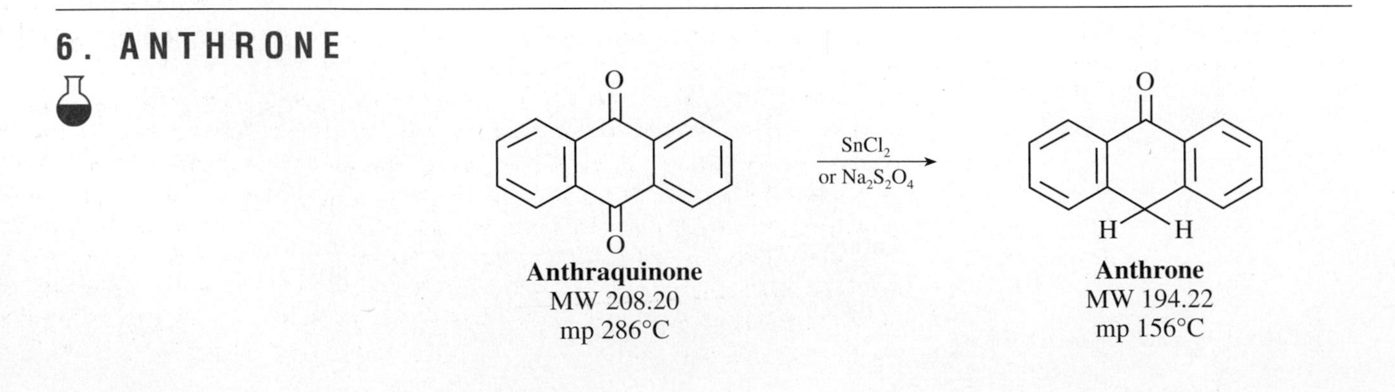

Procedure 1: Tin(II) Chloride Reduction

$SnCl_2 \cdot 2H_2O$

Tin(II) chloride dehydrate

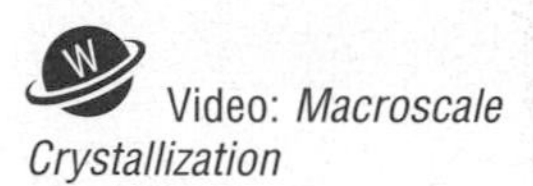
Video: *Macroscale Crystallization*

In a 50-mL round-bottomed flask provided with a reflux condenser, put 2.0 g of anthraquinone, 16 mL of acetic acid, and a solution made by warming 5.3 g of tin(II) chloride dihydrate with 5.3 mL of concentrated hydrochloric acid. Add a boiling stone to the reaction mixture, note the time, and reflux gently until crystals of anthraquinone have completely disappeared (8–10 minutes); then reflux for an additional 15 minutes and record the total time. Disconnect the flask, heat it on a steam bath, and add water (about 5 mL) in 0.5-mL portions until the solution is saturated. Let the solution stand for crystallization. Collect and dry the product, and determine the melting point. The yield of pale-yellow crystals is about 1.7 g.

Cleaning Up. Dilute the filtrate with a large volume of water and flush the solution down the drain.

Procedure 2: Hydrosulfite Reduction

$Na_2S_2O_4$

Sodium hydrosulfite

Decomposes on storage

Video: *Microscale Filtration on the Hirsch Funnel*

Into a 250-mL round-bottomed flask, which can be heated under reflux, put 2.0 g of anthraquinone, 2.4 g of sodium hydroxide, 6 g of fresh sodium hydrosulfite, and 50 mL of water. Heat over a free flame and swirl for a few minutes to convert the anthraquinone into the deep-red anthrahydroquinone anion. Note that particles of different appearance begin to separate even before all of the anthraquinone has dissolved. Arrange for refluxing and reflux for 45 minutes; cool, filter the product, wash it well with water, and let dry. Note the weight of the product and melting point of the crude material, and crystallize it from 95% ethanol; the solution may require filtration to remove insoluble impurities. Record the approximate volume of solvent used and, if the first crop of crystals recovered is not satisfactory, concentrate the mother liquor and secure a second.

Cleaning Up. The aqueous filtrate is treated with household bleach (sodium hypochlorite) until the reaction is complete. Remove the solid by filtration. It may be placed in the nonhazardous solid waste container. The filtrate can be diluted with water and flushed down the drain.

Comparison of Results

Compare your results with those obtained by other students using an alternative procedure with respect to yield, quality of product, and working time. Which is the better laboratory procedure? Then consider the cost of the three solvents concerned, the cost of the two reducing agents (prices can be found in a current catalog), the relative ease of recovery of the organic solvents, and the prudent disposal of byproducts, and decide which method would be preferred as a manufacturing process.

7. FLUORESCENT ANTHRACENE

Zn dust (Cu)

Anthrone
MW 194.22, mp 156°C

Anthracene
MW 178.22, mp 216°C

IN THIS EXPERIMENT, the ketone anthrone is reduced to the hydrocarbon anthracene using activated zinc dust and sodium hydroxide. The product crystallizes from the reaction mixture, but the excess zinc must be destroyed with hydrochloric acid before the product is isolated. After filtration, the anthracene is recrystallized from toluene.

Photo: *Filtration Using a Pasteur Pipette*; Videos: *Filtration of Crystals Using the Pasteur Pipette; Microscale Filtration on the Hirsch Funnel*

Put 100 mg of zinc dust into a 5-mL round-bottomed flask and activate the dust by adding 0.6 mL of water and 0.1 mL of a 5% aqueous copper(II) sulfate solution, and swirl for a minute or two (or use a magnetic stirrer and stirring bar). Add 40 mg of anthrone, 100 mg of sodium hydroxide, and 1 mL of water, heat to boiling on a sand bath, note the time, and start refluxing the mixture (Fig. 31.4). Anthrone at first dissolves as the yellow anion of anthranol, but anthracene soon begins to separate as a white precipitate.

Hydrogen liberated

Methanol removes water without dissolving much anthracene.

In about 15 minutes, the yellow color initially observed on the walls disappears, but refluxing should be continued for a full 30 minutes. Then remove the heat, rinse down any anthracene in the condenser with a few drops of water, and cool the flask in ice. Remove the solvent with a Pasteur pipette and very carefully add 0.2 mL of concentrated hydrochloric acid dropwise to dissolve the remaining zinc; the hydrogen gas given off causes foaming. Again, cool the mixture and add water to wash down any unreacted zinc. After all the zinc has reacted, remove the acid and wash the product liberally with water. Collect the product on a Hirsch funnel and then wash once with ice-cold methanol to remove as much of the water as possible from the product.

Photos: *Crystallization; Use of the Wilfilter*; Video: *Recrystallization*

The product need not be dried before recrystallization from about 0.5 mL of boiling toluene. Slow cooling of the toluene solution will give thin, colorless, beautiful fluorescent plates. The yield should be about 20 mg. Collect the product on a Wilfilter (Fig. 31.3). In washing the equipment with acetone, you should be able to

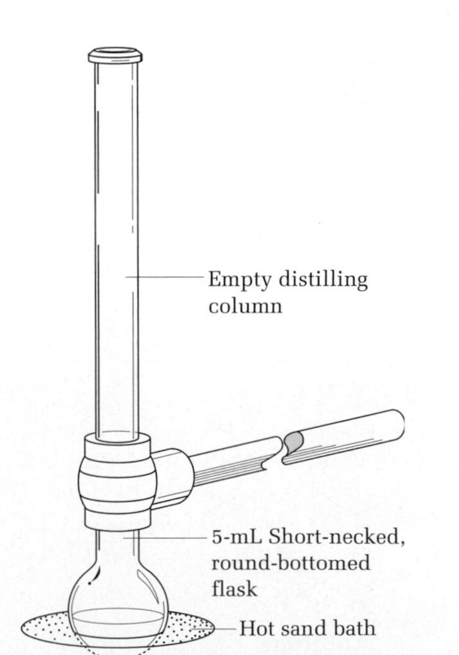

FIG. 31.4
An apparatus for refluxing the reaction mixture in a reaction flask.

observe the striking fluorescence of very dilute solutions. The fluorescence is quenched by a bare trace of impurity.

Anthracene can also be recrystallized from ethanol. *See* Chapter 4, Experiment 2.

Cleaning Up. The aqueous solution is diluted with water, neutralized with sodium carbonate, and filtered under vacuum to remove zinc hydroxide; the filtrate is diluted with water and flushed down the drain. The solid can be placed in the nonhazardous solid waste container. Toluene and acetone are placed in the organic solvents waste container.

8. FLUORESCENT ANTHRACENE

Put 3.8 g of zinc dust into a 125-mL round-bottomed flask and activate the dust by adding 23 mL of water and 0.4 mL of copper(II) sulfate solution, and swirling for a minute or two. Add 1.5 g of anthrone, 3.8 g of sodium hydroxide, and 40 mL of water; attach a reflux condenser, heat to boiling, note the time, and start refluxing the mixture. Anthrone at first dissolves as the yellow anion of anthranol, but anthracene soon begins to separate as a white precipitate. In about 15 minutes, the yellow color initially observed on the walls disappears, but refluxing should be continued for a full 30 minutes. Then stop the heating, use a water wash bottle to rinse down anthracene that has lodged in the condenser, and filter the still-hot mixture on a large Büchner funnel. It usually is possible to decant from, and so remove, a mass of zinc.

Reflux time: 30 minutes

After a liberal washing with water, blow or shake out the gray cake into a 250 mL beaker and rinse the funnel and paper with water. To remove most of the zinc metal and zinc oxide, add 8 mL of concentrated hydrochloric acid and heat with stirring for 20–25 minutes, when initial frothing due to liberated hydrogen should have ceased. Collect the now nearly white precipitate on a Büchner funnel, liberally wash with water, release the suction, rinse the walls of the funnel with methanol, use enough methanol to cover the cake, and then apply suction again. Wash again with enough methanol to cover the cake, and then remove the solvent thoroughly by suction.

Remove all sources of ignition. Hydrogen is liberated.

Video: *Macroscale Crystallization*

The product need not be dried before recrystallization from toluene. Transfer the methanol-moist material to a 50-mL Erlenmeyer flask and add 22 mL of toluene; a liberal excess of solvent is used to avoid crystallization in the funnel. Be sure there are no flames nearby. Heat the mixture on a hot plate to bring the anthracene into solution; there will be a small residue of zinc remaining. Filter by gravity using the usual technique to remove zinc. From the filtrate, anthracene is obtained as thin, colorless, beautiful fluorescent plates. The yield should be about 1 g. In washing the equipment with acetone, you should be able to observe the striking fluorescence of very dilute solutions. The fluorescence is quenched by a bare trace of impurity.

Methanol removes water without dissolving much anthracene.

CAUTION: Toluene is a highly flammable solvent.

Cleaning Up. The aqueous solution is diluted with water, neutralized with sodium carbonate, and filtered under vacuum to remove zinc hydroxide. The filtrate is diluted with water and flushed down the drain. The solid waste can be placed in the nonhazardous solid waste container. Toluene and acetone are placed in the organic solvents waste container.

QUESTIONS

1. Calculate the number of moles of hydrogen chloride liberated in the synthesis of 2-(4-benzoyl)benzoic acid. If this gaseous acid were dissolved in water, hydrochloric acid would be formed. How many milliliters of concentrated hydrochloric acid would be formed in this reaction? The concentrated acid is 12 *M* HCl.
2. Write a mechanism for the formation of anthraquinone from 2-benzoylbenzoic acid, clearly indicating the role of sulfuric acid. What is the name commonly given to this type of reaction?

CHAPTER 32

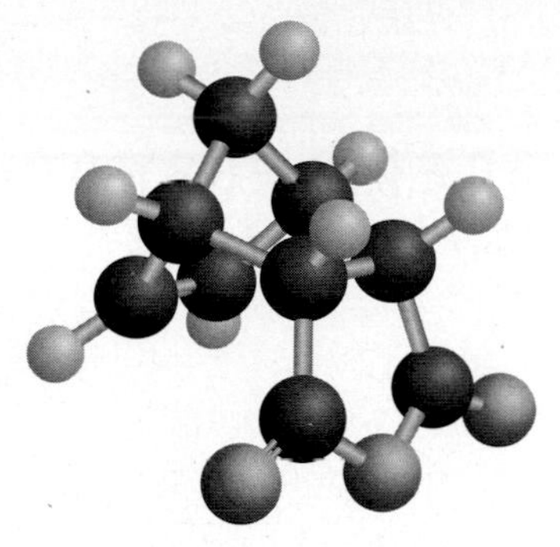

Friedel–Crafts Acylation of Ferrocene: Acetylferrocene

PRELAB EXERCISE: How many possible isomers could exist for diacetylferrocene? Explain. Calculate the volume of 3 *M* aqueous sodium hydroxide needed to neutralize the acetic and phosphoric acids in the synthesis of acetylferrocene.

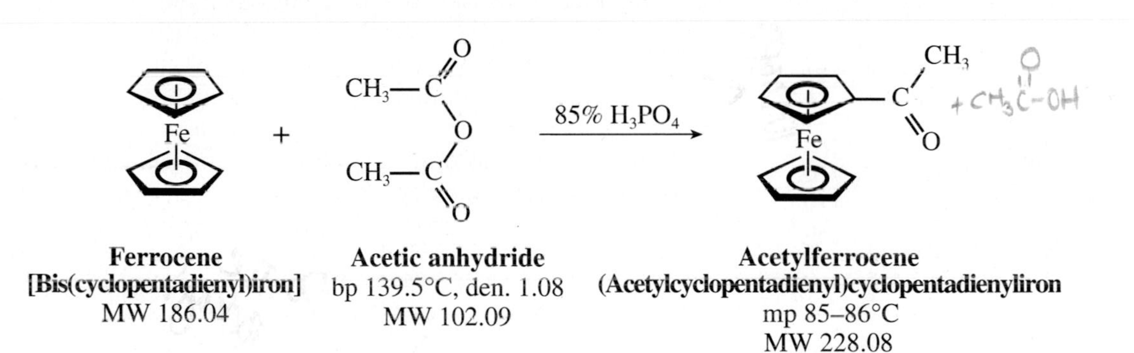

The Friedel–Crafts acylation of benzene requires aluminum chloride as the catalyst. However, ferrocene, with a very high π-electron density that has been referred to as a superaromatic compound, can be acylated under much milder conditions with phosphoric acid as a catalyst. Because the acetyl group is a deactivating substituent, the addition of a second acetyl group, which requires more vigorous conditions, will occur in the nonacetylated cyclopentadienyl ring to give 1,1'-diacetylferrocene. Because ferrocene gives just one monoacetyl derivative and just one diacetyl derivative, it was assigned an unusual sandwich structure.

Acetylferrocene and ferrocene (both highly colored) are easily separated by column chromatography (*see* Chapter 9).

EXPERIMENT

ACETYLFERROCENE

Microscale Procedure

IN THIS EXPERIMENT, ferrocene reacts with acetic anhydride in an acid-catalyzed Friedel-Crafts reaction to give aetylferrocene. Excess acetic anhydride is removed by reaction with water to give acetic acid, and then base is added to react with the acetic acid. The crude product is isolated by filtration and then chromatographed on alumina. There is obvious separation of these components on a chromatography column because the unreacted ferrocene is yellow, and the acetyl product is red-orange.

⚠ Handle acetic anhydride and phosphoric acid with care. Wipe up spills immediately.

To 93 mg of dry sublimed ferrocene (*see* Chapter 49) in a 10 × 100 mm reaction tube, add 0.35 mL (0.38 g) of acetic anhydride followed by 0.1 mL (170 mg) of 85% phosphoric acid. Cap the tube with a septum bearing an empty syringe needle, and warm it on a steam bath or in a beaker of boiling water while agitating the mixture to dissolve the ferrocene. Heat the mixture for an additional 10 minutes, and then cool the tube thoroughly in ice. Carefully add to the solution 0.5 mL of ice-water dropwise with thorough mixing, followed by the dropwise addition of a 3 *M* aqueous sodium hydroxide solution until the mixture is neutral (test with indicator paper and avoid an excess of base). Collect the product on a Hirsch funnel, wash it thoroughly with water, and press it as dry as possible between sheets of filter paper. Save a sample of this material for melting-point and thin-layer chromatographic (TLC) analyses, and purify the remainder by column chromatography.

Photo: *Capping a Reaction Tube with a Septum*; Video: *Microscale Filtration on the Hirsch Funnel*

Column Chromatography

Photo: *Column Chromatography*; Video: *Column Chromatography*

Follow the procedure as given in Chapter 9 for packing a microscale chromatography column and adding the sample (about 90 mg of a ferrocene/acetylferrocene mixture). Use Brockman activity III alumina as the adsorbent.

Elution

⚠ Hexanes are extremely flammable. No flames!

Carefully add hexanes to the column and begin to elute the product from the column. Unreacted ferrocene will move down the column as a yellow band. Collect this in a 10-mL flask. Wash any crystalline material that collects on the tip of the valve into the flask with a few drops of ether. Then elute the column with a 50:50 mixture of hexanes and ether. This will move the acetylferrocene down as an orange-red band. Collect this in a separate 10-mL Erlenmeyer flask. Ferrocene is highly symmetrical and nonpolar, while the acetyl compound is the opposite. These properties govern the order of elution from the chromatography column. Spot a TLC plate with these two solutions as well as the crude acetylferrocene and the ferrocene starting material, and analyze as described in the next section. Any diacetylferrocene will be seen as a dark band at the top of the column. Evaporate the solvents under reduced pressure and determine the weights of the residues (Fig. 32.1).

FIG. 32.1
The evaporation of a solvent under reduced pressure. Heat and swirling motion is supplied by one hand; the vacuum is controlled by the thumb of the other hand.

To aspirator

Flash Chromatography

Equilibrium between the solutes adsorbed on the alumina and the eluting solvent is established rapidly, so it does no harm to increase the flow rate of the solvent through the column. This can be done by applying pressure with a rubber bulb above the solvent. There is, however, a possibility that air will be sucked into the bottom of the column when the bulb is removed, which will destroy the uniform packing of the column. Practice this technique on a column that has been eluted in the usual way using the gravity flow of the solvent. On a macroscale, special fittings are available for applying air pressure to larger columns to speed up the elution process.

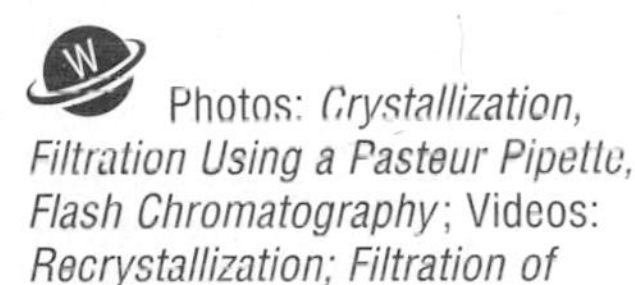
Photos: *Crystallization, Filtration Using a Pasteur Pipette, Flash Chromatography*; Videos: *Recrystallization; Filtration of Crystals Using the Pasteur Pipette*

Recrystallize the acetylferrocene from hot hexanes. Dissolve the residue in a filter flask in the minimum quantity of boiling solvent (about 1 mL), and transfer this hot solution with a Pasteur pipette to a reaction tube. Allow the tube to cool slowly to room temperature and then cool it in ice for at least 10 minutes. Acetylferrocene crystallizes as dark-red rosettes of needles. Remove the solvent with a Pasteur pipette and then, under vacuum, scrape out the product onto filter paper. After it is dry, determine the weight, calculate the percent yield, and determine the melting point.

Perform a TLC analysis on your product. Dissolve very small samples of pure ferrocene, the crude reaction mixture, and recrystallized acetylferrocene, each in a few drops of toluene. Spot the three solutions with microcapillaries on silica gel plates, and develop the chromatogram with 30:1 toluene and an absolute ethanol mixture. Visualize the spots under an ultraviolet (UV) lamp if the silica gel has a fluorescent indicator, or by adsorption of iodine vapor. Do you detect unreacted ferrocene in the reaction mixture and/or a spot that might be attributed to diacetylferrocene?

Cleaning Up. The reaction mixture filtrate can be flushed down the drain. Unused chromatography, recrystallization, and TLC solvents should be placed in the organic solvents waste container. The alumina from the column should be placed in the hood to allow the hexanes and ether to evaporate from it. Once free of organic solvents, the alumina can be placed in the nonhazardous solid waste container. If local regulations do not allow for the evaporation of solvents, dispose of the wet alumina in a waste container provided for the purpose.

Macroscale Procedure

IN THIS EXPERIMENT, ferrocene reacts with acetic anhydride in an acid-catalyzed Friedel-Crafts reaction to give acetylferrocene. Excess acetic anhydride is removed by reaction with water to give acetic acid, and then base is added to react with the acetic acid. The crude product is isolated by filtration and then purified by crystallization from hexane. There is obvious separation of these components by chromatography because the unreacted ferrocene is yellow, and the acetyl product is red-orange.

Both acetic anhydride and phosphoric acid are corrosive to tissue. Handle both with care and wipe up any spills immediately.

In a 25-mL round-bottomed flask, place 3.0 g of ferrocene, 10.0 mL of acetic anhydride, and 2.0 mL of 85% phosphoric acid. Equip the flask with a reflux condenser and a calcium chloride drying tube. Warm the flask gently on a hot water or steam bath while swirling to dissolve the ferrocene; then heat strongly for an additional 10 minutes. Pour the reaction mixture onto 50 g of crushed ice in a 400 mL beaker and rinse the flask with 10 mL of ice water. Stir the mixture for a few minutes with a glass rod, add 75 mL of 3 *M* sodium hydroxide solution (the solution should still be acidic), and then add solid sodium bicarbonate (be careful of foaming) until the remaining acid has been neutralized. Stir and crush all lumps, allow the mixture to stand for 20 minutes, and then collect the product by suction filtration. Press the crude material as dry as possible between sheets of filter paper, save a few crystals for TLC analysis, transfer the remainder to an Erlenmeyer flask, and add 40 mL of hexanes to the flask. Boil the solvent on a hot water or steam bath for a few minutes and then decant the dark-orange solution into another Erlenmeyer flask, leaving a gummy residue of polymeric material. Treat the solution with decolorizing charcoal and filter it through a fluted filter paper placed in a warm stemless funnel that is seated into an appropriately-sized Erlenmeyer flask. Evaporate the solvent (use an aspirator tube; *see* Fig. 8.11 on page 174) until the volume is about 20 mL. Set the flask aside to cool slowly to room temperature. Beautiful rosettes of dark orange-red needles of acetylferrocene will form. After the product has been cooled in ice, collect it on a Büchner funnel and wash the crystals with a small quantity of cold solvent. Pure acetylferrocene has a melting point of 84°C–85°C. Your yield should be about 1.8 g.

Hexanes are extremely flammable. No flames!

Video: *Macroscale Crystallization*

Perform a TLC analysis on your product. Dissolve very small samples of pure ferrocene, the crude reaction mixture, and recrystallized acetylferrocene, each in a few drops of toluene. Spot the three solutions with microcapillaries on silica gel plates, and develop the chromatogram with 3:1 toluene and an absolute ethanol mixture. Visualize the spots under a UV lamp, if the silica gel has a fluorescent indicator, or by adsorption of iodine vapor. Do you detect unreacted ferrocene in the reaction mixture and/or a spot that might be attributed to diacetylferrocene?

See *Web Links*

Cleaning Up. The reaction mixture filtrate can be flushed down the drain. Unused recrystallization and TLC solvents should be placed in the organic solvents waste container. The decolorizing charcoal, once free of solvent, can be placed in the nonhazardous solid waste container. If local regulations do not allow for the evaporation of solvents, dispose of the wet alumina in the waste container provided.

QUESTIONS

1. What is the structure of the intermediate species that attacks ferrocene to form acetylferrocene? What other organic molecule is formed?
2. Why does the second acetyl group enter the unoccupied ring to form diacetylferrocene?

33 CHAPTER

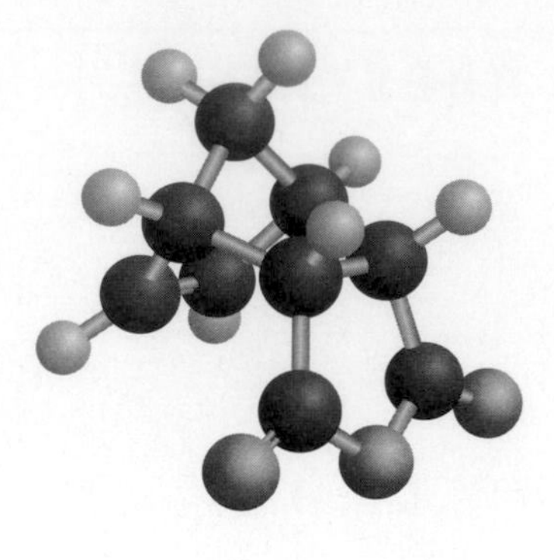

Reactions of Triphenylmethyl Carbocation, Carbanion, and Radical

PRELAB EXERCISE: Draw all of the resonance structures of the triphenylmethyl carbocation.

Triphenylmethanol, which is prepared in Chapter 38, has played an interesting role in the history of organic chemistry. It was converted to the first stable carbocation and the first stable free radical. In the experiments in this chapter, triphenylmethanol is easily converted to the triphenylmethyl (trityl) carbocation, carbanion, and radical. Each of these is stabilized by 10 contributing resonance forms and, consequently, is unusually stable. Because of their long conjugated systems, these forms absorb radiation in the visible region of the spectrum and thus can be detected visually.

Resonance forms contribute to stability.

EXPERIMENTS

THE TRIPHENYLMETHYL (TRITYL) CARBOCATION

The reactions of triphenylmethanol are dominated by the ease with which it dissociates to form the relatively stable triphenylmethyl carbocation. When colorless triphenylmethanol is dissolved in concentrated sulfuric acid, an orange-yellow solution results that gives a fourfold depression of the melting point of sulfuric acid, meaning that 4 moles of ions are produced per mole of triphenylmethanol. If the triphenylmethanol were simply protonated, only 2 moles of ions would result.

$$(C_6H_5)_3COH + 2\,H_2SO_4 \rightleftharpoons (C_6H_5)_3C^+ + H_3O^+ + 2\,HSO_4^-$$

$$(C_6H_5)_3COH + H_2SO_4 \not\rightleftharpoons (C_6H_5)_3COH_2^+ + HSO_4^-$$

The central carbon atom in the carbocation is sp^2 hybridized, and thus the three carbons attached to it are coplanar and disposed at 120° angles:

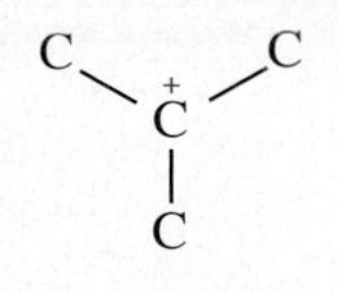

However, the three phenyl groups, because of steric hindrance, cannot lie on one plane. Therefore, the carbocation is propeller shaped:

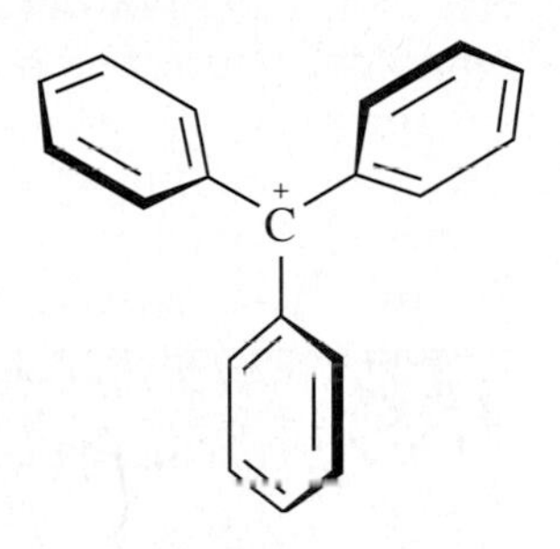

1. Trityl Methyl Ether

Triphenylmethanol is a tertiary alcohol and undergoes, as expected, S_N1 reactions. The intermediate cation is, however, stable enough to be seen in sulfuric acid solution as a red-brown to yellow solution. After dissolution in concentrated sulfuric acid, the hydroxyl anion is protonated; then the OH_2^+ portion is lost as H_2O (which is then itself protonated), leaving the carbocation. The bisulfate ion is a very weak nucleophile and does not compete with methanol in the formation of the product, trityl methyl ether.

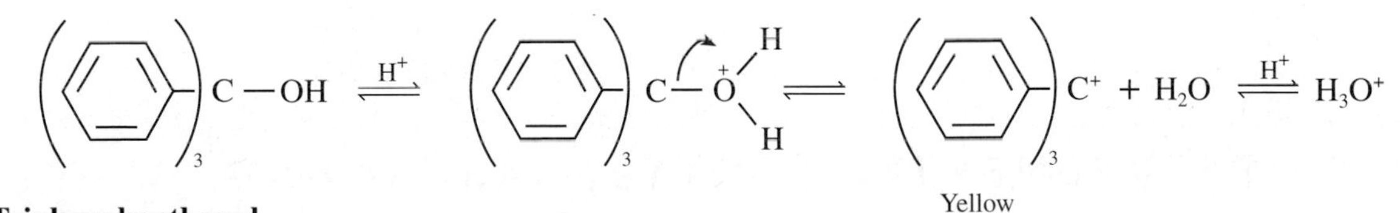

Yellow

Triphenylmethanol
MW 260.34, mp 163°C

$$(C_6H_5)_3C^+ + CH_3OH \rightleftharpoons (C_6H_5)_3C-\overset{+}{O}(H)-CH_3 \xrightarrow{CH_3OH} (C_6H_5)_3C-OCH_3 + CH_3\overset{+}{O}H_2$$

Triphenylmethyl methyl ether
Trityl methyl ether
MW 274.37, mp 96°C

Microscale Procedure

IN THIS EXPERIMENT, triphenylmethanol is dissolved in sulfuric acid, and the resulting ion is allowed to react with methanol to give a crystalline ether. This material is recrystallized from methanol.

Video: *Microscale Filtration on the Hirsch Funnel*

In a 10 × 100 mm reaction tube, place 50 mg of triphenylmethanol, grind the crystals to a fine powder with a glass stirring rod, and add about 0.4–0.5 mL of concentrated sulfuric acid (use the tube calibration to judge the amount). Continue to stir and grind the solid to dissolve all of the alcohol. Note and explain the color. Using a Pasteur pipette, transfer the sulfuric acid solution to 3 mL of ice-cold methanol in another reaction tube. Use some of the cold methanol to rinse out the first tube. If necessary, induce crystallization by cooling the solution, adding a seed crystal, or scratching the tube with a glass stirring rod at the liquid-air interface. Collect the product by filtration on a Hirsch funnel. Do not use filter paper on top of the polyethylene frit because this strongly acidic solution may attack the paper. Wash the crystals well with water and squeeze them between sheets of filter paper to aid drying. Determine the weight of the crude material, calculate the percent yield, and save a sample for a melting-point determination. Recrystallize the product from boiling methanol and determine the melting point of the purified trityl methyl ether. *See* Figure 33.1 for the ^1H NMR spectrum.

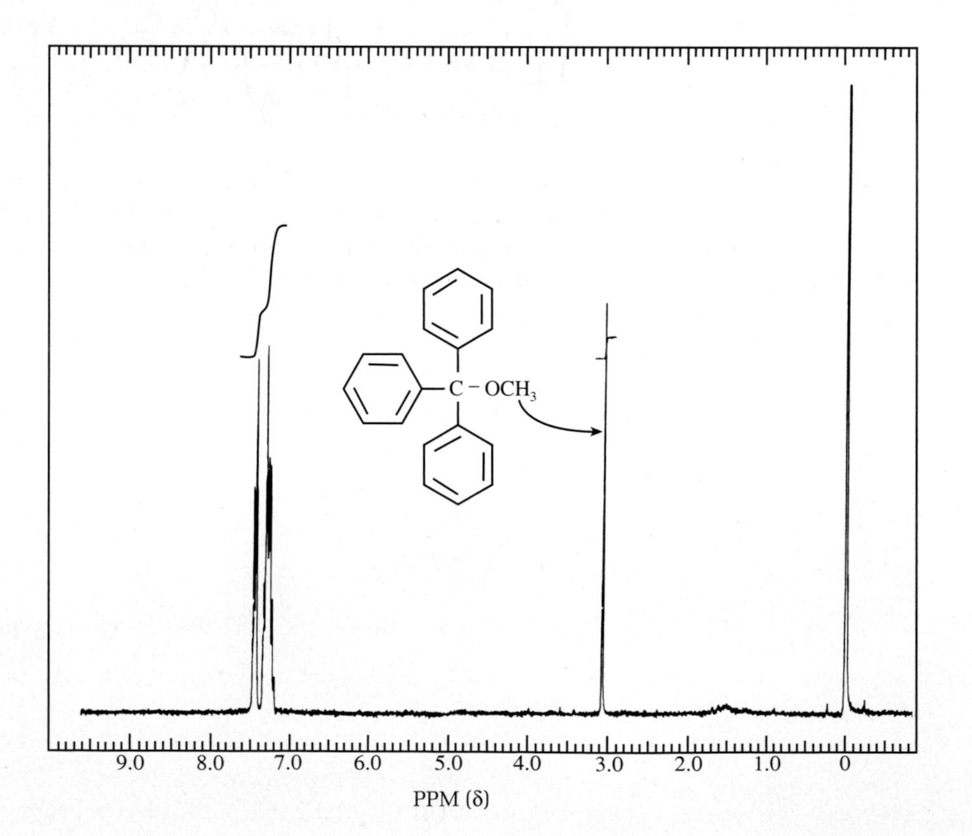

FIG. 33.1
The ^1H NMR spectrum of triphenylmethyl methyl ether, also known as trityl methyl ether (250 MHz).

Cleaning Up. Dilute the filtrate with water, neutralize with sodium carbonate, and flush down the drain.

Macroscale Procedure

Video: *Microscale Filtration on the Hirsch Funnel*

In a 10-mL Erlenmeyer flask, place 0.5 g of triphenylmethanol, grind the crystals to a fine powder with a glass stirring rod, and add 5 mL of concentrated sulfuric acid. Continue to stir the mixture to dissolve all of the alcohol. Using a Pasteur pipette, transfer the sulfuric acid solution to 30 mL of ice-cold methanol in a 50-mL Erlenmeyer flask. Use some of the cold methanol to rinse out the first flask. Induce crystallization (if necessary) by cooling the solution, adding a seed crystal, or scratching the solution with a glass stirring rod at the liquid-air interface. Collect the product by filtration on a Hirsch funnel. Do not use filter paper on top of the polyethylene frit because this strongly acidic solution may attack the paper. Wash the crystals well with water and squeeze them between sheets of filter paper to aid drying. Determine the weight of the crude material, calculate the percent yield, and save a sample for a melting-point determination. Recrystallize the product from boiling methanol and determine the melting point of the purified trityl methyl ether. *See* Figure 33.1 for the ^{1}H NMR spectrum.

Cleaning Up. Dilute the filtrate with water, neutralize with sodium carbonate, and flush down the drain.

2. Triphenylmethyl (Trityl) Bromide

IN THIS EXPERIMENT, an acetic acid solution of triphenylmethanol reacts with hydrobromic acid to give the corresponding bromide that is recrystallized from hexanes.

Reactions that proceed through the carbocation

The triphenylmethanol is dissolved in a good ionizing solvent (acetic acid) and allowed to react with a strong acid and nucleophile (hydrobromic acid). The intermediate carbocation reacts immediately with the bromide ion. Acetate ion, to the extent it is present, is not a good nucleophile. Experiments 2–9 can be run on five times the quantities specified here.

$$(C_6H_5)_3C{-}OH + HBr \rightleftharpoons (C_6H_5)_3C{-}\overset{+}{O}H_2\ Br^- \rightleftharpoons (C_6H_5)_3C^+ + H_2O$$

$$(C_6H_5)_3C^+ \xrightarrow{Br^-} (C_6H_5)_3C{-}Br$$

Triphenylmethyl bromide
Trityl bromide
MW 323.24, mp 154°C

Microscale Procedure

Measure the acid in a 1-mL graduated pipette fitted with a pipette pump.

Dissolve 100 mg of triphenylmethanol in 2 mL of warm acetic acid on a steam or hot water bath, add 0.2 mL of 47% hydrobromic acid, and heat the mixture for

5 minutes on a steam bath or in a beaker of boiling water. Cool it in ice, collect the product on a Hirsch funnel, wash the product with water and hexanes, allow it to dry, determine the weight, and calculate the percent yield. Recrystallize the product from hexanes, and then compare the melting points of the recrystallized and crude materials. The compound crystallizes slowly; allow adequate time for crystals to form. Use the Beilstein test[1] to test for halogen in both the product and starting materials.

Photo: *Crystallization*; Videos: *Microscale Filtration on the Hirsch Funnel; Recrystallization*

Cleaning Up. Dilute the filtrate from the reaction with water, neutralize with sodium carbonate, and flush down the drain. The hexanes mother liquor from the crystallization goes in the organic solvents waste container.

3. For Further Investigation: Triphenylmethyl (Trityl) Iodide?

IN THIS EXPERIMENT, an acetic acid solution of triphenylmethanol is reacted with hydriodic acid to give, by analogy with the previous reactions, a product expected to be the corresponding iodide. Analysis by the Beilstein test will confirm or disprove this.

The iodide is prepared in a reaction very similar to the preparation of the bromide. Bisulfite is added to react with any iodine formed.

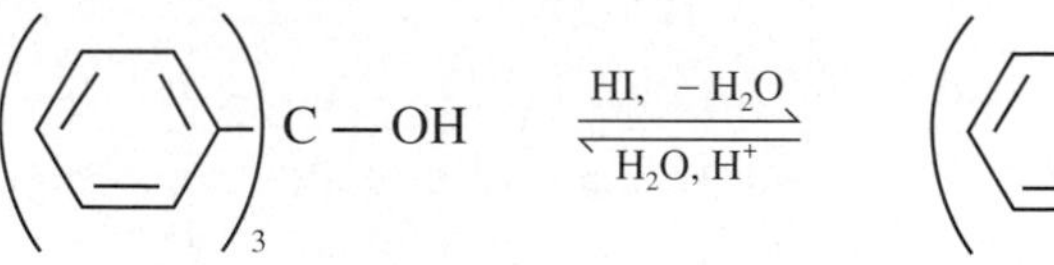

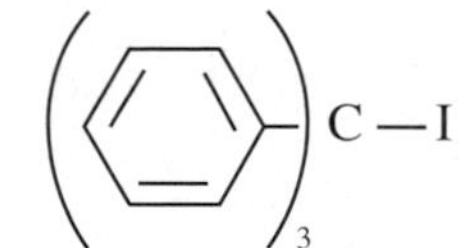

Triphenylmethyl iodide
Trityl iodide
MW 370.22, mp 183°C

Microscale Procedure

Dissolve 100 mg of triphenylmethanol in 2 mL of warm acetic acid, add 0.2 mL of 47% hydriodic acid, heat the mixture for 1 hour on a steam bath, cool it, and add it to a solution of 0.1 g of sodium bisulfite dissolved in 2 mL water in a 10-mL Erlenmeyer flask. Collect the product on a Hirsch funnel, wash it with water, press out as much water as possible, and recrystallize the crude and moist product from methanol (about 3–4 mL). Determine the weight of the dry product, calculate the percent yield, and determine the melting point. Run the Beilstein test.[2] What has been produced and why?

Cleaning Up. Dilute the filtrate from the reaction with water, neutralize with sodium carbonate, and flush down the drain. The methanol from the crystallization should be placed in the halogenated organic solvents waste container.

[1]To carry out the Beilstein test for halogens, heat the tip of a copper wire in a burner flame until no further coloration of the flame is noticed. Allow the wire to cool slightly; then dip the wire into the unknown (solid or liquid) and again heat it in the flame. A green flash is indicative of chlorine, blue-green of bromine, and blue of iodine; fluorine is not detected because copper fluoride is not volatile. The test is very sensitive; halogen-containing impurities may give misleading results. Run the test on a compound known to contain a halogen for comparison to your unknown.

[2]*See* footnote1.

4. Triphenylmethyl Fluoborate and the Tropilium Ion

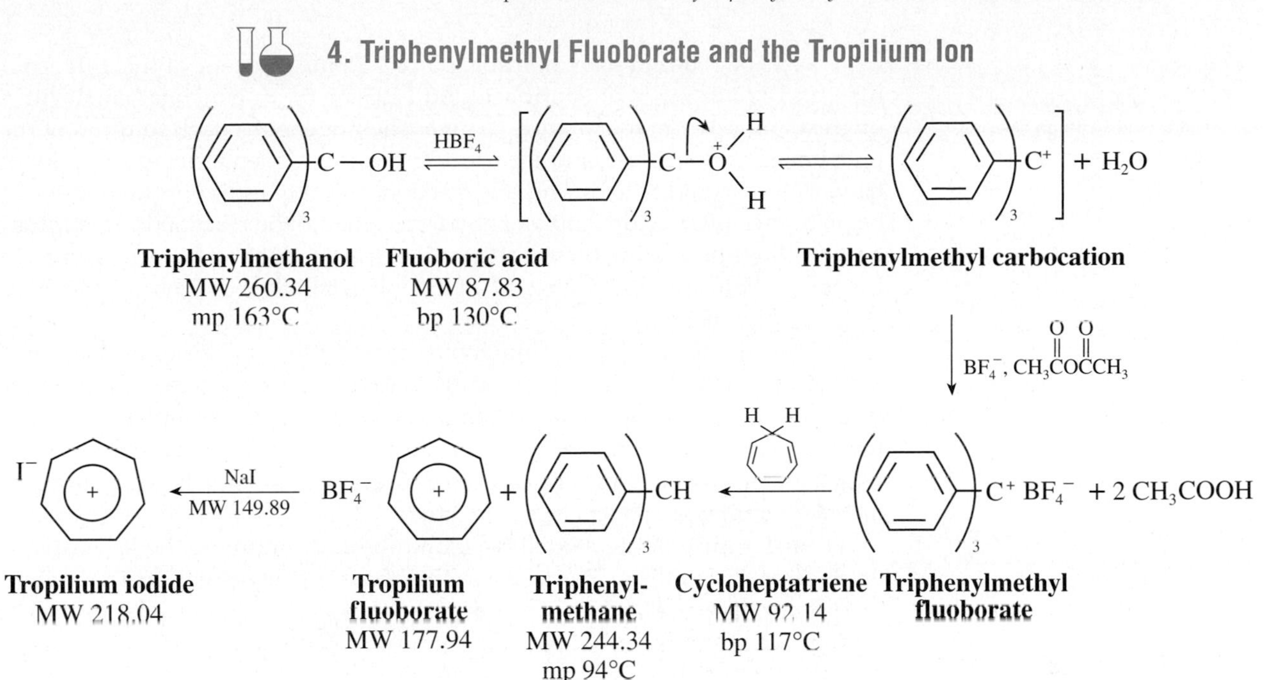

Tropilium ion—an aromatic ion

Reaction of triphenylmethanol with fluoboric acid in the presence of acetic anhydride generates trityl fluoborate, a stable salt. The fluoboric acid protonates the triphenylmethanol, which loses the elements of water in an equilibrium reaction. The water reacts with the acetic anhydride to form acetic acid, and thus drives the reaction to completion.

The salt so formed is the fluoborate of the triphenyl carbocation; it is a powerful base. It can be isolated, but in this case, it will be used *in situ* to react with the hydrocarbon cycloheptatriene.

Cycloheptatriene is more basic than most hydrocarbons, and will lose a proton to the triphenylmethyl carbocation to give triphenylmethane and the cycloheptatrienide carbocation, the tropilium ion. To demonstrate that hydride ion transfer has taken place, isolate triphenylmethane and determine its melting point.

The Hückel Rule: $4n + 2$ π electrons

The tropilium ion has a planar structure—each carbon bears a single proton, and the ion contains 6 π electrons. It is aromatic, a characteristic that can be confirmed by nuclear magnetic resonance (NMR) spectroscopy.

The fluoborate group can be displaced by the iodide ion to prepare tropilium iodide. You can see that the iodide is ionic by watching the reaction of the aqueous solution of this ion with silver ions.

Microscale Procedure

IN THIS EXPERIMENT, triphenylmethanol and fluoboric acid react to form a fluoborate salt that is a powerful base. It can remove a proton from cycloheptatriene to give tropilium ion, a carbocation that is aromatic because it contains 6 π electrons. This ion is isolated by filtration as the fluoborate salt. The filtrate is made basic and extracted with ether. This ether solution is washed, dried, and evaporated to give triphenylmethane. The fluoborate salt is easily converted to the iodide, which can be isolated and crystallized from methanol.

CAUTION: Handle fluoboric acid and acetic anhydride with great care. These reagents are toxic and corrosive. Avoid breathing the vapors or any contact with the skin. In case of skin contact, rinse the affected part under running water for at least 10 minutes. Carry out this experiment in the hood.

Video: *Microscale Filtration on the Hirsch Funnel*

In a 10 × 100-mm reaction tube, place 1.75 mL of acetic anhydride, cool the tube, and add 88 mg of fluoboric acid. Add 195 mg of triphenylmethanol with thorough stirring. Warm the mixture to give a homogeneous dark solution of the triphenylmethyl fluoborate, and then add 78 mg of cycloheptatriene. The color of the trityl cation should fade during this reaction, and the tropilium fluoborate should begin to precipitate. Add 2 mL of anhydrous ether to the reaction tube, stir the contents well while cooling on ice, and collect the product by filtration on a Hirsch funnel. Wash the product with 2 mL of dry ether, and then dry the product between sheets of filter paper.

Photos: *Extraction with Ether; Crystallization*; Videos: *Picking a Solvent; Recrystallization*

To the filtrate, add 3 *M* sodium hydroxide solution and shake the flask to allow all the acetic anhydride and fluorboric acid to react with the base. Test the aqueous layer with indicator paper to ascertain that neutralization is complete; then transfer the mixture to a reaction tube and draw off the aqueous layer. Wash the ether once with 2 mL of water and dry the ether over calcium chloride pellets, adding the drying agent until it no longer clumps together.

Transfer the ether to a tared reaction tube and evaporate the solvent to leave crude triphenylmethane. Remove a sample for a melting-point determination and recrystallize the residue from an appropriate solvent, determined by experimentation. Prove to yourself that the compound isolated is indeed triphenylmethane. Obtain an infrared (IR) spectrum and an NMR spectrum.

Structure proof: NMR

To determine the NMR spectrum of the tropilium fluoborate collected on a Hirsch funnel, dissolve about 50 mg of the product in 0.3 mL of deuterated dimethyl sulfoxide that contains 1% tetramethylsilane as a reference. Compare the NMR spectrum obtained to that of the starting material, cycloheptatriene. The spectrum of the latter can be obtained in deuterochloroform, again using tetramethylsilane as the reference compound.

Photo: *Filtration Using a Pasteur Pipette*; Video: *Filtration of Crystals Using the Pasteur Pipette*

The tropilium fluoborate can be converted to tropilium iodide by dissolving the fluoborate in a few drops of hot, but not boiling, water and adding to this solution 0.25 mL of a saturated solution of sodium iodide. Cool the mixture in ice, remove the solvent with a pipette, and wash the crystals with 0.5 mL of ice-cold methanol. Scrape most of the crystals onto a piece of filter paper and allow them to dry before determining the weight. To the crystalline residue in the reaction tube, add a few drops of water, warm the tube if necessary to dissolve the tropilium iodide, add a drop of 2% aqueous silver nitrate solution, and note the result.

Cleaning Up. Solutions that contain the fluoborate ion (the neutralized filtrate, the NMR sample, the filtrate from the iodide preparation) should be treated with aqueous calcium chloride to precipitate insoluble calcium fluoride, which is then removed by filtration and placed in the nonhazardous solid waste container. The filtrate can be flushed down the drain. Under the hood, allow the ether to evaporate from the calcium chloride and then place the drying agent in the nonhazardous solid waste container. If local regulations do not allow for the evaporation of solvents in a hood, dispose of the wet pellets in the waste container provided. The recrystallization solvent goes in the organic solvents waste container.

THE TRITYL CARBANION

The triphenylmethyl carbanion, the trityl anion, can be generated by reacting triphenylmethane with *n*-butyllithium, a very strong base. The reaction generates the blood-red lithium triphenylmethide and butane. The triphenylmethyl anion reacts much as a Grignard reagent does (*see* Chapter 38). In the present experiment, it reacts with carbon dioxide to give triphenylacetic acid after acidification. Avoid an excess of *n*-butyllithium because on reaction with carbon dioxide, it gives the vile-smelling pentanoic acid.

$$(C_6H_5)_3C{-}H + Li^+\,\bar{C}H_2CH_2CH_2CH_3 \longrightarrow (C_6H_5)_3C{:}^- \; Li^+ + CH_3CH_2CH_2CH_3$$

Triphenylmethane
MW 244.34
mp 94°C

***n*-Butyllithium**
MW 64.06

Trityl carbanion
Lithium triphenylmethide

$\downarrow$ CO_2, H^+

$$(C_6H_5)_3C{-}C(=O)OH$$

Triphenylacetic acid
MW 288.35, mp 270–273°C

5. Synthesis of the Trityl Carbanion and Triphenylacetic Acid

IN THIS EXPERIMENT, triphenylmethane in ether is reacted with the strong base n-butyllithium under nitrogen to give the trityl carbanion. When the solution of this carbanion is poured onto dry ice, it is converted to triphenylacetic acid. This acid is made basic with aqueous sodium hydroxide, and the solution is washed with ether. The triphenylacetic acid crystallizes when the aqueous layer is made acidic.

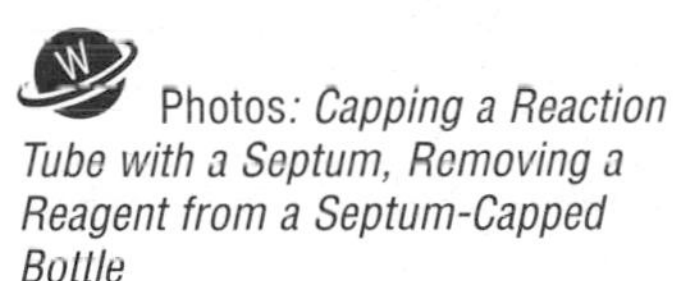

Photos: *Capping a Reaction Tube with a Septum, Removing a Reagent from a Septum-Capped Bottle*

Microscale Procedure

Or use 0.25 mL of a 1.6 *M* solution of *n*-butyllithium.

CAUTION: Use great care in handling the *n*-butyllithium. It reacts avidly with air and violently with water.

To a reaction tube (Fig. 33.2) that has been dried for at least 30 minutes in a 110°C oven and capped with a rubber septum, add 100 mg of triphenylmethane and 3 mL of anhydrous diethyl or *t*-butyl methyl ether. Add an empty syringe needle to the septum; flush out the air in the tube by passing a slow current of nitrogen into the tube for about 30 seconds. Remove the nitrogen inlet needle and, with thorough mixing of the solution, add to the reaction mixture 0.4 mL of a 1.0 *M* solution of *n*-butyllithium in hexane (a commercial product).

Reaction with dry ice

The sodium hydroxide converts the acid to its water-soluble anion.

In a 30-mL beaker, place one or two small pieces of dry ice (solid carbon dioxide) that have been wiped free of any adhering frost. Handle the dry ice with a towel or gloves. Contact with the skin can cause frostbite because the sublimation temperature of dry ice is −78.5°C. Calculate how much carbon dioxide is needed to react with the anion prepared from 100 mg of triphenylmethane; use at least a tenfold excess. Using a Pasteur pipette, transfer the solution of the anion to the dry ice; the reaction is immediate. Allow the unreacted dry ice to sublime; then add 3 mL of 1.5 *M* aqueous sodium hydroxide solution, dissolve as much solid as possible, and transfer the liquid to a reaction tube. Shake the mixture, draw off the ether layer, and wash the aqueous layer with two 2.5-mL portions of ether, which are also discarded. Using 3 *M* hydrochloric acid, acidify the aqueous layer to a pH < 4.

FIG. 33.2
An apparatus for synthesizing the trityl carbanion.

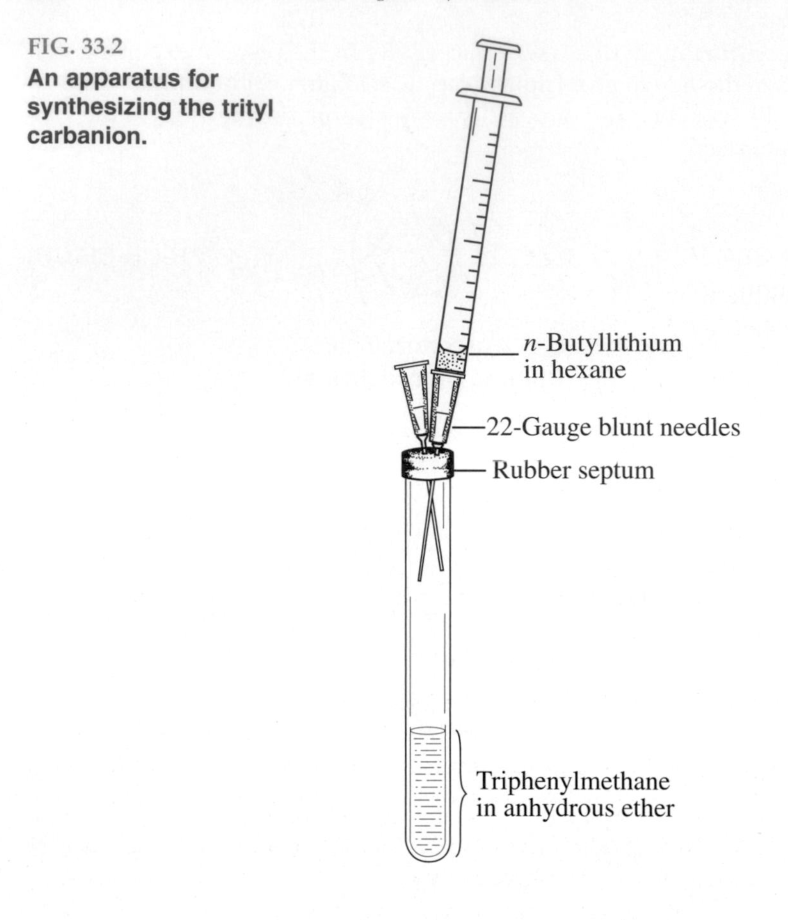

Photo: *Extraction with Ether*

Collect the precipitate, wash it with water, and allow it to air-dry. Take the melting point. When pure, triphenylacetic acid melts at 267°C. The purity and identity of the product can be assessed using thin-layer chromatography (TLC) and IR spectroscopy (using a potassium bromide disk).

Cleaning Up. There should be no excess *n*-butyllithium, but a few drops of the solution in hexane can be destroyed by reaction with *t*-butyl alcohol. Place ether in the organic solvents waste container. Dilute the aqueous layer with water, neutralize with sodium carbonate, and then flush down the drain.

TRIPHENYLMETHYL: A STABLE FREE RADICAL

Gomberg: the first free radical

In the early part of the 19th century, many attempts were made to prepare methyl, ethyl, and similar radicals in a free state, as sodium had been prepared from sodium chloride. Many well-known chemists tried: Gay-Lussac's CN turned out to be cyanogen $(CN)_2$; Bunsen's cacodyl from $(CH_3)_2AsCl$ proved to be $(CH_3)_2As{-}As(CH_3)_2$; and the Kolbe electrolysis gave $CH_3CH_2CH_2CH_3$ instead of CH_3CH_2. Moses Gomberg at the University of Michigan prepared the first free radical in 1900. He had prepared triphenylmethane and chlorinated it to give triphenylmethyl chloride, which he hoped to couple to form hexaphenylethane:

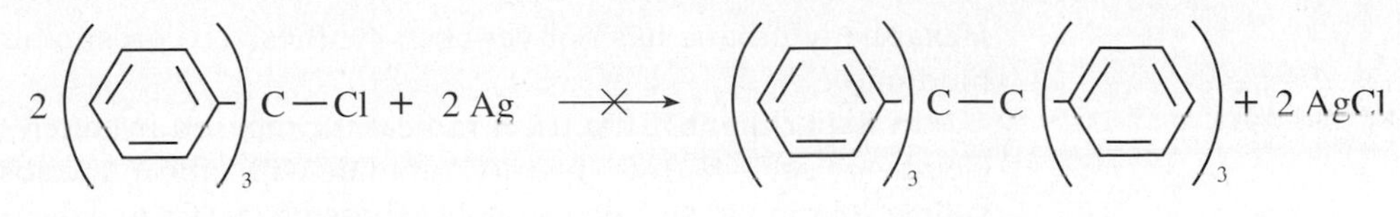

Triphenylmethyl chloride **Hexaphenylethane**

The solid that Gomberg obtained was a high-melting, sparingly soluble, white solid, which turned out on analysis to be not a hydrocarbon, but instead an oxygen-containing compound ($C_{38}H_{30}O_2$). Repeating the experiment in a vacuum, he obtained a yellow solution that, on evaporation, deposited crystals of a colorless hydrocarbon that was remarkably reactive. It readily reacted with oxygen, bromine, chlorine, and iodine. On dissolution, the white hydrocarbon gave a yellow solution in a vacuum; the hydrocarbon was deposited when he evaporated the solution once more. Gomberg interpreted these results as follows:

> The experimental evidence . . . forces me to the conclusion that we have to deal here with a free radical, triphenylmethyl, $(C_6H_5)_3C\cdot$. The action of zinc results, as it seems to me, in a mere abstraction of the halogen:
>
> $$2\,(C_6H_5)_3C{-}Cl + Zn \longrightarrow 2\,(C_6H_5)_3C\cdot + ZnCl_2$$
>
> Now as a result of the removal of the halogen atom from triphenylchloromethane, the fourth valence of the methane is bound either to take up the complicated group $(C_6H_5)_3\cdot$ or remain as such, with carbon as trivalent. Apparently the latter is what happens.[3]

For a long time after Gomberg first carried out this reaction, it was assumed the radical dimerized to form hexaphenylethane:

$$2\,(C_6H_5)_3C\cdot \longrightarrow (C_6H_5)_3C{-}C(C_6H_5)_3$$

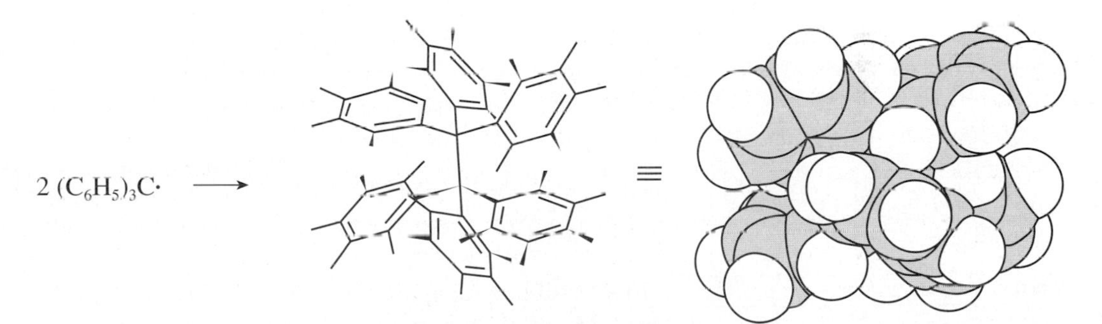

But in 1968, NMR and UV (ultraviolet) evidence showed that the radical is in equilibrium with a different substance:

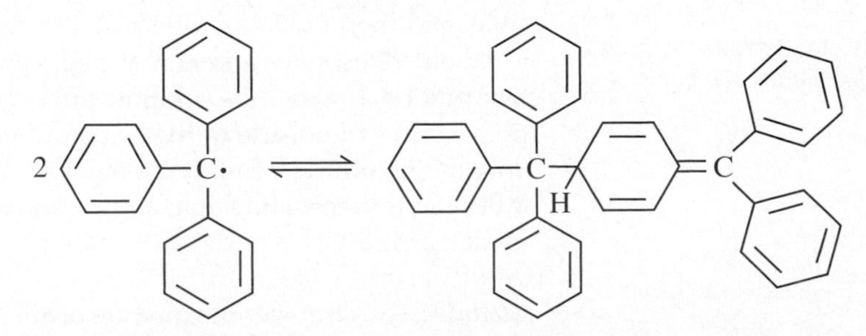

[3]Gomberg MJ. *Am Chem Soc.* **1900**;*22*:757.

Hexaphenylethane has not yet been synthesized, presumably because of steric hindrance.

In Experiment 7, the trityl radical is prepared in much the same fashion as Gomberg used—by reacting trityl bromide with zinc in the absence of oxygen. The yellow solution is then deliberately exposed to air to give the peroxide.

6. Computational Chemistry

With a molecular mechanics program, construct a model of hexaphenylethane; then carry out an energy minimization of the molecule. Note the steric energy. Construct a model of the compound actually formed when the trityl radical dimerizes. Note the steric energy. What is the length of the central C—C bond in your model of hexaphenylethane? What is the C—C bond length in 1,2-diphenylethane? These numbers are probably not correct because valid parameters do not exist for severely sterically hindered molecules of this type, but the space-filling models are worth studying. Do these calculations give any insight as to why hexaphenylethane has not yet been synthesized?

7. Synthesis of Triphenylmethyl Peroxide

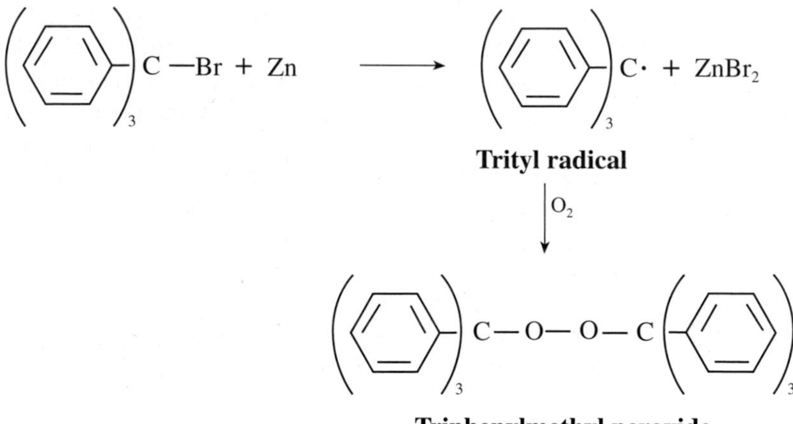

The trityl radical

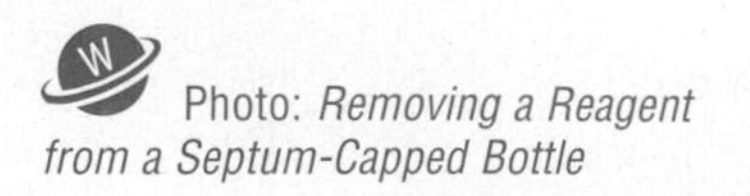
Photo: *Removing a Reagent from a Septum-Capped Bottle*

In a small test tube, dissolve 100 mg of trityl bromide or chloride in 0.5 mL of toluene. Material prepared in Experiment 2 may be used; it need not be recrystallized. Cap the tube with a septum, insert an empty needle, and flush the tube with nitrogen while shaking the contents. Add 0.2 g of fresh zinc dust as quickly as possible, and then flush the tube with nitrogen once more. Shake the tube vigorously for about 10 minutes and note the appearance of the reaction mixture. Using a Pasteur pipette, transfer the solution, but not the zinc, to a 30-mL beaker. Roll the solution around the inside of the beaker to give it maximum exposure to the air and, after a few minutes, collect the peroxide on a Hirsch funnel. Wash the solid with a little cold toluene, allow it to air dry, and determine the melting point. It is reported to melt at 186°C.

Cleaning Up. Transfer the mixture of zinc and zinc bromide to the hazardous waste container.

For Further Investigation

Carry out both of the following reactions. Compare the products formed. Run the necessary solubility and simple qualitative tests. Interpret your results, and propose structures for the compounds produced in each of the reactions.

Perform tests on compounds 1 and 2 to deduce their identities. Use 1:10 ether/hexanes for TLC.

8. Synthesis of Compound 1

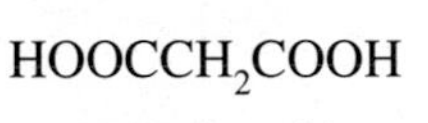

$HOOCCH_2COOH$

Malonic acid

Photo: *Filtration Using a Pasteur Pipette*; Video: *Filtration of Crystals Using the Pasteur Pipette*

Microscale Procedure. In a reaction tube, place 100 mg of triphenylmethanol and 200 mg of malonic acid. Grind the two solids together with a glass stirring rod, and then heat the reaction tube at 150°C for 7 minutes. Use an aspirator tube mounted at the mouth of the tube to carry away undesirable fumes. Allow the tube to cool, and then dissolve the contents in about 0.2 mL toluene. Dilute the solution with 1 mL of 60–80°C hexanes; after cooling the tube in ice, isolate the crystals, wash them with a little cold hexanes, and determine the weight and melting point of the dry crystals.

Cleaning Up. Place the toluene/hexanes mother liquor in the organic solvents waste container.

9. Synthesis of Compound 2

$ClCH_2COOH$

Chloroacetic acid

Microscale Procedure. In a reaction tube, dissolve 100 mg of triphenylmethanol in 2 mL of acetic acid; then add to this 0.4 mL of an acetic acid solution containing 5% of chloroacetic acid and 1% of sulfuric acid. Heat the mixture for 5 minutes, add about 1 mL of water to produce a hot saturated solution, and let the tube cool slowly to room temperature. Collect the product by filtration and wash the crystals with 1:1 methanol-water mixture. Once the crystals are dry, determine the weight and melting point.

Cleaning Up. Dilute the filtrate with water, neutralize with sodium carbonate, and flush down the drain.

QUESTIONS

1. Give the mechanism for the free radical chlorination of triphenylmethane.
2. Is the propeller-shaped triphenylmethyl carbocation a chiral species?
3. Without carrying out the experiments, speculate on the structures of Compounds 1 and 2 synthesized in Experiments 8 and 9.
4. Is the production of triphenylmethyl peroxide a chain reaction?
5. What product would you expect from the reaction of carbon tetrachloride, benzene, and aluminum chloride?

34 CHAPTER

1,2,3,4-Tetraphenyl-naphthalene via Benzyne

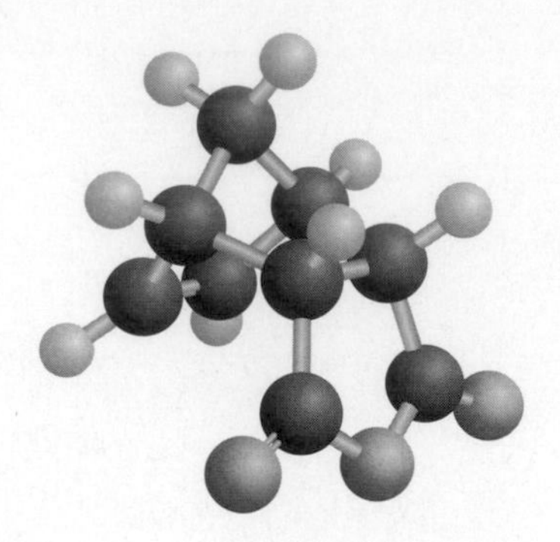

PRELAB EXERCISE: Speculate on the reasons for the stability of the iodonium ion. What type of strain exists in the benzyne molecule?

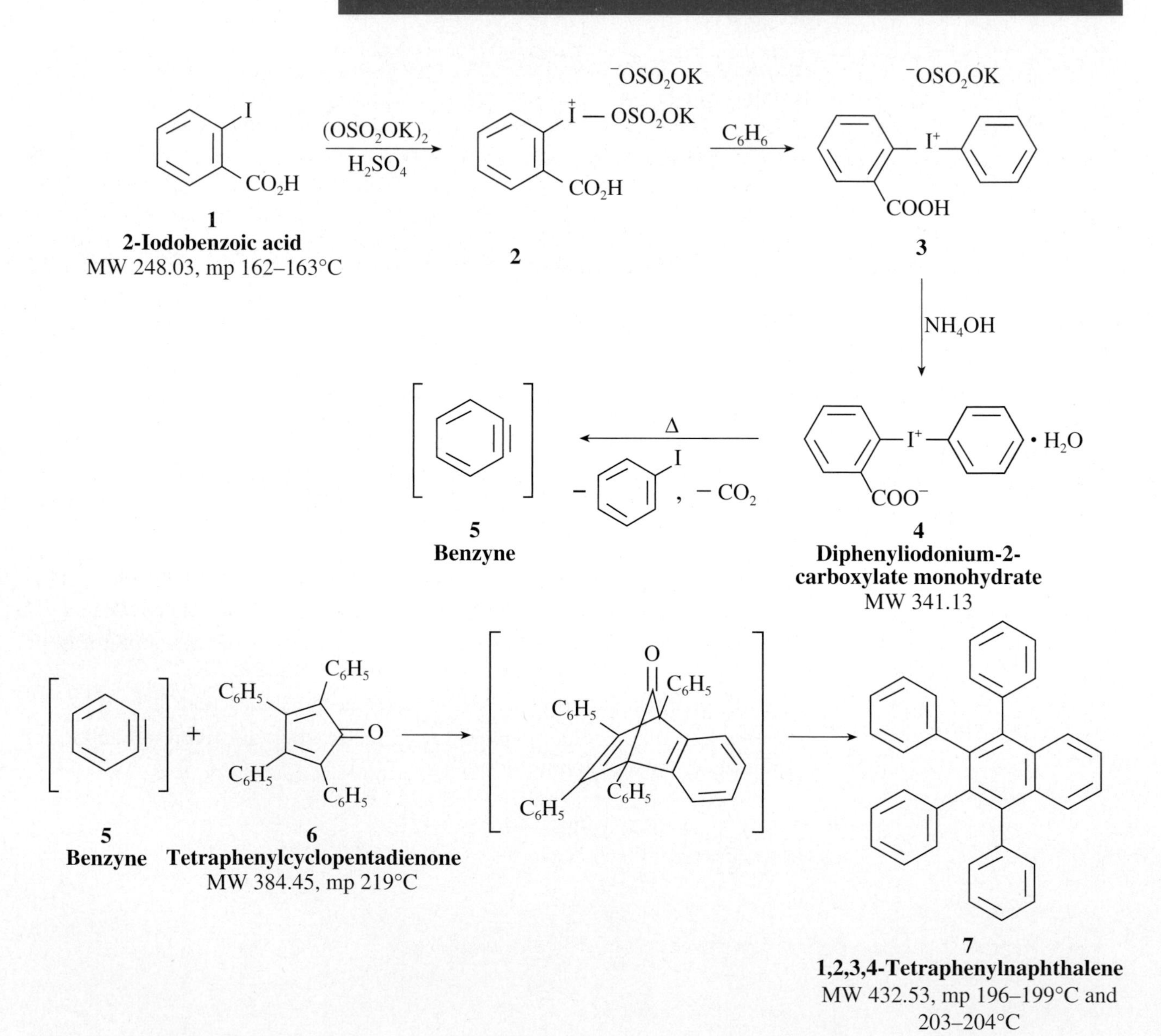

Synthetic use of an intermediate not known as such.

This synthesis of 1,2,3,4-tetraphenylnaphthalene (**7**) demonstrates the transient existence of benzyne (**5**), a hydrocarbon that has not been isolated as such. The precursor, diphenyliodonium-2-carboxylate (**4**), is heated in an inert solvent to a temperature at which it decomposes to benzyne, iodobenzene, and carbon dioxide in the presence of tetraphenylcyclopentadienone (**6**) as the trapping agent. The preparation of the precursor (**4**) illustrates oxidation of a derivative of iodobenzene to an iodonium salt (**2**) and the Friedel–Crafts-like reaction of the substance with benzene to form the diphenyliodonium salt (**3**). Neutralization with ammonium hydroxide then liberates the precursor, the inner salt (**4**), which, when obtained by crystallization from water, is the monohydrate.

EXPERIMENTS

1. DIPHENYLIODONIUM-2-CARBOXYLATE MONOHYDRATE (4)

IN THIS EXPERIMENT, iodobenzoic acid forms a salt with a peroxyacid in sulfuric acid. This salt reacts with benzene to form another salt that on reaction with ammonium hydroxide forms still another salt, which is extracted into dichloromethane. The dichloromethane is dried and evaporated to give the pure salt, diphenyliodonium-2-carboxylate, which is recrystallized from water.

CAUTION: Benzene is a mild carcinogen. Carry out this experiment in the hood. Handle concentrated sulfuric acid with care.

Measure 0.85 mL of concentrated sulfuric acid into a reaction tube and place it in an ice bath to cool. In a 10-mL Erlenmeyer flask that is clamped in an ice bath, place 200 mg of 2-iodobenzoic acid, prepared using the Sandmeyer reaction (*see* Chapter 44), and 260 mg of potassium persulfate.[1] Remove the tube containing the cold sulfuric acid from the ice bath, wipe it dry, and add the acid to the two solids in the flask. Stir the mixture in the ice bath for 4–5 minutes to produce an even suspension; then remove it and note the time. The reaction mixture foams somewhat and acquires a succession of colors. After it has stood at room temperature for 20 minutes, swirl the flask vigorously in an ice bath for 2–3 minutes, add 0.2 mL of benzene, and swirl and cool until the benzene freezes. Then remove and wipe the flask, and note the time at which the benzene melts. Warm the flask in the palm of your hand and swirl frequently at room temperature for 20 minutes to promote completion of reaction in the two-phase mixture. If available, magnetic stirring of the reaction mixture would be convenient, but is not required.

$K_2S_2O_8$
Potassium persulfate (Potassium peroxydisulfate)
MW 270.33
Strong oxidant!

While the reaction is going to completion, place three reaction tubes in an ice bath to chill: one containing 1.9 mL of distilled water, another containing 2.3 mL of concentrated ammonium hydroxide solution, and another containing 4 mL of dichloromethane. At the end of the 20-minute reaction period, chill the reaction mixture thoroughly in an ice bath and add to it dropwise, with vigorous swirling and mixing, the 1.9 mL of ice-cold water that has been chilling. The solid that separates is the potassium bisulfate salt of diphenyliodonium-2-carboxylic acid (**3**).

[1]The fine granular material supplied by Fisher Scientific Co. is satisfactory. Persulfate in the form of large prisms should be finely ground prior to use.

Pour the chilled dichloromethane into the flask that will be used for extracting the product that is liberated when the ammonia is added. While swirling the flask vigorously, add the ammonia dropwise over a 10-minute period. The mixture must be alkaline (pH 9). Test with indicator paper and add more ammonia if necessary. Using a Pasteur pipette, remove the dichloromethane layer, transfer the aqueous layer to a 4-in. test tube, and extract it with two 1.5-mL portions of dichloromethane that are combined with the original dichloromethane extract. Dry the dichloromethane extracts over anhydrous calcium chloride pellets and decant into a tared 10-mL flask. Rinse the drying agent with more solvent; then evaporate the dichloromethane under a stream of air in the hood. Connect the flask to an aspirator, and heat it on the steam bath until the weight is constant. The yield is about 240 mg.

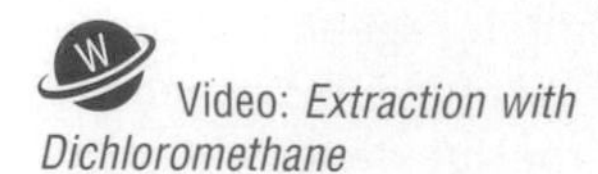
Video: *Extraction with Dichloromethane*

For recrystallization, dislodge the solid with a spatula and dissolve it in 2.8 mL of boiling water. Use a wood boiling stick to prevent bumping of the solution. If desired, the tan solution may be decolorized by adding about 15 pieces of pelletized Norit. On cooling the solution after removal of the Norit, diphenyliodonium-2-carboxylate monohydrate (**4**), the benzyne precursor, separates in colorless, rectangular prisms (mp 219°C–220°C). The yield should be about 200 mg.

Photo: *Crystallization*; Videos: *Recrystallization; Picking a Solvent*

Cleaning Up. Neutralize the aqueous layer with dilute hydrochloric acid. It should then be diluted with water and flushed down the drain. Allow the dichloromethane to evaporate from the drying agent in the hood, and then place the drying agent in the nonhazardous solid waste container along with the Norit. If local regulations do not allow for solvent evaporation in the hood, place the wet drying agent in the waste container provided.

2. DIPHENYLIODONIUM-2-CARBOXYLATE MONOHYDRATE (4)

$K_2S_2O_8$
Potassium persulfate (Potassium peroxydisulfate)
MW 270.33
Strong oxidant!

CAUTION: Benzene is a mild carcinogen. Carry out this experiment in the hood. Handle concentrated sulfuric acid with care.

Measure 8 mL of concentrated sulfuric acid into a 25-mL Erlenmeyer flask and place the flask in an ice bath to cool. Place 2.0 g of *o*-iodobenzoic acid (*see* Chapter 44) and 2.6 g of potassium persulfate[2] in a 125-mL Erlenmeyer flask, swirl the flask containing sulfuric acid vigorously in the ice bath for 2–3 minutes, then remove it and wipe it dry. Place the larger flask in the ice bath and pour the chilled acid down the walls to dislodge any particles of solid. Swirl the flask in the ice bath for 4–5 minutes to produce an even suspension; then remove it and note the time. The reaction mixture foams somewhat and acquires a succession of colors. After it has stood at room temperature for 20 minutes, swirl the flask vigorously in an ice bath for 3–4 minutes, add 2 mL of benzene, and swirl and cool until the benzene freezes. Then remove and wipe the flask and note the time at which the benzene melts. Warm the flask in the palm of your hand and swirl frequently at room temperature for 20 minutes to promote completion of the reaction in the two-phase mixture.

While the reaction is going to completion, place three 50-mL Erlenmeyer flasks in an ice bath to chill: one containing 19 mL of distilled water, another containing 23 mL of 29% ammonium hydroxide solution, and another containing 40 mL of dichloromethane (bp 40.8°C). At the end of the 20-minute reaction period, chill the reaction mixture thoroughly in an ice bath, mount a separatory funnel to deliver into the flask containing the reaction mixture in benzene, and place in the funnel the

[2]The fine granular material supplied by Fisher Scientific Co. is satisfactory. Persulfate in the form of large prisms should be finely ground prior to use.

chilled 19 mL of water. Swirl the reaction flask vigorously while running in the water slowly. The solid that separates is the potassium bisulfate salt of diphenyliodonium-2-carboxylic acid (**3**). Pour the chilled ammonia solution into the funnel, and pour the chilled dichloromethane into the reaction flask so that it will be available for efficient extraction of reaction product **4** as it is liberated from **3** on neutralization. While swirling the flask vigorously in the ice bath, run in the chilled ammonia solution during the course of about 10 minutes. The mixture must be alkaline (pH 9). If not, add more ammonia solution. Pour the mixture into a separatory funnel and rinse the flask with a small amount of fresh dichloromethane, adding this liquid to the separatory funnel. Let the layers separate; then run the lower layer into a tared 125-mL Erlenmeyer flask, through a funnel fitted with a paper containing anhydrous calcium chloride pellets. Extract the aqueous solution with two 10-mL portions of dichloromethane, and run the extracts through the drying agent into the tared flask. Evaporate the dried extracts to dryness on a hot water or steam bath (use an aspirator tube) and remove the solvent from the residual cake of solid by connecting the flask, with a rubber stopper, to the aspirator and heating the flask on a hot water or steam bath until the weight is constant; the yield is about 2.4 g.

Video: *Macoscale Crystallization*

For recrystallization, dislodge the bulk of the solid with a spatula and transfer it onto weighing paper, and then into a 50-mL Erlenmeyer flask. Measure 28 mL of distilled water into a flask and use part of the water to dissolve the residual material in the tared 125-mL flask by heating the mixture to its boiling point over a hot plate or free flame. Pour this solution into the 50-mL flask. Add the remainder of the 28 mL of water to the 50-mL flask and bring the solid into solution at the boiling point. Add a small portion of charcoal for decolorization of the pale-tan solution, swirl, and filter at the boiling point through a funnel, preheated on a steam bath and fitted with moistened filter paper. Diphenyliodonium-2-carboxylate monohydrate (**4**), the benzyne precursor, separates in colorless, rectangular prisms and has a melting point of 219°C–220°C (with decomposition); the yield is about 2.1 g.

Cleaning Up. Neutralize the aqueous layer with dilute hydrochloric acid. It should then be diluted with water and flushed down the drain. Allow the dichloromethane to evaporate from the drying agent in the hood and then place the drying agent in the nonhazardous solid waste container along with the decolorizing charcoal. If local regulations do not allow for solvent evaporation in the hood, place the wet drying agent in a special waste container.

3. PREPARATION OF 1,2,3,4-TETRAPHENYLNAPHTHALENE (7)

Microscale Procedure

Diels–Alder reaction

IN THIS EXPERIMENT, the salt from the previous experiment and a cyclic diene are dissolved in a very high-boiling solvent. At 200°C the salt decomposes to carbon dioxide, iodobenzene, and benzyne. The latter, in a Diels–Alder reaction, reacts with the cyclic diene. This diene decomposes by losing a molecule of carbon monoxide to give the product, tetraphenylnaphthalene, which is crystallized by adding ethanol to the reaction mixture.

Place 100 mg of the diphenyliodonium-2-carboxylate monohydrate just prepared and 100 mg of tetraphenylcyclopentadienone, prepared according to the procedure in Chapter 51, in a reaction tube. Add 0.6 mL of triethylene glycol dimethyl ether (bp 222°C) so that the solvent rinses the walls of the tube. Support the tube vertically, insert a thermometer, and heat it over a hot sand bath. When the temperature reaches 200°C, remove the tube from the heat and note the time. Maintain the mixture at 200°C–205°C by intermittent heating until the purple color is discharged, the evolution of gas (CO_2 and CO) subsides, and a pale-yellow solution is obtained. In case a purple or red color persists after 3 minutes at 200°C–205°C, add additional small amounts of the benzyne precursor and continue to heat until all the solid is dissolved.

Photo: *Filtration Using a Pasteur Pipette*; Videos: *Filtration of Crystals Using the Pasteur Pipette; Microscale Filtration on the Hirsch Funnel*

Cool the hot solution to below 100°C and add to it 0.6 mL of ethanol. Bring the solution to boiling, using a wood stick to prevent bumping. If shiny prisms do not separate at once, add water dropwise at the boiling point of the ethanol until crystals begin to separate. Let crystallization proceed at room temperature and then at 0°C. Collect the product using the Pasteur pipette method or on a Hirsch funnel. Wash the crystals with cold methanol. The yield of dry product should be about 80 mg. The pure hydrocarbon (**7**) exists in two crystalline forms (allotropes) and has a double melting point, the first of which is in the range 196°C–199°C. Let the melting point bath cool to about 195°C, remove the sample, and let the sample solidify. Then determine the second melting point, which, for the pure hydrocarbon, is 203°C–204°C.

Cleaning Up. The filtrate contains ethanol, iodobenzene, triglyme, and probably small amounts of the reactants and the product. Place the mixture in the halogenated organic solvents waste container.

Macroscale Procedure

Diels–Alder reaction

Use the hood; CO is produced.

Place 1.0 g of the diphenyliodonium-2-carboxylate monohydrate just prepared and 1.0 g of tetraphenylcyclopentadienone in a 25 × 150-mm test tube. Add 6 mL of triethylene glycol dimethyl ether (bp 222°C) in a way such that the solvent will rinse the walls of the tube. Support the test tube vertically, insert a thermometer, and heat with a sand bath. When the temperature reaches 200°C, remove from the heat and note the time. Maintain the mixture at 200°C–205°C by intermittent heating until the purple color is discharged, the evolution of gas (CO_2 and CO) subsides, and a pale-yellow solution is obtained. In case a purple or red color persists after 3 minutes at 200°C–205°C, add additional small amounts of the benzyne precursor and continue to heat until all of the solid is dissolved. Let the yellow solution cool to 90°C while heating 6 mL of 95% ethanol to its boiling point on a steam or sand bath. Pour the yellow solution into a 25-mL Erlenmeyer flask, and use a few drops of the hot ethanol to rinse the test tube. Add the remainder of the ethanol to the yellow solution and heat at the boiling point. If shiny prisms do not separate at once, add a few drops of water at the boiling point of the ethanol until prisms begin to separate. Let crystallization proceed at room temperature and then at 0°C. Collect the product via filtration and wash it with methanol. The yield of colorless prisms is 0.8–0.9 g. The pure hydrocarbon (**7**) exists in two crystalline forms (allotropes) and has a double melting point, the first of which is in the range of 196°C–199°C. Let the bath cool to about 195°C, remove the thermometer, and let the sample solidify. Then determine the second melting point, which, for the pure hydrocarbon, is 203°C–204°C.

Cleaning Up. The filtrate contains ethanol, iodobenzene, triglyme, and probably small amounts of the reactants and the product. Place the mixture in the halogenated organic solvents waste container.

QUESTIONS

1. To which general class of compounds does potassium persulfate belong?
2. Calculate the volume of carbon monoxide, at standard temperature and pressure, that is released during the reaction.

35 CHAPTER

Triptycene via Benzyne

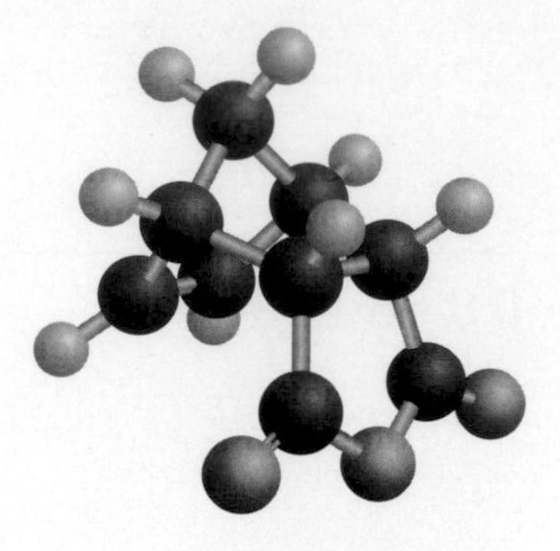

PRELAB EXERCISE: Outline the preparation of triptycene, writing balanced equations for each reaction, including the reactions that are used to remove excess anthracene.

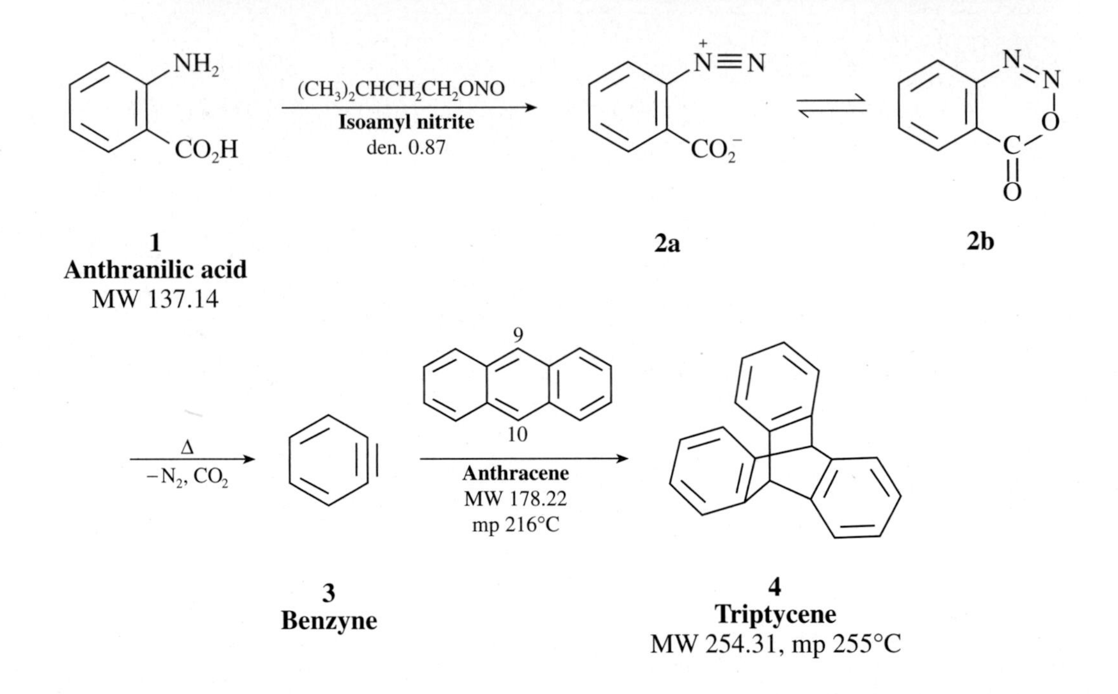

This interesting cage-ring hydrocarbon results from the 9,10-addition of benzyne to anthracene. In one procedure presented in the literature, benzyne is generated under nitrogen from *o*-fluorobromobenzene and magnesium in the presence of anthracene, but the workup is tedious and the yield only 24%. Diazotization of anthranilic acid to benzenediazonium-2-carboxylate (**2a**) or to the covalent form (**2b**) gives another benzyne precursor, but the isolated substance can be kept only at a low temperature and has a tendency to explode. However, isolation of the precursor is not necessary. On slow addition of anthranilic acid to a solution of anthracene and isoamyl nitrite

Aprotic means "no protons"; water (HOH) and alcohols (ROH) are protic.

in an aprotic solvent, the precursor **2a/2b** reacts with anthracene as fast as the precursor is formed. If the anthranilic acid is all present at the start, a side reaction of this substance with benzyne drastically reduces the yield. A low-boiling solvent (CH_2C_{12}, bp 41°C) is used in the reaction reported in the literature, in which the desired reaction goes slowly, and a solution of anthranilic acid is added dropwise over a period of 4 hours. To bring the reaction time into the limits of a laboratory period, the higher boiling solvent 1,2-dimethoxyethane (bp 83°C, water soluble) is specified in this procedure, and a large excess of anthranilic acid and isoamyl nitrite is used. Treatment of the dark reaction mixture with alkali removes acidic byproducts and most of the color, but the crude product inevitably contains anthracene. However, brief heating with maleic anhydride at a suitable temperature leaves the triptycene untouched and converts the anthracene into its maleic anhydride adduct. Treatment of the reaction mixture with alkali converts the adduct into a water-soluble salt and affords colorless, pure triptycene.

Removal of anthracene

EXPERIMENT

TRIPTYCENE

IN THIS EXPERIMENT, anthranilic acid is diazotized with isoamyl nitrite to form an unstable diazonium salt that decomposes to benzyne, which reacts with anthracene also present in the solution. This Diels–Alder reaction gives triptycene. Excess anthracene is then reacted in a very similar reaction with maleic anhydride. This material reacts with base to form a water-soluble salt, leaving the triptycene, which crystallizes from the solvent mixture. It is then recrystallized from a different solvent–methylcyclohexane.

Microscale Procedure

$CH_3OCH_2CH_2OCH_3$
Dimethoxyethane (DME)

Test dimethoxyethane for peroxides. *See* Chapter 2 for the peroxides test.

Carry out the reaction in a hood.

Technique for the slow addition of a solid.

Reaction time: 1 hour

Video: *Microscale Filtration on the Hirsch Funnel*

Place 100 mg of anthracene,[1] 0.1 mL of isoamyl nitrite, 1 mL of dimethoxyethane, and a boiling chip in a 10 × 100 mm reaction tube. In a small filter paper, place 130 mg of anthranilic acid and push this about halfway down the reaction tube. Bring the mixture in the tube to a gentle boil; note that not all of the anthracene dissolves. Add to the filter paper 0.5 mL of dimethoxyethane dropwise to leach the acid into the solution below over a period of not less than 20 minutes. Remove the filter paper, add to the solution 0.1 mL more isoamyl nitrite, add 130 mg more anthranilic acid to the filter paper, replace the paper in the reaction tube, and again leach out the anthranilic acid with 0.5 mL of dimethoxyethane added dropwise over a 20-minute period. Reflux for 10 minutes more; then add 0.5 mL of ethanol and a solution of 0.15 g of sodium hydroxide in 2 mL of water to produce a suspension of solid in a brown alkaline solvent. Cool the mixture thoroughly in ice and also cool a 4:1 methanol-water mixture for rinsing.

Collect the solid on a Hirsch funnel by vacuum filtration, and wash it with the chilled solvent to remove the brown mother liquor. Transfer the moist, nearly colorless solid to a tared, clean reaction tube and evacuate while warming until the

[1]Practical grade anthracene and anthranilic acid are satisfactory.

$CH_3OCH_2CH_2OCH_2$
|
$CH_3OCH_2CH_2OCH_2$

Triethylene glycol dimethyl ether (triglyme)
bp 222°C
Miscible with water

Photo: *Crystallization*; Video: *Recrystallization*

weight is constant; the anthracene-triptycene mixture (mp about 190°C–230°C) weighs about 100 mg.

Add 50 mg of maleic anhydride and 1 mL of triethylene glycol dimethyl ether (triglyme, bp 222°C) and reflux the mixture for 5 minutes. Cool to about 100°C, add 0.5 mL of ethanol and a solution of 150 mg of sodium hydroxide in 2 mL of water; then cool in ice, along with 1.25 mL of a 4:1 methanol-water mixture for rinsing. Triptycene separates as nearly white crystals from the slightly brown alkaline solution. The washed and dried product weighs about 75 mg and melts at 255°C. Recrystallize it from methylcyclohexane (23 mL/g). If an insoluble black impurity is seen in the hot solution, transfer the clear solution to a clean tube using a Pasteur pipette, leaving behind the impurity. Slow cooling will produce flat, rectangular, laminated prisms that can be isolated by removing the solvent with a Pasteur pipette.

Cleaning Up. Dilute the alkaline filtrate from the reaction with water and flush it down the drain. Methylcyclohexane mother liquor from the crystallization goes in the organic solvents waste container.

Macroscale Procedure

$CH_3OCH_2CH_2OCH_3$

1,2-Dimethoxyethane
bp 83°C

Test dimethoxyethane for peroxides. *See* Chapter 2 for the peroxides text.

Carry out the reaction in a hood.

Technique for the slow addition of a solid.

Reaction time: 1 hour

Place 2 g of anthracene,[2] 2 mL of isoamyl nitrite, and 20 mL of 1,2-dimethoxyethane in a 100-mL round-bottomed flask mounted in a hot sand bath and fitted with a short reflux condenser (Fig. 35.1). Insert a filter paper into a 55-mm short stem funnel, moisten the paper with 1,2-dimethoxyethane, and rest the funnel in the top of the condenser. Weigh 2.6 g of anthranilic acid on a folded filter paper, scrape the acid into the funnel with a spatula, and pack it down onto the filter paper. Bring the mixture in the reaction flask to a gentle boil and note that the anthracene does not all dissolve. Measure 20 mL of 1,2-dimethoxyethane into a graduated cylinder, and use a Pasteur pipette to add small portions of the solvent to the anthranilic acid in the funnel—to slowly leach the acid into the reaction flask. If you make sure that the condenser is exactly vertical and the top of the funnel is centered, it should be possible to arrange for each drop to fall free into the flask and not touch the condenser wall. Once dripping from the funnel has started, add fresh batches of solvent to the acid, but only 2–3 drops at a time. Plan to complete leaching the first charge of anthranilic acid in a period of not less than 20 minutes, using about 10 mL of solvent. Then add a second 2.6-g portion of anthranilic acid to the funnel, remove from the heat and run in 2 mL of isoamyl nitrite through the condenser by slightly lifting the funnel. Replace the funnel, resume heating, and leach the anthranilic acid as before in about 20 minutes time. Reflux for an additional 10 minutes; then add 10 mL of 95% ethanol and a solution of 3 g of sodium hydroxide in 40 mL of water to produce a suspension of solid in a brown alkaline liquor. Cool thoroughly in ice; also cool a 4:1 methanol-water mixture for rinsing.

$CH_3OCH_2CH_2OCH_2$
|
$CH_3OCH_2CH_2OCH_2$

Triethylene glycol dimethyl ether (triglyme)
bp 222°C
Miscible with water

Collect the solid on a small Büchner funnel and wash it with the chilled solvent to remove the brown mother liquor. Transfer the moist, nearly colorless solid to a tared 100-mL round-bottomed flask, and evacuate on a steam bath until the weight is constant; the anthracene–triptycene mixture (mp about 190°C–230°C) weighs 2.1 g. Add 1 g of maleic anhydride and 20 mL of triethylene glycol dimethyl ether (triglyme, bp 222°C), heat the mixture to the boiling point under reflux (*see* Fig. 35.1 for the apparatus), and reflux for 5 minutes. Cool to about 100°C, add 10 mL of 95% ethanol and a solution of 3 g of sodium hydroxide in 40 mL of water; then cool in ice, along with 25 mL of a 4:1 methanol-water mixture for rinsing.

[2]Practical grade anthracene and anthranilic acid are satisfactory.

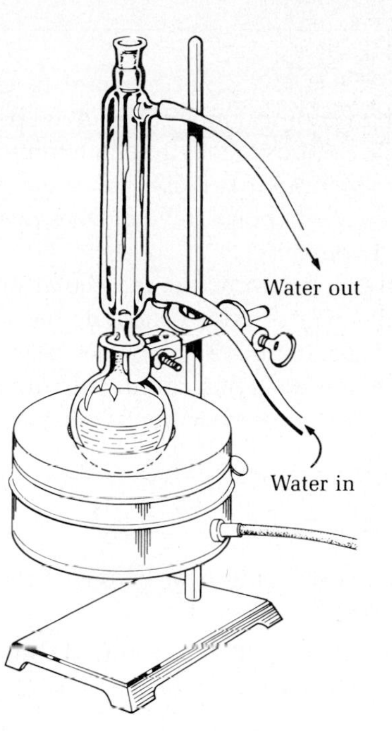

FIG. 35.1
Refluxing a reaction mixture.

Triptycene separates as nearly white crystals from the slightly brown alkaline liquor. The washed and dried product weighs 1.5 g and melts at 255°C. It will appear to be colorless, but it contains a trace of black insoluble material.

Dissolve the product in methylcyclohexane (solubility is 23 mL/g), decant from the specks of black material, and allow the solution to cool slowly. The product crystallizes as flat, rectangular, laminated prisms.

Cleaning Up. Dilute the alkaline filtrate from the reaction with water, and flush it down the drain. Methylcyclohexane mother liquor from the crystallization goes in the organic solvents waste container.

For Further Investigation

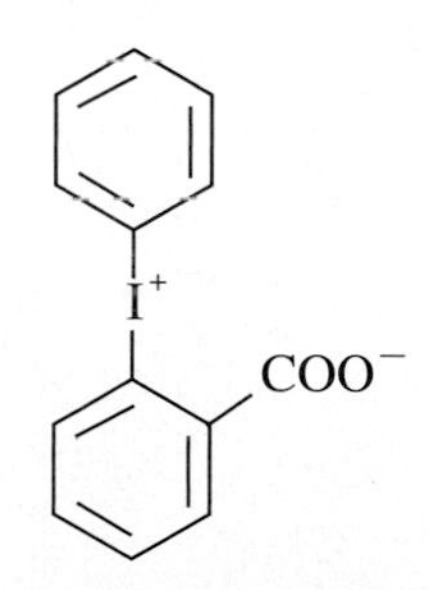

Diphenyliodonium-2-carboxylate

Diphenyliodonium-2-carboxylate is utilized as a benzyne precursor in the synthesis of 1,2,3,4-tetraphenylnaphthalene from tetraphenylcyclopentadienone. For rea sons unknown, the reaction of anthracene with benzyne generated in this way proceeds very poorly and gives only a trace of triptycene. What about the converse proposition? Would benzyne generated by diazotization of anthranilic acid in the presence of tetraphenylcyclopentadienone afford 1,2,3,4-tetraphenylnaphthalene in a satisfactory yield? If you are interested in exploring this possibility, plan and execute a procedure and see what you can discover.

If benzyne is generated from anthranilic acid in the absence of anthracene, as in this experiment, then dibenzocyclobutadiene (also called biphenylene) is formed.[3]

Devise a procedure for isolating this interesting hydrocarbon.

[3]Logullo FM, Seitz AH, Friedman L. *Org Syn*. **1968**;*48*.12, Friedman L, Logullo FM. *J Org Chem*. **1969**;*34*:3089.

COMPUTATIONAL CHEMISTRY

Using the AM1 semiempirical molecular orbital program, calculate the heats of formation of isomeric compounds **2a** and **2b** to determine which might be formed in this reaction. Bear in mind that these will be gas phase heats of formation. Repeat the calculation in water, a polar solvent. Does this change the relative stabilities of the two isomers? Explain.

It is instructive to examine the molecular orbitals of benzyne and dibenzocyclobutadiene. Benzyne is made by joining two *ortho* hydrogens into a new bond. Does the central cyclobutadiene ring, which is antiaromatic, have any effect on the appearance of the orbitals when compared with, say, naphthalene? Compare the bond lengths of the carbon-carbon bonds of benzyne to those of benzene.

QUESTIONS

1. Write equations showing how benzyne might be generated from *o*-fluorobromobenzene and magnesium.
2. Study the mechanism of diazonium salt formation in Chapter 44, and then propose a mechanism for diazotization using isoamyl nitrite.
3. The ^{1}H NMR spectrum of anthracene is given in Figure 35.2. Predict the appearance of the NMR spectrum of triptycene. Which peak in Figure 35.2 will be missing?
4. Draw the structure of the compound formed when maleic anhydride reacts with anthracene. This is a Diels–Alder reaction involving carbons 9 and 10 of anthacene.

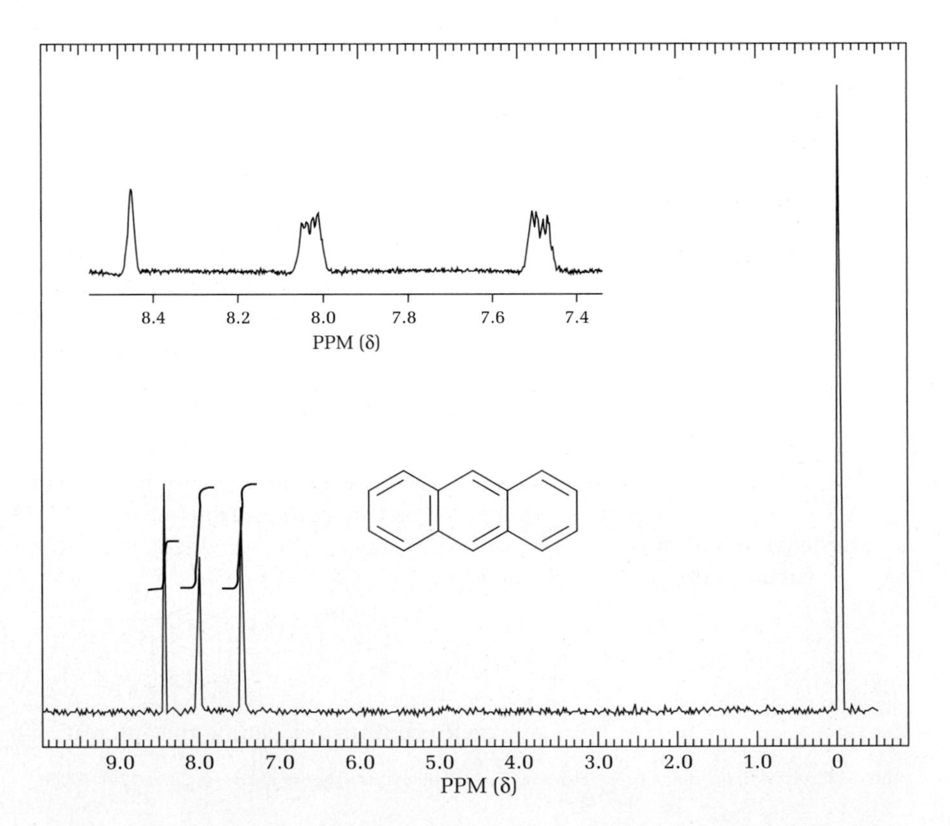

FIG. 35.2
The ^{1}H NMR spectrum of anthracene (250 MHz).

CHAPTER 36

Aldehydes and Ketones

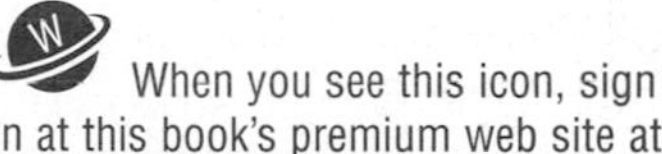

When you see this icon, sign in at this book's premium web site at **www.cengage.com/login** to access videos, Pre-Lab Exercises and other online resources.

PRELAB EXERCISE: Outline a logical series of experiments designed to identify an unknown aldehyde or ketone with the least effort. Consider the time required to complete each identification reaction.

The carbonyl group occupies a central place in organic chemistry. Aldehydes and ketones—compounds such as formaldehyde, acetaldehyde, acetone, and 2-butanone—are very important industrial chemicals used by themselves and as starting materials for a host of other substances. For example, more than 10 billion pounds (4.5 billion kilograms) of formaldehyde-containing plastics are produced in the United States each year.

The carbonyl carbon is sp^2 hybridized, the bond angles between adjacent groups are 120°, and the four atoms R, R′, C, and O lie in one plane:

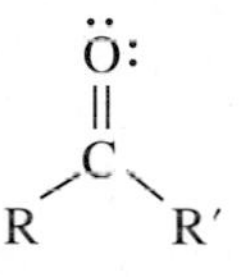

The electronegative oxygen polarizes the carbon-oxygen bond, rendering the carbon electron deficient and hence subject to nucleophilic substitution.

$$\rangle C{=}\ddot{O}: \longleftrightarrow \rangle \overset{+}{C}{-}\ddot{O}:^{-} \quad \text{or} \quad \rangle \overset{\delta+}{C}{=}\overset{\delta-}{\ddot{O}}:$$

Geometry of the carbonyl group

Attack on the sp^2 hybridized carbon occurs via the π-electron cloud above and below the plane of the carbonyl group:

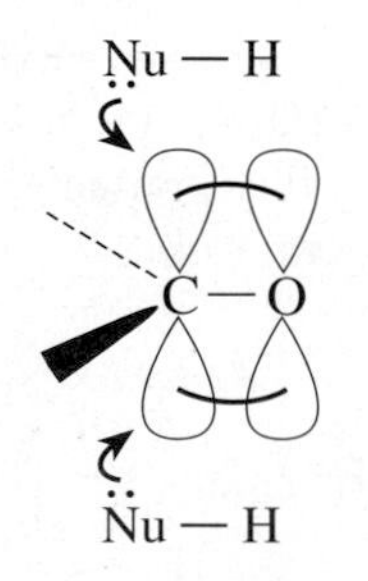

REACTIONS OF THE CARBONYL GROUP

Many reactions of carbonyl groups are acid catalyzed. The acid attacks the electronegative oxygen, which bears a partial negative charge, to create a carbocation that subsequently reacts with the nucleophile:

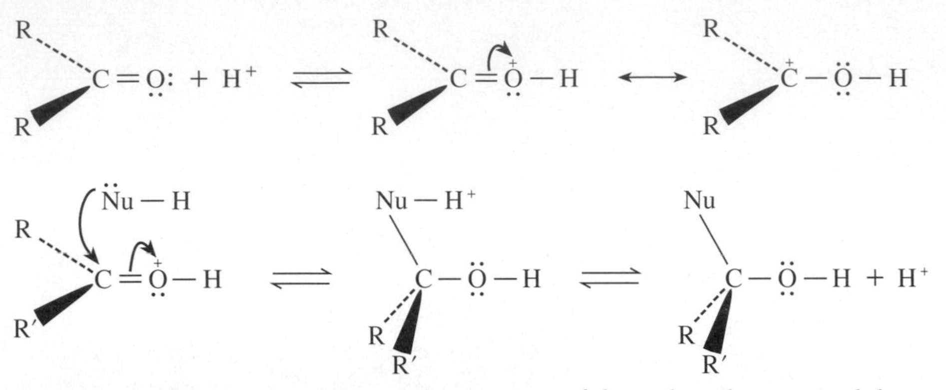

The strength of the nucleophile and the structure of the carbonyl compound determine whether the equilibrium lies on the side of the carbonyl compound or the tetrahedral adduct. Water, a weak nucleophile, does not usually add to the carbonyl group to form a stable compound:

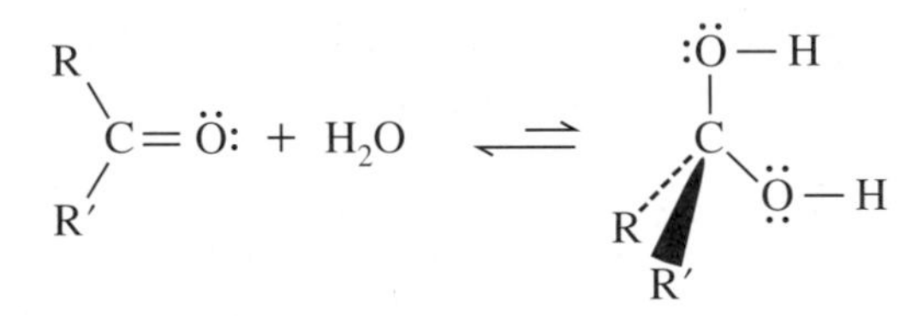

In the special case of trichloroacetaldehyde, however, the electron-withdrawing trichloromethyl group allows a stable hydrate to form:

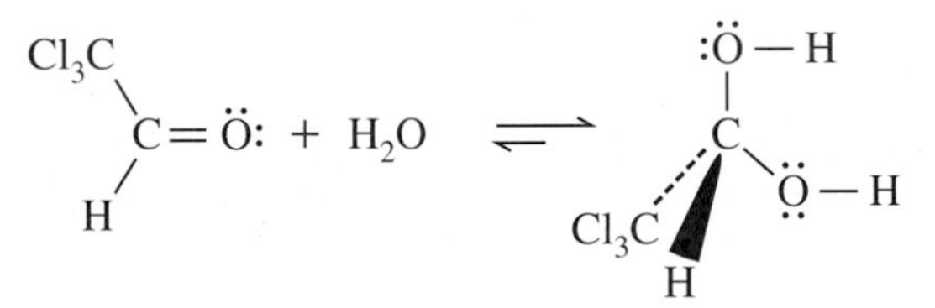

A stable hydrate: chloral hydrate

The compound so formed, chloral hydrate, was discovered by Justus von Liebig in 1832 and was introduced as one of the first sedatives and hypnotics (sleep-inducing substances) in 1869. It is now most commonly encountered in detective fiction as a "Mickey Finn" or "knockout drops."

Reaction with an alcohol; hemiacetals

In an analogous manner, an aldehyde or ketone can react with an alcohol. The product, a hemiacetal or hemiketal, is usually not stable, but in the case of certain cyclic hemiacetals, the product can be isolated. Glucose is an example of a stable hemiacetal.

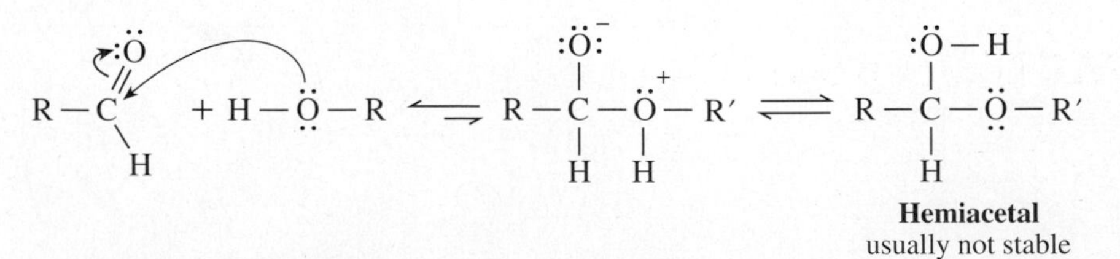

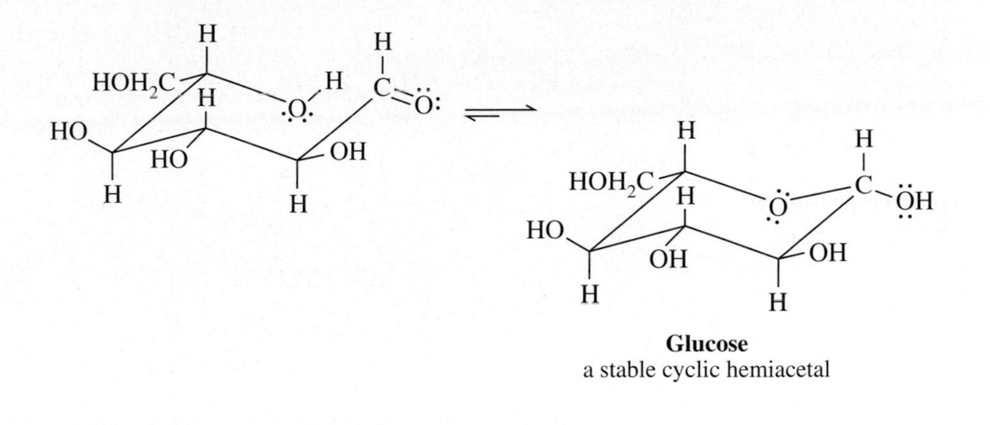

Glucose
a stable cyclic hemiacetal

BISULFITE ADDITION

The bisulfite ion is a strong nucleophile but a weak acid. It will attack the unhindered carbonyl group of an aldehyde or methyl ketone to form an addition product:

$$\mathrm{RHC{=}\ddot{O}{:}} + \mathrm{{:}SO_3H^-\,Na^+} \rightleftharpoons \mathrm{R{-}CH(O{:}^-\,Na^+){-}SO_3H} \longrightarrow \mathrm{R{-}CH(\ddot{O}{-}H){-}SO_3^-\,Na^+}$$

Because these bisulfite addition compounds are ionic water-soluble compounds and can be formed with a maximum 90% yield, they serve as a useful means of separating aldehydes and methyl ketones from mixtures of organic compounds. At high sodium bisulfite concentrations, these adducts crystallize and can be isolated by filtration. The aldehyde or ketone can be regenerated by adding either a strong acid or base:

$$\mathrm{C_6H_5{-}CH(\ddot{O}{-}H){-}SO_3^-\,Na^+} + \mathrm{HCl} \longrightarrow \mathrm{C_6H_5{-}CH{=}\ddot{O}{:}} + \mathrm{SO_2} + \mathrm{H_2O} + \mathrm{NaCl}$$

$$\mathrm{CH_3CH_2{-}C(\ddot{O}{-}H)(CH_3){-}SO_3^-\,Na^+} + \mathrm{NaOH} \longrightarrow \mathrm{CH_3CH_2{-}C({=}\ddot{O}{:}){-}CH_3} + \mathrm{Na_2SO_3} + \mathrm{H_2O}$$

CYANIDE ADDITION

A similar reaction occurs between aldehydes and ketones and hydrogen cyanide, which, like bisulfite, is a weak acid but a strong nucleophile. The reaction is hazardous to carry out because of the toxicity of cyanide, but the cyanohydrins are useful synthetic intermediates:

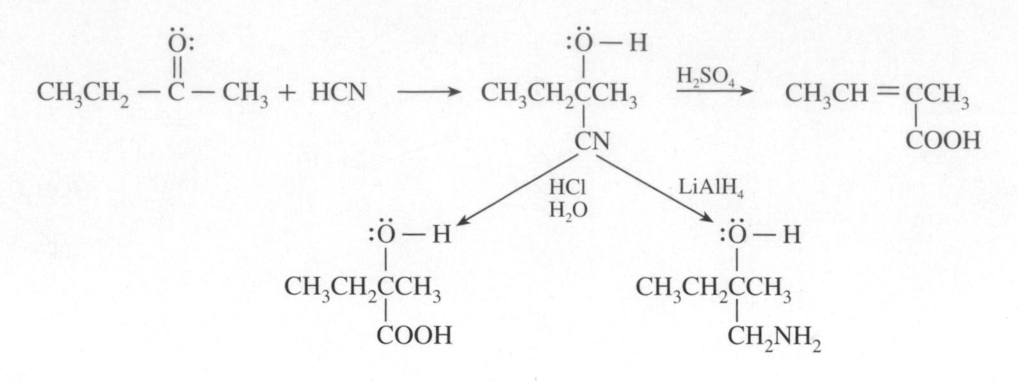

Cyanohydrin formation and reactions

Amines are good nucleophiles and readily add to the carbonyl group:

$$R-CH=\ddot{O} + H_2\ddot{N}R' \rightleftharpoons R-CH(\ddot{O}^-)-\overset{+}{N}H_2R' \rightleftharpoons R-CH(\ddot{O}H)-\ddot{N}HR'$$

The reaction is strongly dependent on the pH. In acid, the amine is protonated (RN^+H_3) and is no longer a nucleophile. In strong base, there are no protons available to catalyze the reaction. But in weak acid solution (pH 4–6), the equilibrium between acid and base (**a**) is such that protons are available to protonate the carbonyl (**b**), and yet there is free amine present to react with the protonated carbonyl (**c**):

(a) $$CH_3\ddot{N}H_2 + HCl \rightleftharpoons CH_3\overset{+}{N}H_3 + Cl^-$$

(b) $$CH_3-CH=\ddot{O} + HCl \rightleftharpoons \left[CH_3-CH=\overset{+}{O}-H \longleftrightarrow CH_3-\overset{+}{C}H-\ddot{O}-H\right] + Cl^-$$

(c) $$CH_3-CH=\overset{+}{O}-H + CH_3\ddot{N}H_2 \rightleftharpoons CH_3-CH(\ddot{O}H)-\overset{+}{N}H_2CH_3 \xrightleftharpoons{-H^+} CH_3-CH(\ddot{O}H)-\ddot{N}H-CH_3$$

SCHIFF BASES

The intermediate hydroxyamino form of the adduct is not stable and spontaneously dehydrates under the mildly acidic conditions of the reaction to give an imine, commonly referred to as a *Schiff base*:

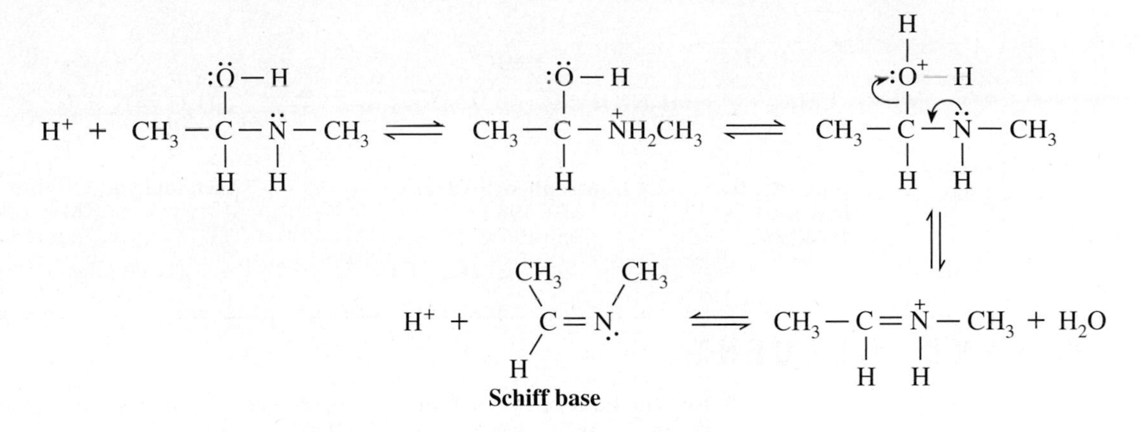

Imine or Schiff base formation

The biosynthesis of most amino acids proceeds through Schiff base intermediates.

OXIMES, SEMICARBAZONES, AND 2,4-DINITROPHENYLHYDRAZONES

Three rather special amines form useful stable imines:

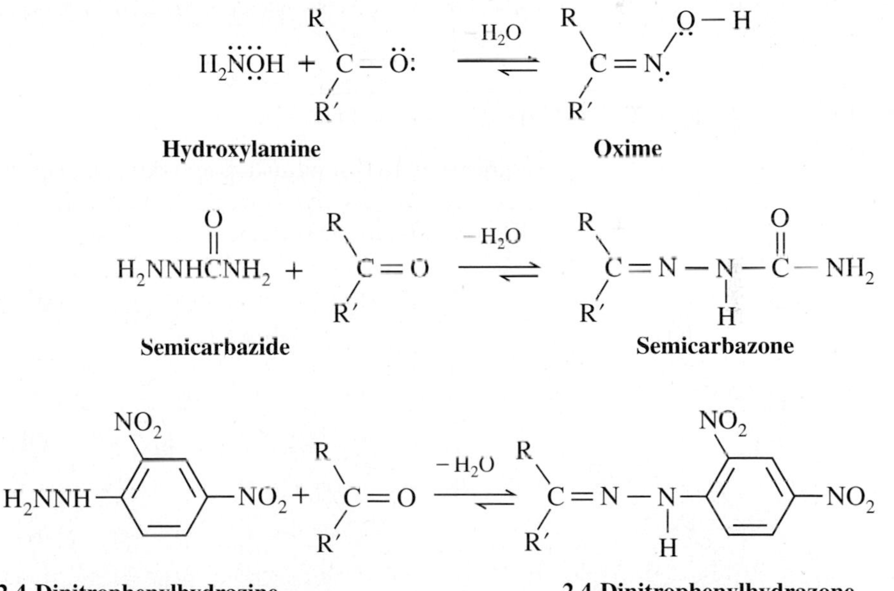

These imines are solids and are useful for the characterization of aldehydes and ketones. For example, IR (infrared) and NMR (nuclear magnetic resonance) spectroscopies may indicate that a certain unknown is acetaldehyde. It is difficult to determine the boiling point of a few milligrams of a liquid, but if it can be converted to a solid derivative, the melting point *can* be determined with that amount. The 2,4-dinitrophenylhydrazones are usually the derivatives of choice because they are crystalline compounds with well-defined melting or decomposition points, and they increase the molecular weight by 180. Ten milligrams of acetaldehyde will give 51 mg of the 2,4-dinitrophenylhydrazone.

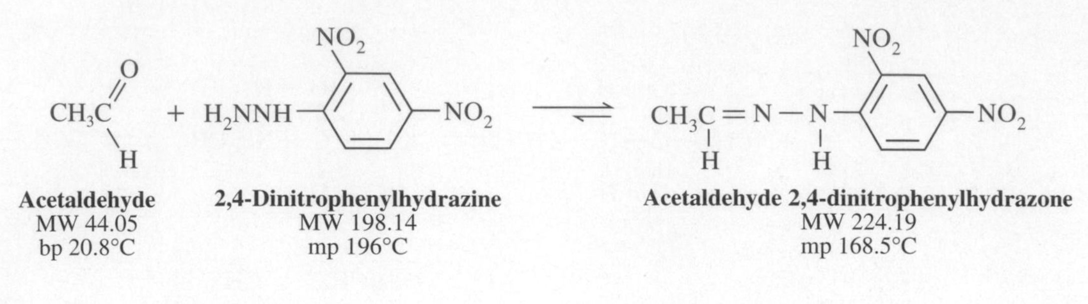

TOLLENS' REAGENT

Before the advent of NMR and IR spectroscopy and mass spectrometry, the chemist was often called on to identify aldehydes and ketones by purely chemical means. Aldehydes can be distinguished chemically from ketones by their ease of oxidation to carboxylic acids. The oxidizing agent, an ammoniacal solution of silver nitrate—Tollens' reagent—is reduced to metallic silver, which is deposited on the inside of a test tube as a silver mirror.

$$2\,Ag(NH_3)_2OH + R{-}C(=O)H \longrightarrow 2\,Ag + R{-}C(=O)O^- \quad NH_4^+ + H_2O + 3\,NH_3$$

SCHIFF'S REAGENT

Another way to distinguish aldehydes from ketones is to use Schiff's reagent. This is a solution of the red dye Basic Fuchsin, which is rendered colorless on treatment with sulfur dioxide. In the presence of an aldehyde, the colorless solution turns magenta.

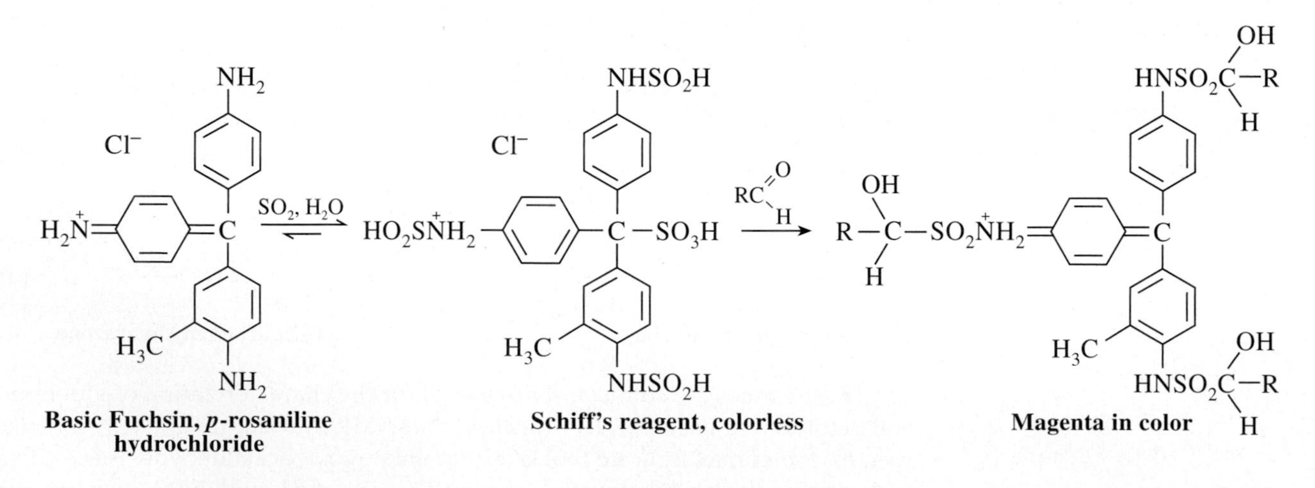

Basic Fuchsin, *p*-rosaniline hydrochloride **Schiff's reagent, colorless** **Magenta in color**

IODOFORM TEST

A test for methyl ketones

Methyl ketones can be distinguished from other ketones by the iodoform test. The methyl ketone is treated with iodine in a basic solution. Introduction of the first iodine atom increases the acidity of the remaining methyl protons, so halogenation

$$\underset{\textbf{Methyl ketone}}{CH_3\overset{O}{\overset{\|}{C}}—}$$

stops only when the triiodo compound has been produced. The base then allows the relatively stable triiodomethyl carbanion to leave, and a subsequent proton transfer gives iodoform, a yellow crystalline solid with a melting point of 119°C–123°C. The test is also positive for fragments or compounds easily oxidized to methyl ketones, such as the fragment $CH_3\overset{|}{C}HOH$ or the compound ethanol. Acetaldehyde also gives a positive test because it is both a methyl ketone and an aldehyde.

$$R—\overset{\ddot{O}:}{\overset{\|}{C}}—CH_3 + OH^- \rightleftharpoons R—\overset{\ddot{O}:}{\overset{\|}{C}}—\bar{\ddot{C}}H_2 + H_2O$$

$$R—\overset{\ddot{O}:}{\overset{\|}{C}}—\bar{\ddot{C}}H_2 + I_2 \longrightarrow R—\overset{O}{\overset{\|}{C}}—CH_2I + I^- \xrightarrow{OH^-} \xrightarrow{I_2} \xrightarrow{OH^-} \xrightarrow{I_2} R—\overset{\ddot{O}:}{\overset{\|}{C}}—CI_3$$

$$R—\overset{\ddot{O}:}{\overset{\|}{C}}—CI_3 + OH^- \longrightarrow R—\overset{:\ddot{O}:^-}{\underset{:\ddot{O}H}{C}}—CI_3 \longrightarrow R—C(=\ddot{O}:)—\ddot{O}—H + :\bar{C}I_3 \longrightarrow R—C(=\ddot{O}:)—\ddot{O}:^- + \underset{\substack{\textbf{Iodoform} \\ \text{mp 123°C}}}{CHI_3}$$

EXPERIMENTS

1. UNKNOWNS

Carry out three tests:

- Known positive
- Known negative
- Unknown

You will be given an unknown that may be any of the aldehydes or ketones listed in Table 36.1. At least one derivative of the unknown is to be submitted to your instructor, but if you first do the bisulfite and iodoform characterizing tests, the results may suggest derivatives whose melting points will be particularly revealing.

You can further characterize the unknown by determining its boiling point, which is best done with a digital thermometer and a reaction tube (*see* Chapter 5). The boiling points of the unknowns are given on this book's web site.

In conducting the following tests, you should perform three tests simultaneously: on a compound known to give a positive test, on a compound known to give a negative test, and on the unknown. In this way you will be able to determine whether the reagents are working as they should, as well as interpret a positive or a negative test.

2. 2,4-DINITROPHENYLHYDRAZONES

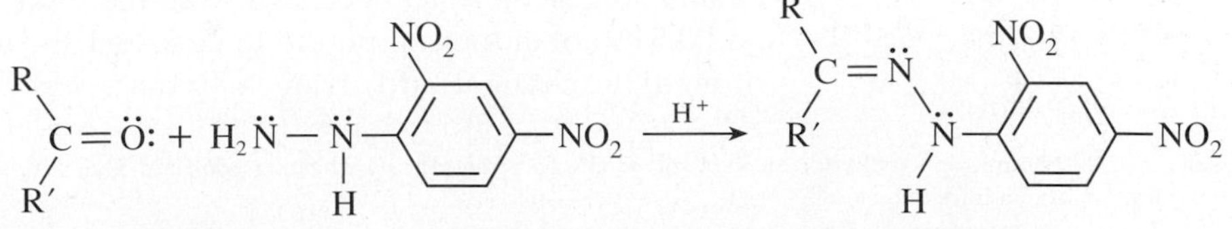

TABLE 36.1 *Melting Points of Derivatives of Some Aldehydes and Ketones*

Compound[a]	*Formula*	n_D^{20}	*MW*	*Water Solubility*	Melting Points (°C) Phenyl-hydrazone	2,4-dinitro-phenyl-hydrazone	Semi-carbazone
Acetone	CH_3COCH_3	1.3590	58.08		42	126	187
n-Butanal	$CH_3CH_2CH_2CHO$	1.3790	72.10	4 g/100 g	Oil	123	95 (106)[b]
3-Pentanone (diethyl ketone)	$CH_3CH_2COCH_2CH_3$	1.3920	86.13	4.7 g/100 g	Oil	156	138
2-Furaldehyde (furfural)	$C_4H_3O\text{-}CHO$	1.5260	96.08	9 g/100 g	97	212 (230)[b]	202
Benzaldehyde	C_6H_5CHO	1.5450	106.12	Insol.	158	237	222
Hexane-2,5-dione	$CH_3COCH_2CH_2COCH_3$	1.4260	114.14	∞	120[c]	257[c]	224[c]
2-Heptanone	$CH_3(CH_2)_4COCH_3$	1.4080	114.18	Insol.	Oil	89	123
3-Heptanone	$CH_3(CH_2)_3COCH_2CH_3$	1.4080	114.18	Insol.	Oil	81	101
n-Heptanal	$n\text{-}C_6H_{13}CHO$	1.4125	114.18	Insol.	Oil	108	109
Acetophenone	$C_6H_5COCH_3$	1.5325	120.66	Insol.	105	238	198
2-Octanone	$CH_3(CH_2)_5COCH_3$	1.4150	128.21	Insol.	Oil	58	122
Cinnamaldehyde	$C_6H_5CH{=}CHCHO$	1.6220	132.15	Insol.	168	255	215
Propiophenone	$C_6H_5COCH_2CH_3$	1.5260	134.17	Insol.	About 48	191	182

[a]Visit this book's web site for data on additional aldehydes and ketones.

[b]Both melting points have been found, depending on crystalline form of derivative.

[c]Monoderivative or diderivative.

An easily prepared derivative of aldehydes and ketones.

Video: *Microscale Crystallization*

To 5 mL of the stock solution[1] of 2,4-dinitrophenylhydrazine in phosphoric acid, add about 0.05 g of the compound to be tested. Five milliliters of the 0.1 *M* solution contains 0.5 mmol (0.0005 mol) of the reagent. If the compound to be tested has a molecular weight of 100, then 0.05 g is 0.5 mmol. Warm the reaction mixture for a few minutes in a water bath, and then let crystallization proceed. Collect the product by suction filtration (Fig. 36.1), wash the crystals with a large amount of water to remove all of the phosphoric acid, press a piece of moist litmus paper onto the crystals, and wash them with more water if they are acidic. Press the product between sheets of filter paper until it is as dry as possible and recrystallize from ethanol. Occasionally, a high molecular weight derivative will not dissolve in a reasonable quantity (10 mL) of ethanol. In that case, cool the hot suspension and isolate the crystals by suction filtration. The boiling ethanol treatment removes impurities so that an accurate melting point can be obtained on the isolated material.

An alternative procedure is applicable when the 2,4-dinitrophenylhydrazone is known to be sparingly soluble in ethanol. Measure 0.5 mmol of crystalline 2,4-dinitrophenylhydrazine into a 50-mL Erlenmeyer flask, add 15 mL of 95% ethanol, digest on a steam bath until all the solid particles are dissolved, and then add 0.5 mmol of the compound to be tested and continue warming. If there is no immediate change, add, from a Pasteur pipette, 3–4 drops of concentrated

[1]Dissolve 2.0 g of 2,4-dinitrophenylhydrazine in 50 mL of 85% phosphoric acid by heating, cool, add 50 mL of 95% ethanol, cool again, and clarify by suction filtration from a trace of residue.

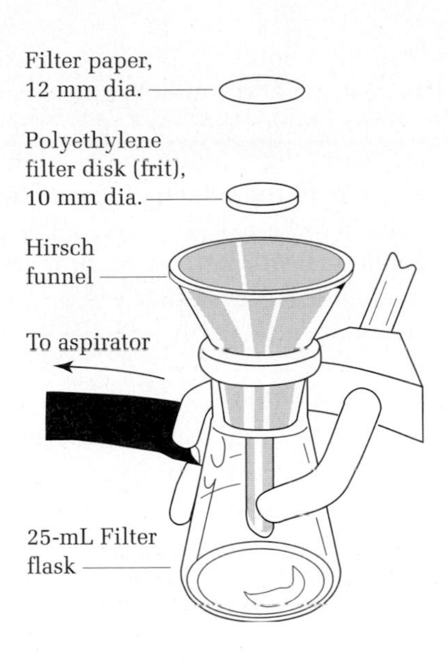

FIG. 36.1
A Hirsch funnel filtration apparatus.

hydrochloric acid as a catalyst and note the result. Warm for a few minutes, then cool and collect the product. This procedure would be used for an aldehyde like cinnamaldehyde ($C_6H_5CH{=}CHCHO$).

The alternative procedure strikingly demonstrates the catalytic effect of hydrochloric acid, but it is not applicable to a substance like diethyl ketone, whose 2,4-dinitrophenylhydrazone is much too soluble to crystallize from the large volume of ethanol. The first procedure is obviously the one to use for an unknown.

Cleaning Up. The filtrate from the preparation of the 2,4-dinitrophenylhydrazone should have very little 2,4-dinitrophenylhydrazine in it, so after dilution with water and neutralization with sodium carbonate, it can be flushed down the drain. Similarly, the mother liquor from crystallization of the phenylhydrazone should have very little product in it, and so should be diluted and flushed down the drain. If solid material is detected, it should be collected by suction filtration, the filtrate flushed down the drain, and the filter paper placed in the solid hazardous waste.

3. SEMICARBAZONES

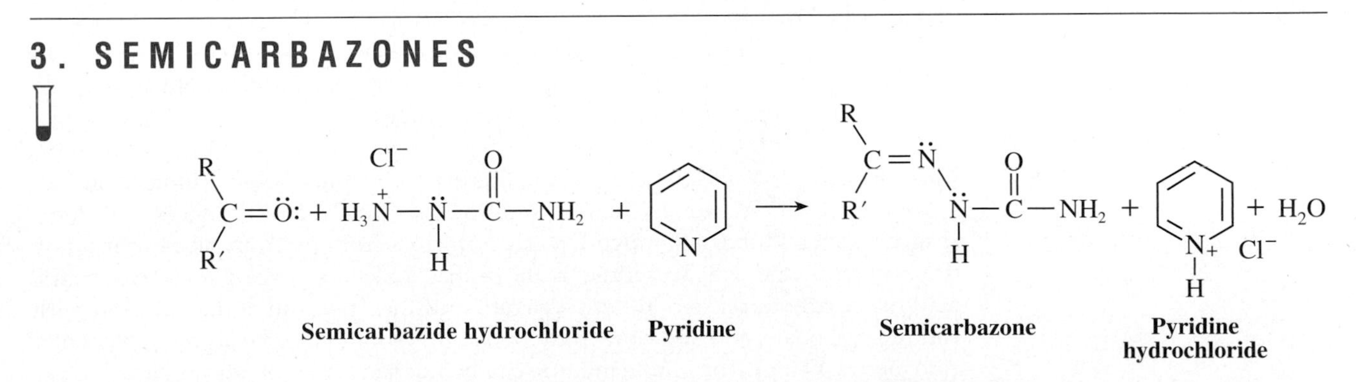

Semicarbazide (mp 96°C) is not very stable in the free form and is used as the crystalline hydrochloride (mp 173°C). Because this salt is insoluble in methanol or ethanol and does not react readily with typical carbonyl compounds in alcohol-water mixtures, pyridine, a basic reagent, is added to liberate free semicarbazide.

Photo: *Use of the Wilfilter*

To 0.5 mL of the stock solution[2] of semicarbazide hydrochloride, which contains 1 mmol of the reagent, add 1 mmol of the compound to be tested and enough methanol (1 mL) to produce a clear solution; then add 10 drops of pyridine (a twofold excess) and warm the solution gently on the steam bath for a few minutes. Cool the solution slowly to room temperature. It may be necessary to scratch the inside of the test tube in order to induce crystallization. Cool the tube in ice, collect the product by suction filtration, and wash it with water followed by a small amount of cold methanol. Recrystallize the product from methanol, ethanol, or ethanol/water. The product can easily be collected on a Wilfilter.

Cleaning Up. Combine the filtrate from the reaction and the mother liquor from the crystallization, dilute with water, make very slightly acidic with dilute hydrochloric acid, and flush the mixture down the drain.

4. TOLLENS' TEST

Test for aldehydes

$$R{-}C(=O)H + 2\ Ag(NH_3)_2OH \longrightarrow 2\ Ag + RCOO^-NH_4^+ + H_2O + 3\ NH_3$$

Clean four or five test tubes by adding a few milliliters of 3 *M* sodium hydroxide solution to each and heating them in a water bath while preparing the Tollens' reagent.

To 2.0 mL of 0.03 *M* silver nitrate solution, add 1.0 mL of 3 *M* sodium hydroxide in a test tube. To the gray precipitate of silver oxide, Ag_2O, add 0.5 mL of a 2.8% aqueous ammonia solution (10 mL of concentrated ammonium hydroxide diluted to 100 mL). Stopper the tube and shake it. Repeat the process until *almost* all of the precipitate dissolves (3.0 mL of ammonia at most); then dilute the solution with water to 10 mL. Empty the test tubes of sodium hydroxide solution, rinse them, and add 1 mL of Tollens' reagent to each. Add 1 drop (no more) of the substance to be tested by allowing it to run down the inside of the inclined test tube. Set the tubes aside for a few minutes without agitating the contents. If no reaction occurs, warm the mixture briefly on a water bath. As a known aldehyde, try 1 drop of a 0.1 *M* solution of glucose. A more typical aldehyde to test is benzaldehyde.

At the end of the reaction, promptly destroy any excess Tollens' reagent with nitric acid because it can form an explosive fulminate on standing. Nitric acid can also be used to remove silver mirrors from test tubes.

Cleaning Up. Place all solutions used in this experiment in a beaker (unused ammonium hydroxide, sodium hydroxide solution used to clean out the tubes, Tollens' reagent from all tubes). Remove any silver mirrors from reaction tubes with a few drops of nitric acid, which is added to the beaker. Make the mixture acidic with nitric acid to destroy unreacted Tollens' reagent and then neutralize the solution with sodium carbonate, and add some sodium chloride solution to precipitate silver chloride (about 40 mg). The whole mixture can be flushed down the drain, or the silver chloride can be collected by suction filtration, and the filtrate flushed down the drain. The silver chloride would go in the nonhazardous solid waste container.

[2]Prepare a stock solution by dissolving 1.11 g of semicarbazide hydrochloride in 5 mL of water; 0.5 mL of this solution contains 1 mmol of reagent.

5. SCHIFF'S TEST

Very sensitive test for aldehydes.

Add 3 drops of the unknown to 2 mL of Schiff's reagent.[3] A magenta color will appear within 10 minutes if aldehydes are present. As in all of these tests, compare the colors produced by a known aldehyde, a known ketone, and the unknown compound.

Cleaning Up. Neutralize the solution with sodium carbonate and flush it down the drain. The amount of *p*-rosaniline in this mixture is negligible (1 mg).

6. IODOFORM TEST

$$R-\overset{\overset{\displaystyle :\ddot{O}}{\|}}{C}-CH_3 + 3\,I_2 + 4\,OH^- \longrightarrow R-C\!\begin{smallmatrix}\nearrow :\ddot{O} \\ \searrow \ddot{O}:^-\end{smallmatrix} + CHI_3 + 3\,:\ddot{I}:^- + 3\,H_2O$$

A methyl ketone **Iodoform** mp 119–123°C

Test for methyl ketones

The reagent contains iodine in potassium iodide solution[4] at a concentration such that 2 mL of solution, on reaction with excess methyl ketone, will yield 174 mg of iodoform. If the substance to be tested is water soluble, dissolve 4 drops of a liquid or an estimated 50 mg of a solid in 2 mL of water in a 20 × 150-mm test tube; add 2 mL of 3 *M* sodium hydroxide, and then slowly add 3 mL of the iodine solution. In a positive test, the brown color of the reagent disappears, and yellow iodoform separates. If the substance to be tested is insoluble in water, dissolve it in 2 mL of 1,2-dimethoxyethane, proceed as above, and at the end, dilute with 10 mL of water.

Suggested test substances are hexane-2,5-dione (water soluble), *n*-butyraldehyde (water soluble), and acetophenone (water insoluble).

Videos: *Microscale Filtration on the Hirsch Funnel, Extraction with Dichloromethane*: Photo: *Use of the Wilfilter*

Iodoform can be recognized by its odor and yellow color and, more securely, from its melting point (119°C–123°C). The substance can be isolated by suction filtration of the test suspension or by adding 0.5 mL of dichloromethane, shaking the stoppered test tube to extract the iodoform into the small lower layer, withdrawing the clear part of this layer with a Pasteur pipette, and evaporating it in a small test tube on a steam bath. The crude solid is crystallized from a methanol-water mixture (*see* Chapter 4). It can be collected on a Wilfilter.

Cleaning Up. Combine all reaction mixtures in a beaker, add a few drops of acetone to destroy any unreacted iodine in the potassium iodide reagent, remove the iodoform by suction filtration, and place the iodoform in the halogenated organic waste container. The filtrate can be flushed down the drain after neutralization (if necessary).

[3]Schiff's reagent is prepared by dissolving 0.1 g Basic Fuchsin (*p*-rosaniline hydrochloride) in 100 mL of water and then adding 4 mL of a saturated aqueous solution of sodium bisulfite. After 1 hour, add 2 mL of concentrated hydrochloric acid.

[4]Dissolve 25 g of iodine in a solution of 50 g of potassium iodide in 200 mL of water.

7. BISULFITE TEST

Forms with unhindered carbonyls

$$\mathrm{R(H)C{=}\ddot{O}\!:} + \mathrm{Na^{+}SO_{3}H^{-}} \rightleftharpoons \mathrm{R{-}C(H)(\ddot{O}H){-}SO_{3}^{-}Na^{+}}$$

Put 1 mL of the stock solution[5] into a 13 × 100-mm test tube and add 5 drops of the substance to be tested. Shake each tube during the next 10 minutes, and note the results. A positive test will result from aldehydes, unhindered cyclic ketones such as cyclohexanone, and unhindered methyl ketones.

If the bisulfite test is applied to a liquid or solid that is very sparingly soluble in water, formation of the addition product is facilitated by adding a small amount of methanol before the addition to the bisulfite solution.

Cleaning Up. Dilute the bisulfite solution or any bisulfite addition products (they will dissociate) with a large volume of water, and flush the mixture down the drain. The amount of organic material being discarded is negligible.

8. IR AND NMR SPECTROSCOPY

IR spectroscopy is extremely useful in analyzing all carbonyl-containing compounds, including aldehydes and ketones (Fig. 36.4 and Fig. 36.11). Refer to the extensive discussion in Chapter 11. In the modern laboratory, spectroscopy has almost completely supplanted the qualitative tests described in this chapter. Figures 36.4 through 36.11 present IR and NMR spectra of typical aldehydes and ketones. Compare these spectra with those of your unknown. Also compare the IR and ^{1}H NMR spectra of the hydrocarbon fluorene (Fig. 36.2 and Fig. 36.3) with those of the ketone derivative, fluorenone (Fig. 36.4 and Fig. 36.5).

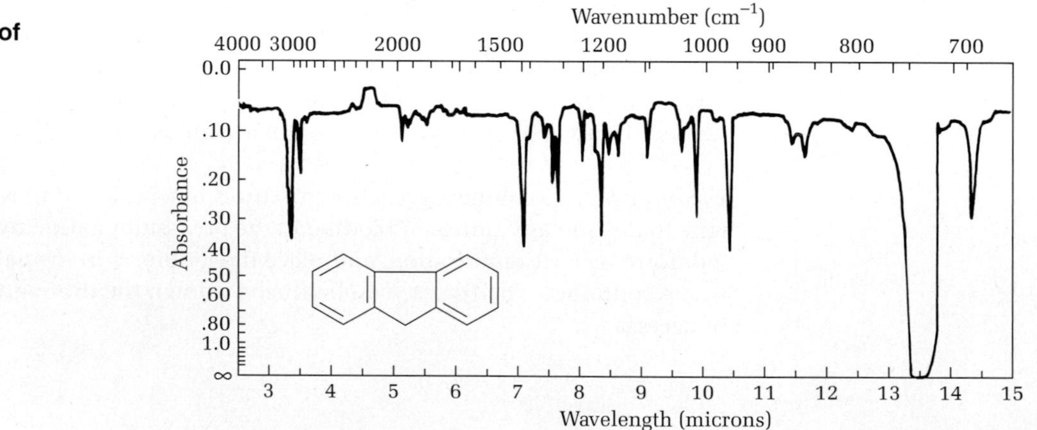

FIG. 36.2
The IR spectrum of fluorene in CS_2.

[5]Prepare a stock solution from 50 g of sodium bisulfite dissolved in 200 mL of water with brief swirling.

FIG. 36.3
The ^{1}H NMR spectrum of fluorene (250 MHz).

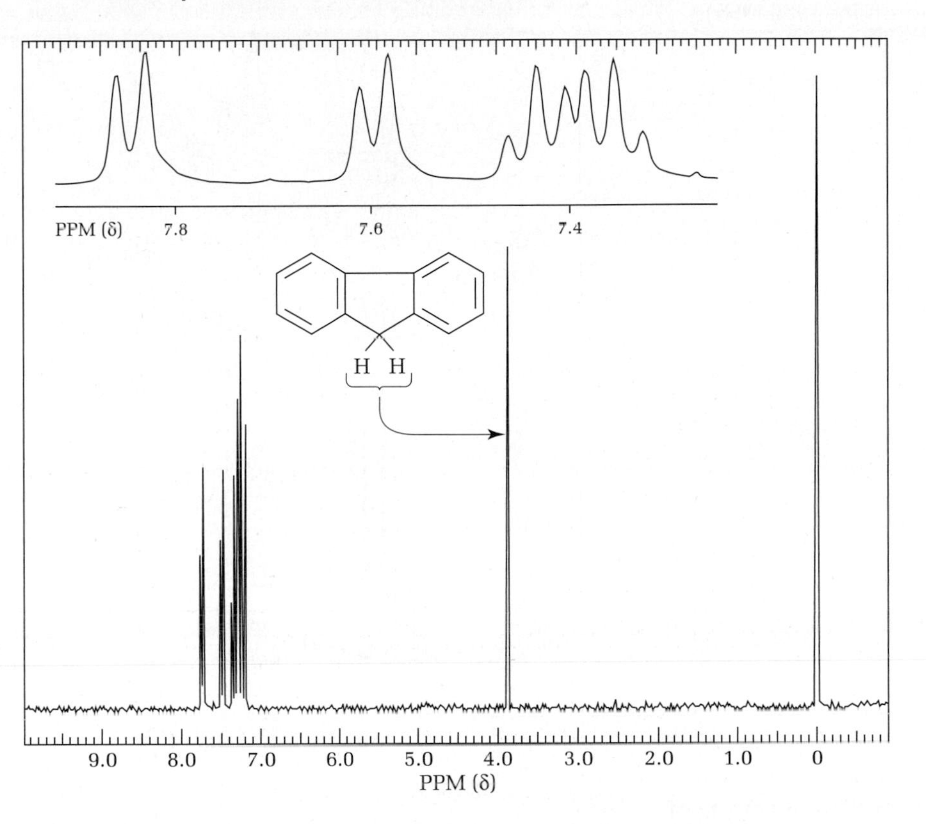

FIG. 36.4
The IR spectrum of fluorenone (KBr disk).

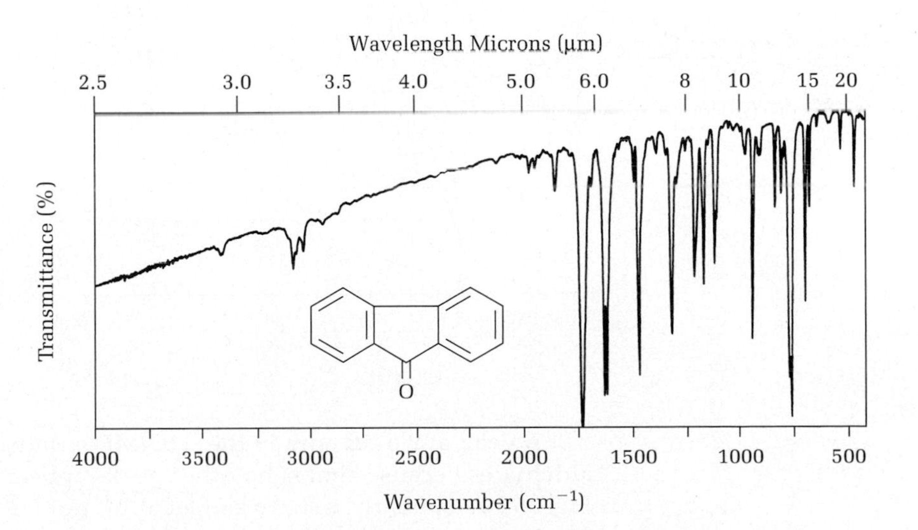

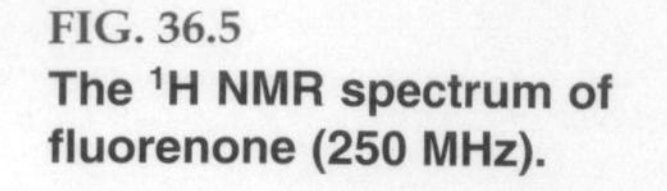

FIG. 36.5
The ^1H NMR spectrum of fluorenone (250 MHz).

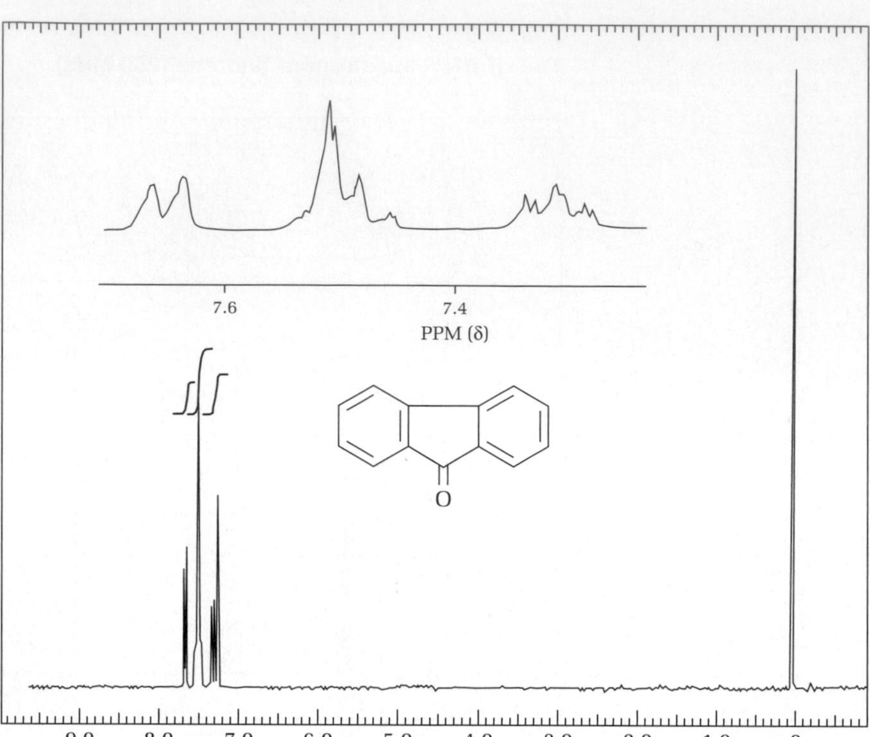

FIG. 36.6
The ^1H NMR spectrum of 2-butanone, $CH_3COCH_2CH_3$ (400 MHz).

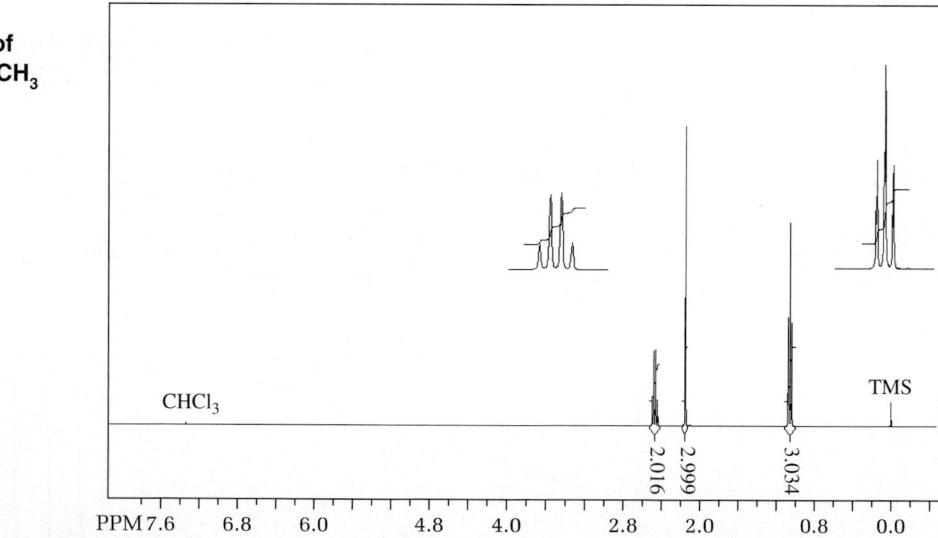

A peak at 9.6–10 ppm in the ^1H NMR spectrum is highly characteristic of aldehydes because almost no other peaks appear in this region (Fig. 36.7 and Fig. 36.10). Similarly, a sharp singlet at 2.2 ppm is very characteristic of methyl ketones; beware of contamination of the sample by acetone, which is often used to clean glassware.

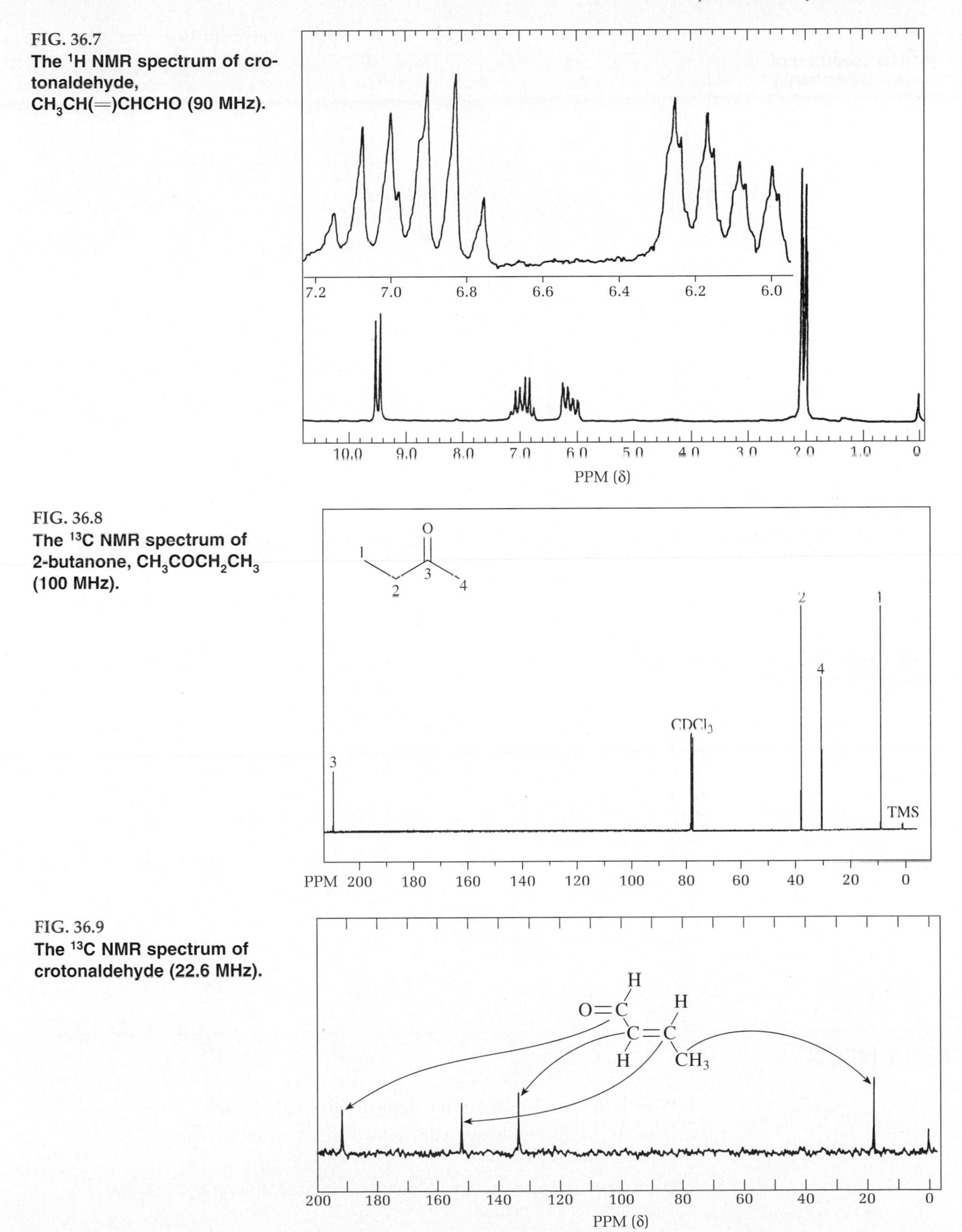

FIG. 36.7
The ¹H NMR spectrum of crotonaldehyde, $CH_3CH(=)CHCHO$ (90 MHz).

FIG. 36.8
The ¹³C NMR spectrum of 2-butanone, $CH_3COCH_2CH_3$ (100 MHz).

FIG. 36.9
The ¹³C NMR spectrum of crotonaldehyde (22.6 MHz).

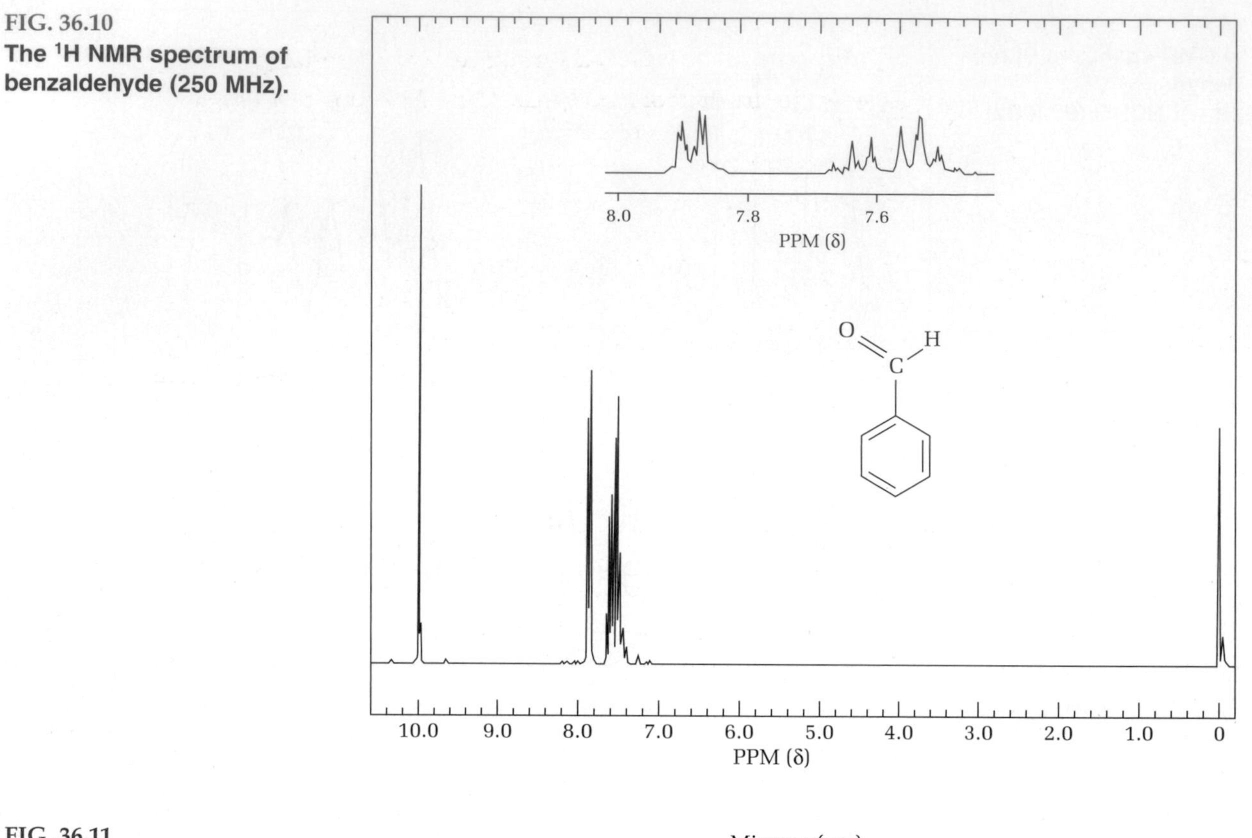

FIG. 36.10
The ^{1}H NMR spectrum of benzaldehyde (250 MHz).

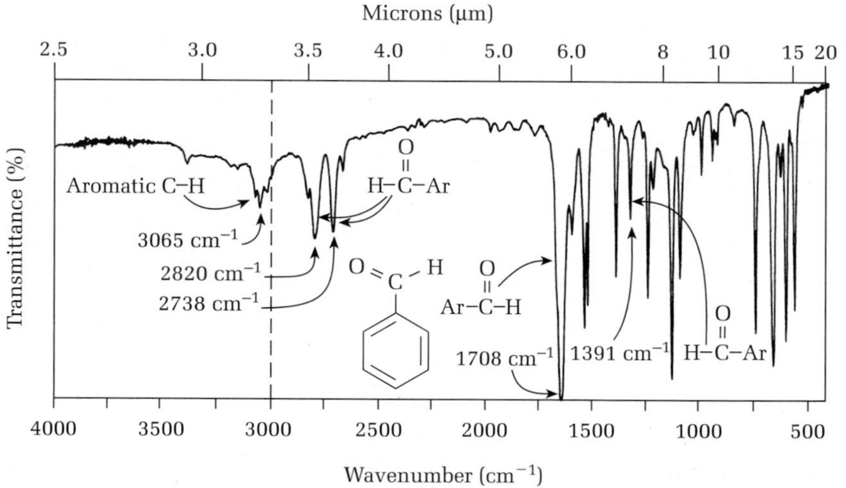

FIG. 36.11
The IR spectrum of benzaldehyde (thin film).

QUESTIONS

1. What is the purpose of making derivatives of unknowns?
2. Why are 2,4-dinitrophenylhydrazones better derivatives than phenylhydrazones?
3. Using chemical tests, how would you distinguish among 2-pentanone, 3-pentanone, and pentanal?

4. Draw the structure of a compound with the empirical formula C_5H_8O that gives a positive iodoform test and does not decolorize permanganate.

5. Draw the structure of a compound with the empirical formula C_5H_8O that gives a positive Tollens' test and does not react with bromine in dichloromethane.

6. Draw the structure of a compound with the empirical formula C_5H_8O that reacts with phenylhydrazine, decolorizes bromine in dichloromethane, and does not give a positive iodoform test.

7. Draw the structure of two geometric isomers with the empirical formula C_5H_8O that give a positive iodoform test.

8. What vibrations cause the peaks at about 3.6 mm (2940 cm^{-1}) in the IR spectrum of fluorene (Fig. 36.2)?

9. Locate the carbonyl peak in Figure 36.4.

10. Assign the various peaks in the ^{1}H NMR spectrum of 2-butanone to specific protons in the molecule (Fig. 36.6).

11. Assign the various peaks in the ^{1}H NMR spectrum of crotonaldehyde to specific protons in the molecule (Fig. 36.7).

37 CHAPTER

Dibenzalacetone by the Aldol Condensation

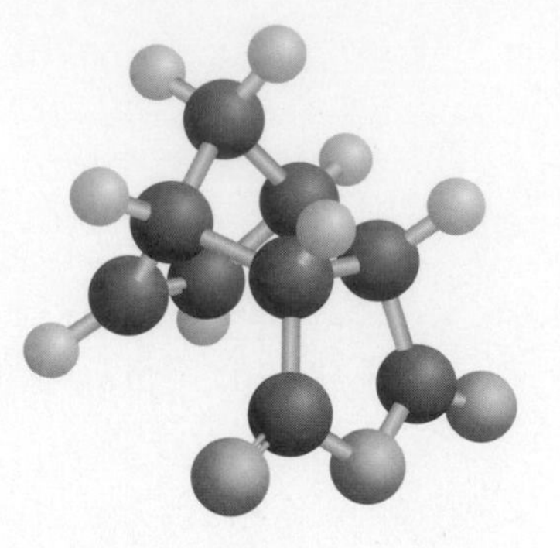

PRELAB EXERCISE: Calculate the volumes of benzaldehyde and acetone needed for the macroscale version of this reaction, taking into account the densities of the liquids and the number of moles of each required.

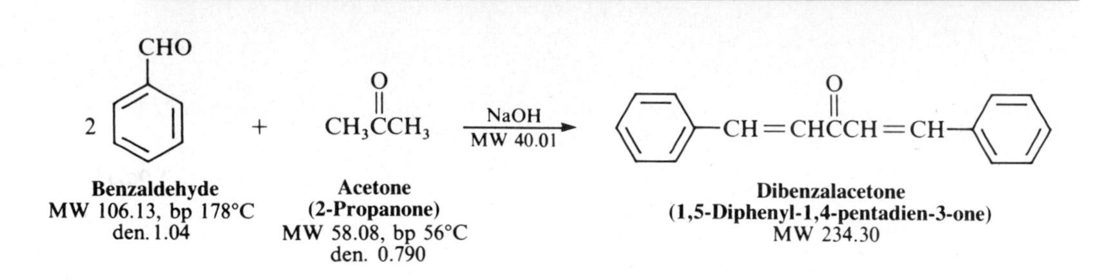

The base-catalyzed reaction of an aldehyde with a ketone is a mixed aldol condensation, known as the Claisen–Schmidt reaction. Dibenzalacetone is readily prepared by the condensation of acetone with two equivalents of benzaldehyde. The aldehyde carbonyl is more reactive than that of the ketone and, therefore, reacts rapidly with the anion of the ketone to give a β-hydroxyketone, which easily undergoes base-catalyzed dehydration. Depending on the relative quantities of the reactants, the reaction can give either mono- or dibenzalacetone.

Dibenzalacetone is innocuous; its spectral properties (ultraviolet [UV] absorbance) indicate why it is used in sunscreens and sunblock preparations. In the present experiment, sufficient ethanol is present as a solvent to readily dissolve the starting material, benzaldehyde, and also the intermediate, benzalacetone. The benzalacetone, once formed, can then easily react with another mole of benzaldehyde to give the product, dibenzalacetone. The mechanism for the formation of benzalacetone is as follows:

$$CH_3\overset{O}{\overset{\|}{C}}CH_3 + OH^- \longrightarrow \left[CH_3\overset{O}{\overset{\|}{C}}-CH_2^- \leftrightarrow CH_3\overset{O^-}{\overset{|}{C}}=CH_2 \right] + H_2O$$

$$CH_3\overset{O}{\overset{\|}{C}}CH_2^- + \overset{H}{C}(=O)-C_6H_5 \rightleftharpoons CH_3\overset{O}{\overset{\|}{C}}CH_2\overset{H}{\underset{O^-}{C}}-C_6H_5 \overset{H_2O}{\rightleftharpoons} CH_3\overset{O}{\overset{\|}{C}}CH_2\overset{H}{\underset{OH}{C}}-C_6H_5 + OH^-$$

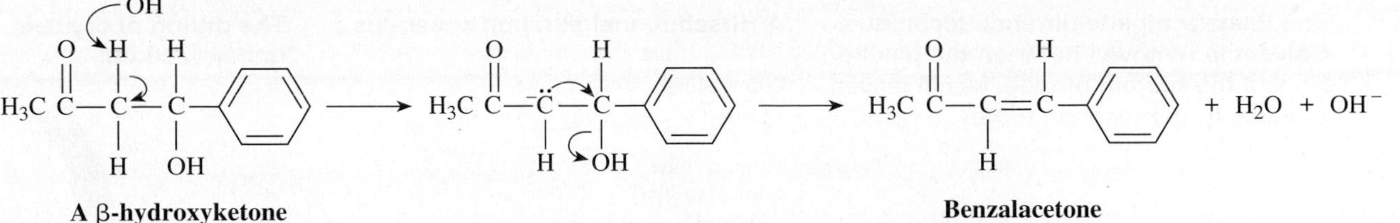

A β-hydroxyketone

Benzalacetone
(4-Phenyl-3-butene-2-one)
mp 42°C

EXPERIMENT

SYNTHESIS OF DIBENZALACETONE[1]

Microscale Procedure

IN THIS EXPERIMENT, an ethanolic solution of acetone and benzaldehyde is added to aqueous sodium hydroxide. The product, dibenzalacetone, crystallizes after a few minutes. The product is filtered from the mixture, washed, pressed dry, and recrystallized from an ethanol-water mixture. This very important reaction is easily carried out.

If sodium hydroxide gets on the skin, wash until the skin no longer has a soapy feeling.

Into a 10 × 100-mm reaction tube, place 2 mL of 3 *M* sodium hydroxide solution. To this solution, add 1.6 mL of 95% ethanol and 0.212 g of benzaldehyde.[2] Then add 0.058 g of acetone to the reaction mixture. Alternatively, your instructor may provide a solution that contains 58 mg of acetone in 1.6 mL of ethanol. Cap the tube immediately and shake the mixture vigorously. The benzaldehyde, initially insoluble, goes into solution, resulting in a clear, pale-yellow solution. After a minute or so, it suddenly becomes cloudy and a yellow precipitate of the product forms. Continue to shake the tube from time to time for the next 30 minutes. If the product fails to crystallize, open the tube and scratch the inside of the tube with a glass rod. Remove the liquid from the tube using a Pasteur pipette by squeezing the bulb of the pipette, lowering the pipette into the liquid and pressing the tip against the bottom of the tube, and then aspirating the liquid into the pipette, leaving the crystals in the tube (Fig. 37.1). Add 3 mL of water, cap, and shake the tube vigorously. Remove the wash liquid as before, and wash the crystals twice more with 3-mL portions of water.

Videos: *Microscale Filtration on the Hirsch Funnel; Formation of an Oil Instead of Crystals; Filtration of Crystals Using the Pasteur Pipette; Microscale Crystallization*

After the final washing, add 3 mL of water to the tube and collect the crystals on a Hirsch funnel using vacuum filtration. Use the filtrate to complete the transfer of the crystals. Squeeze the product between sheets of filter paper to dry it, and then recrystallize the crude dibenzalacetone from a 70:30 ethanol-water mixture. During the recrystallization process (*see* Chapter 4), insert a wooden boiling stick to

[1]Ask your instructor for an alternative procedure for carrying out this experiment, which is found in the *Instructor's Guide*.
[2]The benzaldehyde should be pure. Because it is easily air oxidized to benzoic acid on standing in the lab, it should be freshly purified or from a newly purchased bottle.

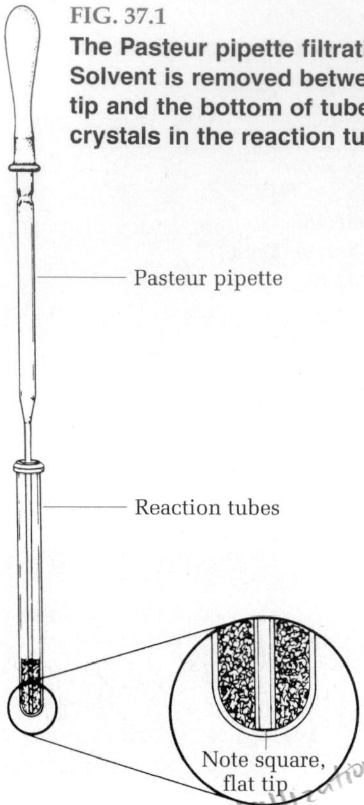

FIG. 37.1
The Pasteur pipette filtration technique. Solvent is removed between the pipette tip and the bottom of tube, which leaves crystals in the reaction tube.

Pasteur pipette

Reaction tubes

Note square, flat tip

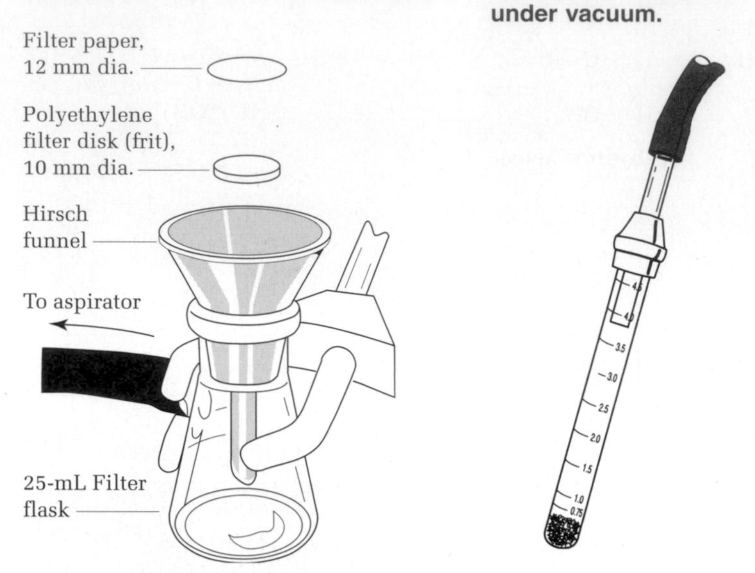

FIG. 37.2
A Hirsch funnel filtration apparatus.

FIG. 37.3
The drying of crystals under vacuum.

after recrystallization, collect on hirsch funnel. - could wash one last time w/0.5mL of cold ethanol.

promote even boiling when heating the solute in the solvent. Remove the tube from the hot sand bath, remove the boiling stick and place the tube in an insulated container to cool slowly to room temperature. Should the product separate as an oil, obtain a seed crystal, heat the solution to dissolve the oil, and add the seed crystal as the solution cools. If the product continues to release oil, add a bit more ethanol. After cooling the tube for several minutes in ice, collect the product by removing the solvent with a pipette (Fig. 37.1) and washing once with about 0.5 mL of ice-cold 70% ethanol while the tube is in ice. Alternatively, collect the crystals on a Hirsch funnel (Fig. 37.2) or Wilfilter and wash once with ice-cold 70% ethanol. Dry the product under vacuum by attaching the tube to an aspirator (Fig. 37.3). Determine the weight of the dibenzalacetone and its melting point, and calculate the percent yield. In a typical experiment, the yield will be 0.10 g (mp 110.5°C–112°C).

Cleaning Up. Dilute the filtrate from the reaction mixture with water, and neutralize it with dilute hydrochloric acid before flushing down the drain. The ethanol filtrate from the recrystallization should be placed in the organic solvents waste container.

Macroscale Procedure

If sodium hydroxide gets on the skin, wash until the skin no longer has a soapy feeling.

Video: *Macroscale Crystallization*

In a 125-mL Erlenmeyer flask containing a magnetic stirring bar, mix 0.05 mol of benzaldehyde with the theoretical quantity of acetone, and add one-half the mixture to a solution of 5 g of sodium hydroxide dissolved in 50 mL of water and 40 mL of ethanol at room temperature (<25°C). After 15 minutes of stirring, add the remainder of the aldehyde-ketone mixture and rinse the container with a little ethanol to complete the transfer. Stir the mixture for 30 minutes; then collect the product by suction filtration on a Büchner funnel. Break the suction and carefully pour 100 mL of water on the product. Reapply the vacuum. Repeat this process three times in order to remove all traces of sodium hydroxide. Finally, press the product as dry as possible on the filter using a cork, and then press it between sheets of filter paper to remove as much water as possible. Save a small sample for a melting-point determination; then recrystallize the product from ethanol using about 10 mL of ethanol for each 4 g of dibenzalacetone. The yield after recrystallization should be about 4 g.

Cleaning Up. Dilute the filtrate from the reaction mixture with water, and neutralize it with dilute hydrochloric acid before flushing down the drain. The ethanol filtrate from the recrystallization should be placed in the organic solvents waste container.

MOLECULAR MODELING

The name *dibenzalacetone* does not completely characterize the molecule made in this experiment. There are actually three isomeric dibenzalacetones: one melting at 110°C–111°C, λ_{max} = 330 nm, ε = 34,300; another melting at 60°C, λ_{max} = 295 nm, ε = 20,000; and a third, a liquid with λ_{max} = 287 nm, ε = 11,000.

Both the melting points and the UV spectral data give some hints regarding the structures of these molecules. The first dibenzalacetone is very symmetrical and can pack well into a crystal lattice. The long wavelength of the UV light absorption maximum and the high value of the molar absorbance (ε) indicate a long, planar conjugated system (*see* Chapter 14). The other two molecules are increasingly less able to pack nicely into a crystal lattice or to have a planar conjugated system. In the last step of the aldol condensation, the loss of water from the β-hydroxyketone can result in the formation of molecules in which the alkene hydrogen atoms are either *cis* or *trans* to each other. Write the structures of the three geometric isomers of dibenzalacetone, and assign each one to the three molecules described above.

Enter the structures of these three isomers into a molecular modeling program and carry out an energy minimization or semiempirical calculation to find the relative steric energies or heats of formation of each molecule. Note: Once the calculation is complete, the lowest energy conformation of each isomer will be as planar as possible so that there can be maximum overlap of the *p* orbitals on each sp^2 hybridized carbon. To test this idea, calculate the steric energy or heat of formation of benzalacetone (Fig. 37.4a) using the usual energy minimization procedure. The result should be an almost planar molecule. Then deliberately hold the dihedral angle defined by atoms 1, 2, 3, and 4 at 90° (Fig. 37.4b), and again calculate the energy of the molecule. In the latter conformation, the *p* orbitals of the carbonyl group are orthogonal to the *p* orbitals of the alkene.

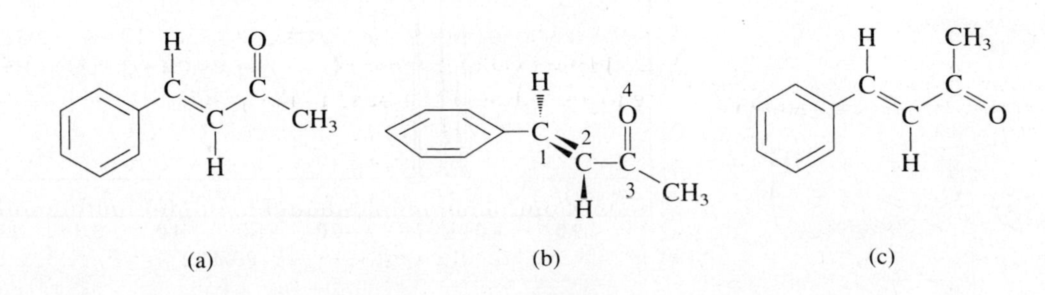

FIG. 37.4
Single bond isomers of benzalacetone.

As you may have discovered in calculating the energies of the three geometric isomers of dibenzalacetone, there is still another form of isomerization entering into the conformations of these molecules: single-bond *cis* and *trans* isomers, exemplified by two of the isomers of benzalacetone (Fig. 37.4a and Fig. 37.4c). Both of these conformers are planar in order to achieve maximum overlap of *p* orbitals, but in 37.4a the carbonyl group is *cis* to the alkene bond, whereas in 37.4c it is *trans*. The barrier to rotation about the single bond is not very high, so these isomers cannot be isolated at room temperature. If your molecular modeling program has a dihedral driver routine, you can calculate the heats of formation of benzalacetone as a function of the dihedral angle defined by atoms 1, 2, 3, and 4 and thus determine the barrier to rotation around this bond in kilocalories per mole.

QUESTIONS

1. Why is it important to maintain equivalent proportions of reagents in this reaction?
2. Which side products do you expect in this reaction? How are they removed?
3. What evidence do you have that your product consists of a single geometric isomer or is a mixture of isomers? Does the melting point give such information?
4. From the ^{1}H NMR spectrum of dibenzalacetone (Fig. 37.5), can you deduce which geometric isomer(s) is (are) formed? (*See* Table 12.1 on page 243.)

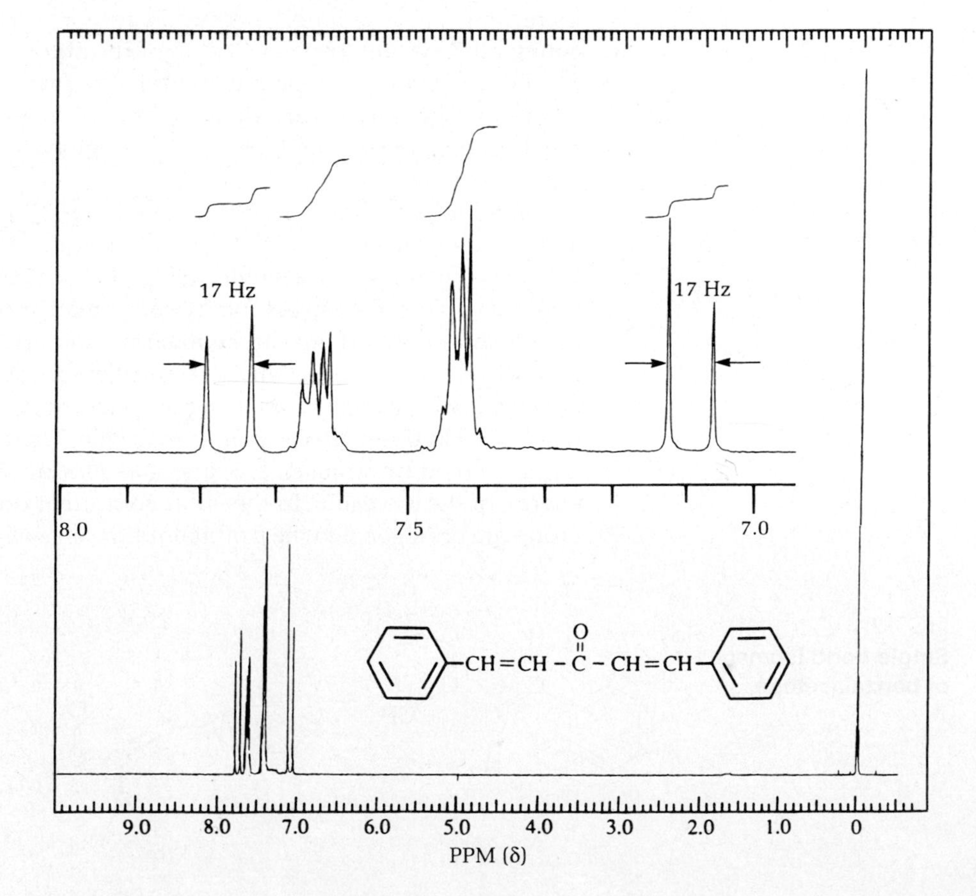

FIG. 37.5
The ^{1}H NMR spectrum of dibenzalacetone (250 MHz). (An expanded spectrum appears above the normal spectrum.)

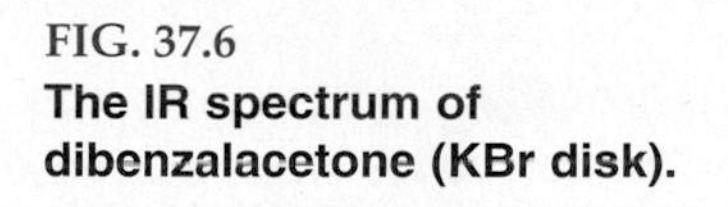

FIG. 37.6
The IR spectrum of dibenzalacetone (KBr disk).

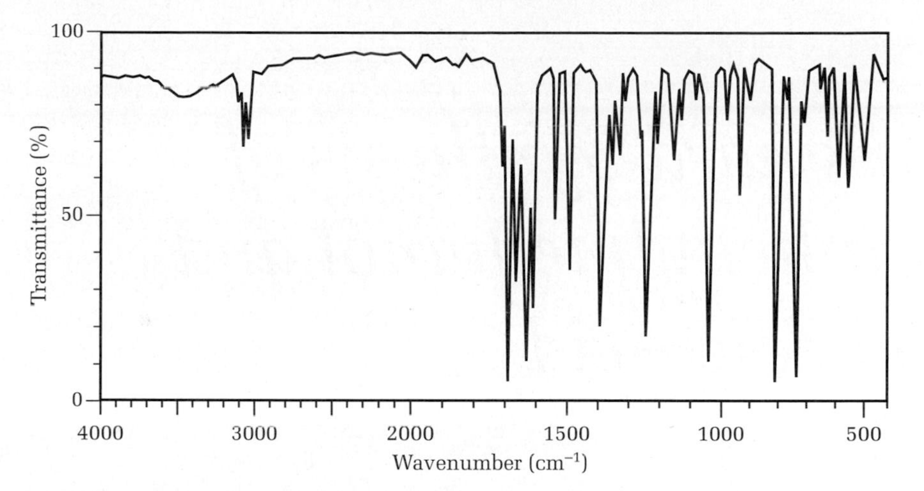

5. How would you change the procedures in this chapter if you wished to synthesize benzalacetone ($C_6H_5CHCHCOCH_3$) or benzalacetophenone ($C_6H_5CHCH{=}COC_6H_5$)?
6. Assign the infrared (IR) peak at 1651 cm^{-1} in Figure 37.6.
7. Draw the structures of the three geometric isomers of dibenzalacetone and assign each one to the three molecules described. Disregard single bond *cis* and *trans* isomerism.
8. Draw the structures of all the single bond *cis* and *trans* isomers for each of the three geometric isomers of dibenzalacetone. There are 10 such single bond isomers. Pick out the one you regard as the most stable and calculate its steric energy. Which three or four are represented by the solid with a melting point of 110–111°C, the solid with a melting point of 60°C, and the liquid?
9. Carry out a molecular mechanics or semiempirical calculation to determine the relative steric energies or heats of formation of 3 of the 10 possible isomeric dibenzalacetones.
10. (a) Calculate the steric energy or heat of formation for one single bond isomer of *trans*-benzalacetone using the usual energy minimization procedure. The result should be a planar molecule. (b) Then deliberately hold the dihedral angle defined by atoms 1, 2, 3, and 4 at 0°, 90°, and 180° and again calculate the energies of the molecule. What is the approximate barrier to rotation about the single bond?
11. Pick out the isomer you regard as the most stable from Question 7 and calculate its steric energy (if this was not done in Question 8).

38 CHAPTER

Grignard Synthesis of Triphenylmethanol and Benzoic Acid

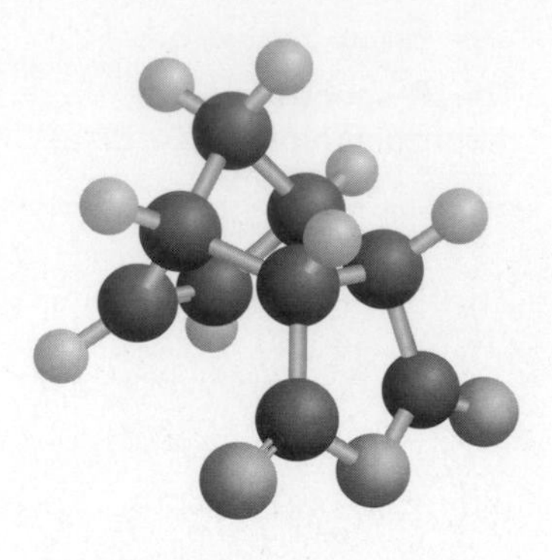

PRELAB EXERCISE: Prepare a flow sheet for the preparation of triphenylmethanol. Using your knowledge of the physical properties of the solvents, reactants, and products, show how the products can be purified. Indicate which layer should contain the product in the liquid/liquid extraction steps.

In 1912, Victor Grignard received the Nobel Prize in Chemistry for his work on the reaction that bears his name, a carbon-carbon bond-forming reaction by which almost any alcohol may be formed from appropriate alkyl halides and carbonyl compounds. The Grignard reagent is easily formed by reacting an alkyl halide, in particular a bromide, with magnesium metal in anhydrous diethyl ether. The reaction can be written and thought of as simply:

$$R—Br + Mg \longrightarrow R—Mg—Br$$

However, the structure of the material in solution is rather more complex. There is evidence that dialkylmagnesium is present:

$$2\ R—Mg—Br \rightleftharpoons R—Mg—R + MgBr_2$$

Structure of the Grignard reagent

The magnesium atoms, which have the capacity to accept two electron pairs from donor molecules to achieve a four-coordinated state, are solvated by the unshared pairs of electrons on diethyl ether:

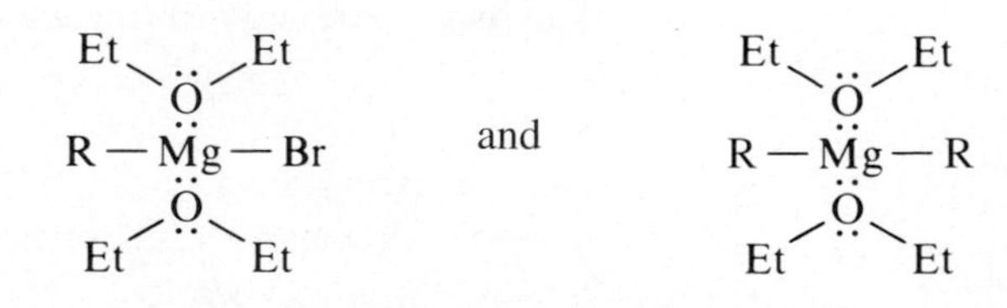

A strong base and strong nucleophile

The Grignard reagent is both a strong base and a strong nucleophile. As a base it will react with all protons that are more acidic than those found on alkenes and alkanes. Thus, Grignard reagents react readily with water, alcohols, amines, thiols, and so on to regenerate an alkane:

$$R\text{—}Mg\text{—}Br + H_2O \longrightarrow R\text{—}H + MgBrOH$$
$$R\text{—}Mg\text{—}Br + R'OH \longrightarrow R\text{—}H + MgBrOR'$$
$$R\text{—}Mg\text{—}Br + R'NH_2 \longrightarrow R\text{—}H + MgBrNHR'$$
$$R\text{—}Mg\text{—}Br + R'C \equiv C\text{—}H \longrightarrow R\text{—}H + R'C \equiv CMgBr$$

(an acetylenic Grignard reagent)

The starting material for preparing the Grignard reagent cannot contain any acidic protons. The reactants and apparatus must be completely and absolutely dry; otherwise the reaction will not start. If proper preparations are made, however, the reaction proceeds smoothly.

Magnesium metal, in the form of a coarse powder, has a coat of oxide on the outside. By crushing the powder under absolutely dry ether in the presence of an organic halide, a fresh surface can be exposed. The reaction will begin at exposed surfaces, as evidenced by a slight turbidity in the solution and evolution of bubbles. Once the exothermic reaction starts, it proceeds easily, the magnesium dissolves, and a solution of the Grignard reagent is formed. The solution is often turbid and gray due to impurities in the magnesium. The reagent is not isolated but reacted immediately with, most often, an appropriate carbonyl compound to give the ether-insoluble salt, magnesium alkoxide, in another exothermic reaction.

$$R\text{—}Mg\text{—}Br + R'\text{—}\overset{\overset{\Large ..}{O}:}{\overset{\|}{C}}\text{—}R'' \longrightarrow R'\text{—}\underset{R}{\underset{|}{\overset{:\ddot{O}:^- MgBr^+}{\overset{|}{C}}}}\text{—}R''$$

In a simple acid-base reaction, this alkoxide is reacted with acidified ice water to give the covalent, ether-soluble alcohol and the ionic water-soluble magnesium salt:

$$R'\text{—}\underset{R}{\underset{|}{\overset{:\ddot{O}:^- MgBr^+}{\overset{|}{C}}}}\text{—}R'' + H^+Cl^- \longrightarrow R'\text{—}\underset{R}{\underset{|}{\overset{:\ddot{O}\text{—}H}{\overset{|}{C}}}}\text{—}R'' + Mg^{2+}Br^-Cl^-$$

A versatile reagent

The great versatility of this reaction lies in the wide range of reactants that undergo it. Thirteen representative reactions are shown in Figure 38.1. In every case except reaction 1, the intermediate alkoxide must be hydrolyzed to give the product. The reaction with oxygen (reaction 2) is usually not a problem because the ether vapor over the reagent protects it from attack by oxygen, but this reaction is one reason why the reagent cannot usually be stored without special precautions. The reaction with solid carbon dioxide (dry ice) occurs readily to produce a carboxylic acid (reaction 3). Reactions 5, 6, and 7 with aldehydes and ketones

FIG. 38.1
The versatility of the Grignard reaction is illustrated by the wide range of reactants that undergo it.

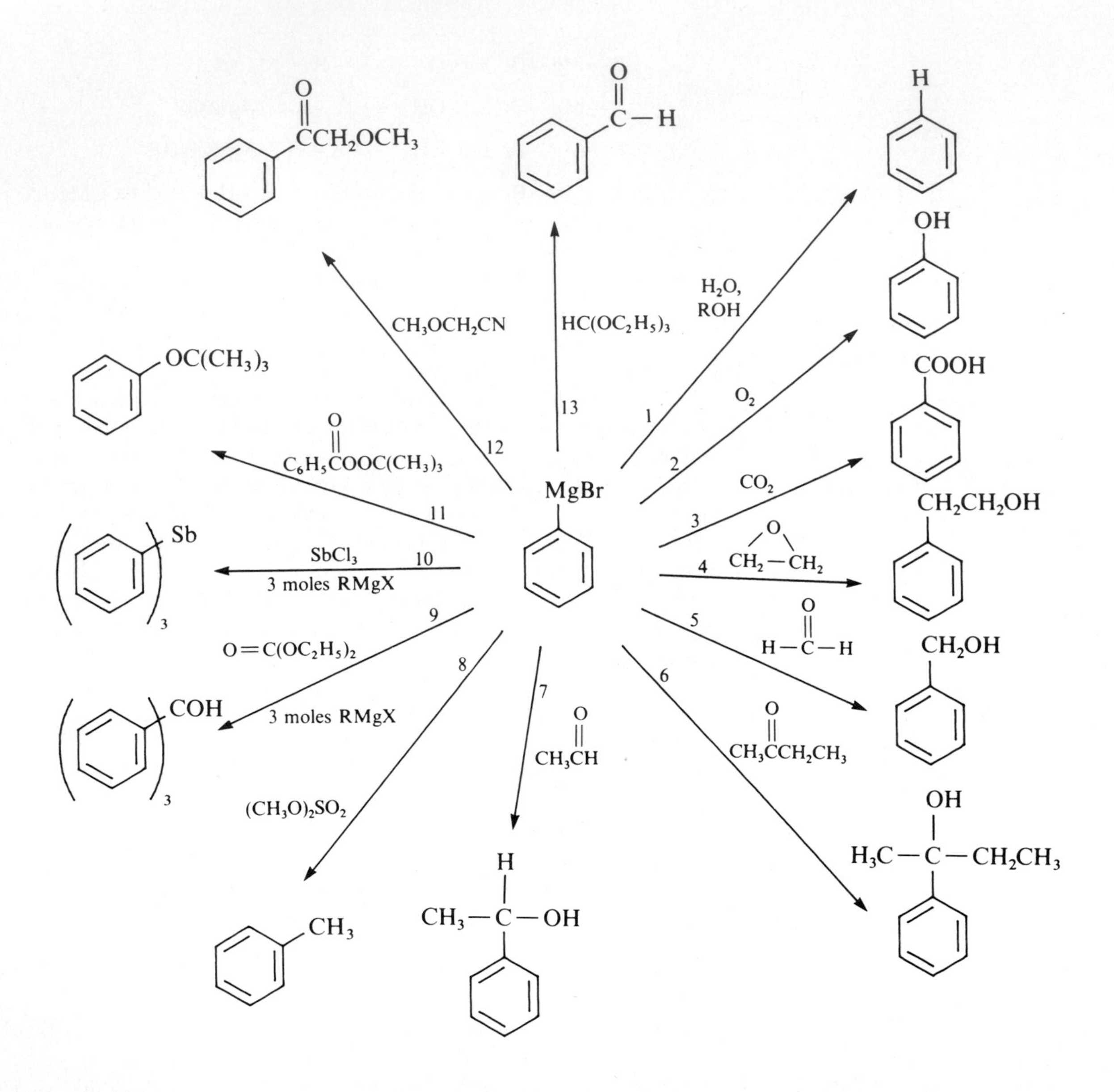

giving, respectively, primary, tertiary, and secondary alcohols are among the most common. The ring-opening of an epoxide via Grignard addition (reaction 4) yields an alcohol as well. Reactions 8–13 are not nearly so common.

In one of the experiments in this chapter, we will carry out another common type of Grignard reaction, the formation of a tertiary alcohol from 2 moles of the reagent and 1 mole of an ester. The ester employed is methyl benzoate, which can be synthesized in Chapter 40. The initially formed product is unstable and decomposes to a ketone, which, being more reactive than an ester, immediately reacts with more Grignard reagent:

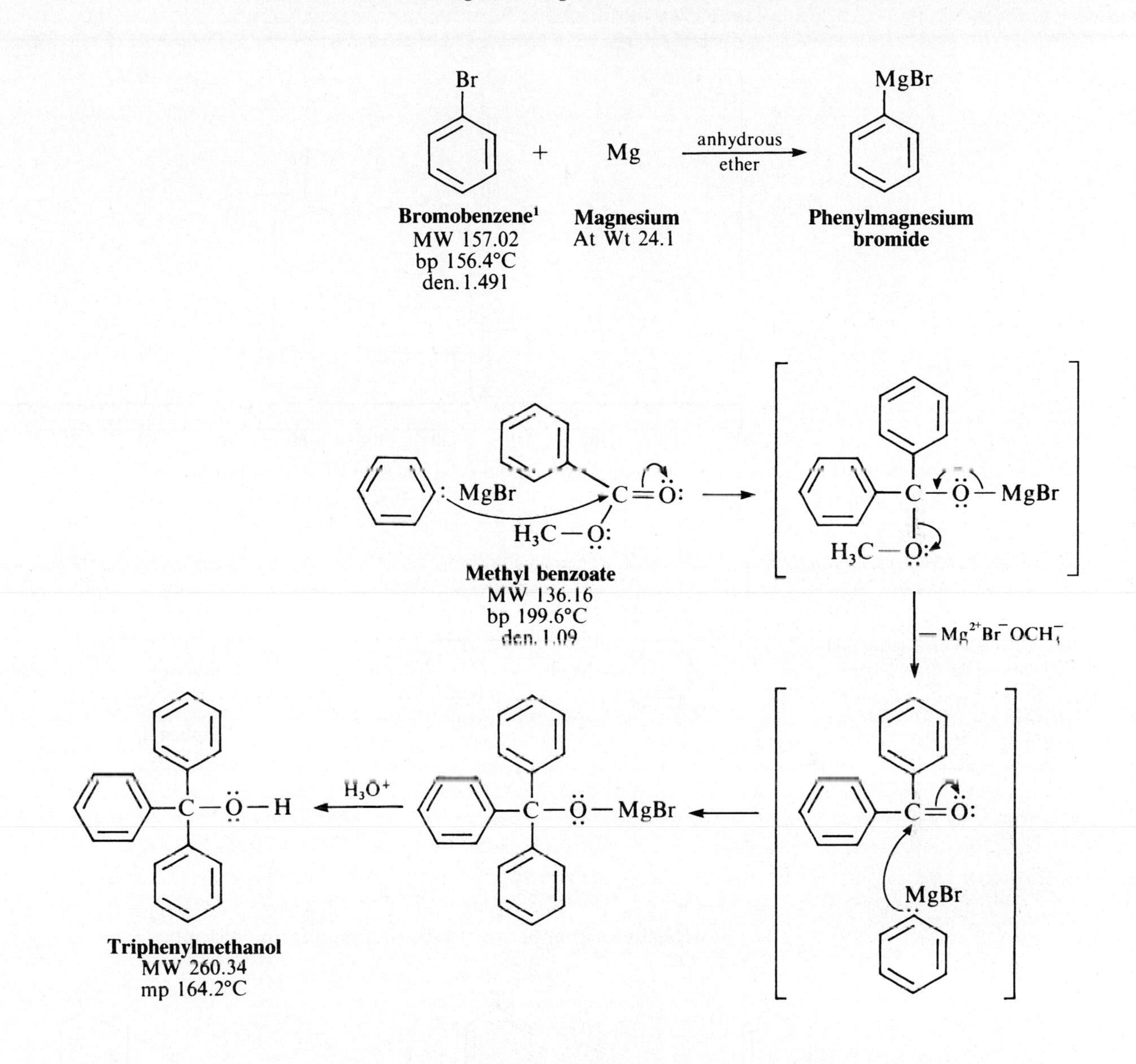

The primary impurity in these experiments is biphenyl, formed by reacting phenylmagnesium bromide with unreacted bromobenzene. (Figure 38.2 shows the ^{13}C NMR spectrum of bromobenzene.) The most effective way to lessen this side reaction is to add the bromobenzene slowly to the reaction mixture so it will react with the magnesium, and not be present in high concentration to react with previously formed Grignard reagent. The impurity is easily eliminated because it is much more soluble in hydrocarbon solvents than triphenylmethanol.

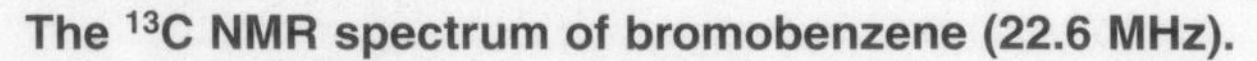

FIG. 38.2
The ^{13}C NMR spectrum of bromobenzene (22.6 MHz).

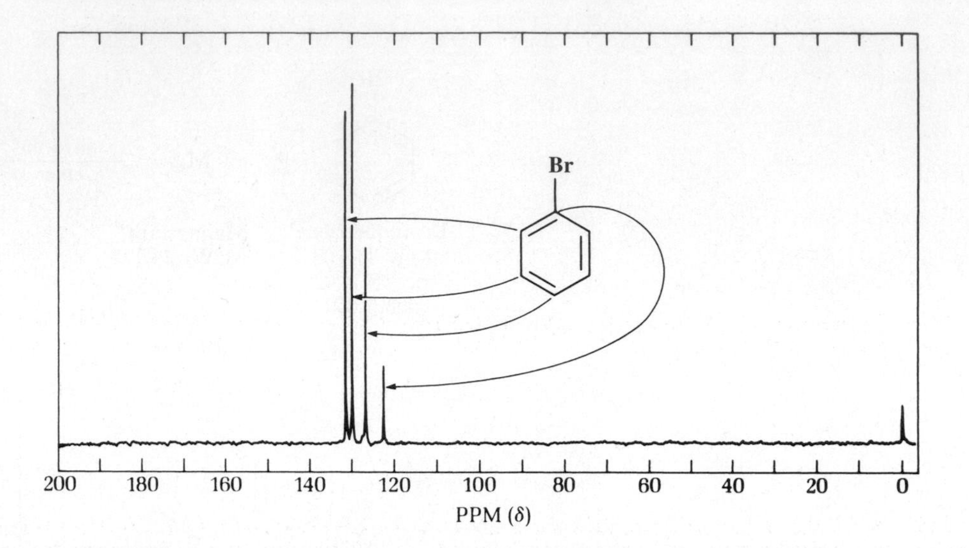

Biphenyl has a characteristic odor. Triphenylmethanol is odorless.

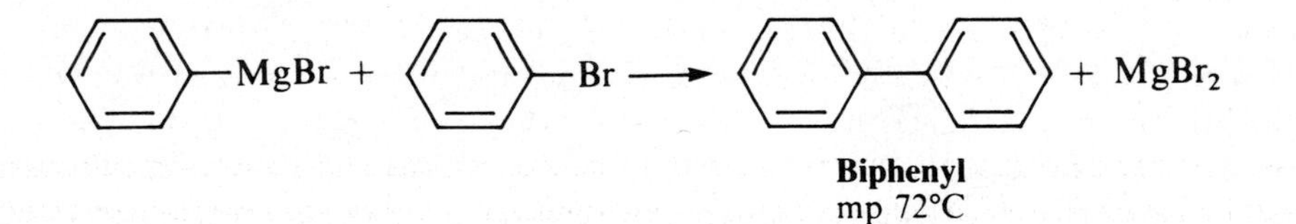

Triphenylmethanol can also be prepared from benzophenone.

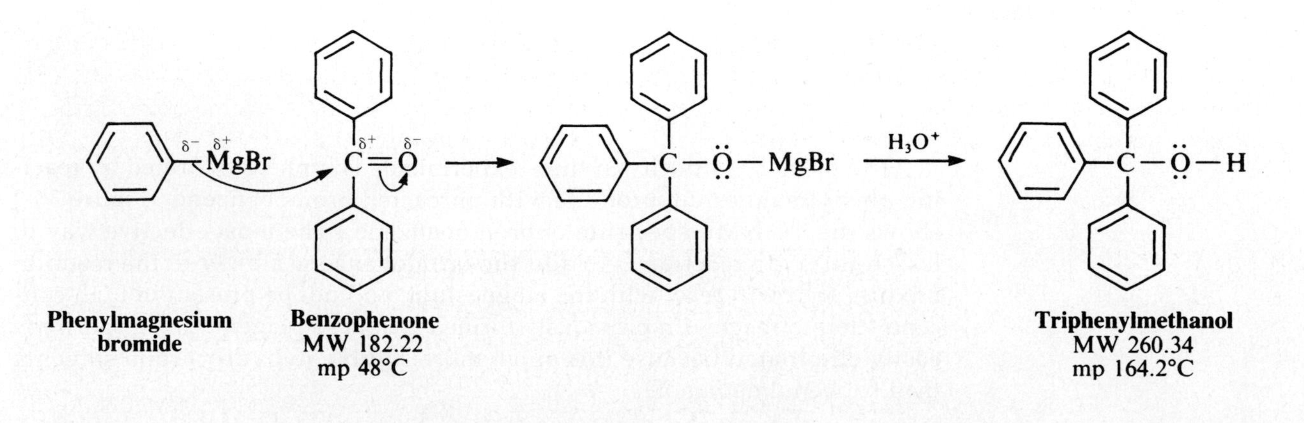

Experiments

1. Phenylmagnesium Bromide (Phenyl Grignard Reagent)

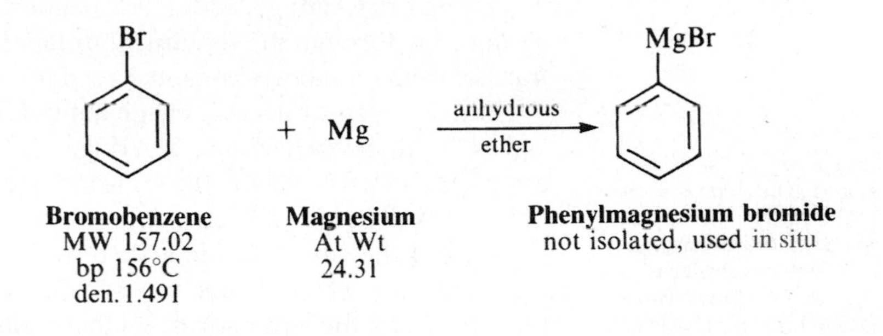

Advance Preparation

It is imperative that all equipment and reagents be absolutely dry. The magnesium and the glassware to be used—two reaction tubes, two 1-dram vials (1 dram = 1.78 mL), and a stirring rod—can be dried in a 110°C oven for at least 30 minutes. Alternatively, if the glassware, syringe, septa, and magnesium appear to be perfectly dry, they can be used without special drying. The plastic and rubber components should be rinsed with acetone if either appears to be dirty or wet with water; then place these components in a desiccator for at least 12 hours. Do not place plastic or rubber components in the oven. New, factory-sealed packages of syringes can be used without prior drying. The ether used throughout this reaction must be absolutely dry (anhydrous ether).

CAUTION: Ether is extremely flammable. Extinguish all flames before using ether.

To prepare the Grignard reagent, anhydrous diethyl ether must be used; elsewhere, ordinary ether (diethyl or *t*-butyl methyl) can be used. For example, ether extractions of aqueous solutions do not need to be carried out with dry ether, which is expensive. It is strongly recommended that *t*-butyl methyl ether be used in all cases except in the preparation of the Grignard reagent itself.

Video: *The Grignard Reaction: Removing a Liquid from a Septum-Capped Bottle with a Syringe*; Photos: *Removing a Reagent from a Septum-Capped Bottle; Polypropylene Syringe Containing Ether*

A very convenient container for anhydrous diethyl ether is a 50-mL septum-capped bottle. This method of dispensing the solvent has three advantages: the ether is kept anhydrous, the exposure to oxygen is minimized, and there is little possibility of its catching fire. Ether is extremely flammable; do not work with this solvent near flames.

To remove ether from a septum-capped bottle, inject a volume of air into the bottle equal to the amount of ether being removed. Pull more ether than needed into the syringe, and then push the excess back into the bottle before removing the syringe. In this way, there will be no air bubbles in the syringe, and it will not dribble (Fig. 38.3).

Procedure

IN THIS EXPERIMENT, magnesium and absolutely dry diethyl ether in a dry, septum-capped tube are reacted with an aromatic halide to give phenylmagnesium bromide. The exothermic reaction is started by crushing the magnesium. If the reaction does not start in 2–3 minutes, it will be necessary to start again, using equipment that was not used in the first attempt.

Remove a reaction tube from the oven, and immediately cap it with a septum. In the operations that follow, keep the tube capped except when it is necessary to open it. After the tube cools to room temperature, add about 2 mmol of magnesium powder. Record the weight of magnesium used to the nearest milligram. The magnesium will become the limiting reagent by using a 5% molar excess of bromobenzene (about 2.1 mmol).

Using a dry syringe, add to the magnesium 0.5 mL of anhydrous diethyl ether by injection through the septum. Your laboratory instructor will demonstrate the transfer from the storage container used in your laboratory.

Diethyl ether can be made and kept anhydrous by storing over Linde 5A molecular sieves. Discard diethyl ether within 90 days because of peroxide formation. *t*-Butyl methyl ether does not have this problem.

Into an oven-dried vial, weigh about 2.1 mmol of dry (stored over molecular sieves) bromobenzene. Using a syringe, add to this vial 0.7 mL anhydrous diethyl ether and *immediately*, with the same syringe, remove all the solution from the vial. This can be done virtually quantitatively so you do not need to rinse the vial. Immediately cap the empty vial to keep it dry for later use. Inject about 0.1 mL of the bromobenzene-ether mixture into the reaction tube and mix the contents by flicking the tube. Pierce the septum with another syringe needle for pressure relief (Fig. 38.4).

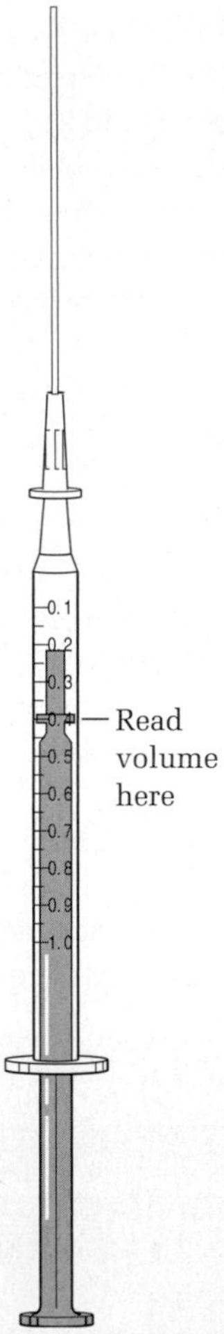

FIG. 38.3
A polypropylene syringe (1 mL, with 0.01-mL graduations). The needle is blunt. When there are no air bubbles in the syringe, it will not dribble ether.

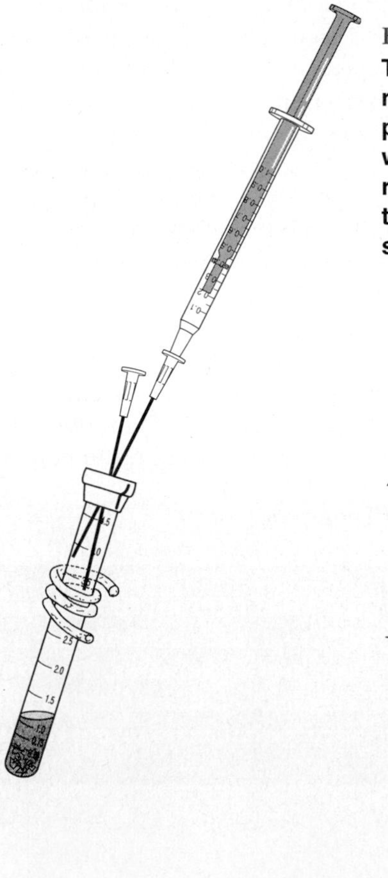

FIG. 38.4
Once the reaction has started, bromobenzene in ether is added slowly from the syringe. The empty needle is for pressure relief, but if condensation is complete (aided by the damp pipe cleaner), it will not be needed. Once the reaction slows down, stir it with a magnetic stirrer and stirring bar.

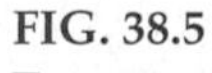

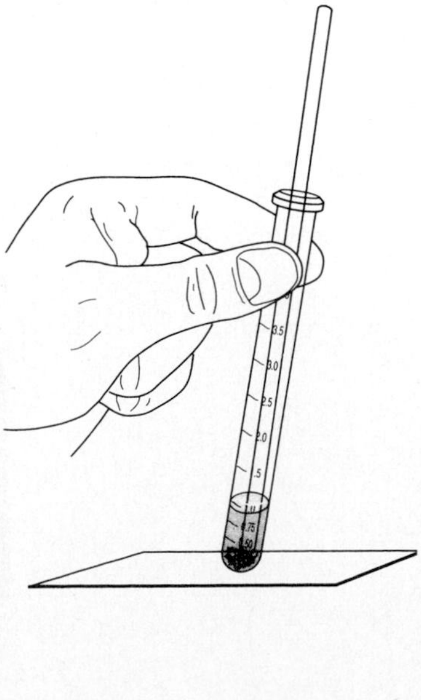

FIG. 38.5
To start the Grignard reaction, remove the septum and apply pressure to the stirring rod while rotating the bottom of the reaction tube on a hard surface that will not scratch the tube, such as a book.

It takes much force to crush the magnesium. Place the tube on a hard surface and bear down with the stirring rod while twisting the reaction tube. Do not pound the magnesium.

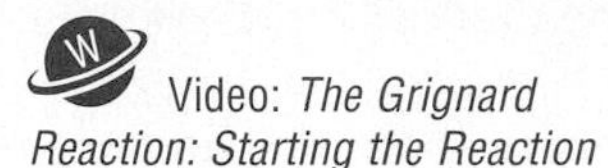

Video: *The Grignard Reaction: Starting the Reaction*

The reaction will not ordinarily start at this point, so remove the septum, syringe, and empty syringe needle and crush the magnesium with a dry stirring rod. You can do this easily in the confines of the 10-mm-diameter reaction tube while it is positioned on a hard surface. There is little danger of poking the stirring rod through the bottom of the tube (Fig. 38.5). Immediately replace the septum, syringe, and empty syringe needle (for pressure relief). The reaction should start within seconds. The formerly clear solution becomes cloudy and soon begins to boil as the magnesium metal reacts with the bromobenzene to form the Grignard reagent—phenylmagnesium bromide.

If the reaction does not start within 1 minute, begin again with completely different, dry equipment (syringe, syringe needle, reaction tube, etc.). Once the Grignard reaction begins, it will continue. To prevent the ether from boiling away, wrap a pipe cleaner around the top part of the reaction tube. Dampen the pipe cleaner with water or, if the room temperature is very hot, with alcohol.

Video: *The Grignard Reaction: Addition of Bromobenzene and Refluxing of Reaction Mixture*

To the refluxing mixture, add slowly and dropwise over a period of several minutes the remainder of the bromobenzene-ether solution at a rate such that the reaction remains under control at all times. After all the bromobenzene solution is added, spontaneous boiling of the diluted mixture may be slow or become slow. At this point, add a magnetic stirring bar to the reaction tube and stir the reaction mixture with a magnetic stirrer. If the rate of reaction is too fast, slow down the stirrer. The reaction is complete when none or a very small quantity of the metal remains. Check to see that the volume of ether has not decreased. If it has, add more anhydrous diethyl ether. Because the solution of the Grignard reagent deteriorates on standing, Experiment 2 should be started at once. The phenylmagnesium bromide can be converted to triphenylmethanol or to benzoic acid.

2. TRIPHENYLMETHANOL

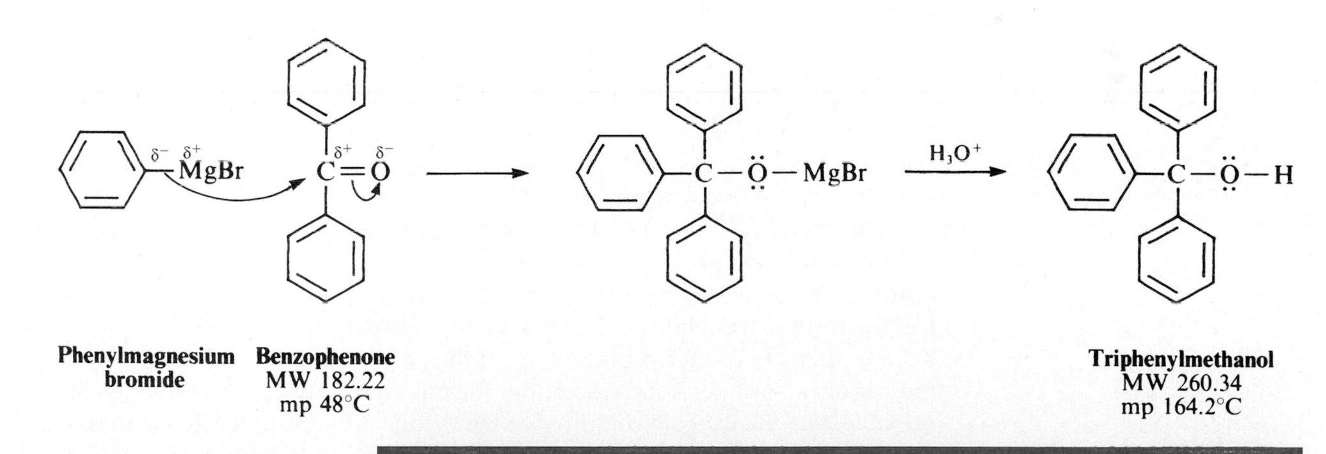

IN THIS EXPERIMENT, the Grignard reagent prepared in Experiment 1 is added to a dry ether solution of benzophenone. Very thorough mixing is required. When the red color disappears, the salt is hydrolyzed by adding hydrochloric acid. More ether is added, the layers separated, and the ether dried and evaporated to give crude product. An impurity, biphenyl, is removed by dissolving it in hexanes, and the product is recrystallized from 2-propanol. The product recrystallizes very slowly, so do not collect the product immediately.

Make t-butyl methyl ether anhydrous by storing it over molecular sieves.

In an oven-dried vial, dissolve 2.0 mmol of benzophenone in 1.0 mL of anhydrous ether by capping the vial and mixing the contents thoroughly. With a dry syringe, remove all the solution from the vial and add it dropwise with *thorough* mixing (magnetic stirring, flicking of the tube, or stirring with a stirring rod near the end of the addition) after each drop to the solution of the Grignard reagent. Add the benzophenone at a rate so as to maintain the ether at a gentle reflux. Rinse the vial with a few drops of anhydrous ether after all the first solution has been added; then add this rinse to the reaction tube.

Mixing the reaction mixture is very important.

After all the benzophenone has been added, the mixture should be homogeneous. If not, mix it again, using a stirring rod if necessary. The syringe can be removed, but leave the pressure-relief needle in place. Allow the reaction mixture to stand at room temperature, and observe its color. The reaction is complete when the red color disappears.

Videos: *The Grignard Reaction: Dissolution of Benzophenone in Ether; Addition to Grignard Reagent; The Grignard Reaction: Mixing Reaction Mixture by Flicking the Reaction Tube; Final Addition of Benzophenone Solution; Color Change at Completion of the Reaction; Addition of Hydrochloric Acid; The Solution Treated with Saturated Sodium Chloride Solution*

At the end of the reaction period, cool the tube in ice and add to it dropwise with stirring (use a glass rod or a spatula) 2 mL of 3 *M* hydrochloric acid. A creamy-white precipitate of triphenylmethanol will separate between the layers. Add more ether (it need not be anhydrous) to the reaction tube and shake the contents to dissolve all the triphenylmethanol. The result should be two perfectly clear layers. Remove a drop of the ether layer for thin-layer chromatographic (TLC) analysis. Any bubbling seen at the interface or in the lower layer is leftover magnesium reacting with the hydrochloric acid. Remove the aqueous layer and shake the ether layer with an equal volume of saturated aqueous sodium chloride solution to remove water and any remaining acid. Carefully remove the entire aqueous layer; then dry the ether layer by adding anhydrous calcium chloride pellets to the reaction tube until the drying agent no longer clumps together. Cork the tube and shake it from time to time over 5–10 minutes to complete the drying.

Using a Pasteur pipette, remove the ether from the drying agent and place it in another tared, dry reaction tube or a centrifuge tube. Use more ether to wash off the drying agent and combine these ether extracts. Evaporate the ether in a hood by blowing nitrogen or air onto the surface of the solution while warming the tube in a beaker of water or in the hand.

After all the solvent has been removed, determine the weight of the crude product. Note the characteristic odor of the biphenyl, which is the product of the side reaction that takes place between bromobenzene and phenylmagnesium bromide during the first reaction.

Trituration (grinding) of the crude product with petroleum ether will remove the biphenyl. Stir the crystals with 0.5 mL of petroleum ether in an ice bath, remove the solvent as thoroughly as possible, add a boiling stick, and recrystallize the residue from boiling 2-propanol (no more than 2 mL). Allow the solution to cool slowly to room temperature, and then cool it thoroughly in ice. Triphenylmethanol crystallizes slowly, so allow the mixture to remain in the ice as long as possible. Stir the ice-cold mixture well and collect the product by vacuum filtration on a Hirsch funnel. Save the filtrate, as concentration may give a second crop of crystals.

Photo: *Filtration Using a Pasteur Pipette*; Videos: *Filtration of Crystals Using the Pasteur Pipette; The Grignard Reaction: Crude Product Triturated and Recrystallized—Pure Triphenylcarbinol Isolated; Microscale Filtration on the Hirsch Funnel*

An alternative method for purifying the triphenylmethanol utilizes a mixed solvent. Dissolve the crystals in the smallest possible quantity of warm ether, and add 1.5 mL of hexanes to the solution. Add a boiling stick to the solution and boil off some of the ether until the solution becomes slightly cloudy, indicating that it is saturated. Allow the solution to cool slowly to room temperature. Triphenylmethanol is deposited slowly as large, thick prisms. Cool the solution in ice; after allowing time for complete crystallization to occur, remove the ether with a Pasteur

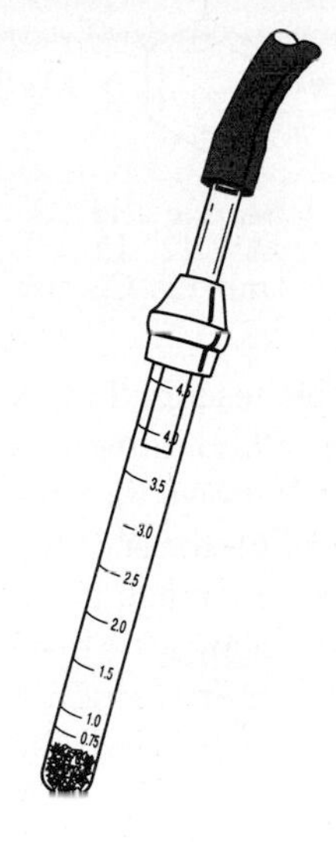

FIG. 38.6
An apparatus for drying crystals in a reaction tube under vacuum.

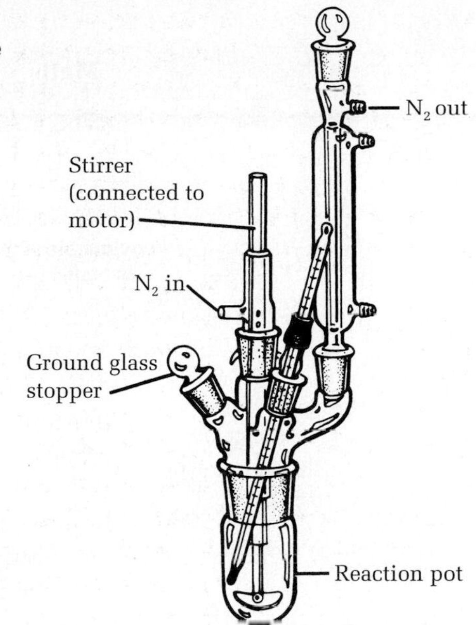

FIG. 38.7
A semimicroscale, research-type apparatus for the Grignard reaction, with provision for a motor-driven stirrer and an inlet and outlet for dry nitrogen.

pipette and wash the crystals once with a few drops of a cold 1:4 ether-hexanes mixture. Dry the crystals in the tube under a vacuum (Fig. 38.6).

Determine the weight, melting point, and percent yield of the triphenylmethanol. Analyze the crude and recrystallized product by TLC on silica gel (*see* Chapter 8), developing the plate with a 1:5 mixture of dichloromethane and petroleum ether. An infrared (IR) spectrum can be determined in chloroform solution or by preparing a mull or KBr disk (*see* Chapter 11). Compare the apparatus used in this experiment with the research-type apparatus shown in Figure 38.7.

Cleaning Up Combine the acidic aqueous layer and saturated sodium chloride layers, dilute with water, neutralize with sodium carbonate, and flush down the drain with excess water. Ether can be allowed to evaporate from the drying agent in the hood, and the drying agent can then be discarded in the nonhazardous solid waste container. If local regulations do not allow for evaporating solvents in a hood, the wet drying agent should be discarded in a special waste container. The petroleum ether and 2-propanol or ether-hexanes mother liquor are placed in the organic solvents waste container.

3. BENZOIC ACID

IN THIS EXPERIMENT, the Grignard reagent prepared in Experiment 1 is squirted onto a piece of dry ice. The resulting white carboxylate salt is hydrolyzed with hydrochloric acid, releasing benzoic acid, which is extracted into the ether layer. The ether solution could be dried and evaporated to give the product, but a better product is obtained by adding base to make the benzoate salt and then adding acid to this basic solution to cause benzoic acid to crystallize. It can be recrystallized from hot water.

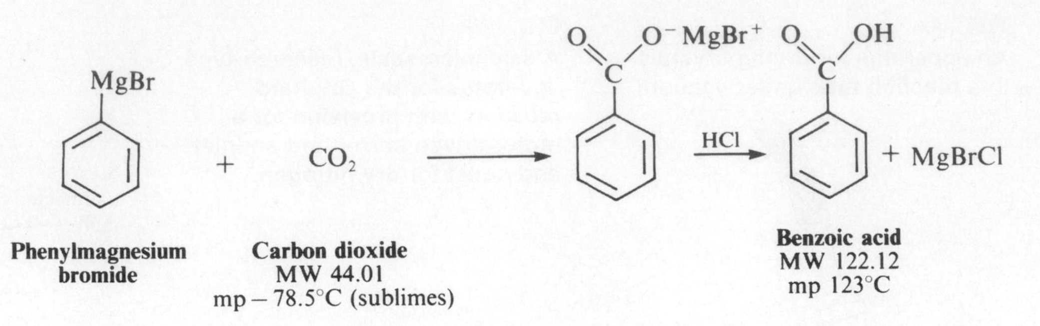

CAUTION: Handle dry ice with a towel or gloves. Contact with the skin can cause frostbite because dry ice sublimes at –78.5°C.

Prepare 2 mmol of phenylmagnesium bromide exactly as described in Experiment 1. Wipe off the surface of a small piece of dry ice (solid carbon dioxide) with a dry towel to remove frost and place it in a dry 30-mL beaker. Remove the pressure-relief needle from the reaction tube; then insert a syringe through the septum, turn the tube upside down, and draw into the syringe as much of the reagent solution as possible. Squirt this solution onto the piece of dry ice; then, using a clean needle, rinse out the reaction tube with 1 mL of anhydrous diethyl ether and squirt this onto the dry ice. Allow excess dry ice to sublime; then hydrolyze the salt by adding 2 mL of 3 *M* hydrochloric acid.

Video: *Extraction with Ether*

Transfer the mixture from the beaker to a reaction tube and shake it thoroughly. Two homogeneous layers should result. Add 1–2 mL of acid or ordinary (not anhydrous) ether if necessary. Remove the aqueous layer and shake the ether layer with 1 mL of water, which is removed and discarded. Then extract the benzoic acid by adding to the ether layer 0.7 mL of 3 *M* sodium hydroxide solution, shaking the mixture thoroughly, and withdrawing the aqueous layer, which is placed in a very small beaker or vial. The extraction is repeated with another 0.5-mL portion of base, and finally with 0.5 mL of water. Now that the extraction is complete, the ether, which can be discarded, contains primarily biphenyl, the byproduct formed during the preparation of phenylmagnesium bromide.

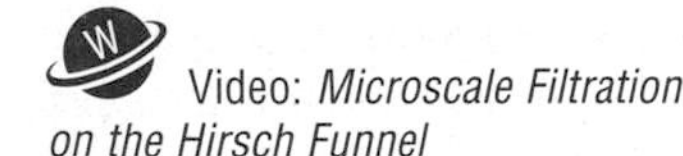
Video: *Microscale Filtration on the Hirsch Funnel*

The combined aqueous extracts are heated briefly to about 50°C to drive off dissolved ether from the aqueous solution, and then made acidic by adding concentrated hydrochloric acid (test with indicator paper). Cool the mixture thoroughly in an ice bath. Collect the benzoic acid on a Hirsch funnel and wash it with about 1 mL of ice water while on the funnel. A few crystals of this crude material are saved for a melting-point determination; the remainder of the product is recrystallized from boiling water.

The solubility of benzoic acid in water is 68 g/L at 95°C and 1.7 g/L at 0°C. Dissolve the acid in very hot water. Let the solution cool slowly to room temperature; then cool it in ice for several minutes before collecting the product by vacuum filtration on a Hirsch funnel. Use the ice-cold filtrate in the filter flask to complete the transfer of benzoic acid from the reaction tube. Turn the product out onto a piece of filter paper, squeeze out any excess water, and allow it to dry thoroughly. Once dry, weigh it, calculate the percent yield, and determine the melting point along with the melting point of the crude material. The product's IR spectrum may be determined as a solution in chloroform (1 g of benzoic acid dissolves in 4.5 mL of chloroform) or as a mull or KBr disk (*see* Chapter 11).

Cleaning Up. Combine all aqueous layers, dilute with a large quantity of water, and flush the slightly acidic solution down the drain.

4. PHENYLMAGNESIUM BROMIDE (PHENYL GRIGNARD REAGENT)

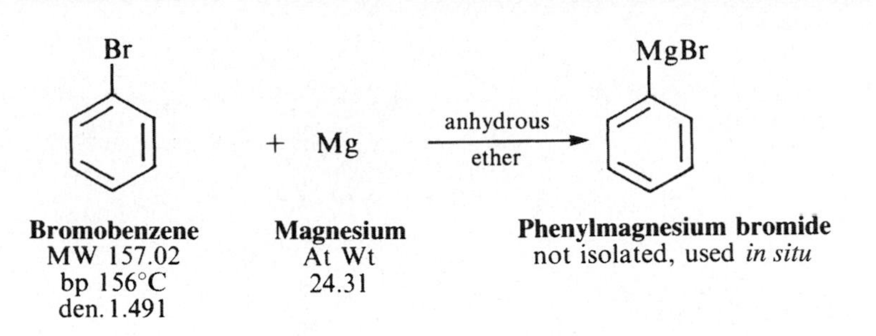

Bromobenzene MW 157.02 bp 156°C den. 1.491

Magnesium At Wt 24.31

Phenylmagnesium bromide not isolated, used *in situ*

Diethyl ether can be kept anhydrous by storing it over Linde 5A molecular sieves. Discard the ether after 90 days because of peroxide formation.

CAUTION: Ether is extremely flammable. Extinguish all flames before using ether.

Specially dried anhydrous diethyl ether is required.

All equipment and reagents must be *absolutely dry.* The Grignard reagent is prepared in a dry, 100-mL round-bottomed flask fitted with a long reflux condenser. A calcium chloride drying tube inserted in a cork that will fit either the flask or the top of the condenser is also made ready (Fig. 38.8a). The flask, condenser, and magnesium (0.082 mol of magnesium turnings) should be as dry as possible to begin with, and then should be further dried in a 110°C oven for at least 35 minutes. Alternatively, the magnesium is placed in the flask, the calcium chloride tube is attached directly, and the flask is heated gently but thoroughly with a cool luminous flame.[1] Do not overheat the magnesium. It will become deactivated through oxidation or, if strongly overheated, can burn. The flask upon cooling pulls dry air through the calcium chloride. Cool to room temperature before proceeding! Extinguish all flames! Ether vapor is denser than air and can travel along bench tops and into sinks. Use care.

Prepare an ice bath in case control of the reaction becomes necessary, although this is usually not the case. Remove the drying tube and fit it to the top of the condenser. Then pour into the flask through the condenser 15 mL of anhydrous diethyl ether and 9 mL (0.086 mol) of bromobenzene. Be sure the graduated cylinders used to measure the ether and bromobenzene are absolutely dry. (More ether is to be added as soon as the reaction starts, but the concentration of bromobenzene is kept high at the outset to promote easy starting.) If there is no immediate sign of reaction, insert a *dry* stirring rod with a flattened end and crush a piece of magnesium firmly against the bottom of the flask under the surface of the liquid, giving a twisting motion to the rod. When this is done properly, the liquid becomes slightly cloudy, and ebullition commences at the surface of the compressed metal. Be careful not to punch a hole in the bottom of the flask. Do not pound the magnesium. Attach the condenser at once, and swirl the flask to provide fresh surfaces for contact. As soon as you are sure that the reaction has started, add an additional 25 mL of anhydrous ether through the top of the condenser before spontaneous boiling becomes too vigorous (replace the drying tube). Note the volume of ether in the flask. Cool in ice if necessary to slow the reaction, but do not overcool the mixture because the reaction can be stopped by too much cooling. Any difficulty in initiating the reaction can be dealt with by trying the following prompts in succession.

[1]Alternatively, if nitrogen gas is available, the reaction can be run under nitrogen using a bubbler.

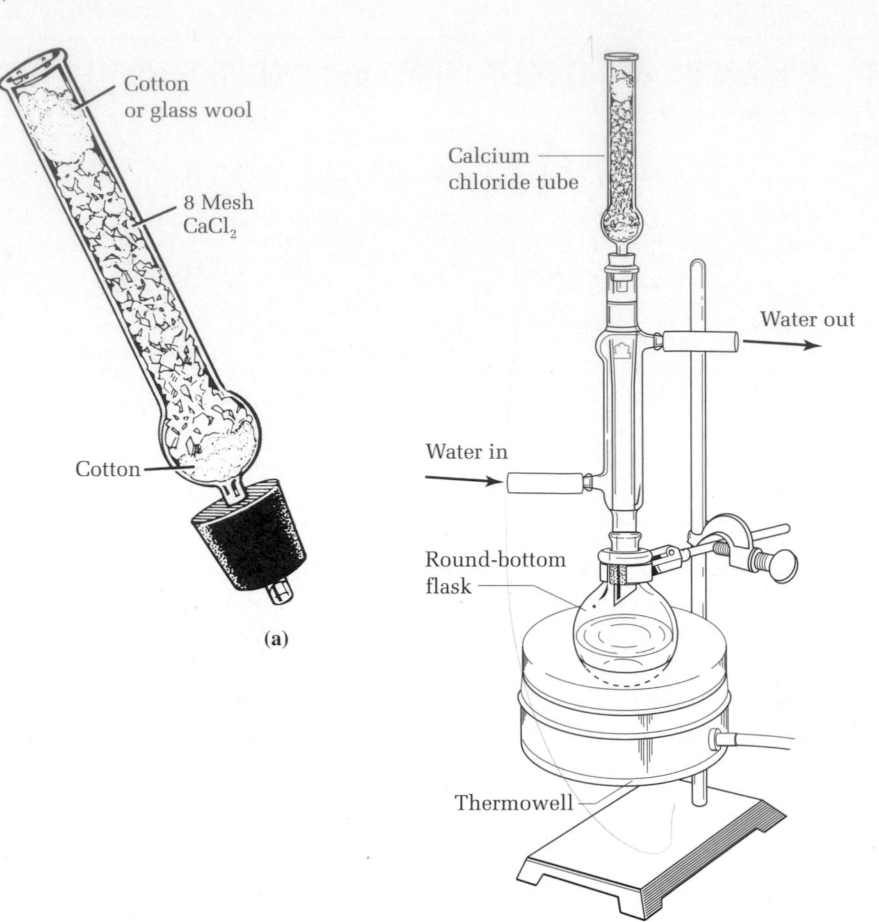

FIG. 38.8
(a) A calcium chloride drying tube fitted with a rubber stopper. Store for future use with a cork in the top and a pipette bulb on the bottom. (b) An apparatus for refluxing the Grignard reaction.

Starting the Grignard reaction

1. Warm the flask with your hands or in a beaker of warm water. Then see if boiling continues when the flask (condenser attached) is removed from the warmth.
2. Try further mashing of the metal with a dry stirring rod.
3. Add a tiny crystal of iodine as a starter (in this case, the ethereal solution of the final reaction product should be washed with sodium bisulfite solution to remove the yellow color).
4. Add a few drops of a solution of phenylmagnesium bromide or methylmagnesium iodide (which can be made in a test tube).
5. Start afresh, taking greater care with the dryness of apparatus, measuring tools, and reagents, and sublime a crystal or two of iodine on the surface of the magnesium to generate Gattermann's activated magnesium before beginning the reaction again.

Use caution when heating to avoid condensation on the outside of the condenser.

Once the reaction begins, spontaneous boiling in the diluted mixture may be slow or become slow. If so, mount the flask and condenser in a heating mantle or Thermowell (using one clamp to support the condenser suffices; Fig. 38.8b) and reflux gently until the magnesium has disintegrated and the solution has acquired a cloudy or brownish appearance. The reaction is complete when only a few remnants of metal (or metal contaminants) remain. Check to see that the volume of ether has not decreased. If it has, add more anhydrous ether. Because the solution of Grignard reagent deteriorates on standing, Experiment 5 should be started at once.

5. TRIPHENYLMETHANOL FROM METHYL BENZOATE

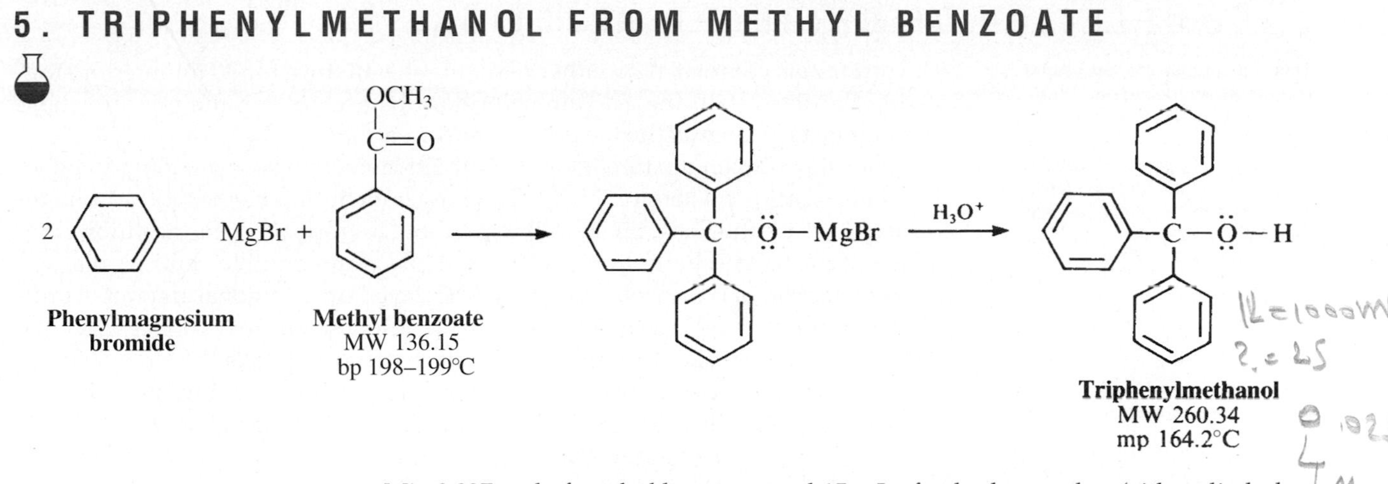

Mix 0.037 mol of methyl benzoate and 15 mL of anhydrous ether (either diethyl or *t*-butyl methyl ether) in a separatory funnel, cool the flask containing the phenylmagnesium bromide solution briefly in an ice bath, remove the drying tube, and insert the stem of a separatory funnel into the top of the condenser. Run in the methyl benzoate solution *slowly* with only such cooling as is required to control the mildly exothermic reaction, which affords an intermediate salt that separates as a white solid. Replace the calcium chloride tube; swirl the flask until it is at room temperature and the reaction has subsided. Go to Experiment 7.

6. TRIPHENYLMETHANOL FROM BENZOPHENONE

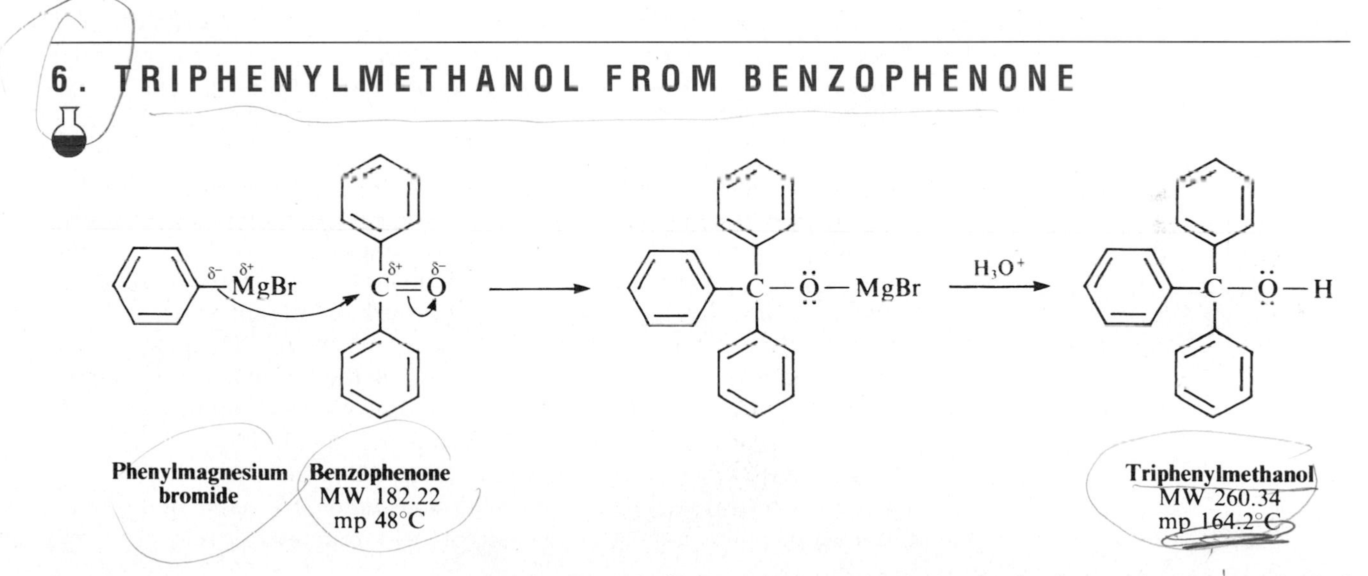

Dissolve 0.037 mol of benzophenone in 25 mL of anhydrous ether (either diethyl or *t*-butyl methyl ether) in a separatory funnel and cool the flask containing *half* the phenylmagnesium bromide solution (0.041 mol) briefly in an ice bath. (The other half can be used to make benzoic acid.) Remove the drying tube and insert the stem of the separatory funnel into the top of the condenser. Add the benzophenone solution *slowly* with swirling, and only such cooling as is required to control the mildly exothermic reaction, which gives a bright-red solution and then precipitates a white salt. Replace the calcium chloride tube; swirl the flask until it is at room temperature and the reaction has subsided. Go to Experiment 7.

7. COMPLETION OF GRIGNARD REACTION

This is a suitable stopping point.

The reaction is completed by either refluxing the mixture for 30 minutes or stoppering the flask with the calcium chloride tube and letting the mixture stand overnight (subsequent refluxing is then unnecessary).[2]

In this part of the experiment, ordinary (not anhydrous) diethyl ether or *t*-butyl methyl ether may be used.

Pour the reaction mixture into a 250-mL Erlenmeyer flask containing 50 mL of 10% sulfuric acid and about 25 g of ice, and use both ordinary ether and 10% sulfuric acid to rinse the flask. Swirl well to promote hydrolysis of the addition compound; basic magnesium salts are converted into water-soluble neutral salts, and triphenylmethanol is distributed into the ether layer. An additional amount of ordinary ether may be required. Pour the mixture into a separatory funnel (rinse the flask with ether), shake, and draw off the aqueous layer. Shake the ether solution with 10% sulfuric acid to further remove magnesium salts, and wash with saturated sodium chloride solution to remove water that has dissolved in the ether. The amounts of liquid used in these washing operations are not critical. In general, an amount of wash liquid equal to one-third of the ether volume is adequate.

Saturated aqueous sodium chloride solution removes water from ether.

To effect final drying of the ether solution, pour the ether layer out of the neck of the separatory funnel into an Erlenmeyer flask, add about 5 g of calcium chloride pellets, swirl the flask intermittently, and after 5 minutes remove the drying agent by gravity filtration (using a filter paper and funnel) into a tared Erlenmeyer flask. Rinse the drying agent with a small amount of ether. Add 25 mL of 66°C–77°C hexanes and concentrate the ether-hexanes solutions (with a steam bath or hot plate) in an Erlenmeyer flask under an aspirator tube (*see* Fig. 9.6 on page 196). Evaporate slowly until crystals of triphenylmethanol just begin to separate; then let crystallization proceed, first at room temperature and then at 0°C. The product should be colorless and should melt at no lower than 160°C. Concentration of the mother liquor may yield a second crop of crystals. A typical student yield is 5.0 g. Evaporate the mother liquors to dryness and save the residue for later isolation of the components by chromatography.

Perform a TLC analysis of the first crop of triphenylmethanol and the residue from the evaporation of the mother liquors. Dissolve equal quantities of the two solids (a few crystals) and also biphenyl in equal quantities of dichloromethane (1 or 2 drops). Using a microcapillary, spot equal quantities of material on silica gel TLC plates and develop the plates in an appropriate solvent system. Try a 1:3 dichloromethane–hexanes mixture first, and adjust the relative quantities of solvent as needed. The spots can be seen by examining the TLC plate under a fluorescent lamp or by treating the TLC plate with iodine vapor. From this analysis, decide how pure each of the solids is and whether it would be worthwhile to attempt to isolate more triphenylmethanol from the mother liquors.

Turn in the product in a vial labeled with your name, the name of the compound, its melting point, and the overall percent yield from benzoic acid.

Cleaning Up. Combine the acidic aqueous layer and saturated sodium chloride layers, dilute with water, neutralize with sodium carbonate, and flush down the drain with excess water. Ether can be allowed to evaporate from the drying agent in the hood, and the drying agent can then be discarded in the nonhazardous solid waste container. If local regulations do not allow for evaporating solvents in a hood, the wet drying agent should be discarded in a special waste container. The ether-hexanes mother liquor is placed in the organic solvents waste container.

Dispose of recovered and waste solvents in the appropriate containers.

[2]A rule of thumb for organic reactions: A 10°C rise in temperature will double the rate of the reaction.

8. BENZOIC ACID

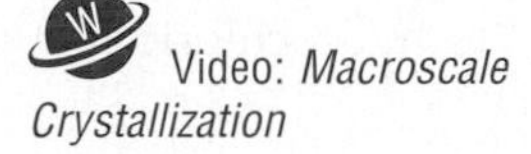

Video: *Macroscale Crystallization*

Wipe the frost from a piece of dry ice, transfer the ice to a cloth towel, and crush it with a hammer. Without delay (so moisture will not condense on the cold solid), transfer about 10 g of dry ice to a 250-mL beaker. Cautiously pour one-half of the solution of phenylmagnesium bromide prepared in Experiment 1 onto the dry ice. A vigorous reaction will ensue. Allow the mixture to warm up, and stir it until the dry ice has evaporated. To the beaker, add 20 mL of 3 *M* hydrochloric acid; then heat the mixture over a steam bath in the hood to boil off the ether. Cool the beaker thoroughly in an ice bath, and collect the solid product by vacuum filtration on a Büchner funnel.

Transfer the solid back to the beaker and dissolve it in a minimum quantity of saturated sodium bicarbonate solution (2.8 *M*). Note that a small quantity of a byproduct remains suspended and floating on the surface of the solution. Note the odor of the mixture. Transfer it to a separatory funnel and shake it briefly with about 15 mL of ether. Discard the ether layer, place the clear aqueous layer in the beaker, and heat it briefly to drive off dissolved ether. Carefully add 3 *M* hydrochloric acid to the mixture until the solution tests acidic to pH paper. Cool the mixture in ice and collect the product on a Büchner funnel. Recrystallize it from a minimum quantity of hot water and isolate it in the usual manner. Determine the melting point and the weight of the benzoic acid; calculate its yield based on the weight of magnesium used to prepare the Grignard reagent.

Cleaning Up. Combine all aqueous layers, dilute with a large quantity of water, and flush the slightly acidic solution down the drain. The ether-hexanes mother liquor from the recrystallization goes in the organic solvents waste container. The TLC developer, which contains dichloromethane, is placed in the halogenated organic waste container. Calcium chloride from the drying tube should be dissolved in water and flushed down the drain.

QUESTIONS

1. Triphenylmethanol can also be prepared by reacting ethyl benzoate with phenylmagnesium bromide, and by reacting diethylcarbonate with phenylmagnesium bromide. Write stepwise reaction mechanisms for these two reactions.

$$C_2H_5O\overset{\overset{\large O}{\|}}{C}OC_2H_5$$

2. If the ethyl benzoate used to prepare triphenylmethanol is wet, what byproduct is formed?
3. Exactly what weight of dry ice is needed to react with 2 mmol of phenylmagnesium bromide? Would an excess of dry ice be harmful?
4. In the synthesis of benzoic acid, benzene is often detected as an impurity. How does this come about?
5. In Experiment 3, the benzoic acid could have been extracted from the ether layer using sodium bicarbonate solution. Write equations showing how this might be done and how the benzoic acid would be regenerated. What practical reason makes this extraction method less desirable than sodium hydroxide extraction?

6. What is the weight of frost (ice) on the dry ice that will react with all of the Grignard reagent used in Experiment 8?
7. How many moles of carbon dioxide are contained in 10 g of dry ice?
8. In Experiment 8 just after the dry ice has evaporated from the beaker, what is the white solid remaining?
9. Write an equation for the reaction of the white solid with 3 *M* hydrochloric acid in Experiment 8.
10. Write an equation for the reaction of the product with sodium bicarbonate.
11. Would you expect sodium benzoate to have an odor? Why or why not?
12. What odor do you detect after the product has dissolved in sodium bicarbonate solution?
13. What is the purpose of the ether extraction?

See *Web Links*

14. "Isolate the product in the usual way." What is the meaning of this sentence?

CHAPTER 39

The Wittig and Wittig-Horner Reactions

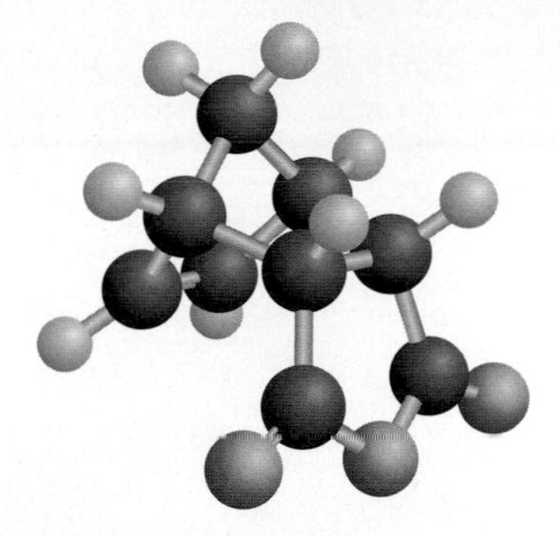

PRELAB EXERCISE: Account for the fact that the Wittig-Horner reaction of cinnamaldehyde gives almost exclusively the *E,E*-butadiene with very little contaminating *E,Z*-product.

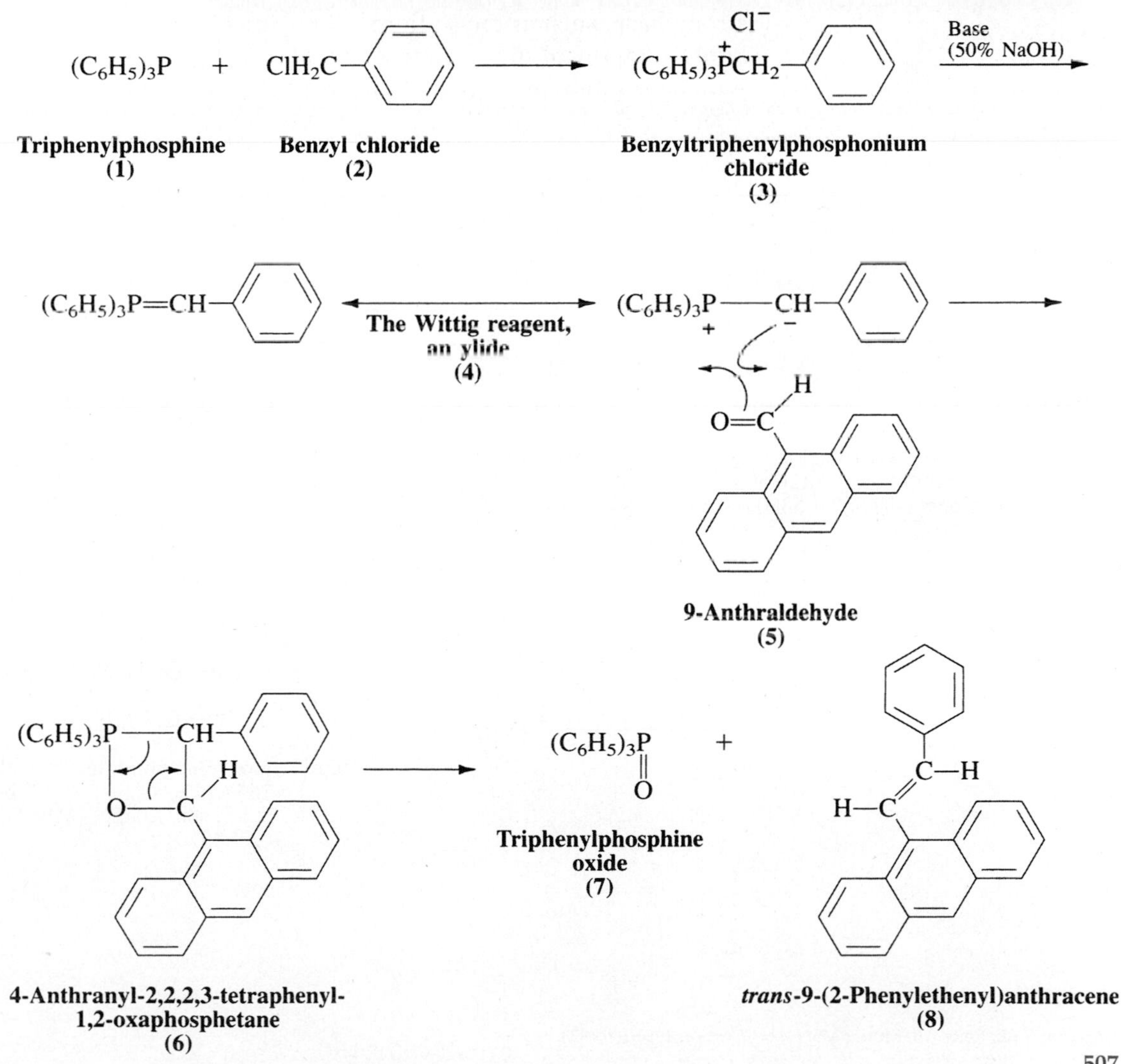

The Wittig reaction affords an invaluable method for converting a carbonyl compound to an olefin; for example, the conversion of 9-anthraldehyde **(5)** and benzyl chloride **(2)** by means of the Wittig reagent **(4)** through a four-centered intermediate to the product **(8)**. Because the active reagent, an ylide, is unstable, it is generated in the presence of a carbonyl compound by dehydro-halogenation of the phosphonium chloride **(3)** with base. Usually a very strong base, such as phenyllithium in dry ether, is employed, but in the experiments in this chapter, 50% sodium hydroxide is employed. The existence of the four-membered ring intermediate was proved by nuclear magnetic resonance (NMR) in 1984.[1] The phenylethenylanthracene product produces a green fluorescence in a product called Cyalume, a chemiluminescent substance, in the light stick reaction (*see* Chapter 61).

The light stick reaction

Simplified Wittig reaction

When the halogen compound employed in the first step has an activated halogen atom ($RCH{=}CHCH_2X$, $C_6H_5CH_2X$, XCH_2CO_2H), a simpler procedure known as the *Horner phosphonate modification* of the Wittig reaction is applicable. When benzyl chloride is heated with triethyl phosphite, chloroethane is eliminated from the initially formed phosphonium chloride with the production of diethyl benzylphosphonate. This phosphonate is stable, but in the presence of a strong base, such as the sodium methoxide used here, it condenses with a carbonyl component in the same way that a Wittig ylide condenses. Thus it reacts with benzaldehyde to give *E*-stilbene, and with cinnamaldehyde to give *E*,*E*-1, 4-diphenyl-1,3-butadiene.

C_6H_5–CH_2Cl + $(CH_3CH_2O)_3P$ → C_6H_5–$CH_2P^+(OCH_2CH_3)_3$ Cl^-

Benzyl chloride
bp 179°C, MW 126.59
den. 1.10, n_D^{20} 1.5380

Triethyl phosphite
bp 156°C, MW 166.16
den. 0.94, n_D^{20} 1.4130

Benzyltriethylphosphonium chloride

↓

C_6H_5–$CH_2P^+(O^-)(OCH_2CH_3)$–O–CH_2CH_3 + $ClCH_2CH_3$

Diethyl benzylphosphonate
bp 156°C/9 torr
MW 228.23, den. 1.045
n_D^{20} 1.4970

Chloroethane
bp 12.3°C
MW 64.52, den. 0.891

[1]Maryanoff BE, Reitz AB, Mutter MS. *J Am Chem Soc.* **1984**;*106*:1873.

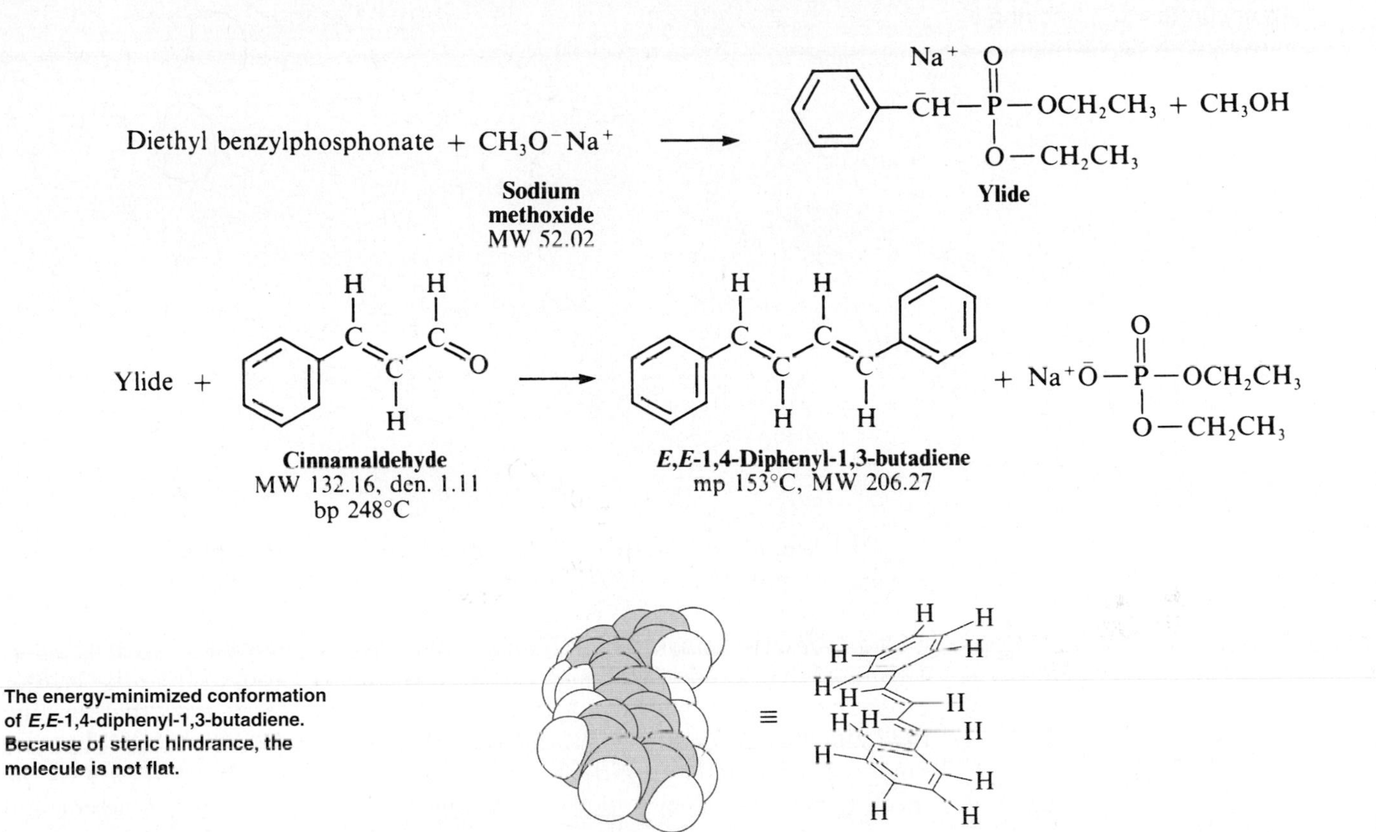

The energy-minimized conformation of *E,E*-1,4-diphenyl-1,3-butadiene. Because of steric hindrance, the molecule is not flat.

EXPERIMENTS

1. SYNTHESIS OF *trans*-9-(2-PHENYLETHENYL) ANTHRACENE[2]

Microscale Procedure

IN THIS EXPERIMENT, a phosphonium chloride is reacted with base to form the Wittig reagent. An aldehyde in the same solution reacts with the Wittig reagent to give the alkene product. The solvent, dichloromethane, is dried and evaporated to give the product that is, in turn, recrystallized from 1-propanol.

[2]For macroscale syntheses of *trans*-9-(2-phenylethenyl)anthracene, see Becker HD, Andersson KJ. *Org Chem.* **1983**;*48*:4552. Merkl G, Merz A. *Synthesis.* **1973**;295. Silversmith EF. *J Chem Educ.* **1986**;*63*:645.

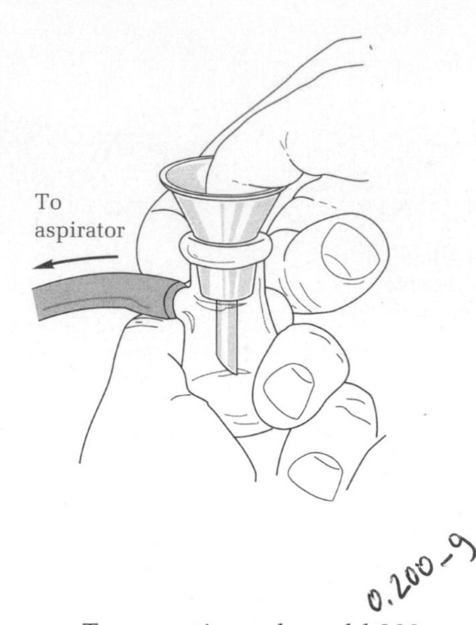

FIG. 39.1
An apparatus for removing a solvent under a vacuum.

To a reaction tube, add 200 mg of benzyltriphenylphosphonium chloride, 115 mg of 9-anthraldehyde, 0.6 mL of dichloromethane, and a magnetic stirring bar. With rapid magnetic stirring, 0.26 mL of 50% sodium hydroxide solution is added dropwise from a Pasteur pipette. After stirring vigorously for 30 minutes, 1.5 mL of dichloromethane and 1.5 mL of water are added; then the tube is capped and shaken. The organic layer is removed and placed in another reaction tube, and the aqueous layer is extracted with 1 mL of dichloromethane. The combined dichloromethane extracts are dried over calcium chloride pellets, the dichloromethane is removed, the drying agent is washed with more solvent, and the solvent is removed under vacuum in the filter flask (Fig. 39.1). To the solid remaining in the filter flask, add 3 mL of 1-propanol, heat and transfer the hot solution to an Erlenmeyer flask to crystallize. After cooling spontaneously to room temperature, cool the flask in ice and collect the product on a Hirsch funnel. The triphenylphosphine oxide remains in the propanol solution. The reported melting point of the product is 131°C–132°C.

Macroscale Procedure

Into a 10-mL round-bottomed flask, place 0.97 g of benzyltriphenylphosphonium chloride, 0.57 g of 9-anthraldehyde, 3 mL of dichloromethane, and a stirring bar. Clamp the flask over a magnetic stirrer and stir the mixture at high speed while adding 1.3 mL of 50% sodium hydroxide solution dropwise from a Pasteur pipette. After the addition is complete, continue the stirring for 30 minutes; then transfer the contents of the flask to a 50-mL separatory funnel using 10 mL of water and 10 mL of dichloromethane to complete the transfer. Shake the mixture, remove the organic layer, and then extract the aqueous layer once more with 5 mL of dichloromethane. Dry the combined organic layers with anhydrous calcium chloride pellets, transfer the solution to a 50-mL Erlenmeyer flask, evaporate it to dryness on a steam bath, and recrystallize the yellow partially crystalline residue from 15 mL of 2-propanol. The product recrystallizes as thin yellow plates (mp 131°C–132°C). Save the product for use as a fluorescer in the Cyalume chemiluminescence experiment (*see* Chapter 61). Examine a dilute ethanol solution of the product under an ultraviolet (UV) lamp.

2. SYNTHESIS OF 1,4-DIPHENYL-1,3-BUTADIENE

IN THIS EXPERIMENT, benzyl chloride and triethyl phosphite are reacted with each other to give a phosphonate salt. This salt, on reaction with the strong base, sodium methoxide, gives a Wittig reagent-like ylide that reacts with an aldehyde to give the product alkene, diphenylbutadiene. It is isolated by filtration and can be recrystallized, if necessary, from methylcyclohexane.

CAUTION: Take care to keep organophosphorus compounds off the skin.

Use freshly prepared or opened sodium methoxide, keep the bottle closed to minimize contact with moist air.

CAUTION: Handle benzyl chloride in the hood. It is a severe lachrymator (tear producer) and irritant of the respiratory tract.

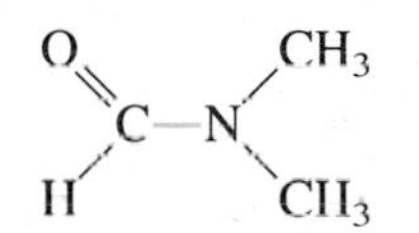

***N*,*N*-Dimethylformamide (DMF)**
MW 73.10, bp 153°C
n_D^{20} 1.4310

A highly polar solvent capable of dissolving ionic compounds such as sodium methoxide yet miscible with water.

CAUTION: Avoid skin contact with dimethylformamide.

The success of this reaction is strongly dependent on the purity of the starting materials.[3] Benzyl chloride and triethyl phosphite are usually pure enough as received from the supplier. Cinnamaldehyde from a new, previously unopened bottle should be satisfactory, but because it air oxidizes extremely rapidly it should be distilled, preferably under nitrogen or at reduced pressure, if there is any doubt about its quality. Sodium methoxide, direct from reputable suppliers, has been found on occasion to be completely inactive. No easy method exists for determining its activity; therefore, it is best prepared using the procedure given in *Organic Syntheses*.[4]

To a 10 mm × 100-mm reaction tube, add 316 mg of benzyl chloride (α-chlorotoluene), 415 mg of triethyl phosphite, and a boiling chip (Fig. 39.2). Because the triethyl phosphite has an offensive odor, obtain this material from a dispenser that has been calibrated to deliver the correct quantity. Place the open reaction tube to a depth of about 1 cm in a sand bath that has been preheated to 210°C to 220°C. Reflux the reaction mixture for 1 hour. The vapors condense on the cool upper portion of the reaction tube, so a condenser is not necessary. At an internal temperature of about 140°C (which you need not monitor), ethyl chloride is evolved; by the end of the reflux period, the temperature of the reaction mixture will be 200 to 220°C. At the end of the reaction period, remove the tube from the sand bath, cool it to room temperature, and then add the contents to 160 mg of sodium methoxide in a 10-mL Erlenmeyer flask using 1 mL of dry dimethylformamide (DMF) to complete the transfer. Cool the mixture in ice, mix the contents of the flask thoroughly, then add 330 mg of freshly distilled cinnamaldehyde in 1 mL of dry DMF dropwise with thorough mixing in the ice bath. Allow the mixture to come to room temperature. Note the changes that take place over the next few minutes.

After a *minimum* of 10 minutes, add 2 mL of methanol to the Erlenmeyer flask with thorough stirring. Add water to the flask until it is almost full, and stir the mixture until it is homogeneous in color. The hydrocarbon precipitates from the reaction mixture after the methanol and water have been added. Collect the product by vacuum filtration on a Hirsch funnel (Fig. 39.3) and wash the crystals first with water to remove the red color, and then with ice-cold methanol to remove the yellow color. The hydrocarbon is completely insoluble in water and only sparingly

[3]The instructor should try out this experiment before assigning it to a class.

[4]For enough sodium methoxide for 50 microscale or 2 macroscale reactions, add 3.5 g of sodium spheres to 50 mL of anhydrous methanol in a 100-mL flask equipped with a condenser. Add the sodium a few pieces at a time through the top of the condenser at a rate so as to keep the reaction under control. It is safest to wait for the complete reaction of one portion of sodium before adding the next. After all the sodium has reacted, remove the methanol on a rotary evaporator, first over hot water and then over a bath heated to 150°C. The purified sodium methoxide will be a free-flowing white powder that should keep for several weeks in a desiccator. The yield should be 8.2 g. Source: *Org Synth.* Col. Vol. IV, **1963**:651.

FIG. 39.2
A microscale reflux apparatus.

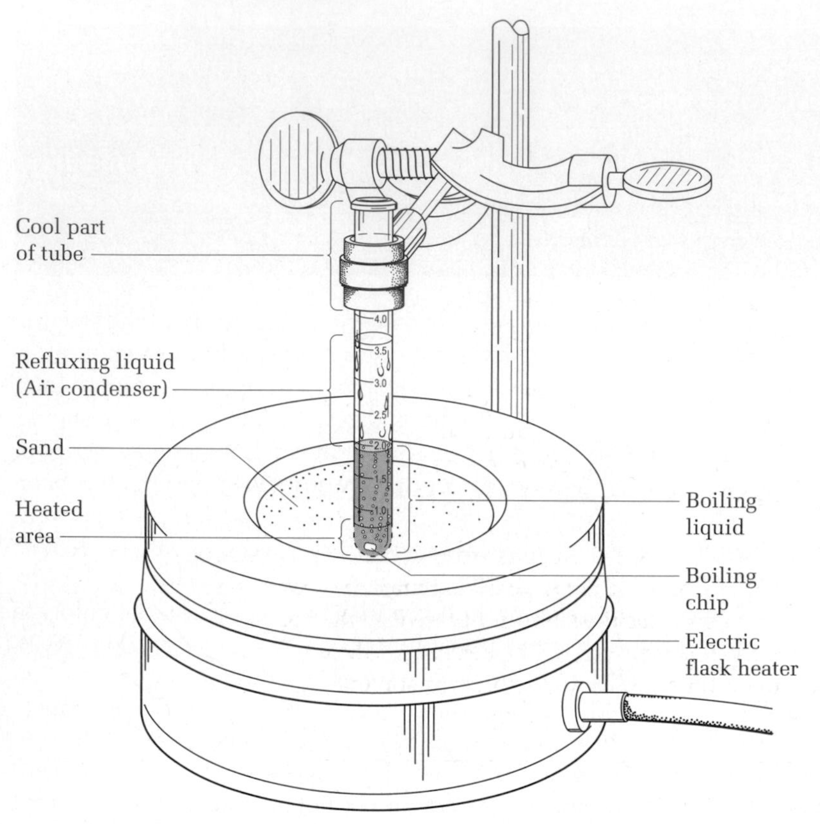

FIG. 39.3
A Hirsch funnel for filtration.

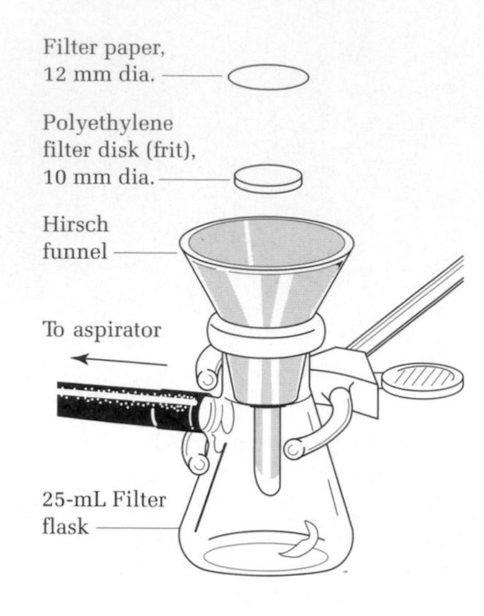

Video: *Microscale Filtration on the Hirsch Funnel*

soluble in methanol. Weigh the dry product, which should be faint yellow in color, determine its melting point, and calculate the crude yield. If the melting point is below 150°C, recrystallize the product from methylcyclohexane (10 mL/g), and again determine the melting point and yield. Turn in the product in a labeled vial, giving crude and recrystallized weights, melting points, and yields.

Cleaning Up. The filtrate and washings from this reaction are dark, oily, and smell bad. The mixture contains DMF and could contain traces of all starting materials. Keep the volume as small as possible, and place it in the hazardous waste container for organophosphorus compounds. If methylcyclohexane was used for recrystallization, place the mother liquor in the organic solvents waste container.

3. SYNTHESIS OF 1,4-DIPHENYL-1,3-BUTADIENE

The success of this reaction is strongly dependent on the purity of the starting materials.

With the aid of pipettes and a pipetter, measure into a 25 × 150-mm test tube 5 mL of benzyl chloride (α-chlorotoluene) and 7.7 mL of triethyl phosphite. Add a boiling stone, insert a cold finger condenser, and gently reflux the liquid with a flask heater for 1 hour. Alternatively, carry out the reaction in a 25-mL round-bottomed flask equipped with a reflux condenser. (Elimination of ethyl chloride

CAUTION: Keep organophosphorus compounds off the skin. Handle benzyl chloride in the hood. It is a severe lachrymator (tear producer) and respiratory tract irritant.

Use freshly prepared or opened sodium methoxide; keep the bottle closed to minimize oxidation.

Avoid skin contact with dimethylformamide.

***N,N*-Dimethylformamide (DMF)**
MW 73.10, bp 153°C
n_D^{20} 1.4310

A highly polar solvent capable of dissolving ionic compounds such as sodium methoxide, yet miscible with water.

starts at about 130°C and, in the time period specified, the temperature of the liquid rises to 190°C–200°C.) Let the phosphonate ester cool to room temperature, pour it into a 125-mL Erlenmeyer flask containing 2.4 g of sodium methoxide, and add 40 mL of dimethylformamide (DMF), using a part of this solvent to rinse the test tube. Swirl the flask vigorously in a water-ice bath to thoroughly chill the contents and continue swirling while running in 5 mL of cinnamaldehyde by pipette. The mixture soon turns deep red; then crystalline hydrocarbon starts to separate. When there is no further change (about 2 minutes), remove the flask from the cooling bath and let it stand at room temperature for about 10 minutes. Then add 20 mL of water and 10 mL of methanol, swirl vigorously to dislodge crystals, and finally collect the product on a suction funnel using the red mother liquor to wash the flask. Wash the product with water until the red color of the product is completely replaced by a yellow color. Then wash with methanol to remove the yellow impurity, and continue until the wash liquor is colorless. The yield of the crude, faintly yellow hydrocarbon (mp 150°C–151°C) should be about 5.7 g. This material is satisfactory for use in Chapter 50 (1.5 g required). A good solvent for recrystallization of the remaining product is methylcyclohexane (bp 101°C, 10 mL/g; use more if the solution requires filtration). Pure *E,E*-1,4-diphenyl-1,3-butadiene melts at 153°C.

Cleaning Up. The filtrate and washings from this reaction are dark, oily, and smell bad. The mixture contains DMF and could contain traces of all starting materials. Keep the volume as small as possible and place it in the hazardous waste container for organophosphorus compounds. If methylcyclohexane was used for recrystallization, place the mother liquor in the organic solvents waste container.

QUESTIONS

1. Show how 1,4-diphenyl-1,3-butadiene might be synthesized from benzaldehyde and an appropriate halogenated compound.
2. Explain why the methyl groups of trimethyl phosphite give two peaks in the ^{1}H NMR spectrum (Fig. 39.4).

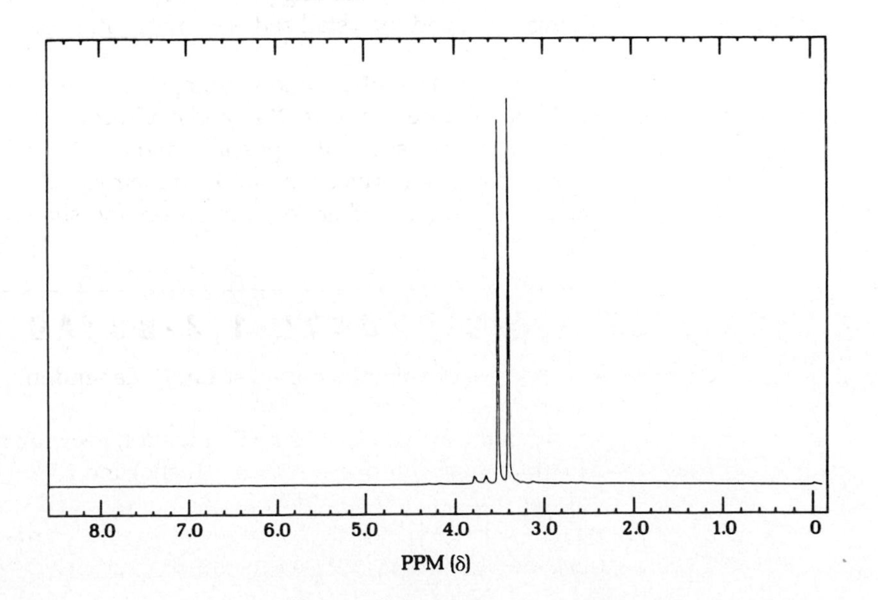

FIG. 39.4
The ^{1}H NMR spectrum of trimethyl phosphite (90 MHz).

3. Write the equation for the reaction between sodium methoxide and moist air.
4. The Wittig reaction usually gives a mixture of *cis* and *trans* isomers. Using a molecular mechanics program, calculate the steric energies or heats of formation of both possible products in Experiment 1 or 2. Are these compounds planar? Are the most stable molecules produced in these two experiments? If not, why not?

CHAPTER 40

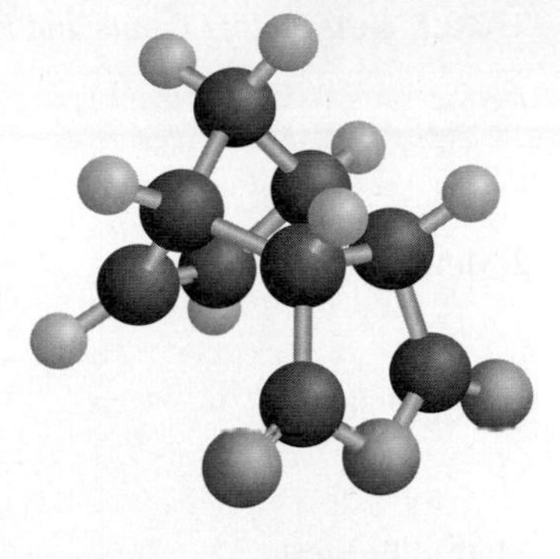

Esterification and Hydrolysis

When you see this icon, sign in at this book's premium web site at **www.cengage.com/login** to access videos, Pre-Lab Exercises and other online resources.

PRELAB EXERCISE: Write the detailed mechanism for the acid-catalyzed hydrolysis of methyl benzoate.

$$R-\overset{\overset{\displaystyle O}{\|}}{C}-O-R'$$

The ester group

The ester group is an important functional group that can be synthesized in a variety of ways. The low molecular weight esters have very pleasant odors, and indeed comprise the major flavor and odor components of a number of fruits. Although a natural flavor may contain nearly 100 different compounds, single esters approximate natural odors and are often used in the food industry for artificial flavors and fragrances (Table 40.1).

Flavors and fragrances

Esters can be prepared by reacting a carboxylic acid with an alcohol in the presence of a catalyst such as concentrated sulfuric acid, hydrogen chloride, *p*-toluenesulfonic acid, or the acid form of an ion exchange resin. For example, methyl acetate can be prepared as follows:

$$\underset{\textbf{Acetic acid}}{CH_3\overset{\overset{\displaystyle O}{\|}}{C}-OH} + \underset{\textbf{Methanol}}{CH_3OH} \overset{H^+}{\rightleftharpoons} \underset{\textbf{Methyl acetate}}{CH_3\overset{\overset{\displaystyle O}{\|}}{C}-OCH_3} + H_2O$$

Fischer esterification

The Fischer esterification reaction reaches equilibrium after a few hours of refluxing. The position of the equilibrium can be shifted by adding more of the acid or of the alcohol, depending on cost or availability. The mechanism of the reaction involves initial protonation of the carboxyl group, attack by the nucleophilic hydroxyl, a proton transfer, and loss of water followed by loss of the catalyzing proton to give the ester. Each of these steps is completely reversible, so this process is also, in reverse, the mechanism for the hydrolysis of an ester:

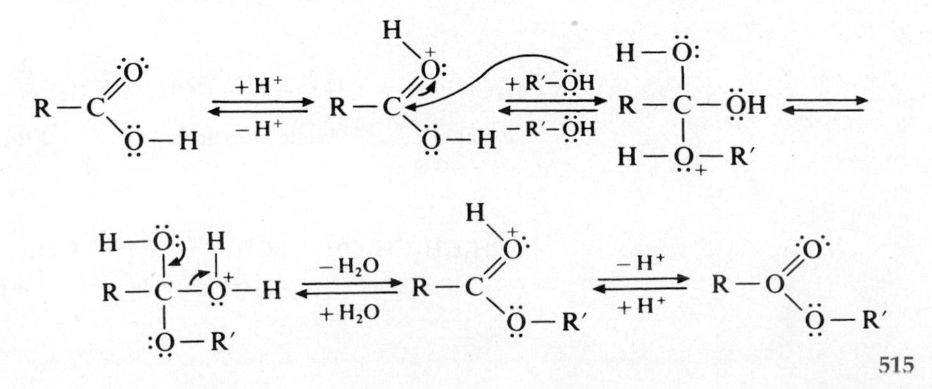

TABLE 40.1 *Boiling Points and Fragrances of Esters*

Ester		Formula	bp (°C)	Fragrance
2-Methylpropyl formate	(a)	$HC(=O)-OCH_2CH(CH_3)CH_3$	98.4	Raspberry
1-Propyl acetate	(b)	$CH_3C(=O)-OCH_2CH_2CH_3$	101.7	Pear
Methyl butyrate	(c)	$CH_3CH_2CH_2C(=O)-OCH_3$	102.3	Apple
Ethyl butyrate	(d)	$CH_3CH_2CH_2C(=O)-OCH_2CH_3$	121	Pineapple
2-Methylpropyl propionate	(e)	$CH_3CH_2C(=O)-OCH_2CH(CH_3)CH_3$	136.8	Rum
3-Methylbutyl acetate	(f)	$CH_3C(=O)-OCH_2CH_2CH(CH_3)CH_3$	142	Banana
Benzyl acetate	(g)	$CH_3C(=O)-OCH_2-C_6H_5$	213.5	Peach
Octyl acetate	(h)	$CH_3C(=O)-OCH_2(CH_2)_6CH_3$	210	Orange
Methyl salicylate	(i)	$o-HOC_6H_4-C(=O)-OCH_3$	222	Wintergreen

Other methods are available for synthesizing esters, most of which are more expensive but readily carried out on a small scale. For example, alcohols react with anhydrides and with acid chlorides:

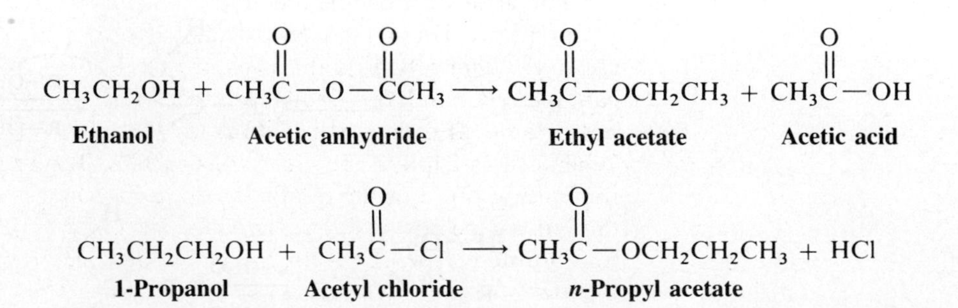

In the latter reaction, an organic base such as pyridine is usually added to react with the hydrogen chloride.

Other ester syntheses

Other methods can also be used to synthesize the ester group. Among these are the addition of 2-methylpropene to an acid to form *t*-butyl esters, the addition of ketene to make acetates, and the reaction of a silver salt with an alkyl halide:

$$CH_2{=}C(CH_3)CH_3 + CH_3CH_2C(=O){-}OH \xrightarrow{H^+} CH_3CH_2C(=O){-}OC(CH_3)_2CH_3$$

2-Methylpropene (isobutylene) **Propionic acid** ***t*-Butyl propionate**

$$CH_2{=}C{=}O + HOCH_2{-}C_6H_5 \longrightarrow CH_3C(=O){-}OCH_2{-}C_6H_5$$

Ketene **Benzyl alcohol** **Benzyl acetate**

$$CH_3C(=O){-}OAg + BrCH_2CH_2CH(CH_3)CH_3 \longrightarrow CH_3C(=O){-}OCH_2CH_2CH(CH_3)CH_3$$

Silver acetate **1-Bromo-3-methylbutane** **3-Methylbutyl acetate**

As noted previously, Fischer esterification is an equilibrium process. Consider the reaction of acetic acid with 1-butanol to give *n*-butyl acetate:

$$CH_3C(=O){-}OH + HOCH_2CH_2CH_2CH_3 \overset{H^+}{\rightleftharpoons} CH_3C(=O){-}OCH_2CH_2CH_2CH_3 + H_2O$$

The equilibrium constant is as follows:

$$K_{eq} = \frac{[n\text{-BuOAc}][H_2O]}{[n\text{-BuOH}]\,[HOAc]}$$

For primary alcohols reacting with unhindered carboxylic acids, $K_{eq} \approx 4$. If equal quantities of 1-butanol and acetic acid are allowed to react at equilibrium, the theoretical yield of ester is only 67%. To upset the equilibrium we can, by Le Chatelier's principle, increase the concentration of either the alcohol or acid. If either one is doubled, the theoretical yield increases to 85%. When one is tripled, the yield goes to 90%. But note that in the example cited, the boiling point of the relatively nonpolar ester is only about 8°C higher than the boiling points of the polar acetic acid and 1-butanol, so a difficult separation problem exists if the product must be isolated by distillation after the starting materials are increased in concentration.

TABLE 40.2 *The Ternary Azeotrope of Boiling Point 90.7°C*

		Percentage Composition of Azeotrope		
Compound	*Boiling Point of Pure Compound (°C)*	**Vapor Phase**	**Upper Layer**	**Lower Layer**
1-Butanol	117.7	8.0	11.0	2.0
n-Butyl acetate	126.7	63.0	86.0	1.0
Water	100.0	29.0	3.0	97.0

Another way to upset the equilibrium is to remove water. This can be done by adding to the reaction mixture molecular sieves (an artificial zeolite), which preferentially adsorb water. Most other drying agents, such as anhydrous sodium sulfate or calcium chloride pellets, will not remove water at the temperatures used to make esters.

Azeotropic distillation

A third way to upset the equilibrium is to preferentially remove the water as an azeotrope (a constant-boiling mixture of water and an organic liquid). The information in Table 40.2 can be found in a chemistry handbook table of ternary (three-component) azeotropes. These data tell us that vapor that distills from a mixture of 1-butanol, *n*-butyl acetate, and water will boil at 90.7°C and that the vapor contains 8% alcohol, 63% ester, and 29% water. The vapor is homogeneous, but when it condenses, it separates into two layers. The upper layer is composed of 11% alcohol, 86% ester, and 3% water, but the lower layer consists of 97% water with only traces of alcohol and ester. If some ingenious way can be devised to remove the lower layer from the condensate and still return the upper layer to the reaction mixture, then the equilibrium can be upset, and nearly 100% of the ester can be produced in the reaction flask.

The apparatus shown in Figure 40.1, modeled after that of Ernest W. Dean and David D. Stark, achieves the desired separation of the two layers. The mixture of equimolar quantities of 1-butanol and acetic acid is placed in the flask along with an acid catalyst. Stirring reduces bumping. The vapor, the temperature of which is 90.7°C, condenses and runs down to the sidearm, which is closed with a cork. The layers separate, with the denser water layer remaining in the sidearm, and the lighter ester plus alcohol layer running down into the reaction flask. As soon as the theoretical quantity of water has collected, the reaction is over, and the product in the flask should contain an ester of high purity. The macroscale apparatus is illustrated in Figure 40.2.

Esterification using a carboxylic acid and an alcohol requires an acid catalyst. In the first experiment, the acid form of an ion-exchange resin is used. This resin, in the form of small beads, is a cross-linked polystyrene that bears sulfonic acid groups on some of the phenyl groups. It is essentially an immobilized form of *p*-toluenesulfonic acid, an organic-substituted sulfuric acid.

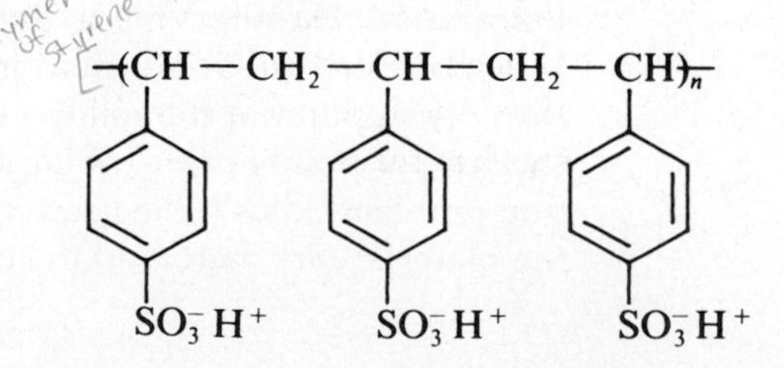

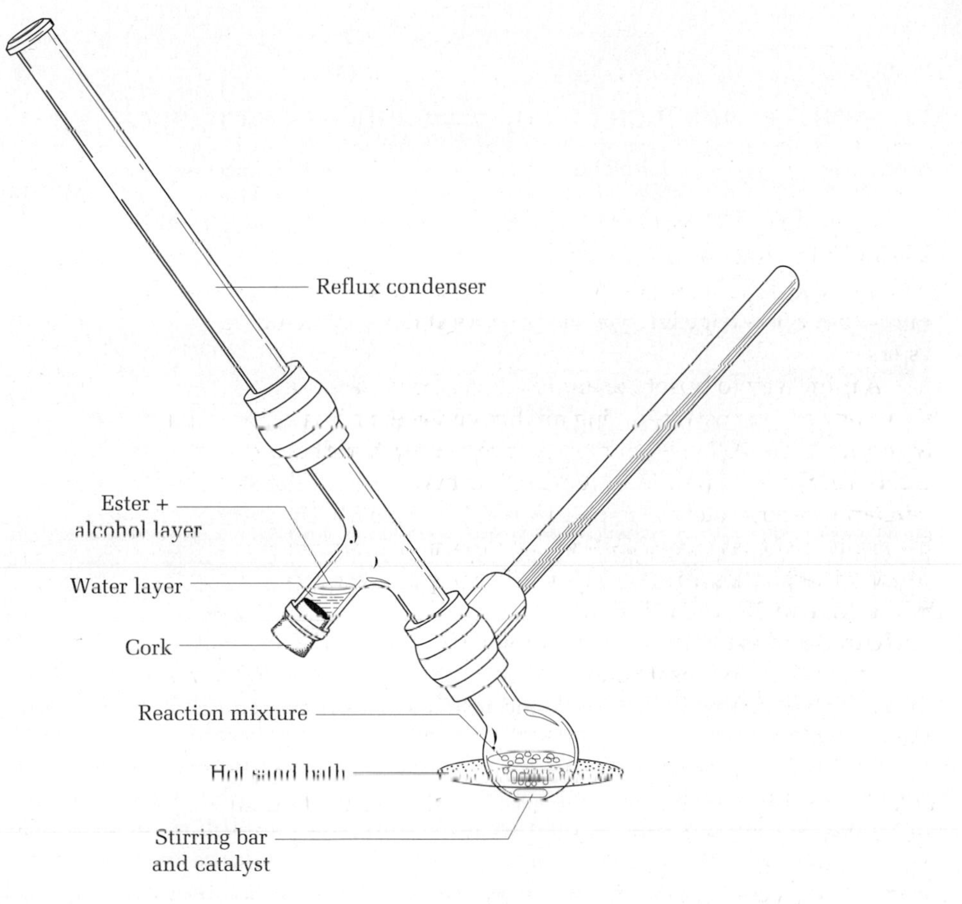

FIG. 40.1
A microscale Dean-Stark azeotropic esterification apparatus. A cork is used instead of a septum so that layer separation can be observed clearly.

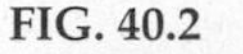

FIG. 40.2
A macroscale azeotropic distillation apparatus, with a Dean-Stark trap where water collects.

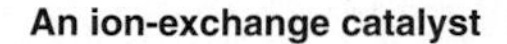

An ion-exchange catalyst

This catalyst has the distinct advantage that at the end of the reaction it can be easily removed by filtration. Immobilized catalysts of this type are becoming more and more common in organic synthesis.

If concentrated sulfuric acid were used as the catalyst, it would be necessary to dilute the reaction mixture with ether; wash the ether layer successively with water, sodium carbonate solution, and saturated sodium chloride solution; and then dry the ether layer with anhydrous calcium chloride pellets before evaporating the ether.

EXPERIMENTS

1. n-BUTYL ACETATE BY AZEOTROPIC DISTILLATION OF WATER

$$CH_3\overset{\overset{\displaystyle O}{\|}}{C}{-}OH + HOCH_2CH_2CH_2CH_3 \underset{}{\overset{H^+}{\rightleftharpoons}} CH_3\overset{\overset{\displaystyle O}{\|}}{C}{-}OCH_2CH_2CH_2CH_3 + H_2O$$

Acetic acid
MW 60.05
bp 117.9°C, den 1.049
n_D^{20} 1.3720

1-Butanol
MW 74.12
bp 117.7°C, den 0.810
n_D^{20} 1.3990

***n*-Butyl acetate**
MW 116.16
bp 126.5°C, den 0.882
n_D^{20} 1.3940

MW 18

IN THIS EXPERIMENT, acetic acid is esterified with butanol in the presence of an acid catalyst attached to very small polystyrene beads. Because esterification is an equilibrium reaction, the equilibrium is upset by removing the water formed in a unique apparatus patterned after the Dean-Stark water separator. The product is isolated simply by withdrawing it from the reaction flask with a Pasteur pipette, leaving the catalyst behind. It can be further purified by simple distillation.

Ion-exchange resin catalyst

A cork instead of a septum allows the accumulation of water to be observed.

Video: *Ester Synthesis; Using Dean-Stark Apparatus*

In a 5-mL short-necked, round-bottomed flask, place 0.2 g of Dowex 50X2-100 ion-exchange resin,[1] 0.60 g of acetic acid, 0.74 g of 1-butanol, and a stirring bar. Attach the addition port with the sidearm corked and an empty distilling column, as shown in Figure 40.1, and clamp the apparatus at the angle shown. Heat the flask while stirring on a hot sand bath, and bring the reaction mixture to a boil. Stirring the mixture prevents it from bumping. As an option, you might hold a thermometer just above the boiling liquid and note when a temperature of about 91°C has been reached. Remove the thermometer and allow the reaction mixture to reflux in a manner such that the vapors condense about one-third of the way up the empty distilling column, which is functioning as an air condenser. Note that the material that condenses is not homogeneous, as droplets of water begin to collect in the upper part of the apparatus. As the sidearm fills with condensate, it is cloudy at first and then two layers separate. When the volume of the lower aqueous layer does not appear to increase, the reaction is over. This will take about 20–30 minutes. Carefully remove the apparatus from the heat, allow it to cool, and then tip the apparatus very carefully to allow all of the upper layer in the sidearm to run back into the reaction flask. Disconnect the apparatus.

Remove the product from the reaction flask with a Pasteur pipette, determine its weight and boiling point, and assess its purity by thin-layer chromatographic (TLC) analysis and infrared (IR) spectroscopy. The product can be analyzed by gas chromatography on a 10-ft (3-m) Carbowax column. At 152°C, the 1-butanol has a retention time of 2.1 minutes, the *n*-butyl acetate 2.5 minutes, and the acetic acid

[1]The Dowex resin as received should be washed with water by decantation to remove much of its yellow color. It is then collected by vacuum filtration on a Büchner funnel and returned, in a slightly damp state, to the reagent bottle.

7.5 minutes. Look for the presence of hydroxyl and carboxyl absorption bands in the IR spectrum. The product can easily be purified by simple distillation (*see* Figure 5.7 on page 94).

Cleaning Up. Place the catalyst in the solid hazardous waste container.

2. ISOBUTYL PROPIONATE (2-METHYLPROPYL PROPANOATE) BY FISCHER ESTERIFICATION

$$CH_3CH_2\overset{O}{\overset{\|}{C}}OH + HOCH_2\overset{CH_3}{\overset{|}{C}}HCH_3 \overset{H^+}{\rightleftharpoons} CH_3CH_2\overset{O}{\overset{\|}{C}}OCH_2\overset{CH_3}{\overset{|}{C}}HCH_3 + H_2O$$

Propanoic acid	**2-Methyl-1-propanol**	**Isobutyl propionate**
MW 74.08	MW 74.12	MW 130.19
bp 141°C	bp 108°C	bp 136.8°C

IN THIS EXPERIMENT, a propionate ester is prepared by reacting excess propanoic acid with an alcohol in the presence of an acid catalyst on polystyrene beads (Dowex 50X). The excess acid serves to drive the equilibrium reaction toward formation of the product. The excess propanoic acid is removed by reaction with potassium carbonate, and the water is adsorbed by silica gel when the reaction mixture is chromatographed.

Procedure

To a reaction tube, add 112 mg of 2-methyl-1-propanol (isobutyl alcohol), 148 mg of propanoic acid (propionic acid), 50 mg of Dowex 50X2-100 ion-exchange resin, and a boiling chip. Attach the empty distilling column as an air condenser (Fig. 40.3). Reflux the resulting mixture for 1 hour or more, cool it to room temperature, remove the product mixture from the resin with a Pasteur pipette, and chromatograph the product (2-methyl-1-propyl propanoate) on a silica gel column.

Video: *Filtration of Crystals Using the Pasteur Pipette*

Chromatography Procedure

CAUTION: Carry out this part of the experiment in a hood. Dichloromethane is a suspected carcinogen.

Assemble the column as depicted in Figure 40.4, being sure that it is clamped in a vertical position. Close the valve and fill the entire glass column with dichloromethane to the bottom of the funnel. In a small beaker, prepare a slurry of 1 g of silica gel in 4 mL of dichloromethane. Stir the slurry gently to remove air bubbles and gently swirl, pour, and scrape the slurry into the funnel, which has a capacity of 10 mL. After some of the silica gel has been added to the column, open the stopcock and allow solvent to drain slowly into an Erlenmeyer flask. Use this dichloromethane to rinse the beaker containing the silica gel. As the silica gel is being added, tap the column with a glass rod or pencil so the adsorbent will pack tightly into the column. Continue to tap the column while cycling the dichloromethane through the column once more; then add 1 g of anhydrous

Photo: *Column Chromatography*; Video: *Column Chromatography*

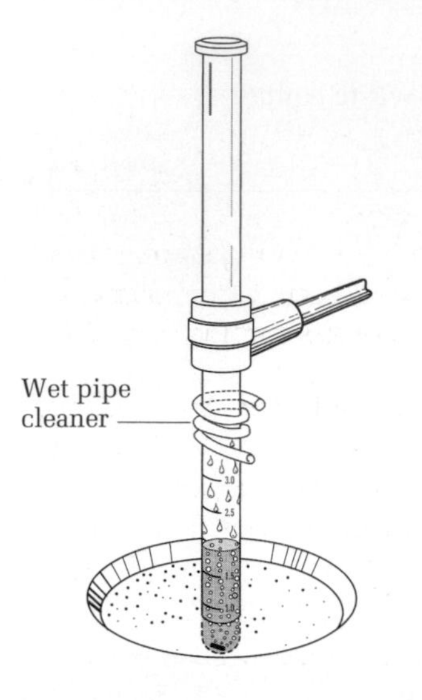

FIG. 40.3
An esterification apparatus.

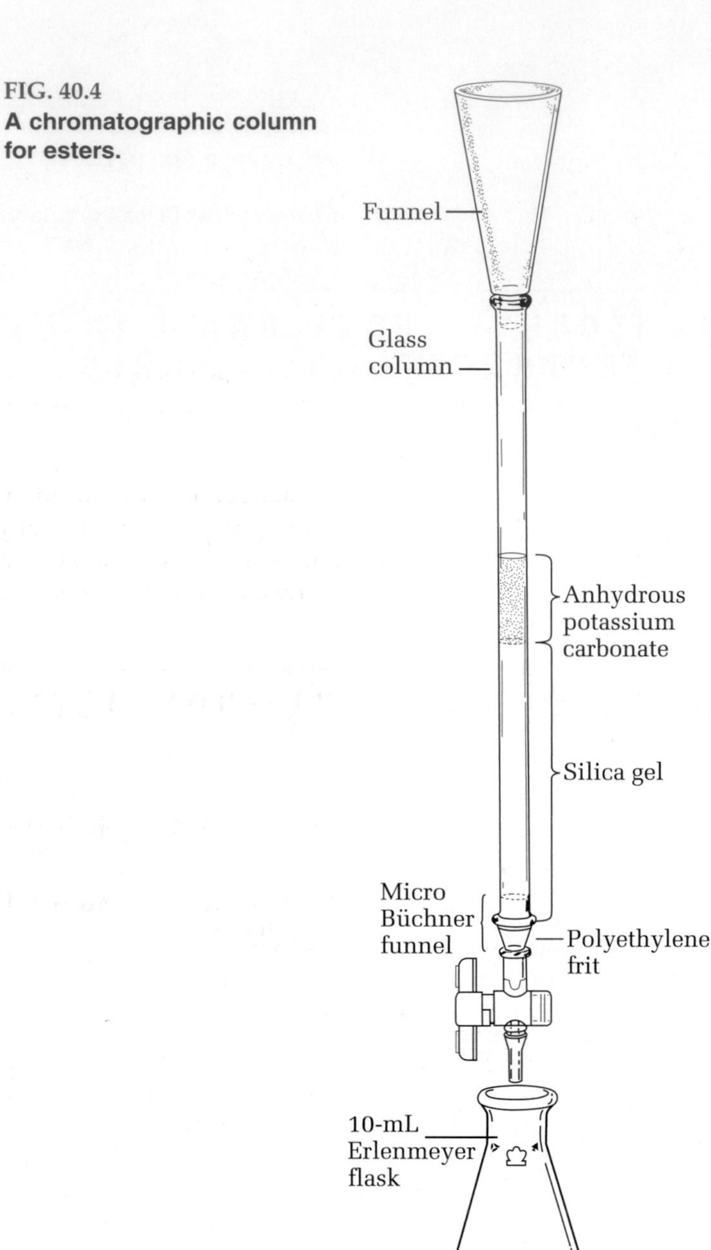

FIG. 40.4
A chromatographic column for esters.

potassium carbonate to the top of the silica gel. The potassium carbonate will remove water from the esterification mixture as well as react with any carboxylic acid present. Allow the solvent to flow until it reaches the top of the potassium carbonate layer.

Adding the Sample

The solvent is drained just to the surface of the potassium carbonate. Using a Pasteur pipette, add the sample to the column and let it run into the adsorbent, stopping when the solution reaches the top of the potassium carbonate. The flask and ion-exchange resin are rinsed twice with 0.5-mL portions of dichloromethane that are run into the column, with the eluent being collected in a tared reaction tube. The elution is completed with 1 mL more of dichloromethane.

Analysis of the Product

Evaporate the dichloromethane under a stream of air or nitrogen in the hood and remove the last traces by connecting the reaction tube to a water aspirator. Because the dichloromethane boils at 40°C and the product boils at 137°C, separation of the two is easily accomplished. Determine the weight of the product and its boiling point and calculate the yield. The ester should be a perfectly clear, homogeneous liquid. Obtain an IR spectrum and analyze it for the presence of unreacted alcohol and carboxylic acid. Check the purity of the product by TLC and/or gas chromatography.

Cleaning Up. Place the catalyst in the hazardous waste container. Any dichloromethane should be placed in the halogenated organic waste container. The contents of the chromatography column, if free of solvent, can be placed in the nonhazardous solid waste container; otherwise, this material is classified as a hazardous waste and must go in a container so designated.

3. BENZYL ACETATE FROM ACETIC ANHYDRIDE

$$C_6H_5CH_2OH + CH_3C(=O){-}O{-}C(=O)CH_3 \longrightarrow C_6H_5CH_2OC(=O)CH_3 + CH_3C(=O)OH$$

Benzyl alcohol	Acetic anhydride	Benzyl acetate	Acetic acid
MW 108.14	MW 102.09	MW 150.18	MW 60.06
bp 205°C	bp 138–140°C	bp 213.5°C	bp 117–118°C
n_D^{20} 1.5400	n_D^{20} 1.3900	n_D^{20} 1.5020	n_D^{20} 1.3720

To a reaction tube, add 108 mg of benzyl alcohol, 102 mg of acetic anhydride, and a boiling chip. Reflux the mixture for at least 1 hour, cool the mixture to room temperature, and chromatograph the liquid in exactly the same manner described in Experiment 2. Analyze the product by TLC and by IR spectroscopy as a thin film between salt or silver chloride plates. Note the presence or absence of hydroxyl and carboxyl bands.

This ester cannot be prepared by Fischer esterification using Dowex because polymerization seems to occur when this ion-exchange resin is used.

Cleaning Up. Place the catalyst in the hazardous waste container. Any dichloromethane should be placed in the halogenated organic waste container. The contents of the chromatography column, if free of solvent, can be placed in the nonhazardous solid waste container. Otherwise, this material is classified as a hazardous waste and must go in the container so designated.

Other Esterifications

This experiment lends itself to wide-ranging experimentation. All three methods of esterification previously described can, in principle, be applied to any unhindered primary or secondary alcohol. These methods work well for most of the esters in Table 40.1, and for hundreds of others as well.

4. METHYL BENZOATE BY FISCHER ESTERIFICATION

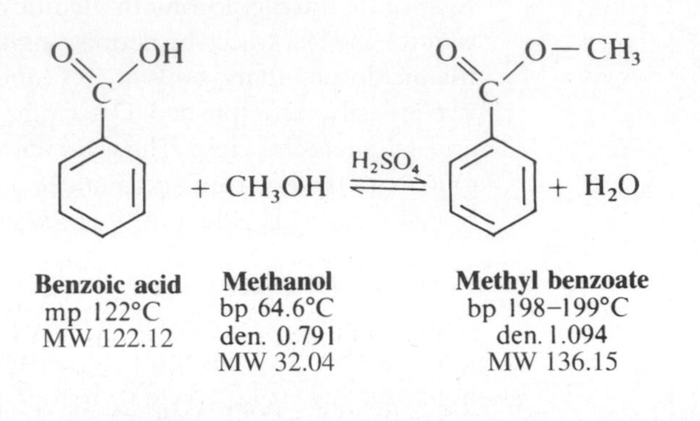

Benzoic acid	Methanol	Methyl benzoate
mp 122°C	bp 64.6°C	bp 198–199°C
MW 122.12	den. 0.791	den. 1.094
	MW 32.04	MW 136.15

IN THIS EXPERIMENT, benzoic acid and excess methanol are refluxed in the presence of a catalytic amount of sulfuric acid. Water and ether are added; the water layer is withdrawn; and the ether layer is washed with water, bicarbonate, and saturated salt solutions. After drying over calcium chloride, the ether is evaporated. Finally, the product, ethyl benzoate, is distilled.

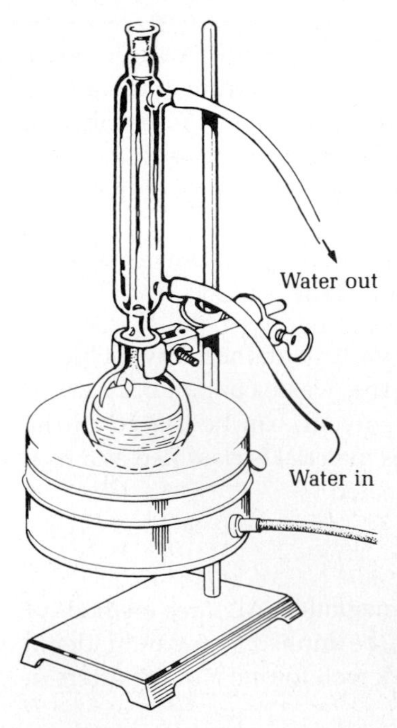

FIG. 40.5
An apparatus for refluxing a reaction mixture.

Place 10.0 g of benzoic acid and 25 mL of methanol in a 125-mL round-bottomed flask, cool the mixture in ice, pour 3 mL of concentrated sulfuric acid *slowly and carefully down the walls of the flask*, then swirl to mix the components. Attach a reflux condenser, add a boiling chip, and reflux the mixture gently for 1 hour. (Fig. 40.5 illustrates this apparatus.) Cool the solution, decant it into a separatory funnel containing 50 mL of water, and rinse the flask with 35 mL of ether. Add this ether to the separatory funnel, shake thoroughly, and drain off the water layer, which contains the sulfuric acid and the bulk of the methanol. Wash the ether in the separatory funnel with 25 mL of water, followed by 25 mL of 0.5 *M* sodium bicarbonate to remove unreacted benzoic acid. Again shake, with frequent release of pressure by inverting the separatory funnel and opening the stopcock, until no further reaction is apparent; then drain off the bicarbonate layer into a beaker. If this aqueous material is made strongly acidic with hydrochloric acid, unreacted benzoic acid may be observed. Wash the ether layer in the separatory funnel with saturated sodium chloride solution, and dry the solution over anhydrous calcium chloride in an Erlenmeyer flask. Add sufficient calcium chloride pellets so that they no longer clump together on the bottom of the flask. After 10 minutes, decant the dry ether solution into a flask, wash the drying agent with an additional 5 mL of ether, and decant again.

Remove the ether by simple distillation or by evaporation on a steam bath under an aspirator tube. Use the methods illustrated in Fig. 4.22 (on page 77) or Fig. 5.10 (on page 97), or use a rotary evaporator (*see* Fig. 9.8 on page 197). When evaporation ceases, add 2–3 g of anhydrous calcium chloride pellets to the residual oil and heat for about 5 minutes longer. Then decant the methyl benzoate into a 50-mL round-bottomed flask, attach a stillhead, dry out the ordinary condenser and use it without water circulating in the jacket, and distill. The boiling point of the ester is so high (199°C) that a water-cooled condenser is liable to crack. Use a tared 25-mL Erlenmeyer flask as the receiver to collect material boiling above 190°C. A typical student yield is about 7 g. *See* Chapter 28 for the nitration of methyl benzoate, and Chapter 38 for its use in the Grignard synthesis of triphenylmethanol.

IR and nuclear magnetic resonance (NMR) spectra of benzoic acid and methyl benzoate are found at the end of the chapter (Figs. 40.6 to 40.11).

Cleaning Up. Pour the sulfuric acid layer into water, combining it with the bicarbonate layer, neutralize it with sodium carbonate, and flush the solution down the drain with excess water. The saturated sodium chloride layer can also be flushed down the drain. If the calcium chloride is free of ether and methyl benzoate, it can be placed in the nonhazardous solid waste container; otherwise it must go into the hazardous waste container. Ether goes into the organic solvents waste container, along with the pot residues from the final distillation.

5. HYDROLYSIS (SAPONIFICATION): THE PREPARATION OF SOAP

In general, the reversal of esterification is called *hydrolysis*. In the case of the hydrolysis of a fatty acid ester, it is called saponification (soap making). In this experiment, the saturated fat made from hydrogenated olive oil in Chapter 25 will be saponified to give a soap, which, in this case, will be primarily sodium stearate.

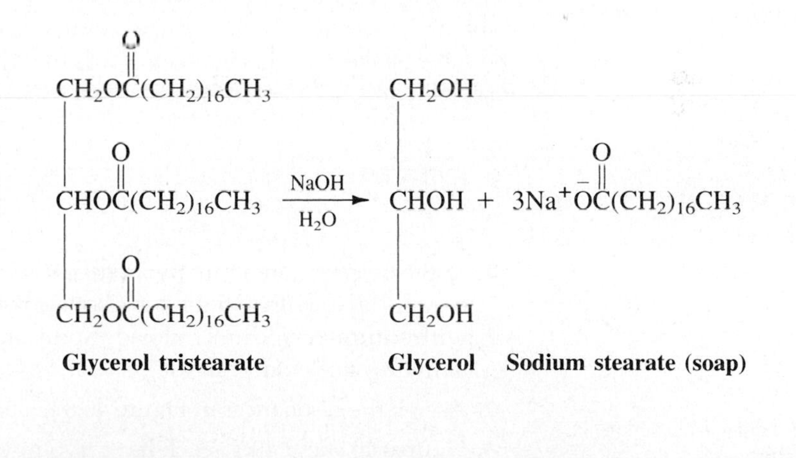

Microscale Procedure

IN THIS EXPERIMENT, a saturated fat is heated with sodium hydroxide in a water-ethanol mixture. The resulting soap is precipitated in a salt solution and collected on a Hirsch funnel, where it is washed free of base and salt.

Place 0.18 g of the saturated triglyceride prepared in Chapter 25, Experiment 4, in a 5-mL short-necked, round-bottomed flask. Add 1.5 mL of a 50:50 water-ethanol solution that contains 0.18 g of solid sodium hydroxide (weigh this quickly). Add an air condenser and gently reflux the mixture on a sand bath for 30 minutes, taking care not to boil away the ethanol. At the end of the reaction period, some of the soap will have precipitated. Transfer the mixture to a 10-mL Erlenmeyer flask containing a solution of 0.8 g of sodium chloride in 3 mL of water. Collect the precipitated soap on a Hirsch funnel and wash it free of excess sodium hydroxide and salt using 4 mL of distilled ice water.

Video: *Microscale Filtration on the Hirsch Funnel*

Test the soap by adding a very small piece (about 5–15 mg) to a centrifuge or test tube along with 3–4 mL of distilled water. Cap the tube and shake it vigorously. Note the height and stability of the bubbles. Add a crystal of magnesium chloride or calcium chloride to the tube. Shake the tube again and note the results. For comparison, do these same tests with a few grains of a detergent instead of the soap.

Macroscale Procedure

In a 100-mL round-bottomed flask, place 5 g of hydrogenated olive oil (Chapter 25) or lard or solid shortening (e.g., Crisco). Add 20 mL of ethanol and a hot solution of 5 g of sodium hydroxide in 20 mL of water. This solution, prepared in a beaker, will become very hot as the sodium hydroxide dissolves. Fit the flask with a water-cooled condenser, and reflux the mixture on a sand bath for 30 minutes. At the end of the reaction period, some of the soap may precipitate from the reaction mixture. Transfer the mixture to a 250-mL Erlenmeyer flask containing an ice-cold solution of 25 g of sodium chloride in 90 mL of distilled water. Collect the precipitated soap on a Büchner funnel and wash it free of excess sodium hydroxide and salt using no more than 100 mL of distilled ice water.

Test the soap by adding a small piece to a test tube along with about 5 mL of water. Cap the tube and shake it vigorously. Note the height and stability of the bubbles formed. Add a few crystals of magnesium chloride or calcium chloride to the tube. Shake the tube again and note the results. For comparison, do these same tests with an amount of detergent equal to that of the soap used.

QUESTIONS

1. In the preparation of methyl benzoate, what is the purpose of (a) washing the organic layer with sodium bicarbonate solution? (b) washing the organic layer with saturated sodium chloride solution? (c) treating the organic layer with anhydrous calcium chloride pellets?
2. Assign the resonances in Figure 40.8 to specific protons in methyl benzoate.
3. Figures 40.9 and 40.10 each have two resonances that are very small. What do the carbons causing these peaks have in common?

For Additional Experiments, sign in at this book's premium website at **www.cengage.com/login**.

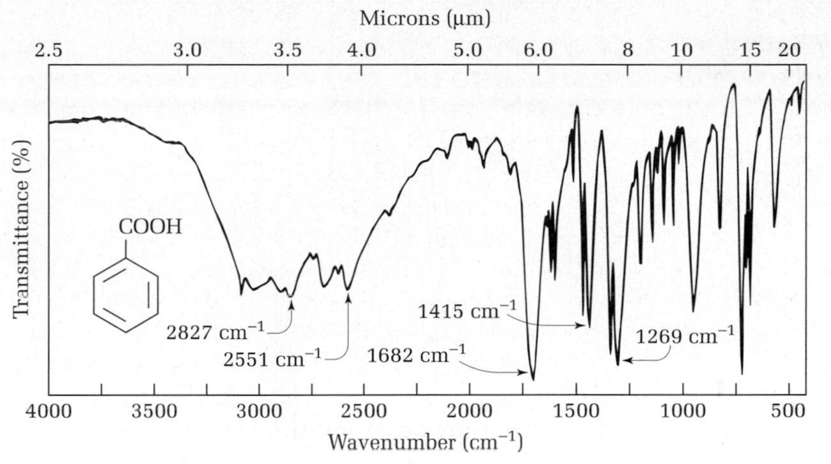

FIG. 40.6
The IR spectrum of benzoic acid (KBr disk).

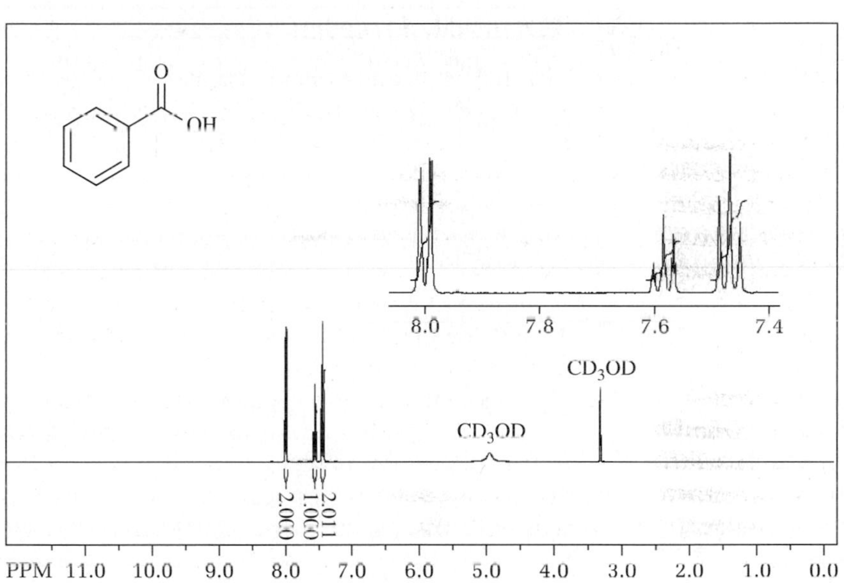

FIG. 40.7
The ^{1}H NMR spectrum of benzoic acid (250 MHz).

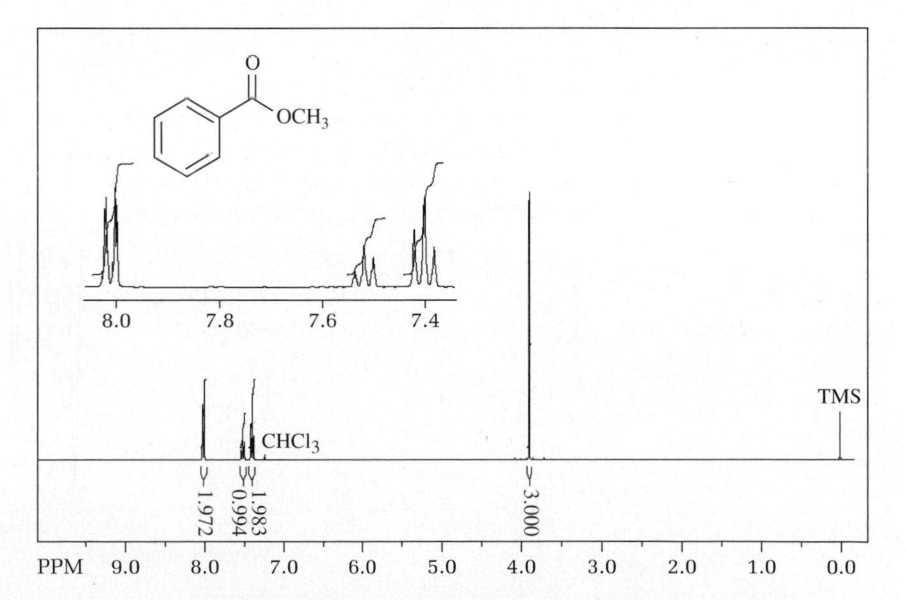

FIG. 40.8
The ^{1}H NMR spectrum of methyl benzoate (400 MHz).

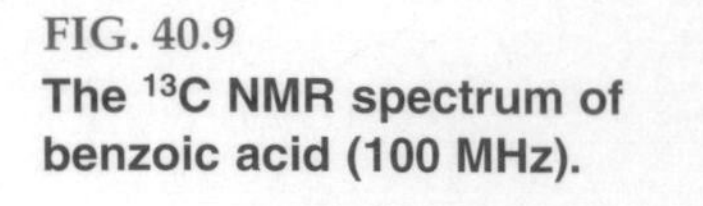

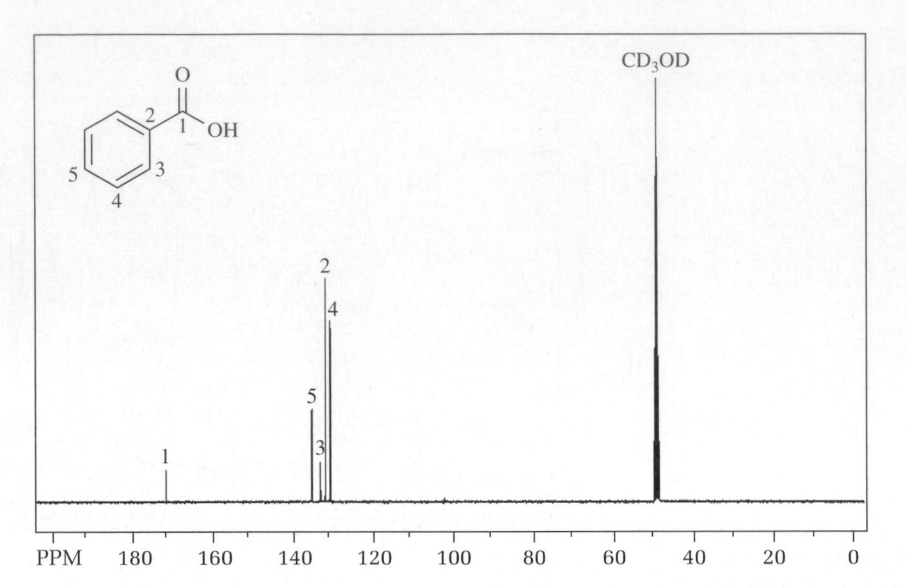

FIG. 40.9
The ^{13}C NMR spectrum of benzoic acid (100 MHz).

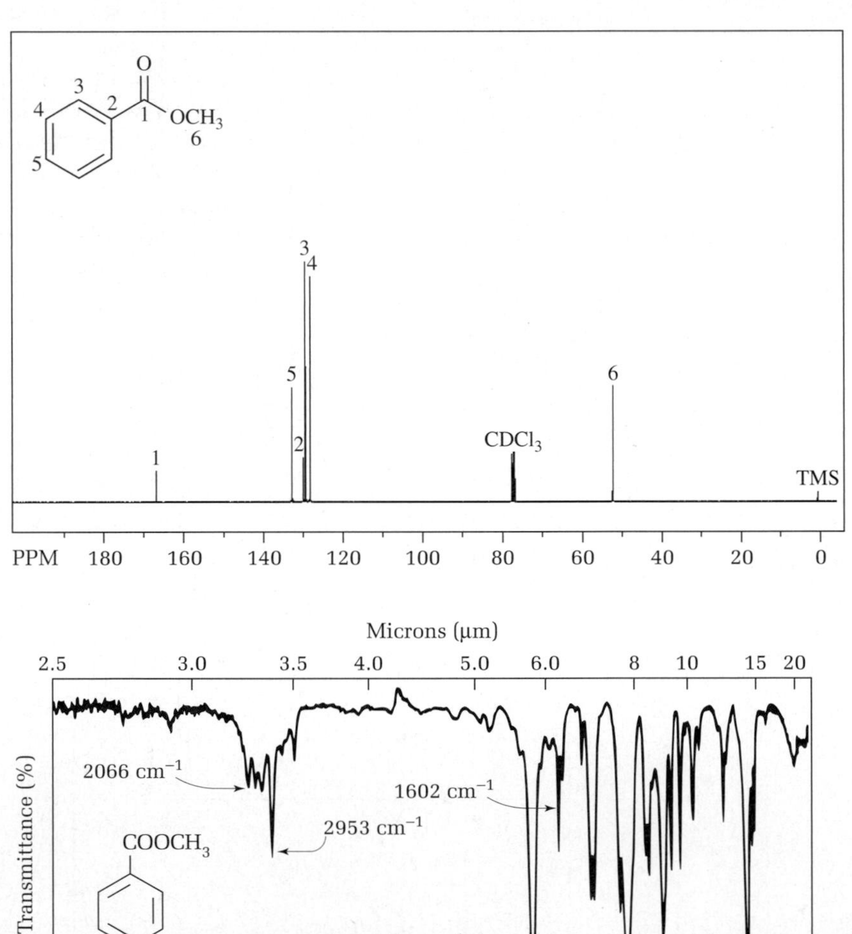

FIG. 40.10
The ^{13}C NMR spectrum of methyl benzoate (100 MHz).

FIG. 40.11
The IR spectrum of methyl benzoate.

CHAPTER 41

Acetylsalicylic Acid (Aspirin)

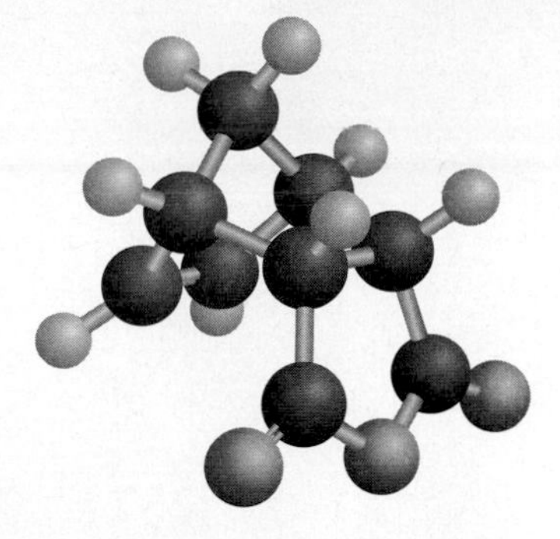

PRELAB EXERCISE: Write detailed mechanisms showing how pyridine and sulfuric acid catalyze the formation of acetylsalicylic acid.

Aspirin is among the most fascinating and versatile drugs known to medicine; it is also among the oldest. The first known use of an aspirin-like preparation can be traced to ancient Greece and Rome. Salicin, an extract obtained from the bark of willow and poplar trees, has been used as a pain reliever (analgesic) for centuries. In the middle of the 19th century, it was found that salicin is a glycoside formed from a molecule of salicylic acid and a sugar molecule. The sodium salt of salicylic acid, sodium salicylate, is easily synthesized on a large scale by heating sodium phenoxide with carbon dioxide at 150°C under slight pressure (the Kolbe synthesis):

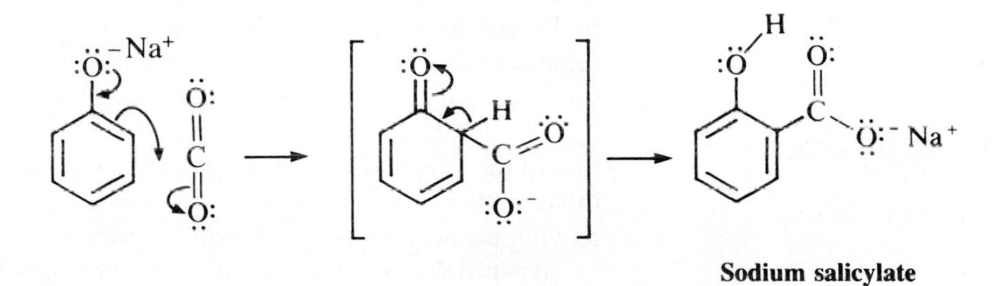

Unfortunately, however, salicylic acid attacks the mucous membranes of the mouth and esophagus, and causes gastric pain that may be worse than the discomfort it was intended to cure. Felix Hoffmann, a chemist for Friedrich Bayer, a German dye company, reasoned that the corrosive nature of salicylic acid could be altered by adding an acetyl group. In 1893 the Bayer Company obtained a patent on acetylsalicylic acid, despite the fact that it had been synthesized some 40 years previously by Charles Gerhardt. Bayer coined the name *Aspirin* for its new product to reflect its acetyl nature and its natural occurrence in the spirea plant. Over the years, Bayer has allowed the term *aspirin* to fall into the public domain so it is no longer capitalized. The manufacturers of Coke (Coca-Cola) and Sanka are working hard to prevent a similar fate befalling their trademarks.

In 1904 Carl Duisberg, the head of Bayer, decided to emulate John D. Rockefeller's Standard Oil Company and formed an *interessen gemeinschaft* (I.G.) of the dye industry (*Farbenindustrie*). This cartel completely dominated the world dye industry before World War I, and it continued to prosper between the wars, even though some of its assets were seized and sold after World War I. Soon after, an American company, Sterling Drug, bought the rights to aspirin for $5.3 million. Sterling was bought by Eastman Kodak in 1988, which was then sold to Smith-Kline Beecham in 1995, which combined in 2000 with Glaxo Wellcome to become GlaxoSmithKline.

Because of their involvement at Auschwitz, the top management of I.G. Farbenindustrie was tried and convicted at the Nuremberg trials after World War II, and the cartel was broken into three large branches—Bayer, Hoechst, and BASF (*Badische Anilin und Sodafabrik*). Each of these companies does more business than DuPont, the largest American chemical company. In 1997, the American rights to the Bayer name and trademark were sold back to Bayer A.G. (Aktiengesellschaft) for $1 billion. At present, virtually all aspirin is made by Rhodia in France and Thailand; none is made in the United States, although there is a $2.5 billion dollar market for over-the-counter pain killers in North America.

The gross profit margin on Bayer aspirin is 80%. After advertising costs, the profit margin is 60%. Aspirin made in bulk by Rhodia costs about $4 per pound.

By law, all drugs sold in the United States must meet the purity standards set by the U.S. Food and Drug Administration (FDA), and so all aspirin is essentially the same. Each tablet contains 0.325 g (325 mg) of acetylsalicylic acid held together with a binder. The remarkable difference in price for aspirin is primarily a reflection of the advertising budget of the company that sells it. The Bayer brand has 5% of the painkiller market; lower-priced generic aspirin has 18%.

Aspirin is an analgesic, an antipyretic (fever reducer), and an anti-inflammatory agent. It is the premier drug for reducing fever, a role for which it is uniquely suited. As an anti-inflammatory, it has become the most widely effective treatment for arthritis. Unfortunately, people suffering from arthritis must take so much aspirin (several grams per day) that gastric problems may result. For this reason, aspirin is often combined with a buffering agent. Bufferin™ is an example of such a preparation.

The ability of aspirin to diminish inflammation is apparently due to its inhibition of the synthesis of prostaglandins, a group of C_{20} molecules that exacerbate inflammation. Aspirin alters the oxygenase activity of prostaglandin synthetase by moving the acetyl group to a terminal amine group of the enzyme.

If aspirin were a new invention, the FDA would place many hurdles in the path of its approval. It has been implicated, for example, in Reye's syndrome, a brain disorder that strikes children and young persons under age 18 who take aspirin after contracting flu or chickenpox. It has an effect on platelets, which play a vital role in blood clotting. In newborn babies and their mothers, aspirin can lead to uncontrolled bleeding and circulation problems for the baby—even brain hemorrhages in extreme cases. This same effect can be turned into an advantage, however. Heart specialists urge potential stroke victims to take aspirin regularly to inhibit clotting in their arteries, and it has been shown that taking aspirin will help prevent first heart attacks from occurring.

Over 140 common medications contain aspirin, including, for example, Alka-Seltzer, Anacin, Coricidin, Excedrin, Midol, and Vanquish. Despite its side effects, aspirin is one of the safest, cheapest, and efficacious nonprescription drugs, although acetaminophen (Tylenol, etc.) has 40% and ibuprofen (Advil, etc.) has 26% of the painkiller market in dollar volume; naproxen (Aleve) has 6% of the market. Aspirin is made commercially employing the same synthesis presented in this chapter.

The mechanism for the acetylation of salicylic acid is as follows:

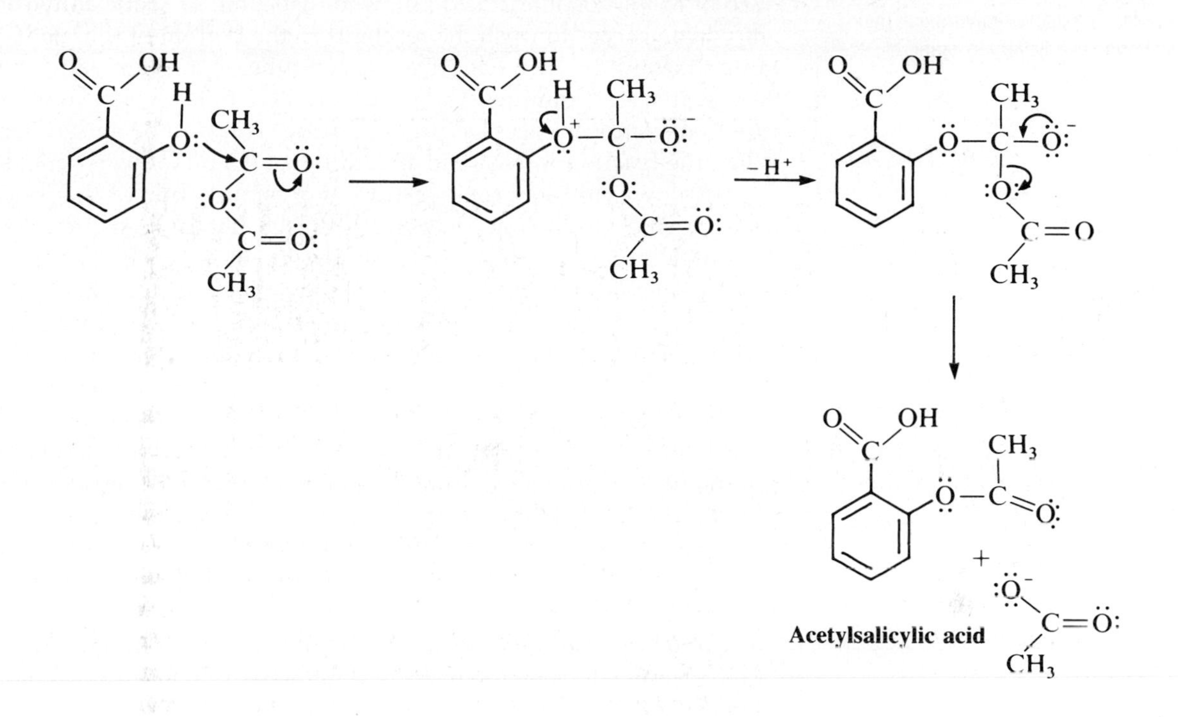

EXPERIMENT

SYNTHESIS OF ACETYLSALICYLIC ACID (ASPIRIN)

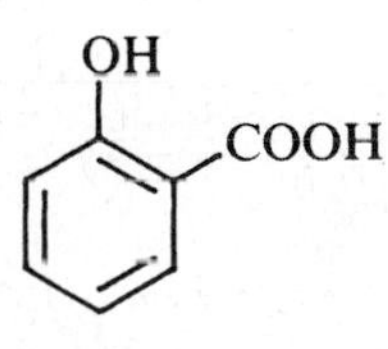

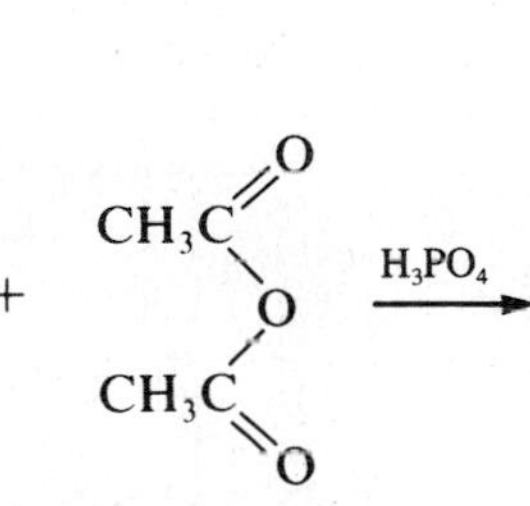

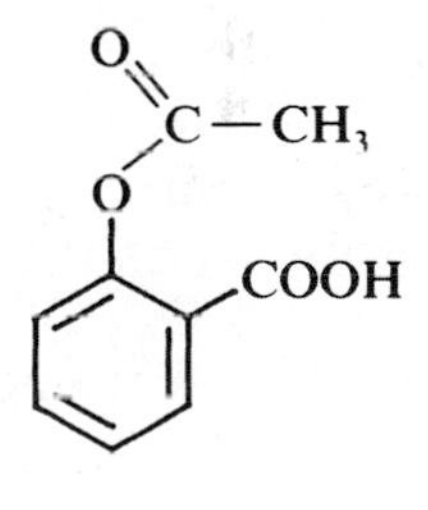

Salicylic acid	**Acetic anhydride**	**Acetylsalicylic acid**
MW 138.12, mp 159°C	MW 102.09, bp 140°C	MW 180.15, mp 128–137°C

Microscale Procedure

IN THIS EXPERIMENT, salicylic acid, acetic anhydride, and an acid catalyst are heated briefly. Water is added to convert excess acetic anhydride to acetic acid, and also serves as the crystallization solvent. After the water is removed from the crystals, they are washed, collected, and dried to give pure aspirin.

Measure acetic anhydride in the hood, as it is very irritating to breathe.

To a reaction tube, add 138 mg of salicylic acid, a boiling chip, and one small drop of 85% phosphoric acid followed by 0.3 mL of acetic anhydride (the acetic anhydride serves to wash the reactants to the bottom of the tube). Mix the reactants thoroughly; then heat the reaction tube on a steam bath or in a beaker of 90°C water for 5 minutes. Cautiously add 0.2 mL of water to the reaction mixture to decompose excess acetic anhydride. This will be an exothermic reaction. When the reaction is over, add an additional 0.3 mL of water and allow the tube to cool slowly to room temperature. If crystallization of the product does not occur during the cooling process, add a seed crystal or scratch the inside of the tube with a glass stirring rod. Cool the tube in ice until crystallization is complete (at least 10 minutes) and then remove the solvent with a Pasteur pipette. If the crystals are too fine for this procedure, collect the product by vacuum filtration on a Hirsch funnel. Complete the transfer of the product to the funnel using a very small quantity of ice water. In either case, turn the product out onto a piece of filter paper and squeeze the crystals between sheets of filter paper to absorb excess water. Allow the product to air-dry thoroughly before determining the weight and calculating the percent yield. Determine the melting point and the infrared (IR) spectrum in chloroform solution. Figure 41.1 shows the IR spectrum of acetylsalicylic acid. Aspirin is hydrolyzed by boiling water, but the reaction is not rapid; if desired, the product may be recrystallized from a small quantity of very hot water.

Photo: *Filtration Using a Pasteur Pipette*; Videos: *Filtration of Crystals Using the Pasteur Pipette; Microscale Filtration on the Hirsch Funnel*

Compare a tablet of commercial aspirin with your sample. Test the solubility of the commercial tablet in water and in toluene, and observe if it dissolves completely. Compare its behavior when heated in a melting-point capillary to the behavior of your sample. If an impurity is found, it is probably a substance used as binder for the tablets. Is the binder organic or inorganic?

Cleaning Up. Dilute the filtrate with water and flush it down the drain.

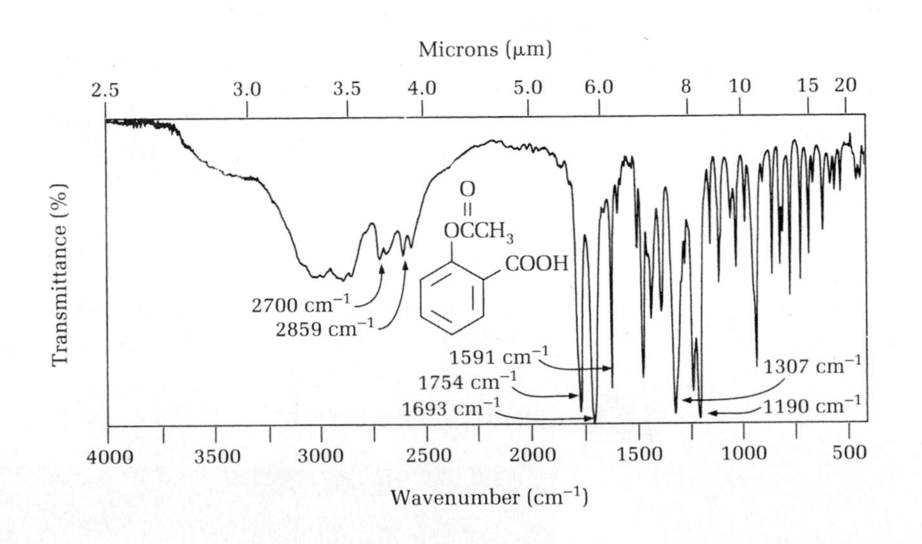

FIG. 41.1
The IR spectrum of acetylsalicylic acid in chloroform.

Macroscale Procedure

Measure acetic anhydride in the hood, as it is very irritating to breathe.

Acetylation catalysts:
1. CH_3COO^- Na^+
2.

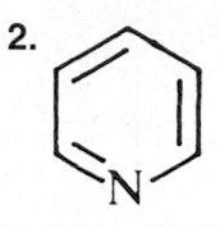

3. $BF_3 \cdot (C_2H_5)_2O$
4. H_2SO_4

Pyridine has a bad odor, and boron trifluoride etherate and sulfuric acid are corrosive; work in the hood. Handle with care.

Video: *Macroscale Crystallization*

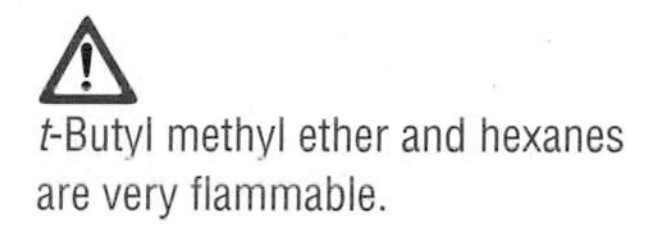
t-Butyl methyl ether and hexanes are very flammable.

Propose mechanisms to interpret your results.

Place 1 g of salicylic acid in each of four 13 × 100-mm test tubes and add to each tube 2 mL of acetic anhydride. To the first tube, add 0.2 g of anhydrous sodium acetate, note the time, stir the mixture with a glass stirring rod, replace the rod with a thermometer, and record the time required for a 4°C rise in temperature. Continue to stir occasionally while starting the next acetylation. Obtain a clean thermometer, put it in the second test tube, add 5 drops of pyridine, stir, observe as before, and compare it to the first results. To the third and fourth test tubes, add 5 drops of boron trifluoride etherate[1] and 5 drops of concentrated sulfuric acid, respectively. What is the order of activity of the four catalysts as judged by the rates of the reactions?

Put all of the tubes in a beaker of hot water for 5 minutes to dissolve solid material and complete the reactions; then pour all of the solutions into a 125-mL Erlenmeyer flask containing 50 mL of water and rinse the tubes with water. Swirl to aid hydrolysis of the excess acetic anhydride; then cool thoroughly in ice, scratch the side of the flask with a stirring rod to induce crystallization, and collect the crystalline solid; the yield is about 4 g.

Acetylsalicylic acid melts with decomposition at temperatures reported between 128°C to 137°C. It can be crystallized by dissolving it in *t*-butyl methyl ether, adding an equal volume of hexanes, and letting the solution stand undisturbed in an ice bath.

Test the solubility of your sample in toluene and in hot water, and note the peculiar character of the aqueous solution when it is cooled and when it is then rubbed against the tube with a stirring rod. Note also that the substance dissolves in cold sodium bicarbonate solution and is precipitated by adding concentrated hydrochloric acid.

Compare a tablet of commercial aspirin to your sample. Test the solubility of the commercial tablet in water and in toluene, and observe if it dissolves completely. Compare its behavior when heated in a melting-point capillary to the behavior of your sample. If an impurity is found, it is probably a substance used as binder for the tablets. Is the binder organic or inorganic? To interpret your results, consider the mechanism whereby salicylic acid is acetylated.

Note that acetic acid is eliminated during the reaction. What effect would sodium acetate have? How might boron trifluoride etherate or sulfuric acid affect the nucleophilic attack of the phenolic oxygen on acetic anhydride? With what might the pyridine, a base, associate?

The IR and ^1H NMR (nuclear magnetic resonance) spectra of acetylsalicylic acid are presented in Figure 41.1 (on page 532) and Figure 41.2.

Cleaning Up. Combine the aqueous filtrates, dilute with water, and flush the solution down the drain.

[1]*To the instructor*: Commercial reagent, if dark, should be redistilled (bp 126°C, water-white).

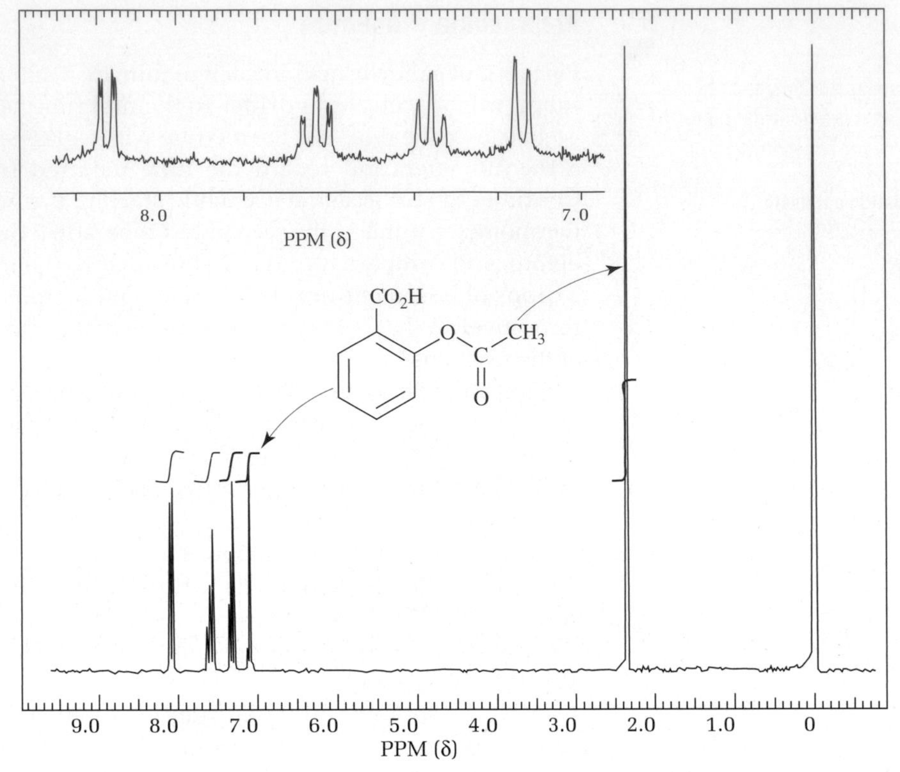

FIG. 41.2
The ¹H NMR spectrum of acetylsalicylic acid (250 MHz). The COOH proton does not appear on this spectrum.

QUESTIONS

1. Hydrochloric acid is about as strong a mineral acid as sulfuric acid. Why would HCl not be a satisfactory catalyst in this reaction?
2. How do you account for the smell of vinegar when an old bottle of aspirin is opened?

See *Web Links*

CHAPTER 42

Malonic Ester of a Barbiturate

PRELAB EXERCISE: Which barbiturate discussed in this chapter cannot be synthesized by the acetoacetic ester reaction?

Barbiturates are central nervous system (CNS) depressants used as hypnotic drugs and anesthetics. They are all derivatives of barbituric acid (R = R' = H), which has no sedative properties. It is called an *acid* because the carbonyl groups render the imide hydrogens acidic:

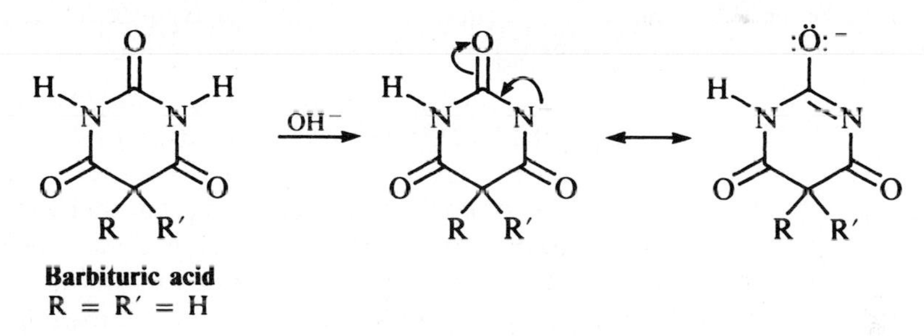

Barbituric acid
R = R′ = H

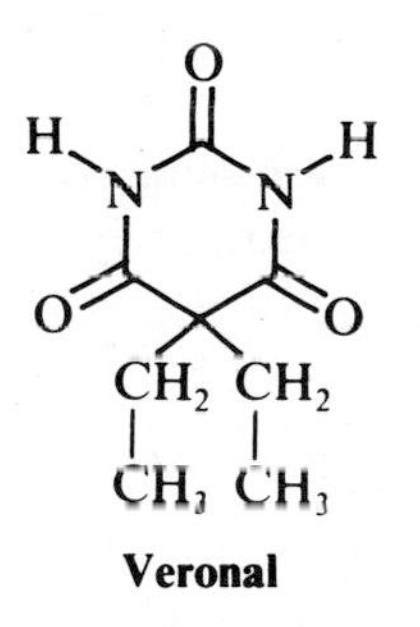

Veronal

Barbituric acid, derived from urea, was first synthesized in 1864 by Adolph von Baeyer. Apparently it was named at a tavern on St. Barbara's day. At the beginning of the 20th century, the great chemist Emil Fischer synthesized the first hypnotic (sleep-inducing) barbiturate, the 5,5-diethyl derivative of barbituric acid, at the direction of Joseph von Mering. Von Mering, who made the seminal discovery that removing the pancreas causes diabetes, named the new derivative of barbituric acid Veronal because he regarded Verona as the most restful city on Earth.

Barbiturates are the most widely used sleeping pills, and are classified according to their duration of action. Because it is quite easy to put almost any conceivable R group onto diethyl malonate, several thousand derivatives of barbituric acid have been synthesized. Studies of these derivatives have shown that as the alkyl chain (R) gets longer or as double bonds are introduced into the chain, the duration of action and the time of onset of action decreases. The maximum sedation occurs when the alkyl chains contain five or six carbons, as found in amobarbital (R = ethyl, R' = 3-methylbutyl) and pentobarbital (R = ethyl, R' = 1-methylbutyl).

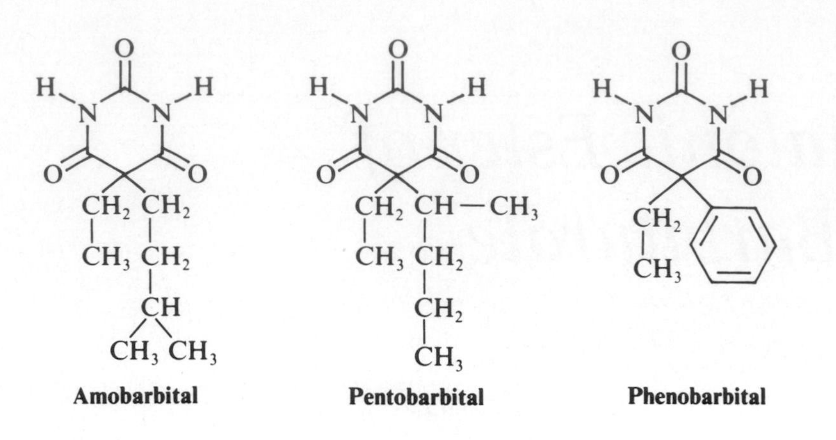

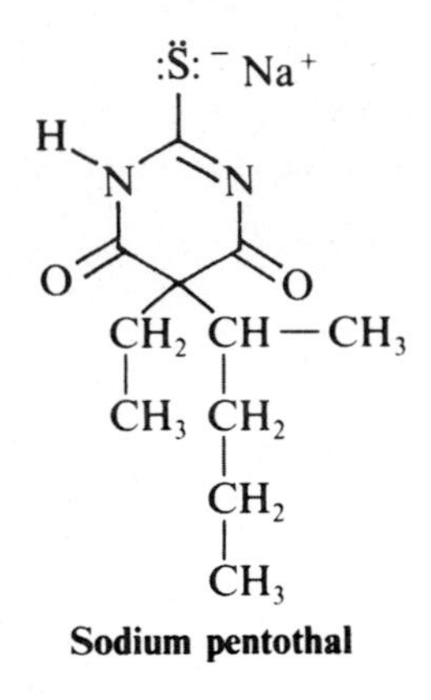

These two molecules illustrate how subtle changes in molecular structure can affect action. Amobarbital requires 30 minutes to take effect and sedation lasts for 5–6 hours, whereas pentobarbital takes effect in 15 minutes and sedation lasts only 2–3 hours. Phenobarbital (R = ethyl, R' = phenyl), on the other hand, requires over 1 hour to take effect, but sedation lasts for 6–10 hours. When the alkyl chains are made much longer, the sedative properties decrease, and the substances become anticonvulsants, which are used to treat epileptic seizures. If the alkyl group is too long or is substituted at one of the two nitrogens, convulsants are produced.

The usual dosage is 10–50 mg per pill. The continuous use of barbiturates leads to physiologic dependence (addiction), and withdrawal symptoms are identical to those experienced by a heroin addict. Replacing the oxygen atom at C-2 with sulfur results in the compound pentothal. Administered intravenously as the sodium salt, pentothal is a fast-acting general anesthetic. Like all general anesthetics, the details of its mode of action are unknown. In low, subanesthetic doses, sodium pentothal reduces inhibitions and the will to resist and thus functions as the so-called truth serum, apparently because it takes less mental effort to tell the truth than to prevaricate.

Most barbiturates are made from diethyl malonate. The methylene protons between the two carbonyl groups are acidic ($pK_a = 13.3$), and resonance will give a highly stabilized enolate anion:

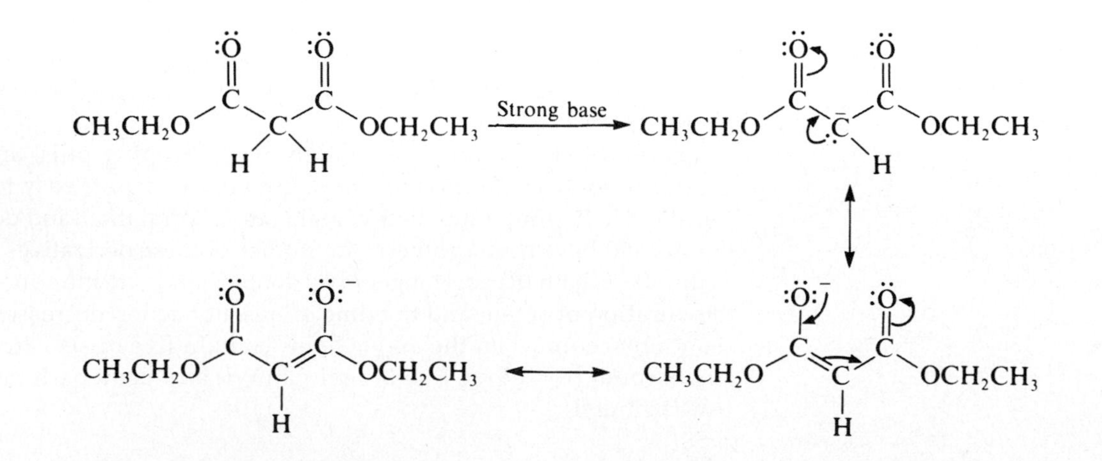

The acidic protons can be removed with a strong base, most often sodium ethoxide in dry ethanol. In Experiment 1, carbonate functions as the strong base because, in the absence of water, it is not solvated. In association with the phase-transfer catalyst tricaprylmethylammonium ion, the carbonate is soluble in the organic phase and can react with the diethyl malonate.

$$K_2CO_{3(s)} + 2\ H_3C\overset{+}{N}(-(CH_2)_7CH_3)_3\ Cl^- \rightleftharpoons \left(H_3C\overset{+}{N}(-(CH_2)_7CH_3)_3\right)_2 CO_3^{2-} + 2\ KCl$$

Tricaprylmethylammonium chloride
Phase transfer catalyst

Soluble in organic phase

The enolate anion of diethyl malonate can be alkylated by an S_N2 displacement of bromide to give diethyl *n*-butylmalonate:

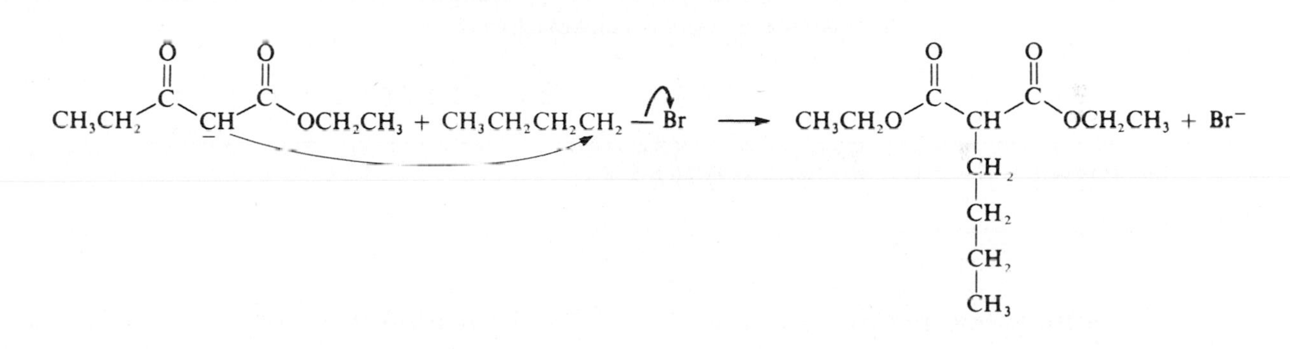

Also, because the product still contains one acidic proton, the process can be repeated using the same or a different alkyl halide. The previous product could be hydrolyzed and decarboxylated to give a carboxylic acid:

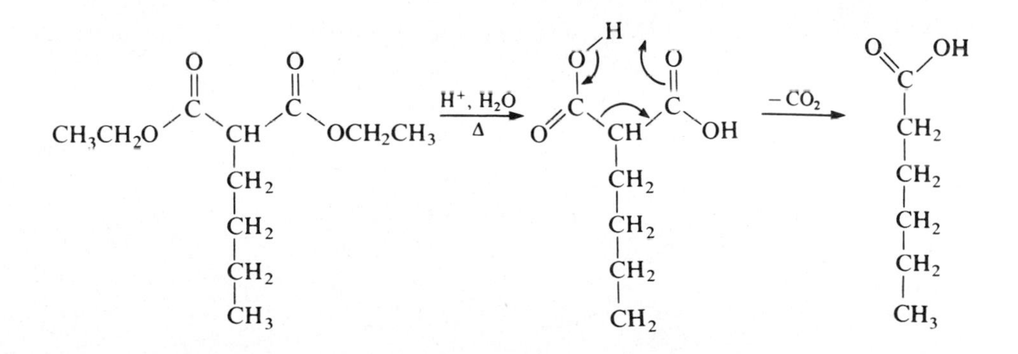

In Experiment 3, the diethyl *n*-butylmalonate is allowed to react with urea in the presence of a strong base, sodium ethoxide, to give the barbiturate:

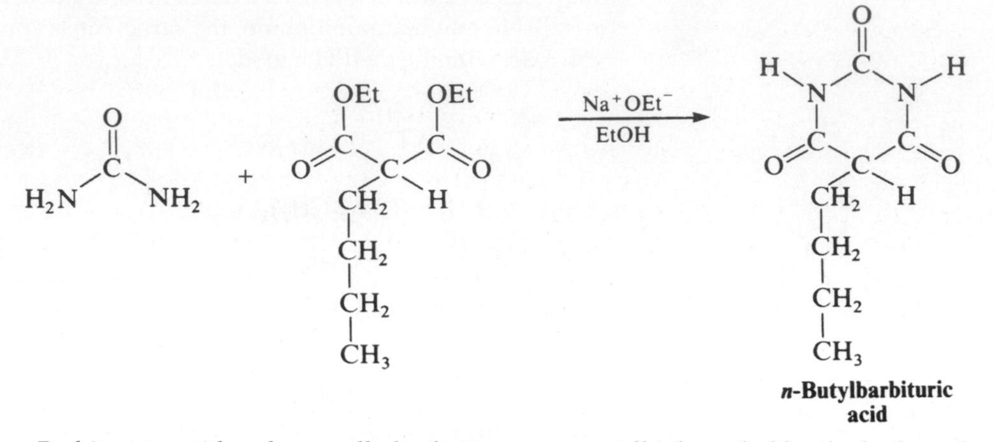

***n*-Butylbarbituric acid**

Barbiturates with only one alkyl substituent are rapidly degraded by the body and are therefore physiologically inactive.

EXPERIMENTS

1. DIETHYL *n*-BUTYLMALONATE

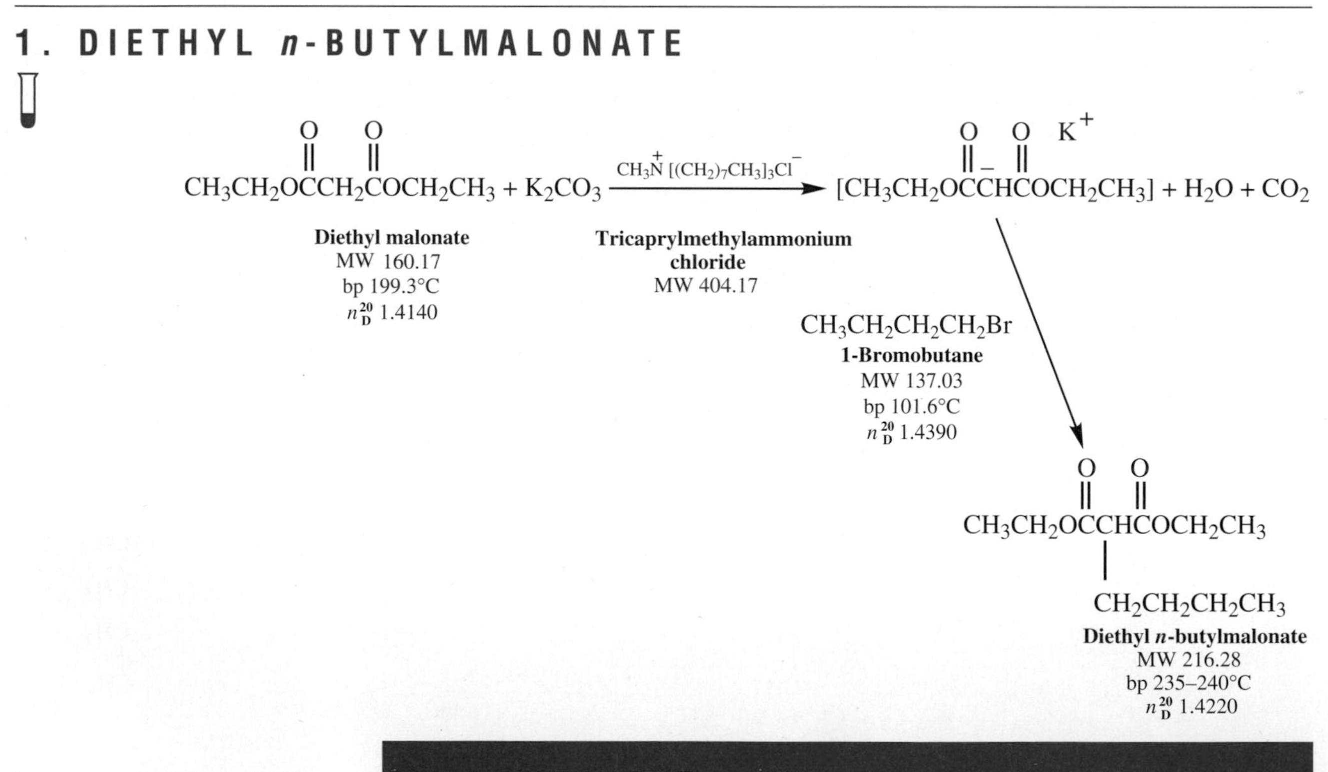

To check reagents, the instructor should run this experiment before assigning it to the class.

IN THIS EXPERIMENT, potassium carbonate combines with a quaternary ammonium salt to form a strong base that forms an anion from an ester of malonic acid. This anion reacts with bromobutane, also present in the reaction mixture, to give the product, a butyl derivative of the malonic ester. This ester is isolated by adding the reaction mixture to water and extracting with ether. The ether is dried and evaporated to give the product.

CAUTION: Tricaprylmethylammonium chloride (Aliquat 336) is a toxic irritant. Handle with care.

Into a 5-mL long-necked, round-bottomed flask, weigh 40 mg of tricaprylmethylammonium chloride, 400 mg of diethyl malonate, 350 mg of 1-bromobutane, and 415 mg of anhydrous potassium carbonate. Attach the empty distillation column as an air condenser and reflux the mixture for 1.5 hours (Fig. 42.1). Allow the mixture to cool somewhat; then cool further in ice. Using 2.5 mL of water, transfer the reaction mixture to a reaction tube, shake the mixture, draw off the organic layer, and extract the aqueous layer with two 1.5-mL portions of ether. Dry the combined organic extracts over anhydrous calcium chloride pellets or sodium sulfate, remove the ether solution from the drying agent, and complete the drying of the ether over Dri-Rite (commercial anhydrous calcium sulfate) or molecular sieves. Place the dry ether solutions in a *dry*, tared 5-mL round-bottomed, long-necked flask and remove the ether by distillation or evaporation in the hood. Complete the removal of the ether by connecting the flask to an aspirator. Determine the weight and calculate the yield of product. Analyze the product by thin-layer chromatography (TLC) on silica gel using dichloromethane as the eluent. Compare the product to a sample of authentic, pure diethyl *n*-butylmalonate, and to the starting materials diethyl malonate and bromobutane. The sample must be pure to be used as a starting material in the next reaction. It must be perfectly clear in appearance and absolutely dry. It can also be analyzed by gas chromatography using a Carbowax column at 160°C.

Video: *Extraction with Ether*

Cleaning Up. The aqueous layer after dilution with water can be flushed down the drain. The drying agents, once free of ether, can be placed in the nonhazardous solid waste container. Ether recovered from the distillation is placed in the organic solvents waste container, and the dichloromethane goes in the halogenated organic solvents waste container.

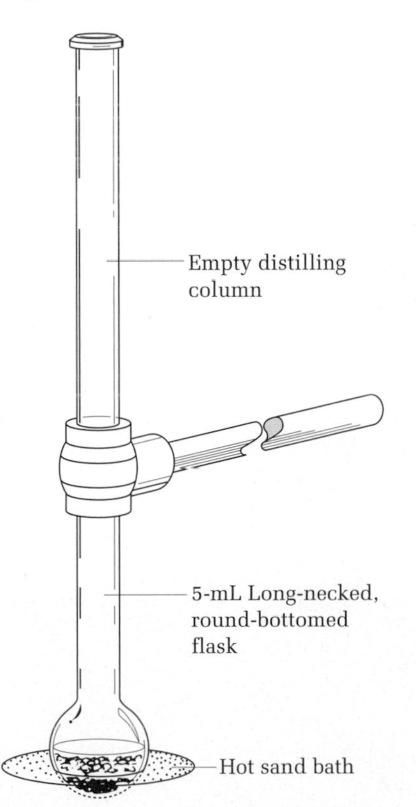

FIG. 42.1
A microscale reflux apparatus.

2. DIETHYL BENZYLMALONATE

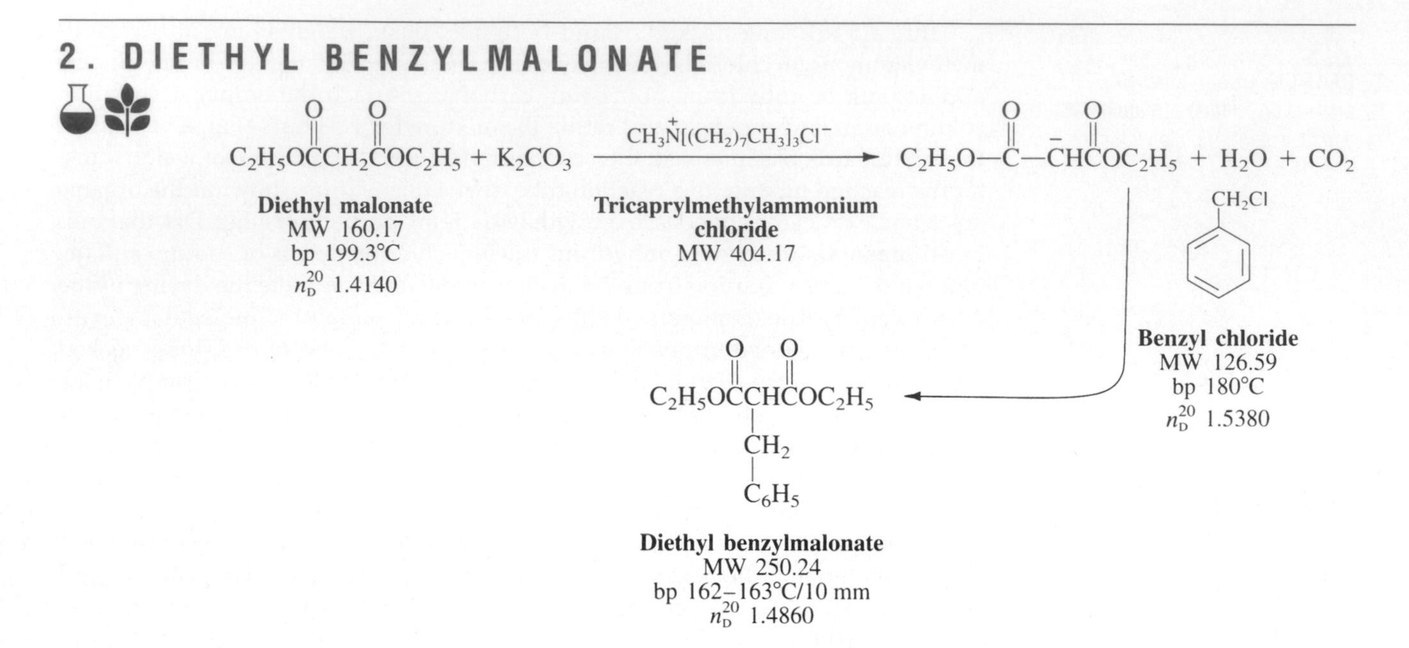

CAUTION: Tricaprylmethylammonium chloride (Aliquat 336) is a toxic irritant. Handle with care.

Into a 50-mL round-bottomed flask, weigh 0.8 g of tricaprylmethylammonium chloride, 8.00 g of diethyl malonate, 13.8 g of benzyl chloride, and 8.3 g of anhydrous potassium carbonate. Attach a water-cooled reflux condenser and reflux the reaction mixture for 1.5 hours (Fig. 42.2). Allow the mixture to cool somewhat and then transfer it, using 30 mL of water to rinse out the flask, to a small separatory funnel. Shake the mixture thoroughly and draw off the aqueous layer after the layers have separated. Place the organic layer in a 50-mL Erlenmeyer flask and extract

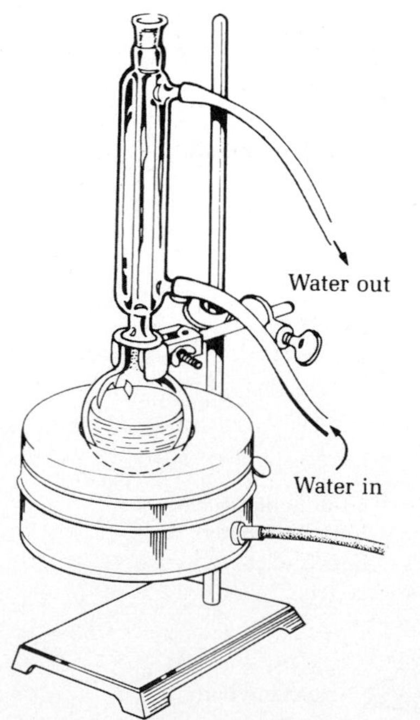

FIG. 42.2
A reflux apparatus for macroscale experiments.

To check reagents, the instructor should run this experiment before assigning it to the class.

the aqueous layer with two 10-mL portions of ether. Dry the combined extracts over anhydrous calcium chloride pellets (about 10 g), decant the liquid into a 100-mL round-bottomed flask, and evaporate the solvent on a rotary evaporator. Distill the residue from a 25-mL round-bottomed flask, collecting the material that boils from about 150°C to 165°C at 10 mm pressure. Use the apparatus depicted in Figure 6.7 (on page 116). The product, diethyl benzylmalonate, is reported to boil at 162–163°C/10 mm. See the nomograph (Fig. 6.19 on page 124) to determine the boiling point at a different pressure.

Cleaning Up. The aqueous layer after dilution with water can be flushed down the drain. Calcium chloride, once free of the ether (evaporate in the hood), can be placed in the nonhazardous solid waste container. Ether recovered from the distillation is placed in the organic solvents waste container. If local regulations do not allow for the evaporation of solvents in a hood, the wet solid should be disposed in a special waste container.

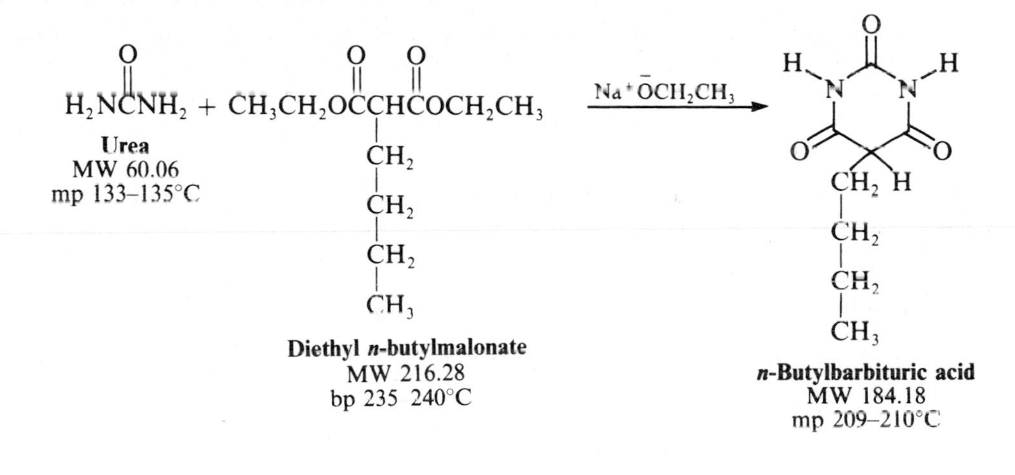

3. *n*-BUTYLBARBITURIC ACID

Experimental Considerations

CAUTION: Handle sodium metal with care. It will react violently with water and can cause a fire or an explosion. Avoid all contact with moisture in any form.

In this experiment, absolute (anhydrous, 100%) ethanol is reacted with sodium metal to prepare a solution of sodium ethoxide in a 5-mL flask. (Commercially prepared sodium ethoxide is often not active.) The empty distilling column functions as an air condenser, and moisture is prevented from diffusing into the apparatus by capping the distilling column with a septum and an empty syringe needle.

The sodium metal may come in the form of small spheres or a large block stored under mineral oil. If in block form, it must be shaved off with a knife and transferred to a beaker containing light mineral oil or hexanes. Use tweezers to handle the sodium, blot it off, and then transfer it to a dry beaker for weighing. Blot off each piece as it is added to the reaction flask down through the condenser.

Be extremely careful to have no water near the sodium. If a piece of sodium is accidentally dropped, place it in a beaker containing 1-propanol (see "Cleaning Up" on page 542). Do not dispose of it in a trash can because it can easily start a fire due to oxidation.

Microscale Procedure

IN THIS EXPERIMENT, sodium ethoxide is prepared from sodium and ethanol, which is then reacted with *n*-butylmalonate and urea to produce a barbiturate. The product is isolated by acidification and evaporation until crystallization occurs. The solid is collected on a Hirsch funnel, washed with hexanes, and recrystallized from water.

To a perfectly dry (oven- or flame-dried) 5-mL round-bottomed, long-necked flask, add 2.5 mL of absolute ethanol. Attach to the flask the empty distilling column capped with a septum and an empty syringe needle. Add to the ethanol 23 mg of sodium metal. Wait for all of the sodium to react; do not proceed until all traces of the metal have disappeared. To the sodium ethoxide solution, add 216 mg of dry *n*-butylmalonate (synthesized in Experiment 1 or purchased commercially[1]). Immediately add 60 mg of urea that has been dissolved in 1.1 mL of absolute ethanol. Reflux the resulting mixture on a warm sand bath or a steam bath for 2 hours. Which solid separates during this reaction? It may cause the reaction mixture to bump during the first part of the refluxing period, so be sure to clamp the apparatus firmly, as always.

Video: *Microscale Filtration on the Hirsch Funnel*

At the completion of the reaction, acidify the solution with 1 mL of 3 *M* hydrochloric acid (check pH with indicator paper), and then reduce the volume in the flask to one-half by distilling the ethanol or by removing the solvent on a rotary evaporator. Cool the solution on ice until no more product crystallizes; then collect it by vacuum filtration on a Hirsch funnel. Empty the filter flask; then wash unreacted diethyl *n*-butylmalonate (detected by its odor) from the crystals with hexanes, press the crystals dry, and then recrystallize the product from boiling water (20 mL of water per gram of product). The acid crystallizes slowly. Cool the solution for at least 30 minutes before collecting the product by vacuum filtration. Dry the barbituric acid on filter paper until the next laboratory period; then weigh it, calculate the percent yield, and determine the melting point. Pure *n*-butylbarbituric acid melts at 209°C–210°C.

Wet all paper that has come in contact with sodium metal with alcohol before disposal. This will decompose traces of the sodium metal.

Cleaning Up. Add scrap sodium metal to 1-propanol in a beaker. After all traces of the sodium have disappeared, the alcohol can be diluted with water, neutralized with dilute hydrochloric acid, and flushed down the drain. Combine aqueous filtrates and mother liquors from the crystallization, neutralize with sodium carbonate, and flush down the drain. The ligroin wash goes in the organic solvents waste container.

4. BENZYLBARBITURIC ACID

Experimental Considerations

In this experiment, absolute (anhydrous, 100%) ethanol is reacted with sodium metal to prepare a solution of sodium ethoxide in the 250-mL flask. (Commercially prepared sodium ethoxide is often not active.) Moisture is prevented from diffusing

[1] *n*-Butylmalonate is available commercially from the Aldrich Chemical Company.

into the apparatus by capping the condenser with a drying tube. The drying tube is prepared by first placing a piece of cotton in the tube, followed by calcium chloride and another firmly placed piece of cotton. At the end of the experiment, dispose of the calcium chloride or cap each end of the drying tube very carefully with corks to save it for later use. The calcium chloride is highly hygroscopic.

The sodium metal may come in the form of small spheres or a large block stored under mineral oil. If in block form, it must be shaved off with a knife and transferred to a beaker containing light mineral oil or hexanes. Use tweezers to handle the sodium, blot it off, and then transfer it to a dry beaker for weighing. Blot off each piece as it is added to the reaction flask down through the condenser.

Be extremely careful to have no water near the sodium. If a piece of sodium is accidentally dropped, place it in a beaker containing 1-propanol (see "Cleaning Up" on page 544). Do not dispose of it in a trash can because it can easily start a fire due to oxidation.

Procedure

To a perfectly dry 250-mL round-bottomed flask, add 50 mL of absolute ethanol; then attach a water cooled condenser and a calcium chloride drying tube (Fig. 42.3). Add to the ethanol 0.46 g of sodium metal. Warm the flask on a steam bath to dissolve all of the sodium; do not proceed until all traces of the metal have disappeared. To the sodium ethoxide solution, add 5.00 g of dry *n*-benzylmalonate (synthesized in Experiment 2 or purchased commercially[2]). Immediately add a hot

FIG. 42.3
A reflux apparatus with a calcium chloride drying tube.

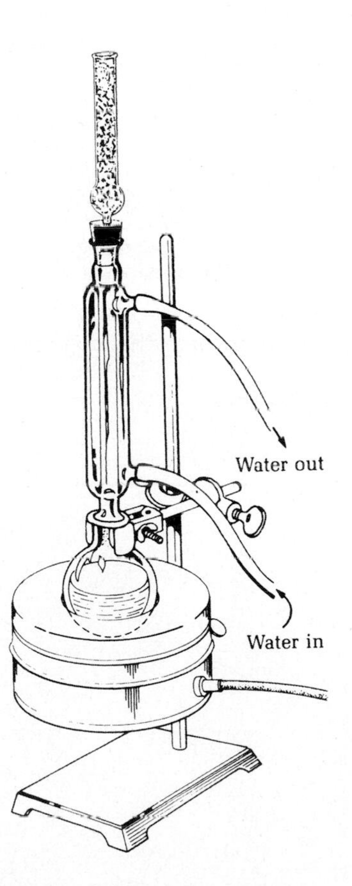

[2]*n*-Benzylmalonate is available from the Aldrich Chemical Company.

solution of 1.2 g of dry urea that has been dissolved in 22 mL of absolute ethanol. Reflux the resulting mixture on a steam bath or a heating mantle for 2 hours. Which solid separates during this reaction? It may cause the reaction mixture to bump during the first part of the refluxing period, so be sure to clamp the apparatus firmly, as always.

At the completion of the reaction, acidify the solution with 20.0 mL of 3 *M* hydrochloric acid (check pH with indicator paper), and then reduce the volume to one-half by removing the ethanol either by distillation or by evaporation on a rotary evaporator. Cool the solution in an ice bath until no more product crystallizes; then collect it by vacuum filtration on a Büchner funnel. Empty the filter flask; then wash unreacted diethyl benzylmalonate (detected by its odor) from the crystals with hexanes, press the crystals dry, and recrystallize the product from boiling water. The acid crystallizes slowly. Cool the solution for at least 30 minutes before collecting the product by vacuum filtration. Dry the barbituric acid on filter paper until the next laboratory period; then weigh it, calculate the percent yield, and determine the melting point. Pure benzylbarbituric acid melts at 206°C.

Cleaning Up. Add scrap sodium metal to 1-propanol in a beaker. After all traces of sodium have disappeared, the alcohol can be diluted with water, neutralized with dilute hydrochloric acid, and flushed down the drain. Combine aqueous filtrates and mother liquors from the crystallization, neutralize with sodium carbonate, and flush down the drain. The hexanes wash goes in the organic solvents waste container.

QUESTIONS

For Additional Experiments, sign in at this book's premium website at **www.cengage.com/login.**

1. The anion of diethyl benzylmalonate is often made by reacting the diester with sodium methoxide. What weight of sodium would be required in Experiment 2 if that method were employed?
2. Outline all of the steps in the synthesis of pentobarbital.
3. What problem might be encountered if 1-bromo-2,2-dimethylpropane, $BrCH_2C(CH_3)_3$, were used as the halide in this synthesis?

CHAPTER 43

Amines

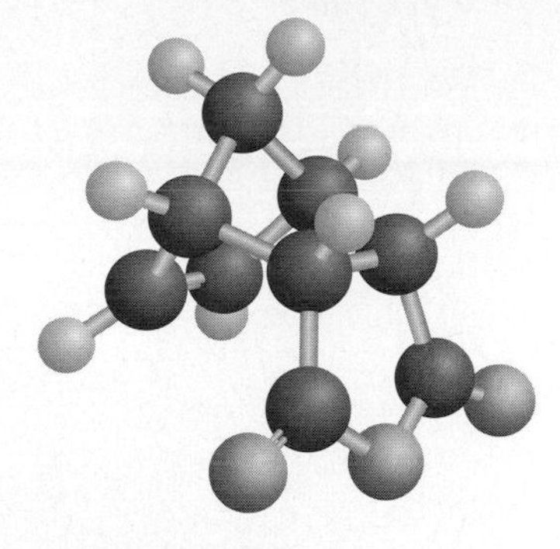

PRELAB EXERCISE: Explain why an ordinary amide from a primary amine does not dissolve in alkali, whereas the corresponding sulfonamide of the amine does dissolve.

Amines are weak bases, and may be regarded as organic substitution products of ammonia. Just as ammonia reacts with acids to form the ammonium ion, so amines react with acid to form the organoammonium ion:

$$\ddot{N}H_3 + H_3\ddot{O}^+ \rightleftharpoons \overset{+}{N}H_4 + H_2\ddot{O}:$$

Ammonia **Ammonium ion**

$$CH_3\ddot{N}H_2 + H_3\ddot{O}^+ \rightleftharpoons CH_3\overset{+}{N}H_3 + H_2\ddot{O}:$$

Methylamine **Methylammonium ion**

When an amine dissolves in water, the following equilibrium is established:

$$R\ddot{N}H_2 + H_2O \rightleftharpoons R\overset{+}{N}H_3 + OH^-$$

and from this the basicity constant can be defined as

$$K_b = \frac{[RNH_3^+][OH^-]}{[RNH_2]}$$

Strong bases have larger values of K_b, meaning the amine has a greater tendency to accept a proton from water to increase the concentration of RNH_3^+ and OH^-. The basicity constant (K_b) for ammonia is 1.8×10^{-5}.

$$\ddot{N}H_3 + H_2O \rightleftharpoons NH_4^+ + OH^-$$

$$K_b = 1.8 \times 10^{-5} = \frac{[NH_4^+][OH^-]}{[NH_3]}$$

Alkyl amines such as methylamine (CH_3NH_2) or *n*-propylamine ($CH_3CH_2CH_2NH_2$) are stronger bases than ammonia because the alkyl groups are electron donors and increase the effective electron density on nitrogen. Aromatic amines, on the other hand, are weaker bases than ammonia. Delocalization of the unshared pair of electrons on nitrogen onto the aromatic ring means that the electrons cannot be shared with an acidic proton.

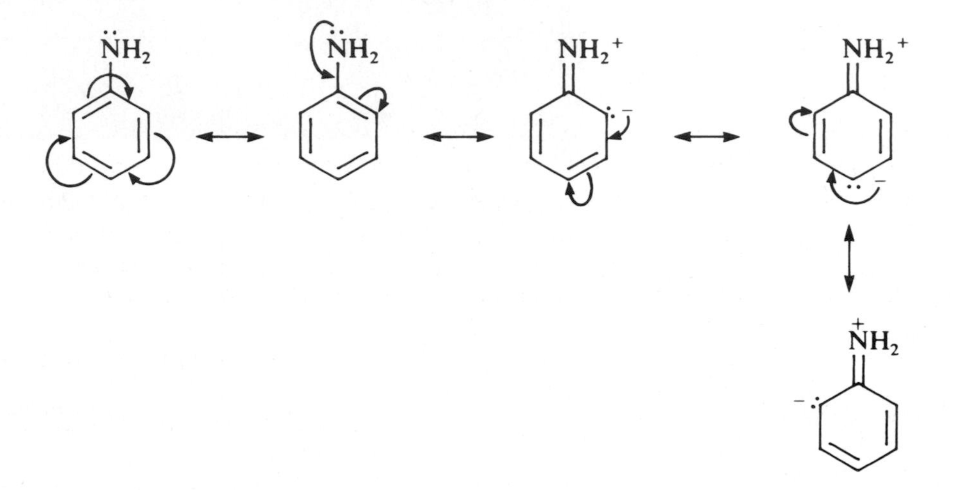

Low molecular weight amines have a powerful fishy odor. Slightly higher molecular weight diamines have names suggestive of their odors: $H_2N(CH_2)_5NH_2$ is cadaverine; and $H_2N(CH_2)_4NH_2$ is putrescine. The lower molecular weight amines with up to about five carbon atoms are soluble in water. Many amines, especially the liquid aromatic amines, undergo light-catalyzed free-radical air oxidation reactions to give a large variety of highly colored decomposition products. The higher molecular weight amines that are insoluble in water will dissolve in acid to form ionic amine salts. This reaction is useful for both characterization and separation purposes. The ionic amine salt on treatment with base will regenerate the amine.

Nitriles and amides are not basic.

Nitriles, R—C≡N, are not basic and neither are amides or substituted amides. The adjacent electron-withdrawing carbonyl oxygen effectively removes the unshared pair of electrons on nitrogen:

$$R-\overset{:\ddot{O}}{\overset{\|}{C}}-\ddot{N}H_2 \longleftrightarrow R-\overset{:\ddot{O}:^-}{\overset{|}{C}}=\overset{+}{N}H_2$$

The object of the experiments in this chapter is to identify an unknown amine or amine salt. Procedures for solubility tests are given in Experiment 1. Experiment 2 presents the Hinsberg test for distinguishing between primary, secondary, and tertiary amines; Experiment 3 gives procedures for preparing solid derivatives for melting-point characterizations; and Experiment 4 gives spectral characteristics. You will apply the procedures to known substances along with an unknown substance.

EXPERIMENTS

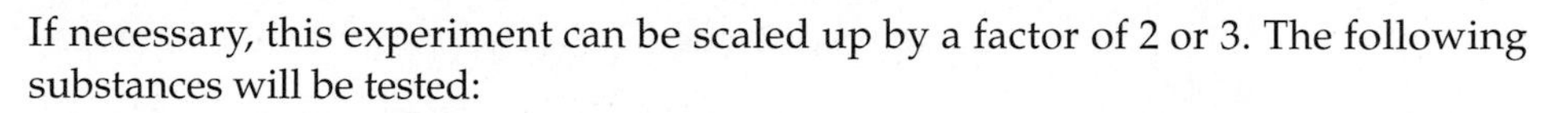

1. SOLUBILITY

If necessary, this experiment can be scaled up by a factor of 2 or 3. The following substances will be tested:

Aniline, $C_6H_5NH_2$ (bp 184°C)
p-Toluidine, $CH_3C_6H_4NH_2$ (mp 43°C)
Pyridine, C_5H_5N (bp 115°C; a tertiary amine)
Methylamine hydrochloride, dec. (salt of CH_3NH_2; bp 6.7°C)
Aniline hydrochloride, dec. (salt of $C_6H_5NH_2$)
Aniline sulfate, dec.

Pyridine

CH_3 / NH_2 — *p*-Toluidine

Video: *Instant Microscale Distillation*

Use caution in smelling unknowns and do so only once.

Amines are often contaminated with dark oxidation products. Amines can be easily purified by a small-scale distillation (Fig. 43.1).

First, determine if the substance has a fishy, ammonia-like odor; if so, it probably is an amine of low molecular weight. Then test its solubility in water by putting one drop, if the unknown is a liquid, or an estimated 10 mg if a solid, into a reaction tube or a small glass centrifuge tube, adding 0.1 mL of water, and seeing if the substance dissolves when cooled. If the substance is a solid, rub it well with a stirring rod and break up any lumps before drawing a conclusion.

If the substance is *readily soluble in cold water* and if the odor is suggestive of an amine, test the solution with pH paper and further determine if the odor disappears by adding a few drops of 3 *M* hydrochloric acid. If the properties are more like those of a salt, add a few drops of 3 *M* sodium hydroxide solution. If the solution remains clear, adding a small amount of sodium chloride may cause separation of a liquid or solid amine.

If the substance is not soluble in cold water, see if it will dissolve on heating; be careful not to mistake the melting of a substance for dissolving. If it dissolves in hot water, add a few drops of 3 M alkali and see if an amine precipitates. (If you are in doubt as to whether a salt has dissolved partially or not at all, pour off the supernatant liquid and make it basic; then add the liquid back to the tube and note that the salt should completely dissolve.)

If the substance is insoluble in hot water, add 3 *M* hydrochloric acid, heat if necessary, and see if it dissolves. If so, make the solution basic and see if an amine precipitates.

Cleaning Up. All solutions should be combined, made basic to free the amines, and extracted with an organic solvent such as hexanes. The aqueous layer is flushed down the drain, and the organic layer is placed in the organic solvents waste container.

2. HINSBERG TEST

CAUTION: Handle amines with care. Many are toxic, particularly the aromatic amines.

If necessary, this experiment can be doubled in scale.

The procedure for distinguishing amines with benzenesulfonyl chloride is to run them in parallel on the following substances:

Aniline, $C_6H_5NH_2$ (bp 184°C)
N-Methylaniline, $C_6H_5NHCH_3$ (bp 194°C)
Triethylamine, $(CH_3CH_2)_3N$ (bp 90°C)

Primary and secondary amines react in the presence of alkali with benzenesulfonyl chloride ($C_6H_5SO_2Cl$) to give sulfonamides.

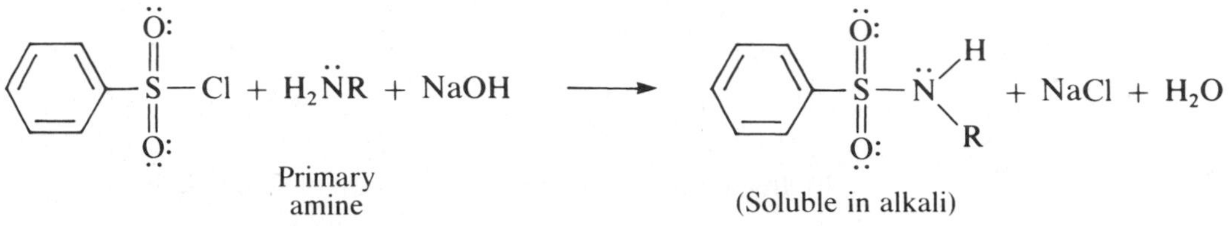

$$C_6H_5SO_2Cl + H_2NR + NaOH \longrightarrow C_6H_5SO_2NHR + NaCl + H_2O \qquad 1$$

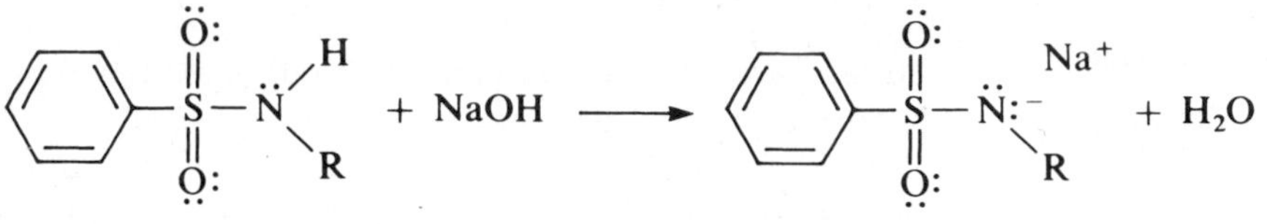

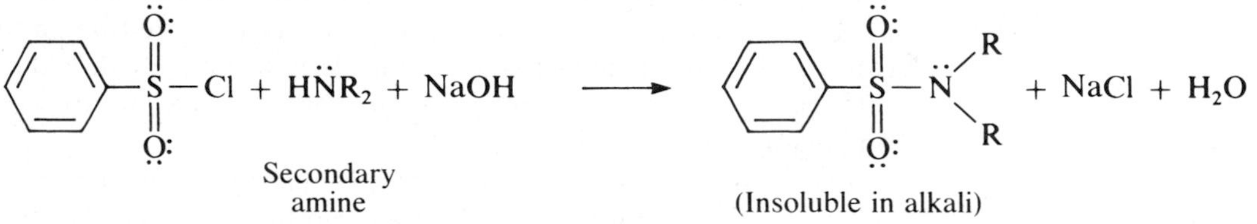

Primary amine; (Soluble in alkali)

$$C_6H_5SO_2NHR + NaOH \longrightarrow C_6H_5SO_2N^{-}R\ Na^{+} + H_2O$$

$$C_6H_5SO_2Cl + HNR_2 + NaOH \longrightarrow C_6H_5SO_2NR_2 + NaCl + H_2O \qquad 2$$

Secondary amine; (Insoluble in alkali)

The sulfonamides are distinguishable because the derivative from a primary amine has an acidic hydrogen, which renders the product soluble in alkali (reaction 1); whereas the sulfonamide from a secondary amine is insoluble (reaction 2). Tertiary amines lack the necessary acidic hydrogen for the formation of benzenesulfonyl derivatives.

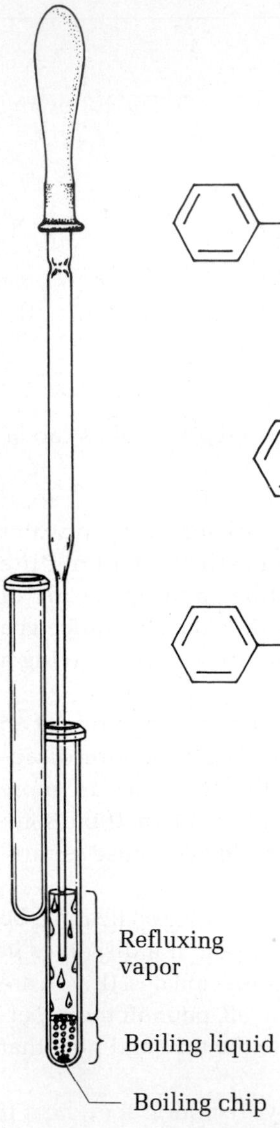

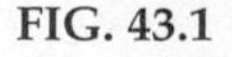

FIG. 43.1
The apparatus for instant microscale distillation. When the refluxing vapor is drawn into the pipette, it will condense to a clear liquid, which is then squirted into the empty reaction tube. Both tubes can be held in one hand.

Procedure

To a reaction tube or small centrifuge tube, add 50 mg of the amine, 200 mg of benzenesulfonyl chloride, and 1 mL of methanol. Over a hot sand bath or a steam bath, heat the mixture to just below the boiling point, cool, and add 2 mL of 6 *M* sodium hydroxide. Shake the mixture for 5 minutes; then allow the tube to stand for an additional 10 minutes with intermittent shaking. If the odor of benzenesulfonyl chloride is detected, warm the mixture to hydrolyze it (*see* Question 7). Cool the mixture and acidify it by adding 6 *M* hydrochloric acid dropwise while stirring. If a precipitate is observed at this point, the amine is either primary or secondary. If no precipitate is observed, the amine is tertiary.

If a precipitate is present, remove it by filtration on a Hirsch funnel, wash it with 2 mL of water, and transfer it to a reaction tube. Add 2.5 mL of 1.5 *M* sodium hydroxide solution and warm the mixture to 50°C. Shake the tube vigorously for

Low molecular weight amines have very bad odors; work in the hood.

CAUTION: Do not allow benzenesulfonyl chloride to come in contact with the skin.

Distinguishing among primary (1°), secondary (2°), and tertiary (3°) amines

2 minutes. If the precipitate dissolves, the amine is primary. If it does not dissolve, it is secondary. The sulfonamide of a primary amine can be recovered by acidifying the alkaline solution. Once dry, these sulfonamides can be used to characterize the amine by their melting points.

Cleaning Up. The slightly acidic filtrate can be diluted and then flushed down the drain if the unknown is primary or secondary. If the unknown is tertiary, make the solution basic, and extract the amine with hexanes. Place the hexanes layer in the organic solvents waste container and flush the aqueous layer down the drain.

3. SOLID DERIVATIVES

Video: *Microscale Filtration on the Hirsch Funnel*

Acetyl ($CH_3C{=}O$) derivatives of primary and secondary amines are usually solids suitable for melting-point characterization and are readily prepared by reaction with acetic anhydride, even in the presence of water. Benzoyl ($C_6H_5C{=}O$) and benzenesulfonyl ($C_6H_5SO_2$) derivatives are made by reacting the amine with the appropriate acid chloride in the presence of alkali, as in Experiment 2 (the benzenesulfonamides of aniline and of *N*-methylaniline melt at 110°C and 79°C, respectively).

Solid derivatives suitable for the characterization of tertiary amines are the methiodides and picrates:

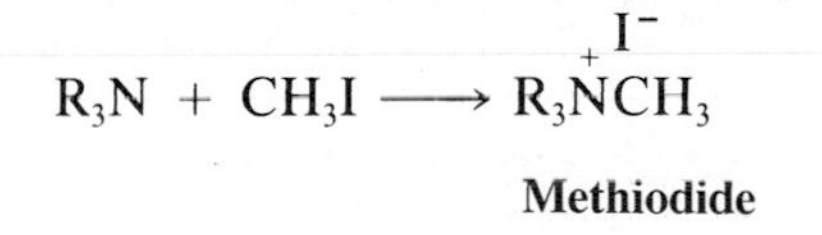

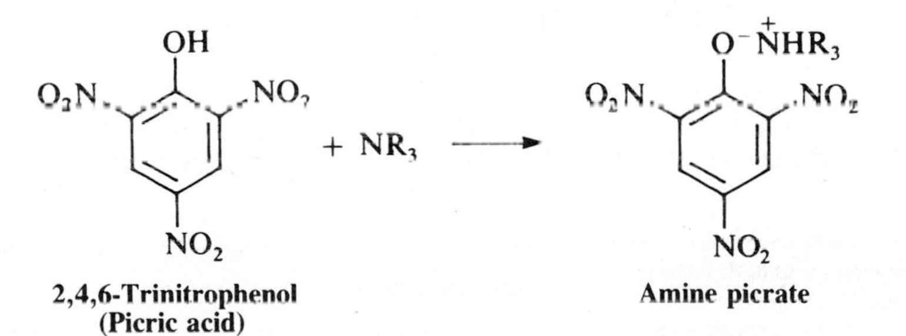

Typical derivatives are to be prepared and, although a determination of melting points is not necessary because the values are given, the products should be saved for possible identification of unknowns.

Acetylation of Aniline with Acetic Anhydride

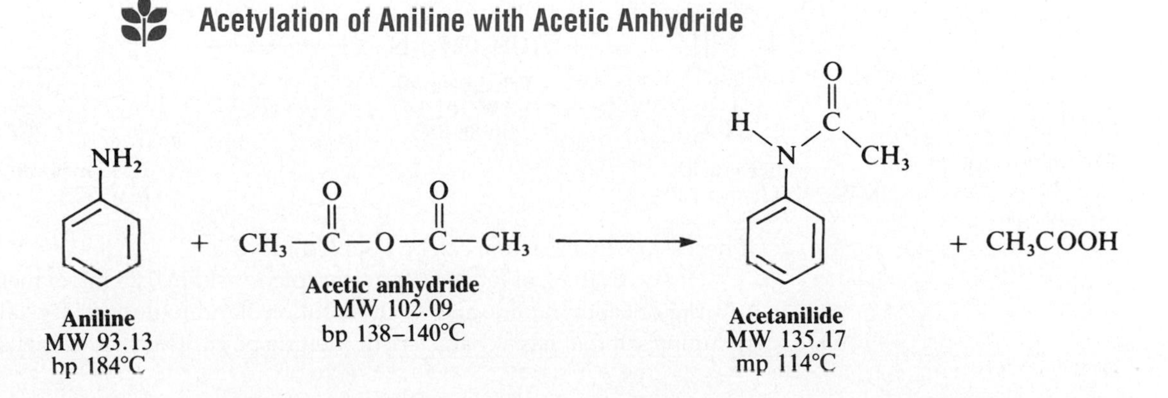

With patience, this reaction and the associated recrystallizations can be carried out in a melting point tube with one-tenth the indicated quantities of material using essentially the same techniques; if deemed necessary, this reaction and associated recrystallizations can also be carried out on 10–20 times as much material.

In a dry, small test tube, place 46 mg (or 5 drops) of freshly distilled aniline, 51 mg (or 5 drops) of pure acetic anhydride, and a small boiling chip. Reflux the mixture for about 8 minutes, and then add water to the mixture dropwise with heating until all the product is in solution. This will require very little water. (Note: On a larger scale, exercise caution when adding water to the reaction mixture because the exothermic hydrolysis of acetic anhydride is slow at room temperature but can get out of control at higher temperatures.) Allow the mixture to cool spontaneously to room temperature; then cool the mixture in ice. Remove the solvent with a Pasteur pipette and recrystallize the crude material again from boiling water. Isolate the crystals in the same way: cool the tube in ice water until crystals form, decant the water, and then wash the crystals with 1–2 drops of ice-cold acetone. Remove the acetone with a Pasteur pipette and dry the product in the test tube. Once dry, determine the weight, calculate the percent yield, and determine the melting point.

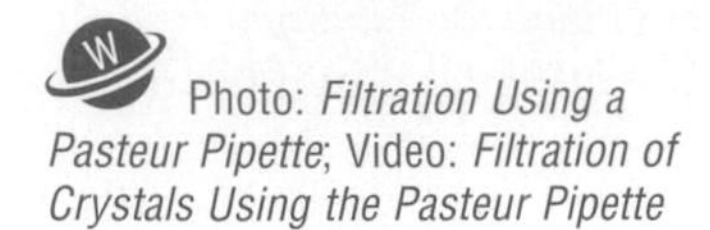

Cleaning Up. The slightly acidic filtrate and washings can be diluted and then flushed down the drain.

Acetylation of Aniline Hydrochloride with Acetic Anhydride

If necessary, this experiment can be carried out on 10–20 times as much material.

Dissolve 0.26 g of aniline hydrochloride in 2.5 mL of water; then add 0.21 g of acetic anhydride followed immediately by 0.25 g of sodium acetate. Allow the reaction mixture to stir (using a magnetic stir bar and magnetic stirrer) for 10 minutes; then cool it in ice. Recrystallize the product from water and wash it with acetone as described in the previous acetylation. See Chapter 45 for the mechanism.

Cleaning Up. The slightly acidic filtrate can be diluted and then flushed down the drain if the unknown is a primary or secondary amine. If the unknown is a tertiary amine, make the solution basic and extract the amine with hexanes. Place the hexanes layer in the organic solvents waste container and flush the aqueous layer down the drain.

NOTE: It is preferable to use the moist reagent of picric acid (35% water) and to not allow this reagent to dry out. Your instructor may provide a stock solution containing 3 g of moist picric acid in 25 mL of methanol.

Formation of an Amine Picrate from Picric Acid and Triethylamine

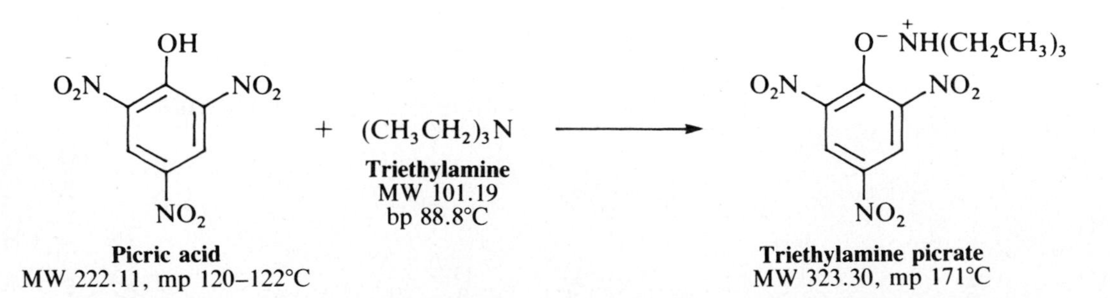

If necessary, this reaction can be doubled in scale.

Dissolve 30 mg of moist (35% water) picric acid in 0.25 mL of methanol; then add 10.1 mg of triethylamine and let the solution stand to deposit crystals of the picrate, an amine salt that has a characteristic melting point. This picrate melts at 171°C.

Cleaning Up. The filtrate and the product from this reaction can be disposed of by dilution with a large volume of water and flushing down the drain. Larger quantities of moist picric acid (1 g) can be reduced with tin and hydrochloric acid to the corresponding triaminophenol.[1] Dry picric acid is hazardous and should not be handled.

4. THE NMR AND IR SPECTRA OF AMINES

The proton bound to nitrogen can appear between 0.6 ppm and 7.0 ppm on the nuclear magnetic resonance (NMR) spectrum, with the position depending on the solvent, the concentration, and the structure of the amine. The peak is sometimes extremely broad, owing to slow exchange and interaction of the proton with the electric quadrupole of the nitrogen. If adding a drop of D_2O (heavy water) to the sample causes the peak to disappear, this is evidence for an amine hydrogen; but alcohols, phenols, and enols will also exhibit this exchange behavior. Figure 43.2 illustrates the ^1H NMR spectrum of aniline, in which the amine hydrogens appear as a sharp peak at 3.5 ppm. Infrared (IR) spectroscopy can also be very useful for identification purposes. Primary amines, both aromatic and aliphatic, show a weak doublet between 3300 and 3500 cm^{-1} (N—H stretching) and a strong absorption between 1560 and 1640 cm^{-1} due to N—H bending (Fig. 43.3). Secondary amines show a single peak between 3310 and 3450 cm^{-1}. Tertiary amines have no useful IR absorptions. The characteristic ultraviolet (UV) absorption shifts of aromatic amines in the presence and absence of acids are discussed in Chapter 14.

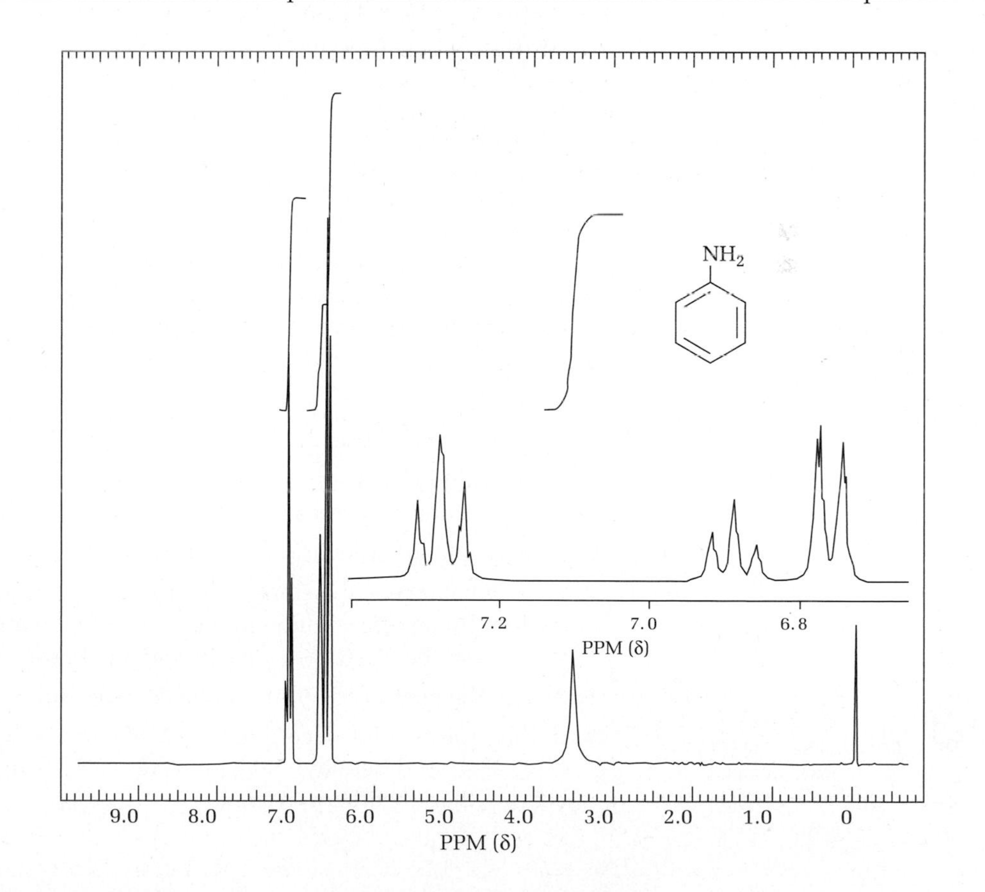

FIG. 43.2
The ^1H NMR spectrum of aniline (250 MHz).

[1]Armour MA. *Hazardous Laboratory Chemicals Disposal Guide;* Lewis Publishers: Edmonton, Alberta, Canada, 1996.

FIG. 43.3
The IR spectrum of aniline.

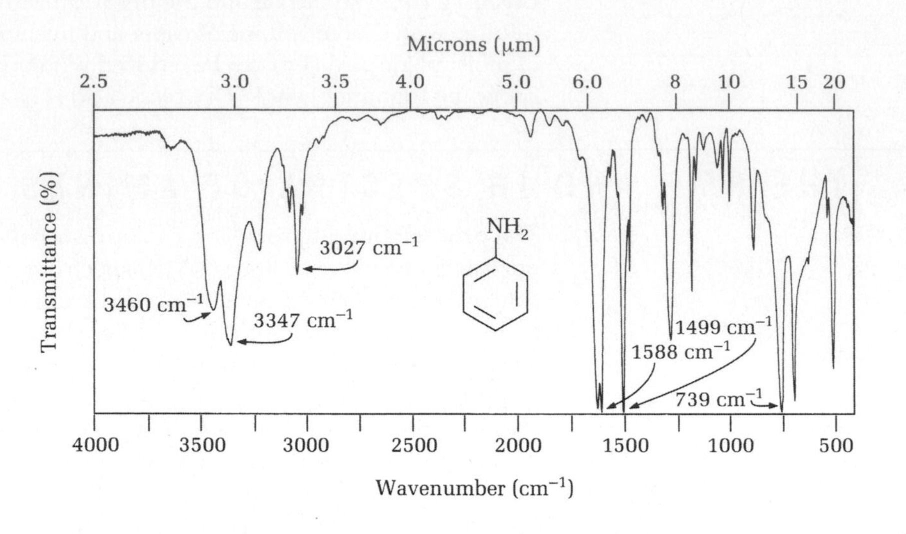

5. UNKNOWNS[2]

Determine first if the unknown is an amine or an amine salt; then determine whether the amine is primary, secondary, or tertiary. Complete identification of your unknown may be required.

QUESTIONS

1. How could you most easily distinguish between samples of 2-aminonaphthalene and acetanilide?
2. Would you expect the reaction product from benzenesulfonyl chloride and ammonia to be soluble or insoluble in alkali?
3. Is it safe to conclude that a substance is a tertiary amine because it forms a picrate?
4. Why is it usually true that amines that are insoluble in water are odorless?
5. Low quality (technical grade) dimethylaniline contains traces of aniline and methylaniline. Suggest a method for eliminating these impurities.
6. How would you prepare aniline from aniline hydrochloride?
7. Write a balanced equation for the reaction of benzenesulfonyl chloride with sodium hydroxide solution in the absence of an amine. What solubility would you expect the product to have in acid and base?
8. Assign the peak at 3347 cm^{-1} in the IR spectrum of aniline (Fig. 43.3).
9. In the proton noise-decoupled ^{13}C NMR spectra of benzenesulfonyl chloride (Fig. 43.4) and benzoyl chloride (Fig. 43.5), why is the peak of the carbon bearing the substituent so small?

For Additional Experiments, sign in at this book's premium website at **www.cengage.com/login.**

[2]Visit this book's website for tables of physical properties of selected primary, secondary, and tertiary amines.

FIG. 43.4
The ^{13}C NMR spectrum of benzenesulfonyl chloride (22.6 MHz)

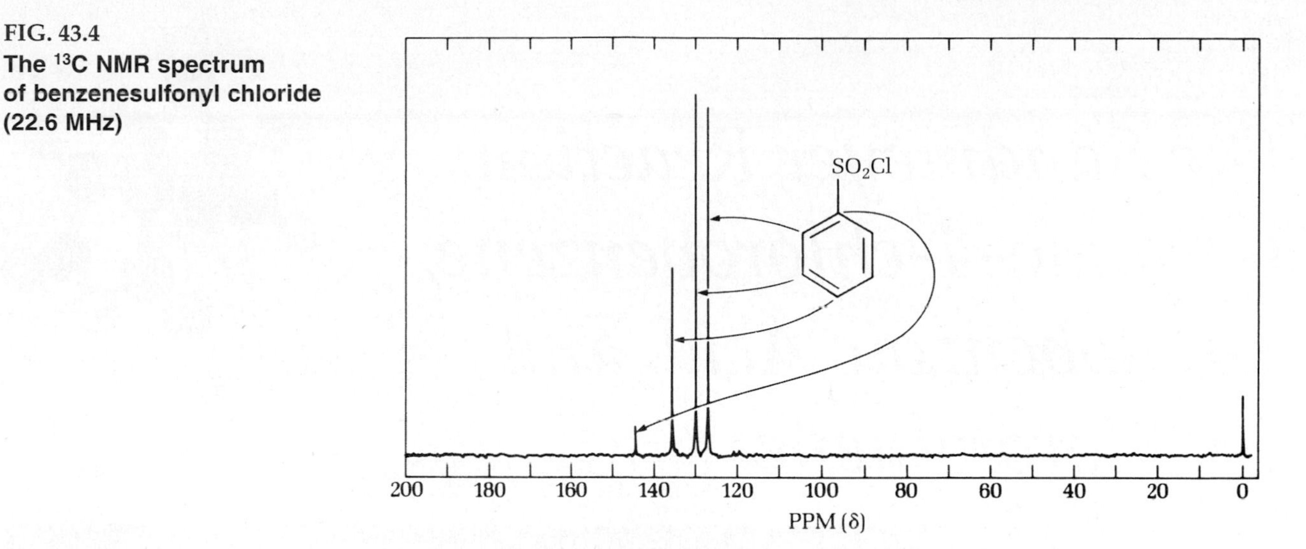

FIG. 43.5
The ^{13}C NMR spectrum of benzoyl chloride (22.6 MHz).

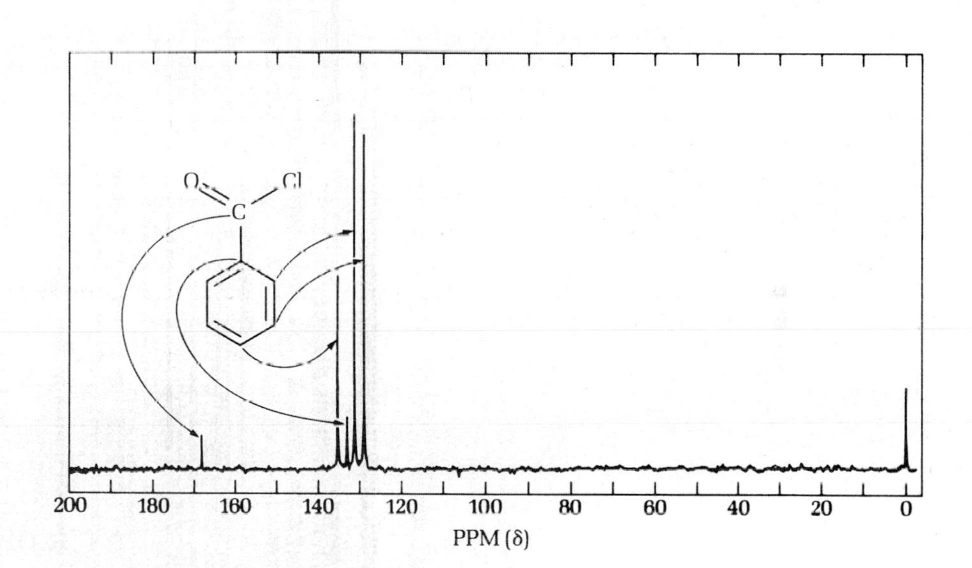

44 CHAPTER

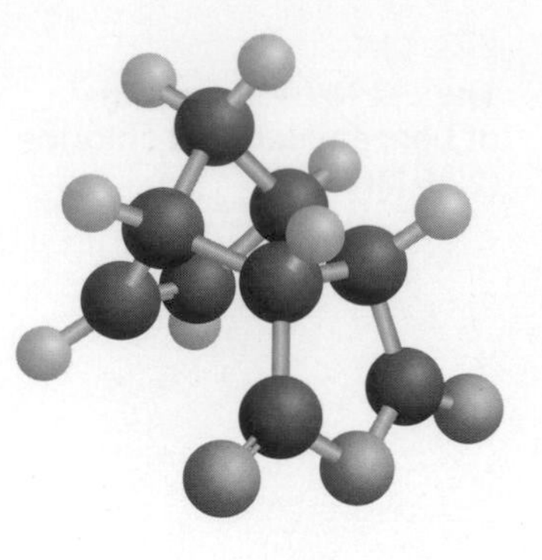

The Sandmeyer Reaction: 1-Bromo-4-chlorobenzene, 2-Iodobenzoic Acid, and 4-Chlorotoluene

PRELAB EXERCISE: Using the Sandmeyer reaction as one step, outline the steps necessary to prepare 4-bromotoluene, 4-iodotoluene, and 4-fluorotoluene from benzene.

The Sandmeyer reaction is a versatile means of replacing the amine group of a primary aromatic amine with a variety of different substituents:

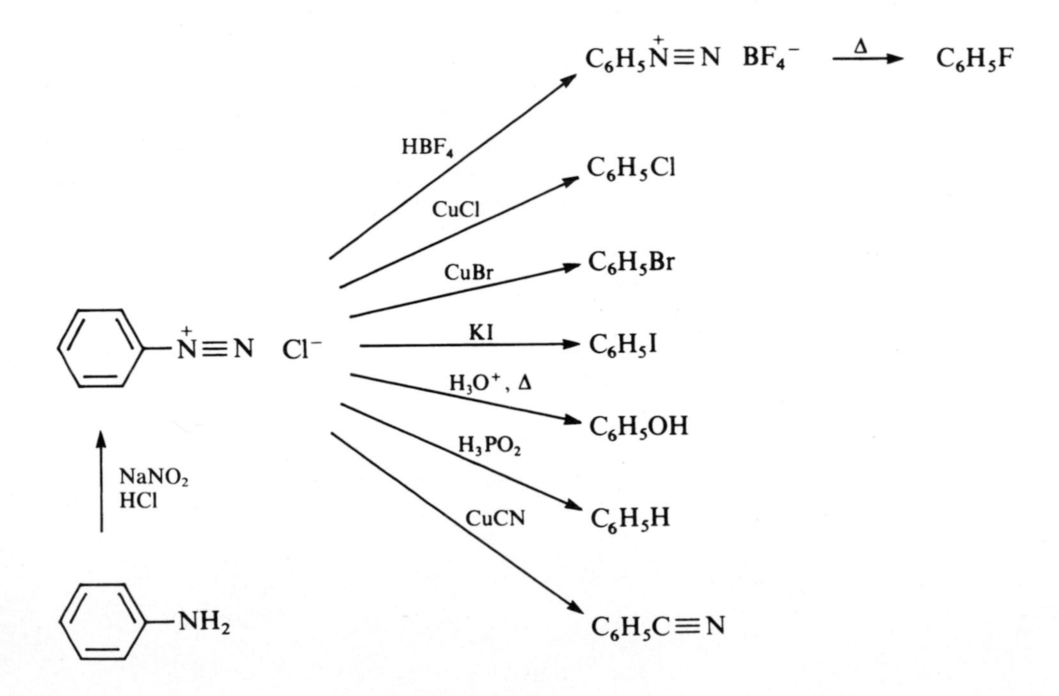

The benzene diazonium salt is formed by reacting nitrous acid with the amine in acid solution. Nitrous acid is not stable and must be prepared in situ; in strong acid, it dissociates to form nitroso ions, $:\overset{+}{N}{=}O$ which attack the nitrogen of the amine. The intermediate so formed loses a proton, rearranges, and finally loses water to form the resonance-stabilized diazonium ion.

$$\underset{\textbf{Sodium nitrite}}{NaNO_2} + HCl \rightleftharpoons \underset{\textbf{Nitrous acid}}{HO\ddot{N}O} + Na^+Cl^-$$

$$H_3O^+ + HO\ddot{N}O \rightleftharpoons H_2O + H_2\overset{+}{\ddot{O}}\ddot{N}O \rightleftharpoons 2\,H_2O + :\overset{+}{N}{=}O$$

$$\underset{\textbf{Primary aromatic amine}}{C_6H_5\ddot{N}H_2} + :\overset{+}{N}{=}O \longrightarrow C_6H_5\overset{+}{N}H_2{-}\ddot{N}{=}O \xrightarrow{-H^+} C_6H_5\ddot{N}H{-}\ddot{N}{=}O \longrightarrow$$

$$C_6H_5{-}\ddot{N}{=}\ddot{N}{-}OH \overset{H+}{\rightleftharpoons} C_6H_5{-}\ddot{N}{=}\ddot{N}{-}\overset{+}{O}H_2 \overset{-H_2O}{\rightleftharpoons} \underset{\textbf{Diazonium ion}}{[C_6H_5{-}\overset{+}{N}{\equiv}N \leftrightarrow C_6H_5{-}\ddot{N}{=}\overset{+}{\ddot{N}}]}$$

The benzene diazonium ion is reasonably stable in aqueous solution at 0°C; on warming it will form the phenol. A versatile functional group, it will undergo all the reactions depicted on the previous page as well as couple to aromatic rings activated with substituents such as amino and hydroxyl groups to form the huge class of azo dyes (*see* Chapter 46).

$$\underset{\textbf{Benzenediazonium chloride}}{C_6H_5{-}\overset{+}{N}{\equiv}N\;Cl^-} + \underset{\textbf{\textit{N,N}-Dimethylaniline}}{C_6H_5{-}N(CH_3)_2} \longrightarrow \underset{\textbf{A diazo compound}}{C_6H_5{-}N{=}N{-}C_6H_4{-}N(CH_3)_2}$$

Diazonium salts are not ordinarily isolated because the dry solid is explosive.

THREE SANDMEYER REACTIONS

Experiments 1, 2, and 3

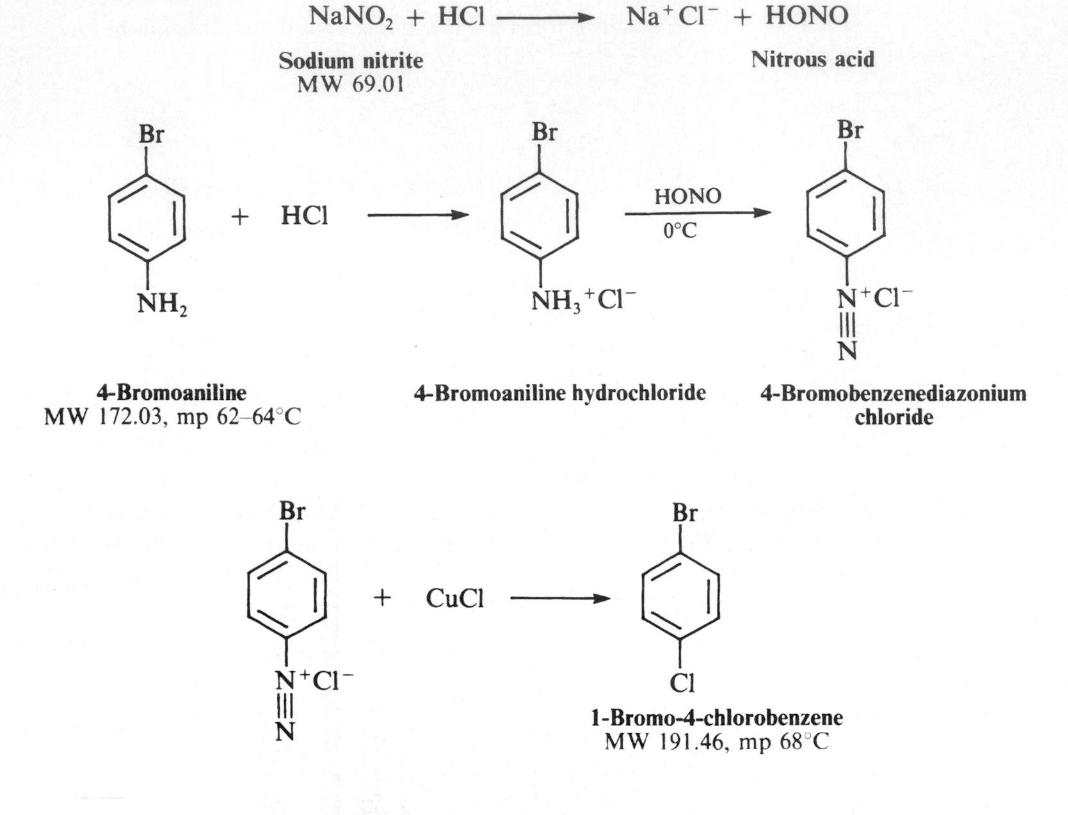

In the first microscale procedure (Experiments 1, 2, and 3), after the copper(I) chloride is prepared, 4-bromoaniline is dissolved in the required amount of hydrochloric acid, two more equivalents of acid are added, and the mixture is cooled in ice to produce a paste of the crystalline amine hydrochloride. When this salt is treated with one equivalent of sodium nitrite at 0–5°C, nitrous acid is generated and reacts to produce the diazonium salt. The excess hydrochloric acid (beyond the two equivalents required to form the amine hydrochloride and react with sodium nitrite) maintains sufficient acidity to prevent formation of the diazo-amino compound and rearrangement of the diazonium salt. The diazonium salt reacts with copper(I) chloride to give the easily sublimed 1-bromo-4-chlorobenzene.

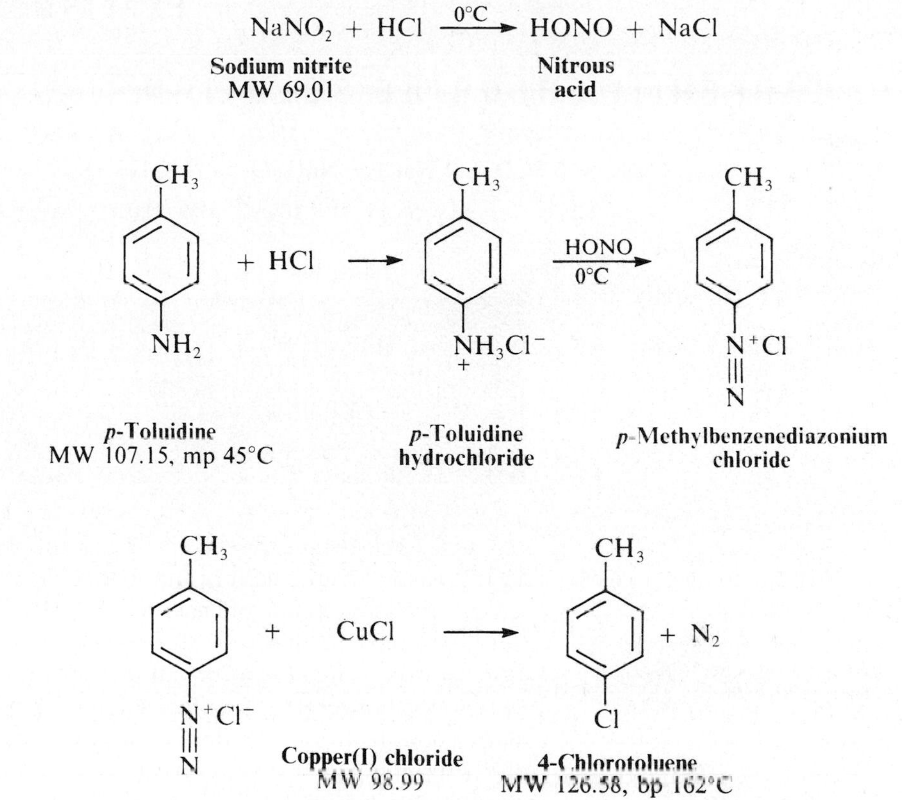

Experiments 6, 7, and 8

Experiments 4 and 5

CAUTION: *p*-Toluidine, like many aromatic amines, is highly toxic. *o*-Toluidine is not only highly toxic, but is also a suspected carcinogen.

The second procedure (Experiment 4 microscale, Experiment 5 macroscale) employs potassium iodide, which reacts with a diazonium salt to give an iodobenzoic acid.

The third procedure is a macroscale synthesis of 4-chlorotoluene (Experiments 6, 7, and 8). *p*-Toluidine is dissolved in the required amount of hydrochloric acid, two more equivalents of acid are added, and the mixture is cooled in ice to produce a paste of the crystalline amine hydrochloride. When this salt is treated with one equivalent of sodium nitrite at 0–5°C, nitrous acid is liberated and reacts to produce the diazonium salt. The excess hydrochloric acid, beyond the two equivalents required to form the amine hydrochloride and react with sodium nitrite, maintains sufficient acidity to prevent formation of the diazo-amino compound and rearrangement of the diazonium salt. The diazonium salt reacts with copper(I) chloride to give liquid 4-chlorotoluene.

Copper(I) chloride is made by reducing copper(II) sulfate with sodium sulfite (which is produced as required from the cheaper sodium bisulfite). The white solid is left covered with the reducing solution for protection against air oxidation until it is to be used and then dissolved in hydrochloric acid. When the diazonium salt solution is added, a complex forms and rapidly decomposes to give *p*-chlorotoluene and nitrogen. The mixture is very discolored, but steam distillation leaves behind most of the impurities and all salts, and yields material substantially pure except for the presence of a trace of yellow pigment, which can be eliminated by distilling the dried oil.

EXPERIMENTS

1. COPPER(I) CHLORIDE SOLUTION

$$2\ CuSO_4 \cdot 5\ H_2O + 4\ NaCl + NaHSO_3 + NaOH \rightarrow 2\ CuCl + 3\ Na_2SO_4 + 2\ HCl + 10\ H_2O$$

MW 249.71 MW 58.45 MW 104.97 MW 40.01 MW 99.02

IN THIS EXPERIMENT, an aqueous solution of copper sulfate, sodium chloride, sodium bisulfite, and sodium hydroxide react to give a precipitate of copper(I) chloride. Immediately before use, this white salt is dissolved in hydrochloric acid.

In a reaction tube, dissolve 0.30 g of copper(II) sulfate crystals ($CuSO_4 \cdot 5\ H_2O$) in 1.0 mL of water by boiling and then add 140 mg of sodium chloride, which may give a small precipitate of basic copper(II) chloride. Prepare a solution of sodium sulfite from 70 mg of sodium bisulfite and 0.4 mL of 3 *M* sodium hydroxide solution (or use 57 mg of sodium sulfite in 0.4 mL of water). Add this solution dropwise to the hot copper(II) sulfate solution. Rinse the tube that contained the sulfite with a couple drops of water and add it to the copper(I) chloride mixture. Shake the reaction mixture well; then allow it to cool in ice while preparing the diazonium salt (Experiment 2). When you are ready to use the copper(I) chloride (in Experiment 3), remove the water above the solid, wash the white solid once with water, remove the water, and dissolve the solid in 0.45 mL of concentrated hydrochloric acid. This solution is susceptible to air oxidation and should not stand for an appreciable time before use.

$NaHSO_3$, sodium bisulfite (sodium hydrogen sulfite), not $Na_2S_2O_4$, sodium hydrosulfite

Cleaning Up. Combine the filtrate and washings, neutralize with sodium carbonate, and flush down the drain.

2. DIAZOTIZATION OF 4-BROMOANILINE

IN THIS EXPERIMENT, an ice-cold acidic solution of an aromatic amine is treated with solid sodium nitrite to give a solution of the diazonium salt.

In a 10-mL Erlenmeyer flask, place 172 mg of 4-bromoaniline and 0.50 mL of 3 *M* hydrochloric acid. Warm the mixture on a sand bath to dissolve the amine and ensure transformation into hydrochloride. Cool the flask in ice (the hydrochloride will crystallize), and add an ice-cold solution of 70 mg of fresh sodium nitrite in 0.2 mL of water. Use a drop of water to complete the transfer of the nitrite solution. Mix the solid and the sodium sulfite thoroughly. The result should be a homogeneous yellow solid/liquid mixture. Do not allow this mixture to warm above 0°C.

3. SANDMEYER REACTION: SYNTHESIS OF 1-BROMO-4-CHLOROBENZENE

IN THIS EXPERIMENT, the cold copper(I) chloride solution is combined with the cold diazonium salt solution made from 4-bromoaniline in Experiment 2. Nitrogen gas bubbles off, and the diazonium ion is replaced with a chlorine atom to give the product, 1-bromo-4-chlorobenzene.

Photo: *Sublimation Apparatus*; Video: *Microscale Filtration on the Hirsch Funnel.*

Cool the copper(I) chloride solution (from Experiment 1) in ice, and add it to the diazonium chloride solution dropwise with a Pasteur pipette and with thorough mixing. Rinse the tube that contained the copper(I) chloride with a drop of water. Allow the reaction mixture to warm to room temperature and observe the reaction mixture closely. It bubbles as nitrogen gas is evolved. Heat the mixture on a steam or a sand bath to complete the reaction. Cool the mixture thoroughly in ice, and collect the dark solid on a Hirsch funnel. Squeeze the product between sheets of filter paper to dry it, and then sublime the bromochlorobenzene on a steam or sand bath. After initial sublimation, scrape the residue in the filter flask to sublime more product.

The extraction can be carried out in a microscale separatory funnel (*see* Fig. 7.7 on page 139).

Another way to isolate the product is to extract the product with three 0.5-mL portions of ether and wash the ether extract with 0.2 mL of 3 *M* sodium hydroxide solution, which will remove any 4-bromophenol, and then with 0.5 mL of saturated sodium chloride solution. Dry the ether solution over anhydrous calcium chloride pellets. Add sufficient drying agent so that it no longer clumps together. After 5–10 minutes, remove the ether with a Pasteur pipette, wash the drying agent with ether, and evaporate the ether solutions in a small, tared filter flask. Use care in this evaporation because the product is volatile. Determine the weight of the crude product; then purify it by sublimation at atmospheric pressure.

Photo: *Filtration Using a Pasteur Pipette*; Videos: *Extraction with Ether, Filtration of Crystals Using the Pasteur Pipette*

The 15-mL centrifuge tube is filled with ice water while the filter flask is heated on a sand or steam bath (refer to the sublimation apparatus in Fig. 6.21 on page 126). Close the sidearm with a rubber pipette bulb. Roll the flask in the sand and/or use a heat gun to drive product from the sides of the flask to the centrifuge tube. Replace the ice water in the centrifuge tube with room-temperature water before removing it so that moisture will not condense on the tube. Scrape the product onto a piece of weighing paper, determine the weight and the melting point, and analyze it for purity by thin-layer chromatography (TLC) and infrared (IR) spectroscopy in a chloroform solution.

Photo: *Sublimation Apparatus*

Cleaning Up. Combine all aqueous filtrates and washings, neutralize with sodium carbonate, and flush down the drain. Allow the ether to evaporate from the drying agent in the hood before the drying agent is placed in the nonhazardous solid waste container. If local regulations do not allow for the evaporation of solvents in a hood, the wet solid should be disposed in a special waste container.

4. SANDMEYER REACTION: SYNTHESIS OF 2-IODOBENZOIC ACID

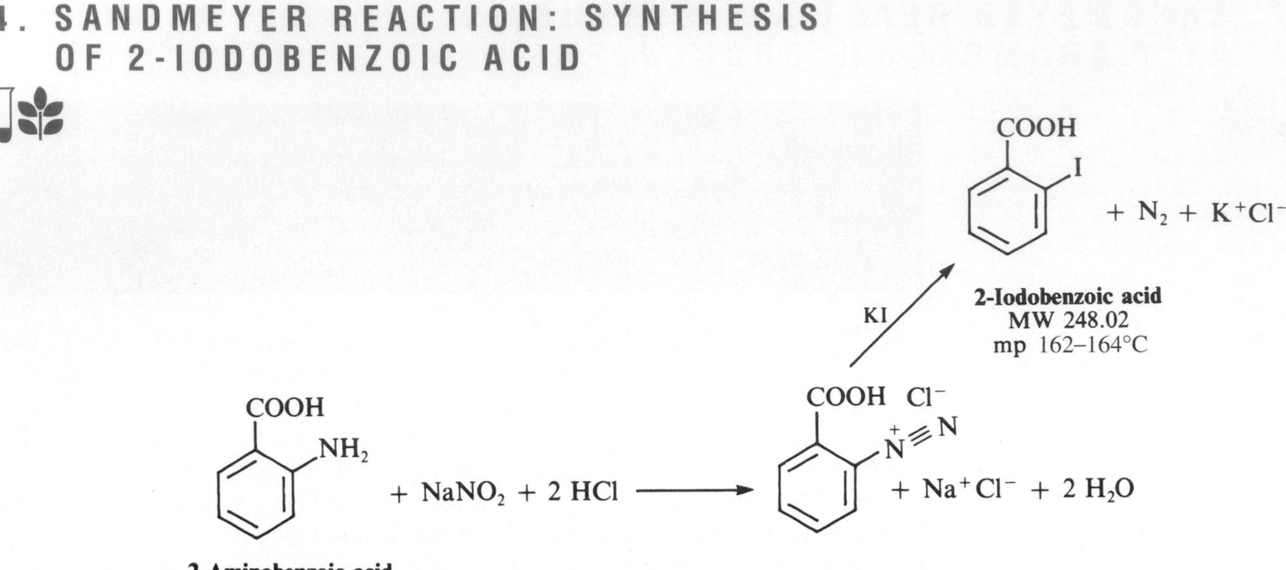

CAUTION: Do not measure hydrochloric acid in a syringe that has a metal needle.

To a reaction tube containing 137 mg of 2-aminobenzoic acid (anthranilic acid), add 1 mL of water and 0.25 mL of concentrated hydrochloric acid. Heat to dissolve the amino acid and form the hydrochloride; then cool the tube in ice. Cap the tube with a septum; then, with the tube in an ice bath, add a solution of 75 mg of sodium nitrite dissolved in 0.3 mL of water using a syringe equipped with a needle. This addition should be made dropwise with thorough agitation or magnetic stirring of the reaction. Rinse the syringe with 0.1 mL of water, which is also injected into the reaction mixture. After 5 minutes, a solution of 0.17 g of potassium iodide in 0.25 mL of water is added when a brown complex partially separates.

An empty distilling column is added to the top of the reaction tube to catch any foam. The mixture is allowed to stand without cooling for 5 minutes, and then cautiously warmed to 40°C. At this point, a vigorous reaction begins, with nitrogen gas evolution, foaming, and separation of a tan-colored solid. After reacting for 10 minutes, the mixture is heated in a beaker of boiling water for 10 minutes and then cooled in ice. A few milligrams of sodium sulfite are added to destroy any iodine present, and the granular tan-colored product is collected and washed with water.

Video: *Microscale Crystallization*

The still-moist product is dissolved in 0.7 mL of ethanol; 0.35 mL of water is added; and the hot, brown solution is treated with enough granular Norit to remove most of the color. The solution is transferred to another reaction tube to remove the decolorizing charcoal, diluted at the boiling point with 0.35–0.40 mL of water, and allowed to stand. 2-Iodobenzoic acid separates in large, slightly yellow needles. The yield should be about 150 mg, with a melting point near 164°C.

Cleaning Up. The reaction mixture filtrate and mother liquor from the crystallization are combined, neutralized with sodium carbonate, and flushed down the drain. Norit is placed in the nonhazardous solid waste container.

5. SANDMEYER REACTION: SYNTHESIS OF 2-IODOBENZOIC ACID

COOH, NH_2 $\xrightarrow[\text{HCl, 0°C}]{NaNO_2}$ COOH, $\overset{+}{N_2}Cl^-$

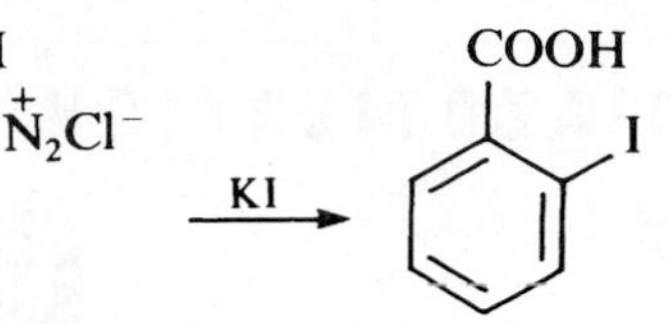

2-Aminobenzoic acid
(Anthranilic acid)
MW 137.14, mp 144–148°C

2-Iodobenzoic acid
MW 248.02, mp 162–164°C

A 100-mL round-bottomed flask containing 3.4 g of anthranilic acid, 25 mL of water, and 6 mL of concentrated hydrochloric acid is heated until the solid is dissolved. The mixture is then cooled in ice while bubbling in nitrogen to displace the air. When the temperature reaches 0–5°C, a solution of 1.8 g of sodium nitrite in 6 mL of water is added slowly. After 5 minutes, a solution of 4.25 g of potassium iodide in 6 mL of water is added when a brown complex partially separates. This mixture is allowed to stand without cooling for 5 minutes (under nitrogen) and then warmed to 40°C, at which point a vigorous reaction ensues (gas evolution and separation of a tan-colored solid). After reacting for 10 minutes, the mixture is heated on a steam or sand bath for 10 minutes and then cooled in ice. A pinch of sodium bisulfite is added to destroy any iodine present, and the granular tan-colored product is collected and washed with water. The still-moist product is dissolved in 18 mL of 95% ethanol; then 9 mL of hot water is added, and the brown solution is treated with decolorizing charcoal, filtered, diluted at the boiling point with an additional 9–10 mL of water, and allowed to stand. 2-Iodobenzoic acid separates in large, slightly yellow needles (mp 164°C); the yield is approximately 4.3 g (72%).

Video: *Macroscale Crystallization*

Cleaning Up. The reaction mixture, filtrate, and mother liquor from the crystallization are combined, neutralized with sodium carbonate, and flushed down the drain with a large excess of water. Decolorizing charcoal is placed in the non-hazardous solid waste container.

6. COPPER(I) CHLORIDE SOLUTION

$$2\ CuSO_4 \cdot 5\ H_2O + 4\ NaCl + NaHSO_3 + NaOH \rightarrow 2\ CuCl + 3\ Na_2SO_4 + 2\ HCl + 10\ H_2O$$

MW 249.71 MW 58.45 MW 104.97 MW 40.01 MW 99.02

In a 250-mL round-bottomed flask (to be used later for steam distillation), dissolve 15 g of copper(II) sulfate crystals ($CuSO_4 \cdot 5\ H_2O$) in 50 mL of water by boiling; then add 7 g of sodium chloride, which may give a small precipitate of basic copper(II) chloride. Prepare a solution of sodium sulfite from 3.5 g of sodium bisulfite, 2.25 g of sodium hydroxide, and 25 mL of water; add this, not too rapidly, to the hot copper(II) sulfate solution (rinse flask and neck). Shake well, put the flask in a pan of cold water in a slanting position favorable for decantation, and let the mixture stand to cool and settle during the diazotization. When you are ready to use the

$NaHSO_3$, sodium bisulfite (sodium hydrogen sulfite), not $Na_2S_2O_4$, sodium hydrosulfite

One may stop here.

copper(I) chloride for Experiment 8, decant the supernatant liquid, wash the white solid once with water by decantation, and dissolve the solid in 23 mL of concentrated hydrochloric acid. This solution is susceptible to air oxidation, and should not stand for an appreciable time before use.

7. DIAZOTIZATION OF *p*-TOLUIDINE

IN THIS EXPERIMENT, a solid amine is dissolved in hydrochloric acid. The ice-cold solution is treated with solid sodium nitrite to diazotize the amine.

CAUTION: *p*-Toluidine is toxic. Handle with care.

Put 5.5 g of *p*-toluidine and 7.5 mL of water in a 50-mL Erlenmeyer flask. Measure 12.5 mL of concentrated hydrochloric acid, and add 5 mL of it to the flask. Heat over a hot plate and swirl to dissolve the amine to ensure that it is all converted into the hydrochloride. Add the remaining acid, cool thoroughly in an ice bath, and let the flask stand in the bath while preparing a solution of 3.5 g of sodium nitrite in 10 mL of water. To maintain a temperature of 0–5°C during diazotization, add a few pieces of ice to the amine hydrochloride suspension, and add more later as the first ones melt. Pour in the nitrite solution in portions during 5 minutes with swirling in the ice bath. The solid should dissolve, and a clear solution of the diazonium salt will form. After 3–4 minutes, test for excess nitrous acid: Dip a stirring rod in the solution, touch off the drop on the wall of the flask to remove it, put the rod in a small test tube, and add a few drops of water. Then insert a strip of starch-iodide paper – an instantaneous deep-blue color due to a starch-iodine complex indicates the desirable presence of a slight excess of nitrous acid. (The sample tested is diluted with water because strong hydrochloric acid alone produces the same color on starch-iodide paper, after a slight induction period.) Leave the solution in an ice bath.

8. SANDMEYER REACTION: SYNTHESIS OF 4-CHLOROTOLUENE

IN THIS EXPERIMENT, the ice-cold diazonium salt solution from Experiment 7 is poured into the copper(I) chloride solution prepared in Experiment 6. Nitrogen gas is evolved, and the product, 4-chlorotoluene, separates as an oil. It is isolated by steam distillation—the distillate is extracted with ether, and the ether is washed, dried, and evaporated to leave crude product that is purified by simple distillation.

Complete the preparation of copper(I) chloride solution from Experiment 6, cool it in an ice bath, pour in the solution of diazonium chloride through a long-stemmed funnel, and rinse the flask. Swirl occasionally at room temperature for 10 minutes. You should observe initial separation of a complex of the two components followed by its decomposition, with the liberation of nitrogen and the separation of an oil. Arrange for steam distillation (see Fig. 6.4 on page 108) or generate steam in situ simply by boiling the flask contents with a Thermowell using an apparatus for simple distillation (see Fig. 5.10 on page 100; add more water during the distillation). Do not start the distillation until bubbling in the mixture has practically ceased and an oily layer has separated. Then steam distill and note that *p*-chlorotoluene, although lighter than the solution of inorganic salts in which it was produced, is heavier (den. 1.07) than water. Extract the distillate with a little ether, wash the extract with 3 *M* sodium hydroxide solution to remove any *p*-cresol ($CH_3C_6H_4OH$) present, and then wash with a saturated sodium chloride solution. Dry the ether solution over about 2.5 g of anhydrous calcium chloride pellets and filter or decant it into a tared flask, evaporate the ether, and determine the yield and percent yield of product (your crude yield should be about 4.5 g). Pure *p*-chlorotoluene, the IR and nuclear magnetic resonance (NMR) spectra of which are shown in Figures 44.1 through 44.3, is obtained after simple distillation of this crude product. Analyze your crude and distilled product by TLC (using a 3:1 mixture of hexane and dichloromethane as the eluent), IR spectroscopy, and gas chromatography using a Carbowax column. Is it pure?

CAUTION: Use an aspirator tube.

Cleaning Up. Combine the pot residue from the steam distillation with the aqueous washings, neutralize with sodium carbonate, dilute with water, and flush down the drain. Allow the ether to evaporate from the drying agent in the hood; then place the drying agent in the nonhazardous solid waste container.

Dispose of copper salts and solutions in the waste container provided.

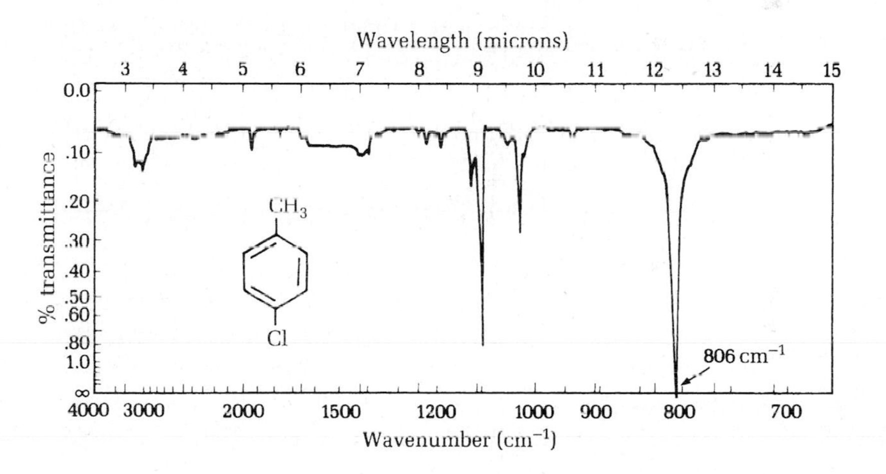

FIG. 44.1
The IR spectrum of *p*-chlorotoluene in carbon disulfide.

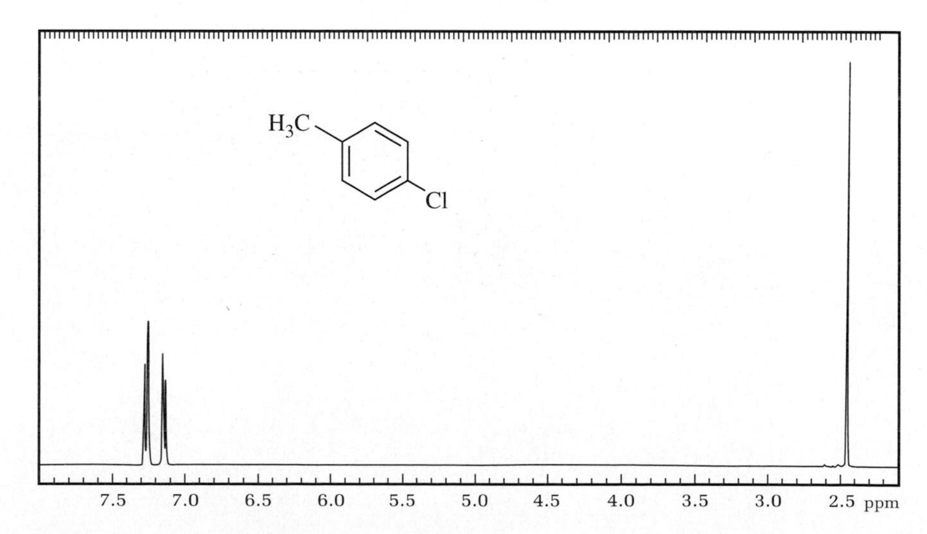

FIG. 44.2
The ^{1}H NMR spectrum of *p*-chlorotoluene (90 MHz).

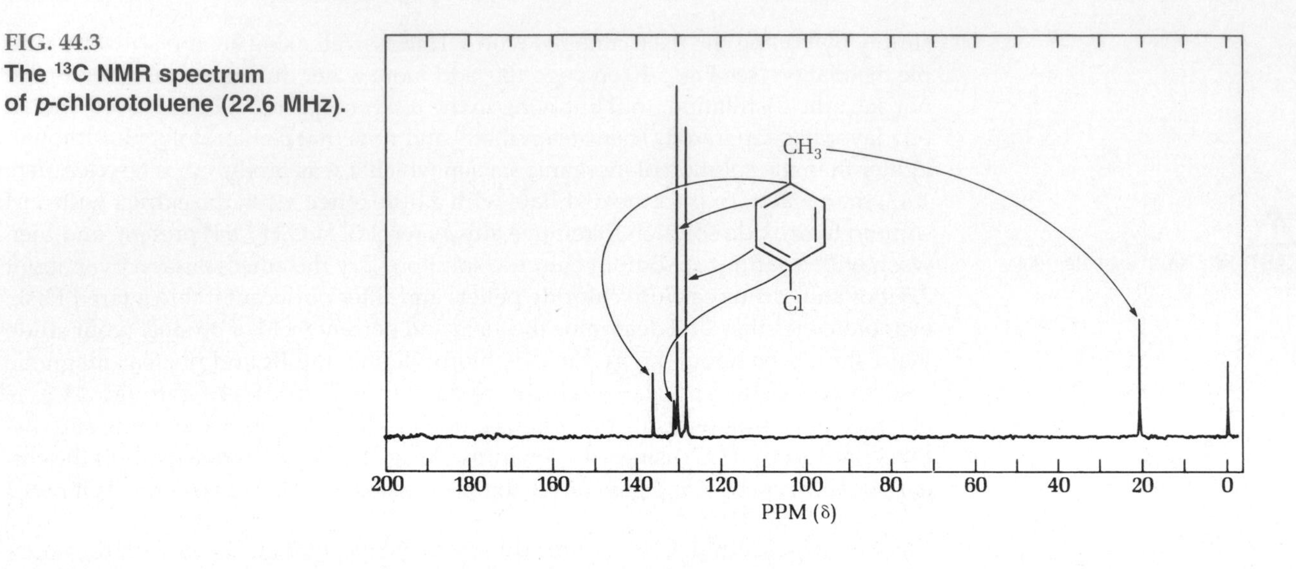

FIG. 44.3
The ^{13}C NMR spectrum of *p*-chlorotoluene (22.6 MHz).

QUESTIONS

1. Nitric acid is generated by the action of sulfuric acid on sodium nitrate. Nitrous acid is prepared by the action of hydrochloric acid on sodium nitrite. Why is nitrous acid prepared in situ rather than obtained from the reagent shelf?
2. Which byproduct would be obtained in high yield if the diazotization of *p*-toluidine were carried out at 30°C instead of 0–5°C?
3. How would 4-bromoaniline be prepared from benzene?

CHAPTER 45

Synthesis and Bioassay of Sulfanilamide and Derivatives

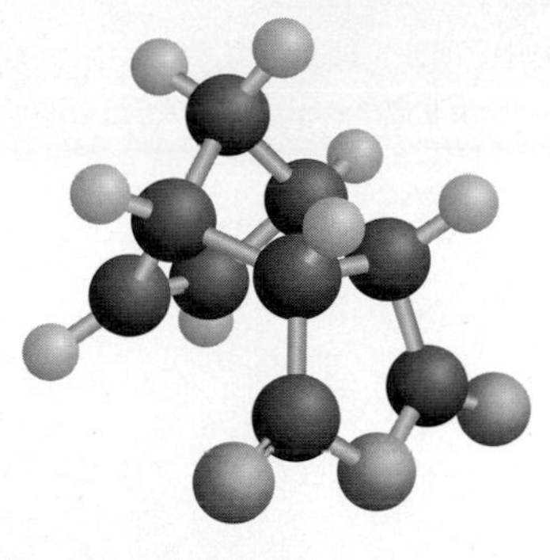

PRELAB EXERCISE: Write balanced equations for each step in the reaction sequence of the synthesis of sulfanilamide. Prepare detailed flow sheets for the three reactions. Look up the solubilities of all reagents and indicate—in separation steps—in which layer you would expect to find the desired product. Calculate the theoretical volume of ammonium hydroxide solution (usually sold as a 28% by weight aqueous solution) needed to react with 5 g of *p*-acetaminobenzenesulfonyl chloride to form sulfanilamide.

Paul Ehrlich, the father of immunology and chemotherapy, discovered Salvarsan, an arsenical "magic bullet" (a favorite phrase of his) used to treat syphilis. He hypothesized at the beginning of the 20th century that it might be possible to find a dye that would selectively stain (or dye) a bacterial cell and thus destroy it. In 1932, I.G. Farbenindustrie patented Prontosil, a new azo dye that they put through routine testing for chemotherapeutic activity when it was noted that it had a particular affinity for protein fibers like silk.

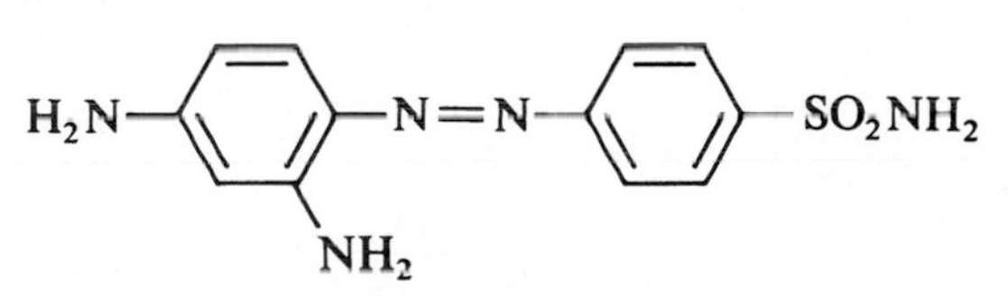

Prontosil

Prontosil was found to be effective against streptococcal infections in mice, but somewhat surprisingly, it was ineffectual in vitro (in an artificial environment outside of the living animal). A number of other dyes were tested, but only those having the group

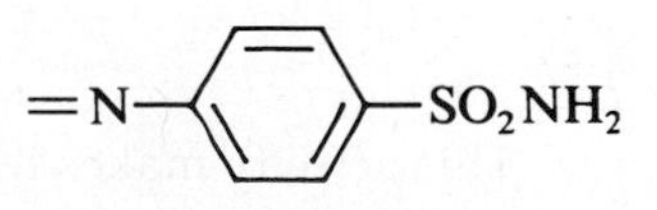

were effective. French workers hypothesized that the antibacterial activity had nothing to do with the identity of the compounds as dyes, but rather with the reduction of the dyes in the body to *p*-aminobenzenesulfonamide, known commonly as sulfanilamide.

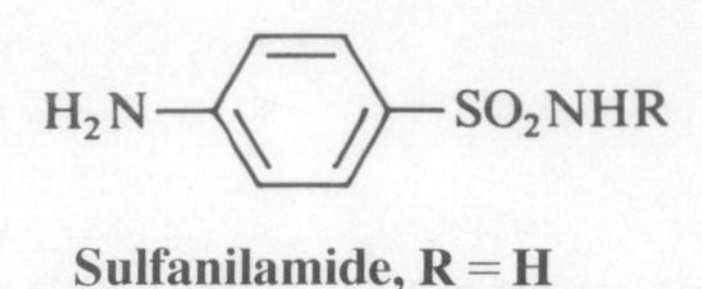

Sulfanilamide, R = H

On the basis of this hypothesis, sulfanilamide was tested and found to be the active substance.

Because sulfanilamide had been synthesized in 1908, its manufacture was not protected by patents, so the new drug and thousands of its derivatives were rapidly synthesized and tested. When the R group in sulfanilamide is replaced with a heterocyclic ring system—for example, pyridine, thiazole, diazine, merazine, and so on—the sulfa drug so produced is often faster acting or less toxic than sulfanilamide. Although they have been supplanted for the most part by antibiotics of microbial origin, these drugs still find wide application in chemotherapy.

Unlike that of most drugs, the mode of action of the sulfa drugs is now completely understood. Bacteria must synthesize folic acid for growth. Higher animals, like humans, do not synthesize folic acid and hence must acquire it in their food. Sulfanilamide inhibits the formation of folic acid, stopping the growth of bacteria. This process is highly selective because the synthesis of folic acid does not occur in humans, so only bacteria are affected.

A closer look at these events reveals that bacteria synthesize folic acid using several enzymes, including one called dihydropteroate synthetase, which catalyzes the attachment of *p*-aminobenzoic acid to a pteridine ring system. When sulfanilamide is present, it competes with the *p*-aminobenzoic acid (note the structural similarity) for the active site on the enzyme.

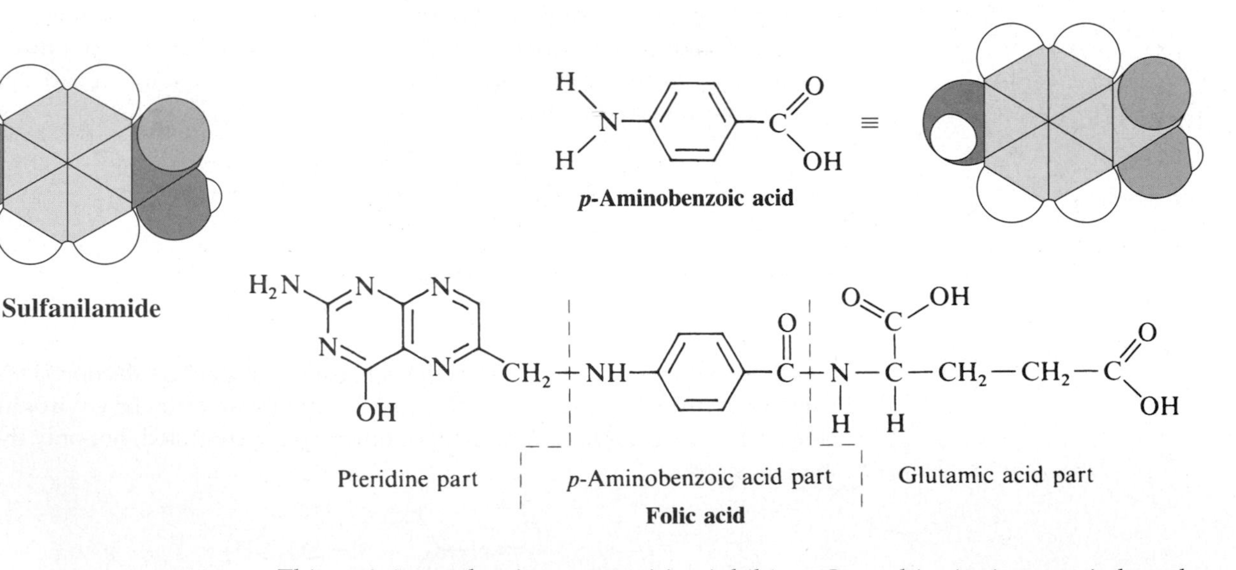

This activity makes it a competitive inhibitor. Once this site is occupied on the enzyme by sulfanilamide, folic acid synthesis ceases and bacterial growth stops. Folic acid can also be synthesized in the laboratory.[1]

[1]Plante LT, Williamson KL, Pastore EJ. In: *Methods in Enzymology*; McCormick DB, Wright LD, Eds; Academic Press: New York, 1990; 66:533.

PART 1: THE SYNTHESIS AND BIOASSAY OF SULFANILAMIDE

Part 1 of this chapter represents a multistep synthesis of sulfanilamide, starting with nitrobenzene. Nitrobenzene is reduced with tin and hydrochloric acid to give the anilinium ion, which is converted to aniline with base:

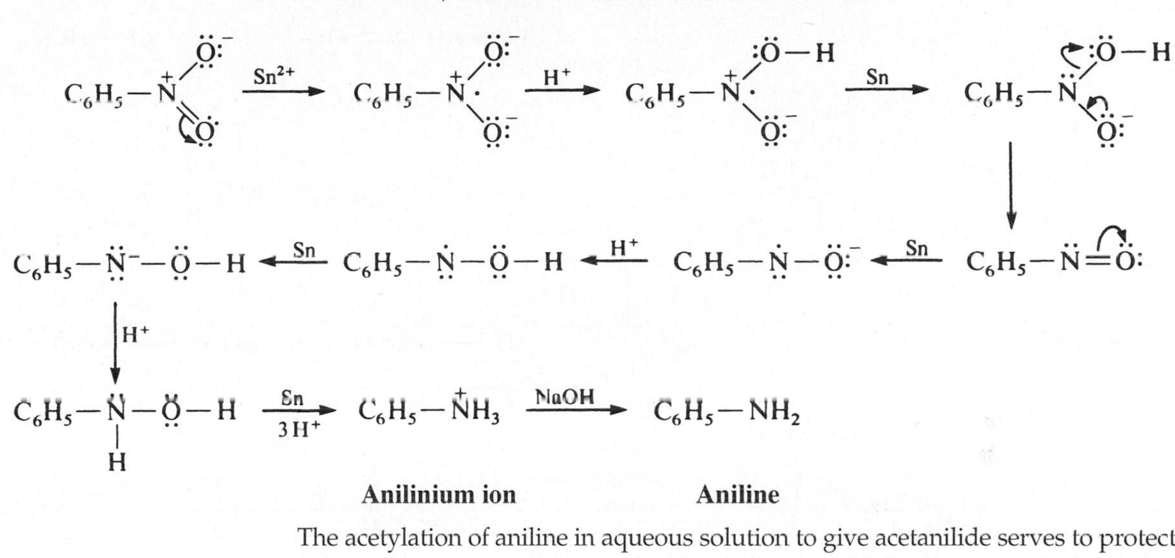

The acetylation of aniline in aqueous solution to give acetanilide serves to protect the amine group from reaction with chlorosulfonic acid. Furthermore, if the amine group were protonated, it would be a *meta* director to substitution. The resulting *meta*-substituted product does not have antibacterial activity. The generation of acetanilide is not difficult, with the acetylation taking place readily in aqueous solution. Aniline reacts with acid to give the water-soluble anilinium ion. The anilinium ion reacts with the acetate ion to set up an equilibrium that liberates a small quantity of aniline:

$$\underset{\textbf{Aniline}}{C_6H_5-\ddot{N}H_2} + H^+ \longrightarrow \underset{\textbf{Anilinium ion}}{C_6H_5-\overset{+}{N}H_3}$$

$$C_6H_5-\overset{+}{N}H_3 + CH_3-C(=\ddot{O}{:})-\ddot{O}{:}^- \rightleftharpoons C_6H_5-\ddot{N}H_2 + CH_3-C(=\ddot{O}{:})-\ddot{O}-H$$

This aniline reacts with acetic anhydride to give water-insoluble acetanilide:

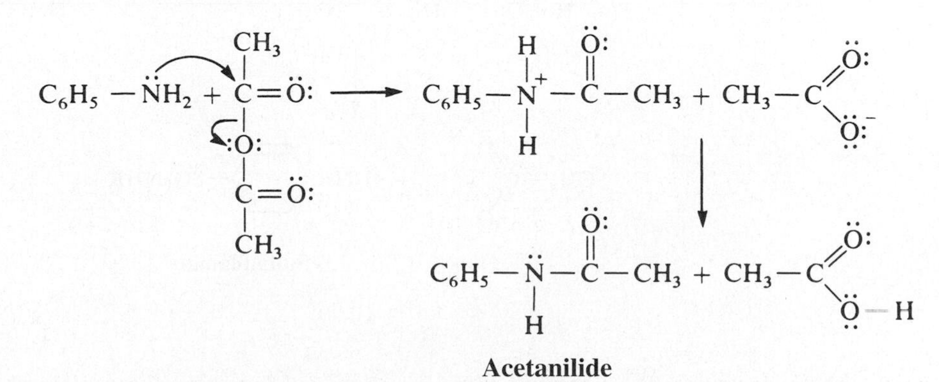

This upsets the equilibrium, releasing more aniline, which can then react with acetic anhydride to give more acetanilide.

Acetanilide reacts with chlorosulfonic acid in an electrophilic aromatic substitution:

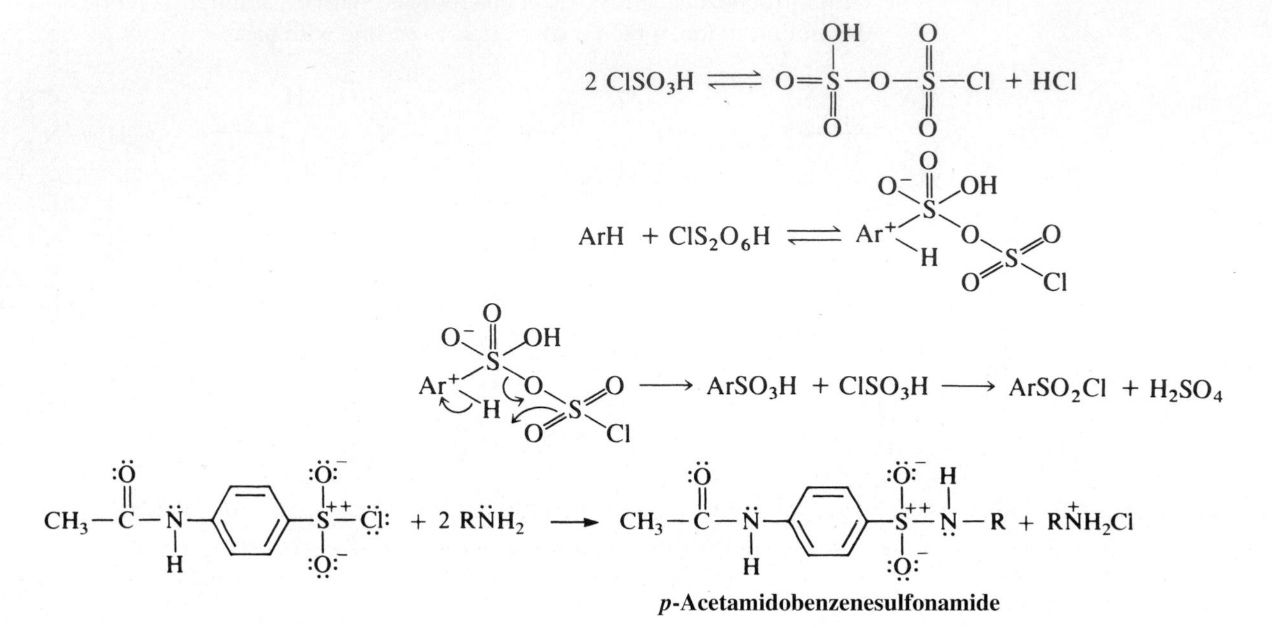

The protecting amide group is removed from the *p*-acetamidobenzenesulfonamide by acid hydrolysis. The amide group is more easily hydrolyzed than the sulfonamide group:

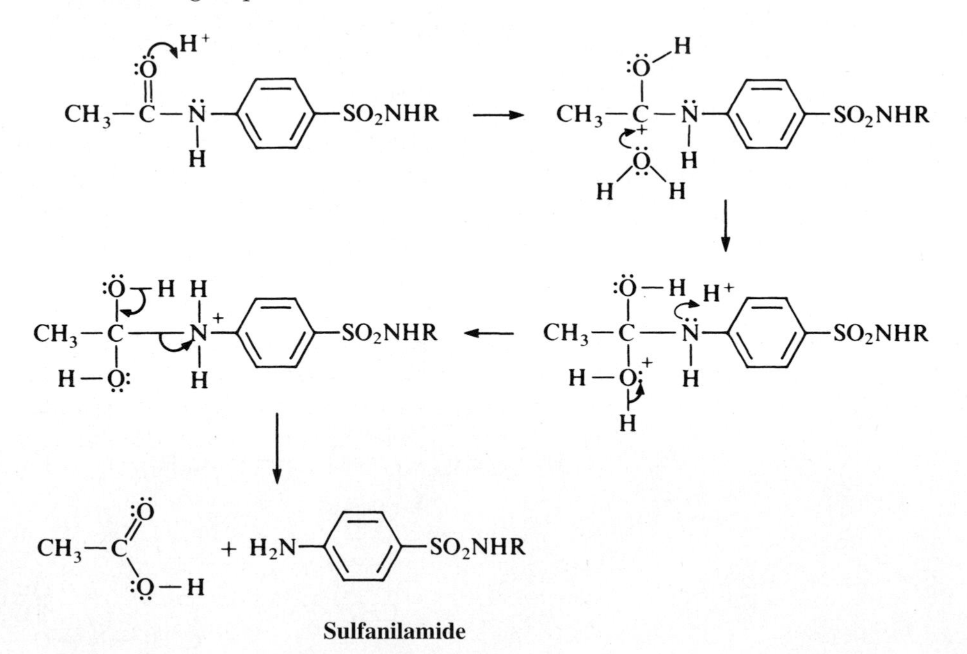

THE SYNTHETIC SEQUENCE OF SULFANILAMIDE

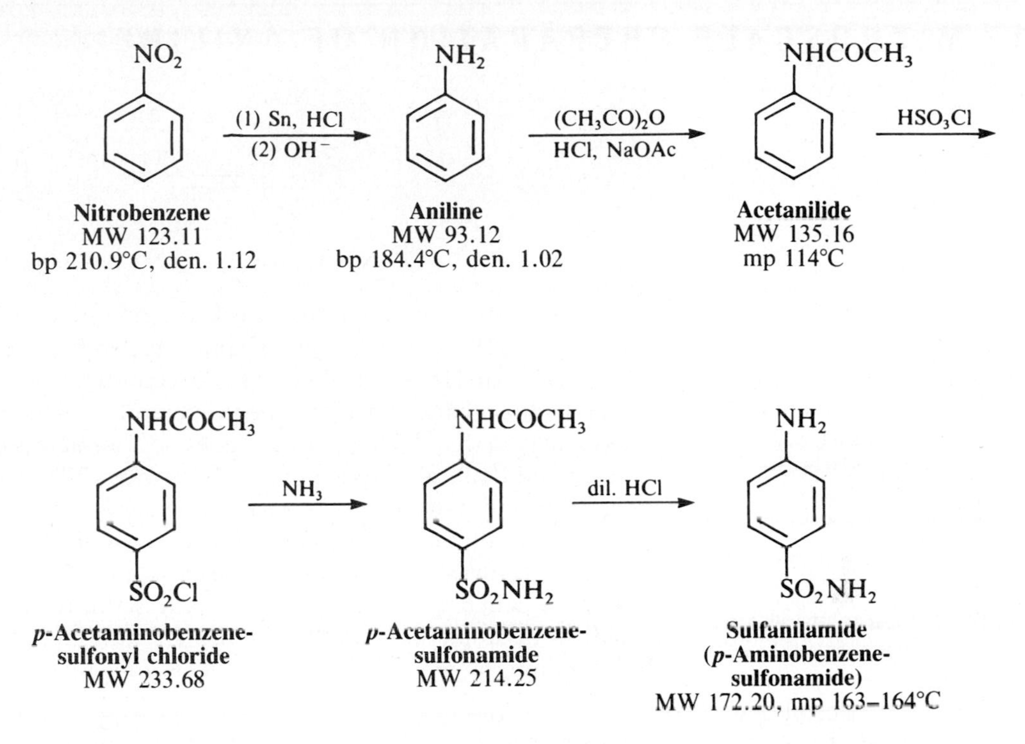

Nitrobenzene is reduced to aniline by tin and hydrochloric acid. A double salt with tin having the formula $(C_6H_5NH_3)_2SnCl_4$ separates partially during the reaction, and at the end it is decomposed by adding excess alkali, which converts the tin into water soluble stannite or stannate (Na_2SnO_2 or Na_2SnO_3) and also liberates aniline.

The aniline is separated from inorganic salts, and the insoluble impurities derived from the tin by steam distillation (*see* Chapter 6). The aniline layer from the steam distillate is then dried, distilled, and acetylated with acetic anhydride in aqueous solution to give acetanilide. Treatment of the acetanilide with excess chlorosulfonic acid affects substitution of the chlorosulfonyl group, and affords the *para*-substituted product, *p*-acetaminobenzenesulfonyl chloride. The amino group must be protected by acetylation to permit the acid chloride group to be added to the ring. If the amine group were protonated, substitution would occur at the *meta* position. The amide nitrogen of acetanilide is not basic enough to react with chlorosulfonic acid. The next step in the synthesis is ammonolysis of the sulfonyl chloride to give *p*-acetaminobenzenesulfonamide, and the terminal step is removal of the protective acetyl group to give sulfanilamide, which can be tested for antibacterial activity.

Use the total product obtained at each step as the starting material for the next step, and adjust the amounts of reagents accordingly. Keep an accurate record of your working time. Aim for a high overall yield of pure final product in the shortest possible time. Study the procedures carefully before coming to the laboratory so that your work will be efficient. A combination of consecutive steps that avoids a needless isolation saves time and increases the yield.

Experiments

1. Microscale Preparation of Aniline

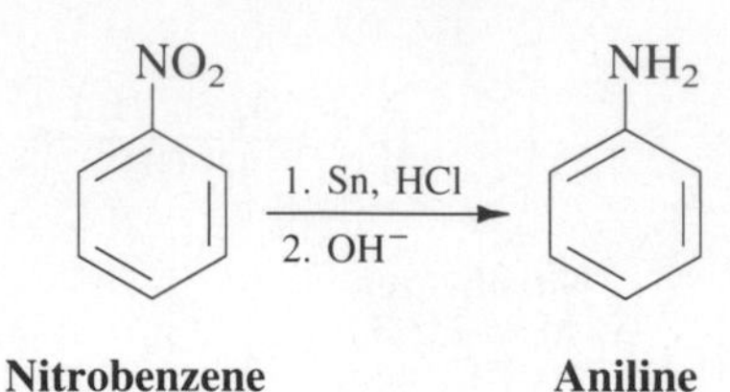

CAUTION: Measure nitrobenzene (toxic) in the hood. Do not breathe the vapors. Use a nonmercury thermometer and wear gloves.

Reaction time: 0.5 hours

CAUTION: Handle aniline with care. It may be a carcinogen.

In a 5-mL round-bottomed, long-necked flask, place 1.25 g of granulated tin, 600 mg of nitrobenzene, 2.75 mL of concentrated hydrochloric acid, and a magnetic stirring bar. Have an ice bath ready to control the reaction, insert a thermometer, and let the mixture react until the temperature reaches 60°C; then cool it briefly, if necessary, so that the temperature does not rise above 62°C. Swirl the reaction mixture and maintain the temperature between 55°C and 60°C for 15 minutes. Remove the thermometer, rinse it with water, and fit the flask with an empty fractionating column that will function as an air condenser. Additional cooling can, if necessary, be furnished by a damp pipe cleaner (Fig. 45.1). Heat the reaction mixture on a steam bath or a boiling water bath with frequent swirling or magnetic stirring until droplets of nitrobenzene are absent from the condenser, and the color due to an intermediate reduction product is gone (about 15 minutes). During this

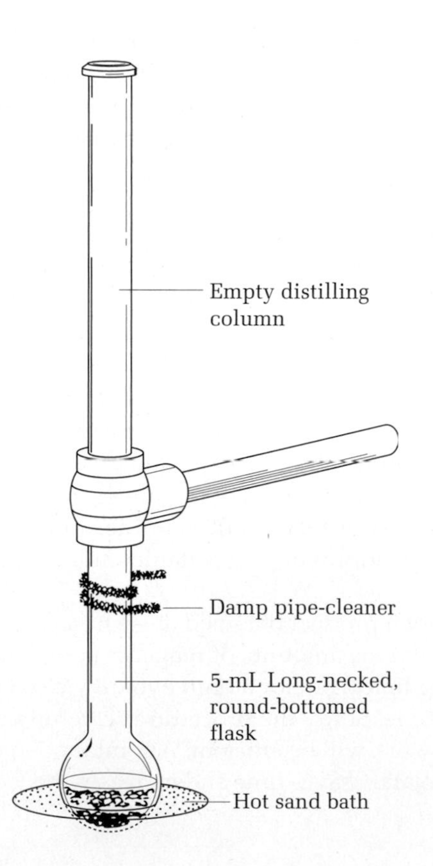

FIG. 45.1
A microscale apparatus for refluxing a reaction mixture. Use a black Viton connector.

CAUTION: Handle the hot solution of sodium hydroxide with care; it is very corrosive.

Videos: *Extraction with Ether, Steam Distillation Apparatus*

The aniline can be isolated (*see* Experiment 2) or converted directly to acetanilide (*see* Experiment 3).

period, dissolve 2 g of sodium hydroxide in 5 mL of water and cool the solution to room temperature.

At the end of the reduction period, transfer the acid solution to a 15-mL centrifuge tube, cool the mixture in ice, and add the sodium hydroxide solution dropwise with stirring and ice cooling. The alkali neutralizes the aniline hydrochloride, releasing aniline. Gently extract the mixture with three 1.5-mL portions of ether; place the ether extracts in the 5-mL short-necked, round-bottomed flask, which need not be clean or dry. If intractable emulsions form, centrifuge the tube for 1 minute to cause the layers to separate. Transfer any emulsion layer to the flask along with the ether layer.

Construct the apparatus shown in Figure 45.2, add a boiling chip, and remove the ether (bp 55°C) by distillation. To the dirty residue of aniline and water, add 2 mL of water with a syringe through the addition port. Heat the contents of the flask to boiling. A mixture of water and aniline will steam distill (*see* Chapter 6), and the distillate will be cloudy as the two layers, aniline and water, separate. As the distillation proceeds, add water with a syringe through the addition port to keep the volume in the distilling flask constant. Because aniline is fairly soluble in

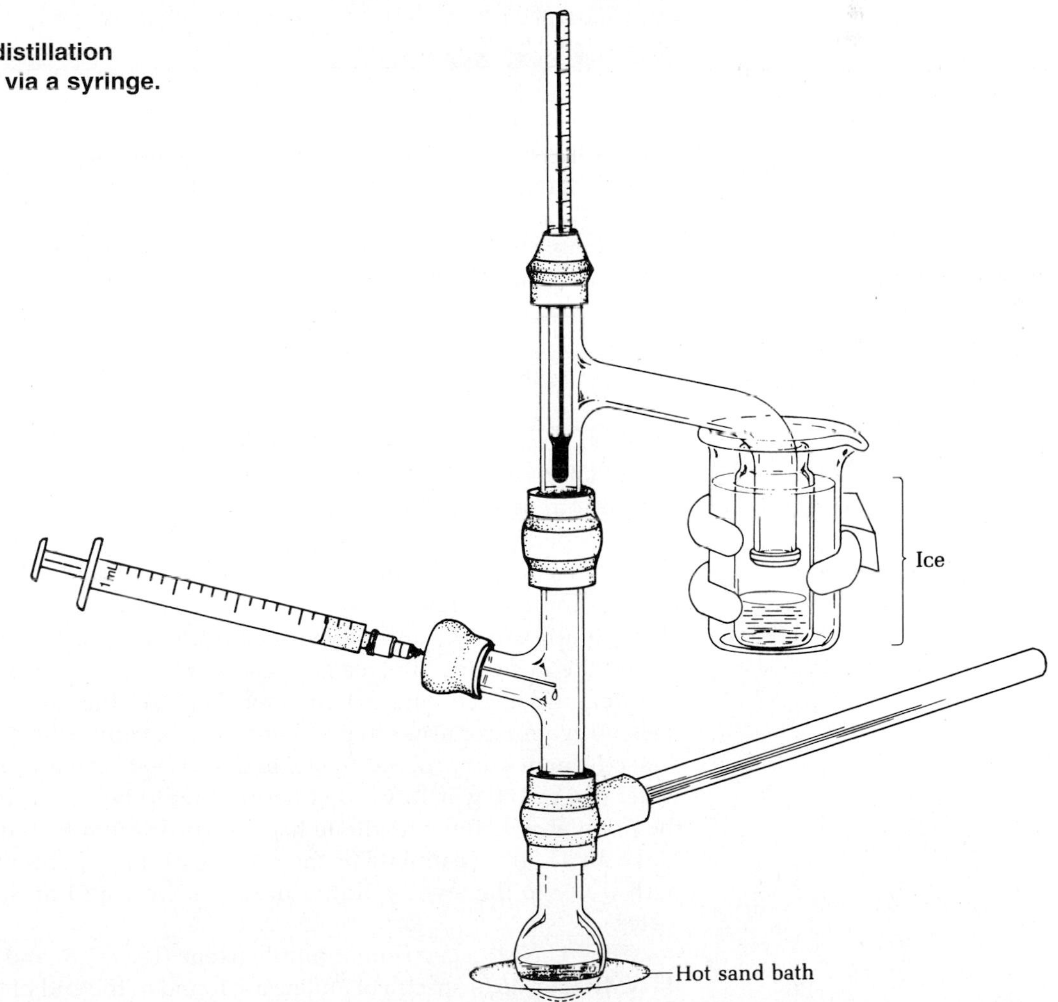

FIG. 45.2
A small-scale steam distillation apparatus. Add water via a syringe.

water (3.6 g/100 g of water at 18°C), distillation should be continued somewhat beyond the point at which the distillate has lost its original turbidity (2.5–3 mL more). Make an accurate estimate of the volume of the distillate, for example, by pouring it into a 10-mL graduated cylinder.

The ^{13}C NMR (nuclear magnetic resonance) spectrum of nitrobenzene (Fig. 45.6) and the infrared (IR; Fig. 45.7) and ^{1}H NMR spectra of aniline (Fig. 45.8) are found at the end of the chapter.

Cleaning Up. Filter the pot residue from the steam distillation through a layer of Celite filter aid on a Büchner funnel. Discard the solid in the nonhazardous solid waste container. The filtrate, after neutralizing with hydrochloric acid, can be flushed down the drain.

2. MACROSCALE PREPARATION OF ANILINE

IN THIS EXPERIMENT, nitrobenzene is reduced with tin and hydrochloric acid. The reaction mixture is neutralized with strong base, and the aniline is isolated by steam distillation.

CAUTION: Measure nitrobenzene and dichloromethane (both are toxic) in the hood. Do not breathe the vapors.

CAUTION: Handle aniline with care. It may be a carcinogen.

Reaction time: 0.5 hours

CAUTION: Handle the hot solution of sodium hydroxide with care; it is very corrosive.

Suitable point of interruption

The reduction of the nitrobenzene is carried out in a 500-mL round-bottomed flask suitable for steam distillation of the reaction product. Put 25 g of granulated tin and 12.0 g of nitrobenzene in the flask, make an ice-water bath ready, add 55 mL of concentrated hydrochloric acid, insert a thermometer, and swirl well to promote reaction in the three-phase system. Let the mixture react until the temperature reaches 60°C; then cool briefly in ice just long enough to prevent a rise above 60°C, so the reaction will not get out of hand. Continue to swirl, cool as required, and maintain the temperature between 55°C and 60°C for 15 minutes. Remove the thermometer, rinse it with water, fit the flask with a reflux condenser, and heat on a steam bath or heating mantle with frequent swirling until droplets of nitrobenzene are absent from the condenser and the color, due to the formation of an intermediate reduction product, is gone (about 15 minutes). During this period, dissolve 40 g of sodium hydroxide in 100 mL of water and cool to room temperature.

At the end of the reduction reaction, cool the acid solution in ice (to prevent volatilization of aniline) during gradual addition of the solution of alkali. This alkali neutralizes the aniline hydrochloride, releasing aniline, which will now be volatile in steam. Attach a stillhead with a steam-inlet tube, a condenser, an adapter, and a receiving Erlenmeyer flask (*see* Fig. 6.4 on page 108); heat the flask with a microburner to prevent the flask from filling with water from condensed steam and proceed to steam distill. Because aniline is fairly soluble in water (3.6 g/100 g at 18°C), distillation should be continued somewhat beyond the point at which the distillate has lost its original turbidity (50–60 mL more). Make an accurate estimate of the volume of distillate by filling a second flask with water to the level of liquid in the receiver and measuring the volume of water.

The ^{13}C NMR spectrum of nitrobenzene (Fig. 45.6) and the IR (Fig. 45.7) and ^{1}H NMR (Fig. 45.8) spectra of aniline are found at the end of the chapter.

3. ISOLATION OF ANILINE—ALTERNATIVE CHOICES

Salting out aniline

At this point, aniline can be isolated. You could reduce the solubility of aniline by dissolving in the steam distillate 0.2 g of sodium chloride per milliliter, extracting the aniline with three 50-mL portions of dichloromethane, drying the extract, distilling the dichloromethane (bp 41°C), and then distilling the aniline (bp 184°C). Alternatively, the aniline can be converted directly to acetanilide. The procedure calls for pure aniline, but note that the first step is to dissolve the aniline in water and hydrochloric acid. Your steam distillate is a mixture of aniline and water, both of which have been distilled. Are they not both water-white and presumably pure? Hence, an attractive procedure would be to assume that the steam distillate contains the theoretical amount of aniline and to add to it, in turn, appropriate amounts of hydrochloric acid, acetic anhydride, and sodium acetate, calculated from the quantities given in Experiment 2.

Cleaning Up. Filter the pot residue from the steam distillation through a layer of Celite® filter aid on a Büchner funnel. Discard the tin salts in the nonhazardous solid waste container. The filtrate, after neutralizing it with hydrochloric acid, can be flushed down the drain.

4. ACETANILIDE: ACETYLATION IN AQUEOUS SOLUTION

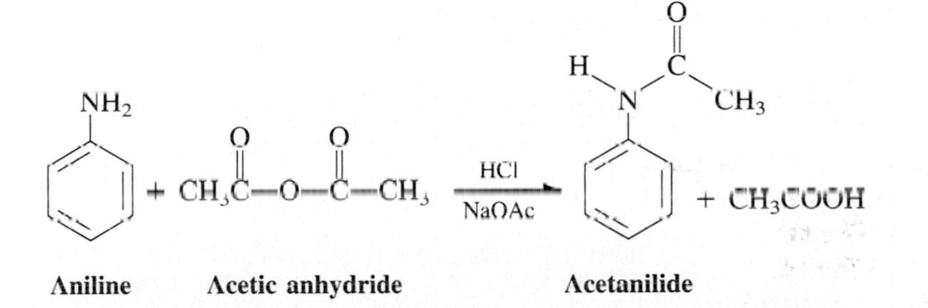

IN THIS EXPERIMENT, the steam distillate, a mixture of aniline and water, is reacted with acetic anhydride to form crystalline acetanilide that is isolated by filtration. It is carefully dried for the next experiment.

From the volumes of reagents given for this experiment, calculate the numbers of grams, the numbers of moles, and the mole ratios. What is the limiting reagent? Why is sodium acetate found in this mixture?

CAUTION: Handle aniline and acetic anhydride in the hood; avoid contact with the skin by wearing gloves.

Add to the steam distillate, which consists of a mixture of aniline and water, 0.43 mL of concentrated hydrochloric acid and make up the volume to 13 mL. Add 0.620 mL of acetic anhydride; also prepare a solution of 0.53 g of anhydrous sodium acetate in 3 mL of water. Add the acetic anhydride to the solution of aniline hydrochloride with stirring, and then immediately add all of the sodium acetate solution. Stir, cool in ice, and collect the product by suction filtration on a Hirsch funnel. It should be colorless; when dry, the melting point should be close to 114°C.

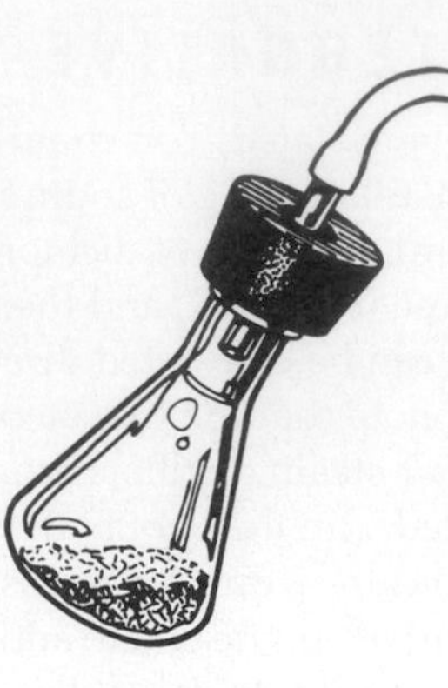

FIG. 45.3
Drying of acetanilide under reduced pressure. Heat the flask on a steam bath or a sand bath.

Because the acetanilide *must be completely dry* for use in the next step, it is advisable to put the material in a tared 25-mL Erlenmeyer flask and to heat it on a steam or sand bath under evacuation until the weight is constant (Fig. 45.3). It could then be used to make sulfanilamide.

Cleaning Up. The aqueous filtrate can be flushed down the drain.

5. ACETANILIDE: ACETYLATION IN AQUEOUS SOLUTION

CAUTION: Handle aniline and acetic anhydride in the hood; avoid contact with the skin.

Dissolve 5.0 g of aniline in 135 mL of water and 4.5 mL of concentrated hydrochloric acid; if the solution is colored, filter it by suction through a pad of decolorizing charcoal. Measure out 6.2 mL of acetic anhydride; also prepare a solution of 5.3 g of anhydrous sodium acetate in 30 mL of water. Add the acetic anhydride to the solution of aniline hydrochloride with stirring, and immediately add all of the sodium acetate solution. Stir, cool in ice, and collect the product. It should be colorless, with a melting point close to 114°C. Because the acetanilide *must be completely dry* for use in the next step, it is advisable to put the material in a tared 125-mL Erlenmeyer flask and to heat this on a steam or sand bath under evacuation until the weight is constant (Fig. 45.3).

^{1}H NMR (Fig. 45.9) and IR (Fig. 45.10) spectra of acetanilide are found at the end of the chapter.

Cleaning Up. The aqueous filtrate, after neutralizing it with hydrochloric acid, can be flushed down the drain with a large excess of water.

6. SULFANILAMIDE

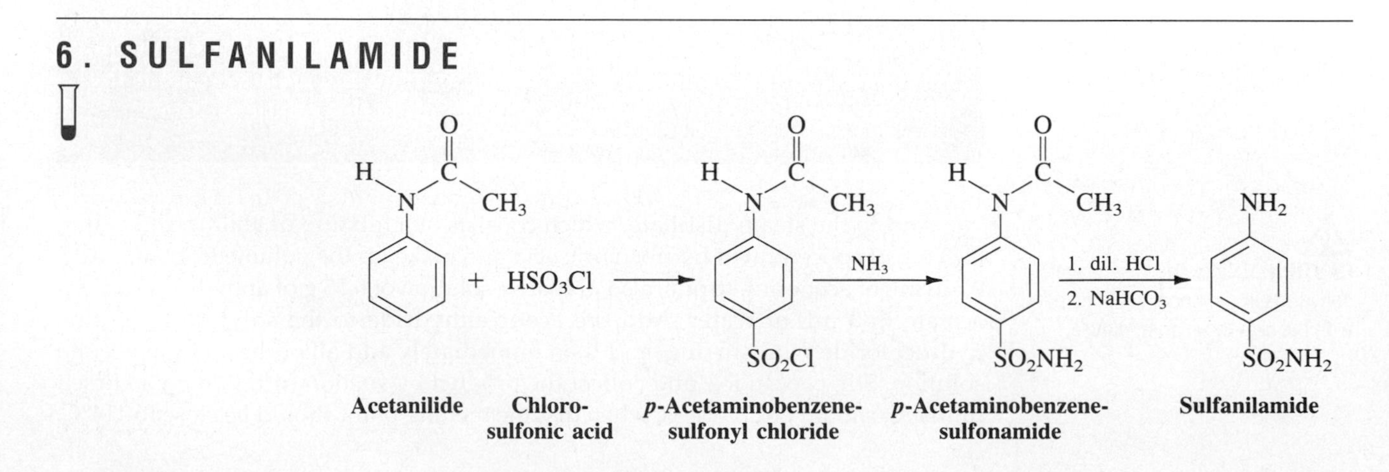

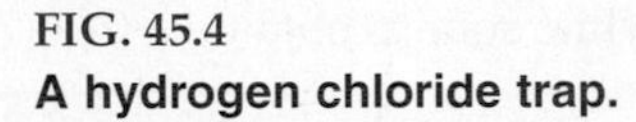

FIG. 45.4
A hydrogen chloride trap.

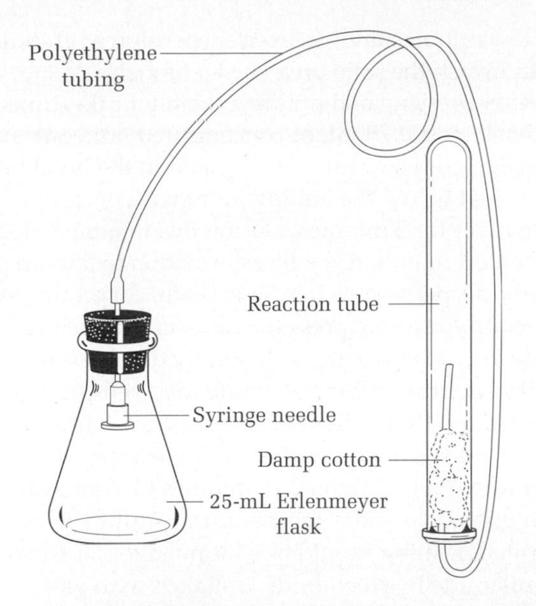

IN THIS EXPERIMENT, perfectly dry acetanilide is treated with chlorosulfonic acid, a highly reactive reagent. The hydrogen chloride evolved is trapped; the reaction mixture is added carefully to water; and the product, *p*-acetaminobenzenesulfonyl chloride, is isolated by filtration. This solid is added to aqueous ammonia to form *p*-acetaminobenzenesulfonamide, which is hydrolyzed with hot hydrochloric acid. The resulting solution is neutralized with sodium bicarbonate to give sulfanilamide.

HSO_3Cl
Chlorosulfonic acid

CAUTION: Chlorosulfonic acid is a corrosive chemical and reacts violently with water. Withdraw with a pipette and pipette bulb. Neutralize any spills and drips immediately. The wearing of gloves and a face shield is advised. Do not breathe any fumes of this chemical, as it burns the mucous membranes of the nose and throat; work within a hood.

Photos: *Capping a Reaction Tube with a Septum, Placing a Polyethylene Tube through a Septum*

The chlorosulfonation of acetanilide in the preparation of sulfanilamide is conducted without solvent in the 25-mL Erlenmeyer flask used for drying the precipitated acetanilide from Experiment 3. Alternatively, commercial acetanilide may be employed.

Fit the Erlenmeyer flask with a septum connected to a short length of polyethylene tubing leading into a reaction tube that contains a small piece of damp cotton to trap hydrogen chloride vapors (Fig. 45.4). Working in the hood, add 0.625 mL of chlorosulfonic acid, a few drops at a time, to 0.25 g of acetanilide from a capped vial using a Pasteur pipette (not a syringe with metal needle). Connect the flask to the gas trap between additions. The reaction subsides in 5–10 minutes, with only a few small pieces of acetanilide remaining undissolved.

$$HSO_3Cl + H_2O \longrightarrow H_2SO_4 + HCl$$

When this point has been reached, heat the mixture on a steam bath for 10 minutes to complete the reaction, cool the flask in ice, and deliver the oily product of *p*-acetaminobenzenesulfonyl chloride by drops with a Pasteur pipette while stirring it into 3.5 mL of ice water contained in a 10-mL Erlenmeyer flask in the hood. If the material in the flask solidifies (it usually does not), add 3.5 g of ice to the reaction flask. Use extreme caution when adding the oil to the ice water, and when rinsing out any containers that have held chlorosulfonic acid. Rinse the Erlenmeyer flask with cold water, and stir the precipitated *p*-acetaminobenzenesulfonyl chloride

Video: *Microscale Filtration on the Hirsch Funnel*

for a few minutes until an even suspension of granular white solid is obtained. Collect and wash the solid on a Hirsch funnel.

Do not let the mixture stand before adding ammonium hydroxide solution.

After pressing and draining the filter cake, transfer the solid to the rinsed reaction flask, add 0.75 mL of concentrated aqueous ammonia solution (ammonium hydroxide) and 0.75 mL of water, and in the hood heat the mixture over a hot sand bath to just below the boiling point with occasional swirling. Heat the mixture in this manner for 5 minutes. During this treatment, a change can be noted as the sulfonyl chloride undergoes transformation to a more pasty suspension of the amide. Cool the suspension well in an ice bath, collect the *p*-acetaminobenzenesulfonamide by suction filtration, press the cake on a Hirsch funnel, and allow it to drain thoroughly. Any excess water will unduly dilute the acid used in the next step.

Transfer the still-moist amide to a well-drained reaction flask, add 0.25 mL of concentrated hydrochloric acid and 0.5 mL of water, boil the mixture gently until all the solid has dissolved (5–10 minutes), and then continue the heating at the boiling point for an additional 10 minutes (do not evaporate to dryness). The solution, when cooled to room temperature, should deposit no solid amide, but if it is deposited, heating should be continued for a further period. The cooled solution of sulfanilamide hydrochloride is shaken with granulated decolorizing charcoal and filtered by removing the solution with a Pasteur pipette.

Place the solution in a 30-mL beaker and cautiously add an aqueous solution of 0.25 g of sodium bicarbonate while stirring to neutralize the hydrochloride. After the foam has subsided, test the suspension with indicator paper; if it is still acidic, add more bicarbonate until the neutral point is reached. Cool thoroughly in ice and collect the granular, white precipitate of sulfanilamide. The crude product (mp 161–163°C) on crystallization from alcohol or water affords pure sulfanilamide (mp 163–164°C) with about 90% recovery. Save a small amount (~10 mg) of sample for the bioassay.

The IR spectrum (Fig. 45.11) of sulfanilamide is found at the end of this chapter.

HSO_3Cl
Chlorosulfonic acid

Cleaning Up. Rinse the cotton in the trap with water, add this rinse to the combined aqueous filtrates from all reactions, and depending on the pH, neutralize the solution by adding either 3 *M* hydrochloric acid or sodium carbonate. Flush the neutral solution down the drain. Any spilled drops of chlorosulfonic acid should be covered with sodium carbonate; the resulting powder should be collected in a beaker, dissolved in water, and then flushed down the drain.

7. SULFANILAMIDE

CAUTION: Chlorosulfonic acid is a corrosive chemical and reacts violently with water. Withdraw with a pipette and pipette bulb. Neutralize any spills and drips immediately. The wearing of gloves and a face shield is advised. Do not breathe any fumes of this chemical, as it burns the mucous membranes of the nose and throat; work within a hood.

The chlorosulfonation of acetanilide in the preparation of sulfanilamide is conducted without solvent in a 125-mL Erlenmeyer flask used for drying the precipitated acetanilide from Experiment 2. Because the reaction is most easily controlled when the acetanilide is in the form of a hard cake, the dried solid is melted by heating the flask over a hot plate; as the melt cools, the flask is swirled to distribute the material as it solidifies over the lower walls of the flask. Let the flask cool while making provision for trapping the hydrogen chloride evolved in the chlorosulfonation. Fit the Erlenmeryer flask with a hydrogen chloride gas trap as shown in Fig. 45.5. The tube in the filter flask should be about 1 cm above the surface of the water, and *must not dip into the water*.

$$HSO_3Cl + H_2O \longrightarrow H_2SO_4 + HCl$$

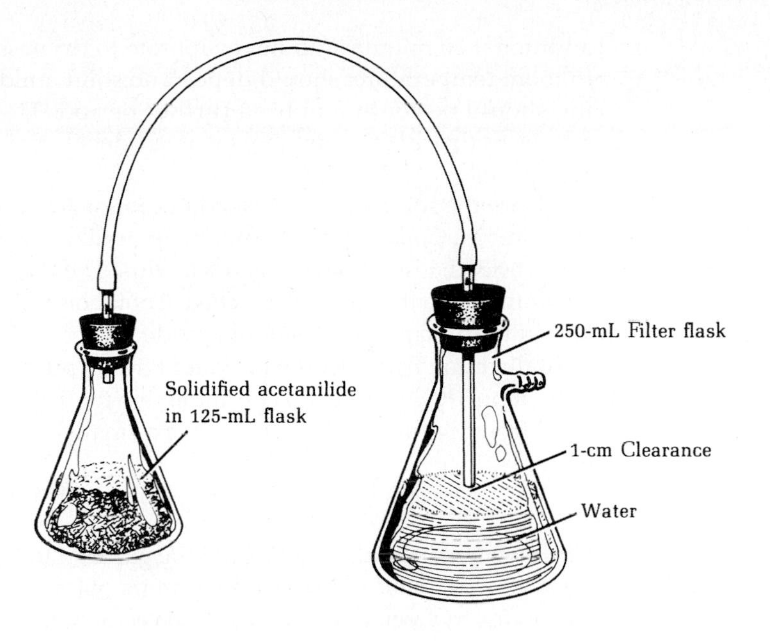

FIG. 45.5
The chlorosulfonation apparatus fitted with HCl gas trap.

Cool the flask containing the acetanilide thoroughly in an ice-water bath. For each 5.0 g of acetanilide, measure 12.5 mL of chlorosulfonic acid in a graduated cylinder (supplied with the reagent and kept away from water). Add this reagent to the flask in 1–2 mL portions with a capillary dropping tube, and connect the flask to the gas trap. The flask is now removed from the ice bath and swirled, until a portion of the solid has dissolved and the evolution of hydrogen chloride is proceeding rapidly. Occasional cooling in ice may be required to prevent too brisk a reaction.

The reaction subsides in 5–10 minutes, with only a few lumps of acetanilide remaining undissolved. When this point has been reached, heat the mixture on a steam or sand bath for 10 minutes to complete the reaction, cool the flask under the tap, and working in the hood, deliver the oil by drops with a capillary dropper while stirring it into 75 mL of ice water contained in a beaker cooled in an ice bath. Use extreme caution when adding the oily *p*-acetaminobenzenesulfonyl chloride to the ice water, and when rinsing out any containers that have held chlorosulfonic acid. Rinse the Erlenmeyer flask with cold water and stir the precipitated *p*-acetaminobenzenesulfonyl chloride for a few minutes until an even suspension of granular white solid is obtained. Collect and wash the solid on a Büchner funnel.

Do not let the mixture stand before adding ammonia.

After pressing and draining the filter cake, transfer the solid to the rinsed reaction flask, add (for each 5 g of acetanilide) 15 mL of concentrated aqueous ammonia solution and 15 mL of water, and heat the mixture over a flame or sand bath with occasional swirling (in the hood). Maintain the temperature of the mixture just below the boiling point for 5 minutes. During this treatment, a change can be noted as the sulfonyl chloride undergoes transformation to a more pasty suspension of the amide. Cool the suspension well in an ice bath, collect the *p*-acetaminobenzenesulfonamide by suction filtration, press the cake on the funnel, and allow it to drain thoroughly. Any excess water will unduly dilute the acid used in the next step.

Transfer the still-moist amide to the well-drained reaction flask, add 5 mL of concentrated hydrochloric acid and 10 mL of water (for each 5 g of *p*-acetamidobenzenesulfonamide), and boil the mixture gently until the solid has all dissolved (5–10 minutes); then continue the heating at the boiling point for an additional 10 minutes (do not evaporate to dryness). The solution, when cooled to room temperature, should deposit no solid amide, but if it is deposited, heating should be continued for a further period. The cooled solution of sulfanilamide hydrochloride is shaken with decolorizing charcoal and filtered by suction.

Place the solution in a beaker and cautiously add an aqueous solution of 5 g of sodium bicarbonate with stirring to neutralize the hydrochloride. After the foam has subsided, test the suspension with litmus; if it is still acidic, add more bicarbonate until the neutral point is reached. Cool thoroughly in ice and collect the granular, white precipitate of sulfanilamide. The crude product (mp 161–163°C) on crystallization from alcohol or water affords pure sulfanilamide (mp 163–164°C) with about 90% recovery. Determine the melting point and calculate the overall yield from the starting material. Save a small amount (~10 mg) of sample for the bioassay.

The IR spectrum (Fig. 45.11) of sulfanilamide is found at the end of the chapter.

Cleaning Up. Add the water from the gas trap to the combined aqueous filtrates from all reactions, and depending on its pH, neutralize the solution by adding either 10% hydrochloric acid or sodium carbonate. Flush the neutral solution down the drain with a large excess of water. Any spilled drops of chlorosulfonic acid should be covered with sodium carbonate; the resulting powder should be collected in a beaker, dissolved in water, and then flushed down the drain.

8. BIOASSAY OF SULFANILAMIDE

You may bioassay your synthesized sulfanilamide and commercial sulfanilamide with *Bacillus cereus* or *E. coli* to test for its antibiotic behavior. There are many bacterial cultures commercially available, but *Bacillus cereus* is preferred because it is nonpathogenic, requires no special medium for growth, and grows at room temperature in a short time.[2] Cultures of *E. coli* not only have to be handled with care due to their potential pathogenicity, but their optimal growth is at 37°C, which requires an incubator. In addition, *E. coli* is a gram-negative bacteria unlike *Bacillus cereus*, which is gram-positive. It would be interesting to test your compound with both bacteria and to compare the two results.

Agar culture plates are prepared in sterile Petri dishes (equipped with caps) or can be purchased. These plates are stored in the refrigerator, preferably in plastic bags to keep the plates from drying out. Whenever you handle the sterile agar Petri dishes, try to minimize any exposure to random forms of bacteria or dirt in the laboratory. Always wear gloves when handling the plates and the bacteria because your hands can easily contaminate the agar. Open the cover of the dish a few inches vertically to have just enough room to swab the plate with a known strain of bacteria using a sterilized copper wire. Once you have your sulfanilamide sample ready, you will add your sample to the plate.

After preparing your assay, you will observe the growth of bacteria on the plate. If the bacteria do not grow in the areas where sulfanilamide is present, it is because this chemical is exhibiting antibiotic behavior.

[2]Other nonpathogens for acceptable use are *Staphylococcus epidermis* and the following strains of *E. coli*: K-12, 1776B, and 1776C.

Bioassay Using Agar Medium Bacterial Growth Plates

All students must be wearing gloves during this entire procedure. Obtain a prepared, sterile agar plate. Insert a sterile swab into the culture tube containing the bacteria and let the cotton absorb the liquid medium. Remove the swab from the culture tube, and then tilt the cover of the agar plate at an angle to keep exposure to unwanted bacteria at a minimum. Do not lay the top of the plate on the benchtop, and do not allow anyone to breathe or come in close contact with the uncovered plate. The moist swab is brought in contact with the agar and then rubbed across its surface. Start at the center of the plate and very gently rub back and forth, moving down toward the edge of the plate. Turn the plate 90 degrees and repeat the rubbing of the swab. Turn the plate 135 degrees and rub the remaining section of the plate. The plate should now be evenly coated with the bacterial culture. Replace the cover on the culture dish. Dispose of the used sterile swab in the appropriate biohazard waste container.

Next, the isolated product will be placed in contact with the inoculated culture. Add the required volume of acetone to your sample to make a 5 to 10% (wt/vol) solution of your product in acetone. Dip the tweezers into alcohol to sterilize. Remove the tweezers from the alcohol and fan back and forth to evaporate the alcohol. Pick up one sterile sample disk using the sterile tweezers, and dip it in the solution of your product. Allow the solvent to dry, and then transfer the disk to the culture dish by lifting the lid slightly and placing it in the agar.

Next, transfer a control disk to the culture dish. The control disk is made sterile by dipping it in ethanol and then allowing it to dry. Wrap the edges of the dish with Parafilm to keep the agar from drying out and cracking. Allow the dish to sit on the benchtop for about 10 minutes to seat the disks in the agar. Invert the culture dish so that the bottom is facing up. Place the dish either at room temperature in a secure location, or in an incubator oven. The results of the assay can be assessed after 24 hours or longer. *See* Chapter 6 for an example of an agar plate.

CAUTION: Wash hands after handling bacterial plates.

Cleaning Up. For glass petri plates that need to be cleaned for the next use, if an autoclave is available, then the agar plates and contaminated materials should be autoclaved for at least 1 hour. If an autoclave is not available, soak the agar plates and contaminated materials in bleach for several days. For plastic disposable plates, tape each plate shut with masking tape and put them into the appropriate biohazard solid waste disposal container.

PART 2: SYNTHESIS AND BIOASSAY OF SULFANILAMIDE DERIVATIVES[3]

As mentioned at the beginning of this chapter, sulfanilamide and its derivatives can exhibit biological activity. For example, when the R group in sulfanilamide is replaced with a heterocyclic ring system, the sulfa drug so formed is often faster acting or less toxic to humans than sulfanilamide. What if the R group is an alkyl group? What about derivatizing sulfanilamide such that the primary amine is functionalized to an amide? Will these derivatives be more or less active than sulfanilamide? You will be able to answer these questions with the following experiments.

[3]Note: Some people are allergic to sulfanilamide, so be aware of any individual sensitivity.

EXPERIMENTS

1. SYNTHESIS OF 4-AMINO-*N*-METHYLBENZENESULFONAMIDE (*N*-1-METHYLSULFANILAMIDE)

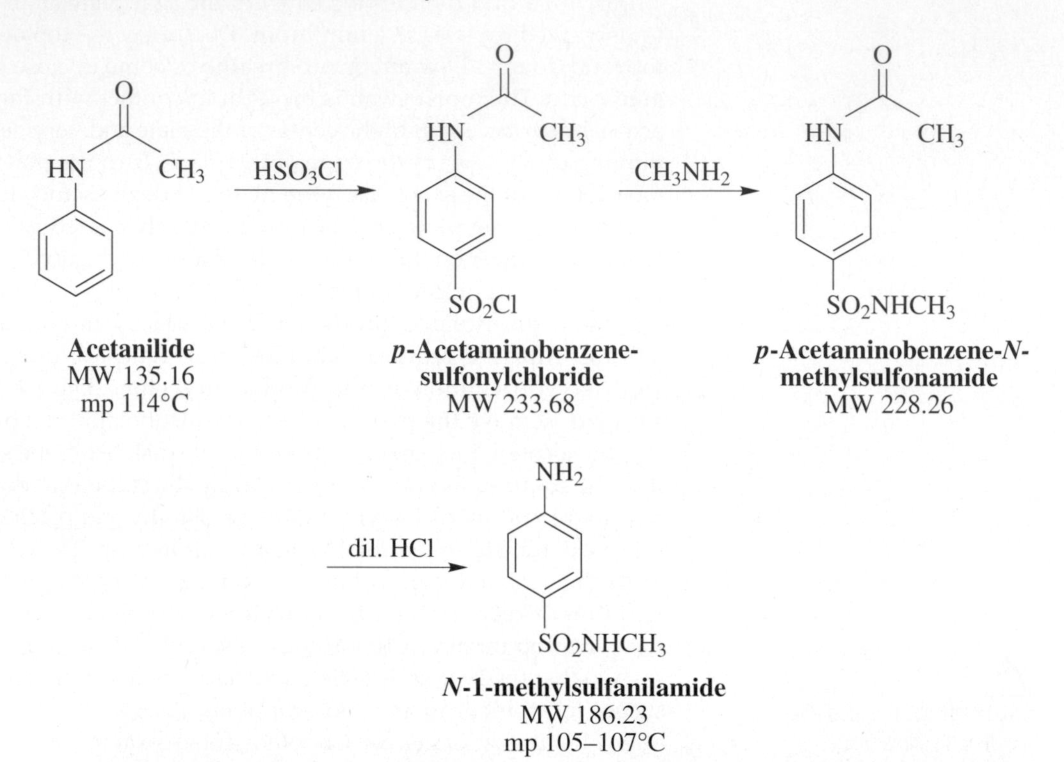

IN THIS EXPERIMENT, perfectly dry acetanilide is treated with chlorosulfonic acid, a highly reactive reagent. The hydrogen chloride evolved is trapped, the reaction mixture is added carefully to water, and the product (*p*-acetaminobenzenesulfonyl chloride) is isolated by filtration. This solid is added to a solution of methylamine to form *p*-acetamino-*N*-methylbenzenesulfonamide, which is hydrolyzed with hot hydrochloric acid. The resulting solution is neutralized with sodium bicarbonate to give *N*-1-methylsulfanilamide.

The chlorosulfonation of acetanilide in the preparation of *N*-1-methylsulfanilamide is conducted without solvent in the 25-mL Erlenmeyer flask used for drying the precipitated acetanilide from Experiment 4 in Part 1. Alternatively, commercial acetanilide may be employed.

Fit the Erlenmeyer flask with a septum connected to a short length of polyethylene tubing leading into an inverted reaction tube that contains a small piece of damp cotton to trap hydrogen chloride vapors (*see* Fig. 45.4 on page 575). Add 0.625 mL of chlorosulfonic acid, a few drops at a time, to 0.25 g of acetanilide from a capped vial in the hood using a Pasteur pipette (not a syringe with metal needle). Connect the

CAUTION: Chlorosulfonic acid is a corrosive chemical that reacts violently with water. Withdraw with a pipette and pipette bulb. Neutralize any spills and drips immediately. The wearing of gloves and a face shield is advised. Do not breathe any fumes of this chemical, as it burns the mucous membranes of the nose and throat; work within a hood.

Photos: *Capping a Reaction Tube with a Septum, Placing a Polyethylene Tube through a Septum*

Do not let the mixture stand before addition of methylamine.

Erlenmeyer flask to the gas trap between additions. In 5–10 minutes, the reaction subsides, with only a few small pieces of acetanilide remaining undissolved.

When this point has been reached, heat the mixture on a steam or sand bath for 10 minutes to complete the reaction, cool the flask in ice, and deliver the oily product dropwise with a Pasteur pipette while stirring it into 3.5 mL of ice water contained in a 10-mL Erlenmeyer flask in the hood. If the material in the flask solidifies (it usually does not), add 3.5 g of ice to the reaction flask. Use extreme caution when adding the oily *p*-acetaminobenzenesulfonyl chloride to the ice water, and when rinsing out any containers that have held chlorosulfonic acid. Rinse the flask with cold water and stir the precipitated *p*-acetaminobenzenesulfonyl chloride for a few minutes until an even suspension of granular white solid is obtained. Collect and wash the solid on a Hirsch funnel.

After pressing and draining the filter cake, transfer the solid to the rinsed reaction flask, add 0.02 mol of methylamine (2.5 mL of a 33% solution of methylamine in absolute ethanol),[4] and heat the mixture over a hot sand bath to just below the boiling point with occasional swirling in the hood. Heat the mixture in this manner for approximately 15 minutes. Monitor the reaction by thin-layer chromatography (TLC) using ethyl acetate. When complete, remove all excess solvent with a stream of nitrogen gas to leave behind a white solid.

Transfer the still-moist amide to the well-drained reaction flask, add 0.25 mL of concentrated hydrochloric acid and 0.5 mL of water, and boil the mixture gently until all of the solid has dissolved (5–10 minutes); then continue heating at the boiling point for an additional 10 minutes (do not evaporate to dryness). The solution, when cooled to room temperature, should deposit no solid amide; if a solid is deposited, heating should be continued for a further period.

Place the solution in a 30-mL beaker, and cautiously add an aqueous solution of 0.25 g of sodium bicarbonate while stirring to neutralize the hydrochloride. After the foam has subsided, test the suspension with indicator paper; if it is still acidic, add more bicarbonate until the neutral point is reached. Cool thoroughly in ice and collect the granular, white precipitate of *N*-1-methylsulfanilamide. The crude product can be recrystallized in water to afford pure *N*-1-methylsulfanilamide (mp 105–107°C) with about 65% recovery. Save a small amount of sample (~10 mg) for the bioassay.

Cleaning Up. Rinse the cotton in the trap with water, add this rinse to the combined aqueous filtrates from all reactions, and depending on its pH, neutralize the solution by adding either 3 *M* hydrochloric acid or sodium carbonate. Flush the neutral solution down the drain. Any spilled drops of chlorosulfonic acid should be covered with sodium carbonate; the resulting powder should be collected in a beaker, dissolved in water, and then flushed down the drain.

2. 4-AMINO-*N*-METHYLBENZENESULFONAMIDE (*N*-1-METHYLSULFANILAMIDE)

The chlorosulfonation of acetanilide in the preparation of *N*-1-methylsulfanilamide is conducted without solvent in the 125-mL Erlenmeyer flask used for drying the precipitated acetanilide from Experiment 5 in Part 1 (commercial acetanilide can also be used). Because the reaction is most easily controlled when the acetanilide

[4]The 33% solution of methylamine in absolute ethanol can be purchased from Sigma-Aldrichcatalog, item no. 534102–Methylamine solution purum, 33% in absolute ethanol (~8M).

is in the form of a hard cake, the dried solid is melted by heating the flask over a hot plate. As the melt cools, the flask is swirled to distribute the material over the lower walls of the flask as it solidifies. Let the flask cool while making provision for trapping the hydrogen chloride evolved in the chlorosulfonation. Fit the Erlenmeyer flask with a hydrogen chloride gas trap as shown in Fig. 45.5. The tube in the filter flask should be about 1 cm above the surface of the water and *must not dip into the water*.

CAUTION: Chlorosulfonic acid is a corrosive chemical that reacts violently with water. Withdraw with pipette and a pipette bulb. Neutralize any spills and drips immediately. The wearing of gloves and a face shield is advised. Do not breathe any fumes of this chemical, as it burns the mucous membranes of the nose and throat; work within a hood.

Cool the flask containing the acetanilide thoroughly in an ice-water bath. For each 5.0 g of acetanilide, measure 12.5 mL of chlorosulfonic acid in the graduated cylinder (supplied with the reagent and kept away from water). Add the reagent to the flask in 1–2 mL portions with a Pasteur pipette, and connect the flask to the gas trap. Remove the flask from the ice bath and swirl until a portion of the solid has dissolved and the evolution of hydrogen chloride is proceeding rapidly. Occasional cooling in ice may be required to prevent too brisk a reaction.

The reaction subsides in 5–10 minutes, with only a few lumps of acetanilide remaining undissolved. When this point has been reached, heat the mixture on a sand bath for 10 minutes to complete the reaction, cool the flask under the tap, and deliver the oil dropwise with a Pasteur pipette while stirring it into 75 mL of ice water in a beaker cooled in an ice bath (in the hood). Use extreme caution when adding the oily *p*-acetaminobenzenesulfonyl chloride to the ice water, and when rinsing out any containers that have held chlorosulfonic acid. Rinse the flask with cold water and stir the precipitated *p*-acetaminobenzenesulfonyl chloride for a few minutes until an even suspension of granular white solid is obtained. Collect and wash the solid on a Büchner funnel. Save a small sample (~6 mg) for bioassay.

Do not let the mixture stand before addition of methylamine.

After pressing and draining the filter cake, transfer the solid to the rinsed reaction flask, add 0.42 mol of methylamine (52 mL of a 33% solution of methylamine in absolute ethanol),[5] for each 5 g of acetanilide, and heat the mixture over a flame with occasional swirling (in the hood). Maintain the temperature of the mixture just below the boiling point for approximately 30 minutes. Monitor the reaction by TLC using ethyl acetate. When complete, remove all excess solvent with a nitrogen gas line to leave behind a white solid. Save a small sample (~6 mg) for bioassay.

Transfer the still-moist amide to the well-drained reaction flask, add 5 mL of concentrated hydrochloric acid and 10 mL of water (for each 5 g of acetanilide), and boil the mixture gently until the solid has all dissolved (5–10 minutes); then continue the heating at the boiling point for an additional 10 minutes (do not evaporate to dryness). The solution, when cooled to room temperature, should deposit no solid amide; if a solid is deposited, heating should be continued for a further period.

Place the solution in a beaker, and cautiously add an aqueous solution of 5 g of sodium bicarbonate with stirring to neutralize the hydrochloride. After the foam has subsided, test the suspension with indicator paper; if it is still acidic, add more bicarbonate until the neutral point is reached. Cool thoroughly in ice and collect the granular, white precipitate of *N*-1-methylsulfanilamide. The crude product can be recrystallized in water to afford pure *N*-1-methylsulfanilamide

[5]*See* footnote 4.

(mp 105–107°C) with about 65% recovery. Determine the melting point, and calculate the overall yield from the starting material. Save a small amount (~10 mg) for the bioassay.

Cleaning Up. Add the water from the gas trap to the combined aqueous filtrates from all reactions, and depending on its pH, neutralize the solution by adding either 10% hydrochloric acid or sodium carbonate. Flush the neutral solution down the drain with a large excess of water. Any spilled drops of chlorosulfonic acid should be covered with sodium carbonate; the resulting powder should be collected in a beaker, dissolved in water, and then flushed down the drain.

3. SYNTHESIS OF ACYLATED SULFANILAMIDE DERIVATIVES

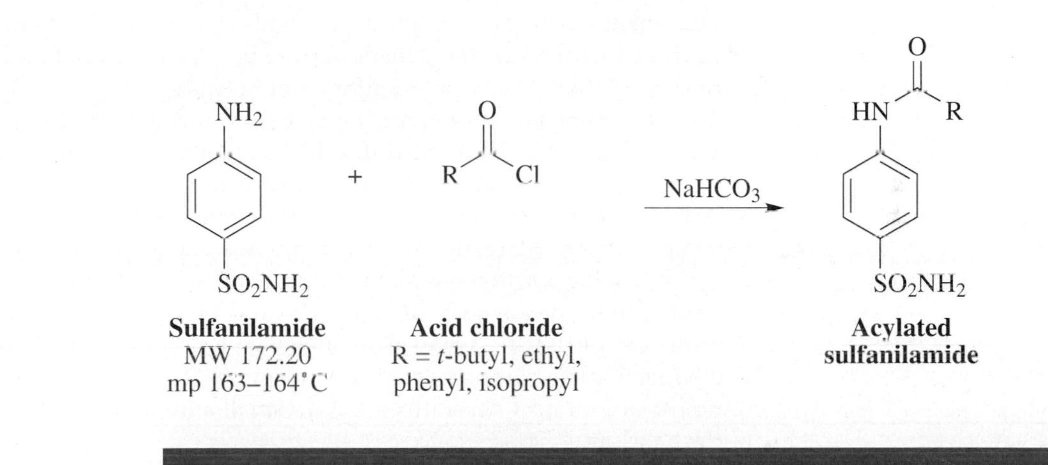

IN THIS EXPERIMENT, sulfanilamide is treated with an acid chloride to produce an acylated sulfanilamide. The resulting solution is neutralized by sodium bicarbonate.

CAUTION: Tetrahydrofuran easily forms peroxides. Discard material that has been exposed to oxygen after 30 days.

The acylation of sulfanilamide to prepare the acylated derivative is conducted in a 25-mL round-bottom flask equipped with a magnetic stirring bar. Transfer 3 mL of tetrahydrofuran to the flask, and then add 0.25 g of sulfanilamide (from Experiment 6 or Experiment 7 in Part 1; alternatively, commercial sulfanilamide may be employed) and 0.305 g of sodium bicarbonate. While the mixture is stirring, add 3.63 mol of the desired acid chloride. Attach a condenser to the flask, and heat at reflux until the reaction goes to completion. Monitor progress of the reaction by thin-layer chromatography every 15 minutes using a 20/30/50 mixture of dichloromethane, ethanol, and hexanes to elute.

When this point has been reached, add 15 mL of cold water to cause complete precipitation of the acylated derivative. Collect the granular white solid, and wash it with ether on a Hirsch funnel. The crude product can be recrystallized in alcohol to afford the pure acylated sulfanilamide. Save a small amount of pure sample (~10 mg) for the bioassay.

Yield and melting points of four acyl derivatives are displayed in Table 45.1.

TABLE 45.1 *Physical Properties of Acylated Sulfanilamide Derivatives*

Derivative	*R*	*% Yield*[a]	*mp (°C)*
4-Pivaloylamido benzenesulfonamide	$(CH_3)_3C$	60%–67%	231–233
4-Propionamido benzenesulfonamide	CH_3CH_2	66%–76%	225–227
4-Benzoylamino benzenesulfonamide	C_6H_5	60%–86%	288–290
4-Isobutanamido benzenesulfonamide	$(CH_3)_2CH$	59%–79%	249–251

[a]Percent yield after recrystallization.

4. SYNTHESIS OF ACYLATED SULFANILAMIDE DERIVATIVES

CAUTION: Tetrahydrofuran easily forms peroxides. Discard material that has been exposed to oxygen after 30 days.

The acylation of sulfanilamide is conducted in a 50 mL round-bottom flask equipped with a 1-inch magnetic stirring bar. Add 12 mL of tetrahydrofuran to the flask, and then add 1.0 g of sulfanilamide (from Experiment 7 in Part 1; alternatively, commercial sulfanilamide may be employed) and 1.3 g of sodium bicarbonate. While the mixture is stirring, add 14.5 mol of the desired acid chloride. Attach a condenser to the flask, and heat at reflux until the reaction goes to completion. Monitor progress of the reaction by thin-layer chromatography every 15 minutes using a 20/30/50 mixture of dichloromethane, ethanol, and hexanes to elute.

When the reaction is complete, add 30 mL of cold water to cause complete precipitation of the acylated derivative. Collect the granular white solid, and wash with ether on a fritted glass funnel. The crude product can be recrystallized from ethanol to afford pure acylated sulfanilamide. *See* Table 45.1 for yields and melting points of four acyl derivatives. Save a small amount of pure sample (~10 mg) for the bioassay.

5. BIOASSAY OF SULFANILAMIDE AND DERIVATIVES

IN THIS EXPERIMENT, sulfanilamide derivatives are tested for antibacterial activity and compared to the activity of sulfanilamide.

The bioassay procedure is described in Experiment 8 of Part 1 of this chapter. You will need approximately 6 mg of each derivative and of sulfanilamide to test for bioactivity. Be sure to label carefully the agar plates so it is clear what compounds are being tested. The compounds to be tested can be placed directly on the agar/bacteria in a small pile, where they will slowly be absorbed and dispersed into the agar. If a derivative is active toward bacteria, there will be no bacterial growth in the immediate area where the derivative was placed on the plate. Observe the results from the bioassay and assess whether any of the derivatives are active. Compare the activities of the derivatives with each other and with the sulfanilamide. From these results, what conclusions can you draw about the activity of sulfanilamide derivatives? Which derivatives are antibacterial? Does the primary aromatic amine need to be in the "free" form (NH_2) to be antibacterial? Is *N*-methylsulfanilamide more toxic or less toxic toward bacteria than sulfanilamide?

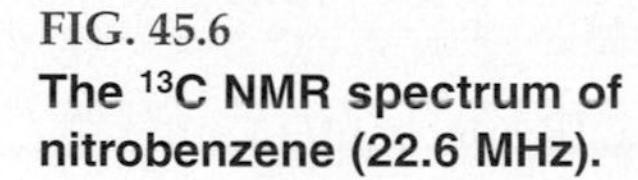

FIG. 45.6
The ^{13}C NMR spectrum of nitrobenzene (22.6 MHz).

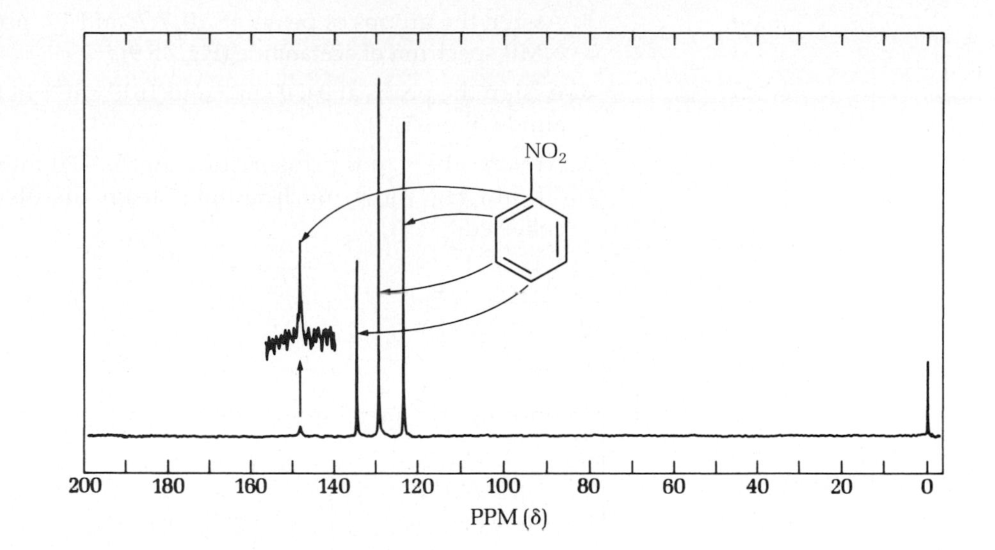

FIG. 45.7
The IR spectrum of aniline.

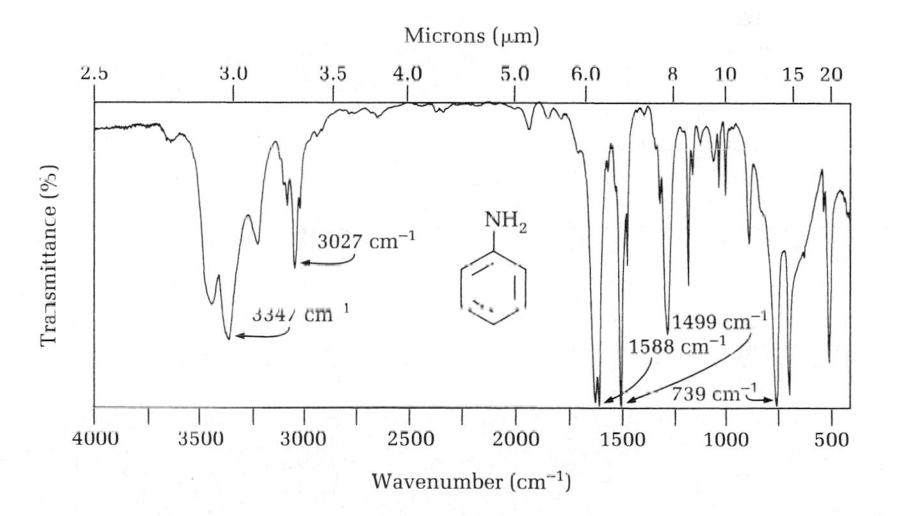

QUESTIONS

1. Why is an acetyl group added to aniline (making acetanilide) and then removed to regenerate the amine group in sulfanilamide?
2. What happens when chlorosulfonic acid comes in contact with water?
3. Acetic anhydride, like any anhydride, reacts with water to form a carboxylic acid. How then is it possible to carry out an acetylation in aqueous solution? What is the purpose of the hydrochloric acid and the sodium acetate in this reaction?
4. What happens when *p*-acetaminobenzenesulfonyl chloride is allowed to stand for some time in contact with water?

5. Assign the groups of peaks at 7.0, 7.2, and 7.4 ppm to specific protons in the NMR spectrum of acetanilide (Fig. 45.9).
6. Assign the peaks at 3470 cm^{-1} and 1619 cm^{-1} in the IR spectrum of sulfanilamide (Fig. 45.11).
7. At 98°C the vapor pressure of water is 720 mm Hg, and that of aniline is 40 mm Hg. How much aniline steam distills with each gram of water collected?

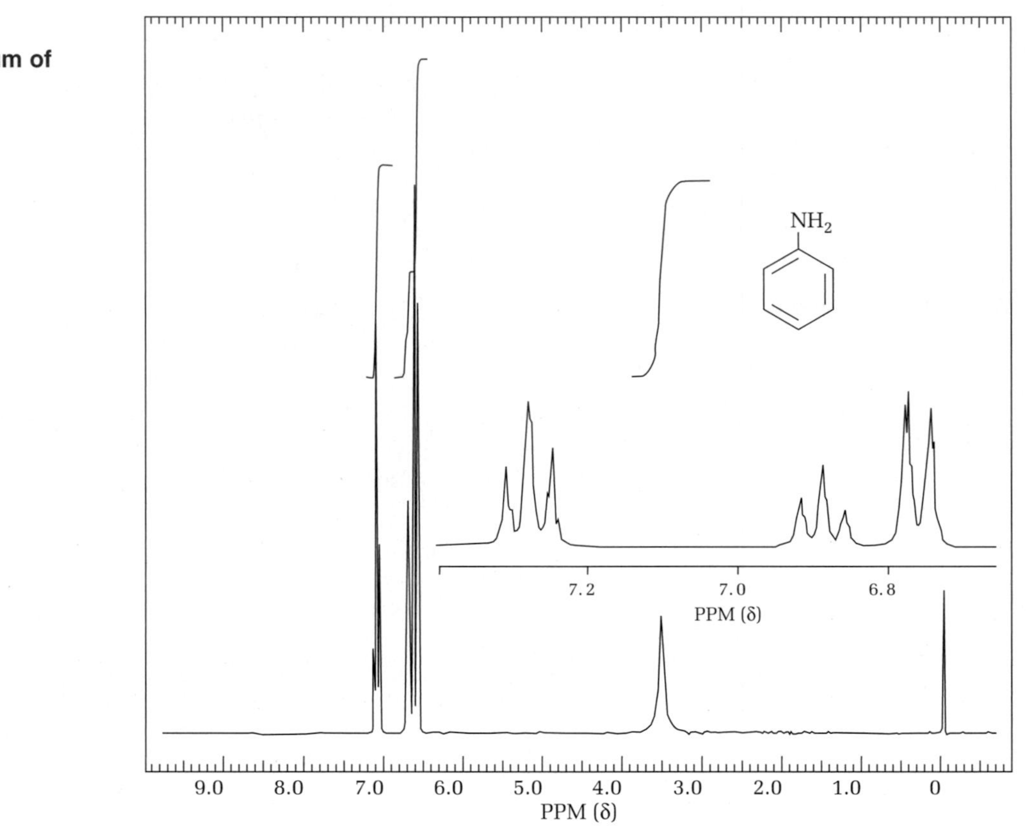

FIG. 45.8
The ^1H NMR spectrum of aniline (250 MHz).

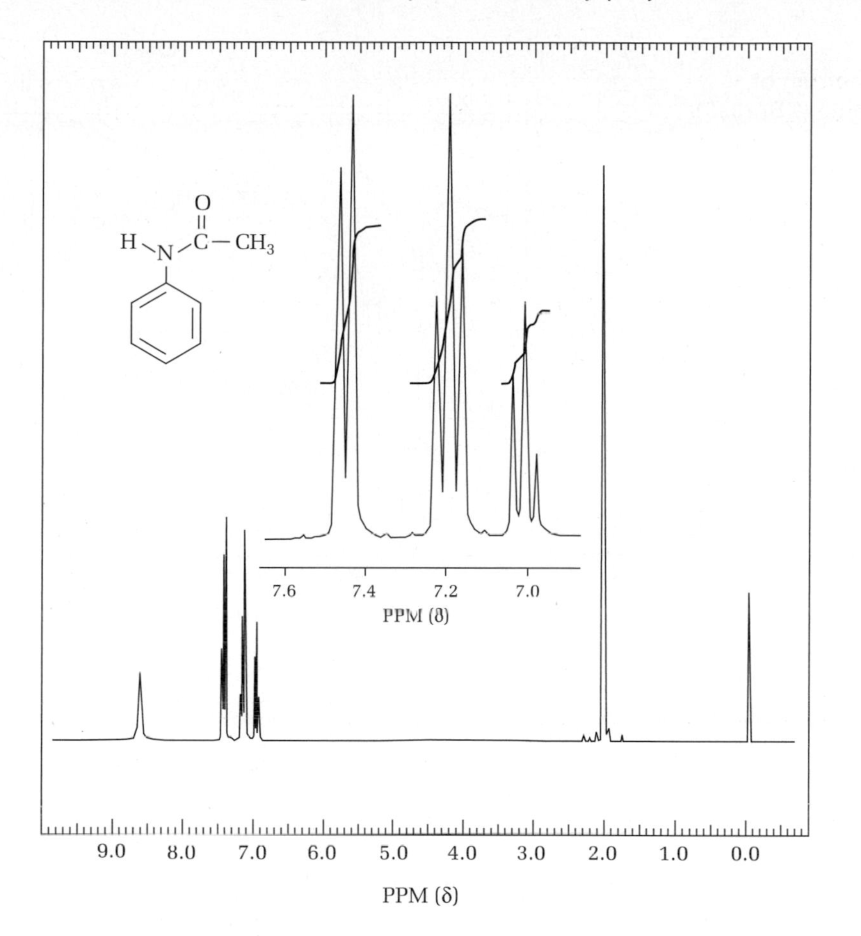

FIG. 45.9
The ^{1}H NMR spectrum of acetanilide (250 MHz). The amide proton shows a characteristically broad peak.

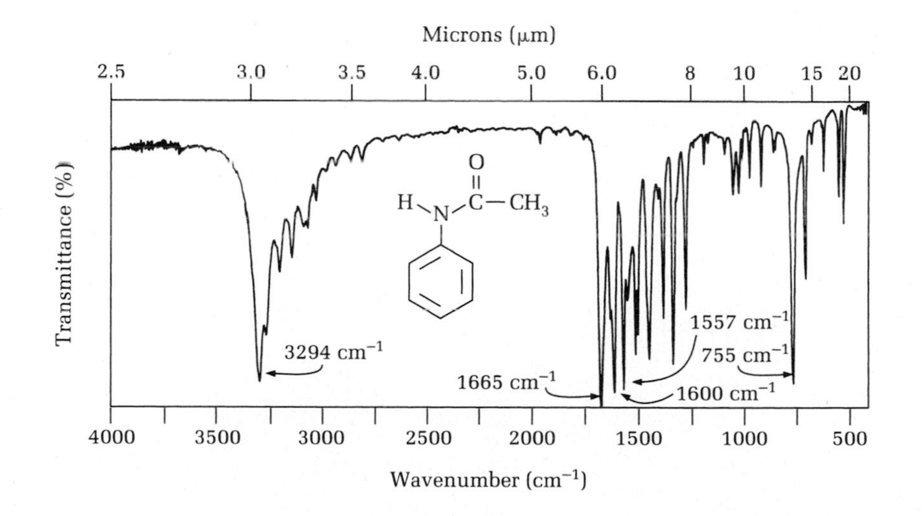

FIG. 45.10
The IR spectrum of acetanilide (KBr disk).

FIG. 45.11
The IR spectrum of sulfanilamide (KBr disk).

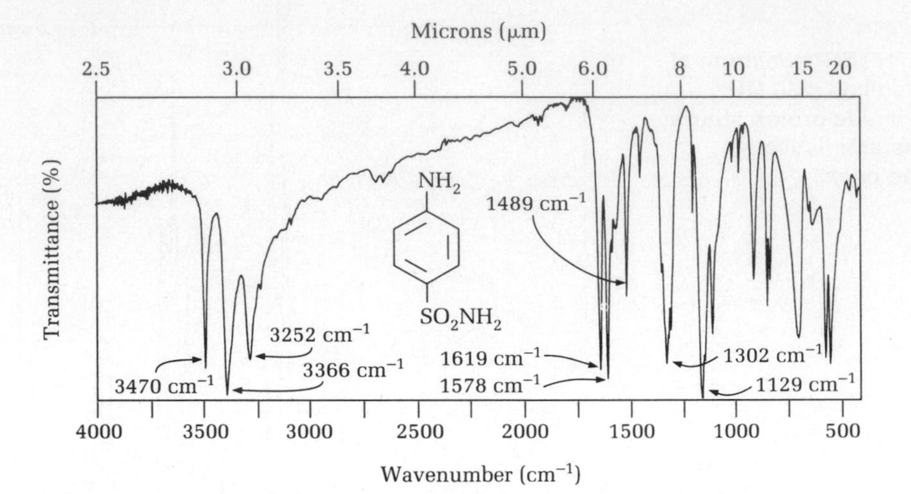

CHAPTER 46

Dyes and Dyeing[1]

PRELAB EXERCISE: Operating on the simple hypothesis that the intensity of a dye on a fiber will depend on the number of strongly polar or ionic groups in the fiber molecule, predict the relative intensities produced by Methyl Orange when it is used to dye a variety of different fibers, such as are found in Multifiber Fabric. Examine the structures of the fiber molecules and try to order them in terms of polarity. Wool is a protein.

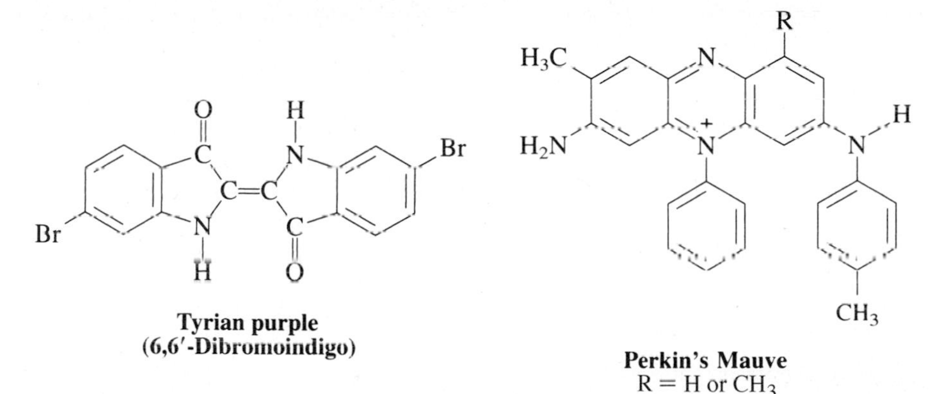

Tyrian purple (6,6′-Dibromoindigo)

Perkin's Mauve
R = H or CH_3

Since prehistoric times, humans have been dyeing cloth. The "wearing of the purple" has long been synonymous with royalty, attesting to the cost and rarity of Tyrian purple, a dye derived from the sea snail *Murex brandaris*. The organic chemical industry originated in 1856 with William Henry Perkin's discovery of the first synthetic dye, Perkin's Mauve. The structure of this dye has now been elucidated.[2]

A fiber usually absorbs dyes from an aqueous solution. A natural fiber such as cotton has a surface area of about 4.4 hectares per kilogram (5 acres per pound). The dye penetrates the pores in the fiber and is bound to the fiber by electrostatic

[1]For a detailed discussion of the chemistry of dyes and dyeing, see Fieser LF, Fieser M. *Topics in Organic Chemistry*. Reinhold Publishing Corp: New York, **1963**; 357–417.
[2]Meth-Cohn O, Smith MJ. *Chem Soc.*, Perkin Trans. 1, **1994**, **5**.

forces, van der Waals attraction, hydrogen bonding, and covalent bonds (in the case of fiber-reactive dyes). A good dye must be fast to light, heat, and washing; that is, it must not fade, sublime away, or come off during washing.

Among the newest of the dyes are the fiber-reactive compounds, which form a covalent link to the hydroxyl groups of cellulose. This reaction involves an amazing and little-understood nucleophilic displacement of a chloride ion from the triazine part of the dye molecule by the hydroxyl groups of cellulose, yet the reaction occurs in aqueous solution.

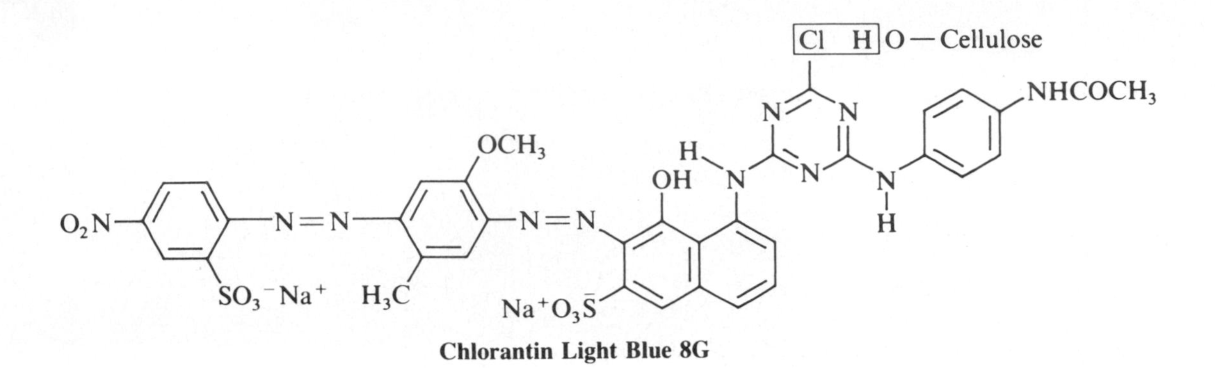

Chlorantin Light Blue 8G

The syntheses of several dyes are described in this chapter; these and other dyes are used to dye a representative group of natural and synthetic fibers. You will receive several pieces of Multifiber Fabric (Fig. 46.1), which has 13 strips of different fibers woven into it. Below the black thread at the top, the fibers are acetate rayon, which is cellulose from any source, that is acetylated on about two of the hydroxyl groups. SEF, Solutia's Self-Extinguishing Fiber, is a copolymer of mostly acrylonitrile with vinyl chloride or bromide. It chars, rather than melts or burns. Fake fur is often made of this fiber. Polyester, bright filament, is a polyester such as polyethylene glycol terephthalate (Dacron) spun to give a bright appearance. Cotton is pure cellulose, and Creslan 61 is polyacrylonitrile. Disperse-dyeable polyester can be dyed with disperse dyes (*see* Experiment 6). Cationic dyeable polyester can be dyed with a basic dye such as Malachite Green or Crystal Violet (*see* Experiment 3). Polyamide is best exemplified by nylon 6.6, which is a polymer of adipic acid and hexamethylenediamine. Polyacrylonitrile, also called polyacrylic, was invented at DuPont and called Orlon. Silk is a polypeptide. Polypropylene is a pure hydrocarbon polymer of propylene and is often used to make women's hosiery. Viscose rayon is regenerated cellulose. Worsted wool is wool that is spun of long, parallel fibers. Like silk, wool is a polyamide.

FIG. 46.1
Multifiber #43 Fabric.

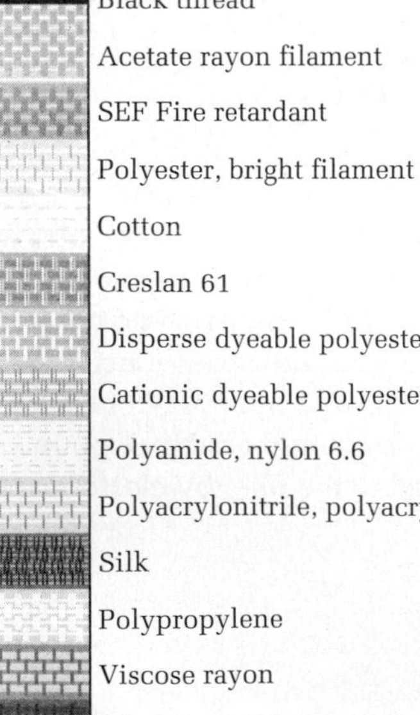

These fibers take up dyes in remarkably different ways. As you will see, the same dye will produce a variety of colors, depending on the fiber. Acetate rayon, having a smaller number of hydroxyl groups compared to cotton, is more difficult to dye with a direct dye than cotton. Fibers with few or no functional groups on the chain accept dyes poorly, for example, polypropylene, which is just a hydrocarbon. Except at the ends of its polymer chains, nylon has few dyeable functional groups. Similarly, Creslan 61, a polyester made by polymerizing ethylene glycol and terephthalic acid, has few polar centers within the polymer and is also difficult to dye. Even more difficult to dye is polyacrylonitrile. Wool and silk are polypeptides that are cross-linked with disulfide bridges. The acidic and basic amino acids (e.g., glutamic acid and lysine) in wool and silk provide many polar groups to which a dye can bind, making these fabrics easy to dye.

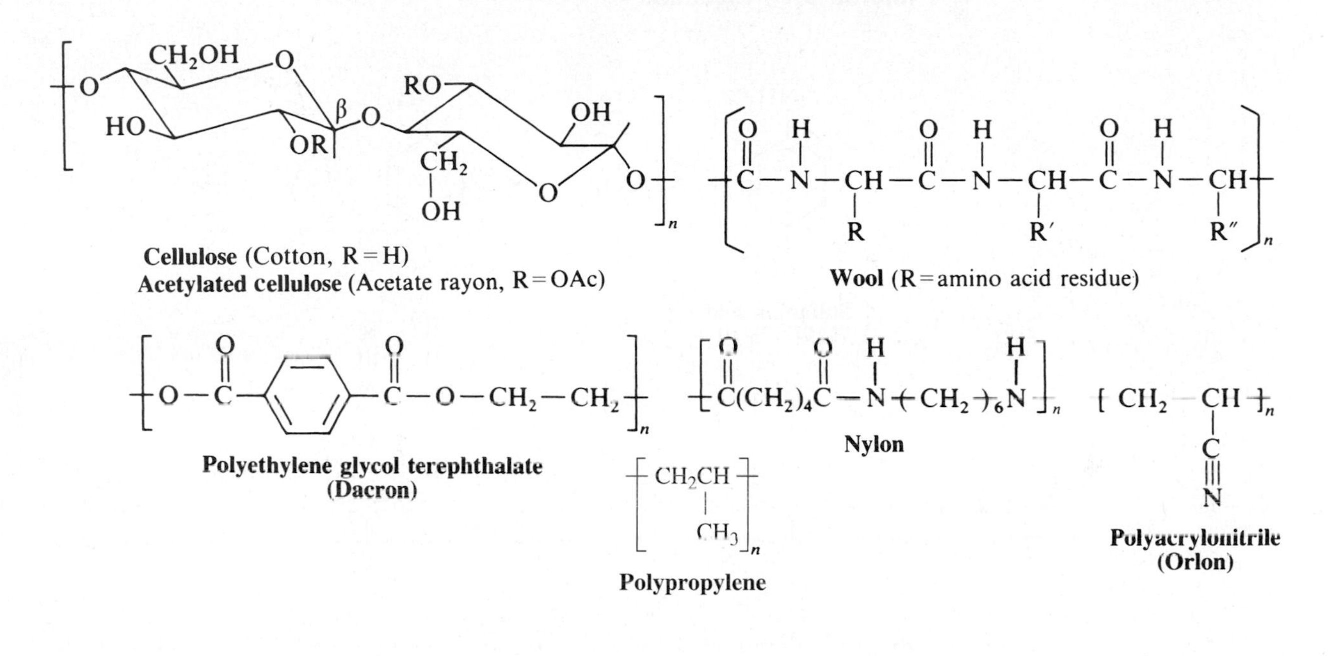

PART 1: DYES

Azo group

The most common dyes are the azo dyes, formed by coupling diazotized amines to phenols. The dye can be made in bulk, and as we shall see, the dye molecule can also be developed on and in the fiber by combining the reactants in the presence of the fiber.

One dye, Orange II, is made by coupling diazotized sulfanilic acid with 2-naphthol in an alkaline solution. Another, Methyl Orange, is prepared by coupling the same diazonium salt with *N*,*N*-dimethylaniline in a weakly acidic solution. Methyl Orange is used as an indicator because it changes color at pH 3.2–4.4. The change in color is due to the transition from one chromophore (azo group) to another (quinonoid system).

You will prepare one of these two dyes and then exchange samples with a classmate and do the tests with both dyes. Both substances dye wool, silk, and skin, so you must work carefully to avoid getting these on your hands or clothes. The dye will eventually wear off your hands; alternatively, you can clean your hands by soaking them for a few seconds in a warm, slightly acidic (H_2SO_4) permanganate solution until heavily stained with manganese dioxide and then removing the stain in a bath of warm, dilute bisulfite solution. (Follow the same procedure if dye is accidentally spilled on your clothing.)

EXPERIMENTS

1. DIAZOTIZATION OF SULFANILIC ACID

Microscale Procedure

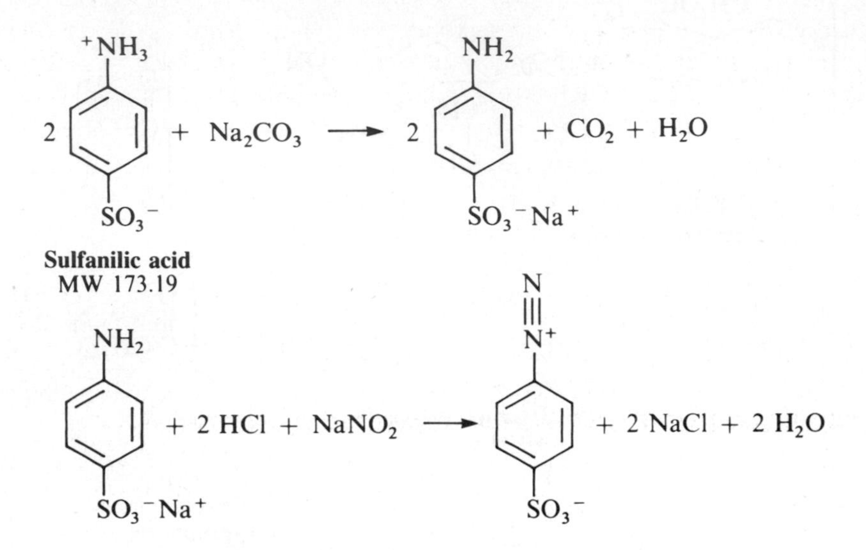

In a 10 × 100-mm reaction tube (Fig. 46.2) dissolve, by boiling, 120 mg of sulfanilic acid monohydrate in 1.25 mL of 2.5% sodium carbonate solution (or use 35 mg of anhydrous sodium carbonate and 1.25 mL of water). Cool the solution to room temperature, add 50 mg of sodium nitrite, and stir until it is dissolved. Cool the tube in ice, and add to it with thorough stirring a mixture of 0.75 g of ice and 0.125 mL of concentrated hydrochloric acid. A powdery white precipitate of the diazonium salt should separate in 1–2 minutes; the material is then ready for use. The product is not collected, but is used in the preparation of Orange II and/or Methyl Orange while in suspension. It is more stable than most diazonium salts and will keep for a few hours.

CAUTION: Avoid skin contact with diazonium salts. Some diazonium salts are explosive when dry. Always use in solution.

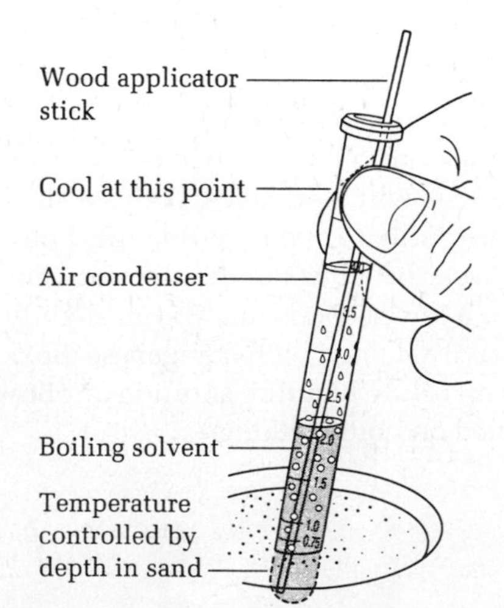

FIG. 46.2
A reaction tube with a boiling stick to promote even boiling.

Macroscale Procedure

In a 50-mL Erlenmeyer flask, dissolve, if necessary by boiling, 2.4 g of sulfanilic acid monohydrate in 25 mL of 2.5% sodium carbonate solution (or use 0.66 g of anhydrous sodium carbonate and 25 mL of water). Cool the solution under the tap, add 0.95 g of sodium nitrite, and stir until it is dissolved. Pour the solution into a flask containing about 15 g of ice and 2.5 mL of concentrated hydrochloric acid. A powdery white precipitate of the diazonium salt should separate in 1–2 minutes; the material is then ready for use. The product is not collected but is used in the preparation of Orange II and/or Methyl Orange while in suspension. It is more stable than most diazonium salts and will keep for a few hours.

2. ORANGE II (1-*p*-SULFOBENZENEAZO-2-NAPHTHOL, SODIUM SALT)

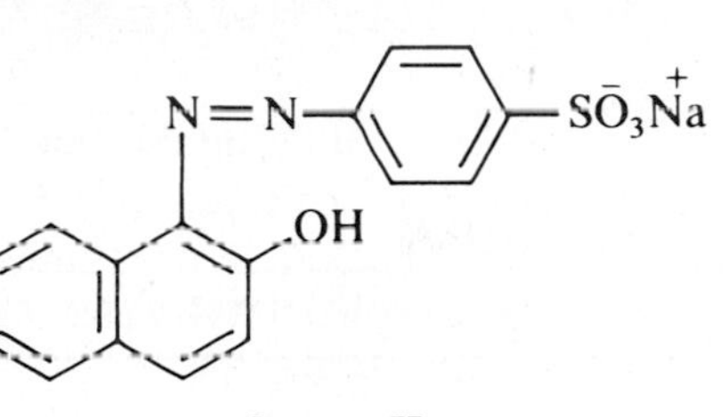

Orange II

⚠ **CAUTION:** Because 2-naphthol is a carcinogen, handle it with care, in the hood. Do not breathe the dust or allow skin contact.

In a 10-mL Erlenmeyer flask, dissolve 90 mg of 2-naphthol in 0.5 mL of 3 *M* sodium hydroxide solution and transfer to this solution—with *thorough* stirring—the suspension of diazotized sulfanilic acid prepared in the microscale procedure of Experiment 1. Rinse all of the diazonium salt into the naphthol solution with a few drops of cold water. Coupling occurs very rapidly, and the dye, being a sodium salt, separates easily from the solution because a considerable excess of sodium ion from the carbonate, the nitrite, and the alkali is present. Stir the crystalline paste thoroughly to effect good mixing and, after 5–10 minutes, heat the mixture in a beaker of boiling water until the solid dissolves. Add 0.25 g of sodium chloride to further decrease the solubility of the product and bring this all into solution by heating and stirring. Allow the flask to cool to near room temperature undisturbed; then cool it in ice. Collect the product on a Hirsch funnel bearing a filter paper (Fig. 46.3). Use a saturated sodium chloride solution rather than water to rinse out the flask and to wash the filter cake free of the dark-colored mother liquor. The filtration is somewhat slow, and the transfer to a Hirsch funnel is difficult because it is a paste.

Video: *Microscale Filtration on the Hirsch Funnel*

Photo: *Vacuum Filtration into a Reaction Tube through a Hirsch Funnel*

The product dries slowly and contains about 20% sodium chloride. Thus, the crude yield is not significant, and the material need not be dried before being purified. This particular azo dye is too soluble to be crystallized from water; it can be obtained in a fairly satisfactory form by adding saturated sodium chloride solution to a hot, filtered solution in water and then cooling it, but the best crystals are obtained from aqueous ethanol.

Transfer the filter cake to a reaction tube, wash the material from the filter paper and funnel with 1 mL of water, and bring all of the solid into solution at the boiling point. It may be necessary to add another 0.25 mL (no more) of water in the

FIG. 46.3
A filter flask with a Hirsch funnel, a polyethylene frit, and filter paper.

course of filtering this hot solution through a Hirsch funnel equipped with a polyethylene frit (Fig. 46.3). Collect the filtrate in a reaction tube, add 2.5–3 mL of ethanol, and allow crystallization to proceed as the tube cools slowly to room temperature. Cool well in ice before collecting the product. Rinse the tube with some of the filtrate and finally, wash the product with a small quantity of ethanol. The yield of pure crystalline material should be about 150 mg. Orange II separates and crystallizes from aqueous alcohol as the dihydrate, containing two molecules of water; allowance for this should be made in calculating the yield. If the water of hydration is eliminated by drying at 120°C, the material will become fiery red.

Cleaning Up. The filtrate from the reaction, although highly colored, contains little dye but is highly soluble in water. It can be diluted with a large quantity of water and flushed down the drain. Alternatively, with the volume kept as small as possible, the filtrate can be placed in the aromatic amines hazardous waste container, or it can be reduced with tin(II) chloride (see Experiment 5). The crystallization filtrate should go into the organic solvents waste container. Wash the Hirsch funnel *thoroughly.*

3. ORANGE II (1-*p*-SULFOBENZENEAZO-2-NAPHTHOL, SODIUM SALT)

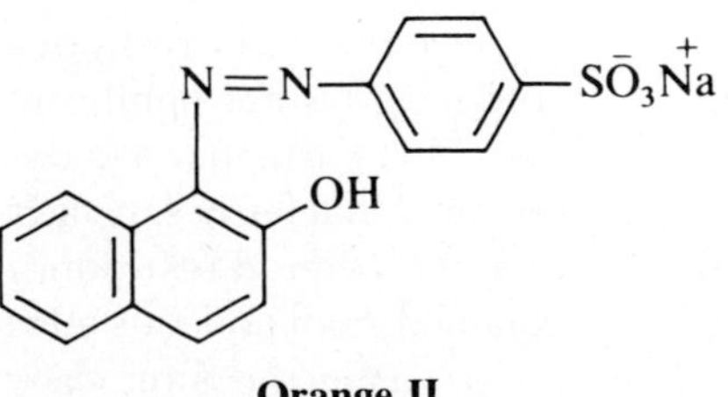

Orange II

CAUTION: Because 2-naphthol is a carcinogen, handle it with care, in the hood. Do not breathe the dust or allow skin contact.

In a 250-mL beaker, dissolve 1.8 g of 2-naphthol in 10 mL of cold 3 *M* sodium hydroxide solution; pour into this solution, with stirring, the suspension of diazotized sulfanilic acid prepared in the macroscale procedure of Experiment 1. Rinse the Erlenmeyer flask with a small amount of water and add it to the beaker. Coupling occurs very rapidly, and the dye, being a sodium salt, separates easily from the solution because a considerable excess of sodium ion from the carbonate, the nitrite, and the alkali is present. Stir the crystalline paste thoroughly to effect good mixing and, after 5–10 minutes, heat the mixture until the solid dissolves. Add 5 g of sodium chloride to further decrease the solubility of the product, bring this all into solution by heating and stirring, set the beaker in a pan of ice and water, and let the solution cool undisturbed. When near room temperature, cool further by stirring and collect the product on a Büchner funnel. Use saturated sodium chloride solution rather than water for rinsing the material from the beaker and for washing the filter cake free of the dark-colored mother liquor. The filtration is somewhat slow.[3]

The product dries slowly; it contains about 20% of sodium chloride. Thus, the crude yield is not significant, and the material need not be dried before being purified. This particular azo dye is too soluble to be crystallized from water; it can be obtained in a fairly satisfactory form by adding saturated sodium chloride solution to a hot, filtered solution in water and cooling, but the best crystals are obtained from aqueous ethanol.

Transfer the filter cake to a beaker, wash the material from the filter paper and funnel with water, and bring the cake into solution at the boiling point. Avoid a large excess of water, but use enough to prevent separation of the solid during filtration (use about 25 mL). Filter by suction through a Büchner funnel that has been preheated on a steam bath. Pour the filtrate into an Erlenmeyer flask, rinse the filter flask with a small quantity of water, add it to the flask, estimate the volume, and if greater than 30 mL, evaporate by boiling. Cool to 80°C, add 50–60 mL of ethanol, and allow crystallization to proceed. Cool the solution well before collecting the product. Rinse the beaker with the mother liquor, and wash finally with a little ethanol. The yield of pure crystalline material is about 3.4 g. Orange II separates and crystallizes from aqueous alcohol as the dihydrate, with two molecules of water; allowance for this should be made in calculating the yield. If the water of hydration is eliminated by drying at 120°C, the material will become fiery red.

Cleaning Up. The filtrate from the reaction, although highly colored, contains little dye but is highly soluble in water. It can be diluted with a large quantity of water and flushed down the drain. Alternatively, with the volume kept as small as possible, the filtrate can be placed in the aromatic amines hazardous waste container, or it can be reduced with tin(II) chloride (*see* Experiment 6). The crystallization filtrate should go into the organic solvents waste container.

[3]If the filtration must be interrupted, fill the funnel with product and close the rubber suction tubing (while the aspirator is still running) with a screw pinchclamp placed near the filter flask. Then disconnect the tubing from the trap and set the unit aside. Thus, suction will be maintained, and filtration will continue.

4. METHYL ORANGE (*p*-SULFOBENZENEAZO-4-DIMETHYLANILINE, SODIUM SALT)

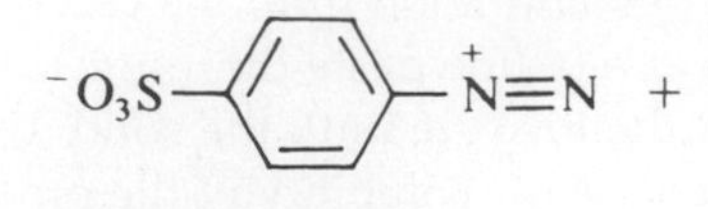

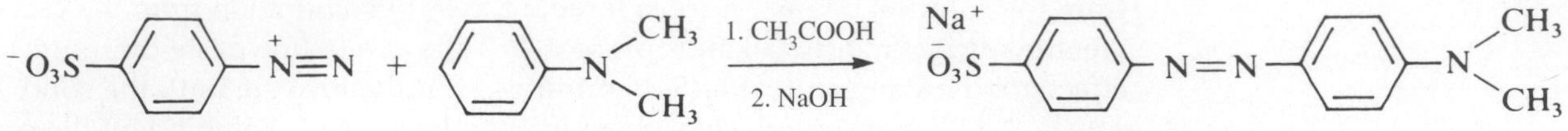

Diazotized sulfanilic acid
4-Diazobenzenesulfonic acid

+

***N,N*-Dimethylaniline**
MW 121.18
bp 194°C

1. CH_3COOH
2. NaOH

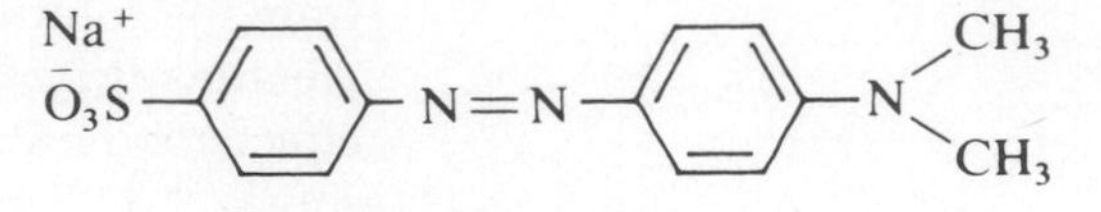

Methyl Orange
4-[4-(Dimethylamino)phenylazo]benzenesulfonic acid, sodium salt
MW 327.34
λ_{max} 507 nm

Methyl Orange is an acid-base indicator.

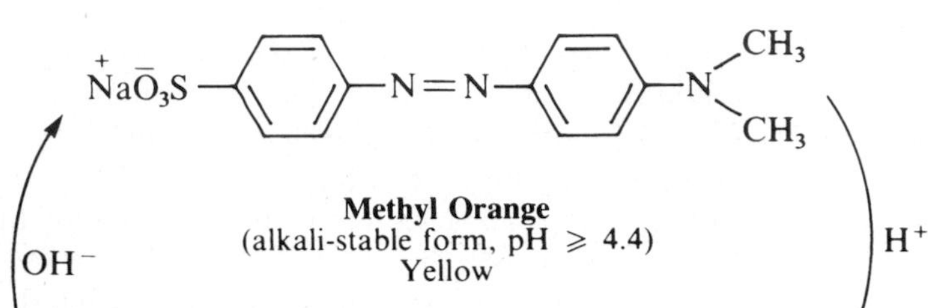

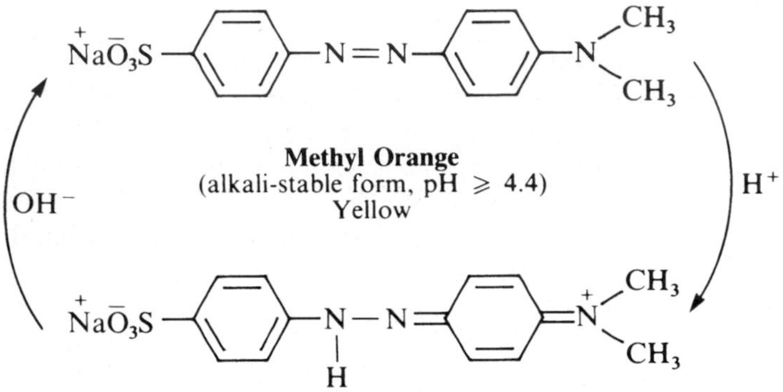

Methyl Orange
(alkali-stable form, pH ⩾ 4.4)
Yellow

OH^- ⇄ H^+

Methyl Orange
(acid-stable form, pH ⩽ 3.2)
Red

Microscale Procedure

In a 10-mL Erlenmeyer flask, mix 75 mg of dimethylaniline and 65 mg of acetic acid. Add the suspension of diazotized sulfanilic acid prepared in the microscale procedure in Experiment 1, with stirring, to the solution of dimethylaniline acetate, rinsing out the last portions with a few drops of water. Stir the mixture thoroughly and, within a few minutes, the red acid-stable form of the dye should separate. A stiff paste should result in 5–10 minutes, at which time, 1 mL of 3 *M* sodium hydroxide is added to produce the orange sodium salt. Heat the mixture to the boiling point with constant stirring (to avoid bumping) in order to dissolve the dye. A better way to obtain crystals would be to place the flask in a small beaker of boiling water. Remove the flask after the dye has been dissolved, and allow it to cool slowly to room temperature; then cool it thoroughly in ice before collecting the product by vacuum filtration on a Hirsch funnel bearing a piece of filter paper. Use saturated sodium chloride solution rather than water to rinse the flask, and to wash the dark mother liquor from the filter cake.

Video: *Microscale Filtration on the Hirsch Funnel*

The crude product does not need to be dried; it can be crystallized from water after performing preliminary solubility tests to determine the proper conditions. The yield should be between 125 mg and 150 mg. Recrystallized material need not necessarily be used for the dyeing process.

Cleaning Up. The highly colored filtrates from the reaction and crystallization are very water soluble. After dilution with a large quantity of water, they can be flushed down the drain because the amount of solid is small. Alternatively, the combined filtrates should be placed in the hazardous waste container, or the mixture can be reduced with tin(II) chloride (*see* Experiment 5). Wash the Hirsch funnel thoroughly.

Macroscale Procedure

In a test tube, thoroughly mix 1.6 mL of dimethylaniline and 1.25 mL of glacial acetic acid. To the suspension of diazotized sulfanilic acid prepared in the macroscale procedure in Experiment 1 contained in a 250-mL beaker, add, with stirring, the solution of dimethylaniline acetate. Rinse the test tube with a small quantity of water and add it to the beaker. Stir and mix thoroughly; within a few minutes, the red, acid-stable form of the dye should separate. A stiff paste should result in 5–10 minutes; 18 mL of 3 M sodium hydroxide solution is then added to produce the orange sodium salt. Stir well and heat the mixture to the boiling point, when a large part of the dye should dissolve. Place the beaker in a pan of ice and water, and allow the solution to cool undisturbed. When cooled thoroughly, collect the product on a Büchner funnel, using saturated sodium chloride solution rather than water to rinse the flask and to wash the dark mother liquor from the filter cake.

The crude product does not need to be dried; it can be crystallized from water after making preliminary solubility tests to determine the proper conditions. The yield is about 2.5–3 g.

Cleaning Up. The highly colored filtrates from the reaction and crystallization are very water soluble. After dilution with a large quantity of water, they can be flushed down the drain because the amount of solid is small. Alternatively, the combined filtrates should be placed in the hazardous waste container, or the mixture can be reduced with tin(II) chloride (*see* Experiment 5).

5. FOR FURTHER INVESTIGATION

Following the procedure for the synthesis of Orange II, couple diazotized sulfanilic acid with one or more of the following naphthols: 4-amino-5-hydroxy-2,7-naphthalenedisulfonic acid, monosodium salt; 6-amino-4-hydroxy-2-naphthalenesulfonic acid; 7-amino-4-hydroxy-2-naphthalenesulfonic acid; 4-hydroxy-1-naphthalenesulfonic acid, sodium salt; or 6-hydroxy-2-naphthalenesulfonic acid, sodium salt. Despite the apparent complexity of these molecules, they are not expensive; they are used for the commercial synthesis of dyes. The many sulfonic acid groups have been added to make the dyes water soluble; the addition of solid sodium chloride "salts out" the dye, rendering it insoluble.

6. TESTS

Solubility and Color

Compare the solubility in water of Orange II to Methyl Orange, and account for the difference in terms of structure. Treat the first solution with alkali and note the change in shade due to salt formation; to the other solution, alternately add acid and alkali.

Reduction

A characteristic of an azo compound is the ease with which the molecule is cleaved at the double bond by reducing agents to give two amines. Because amines are colorless, the reaction is easily followed by the color change. This reaction is useful in preparing hydroxyamino and similar compounds in order to analyze azo dyes by titration with a reducing agent, and in identifying azo compounds from an examination of the cleavage products.

This reaction can, if necessary, be run on five times the indicated scale. Dissolve about 0.1 g of tin(II) chloride in 0.2 mL of concentrated hydrochloric acid, add a small quantity of the azo compound (20 mg), and heat. A colorless solution should result, and no precipitate should form upon adding water. The aminophenol or the diamine products are present as the soluble hydrochlorides; the other product of cleavage—sulfanilic acid—is sufficiently soluble to remain in solution.

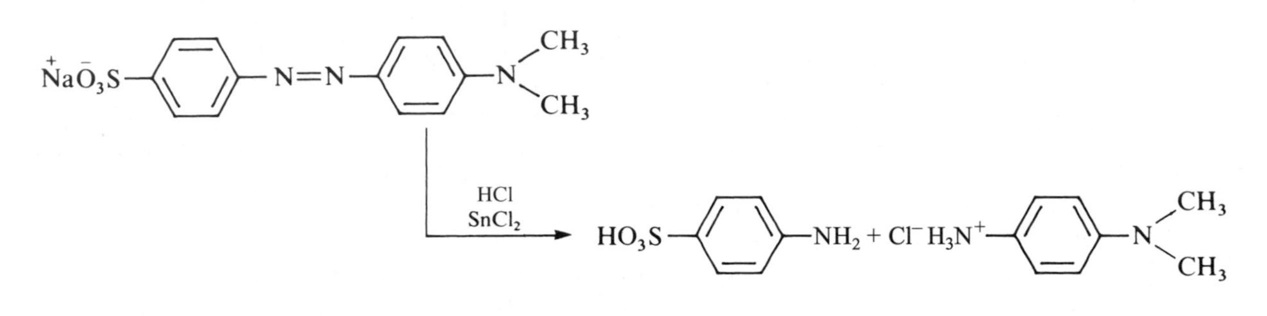

Cleaning Up. Dilute the reaction mixture with water, neutralize with sodium carbonate, and remove the solids by vacuum filtration. The solids go in the aromatic amines hazardous waste container, and the filtrate can be flushed down the drain.

PART 2: DYEING EXPERIMENTS

Using good laboratory techniques, your hands will not be dyed; use care.

The following experiments can, if necessary, be run on up to 10 times the indicated amounts. Wearing disposable gloves (polyethylene or vinyl) might be advisable, although if you use good laboratory techniques, your hands will not be dyed; use care.

1. DIRECT DYES

The sulfonate groups on the Methyl Orange and Orange II molecules are polar, and thus enable these dyes to combine with polar sites in fibers. Wool and silk have many polar sites on their polypeptide chains, and hence bind strongly to a dye of this type. Martius Yellow, picric acid, and eosin are also highly polar dyes, and thus dye directly to wool and silk.

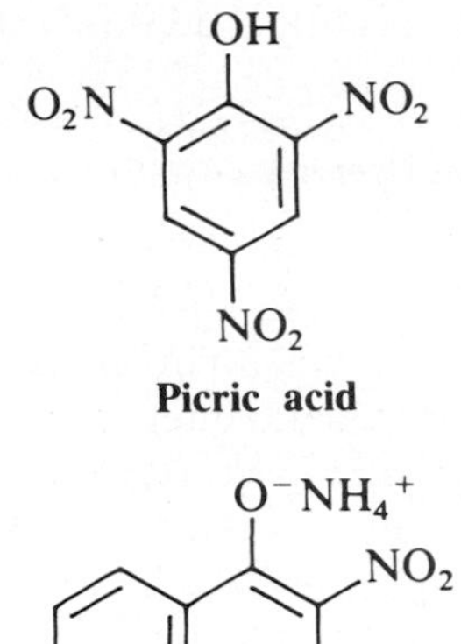

Picric acid

$O^-NH_4^+$
NO_2
NO_2
Martius Yellow

Orange II or Methyl Orange

The dye bath is prepared from 50 mg of Orange II or Methyl Orange, 0.5 mL of 3 *M* sodium sulfate solution, 15 mL of water, and 5 drops of 3 *M* sulfuric acid in a 30-mL beaker. Place a piece of test fabric, a strip 3/4-in. wide, in the bath for 5 minutes at a temperature near the boiling point. Remove the fabric from the dye bath, allow it to cool, and then wash it thoroughly with soap under running water before drying it.

Dye untreated test fabric and one or more of the pieces of test fabric that have been treated with a mordant following this same procedure. *See* Experiment 3 in this part for the application of mordants.

Picric Acid or Martius Yellow

Dissolve 50 mg of one of these acidic dyes in 15 mL of hot water to which a few drops of dilute sulfuric acid have been added. Heat a piece of test fabric in this bath for 1 minute; then remove it with a stirring rod, rinse well, scrub with soap and water, and dry. Describe the results.

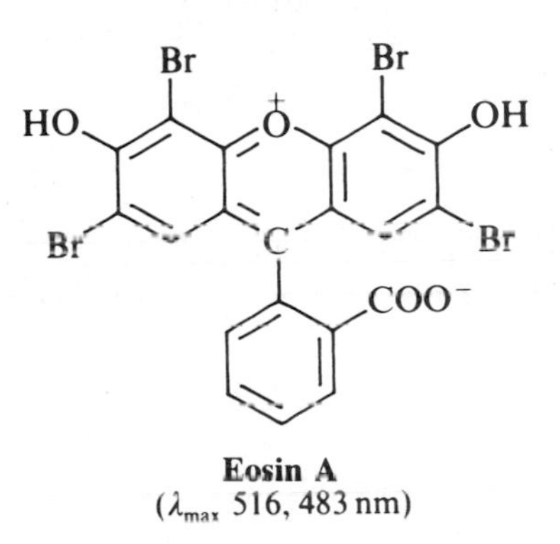

Eosin A
(λ_{max} 516, 483 nm)

Eosin

Dissolve 10 mg of sodium eosin in 20 mL of water, and dye a piece of test fabric by heating it with the solution for about 10 minutes. Eosin is the dye used in red ink. Also dye pieces of mordanted cloth in eosin (*see* Experiment 3 in this part). Wash and rinse the dyed cloth in the usual way, as described above.

Cleaning Up. In each case, dilute the dye bath with a large quantity of water and flush it down the drain.

2. SUBSTANTIVE DYES: CONGO RED

Cotton and rayons do not have the anionic and cationic carboxyl and amine groups of wool and silk, and hence do not dye well with direct dyes. However, cotton and rayons can be dyed with substances of rather high molecular weight showing colloidal properties. Such dyes probably become fixed to the fibers by hydrogen bonding: Congo Red is a substantive dye.

Dissolve 10 mg of Congo Red in 40 mL of water, add about 0.1 mL each of 3 *M* solutions of sodium carbonate and sodium sulfate, heat to a temperature just below the boiling point, and introduce a piece of test fabric. At the end of 10 minutes, remove the fabric and wash in warm water with soap as long as the dye continues to be removed. Place pieces of the dyed material in very dilute hydrochloric acid solution, and observe the result. Rinse and wash the material with soap.

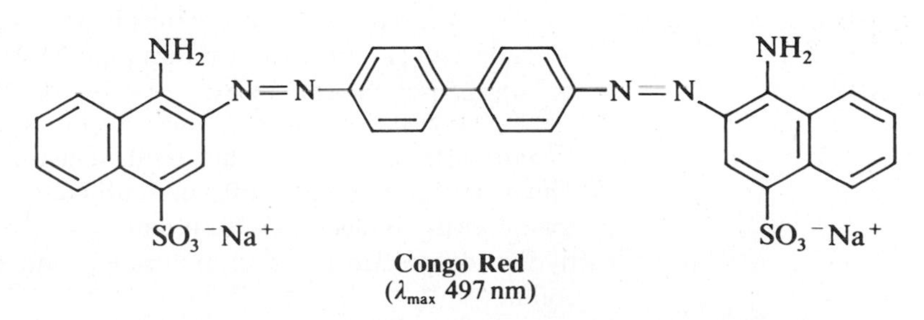

Congo Red
(λ_{max} 497 nm)

Cleaning Up. The dye bath should be diluted with water and flushed down the drain.

3. MORDANT DYES

One of the oldest known methods of producing wash-fast dyes involves using metallic hydroxides, which form a link, or mordant (Latin *mordere*, to bite), between the fabric and the dye. Other substances, such as tannic acid, also function as mordants. The color of the final product depends on both the dye used and the mordant. For instance, the dye Turkey Red (alizarin) appears red with an aluminum mordant, violet with an iron mordant, and brownish-red with a chromium mordant. Some commercially important mordant dyes possess a structure based on triphenylmethane, as well as Crystal Violet and Malachite Green.

Chromium functioning as a mordant

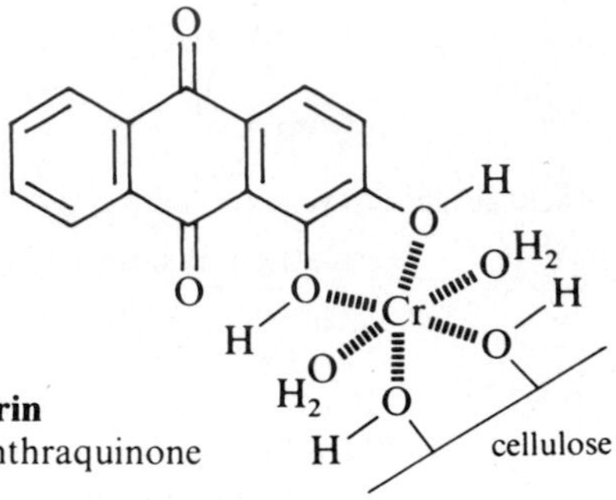

Alizarin
1,2-Dihydroxyanthraquinone

cellulose

Applying Mordants—Tannic Acid, Iron, Tin, Chromium, Copper, and Aluminum

CAUTION: $K(SbO)C_4H_4O_6$ is toxic, and the dust is a suspected carcinogen.

Mordant pieces of test fabric or wool yarn by allowing them to stand in a hot (nearly boiling) solution of 0.1 g of tannic acid in 50 mL of water for 30 minutes. The tannic acid mordant must now be fixed to the cloth; otherwise it would wash out. In order to do this, transfer the cloth or yarn to a hot bath made from 20 mg of the fixing agent potassium antimonyl tartrate [$K(SbO)C_4H_4O_6$; tartar emetic] in 20 mL of water. After 5 minutes, wring or blot the cloth and dry it as much as possible over a warm hot plate.

Mordant 1/2-in. strips of test cloth or yarn in the following mordants, which are 0.1 *M* solutions of the indicated salts: ferrous sulfate, stannous chloride, potassium dichromate, copper sulfate, and potassium aluminum sulfate (alum). (The alum and dichromate solutions should also contain 0.05 *M* oxalic acid.) Immerse pieces of cloth in the solutions, which are kept near the boiling point, for about 15–20 minutes or longer. These mordanted pieces of cloth or yarn can then be dyed with alizarin (1,2-dihydroxyanthraquinone, Turkey Red) and either Methyl Orange or Orange II in the usual way.

Cleaning Up. Mix the mordant baths. The Fe^{2+} and Sn^{2+} will reduce the Cr^{6+} to Cr^{3+}. The mixture can then be diluted with water and flushed down the drain because the quantity of metal ions is extremely small. Alternatively, precipitate the ions as hydroxides, collect by vacuum filtration, and place the solid in the hazardous waste container.

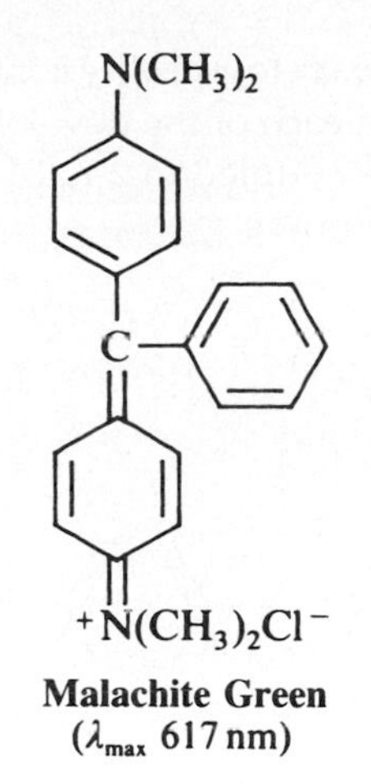

Malachite Green
(λ_{max} 617 nm)

Dyeing with a Triphenylmethane Dye: Crystal Violet or Malachite Green

A dye bath is prepared by dissolving 10 mg of either Crystal Violet or Malachite Green in 20 mL of boiling water. These are basic or cationic dyes. Dye the mordanted cloth or yarn in this bath for 5–10 minutes at a temperature just below the boiling point. Dye another piece of cloth that has not been mordanted and compare the two. In each case, allow as much of the dye to drain back into the beaker as possible and then, using glass rods, wash the dyed cloth or yarn under running water with soap, blot, and dry.

Cleaning Up. The stains on glass produced by triphenylmethane dyes can be removed with a few drops of concentrated hydrochloric acid and washing with water, as HCl forms a di- or trihydrochloride that is more soluble in water than the original monosalt.

The dye bath and acid washings are diluted with water and flushed down the drain because the quantity of dye is extremely small.

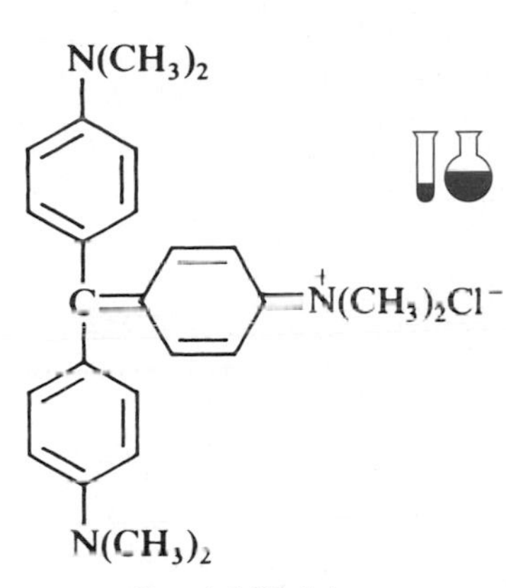

Crystal Violet
(λ_{max} 591, 540 nm)

4. DEVELOPED DYES

A superior method of applying azo dyes to cotton, patented in England in 1880, is that in which cotton is soaked in an alkaline solution of a phenol, and then in an ice-cold solution of a diazonium salt; the azo dye is developed directly on the fiber. The reverse process (ingrain dyeing) of impregnating cotton with an amine, which is then diazotized and developed by immersion in a solution of the phenol, was introduced in 1887. The first ingrain dye was Primuline Red, obtained by coupling the sulfur dye Primuline, after application to the cloth and diazotization, with 2-naphthol. Primuline (substantive to cotton) is a complex thiazole prepared by heating *p*-toluidine with sulfur, and then introducing a solubilizing sulfonic acid group.

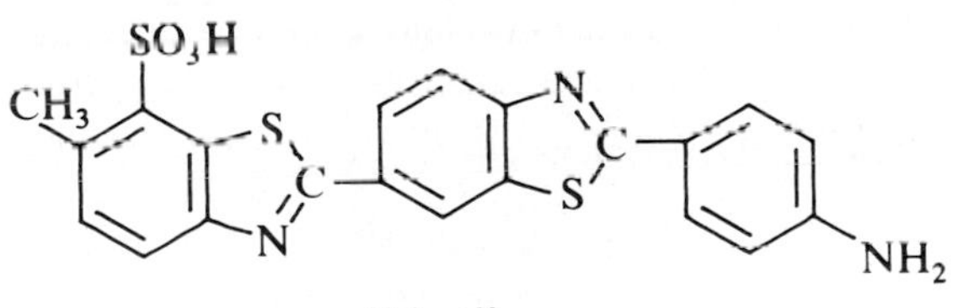

Primuline

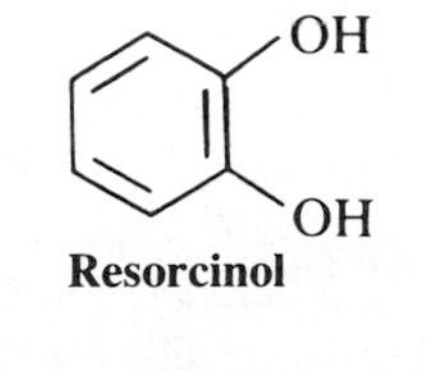

Resorcinol

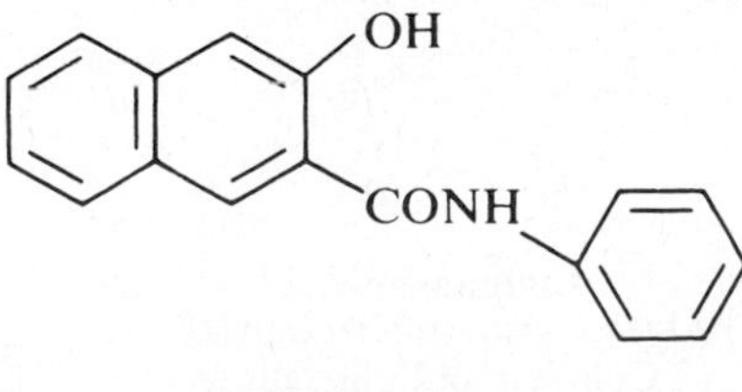

Naphthol AS

Primuline Red

Dye three pieces of cotton cloth in a solution of 20 mg of Primuline and 0.5 mL of sodium carbonate solution in 50 mL of water, at a temperature just below the boiling point for 15 minutes. Then wash the pieces of cloth twice in about 50 mL of water. Prepare a diazotizing bath by dissolving 20 mg of sodium nitrite in 50 mL of water containing a little ice and, just before using the bath, add 0.5 mL of concentrated hydrochloric acid. Allow the pieces of cloth dyed with Primuline to stay in this diazotizing bath for about 5 minutes. Next, prepare three baths for the coupling reaction. Dissolve 10 mg of 2-naphthol in 0.2 mL of 1.5 *M* sodium hydroxide solution and dilute with 10 mL of water; prepare similar baths from phenol, resorcinol, Naphthol AS, or other phenolic substances.

Transfer the three pieces of cloth from the diazotizing bath to a beaker containing about 50 mL of water and stir. Put one piece of cloth in each of the developing baths and allow them to remain for 5 minutes. Primuline coupled to 2-naphthol gives the dye called Primuline Red. Draw the structure of the dye.

Para Red, an Ingrain Color

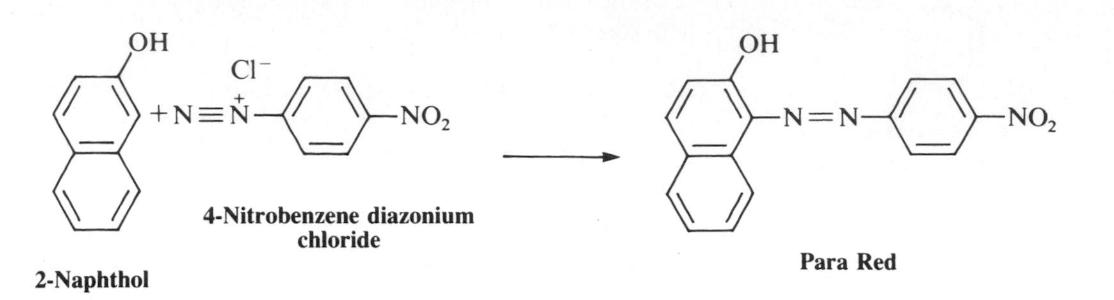

A solution is prepared by suspending 300 mg of 2-naphthol in 10 mL of water, stirring well, and adding 3 *M* sodium hydroxide solution, a drop at a time, until the naphthol just dissolves. Do not add excess alkali. The material to be dyed is first soaked in or painted with this solution, and then dried, preferably in an oven.

CAUTION: 4-Nitroaniline is toxic. Handle with care.

Para Red is the red dye used for the American flag.

Prepare a solution of 4-nitrobenzenediazonium chloride as follows: Dissolve 140 mg of 4-nitroaniline in a mixture of 3 mL of water and 0.6 mL of 3 *M* hydrochloric acid by heating. Cool the solution in ice (the hydrochloride of the amine may crystallize), add all at once a solution of 80 mg of sodium nitrite in about 0.5 mL of water, and stir. In about 10 minutes, a clear solution of the diazonium salt will be obtained. Just before developing the dye on the cloth, add a solution of 80 mg of sodium acetate to 0.5 mL of cold water. Stir in the acetate well, add 30 mL of water, and then immediately add the cloth. The diazonium chloride solution may also be painted onto the cloth.

Good results can be obtained by substituting Naphthol AS for 2-naphthol; in this case, it is necessary to warm the Naphthol AS with alkali and to break the lumps with a flattened stirring rod to bring the naphthol into solution.

Cleaning Up. The dye baths should be diluted with water and flushed down the drain because the quantity of dye is extremely small.

5. DYEING WITH INDIGO: THE DENIM DYE

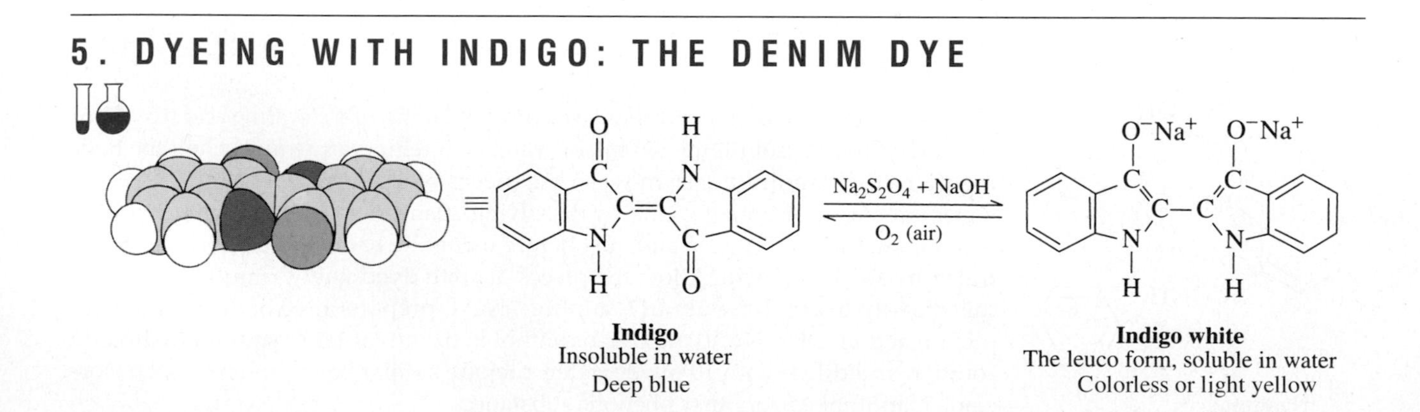

Indigo
Insoluble in water
Deep blue

Indigo white
The leuco form, soluble in water
Colorless or light yellow

Look around you. No matter where in the world you are carrying out these experiments, you will find people wearing cotton clothes dyed with indigo. Indigo, "the king of dyes," has been used for more than 3500 years, first in India and then in ancient Egypt. It is a unique dye in that it can be abraded from the surface of a fabric, where other dyes penetrate the fiber. This explains why the knees and other parts of blue jeans (dyed exclusively with indigo) will gradually turn white. An advantage of this fact has been taken by manufacturers who sell "stonewashed" denim blue jeans that have been tumbled with pumice to wear off some of the dye (which can then be reused).

Denim, de Nimes, from Nimes, in southern France

This dye originally came from the leaves of the indigo plant, but it is now produced by chemical synthesis. It is one of the *vat dyes*, a term applied to dyes that are reduced to a colorless (leuco) form that is then oxidized on the surface of the fiber. Formerly, the reduction was carried out by fermenting plant leaves; now, chemical reducing agents, most commonly sodium hydrosulfite, are used.

Experimental Considerations

Indigo is a dark-blue powder that is completely insoluble in water and is not easily wet by water. To disperse the dye in water, a bit of soap is added to the dye bath. The liquid will appear blue, but the dye is not dissolved; it is merely suspended. Sodium hydrosulfite ($Na_2S_2O_4$) is then added to the hot dye bath. Because this reducing agent decomposes on storage, it is difficult to state exactly how much should be used. Add the required amount, stir, and then look through the side of the beaker. If the solution is not transparent and light yellow in color, add more of the reducing agent. From the top, the liquid will still appear to be dark blue because of a film of the oxidized dye on the surface of the solution.

Procedure

Use 100 mg of indigo, and a drop of detergent or a pinch of soap powder. Boil the dye with 50 mL of water, 2.5 mL of 3 *M* sodium hydroxide solution, and about 0.5 g of sodium hydrosulfite until the dye is reduced. At this point, a clear solution will be seen through the side of the beaker. Add more sodium hydrosulfite if necessary to produce a clear solution. Introduce a piece of cloth, and boil the solution gently for 10 minutes. Rinse the cloth well in water, and then allow it to dry. To increase the intensity of the dye, repeat the process several times with no drying. Describe what happens during the drying process.

Other dyes that can be used in this procedure are Indanthrene Brilliant Violet and Indanthrene Yellow.

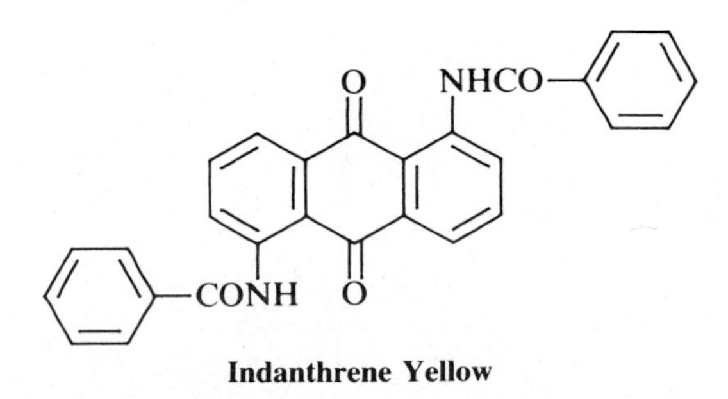

Indanthrene Yellow

Cleaning Up. Add household bleach (5.25% sodium hypochlorite solution) to the dye bath to oxidize it to the starting material. The mixture can be diluted with water and flushed down the drain; alternatively, the small amount of solid can be removed by filtration, with the solid placed in the aromatic amines hazardous waste container.

6. DISPERSE DYES: DISPERSE RED

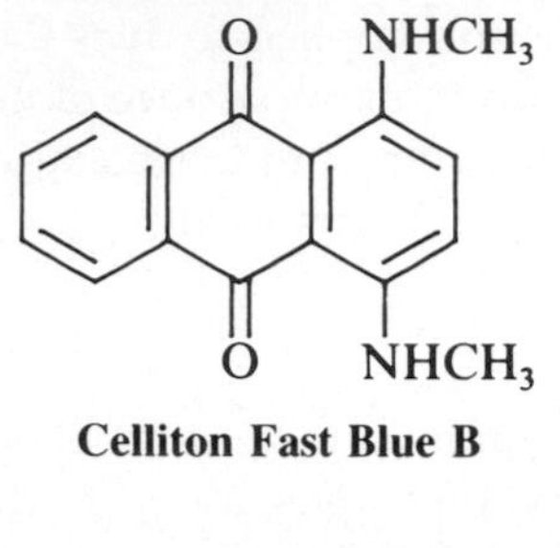

Celliton Fast Blue B

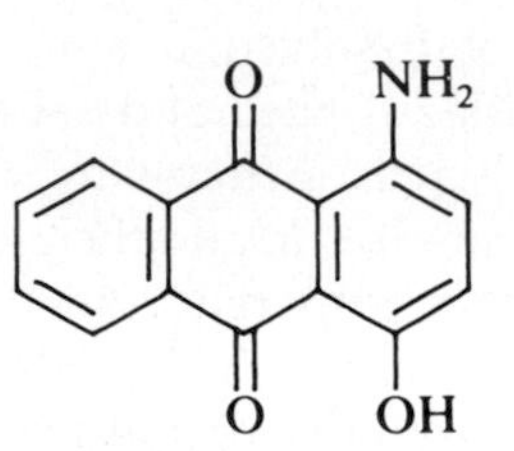

Celliton Fast Pink B

Fibers such as polyester, acetate rayon, nylon, and polypropylene are difficult to dye with conventional dyes because they contain so few polar groups. These fibers are dyed with substances that are insoluble in water but, when subjected to elevated temperatures (pressure vessels), are soluble in the fiber as true solutions. They are applied to the fiber in the form of a dispersion of finely divided dye (hence the name). The Cellitons are typical disperse dyes.

Disperse Red, a brilliant red dye used commercially, is synthesized in this experiment.

Diazotization of 2-Amino-6-methoxybenzothiazole

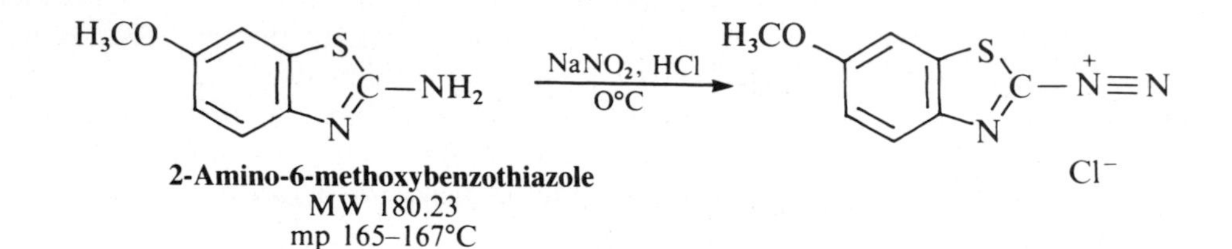

To 135 mg (0.75 mmol) of 2-amino-6-methoxybenzothiazole in 1.5 mL of water in a test tube, add 0.175 mL of concentrated hydrochloric acid; then cool the solution to 0–5°C. To this mixture add, dropwise, an *ice-cold* solution of 55 mg of sodium nitrite that has been dissolved in 0.75 mL of water. The reaction mixture changes color, and some of the diazonium salt crystallizes out, but it should not foam. Foaming, caused by the evolution of nitrogen, is an indication that the mixture is too warm. Keep the mixture ice-cold until used in the coupling reaction.

Disperse Red

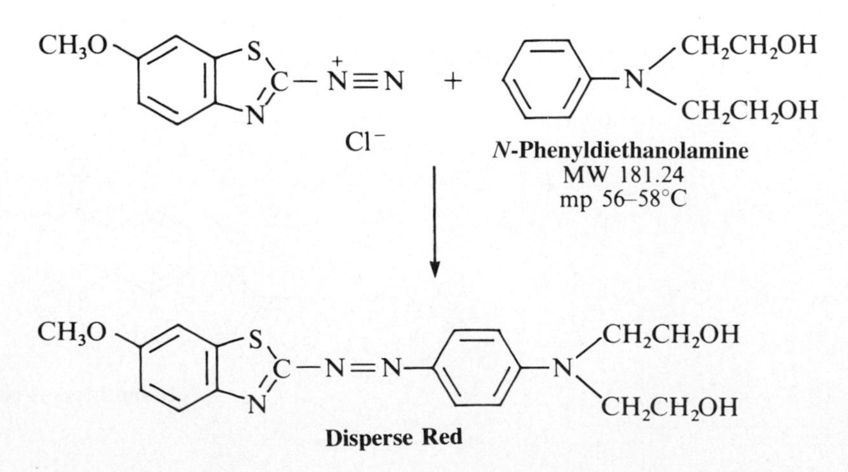

To 135 mg (0.75 mmol) of *N*-phenyldiethanolamine in 0.75 mL of hot water, add just enough 3 *M* hydrochloric acid to bring the amine into solution. This amount is less than 0.5 mL. Cool the resulting solution to 0°C in ice, and add to it, dropwise and with *very thorough mixing*, the diazonium chloride solution. Mix the solution well by drawing it into a Pasteur pipette and expelling it into a cold reaction tube. Allow the mixture to come to room temperature over a period of 10 minutes; then add 225 mg of sodium chloride and heat the mixture to boiling. The sodium chloride decreases the solubility of the product in water. Allowing the hot solution to cool slowly to room temperature should afford easily filterable crystals. Collect the dye on a Hirsch funnel, wash it with a few drops of saturated sodium chloride solution, and press it dry on the funnel.

Video: *Microscale Filtration on the Hirsch Funnel*

This reaction often produces a noncrystalline product that looks like purple tar, which is the dye. It can be used to dye the Multifiber Fabric, so do not discard the reaction mixture.

Save the filtrate. Add the material on the filter paper to 50 mL of boiling water, and dye a piece of test cloth for 5 minutes. Do the same with the filtrate. Wash the pieces of cloth with soap and water, rinse and dry them, and compare the results.

Cleaning Up. The dye baths should be diluted with water and flushed down the drain because the quantity of dye is extremely small.

7. TEST IDENTIFICATION STAIN

In the dye industry, a proprietary mixture of three dyes called *Test Identification Stain*[4] is used to dye cloth. Prepare a dye bath by dissolving 25 mg of this dye in 25 mL of water, adding 1 drop of acetic acid, and bringing the mixture to a boil. Dye a piece of test cloth in the mixture for 5 minutes, remove it, immediately wash it under running water, and then scrub it *thoroughly* with soap and water before rinsing and drying. Describe the rather extraordinary result and suggest how this mixture could be used industrially. Several pieces of test cloth can be dyed in this one dye bath. Analyze the dye mixture by thin-layer chromatography, using a 40:60 mixture of ethanol and hexane as the eluent.

Cleaning Up. Dilute the dye bath with copious amounts of water and flush it down the drain. Dyeing cloth removes much of the dye from the dye bath.

8. FLUORESCEIN AND EOSIN

Fluorescein, as its name implies, is fluorescent; in fact, it is so fluorescent that the sodium salt in water can be detected under ultraviolet (UV) light at concentrations of 0.02 ppm. Because of this property, fluorescein is used to trace the paths of underground rivers such as those found in Mammoth Cave in Kentucky, to pinpoint sources of contamination of drinking water, to find leaks in the huge condensers found in power plants, and so on. It is also used to visualize scratches on the cornea of the eye. The tetrabromo derivative is eosin, a dye used in lipstick, nail polish, red ink, and as a biological stain.

[4]The Test Identification Stain is available from Kontes.

The synthesis of fluorescein is very straightforward; however, in this experiment, we have deliberately omitted the equations for the synthesis, the mechanism of the reaction, the structures of the molecules in acid and base, and the structures of the tetrabromo derivatives. These are left for you to solve, or look up in the library.[5]

Synthesis of Fluorescein

Videos: *Filtration of Crystals Using the Pasteur Pipette, Microscale Filtration on the Hirsch Funnel*

In a reaction tube, heat 100 mg of zinc chloride in a hot sand bath until no more moisture comes from the tube; then add 200 mg of resorcinol and 140 mg of phthalic anhydride. Place a thermometer into the mixture and heat the mixture to 180°C. Stir with a stirring rod and continue heating for 10 minutes, at which time the mixture should cease bubbling and become stiff. Cool the mixture to below 100°C, add 2 mL of water and 0.2 mL of concentrated hydrochloric acid, then stir and grind the solid (this process is called *trituration*) while heating it on a steam bath or boiling water bath (but not on a sand bath because it will bump). Remove the liquid with a Pasteur pipette. After again triturating with another similar portion of water and acid, collect the solid on a Hirsch funnel, wash it well with water, and allow it to dry. A typical yield is 240 mg of fluorescein.

Fluorescence

Add 50 mg of fluorescein to 0.5 mL of 3 *M* sodium hydroxide and 0.5 mL of water; then dilute the mixture to 100 mL. Examine the solution by transmitted light, and then by reflected sunlight or UV light. Make the solution acidic and again examine it under sunlight or UV light.

Tetrabromofluorescein, Eosin: The Red Ink Dye

In a reaction tube, place 100 mg of fluorescein and 0.4 mL of ethanol. To this mixture, add 215 mg of bromine in 0.5 mL of carbon tetrachloride, dropwise, mixing thoroughly after each drop. After half the bromine is added, a solution of soluble dibromofluorescein results. When all the bromine has been added, the tetrabromo compound will precipitate. After about 15 minutes, cool the reaction mixture in ice and remove the solvent with a Pasteur pipette. Wash the product in the reaction tube with a few drops of ice-cold ethanol; then dry it under vacuum. The yield should be about 160 mg.

Prepare the ammonium salt of eosin by exposing the salt to ammonia vapors. Moisten about 80 mg of fluorescein with a few drops of ethanol on a small filter paper. Fold up the paper and push it into a reaction tube above 0.5 mL of ammonium hydroxide. Cap the tube with a septum. After about an hour, a test portion of the derivative should be completely soluble in water. Dye a piece of test cloth with this dye. Eosin is used commercially to dye silk.

Answer the following questions. The answers for some will come from your experimental observations:

1. Write a balanced equation for the reaction of resorcinol with phthalic anhydride.

[5]If you would like to read about the original synthesis, see *J Prakt Chem.* **1922**; 104:123, which will require a reading knowledge of German, and the structure proof in *Comp Rendu.* **1937**; 205:864, which will require a reading knowledge of French. Further searching in the library may disclose other papers in English.

2. Calculate the number of moles of starting material, determine the limiting reagent, and calculate the theoretical yield of fluorescein.
3. In view of the balanced equation, why is it necessary to dehydrate zinc chloride, a very hygroscopic substance?
4. To which general category of substances does zinc chloride belong that is relevant to its use in this reaction? What two functions does zinc chloride serve?
5. Write a possible mechanism for the synthesis of fluorescein.
6. Write an equation showing the reaction of fluorescein with sodium hydroxide. Hint: A quinoid structure is involved.
7. How do you account for the changes in fluorescence in going from an acidic to a basic medium?
8. Compare the color and fluorescence changes of fluorescein to those of the related compound, phenolphthalein.
9. Write a balanced equation for the bromination of fluorescein to give first the dibromo, and then the tetrabromo compound. Explain your choices for the positions of bromination.
10. Write a balanced equation for the reaction of tetrabromofluorescein with ammonia.

9. OPTICAL BRIGHTENERS: FLUORESCENT WHITE DYES

Most modern detergents contain a blue-white fluorescent dye that is adsorbed on the cloth during the washing process. These dyes fluoresce; that is, absorb UV light and reemit light in the visible blue region of the spectrum. This blue color counteracts the pale-yellow color of white goods, which develops because of a buildup of insoluble lipids. The modern-day usage of optical brighteners has replaced a past custom of using bluing (ferriferrocyanide).

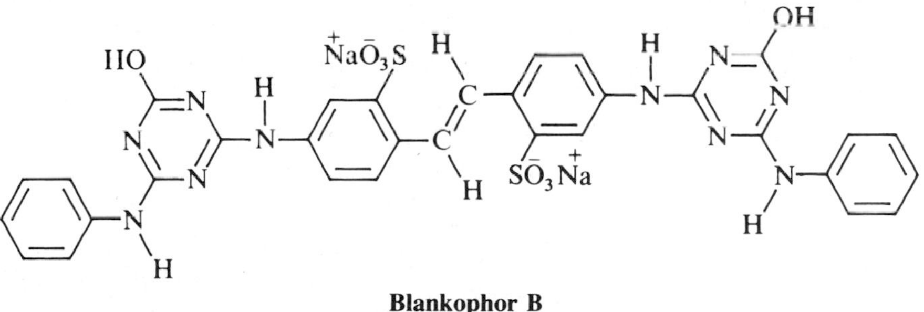

Blankophor B
An optical brightener

Dyeing with Detergents

Immerse a piece of test fabric in a hot solution (0.5 g of detergent, 200 mL of water) of a commercial laundry detergent that you suspect may contain an optical brightener (e.g., Tide and Cheer) for 15 minutes. Rinse the fabric thoroughly, dry, and compare it to an untreated fabric sample under a UV lamp.

See *Web Links*

Cleaning Up. Dilute the solution with water and flush down the drain.

QUESTIONS

1. Write reactions showing how nylon can be synthesized so that it will react with (a) basic dyes and (b) acidic dyes.
2. Draw the resonance form of dimethylaniline that is most prone to react with diazotized sulfanilic acid.
3. Draw a resonance form of indigo that would be present in base.
4. Draw a resonance form of indigo that has been reduced and is therefore colorless.

CHAPTER 47

Martius Yellow

PRELAB EXERCISE: Prepare a time line for this experiment, clearly indicating which experiments can be carried out simultaneously.

The experiments in this chapter were introduced by Louis Fieser of Harvard University over 70 years ago. Although the initial procedure starts with macroscale quantities of materials, the later steps often become microscale experiments and benefit greatly from using microscale equipment. It has long been the basis for an interesting laboratory competition based on speed and skill. Ken Williamson, one of the present authors, was a winner over 50 years ago.

Starting with 5 g of 1-naphthol, a skilled chemist familiar with the procedures can prepare pure samples of the seven compounds in this chapter in 3–4 hours. In a first trial of the experiments here, a particularly competent student who plans his or her work in advance can complete the program in two laboratory periods (6 hours).

The first compound of the series, Martius Yellow, a mothproofing dye for wool (1 g of Martius Yellow will dye 200 g of wool) discovered in 1868 by Karl Alexander von Martius, is the ammonium salt of 2,4-dinitro-1-naphthol (1). This compound is obtained by sulfonating 1-naphthol with sulfuric acid, and treating the resulting disulfonic acid with nitric acid in an aqueous medium. The exchange of groups occurs with remarkable ease, and it is not necessary to isolate the disulfonic acid. The advantage of introducing the nitro groups in this indirect way is that 1-naphthol is highly sensitive to oxidation, and would be partially destroyed by direct nitration. Martius Yellow is prepared by reacting the acidic phenolic group of compound **1** with ammonia to form the ammonium salt. A small portion of this salt (Martius Yellow) is converted by acidification and crystallization into pure 2,4-dinitro-1-naphthol, a sample of which is saved. The remainder is suspended in water and reduced to diaminonaphthol with sodium hydrosulfite according to the following equation:

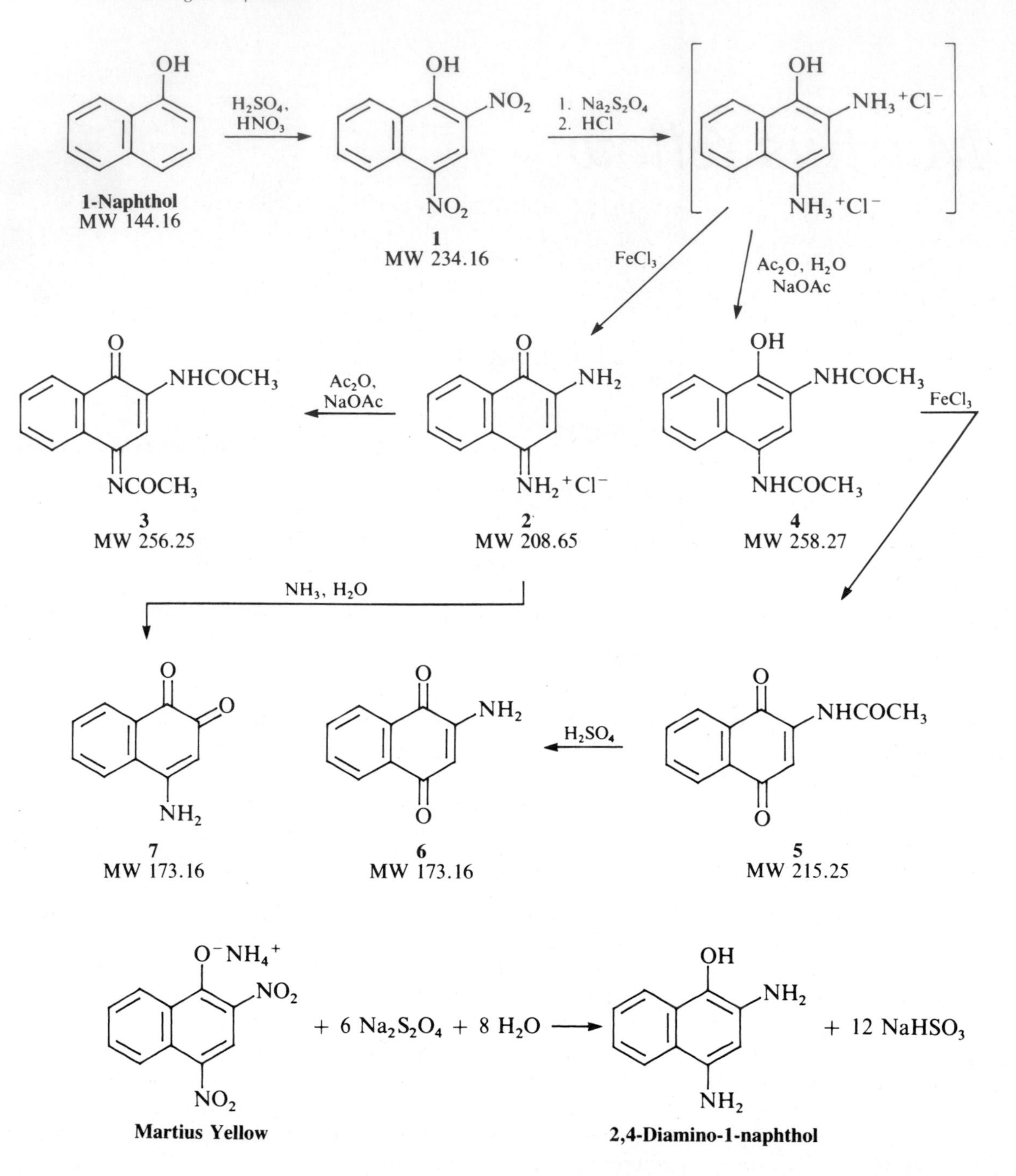

The diaminonaphthol separates in the free condition, rather than as an ammonium salt. The diamine, unlike the dinitro compound, is a very weak acidic substance that does not readily form salts.

Because 2,4-diamino-1-naphthol is exceedingly sensitive to air oxidation as a free base, it is at once dissolved in dilute hydrochloric acid. The solution of diaminonaphthol dihydrochloride is clarified with decolorizing charcoal and divided into two equal parts. One part on oxidation with iron(III) chloride affords the fiery red 2-amino-1,4-naphthoquinonimine hydrochloride (**2**). This substance, like many other salts, has no melting point; it is therefore converted for identification to the yellow diacetate (**3**). Compound **2** is remarkable in that it is stable enough to be isolated. On hydrolysis, it affords the orange 4-amino-1,2-naphthoquinone (**7**).

The other part of the diaminonaphthol dihydrochloride solution is treated with acetic anhydride and then sodium acetate; the reaction in aqueous solution effects selective acetylation of the amino groups and affords 2,4-diacetylamino-1-naphthol (**4**). Oxidation of compound **4** by Fe^{3+} and oxygen from the air results in the cleavage of the acetylamino group at the 4-position, and the product is 2-acetylamino-1,4-naphthoquinone (**5**). This yellow substance is hydrolyzed by sulfuric acid to the red 2-amino-1,4-naphthoquinone (**6**), the last member of the series. The reaction periods are brief, and the yields high; however, remember to scale down the quantities of the reagents and the solvents if the quantity of the starting material is less than that called for.[1]

EXPERIMENTS

If necessary, the quantities in the following seven experiments can be doubled.

1. PREPARATION OF 2,4-DINITRO-1-NAPHTHOL (1)

Use care when working with hot concentrated sulfuric and nitric acids.

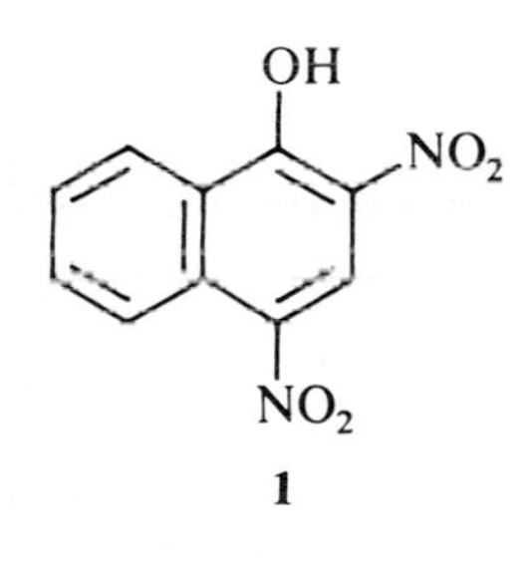

Avoid contact of the yellow product and its orange ammonium salt with the skin.

Place 2.5 g of pure 1-naphthol[2] in a 50-mL Erlenmeyer flask, add 5 mL of concentrated sulfuric acid, and heat the mixture with swirling on a steam bath or a hot plate for 5 minutes, at which time the solid should have dissolved and an initial red color should be discharged. Cool in an ice bath, add 13 mL of water, and cool the solution rapidly to 15°C. Measure 3 mL of concentrated nitric acid into a test tube, and transfer it with a Pasteur pipette in small portions (0.25 mL each) to the chilled aqueous solution while maintaining the temperature between 15°C and 20°C by swirling the flask vigorously in the ice bath. When the addition is complete and the exothermic reaction has subsided (1–2 minutes), warm the mixture gently to 50°C (1 minute), whereupon the nitration product should separate as a stiff yellow paste. Apply heat (the full heat of the steam bath for an additional 1 minute if using a steam bath), fill the flask with water, break up the lumps and stir to an even paste, collect the product (**1**) on a small Büchner funnel, wash it well with water, and then wash it into a 250-mL beaker with water (50 mL). Add 75 mL of hot water and 2.5 mL of concentrated ammonium hydroxide solution (den. 0.90), heat to the boiling point, and stir to dissolve the solid. Filter the hot solution by suction if it is dirty, add 5 g of ammonium chloride to the filtrate to salt out the ammonium salt (Martius Yellow), cool in an ice bath, collect the orange salt, and wash it with water containing 1%–2% of ammonium chloride. The salt does not have to be dried (the dry weight yield is 3.8 g—a percent yield of 88%).

[1]This series of reactions lends itself to a laboratory competition, the rules for which might be as follows: (a) No practice or advance preparation is allowed except the collection of reagents not available at the contestant's bench [ammonium chloride, sodium hydrosulfite, iron(III) chloride solution, and acetic anhydride]. (b) The time scored is the actual working time, including that required for cleaning the apparatus and bench; labels for ziplock bags can be prepared outside of the working period. (c) Time is not charged during an interim period (overnight) such as when solutions are crystallizing or solids are drying, on the condition that during this period no adjustments are made and no cleaning or other work is done. (d) Melting-point and color-test characterizations are omitted. (e) Successful completion of the contest requires preparing authentic and macroscopically crystalline samples of all seven compounds. (f) Judgment of the winners among the successful contestants is based on quality and quantity of samples, technique and neatness, and working time. (Superior performance is 3–4 hours.)

[2]If the 1-naphthol is dark, it can be purified by distillation at atmospheric pressure in the hood. The colorless distillate is most easily pulverized (also in the hood) before it has completely cooled and hardened.

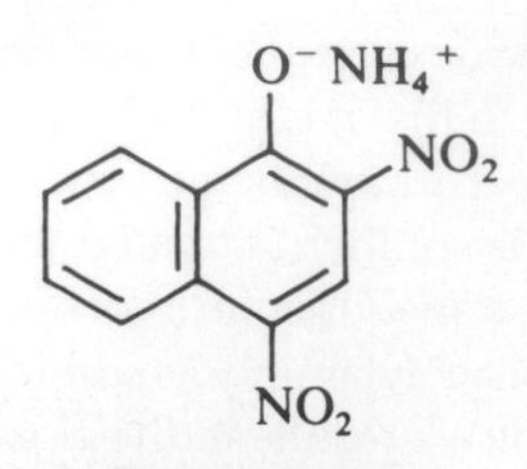

Martius Yellow

Set aside an estimated 150 mg of the moist ammonium salt. This sample is to be dissolved in hot water, the solution is acidified with HCl, and the free 2,4-dinitro-1-naphthol (**1**) is crystallized from methanol or ethanol (use decolorizing charcoal if necessary); it forms yellow needles (mp 138°C).

The proton nuclear magnetic resonance (^{1}H NMR) and infrared (IR) spectra of 1-naphthol are shown in Figures 47.1 and 47.2.

FIG. 47.1
The ^{1}H NMR spectrum of 1-naphthol (250 MHz).

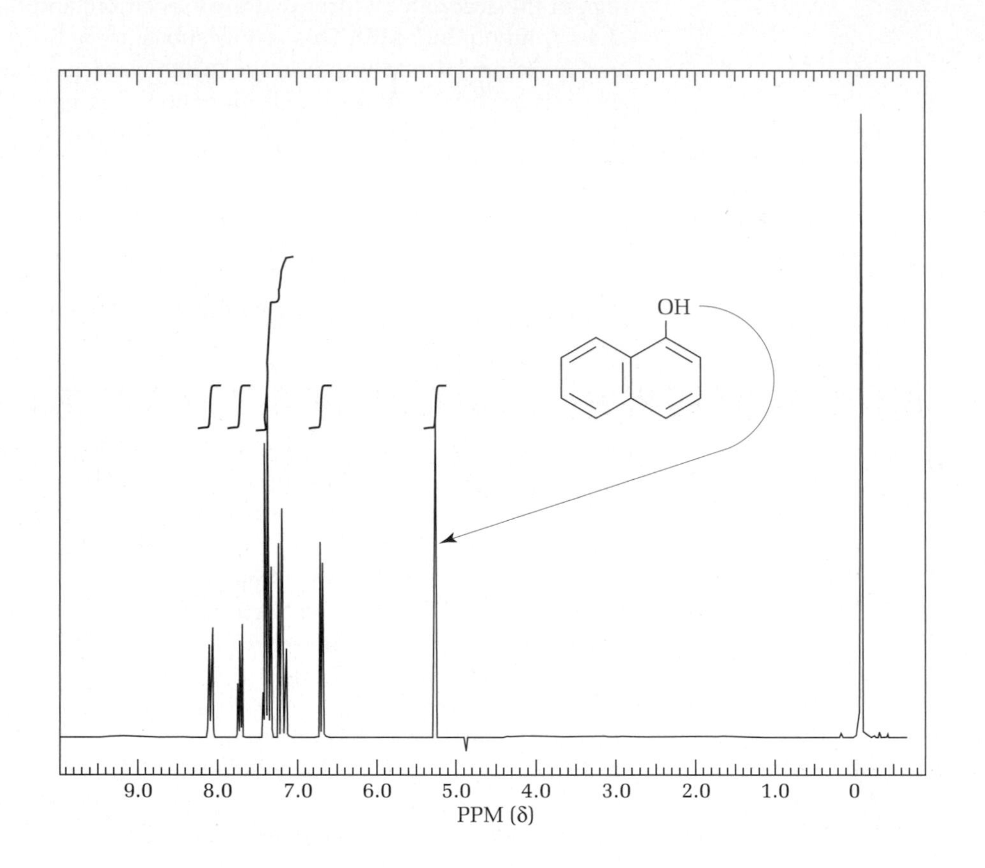

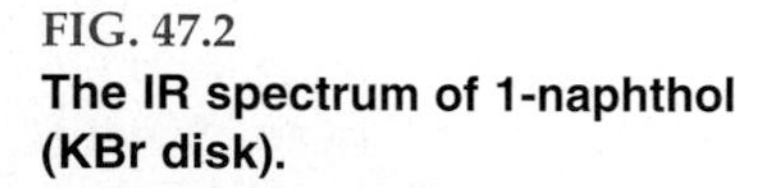
FIG. 47.2
The IR spectrum of 1-naphthol (KBr disk).

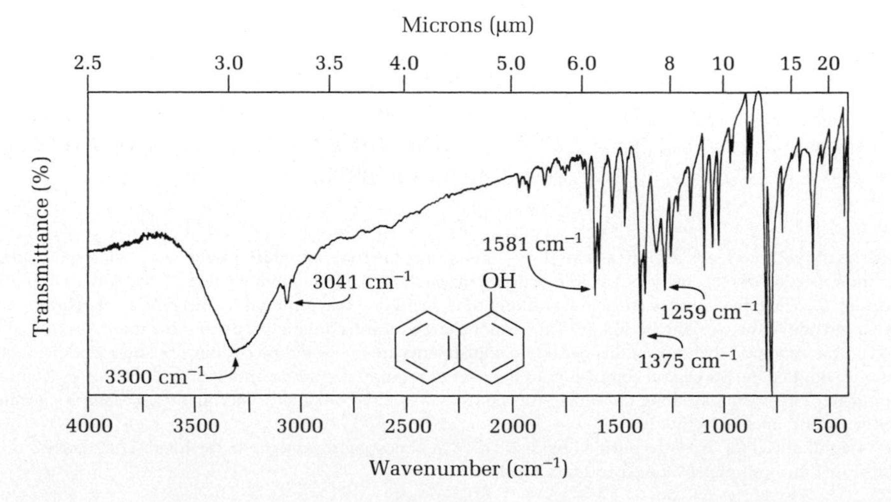

Cleaning Up. Combine aqueous filtrates, dilute with water, neutralize with sodium carbonate, and flush the solution down the drain. Recrystallization solvents go in the organic solvents waste container.

Preparation of Unstable 2,4-Diamino-1-Naphthol

$Na_2S_2O_4$
Sodium hydrosulfite

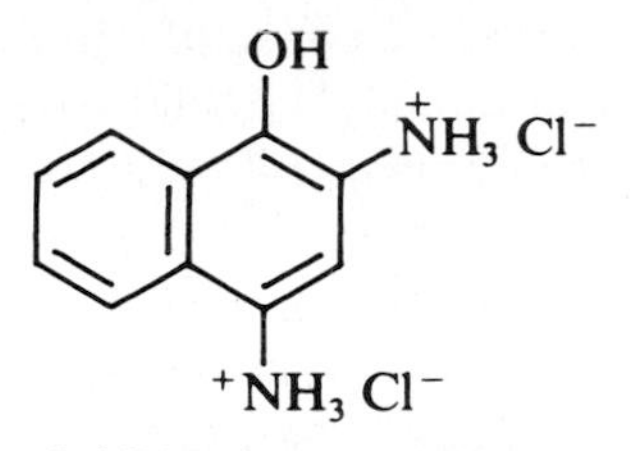

2,4-Diamino-1-naphthol dihydrochloride

Wash the remaining ammonium salt into a beaker with about 100 mL of water, add 20 g of sodium hydrosulfite, stir until the original orange color has disappeared and a crystalline tan precipitate has formed (5–10 minutes), and then cool in ice. Prepare two separate solutions: (1) 1 g of sodium hydrosulfite in 50 mL of water for use in washing and (2) a 250-mL beaker containing 3 mL of concentrated hydrochloric acid and 12 mL of water. When collecting the precipitate by suction filtration on a small Büchner funnel, use the hydrosulfite solution for rinsing and washing, be sure to avoid even briefly sucking air through the cake after the reducing agent has been drained away, and immediately wash the solid into the beaker containing the dilute hydrochloric acid and stir to convert all the diamine to the dihydrochloride.

The acid solution, often containing suspended sulfur and filter paper, is clarified using suction filtration through a moist charcoal bed made by shaking 1 g of powdered decolorizing charcoal (carbon) with 13 mL of water in a stoppered flask to produce a slurry, and pouring this on filter paper placed in a 50-mm Büchner funnel. Pour out the water from the filter flask, and then filter the acid solution of dihydrochloride. Divide the pink or colorless filtrate into approximately two equal parts, and immediately add the reagents for converting one part to compound **2** (Experiment 2) and the other part to compound **4** (Experiment 4).

Cleaning Up. Neutralize the filtrate with sodium carbonate and dilute with water. In the hood, cautiously add household bleach (aqueous sodium hypochlorite solution) to the mixture until a test with 5% silver nitrate proves that no more hydrosulfite is present (the absence of a black precipitate). Neutralize the solution and filter it through Celite to remove any suspended solids. Dilute the filtrate with water and flush it down the drain. The solid residue goes in the nonhazardous solid waste container.

2. PREPARATION OF 2-AMINO-1,4-NAPHTHOQUINONIMINE HYDROCHLORIDE (2)

O
NH_2
$^+\overset{\|}{N}H_2\ Cl^-$
2

To one-half of the diamine dihydrochloride solution (Experiment 1), add 12.5 mL of 1.3 *M* iron(III) chloride solution,[3] cool in ice, and, if necessary, initiate crystallization by scratching the inside of the test tube. Rub the liquid film with a glass stirring rod at a single spot slightly above the surface of the liquid. If efforts to induce crystallization are unsuccessful, add more hydrochloric acid until crystallization occurs. Collect the red product and wash with 3 *M* HCl (aq). The dry weight yield is 1.2–1.35 g.

[3]*Preparation of iron(III) chloride solution*: Dissolve 45 g of $FeCl_3 \cdot 6\ H_2O$ (MW 270.32) in 50 mL of water and 50 mL of concentrated hydrochloric acid by warming, cooling, and filtration (produces 124 mL of solution).

Divide the moist product into three equal parts; then spread out one part to dry for conversion to compound **3** (Experiment 3). The other two parts can be used while still moist for conversion to compound **7** (Experiment 7) and for recrystallization. Dissolve one part by warming in a little water containing 2–3 drops of hydrochloric acid, shake for a minute or two with decolorizing charcoal, filter by suction, and add concentrated hydrochloric acid to decrease the solubility. Collect the product by suction filtration.

Cleaning Up. Neutralize the filtrate with sodium carbonate and collect the iron hydroxide by vacuum filtration through Celite on a Büchner funnel. The solid goes into the nonhazardous solid waste container, whereas the filtrate is diluted with water and flushed down the drain.

3. PREPARATION OF 2-AMINO-1,4-NAPHTHOQUINONIMINE DIACETATE (3)

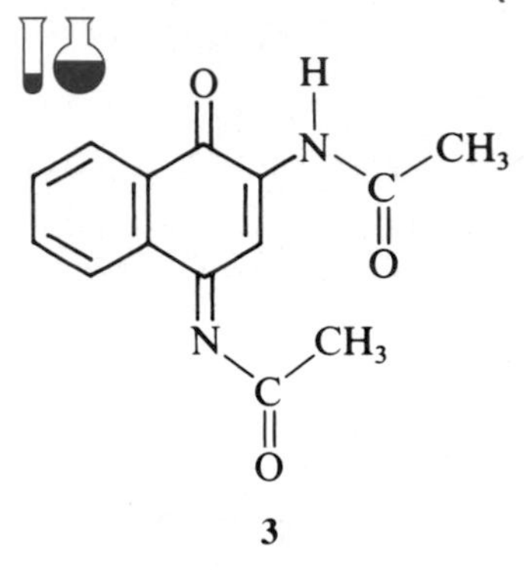

A mixture of 0.25 g of the dry quinonimine hydrochloride (**2**; Experiment 2), 0.25 g of sodium acetate (anhydrous), and 1.5 mL of acetic anhydride is stirred in a reaction tube and warmed gently on a sand or steam bath. With thorough stirring, the red salt should soon change into yellow crystals of the diacetate. The solution may appear red, but as soon as particles of red solid have disappeared, the mixture can be poured into about 5 mL of water. Stir until the excess acetic anhydride has either dissolved or become hydrolyzed, collect and wash the product (the dry weight yield is 250 mg) with water, and (drying is unnecessary) crystallize it from ethanol or methanol; yellow needles result (mp 189°C).

Cleaning Up. The filtrate should be diluted with water and flushed down the drain. The crystallization solvent goes in the organic solvents waste container.

4. PREPARATION OF 2,4-DIACETYLAMINO-1-NAPHTHOL (4)

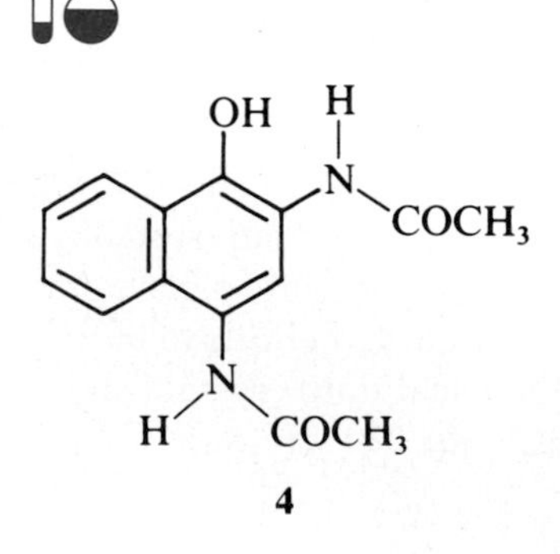

To one-half of the diaminonaphthol dihydrochloride solution saved from Experiment 1, add 1.5 mL of acetic anhydride, stir vigorously, and then add a solution of 1.5 g of sodium acetate (anhydrous) and about 50 mg of sodium hydrosulfite in 10–15 mL of water. The diacetate may precipitate as a white powder, or it may separate as an oil that solidifies when the solution chilled in ice and rubbed with a stirring rod. Collect the product and, to hydrolyze any triacetate present, dissolve it in 2.5 mL of 3 *M* sodium hydroxide and 25 mL of water by stirring at room temperature. If the solution is colored, a few additional milligrams of sodium hydrosulfite may bleach it. Filter by suction and acidify by gradual addition of well-diluted hydrochloric acid (1 mL of concentrated acid in 19 mL of water). The diacetate tends to remain in a supersaturated solution; hence, either to initiate crystallization or to ensure maximum separation, it is advisable to stir well, rub the walls with a stirring rod, and cool in ice. Collect the product, wash it with water, and divide it into thirds (the dry weight yield is 1–1.3 g).

Two-thirds of the material can be converted without drying into compound **5**, and the other third can be used to prepare a crystalline sample. Dissolve the one-third reserved for crystallization (moist or dry) in enough hot acetic acid to bring about solution, add a solution of a small crystal of tin(II) chloride in a few drops of 10% HCl (aq) solution to inhibit oxidation, and dilute gradually with 5 to 6 volumes of water at the boiling point. Crystallization may be slow, and cooling and scratching may be necessary. The pure diacetate forms colorless prisms (mp 224°C, dec.).

Cleaning Up. Combine all filtrates—including the acetic acid used to crystallize the product—dilute with water, neutralize with sodium carbonate, and flush the solution down the drain.

5. PREPARATION OF 2-ACETYLAMINO-1,4-NAPTHOQUINONE (5)

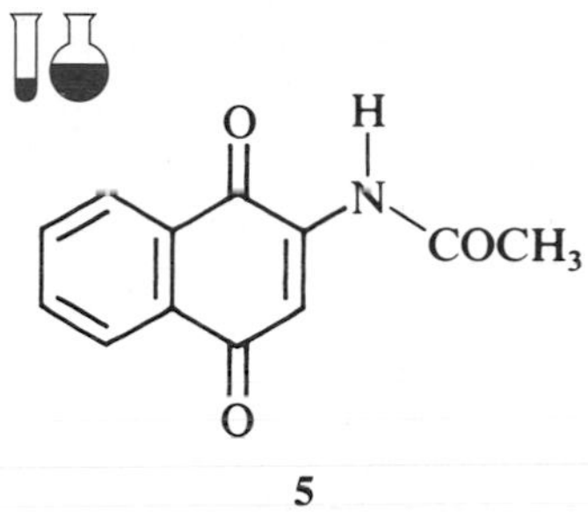

Dissolve 0.75 g of the moist diacetylaminonaphthol (compound **4**; Experiment 4) in 5 mL of acetic acid (hot), dilute with 10 mL of hot water, and add 5 mL of 0.13 *M* iron(III) chloride solution. The product separates promptly into flat, yellow needles, which are collected (after cooling) and washed with a little alcohol; the yield is usually 0.6 g. Dry one-half of the product for conversion to compound **6** and crystallize the remaining half from 95% ethanol (mp 204°C).

Cleaning Up. Dilute the filtrate with water, neutralize it with sodium carbonate, and flush it down the drain. A negligible quantity of iron is disposed of in this way.

6. PREPARATION OF 2-AMINO-1,4-NAPTHOQUINONE (6)

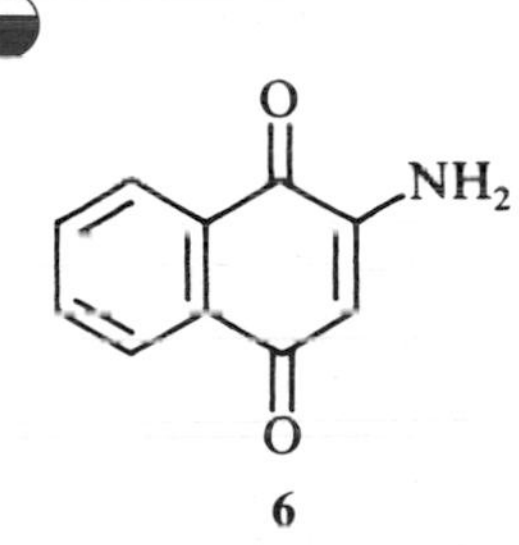

To 0.25 g of dry 2-acetylamino-1,4-naphthoquinone (**5**) contained in a reaction tube, add 1 mL of concentrated sulfuric acid and heat the mixture on a steam or sand bath with swirling to promote rapid solution (1–2 minutes). After an additional 5 minutes, cool the deep red solution, dilute it with a large amount of water, and collect the precipitated product. Wash the crude material with water and recrystallize it from alcohol or an alcohol:water mixture[4] while it is still moist. The yield of red needles of the amino quinone (mp 206°C) will be about 200 mg.

Cleaning Up. The filtrate is neutralized with sodium carbonate, diluted with water, and flushed down the drain.

7. PREPARATION OF 4-AMINO-1,2-NAPHTHOQUINONE (7)

Compound 2 can be moist.

Dissolve 0.5 g of the aminonaphthoquinonimine hydrochloride (**2**) reserved from Experiment 2 in 12 mL of water, add 1 mL of concentrated ammonium hydroxide solution (den 0.90), and boil the mixture for 5 minutes. The free quinonimine that initially precipitated is hydrolyzed to a mixture of the aminoquinone (**7**) and the isomeric compound **6**. Cool, collect the precipitate, suspend it in about 25 mL of water, and add 12.5 mL of 3 *M* sodium hydroxide solution. Stir well, remove the small amount of residual 2-amino-1,4-naphthoquinone (**6**) by filtration, and acidify

[4]The exact ratio of alcohol:water is determined experimentally. Try different ratios, such as 50:50, 70:30, or 80:20. *See* Chapter 4 on how to choose a recrystallization solvent.

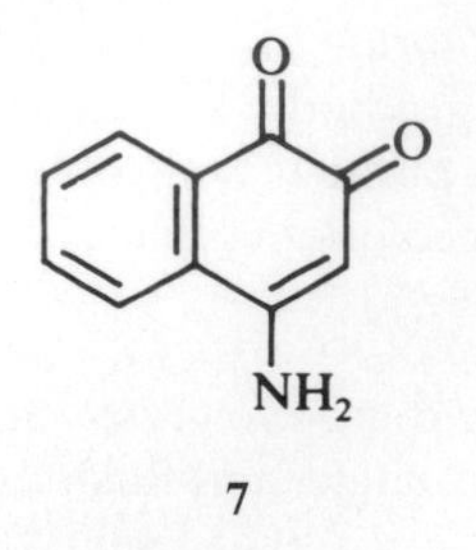

the filtrate with acetic acid. The orange precipitate of compound **7** is collected, washed, and crystallized while still wet from the addition of 250–300 mL of hot water (the separation is slow). The yield of orange needles (dec. about 270°C) is about 200 mg.

Cleaning Up. The filtrate is neutralized with sodium carbonate, diluted with water, and flushed down the drain.

QUESTIONS

1. Write a balanced equation for the preparation of 2-amino-1,4-naphthoquinonimine hydrochloride (**2**).
2. What part of the molecule accounts for the peak at 3300 cm^{-1} in the IR spectrum of 1-naphthol (Fig. 47.2)?

CHAPTER 48

Diels–Alder Reaction

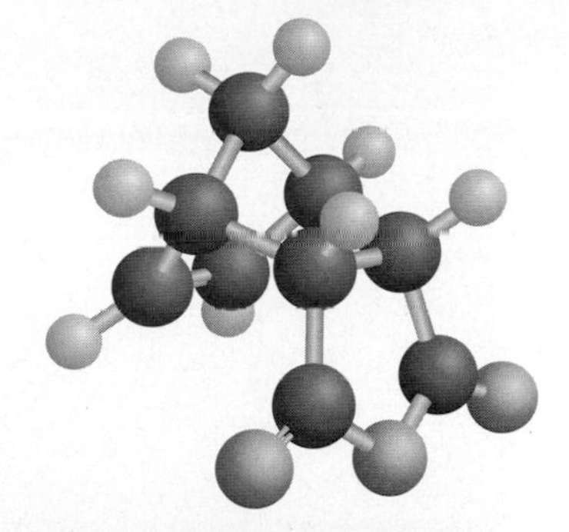

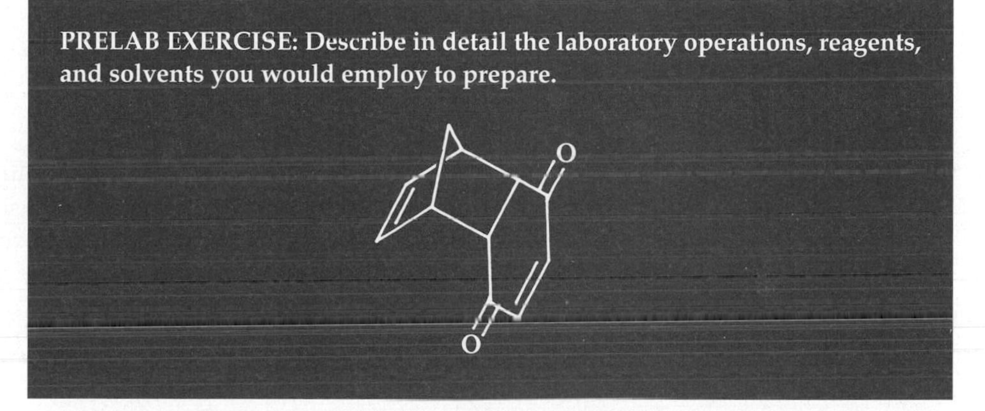

PRELAB EXERCISE: Describe in detail the laboratory operations, reagents, and solvents you would employ to prepare.

Otto Diels and his pupil, Kurt Alder, received the Nobel Prize in Chemistry in 1950 for their discovery and work on the reaction that bears their names. Its great usefulness lies in its high yield and high stereospecificity. A cycloaddition reaction, the Diels–Alder reaction involves the 1,4-addition of a conjugated diene in the s-*cis* conformation to an alkene in which two new σ (sigma) bonds are formed from two π (pi) bonds.

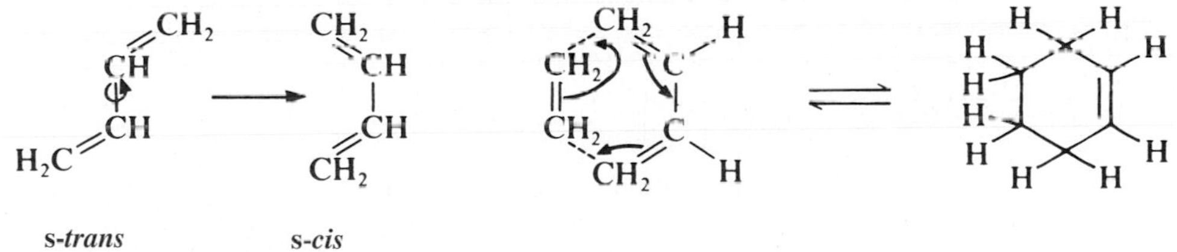

The adduct is a six-membered ring alkene. The diene can have the two conjugated bonds contained within a ring system, as with cyclopentadiene or cyclohexadiene, or the molecule can be an acyclic diene that must be in the *cis* conformation about the single bond before reaction can occur.

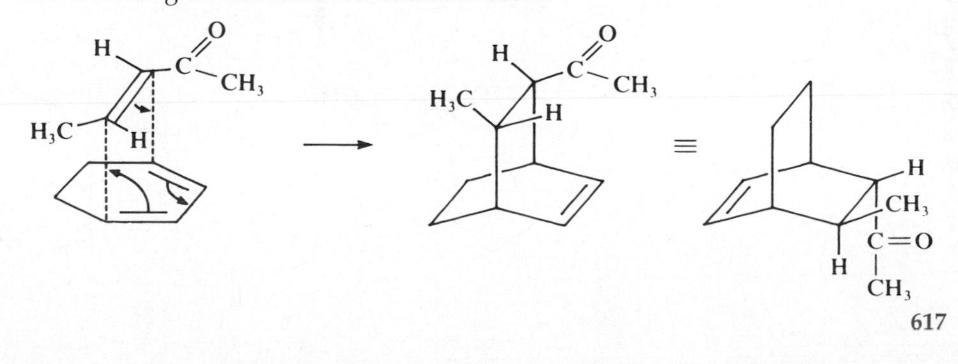

The reaction works best when there is a marked difference between the electron densities in the diene and the alkene with which it reacts—the dienophile. Usually, the dienophile has electron-attracting groups attached to it, whereas the diene is electron rich, for example, as in the reaction of methyl vinyl ketone with 1,3-butadiene.

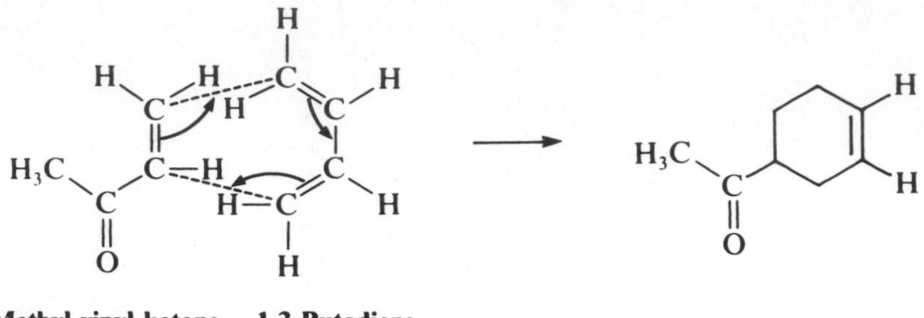

Methyl vinyl ketone 1,3-Butadiene

Retention of the configurations of the reactants in the products implies that both new σ bonds are formed almost simultaneously. If not, then the intermediate with a single new bond could rotate about that bond before the second σ bond is formed, thus destroying the stereospecificity of the reaction.

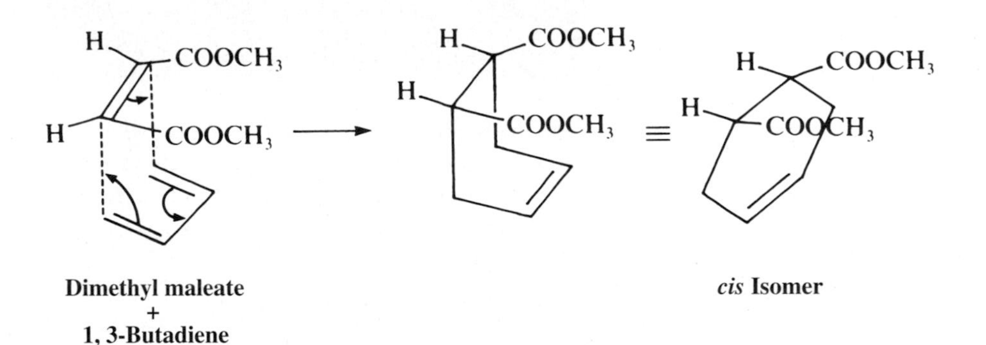

Dimethyl maleate + 1, 3-Butadiene

***cis* Isomer**

The following does not happen:

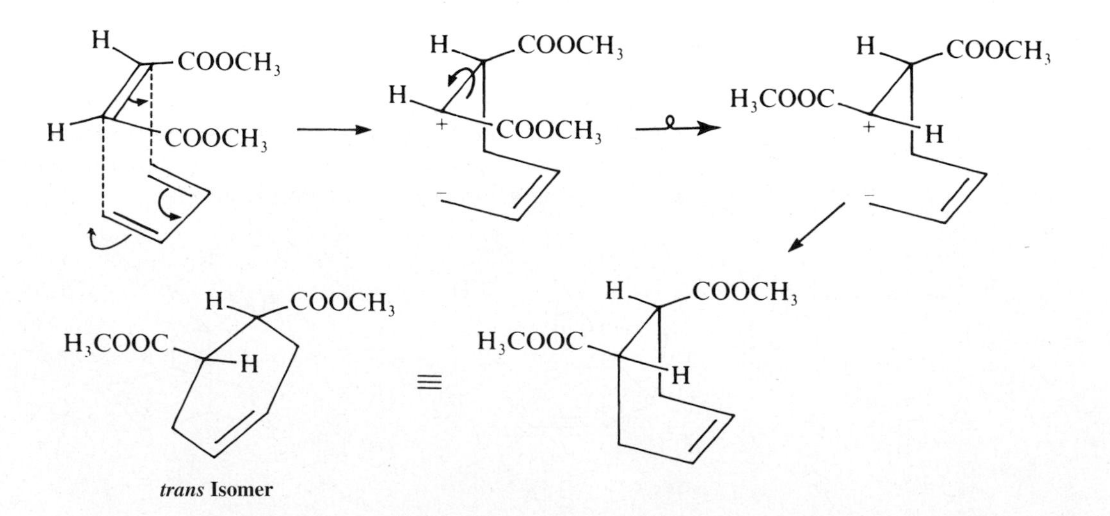

***trans* Isomer**

A highly stereospecific reaction

This reaction is not polar in that no charged intermediates are formed. Neither is it a radical reaction, because no unpaired electrons are involved. It is instead known as a *concerted reaction*, or one in which several bonds in the transition state are simultaneously made and broken. When a cyclic diene and a cyclic dienophile react with each other, more than one stereoisomer may be formed. The isomer that predominates is the one that involves maximum overlap of π electrons in the transition state. The transition state for the formation of the *endo* isomer involves a sandwich with the diene directly above the dienophile. To form the *exo* isomer, the diene and dienophile would need to be arranged in a stair-step fashion. Robert B. Woodward and Roald Hoffmann formulated the theoretical rules involving the correlation of orbital symmetry, which govern the Diels–Alder and other electrocyclic reactions.

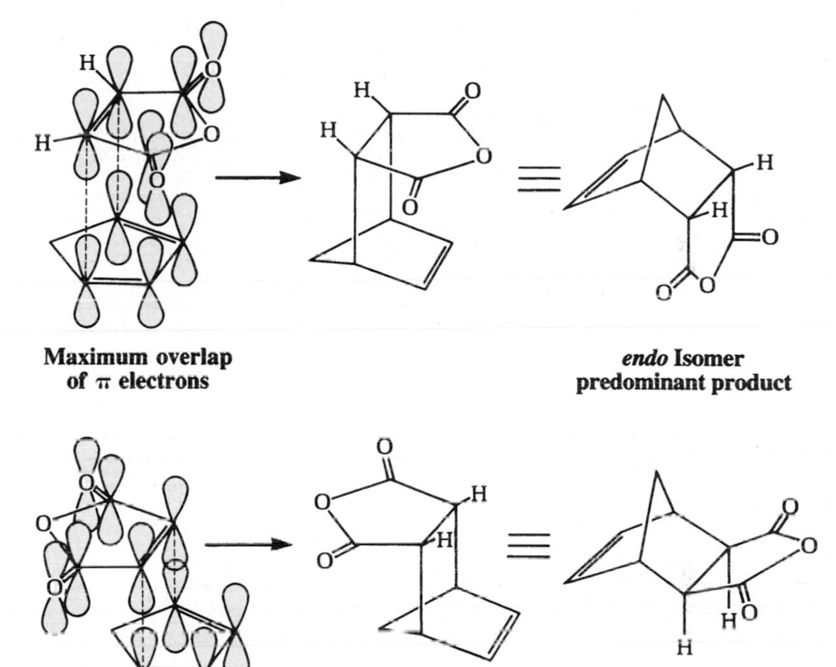

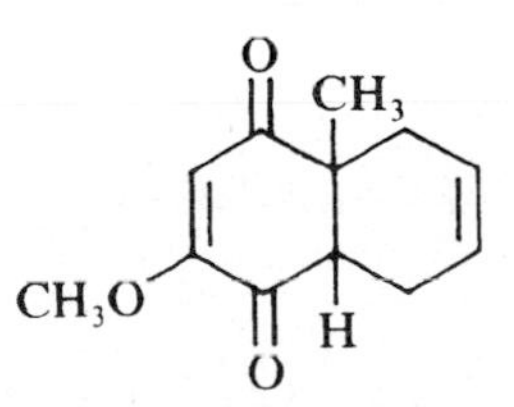

Woodward's Diels–Alder adduct

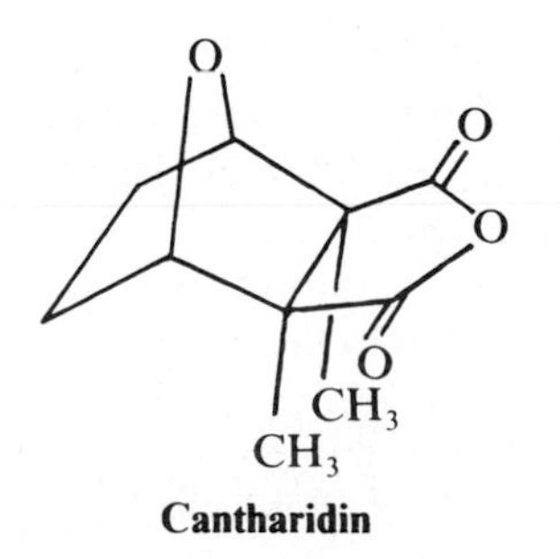

Cantharidin

The Diels–Alder reaction has been used extensively in the synthesis of complex natural products because it is possible to exploit the formation of a number of chiral centers in one reaction, and also because of the regioselectivity of the reaction. For example, the first step in Woodward's synthesis of cortisone was the formation of a Diels–Alder adduct.

But the reaction is also subject to steric hindrance, especially when the difference between the electron-withdrawing and electron-donating characteristics of the two reactants is not great. Cantharidin is a terpenoid that is the active ingredient in the poison secreted by the Spanish fly (*Lytta vesicatoria*), and it is a powerful vesicant (blister former). When Woodward tried to synthesize cantharidin by the Diels–Alder condensation of furan with dimethylmaleic anhydride, the reaction did not work. The reaction possesses $-\Delta V^*$ (it proceeds with a net decrease in volume). High pressure should overcome this problem, but when attempted, this reaction did not proceed even at 600,000 lb/in.^{2} (4.1 × 10^{10} dynes/cm^2). A closely related reaction will proceed at 300,000 lb/in.^{2} and has been used to synthesize this molecule.[1]

Cyclopentadiene is obtained from the light oil that is produced from coal tar distillation, but exists as the stable dimer, dicyclopentadiene, which is the Diels–Alder adduct from two molecules of the diene. Thus, generation of cyclopentadiene by pyrolysis of the dimer represents a reverse Diels–Alder reaction. *See* Figures 48.5 to 48.7 at the end of the chapter for nuclear magnetic resonance (^{1}H and ^{13}C NMR) and infrared (IR) spectra, respectively, of dicyclopentadiene.

In the Diels–Alder addition of cyclopentadiene and maleic anhydride, the two molecules approach each other in the orientation shown in the top drawing on the previous page because this orientation provides maximal overlap of π bonds of the two reactants, and favors formation of an initial π complex and then the final *endo* product. Dicyclopentadiene also has the endo configuration.

EXPERIMENTS

1. CRACKING OF DICYCLOPENTADIENE

⚠ Check dicyclopentadiene for peroxides. Commercially available test strips can test for the presence of peroxides. If peroxides are present, the bottle should be discarded. If the bottle is over 12 months old, test for peroxides and/or discard bottle. The presence of peroxides during a distillation can be very dangerous because an explosion may occur.

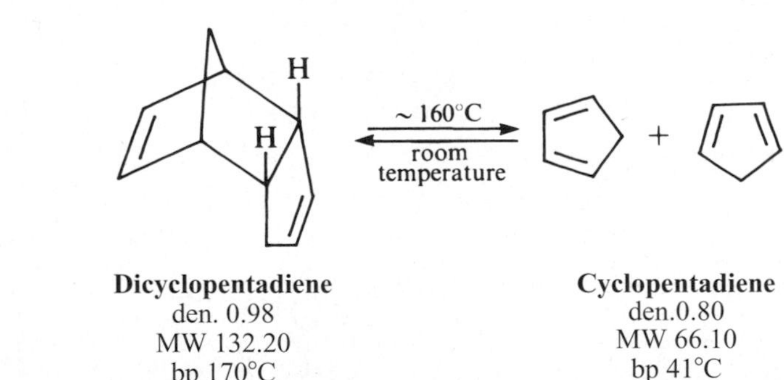

Dicyclopentadiene
den. 0.98
MW 132.20
bp 170°C
n_D^{20} 1.5100

Cyclopentadiene
den.0.80
MW 66.10
bp 41°C

IN THIS EXPERIMENT, the dimer of cyclopentadiene is cracked into two molecules of the monomer by dripping it onto very hot mineral oil. The product, which boils at 41°C, is collected in an ice-cooled receiver.

Microscale Procedure

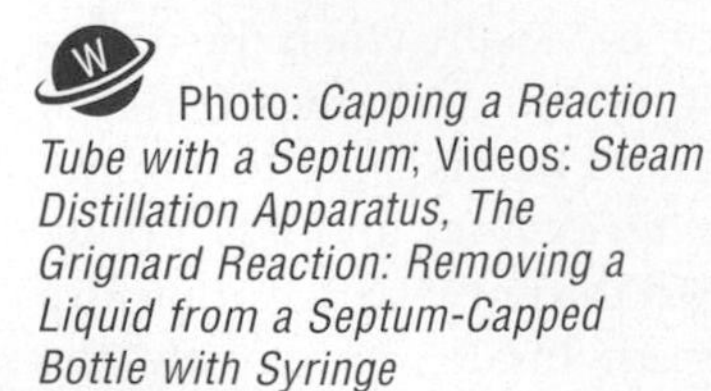
Photo: *Capping a Reaction Tube with a Septum*; Videos: *Steam Distillation Apparatus, The Grignard Reaction: Removing a Liquid from a Septum-Capped Bottle with Syringe*

Half fill with mineral oil the 5-mL, short-necked, round-bottomed flask equipped with an addition port bearing a septum on the sidearm, and topped with a distillation head and thermometer (Fig. 48.1). Set the controller on the flask heater at half maximum. The temperature can be controlled by piling up or scraping away sand from the flask. You can either start with the thermometer down in the oil and raise it as the temperature approaches 250°C, or leave the thermometer in the high position and simply wait until you judge the oil is hot enough.

[1]Dauben WG, Kessel CR, Takemura KH. *J Am Chem Soc.* **1980;***102*:6893.

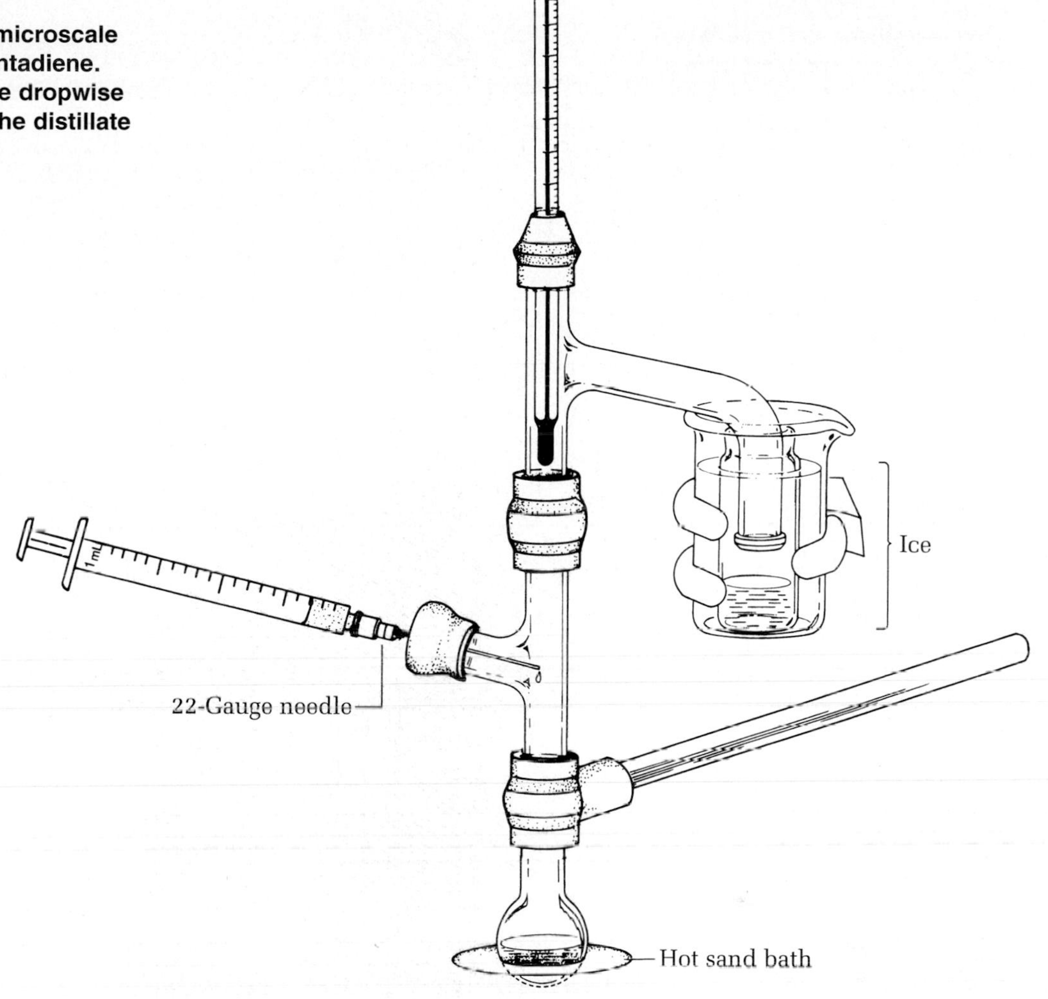

FIG. 48.1
An apparatus for the microscale cracking of dicyclopentadiene. Add dicyclopentadiene dropwise via a syringe so that the distillate temperature does not exceed 45°C.

Place a small, tared collection vial in an ice-filled 30-mL beaker at the end of the distilling head, taking care to keep water out of the vial. Using a syringe, draw 0.6 mL of dicyclopentadiene (Fig. 48.2) from a septum-capped storage container after first injecting 0.6 mL of air into the container to overcome the vacuum. Insert the needle of the filled syringe into a rubber stopper or cork to avoid loss of the contents until they are used.

When the mineral oil is judged to be hot enough (250°C), inject the dicyclopentadiene through the septum on the addition port. Add it dropwise at a rate such that the temperature of the thermometer never exceeds 45°C. The boiling point of cyclopentadiene is 41°C. Add the dimer over a 10-minute period. If the dimer is added too slowly, the yield will be lower. Once all the dicyclopentadiene has been added, remove the syringe, remove the collection vial and quickly cap it, and then weigh the vial to determine the product yield. Rinse the syringe with acetone in the hood because the dimer has a very disagreeable odor. Calculate the percent yield of cyclopentadiene. If the product is cloudy, add a small quantity of anhydrous

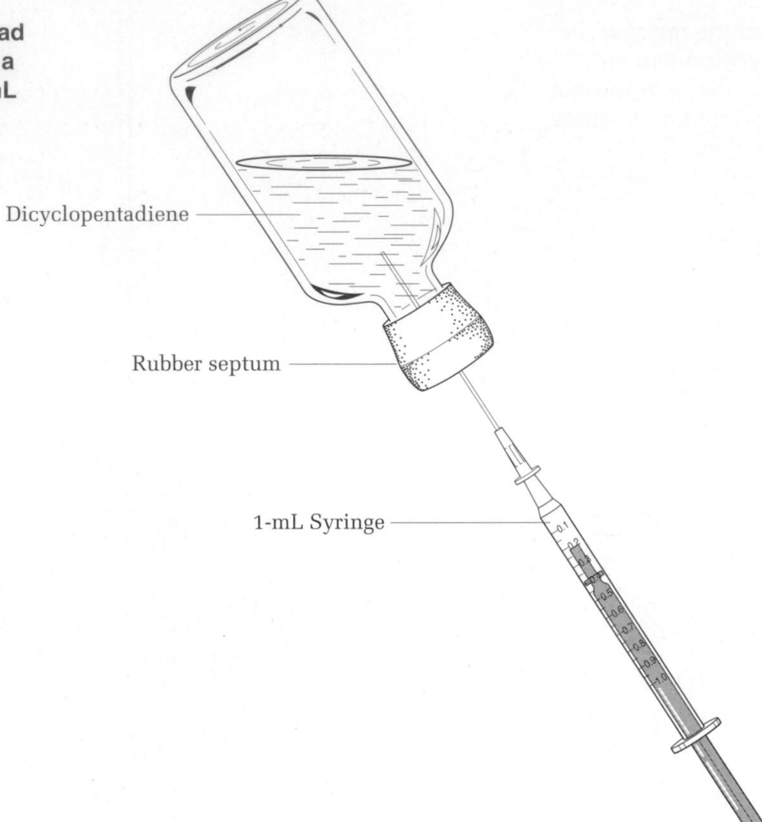

FIG. 48.2
Dicyclopentadiene has a very bad odor, and so is dispensed from a closed container. Remove 0.6 mL from the septum-capped bottle after injecting 0.6 mL of air.

Don't disassemble the apparatus until you are sure you have 0.3 mL of product.

calcium chloride pellets to dry it if the maleic anhydride experiment (Experiment 2) is being done next. It need not be dry to make ferrocene. Keep this cyclopentadiene on ice and use it the same day it is prepared for Experiment 2.

The ^{1}H NMR, IR, and ^{13}C NMR spectra of dicyclopentadiene appear in Figures 48.5, 48.6, and 48.7, respectively, at the end of the chapter.

Photo: *Reverse Diels–Alder*, Video: *Macroscale Cracking of Dicyclopentadiene*

Cleaning Up. Place the mineral oil from the reaction flask and any unused dicyclopentadiene in the organic solvents waste container. If calcium chloride was used, free it of cyclopentadiene by evaporation in the hood, and then place it in the nonhazardous solid waste container.

Check dicyclopentadiene for peroxides. Commercially available test strips can test for the presence of peroxides. If peroxides are present, the bottle should be discarded. If the bottle is over 12 months old, test for peroxides and/or discard bottle. The presence of peroxides during a distillation can be very dangerous because an explosion may occur.

Macroscale Procedure

Measure 20 mL of dicyclopentadiene into a 100-mL flask, and arrange for fractional distillation into an ice-cooled receiver (*see* Fig. 19.2 on page 337). Heat the dimer with an electric flask heater until it refluxes briskly and at a rate such that the monomeric diene begins to distill in about 5 minutes, and soon reaches a steady boiling point between 40°C and 42°C. Apply heat continuously to promote rapid distillation without exceeding the boiling point of 42°C. Distillation for 45 minutes should provide the 12 mL of cyclopentadiene required for two preparations of the adduct; continued distillation for another half hour produces a total of about 20 mL of monomer.

The ^{1}H NMR, IR, and ^{13}C NMR spectra of dicyclopentadiene appear in Figures 48.5, 48.6, and 48.7, respectively, at the end of the chapter.

Cleaning Up. Pour the pot residue of dicyclopentadiene and any unused cyclopentadiene into the recovered dicyclopentadiene container. This recovered material can, despite its appearance, be cracked in the future to give cyclopentadiene, but test it for peroxides. If the pot residue is not to be recycled, place it in the organic solvents waste container.

2. SYNTHESIS OF *cis*-NORBORNENE-5,6-*endo*-DICARBOXYLIC ANHYDRIDE

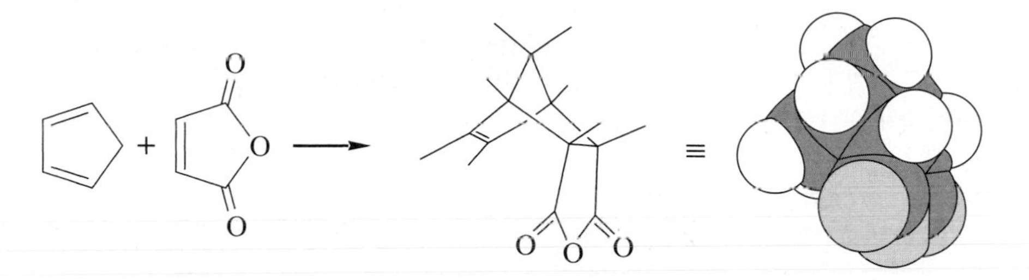

Maleic anhydride
mp 53°C, MW 98.06

***cis*-Norbornene-5,6-*endo*-dicarboxylic anhydride**
mp 165°C, MW 164.16

IN THIS EXPERIMENT, maleic anhydride (the dienophile) is reacted with cyclopentadiene (the diene) in a mixture of solvents that will dissolve the starting material but not the product, which crystallizes. The product is collected by Pasteur pipette filtration.

Microscale Procedure

Mixing of the reactants is very important. Pull the reaction mixture into a pipette and then expel it into the reaction tube.

Dissolve 0.20 g of powdered maleic anhydride in 1 mL of ethyl acetate in a tared 10 × 100-mm reaction tube; then add 1 mL hexanes (bp 60–80°C). This balanced combination of solvents is used because the product is too soluble in pure ethyl acetate, but not soluble enough in pure hexanes. To the solution of maleic anhydride, add 0.20 mL (0.160 g) of dry cyclopentadiene, mix the reactants, and observe the reaction. Allow the tube to cool to room temperature, during which time crystallization of the product should occur. If crystallization does not occur, scratch the inside of the test tube with a stirring rod at the liquid-air interface. The scratch marks on the inside of the tube often form the nuclei upon which crystallization starts. Should crystallization occur very rapidly at room temperature, the crystals will be very small. If so, save a seed crystal, heat the mixture in the reaction tube until the product dissolves, seed it, and allow it to cool slowly to room temperature. You will be rewarded with large plate-like crystals. Remove the solvent from the crystals with a Pasteur pipette that is forced to the bottom of the tube, wash the crystals with one portion of cold

Video: *Filtration of Crystals Using the Pasteur Pipette*

hexanes, and remove the solvent (*see* Fig. 4.12 on page 72). Scrape the product onto a piece of filter paper, allow the crystals to air-dry, determine their weight, and calculate the yield of the product. Determine the melting point of the product, and turn in any material not used in the next experiment. Thin-layer chromatography (TLC) of the product is not necessary because the product is quite pure. The IR spectrum of the anhydride adduct is in Figure 48.8, the ^{1}H NMR spectrum is in Figure 48.9, and the ^{13}C NMR spectrum is in Figure 48.10—all at the end of the chapter.

Cleaning Up. Place the crystallization solvent mixture in the organic solvents waste container. It contains a very small quantity of the product.

Macroscale Procedure

Place 6 g of maleic anhydride in a 125-mL Erlenmeyer flask, and dissolve the anhydride in 16 mL of ethyl acetate by heating on a hot plate or a steam bath. Add 16 mL of hexane (bp 60–80°C), cool the solution thoroughly in an ice-water bath, and leave it in the bath (some anhydride may crystallize).

CAUTION: Cyclopentadiene is flammable.

The freshly distilled cyclopentadiene may be slightly cloudy because of moisture condensation in the cooled receiver and water in the starting material. Add about 1 g of calcium chloride pellets to remove the moisture. Be sure to use the cyclopentadiene immediately because it will dimerize within a few hours of its preparation. Measure 6 mL of dry cyclopentadiene, and add it to the ice-cold solution of maleic anhydride. Swirl the solution in an ice bath for a few minutes until the exothermic reaction ceases and the adduct separates as a white solid. Then heat the mixture on a hot plate or a steam bath until the solid is completely dissolved.[2] If you allow the solution to stand undisturbed, you will be rewarded with a beautiful display of crystal formation. The anhydride crystallizes in long spars (mp 164–165°C); a typical yield is 8.2 g.[3] The IR spectrum of the anhydride adduct is in Figure 48.8, the ^{1}H NMR spectrum is in Figure 48.9, and the ^{13}C NMR spectrum is in Figure 48.10—all at the end of the chapter.

Rapid addition at 0°C

Cleaning Up. Place the crystallization solvent mixture in the organic solvents waste container. It contains a very small quantity of the product. Allow the organic material to evaporate from the drying agent (in the hood); then place it in the nonhazardous solid waste container.

[2]In case moisture has gotten into the system, a little of the corresponding diacid may remain undissolved at this point and should be removed by filtering the hot solution.

[3]The student need not work up the mother liquor, but may be interested in learning the result of this procedure. Concentration of the solution to a small volume is not satisfactory because of the presence of dicyclopentadiene, formed by the dimerization of excess monomer; the dimer has high solvent power. Hence, the bulk of the solvent is evaporated on a steam bath, and the flask is connected to a water pump with a rubber stopper and glass tube, and then heated under vacuum on a steam bath until dicyclopentadiene is removed and the residue solidifies. Crystallization from 1:1 ethyl acetate–ligroin affords 1.3 g of adduct (mp 156–158°C); the total yield is 95%.

3. *cis*-NORBORNENE-5,6-*endo*-DICARBOXYLIC ACID

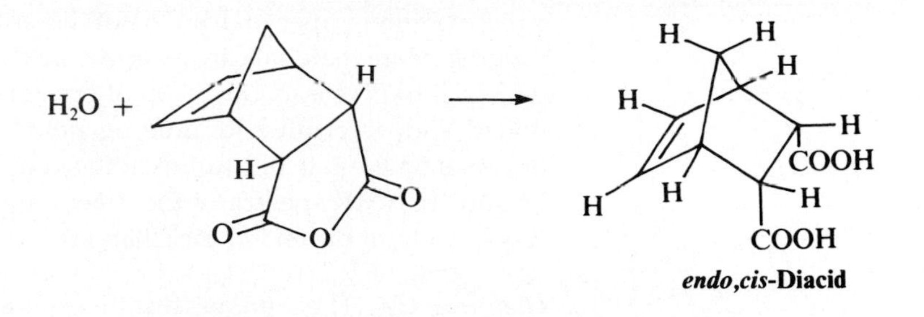

IN THIS EXPERIMENT, the product from Experiment 2 is boiled with water to hydrolyze the anhydride to the corresponding dicarboxylic acid. On cooling the aqueous solution, the product crystallizes.

Microscale Procedure

To 0.2 g (200 mg) of the anhydride from Experiment 2, add 2.5 mL of water and a boiling stick in a 10 × 100-mm reaction tube. Heat the mixture to boiling by immersing the tube in a hot sand bath (Fig. 48.3). The anhydride may appear to melt and form globules on the bottom of the tube. As the reaction proceeds, the anhydride will react with the water, and the diacid, which is soluble in boiling water, will be formed. Continue to heat for about 2 minutes after the last globule disappears. Remove the boiling stick from the hot solution, and allow the mixture to cool to room temperature.

If crystallization of the diacid does not occur, follow the same procedure used for the anhydride (Experiment 2). On slow cooling with simultaneous crystal growth, the solution will deposit long, needlelike crystals. Again, cool the mixture

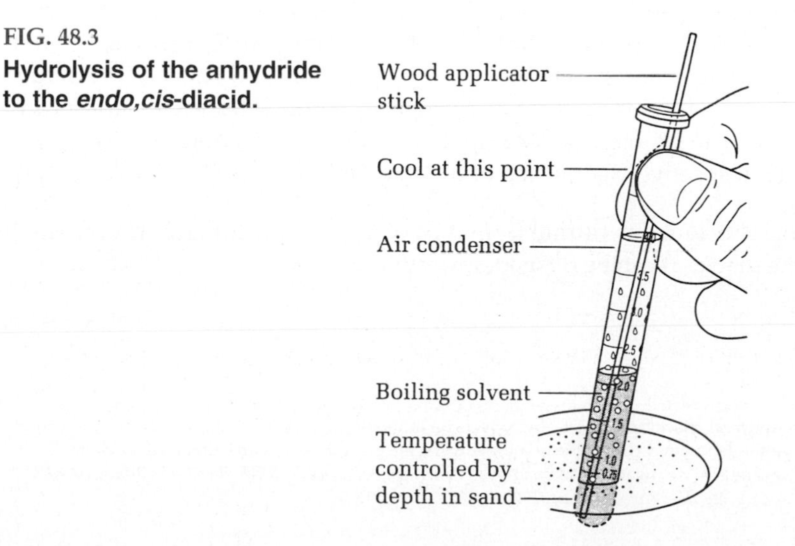

FIG. 48.3
Hydrolysis of the anhydride to the *endo,cis*-diacid.

in ice, allow sufficient time for crystal growth to occur, and then collect the product by filtration on a Hirsch funnel. Use the ice-cold filtrate to complete the transfer. Wash the crystals once with a small quantity of ice water, and place the product on a piece of filter paper to dry. Do not discard the filtrate until you have weighed the product. More material can be recovered by concentrating the filtrate, and allowing it to cool to give a second crop of crystals. This is a general strategy. Weigh the diacid and determine its melting point and percent yield. The melting point depends on the rate of heating as the anhydride reforms and water splits out. The IR and ^1H NMR spectra of the diacid are presented in Figures 48.11 and 48.12, respectively, at the end of the Chapter.[4]

Cleaning Up. The aqueous filtrate from the crystallization contains a very small quantity of the diacid. It can be flushed down the drain.

Macroscale Procedure

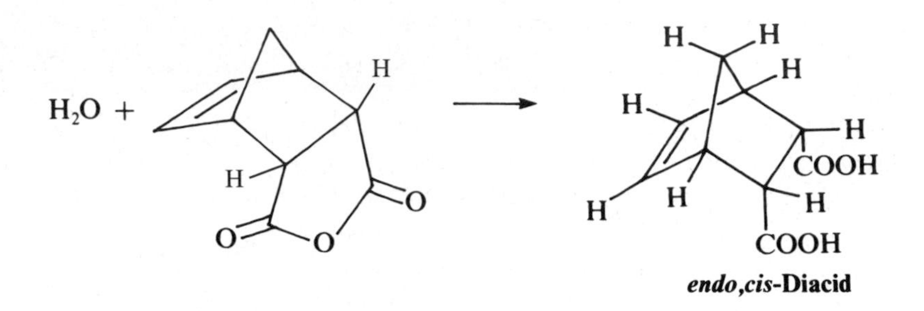

For preparation of the *endo,cis*-diacid, place 4.0 g of the anhydride from Experiment 2 and 50 mL of distilled water in a 125-mL Erlenmeyer flask, grasp with a clamp, swirl over a hot plate, and bring the contents to the boiling point, at which time the solid partially dissolves and partly melts. Continue to heat until all the oil is dissolved; then let the solution stand undisturbed. Because the diacid has a strong tendency to remain in a supersaturated solution, allow 30 minutes or more if necessary for the solution to cool to room temperature; then drop in a boiling stone or touch the surface of the liquid once or twice with a stirring rod. Observe the stone and its surroundings carefully, waiting several minutes before applying the more effective method of making one scratch with a stirring rod on the inner wall of the flask at the air-liquid interface. Let crystallization proceed spontaneously to give large needles; then cool the solution in ice and collect the product. The melting point depends on the rate of heating as the anhydride reforms and water splits out. The IR and ^1H NMR spectra of the diacid are found in Figures 48.11 and 48.12, respectively at the end of the chapter.[5]

The temperature of decomposition is variable.

Video: *Macroscale Crystallization*

Cleaning Up. The aqueous filtrate from the crystallization contains a very small quantity of the diacid. It can be flushed down the drain.

[4]The *endo,cis*-diacid is stable to alkali, but can be isomerized to the *trans*-diacid (mp 192°C) by conversion to the dimethyl ester (3 g acid, 10 mL methanol, 0.5 mL concentrated H_2SO_4; reflux for 1 hour). This ester is equilibrated with sodium ethoxide in refluxing ethanol for 3 days and saponified. For an account of a related epimerization and discussion of the mechanism, see Meinwald J, Gassman PG. *J Am Chem Soc.* **1960**;*82*:5445. *See also* Williamson KL, Li YF, Lacko R, Youn CH. *J Am Chem Soc.* **1969**;*91*:6129; Williamson KL, Li YF. *J Am Chem Soc.* **1970**;*92*:7654.
[5]*See* footnote 4.

4. SYNTHESIS OF COMPOUND X[6]

IN THIS EXPERIMENT, the diacid from Experiment 3 is dissolved in concentrated sulfuric acid and warmed briefly. Water is added very carefully to the mixture, which causes the product to crystallize. A seed crystal is saved, and the product is recrystallized from the aqueous acid mixture. Allow at least 10 minutes for compound X to crystallize.

Microscale Procedure

General rule: If you do not have the necessary quantity of diacid, scale down the amounts of reactants and solvents to match the lesser quantity of starting material.

To a tared 10 × 100-mm reaction tube, add 0.15 g of the *endo,cis*-diacid from Experiment 3, followed by 0.25 mL of concentrated sulfuric acid. Warm the mixture on a steam bath or in a beaker of boiling water for about 2 minutes to allow the anhydride to dissolve/react, cool; then *cautiously* add 0.70 mL of water to the test tube. This should be done dropwise with vigorous mixing of the contents after the addition of each drop. The product will crystallize as a fine powder, often gray in color.

Do not put a wood boiling stick into this solution. The sulfuric acid will attack the wood.

Video: *Microscale Filtration on the Hirsch Funnel*; Photo: *Use of the Wilfilter*

Save a seed crystal and heat the tube on a hot sand bath until the crystals redissolve. Seed the solution and allow it to cool slowly to room temperature. Compound X will crystallize in platelike crystals. The crystallization process for this compound is fairly slow; allow at least 10 minutes for the solution to come to room temperature, and a further 10 minutes in an ice bath before collecting the crystals on a Hirsch funnel or a Wilfilter (Fig. 48.4). The longer you wait to collect the product, the higher your yield will be. Wash the crystals with one small portion of ice water, and scrape the product onto a piece of filter paper. Squeeze the crystals between sheets of filter paper to complete the drying process; then determine the weight, yield, and melting point of the product.

FIG. 48.4
The Wilfilter filtration apparatus.

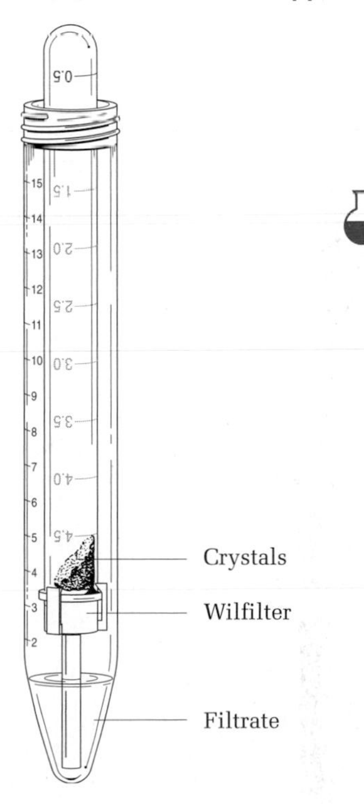

Cleaning Up. Dilute the aqueous filtrate with water, neutralize it with sodium carbonate, and flush the resulting solution down the drain. It contains an extremely small quantity of compound X.

Macroscale Procedure

To prepare compound X, place 1 g of the *endo,cis*-diacid and 5 mL of concentrated sulfuric acid in a 50-mL Erlenmeyer flask, and heat gently on a hot plate for a minute or two until all the crystals are dissolved. Then cool in an ice bath, add a small piece of ice, swirl to dissolve, and add further ice until the volume is about 20 mL. Heat to the boiling point, and let the solution simmer on the hot plate for 5 minutes. Cool well in ice, scratch the flask to induce recrystallization, and allow for some delay in complete separation. The crystals will often be gray in color. Collect, wash with water, and recrystallize from water. Compound X (about 0.7 g) forms large prisms (mp 203°C). To run an NMR spectrum of compound X, the compound must be dissolved in deuterodimethyl sulfoxide (DMSO-d_6).

[6]The synthesis of compound X was introduced by James A. Deyrup.

Cleaning Up. Dilute the aqueous filtrate with water, neutralize it with sodium carbonate, and flush the resulting solution down the drain. It contains a small quantity of compound X.

5. STRUCTURE DETERMINATION OF COMPOUND X

To determine the formula for compound X, which is an isomer of the diacid, try to answer the following questions: What is meant by "isomer of the starting material"? How many functional groups are in the starting material? What happens when each functional group reacts with a strong acid? What ionic intermediate is formed when the diacid dissolves in concentrated sulfuric acid? Why is the ^{1}H NMR spectrum of X (Fig. 48.13) so much more complex than the spectra of the anhydride and diacid (Fig. 48.9 and Fig. 48.12)? What functional group is missing from compound X (which is seen in Fig. 48.9 and Fig. 48.12)? Write formulas for the possible structures of compound X, and devise tests to distinguish among them. What functional group is present in compound X that is not found in the diacid, as determined by an analysis of the IR spectrum of compound X (Fig. 48.14)?

Computational Chemistry

Would you predict that compound X is more or less stable than the isomeric diacid from which it is formed? Test your prediction by carrying out a heat of formation calculation on each of the isomers. To do this, first find the conformation of each isomer with the minimum steric energy using a molecular mechanics program. Then submit each of these to a heat of formation calculation at the AM1 or higher level of calculation.

QUESTIONS

For Additional Experiments, sign in at this book's premium website at **www.cengage.com/login.**

1. In the cracking of dicyclopentadiene, why is it necessary to distill the product very slowly?
2. Draw the products of the following reactions:

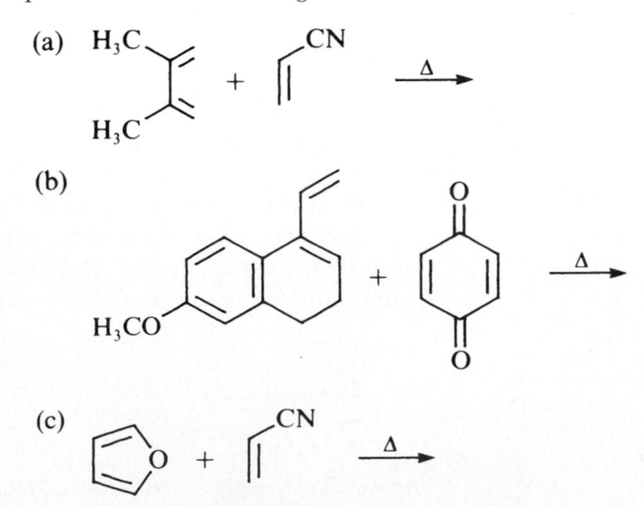

3. What starting material would be necessary to prepare the following compound by the Diels–Alder reaction?

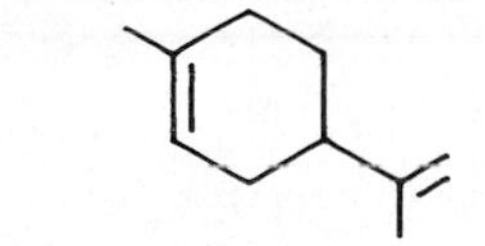

4. If the Diels–Alder reaction between dimethylmaleic anhydride and furan had worked, would cantharidin have been formed?
5. Determine the heats of formation or the steric energies of the *exo*- and *endo*-anhydride adducts using a molecular mechanics program. What do these energies tell you about the mechanism of the reaction?
6. Which molecule, norbornene dicarboxylic acid or compound X, would you predict to be more stable? Why?

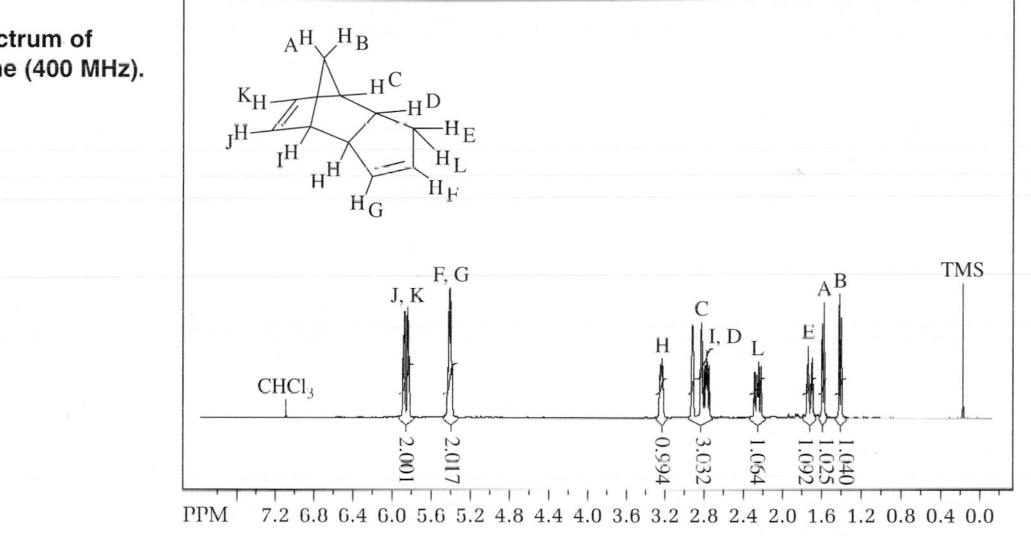

FIG. 48.5
The ^1H NMR spectrum of dicyclopentadiene (400 MHz).

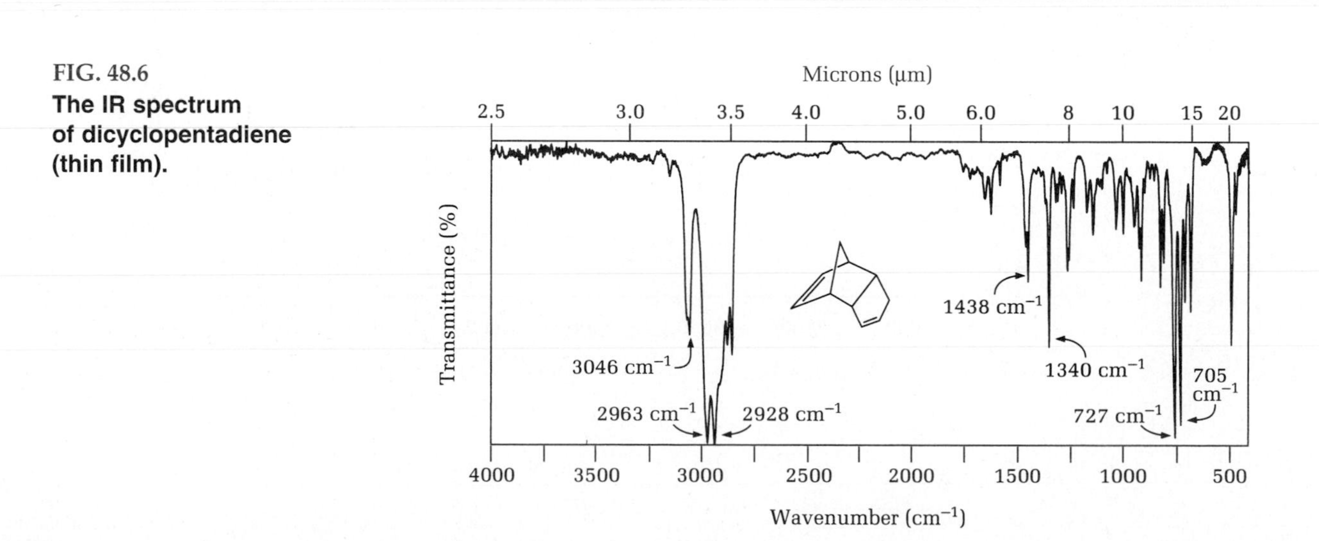

FIG. 48.6
The IR spectrum of dicyclopentadiene (thin film).

FIG. 48.7

The ^{13}C NMR spectrum of dicyclopentadiene (100 MHz).

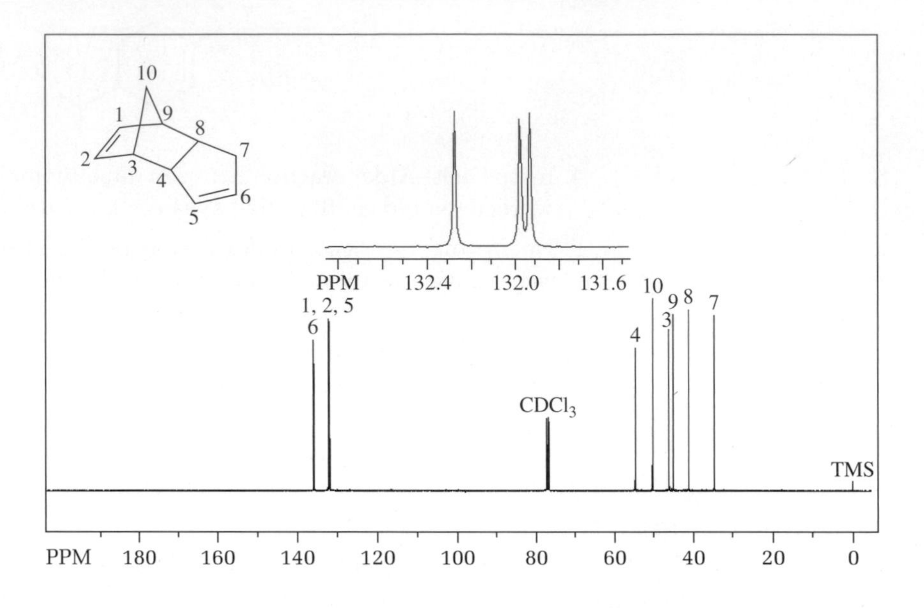

FIG. 48.8

The IR spectrum of *cis*-norbornene- 5,6-*endo*-dicarboxylic anhydride.

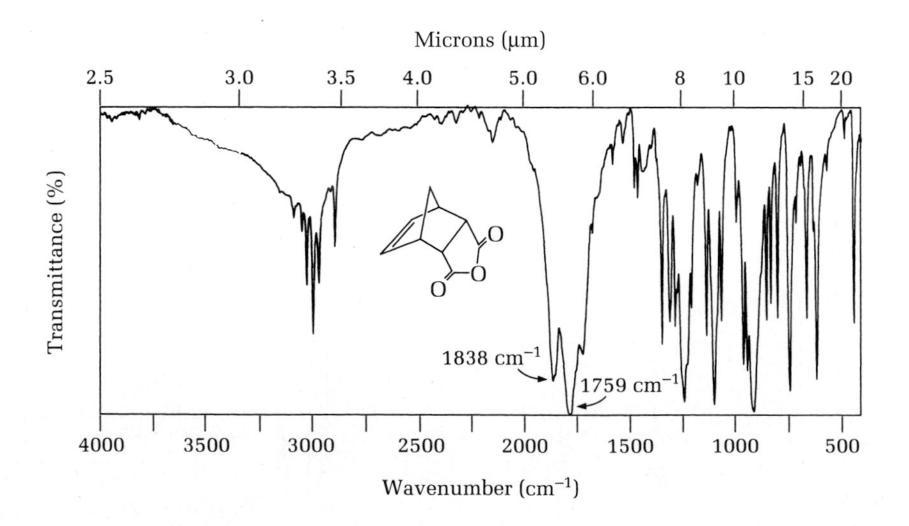

FIG. 48.9

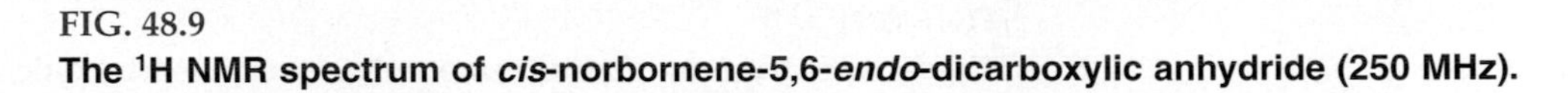

The ^{1}H NMR spectrum of *cis*-norbornene-5,6-*endo*-dicarboxylic anhydride (250 MHz).

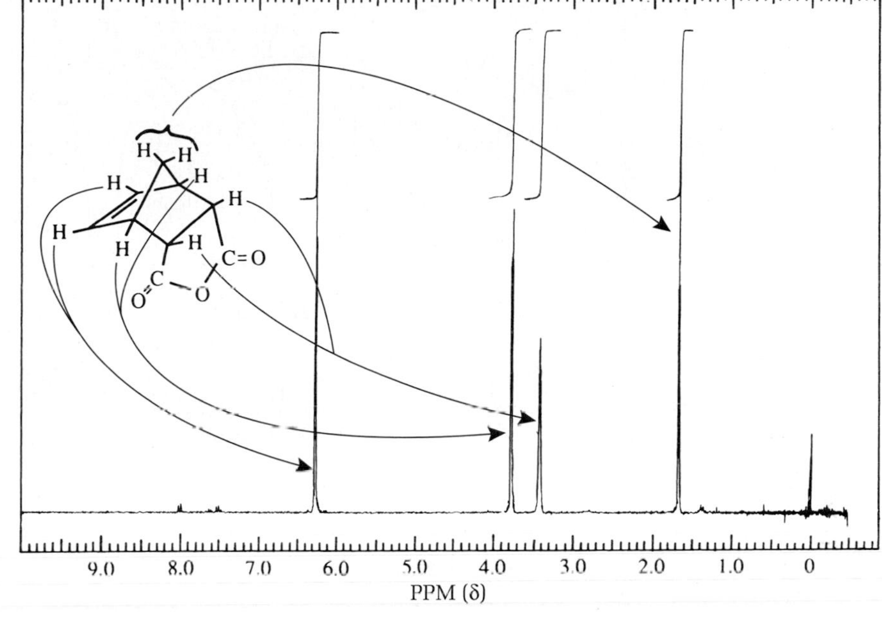

FIG. 48.10

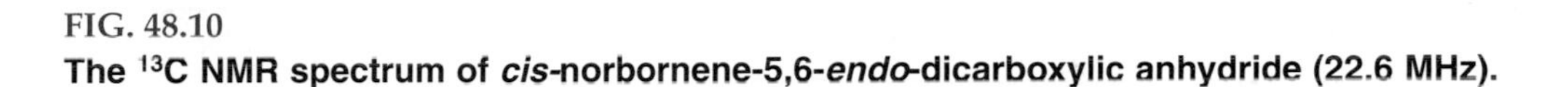

The ^{13}C NMR spectrum of *cis*-norbornene-5,6-*endo*-dicarboxylic anhydride (22.6 MHz).

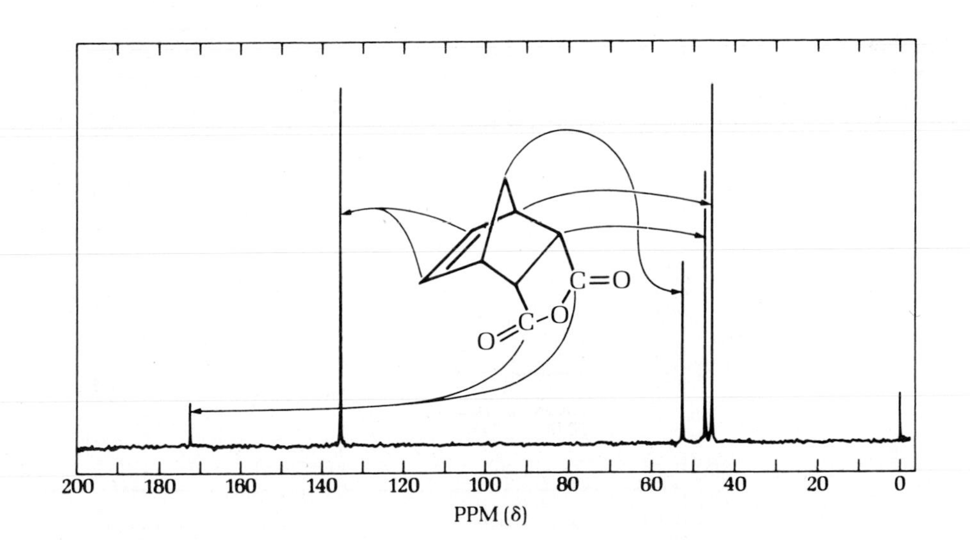

FIG. 48.11
The IR spectrum of *cis*-norbornene-5,6-*endo*-dicarboxylic acid (KBr disk).

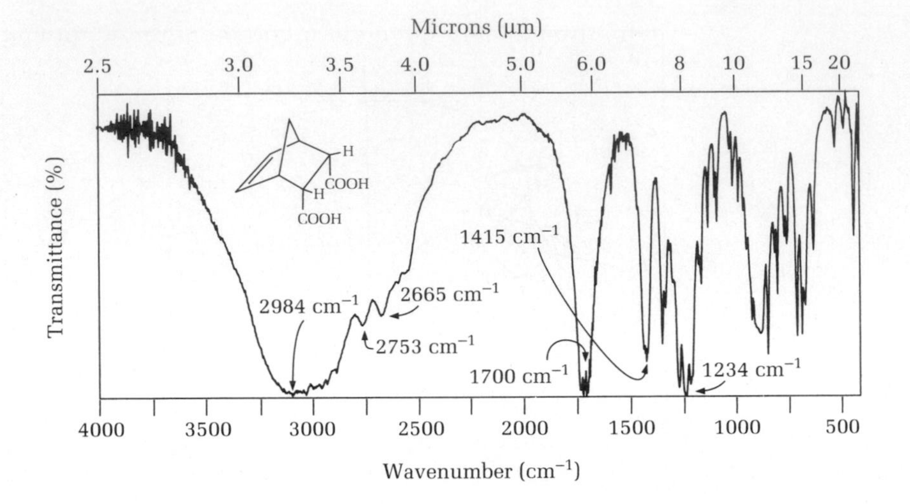

FIG. 48.12
The ^{1}H NMR spectrum of *cis*-norbornene-5,6-*endo*-dicarboxylic acid (250 MHz).

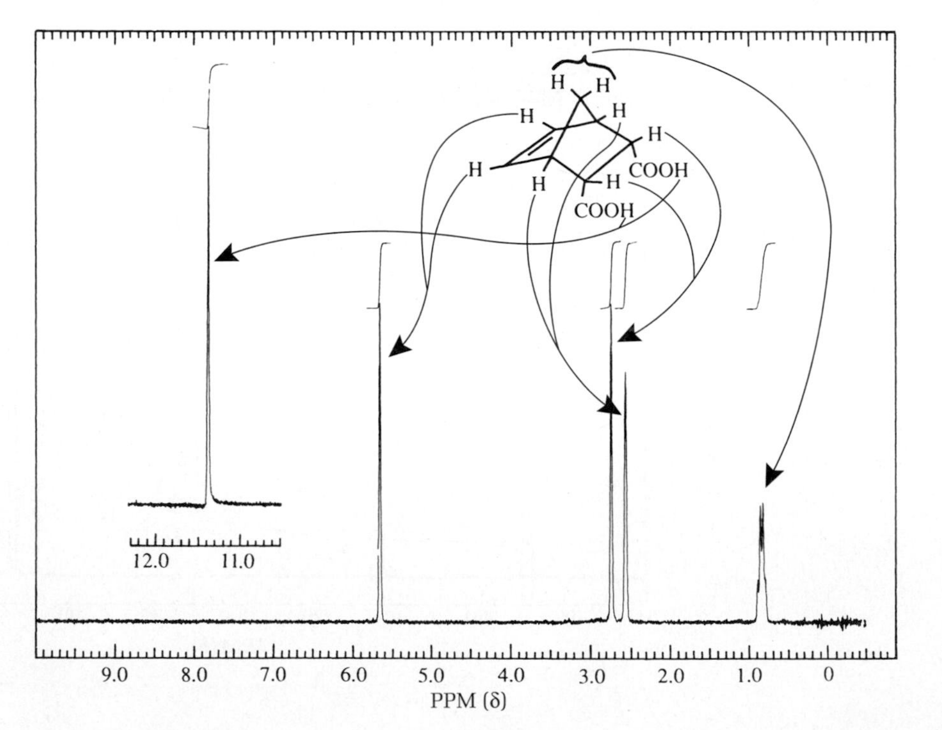

FIG. 48.13

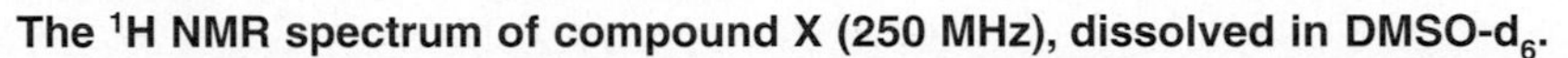

The ^{1}H NMR spectrum of compound X (250 MHz), dissolved in DMSO-d_6.

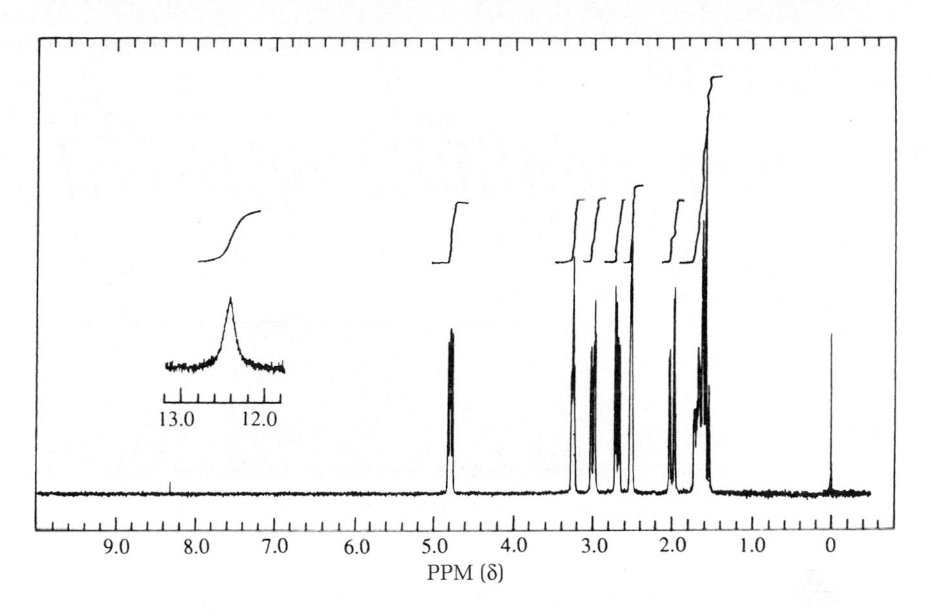

FIG. 48.14

The IR spectrum of compound X (KBr disk).

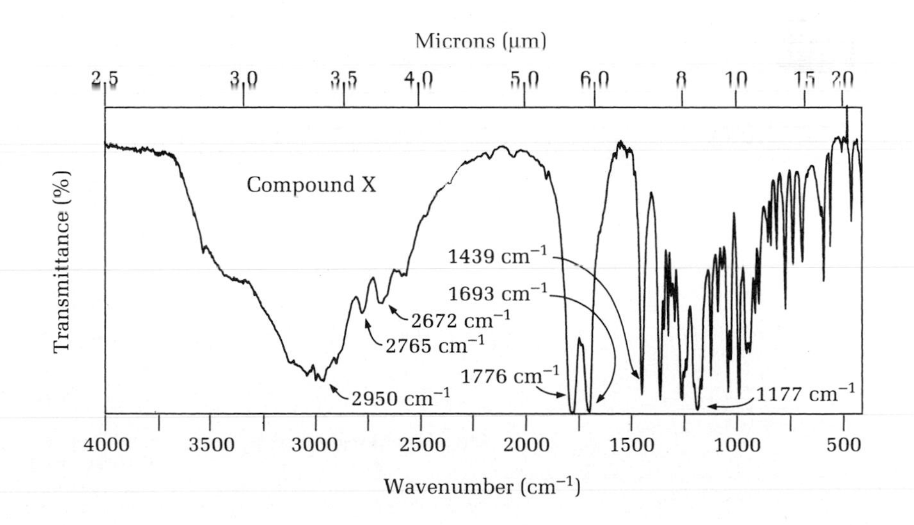

49 CHAPTER

Ferrocene [Bis(cyclopentadienyl)iron]

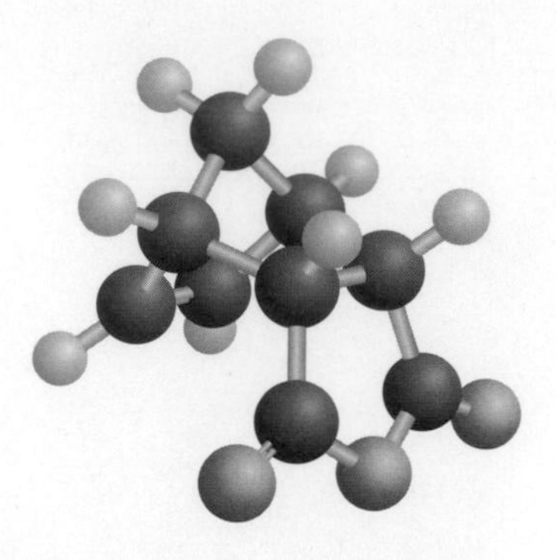

PRELAB EXERCISE: Propose a detailed outline of the procedure for synthesizing ferrocene, paying particular attention to the time required for each step.

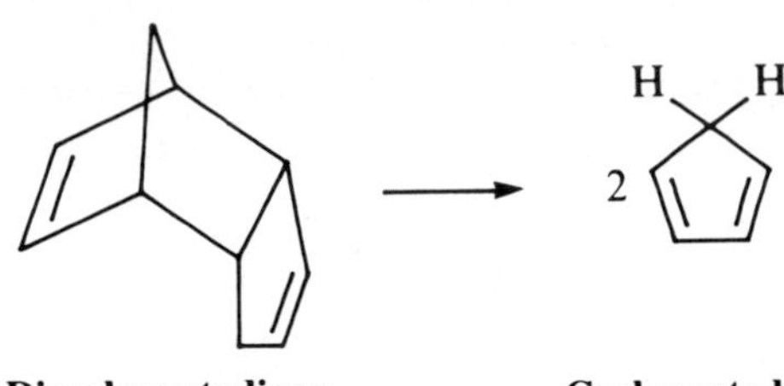

Dicyclopentadiene
den. 0.98, MW 132.20

Cyclopentadiene
bp 41°C, den. 0.80
MW 66.10

Potassium cyclopentadienide

$$2\ C_5H_5^-K^+ + FeCl_2 \cdot 4\ H_2O \longrightarrow Fe(C_5H_5)_2 + 2\ KCl + 4\ H_2O$$

Iron(II) chloride tetrahydrate
MW 198.81

Ferrocene
Bis(cyclopentadienyl)iron
MW 186.04, mp 172–174°C

The Grignard reagent is a classic organometallic compound. The magnesium ion, in Group IIA of the periodic table, needs to lose two—and only two—electrons to achieve the inert gas configuration. This metal has a strong tendency to form ionic bonds by electron transfer:

$$RBr + Mg \longrightarrow \overset{\delta-}{R} - \overset{\delta+}{MgBr}$$

In the transition metals, the situation is not so simple. Consider the bonding between iron and carbon monoxide in $Fe(CO)_5$:

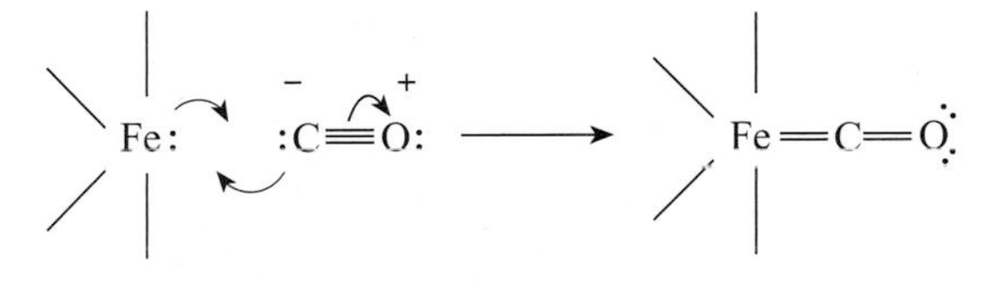

The pair of electrons on the carbon atom is shared with iron to form a σ(sigma) bond between the carbon and iron. The π (pi) bond between iron and carbon is formed from a pair of electrons in the *d* orbital of iron. The π bond is thus formed by the overlap of a *d* orbital of iron with the p-π bond of the carbonyl group. This mutual sharing of electrons results in a relatively nonpolar bond.

Iron has six electrons in the 3*d* orbital, two in the 4*s*, and none in the 4*p* orbital. The inert gas configuration requires 18 electrons—ten 3*d*, two 4*s*, and six 4*p* electrons. Iron pentacarbonyl enters this configuration by accepting two electrons from each of the five carbonyl groups, for a total of 18 electrons. Back-bonding of the *d*-π type distributes the excess electrons among the five carbon monoxide molecules.

Early attempts to form σ-bonded derivatives linking alkyl carbon atoms to iron were unsuccessful, but P. L. Pauson in 1951 succeeded in preparing a very stable substance, ferrocene, $C_{10}H_{10}Fe$, by reacting 2 mol of cyclopentadienylmagnesium bromide with anhydrous ferrous chloride. Another group of chemists—Geoffrey Wilkinson, Myron Rosenblum, Mark Whiting, and Robert B. Woodward—recognized that the properties of ferrocene (its remarkable stability when exposed to water, acids, and air and its ease of sublimation) could be explained only if it had the structure depicted previously, and that the bonding of the ferrous iron with its six electrons must involve all 12 of the π electrons on the two cyclopentadiene rings, with a stable 18-electron inert gas structure as the result.

In this chapter, ferrocene is prepared by reacting the anion of cyclopentadiene with iron(II) chloride. Abstraction of one of the acidic allylic protons of cyclopentadiene with base gives the aromatic cyclopentadienyl anion. It is considered aromatic because it conforms to the Hückel rule in having $4n + 2\pi$ electrons (where n is 1). Two molecules of this anion will react with iron(II) chloride to give ferrocene, the most common member of the class of metal organic compounds referred to as metallocenes. In this centrosymmetric sandwich-type π complex, all carbon atoms are equidistant from the iron atom, and the two cyclopentadienyl rings rotate more or less freely with respect to each other. The extraordinary stability of ferrocene (stable to 500°C) can be attributed to the sharing of the 12 π electrons of the two cyclopentadienyl rings with the 6 outer shell electrons of iron(II) to give the iron a stable 18-electron inert gas configuration. Ferrocene is soluble in organic solvents, can be dissolved in concentrated sulfuric acid and recovered unchanged, and is resistant to other acids and bases as well (in the absence of oxygen). This behavior is consistent with that of an aromatic compound; ferrocene can undergo electrophilic aromatic substitution reactions with ease.

Ferrocene: soluble in organic solvents, stable to 500°C

Cyclopentadiene readily dimerizes at room temperature by a Diels–Alder reaction to give dicyclopentadiene. This dimer can be "cracked" by heating (an example of the reversibility of the Diels–Alder reaction) to give low-boiling cyclopentadiene. In most syntheses of ferrocene, the anion of cyclopentadiene is prepared by reacting the diene with metallic sodium. Subsequently, this anion is allowed to react with anhydrous iron(II) chloride. Here, the anion is generated using powdered potassium hydroxide, which functions as both a base and a dehydrating agent.

The anion of cyclopentadiene decomposes rapidly in air and iron(II) chloride; although reasonably stable in the solid state, it is readily oxidized to the iron(III) (ferric) state in solution. Consequently, this reaction must be carried out in the absence of oxygen, which is accomplished by bubbling nitrogen gas through any solutions used in order to displace dissolved oxygen and to flush air from the apparatus. Rather elaborate apparatus is used in research laboratories to carry out experiments in the absence of oxygen. A very simple apparatus is used in this chapter because no gases are evolved, no heating is necessary, and the reaction is only mildly exothermic.

EXPERIMENTS

1. MICROSCALE SYNTHESIS OF FERROCENE

IN THIS EXPERIMENT, one of the reactants and the ionic intermediate will react with atmospheric oxygen; therefore, the reaction is run under a nitrogen atmosphere. Potassium hydroxide is partially dissolved in a solvent, cyclopentadiene is added, and, on shaking and stirring, is converted to the colored anion. To this solution of the cyclopentadienyl anion is added a solution of iron(II) chloride to produce ferrocene. The mixture is poured onto ice-cold acid, and the crystalline product is collected on a Hirsch funnel. Once dry, the product is purified by sublimation.

To a 5-mL short-necked, round-bottomed flask, add a magnetic stirring bar and then quickly add 0.75 g of finely powdered potassium hydroxide,[1] followed by 1.25 mL of dimethoxyethane (DME). The funnel that is a part of the chromatography column makes a convenient addition funnel. Cap the flask with a good septum and

[1]Potassium hydroxide is easily ground to a fine powder in 25-g batches in 1 minute, employing an ordinary food blender (e.g., Waring, Osterizer brands). The finely powdered base is transferred in a hood to a bottle with a tightly fitting cap. Alternatively, grind about 1 g potassium hydroxide in a mortar and transfer it rapidly to the reaction flask.

FIG. 49.1
The apparatus for flushing air from the reaction flask. Check the nitrogen flow rate by allowing it to bubble through an organic solvent before inserting the needle through the septum.

Nitrogen
Rubber tube
Empty syringe needle
Rubber septum

$CH_3OCH_2CH_2OCH_3$

1,2-Dimethoxyethane
(Ethylene glycol dimethyl ether, monoglyme), bp 85°C
Completely miscible with water

Video: *Ferrocene Synthesis*

CAUTION: Potassium hydroxide is extremely corrosive and hygroscopic. Immediately wash any spilled powder or solutions from the skin and wipe up all spills. Keep containers tightly closed. Work in the hood.

CAUTION: Dimethyl sulfoxide is rapidly absorbed through the skin. Wash off spills with water. Wear disposable gloves when shaking the apparatus.

$CH_3\overset{O}{\overset{\|}{S}}CH_3$

Dimethyl sulfoxide, DMSO
(Methyl sulfoxide), bp 189°C
Completely miscible with water

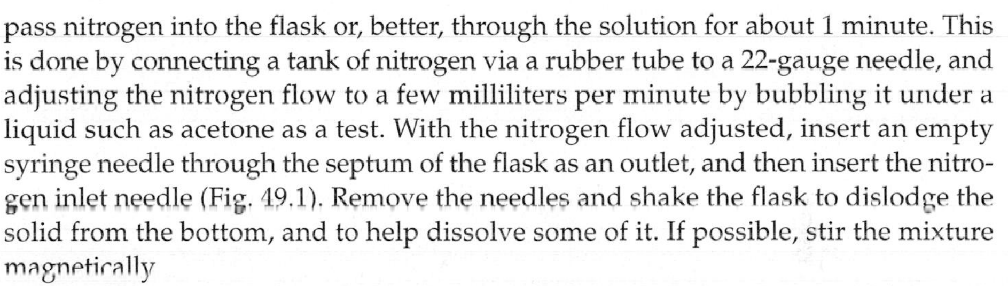

pass nitrogen into the flask or, better, through the solution for about 1 minute. This is done by connecting a tank of nitrogen via a rubber tube to a 22-gauge needle, and adjusting the nitrogen flow to a few milliliters per minute by bubbling it under a liquid such as acetone as a test. With the nitrogen flow adjusted, insert an empty syringe needle through the septum of the flask as an outlet, and then insert the nitrogen inlet needle (Fig. 49.1). Remove the needles and shake the flask to dislodge the solid from the bottom, and to help dissolve some of it. If possible, stir the mixture magnetically.

To a 10 × 100-mm reaction tube, add 0.35 g of finely powdered green iron(II) chloride tetrahydrate and 1.5 mL of dimethyl sulfoxide (DMSO). Cap the tube with a good rubber septum, insert an empty syringe needle through the septum, and pass nitrogen into the tube for about 1 minute to displace the oxygen present. Remove the needles, and then shake the vial vigorously to dissolve all the iron chloride. Some warming may be needed.

Using an accurate syringe, inject 0.300 mL of freshly prepared cyclopentadiene (see Chapter 48 for the preparation of cyclopentadiene) into the flask containing the potassium hydroxide. Do not grasp the body of the syringe because the heat of your hand will cause the cyclopentadiene to volatilize; hold the syringe at the top. Stir the mixture vigorously and note the color change as the potassium cyclopentadienide is formed. After waiting about 5 minutes for the anion to form, pierce the septum with an empty needle for pressure relief and inject the iron(II) chloride solution contained in the reaction tube in six 0.25-mL portions over a 10-minute period. Stir the mixture well with a magnetic stirrer. If the stirring bar is immobilized, then between injections, remove both needles from the septum and shake the flask vigorously. After all the iron(II) chloride solution has been added, rinse the reaction tube with an additional 0.25 mL of DMSO and add this to the flask. Continue to stir or shake the solution for about 15 minutes to complete the reaction.

Isolation of ferrocene

To isolate the ferrocene, pour the dark slurry onto a mixture of 4.5 mL of 6 *M* hydrochloric acid and 5 g of ice in a 30-mL beaker. Stir the contents of the beaker thoroughly to dissolve and neutralize all the potassium hydroxide. Collect the crystalline orange ferrocene on a Hirsch funnel, wash the crystals well with water, press out excess water, squeeze the product between sheets of filter paper to complete the drying, and then purify the ferrocene by sublimation. The filtrate is blue because of dissolved ferrocinium ion. It can be reduced with a mild reducing agent, such as ascorbic acid, to regenerate ferrocene. The amount of ferrocene produced is usually negligible.

Purification by sublimation

To sublime the ferrocene, add the crude dry product to a 25-mL filter flask equipped with a neoprene filter adapter (Pluro stopper) and a 15-mL centrifuge tube that is pushed to within 5 mm of the bottom of the flask (Fig. 49.2). Put a rubber bulb on the side arm of the flask to cap it off; then add ice to the centrifuge tube and heat the flask on a sand bath to sublime the product. Tilting and rolling the filter flask in the hot sand will help drive ferrocene onto the centrifuge tube. Remove the flask from the sand bath and use a heat gun to drive the last of the ferrocene from the sides of the flask to the centrifuge tube. Ferrocene sublimes nicely at atmospheric pressure. Vacuum sublimation is not needed.

When the sublimation is complete, cool the flask, remove the ice water from the centrifuge tube, and replace it with room temperature water (to prevent moisture from collecting on the tube). Transfer the product to a tared, stoppered vial, determine the weight, and calculate the percent yield. Determine the melting point in an evacuated capillary that will contain the ferrocene, because the product sublimes at the melting point (Fig. 49.3) (*see* Chapter 3 for this technique).

Cleaning Up. The filtrate should be slightly acidic. Neutralize it with sodium carbonate, dilute it with water, and flush it down the drain. Place any unused

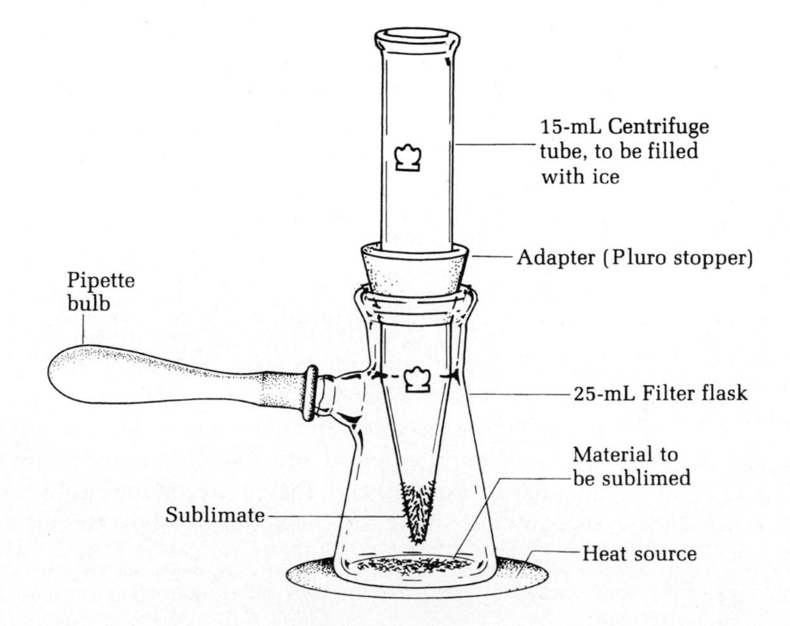

FIG. 49.2
The apparatus for the sublimation of ferrocene.

cyclopentadiene in the recovered dicyclopentadiene or the organic solvents waste container. Add 0.4 mL concentrated nitric acid to the sublimation flask to clean it. After at least 24 hours, dilute the acid, neutralize it with sodium carbonate, and flush the solution down the drain.

FIG. 49.3
Evacuation of a melting point capillary prior to sealing.

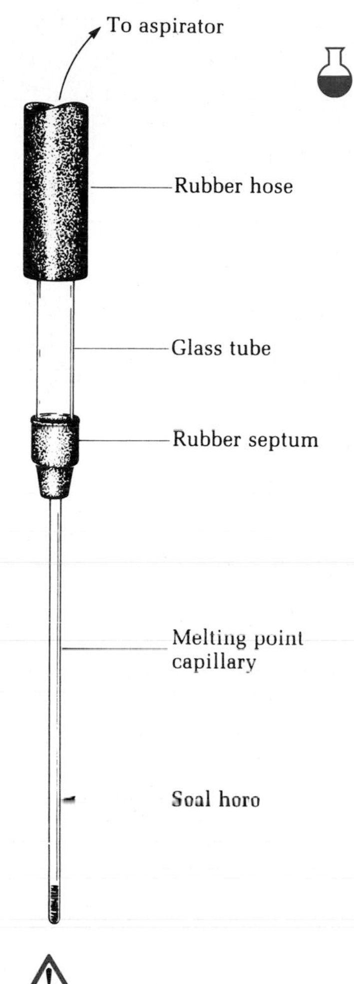

2. MACROSCALE SYNTHESIS OF FERROCENE

Following the procedure described in Experiment 1 of Chapter 48, prepare 3 mL of cyclopentadiene. It need not be dry. While this distillation is taking place, rapidly weigh 12.5 g of finely powdered potassium hydroxide[2] into a 50-mL Erlenmeyer flask, add 30 mL of dimethoxyethane (DME), and immediately cool the mixture in an ice bath. Swirl the mixture in the ice bath for a minute or two; then bubble nitrogen through the solution for about 2 minutes. Quickly stopper the flask and shake the mixture to dislodge the cake of potassium hydroxide from the bottom of the flask, and to dissolve as much of the base as possible (much will remain undissolved).

Grind 3.5 g of iron(II) chloride tetrahydrate to a fine powder, then add 3.5 g of the green salt to 12.5 mL of dimethyl sulfoxide (DMSO) in a 25-mL Erlenmeyer flask. Pass nitrogen through the DMSO mixture for about 2 minutes, stopper the flask, and shake it vigorously to dissolve all the iron(II) chloride. Gentle warming of the flask on a steam bath may be necessary to dissolve the last traces of iron(II) chloride. Transfer the solution rapidly to a 60-mL separatory funnel (which serves as an addition funnel) equipped with a cork or stopper to fit into the 50-mL Erlenmeyer flask, flush air from the funnel with a stream of nitrogen, and stopper it.

Transfer 3 mL of the freshly distilled cyclopentadiene to the slurry of potassium hydroxide in dimethoxyethane. Shake the flask vigorously and note the color change as the potassium cyclopentadienide is formed. After waiting about 5 minutes for the anion to form, quickly replace the cork or stopper on the Erlenmeyer flask that is fitted with the separatory funnel in order to avoid admission of air to the flask; see Fig. 49.4. Figure 49.5 depicts a research-quality apparatus that would be used for this experiment.

Add the iron(II) chloride solution to the base dropwise over a period of 20 minutes with vigorous swirling and shaking. Dislodge the potassium hydroxide should it cake on the bottom of the flask. The shaking will allow nitrogen to pass from the Erlenmeyer flask into the separatory funnel as the solution leaves the funnel.[3] Continue to shake and swirl the solution for 10 minutes after all the iron(II) chloride is added; then pour the dark slurry onto a mixture of 45 mL of 6 *M* hydrochloric acid and 50 g of ice in a 250-mL beaker. Stir the contents of the beaker thoroughly to dissolve and neutralize all the potassium hydroxide. Collect the crystalline orange ferrocene on a Büchner funnel, wash the crystals with water, press out excess water, and allow the product to dry overnight on a watch glass.

CAUTION: Potassium hydroxide is extremely corrosive and hygroscopic. Immediately wash any spilled powder or solutions from the skin, and wipe up all spills. Keep containers tightly closed. Work in the hood.

[2]Potassium hydroxide is easily ground to a fine powder in 75-g batches in 1 minute employing an ordinary food blender (e.g., Waring, Osterizer brands). The finely powdered base is transferred in a hood to a bottle with a tightly fitting cap. If a blender is not available, crush and then grind 27 g of potassium hydroxide pellets in a large mortar and quickly weigh 25 g of the resulting powder into a 125-mL Erlenmeyer flask.

[3]If the particular separatory funnel being used does not allow nitrogen to pass from the flask to the funnel, connect the two with a rubber tube leading to a glass tube and stopper at the top of the separatory funnel and to a syringe needle that pierces the rubber stopper in the flask (suggestion of D. L. Fishel).

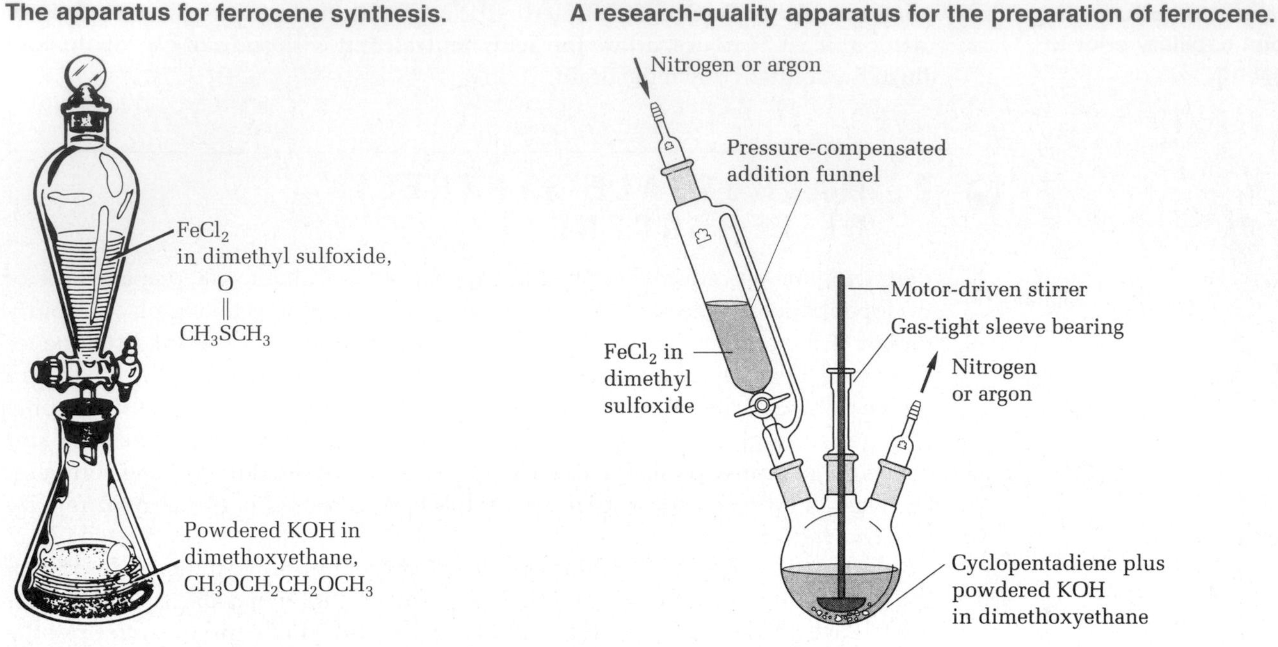

FIG. 49.4
The apparatus for ferrocene synthesis.

FIG. 49.5
A research-quality apparatus for the preparation of ferrocene.

CAUTION: Dimethoxyethane can form peroxides. Discard 90 days after opening the bottle because of peroxide formation and potential for an explosion.

CAUTION: Dimethyl sulfoxide is rapidly absorbed through the skin. Wash off spills with water.

Recrystallize the ferrocene from methanol, or better, hexanes. It is also very easily sublimed. In a hood, place about 0.5 g of crude ferrocene on a watch glass on a hot plate set to about 150°C. Invert a glass funnel over the watch glass. Ferrocene will sublime in about 1 hour, leaving nonvolatile impurities behind. Pure ferrocene melts at 172–174°C. Determine the melting point in an evacuated capillary that will contain the ferrocene (Fig. 49.3), because the product sublimes at the melting point. Compare the melting points of your sublimed and recrystallized materials.

Cleaning Up. The filtrate from the reaction mixture should be slightly acidic. Neutralize it with sodium carbonate, dilute it with water, and flush it down the drain. Place any unused cyclopentadiene in the recovered dicyclopentadiene or the organic solvents waste container. If the ferrocene has been crystallized from methanol or hexanes, place the mother liquor in the organic solvents waste container.

QUESTIONS

1. If ferrocene and all of the reagents in this experiment are stable in air before the reaction begins, why must air be so carefully excluded during the reaction?
2. What special properties do the solvents dimethoxyethane and dimethyl sulfoxide have compared to diethyl ether, for example, that make them particularly suited for this reaction?
3. What is it about ferrocene that allows it to sublime easily whereas many other compounds do not?

CHAPTER 50

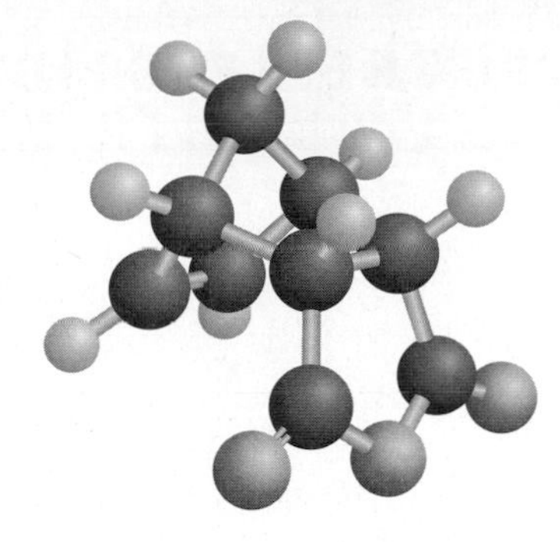

A Diels–Alder Reaction Puzzle: The Reaction of 2,4-Hexadien-1-ol with Maleic Anhydride

PRELAB EXERCISE: Predict the approximate frequencies and chemical shifts for important peaks to be found in the IR and NMR spectra of the expected product. Determine all of the electrophilic and nucleophilic sites in the expected product. This exercise may help in determining how the actual product forms.

2,4-Hexadien-1-ol can be shown by simple molecular mechanics calculations (*see* Chapter 15) to exist predominantly in the s-*trans* configuration in which there is overlap of the π-electrons of the alkenes. There is free rotation around the 3–4 single bond and again, overlap of π-electrons so that the s-*cis* configuration also makes a contribution to the possible conformers. It is in this s-*cis* configuration that the diene can enter into the Diels–Alder reaction (*see* the first page of the discussion of the Diels–Alder reaction in Chapter 48).

The Diels–Alder reaction is highly stereospecific. Note that four chiral centers, all *cis* to one another, are formed in this one reaction from starting materials having no chiral centers (*see* Chapter 48).

One of the goals of modern chemistry is to have a minimum impact on the environment. Microscale experiments are designed to do just this, but it is possible to carry this idea one step further: A solvent, no matter how small the quantity, must be collected and disposed of. So let's try eliminating the solvent! Goodwin and Rogers[1] accomplished this in the following experiment.

[1]See Instructor's Guide for reference to the original literature.

1. THE REACTION OF 2,4-HEXADIEN-1-OL WITH MALEIC ANHYDRIDE

IN THIS EXPERIMENT, the s-*cis* conformation of a dienol reacts with maleic anhydride in the absence of a solvent, to give a pure solid product, but not the one expected. Analysis of IR and NMR spectra are used to deduce the structure.

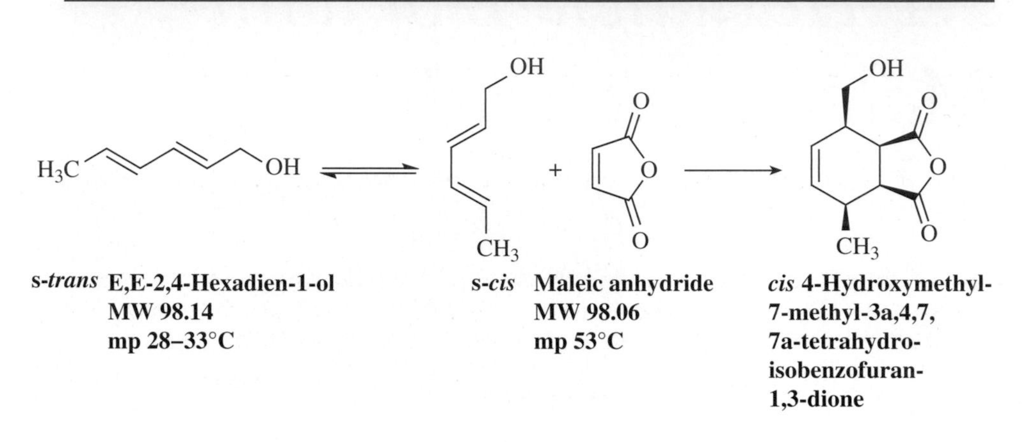

Weigh into a very small vial 147 mg of *E,E*-2,4-hexadien-1-ol. Since this is a low-melting solid, this will not be easy. Using a heat gun or similar source of heat, melt the material so it flows to the bottom of the vial. Cool the vial until the dieneol re-solidifies. Then add 147 mg of finely powdered maleic anhydride. The two materials happen to have almost the same molecular weights; the exact weight of the alcohol does not matter as long as an exactly equal weight of maleic anhydride is added because this is a 1:1 reaction.

Using the pointed end of a metal spatula, stir the reaction mixture thoroughly and vigorously until complete liquefaction and resolidification occur. Note carefully the changes in physical appearance and properties as the reaction proceeds. After about 10–15 minutes, a white solid will result. Determine the melting point range, which is reported to be 159–161°C. If the melting point is much lower, then the product is contaminated with one of the two starting materials. It can be recrystallized from boiling toluene, but an alternate technique may suffice for purification: trituration.

Add two or three drops of toluene to the solid and grind it thoroughly with the solvent. This is the process of trituration. The toluene will dissolve any small amount of maleic anhydride or dieneol that is in excess, leaving the less soluble product behind. Scrape the damp material onto a piece of filter paper, squeeze it to remove the solvent, allow it to dry and again determine the melting point. Once you are satisfied with the purity of the product, obtain IR and ^{1}H NMR spectra (from 0 to 13 ppm) in either deuterated acetone or chloroform.

In analyzing the IR spectrum, note which peaks are absent from the expected structure and which new peaks are present. As noted in Chapter 11, infrared spectroscopy is especially useful for detecting and distinguishing among all carbonyl-containing compounds. The IR spectra of a number of Diels–Alder adducts of maleic anhydride are shown in Chapter 12. As you will soon discover, the product is not the expected one. But in the course of the synthesis, it should have been obvious that it did not react with some other material or give off some molecule, such as water. The product you have isolated is an isomer of the expected product.

Once you have determined all the electrophilic and nucleophilic sites in the expected product, can you predict what intramolecular reaction might occur?

The proton NMR spectrum is very well resolved, but will reveal little to help you except for the chemical shift of one peak.

Cleaning Up. Except for the disposal of the product in the waste container provided, there is no chemical waste from this experiment.

QUESTION

1. What simple chemical test might be used to confirm the correct structure of this molecule?

51 CHAPTER

Tetraphenylcyclopentadienone

PRELAB EXERCISE: Write a detailed mechanism for the formation of tetraphenylcyclopentadienone from benzil and 1,3-diphenylacetone. To which general class of reactions does this condensation belong?

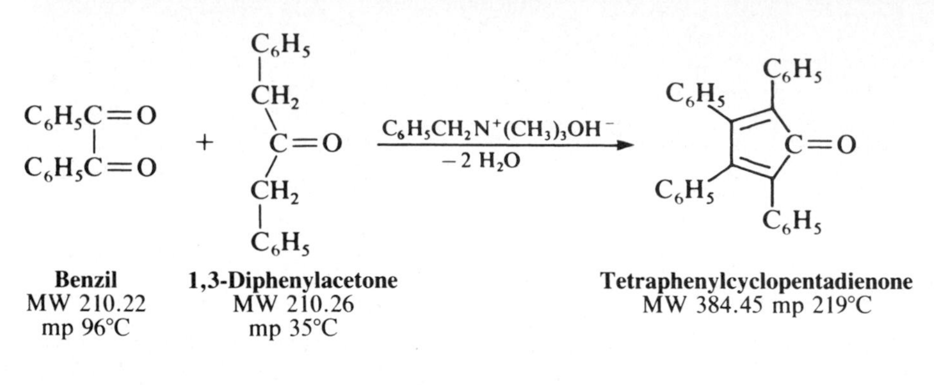

Cyclopentadienone

Cyclopentadienone is an elusive compound that has been sought for many years but with little success. Molecular orbital calculations predict that it should be highly reactive, and so it is—it exists only as the dimer. The tetraphenyl derivative of this compound is synthesized in this chapter. This derivative is stable and reacts readily with dienophiles. It is used not only for the synthesis of highly aromatic, highly arylated compounds, but also for examination of the mechanism of the Diels–Alder reaction itself. Tetraphenylcyclopentadienone has been carefully studied by means of molecular orbital methodology in attempts to understand its unusual reactivity, color, and dipole moment.

Louis Fieser introduced the idea of using very high-boiling solvents to speed up this and many other reactions.

The literature procedure for the condensation of benzil with 1,3-diphenylacetone in ethanol, with potassium hydroxide as a basic catalyst, is not optimal because of the low boiling point of the alcohol and the limited solubility of both potassium hydroxide and the reaction product in this solvent. Triethylene glycol is a better solvent and permits operation at a higher temperature. In the procedure that follows, the glycol is used with benzyltrimethylammonium hydroxide, a strong base readily soluble in organic solvents, which serves as a catalyst.

In Chapter 52, tetraphenylcyclopentadienone will be used to synthesize hexaphenylbenzene and dimethyl tetraphenylphthalate. The mechanism for the formation of tetraphenylcyclopentadienone is as follows:

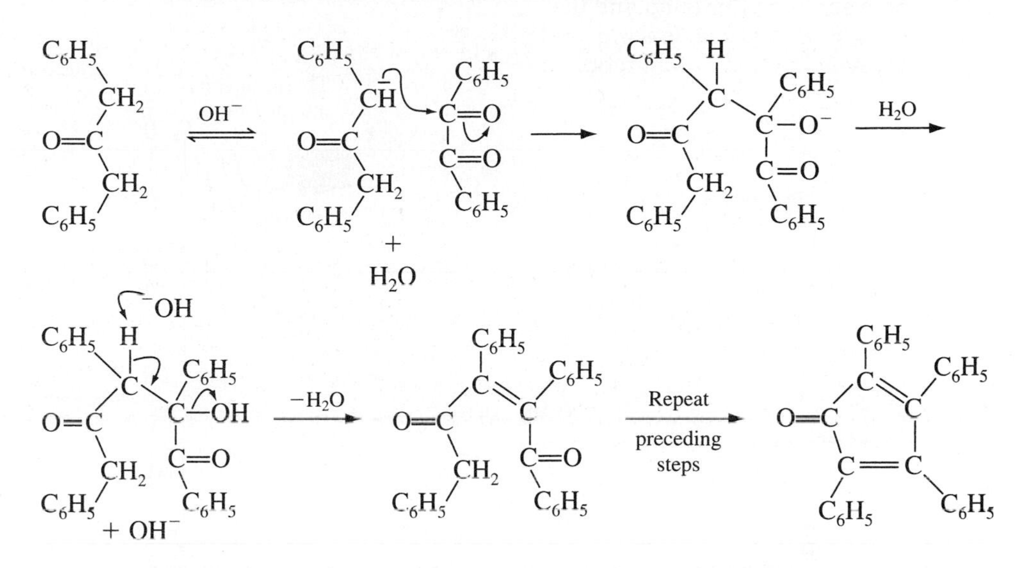

EXPERIMENTS

1. TETRAPHENYLCYCLOPENTADIENONE

IN THIS EXPERIMENT, benzil and diphenyl acetone are dissolved in a very high-boiling solvent and treated with a strong organic base. The product crystallizes and is collected and washed with methanol. It can, if necessary, be recrystallized from a high-boiling solvent.

CAUTION: Triton B is toxic and corrosive.

See Chapter 54 for the preparation of benzil. Into a 10 × 100-mm reaction tube, place 42 mg of pure benzil (free of benzoin), 42 mg of 1,3-diphenylacetone, and 0.4 mL of triethylene glycol, using the solvent to wash the walls of the tube. Clamp the tube over a hot sand bath, insert a thermometer, and heat the solution until the benzil is dissolved. Remove the tube from the heat; then, using a 1-mL syringe, add to the solution 0.20 mL of a 40% solution of benzyltrimethylammonium hydroxide in methanol (Triton B) when the temperature of the solution reaches exactly 100°C. Stir once to mix. Crystallization usually starts in 10–20 seconds. Let the mixture cool until it is near room temperature; then cool it in cold water. Add 0.5 mL of methanol, cool the tube in ice, and collect the product.

Short reaction period

If the crystals are large enough, collection can be done by inserting a Pasteur pipette into the tube and removing the solvent between the tip of the pipette and the bottom of the tube (Fig. 51.1). If the crystals are small, collect them on a Wilfilter, a Hirsch funnel, or a microscale Büchner funnel (Fig. 51.2). In any case, wash the crystals with cold methanol until the washings are purple-pink, not

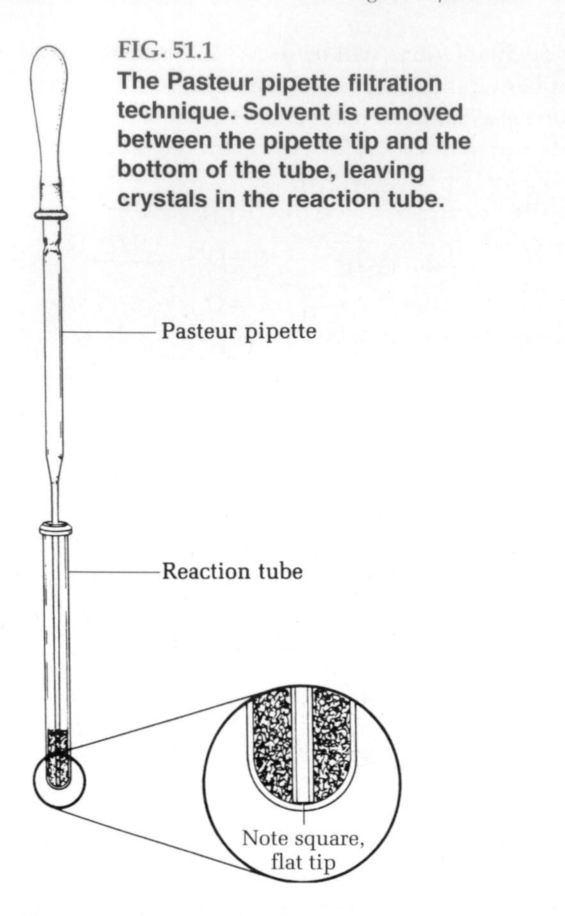

FIG. 51.1
The Pasteur pipette filtration technique. Solvent is removed between the pipette tip and the bottom of the tube, leaving crystals in the reaction tube.

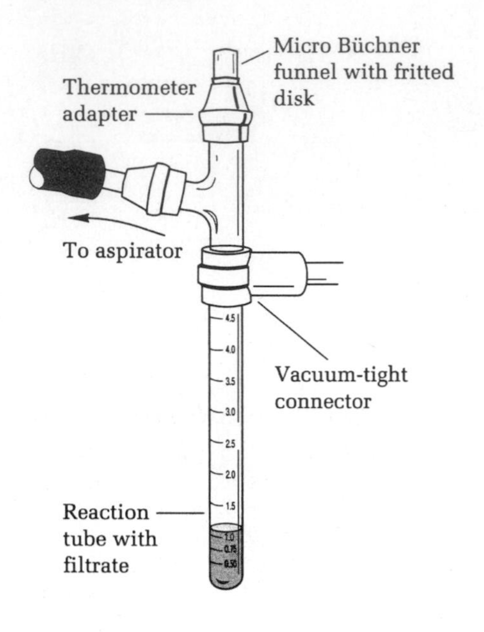

FIG. 51.2
An apparatus for filtration on a microscale Büchner funnel.

Photos: *Filtration Using a Pasteur Pipette, Micro Büchner Funnel, Use of the Wilfilter*, Videos: *Filtration of Crystals Using the Pasteur Pipette, Microscale Crystallization*

brown. The yield of deep-purple crystals is about 60 mg. If either the crystals are not well formed or the melting point is low, place the material in a reaction tube, add 0.6 mL of triethylene glycol, stir with a stirring rod, and raise the temperature to 220°C to bring the solid into solution. Let it stand for crystallization. (If the solution is initially pure and then recrystallized, the recovery is about 90%.)

Cleaning Up. Because the filtrate and washings from the reaction contain Triton B, place them in the hazardous waste container. Dilute the crystallization solvent with water and flush it down the drain.

2. TETRAPHENYLCYCLOPENTADIENONE

CAUTION: Triton B is toxic and corrosive.

Short reaction period

See Chapter 54 for the preparation of benzil. Measure into a 25 × 150-mm test tube 2.1 g of benzil, 2.1 g of 1,3-diphenylacetone, and 10 mL of triethylene glycol, using the solvent to wash the walls of the test tube. Support the test tube in a hot sand bath, stir the mixture with a stirring rod, heat until the benzil is dissolved, and then remove it from the sand. Measure 1 mL of a commercially available 40% solution of benzyltrimethylammonium hydroxide (Triton B) in methanol into a 10 × 75-mm test tube, adjust the temperature of the solution to exactly 100°C, remove from the heat, add the catalyst, and stir once to mix. Crystallization usually starts in 10–20 seconds. Let the temperature drop to about 80°C; then cool under the tap, add 10 mL

Video: *Macroscale Crystallization*

of methanol, stir until a thin crystal slurry forms, collect the product, and wash it with methanol until the filtrate is purple-pink, not brown. The yield of deep-purple crystals is 3.3–3.7 g. If either the crystals are not well formed or the melting point is low, place 1 g of material and 10 mL of triethylene glycol in a vertically supported test tube, stir with a stirring rod, raise the temperature to 220°C to bring the solid into solution, and let stand for crystallization. (If the solution is initially pure and then recrystallized, the recovery is about 90%.)

Cleaning Up. Because the filtrate and washings from the reaction contain Triton B, place them in the hazardous waste container. Dilute the crystallization solvent with water and flush it down the drain.

QUESTION

1. Draw the structure of the dimer of cyclopentadienone. Why doesn't tetraphenylcyclopentadienone undergo dimerization? This will become much clearer if an energy-minimized structure is generated using a molecular mechanics program.

52 CHAPTER

Hexaphenylbenzene and Dimethyl Tetraphenylphthalate

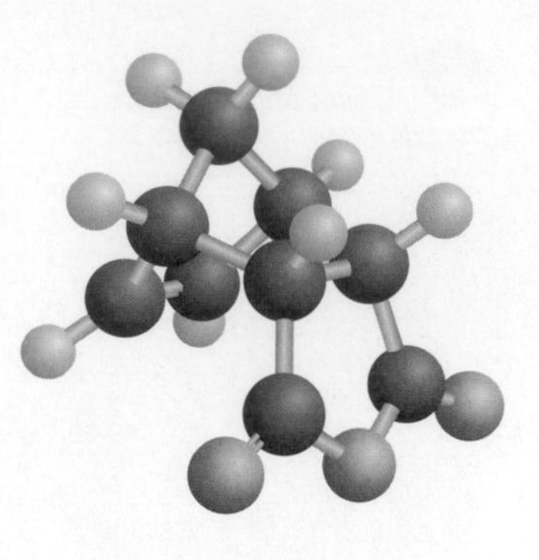

PRELAB EXERCISE: Explain the driving force behind the loss of carbon monoxide from the intermediates formed in the two reactions in this chapter.

This experiment illustrates two examples of the Diels–Alder reaction, a means of synthesizing molecules that would be extremely difficult to synthesize in any other way. Both reactions employ as the diene the tetraphenylcyclopentadienone prepared in Chapter 51. Although the Diels–Alder reaction is reversible, the intermediate in each of these reactions spontaneously loses carbon monoxide (why?) to form the products.

In the first experiment, the dienone is condensed with dimethyl acetylenedicarboxylate using as the solvent 1,2-dichlorobenzene. This solvent is chosen for its solvent properties as well as its high boiling point, which guarantees that the reaction is completed in 1–2 minutes.

In the second experiment, the dienone is condensed with diphenylacetylene to produce hexaphenylbenzene. Hexaphenylbenzene melts at 465°C without decomposition. Few completely covalent organic molecules have higher melting points. For comparison, lead melts at 327.5°C. As is often the case, a high melting point also means limited solubility. The solvent used to recrystallize hexaphenylbenzene, diphenyl ether, has a very high boiling point (bp 259°C) and has superior solvent power.

EXPERIMENTS

1. DIMETHYL TETRAPHENYLPHTHALATE

IN THIS EXPERIMENT, purple tetraphenylcyclopentadienone undergoes a Diels–Alder reaction with an acetylene diester in a high-boiling solvent. The intermediate adduct spontaneously loses carbon monoxide to give the high-melting product that is purified by crystallization.

Microscale Procedure

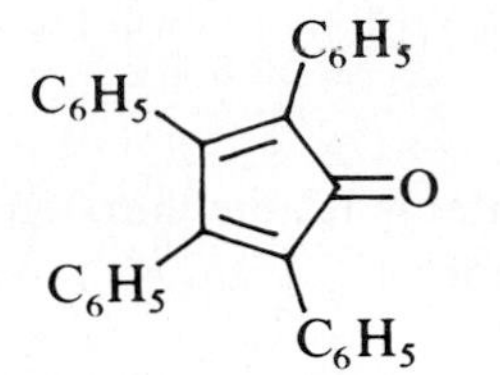

Tetraphenylcyclopentadienone
MW 384.45, mp 219°C

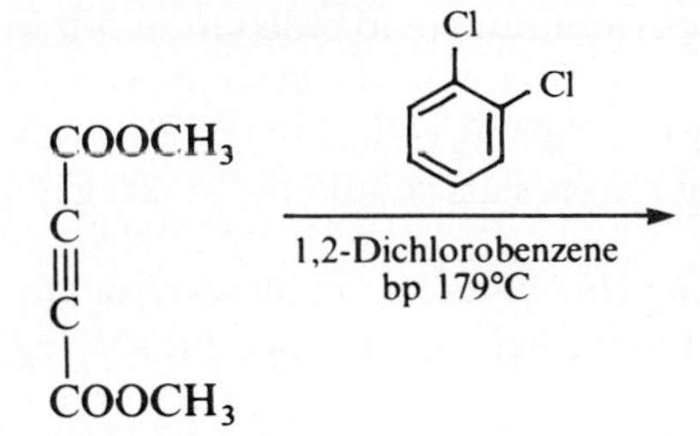

Dimethyl acetylenedicarboxylate
MW 142.11, bp ~300°C

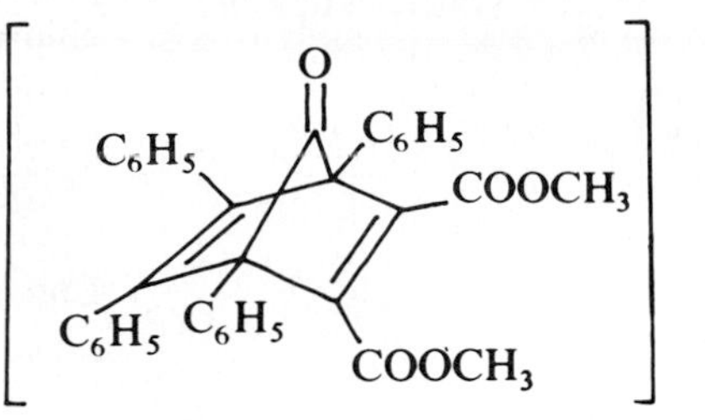

−CO

$COOCH_3$

$COOCH_3$

Dimethyl tetraphenylphthalate
MW 498.58, mp 258°C

CAUTION: Work in a hood. Dimethyl acetylenedicarboxylate is corrosive and a lachrymator (tear producer). 1,2-Dichlorobenzene is a toxic irritant.

Photos: *Filtration Using a Pasteur Pipette' Use of the Wilfilter,* Videos: *Filtration of Crystals Using the Pasteur Pipette, Microscale Filtration on the Hirsch Funnel*

Measure into a reaction tube 50 mg of tetraphenylcyclopentadienone (prepared in Chapter 51), 0.4 mL of 1,2-dichlorobenzene, and 27.5 mg of dimethyl acetylenedicarboxylate. Clamp the tube over a hot sand bath, insert a thermometer, and raise the temperature to the boiling point (180–185°C). Boil gently until there is no further color change, and let the rim of condensate rise just high enough to wash the walls of the tube. The pure adduct is colorless and, if the starting ketone is adequately pure, the color changes from purple to pale tan. A 5-minute boiling period should be sufficient. Cool the tube to 100°C, slowly stir in 0.6 mL of 95% ethanol, and let crystallization proceed. After the mixture has cooled to near room temperature, cool it in ice. Then collect the product, either by removing the solvent with a Pasteur pipette or, if the crystals are too small for this technique, on a Hirsch funnel or a Wilfilter. Wash the crystals with cold methanol. The yield of dimethyl tetraphenylphthalate will be about 50 mg.

Cleaning Up. Because the filtrate from this reaction contains 1,2-dichlorobenzene, place it in the halogenated organic solvents waste container.

Macroscale Procedure

CAUTION: Work in a hood. Dimethyl acetylenedicarboxylate is corrosive and a lachrymator (tear producer). 1,2-Dichlorobenzene is a toxic irritant.

Measure into a 25 × 150-mm test tube 2 g of tetraphenylcyclopentadienone, 10 mL of 1,2-dichlorobenzene, and 1 mL (1.1 g) of dimethyl acetylenedicarboxylate. Clamp the test tube in a hot sand bath, insert a thermometer, and raise the temperature to the boiling point (180–185°C). Boil gently until there is no further color change and

Reaction time: about 5 minutes

Video: *Macroscale Crystallization*

let the rim of condensate rise just high enough to wash the walls of the tube. The pure adduct is colorless and, if the starting ketone is adequately pure, the color changes from purple to pale tan. A 5-minute boiling period should be sufficient. Cool to 100°C, slowly stir in 15 mL of 95% ethanol, and let crystallization proceed. Cool under the tap, collect the product, and rinse the tube with methanol. The yield of colorless crystals should be 2.1–2.2 g.

Cleaning Up. Because the filtrate from this reaction contains 1,2-dichlorobenzene, place it in the halogenated organic solvents waste container.

2. HEXAPHENYLBENZENE

IN THIS EXPERIMENT, tetraphenylcyclopentadienone is heated over a flame with excess diphenylacetylene. Some of the diphenylacetylene is removed with a pipette to allow the very hot reaction mixture to become even hotter to complete the reaction. The product, which melts above the melting point of lead, is recrystallized from diphenyl ether, a high-boiling point solvent, and is collected by filtration.

Diphenylacetylene is a less reactive dienophile than dimethylacetylenedicarboxylate; but when heated with tetraphenylcyclopentadienone without solvent to a high temperature (ca. 380–400°C), the reaction will proceed. In the following procedure, a large excess of dienophile will serve as the solvent. Because refluxing diphenylacetylene (bp about 300°C) keeps the temperature below the melting point of the product, removal of the diphenylacetylene causes the reaction mixture to melt, which ensures completion of the reaction.

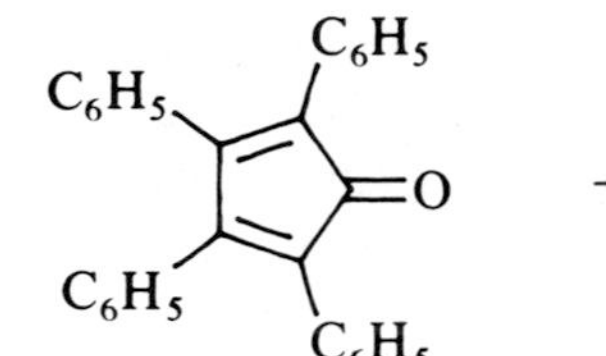

Tetraphenylcyclopentadienone
MW 384.45, mp 219°C

+

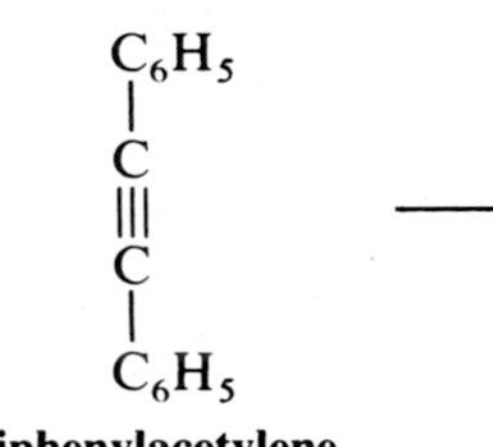

Diphenylacetylene
MW 178.22, mp 61°C

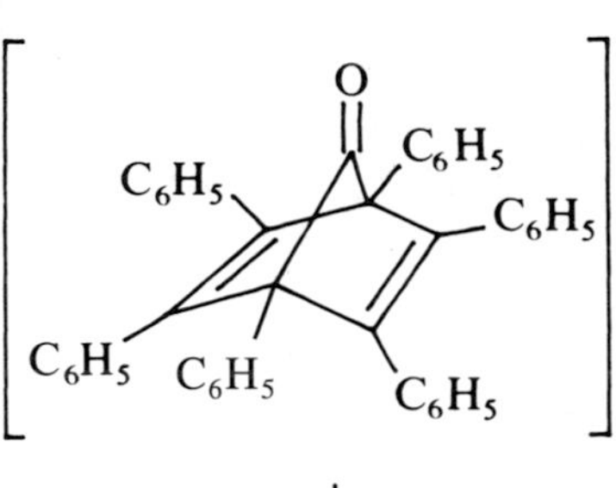

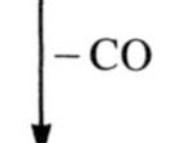

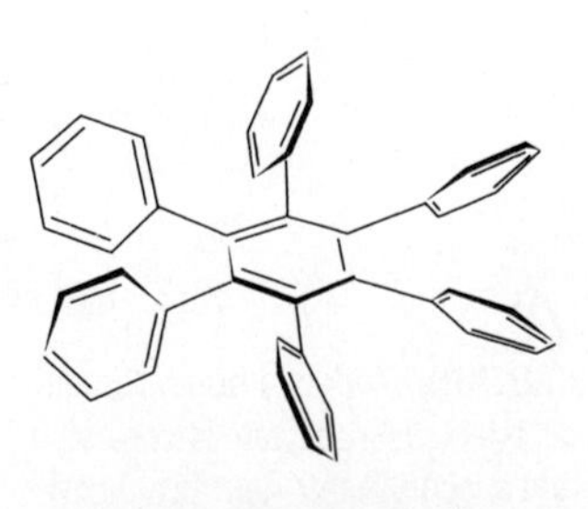

Hexaphenylbenzene
MW 534.66
mp 465°C
Energy-minimized conformation

Microscale Procedure

Tetraphenylcyclopentadienone is a hydrocarbon with a very high melting point. Although the small amount of carbon monoxide produced probably presents no hazard, work in a hood.

Videos: *Instant Microscale Distillation, Microscale Filtration on the Hirsch Funnel, Filtration of Crystals Using the Pasteur Pipette*; Photo: *Filtration Using a Pasteur Pipette*

This is the only experiment in this text that requires a flame to conduct a microscale reaction. Use care to ensure that no flammable solvents are nearby.

Place 50 mg each of tetraphenylcyclopentadienone (prepared in Chapter 51) and diphenylacetylene (which can be prepared in Chapter 58) in a reaction tube, clamp it upright, and heat the mixture with the flame of a microburner. This is the only experiment in this text that requires a flame to conduct a microscale reaction. Use care to ensure that no flammable solvents are nearby. Soon after the reactants have melted with strong bubbling, the white product becomes visible. Let the diphenylacetylene reflux briefly on the walls of the tube; then remove some of this excess diphenylacetylene by inserting a Pasteur pipette into the vapors above the solid and drawing the hot vapors into the pipette. Repeat this operation until it is possible, through strong heating with the flame, to melt the mixture completely. Then let the melt cool and solidify. Add 0.25 mL of diphenyl ether (bp 259°C), heat the mixture *carefully* to dissolve the solid, then let the product crystallize slowly. Extinguish the flame and, when the tube is cold, add 0.5 mL of toluene to dilute the mixture. Collect the product using the Pasteur pipette method or by vacuum filtration on a Hirsch funnel and wash it with toluene.

The yield of colorless plates is about 60 mg. Using a Mel-Temp apparatus, *equipped with a 500°C thermometer*, determine the melting point. To avoid oxidation, seal the sample in an evacuated capillary tube. The product should melt at 465°C.

Cleaning Up. Place all filtrates in the organic solvents waste container.

Macroscale Procedure

Tetraphenylcyclopentadienone is a hydrocarbon with a very high melting point. Although the small amount of carbon monoxide produced probably presents no hazard, work in a hood.

Place 0.5 g each of tetraphenylcyclopentadienone (prepared in Chapter 51) and diphenylacetylene (which can be prepared in Chapter 58) in a 25 × 150-mm test tube supported by a clamp, and heat the mixture strongly with the free flame of a microburner held in the hand (do not insert a thermometer into the test tube; the temperature will be too high). Soon after the reactants have melted with strong bubbling, white masses of the product become visible. Let the diphenylacetylene reflux briefly on the walls of the tube; then remove some of the diphenylacetylene by letting it condense for a minute or two on a cold finger condenser filled with water but without fresh water running through it (Fig. 52.1), minus the paper thimble. Remove the flame, withdraw the cold finger, and wipe it with a towel. Repeat the operation until you are able, by strong heating, to melt the mixture completely. Then let the melt cool and solidify. Add 10 mL of diphenyl ether (bp 259°C), using it to rinse the walls. Heat *carefully* over a free flame to dissolve the solid; then let the product crystallize. When cold, add 10 mL of toluene to thin the mixture, collect the product, and wash with fresh toluene.

In case the hexaphenylbenzene is contaminated with insoluble material, crystallization from a filtered solution can be accomplished as follows: Place 10 mL of diphenyl ether in a 25 × 150-mm test tube, pack the sample of hexaphenylbenzene into a 10-mm extraction thimble, and suspend this in a test tube with two nichrome wires, as shown in Figure 52.1. Insert a cold finger condenser supported by an inverted filter adapter and adjust the condenser and the wires so that condensing liquid will drop into the thimble. Let the diphenyl ether reflux until the hexaphenylbenzene in the thimble is dissolved; then let the product crystallize, add toluene, collect the product, and wash with toluene (as described previously). A larger version of this apparatus is shown in Figure 52.2.

The yield of colorless plates is 0.6–0.7 g. The melting point of the product can be determined with a Mel-Temp apparatus and a 500°C thermometer. To avoid oxidation, seal the sample in an evacuated capillary tube. The product should melt at 465°C.

FIG. 52.1
A Soxhlet-type extractor.

Cold finger condenser
Filter adapter
Wire
Paper thimble

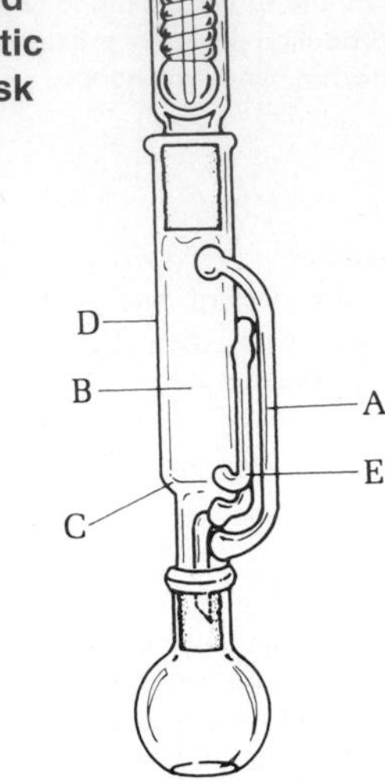

FIG. 52.2
A research-quality Soxhlet extractor. Solvent vapors from the flask rise through A and up into the condenser. As they condense, the liquid just formed returns to B, where the sample to be extracted is placed. (The bottom of B is sealed at C.) Liquid rises in B to level D, at which time the automatic siphon, E, starts. Extracted material accumulates in the flask as more pure liquid vaporizes to repeat the process.

Cleaning Up. Place all filtrates in the organic solvents waste container.

COMPUTATIONAL CHEMISTRY

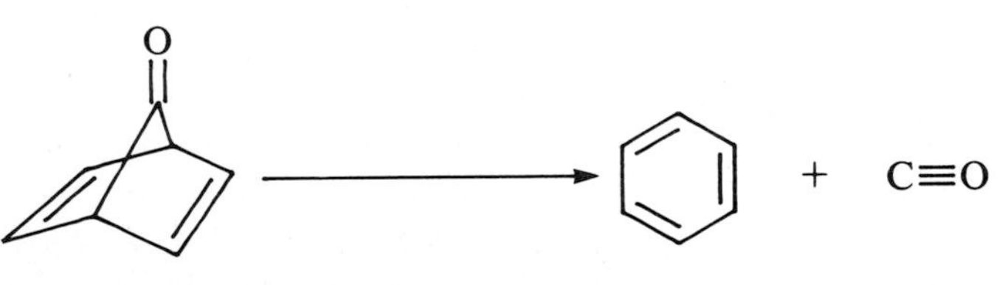

Bicyclo[2.2.1]hepta-2,5-dien-7-one **Benzene** **Carbon monoxide**

Calculate, using the AM1 semiempirical method, the heats of formation of bicyclo[2.2.1]hepta-2,5-dien-7-one and of benzene and carbon monoxide in order to confirm theoretically that a reaction might occur.

QUESTIONS

1. Which two factors probably contribute to the very high melting points of these two hexasubstituted benzenes?
2. What volume of carbon monoxide, measured at standard temperature and pressure (STP), is produced by the decomposition of 38 mg of tetraphenylcyclopentadienone?
3. Is the energy-minimized conformation of hexaphenylbenzene a chiral molecule? Explain why or why not.

Derivatives of 1,2-Diphenylethane: A Multistep Synthesis[1]

The next eight chapters provide various procedures for the rapid preparation of small samples of twelve related compounds, using benzaldehyde and phenylacetic acid as the initial reactants (see synthetic scheme on the next page). The quantities of reagents specified in the procedures will often provide somewhat more of each intermediate than is required for completing subsequent steps in the sequence of reactions. If the experiments are dovetailed, the entire series of preparations can be completed in a very short working time. For example, one can start the preparation of benzoin (record the time of starting; do not rely on memory) and during the reaction period, start the preparation of phenylcinnamic acid. The preparation of phenylcinnamic acid requires refluxing for 35 minutes and, while it is proceeding, the benzoin preparation can be stopped when the time is up and the product allowed to crystallize. The phenylcinnamic acid mixture can be allowed to stand (and cooled) until one is ready to isolate it. Also, you may want to observe the crystals occasionally during a crystallization process; however, you should utilize most of your time for other operations.

Specific points related to stereochemistry and reaction mechanisms are discussed at the beginning of each individual chapter. Because several of the compounds have characteristic ultraviolet or infrared spectra, pertinent spectroscopic constants are recorded, and brief interpretations of the data are presented. Molecular mechanics calculations give insight into the conformations of several of these compounds.

QUESTIONS

1. Starting with 150 mg of benzaldehyde and assuming an 80% yield for each step, what yield of diphenylacetylene, in milligrams, might you expect in the synthetic sequence on page 654?
2. From the information given and assuming a yield of 80% for the last two reactions, what yield of stilbene dibromide would you expect when starting with 150 mg of benzaldehyde?

[1]If your work is well organized and no setbacks occur, the experiments can be completed in about four laboratory periods. Your instructor may elect to name a certain number of periods in which you are to make as many of the compounds as possible; your instructor may also decide to require only the submission of the end products in each series.

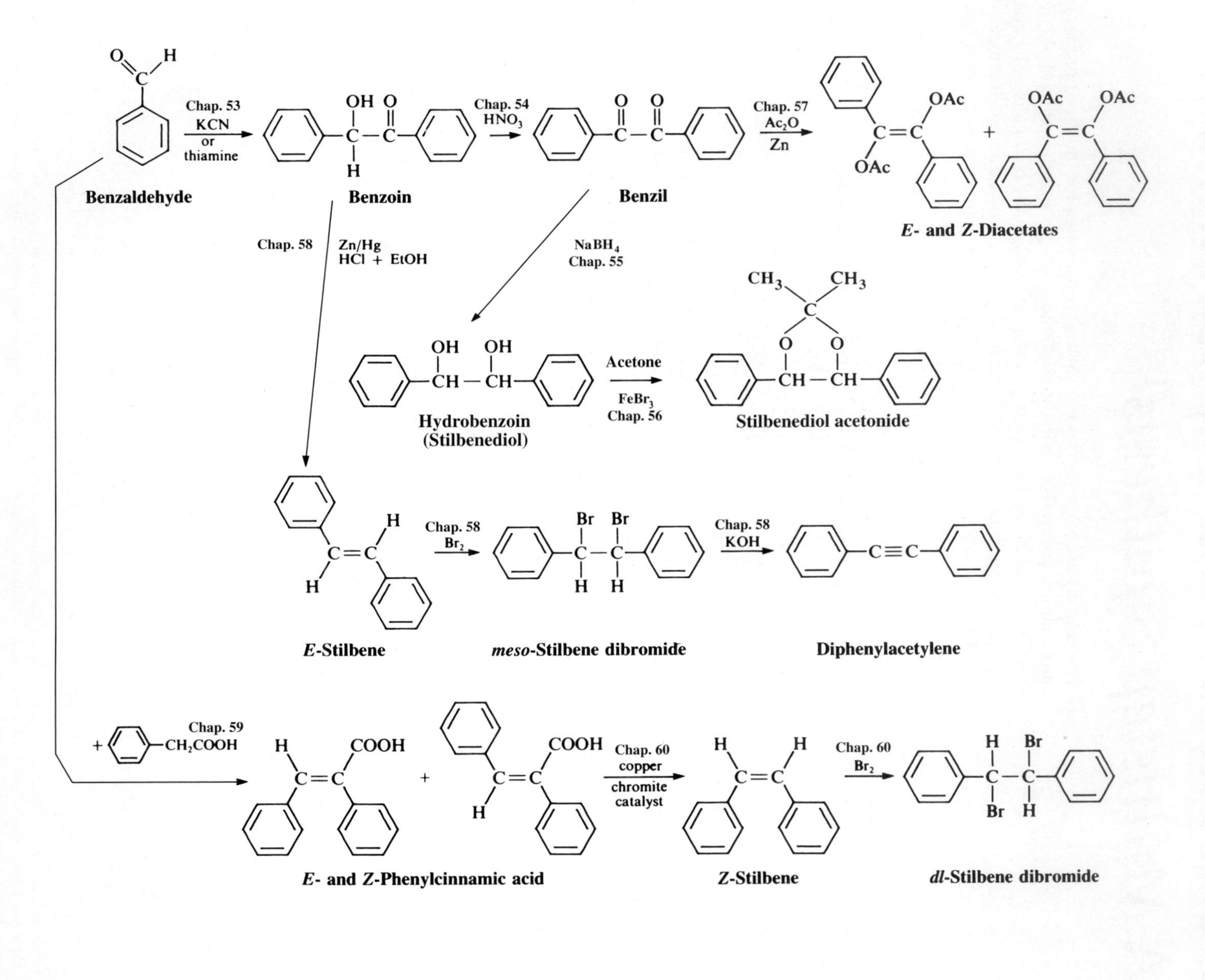

Benzaldehyde
Chap. 53
KCN
or
thiamine
Benzoin
Chap. 54
HNO3
Benzil
Chap. 57
Ac2O
Zn
E- and Z-Diacetates
Chap. 58
Zn/Hg
HCl + EtOH
NaBH4
Chap. 55
Hydrobenzoin
(Stilbenediol)
Acetone
FeBr3
Chap. 56
Stilbenediol acetonide
E-Stilbene
Chap. 58
Br2
meso-Stilbene dibromide
Chap. 58
KOH
Diphenylacetylene
Chap. 59
CH2COOH
E- and Z-Phenylcinnamic acid
Chap. 60
copper
chromite
catalyst
Z-Stilbene
Chap. 60
Br2
dl-Stilbene dibromide

CHAPTER 53

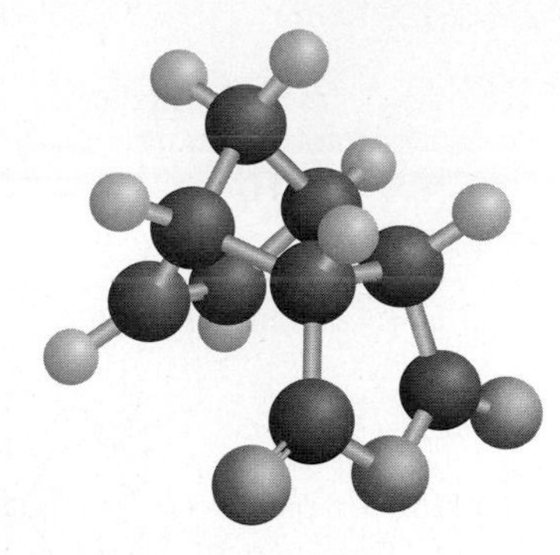

The Benzoin Condensation: Catalysis by the Cyanide Ion and Thiamine

When you see this icon, sign in at this book's premium website at **www.cengage.com/login** to access videos, Pre-Lab Exercises and other online resources.

PRELAB EXERCISE: What purposes does sodium hydroxide serve in the thiamine-catalyzed benzoin condensation?

The reaction of 2 mol of benzaldehyde to form a new carbon-carbon bond is known as the *benzoin condensation*. It is catalyzed by two rather different catalysts—the cyanide ion and thiamine (a B-complex vitamin)—which, on close examination, seem to function in exactly the same way.

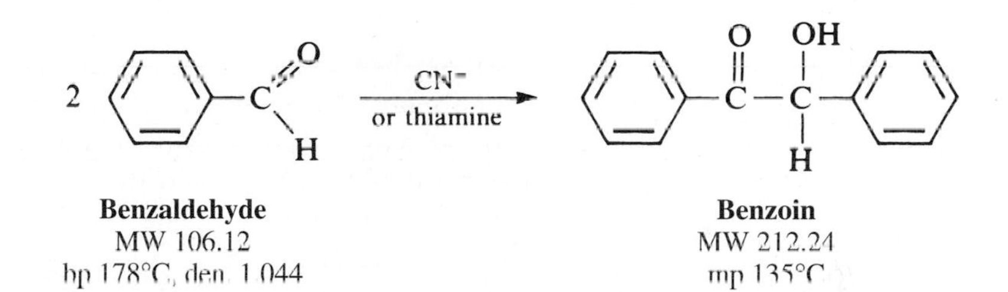

Consider the reaction catalyzed by the cyanide ion. The cyanide ion attacks the carbonyl oxygen to form a stable cyanohydrin, mandelonitrile, a liquid with a boiling point of 170°C that under the basic conditions of the reaction, loses a proton to give a resonance-stabilized carbanion (A). The carbanion attacks another molecule of benzaldehyde to give compound **B**, which undergoes a proton transfer and loses cyanide to give benzoin. Evidence for this mechanism lies in the failure of 4-nitrobenzaldehyde to undergo the reaction because the nitro group reduces the nucleophilicity of the anion in compound **A**. On the other hand, a strong electron-donating group in the 4-position of the phenyl ring makes the loss of the proton from the cyanohydrin very difficult; thus 4-dimethylaminobenzaldehyde also does not undergo the benzoin condensation with itself.

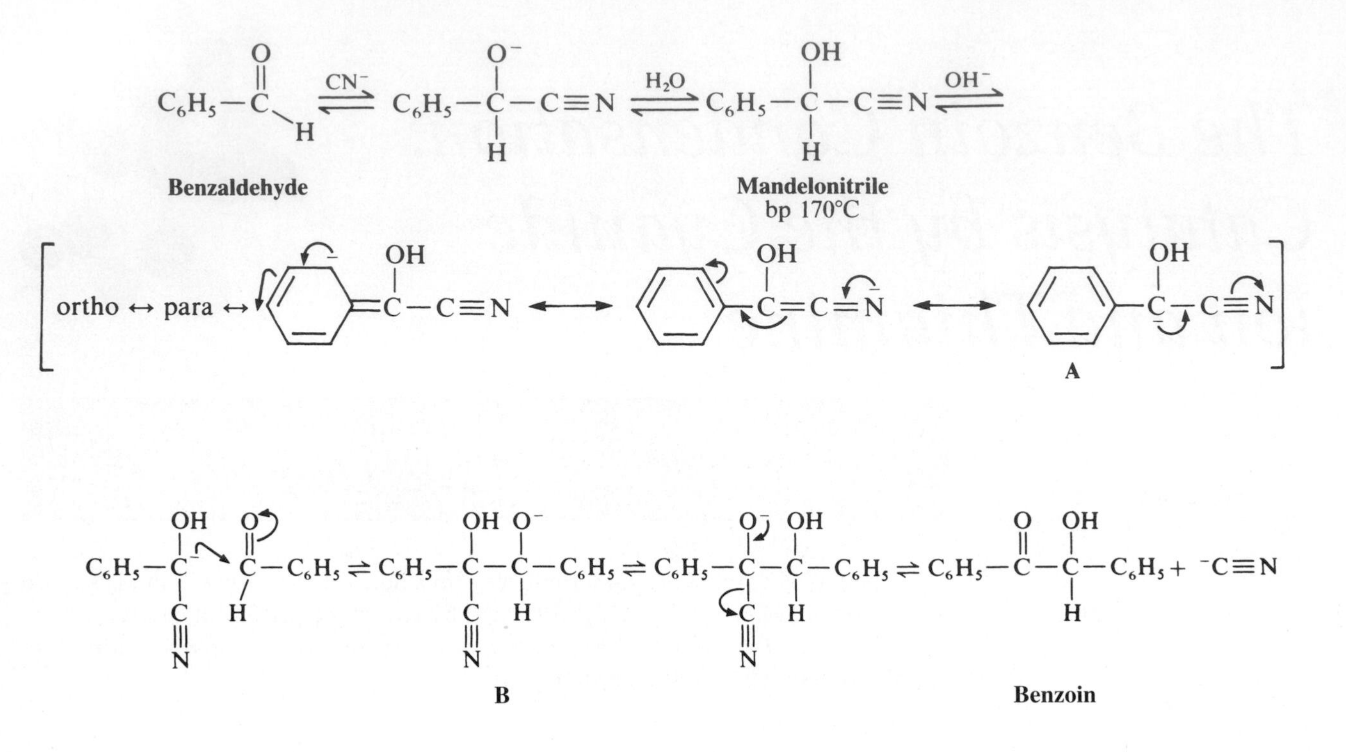

Several biochemical reactions bear a close resemblance to the benzoin condensation, but are not obviously catalyzed by the highly toxic cyanide ion. Approximately 30 years ago, Ronald Breslow proposed that vitamin B_1 (also known as thiamine hydrochloride), in the form of the coenzyme thiamine pyrophosphate, can function in a manner completely analogous to the cyanide ion in promoting reactions like the benzoin condensation. The resonance-stabilized conjugate base of the thiazolium ion, thiamine, and the resonance-stabilized carbanion (C), which it forms, are the keys to the reaction. Like the cyanide ion, the thiazolium ion has just the right balance of nucleophilicity, the ability to stabilize the intermediate anion, and good leaving group qualities.

The cyanide ion binds irreversibly to hemoglobin, rendering it useless as a carrier of oxygen.

In the reactions that follow, the cyanide ion functions as a fast and efficient catalyst, although in large quantities it is highly toxic. The amount of potassium cyanide used in the first experiment (15 mg) is about eight times lower than the average fatal dose, a difference that underlines the advantage of carrying out organic experiments on a microscale.

The importance of thiamine is evident in that it is a vitamin, an essential substance that must be provided in the diet to prevent beriberi, a nervous system disease. The thiamine-catalyzed reaction is much slower, but the catalyst is completely nontoxic.

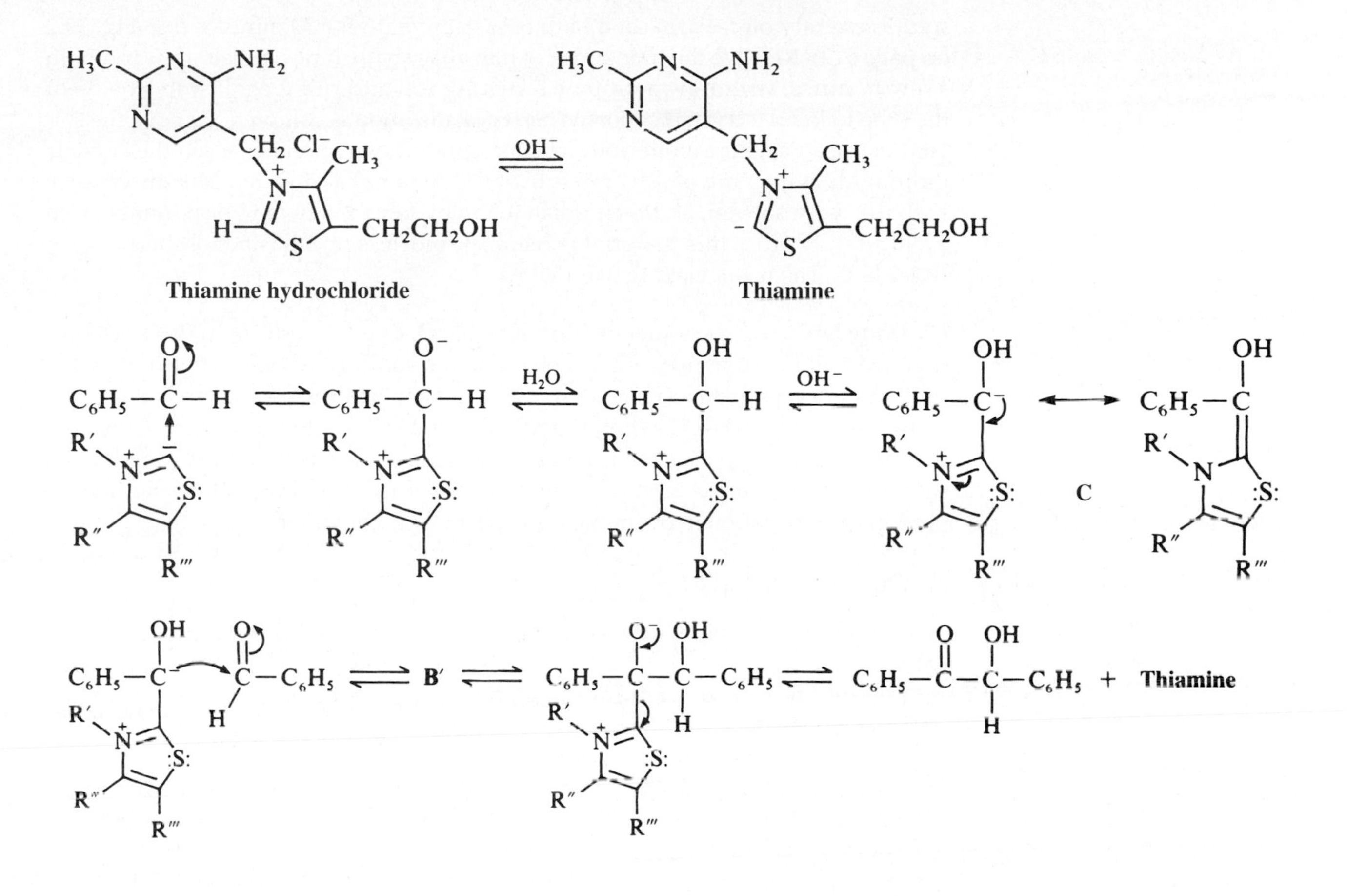

EXPERIMENTS

1. CYANIDE-CATALYZED BENZOIN CONDENSATION

Reaction time: 30 minutes

IN THIS EXPERIMENT, a tiny amount of potassium cyanide catalyzes the condensation of two molecules of benzaldehyde to a molecule of benzoin in a 30-minute reaction. The product crystallizes from the reaction mixture and is isolated by filtration.

CAUTION: Potassium cyanide is poisonous. Do not handle if you have open cuts on your hands. Never acidify a cyanide solution because HCN gas is evolved. Wash your hands after handling cyanide.

Microscale Procedure

In a reaction tube, place 15 mg of potassium cyanide (poison!); dissolve it in 0.15 mL of water; add 0.30 mL of 95% ethanol and from a small, accurate syringe add 0.15 mL (157 mg) of pure benzaldehyde.[1] Introduce a boiling chip and reflux the

[1]Commercial benzaldehyde is inhibited against autoxidation with 0.1% hydroquinone. If the material available is yellow or contains benzoic acid crystals, it should be purified. *See also* footnote 2.

solution gently on a warm sand bath or a steam bath for 30 minutes (see Fig. 39.2 on page 512). Remove the tube and cool it in an ice bath; if no crystals appear within a few minutes, withdraw a drop on a stirring rod and rub it against the inside of the tube to induce crystallization. When crystallization is complete, remove the solvent using a Pasteur pipette and, while keeping the tube on ice, wash the crystals thoroughly with 1 mL of a 1:1 mixture of 95% ethanol and water. Mix the crystals with the wash solvent, and then isolate them by filtration on a Hirsch funnel or on a Wilfilter. If pure, this material is usually colorless and has a melting point of 134–135°C. The usual yield is 100–120 mg.

Photos: *Filtration Using a Pasteur Pipette, Use of the Wilfilter*; Videos: *Filtration of Crystals Using the Pasteur Pipette, Microscale Filtration on the Hirsch Funnel*

Cleaning Up. Add the aqueous filtrate to 10 mL of a 1% sodium hydroxide solution. Add 10 mL of household bleach (aqueous sodium hypochlorite solution) to oxidize the cyanide ion. The resulting solution can be tested for cyanide using the Prussian blue test described on the website under Tests for Nitrogen. (Continue this process until no cyanide is present.) When cyanide is no longer present, the solution can be diluted with water and flushed down the drain. Place the ethanol used in crystallization in the organic solvents waste container.

Macroscale Procedure

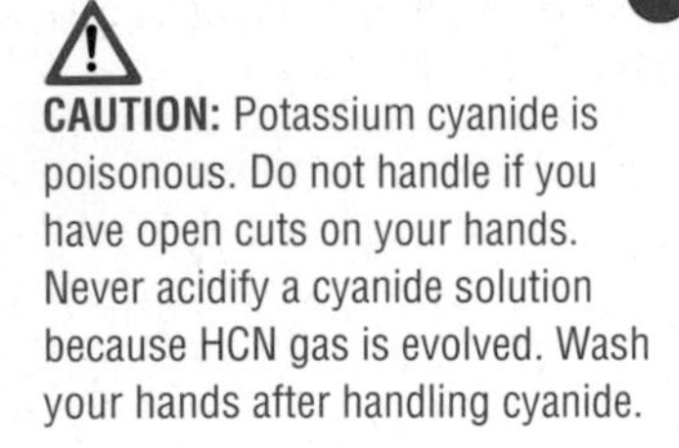

CAUTION: Potassium cyanide is poisonous. Do not handle if you have open cuts on your hands. Never acidify a cyanide solution because HCN gas is evolved. Wash your hands after handling cyanide.

Place 0.75 g of potassium cyanide (poison!) in a 50-mL round-bottomed flask, dissolve it in 7.5 mL of water, add 15 mL of 95% ethanol and 7.5 mL of pure benzaldehyde,[2] introduce a boiling stone, attach a short condenser, and reflux the solution gently on a flask heater for 30 minutes (Fig. 53.1). Remove the flask and cool it in an ice bath; if no crystals appear within a few minutes, withdraw a drop on a stirring rod and rub it against the neck of the flask to induce crystallization. When crystallization is complete, collect the product and wash it free of the yellow mother liquor with a 1:1 mixture of 95% ethanol and water. Usually this first-crop material is colorless and of satisfactory melting point (134–135°C); the usual yield 5–6 g.[3]

Video: *Macroscale Crystallization*

Cleaning Up. Add the aqueous filtrate to 5 mL of a 1% sodium hydroxide solution. Add 25 mL of household bleach (aqueous sodium hypochlorite solution) to oxidize the cyanide ion. The resulting solution can be tested for cyanide using the Prussian blue test described on the website under Tests for Nitrogen. Continue this process until no cyanide is present. When cyanide is no longer present, the solution can be diluted with water and flushed down the drain. Place the ethanol used in crystallization in the organic solvents waste container.

[2]Commercial benzaldehyde inhibited against autoxidation with 0.1% hydroquinone is usually satisfactory. If the material available is yellow or contains benzoic acid crystals, it should be shaken with equal volumes of a 5% sodium carbonate solution until carbon dioxide is no longer evolved; then dry the upper layer over calcium chloride and distill (bp 178–180°C). Avoid exposing the hot liquid to air. The distillation step can be omitted if the benzaldehyde is colorless.

[3]Concentration of the mother liquor to a volume of 10 mL gives a second crop (1.4 g, mp 133–134.5°C); the best total yield 6.8 g (87%). Recrystallization can be accomplished with either methanol (11 mL/g) or 95% ethanol (7 mL/g) with 90% recovery in the first crop.

2. THIAMINE-CATALYZED BENZOIN CONDENSATION

Microscale Procedure

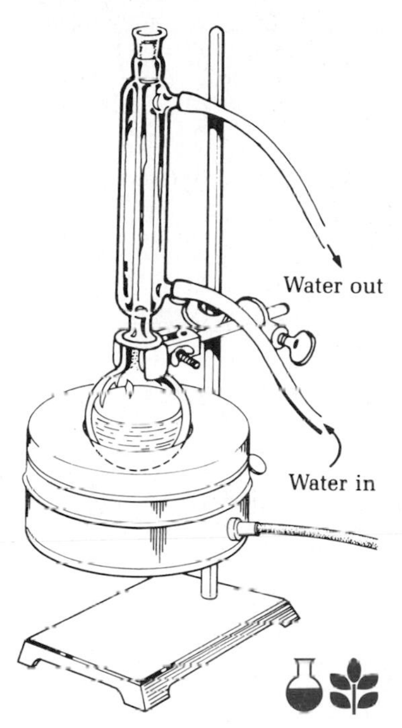

FIG. 53.1
A macroscale reflux apparatus.

In a reaction tube, place 26 mg of thiamine hydrochloride, dissolve it in a drop of water, add 0.30 mL of 95% ethanol, and cool the solution in an ice bath. Add 0.05 mL of 3 *M* sodium hydroxide followed by 0.15 mL (157 mg) of pure benzaldehyde,[4] stir well, add the distilling column as an air condenser, and heat the mixture in a water bath at 60°C for 1–1.5 hours. The progress of the reaction should be followed by thin-layer chromatography (TLC). Use benzaldehyde in one lane, benzoin in another, and the reaction mixture in the center. Elute with a 50:50 mixture of petroleum ether and ethyl acetate. Alternatively, the reaction mixture can be stored at room temperature for at least 24 hours, although a week would do no harm. (The rate of most organic reactions doubles for each 10°C rise in temperature.)

Cool the reaction mixture in an ice bath. If crystallization does not occur, withdraw a drop of solution on a stirring rod and rub it against the inside surface of the tube to induce crystallization. Remove the solvent using a Pasteur pipette and, while keeping the mixture on ice, wash the crystals with a 1:1 ice-cold mixture of 95% ethanol and water. The product should be colorless and of sufficient purity (mp 134–135°C) for use in subsequent reactions; the usual yield is 100–120 mg. If desired, the moist product can be recrystallized from 95% ethanol (7 mL/g) or methanol (11 mL/g) with 90% recovery.

Cleaning Up. The aqueous filtrate, after neutralization with dilute hydrochloric acid, is diluted with water and flushed down the drain. Ethanol used in crystallization should be placed in the organic solvents waste container.

Macroscale Procedure

Place 1.3 g of thiamine hydrochloride in a 50-mL Erlenmeyer flask, dissolve it in 4 mL of water, add 15 mL of 95% ethanol, and cool the solution in an ice bath. Add 2.5 mL of 3 *M* sodium hydroxide dropwise with swirling such that the temperature of the solution does not rise above 20°C. To the yellow solution, add 7.5 mL of pure benzaldehyde[5] and heat the mixture at 60°C for 1–1.5 hours. The progress of the reaction can be followed by thin-layer chromatography (TLC). Alternatively, the reaction mixture may be stored at room temperature for at least 24 hours. (The rate of most organic reactions doubles for each 10°C rise in temperature.)

Cool the reaction mixture in an ice bath. If crystallization does not occur, withdraw a drop of solution on a stirring rod and rub it against the inside surface of the flask to induce crystallization. Collect the product by suction filtration and wash it free of the yellow mother liquor with a 1:1 mixture of 95% ethanol and water. The product should be colorless and of sufficient purity (mp 134–135°C) for use in subsequent reactions; the usual yield is 5–6 g. If desired, the moist product can be recrystallized from 95% ethanol (8 mL/g).

Video: *Macroscale Crystallization*

[4] *See* footnote 1.
[5] *See* footnote 1.

Cleaning Up. The aqueous filtrate, after neutralization with dilute hydrochloric acid, is diluted with water and flushed down the drain. Ethanol used in crystallization should be placed in the organic solvents waste container.

QUESTIONS

1. Speculate on the structure of the compound formed when 4-dimethylaminobenzaldehyde is condensed with 4-chlorobenzaldehyde.
2. Why might the presence of benzoic acid be deleterious to the benzoin condensation?
3. How many *p* electrons are in the thiazoline ring of thiamine hydrochloride? of thiamine?
4. Locate the CH and OH protons in the proton nuclear magnetic resonance (^1H NMR) spectrum of benzoin (Fig. 53.2).

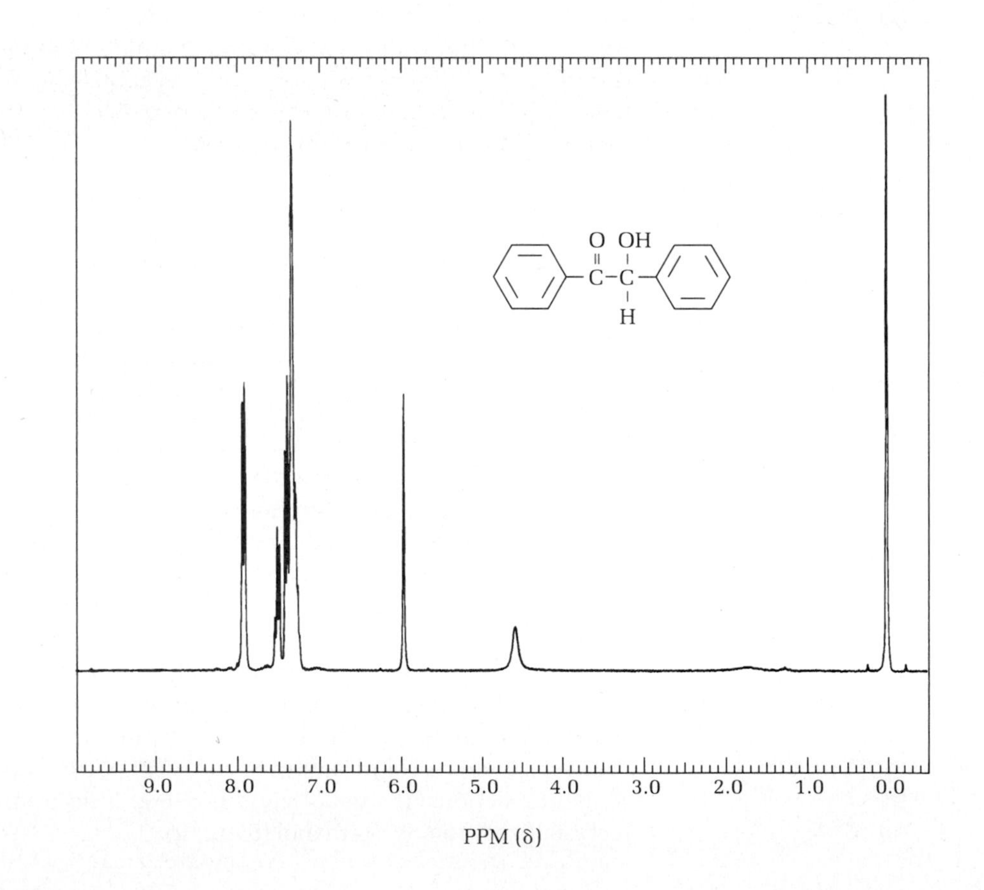

FIG. 53.2
The ^1H NMR spectrum of benzoin.

CHAPTER 54

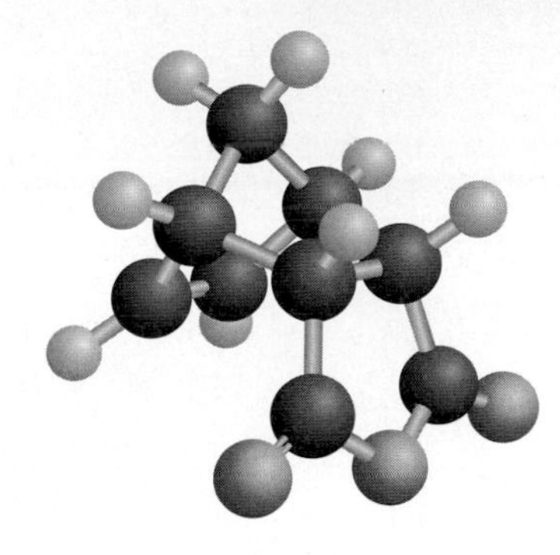

Nitric Acid Oxidation; Preparation of Benzil from Benzoin; and Synthesis of a Heterocycle: Diphenylquinoxaline

PRELAB EXERCISE: Write a detailed mechanism for the formation of 2,3-dimethylquinoxaline.

Benzoin can be oxidized to benzil, an α-diketone, very efficiently by nitric acid or by copper(II) sulfate in pyridine. When oxidized with sodium dichromate in acetic acid, the yield is lower because some of the material is converted into benzaldehyde by cleavage of the bond between two oxidized carbon atoms that is activated by both phenyl groups (a). Similarly, when hydrobenzoin is oxidized with dichromate or permanganate, it yields chiefly benzaldehyde and only a trace of benzil (b).

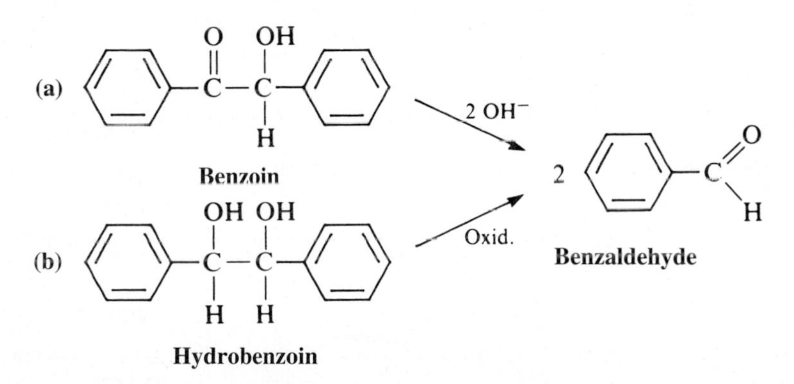

Ultraviolet (UV) spectroscopy is used to help characterize aromatic molecules such as benzoin. The absorption band at 247 nm in Figure 54.1 is attributable to the presence of the phenyl ketone group,

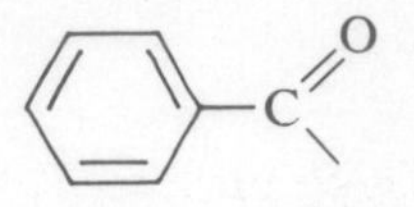

in which the carbonyl group is conjugated with a benzene ring. Aliphatic αβ-unsaturated ketones, R—CH=CH—C=O, show selective absorption of UV light of comparable wavelength. See the infrared (IR) spectrum of benzoin in Figure 54.2.

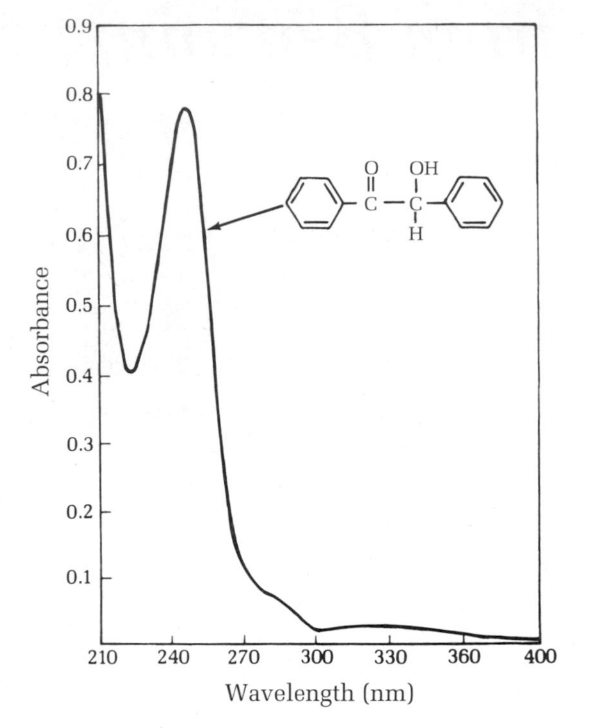

FIG. 54.1
The UV spectrum of benzoin. λ_{max}^{EtOH} = 247 nm (ε = 13,200). Concentration: 12.56 mg/L = 5.92 × 10^{-5} mol/L. *See* Chapter 14 for the relationship between the extinction coefficient (ε), absorbance (*A*), and concentration (*C*).

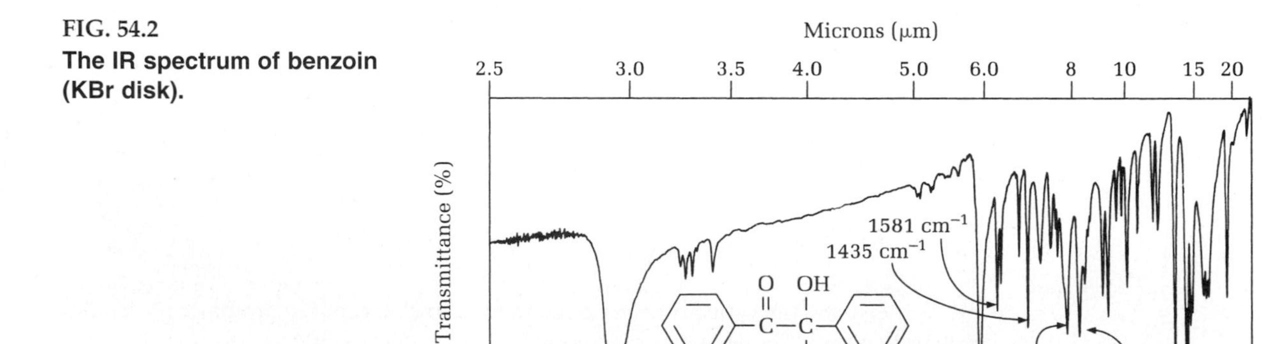

FIG. 54.2
The IR spectrum of benzoin (KBr disk).

EXPERIMENTS

1. NITRIC ACID OXIDATION OF BENZOIN

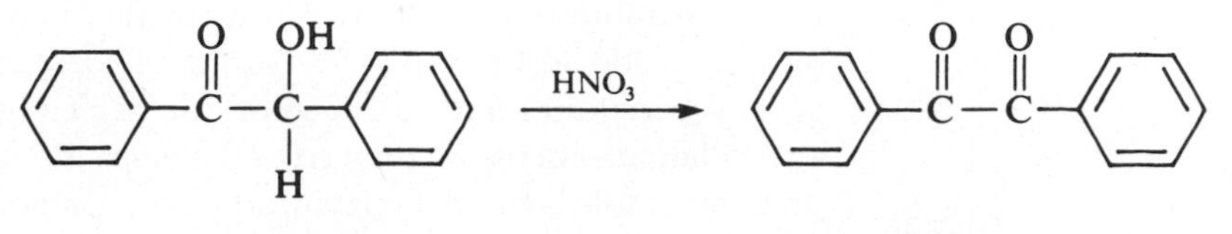

Benzoin
MW 212.24, mp 135°C

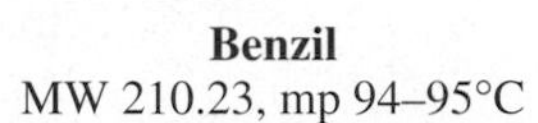

Benzil
MW 210.23, mp 94–95°C

CAUTION: Handle nitric acid with great care. It is highly corrosive to tissue and a very strong oxidant.

IN THIS EXPERIMENT, the hydroxyl group of benzoin is oxidized with hot nitric acid to a carbonyl group to give benzil, which is yellow in color. The product is collected on a Hirsch funnel, washed with water, and recrystallized from ethanol.

Microscale Procedure

Reaction time: 11 minutes

Photos: *Filtration Using a Pasteur Pipette, Use of the Wilfilter,* Videos: *Filtration of Crystals, Using the Pasteur Pipette, Microscale Filtration on the Hirsch Funnel*

Heat a mixture of 100 mg of benzoin and 0.35 mL of concentrated nitric acid on a steam bath or in a small beaker of boiling water for about 11 minutes. Carry out the reaction in the hood or use an aspirator tube near the top of the tube to remove nitrogen oxides. Be sure all of the benzoin is washed down inside the tube and is oxidized. Add 2 mL of water to the reaction mixture, cool to room temperature, and stir the mixture for 1–2 minutes to coagulate the precipitated product. Remove the solvent with a Pasteur pipette and wash the solid with an additional 2 mL of water. Dissolve the solid in 0.5 mL of hot ethanol and add water dropwise to the hot solution until the solution appears to be cloudy, indicating that it is saturated. Heat to bring the product completely into solution and allow it to cool slowly to room temperature. Cool the tube in ice and isolate the product using a Wilfilter or a Hirsch funnel. Scrape the benzil onto a piece of filter paper, squeeze out excess solvent, and allow the solid to dry. Record the percent yield, the crystalline form, the color, and the melting point of the product.

Cleaning Up. The aqueous filtrate should be neutralized with sodium carbonate, diluted with water, and flushed down the drain. Place the ethanol used in crystallization in the organic solvents waste container.

Macroscale Procedure

Reaction time: 11 minutes

Video: *Macroscale Crystallization*

Heat a mixture of 4 g of benzoin and 14 mL of concentrated nitric acid on a steam bath for about 11 minutes. Carry out the reaction in a hood or use an aspirator tube near the top of the flask to remove nitrogen oxides. Add 75 mL of water to the reaction mixture, cool to room temperature, and swirl for 1–2 minutes to coagulate the precipitated product. Collect and wash the yellow solid on a Hirsch funnel, pressing the solid on the filter to squeeze out the water. This crude product (the dry weight yield is 3.7–3.9 g) need not be dried, but can be crystallized at once from ethanol. Dissolve the product in 10 mL of hot ethanol, add water dropwise to the cloud point, and set aside to crystallize. Record the yield, the crystalline form, the color, and the melting point of the purified benzil. The IR and ^{1}H NMR spectra are found at the end of the chapter (Fig. 54.5 and Fig. 54.6).

A Test for the Presence of Unoxidized Benzoin

Dissolve about 0.5 mg of crude or purified benzil in 0.5 mL of 95% ethanol or methanol, and add 1 drop of 3 *M* sodium hydroxide. If benzoin is present, the solution soon acquires a purplish color due to a complex of benzil with a product of autoxidation of benzoin. If no color develops in 2–3 minutes, which is an indication that the sample is free of benzoin, add a small amount of benzoin. Observe the color that develops, and note that if the test tube is stoppered and shaken vigorously, the color momentarily disappears. When the solution is allowed to stand, the color reappears.

Cleaning Up. The aqueous filtrate should be neutralized with sodium carbonate, diluted with water, and flushed down the drain. Ethanol used in crystallization should be placed in the organic solvents waste container.

2. PREPARATION OF BENZIL QUINOXALINE (DIPHENYLQUINOXALINE)

A reaction that characterizes benzil as an α-diketone is a condensation reaction of 1,2-phenylenediamine to the quinoxaline derivative. The aromatic heterocyclic ring formed in the condensation is fused to a benzene ring to give a bicyclic system analogous to naphthalene. The word *heterocyclic* refers to the fact that the ring contains an atom or atoms other than carbon.

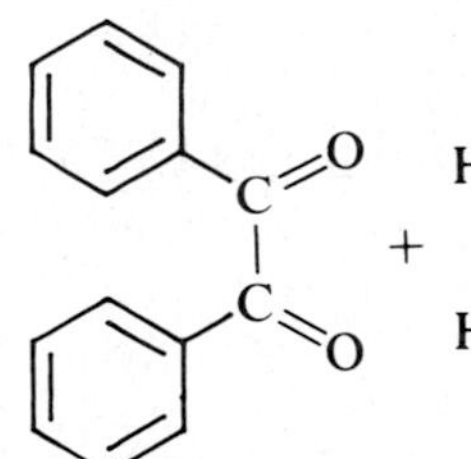

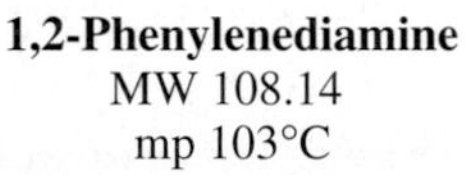

Benzil
MW 210.23
mp 94–95°C

1,2-Phenylenediamine
MW 108.14
mp 103°C

2,3-Diphenylquinoxaline
MW 282.33
mp 126°C

CAUTION: Handle *o*-phenylenediamine with care. It is classified as a suspected carcinogen. Similar compounds (hair dyes) are low-risk carcinogens. Carry out the reaction in the hood.

IN THIS EXPERIMENT, the two carbonyl groups of benzil are condensed with two amine groups to give an aromatic compound with two nitrogen atoms in the ring. The diamine must be purified by sublimation before being heated with the benzil (no solvent). The product is crystallized from methanol or a mixture of methanol and water. (See Chapter 4.)

Microscale Procedure

Commercial 1,2-phenylenediamine (also known as *o*-phenylenediamine) is usually badly discolored (due to air oxidation) and gives a poor result unless it is purified as follows. Place 100 mg of material in a reaction tube, evacuate the tube at full aspirator suction, clamp it, and heat the bottom of the tube with a hot sand bath to distill or sublime colorless 1,2-phenylenediamine from the dark residue into the upper half of the tube (Fig. 54.3). Let the tube cool until the melt has solidified; then scrape out the white solid.

Video: *Microscale Filtration on the Hirsch Funnel*

Reaction time: 11 minutes

Weigh 105 mg of benzil and 54 mg of your purified 1,2-phenylenediamine into a reaction tube and heat on a steam bath for 10 minutes, which changes the initially molten mixture to a light tan solid. Dissolve the solid in hot methanol (about 2.5 mL) and let the solution stand undisturbed. If crystallization does not occur within 10 minutes, reheat the solution and dilute it with a little water to the point of saturation. The crystals should be filtered on a Hirsch funnel as soon as they are formed because brown oxidation products accumulate on standing. The quinoxaline forms colorless needles (mp 125–126°C); the yield is about 90 mg.

Cleaning Up. The residues from the distillation (sublimation) of 1,2-phenylenediamine (the tube can be rinsed with acetone) and the solvent from the reaction should be placed in the aromatic amines hazardous waste container because the diamine may be a carcinogen.

Macroscale Procedure

Commercial *o*-phenylenediamine (also known as 1,2-phenylenediamine) is usually badly discolored (due to air oxidation) and gives a poor result unless it is purified as follows. Place 200 mg of material in a 20 × 150-mm test tube, evacuate the tube at full aspirator suction, clamp it in a nearly horizontal position, and heat the bottom of the tube in a hot sand bath to sublime colorless *o*-phenylenediamine from the dark residue into the upper half of the tube. Let the tube cool in position until the melt has solidified; then scrape out the white solid.

Reaction time: 11 minutes

Weigh 0.20 g of benzil (theoretical = 210 mg) and 0.10 g of your purified *o*-phenylenediamine (theoretical = 108 mg) into 20 × 150-mm test tube and heat it in a steam bath for 10 minutes, which changes the initially molten mixture to a light tan solid. Dissolve the solid in hot methanol (about 5 mL) and let the solution stand undisturbed. If crystallization does not occur within 10 minutes, reheat the solution and dilute it with a little water to the point of saturation. The crystals should be filtered as soon as they are formed because brown oxidation products accumulate on standing. The quinoxaline forms colorless needles (mp 125–126°C); the yield is 185 mg. The UV spectrum is shown in Figure 54.4.

Cleaning Up. Place all the residues from the distillation of 1,2-phenylenediamine and the solvent from the reaction in the aromatic amines hazardous waste container because the diamine may be a carcinogen.

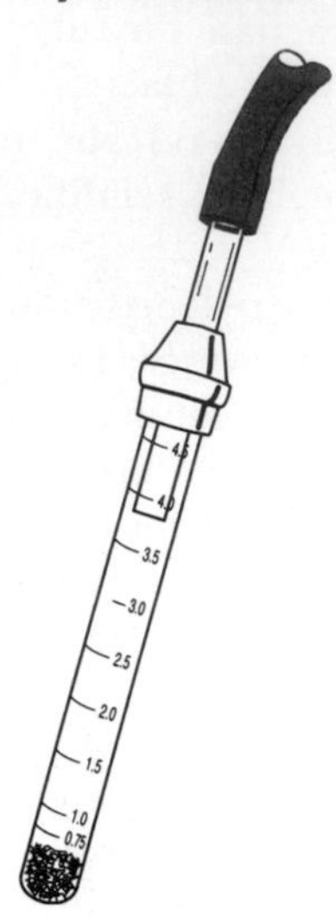

FIG. 54.3
An apparatus for the vacuum sublimation of 1,2-phenylenediamine.

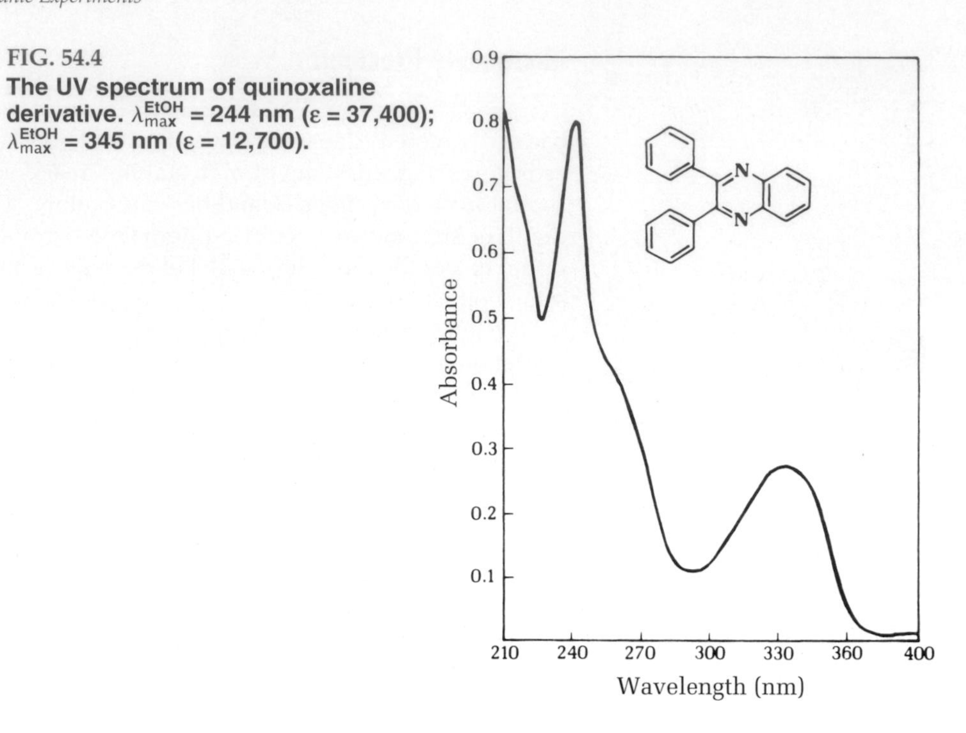

FIG. 54.4
The UV spectrum of quinoxaline derivative. λ_{max}^{EtOH} = 244 nm (ε = 37,400); λ_{max}^{EtOH} = 345 nm (ε = 12,700).

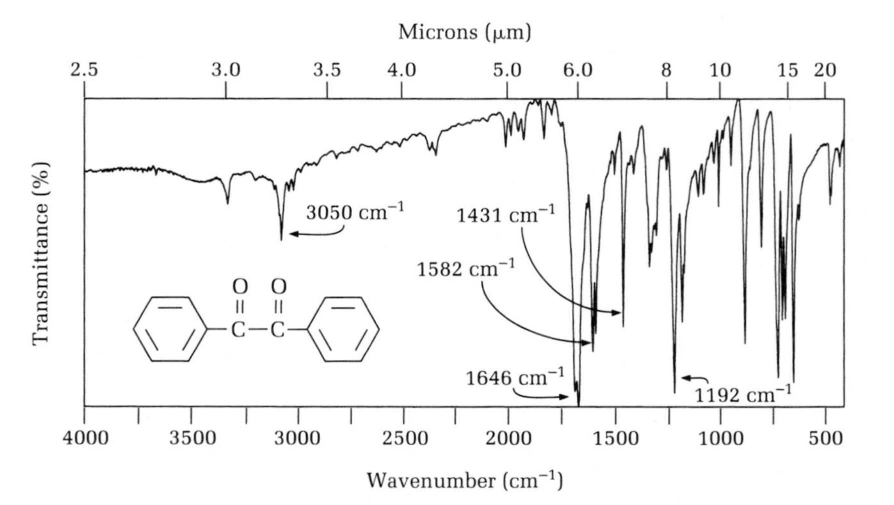

FIG. 54.5
The IR spectrum of benzil (KBr disk).

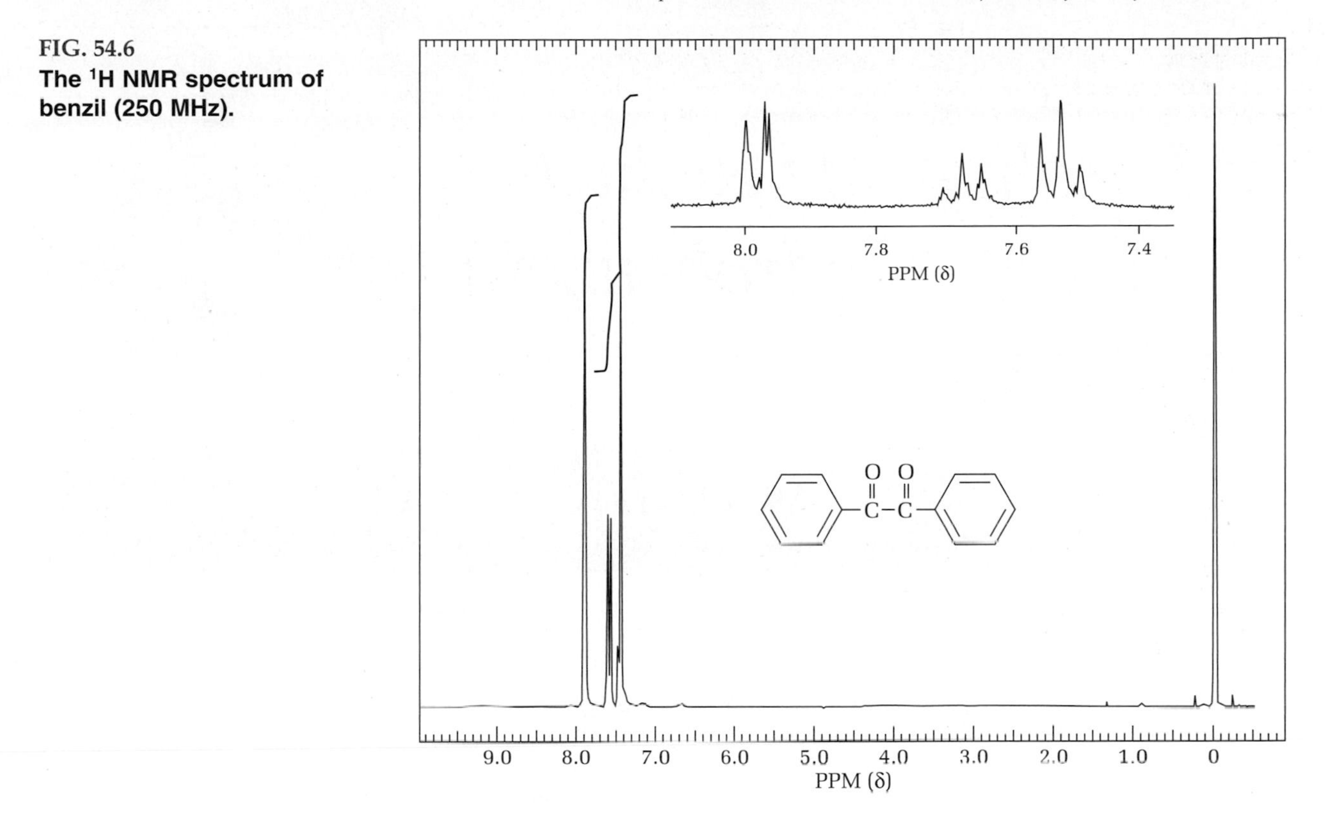

FIG. 54.6
The ^{1}H NMR spectrum of benzil (250 MHz).

QUESTIONS

1. Assign the peak at 1646 cm^{-1} in the IR spectrum of benzil (Fig. 54.5).
2. Assign the peaks at 3410 cm^{-1} and 1664 cm^{-1} in the IR spectrum of benzoin (Fig. 54.2).

55 CHAPTER

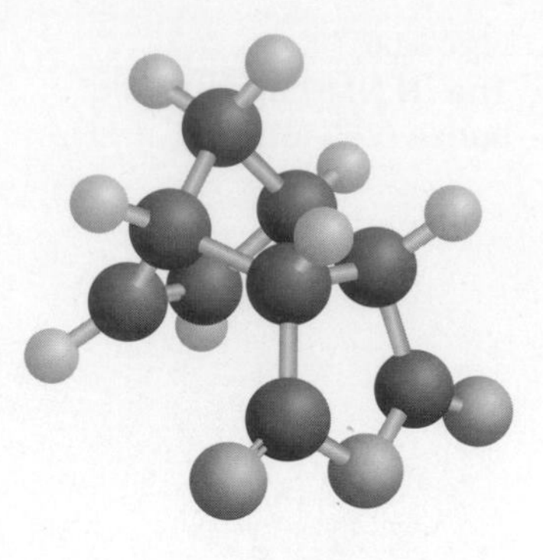

The Borohydride Reduction of a Ketone: Hydrobenzoin from Benzil

PRELAB EXERCISE: Compare the reducing capabilities of lithium aluminum hydride to those of sodium borohydride.

In 1943, H. I. Schlesinger and Herbert C. Brown discovered sodium borohydride. Brown devoted his entire scientific career to this reagent, making it and other hydrides the most useful and versatile of reducing reagents. He received a Nobel Prize in Chemistry in 1979 for his work.

Considering the extreme reactivity of most hydrides (such as sodium hydride and lithium aluminum hydride) toward water, sodium borohydride is somewhat surprisingly solid as a stabilized aqueous solution in 14 *M* sodium hydroxide containing 12% sodium borohydride. Unlike lithium aluminum hydride, sodium borohydride is insoluble in ether and soluble in methanol and ethanol.

Sodium borohydride is a mild and selective reducing reagent. In ethanol solution, it reduces aldehydes and ketones rapidly at 25°C and esters very slowly, and it is inert toward functional groups that are readily reduced by lithium aluminum hydride, including carboxylic acids, epoxides, lactones, nitro groups, nitriles, azides, amides, and acid chlorides.

The present experiment is a typical sodium borohydride reduction. These same conditions and isolation procedures could be applied to hundreds of other ketones and aldehydes.

EXPERIMENT

SODIUM BOROHYDRIDE REDUCTION OF BENZIL

The addition of two atoms of hydrogen to benzoin or four atoms of hydrogen to benzil gives a mixture of stereoisomeric diols, of which the predominant isomer is the nonresolvable (1*R*,2*S*)-hydrobenzoin, the *meso* isomer, accompanied by the enantiomeric (1*R*,2*R*) and (1*S*,2*S*) compounds.

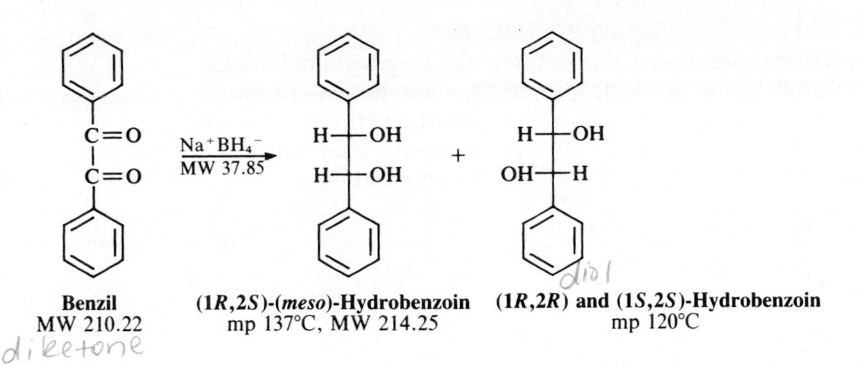

The reaction proceeds rapidly at room temperature; the intermediate borate ester is hydrolyzed with water to give the product alcohol.

$$4\ R_2C{=}O + Na^+BH_4^- \longrightarrow (R_2CHO)_4B^-Na^+$$

$$(R_2CHO)_4B^-Na^+ + 2\ H_2O \longrightarrow 4\ R_2CHOH + Na^+BO_2^-$$

The procedure that follows specifies the use of benzil rather than benzoin, because you can then follow the progress of the reduction by the discharge of the yellow color of the benzil.

Chapter 54 contains the infrared (IR) and nuclear magnetic resonance (NMR) spectra for benzil (Fig. 54.5 and Fig. 54.6, respectively). The ultraviolet (UV) spectrum is shown in Figure 55.2 at the end of this chapter.

Microscale Procedure

IN THIS EXPERIMENT, a solid diketone dissolved in ethanol is reduced with solid sodium borohydride to a dialcohol that crystallizes from the reaction mixture and is isolated by filtration. It is easy to follow the progress of the reaction—the diketone is yellow, and the dialcohol is colorless in solution.

Photos: *Filtration Using a Pasteur Pipette, Use of the Wilfilter,* Videos: *Filtration of Crystals Using the Pasteur Pipette, Microscale Crystallization*

In a 10 × 100-mm reaction tube, dissolve 50 mg of benzil in 0.5 mL of 95% ethanol; then cool the solution in ice to produce a fine suspension. Add to this suspension 10 mg of sodium borohydride (a large excess). The benzil dissolves, the mixture warms up, and the yellow color disappears in 2–3 minutes. After a total of 10 minutes, add 0.5 mL of water, heat the solution to the boiling point, filter the solution in case it is not clear (usually not necessary), and dilute the hot solution with hot water to the point of saturation (cloudiness)—a process that requires about 1 mL of water. (1*R*,2*S*)-Hydrobenzoin separates in lustrous thin plates (mp 136–137°C) and is best isolated by withdrawing some of the solvent using a Pasteur pipette and then using a Wilfilter (Fig. 55.1). It also can be collected on a microscale Büchner funnel. The yield is about 35 mg.

Cleaning Up. Dilute the aqueous filtrate with water, neutralize with acetic acid (to destroy the borohydride), and then flush the mixture down the drain.

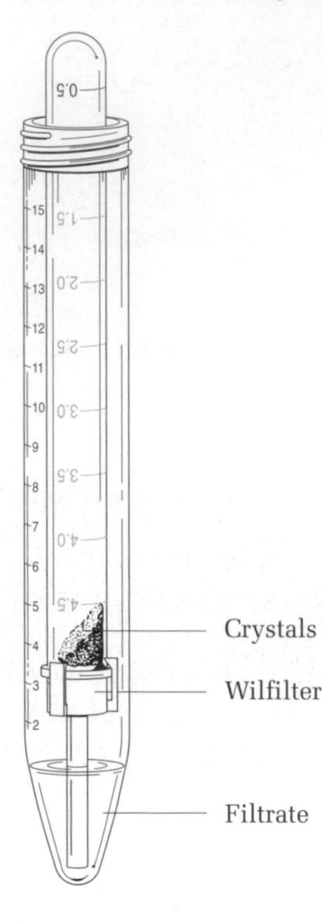

FIG. 55.1
The Wilfilter filtration apparatus. *See* Chapter 4.

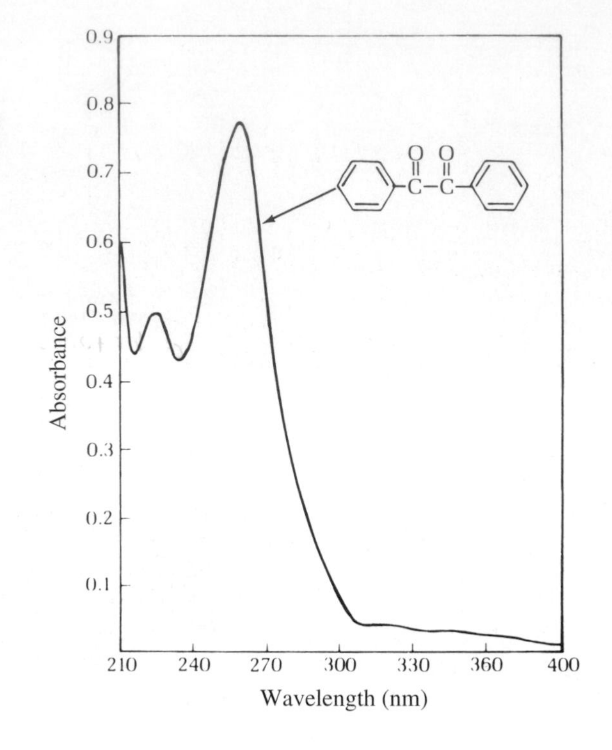

FIG. 55.2
The UV spectrum of benzil. λ_{max}^{EtOH} = 260 nm (ε = 19,800). One-centimeter cells and 95% ethanol have been employed for this spectrum.

Macroscale Procedure

In a 50-mL Erlenmeyer flask, dissolve 0.5 g of benzil in 5 mL of 95% ethanol, and cool the solution under the tap to produce a fine suspension. Then add 0.1 g of sodium borohydride (large excess). The benzil dissolves, the mixture warms up, and the yellow color disappears in 2–3 minutes. After a total of 10 minutes, add 5 mL of water, heat to the boiling point, filter in case the solution is not clear, dilute to the point of saturation with more water (10 mL), and set the solution aside to crystallize. *meso*-Hydrobenzoin separates in lustrous thin plates (mp 136–137°C); the yield is about 0.35 g. A second crop of material can be obtained by adding solid sodium chloride to the filtrate until no more of the salt will dissolve. Collect the second crop of crystals via filtration and wash the crystals with water.

Cleaning Up. Dilute the aqueous filtrate with water, neutralize it with acetic acid (to destroy the borohydride), and flush the mixture down the drain.

QUESTIONS

1. Using the *R*,*S* system of nomenclature, draw and name all of the isomers of hydrobenzoin.
2. Calculate the theoretical weight of sodium borohydride needed to reduce 50 mg of benzil.

CHAPTER 56

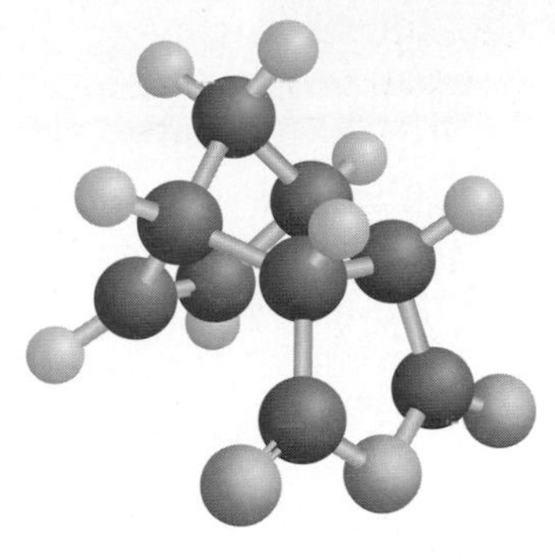

The Synthesis of 2,2-Dimethyl-1,5-Dioxolane; The Acetonide Derivative of a Vicinal Diol

PRELAB EXERCISE: Draw the most stable structures of *meso*- and *racemic*-1,2-stilbenediol using Newman projections. Draw the same two isomers with the hydroxyl groups eclipsed.

In Chapter 55, yellow benzil was reduced with sodium borohydride to the colorless diol. The melting point of the product (137°C) indicates that (1*R*,2*S*)-(*meso*)-hydrobenzoin has been produced. But how do we know that this is the isomer produced? The object of this experiment is to demonstrate, using logic based on nuclear magnetic resonance (NMR) spectral data, that the product is indeed the *meso* isomer and not the *racemic* isomer.

The diol is reacted with acetone to form a crystalline acetonide (a ketal) that will have either structure 1 or structure 2. An examination of the NMR spectrum of this derivative will reveal whether or not the two methyl groups of the acetonide are equivalent, and thus whether the diol is *meso* or *racemic*. Rather than simply explaining each step in this process and the reasoning behind it, you will discover them for yourself by answering the questions at the end of this chapter. By considering each question in turn, you will see how it is possible to determine the stereochemistry of the starting diol.

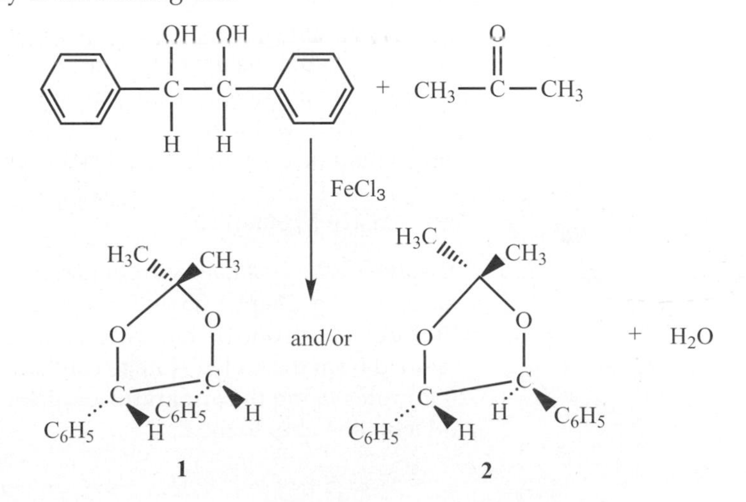

EXPERIMENT

IN THIS EXPERIMENT, a solid diol is dissolved in acetone that is both a solvent and a reactant. The addition of a catalyst, the Lewis acid iron(III) chloride, and heating of the solution causes the formation of a cyclic acetonide. At the end of the reaction, the mixture is poured into water containing a base to react with the acid, and the product is extracted into dichloromethane. This solution is dried and then evaporated to give the solid product, which is then crystallized from hexane.

Microscale Procedure

Dissolve 100 mg of the diol in 3 mL of acetone and add 30 mg of anhydrous iron(III) chloride. The iron(III) chloride is very hygroscopic and should therefore be weighed from the sealed storage container into a closed vial. Reflux the mixture for 20 minutes using the apparatus depicted in Figure 56.1.

Video: *Extraction with Dichloromethane*; Photo: *Extraction with Dichloromethane*

Transfer the reaction mixture to a 15-mL centrifuge tube containing 5 mL of a 1 *M* potassium carbonate solution. Extract the product three times with 1-mL portions of diethyl ether that are transferred to a reaction tube. Wash the combined extracts with 2.5 mL of water and then dry the solution over a few pieces of granular anhydrous calcium chloride. Transfer the dry solution to a clean reaction tube and carefully evaporate it to dryness.

Video: *Filtration of Crystals Using the Pasteur Pipette*; Photos: *Filtration Using a Pasteur Pipette, Use of the Wilfilter*

Dissolve the crude product in 0.3–0.5 mL of hexanes and stir the resulting slurry for a few minutes. The starting diol will not be soluble in hexanes, thus the product can be isolated in pure form by the filtration of the slurry using a Wilfilter (Fig. 56.2). After filtration, concentrate the resulting filtrate and dry the product under vacuum, if necessary; then record the yield, the melting point (reported melting point is 57–59°C), and the infrared (IR) spectrum of the product. Record the ^{1}H NMR spectrum of a solution in deuterochloroform ($CDCl_3$) and assign the stereochemistry of the product and thus of the diol. Store the acetonide in a refrigerator before analysis; it slowly decomposes at room temperature.

Cleaning Up. Place any excess or recovered dichloromethane in the halogenated waste container. Other organic solutions go in the organic solvents waste container. The aqueous phases from the extractions can, after dilution with water, be washed down the drain. The drying agent (the calcium chloride pellets), after removal of the solvent in the hood, can be placed in the nonhazardous solid waste container. If local regulations do not allow for the evaporation of solvents in a hood, the wet solid should be disposed in a special waste container.

Macroscale Procedure

Dissolve 0.30 g of *meso*-hydrobenzoin (*meso*-1,2-stilbenediol), the product of the experiment in Chapter 55, in 9 mL of acetone and add 0.1 g of anhydrous iron(III) chloride. The iron(III) chloride is very hygroscopic and should therefore be weighed from the sealed storage container into a closed vial. Reflux the mixture for 20 minutes using the apparatus depicted in Figure 25.3 (on page 380). Be sure to add a boiling chip to the flask.

FIG. 56.1
The esterification apparatus.

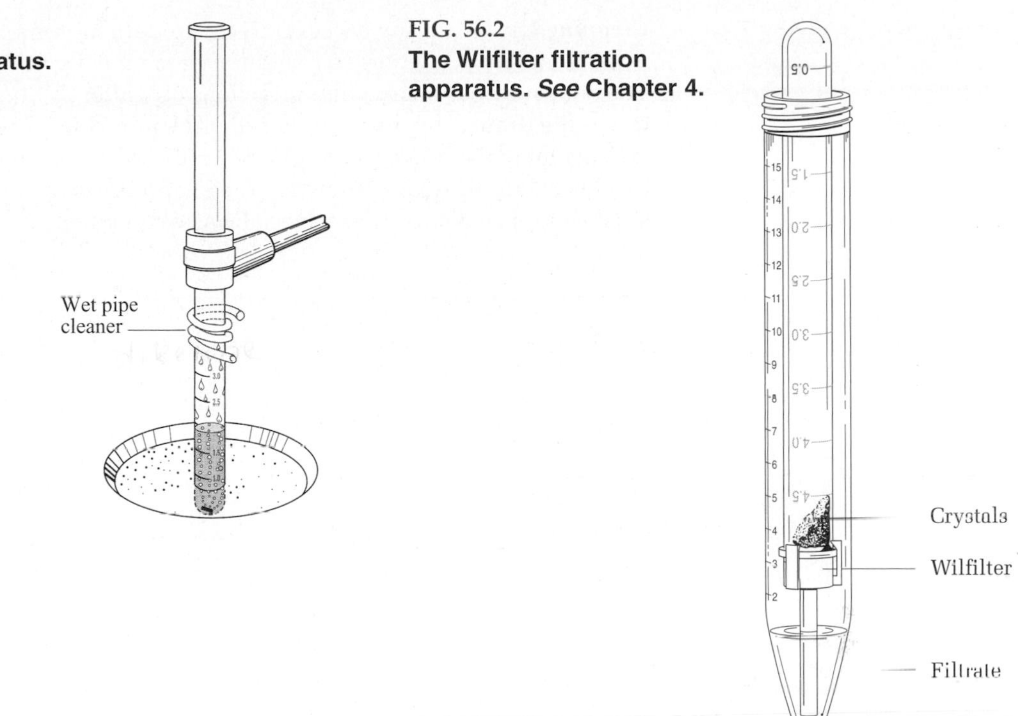

FIG. 56.2
The Wilfilter filtration apparatus. *See* Chapter 4.

Transfer the reaction mixture to a small separatory funnel that contains 30 mL of a 1 *M* potassium carbonate solution. Extract the product three times with 3-mL portions of diethyl ether that are transferred to a 10-mL Erlenmeyer flask. If you are uncertain which layer is the diethyl ether layer, choose a drop from one layer and add a few drops of water. If two phases can be seen, the layer you chose is the diethyl ether layer. Wash the combined extracts with 8 mL of water, and then dry the diethyl ether solution over granular anhydrous calcium chloride pellets. Transfer the dry solution to a clean 25-mL filter flask using more diethyl ether to wash the tube and drying agent. Evaporate the solution to dryness under vacuum (*see* Fig. 25.9 on page 388). Warm the filter flask with your hand to speed the process. The product should crystallize in the filter flask. It may be necessary to scratch or seed the thick liquid that sometimes results to initiate crystallization.

At this point, the crude product could be used for NMR analysis. Dissolve about 50 mg of the crystals or oil in deuterochloroform and transfer the solution to a clean, dry NMR tube. Bring the level of the deuterochloroform in the tube to 5.5 cm, cap the tube, and invert it several times to mix the solution.

To purify the crude product, dissolve it in 10–20 mL of hexanes and stir the resulting slurry for a few minutes. The starting diol will not be soluble in hexanes, thus the product can be isolated in pure form by filtration of the slurry. After filtration, concentrate the resulting filtrate and dry it under vacuum, if necessary (*see* Fig. 4.22 on page 77); then record the yield, the melting point (reported melting point is 57–59°C), and the IR spectrum of the product.

Record the ^{1}H NMR spectrum of a solution in deuterochloroform and assign the stereochemistry of the product and thus of the diol. Store the acetonide in a refrigerator before analysis; it slowly decomposes at room temperature.

Cleaning Up. Place any excess or recovered dichloromethane in the halogenated waste container. Other organic solutions go in the organic solvents waste container. The aqueous phases from the extractions can, after dilution with water, be washed down the drain. The drying agent (the calcium chloride pellets), after removal of the solvent in the hood, can be placed in the nonhazardous solid waste container. If local regulations do not allow for the evaporation of solvents in a hood, the wet solid should be disposed in a special waste container.

QUESTIONS

1. Write the mechanism and discuss the stereochemistry of the sodium borohydride reduction of diphenylethanedione (benzil).
2. Give the mechanism of acetonide formation. What is the role of iron(III) chloride?
3. Carefully draw the structures, using Newman projections, of (1*R*,2*S*)-(*meso*)-hydrobenzoin and (1*R*,2*R*)- and (1*S*,2*S*)-(*racemic*)-hydrobenzoin with the hydroxyl groups eclipsed. Then draw the corresponding acetonides of these two isomers.
4. Is acetonide 1 (on page 671) from the *meso* or the *racemic* diol?
5. Consider the symmetry of the two different acetonides formed. How many methyl peaks would you expect to observe in the ^1H NMR spectrum for the acetonide of the *meso* isomer? for the *racemic* isomer?
6. Compare and contrast the IR spectra of the dione, the diol, and the acetonide. Assign the major peaks to functional groups.

CHAPTER 57

1,4-Addition: Reductive Acetylation of Benzil

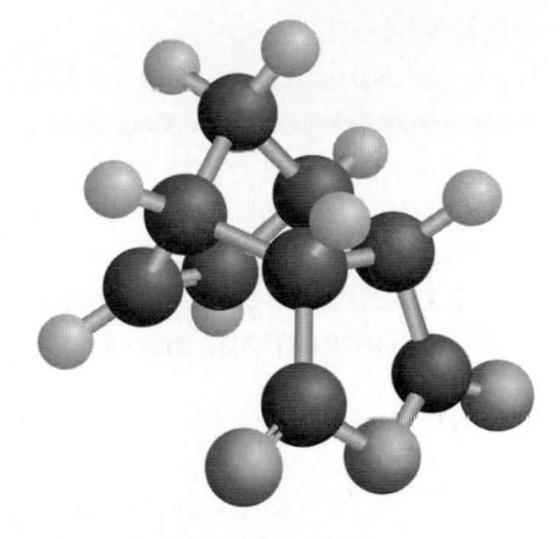

PRELAB EXERCISE: Write the complete mechanism for the reaction of acetic anhydride with an alcohol.

AC = acetyl group

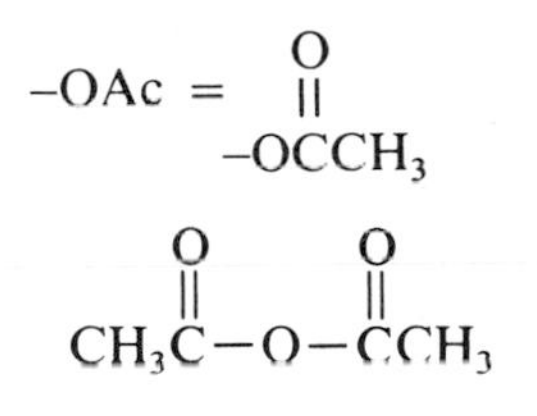

Ac_2O *is acetic anhydride*

In one of the first demonstrations of the phenomenon of 1,4-addition, Johannes Thiele (1899) established that the reduction of benzil with zinc dust in a mixture of acetic anhydride and sulfuric acid involves the 1,4-addition of hydrogen to the α-diketone grouping, and acetylation of the resulting enediol before it can undergo ketonization to benzoin. The process of reductive acetylation results in a mixture of the *E*- and *Z*-isomers (1 and 2). Thiele and subsequent investigators isolated the more soluble, lower-melting *Z*-stilbenediol diacetate (2) only in impure form (mp 110°C). Separation of the two isomers by chromatography is not feasible because they are equally adsorbable on alumina. However, separation is possible by fractional crystallization (described in the following procedures), and both isomers can be isolated in pure condition. In the method prescribed here for the preparation of the isomer mixture, hydrochloric acid is substituted for sulfuric acid because sulfuric acid gives rise to colored impurities and is reduced to sulfur and hydrogen sulfide.[1]

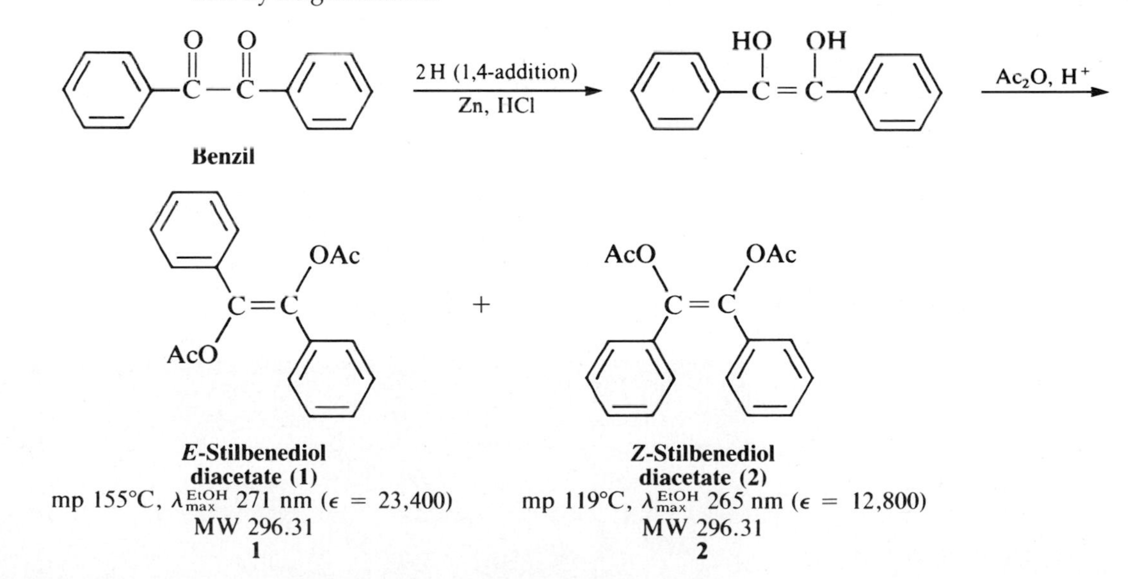

***E*-Stilbenediol diacetate (1)**
mp 155°C, λ_{max}^{EtOH} 271 nm (ϵ = 23,400)
MW 296.31
1

***Z*-Stilbenediol diacetate (2)**
mp 119°C, λ_{max}^{EtOH} 265 nm (ϵ = 12,800)
MW 296.31
2

[1]If acetyl chloride (2 mL) is substituted for the hydrochloric acid–acetic anhydride mixture in the procedure, the Z-isomer is the sole product.

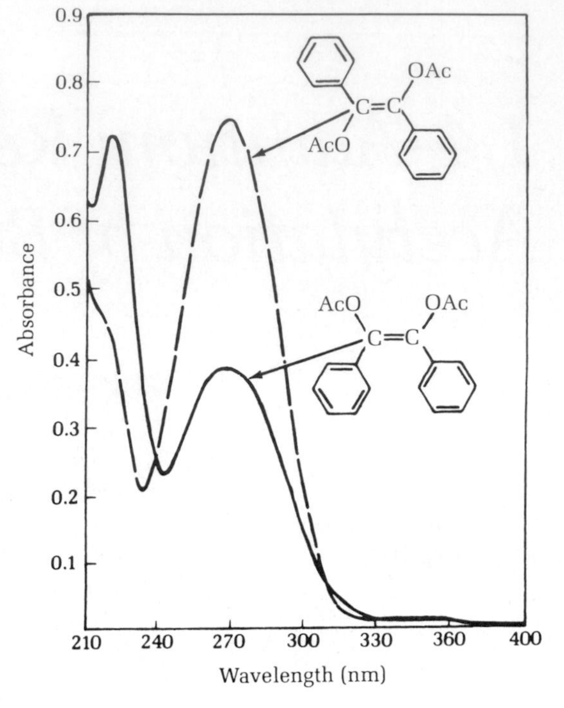

FIG. 57.1
The UV spectra of *Z*- and *E*-stilbene diacetate. *Z*: λ_{max}^{EtOH} = 223 nm (ε = 20,500) and 269 nm (ε = 10,800); *E*: λ_{max}^{EtOH} = 272 nm (ε = 20,800). The two phenyl rings cannot be coplanar in the *Z*-diacetate, which prevents overlap of the *p* orbitals of the phenyl groups with those of the central double bond. This steric inhibition of resonance accounts for the diminished intensity of the *Z*-isomer relative to the *E*-isomer. Both spectra were run at the same concentration.

Assignment of configuration

The configurations of this pair of geometrical isomers remained unestablished for over 50 years, but the tentative inference that the higher melting isomer has the more symmetrical *E* configuration eventually was found to be correct. Evidence of infrared (IR) spectroscopy is of no avail; the spectra are nearly identical in the interpretable region (4000–1200 cm^{-1}) characterizing the acetoxyl groups, but differ in the fingerprint region (1200–400 cm^{-1}). However, the isomers differ markedly in ultraviolet (UV) absorption (Fig. 57.1). In analogy to *E*- and *Z*-stilbene, the conclusion is justified that the higher melting isomer, because it has an absorption band at a longer wavelength and a higher intensity than its isomer, does indeed have configuration 1.

Experiments

1. Reductive Acetylation of Benzil

IN THIS EXPERIMENT, an ice-cold liquid anhydride reacts with an ice-cold solution of the dienol of a diketone (made by the zinc dust reduction of the diketone). The product, a mixture of solid diacetate isomers, is isolated by filtration. Because the solid mixture is contaminated with zinc dust, it is dissolved in ether to leave the zinc dust behind. The *Z*- and *E*-isomers are separated by fractional crystallization. As the ether is evaporated, the *E*-isomer is the first to crystallize out.

Microscale Procedure

Handle acetic anhydride and hydrochloric acid with care.

Place one reaction tube containing 0.7 mL of acetic anhydride and another containing 0.1 mL of concentrated hydrochloric acid in an ice bath and, when both are thoroughly chilled, transfer the acid to the anhydride dropwise by means of a Pasteur pipette. Transfer the acidic anhydride solution into a reaction tube containing 100 mg of pure benzil (prepared in Chapter 54) and 100 mg of zinc dust, and stir the mixture for 2–3 minutes in an ice bath. The reaction will give off heat. Remove the tube and allow it to warm up slowly, cooling in ice if the tube becomes too hot to touch.

Use fresh zinc dust.

When there is no further exothermic effect, let the mixture stand for 5 minutes; then add 2.5 mL of water. Swirl, break up any lumps of product, and allow a few minutes for the hydrolysis of excess acetic anhydride. Next, collect the mixture of product and zinc dust on a Hirsch funnel, wash with water, and press and apply suction to the filter cake until there is no further drip. Digest the solid in a 25-mL Erlenmeyer flask (drying is not necessary) with 7.0 mL of diethyl ether to dissolve the organic material; add about 1 g of anhydrous calcium chloride pellets, and swirl briefly. Using a Pasteur pipette, transfer the ether solution into another 10-mL Erlenmeyer flask, wash the drying agent, and concentrate the ether solution (using a steam bath, a boiling stone, and a water aspirator) to a volume of approximately 1.5 mL. Transfer to a reaction tube, cork the tube, and let the tube stand undisturbed.

Video: *Microscale Filtration on the Hirsch Funnel*

The *E*-diacetate (1) soon begins to separate in the form of prismatic needles; after 20–25 minutes, crystallization appears to stop. Remove the mother liquor from the reaction tube with a Pasteur pipette, or use a Wilfilter (*see* Fig. 4.14 on page 73) or the microscale filtration apparatus shown in Figure 57.2, and evaporate it to dryness in another reaction tube.

Photos: *Filtration Using a Pasteur Pipette, Use of the Wilfilter,* Videos: *Filtration of Crystals Using the Pasteur Pipette, Microscale Crystallization*

Dissolve the white solid remaining in the second reaction tube in 1.0 mL of methanol, let the solution stand undisturbed for about 10 minutes, and drop in a tiny crystal of the *E*-diacetate. This should give rise, in 20–30 minutes, to a second crop of the *E*-diacetate. Remove the mother liquor from these crystals and concentrate it to 0.7 mL, let it cool to room temperature as before, and again seed with a crystal of the *E*-diacetate; this usually affords a third crop of this diacetate. A typical yield for the three crops might be 30 mg (mp 154–156°C), 5 mg (mp 153–154°C), and 5 mg (mp 153–155°C).

At this point, the mother liquor should be rich enough in the more soluble *Z*-diacetate (2) for its isolation. Concentrate the methanol mother liquor and the washings from the third crop of the *E*-diacetate to a volume of 0.4 mL, stopper the tube, and let the solution stand undisturbed overnight. The *Z*-diacetate sometimes separates spontaneously in large rectangular prisms of great beauty. If the solution remains supersaturated, adding a seed crystal of the *Z*-diacetate causes prompt separation of this diacetate in a paste of small crystals [e.g., 20 mg (mp 118–119°C) and then a second crop of 7 mg (mp 116–117°C)]. This experiment illustrates the great convenience of column chromatography in isomer separations when it can be employed. Separating compounds by fractional crystallization is rarely necessary.

Cleaning Up. Spread the zinc out on a watch glass for about 20 minutes before wetting it and placing it in the nonhazardous solid waste container. Sometimes zinc dust from a reaction like this is pyrophoric (spontaneously flammable in air) because it is so finely divided and has such a large, clean surface area that is able to react with air. Neutralize the aqueous filtrate with sodium carbonate, remove the

FIG. 57.2
A microscale filtration assembly.

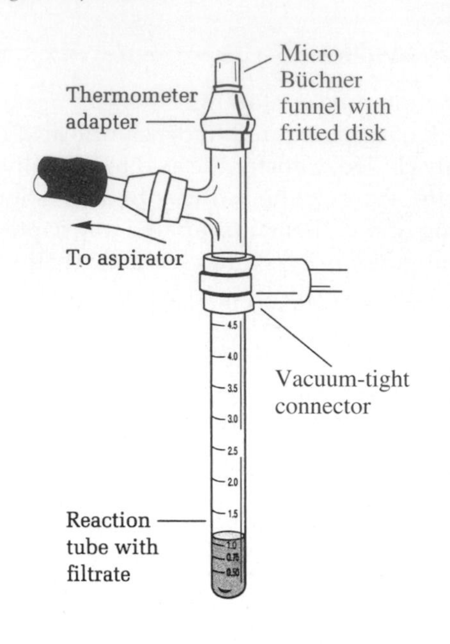

zinc salts by vacuum filtration, dilute the filtrate with water, and flush it down the drain. The zinc salts are placed in the nonhazardous solid waste container. Place the filtrate from the fractional crystallization in the organic solvents waste container. Allow the solvent to evaporate from the calcium chloride pellets in the hood; then place the drying agent in the nonhazardous solid waste container. If local regulations do not allow for the evaporation of solvents in a hood, the wet solid should be disposed in a special waste container.

Macroscale Procedure

Handle acetic anhydride and hydrochloric acid with care.

Reaction time: 10 minutes

Carry out the reaction in the hood.

Use fresh zinc dust.

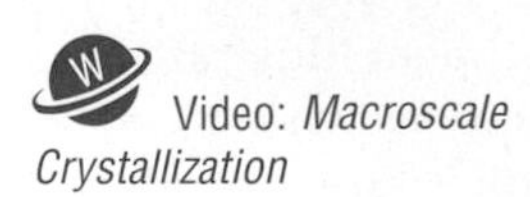
Video: *Macroscale Crystallization*

Place one test tube (20 × 150 mm) containing 7 mL of acetic anhydride and another test tube (13 × 100 mm) containing 1 mL of concentrated hydrochloric acid in an ice bath. When both are thoroughly chilled, transfer the acid to the anhydride dropwise in not less than 1 minute by means of a Pasteur pipette. Wipe the test tube dry, pour the chilled solution into a 50-mL Erlenmeyer flask containing 1 g of pure benzil and 1 g of zinc dust, and swirl for 2–3 minutes in an ice bath. Remove the flask and hold it in the palm of your hand; if it begins to warm up, cool further in ice. When there is no further exothermic effect, let the mixture stand for 5 minutes; then add 25 mL of water. Swirl, break up any lumps of product, and allow a few minutes for the hydrolysis of excess acetic anhydride. Then collect the mixture of product and zinc dust by vacuum filtration, wash with water, and press and apply suction to the cake until there is no further drip. Digest the solid (drying is not necessary) with 70 mL of *t*-butyl methyl ether to dissolve the organic material, add about 4 g of calcium chloride pellets, and swirl briefly. Filter the solution into a 125-mL Erlenmeyer flask, concentrate the filtrate (using a steam bath, a boiling stone, and a water aspirator) to a volume of approximately 15 mL,[2] cork the flask, and let the flask stand undisturbed.

[2]Measure 15 mL of a solvent into a second flask of the same size and compare the levels in the two flasks in order to gauge the approximate amount.

The *E*-diacetate (1) soon begins to separate in prismatic needles; after 20–25 minutes, crystallization appears to stop. Remove the crystals by filtration on a Hirsch funnel. Wash them once with ether and then evaporate the ether to dryness. Dissolve the residue in 10 mL of methanol and transfer the hot solution to a 25-mL Erlenmeyer flask.

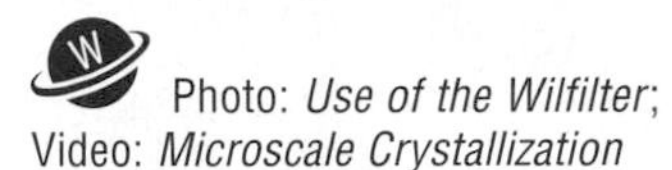
Photo: *Use of the Wilfilter*; Video: *Microscale Crystallization*

Let the solution stand undisturbed for about 10 minutes, then drop in one tiny crystal of the *E*-diacetate. This should give rise, in 20–30 minutes, to a second crop of the *E*-diacetate. Then concentrate the mother liquor and washings to a volume of 7–8 mL, let cool to room temperature as before, and again seed with a crystal of the *E*-diacetate; this usually affords a third crop of this diacetate. A typical yield for the three crops might be 300 mg (mp 154–156°C), 50 mg (mp 153–156°C), and 50 mg (mp 153–155°C). The products can also be collected using a Wilfilter or a microscale Büchner funnel.

At this point, the mother liquor should be rich enough in the more soluble *Z*-diacetate (2) for its isolation. Concentrate the methanol mother liquor and washings from the third crop of the *E*-diacetate to a volume of 4–5 mL, stopper the flask, and let the solution stand undisturbed overnight. The *Z*-diacetate sometimes separates spontaneously in large rectangular prisms of great beauty. If the solution remains supersaturated, adding a seed crystal of the *Z*-diacetate causes prompt separation of this diacetate in a paste of small crystals [e.g., 215 mg (mp 118–119°C) and then a second crop of 70 mg (mp 116–117°C)].

Cleaning Up. Spread the zinc out on a watch glass for about 20 minutes before wetting it and placing it in the nonhazardous solid waste container. Sometimes zinc dust from a reaction like this is pyrophoric (spontaneously flammable in air) because it is so finely divided and has such a large, clean surface area that is able to react with air. Neutralize the aqueous filtrate with sodium carbonate, remove the zinc salts by vacuum filtration, dilute the filtrate with water, and flush it down the drain. The zinc salts are placed in the nonhazardous solid waste container. Place the filtrate from the fractional crystallization in the organic solvents waste container. Allow the solvent to evaporate from the calcium chloride pellets in the hood, and then place the drying agent in the nonhazardous solid waste container. If local regulations do not allow for the evaporation of solvents in a hood, the wet solid should be disposed in a special waste container.

COMPUTATIONAL CHEMISTRY

Construct *Z*- and *E*-stilbene diacetate and carry out an energy minimization on the resulting molecules using a molecular mechanics program. Repeat the calculation using the semiempirical molecular orbital program AM1. From the resulting conformations, can you explain why the UV extinction coefficient ε for the peak at 269 nm is smaller for the *Z*-isomer than for the *E*-isomer?

58 CHAPTER

The Synthesis of an Alkyne from an Alkene; Bromination and Dehydrobromination: Stilbene and Diphenylacetylene

PRELAB EXERCISE: Calculate the theoretical quantities of thionyl chloride and sodium borohydride needed to convert benzoin to *E*-stilbene.

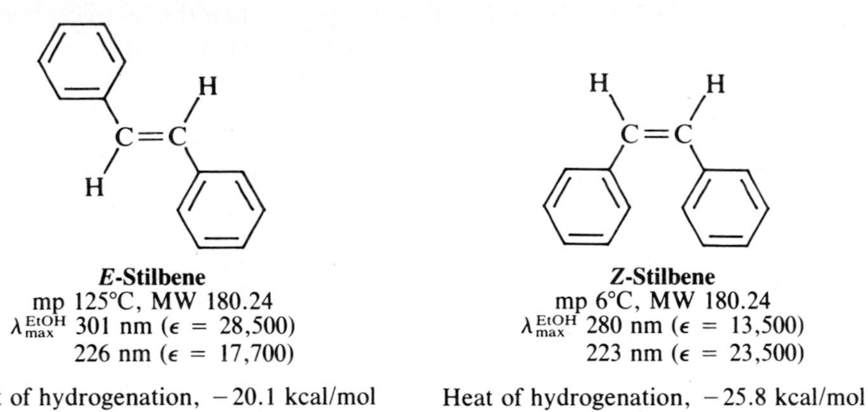

In this chapter's experiments, benzoin, which was prepared in Chapter 53, is converted to the alkene *trans*-stilbene (*E*-stilbene), which is in turn brominated and dehydrobrominated to form diphenylacetylene, an alkyne.

One method of preparing *E*-stilbene is the reduction of benzoin with zinc amalgam in a mixture of ethanol and hydrochloric acid, presumably through an intermediate:

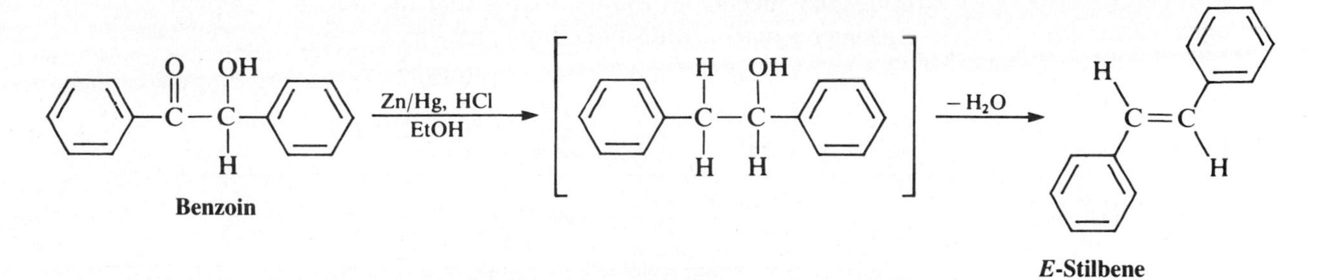

The procedure that follows is quick and affords very pure hydrocarbon. It involves three steps: (1) replacing the hydroxyl group of benzoin by chlorine to form desyl chloride, (2) reducing the keto group with sodium borohydride to give what appears to be a mixture of the two diastereoisomeric chlorohydrins, and (3) eliminating the elements of hypochlorous acid with zinc and acetic acid. The last step is analogous to the debromination of an olefin dibromide.

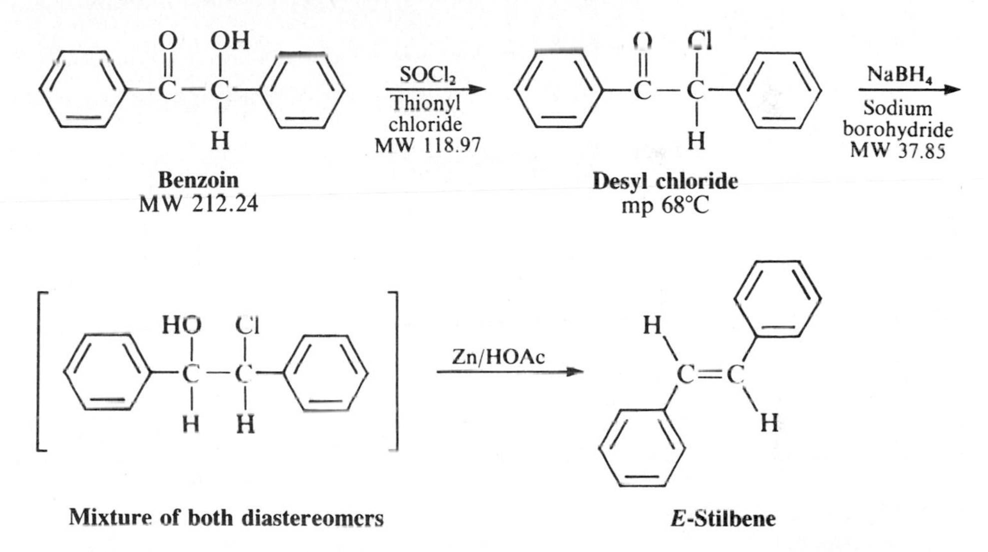

The minimum energy conformations of *E*- and *Z*-stilbene are shown in Figures 58.1 and 58.2, respectively. These have been calculated using Spartan's

FIG. 58.1
The minimum energy conformation of *E*-stilbene.

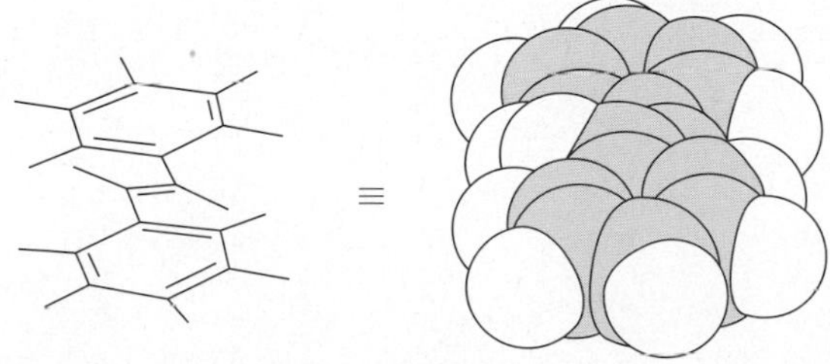

FIG. 58.2
The minimum energy conformation of *Z*-stilbene.

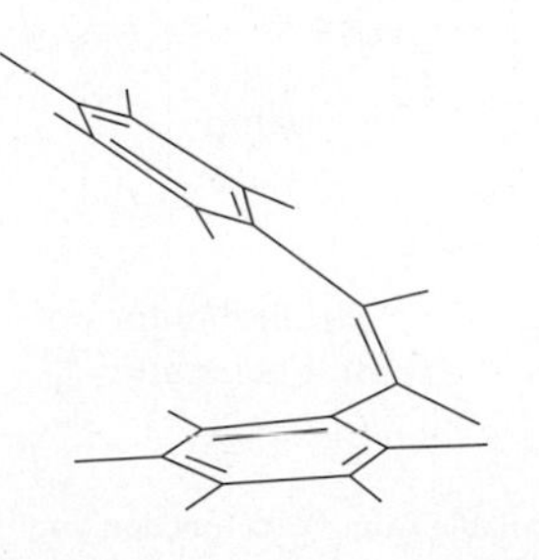

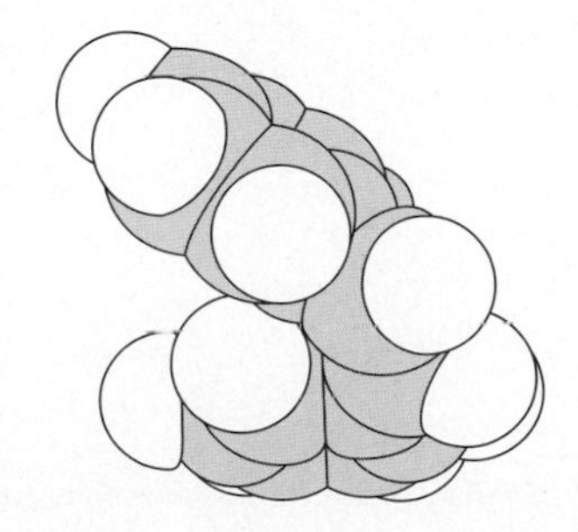

molecular mechanics routine.[1] Note that the *E*-isomer is planar, whereas the *Z*-isomer is markedly distorted from planarity.

Experiments

1. Stilbene

IN THIS EXPERIMENT, the hydroxyl group of solid benzoin is converted to a chloride by heating with liquid thionyl chloride, which acts as both a solvent and a reactant. The hydrogen chloride that is evolved is captured in a gas trap. The thionyl chloride is evaporated under vacuum to leave an oily product—a keto chloride. This material is dissolved in ethanol and reduced to a mixture of isomeric hydroxy chlorides (chlorohydrins) with sodium borohydride. Without isolation, this solution of chlorohydrins is reduced again, this time with zinc dust and acetic acid. This process removes the elements of hypochlorus acid (HOCl) to give the product—stilbene.

Microscale Procedure

Photos: *Placing a Polyethylene Tube through a Septum, Gas Trap*

Place 100 mg of benzoin (crushed to a powder) in a reaction tube, cover it with 0.15 mL of thionyl chloride, and add a boiling chip and a rubber septum bearing a polyethylene tube leading to another reaction tube bearing a plug of damp cotton (Fig. 58.3). This will serve to trap the hydrogen chloride and sulfur dioxide evolved

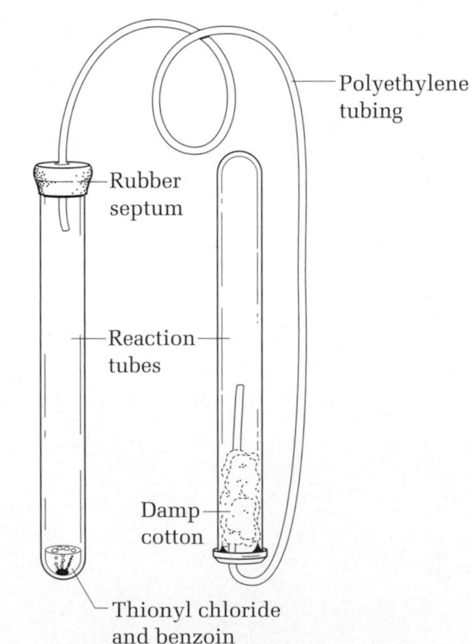

FIG. 58.3
A trap for hydrogen chloride and sulfur dioxide. Warm the tubing in a steam bath to bend it permanently.

[1]Spartan molecular modeling software is available from Wavefunction, Inc., Irvine, California.

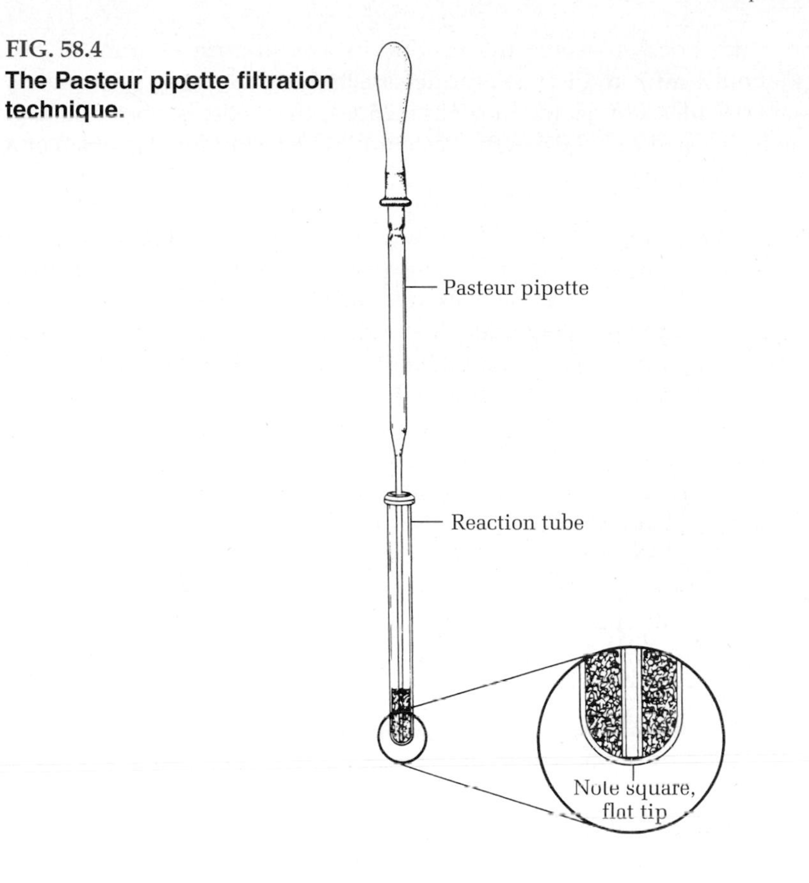

FIG. 58.4
The Pasteur pipette filtration technique.

in this reaction. Warm the reaction mixture gently on a steam bath or on a warm sand bath until the solid has dissolved, and then more strongly for 5 minutes.

Caution: If the mixture of benzoin and thionyl chloride is allowed to stand at room temperature for an appreciable time before being heated, an undesired reaction intervenes,[2] and the synthesis of *E*-stilbene is spoiled.

Photos: *Extraction with Ether, Filtration Using a Pasteur Pipette, Use of the Wilfilter;* Videos: *Extraction with Ether, Filtration of Crystals Using the Pasteur Pipette*

To remove excess thionyl chloride (bp 77°C), connect the reaction tube to an aspirator for a few minutes, add 0.5 mL of hexanes, boil it off, and evacuate again. Desyl chloride is thus obtained as a viscous, pale-yellow oil (it will solidify on standing). Dissolve it in 1 mL of 95% ethanol, cool under the tap, and add 9 mg of sodium borohydride (an excess is harmful to the success of the reaction). Stir to break up any lumps of the borohydride; after 10 minutes, add to the solution of chlorohydrins 60 mg of zinc dust and 0.15 mL of acetic acid; then reflux for 1 hour and cool under the tap. When white crystals separate, add 2 mL of *t*-butyl methyl ether, wash the ether solution once with an equal volume of water containing 1 drop of concentrated hydrochloric acid (to dissolve basic zinc salts), and then wash with water. Dry the ether over calcium chloride pellets, which are added until they no longer clump together. Remove the ether from the drying agent with a Pasteur pipette (Fig. 58.4), wash the drying agent with ether, evaporate the ether to

[2]
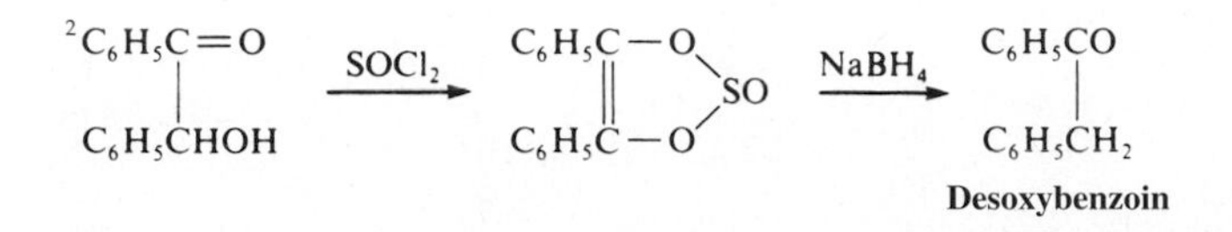

dryness under the hood, dissolve the residue in a minimum amount of hot 95% ethanol (about 1 mL), and let the product crystallize. *E*-Stilbene separates as diamond-shaped iridescent plates (mp 124–125°C); the yield is about 50 mg. *E*-Stilbene can be isolated easily using the Pasteur pipette technique (Fig. 58.4) or a Wilfilter (*see* Fig. 4.14 on page 73).

Sometimes zinc dust from a reaction like this is pyrophoric (spontaneously flammable in air) because it is so finely divided and has such a large, clean surface area able to react with air.

Cleaning Up. Combine the washings from the cotton in the trap and all aqueous layers, neutralize with sodium carbonate, remove zinc salts by vacuum filtration, and flush the filtrate down the drain with excess water. The zinc salts are placed in the nonhazardous solid waste container. Allow the ether to evaporate from the calcium chloride pellets in the hood, and then place the drying agent in the nonhazardous solid waste container. If local regulations do not allow for the evaporation of solvents in a hood, the wet solid should be disposed in a special waste container. Ethanol mother liquor goes in the organic solvents waste container. Spread out any isolated zinc on a watch glass to dry and air oxidize. Wet it with water, and place it in the nonhazardous solid waste container.

Macroscale Procedure

Carry out the procedure in the hood.

Place 2 g of benzoin (crushed to a powder) in a 50-mL round-bottomed flask, cover it with 4 mL of thionyl chloride,[3] warm gently on a steam bath (in the hood) until all the solid has dissolved, and then heat more strongly for 5 minutes.

Caution: If the mixture of benzoin and thionyl chloride is left standing at room temperature for an appreciable time before being heated, an undesired reaction intervenes,[4] and the synthesis of *E*-stilbene is spoiled.

To remove excess thionyl chloride (bp 77°C), evacuate on an aspirator for a few minutes, add 5 mL of hexanes, boil it off (in the hood), and again evacuate. Desyl chloride is thus obtained as a viscous, pale-yellow oil, that will solidify on standing. Dissolve it in 20 mL of 95% ethanol, cool under the tap, and add 180 mg of sodium borohydride (an excess is harmful to the success of the reaction). Stir to break up any lumps of the borohydride; after 10 minutes, add to the solution of chlorohydrins 1 g of zinc dust and 2 mL of acetic acid, and reflux for 1 hour. Then cool under the tap. When white crystals separate, add 25 mL of *t*-butyl methyl ether and decant the solution from the bulk of the zinc into a separatory funnel. Wash the solution twice with an equal volume of water containing 0.5–1 mL of concentrated hydrochloric acid (to dissolve basic zinc salts) and then, in turn, with a sodium carbonate solution and a saturated sodium chloride solution. Dry the ether over anhydrous calcium chloride pellets (2 g), filter to remove the drying agent, evaporate the filtrate to dryness under the hood, dissolve the residue in a minimum amount of hot 95% ethanol (15–20 mL), and let the product crystallize. *E*-Stilbene separates as diamond-shaped iridescent plates (mp 124–125°C); the yield is about 1 g. The UV spectrum is shown in Figure 58.5.

[3]The reagent can be dispensed from a burette or measured by pipette; in the latter case, the liquid should be drawn into the pipette with a pipette bulb, not by mouth.

[4] $C_6H_5C{=}O$ / C_6H_5CHOH $\xrightarrow{SOCl_2}$ $C_6H_5C{-}O$ / $C_6H_5C{-}O$ (cyclic, SO) $\xrightarrow{NaBH_4}$ C_6H_5CO / $C_6H_5CH_2$

Desoxybenzoin

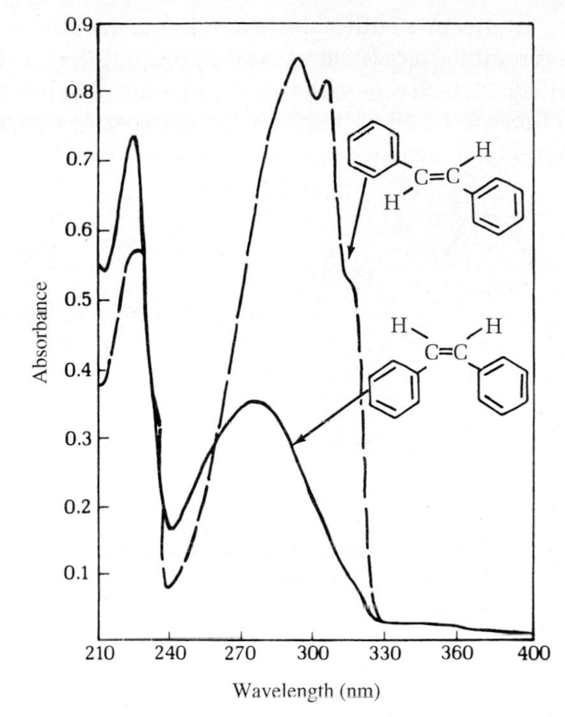

FIG. 58.5
The UV spectra of *Z*- and *E*-stilbene. *Z*: λ_{max}^{EtOH} = 224 nm (ε = 23,300) and 279 nm (ε = 11,100); *E*: λ_{max}^{EtOH} = 226 nm (ε = 18,300) and 295 nm (ε = 27,500). As in the diacetates, steric hindrance and the lack of coplanarity in these hydrocarbons cause the long-wavelength absorption of the *Z*-isomer to be of diminished intensity relative to the *E*-isomer.

Sometimes zinc dust from a reaction like this is pyrophoric (spontaneously flammable in air) because it is so finely divided and has such a large, clean surface area able to react with air.

Cleaning Up. Combine the washings from the cotton in the trap and all aqueous layers, neutralize with sodium carbonate, remove zinc salts by vacuum filtration, and flush the filtrate down the drain with excess water. The zinc salts are placed in the nonhazardous solid waste container. Allow the ether to evaporate from the calcium chloride in the hood, and then place it in the nonhazardous solid waste container. If local regulations do not allow for the evaporation of solvents in a hood, the wet solid should be disposed in a special waste container. Ethanol mother liquor goes in the organic solvents waste container. Any zinc isolated should be spread out on a watch glass to dry and air oxidize. It should then be wetted with water, and placed in the nonhazardous solid waste container.

2. *meso*-STILBENE DIBROMIDE

IN THIS EXPERIMENT, an acetic acid solution of stilbene, which is an alkene, is brominated with a solid bromine donor (which eliminates handling hazardous liquid bromine) to give a dibromide that crystallizes and is isolated by filtration.

E-Stilbene reacts with bromine predominantly by the usual process of *trans*-addition, and affords the optically inactive, nonresolvable *meso*-dibromide; the much lower melting enantiomeric mixture of dibromides is a very minor product of the reaction.

Pyridine + HBr + Br_2

↓

Pyridinium hydrobromide perbromide

In this procedure, the brominating agent will be pyridinium hydrobromide perbromide,[5] a crystalline, nonvolatile, odorless complex of high molecular weight (319.86), which dissociates, in the presence of a bromine acceptor such as an alkene, to liberate 1 mol of bromine. For microscale experiments, the perbromide is far more convenient and agreeable to measure and use than free bromine.

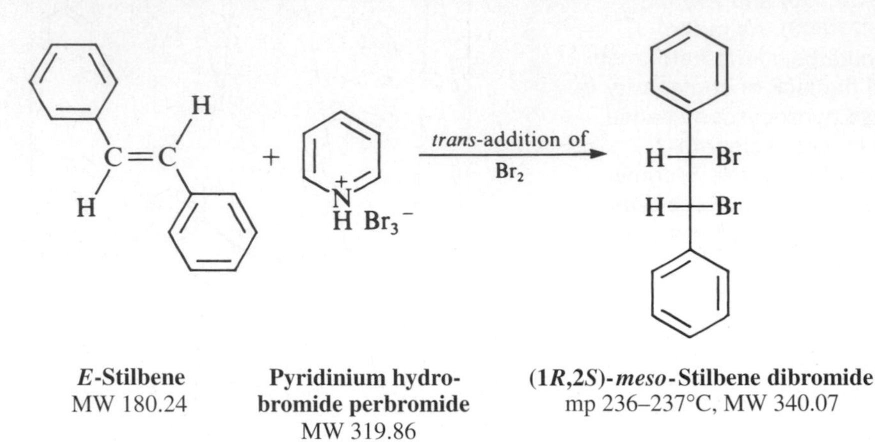

CAUTION: Pyridinium hydrobromide perbromide is corrosive and a lachrymator (tear producer).

Carry out the procedure in the hood.

Photo: *Use of the Wilfilter*, Video: *Microscale Filtration on the Hirsch Funnel*

Microscale Procedure

In a reaction tube, dissolve 50 mg of *E*-stilbene in 1 mL of acetic acid by heating on a steam bath or a hot water bath; then add 100 mg of pyridinium hydrobromide perbromide. Mix by swirling; if necessary, rinse crystals of the reagent down the walls of the flask with a little acetic acid and continue the heating for an additional 1–2 minutes. The dibromide separates as small plates. Cool this mixture under the tap, collect the product on a Wilfilter (*see* Fig. 4.14 on page 73) or on a Hirsch funnel, and wash it with methanol to remove any color; the yield of colorless crystals (mp 236–237°C) is 80 mg. Use this material to prepare the diphenylacetylene after determining the percent yield and the melting point.

Cleaning Up. To the filtrate, add sodium bisulfite (until a negative test with starch-iodide paper is observed) to destroy any remaining perbromide, neutralize with sodium carbonate, and extract the pyridine released with ether, which goes in the organic solvents waste container. The aqueous layer can then be diluted with water and flushed down the drain.

Total time required: 10 minutes

Macroscale Procedure

CAUTION: Pyridium hydrobromide perbromide is corrosive and a lachrymator (tear producer).

In a 50-mL Erlenmeyer flask, dissolve 1 g of *E*-stilbene in 20 mL of acetic acid by heating on a steam bath; then add 2 g of pyridinium hydrobromide perbromide.[6] Mix by swirling; if necessary, rinse crystals of the reagent down the walls of the flask with a little acetic acid; continue the heating for an additional 1–2 minutes.

[5]Crystalline pyridinium hydrobromide perbromide suitable for small-scale experiments is available from the Aldrich Chemical Company. Massive crystals commercially available should be recrystallized from acetic acid (4 mL/g). Pyridinium hydrobromide perbromide can also be prepared as follows: Mix 15 mL of pyridine with 30 mL of 48% hydrobromic acid, and cool. Add 25 g of bromine gradually with swirling, cool, and collect the product (use acetic acid for rinsing and washing). Without drying the solid, crystallize it from 100 mL of acetic acid. The yield of orange needles should be 33 g (69%).
[6]*See* footnote 5.

Carry out the procedure in the hood.

Total time required: 10 minutes

The dibromide separates almost at once as small plates. Cool the mixture under the tap, collect the product, and wash it with methanol; the yield of colorless crystals (mp 236–237°C) is about 1.6 g. Use 0.5 g of this material to prepare the diphenylacetylene, and turn in the remainder to your instructor.

Cleaning Up. To the filtrate, add sodium bisulfite (until a negative test with starch-iodide paper is observed) to destroy any remaining perbromide, neutralize with sodium carbonate, and extract the pyridine released during the reaction with ether, which goes in the organic solvents waste container. The aqueous layer can then be diluted with water and flushed down the drain.

3. SYNTHESIS OF DIPHENYLACETYLENE

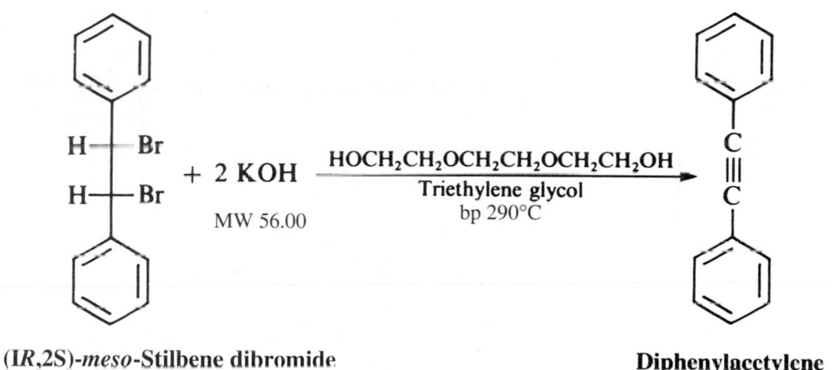

One method for the preparation of diphenylacetylene involves the oxidation of benzil dihydrazone with mercuric oxide; the intermediate diazo compound loses nitrogen as it becomes the hydrocarbon:

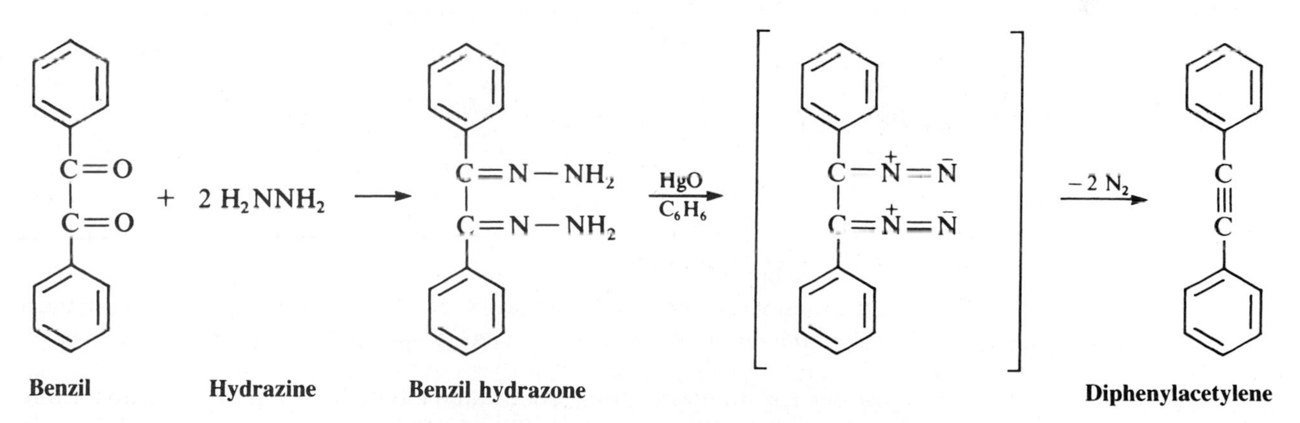

The method used in the following procedure involves the dehydrohalogenation of *meso*-stilbene dibromide. An earlier procedure in the chemical literature called for refluxing the dibromide with 43% ethanolic potassium hydroxide in an oil bath at 140°C for 24 hours. In the following procedure, the reaction time is reduced to a few minutes by using high-boiling triethylene glycol as the solvent to permit operation at a higher reaction temperature, a technique introduced by Louis Fieser.

IN THIS EXPERIMENT, a solution of stilbene dibromide in a very high-boiling solvent reacts with potassium hydroxide to remove 2 mol of HBr and form an alkyne. After the reaction mixture is diluted with water, the product crystallizes, is isolated by filtration, and is recrystallized from ethanol to give pure diphenylacetylene.

Microscale Procedure

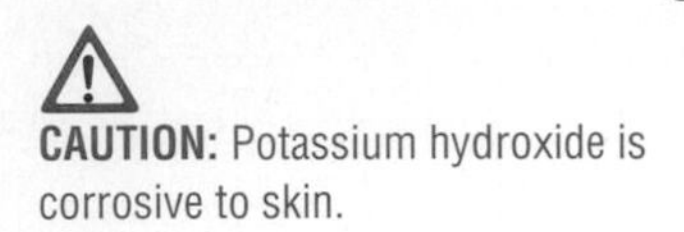

CAUTION: Potassium hydroxide is corrosive to skin.

Reaction time: 5 minutes

Photo: *Filtration Using a Pasteur Pipette*; Video: *Filtration of Crystals Using the Pasteur Pipette*

In a reaction tube, place 80 mg of *meso*-stilbene dibromide, 40 mg of potassium hydroxide,[7] and 0.5 mL of triethylene glycol. Heat the mixture on a hot sand bath to 160°C, at which point, potassium bromide begins to separate. By intermittent heating, keep the mixture at 160–170°C for an additional 5 minutes; then cool to room temperature, remove the thermometer, and add 2 mL of water. The diphenylacetylene that separates as a nearly colorless, granular solid is collected with a Pasteur pipette. The crude product need not be dried, but can be crystallized directly from 95% ethanol. Let the ethanol solution stand undisturbed to observe the formation of beautiful, very large spars of colorless crystals. After a first crop has been collected, the mother liquor can be concentrated to afford a second crop of pure product; the total yield is 35 mg (mp 60–61°C).

Cleaning Up. Combine the crystallization mother liquor with the filtrate from the reaction, dilute with water, neutralize with 3 *M* hydrochloric acid, and flush down the drain.

Macroscale Procedure

CAUTION: Potassium hydroxide is corrosive to skin.

Reaction time: 5 minutes

Photo: *Filtration Using a Pasteur Pipette*; Videos: *Filtration of Crystals Using the Pasteur Pipette, Microscale Filtration on the Hirsch Funnel*

In a 20 × 150-mm test tube, place 0.5 g of *meso*-stilbene dibromide, 3 pellets of potassium hydroxide[8] (250 mg), and 2 mL of triethylene glycol. Insert a thermometer into a 10 × 75-mm test tube containing enough triethylene glycol to cover the bulb, and then slip this assembly into the larger tube. Clamp the tube in a vertical position in a hot sand bath and heat the mixture to 160°C, at which point potassium bromide begins to separate. By intermittent heating, keep the mixture at 160–170°C for an additional 5 minutes; then cool to room temperature, remove the thermometer and small tube, and add 10 mL of water. The diphenylacetylene that separates as a nearly colorless, granular solid is collected by suction filtration. The crude product need not be dried, but can be crystallized directly from 95% ethanol. Let the ethanol solution stand undisturbed to observe the formation of beautiful, very large spars of colorless crystals. After a first crop has been collected, the mother liquor can be concentrated to afford a second crop of pure product; the total yield is about 0.23 g (mp 60–61°C). The UV spectrum is shown in Figure 58.6.

Cleaning Up. Combine the crystallization mother liquor with the filtrate from the reaction, dilute with water, neutralize with 3 *M* hydrochloric acid, and flush down the drain.

[7]Potassium hydroxide pellets consist of 85% KOH and 15% water.
[8]*See* footnote 7.

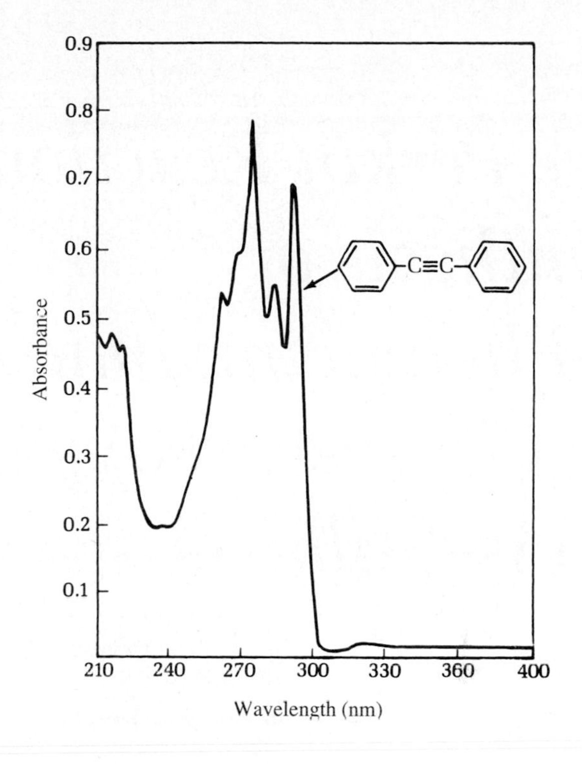

FIG. 58.6
The UV spectrum of diphenylacetylene. λ_{max}^{EtOH} = 279 nm (ε = 31,400). This spectrum is characterized by considerable fine structure (multiplicity of bands) and a high extinction coefficient.

QUESTIONS

1. Using a molecular mechanics program, calculate the energy difference or the difference in heats of formation between *Z*- and *E*-stilbene. How do your computed values compare to the difference in heats of hydrogenation of the two isomers (refer to the structures on the first page of this chapter)?
2. Why is *Z*-stilbene not planar? Explain.
3. How do you account for the large difference in the extinction coefficients (ε) of the long-wavelength peaks for *Z*- and *E*-stilbene (*see* Fig. 58.5 on page 685)?

59 CHAPTER

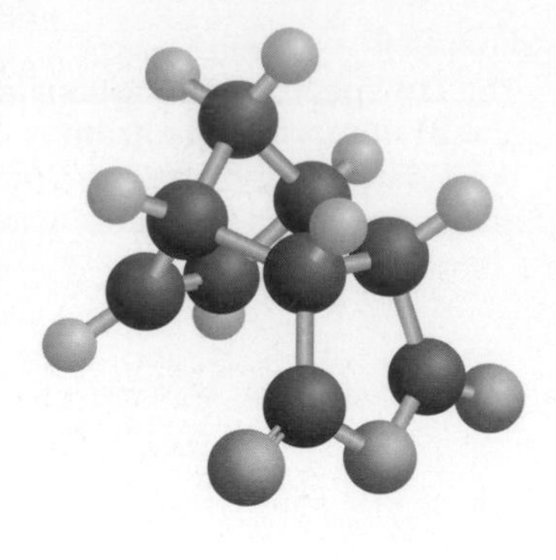

The Perkin Reaction: Synthesis of α-Phenylcinnamic Acid and Its Decarboxylation to cis-Stilbene

PRELAB EXERCISE: At the end of the Perkin reaction, the products are present as mixed anhydrides. What are these mixed anhydrides, how are they formed, and how are they converted to the product?

The reaction of benzaldehyde with phenylacetic acid to produce a mixture of the α-carboxylic acid derivatives of *Z*- and *E*-stilbene, a form of an aldol condensation known as the *Perkin reaction*, is effected by heating a mixture of the components with acetic anhydride and triethylamine. In the course of the reaction, the phenylacetic acid is probably present both as the anion and as the mixed anhydride resulting from equilibration with acetic anhydride. A 5-hour reflux period specified in an early procedure has been shortened by a factor of 10 by restricting the amount of the volatile acetic anhydride, using an excess of the less expensive high-boiling aldehyde component, and using a condenser that permits some evaporation and consequent elevation of the reflux temperature.

E-Stilbene is a byproduct of the condensation, but it has been shown that neither the *E*- nor *Z*-acid undergoes decarboxylation under the conditions of the condensation reaction.

At the end of the reaction, the α-phenylcinnamic acids are present in part as the neutral mixed anhydrides, but these can be hydrolyzed by adding excess hydrochloric acid. The organic material is taken up in ether, and the acids are extracted with alkali. Neutralization with acetic acid ($pK_a = 4.76$) then causes precipitation of only the less acidic *E*-acid (refer to the pK_a values given the structural formulas on the next page); the *Z*-acid separates when hydrochloric acid is added.

Although *Z*-stilbene is less stable and has a lower melting point than *E*-stilbene, the reverse is true of the α-carboxylic acids, and in this preparation the more stable, higher melting *E*-acid is the predominant product. Evidently, the steric interference between the carboxyl and phenyl groups in the *Z*-acid is greater than that between

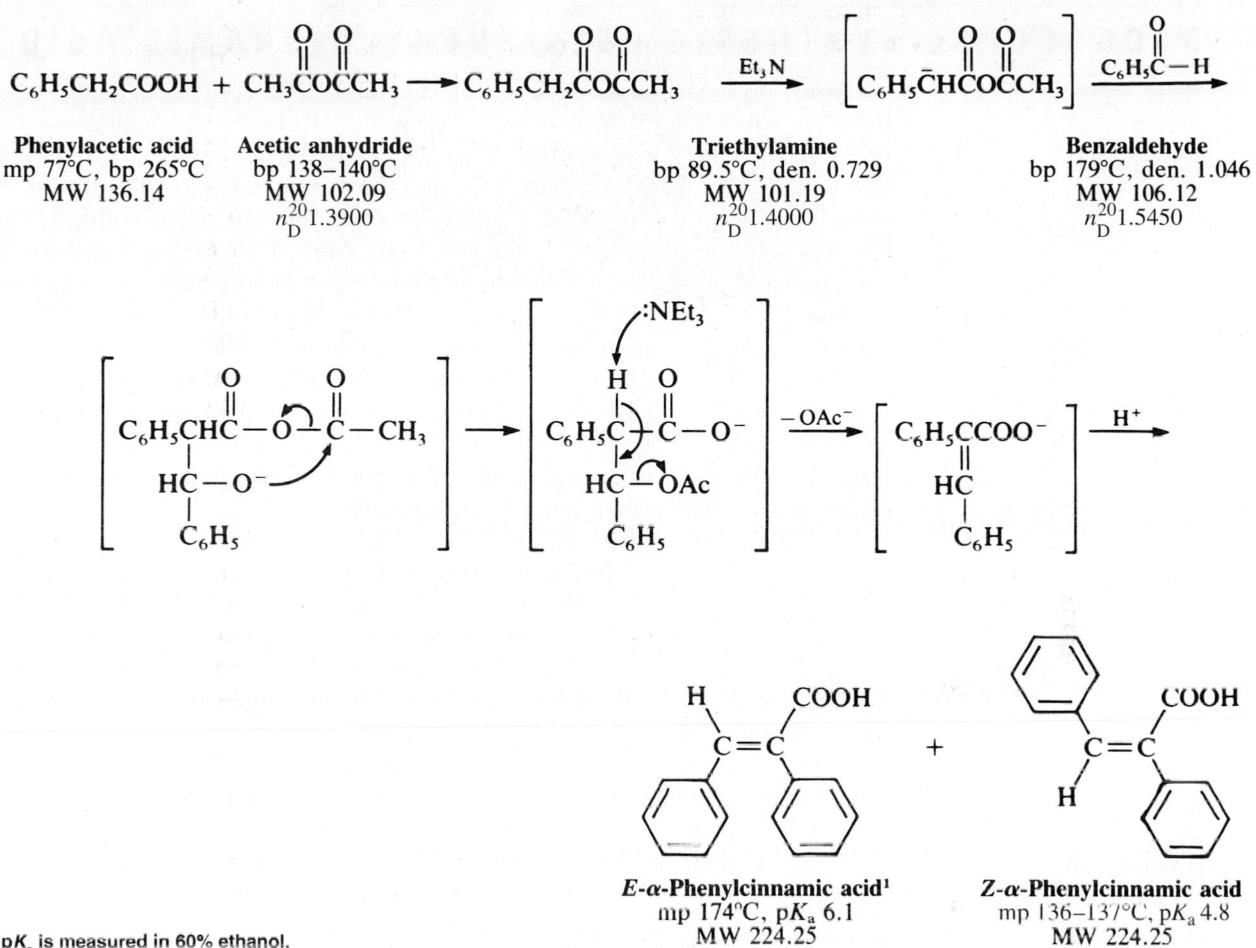

The pK_a is measured in 60% ethanol.

the two phenyl groups in the *E*-acid, a statement you can check by calculation. Steric hindrance is also evident from the fact that the *Z*-acid is not subject to Fischer esterification (ethanol and an acid catalyst), whereas the *E*-acid is.

Experiments: Part 1

IN THIS EXPERIMENT, four reagents react in an aldol-type reaction (the Perkin reaction) between (1) a solution of phenylacetic acid and acetic anhydride to give a mixed anhydride that (2) reacts with an amine to give an anion that then (3) condenses with benzaldehyde and eliminates a molecule of acetic acid to give the product (4), which is a mixture of isomeric phenylacetic acids. The product is isolated by adding hydrochloric acid, which causes the isomeric products to crystallize. The crystals are dissolved in ether and are extracted with aqueous base. The basic solution on acidification with acetic acid deposits the *E*-acid, which is isolated by filtration and recrystallized. The aqueous filtrate on acidification with hydrochloric acid causes the *Z*-acid to crystallize. It is also isolated by filtration.

1. MICROSCALE SYNTHESIS OF α-PHENYLCINNAMIC ACID

Reflux time: 35 minutes

Photos: *Extraction with Ether, Use of the Wilfilter,* Videos: *Extraction with Ether, Microscale Filtration on the Hirsch Funnel*

Measure into a reaction tube 250 mg of phenylacetic acid, 0.30 mL of benzaldehyde, 0.2 mL of triethylamine, and 0.2 mL of acetic anhydride. Insert a boiling stone and reflux the mixture for 35 minutes (Fig. 59.1). Cool the yellow melt, add 0.4 mL of concentrated hydrochloric acid, and mix; the mixture sets to a stiff paste. Add 2.5 mL of *t*-butyl methyl ether, warm to dissolve the bulk of the solid, wash the ethereal solution twice with water, and then extract it five times with 2-mL portions of 0.5 *M* potassium hydroxide solution.[1] Set aside the dark-colored ethereal solution (discard it after the product is in hand).[2] Acidify the combined, colorless alkaline extract to pH 6 by adding 0.4 mL of acetic acid, collect the *E*-acid that precipitates, and save the filtrate and washings. The yield of *E*-acid (mp 163–166°C) is usually about 290 mg. Crystallize 30 mg of the material by dissolving it in 0.6 mL of *t*-butyl methyl ether, adding 0.8 mL of hexanes), heating briefly to the boiling point, and letting the solution stand. Silken needles form (mp 173–174°C). These can be easily collected on a Wilfilter (*see* Fig. 4.14 on page 73).

Adding 0.5 mL of concentrated hydrochloric acid to the aqueous filtrate from the precipitation of the *E*-acid produces a cloudy emulsion, which on standing for about 1/2 hour, coagulates to crystals of the *Z*-acid; the yield is 30 mg (mp 136–137°C). The *E*-acid can be recrystallized by dissolving 30 mg in 0.5 mL of *t*-butyl methyl ether, filtering (if necessary) from a trace of sodium chloride, adding 1 mL of hexanes, and evaporating to a volume of 0.5 mL; the acid separates as a hard crust of prisms (mp 138–139°C).

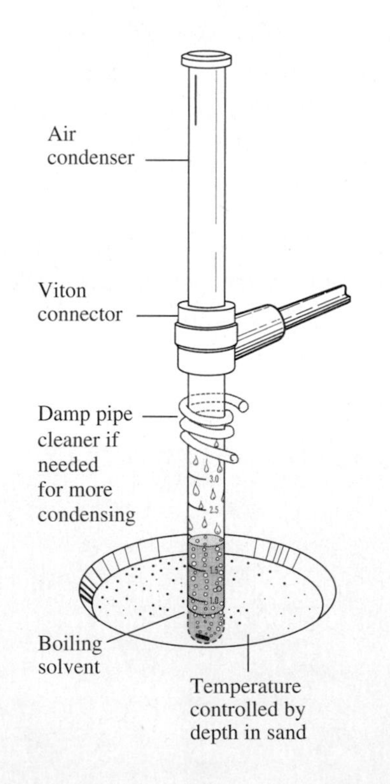

FIG. 59.1
A microscale reflux apparatus.

[1]If stronger alkali is added, the potassium salt may separate.
[2]To isolate the stilbene, wash the ethereal solution with saturated sodium bisulfite solution to remove the benzaldehyde, dry, evaporate, and crystallize the residue from methanol. Large, slightly yellow spars (mp 122–124°C) separate (yield = 9 mg).

Cleaning Up. The dark-colored ether solution and the mother liquor from the crystallization are placed in the organic solvents waste container. After the *Z*-acid has been removed from the reaction mixture, the acidic solution is diluted with water and flushed down the drain.

2. MACROSCALE SYNTHESIS OF α-PHENYLCINNAMIC ACID

Reflux time: 35 minutes

Measure into a 25-mL round-bottomed flask 2.5 g of phenylacetic acid, 3 mL of benzaldehyde, 2 mL of triethylamine, and 2 mL of acetic anhydride. Add a water-cooled condenser and a boiling stone, and reflux the mixture for 35 minutes (Fig. 59.2). Cool the yellow melt, add 4 mL of concentrated hydrochloric acid, and swirl; the mixture sets to a stiff paste. Add *t*-butyl methyl ether, warm to dissolve the bulk of the solid, and transfer to a separatory funnel by using more ether. Wash the ethereal solution twice with water; then extract it with a mixture of 25 mL of water and 5 mL of 3 *M* potassium hydroxide solution.[3] Repeat the extraction two more times and discard the dark-colored ethereal solution (discard it after the product is in hand).[4] Acidify the combined, colorless, alkaline extract to pH 6 by adding 5 mL of acetic acid, collect the *E*-acid that precipitates, and save the filtrate and washings. The yield of *E*-acid (mp 163–166°C) is usually about 2.9 g. Crystallize 0.3 g of material by dissolving it in 8 mL of *t*-butyl methyl ether, adding 8 mL of hexanes or pentane, heating briefly to the boiling point, and letting the solution stand. Silken needles form (mp 173–174°C).

FIG. 59.2
A macroscale reflux apparatus.

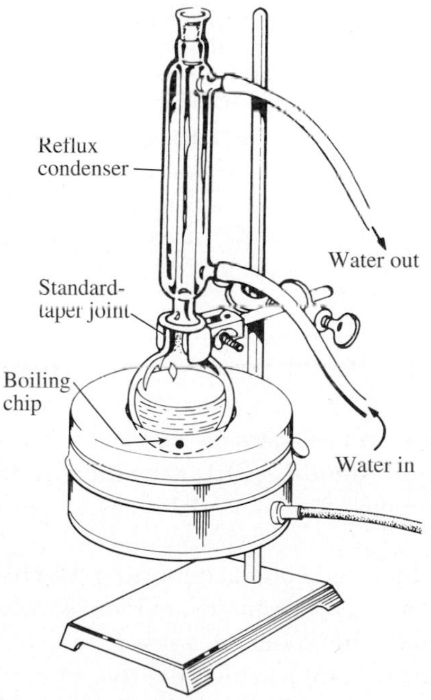

Reflux condenser
Water out
Standard-taper joint
Boiling chip
Water in

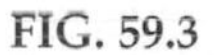

FIG. 59.3
The UV spectra of *Z*- and *E*-α-phenylcinnamic acid, run at identical concentrations. *E*: λ_{max}^{EtOH} = 222 nm (ε = 18,000), 282 nm (ε = 14,000); *Z*: λ_{max}^{EtOH} = 222 nm (ε = 15,500), 292 nm (ε = 22,300).

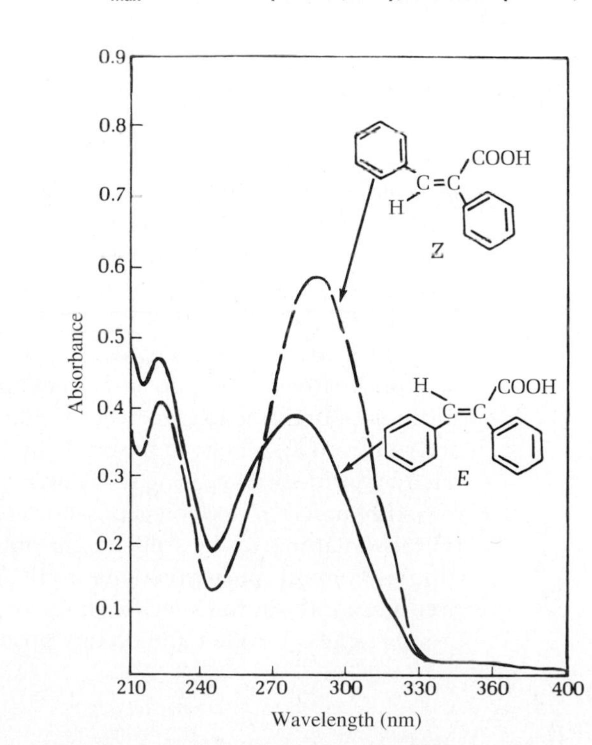

[3] *See* footnote 1.
[4] *See* footnote 2.

Adding 5 mL of concentrated hydrochloric acid to the aqueous filtrate from the precipitation of the *E*-acid produces a cloudy emulsion, which, on standing for about 1/2 hour, coagulates to crystals of *Z*-acid; the yield is about 0.3 g (mp 136–137°C). The *E*-acid can be recrystallized by dissolving 30 mg in 0.5 mL of *t*-butyl methyl ether, filtering (if necessary) from a trace of sodium chloride, adding 1 mL of hexanes and evaporating to a volume of 0.5 mL; the acid separates as a hard crust of prisms (mp 138–139°C).

Cleaning Up. The dark-colored ether solution and the mother liquor from the crystallization are placed in the organic solvents waste container. After the *Z*-acid has been removed from the reaction mixture, the acidic solution is diluted with water and flushed down the drain.

DECARBOXYLATION: SYNTHESIS OF *cis*-STILBENE

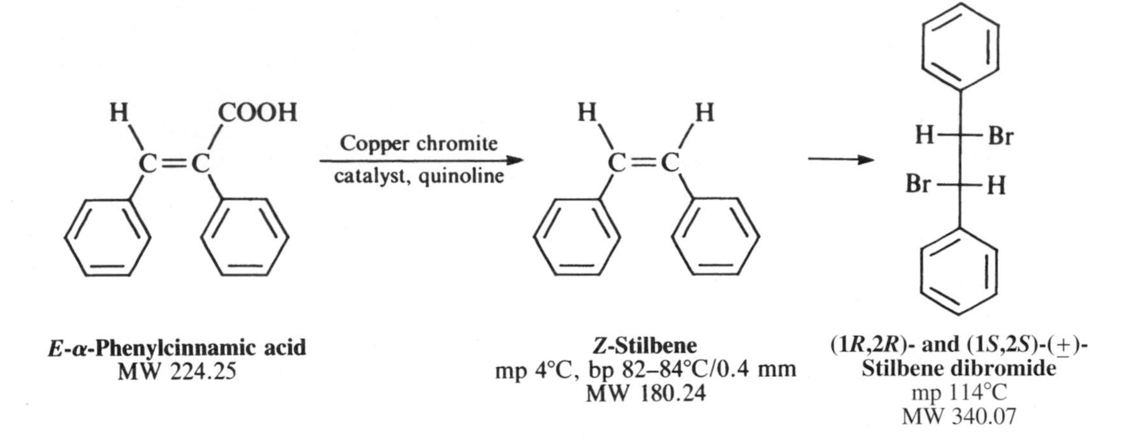

The catalyst for this reaction is copper chromite, 2 CuO · Cr_2O_3, a relatively inexpensive, commercially available catalyst used for both hydrogenation and dehydrogenation as well as decarboxylation.

Decarboxylation of *E*-α-phenylcinnamic acid is effected by refluxing the acid in quinoline in the presence of a trace of copper chromite catalyst; both the basic properties and boiling point (237°C) of quinoline make it a particularly favorable solvent. *Z*-Stilbene, a liquid at room temperature, can be characterized by the *trans* addition of bromine to give the crystalline (±)-dibromide. A little *meso*-dibromide derived from *E*-stilbene in the crude hydrocarbon starting material is easily separated by virtue of its sparing solubility.

Although free rotation is possible around the single bond connecting the chiral carbon atoms of the stilbene dibromides and hydrobenzoins, evidence from dipole-moment measurements indicates that the molecules tend to exist predominantly in the specific shape or conformation in which the two phenyl groups repel each other and occupy positions as far apart as possible. The optimal

FIG. 59.4
The favored conformations of stilbene dibromide.

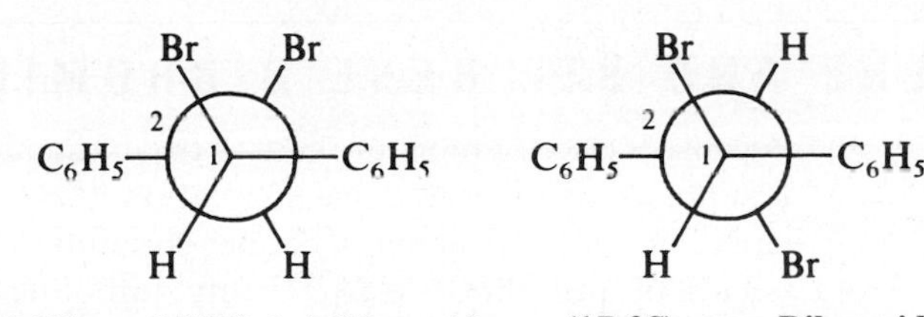

conformations of the (±)-dibromide and the *meso*-dibromide are represented in Figure 59.4 by Newman projection formulas, in which the molecules are viewed along the axis of the bond connecting the two chiral carbon atoms. In the *meso*-dibromide, the two repelling phenyl groups are on opposite sides of the molecule and so are the two large bromine atoms. Hence, the structure is far more symmetrical than that of the (±)-dibromide. X-ray diffraction measurements of the dibromides in the solid state confirm the conformations indicated in Figure 59.4. The Br-Br distances found are meso-dibromide, 4.50 nm and (±)-dibromide, 3.85 nm. The difference in symmetry of the two optically inactive isomers accounts for the marked contrast in properties:

	mp (°C)	Solubility in Ether (18°C)
±-Dibromide	114	1 part in 3.7 parts
meso-Dibromide	237	1 part in 1025 parts

Experiments: Part 2

IN THIS EXPERIMENT, phenylcinnamic acid is dissolved in the high-boiling solvent quinoline and treated with the catalyst copper chromite. This mixture is heated under vacuum to remove traces of water, and then at high temperature to cause decarboxylation and formation of *cis*-stilbene. The solid catalyst is removed by filtration, the solution is diluted with ether, and the basic quinoline is removed by extraction with aqueous hydrochloric acid. The ether solution is dried and evaporated to give crude stilbene, which is dissolved in acetic acid and brominated (with solid pyridinium hydrobromide perbromide) on mild warming. As the solution cools, a small amount of *meso*-stilbene dibromide crystallizes and is removed by filtration. The acetic acid solution is diluted with water and then extracted with ether. The ether extracts are washed with water and then alkali to remove the acetic acid. The ether is dried and evaporated to give a dark oil that crystallizes. The crystals are purified by recrystallization from methanol to give pure (*R*)- and (*S*)-stilbene dibromide.

3. *cis*-STILBENE AND STILBENE DIBROMIDE

$2CuO \cdot Cr_2O_3$
Copper chromite

Copper chromite is toxic; weigh it in the hood.

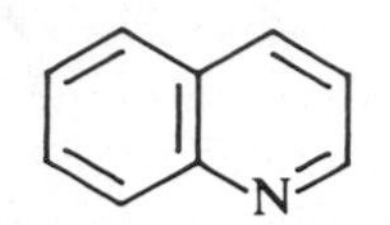

Quinoline, bp 237°C

Reaction time in the first step: 10 minutes

Carry out the procedure in the hood.

Because a trace of moisture causes troublesome spattering, the reactants and catalyst are dried prior to decarboxylation. Place 250 mg of crude, dry *E*-α-phenylcinnamic acid and 40 mg of copper chromite catalyst into a reaction tube, and add 0.3 mL of quinoline (bp 237°C; any quinoline that has darkened during storage should be redistilled over a little zinc dust) to wash down the solids. Connect to an aspirator and turn it on full force. Be sure that you have a good vacuum (use a pressure gauge) and heat the tube strongly on a steam bath with most of the rings removed, or heat on a sand bath. Heat and evacuate for 5–10 minutes to remove all traces of moisture. Then wipe the outside walls of the tube dry, insert a thermometer, clamp the tube over a hot sand bath, raise the temperature to 230°C, and note the time. Next, maintain a temperature close to 230°C for 10 minutes. Cool the yellow solution containing the suspended catalyst to 25°C, add 3 mL of *t*-butyl methyl ether, and filter the solution by gravity (use more ether for rinsing). Transfer the solution to a reaction tube and remove the quinoline by extraction twice with about 1.5 mL portions of water containing 0.35 mL of concentrated hydrochloric acid. Then shake the ethereal solution well with water containing a little sodium hydroxide solution, draw off the alkaline liquor, and acidify it. A substantial precipitate will show that decarboxylation was incomplete, in which case the starting material can be recovered and the reaction repeated. If there is only a trace of precipitate, shake the ethereal solution with saturated sodium chloride solution for preliminary drying, dry the ethereal solution over calcium chloride pellets, remove the ether from the drying agent with a Pasteur pipette, and evaporate the ether. The residual brownish oil (about 150 mg) is crude *Z*-stilbene, which contains a small amount of the *E*-isomer formed by rearrangement during heating.

The second step requires about 30 minutes.

Video: *Extraction with Ether*; Photos: *Extraction with Ether, Use of the Wilfilter*

Dissolve the crude *Z*-stilbene (e.g., 150 mg) in 1 mL of acetic acid and, in subdued light, add double the weight of pyridinium hydrobromide perbromide (e.g., 300 mg). Warm on a steam or sand bath until the reagent is dissolved; then cool under the tap and scratch to effect separation of a small crop of plates of the *meso*-dibromide. Filter the solution by suction if necessary, dilute extensively with water, and extract with ether. Wash the solution twice with water and then with saturated sodium bicarbonate solution (1.3 *M*) until neutral; shake with saturated sodium chloride solution, dry over calcium chloride pellets, and evaporate to a volume of about 1 mL. If a little more of the sparingly soluble *meso*-dibromide separates, remove it and then evaporate the remainder of the solvent. The residual (±)-dibromide is obtained as a dark oil that readily solidifies when rubbed with a stirring rod. Dissolve it in a small amount of methanol and let the solution stand to crystallize. Isolate the product on a Wilfilter. The (±)-dibromide separates as colorless prismatic plates (mp 113–114°C); the yield is about 60 mg.

Cleaning Up. Place the catalyst removed by filtration in the nonhazardous solid waste container. Treat the combined aqueous layers from all parts of the experiment containing quinoline and pyridine with a small quantity of bisulfite to destroy any bromine, and neutralize with sodium carbonate. Extract the quinoline and pyridine that are released into ligroin, and place the hexanes in the organic solvents waste container. Dilute the aqueous layer with water and flush it down the drain. After ether is allowed to evaporate from the calcium chloride pellets in the hood, the pellets can be placed in the nonhazardous solid waste container. If local regulations do not allow for the evaporation of solvents in a hood, the wet solid should be disposed in a special waste container. Methanol from the crystallization goes in the organic solvents waste container.

4. *cis*-STILBENE AND STILBENE DIBROMIDE

$2CuO \cdot Cr_2O_3$
Copper chromite

Copper chromite is toxic; weigh it in the hood.

Reaction time in the first step: 10 minutes

Because a trace of moisture causes troublesome spattering, the reactants and catalyst are dried prior to decarboxylation. The two preparations of copper chromite described in the *Journal of the American Chemical Society*[5] are both satisfactory. Place 2.5 g of crude, dry *E*-α-phenylcinnamic acid and 0.2 g of copper chromite catalyst into a 20 × 150-mm test tube, and add 3 mL of quinoline (bp 237°C; any quinoline that has darkened during storage should be redistilled over a little zinc dust) to wash down the solids. Connect with a rubber stopper to an aspirator and turn it on full force. Be sure that you have a good vacuum (check with a pressure gauge) and heat the tube strongly on a hot water bath or a sand bath. Heat and evacuate for 5–10 minutes to remove all traces of moisture. Then wipe the outside walls of the test tube dry, insert a thermometer, clamp the tube over a microburner, raise the temperature to 230°C, and note the time. Next, maintain a temperature close to 230°C for 10 minutes. Cool the yellow solution containing suspended catalyst to 25°C, add 30 mL of *t*-butyl methyl ether, and filter the solution by gravity (use more ether for rinsing). Transfer the solution to a separatory funnel and remove the quinoline by extraction twice with about 15-mL portions of water containing 3–4 mL of concentrated hydrochloric acid. Then shake the ethereal solution well with water containing a little sodium hydroxide solution, draw off the alkaline liquor, and acidify it. A substantial precipitate will show that decarboxylation was incomplete, in which case the starting material can be recovered and the reaction repeated. If there is only a trace of precipitate, shake the ethereal solution with saturated sodium chloride solution for preliminary drying, dry the ethereal solution over anhydrous calcium chloride pellets, remove the drying agent by filtration, and evaporate the ether. The residual brownish oil (1.3–1.8 g) is crude *Z*-stilbene, which contains a small amount of the *E*-isomer formed by rearrangement during heating.

Carry out the procedure in the hood.

The second step requires about 30 minutes.

Dissolve the crude *Z*-stilbene (e.g., 1.5 g) in 10 mL of acetic acid and, in subdued light, add double the weight of pyridinium hydrobromide perbromide (e.g., 3.0 g). Warm on a hot water or sand bath until the reagent is dissolved; then cool under the tap and scratch to effect separation of a small crop of plates of the *meso*-dibromide (10–20 mg). Filter the solution by suction, dilute extensively with water, and extract with *t*-butyl methyl ether. Wash the solution twice with water, and then with 5% sodium bicarbonate solution until neutral; shake with saturated sodium chloride solution, dry over calcium chloride pellets, and evaporate to a volume of about 10 mL. If a little more of the sparingly soluble *meso*-dibromide separates, remove it by gravity filtration and then evaporate the remainder of the solvent. The residual DL-dibromide is obtained as a dark oil that readily solidifies when rubbed with a stirring rod. Dissolve it in a small amount of methanol and let the solution stand to crystallize. The DL-dibromide separates as colorless prismatic plates (mp 113–114°C); the yield is about 0.6 g.

Cleaning Up. Place the catalyst removed by filtration in the nonhazardous solid waste container. Treat the combined aqueous layers from all parts of the experiment containing quinoline and pyridine with a small quantity of bisulfite to destroy any bromine and neutralize with sodium carbonate. Extract the quinoline and pyridine that are released into hexanes and place the hexanes in the organic solvents waste container. Dilute the aqueous layer with water and flush it down the drain. After the ether is allowed to evaporate from the calcium chloride pellets in the

[5] *J Am Chem Soc.* **1932**; 54:1138; *J Am Chem Soc.* **1950**; 72:2626.

hood, the pellets can be placed in the nonhazardous solid waste container. If local regulations do not allow for the evaporation of solvents in a hood, the wet solid should be disposed in a special waste container. Methanol from the crystallization goes in the organic solvents waste container.

COMPUTATIONAL CHEMISTRY

Construct *Z*- and *E*-α-phenylcinnamic acids and carry out energy minimizations. Be sure to arrange the carboxyl group to give the lowest steric energy, which you should note. Then carry out AM1 semiempirical molecular orbital heats of formation calculations for the two isomers. Do the calculations confirm that the *Z*-isomer is more stable than the *E*-isomer? Would simple inspection of the space-filling model have led to the same conclusion? Can you rationalize the absorbances in the ultraviolet (UV) spectra of the two isomers (Fig. 59.3)?

The Br-Br distances in (1*R*,2*R*)- or (1*S*,2*S*)-stilbene dibromide and (1*R*,2*S*)-stilbene dibromide are 4.50 and 3.85 nm, respectively, as determined by X-ray diffraction of the crystals. Do simple molecular mechanics energy-minimized structures give these same intramolecular distances once the global minimum is found?

QUESTIONS

1. Draw the mechanism that shows how the bromination of *E*-stilbene produces *meso*-stilbene dibromide.
2. Can nuclear magnetic resonance (^{1}H NMR) spectroscopy be used to distinguish between *meso*- and (±)-stilbene dibromide? Explain why or why not.
3. Why is *Z*-stilbene not planar?
4. Draw a Newman projection of (1*R*,2*S*)-stilbene dibromide (the *meso*-isomer).
5. Calculate the volume (measured at standard temperature and pressure) of the carbon dioxide evolved in this decarboxylation reaction.

See *Web Links*

CHAPTER 60

Multicomponent Reactions: The Aqueous Passerini Reaction

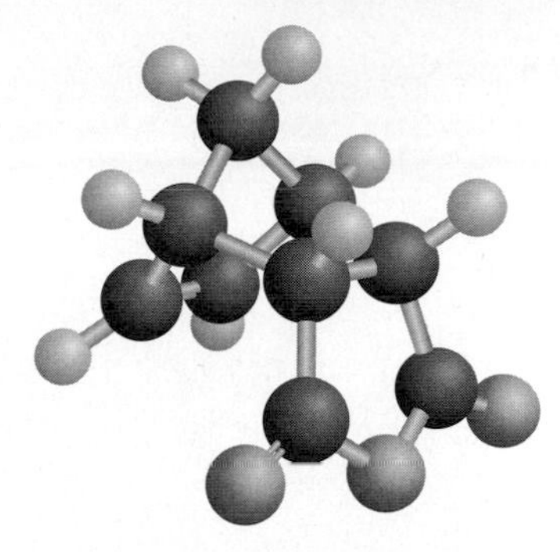

PRELAB EXERCISE: The Ugi reaction is similar to the Passerini reaction in that it also yields a product with multifunctionality and is multicomponent. What functional groups are present in an Ugi reaction product, and what starting materials are used to make this product?

One of the primary objectives of pharmaceutical organic chemistry is the synthesis of drugs useful in combatting disease. In recent years, high throughput screening, often carried out by laboratory robots, allows the rapid analysis of new chemical compounds for effectiveness against a wide variety of, for example, bacteria if a new antibiotic is sought. But the ability to analyze thousands of compounds for biological activity in a short time would be for naught if the compounds for analysis were not available.

Most of the reactions in this text involve reacting A with B to produce C. A further reaction might involve reacting C with D to give E. This is a laborious process if one wants to produce, say, 1000 different E molecules using 10 different A, B, and D molecules. A few reactions have been discovered that allow A, B, and D to be mixed together and E isolated by filtration, and where A, B, and D can have a wide variety of structures. The Passerini reaction is one of these.

Discovered in 1921 by Mario Passerini, of the University of Florence, the reaction remained unutilized for many years. It is ideally suited for the production of large numbers of molecules having different functional groups, and therefore possibly different biological activities.

A closely related reaction, the Ugi reaction, utilizes four components. Thousands of different compounds were synthesized using the reaction and one was utilized in the synthesis of Crixivan, employed as a part of the therapy for the treatment of HIV infection and AIDS. The Food and Drug Administration approved the drug a record 42 days after the application was received; it is now a $500 million-per-year product of Merck & Co.

The Passerini reaction involves the reaction between a carboxylic acid, an aldehyde or ketone, and an isocyanide. A very wide variety of these four functional groups can be used to synthesize thousands of different molecules, which with a peptide-like structure, are potentially biologically active or can serve as building blocks to make active molecules.

A few years ago, Michael Pirrung[1] discovered that this reaction could be run in water. This is very surprising, because the reactants are not completely soluble in water. He found the rate of the reaction was increased 50 times over the use of organic solvents and the yields were all greater than 90%! The exact reason for the rate enhancement is not understood.

This reaction is the epitome of green chemistry. Like the Diels-Alder reactions, all the atoms in the starting material end up in the product, so there is no waste and the solvent could not be more benign: water.

The mechanism of the Passerini reaction has been the subject of some debate over the years; it is not simple. The mechanism given below involves ionic intermediates, which are likely to form in aqueous conditions. An alternate mechanism has been proposed for aprotic conditions; it is a concerted mechanism.

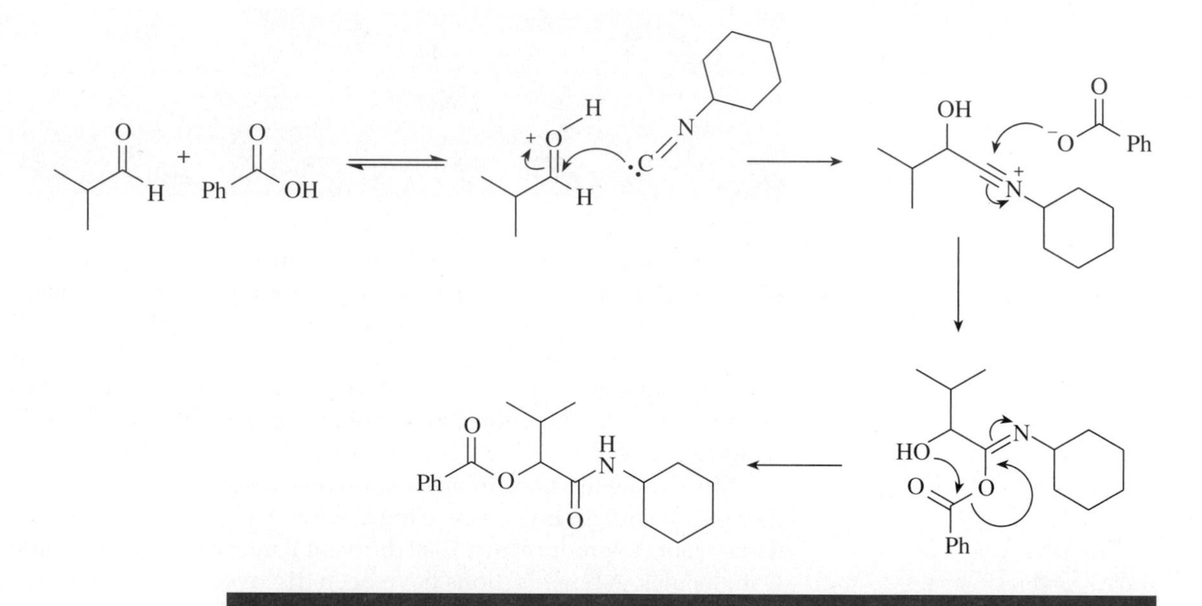

IN THIS EXPERIMENT, benzoic acid, benzaldehyde and *t*-butyl isocyanide are stirred together in aqueous solution for 1/2 hour. The product crystallizes and is isolated by filtration before recrystallization from ethanol.

THE AQUEOUS PASSERINI REACTION

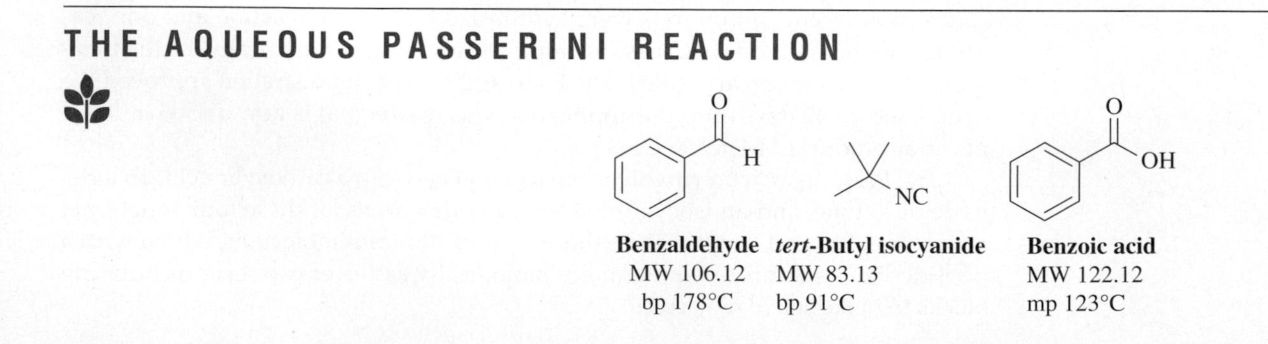

[1]a) Pirrung MC, Das Sarma K. *J Am Chem Soc.* **2003;***126*:444–445. b) Pirrung MC, Das Sarma K. *Tetrahedron.* **2005;***61*:11456–11472.

To a reaction tube, add 0.5 mmole of *t*-butyl isocyanide. This must be done in a good, working hood; the isocyanide has a particularly foul odor. Then add 0.5 mmole of benzaldehyde followed by 0.5 mmole of benzoic acid, 2 mL of water and a magnetic stirring bar. Note the appearance of the reaction mixture at initiation and during the next 25 minutes while the mixture is stirred at the highest possible speed. The product should crystallize as the reaction takes place. Isolate the ester amide by filtration on the Hirsch funnel or, better, by employing the Wilfilter.

Dissolve the product in about 3 mL of boiling ethanol, add water dropwise until cloudiness develops and crystals begin to form. Cool the mixture in ice for 30–60 minutes and then collect the product on the Hirsch funnel. Once dry, determine the melting point. It is reported to be 151–153°C. In addition, obtain an IR spectrum and ^{1}H NMR spectrum of the product and annotate each spectrum (see questions below).

This experiment lends itself to many individual research projects by simply varying the chemical nature of the three components. Apparently, water solubility slows down the reaction, so that cinnamic acid reacts more rapidly than benzoic acid.

QUESTIONS

1. In the IR spectrum, which peak corresponds to the carbonyl adjacent to the phenyl group and which to the carbonyl adjacent to the amide group?
2. In the ^{1}H NMR spectrum, identify the peaks corresponding to the *tert*-butyl group, the phenyl groups, and the proton adjacent to the carbonyl group.

61 CHAPTER

Chemiluminescence: Syntheses of Cyalume and Luminol

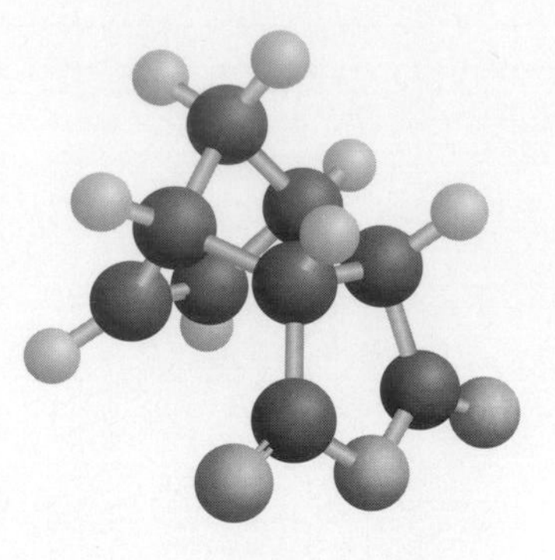

When you see this icon, sign in at this book's premium web site at **www.cengage.com/login** to access videos, Pre-Lab Exercises, and other online resources.

PRELAB EXERCISE: Write a balanced equation for the reduction of nitrophthalhydrazide to aminophthalhydrazide using sodium hydrosulfite, which is oxidized to bisulfite.

Chemiluminescence is a process whereby light is produced via a chemical reaction, with the evolution of little or no heat. The periodic flashes of the male firefly in its quest for a mate and the glow of light seen in the wake of a boat, under a rotten log, or among the many organisms found at great depths in the ocean are examples of natural chemiluminescence. The interaction of luciferin from the firefly, the enzyme luciferase, adenosine triphosphate (ATP), and molecular oxygen is the most carefully studied of these reactions. In this chapter, both Cyalume, an invention of the American Cyanamid Company, and luminol are synthesized.

"Light sticks" sold at entertainment events and as toys are solutions of Cyalume. Bending the light stick activates it (by mixing hydrogen peroxide with the Cyalume) to give off a bright glow in a variety of colors. These sticks are also used as emergency lights on life vests, as roadside markers, for night diving, on fishing lures to catch swordfish, and even in golf balls that light up at night! They are invaluable in an emergency wherever a sparkless source of light is needed, A variety of colors can be produced by allowing synthetic Cyalume to interact with a variety of fluorescers, including one you can prepare by the Wittig reaction. Luminol is not a commercial product. Luminol's chemiluminescence, the mechanism of the reaction, and peroxyoxalate chemistry are presented in several journal articles.[1]

[1]For the chemiluminescence of luminol, *see* Huntress EH, Stanley LN, Parker AS. *J Chem Educ.* **1934**;*11*:142. The mechanism of the reaction has been investigated by White EH, et al. *J Am Chem Soc.* **1964**;*86*:940–42. For the discovery of Cyalume, see Rauhut M.M. *Accnts Chem Res.* **1969**;**2**(3):80–87. Peroxyoxalate chemistry is discussed by Mohan AG, Turro NJ. *J Chem Educ.* **1974**;*51*:528.

CYALUME

Cyalumes are the most efficient nonenzyme fluorescing substances known, with quantum efficiencies of up to 26%.

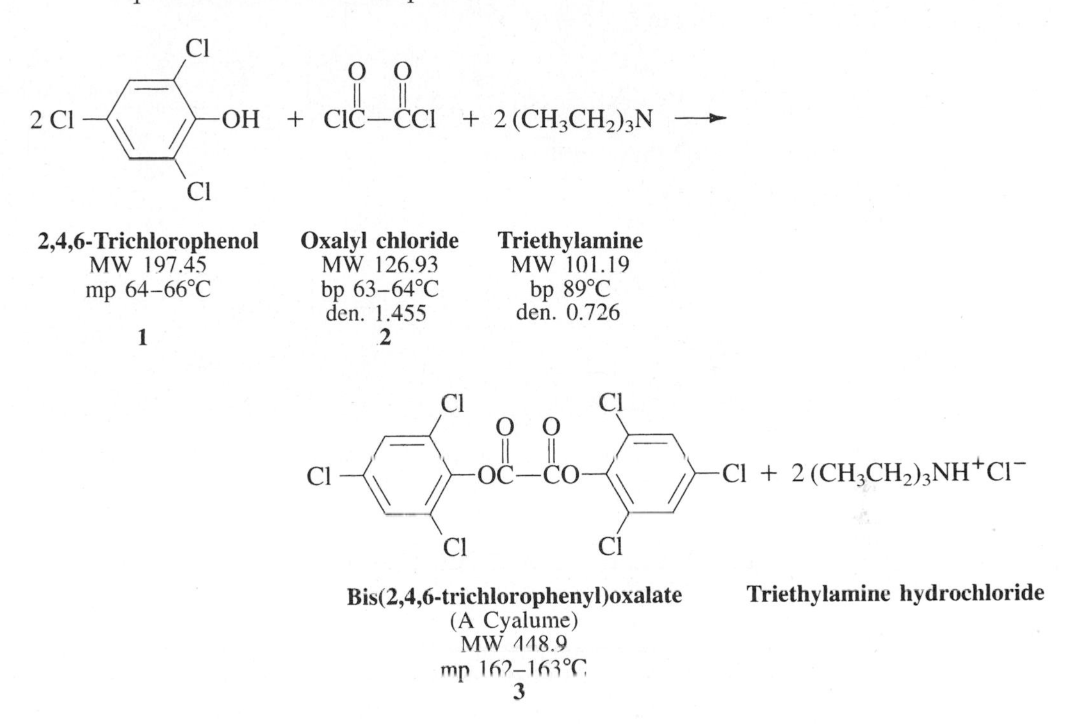

2,4,6-Trichlorophenol (1) reacts with the diacid chloride, oxalyl chloride (2), in the presence of triethylamine to give the desired oxalic acid ester (3) and the amine hydrochloride. Several different oxalic acid esters will work in this luminescent reaction.

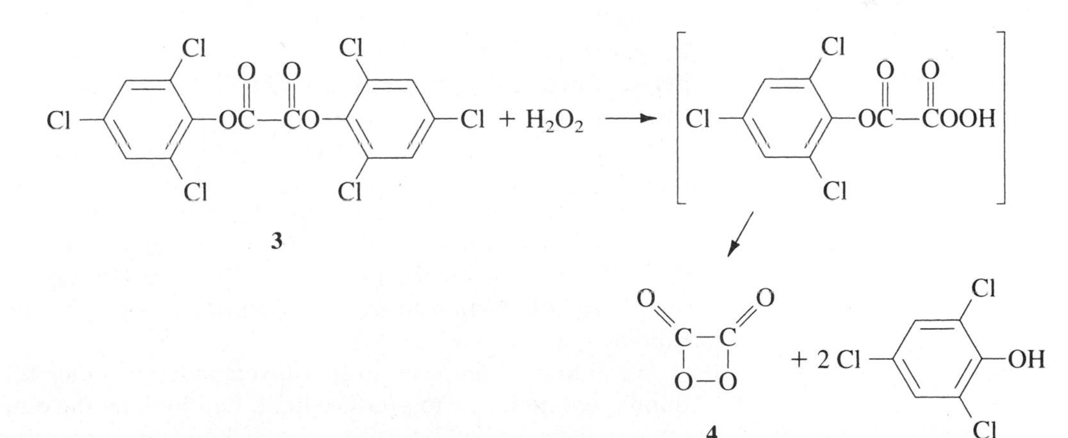

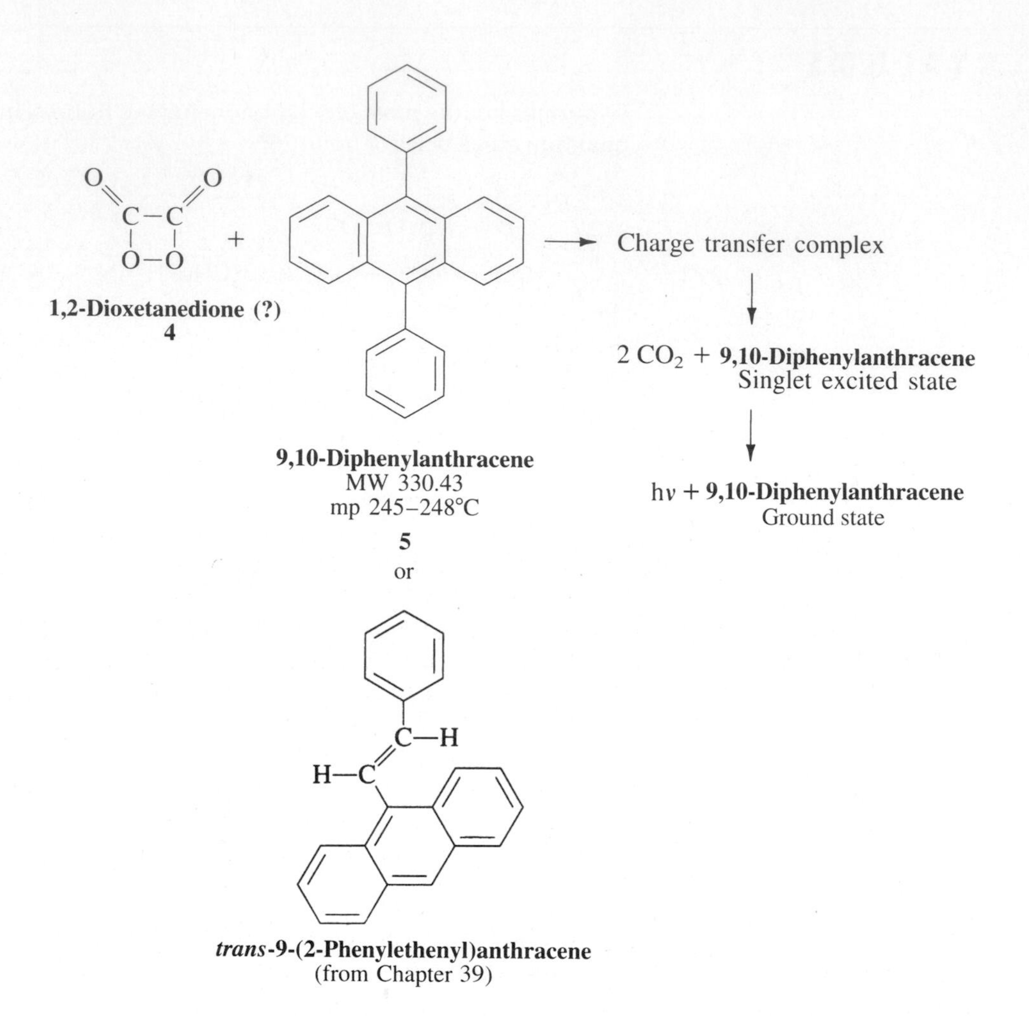

In this light-producing reaction, the ester undergoes nucleophilic attack by hydrogen peroxide to produce a peroxyoxalic acid, which, by an intramolecular displacement reaction, generates a small intermediate molecule, 1,2-dioxetanedione (4). This very unstable molecule transmits its energy to a fluorescer, such as 9,10-diphenylanthracene (5), through a charge-transfer complex that produces the excited singlet state of the fluorescer when it decomposes to two molecules of carbon dioxide. *trans*-9-(2-Phenylethenyl)anthracene, synthesized by the Wittig reaction in Chapter 39, is also an excellent fluorescer. The observed color depends on the structure of the fluorescer. Other fluorescers are 9,10-bis(phenylethynyl)anthracene that emits green light, and Rhodamine B and rubrene that both emit red light.

Cyalume decomposes to 1,2-dioxetanedione, which transfers its energy to a fluorescent molecule to produce light. Luminol, on the other hand, is itself fluorescent and gives off light from the excited state. In addition, the luminol reaction is complete in a few minutes, but the peroxyoxalate reaction can last for a day or more.

LUMINOL

The oxidation of luminol produces a striking emission of blue-green light. An alkaline solution of the compound is allowed to react with a mixture of hydrogen peroxide and potassium ferricyanide. The dianion (7) is oxidized to the triplet excited state (defined as two unpaired electrons of like spin) of the amino phthalate ion. This slowly undergoes intersystem crossing to the singlet excited state (defined as two unpaired electrons of opposite spin; (9), which decays to the ground state ion (8) with the emission of one quantum of light (a photon) per molecule. (*See* Chapter 62 for a more detailed discussion of photochemistry.)

Luminol (6) is prepared by reducing the nitro derivative (13) formed upon the thermal dehydration of a mixture of 3-nitrophthalic acid (11) and hydrazine (12). An earlier procedure for effecting the first step called for adding hydrazine sulfate to an alkaline solution of the acid, evaporating to dryness, and baking the resulting mixture of the hydrazine salt and sodium sulfate at 165°C; this process required 4.5 hours for completion. Louis Fieser reduced this working time drastically by adding high-boiling triethylene glycol (bp 290°C) to an aqueous solution of the hydrazine salt, distilling the excess water, and raising the temperature to a point where dehydration to intermediate 13 is complete within a few minutes. Nitrophthalhydrazide (3) is insoluble in dilute acid, but soluble in alkali by virtue of enolization; it is conveniently reduced to luminol (6) by sodium hydrosulfite (sodium dithionite) in an alkaline solution. In dilute, weakly acidic or neutral solutions, luminol exists largely as the dipolar ion (14), which exhibits beautiful blue fluorescence.

An alkaline solution of luminol contains the doubly enolized anion (15), and displays particularly marked chemiluminescence when oxidized with a combination of hydrogen peroxide and potassium ferricyanide.

CHEMILUMINESCENCE OF LUMINOL

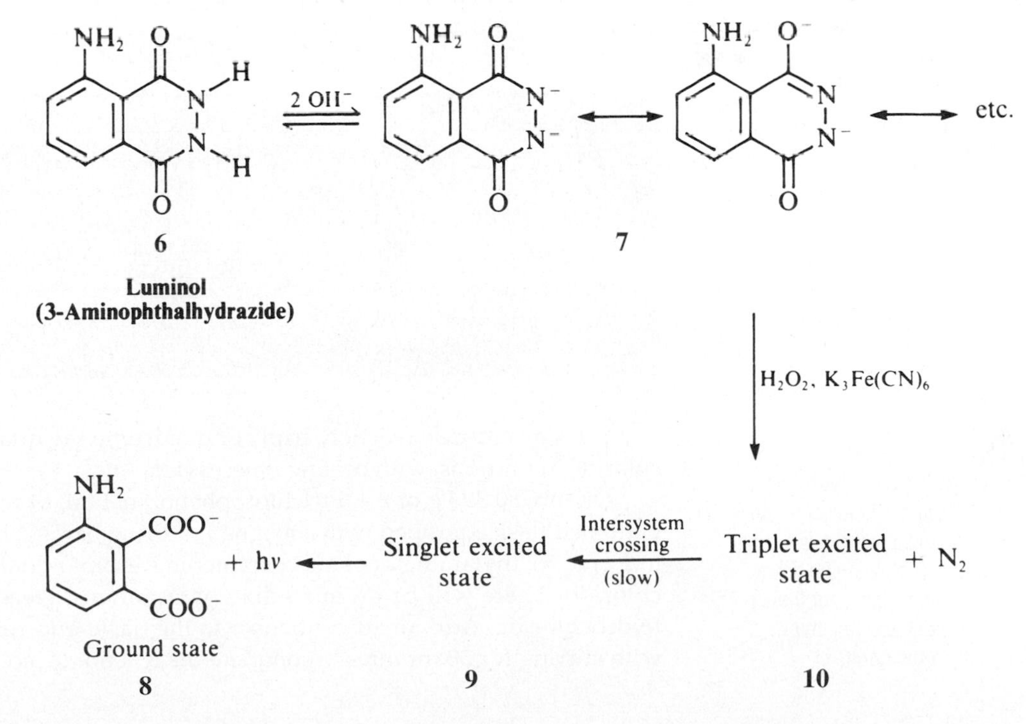

SYNTHESIS OF LUMINOL

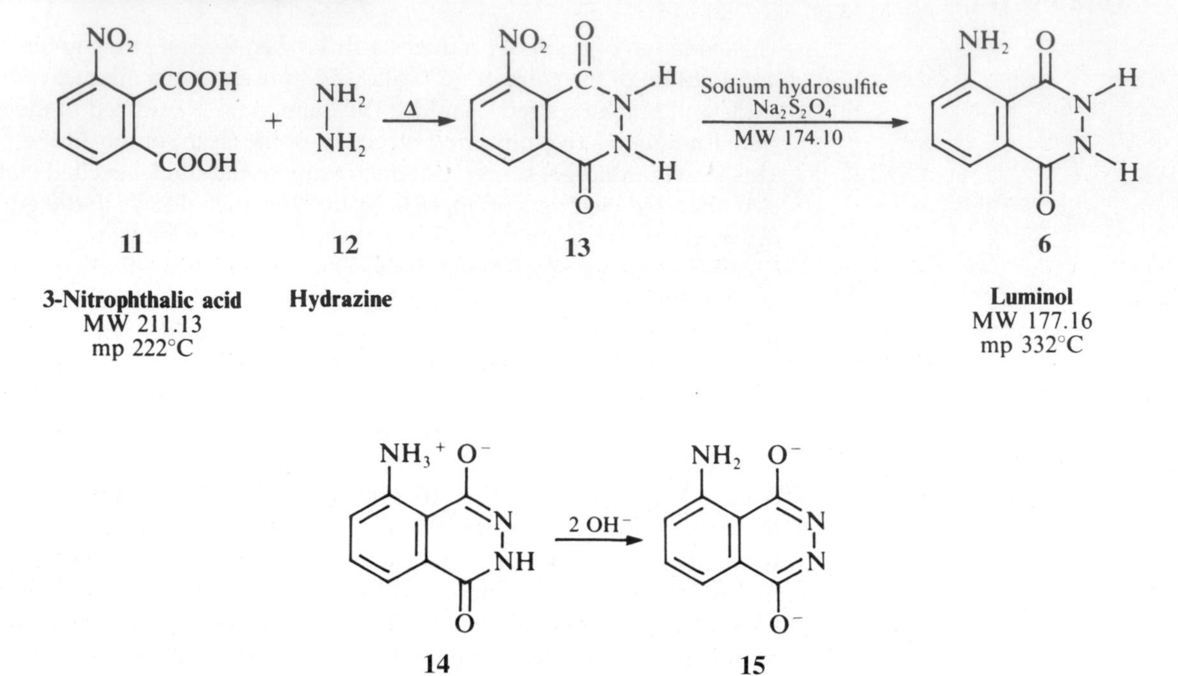

EXPERIMENTS

CYALUME

1. Synthesis of Bis(2,4,6-Trichlorophenyl)oxalate

IN THIS EXPERIMENT, a phenol dissolved in toluene is treated with the diacid chloride oxalyl chloride in the presence of an amine to give a diester. Upon cooling of the hot reaction mixture, the product (along with solid amine hydrochloride) crystallizes. The mixture of solids obtained by filtration is washed with water to remove the amine hydrochloride. The solid diester, a Cyalume, is collected by filtration, dried, and then recrystallized from toluene.

CAUTION: 2,4,6-Trichlorophenol is a carcinogen. Oxalyl chloride is corrosive and a lachrymator (tear producer). Triethylamine has a very bad odor. Carry out the measurement and transfer of reagents in the hood.

For a macroscale reaction, triple or quadruple the quantities. Add the thionyl chloride in portions, with cooling if necessary.

Dissolve 0.395 g of 2,4,6-trichlorophenol in 3 mL of toluene in a 5-mL round-bottomed flask equipped with a magnetic stirring bar. Add to this solution 0.28 mL (0.2 g) of triethylamine, cool the container in ice, and add 0.10 mL (0.14 g) of oxalyl chloride. There will be an immediate voluminous precipitate of triethylamine hydrochloride. Add an air condenser to the flask, and reflux the mixture gently with stirring for 30 minutes to complete the reaction. Cool the mixture well in ice,

Video: *Microscale Filtration on the Hirsch Funnel*

Magnetic stirring of this reaction mixture is desirable, but not mandatory. The precipiate is viscous and does not stir well.

and collect the solid on a Hirsch funnel. Complete the transfer of material with a small volume of hexane(s), which is also used to wash the solid on the filter.

Press the material as dry as possible on the filter; then suspend it in about 5–6 mL of water in a 10-mL Erlenmeyer flask. Stir the solid well with the water, using a spatula to break up any lumps; then cap the flask and shake it vigorously. In this process, the amine hydrochloride dissolves in water and leaves the product as a suspension. Empty the filtrate in the previously used filter flask into the flammable waste container; then collect the solid product from the aqueous solution on a Hirsch funnel. Wash the solid well with water and press it as dry as possible on the filter. Spread out the product to dry, determine its weight, and then recrystallize it from a minimum volume of boiling toluene. Allow the product to crystallize slowly; then cool the tube in ice. Collect the product on a Hirsch funnel [use hexane(s) to complete the transfer of material] and wash the crystals with a small volume of hexane(s). Determine the weight and melting point of the dry material and calculate the percent yield.

Cleaning Up. Flush the aqueous filtrate down the drain and discard the organic filtrates in the flammable waste container.

2. The Chemiluminescent Reaction

Dissolve 50 mg (or less, if that is all you have) of bis(2,4,6-trichloro)oxalate and about 3 mg of 9-(2-phenylethenyl)anthracene (from Chapter 39) or 9,10-diphenylanthracene in 5 mL of diethyl phthalate by warming on a steam or sand bath. In another container, shake 0.2 mL of 30% hydrogen peroxide (or, less satisfactory, 2 mL of 3% peroxide) with 5 mL of diethyl phthalate. In a darkened room, add the peroxide suspension dropwise to the solution of the oxalate ester and the fluorescer. Try cooling and warming the mixture after the initial mixing, and note the effect of temperature on the rate of a reaction. Also try adding an additional 1–2 mg of the fluorescer (9,10-diphenylanthracene is expensive) and more hydrogen peroxide.

Cleaning Up. Dispose of the mixture in the halogenated waste container.

Computational Chemistry

Using a molecular mechanics program, calculate the energy, after minimization, of the dimer of carbon dioxide, 1,2-dioxetanedione (4), the molecule that is the intermediate in Cyalume chemiluminescence. Note the bond lengths and angles. This molecule, with only six atoms, is small enough to carry out some of the more advanced calculational methods in a reasonable time frame. Using the energy-minimized structure as input, calculate the optimal geometry of 1,2-dioxetanedione using the AM1 semiempirical molecular orbital method. Again note the bond lengths and angles.

On a high-speed Apple Macintosh computer with the program Spartan,[2] this calculation takes less than 8 seconds. Using this AM1 geometry as input, again calculate the optimal geometry and heat of formation with an ab initio program. At the HF/6-31G*[3] level, this calculation can take a few minutes. The energy is given

[2]Spartan molecular modeling software is available from Wavefunction, Inc., Irvine, California.
[3]The asterisk means that the program has been improved.

by the program in hartrees (1 hartree = 365.5 kcal/mol). Note the changes in energy and in geometry as the calculational method supposedly gets better (and more lengthy). Ab initio calculations are, due to this time constraint, limited to molecules with fewer than 100 atoms. The time needed to carry out the calculation increases at more than the fourth power (n^4) of the number of atoms in the molecule.

LUMINOL

1. Synthesis of Luminol

Microscale Procedure

IN THIS EXPERIMENT, nitrophthalic acid is dissolved in an aqueous solution of hydrazine. The high-boiling solvent triethylene glycol is added, and the water is distilled off to allow the reaction to take place at a high temperature. Dilution of the reaction mixture with hot water causes the nitrophthalhydrazide to crystallize. After isolation by filtration, the damp product is reduced with sodium hydrosulfite in a basic solution. After heating to complete the reaction, the base is neutralized with acetic acid. On cooling, the product, luminol (3-aminophthalhydrazide), crystallizes and is isolated by filtration.

CAUTION: Hydrazine is a carcinogen. Handle with care. Wear gloves and carry out this experiment in the hood.

$HO(CH_2CH_2O)_3H$

Triethylene glycol
bp 290°C

Video: *Microscale Filtration on the Hirsch Funnel*

The two-step synthesis of a chemiluminescent substance can be completed in 25 minutes.

CAUTION: Contact of solid sodium hydrosulfite with moist combustible material may cause a fire.

First, heat a tube containing 3 mL of water on a steam or sand bath. Then heat a mixture of 200 mg of 3-nitrophthalic acid and 0.4 mL of an 8% aqueous solution of hydrazine (dilute 3.12 g of a commercial 64% hydrazine solution to a volume of 25 mL; use *caution*!) in a 10 × 100-mm reaction tube over a hot sand bath until the solid is dissolved. Add 0.6 mL of triethylene glycol, and clamp the tube in a vertical position above a hot sand bath. Insert a boiling chip and a thermometer, and boil the solution vigorously to distill the excess water. Intermittently remove the thermometer and replace it with an aspirator tube to facilitate this. There will be a period during which the solution will boil at 110°C and then over a 3–4-minute period, it will rise to 215°C. Lift the tube from the hot sand and, by intermittent gentle heating, maintain a temperature of 215–220°C for 2 minutes. Remove the tube, cool to about 100°C (crystals of the product often appear), add the 3 mL of hot water, cool the tube in cold water, and collect the light-yellow granular nitro compound (13) by vacuum filtration on a Hirsch funnel. The reason for adding hot water and then cooling rather than adding cold water is that the solid is then obtained in a more easily filterable form. The dry weight should be about 140 mg.

The nitro compound need not be dried and can be transferred at once, for reduction, to the uncleaned tube in which it was prepared. Add 1.0 mL of 3 *M* sodium hydroxide solution and stir with a stirring rod. To the resulting deep brown-red solution, add 0.6 g of fresh sodium hydrosulfite dihydrate (not sodium hydrogen sulfite or sodium bisulfite). Wash the solid down the walls with a little water. Heat to the boiling point, stir, and keep the mixture hot for 5 minutes, during which time some of the reduction product may separate. Then add 0.4 mL of acetic acid, cool the tube in a beaker of cold water, and stir; collect the resulting

$Na_2S_2O_4 \cdot 2\ H_2O$
Sodium hydrosulfite dihydrate (Sodium dithionite)
MW 210.15

precipitate of light-yellow luminol (6) by vacuum filtration on a Hirsch funnel. The filtrate on standing overnight usually deposits a further crop of luminol (20–40 mg).

Cleaning Up. Combine the filtrate from the first and second reactions, dilute with a few milliliters of water, neutralize with sodium carbonate, add 3 mL of household bleach (5.25% sodium hypochlorite solution), and heat the mixture to 50°C for 1 hour. This will oxidize any unreacted hydrazine and hydrosulfite. Dilute the mixture and flush it down the drain.

Macroscale Procedure

CAUTION: Hydrazine is a carcinogen. Handle with care. Wear gloves and carry out this experiment in the hood.

The two-step synthesis of a chemiluminescent substance can be completed in 25 minutes.

First, heat a flask containing 15 mL of water on a steam bath. Then heat a mixture of 1 g of 3-nitrophthalic acid and 2 mL of an 8% aqueous solution of hydrazine (dilute 3.12 g of a commercial 64% hydrazine solution to a volume of 25 mL; use caution!) in a 20 × 150-mm test tube over a Thermowell until the solid is dissolved, add 3 mL of triethylene glycol, and clamp the tube in a vertical position in a hot sand bath. Insert a thermometer, a boiling chip, and an aspirator tube connected to an aspirator and boil the solution vigorously to distill the excess water (110–130°C). Let the temperature rise rapidly (3–4 minutes) until it reaches 215°C. Remove the burner, note the time, and, by intermittent gentle heating, maintain a temperature of 215–220°C for 2 minutes. Remove the tube, cool to about 100°C (crystals of the product often appear), add the 15 mL of hot water, cool under the tap, and collect the light-yellow granular nitro compound (13). The dry weight yield is 0.7 g. The reason for adding hot water and then cooling rather than adding cold water is that the solid is then obtained in more easily filterable form.

CAUTION: Contact of solid sodium hydrosulfite with moist combustible material may cause a fire.

Video: *Microscale Filtration on the Hirsch Funnel*

The nitro compound need not be dried and can be transferred at once, for reduction, to the uncleaned test tube in which it was prepared. Add 5 mL of 3 *M* sodium hydroxide solution, and stir with a stirring rod. To the resulting deep brown-red solution, add 3 g of fresh sodium hydrosulfite dihydrate (not sodium hydrogen sulfite or sodium bisulfite). Wash the solid down the walls with a little water. Heat to the boiling point, stir, and keep the mixture hot for 5 minutes, during which time some of the reduction product may separate. Then add 2 mL of acetic acid, cool under the tap, and stir; collect the resulting precipitate of light-yellow luminol (6). The filtrate on standing overnight usually deposits a further crop of luminol (0.1–0.2 g).

Cleaning Up. Combine the filtrate from the first and second reactions, dilute with a few milliliters of water, neutralize with sodium carbonate, add 40 mL of household bleach (aqueous sodium hypochlorite solution), and heat the mixture to 50°C for 1 hour. This will oxidize any unreacted hydrazine and hydrosulfite. Dilute the mixture and flush it down the drain.

2. The Light-Producing Reaction

IN THIS EXPERIMENT, a basic solution of luminol is mixed with a solution of hydrogen peroxide in the presence of an iron catalyst to produce light.

This reaction can be run on a scale five times larger. Dissolve the first crop of moist luminol (dry weight is about 40–60 mg) in 2 mL of 3 *M* sodium hydroxide solution and 18 mL of water; this is stock solution A. Prepare a second stock solution (B) by mixing 4 mL of 3% aqueous potassium ferricyanide, 4 mL of 3% hydrogen peroxide, and 32 mL of water. Next, dilute 5 mL of solution A with 35 mL of water and, in a dark place, pour this solution and solution B simultaneously into an Erlenmeyer flask. Swirl the flask and, to increase the brilliance, gradually add further small quantities of sodium hydroxide and ferricyanide crystals.

Ultrasonic sound also can be used to promote this reaction. Prepare stock solutions A and B again, but omit the hydrogen peroxide. Place the combined solutions in an ultrasonic cleaning bath (sonicator), or immerse an ultrasonic probe into the reaction mixture. Spots of light are seen where the ultrasonic vibrations produce hydroxyl radicals.

The sanguinary minded can mix solutions A and B, omitting the ferricyanide from solution B. Light can be generated by adding blood dropwise to the reaction mixture.

Cleaning Up. Add 2 mL of 3 *M* hydrochloric acid, dilute the solution with water, and flush the mixture down the drain.

CHAPTER 62

Photochemistry: The Synthesis of Benzopinacol

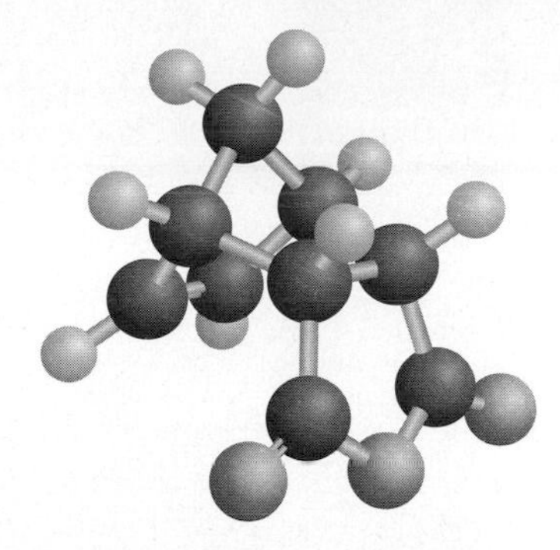

PRELAB EXERCISE: Draw a mechanism for the base-catalyzed cleavage of benzopinacol.

Photochemistry, as the name implies, is the chemistry of reactions initiated by light. A molecule can absorb light energy and then undergo isomerization, fragmentation, rearrangement, dimerization, or hydrogen atom abstraction. The last reaction, hydrogen atom abstraction, is the subject of this chapter. On irradiation with sunlight, benzophenone abstracts a proton from the solvent 2-propanol and becomes reduced to benzopinacol.

$$2\,C_6H_5\overset{\overset{\displaystyle O}{\|}}{C}C_6H_5 \;+\; CH_3\overset{\overset{\displaystyle OH}{|}}{C}HCH_3 \;\xrightarrow{h\nu}\; C_6H_5-\underset{\underset{\displaystyle C_6H_5}{|}}{\overset{\overset{\displaystyle OH}{|}}{C}}-\underset{\underset{\displaystyle C_6H_5}{|}}{\overset{\overset{\displaystyle OH}{|}}{C}}-C_6H_5 \;+\; CH_3\overset{\overset{\displaystyle O}{\|}}{C}CH_3$$

Benzophenone	2-Propanol	Benzopinacol	Acetone
MW 182.21	MW 60.09	MW 366.44	MW 58.08
mp 48°C	bp 82°C	mp 189°C	bp 56°C
	n_D^{20} 1.3770		n_D^{20} 1.3590

The energy of light varies with its frequency according to the following equation:

$$\Delta E = hv = h\left(\frac{c}{\lambda}\right)$$

where h is Planck's constant, v is the frequency of the light, λ is its wavelength, and c is the speed of light. The fact that benzophenone is colorless means it does not absorb visible light, yet irradiation of benzophenone in alcohol results in a chemical change. Pyrex glass is not transparent to ultraviolet (UV) light with a wavelength shorter than 290 nm, so some wavelength between 290 nm and 400 nm (the edge of the visible region) must be responsible. The UV spectrum of benzophenone indicates an absorption band centered at about 355 nm. Light of that wavelength is absorbed by benzophenone.

What, in a general sense, occurs when a molecule absorbs light? A photon is absorbed only if its energy (wavelength) corresponds exactly to the difference between two electronic energy levels in the molecule. In benzophenone, the electrons that are most loosely held and thus most easily excited are the two pairs of nonbonded electrons on the carbonyl oxygen, called the *n*-electrons.

C=Ö: ← *n*-electrons

These electrons have paired spins, C=O ↑↓ ↑↓ and only one is excited by the photon of light. The electron goes into the lowest unoccupied excited energy level, π^*. Each electronic energy level of benzophenone has within it many vibrational and rotational energy levels. An electron can reside in many of these levels in the lower electronic energy level, the ground state S_0, and it can be promoted to many vibrational and rotational energy levels in the first excited state (S_1). Hence, the UV spectrum does not appear as a single sharp peak but as a band composed of many peaks that arise from transitions between these many energy levels (Fig. 62.1).

The spin of the electron cannot change in going from the ground state S_0 to the excited singlet state S_1 (conservation of angular momentum). Once in a higher vibrational or rotational state, it can drop to S_1 by vibrational relaxation, losing some of its energy as heat. It then undergoes either fluorescence or intersystem crossing. As seen in Figure 62.1, the rate for fluorescence, a light-emitting process in which the electron returns to the ground state, is 10^4 times slower than intersystem crossing; so benzophenone does not fluoresce. The electron flips its orientation

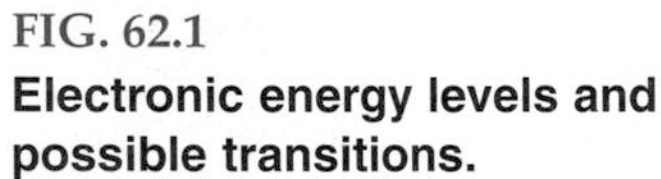
FIG. 62.1
Electronic energy levels and possible transitions.

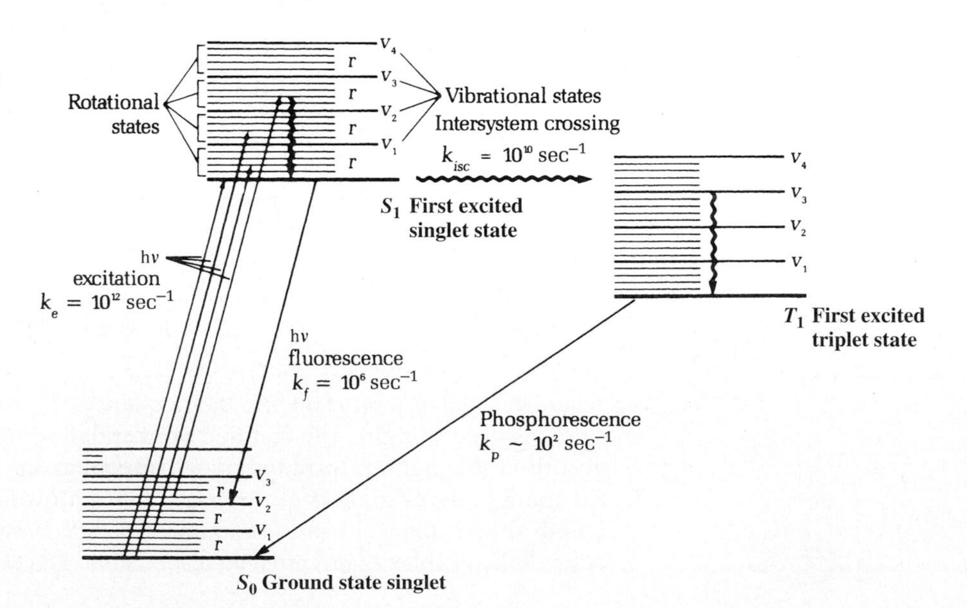

during intersystem crossing so that it has the same orientation as the electron with which it was paired in the ground state. In this new state, called the *triplet*, it can lose energy as light; but this process, phosphorescence, is a slow one. The lifetime of the triplet state is long enough for chemical reactions to take place. The triplet can also lose energy as heat in a radiationless transition, but the probability of this happening is relatively low. This situation can be represented diagrammatically:

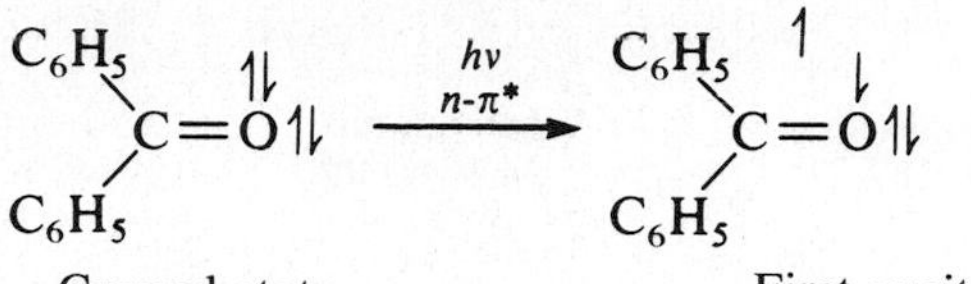

Ground state

First excited singlet state S_1

Intersystem crossing

$(C_6H_5)_2\dot{C}{=}O\cdot \equiv$ [C_6H_5, C_6H_5 / $C{=}O$ with unpaired electrons]

First excited triplet state, T_1

This T_1 state is a diradical and can abstract a methine proton from the solvent and then a hydroxyl proton to give acetone and diphenyl hydroxy radical, which dimerizes to give the product.

Removal of the methine proton

$$(C_6H_5)_2\dot{C}{=}O\cdot + H{-}C(CH_3)_2{-}O{-}H \longrightarrow (C_6H_5)_2\dot{C}{-}OH + \cdot C(CH_3)_2{-}O{-}H$$

2-Propanol

Removal of the hydroxyl proton

$$(C_6H_5)_2\dot{C}{=}O\cdot + H{-}O{-}\dot{C}(CH_3)_2 \longrightarrow (C_6H_5)_2\dot{C}{-}OH + O{=}C(CH_3)_2$$

Acetone

Dimerization

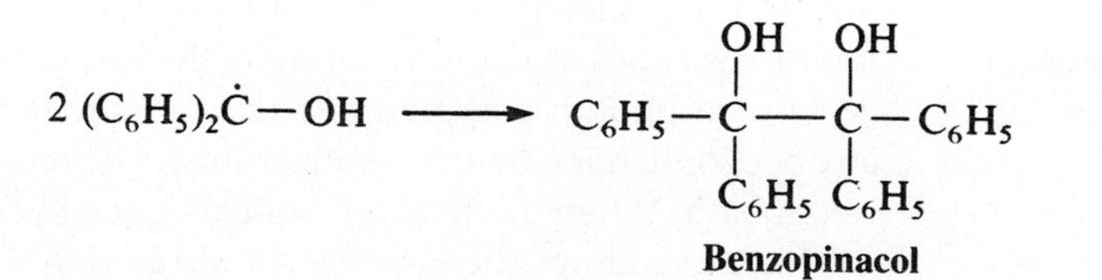

Experiments

1. Benzopinacol

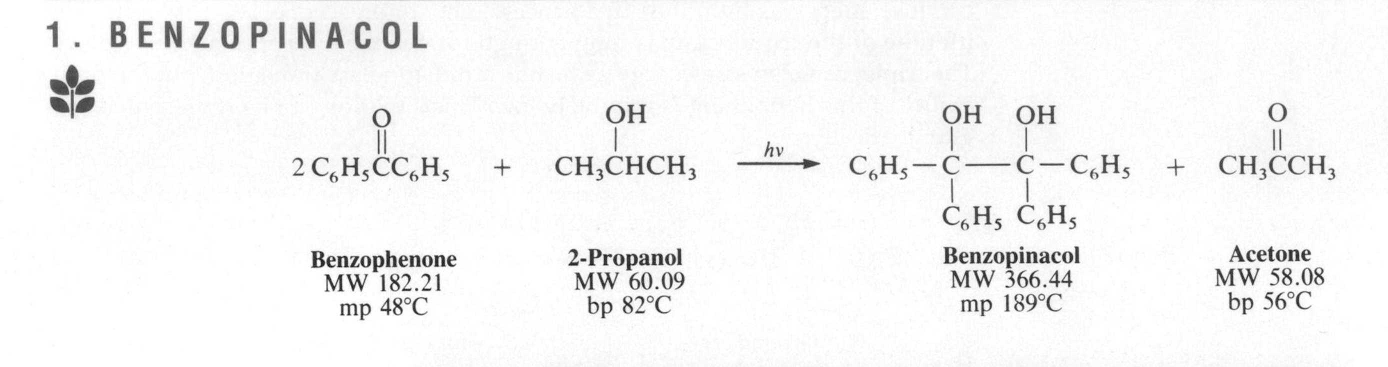

IN THIS EXPERIMENT, a tenth of a gram of benzophenone is dissolved in isopropyl alcohol in the presence of a tiny drop of acetic acid. Irradiation by the sun or some other source of UV light results in the photochemical synthesis of benzopinacol, which crystallizes from the reaction mixture.

This experiment should be done when there is good prospect for several hours of bright sun. The benzopinacol is cleaved by alkali to benzhydrol and benzophenone (Experiment 2), and it is rearranged in acid to benzopinacolone (Experiment 3).

Microscale Procedure

In a 2-mL (1-dram) ampoule, dissolve 100 mg of benzophenone in 1 mL of isopropyl alcohol by warming on a sand bath. Add a microdrop of acetic acid, cool the ampoule in dry ice, and seal the neck of the ampoule in a flame (Fig. 62.2). Set the ampoule outside in bright sunlight. On a larger scale, this experiment can take as long as a week due to absorption of UV radiation by the thick glass walls of a flask, and by self-absorption of the thick layers of solution. On a microscale, the glass of an ampoule is very thin and the thickness of the entire solution is small, so the reaction proceeds very rapidly. A rough parabolic reflector fashioned from aluminum foil can speed up the reaction even more.

Reaction time: 1 day to 1 week

Because benzopinacol is only sparingly soluble in the isopropyl alcohol, its formation is followed by the separation of small, colorless crystals (benzophenone forms large, thick prisms) from around the walls of the ampoule. If the reaction mixture is exposed to direct sunlight, the first crystals separate in less than an hour, and the reaction is complete in a day. In winter, the reaction will take longer, and any benzophenone that crystallizes must be brought into solution by warming on a steam bath. When the reaction appears to be over, chill the tube (if necessary) and collect the product. The material should be pure (mp 188–189°C). If the yield is low, more material can be obtained by further exposure of the mother liquor to sunlight.

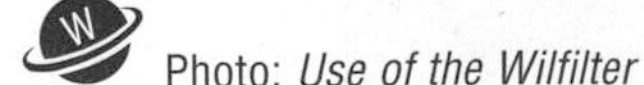
Photo: *Use of the Wilfilter*

Cleaning Up. Dilute the isopropyl alcohol filtrate with water and flush the solution down the drain. Should any unreacted benzophenone precipitate, collect it by vacuum filtration, discard the filtrate down the drain, and place the recovered solid in the nonhazardous solid waste container.

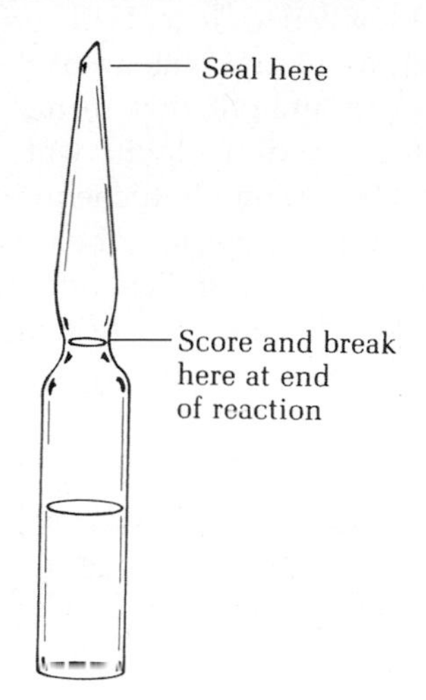

FIG. 62.2
Microscale photolysis in a 2-mL (1-dram) ampoule.

Reaction time: 4 days to 2 weeks

Macroscale Procedure

In a 50-mL round-bottomed flask, dissolve 5 g of benzophenone in 30–35 mL of isopropyl alcohol by warming on a steam bath, fill the flask to the neck with more isopropyl alcohol, and add 1 drop of glacial acetic acid. (If the acid is omitted, enough alkali may be derived from the glass of the flask to destroy the reaction product by the alkaline cleavage described in Experiment 2.) Stopper the flask with a well-rolled, tight-fitting cork or a lightly greased glass stopper, which is then wired in place. Invert the flask in a 100-mL beaker placed where the mixture will be most exposed to direct sunlight for some time. Because benzopinacol is only sparingly soluble in alcohol, its formation can be followed by the separation from around the walls of the flask of small, colorless crystals (benzophenone forms large, thick prisms). If the reaction mixture is exposed to direct sunlight, the first crystals separate in about 5 hours, and the reaction is practically complete (with 95% yield) in 4 days. In winter, the reaction may take as long as 2 weeks, and any benzophenone that crystallizes must be brought into solution by warming on a steam bath. When the reaction appears to be over, chill the flask (if necessary) and collect the product. The material should be pure (mp 188–189°C). If the yield is low, more material can be obtained by further exposure of the mother liquor to sunlight.

Cleaning Up. Dilute the isopropyl alcohol filtrate with water and flush the solution down the drain. Should any unreacted benzophenone precipitate, collect it by vacuum filtration, discard the filtrate down the drain, and place the recovered solid in the nonhazardous solid waste container.

2. ALKALINE CLEAVAGE

IN THIS EXPERIMENT, you can demonstrate that the alkaline cleavage of benzopinacol produces a low melting solid mixture that can be shown to be equal parts of an alcohol and a ketone. Alternatively, irradiation of benzophenone in an alkaline solution of benzophenone can be shown to produce only the alcohol benzhydrol.

Suspend a small test sample of benzopinacol in alcohol and heat to boiling on a steam bath, being sure that the amount of solvent is not sufficient to dissolve the solid. Add 1 drop of sodium hydroxide solution, heat for 1–2 minutes, and observe the result. The solution contains equal parts of benzhydrol and benzophenone, which are formed by the following reaction:

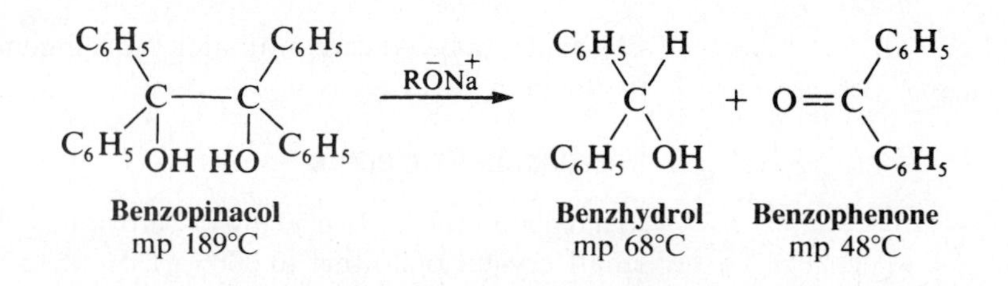

The low-melting products resulting from the cleavage are far more soluble than the starting material. Analyze the mixture by thin-layer chromatography.

Benzophenone can be converted into benzhydrol in nearly quantitative yield by following the procedure outlined for the preparation of benzopinacol (Experiment 1), modified by adding a *very* small piece of sodium (5 mg) instead of the acetic acid. The reaction is complete when, after exposure to sunlight, the greenish-blue color disappears. To obtain the benzhydrol, the solution is diluted with water, acidified, and evaporated. Benzopinacol is produced as before by photochemical reduction, but it is at once cleaved by the sodium alkoxide. The benzophenone formed by cleavage is converted into more benzopinacol, cleaved, and eventually consumed. Figures 62.3 and 62.4, presented at the end of the chapter, show the infrared (IR) and ^{1}H NMR (nuclear magnetic resonance) spectra, respectively, of benzophenone.

3. PINACOLONE REARRANGEMENT

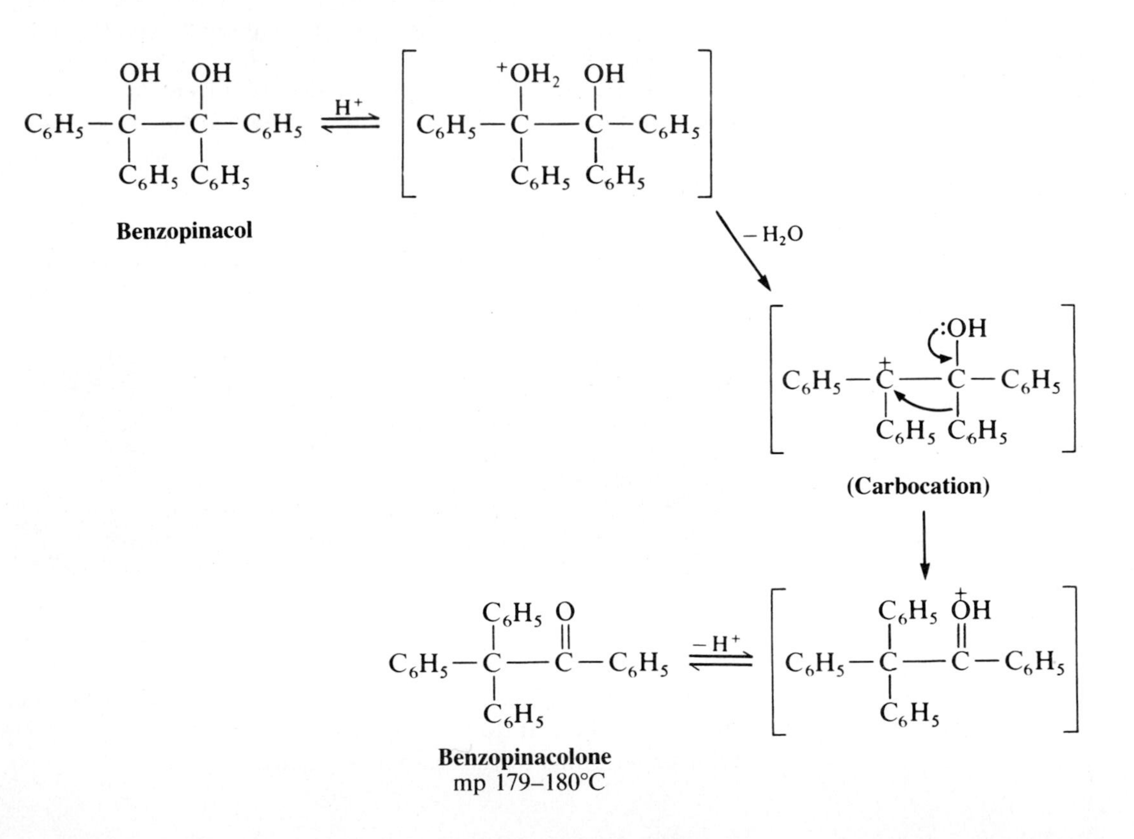

The acid-catalyzed carbonium ion rearrangement is characterized by rapidity and by the high yield.

Microscale Procedure

In a reaction tube, place 50 mg of benzopinacol, 0.25 mL of acetic acid, and one *very* small crystal of iodine (0.0005 g). Heat to boiling for 1–2 minutes on a sand bath until the crystals are dissolved; then reflux the red solution for 5 minutes.

On cooling, the pinacolone separates as a stiff paste. Thin the paste with alcohol, collect the product by vacuum filtration on a Hirsch funnel, and wash it free of iodine with cold 95% ethanol. The material should be pure; the expected yield is 95%.

Cleaning Up. Dilute the filtrate with water, neutralize with sodium carbonate, and flush down the drain.

Macroscale Procedure

In a 100-mL round-bottomed flask, place 5 g of benzopinacol, 25 mL of acetic acid, and two or three very small crystals of iodine (0.05 g). Heat to the boiling point for 1–2 minutes under a reflux condenser until the crystals are dissolved; then reflux the red solution for 5 minutes. On cooling, the pinacolone separates as a stiff paste. Thin the paste with alcohol, collect the product, and wash it free from iodine with alcohol. The material should be pure, with a yield of 95%.

Cleaning Up. Dilute the filtrate with water, neutralize with sodium carbonate, and flush down the drain.

QUESTIONS

1. Would the desired reaction occur if ethanol or *t*-butyl alcohol were used instead of isopropyl alcohol in the attempted photochemical dimerization of benzophenone?
2. Assign the peak at 1639 cm^{-1} in the IR spectrum of benzophenone (Fig. 62.3).

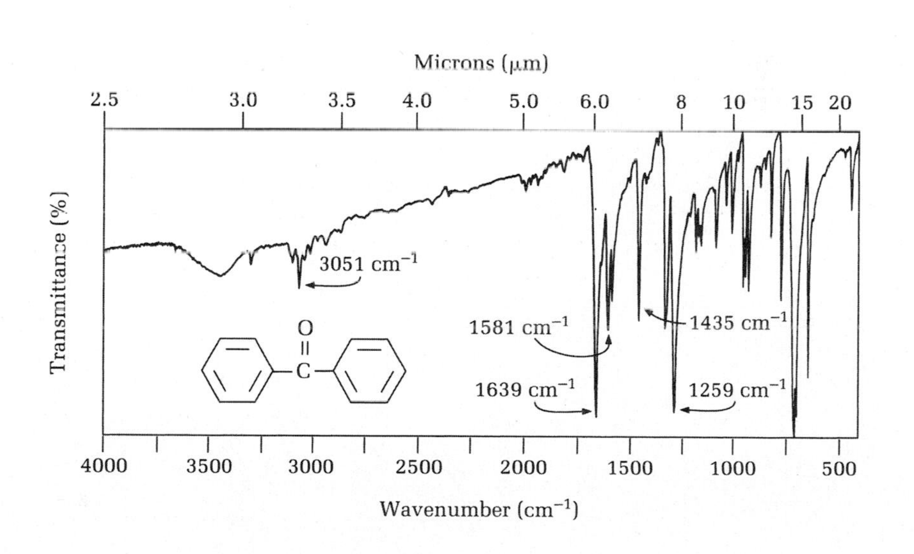

FIG. 62.3
The IR spectrum of benzophenone (KBr disk).

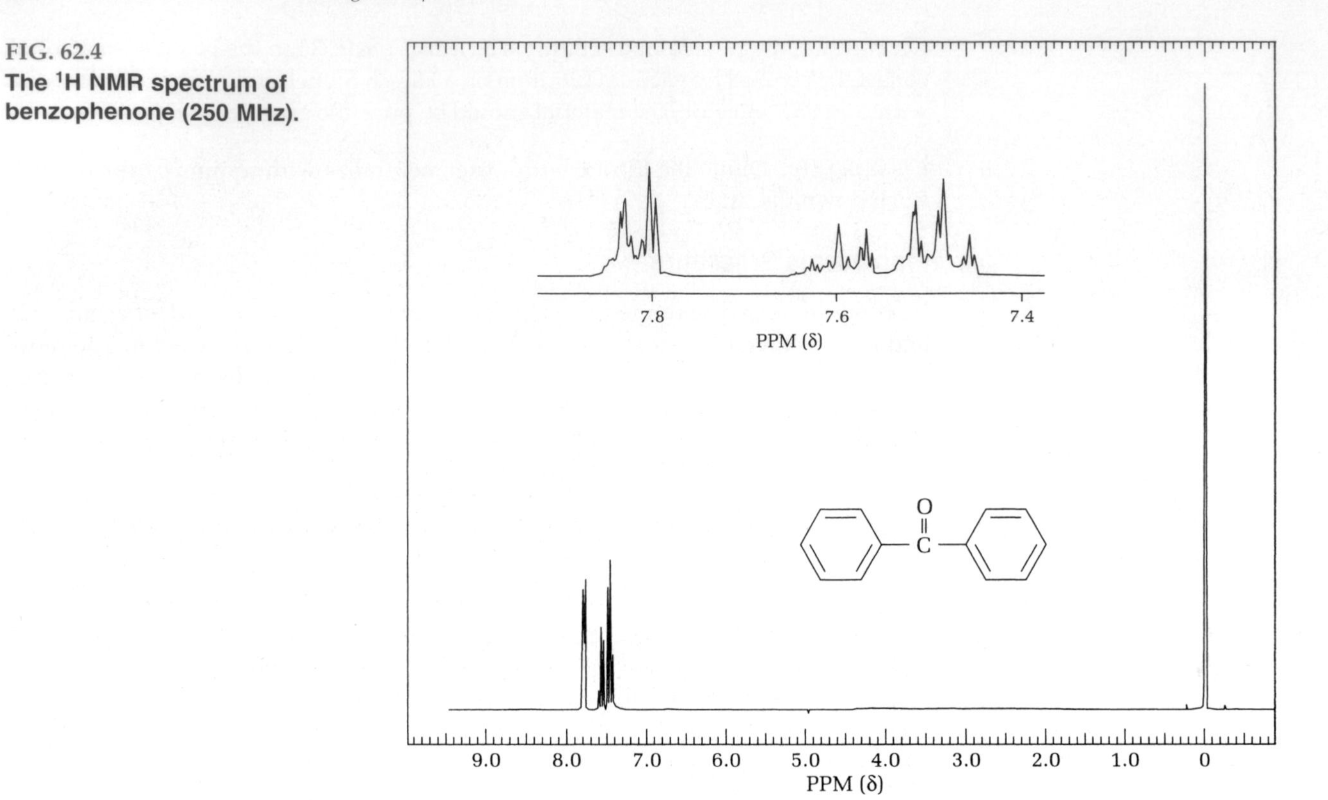

FIG. 62.4
The ¹H NMR spectrum of benzophenone (250 MHz).

CHAPTER 63

Carbohydrates and Sweeteners

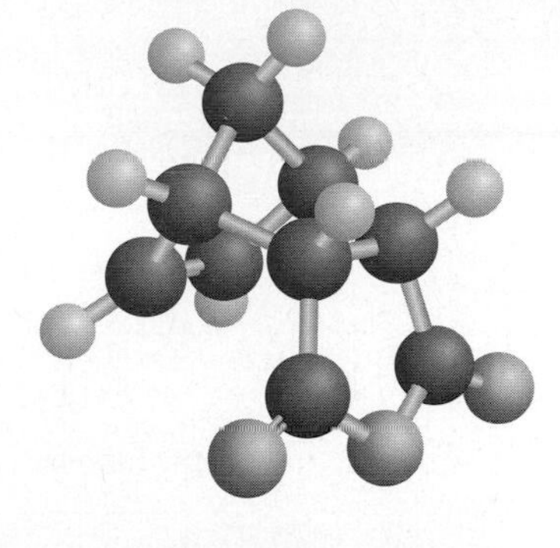

When you see this icon, sign in at this book's premium website at **www.cengage.com/login** to access videos, Pre-Lab Exercises and other online resources.

PRELAB EXERCISE: Draw a flow sheet showing the sequence of tests you would conduct to identify an unknown carbohydrate.

Carbohydrates, the direct product of the photosynthetic combination of carbon dioxide and water, are, by weight, the most common organic compounds on the earth. Because most have the empirical formula $C_nH_{2n}O_n = C_n(H_2O)_n$, it is easy to see why early scientists considered these molecules to be hydrates of carbon. They are polyhydroxyaldehydes and ketones, and exist as hemi- and full acetals and ketals. Carbohydrates are classified as mono-, di-, and polysaccharides.

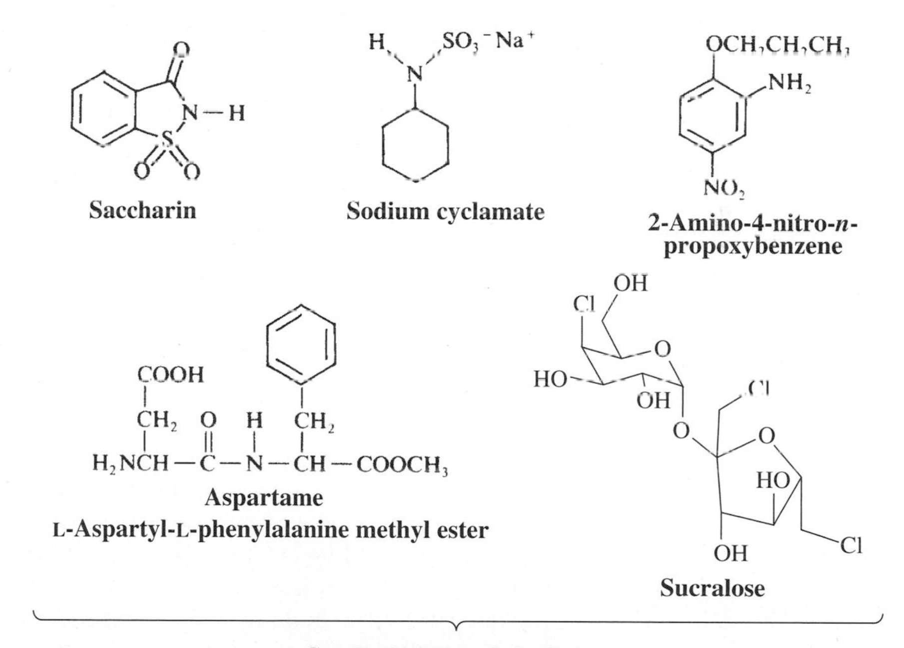

Synthetic Sugar Substitutes

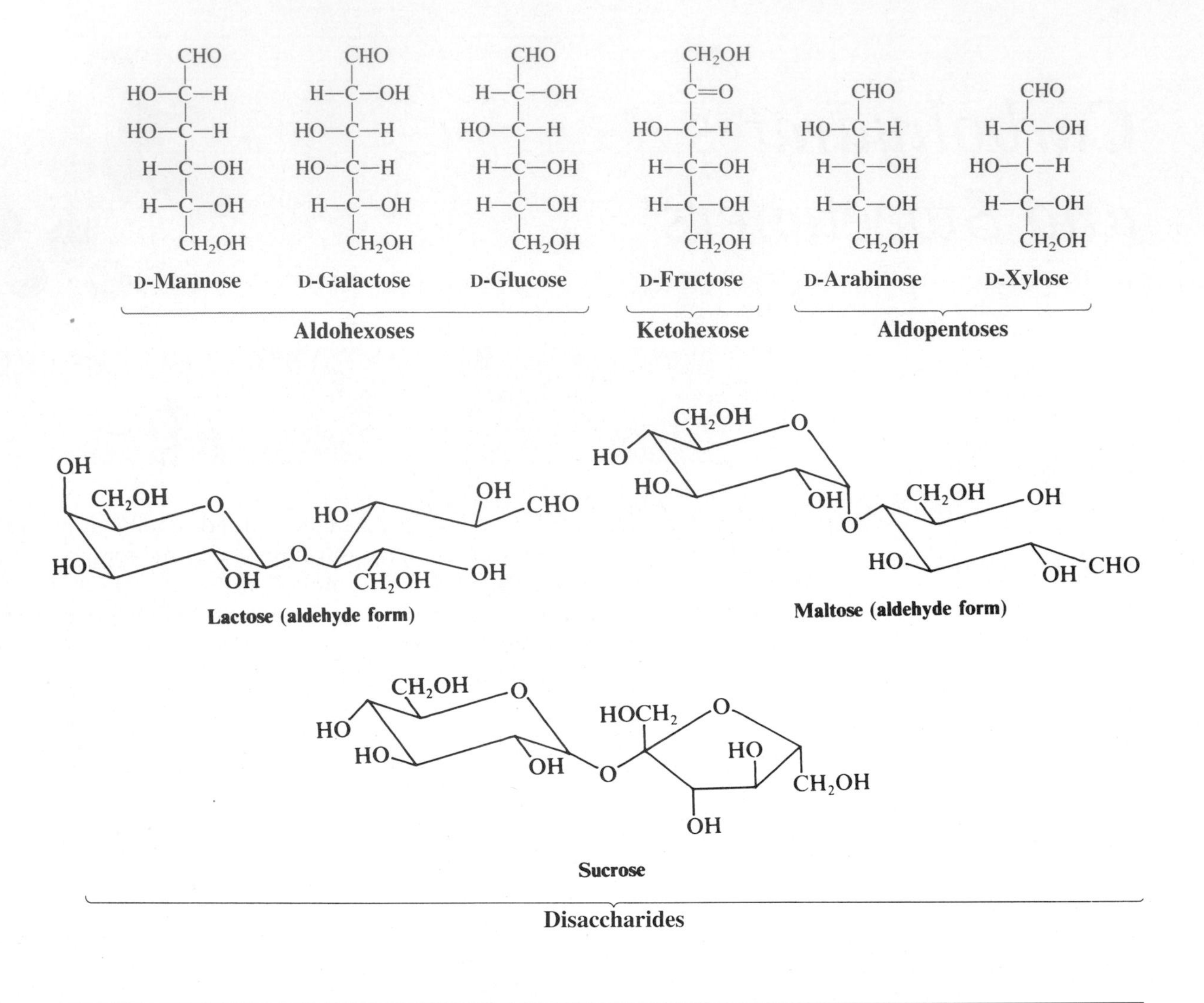

SWEETENERS

The term *sugar* applies to mono-, di-, and oligosaccharides, which are all soluble in water and thereby distinguished from polysaccharides like starch and cellulose, which are not water soluble. Many natural sugars are sweet, but the data in Table 63.1 show that sweetness varies greatly with stereochemical configuration and is exhibited by compounds of widely differing structural type. Note that D-fructose is 50% sweeter than sucrose, which accounts for the presence of high-fructose corn syrup in many commercial foods and drinks.

The synthesis and sale of artificial, synthetic sweeteners constitute a multibillion-dollar industry. Aspartame is one of the most widely used, but it is not stable to heat and does not have a particularly long shelf life in soft drinks. It is the methyl ester of a phenylalanine/aspartic acid dipeptide. The body will metabolize it into phenylalanine, and hence aspartame carries a warning for those suffering from phenylketonuria. Sucralose, about three times sweeter than Aspartame, is stable to heat and can thus be used in baked goods. It is manufactured by selective chlorination of sucrose. Truvia, a naturally occurring sweetener from the stevia plant, is a

TABLE 63.1 *Relative Sweetness of Sugars and Sugar Substitutes*

	Sweetness	
Compound	***To Humans***	***To Bees***
Monosaccharides		
D-Fructose	1.50	
D-Glucose	0.55	
D-Mannose	Sweet, then bitter	–[a]
D-Galactose	0.55	–
D-Arabinose	0.70	–
D-Xylose	Very sweet	
Disaccharides		
Sucrose (glucose, fructose)	1.0	
Maltose (2 glucose)	0.3	
Lactose (glucose, galactose)	0.2	–
Cellobiose (2 glucose)	Indifferent	–
Gentiobiose (2 glucose)	Bitter	–
Synthetic sugar substitutes		
Sodium cyclamate	~40	
Aspartame (Nutraweet)	180	
Steviol (Truvia)	350	
Saccharin (Sweet'N Low)	550	
Sucralose	600	
Perillartine	2,000	
Neotame	~10,000	

[a]–means that it is sweeter to humans than to bees

glycoside of the carboxy alcohol, steviol. It is 350 times sweeter than sugar. Perillartine is the oxime of the terpene called perillaldehyde that can be isolated from a number of plants. It is 2000 times sweeter than sucrose and is, curiously enough, used primarily in Japan. A new sweetener, not widely used yet, is Neotame, which is 7,000–13,000 times as sweet as sucrose.

The four synthetic sugar substitutes in Table 63.1 have been approved by the Food and Drug Administration in the United States, and most of the other countries in the world. But approval has been a contentious process both here and abroad. Sodium cyclamate has been banned in the United States since 1969, although it is allowed in Canada. By feeding rats the equivalent of 350 cans of soda per day in a human, sodium cyclamate caused bladder cancer in rats. Saccharin was removed from the market at one time in the United States before being reinstated; it is still banned in Canada. Aspartame was one of the most studied substances ever to receive FDA approval.

All of these sweeteners were discovered by accident. The chemist working with sodium cyclamate noticed that his cigarette (!) tasted very sweet after he had laid it on his lab bench! This says a lot about how careless chemists have been in the past (or how many undiscovered sweeteners there might be out there!). The exact physiological basis for sweetness, and therefore the relationship between organic structure and sweetness, is not understood.

One gram of sucrose (common table sugar) dissolves in 0.5 mL of water at 25°C and in 0.2 mL at the boiling point, but the substance has marked, atypical crystallizing properties. Despite the high solubility, it can be obtained in beautiful, large crystals (rock candy). More typical sugars are obtainable in crystalline form only with difficulty, particularly in the presence of a trace of impurity, and even then, only small, ill-formed crystals can be acquired. Alcohol is often added to a water solution to decrease solubility and thus to induce crystallization. The amounts of 95% ethanol required to dissolve 1-g samples at 25°C are 170 mL for sucrose, 60 mL for glucose, and 15 mL for fructose. Some sugars have never been obtained in crystalline condition and are known only as *viscous syrups*.

ANALYSIS OF CARBOHYDRATES

In this chapter, unknown carbohydrates will be identified using a series of classification tests. The Molisch test (Experiment 1) is a general test for carbohydrates. If the test is positive and the unknown is water soluble (which distinguishes it from polysaccharides such as cellulose), then it can be tested with iodine, which will give a positive test with the partially water-soluble polysaccharide, starch.

If the unknown is a sugar, it can be a mono-, di-, or, less commonly, an oligosaccharide. If it is a monosaccharide, it usually has either five or six carbons, and its carbonyl group can be either an aldehyde or a ketone.

To distinguish aldehydosugars from ketosugars, take advantage of the reducing ability of the aldehyde group. Either the traditional Fehling's or Benedict's test will give a positive reaction with any reducing sugar; in Experiment 2, we will employ the far more sensitive red tetrazolium (RT) test. If the RT test is positive, then Barfoed's test (Experiment 3) allows us to make a distinction between reducing mono- and disaccharides.

To distinguish between pentoses and hexoses, run Bial's test (Experiment 4). Seliwanov's test (Experiment 5) allows us to make a distinction between aldohexoses and ketohexoses.

Osazones

With phenylhydrazine, many sugars form beautiful crystalline derivatives called osazones (Experiment 6). Osazones are far less soluble in water than their parent sugars because the molecular weight is increased by 178 units and the number of hydroxyl groups is reduced by 1. It is easier to isolate an osazone than to isolate the sugar, and syrupy sugars often give crystalline osazones. Osazones of the more highly hydroxylic disaccharides are notably more soluble than those of monosaccharides.

Some disaccharides do not form osazones, but a test for the formation or nonformation of the osazone is ambiguous because the glycosidic linkage may suffer hydrolysis in a boiling solution of phenylhydrazine and acetic acid, with formation of an osazone derived from a component sugar and not from the disaccharide. If a sugar has reducing properties (a positive RT test), it is also capable of osazone formation; hence, an unknown sugar should be tested for reducing properties before preparation of an osazone is attempted.

EXPERIMENTS

1. MOLISCH TEST FOR CARBOHYDRATES

IN THIS EXPERIMENT, the Molisch reagent is mixed with a dilute solution of a suspected carbohydrate. If the careful introduction of concentrated sulfuric acid produces a purple color at the interface, a carbohydrate is present.

CAUTION: Handle concentrated sulfuric acid with care. It is very corrosive. Do not mix the layers in the tube because the heat of mixing can form steam, thus causing the solution to erupt from the tube. Use great care in pouring the reaction mixture into a beaker for disposal.

Label several reaction tubes or 10 × 75-mm culture tubes. To 1 mL of a 1% solution of the carbohydrates in Table 63.1 (or a representative selection chosen by your instructor), add 1 drop of Molisch reagent (a solution of 1.25 g of 1-naphthol in 25 mL of ethanol) and mix thoroughly. Add 1 mL of water in another tube as a control. Then, while tilting each tube, carefully introduce 1 mL of concentrated sulfuric acid down the inside walls. The dense sulfuric acid will form a lower layer. Note the appearance of each tube. Do not mix the layers. A purple color at the interface between the two layers is a positive test for carbohydrates.

Cleaning Up. Pour the contents of each tube into a large beaker (caution!) and neutralize with sodium bicarbonate. Dilute the neutral solution with water and flush down the drain.

2. RED TETRAZOLIUM

Red tetrazolium,[1] also known as 2,3,5-triphenyl-2H-tetrazolium chloride, is a nearly colorless, water-soluble, light-sensitive substance that oxidizes aldoses and ketoses, as well as other α-ketols, and is thereby reduced. Freshly prepared aqueous solutions should be used in tests. Any unused solution should be acidified and discarded. The reduced form is a water-insoluble, intensely colored pigment—a diformazan.

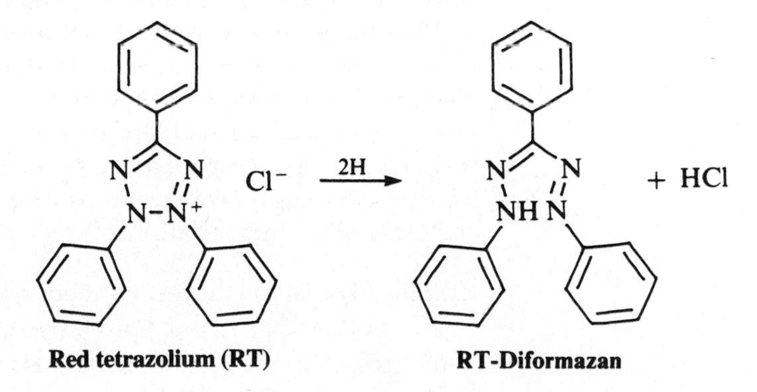

[1]Red tetrazolium is available from Aldrich Chemical Company.

A sensitive test for reducing sugars

RT affords a highly sensitive test for reducing sugars because it distinguishes between α-ketols and simple aldehydes more sharply than Fehling's and Tollens' tests, two older tests that were used for this purpose.

Put 1 drop of each of the 1% test solutions in clean, marked test tubes and, to each tube, add 1 mL of a 0.5% aqueous solution of red tetrazolium (made by dissolving 0.5 g of red tetrazolium in 100 mL of water), and 1 drop of 3 *M* sodium hydroxide solution. Put the tubes in a beaker of hot water, note the time, then note the time at which each tube develops color.

To estimate the sensitivity of the test, use the substance that you regard as the most reactive of the five studied. Dilute 1 mL of the 1% solution with water to a volume of 100 mL and run a test with red tetrazolium on 0.2 mL of the diluted solution.

Cleaning Up. Dilute the test solutions with water and flush down the drain.

3. BARFOED'S TEST

Barfoed's test is used to distinguish between reducing mono- and disaccharides. In clean, labeled reaction tubes or 10 × 75-mm culture tubes, place 1 mL of 1% solutions of a representative selection of reducing mono- and disaccharide solutions chosen from Table 63.1. To each, add 0.5 mL of Barfoed's reagent (made by dissolving 6.7 g of copper(II) acetate and 0.9 mL of acetic acid in 100 mL of water) and mix. Heat the tubes in a hot water bath for 15 minutes. Note the time needed for a red precipitate to appear in each tube.

Cleaning Up. Dilute the test solutions with water and flush down the drain.

4. BIAL'S TEST

Bial's test is used to distinguish between pentoses and hexoses. In clean, labeled reaction tubes or 10 × 75-mm culture tubes, place 1 mL of 1% solutions of a representative selection of pentoses and hexoses chosen from Table 63.1. In one tube, place 1 mL of distilled water to serve as a reference or control solution.

To each tube, add 1 mL of Bial's reagent and a boiling chip. (To make Bial's reagent: dissolve 300 mg of 3,5-dihydroxytoluene in 100 mL of concentrated hydrochloric acid; then add 0.3 mL of a 10% iron(III) chloride solution.) Heat each tube to boiling on a sand bath and note the color that develops. In tubes that have weak or no color, the product of the reaction can be concentrated by diluting the mixture with 2 mL of water, adding 0.5 mL of cyclohexanol, and then extracting the colored product into the alcohol layer by shaking the tube vigorously.

Cleaning Up. Dilute the test solutions with water and flush down the drain.

5. SELIWANOV'S TEST

Seliwanov's test is used to distinguish between aldohexoses and ketohexoses. To a number of reaction tubes or 10 × 75-mm culture tubes, add 1 mL of the Seliwanov reagent, which is prepared by dissolving 50 mg of 1,3-dihydroxybenzene (resorcinol) in a mixture of 33 mL of concentrated hydrochloric acid and 67 mL of distilled water. To each labeled tube, add 5 drops of a 1% solution of each of the sugars to be tested. After mixing, heat the tubes in a boiling water bath for 2 minutes. Examine the tubes. The intensity of the color is proportional to the α-ketol concentration.

Cleaning Up. Dilute the test solutions with water and flush down the drain.

6. OSAZONES

CAUTION: Conduct this procedure in the hood. Avoid skin contact with phenylhydrazine.

This experiment can be carried out on a scale two or three times larger than what is specified here.

Put 0.33-mL portions of phenylhydrazine reagent (made by neutralizing phenylhydrazine (3 mL) with 9 mL of acetic acid in a 50-mL Erlenmeyer flask, adding 15 mL of water, transferring the mixture to a graduated cylinder, and making up the volume to 30 mL) into each of four cleaned, numbered reaction tubes. Add 1-mL portions of 1% solutions of glucose, fructose, lactose, and maltose, and heat the tubes in the beaker of hot water for 20 minutes. Shake or flick the tubes occasionally to relieve supersaturation. Note the times at which the osazones separate. If, after 20 minutes, no product has separated, cool and scratch the test tube to induce crystallization.

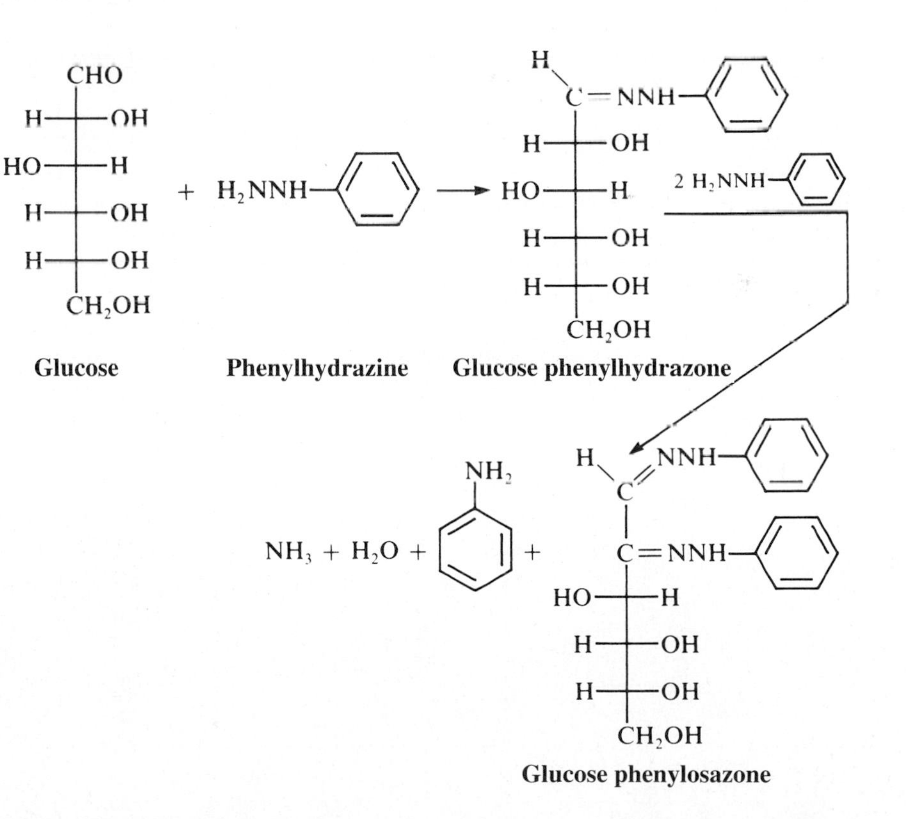

Collect and save the products for possible use in the later identification of unknowns. Because osazones melt with decomposition, heat the bath at a standard rate (0.5°C/s) when determining the melting point.

Cleaning Up. The filtrate can be diluted with water and flushed down the drain. If you must destroy the phenylhydrazine, neutralize the solution and add 2 mL of household bleach (aqueous sodium hypochlorite solution) for each 1 mL of the reagent. Heat the mixture to 45–50°C for 2 hours to oxidize the amine; then cool the mixture and flush it down the drain.

QUESTIONS

For Additional Experiments, sign in at this book's premium website at **www.cengage.com/login**.

1. What is the order of relative reactivity in the **RT** test of the compounds studied?
2. Which test do you regard as the most reliable for distinguishing reducing from nonreducing sugars? Which is most reliable for differentiating an α-ketol from a simple aldehyde?
3. Write a mechanism for the acid-catalyzed hydrolysis of the disaccharide sucrose. Take care to draw the stereochemistry of the products.
4. Will the anomeric forms of glucose give different phenylosazones?

CHAPTER 64

Virstatin, a Possible Treatment for Cholera

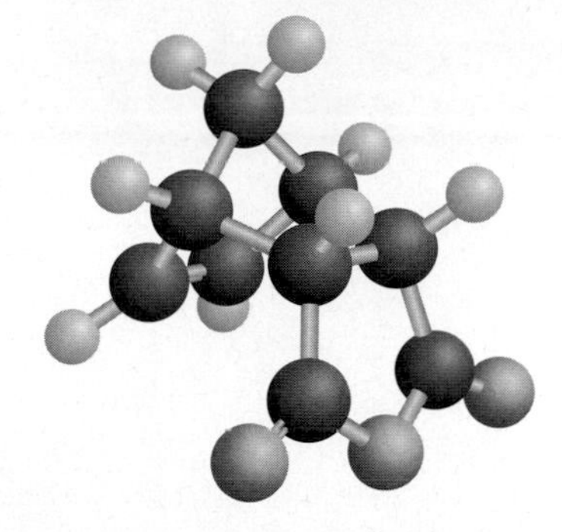

PRELAB EXERCISE: What is the purpose of adding 3 *M* HCl (aq) to the cooled reaction mixture of the hydrolysis step (step 1)?

Cholera is a terrible disease, endemic in much of Africa, and the cause of many millions of deaths in historic times. It is a water-borne disease, the result of poor sanitation and sanitary practices. Eating food and drinking water infected with the gram-negative bacterium *Vibrio cholerae* is the primary means of transmission. No other common disease kills as quickly. Upon infection, death can occur in as little as three hours in severe cases. More commonly, from the onset of exhaustive diarrhea to shock can take 4 to 12 hours, with death in 18 hours to several days when untreated. Loss of fluids can be at the rate of one liter per hour.

The most effective treatment for cholera is by oral rehydration to replace the loss of water and electrolytes; antibiotics merely shorten the course of the disease. It can be prevented by municipal water treatment and the construction of proper water supplies, and so is no longer a health threat in the first world.

A major problem with modern antibiotics is the development of mutations in the bacteria being treated that confers resistance to their action. The most notorious is methicillin resistant *Staphylococcus aureus* (MRSA). Staph bacteria developed resistance to penicillin in 1947, just four years after it was introduced. It is very common in hospitals, and staph is now resistant to a broad spectrum of antibiotics. Platensimycin is one of the antibiotics produced by *Streptomyces platensis*, and it blocks fatty acid biosynthesis which Gram-positive bacteria need to synthesize their cell membrane. It has been shown effective against MRSA in mice.[1] Cholera has developed a similar resistance to antibiotics, so another avenue for stopping the reproduction of this bacterium was explored.

Virstatin is not an antibiotic; it is a small molecule inhibitor. It does not kill the *V. cholerae* bacterium, but instead decreases its virulence. It inhibits the genes that promote colonies of the bacteria to form and that produce the toxin causing debilitating diarrhea. Thus, it can prevent extensive colonization of *V. cholerae* in the intestinal tract and greatly reduce the amount of diarrhea and fluid loss that normally

[1]Wang J, Soisson SM, Young K, et al., Platensimycin is a selective FabF inhibitor with potent antibiotic properties. *Nature*. **2006**;*441*(*7091*):358–361.

occurs. Administered only to cholera patients, Virstatin, a new experimental drug, would not cause bacteria to develop resistance in the same way that broad-spectrum antibiotics do. This is a new approach to antibacterial chemotherapy.[2]

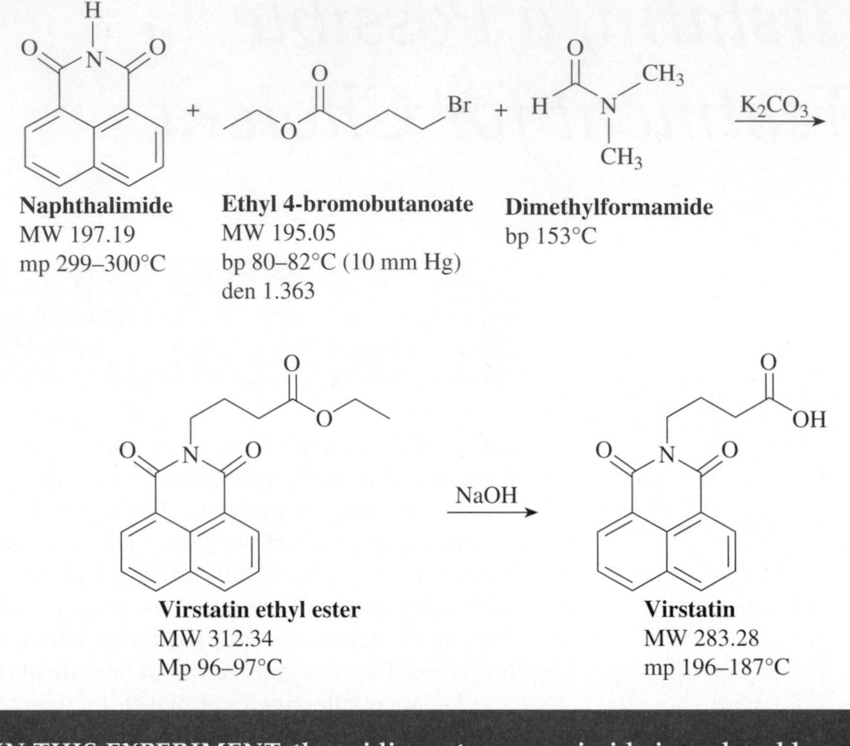

IN THIS EXPERIMENT, the acidic proton on an imide is replaced by an ethyl ester of butanoic acid. Hydrolysis of this ester to the corresponding carboxylic acid gives virstatin.

1. SYNTHESIS OF VIRSTATIN ETHYL ESTER

To a dry 5 mL long-necked flask, add 1.26 mmole of naphthalimide and 0.25 g of anhydrous potassium carbonate followed by 1.3 mL of dimethylformamide (stored over molecular sieves to dry it). Then add 1.9 mmol of dry (molecular sieves) ethyl 4-bromobutanoate. This can be measured volumetrically since it is in excess, but it is more accurate to weigh a small amount of a liquid.

Add a boiling chip and reflux the reaction mixture for 1 hr, making sure that the refluxing vapors stay within the flask. Transfer the cooled solution to a 10 mL Erlenmeyer flask, wash the long-necked flask out with 5 ml of water, and cool the solution in ice for 5 minutes during which time the product will crystallize. Collect the ester on the Hirsch funnel, using water to transfer and wash the product. Place this material in a 25 mL Erlenmeyer flask, and add 10 mL of methanol. Dissolve the product by heating. If any solid material remains, remove it by filtration of the hot solution. Making sure the volume of the solution is 10 mL, add water (about 10 mL)

[2]The present experiment is based on the work of Chriss E. McDonald. A two-step synthesis of virstatin, a virulence inhibitor of *Vibrio cholerae*. *J Chem Ed.*, **2009**;*86*:482–483.

to the boiling solution until a slight cloudiness develops. Allow the solution to cool slowly to room temperature, and then cool the mixture in ice for at least 5 minutes. Collect the resulting crystals on the Hirsch funnel by vacuum filtration. Wash the crystals with 50/50 methanol/water that is ice cold. Press the product between sheets of filter paper to dry it, save samples for mp determination and IR and ^{1}H NMR spectra. Weight the dry product and calculate the yield.

Assign the principal IR and NMR peaks.

2. HYDROLYSIS OF VIRSTATIN ETHYL ESTER TO VIRSTATIN

To a 5 mL long-necked round-bottomed flask containing 300 mg of Virstatin ethyl ester, add 1.5 mL methanol, followed by 0.15 mL 3 *M* aqueous sodium hydroxide solution. Reflux the mixture for 30 minutes, making sure the methanol does not escape. A damp pipe cleaner can be added for further cooling (see Fig. 25.1).

To the cooled reaction mixture, add 3 *M* hydrochloric acid to give a pH of 1, add 3 mL of water and cool the mixture in ice for at least 5 minutes while the product crystallizes. Collect the Virstatin on the Hirsch funnel, wash it with 2 mL of ice water and suck air through the filter until the product is almost dry.

Place the crystals in a 10 mL flask and recrystallize them from 3–5 mL of boiling ethyl acetate. Allow the solution to cool slowly to room temperature, then cool it in ice and collect the crystalline Virstatin on the Hirsch funnel. Wash it with 2 mL of cold hexanes. Weigh the dry product and calculate the yield of product. Determine the mp, IR and ^{1}H NMR spectra.

Assign the principal IR and NMR peaks.

QUESTIONS

1. What is the pKa of the imide proton of naphthalimide? To what do you attribute its acidity?
2. What is the role of DMF in the first step of this experiment, the synthesis of virstatin ethyl ester?

65 CHAPTER

Biosynthesis of Ethanol and Enzymatic Reactions

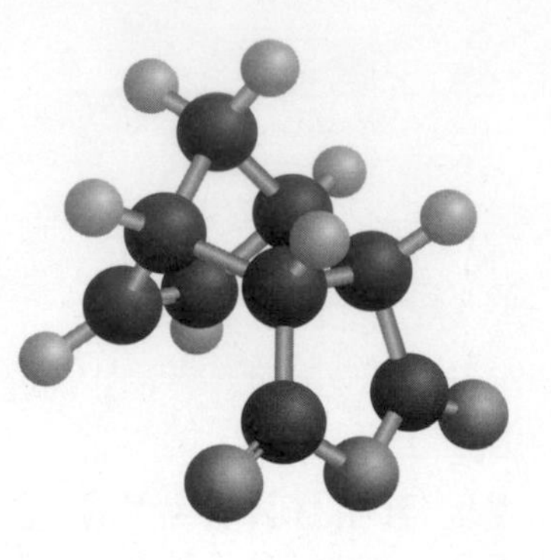

Ethanol and Global Warming

Since 1950, the average temperature of the surface of the earth has risen about 0.7°C. This temperature rise is the largest in recorded history and is projected, at the current rate, to rise 1.4 to 5°C during the current century. The environmental effects of this global warming would be devastating. To cite just one effect: the Greenland and Antarctic ice caps would melt, causing sea levels to rise throughout the world. A two-foot rise would flood 10,000 square miles of the United States and cause some municipal water supplies to turn saline, for example New York City and Philadelphia.

The cause of global warming is an increase of greenhouse gases, primarily carbon dioxide, produced by the burning of fossil fuels. The concentration of carbon dioxide has increased thirty-six times since 1750. These levels are higher than at any time during the last 700,000 years. To mitigate this effect, there has been a world-wide effort to reduce our dependence on these fossil fuels and replace them with other energy sources. In France, 79% of the power generated is nuclear. In Denmark, wind generates 20% of the country's power needs.

In the United States, the world's leader in generation of greenhouse gases, the most obvious attempt at mitigation of carbon dioxide generation has been the use of ethanol as an automobile fuel additive. Auto engines can run without modification with mixtures of up to 10% ethanol.

Ethanol from sugar has been used as an auto fuel in Brazil for more than 30 years. Sugarcane is one of the most efficient photosynthesizers in the plant kingdom, able to convert up to 2% of incident solar energy into biomass. Today, no passenger automobiles in that country run on pure gasoline; more than half the fuel consumed there by automobiles is ethanol. Several million vehicles run on 96% ethanol, twice as many are "flex fuel" automobiles that can run on 20 to 100% ethanol and the remainder run on 20–25% ethanol.

Brazil and the United States account for 70% of the world's ethanol production, and of that, 90% is used as fuel. Sugar cane is the source of the ethanol in Brazil. The cane is pressed to produce a solution of sucrose that is fermented to give ethanol. The bagasse, or dry refuse left after the juice has been extracted from the cane, is used as the fuel for distillation. It is an efficient process—for each unit of energy used to produce the ethanol, about 9 units of energy are produced in the ethanol.

Most of the United States is too cold to grow sugar cane, so within the last few years, much ethanol has been produced from corn, which, of course, does not contain sucrose. Milled corn, i.e., cornstarch, is treated with α-amylase and glucoamylase to give corn syrup, primarily glucose, which can then be fermented to ethanol. For each unit of energy used to produce ethanol in this way, about 1.3 units of energy are used to make the ethanol, a net loss of 30%. This energy input is predominantly fossil-fuel-based and is used to produce the fertilizer and pesticides needed to grow the corn, to plant, maintain, and harvest it, to ferment the glucose, to distill the ethanol, and to transport it to its final destination.

When a mixture of ethanol and water is distilled, the resulting distillate contains 95.5% ethanol and 4.5% water. It is a low-boiling azeotrope, boiling at 78.15°C (*see* Chapter 5, Distillation). Pure ethanol boils at 78.3°C so it is impossible to obtain 100% ethanol by even the most careful distillation. When 95% ethanol is mixed with a hydrocarbon such as gasoline, the mixture will be cloudy because the 5% of water does not dissolve in the gasoline. The two liquids are immiscible (*see* Solubility Tests in Chapter 14). So in order to blend ethanol with gasoline, the ethanol must be dehydrated. This can be accomplished by mixing the 95% ethanol with benzene and distilling the mixture. A low-boiling azeotrope of benzene and water distills first, followed by the higher boiling 100%, or *absolute*, ethanol. Another way to remove the water is to treat the 95% ethanol with molecular sieves (artificial zeolite). *See* page 136, Drying Agents, in Chapter 7, Extraction.

Absolute ethanol is hygroscopic; it will remove moisture from the air. It cannot generally be pumped through pipelines because it picks up water and becomes useless as a gasoline additive. And unlike gas or oil, because of its miscibility with water, it promotes corrosion of pipelines.

The economics of corn-based ethanol are interesting to contemplate. In Brazil, ethanol costs $0.83 per gallon to produce (2007) and in the U.S. $1.14 per gallon (2004). However, there is a U.S. government subsidy of $0.51/gal and an import tariff of $0.54/gal in the U.S. As a consequence of this bounty, farmers in the Midwest turned to corn production to make ethanol. This raised the price of corn, a primary feedstock for animals, and thus the cost of meat. It also raised the cost of high fructose corn syrup. In 2008, 20% of the total U.S. corn crop was used to produce ethanol; it offset 1% of U.S. oil use.

Pure corn syrup is 100% glucose, which is half as sweet as sucrose (sugar) (*see* page 721 in Chapter 63, Carbohydrates and Sweeteners). Enzymatic treatment can convert corn syrup to fructose that is 50% sweeter than sucrose. A mixture of 55% fructose and 45% glucose, which has the same sweetness as sucrose, is sold on a very large scale as high fructose corn syrup. Look at the label of any non-diet soft drink and a large number of other processed foods and it will be clear why the average American consumes more high fructose corn syrup (28 kg/yr) than sucrose (24 kg/yr). It is easier than sugar to work with because the syrup is a liquid, and it is somewhat cheaper than sugar because of the subsidies and tariffs mentioned above. Since 1995, corn growers have received over $40 billion in subsidies.

So if cornstarch is not to be the long-term source of ethanol for fuel in the U.S., what will take its place? The obvious alternative is cellulose, which, like starch, is a polymer of glucose but with a β-linkage between the glucose units. Cellulose can be broken down into glucose by enzymatic treatment. All ruminants, such as sheep and cows, are able to break down fibrous feed such as grass and hay because bacteria in their stomachs generate fermentation enzymes.

A very good source of cellulose is switchgrass, a hardy prairie grass that gives very high yields of cellulose per acre. It is self-seeding so needs no sowing, and is adaptable to a large number of different environments.

Waste cellulose is available from a large number of sources. The problem has been the high cost of the enzymes needed for converting cellulose to glucose ($0.40/gal of ethanol) vs. $0.03/gal ethanol for converting cornstarch to ethanol. A new company, Qteros, claims to have solved this problem.

The use of ethanol to replace fossil fuels will only mitigate the greenhouse gas problem. It cannot possibly reduce it. Consider the entire world being planted with sugar cane. Growing the cane removes carbon dioxide from the air, but subsequent fermentation and combustion releases exactly the same quantity of carbon dioxide that was used to grow it. Overall, heat has been produced from light in a controlled and convenient manner through ethanol, but the carbon footprint from this activity remains in stasis, without a reduction.

PRELAB EXERCISE: List the essential chemical substances, the solvent, and conditions for converting glucose to ethanol.

Humans have been preparing fermented beverages for more than 5000 years. Materials excavated from Egyptian tombs dating to 3000 B.C. demonstrate the operations used in making beer and leavened bread. The history of fermentation, whereby sugar is converted to ethanol by the action of yeast, is also a history of chemistry. The word *gas* was coined by Johannes Baptista van Helmont in 1610 to describe the bubbles produced in fermentation. Antoni van Leeuwenhoek observed and described the cells of yeast in 1680, with his newly invented microscope. In 1754, Joseph Black discovered carbon dioxide and showed it to be a product of fermentation, the burning of charcoal, and respiration. In 1789, Antoine Lavoisier showed that sugar gives ethanol and carbon dioxide, and made quantitative measurements of the amounts consumed and produced.

Fermentation

Glucose $\rightarrow 2\ C_2H_5OH + 2\ CO_2$

Once the mole concept was established (in 1815), Joseph Gay-Lussac could show that 1 mol of fermented glucose gives exactly 2 mol of ethanol and 2 mol of carbon dioxide. But the process of fermentation stumped some great chemists. Little-known Friedrich T. Kützing wrote in 1837, on the basis of microscopic observation, "It is obvious that chemists must now strike yeast off the roll of chemical compounds because it is not a compound but an organized body, an organism." On the other side were chemists such as Jöns Berzelius, who believed that yeast had a catalytic action; and Juston von Liebig, who put forth a "theory of motion of the elements within a compound that caused a disturbance of equilibrium that was communicated to the elements of the substance with which it came in contact thus forming new compounds."

Pasteur's contribution

It remained for Louis Pasteur to show that fermentation was a physiologic action associated with the life processes of yeast. Through his microscope, Pasteur could see the yeast cells that grew naturally on the surface of grapes. He showed that grape juice carefully extracted from the center of a grape and exposed to clean air would not ferment. In his classic paper of 1857, Pasteur described fermentation as the action of a living organism, but because the conversion of glucose to ethanol and carbon dioxide is a balanced equation, other chemists thought a chemical was responsible and disputed his findings. They searched for the substance in yeast that might cause the reaction. The search lasted for 40 years and was eventually ended by a clever experiment by Eduard Büchner. He made a cell-free extract of yeast that would still cause the conversion of sugar to alcohol. This cell-free extract contained the catalysts, which we now call *enzymes*, that were necessary for fermentation—a discovery that earned him the 1907 Nobel Prize in Chemistry. In 1905, Sir Arthur Harden discovered that inorganic phosphate added to the enzymes increased the rate of fermentation and was itself consumed. This result led Harden to eventually isolate fructose 1,6-diphosphate. Clearly, the history of biochemistry is intimately associated with the study of alcoholic fermentation.

Enzymes: fermentation catalysts

Sources of enzymes:

Grape skins

Malt

Saliva

Yeast

Ancient peoples discovered many of the essential reactions of alcoholic fermentation completely by accident. That crushed grapes would soon begin to froth and bubble and produce a pleasant beverage is a discovery lost in time. But what of those who lived in colder climates where the grape did not grow? How did they discover that the starch of wheat or barley could be converted to sugar by the enzymes in malt? When grain germinates, enzymes are produced that turn the starch into sugar. The process of malting involves letting the grain start to germinate and then heating and drying the sprouts to stop the process before the enzymes are used up. The color of the malt depends on the temperature of the drying. The darkest is used for stout and porter; the lighter for brown, amber, and pale ale. Because of a discovery made at least 10,000 years ago that the resulting beverage did not spoil as rapidly if hops were added, we now also have beer.

Other sources exist for the amylases that catalyze the conversion of starch to glucose. The Peruvian campasinos (peasants) make a drink called "chichi" from masticated wheat, which is dried in small cakes. When water, yeast, and more ground wheat are added, the resulting mixture ferments into a beer-like beverage. The enzyme salivary amylase is the catalyst for this starch-to-glucose conversion.

Bakers make use of fermentation by taking advantage of the gas released to leaven bread. In this chapter, baker's yeast is used to convert sucrose, ordinary table sugar, into ethanol and carbon dioxide with the aid of some 14 enzymes as catalysts, in addition to adenosine triphosphate (ATP), phosphate ion, thiamine pyrophosphate, magnesium ion, and reduced nicotinamide adenine dinucleotide (NADH), all present in yeast.

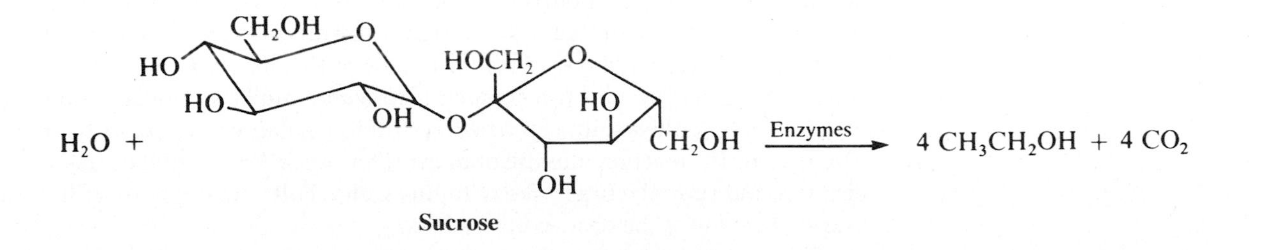

Enzymes are labile.

This experiment involves the fermentation of ordinary cane sugar using baker's yeast. The resulting dilute solution of ethanol, after removing the yeast by filtration, can be distilled according to the procedure of Experiment 2 in Chapter 5 (on page 95).

EXPERIMENTS

1. FERMENTATION OF SUCROSE

Microscale Procedure

IN THIS EXPERIMENT, a warm aqueous solution of sucrose and sodium phosphate in the presence of yeast is allowed to ferment to ethanol and carbon dioxide. The enzyme catalysts from the yeast are rendered inactive when the alcohol concentration reaches 10–12%. Fractional distillation of the reaction mixture will yield 95% pure ethanol, a constant-boiling azeotrope.

Protect the fermenting reaction from exposure to oxygen and contaminating materials. Under aerobic conditions, Acetobacter bacteria from the air can convert ethanol to acetic acid.

Reaction time: 1 week

To a 5-mL round-bottomed, long-necked flask, add 90 mg of dry yeast and 1.25 mL of warm (up to 50°C) water. Shake the mixture thoroughly until it is more or less homogeneous in appearance; then add to it 9 mg of disodium hydrogen phosphate, 1.30 g of sucrose, and an additional 3.75 mL of water, which should be warmed to about 45°C. Shake this mixture to ensure complete mixing, and then fit the neck of the flask with a septum that is connected to a 20-cm length of polyethylene tubing (Fig. 65.1). Lead this tubing beneath the surface of about 2 mL of a saturated aqueous solution of calcium hydroxide (limewater) in a reaction tube. The tube in the limewater will act as a seal to prevent air and unwanted enzymes from entering the flask, but will allow gas to escape. Place the assembly in a warm spot—the optimal temperature for the reaction is 35°C for a week, at which time the evolution of carbon dioxide will have ceased. What is the precipitate in the limewater?

On a small scale, it will be necessary to provide external heat to maintain the fermentation, or to group several reactions closely together in an insulated container. On a larger scale, because fermentation is an exothermic reaction, there is enough heat evolved by the reaction to keep the mixture warm and to promote the biosynthesis of ethanol.

On completion of fermentation, add about 0.25 g of Celite filter aid (diatomite, diatomaceous earth) to the flask and shake it vigorously. Celite is added to make it possible to filter the solution, otherwise the yeast cells would clog the filter paper if they were filtered with that method. Filter the mixture on a Hirsch funnel into a 25-mL filter. Because the small filter flask can tip over easily, clamp it to a ring stand. Moisten a circular filter paper with water, apply gentle suction (have the water supply to the aspirator turned on full force, valve to trap partly open), and slowly pour the reaction mixture onto the filter paper. Wash out the flask with 1 mL of water and rinse the filter cake with this water. Full vacuum in this filtration will evaporate some of the desired ethanol.

The filtrate, which is a dilute solution of ethanol contaminated with a few bits of cellular material and other organic compounds (acetic acid if you are not careful), is saved in a stoppered container until it is distilled following the procedure outlined in Experiment 2 of Chapter 5.

Refer to the "Cleaning Up" instructions after the following macroscale procedure.

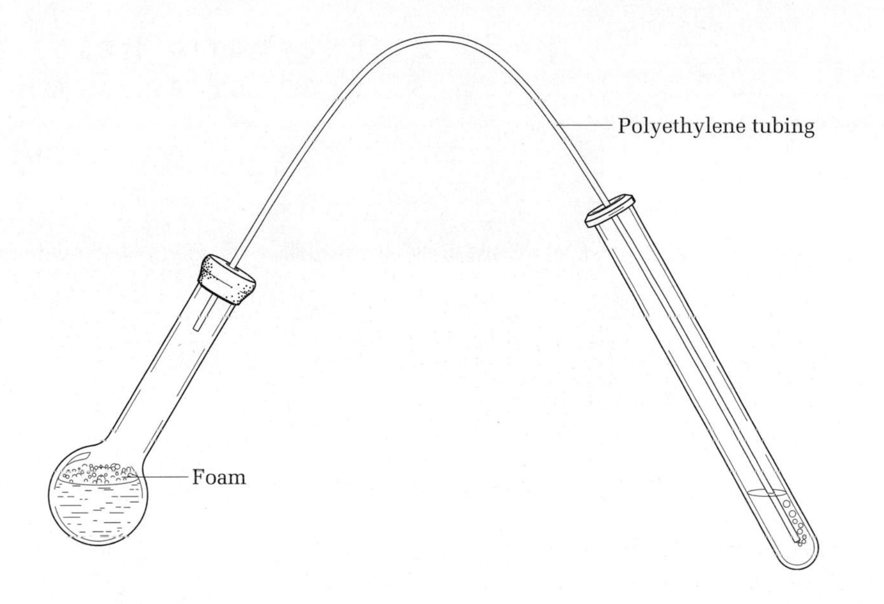

FIG. 65.1
A fermentation apparatus. *See* Figure 18.2 (on page 331) for the technique of threading a polyethylene tube through a septum.

Macroscale Procedure

Protect the fermenting reaction from exposure to oxygen and contaminating materials. Under aerobic conditions, Acetobacter bacteria can convert ethanol to acetic acid.

Reaction time: 1 week

Filter aid

Macerate (grind) one-half cake of yeast or half an envelope of dry yeast in 50-mL of water in a beaker, add 0.35 g of disodium hydrogen phosphate, and transfer this slurry to a 500-mL round-bottomed flask. Add a solution of 51.5 g of sucrose in 150 mL of water and shake to ensure complete mixing. Fit the flask with a one-hole rubber stopper containing a bent glass tube that dips below the surface of a saturated aqueous solution of calcium hydroxide (limewater) in a 15-cm test tube. The tube in limewater will act as a seal to prevent air and unwanted enzymes from entering the flask, but will allow gas to escape. The tube should be about 0.5 cm below the limewater so that limewater will not be sucked back into the flask should the pressure change. Place the assembly in a warm spot on your desk (the optimum temperature for the reaction is 35°C; an example of a warm spot may be a cupboard near a hot water line) for 1 week, at which time the evolution of carbon dioxide will have ceased. What is the precipitate in the limewater?

Upon completion of fermentation, add 10 g of Celite filter aid (diatomite, diatomaceous earth) to the flask, shake vigorously, and filter. Use a 5.5-cm Büchner funnel placed on a neoprene adapter, or Filtervac atop a 500-mL filter flask that is attached to a water aspirator through a trap by vacuum tubing. Celite is added to make it possible to filter the solution. Because the apparatus is top-heavy, clamp the flask to a ring stand. Moisten a circular filter paper with water, apply gentle suction and slowly pour the reaction mixture onto the filter paper. Wash out the flask with a few milliliters of water from your wash bottle, and rinse the filter cake with this water.

The filtrate, which is a dilute solution of ethanol contaminated with bits of cellular material and other organic compounds (acetic acid if you are not careful), is saved in a stoppered flask until it is distilled following the procedure outlined in Experiment 2 of Chapter 5.

Cleaning Up. Because sucrose, yeast, and ethanol are natural products, all solutions produced in this experiment contain biodegradable material and can be flushed down the drain after dilution with water. The limewater can be disposed of in the same way. The Celite filter aid can be placed in the nonhazardous solid waste container.

PART 2. ENZYMATIC REACTIONS: A CHIRAL ALCOHOL FROM A KETONE

PRELAB EXERCISE: Study the biochemistry of the conversion of glucose to ethanol.

ENZYMATIC REDUCTION OF A KETONE TO A CHIRAL ALCOHOL

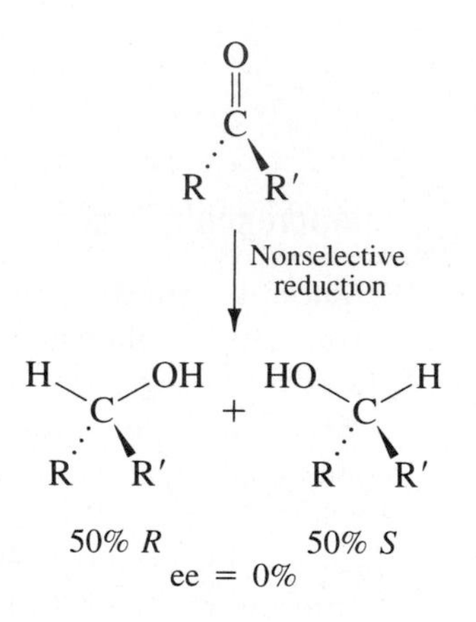

The reduction of an achiral ketone with the usual laboratory reducing agents, such as sodium borohydride or lithium aluminum hydride, will not give a chiral alcohol because the chances for attack on the two sides of the planar carbonyl group are equal. However, if the reducing agent is chiral, a chiral alcohol might be obtained. In recent years, organic chemists have devised a number of such chiral reducing agents, but few are as efficient as those found in nature.

In this part of the chapter, we will study the enzymes found in baker's yeast[1] to reduce ethyl acetoacetate to *S*-(+)-ethyl 3-hydroxybutanoate. This compound is a very useful synthetic building block. Many chiral natural product syntheses are based on this hydroxyester.[2]

$$CH_3\overset{O}{\overset{\|}{C}}CH_2\overset{O}{\overset{\|}{C}}OCH_2CH_3 \xrightarrow{\text{yeast}} (CH_3)(H)(HO)C\text{—}CH_2\overset{O}{\overset{\|}{C}}OCH_2CH_3$$

Ethyl acetoacetate
MW 130.14
bp 181°C, den. 1.021
n_D^{20} 1.4190

***S*-(+)-Ethyl 3-hydroxybutanoate**
MW 132.16
bp 180–182°C (71–73°C/12 mm)
n_D^{20} 1.4210

[1] Servi S. *Synthesis.* **1990**;1–25.
[2] Amstutz R, Hungerbühler E, Seebach D. *Helv Chim Acta.* **1981**;*64*:1796. Seebach D. *Tetrahedron Lett.* **1982**;*159*.

Many different enzymes are present in yeast. The primary ones responsible for converting glucose to ethanol are discussed in Chapter 64. While this fermentation reaction is taking place, certain ketones can be reduced to chiral alcohols.

ee = enantiomeric excess

Whenever a chiral product is produced from an achiral starting material, the chemical yield as well as the optical yield is important—in other words, the stereoselectivity of the reaction is critical. The usual method for recording this is to calculate the enantiomeric excess (ee). A sodium borohydride reduction will produce 50% *R* and 50% *S* alcohol with no enantiomer in excess. If 93% *S*-(+) and 7% ***R***-(–) isomer are produced, then the enantiomeric excess is 86%. In the yeast reduction of ethyl acetoacetate, several authors have reported enantiomeric excesses ranging from 70% to 97%. This optical yield is distinct from the chemical yield, which depends on how much material is isolated from the reaction mixture.

% ee calculation:

$$\frac{([\mathbf{S}] - [\mathbf{R}])}{([\mathbf{S}] + [\mathbf{R}])} \times 100 = \%\text{ee}$$

The use of enzymes to carry out stereospecific chemical reactions is not new, but it is not always possible to predict if an enzymatic reaction (unlike a purely chemical reaction) will occur or how stereospecific it will be if it does take place. Because the yeast reduction of ethyl acetoacetate is easily carried out, it might be an interesting research project to explore the range of possible ketones that yeast will reduce to chiral alcohols. For example, butyrophenone can be reduced to the corresponding chiral alcohol.[3] The following experiments are based on the work of Seebach,[4] Mori,[5] and Ridley.[6] The work of Bucciarelli et al. is also applicable.[7]

EXPERIMENTS

1. ENZYMATIC RESOLUTION

IN THIS EXPERIMENT, ethyl acetoacetate is added to a fermenting sugar solution. After two days, the product formed by the chiral reduction of the methyl ketone of the starting material is extracted with ether. The ether is dried and evaporated to give an optically active hydroxyester.

Microscale Procedure

In a 25-mL flask, dissolve 2.3 g of sucrose and 15 mg of disodium hydrogen phosphate in 8.5 mL of warm (35°C) tap water (*see* Experiment 1). Add 0.5 g of dry yeast and swirl to suspend the yeast throughout the solution. After 15 minutes, while fermentation is progressing vigorously, add 150 mg of ethyl acetoacetate. Store the flask in a warm place, ideally at 30–35°C, for at least 48 hours (a longer time will do no harm). At the end of this time, add 0.5 g of Celite filtration aid and remove the yeast cells by filtration on a Hirsch funnel (*see* Experiment 1). Wash the cells with 1.5 mL of water; then saturate the filtrate with sodium chloride to decrease

Keep the mixture warm; the reaction is slow at low temperatures.

[3]Sih CJ, Rosazza JP. In: *Application of Biochemical Systems in Organic Chemistry*, Part 1. Jones JB, Sih CJ, Perlman D. Wiley: New York, 1976;71–78.
[4]Seebach D. *Tetrahedron Lett.* **1982**;23:159–162.
[5]Mori K. *Tetrahedron.* **1981**;37:1341.
[6]Ridley DD, Stralow M. *Chem Commun.* **1975**;400.
[7]Bucciarelli M, Fomi A, Moretti I, Torre G. *Synthesis.* **1983**;897.

A 15-mL centrifuge tube can be used for the extraction, and the emulsions can be broken in the centrifuge.

CAUTION: Extinguish all flames when working with ether.

Videos: *Microscale Filtration on the Hirsch Funnel, Extraction with Ether*, Photo: *Extraction with Ether*

the solubility of the product. Refer to a handbook for the solubility of sodium chloride in water to determine approximately how much to use. Extract the resulting solution five times with 1.5-mL portions of ether in a test tube, taking care to shake hard enough to mix the layers but not so hard as to form an emulsion between the ether and water. Adding a small amount of methanol may help to break up emulsions. Dry the ether layer by adding anhydrous sodium sulfate or calcium chloride pellets until the drying agent no longer clumps together. After approximately 15 minutes, filter the ether solution into a tared flask and evaporate the filtrate. The remaining residue should weigh about 100 mg. It should, unlike the starting material, give a negative iron(III) chloride test. It can be analyzed by thin-layer chromatography (TLC), using dichloromethane as the solvent, to determine whether unreacted ethyl acetoacetate is present. Infrared (IR) spectroscopy should show the presence of the hydroxyl group, and may show unreduced methyl ketone. The nuclear magnetic resonance (NMR) spectrum of the product is easily distinguished from that of the starting material.

Cleaning Up. Dilute the aqueous layer with water and flush it down the drain. After the ether evaporates from the drying agent in the hood, place the drying agent in the nonhazardous solid waste container. If local regulations do not allow for the evaporation of solvents in a hood, the wet solid should be disposed in a special waste container. Dichloromethane (the TLC solvent) goes in the halogenated organic solvents waste container. Ether distillate goes in the organic solvents waste container.

Macroscale Procedure

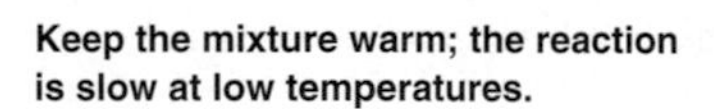

Keep the mixture warm; the reaction is slow at low temperatures.

CAUTION: Extinguish all flames when working with ether.

In a 500-mL flask, dissolve 80 g of sucrose and 0.5 g of disodium hydrogen phosphate in 300 mL of warm (35°C) tap water (*see* Experiment 1). Add two packets (16 g) of dry yeast and swirl to suspend the yeast throughout the solution. After 15 minutes, while fermentation is progressing vigorously, add 5 g of ethyl acetoacetate. Store the flask in a warm place, ideally at 30–35°C, for at least 48 hours (a longer time will do no harm). At the end of this time, add 20 g of Celite® filtration aid and remove the yeast cells by filtration on a 10-cm Büchner funnel (*see* Experiment 1). Wash the cells with 50 mL of water; then saturate the filtrate with sodium chloride to decrease the solubility of the product. Refer to a handbook for the solubility of sodium chloride in water to determine approximately how much to use. Extract the resulting solution five times with 50-mL portions of ether, taking care to shake the separatory funnel hard enough to mix the layers but not so hard as to form an emulsion between the ether and water. Adding a small amount of methanol may help to break up emulsions. Dry the ether layer by adding anhydrous sodium sulfate or calcium chloride pellets until the drying agent no longer clumps together. After approximately 15 minutes, decant the ether solution into a tared distilling flask and remove the ether by distillation or by evaporation. The residue should weigh about 3.5 g. It should, unlike the starting material, give a negative iron(III) chloride test. It can be analyzed by TLC (use dichloromethane as the solvent) to determine whether unreacted ethyl acetoacetate is present. IR spectroscopy should show the presence of the hydroxyl group and may show unreduced methyl ketone. The NMR spectrum of the product is easily distinguished from that of the starting material. NMR and IR spectra of the starting material and *racemic* products are shown in Figures 65.2, 65.3, and 65.4.

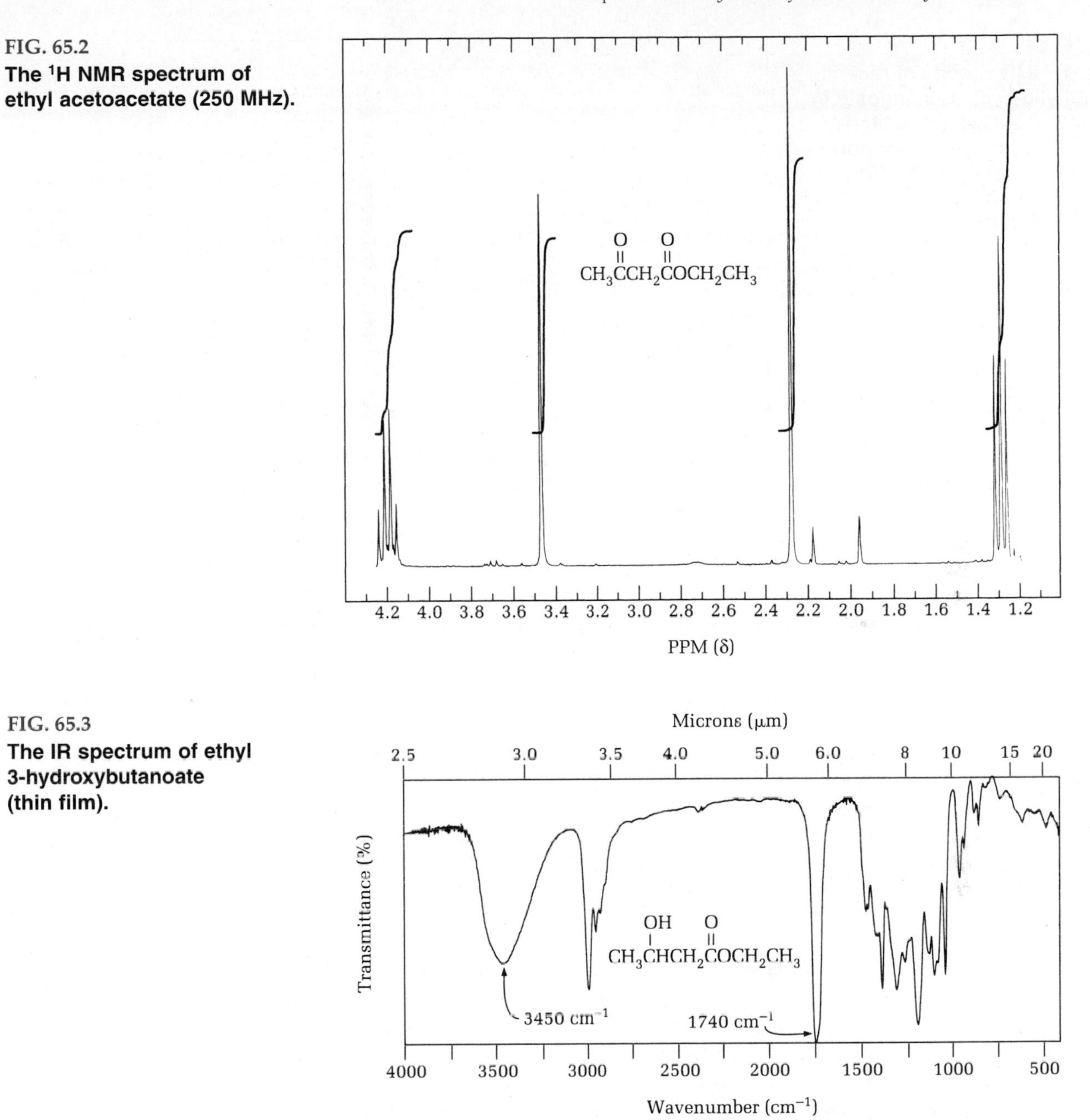

FIG. 65.2
The ^{1}H NMR spectrum of ethyl acetoacetate (250 MHz).

FIG. 65.3
The IR spectrum of ethyl 3-hydroxybutanoate (thin film).

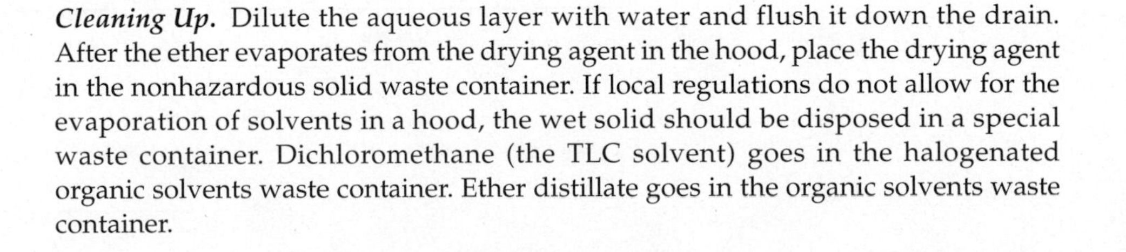

Cleaning Up. Dilute the aqueous layer with water and flush it down the drain. After the ether evaporates from the drying agent in the hood, place the drying agent in the nonhazardous solid waste container. If local regulations do not allow for the evaporation of solvents in a hood, the wet solid should be disposed in a special waste container. Dichloromethane (the TLC solvent) goes in the halogenated organic solvents waste container. Ether distillate goes in the organic solvents waste container.

FIG. 65.4
The ^{1}H NMR spectra of racemic (±)-ethyl 3-hydroxybutanoate in 0.3 mL of carbon tetrachloride and 0.2 mL deuterochloroform (250 MHz). (a) Pure. (b) With 30 mg of shift reagent. (c) With 50 mg of shift reagent.

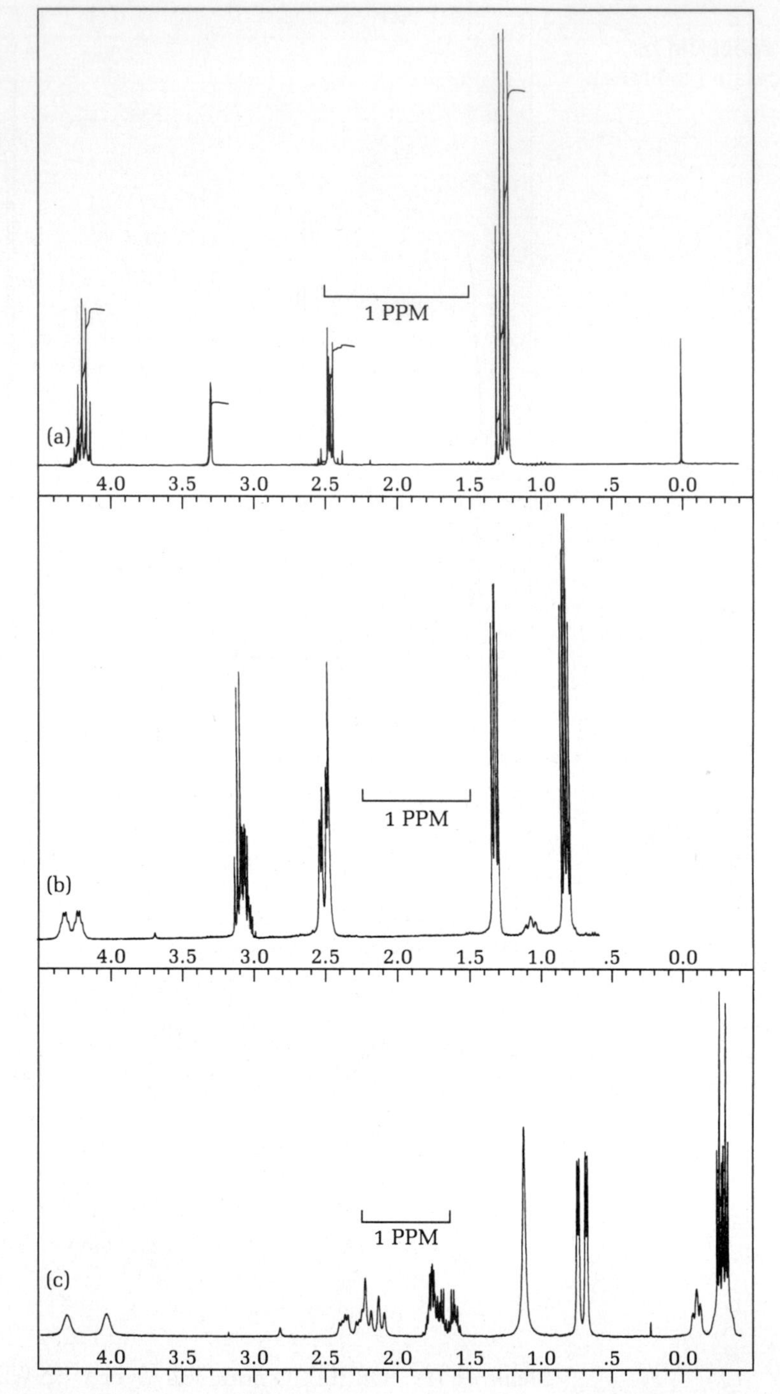

2. DETERMINATION OF OPTICAL PURITY

The optical purity of the product can be determined by using a polarimeter to measure the optical rotation. The specific rotation, hydroxybutanoate has been reported to vary from +31.3° to +41.7° in chloroform. The specific rotation, $[\alpha]_D^{25}$, of *S*(+)-ethyl 3-hydroxybutanoate has been reported to vary from +31.3° to +41.7° in chloroform. The specific rotation, $[\alpha]_D^{25}$, of +37.2° (chloroform) corresponds to an enantiomeric excess of 85%.

Another way to determine optical purity is by gas chromatography using a column packed with a chiral support. The chiral material in the column will interact with the two enantiomers of a chiral substrate differently, causing one enantiomer to have a longer retention time than the other. This method requires much less material than a polarimeter.

A third method for determining the optical purity of a small quantity of material is to use an NMR chiral shift reagent.[8] A chiral shift reagent will complex with a basic center (the hydroxyl group in ethyl 3-hydroxybutanoate) and cause the NMR peaks to shift, usually downfield. Protons nearest the shift reagent/hydroxyl group shift more than those far away. A chiral shift reagent forms diastereomeric complexes so that peaks from one enantiomer shift downfield more than peaks from the other enantiomer. By comparing the areas of the peaks, the ee can be calculated.

CAUTION: Handle carbon tetrachloride in the hood.

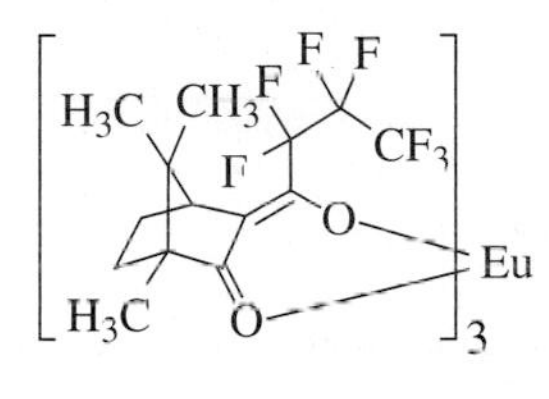

The NMR experiment is easily accomplished by adding 10- to 20-mg increments of europium tris[3-(heptafluoropropylhydroxymethylene)-(+)-camphorate] shift reagent to a solution of 30 mg of pure dry ethyl 3-hydroxybutanoate in a mixture of 0.3 mL of carbon tetrachloride and 0.2 mL of deuterochloroform.[9]

The analysis of the NMR spectra is not completely straightforward. For example, in Figure 65.4(a), can you locate the peaks from the proton on the hydroxyl-bearing carbon atom? (Pay attention to the integrals.) Note that the methylene hydrogens on carbon-2 (δ = 2.45 ppm in Fig. 65.4a) are not magnetically (or chemically) equivalent. They are said to be diastereotopic because diasteriomers could be formed by replacing either one or the other proton with another substituent.

As the optically active shift reagent is added, sets of peaks begin to double as diastereomeric complexes are formed (Fig. 65.4b and Fig. 65.4c). By integrating the pairs of peaks, the enantiomeric excess (ee) of one isomer in relation to the other can be determined. In this experiment, only one set of peaks can be seen if the enzymatic reduction gives only one of the two possible enantiomers of the product. Thus it is a good idea to compare your product to racemic material, either commercially made or prepared by reducing ethyl acetoacetate with sodium borohydride (follow the procedure in Chapter 26). Also bear in mind that the shift reagent itself will contribute peaks to the spectrum.

The best way to be sure of all the peak assignments is to run and integrate a series of spectra where the shift reagent is added in 10-mg increments.

Cleaning Up. Place the contents of the NMR tube in the halogenated solvents waste container.

[8]Lipkowitz KB, Mooney JL. *J Chem Educ.* **1987**;64:985.
[9]Lipkowitz KB, Mooney JL. *J Chem Educ.* **1987**;64:985.

3. PREPARATION OF 3,5-DINITROBENZOATE

3,5-Dinitrobenzoates are common alcohol derivatives. They are easily prepared, the dinitrophenyl group adds considerably to the molecular weight, and they are easily recrystallized.

The 3,5-dinitrobenzoates of a racemic mixture of *R*- and *S*-ethyl 3-hydroxybutanoate, like most racemates, recrystallize together to give crystals that in this case melt at 146°C.

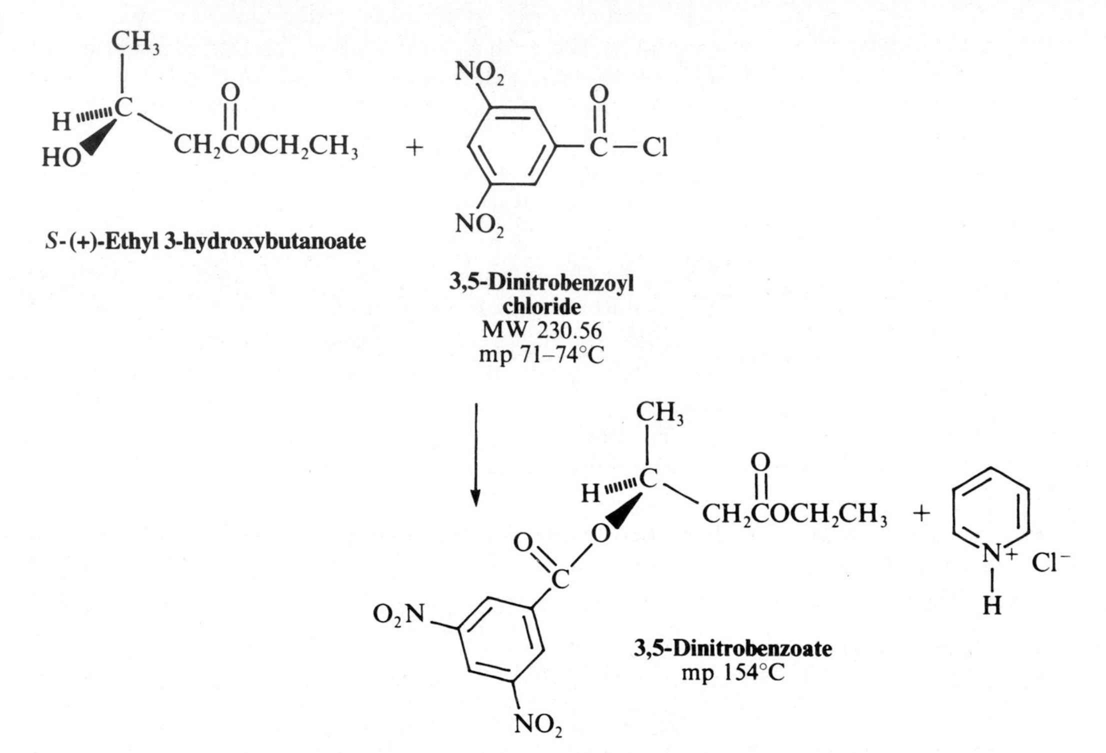

No amount of recrystallization causes one enantiomer to crystallize out while the other remains in solution because they are, after all, mirror images of each other. However, if one enantiomer is in large excess, it is possible to effect a separation by crystallization. In the present case, the *S* enantiomer predominates and it crystallizes out, leaving most of the *R* + *S* racemate in solution. Repeated crystallization increases the melting point and purity to a point at which the optical purity of the crystalline product is almost 100%. At this point, the 3,5-dinitrobenzoate has a melting point of 154°C. Treatment of the derivative with excess acidified ethanol regenerates 100% ee *S*-(+)-ethyl 3-hydroxybutanoate.

Procedure

IN THIS EXPERIMENT, the hydroxyester from Experiment 1 is reacted with an acid chloride in the presence of base to give 3,5-dinitrobenzoate. The reaction mixture is poured into water, any dinitrobenzoic acid is extracted with bicarbonate, and the crystalline product is recrystallized from ethanol. This material, when recrystallized to a constant melting point, is enantiomerically pure.

To 100 mg of the *S*-(+)-ethyl 3-hydroxybutanoate in a reaction tube (Fig. 65.5), add 175 mg of pure[10] 3,5-dinitrobenzoyl chloride and 1 mL of pyridine. Equip the reaction tube with a septum pierced with a syringe needle. Reflux the mixture for 15 minutes and then transfer it with a Pasteur pipette to 3.5 mL of water in another reaction tube. Remove the solvent from the crystals and shake the crystals with 2 mL of 0.5 *M* sodium bicarbonate solution to remove dinitrobenzoic acid. Remove the bicarbonate solution and recrystallize the derivative from ethanol. Collect the product on a Wilfilter. Dry a portion and determine the melting point. If it is not near 154°C, repeat the crystallization. Dry the pure derivative and calculate the percent yield.

Treatment of the derivative with excess acidified ethanol will regenerate 100% ee *S*-(+)-ethyl 3-hydroxybutanoate. If several crops of crude product are pooled, distillation at reduced pressure (*see* Chapter 6) can give chemically pure ethyl 3-hydroxybutanoate (bp 71–73°C/12 mm). The *R* and *S* enantiomers have the same boiling point, so the optical purity will not be changed by distillation.

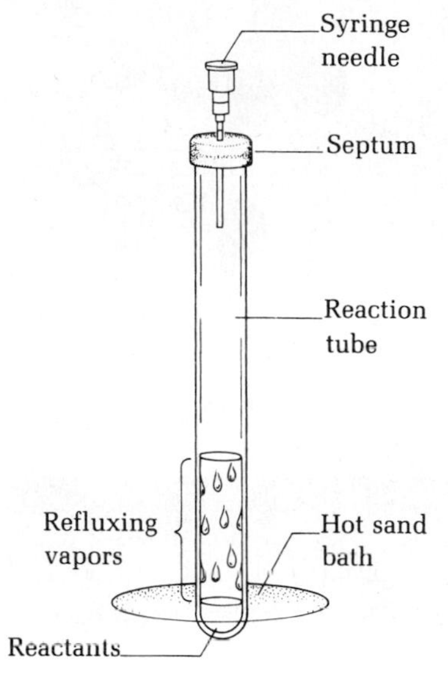

FIG. 65.5
An apparatus for refluxing reaction mixture.

Cleaning Up. The pyridine solvent goes in the organic solvents waste container. Dilute the aqueous layer with water and flush it down the drain. Ethanol from the recrystallization goes in the organic solvents waste container, along with the pot residue if the final product is distilled.

QUESTIONS

1. Using yeast, can glucose be converted to ethanol? Can fructose be converted to ethanol?
2. Write the equation for the formation of the precipitate formed in a test tube containing calcium hydroxide.
3. In this experiment, could 90% ethanol be made by adding more sugar to the fermentation flask?
4. Explain the differences in the ^1H NMR spectra shown in Figure 65.4 (on page 740). Are there signals resulting from the shift reagent? If so, identify them.
5. Look up the prices of racemic ethyl 3-hydroxybutanoate and each of the enantiomers in a chemical catalog (e.g., the Aldrich Chemical Company catalog). How do you explain the price differences?
6. Assign the peaks at 3450 cm^{-1} and 1740 cm^{-1} in the IR spectrum of ethyl 3-hydroxybutanoate (Fig. 65.3 on page 739).

[10]Check the melting point of the 3,5-dinitrobenzoyl chloride. If it is below 70°C, it should be recrystallized from dichloromethane.

66 CHAPTER

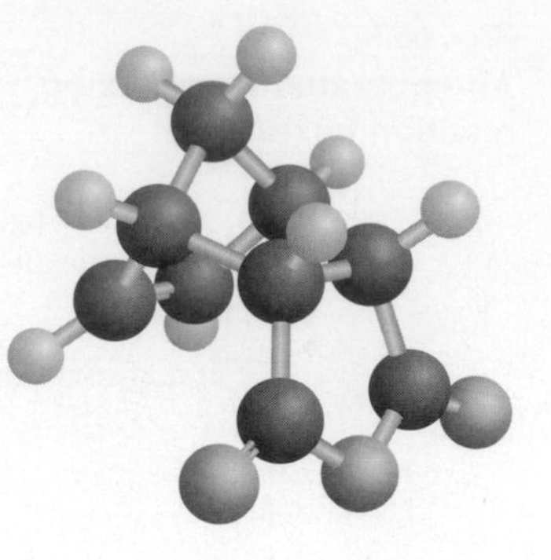

The Synthesis of Natural Products: The Sex Attractant of the Cockroach and Camphor

When you see this icon, sign in at this book's premium website at **www.cengage.com/login** to access videos, Pre-Lab Exercises and other online resources.

PRELAB EXERCISE: Write a balanced equation for the oxidation of 2,5-dimethoxybenzyl-3-methylbutanoate to blattellaquinone.

PART 1: The Synthesis of the Sex Attractant of the German Cockroach

Pheromones are chemical substances animals, generally of the same species, use to communicate. Dogs mark their territory with pheromones, male dogs detect females in estrus, ants lay down trails leading to food, and female insects attract males for mating.

In general pheromones are relatively small compounds that are somewhat volatile, yet complex enough to have some specificity. Consider the trail pheromone of the ant. Ants forage for food in a random manner, but once they find it they put down a trail pheromone as they return to the nest that allows other ants to follow the path to the food. But once the food is exhausted the trail should evaporate. 2-Methyl-4-heptanone, b.p. 158°C meets these requirements.

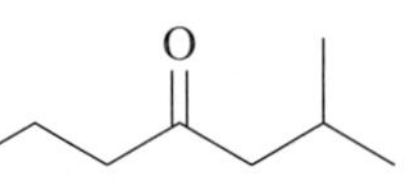

The sex pheromone of the flightless virgin female gypsy moth, *Lymantria dispar,* emitted in picogram amounts, can attract males from two miles away.

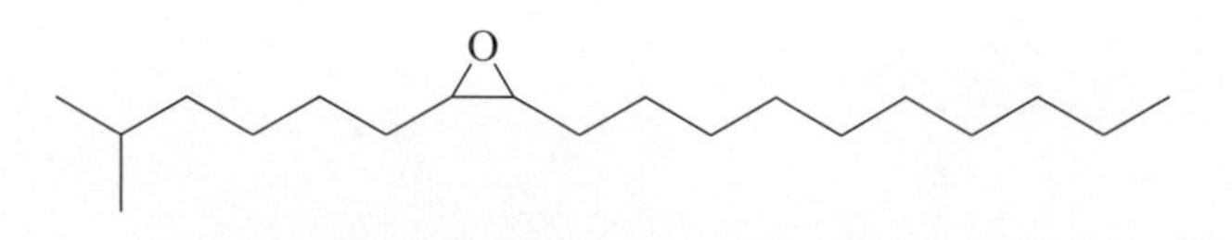

(+) *cis*-7,8-Epoxy-2-methyloctadecane (Disparlure)
Sex attractant of the gypsy moth

The larvae of gypsy moths feed on newly emerged leaves of trees, usually at night, but as their numbers increase they feed day and night and can strip whole hardwood trees of their vegetation in short order. Confined primarily to the northeast United States, they increase in population periodically to become major pests. One approach to their control is to spray infected areas with Disparlure to confuse the males who cannot then find the females. But Disparlure is most effective in traps used to access population densities.

The sex attractant of the common housefly is a hydrocarbon, *cis*-9-tricosene.

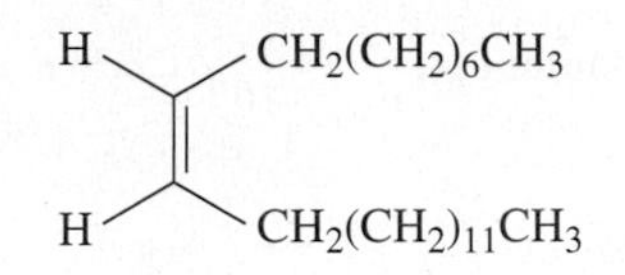

Despite an active search, especially by perfume companies, no human sex pheromone has been positively identified, but an intriguing experiment indicates there is chemical communication between humans. It has been observed that women living in close proximity, as in a dormitory, and not on oral contraceptives, attain menstrual synchrony in about three to four months. Once in synchrony they can detect the odor of β-α-androstenol at very low concentrations.

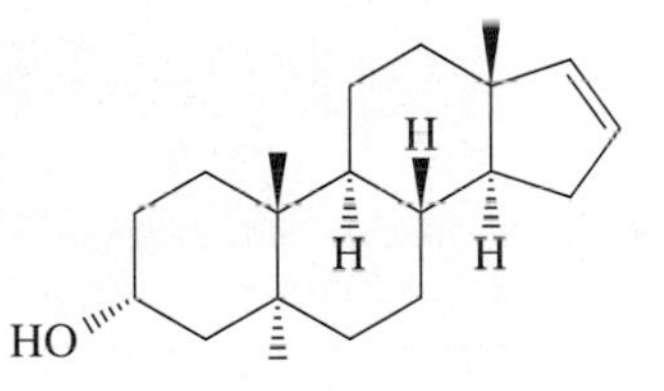

In the present experiment the sex attractant of the German cockroach is synthesized. The cockroach is a major household pest worldwide, spreading allergenic diseases and asthma. The roach has been with us for more than 100 million years; it is remarkably hardy, able to go 45 min without air, 30 min under water, without food for three months and water for one month.

The sex attractant was isolated and by extraction of the rear portion of 15,000 virgin female cockroaches (each one contains only about a nanogram (10^{-9} g) of the attractant) by Nojima, et al.[1] who also determined the structure by NMR and mass spectroscopic methods and synthesized the molecule. The present experiment is taken from the work of P.L. Feist.[2]

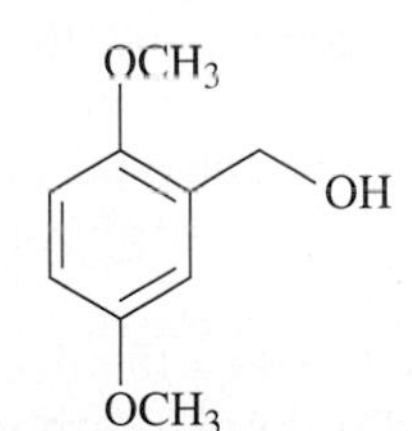

2,5-Dimethoxybenzyl alcohol
MW 168.19
bp 122–125°C (0.1 mm Hg)

O
Cl

3-Methylbutyryl chloride (isovaleryl chloride)
MW 120.58
bp 115–117°C

CH_2CH_3
N
H_3CH_2C CH_2CH_3

Triethylamine
MW 101.19
bp 88.8°C

[1]Nojima, S.; Schal, C.; Webster, F. X.; Santangelo, R. G.; Roelofs, W. L. *Science* **2005**, *307*, 1104–1106.
[2]P. L. Feist, *J. Chem. Ed.*, **2008**, *85*, 1548–1549.

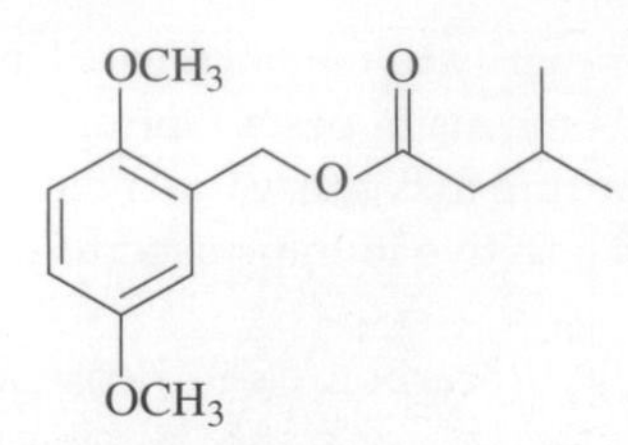

2,5-Dimethoxybenzyl-3-methylbutanoate
MW 252.31

Ceric ammonium nitrate
MW 548.26

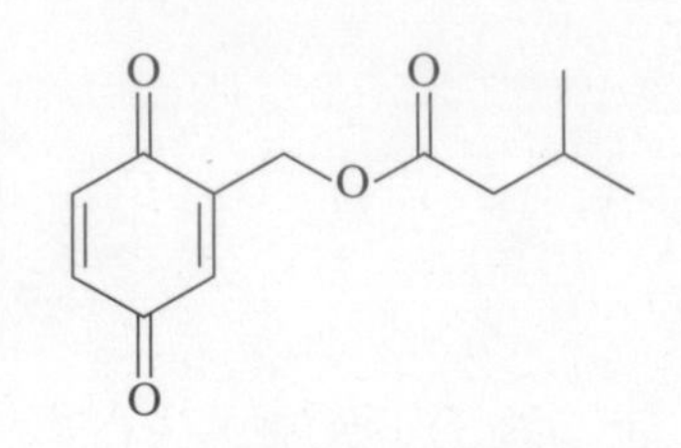

2-Benzyl(3-methylbutanoate)quinone
Gentisyl quinone isovalerate
Blattellaquinone
MW 222.24
mp 56–56.5°C

1. PREPARATION OF 2,5-DIMETHOXYBENZYL-3-METHYLBUTANOATE

Microscale Procedure

IN THIS EXPERIMENT, an alcohol is esterified with an acid chloride in dichloromethane with the aid of triethylamine. After washing the solution with sodium bicarbonate and ammonium chloride solutions and drying the solvent the product is isolated by evaporation of the solvent.

In a 5-mL round-bottomed flask place 1.0 mmol of 2,5-dimethoxybenzyl alcohol in 3.5 mL of dichloromethane. Add a stirring bar and stir until the alcohol is completely dissolved. Add with continued stirring 1.1 mmol of triethylamine followed by the slow dropwise addition of 1.3 mmols of 3-methylbutyryl chloride (isovaleryl chloride). If the solvent begins to boil add the acid chloride more slowly.

After all the butyryl chloride has been added continue stirring for 45 min to complete the reaction. After 20 min and 40 min remove very small samples for TLC analysis on silica gel plates (*see* Chapter 6) using a 50/50 mixture of hexanes and ethyl acetate to develop the plate and a UV lamp to visualize the spots. The starting alcohol has an R_f value of about 0.35 and the product an R_f value of about 0.60.

Transfer the reaction mixture to a 15 mL centrifuge tube and wash the organic layer with 1 mL of saturated sodium bicarbonate solution and then with 2 mL of 10% ammonium chloride solution. Dry the organic layer with anhydrous calcium chloride pellets. Flick the centrifuge tube (*see* Fig. 7.4) and add pellets until they no longer clump together.

Transfer the clear solution to an appropriate round-bottomed flask and remove the solvent by distillation or rotary evaporation. If you have determined the tare weight of the flask you can weigh it to determine the approximate yield of light tan liquid product. Run an infrared spectrum and remove a sample for NMR analysis.

2. 2-BENZYL(3-METHYLBUTANOATE)QUINONE, GENTISYL QUINONE ISOVALERATE, BLATTELLAQUINONE

Microscale Procedure

IN THIS EXPERIMENT, a 1,4-dimethoxybenzene ester is oxidized to a quinone ester, blattellaquinone, with ceric ammonium nitrate in an acetonitrile/water mixture. After extraction of the product into dichloromethane the solution is washed, dried, and evaporated to give the crude crystalline product.

Transfer the crude 2,5-dimethoxybenzyl 3-methylbutanoate to a 5 mL round bottom flask with the aid of 3.5 mL of 50/50 acetonitrile/water. Add a stirring bar and slowly add dropwise 3 mmol of a solution of ceric ammonium nitrate in 1.5 mL of water using a Pasteur pipette. After all oxidizing agent has been added continue to stir the solution for 45 min.

Filter the reaction mixture on the micro Hirsch funnel from any solid material, transfer the filtrate to a 15 mL centrifuge tube and extract it with three 2.5 mL portions of dichloromethane, pulling off the extraction solvent and placing it in a 10 mL Erlenmeyer flask. Save all layers, because on the first extraction the organic layer can be in either the top or bottom layer. If you are in doubt as to whether a separated layer is dichloromethane add a drop of water to it. The water should float on the top.

Wash the combined organic extracts with two 2.5 mL portions of saturated sodium bicarbonate solution. To do this add the aqueous solution to the dichloromethane extracts in a centrifuge tube, cap the tube, shake it, and after the layers separate draw off the bottom layer and place it in an Erlenmeyer flask. Discard the upper layer, return the dichloromethane layer to the centrifuge tube and repeat the process. Dry the organic layer, which is now in the Erlenmeyer flask with anhydrous calcium pellets. Add pellets until they no longer clump together.

Transfer the clear dichloromethane solution to a round-bottomed flask and remove the solvent on the rotary evaporator, or, more conveniently, return it to the filter flask that has been cleaned and dried, place the Hirsch funnel in the top, and attach the flask to the aspirator. By placing your thumb in the Hirsch funnel, the vacuum can be controlled and heat can be applied by holding the flask in the other hand while swirling the contents (Fig 7.9).

If the filter flask is tared the weight of the product can be determined. It should, of course, contain no more than 1.0 mmol of product. If it contains more, then all the solvent has not been removed.

The slightly brown oil in the filter flask will crystallize if cooled, seeded, or scratched with a stirring rod to give tan crystals of blattellaquinone. Remove a sample for melting point determination. The pure product is reported to melt at 56–56.5°C. Raise the temperature very slowly above about 45°C as the mp is being determined. Take IR and NMR spectra to fully characterize the product. This material can be used to determine if it truly will attract male cockroaches.

It can be purified further by column chromatography on alumina (Chapter 9). Use 75/25 hexanes/ethyl acetate to elute the column collecting about 10–15 mL of eluent. Follow the procedure for flash chromatography in Chapter 22. Remove the solvent in the filter flask as before and recrystallize the white residue from 50/50 ether/hexanes before determining the mp.

3. 2,5-DIMETHOXYBENZYL-3-METHYLBUTANOATE

Macroscale Procedure

In a 100 mL round-bottomed flask place 10 mmols of 2,5-dimethoxybenzyl alcohol in 40 mL of dichloromethane. Add a stirring bar and stir until the alcohol is completely dissolved. Add with continued stirring 11 mmol of triethylamine followed by the slow dropwise addition of 13 mmols of 3-methylbutyryl chloride (isovaleryl chloride) from a separatory funnel.

After all the butyryl chloride has been added continue stirring for 45 min to complete the reaction. After 20 min and 40 min remove very small samples for TLC analysis on silica gel plates (*see* Chapter 6) using a 50/50 mixture of hexanes and ethyl acetate to develop the plate and a UV lamp to visualize the spots. The starting alcohol has an R_f value of about 0.35 and the product an R_f value of about 0.60.

Transfer the reaction mixture to a small separatory funnel and wash the organic layer with 10 mL of saturated sodium bicarbonate solution and then with 20 mL of 10% ammonium chloride solution. Dry the organic layer with anhydrous calcium chloride pellets. Add pellets until they no longer clump together.

Transfer the clear solution to an appropriate round-bottomed flask and remove the solvent by distillation or rotary evaporation. If you have determined the tare weight of the flask you can weigh it to determine the approximate yield of light tan liquid product. Run an infrared spectrum and remove a sample for NMR analysis.

4. 2-BENZYL(3-METHYLBUTANOATE)QUINONE, GENTISYL QUINONE ISOVALERATE, BLATTELLAQUINONE

Macroscale Procedure

Transfer the crude 2,5-dimethoxybenzyl 3-methylbutanoate to a 100 mL round bottom flask with the aid of 40 mL of 50/50 acetonitrile/water. Add a stirring bar and slowly add dropwise 30 mmol of a solution of ceric ammonium nitrate in 20 mL of water from a separatory funnel. After all of the oxidizing agent has been added continue to stir the solution for 45 min.

Filter the reaction mixture on the Hirsch funnel from any solid material, saturate the filtrate with solid sodium chloride and transfer it to a separatory funnel. Extract it with it with three 30 mL portions of dichloromethane. Save all layers, because on the first extraction the organic layer can be in either the top or bottom layer. If you are in doubt as to whether a separated layer is dichloromethane add a drop of water to it. The water should float on top.

Wash the combined organic extracts with two 25 mL portions of saturated sodium bicarbonate solution. Transfer the organic layer to an Erlenmeyer flask with anhydrous calcium pellets. Add pellets until they no longer clump together.

Transfer the clear dichloromethane solution to a round-bottomed flask and remove the solvent on the rotary evaporator, or by distillation. If the flask is tared the weight of the product can be determined. It should, of course, contain no more than 10 mmole of product. If it contains more, then all the solvent has not been removed.

The slightly brown oil in the flask will crystallize if cooled, seeded or scratched with a stirring rod to give tan crystals of blattellaquinone. Remove a sample for melting point determination. The pure product is reported to melt at 56–56.5°C. Raise the temperature very slowly above 45°C as the mp is being determined. Take IR and NMR spectra to fully characterize the product. This material can be used to determine if it truly will attract male cockroaches.

It can be purified further by column chromatography on alumina (Chapter 9). Use 75/25 hexanes/ethyl acetate to elute the column. Follow the procedure for flash chromatography in Chapter 22. Remove the solvent in the filter flask as before and recrystallize the white residue from 50/50 ether/hexanes before determining the mp.

Cleaning Up. Aqueous extracts can be flushed down the drain, all waste dichloromethane should be placed in the halogenated solvents waste container, and all solids placed in the hazardous solid waste container.

PART 2: THE CONVERSION OF CAMPHENE TO CAMPHOR

In the following experiments, 2,2-dimethyl-3-methylenebicyclo[2.2.1]heptane, otherwise known as *camphene* (**1**), will be converted to 1,7,7-trimethylbicyclo-[2.2.1]heptan-2-one, which is camphor (**8**), through the intermediate isomeric alcohols borneol (**6**) and isoborneol (**7**) and *endo*- and *exo*-1,7,7-trimethylbicyclo-[2.2.1]heptan-2-ol.

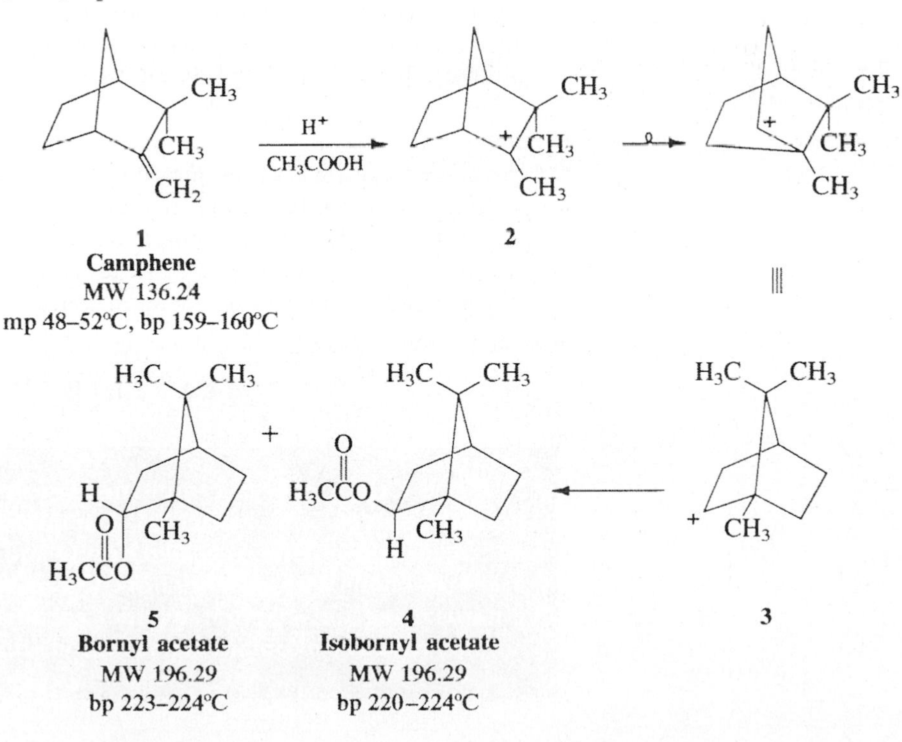

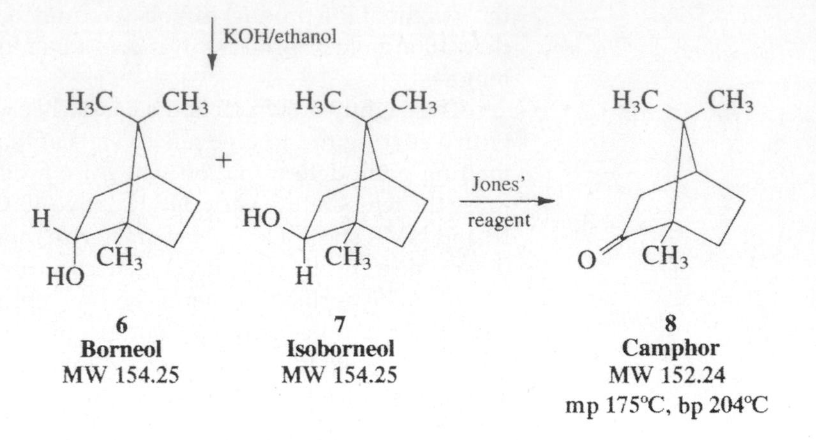

Camphene (**1**), as its odor will indicate, can be isolated from turpentine and a number of other naturally occurring oils. Upon reaction with acetic acid and a catalytic amount of sulfuric acid, it forms a carbocation (**2**) that undergoes rearrangement (Wagner–Meerwein rearrangement) to the secondary carbocation (**3**), with subsequent solvolysis (reaction with the solvent) to form predominantly the *exo*-acetate (**4**) with some of the *endo*-acetate (**5**). Even though **2** is a tertiary carbocation and more stable than **3**, the acetate (**4**) is formed because the acetate ion can more easily attack the unhindered cation (**3**) than the hindered cation (**2**). In Experiment 3, computational chemistry is used to predict and/or confirm that the *exo*-acetate (**4**) will be formed in preference to the *endo*-isomer (**5**).

Isobornyl acetate (**4**), with the acetate group in the *exo* position, is accompanied by a small amount of the isomeric endo compound, bornyl acetate (**5**). These isomers are liquids and not easily purified. They can be hydrolyzed to the corresponding alcohols, borneol (**6**) and isoborneol (**7**), which are solids.

This mixture of isomers can be oxidized to camphor (**8**). All of these compounds have a roughly spherical shape. Intermolecular interactions are relatively small; therefore, the compounds sublime easily. When you look up the physical properties of isoborneol, note the small difference between its melting point and its boiling point.

In carrying out this series of reactions, bear in mind that one of the goals is to prepare enough of the final product to purify and to characterize by thin-layer chromatography (TLC), melting point, and infrared (IR) and nuclear magnetic resonance (NMR) spectroscopy. For characterization you will need about 20 mg. With skill, you should end up with a lot more than that.

EXPERIMENTS

IN THIS EXPERIMENT, camphene is dissolved in acetic acid, which is both the solvent and a reactant. A catalytic amount of sulfuric acid is added, and the mixture is heated briefly and then diluted with ether. Several extractions with water remove the acetic acid and sulfuric acid. The ether is dried and evaporated under vacuum to give primarily isobornyl acetate, which also contains some bornyl acetate.

1. LABORATORY CONVERSION OF CAMPHENE TO CAMPHOR

Microscale Procedure: Camphene (1) to Isobornyl Acetate (4)

In a 5-mL round-bottomed flask, place a stirring bar, 3 mL of acetic acid, and 1.36 g of camphene; to the resulting solution, add a solution of 0.25 g of concentrated sulfuric acid in 0.3 mL of water. Because acetic acid is a solvent and sulfuric acid is a catalyst, the amounts of these reagents need not be measured with great accuracy. Heat the mixture in a boiling water bath for 15 min with stirring. Alternatively, the mixture can simply be heated over a steam bath with frequent swirling of the contents. Do not reflux the mixture directly on a sand bath because the sand bath would be too hot and cause undesired side reactions. The reaction mixture will turn dark in color.

After the heating period, cool the reaction mixture and place it in a 15-mL centrifuge tube, complete the transfer with about 2 mL of *t*-butyl methyl ether, and extract the mixture with two 4-mL portions of water. This extraction will remove most of the acetic acid. The separations are achieved by drawing off the aqueous layer using a 9 in. Pasteur pipette. Wash the ether layer with about 2 mL of 10% sodium carbonate or bicarbonate solution and dry the solution over calcium chloride pellets for about 5 min. Transfer the solution to the tared 25-mL filter flask and remove the ether under reduced pressure (Fig. 66.1). This will take just a minute or so because of the ether's low boiling point. Save about 50 mg of the product for analysis. It will be a brown liquid mixture of isobornyl acetate with some bornyl acetate. In this product isolation, the exact amounts of ether, water, and carbonate solutions are not important, so there is no need to measure them precisely.

Video: *Extraction with Ether*, Photo: *Extraction with Ether*

Cleaning Up. Neutralize the aqueous washes with bicarbonate and then flush the resulting solution down the drain with lots of water; alternatively, combine these washes with the filtrate from the next experiment, neutralize, and flush down the drain. Place the calcium chloride in the nonhazardous waste container.

Microscale Procedure: Isobornyl Acetate (4) [and Bornyl Acetate (5)] to Isoborneol (7) [and Borneol (6)]

IN THIS EXPERIMENT, the acetates made in the previous experiment are hydrolyzed by heating in a solution of potassium hydroxide dissolved in a mixture of water and ethanol. The reaction mixture is poured onto ice, and the crystallized alcohols are collected by filtration.

FIG. 66.1
Evaporation of a solvent under reduced pressure. Heat and swirling motion is supplied by one hand; the vacuum is controlled by thumb of other.

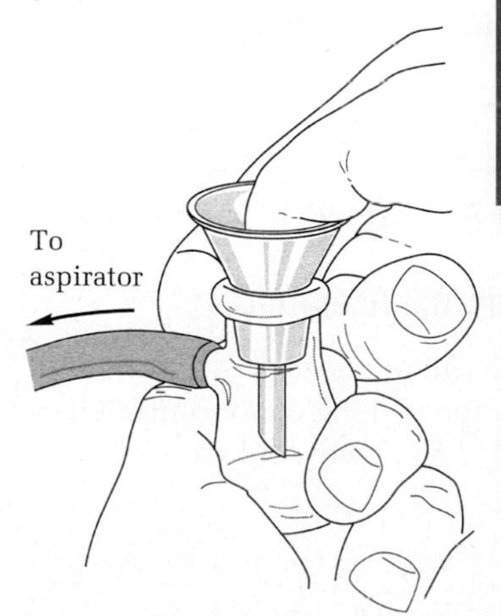

In a 5-mL round-bottomed flask place a boiling chip and a solution of 0.45 g of potassium hydroxide pellets (the pieces are 85% potassium hydroxide and 15% water) dissolved in 0.7 mL water to which is added 2.3 mL of ethanol. To this basic solution add 1.0 g of isobornyl/bornyl acetate, apply an air condenser, and reflux the mixture on a sand bath for 1 h. With careful adjustment of the amount of heat, it is easy to reflux this mixture using an air condenser (*see* Fig. 25.1 on page 379). Just make sure that the top of the condenser is not hot, which would indicate that ethanol may be escaping. Wrap a wet pipe cleaner around the air condenser if more cooling is needed (*see* Fig. 40.3 on page 522). At the end of the hour, pour the

Video: *Microscale Filtration on the Hirsch Funnel*; Photo: *Sublimation Apparatus*

solution onto about 5 g of ice in a 10-mL Erlenmeyer flask. Swirl the mixture until the product is completely solidified; then collect it on a Hirsch funnel. Wash it well with water; then carefully press the product onto the funnel using a spatula. It will have the consistency of light-brown sugar. Try to squeeze out as much of the water as possible. Save about 60 mg of this sample for analysis. It sublimes easily and will give pure white crystals. Determine the melting point of the sublimed product in a sealed melting-point capillary (*see* Fig. 3.6 on page 50 for this technique). Explain the melting point in terms of the expected composition of the product.

Cleaning Up. Neutralize the filtrate with dilute hydrochloric acid or combine it with the washings from the previous experiment, neutralize, and flush down the drain with excess water.

Microscale Procedure: Borneols [(5) and (6)] to Camphor (7)

IN THIS EXPERIMENT the mixture of alcohols from the previous experiment is dissolved in acetone and oxidized with Cr(VI) reagent. On dilution with water, the product (camphor) separates as a solid and is isolated by filtration. It is then purified by sublimation.

CAUTION: Jones' reagent is very corrosive because it is a powerful oxidizing agent prepared in concentrated sulfuric acid. Wash up spills immediately. If spilled on the skin, flush with water until the skin is neutral. Cr^{6+} as a dust is a carcinogen if inhaled. It is unlikely that you would inhale this reagent; however, exercise care when preparing the reagent from solid chromium trioxide.

In a 15-mL centrifuge tube dissolve 0.6 g of the borneols in 1.2 mL of acetone, add a magnetic stirring bar, and then add to the solution 1 mL of Jones' reagent[3] dropwise with thorough stirring and cooling. After all of the reagent is added, the mixture should be homogeneous in appearance. If not, stir it some more. After about 30 min, add water dropwise to the reaction mixture, diluting it to about 12 mL. Collect the solid by filtration on a Hirsch funnel, squeeze it as dry as possible, and then sublime the camphor at atmospheric pressure (Fig. 66.2). Determine the melting point, analyze the product by TLC and IR and NMR spectroscopy, and then submit the product to your instructor in a labeled vial (not in a plastic bag).

Photo: *Sublimation Apparatus*

Cleaning Up. Complete the reduction of any remaining chromic ion in the filtrate by adding solid sodium thiosulfate until the solution becomes cloudy and blue. Neutralize with sodium carbonate and filter the flocculent precipitate of $Cr(OH)_3$ on a Hirsch funnel. Dilute the filtrate with water and flush down the drain. Place the precipitate on the filter paper in the heavy metals hazardous waste container.

2. LABORATORY CONVERSION OF CAMPHENE TO CAMPHOR

Macroscale Procedure: Camphene (1) to Isobornyl Acetate (4)

In a 25-mL round-bottomed flask place a stirring bar, 12 mL of acetic acid, and 5.44 g of camphene; to the resulting solution add a solution of 1.0 g of concentrated sulfuric acid in 1.2 mL of water. Because acetic acid is a solvent and sulfuric acid is

[3] Jones' reagent is prepared by dissolving 13.5 g of chromium trioxide in 11.5 mL of concentrated sulfuric acid and then adding, with great care, enough water to bring the volume up to 50 mL.

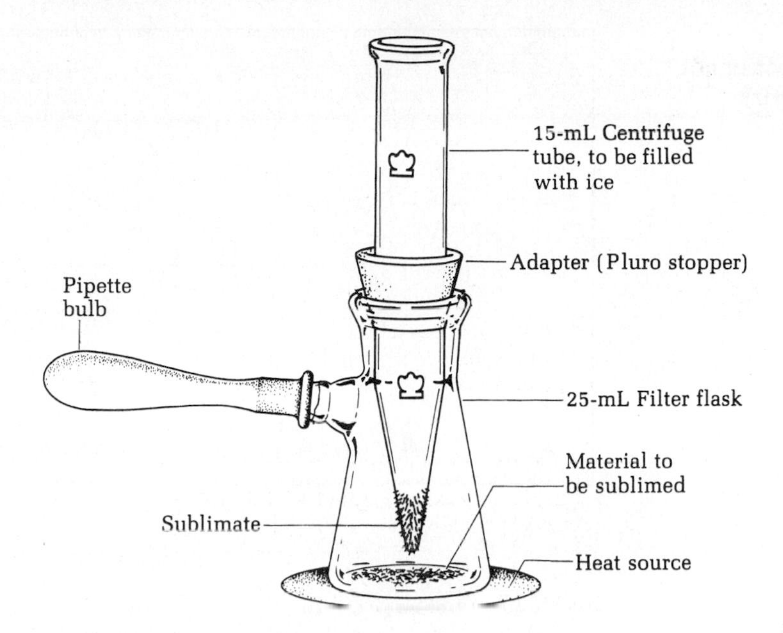

FIG. 66.2
A sublimation apparatus.

a catalyst, the amounts of these reagents need not be measured with great precision. Heat the mixture in a boiling water bath for 15 min with stirring. Alternatively, the mixture can simply be heated over a steam bath with frequent swirling of the contents. Do not reflux the mixture directly on a sand bath because the sand bath would be too hot and cause undesired side reactions. The reaction mixture will turn dark in color.

After the heating period, cool the reaction mixture and place it in a small separatory funnel, complete the transfer with about 8 mL of *t*-butyl methyl ether, and extract the mixture with two 15-mL portions of water. This extraction will remove most of the acetic acid. Wash the ether layer with about 8 mL of 10% sodium carbonate or bicarbonate solution and dry the solution over calcium chloride pellets for about 5 min. Transfer the solution to a tared 100-mL round-bottomed flask and remove the ether with a rotary evaporator. This will take just a minute or so because of ether's low boiling point. Save about 200 mg of the product for analysis. It will be a brown liquid mixture of isobornyl acetate with some bornyl acetate. In this product isolation, the exact amounts of ether, water, and carbonate solutions are not important, so there is no need to measure them precisely.

Cleaning Up. Neutralize the aqueous washes with bicarbonate and then flush the resulting solution down the sink with lots of water; alternatively, combine these washes with the filtrate from the next experiment, neutralize, and flush down the sink. Place the calcium chloride in the nonhazardous waste container.

Macroscale Procedure: Isobornyl Acetate (4) [and Bornyl Acetate (5)] to Isoborneol (7) [and Borneol (6)]

In a 25-mL round-bottomed flask place a boiling chip and a solution of 1.8 g of potassium hydroxide pellets (the pieces are 85% potassium hydroxide and 15% water) dissolved in 2.8 mL water to which is added 9.2 mL of ethanol. To this basic solution add 4.0 g of isobornyl–bornyl acetate, apply a condenser, and reflux the mixture on a sand bath for 1 h. At the end of the hour, pour the solution onto

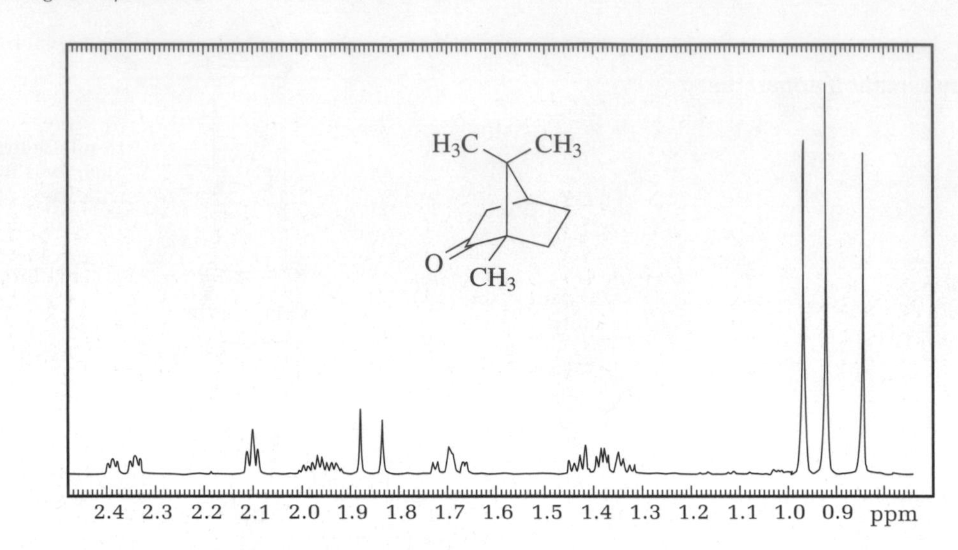

FIG. 66.3
The ^{1}H NMR spectrum of camphor (400 MHz).

about 20 g of ice in a 50-mL Erlenmeyer flask. Swirl the mixture until the product is completely solidified; then collect it on a Hirsch funnel. Wash it well with water, then carefully press the product onto the funnel using a spatula. It will have the consistency of light-brown sugar. Try to squeeze out as much water as possible. Save about 250 mg of this sample for analysis. It sublimes easily and will give pure white crystals. Determine the melting point of the sublimed product in a sealed melting-point capillary. Explain the melting point in terms of the expected composition of the product.

Cleaning Up. Neutralize the filtrate with dilute hydrochloric acid, or combine it with the washings from the previous experiment, neutralize, and flush down the drain with excess water.

CAUTION: Jones' reagent is very corrosive because it is a powerful oxidizing agent prepared in concentrated sulfuric acid. Wash up spills immediately. If spilled on the skin, flush with water until the skin is neutral. Cr^{6+} as a dust is a carcinogen if inhaled. It is unlikely that you would inhale this reagent; however, exercise care when preparing the reagent from solid chromium trioxide.

Macroscale Procedure: Borneols [(5) and (6)] to Camphor (8)

In a 25-mL Erlenmeyer flask, dissolve 2.4 g of the borneols in 4.8 mL of acetone, add a magnetic stirring bar, and then add to the solution 4.0 mL of Jones' reagent[4] dropwise with thorough stirring and cooling. After all of the reagent is added, the mixture should be homogeneous in appearance. If not, stir it some more. After about 30 min, add water dropwise to the reaction mixture, diluting it to about 50 mL. Collect the solid by filtration on a Hirsch funnel, squeeze it as dry as possible, and then sublime the camphor at atmospheric pressure (*see* Fig. 66.2 on page 753). Determine the melting point, analyze the product by TLC and IR and NMR spectroscopy (Fig. 66.3), and then submit the product to your instructor in a labeled vial (not in a plastic bag).

Cleaning Up. Complete the reduction of any remaining chromic ion in the filtrate by adding solid sodium thiosulfate until the solution becomes cloudy and blue. Neutralize with sodium carbonate and then filter the flocculent precipitate of $Cr(OH)_3$ on a Hirsch funnel. Dilute the filtrate with water and flush down the drain. Place the precipitate on the filter paper in the heavy metals hazardous waste container.

[4]Jones' reagent is prepared by dissolving 13.5 g of chromium trioxide in 11.5 mL of concentrated sulfuric acid and then adding, with great care, enough water to bring the volume up to 50 mL.

3. COMPUTATIONAL CHEMISTRY: THE THEORETICAL CONVERSION OF CAMPHENE TO CAMPHOR

Molecular mechanics calculations have been dealt with previously (*see* Chapter 15), but these calculations represent just a part of a much larger topic, computational chemistry—an important new part of organic chemistry. Just a few years ago this area of chemistry was relegated to the specialist, the "theoretical chemist," who laboriously encoded the Cartesian coordinates of all the atoms in a molecule and then submitted the calculation to a large mainframe computer for solution. The data returned in enormous piles of digital printout, which was interpretable only by the expert. In the past few years, however, workstations and even graphically oriented desktop computers can handle these computations, while the input and output can be manipulated and displayed on a computer monitor as a drawing of the molecule. For example, the distribution of charge density can be color encoded and superimposed on the space-filling model of the molecule. Thus computational chemistry has passed from the hands of the specialist to the practicing organic chemist.

In the example presented here, we try to predict by computation which isomer—isobornyl acetate (**4**) or bornyl acetate (**5**)—is favored when camphene is treated with acetic acid and a catalytic amount of sulfuric acid.

As seen in the reaction scheme, camphene (**1**), on protonation, will give a carbocation (**3**), which is attacked by the nucleophile, acetic acid, to give either **4** or **5**, the isomeric acetates. The carbocation, an sp^2-hybridized carbon, is planar and has an empty *p* orbital. We know that the acetic acid must attack this from "above" or "below," along the axis of the *p* orbital. The reactants must come within bonding distance for the product to form, but preventing this is steric hindrance from other parts of the molecule. It is not obvious from scrutiny of the drawing whether there is more steric hindrance from the "top" or the "bottom" of the ion in **3**. Even when molecular models are held in the hand, it looks as if the bridge methyl group is going to get in the way of the acetic acid, so we would predict that attack by acetic acid would take place from the bottom side of the cation.

To approach this problem from a calculational standpoint, we first assemble the framework of the cation on a computer and then minimize its energy. We then proceed to carry out a molecular orbital calculation. In this case we use a semiempirical method because it is fast and will provide the type of information we need. Among the well-known methods are modified neglect of diatomic overlap (MNDO), Austin Method 1 (AM1),[5] and MNDO parameterization method 3 (MNDO/PM3 or simply PM3).

Using a computer program such as Spartan,[6] the lowest unoccupied molecular orbital (LUMO) of the cation (3) can be calculated and superimposed on the space-filling model of the cation. Through color coding, you can see on the computer monitor that there is more of the empty molecular orbital exposed on the top (*exo*) side of the cation than on the bottom (*endo*) side. If you have the facilities, try to reproduce this calculation. This same procedure is used to predict the direction of attack of the borohydride anion on 2-methylcyclohexanone (*see* Chapter 26).

See *Web Links*

[5]Its originator, Michael J. S. Dewar, was a professor at the University of Texas at Austin.
[6]Spartan molecular modeling software is available from Wavefunction, Inc., Irvine, California.

QUESTIONS

1. Blattellaquinone is an example of a parabenzoquinone. Give an example of an orthobenzoquinone.
2. In this experiment, a benzene derivative (an aromatic compound) is converted to a parabenzoquinone. Is a parabenzoquinone aromatic?
3. The molal freezing point depression constant for camphor is 39.7°C. What does this mean, and what effect might it have on your observed melting point for camphor?

CHAPTER 67

Polymers: Synthesis and Recycling

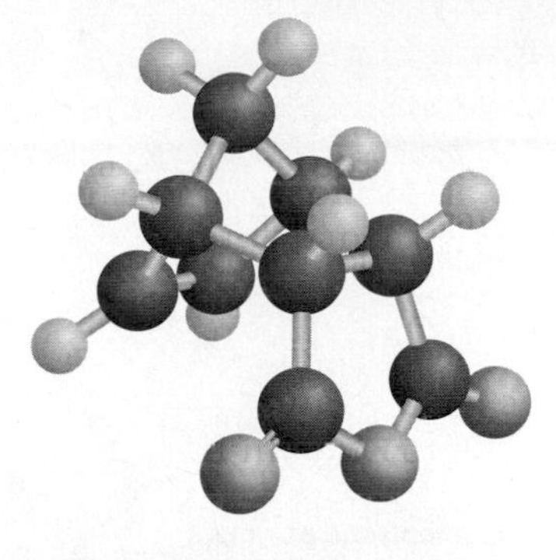

PRELAB EXERCISE: In the preparation of nylon by interfacial polymerization, sebacoyl chloride is synthesized from decanedioic acid and thionyl chloride. What volume of hydrogen chloride is produced in this reaction? What volume of sulfur dioxide is produced?

PART 1: SYNTHESIS OF POLYMERS

Polymers are ubiquitous. Natural polymers such as proteins (polyamino acids), DNA (polynucleotides), and cellulose (polyglucose) are the basic building blocks of plant and animal life. Synthetic organic polymers, or plastics, are now among our most common structural materials. In the United States, we make and use more synthetic polymers than we do steel, aluminum, and copper combined—at present, worth $310 billion annually in an industry employing 1.4 million people.

We use more synthetic polymers than steel, aluminum, and copper combined.

Polymers, from the Greek meaning "many parts," are high molecular weight molecules composed of repeating units of smaller molecules. Most polymers consist of long chains held together by hydrogen bonds, van der Waals forces, and the tangling of the long chains. When heated, the covalent bonds of thermoplastic polymers do not break but will adopt new shapes when the chains slide over one another. The polymer can be reformed upon cooling and become any shape desired, including films, sheets, extrusions, or molded parts in a myriad of forms.

Nitrocellulose

The first synthetic plastic was nitrocellulose, made in 1862 by nitrating cellulose, a natural polymer. Nitrocellulose, when mixed with a plasticizer such as camphor to make it more workable, was originally used as a replacement for ivory in billiard balls and piano keys, and to make celluloid collars. This material, from which the first movie film was made, is notoriously flammable.

Cellulose acetate

Cellulose acetate, made by treating cellulose with acetic acid and acetic anhydride, was originally used as a waterproof varnish to coat the fabric of airplanes during World War I. It later became important as a photographic film base, and as acetate rayon.

Bakelite

The first completely synthetic organic polymer was Bakelite, named for its discoverer Leo Baekeland, a Belgian chemistry professor who invented Velox, the first successful photographic paper. He immigrated to America at age 35 and sold his invention to George Eastman for $1 million in 1899. Baekeland then turned his attention to finding a replacement for shellac, which comes from the Asian lac beetle. At the time, shellac was coming into great demand in the fledgling electrical industry as an insulator. The polymer Baekeland produced (Bakelite) is still used today for electrical plugs, switches, and the black handles and knobs on pots and pans. It has superior electrical insulating properties and very high heat resistance. It is made by the base-catalyzed reaction of excess formaldehyde with phenol. In a low molecular weight form, it is used to glue together layers of plywood or mixed with a filler such as sawdust it can be formed in any desired shape. When it is heated to a high temperature, cross-linking occurs as the polymer "cures."

Thermoplastic polymers

Most polymers are amorphous, linear macromolecules that are thermoplastic and soften at high temperatures. In Bakelite, the polymer cross-links to form a three-dimensional network, while also becoming a dark, insoluble, infusible substance. Such polymers are said to be *thermosetting*. Natural rubber is thermoplastic. It becomes a thermosetting polymer when heated with sulfur, as Charles Goodyear discovered. With 2% sulfur, the rubber becomes cross-linked but is still elastic; at 30% sulfur, the rubber can be made into bowling balls. Some other important thermosetting polymers are urea-formaldehyde resins and melamine-formaldehyde resins. The latter are among the hardest of polymers and take on a high-gloss finish. Melamine is used extensively to manufacture plastic dinnerware.

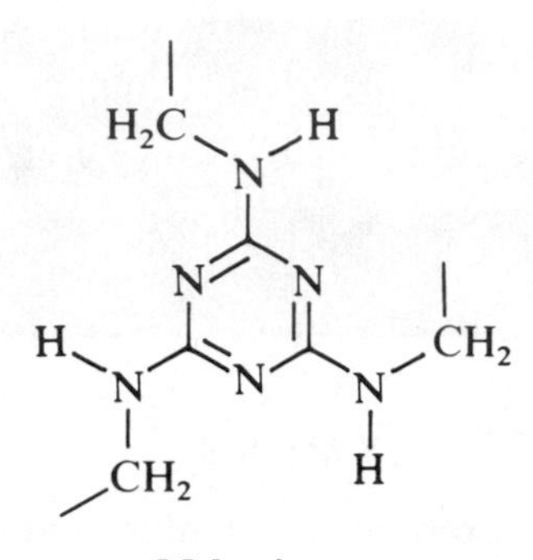

Melamine, a thermosetting polymer

Even though vinyl chloride was discovered in 1835, poly(vinyl chloride) (PVC) was not produced until 1912. It is now one of our most common polymers; production in 1997 was over 20 million metric tons. The monomer is made by the pyrolysis of 1,2-dichloroethane, which is formed by chlorination of ethylene. Free-radical polymerization of PVC follows Markovnikov's rule to give the head-to-tail polymer with high specificity:

$$nCH_2{=}CHCl \longrightarrow —(CH_2CHCl)_n—$$

Vinyl chloride **Polyvinyl chloride**

Plasticizers

Pure PVC is an extremely hard polymer. PVC, as well as some other polymers, can be modified by the addition of plasticizers; the greater the ratio of plasticizer to polymer, the greater the flexibility of the polymer. PVC pipe is rigid and contains little plasticizer, whereas shower curtains contain a large percentage of plasticizers. The most common plasticizer is di(2-ethylhexyl) phthalate, which can be added in concentrations up to 50%.

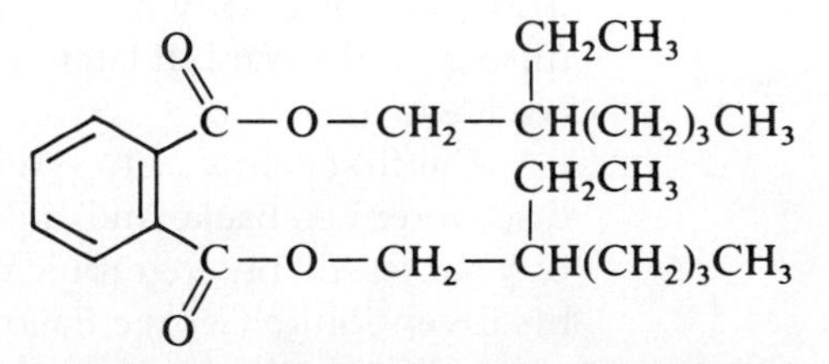

Di(2-ethylhexyl) phthalate (Dioctyl phthalate)

PVC is used for raincoats, house siding, and artificial leather (e.g., handbags, briefcases, and inexpensive shoes). It is found in garden hose, floor covering, swimming pool liners, and automobile upholstery. When vinyl upholstery is exposed to high temperatures, as in the interior of an automobile parked in the sun, the plasticizer distills out. The result is an opaque, difficult-to-remove film on the insides of the windows, and upholstery that is hard and brittle.

Polymerization methods

Monomers can be polymerized in the gas phase, in bulk, as suspensions and as emulsions. The most common method of making PVC is by emulsifying the vinyl chloride monomer in water with surfactants (soaps), water-soluble catalysts, and heat. The monomer is polymerized to solid particles, which are suspended in the aqueous phase. This product can be centrifuged and dried for further processing, or used as such. Chemists can control the average molecular weight of the product, which can become very high in emulsion polymerization. A high molecular weight means a more rigid and stronger polymer, but also one that is more difficult to work with because of greatly increased viscosity. An emulsion of polyvinyl acetate is sold as latex paint. When the vehicle of emulsion (water) evaporates, the polymer is left as a hard film. A thicker emulsion of polyvinyl acetate is an excellent adhesive, the familiar white glue. When vinyl acetate and vinyl chloride are polymerized together, a copolymer results with properties all its own. This copolymer is particularly good for detailed moldings, and was used to make phonograph records.

$—(CH_2CCl_2)_n—$

Polyvinylidene chloride

Poly(vinylidene chloride) is primarily extruded as a film that has low permeability to water vapor and air, and is therefore used as the familiar clinging plastic food wrap, Saran Wrap.

$—(CF_2CF_2)_n—$

Polytetrafluoroethylene

Poly(tetrafluoroethylene), also known as Teflon, is another halogenated polymer and has a number of unique properties. It has a very high melting point (327°C), it does not dissolve in any solvent, and nothing sticks to it. It is also an excellent electrical insulator. Teflon, a product of the DuPont company, is one of the densest polymers and also one of the most expensive. The surface of the polymer must be etched with metallic sodium to form free radicals to which glue can adhere. The polymer has an extremely low coefficient of friction, which makes it useful for bearings. Its chemical inertness makes it an ideal liner for chemical reagent bottle caps, and its no-stick property is ideal for the coating on the inside of frying pans. At 380°C, Teflon is still so viscous that it cannot be injection molded. Instead, it is molded by pressing the powdered polymer at high temperature and pressure, a process called *sintering*.

High-density polyethylene

The polymer produced in highest volume is polyethylene. Invented by the British, who call it *polythene*, and put into production in 1939, for a long time it could only be produced by the oxygen-catalyzed polymerization of ethylene at pressures near 2.81×10^7 kg/m^2 (40,000 lb/in.^{2}). Such pressures are expensive and dangerous to maintain on an industrial scale. The polyethylene produced has a low density and is used primarily to make film for bags of all types—from sandwich bags to trash can liners. The opaque appearance of polyethylene is due to crystallites, regions of order in the polymer that resemble crystals in an otherwise highly branched material. In the 1950s, Karl Ziegler and Giulio Natta developed catalysts composed of titanium chloride ($TiCl_4$), alkyl aluminum, and transition metal halides with which ethylene can be polymerized at pressures of just 3.16×10^5 kg/m^2 (450 lb/in.^{2}). The resulting product has a higher density and a higher softening temperature (20°C) than the low-density material. The catalysts that Ziegler and Natta developed, and for which they received the 1963 Nobel Prize in Chemistry, cause linear polymerization and thus a crystalline product.

$—(CH_2CH_2)_n—$

Polyethylene

High-density polyethylene is as rigid as polystyrene, and yet has high-impact resistance. It is used to mold very large articles such as luggage, the cases for domestic appliances, trash cans, and soft drink crates.

Polystyrene

Polystyrene is a brilliantly clear, high-refractive-index polymer familiar in the form of disposable drinking glasses. It is brittle and produces sharp, jagged edges when fractured. It softens in boiling water, and burns readily with a very smoky flame. But it can be foamed readily and makes a very good insulator—witness the disposable white hot-drink cup. It is used extensively for insulation when properly protected from ignition. The addition of a small quantity (7%) of poly(butadiene) to the styrene makes a polymer that is no longer transparent but has high impact resistance. Blends of acrylonitrile, poly(butadiene), and styrene (ABS) have excellent molding properties, and are used to make automobile bodies. One formulation can be chrome-plated for automobile grills and bumpers.

$CH_2{=}CHCN$	$C_6H_5CH{=}CH_2$	$CH_2{=}CH{-}CH{=}CH_2$
Acrylonitrile	**Styrene**	**1,3-Butadiene**

Rubber

Joseph Priestley, the discoverer of oxygen, named *rubber* for its ability to remove lead pencil marks. Rubber is an *elastomer*, defined as a substance that can be stretched to at least twice its length and return to its original size. The Germans, cut off from a supply of natural rubber, began manufacturing synthetic rubber during World War I. Called "buna" for butadiene and sodium (*Na*), the polymerization catalyst, it was not the ideal substitute. Cars with buna tires had to be jacked up when not in use because their tires would develop flat spots. The addition of about 25% styrene greatly improved the qualities of the product; styrene-butadiene synthetic rubber now dominates the market, a principal outlet being automobile tires. Adding 30% acrylonitrile to butadiene produces nitrile rubber, which is used to make conveyor belts, storage tank liners, rubber hoses, and gaskets.

Synthetic rubber; a styrenebutadiene copolymer

The chemist classifies polymers in several ways. There are thermosetting plastics such as Bakelite and melamine, and the much larger category of thermoplastic materials, which can be molded, blown, and formed after polymerization. There are the arbitrary distinctions made among plastics, elastomers, and fibers. And there are the two broad categories formed by the polymerization reaction itself: (1) *addition polymers* (e.g., vinyl polymerizations), in which a double bond of a monomer is transformed into a single bond between monomers; and (2) *condensation polymers* (e.g., Bakelite), in which a small molecule, such as water or alcohol, is split out as the polymerization reaction occurs.

Addition polymers

Condensation polymers

Nylon

One of the most important condensation polymers is nylon, a name so ingrained into our language that it has lost its trademark status. It was developed by Wallace Carothers, the director of organic chemical research at DuPont, and was the outgrowth of his fundamental research into polymer chemistry. Introduced in 1938, nylon was the first totally synthetic fiber. The most common form of nylon is the polyamide formed by the condensation of hexamethylene diamine and adipic acid:

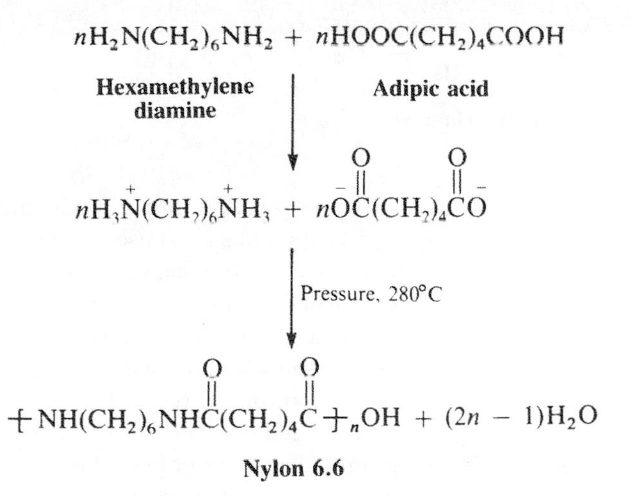

The reactants are mixed together to form a salt that melts at 180°C. This is converted into the polyamide by heating to 280°C under pressure, which eliminates water. Nylon 6.6 is used to make textiles, whereas nylon 6.10, from the 10-carbon diacid, is used for bristles and high-impact sports equipment. Nylon can also be made by interfacial and ring-opening polymerization, both of which are used in the following experiments.

The condensation polymer made by reacting ethylene glycol with 1,4-benzene dicarboxylic acid (terephthalic acid) produces a polymer that is converted in large part into polyester fibers such as Dacron. The polymerization is run as an ester interchange reaction using the methyl ester of terephthalic acid:

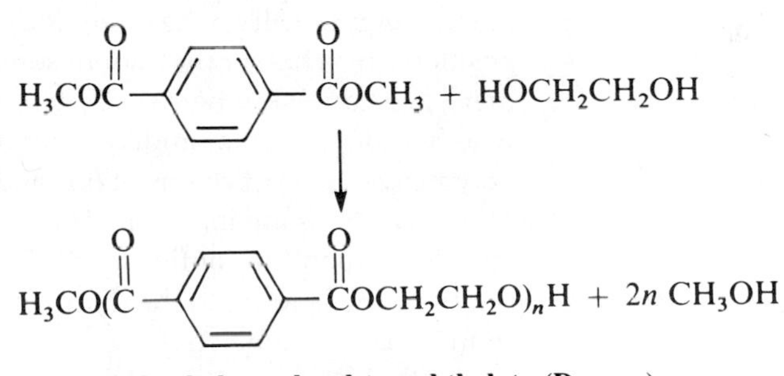

Polyethylene glycol terephthalate (Dacron)

Another condensation polymer, polycarbonate, has received much attention in the last few years. When bisphenol A is reacted with sodium hydroxide followed by phosgene, the polycarbonate condensation polymer is produced. It is hard, clear, and almost shatterproof, and hence used to make such things as CDs and DVD's, eyeglass lenses, laptop computer cases, as well as the lining of containers for canned food and beverages. But controversy has arisen over its use in baby bottles because extremely small amounts can be leached from the bottle, presumably by hydrolysis of the polymer. Most of the bisphenol A so produced is metabolized, but micrograms can be detected in blood and urine.

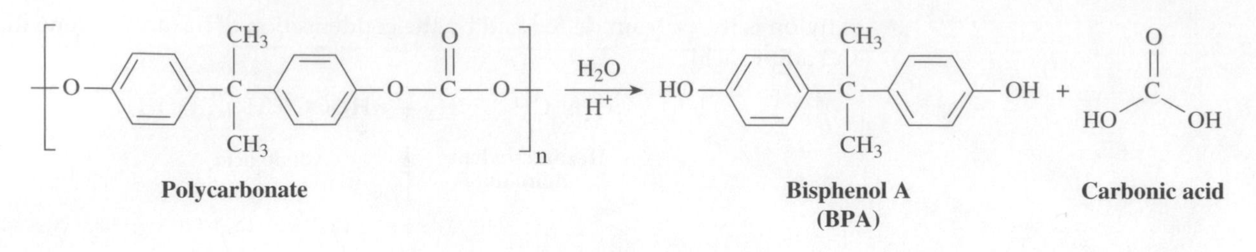

Bayer and GE began producing polycarbonate in 1960, and for the last twenty-five years it has been used as a safer alternative to glass for baby bottles because it is virtually unbreakable. Its relatively high melting point means it can be sterilized or washed in hot water, unlike most other plastics.

Bisphenol A has been carefully studied; it is an endocrine disruptor and has been implicated in a wide variety of problems—neurotoxicity, diabetes, and heart disease among others. It is present in the urine of more than 90% of adults; a formula-fed baby can consume up to 13 micrograms (10^{-6} g) of bisphenol per kilogram of body weight per day. But making a definitive connection between quantities this small and developmental problems is not easy. At present, it is recommended that children up to three years old minimize their exposure to bisphenol A. For others, it is probably not a problem, but media warnings that people should not drink out of "plastic" bottles has caused the withdrawal of all polycarbonate bottles from the shelves of many stores in the United States. Canada has outlawed the sale of polycarbonate baby bottles, but they are considered safe by almost all European nations.

To some extent, this recent attention is a result of greatly improved analytical methods, in particular gas chromatography/mass spectroscopy, which can detect exceedingly small amounts of such substances.

The structure of a polymer is not simple. For example, the polymerization of styrene produces a chiral carbon at each benzyl position. We can ask whether the phenyl rings are all on the same side of the long carbon chain, whether they alternate positions, or whether they adopt some random configuration. In the case of copolymers, we can ask whether the two components alternate (ABABABABAB . . .), whether they adopt a random configuration (AABBBABAAB . . .), or whether they polymerize as short chains of one and then the other (AAAABBBBBAAABBBB . . .). These questions are important because the physical properties of the resulting polymer depend on the configuration. The polymer chemist is concerned with finding the answers, leading to the discovery of catalysts and reaction conditions that can control these parameters.

In the experiments that follow, you will find that the preparation of nylon by interfacial polymerization is a spectacular and reliable experiment that is easily carried out in one afternoon. The synthesis of Bakelite works well, but it takes longer, requiring overnight heating in an oven to complete the polymerization. Nylon by ring-opening polymerization requires skill and care because of the high temperatures involved. The polymerization of styrene also requires care, but is somewhat easier to carry out.

In the recycling experiments in Part 2 of this chapter, you will depolymerize a PET bottle to short-chain polyols, and then use that material to make fiberglass and polyurethane foam.

EXPERIMENTS

1. NYLON BY INTERFACIAL POLYMERIZATION[1]

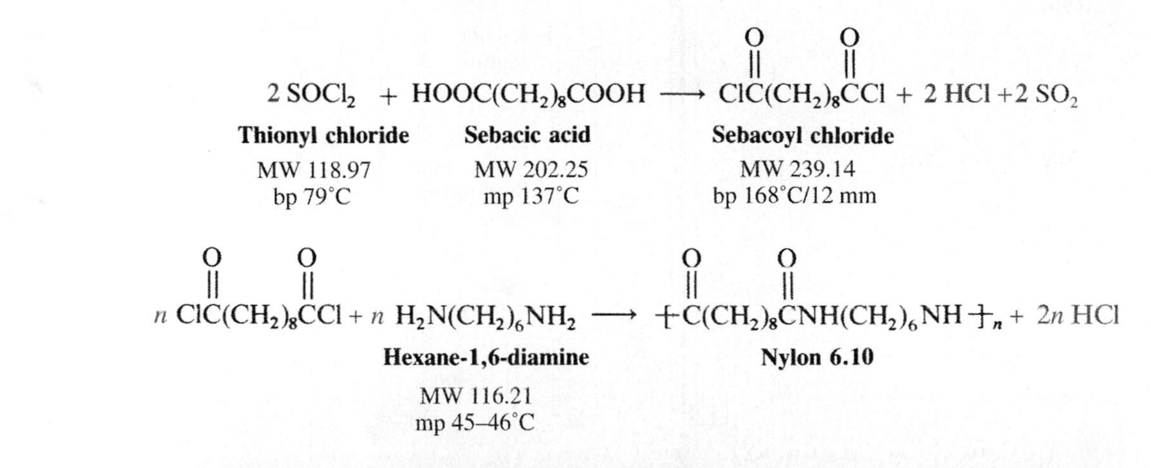

IN THIS EXPERIMENT, a 10-carbon diacid is converted to the corresponding 10-carbon diacid chloride with thionyl chloride. The hydrogen chloride gas given off during the reaction is trapped. The diacid chloride product is dissolved in dichloromethane and transferred to a beaker. An aqueous basic solution of a 6-carbon diamine is floated on top of the dense organic layer. Where the solutions come in contact, nylon 6.10 is formed and can be lifted from the beaker and wrapped on some type of cylinder.

In this experiment, a diamine dissolved in water is carefully floated on top of a solution of a diacid chloride dissolved in an organic solvent. Where the two solutions come in contact (the interface), addition/elimination occurs with formation of a tetrahedral intermediate ultimately giving rise to a film of a polyamide. The reaction stops there unless the polyamide is removed. In the case of nylon 6.10, the product of this reaction, the film is so strong that it can be picked up with a wire hook and continuously removed in the form of a rope.

This reaction works because the diamine is soluble in both water and dichloromethane, the organic solvent used. As the diamine diffuses into the organic layer, a reaction occurs immediately to give the insoluble polymer. The hydrochloric acid produced reacts with the sodium hydroxide in the aqueous layer. The chloride does not hydrolyze before reacting with the amine because it is not very soluble in water. The acid chloride is conveniently prepared using thionyl chloride.

Microscale Procedure

In a reaction tube fitted with a gas trap (Fig. 67.1), place 0.25 g of sebacic acid (1,8-octane dicarboxylic acid, otherwise known as decanedioic acid), 0.25 mL of thionyl chloride and 12 mg (a small drop) of *N,N*-dimethylformamide. Heat the tube

[1]Morgan PW, Kwolek SL. *J Chem Educ.* **1959**;*36*:182.

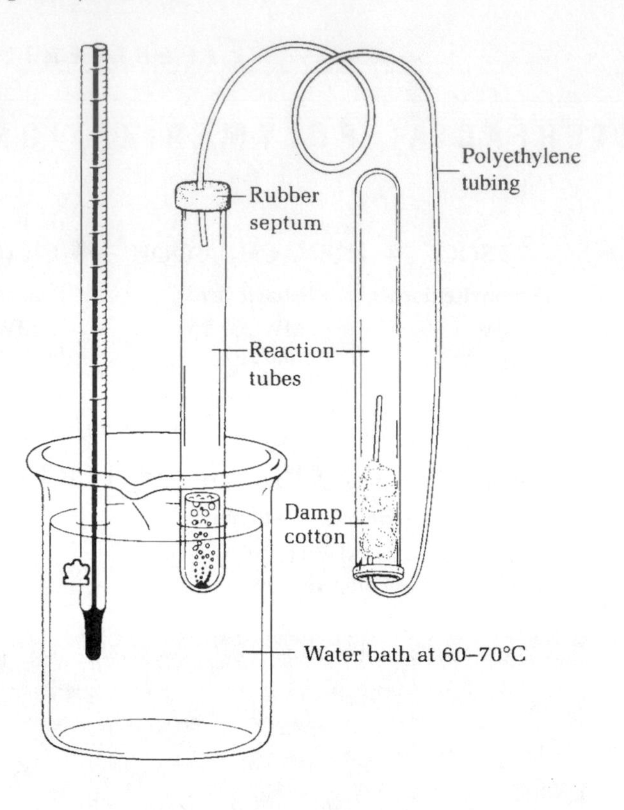

FIG. 67.1
An apparatus for acid chloride synthesis. HCl and SO_2 are trapped in the damp cotton.

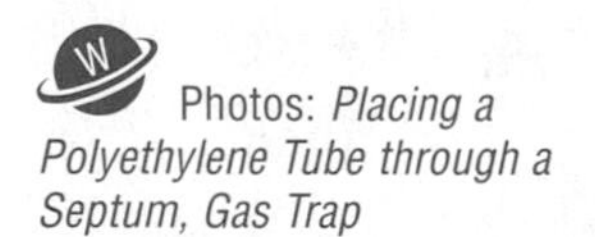
Photos: *Placing a Polyethylene Tube through a Septum, Gas Trap*

to 60–70°C in a water bath in the hood. This is best accomplished by putting very hot water in a beaker and placing the beaker on a steam bath or hot plate to maintain it at 60–70°C. As the reaction proceeds, the product forms a liquid layer on the bottom of the tube. Flick the tube to use this liquid to wash down unreacted acid (if necessary) as the reaction proceeds. When all of the acid has reacted and gas evolution has ceased (about 10–15 minutes), the product should be a clear liquid. Obtain a clean 50-ml beaker, and coat the inside with a thin layer of silicone vacuum grease or silicone oil to prevent the polymer from sticking to the glass. Then add some dichloromethane to the reaction tube and transfer the solution to this 50-mL beaker using a total of 12 mL of dichloromethane. Carefully pour onto the top of this dichloromethane solution 0.25 g of hexane-1,6-diamine (hexamethylenediamine) that has been dissolved in 6 mL of water containing 1 mL of 3 *M* sodium hydroxide. Immediately pick up the polymer film formed between the two layers at its center using a copper wire hook. Wrap the resulting polymer around the outer surface of a bottle, beaker, or graduated cylinder as it is removed. Wind up as much of the polymer as possible, wash it thoroughly in water, and press it as dry as possible. After the polymer has dried, determine its weight and calculate the percent yield. Try dissolving the polymer in two or three solvents. Attach a piece of the polymer to your laboratory report.

Cleaning Up. Add the cotton from the trap to the used reaction mixture, and stir the mixture vigorously to cause the nylon to precipitate. Decant the water and dichloromethane, and then squeeze the solid as dry as possible; place the solid in the nonhazardous solid waste container. Place the dichloromethane in the halogenated organic solvents waste container. After neutralization, dilute the aqueous layer with water and flush down the drain.

Macroscale Procedure[2]

In a 15-mL tapered 14/20 flask, fitted with a condenser and gas trap like the one shown in Fig. 18.3 (on page 332), place 2 g of sebacic acid (1,8-octane dicarboxylic acid, otherwise known as decanedioic acid), 2 mL of thionyl chloride, and 0.1 mL of *N*,*N*-dimethylformamide. Heat the flask to 60–70°C in a water bath in the hood. This is best accomplished by putting very hot water in a beaker and placing the beaker on a steam bath or hot plate to maintain it at 60–70°C. As the reaction proceeds, the product forms a liquid layer on the bottom of the flask. Use this liquid to wash down unreacted acid as the reaction proceeds. When all of the acid has reacted and gas evolution has ceased (about 10–15 minutes), transfer the product, which should be a clear liquid at this point, to a 250-mL beaker using 50 mL of dichloromethane. Carefully pour onto the top of this dichloromethane solution 2 g of hexane-1,6-diamine (hexamethylenediamine) that has been dissolved in 50 mL of water containing 1 g of sodium hydroxide. Pick up the polymer film at the center with a copper wire and wrap it around the outer surface of a bottle, beaker, or graduated cylinder as it is removed. Wind up as much of the polymer as possible, wash it thoroughly in water, and press it as dry as possible. After the polymer has dried, determine its weight and calculate the percent yield. Try dissolving the polymer in two or three solvents. Attach a piece of the polymer to your laboratory report.

Cleaning Up. Add the cotton from the trap to the used reaction mixture and stir the mixture vigorously to cause the nylon to precipitate. Decant the water and dichloromethane, and then squeeze the solid as dry as possible; place the solid in the nonhazardous solid waste container. Place the dichloromethane in the halogenated organic solvents waste container. After neutralization, dilute the aqueous layer with water and flush down the drain.

2. THE CONDENSATION POLYMERIZATION OF PHENOL AND FORMALDEHYDE: BAKELITE

The condensation of phenol with formaldehyde is a base-catalyzed process in which one resonance form of the phenoxide ion attacks formaldehyde, yielding a mixture of methoyl phenols. The resulting dimethoyl phenol is then cross-linked by heat, presumably by dehydration with the intermediate formation of benzylcarbocations. The resulting polymer is Bakelite. Because the cost of phenol is relatively high and the polymer is somewhat brittle, it is common practice to add an extender such as sawdust to the material before cross-linking. The mixture is placed in molds and heated to form the polymer. The resulting polymer, like other thermosetting polymers, is not soluble in any solvent and does not soften when heated. It is the plastic used to make the black handles on kitchen pots and pans that can withstand the heat of an oven.

[2]East GC, Hassell S. *J Chem Educ.* **1983**;*60*:69.

Microscale Procedure

IN THIS EXPERIMENT, phenol and formaldehyde are heated in the presence of ammonium hydroxide to produce a low molecular weight polymer. This material is poured into a disposable tube, separated from the aqueous layer, acidified with acetic acid, and heated overnight to complete the polymerization, producing Bakelite.

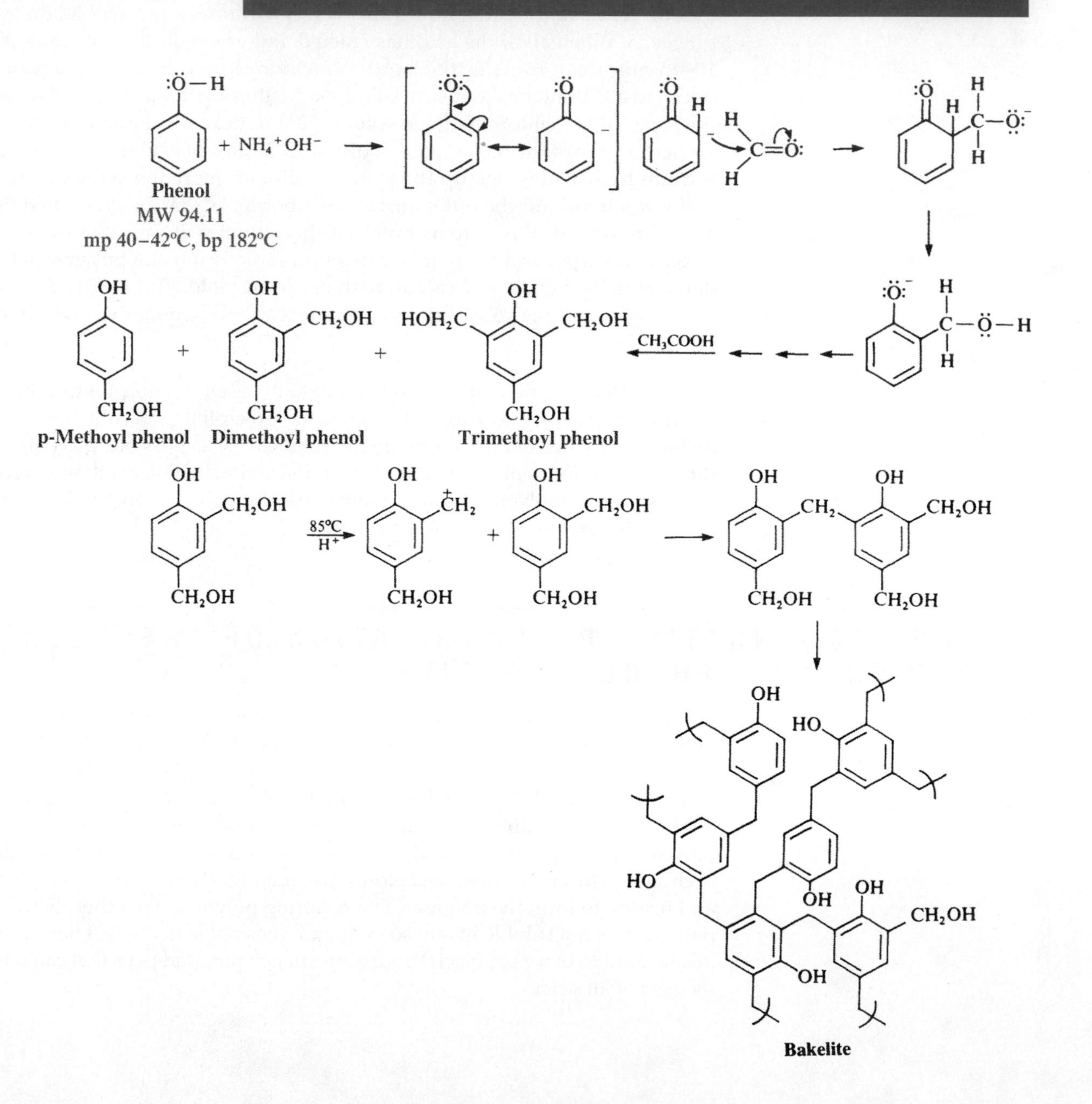

In a 5-mL short-necked, round-bottomed flask, place 0.75 g of phenol and 2.5 mL of 37% by weight aqueous formaldehyde solution. The formaldehyde solution contains 10%–15% methanol, which has been added as a stabilizer to prevent the formaldehyde from polymerizing. Add 0.37 mL of concentrated ammonium hydroxide to the solution, attach an empty distilling column as an air condenser (Fig. 67.2), and reflux it for 5 minutes beyond the point at which the solution turns cloudy; the total reflux time is about 10 minutes. In the hood, pour the warm solution into a small, disposable tube, for example, a 10 × 75-mm soft glass test tube or vial. *Immediately* wash out the empty flask with a small amount of acetone. Heat the tube and contents inside a steam bath or on a sand bath for 10 minutes to complete the reaction. Allow the mixture to stand at room temperature for 10–20 minutes. As the mixture cools, droplets of product will sink to the bottom, giving a clear upper layer that is then poured off. If the upper layer is poured off prematurely, the yield of product will be much lower. Warm the yellow-orange lower layer on a steam bath, and add acetic acid dropwise (0.5 mL maximum) with thorough mixing until the layer is clear. It should remain clear even when the polymer is cooled to room temperature. Heat the tube on a water bath at 60–65°C for 30 minutes; after placing a wood stick in the polymer to use as a handle, leave the tube, labeled with your name, in an 85°C oven overnight or until the next laboratory period. To free the polymer, the tube may need to be broken (Fig. 67.3). A filler such as sawdust or a small piece of cloth can be used to make a stronger polymer. Attach a piece of the polymer to your laboratory report.

Cleaning Up. Place the acetone wash in the organic solvents waste container. Should it be necessary to destroy formaldehyde, add 25 mL of household bleach for each milliliter of the solution. After 20 minutes, flush the solution down the drain. Discard the used glass tube pieces in the nonhazardous solid waste container.

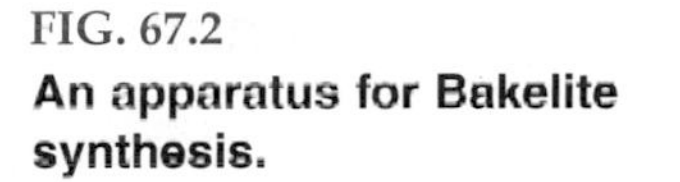

FIG. 67.2
An apparatus for Bakelite synthesis.

FIG. 67.3
To free the polymer, turn the thumbscrew to break the tube. Wear safety glasses. You may wish to wrap the tube in a piece of cloth.

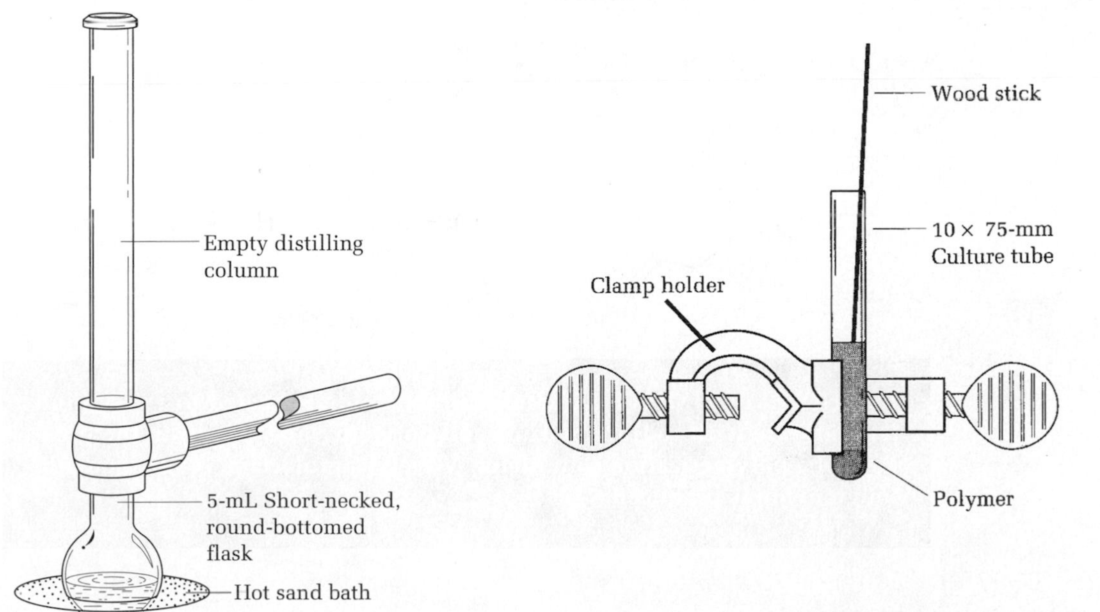

CAUTION: Formaldehyde solution is a suspected carcinogen. Handle the solution with care. Make measurements and carry out the experiment in the hood. The oven should be well ventilated or in a hood. Phenol is very corrosive to the skin.

Macroscale Procedure

In a 25-mL round-bottomed flask, place 3.0 g of phenol and 10 mL of 37% by weight aqueous formaldehyde solution. The formaldehyde solution contains 10%–15% methanol, which has been added as a stabilizer to prevent the formaldehyde from polymerizing. Add 1.5 mL of concentrated ammonium hydroxide to the solution, and reflux it for 5 minutes beyond the point at which the solution turns cloudy; the total reflux time is about 10 minutes. In the hood, pour the warm solution into a disposable test tube or culture tube and draw off the upper layer. Immediately wash out the empty flask with a small amount of acetone. Warm the viscous milky lower layer on a steam or sand bath, and add acetic acid dropwise with thorough mixing until the layer is clear. The layer should remain clear, even when the polymer is cooled to room temperature. Heat the tube on a water bath at 60–65°C for 30 minutes. Then, after placing a wood stick in the polymer to use as a handle, leave the tube, labeled with your name, in an 85°C oven overnight or until the next laboratory period. To free the polymer, the tube may need to be broken. Attach a piece of the polymer to your laboratory report.

Cleaning Up. Place the acetone wash in the organic solvents waste container. Should it be necessary to destroy formaldehyde, add 25 mL of household bleach for each milliliter of the solution. After 20 minutes, flush the solution down the drain. Discard the used glass tube pieces in the nonhazardous solid waste container.

3. NYLON BY RING-OPENING POLYMERIZATION[3]

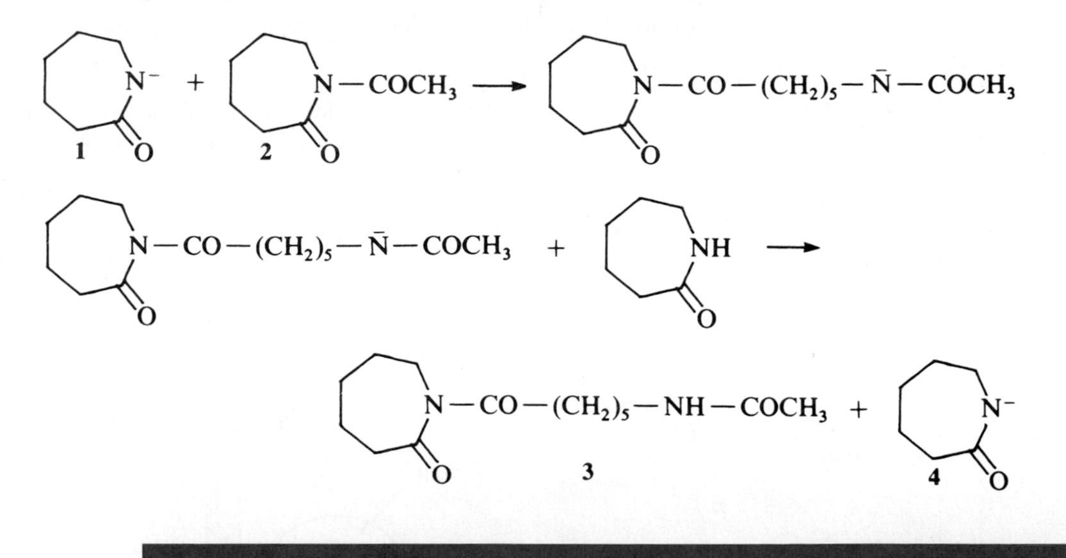

IN THIS EXPERIMENT, a cyclic 6-carbon amide is heated to a very high temperature with sodium hydride and poly(ethylene glycol) in the presence of a polymerization coin itiator, *N*-acetylcaprolactam. A very smooth, tough nylon is produced.

[3]Mathias LJ, Vaidya RA, Canterbury JB. *J Chem Educ.* **1984**;*61*:805.

Interfacial polymerization in the manner described in Experiment 1 is not a commercial process; the ring-opening of caprolactam, however, is a commercial process. The nylon produced, nylon 6.6, is used extensively in automobile tire cords and for gears and bearings in small mechanical devices.

The catalyst used in this reaction is sodium hydride; therefore, this is referred to as an anionic polymerization. The sodium hydride removes the acidic lactam proton to form an anion (1) that attacks the coinitiator, acetylcaprolactam (2), which has an electron-attracting acetyl group attached to the nitrogen. The ring of the acetylcaprolactam is attacked by the anion, and the acetylcaprolactam ring opens, forming a substituted caprolactam (3) that still has an electron-attracting group attached to nitrogen. A proton transfer reaction occurs, generating a new caprolactam anion (4), and 4 then attacks 3, and so on.

Ordinarily, anionic polymerizations must be run in the absence of oxygen, but the addition of poly(ethylene glycol) serves to complex with the sodium ion just as 18-crown-6 (crown ether) does, and enhances the catalytic activity of the sodium hydride.

Microscale Procedure

Set the heater to its maximum setting on your sand bath. The sand must be quite hot before beginning this experiment. Heating can also be done over a small Bunsen burner flame.

Active sodium hydride is gray. If it is white, it has probably decomposed on contact with a slight amount of moisture.

Into a disposable 10 × 75-mm test tube, place 1 g of caprolactam, 60 mg of poly(ethylene glycol), and 1 small drop of *N*-acetylcaprolactam. Heat the mixture. As soon as it has melted, remove it from the heat and add 20 mg of gray (not white) sodium hydride (50% dispersion in mineral oil). Mix the catalyst with the reactants by stirring with a Pasteur pipette, and heat the mixture *rapidly* to boiling (200–230°C). This should take place over a 2-minute period. Polymerization takes place rapidly, as indicated by an increase in viscosity. If polymerization has not occurred within 3 minutes, remove the tube from the heat, cool it somewhat, and add another 15 mg of sodium hydride. When the solution is so viscous that it will barely flow, insert a wood stick and with help from a partner, draw fibers from the melt. After it cools, the nylon 6.6 can usually be removed from the tube as one cylindrical piece. Try dissolving a piece of the nylon or the fibers in various solvents. Test the physical properties of the fibers by stretching them to the breaking point. Describe your observations, and attach a piece of fiber to your laboratory report. Do not forget to turn off the heater.

Cleaning Up. Place the used tube, if coated with polymer, in the nonhazardous solid waste container. Should it be necessary to destroy sodium hydride, add it to excess 1-butanol (38 mL/g of hydride). After the reaction has ceased, cautiously add water; then dilute with more water and flush the solution down the drain.

Macroscale Procedure

Set the heater to its maximum setting on your sand bath. The sand must be quite hot before beginning this experiment. Heating can also be done over a small Bunsen burner flame.

Into a disposable 4-in. test tube, place 4 g of caprolactam, 0.25 g of poly(ethylene glycol), and 2 drops of *N*-acetylcaprolactam. Heat the mixture. As soon as it has melted, remove it from the heat and add 50 mg of gray (not white) sodium hydride (50% dispersion in mineral oil). Mix the catalyst with the reactants by stirring with a Pasteur pipette, and heat the mixture *rapidly* to boiling (200°C–230°C). This should take place over a 2-minute period. Polymerization takes place rapidly, as indicated by an increase in viscosity. If polymerization has not occurred within 3 minute, remove the tube from the heat, cool it somewhat, and add another 50 mg of sodium hydride. When the solution is so viscous that it will barely flow, insert a wood stick and with help from a partner, draw fibers (mono-filament line) several feet long from the melt. After it cools, the nylon-6.6 can usually be removed from the tube as one cylindrical piece. Try dissolving a piece of the nylon or the fibers in various solvents. Test the physical properties of the fibers by stretching them to the breaking point. Describe your observations, and attach a piece of fiber to your laboratory report. Do not forget to turn off the heater.

Cleaning Up. Place the used tube, if coated with polymer, in the nonhazardous solid waste container. Should it be necessary to destroy sodium hydride, add it to excess 1-butanol (38 mL/g of hydride). After the reaction has ceased, cautiously add water; then dilute with more water and flush the solution down the drain.

4. POLYSTYRENE BY FREE-RADICAL POLYMERIZATION

Polystyrene, the familiar crystal-clear brittle plastic used to make disposable drinking glasses and, when foamed, lightweight white cups for hot drinks, is usually made by free-radical polymerization. An initiator is not used commercially because polymerization begins spontaneously at elevated temperatures. At lower temperatures, a variety of initiators could be used [e.g., 2,2′-azobis-(2-methyl-propionitrile), which was used in the free-radical chlorination of 1-chlorobutane]. In this experiment, we use benzoyl peroxide as the initiator. On mild heating, it splits into two benzoyloxy radicals

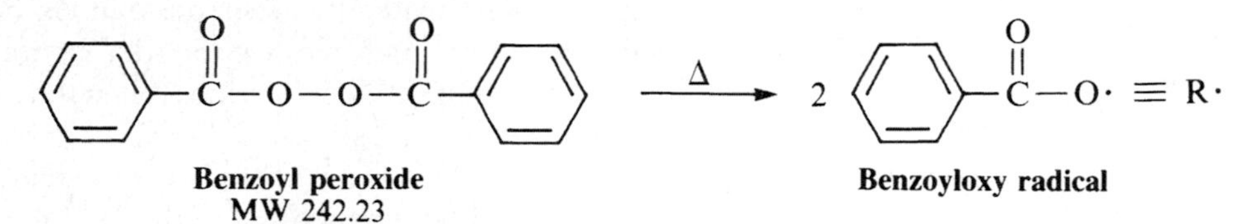

that react with styrene through initiation, propagation, and termination steps to form polystyrene:

Initiation:

$$R\cdot + CH_2{=}CH(C_6H_5) \longrightarrow RCH_2CH(C_6H_5)\cdot$$

Propagation:

$$RCH_2CH(C_6H_5)\cdot + CH_2{=}CH(C_6H_5) \longrightarrow RCH_2CH(C_6H_5)-CH_2CH(C_6H_5)\cdot + CH_2{=}CH(C_6H_5) \longrightarrow RCH_2CH(C_6H_5)-CH_2CH(C_6H_5)-CH_2CH(C_6H_5)\cdot\text{, etc.}$$

Termination:

$$2\,R\cdot \longrightarrow R-R,\quad RCH_2CH(C_6H_5)\cdot + R\cdot \longrightarrow RCH_2CH(C_6H_5)-R,\quad 2\,RCH_2CH(C_6H_5)\cdot \longrightarrow RCH_2CH(C_6H_5)-CH(C_6H_5)CH_2R$$

Styrene
MW 104.15
bp 145–146°C

$$\longrightarrow R-[CH_2-CH(C_6H_5)]_n-CH_2-CH(C_6H_5)-H$$

Polystyrene
MW 300,000–25,000,000

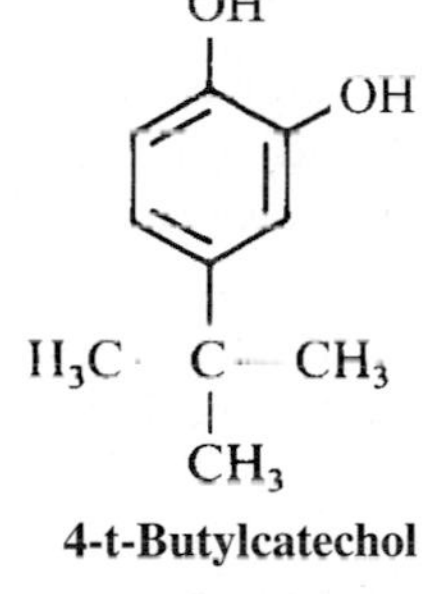

4-t-Butylcatechol

In the termination sequence, the first two reactions are rarely observed; termination occurs almost entirely by the third reaction. The final polymer has about 3000 monomer units in a single chain. The lifetime of the radical is about 1 second.

To prevent styrene from polymerizing in the bottle in which it is sold, the manufacturer adds 10–15 ppm of 4-*tert*-butylcatechol, a radical inhibitor (a particularly good chain terminator). The inhibitor must be removed by passing the styrene through a column of alumina before the styrene can be polymerized.

Synthesis of Polystyrene

Styrene is flammable, an irritant, and has a bad odor. Work with it in the hood.

IN THIS EXPERIMENT, the polymerization inhibitor is removed from commercial styrene by filtering it through alumina into a disposable culture tube. A small amount of peroxide catalyst is added, and the mixture is heated to initiate polymerization. A clear, very brittle plastic is produced.

CAUTION: Benzoyl peroxide is flammable, and may explode on heating or on impact. There is no need for more than a gram or two in the laboratory at any one time. It should be stored in and dispensed from waxed paper containers, not in metal or glass containers with screw-cap lids.

Inside a Pasteur pipette, loosely place a very small piece of cotton; then fill the pipette half full with alumina. Add to the top of the pipette 1.5 mL of styrene, and collect 1 mL in a disposable 10 × 75-mm test tube. Add to the tube 50 mg of benzoyl peroxide and a wood boiling stick; then heat the tube over a hot sand bath. When the temperature reaches about 135°C, polymerization begins; because it is an exothermic process, the temperature rises. Keep the reaction under control by cautious heating. The temperature rises, perhaps to 180°C, well above the boiling point of styrene (145°C); the viscosity also increases. Pull the wood stick from the melt from time to time to form fibers; when a cool fiber is found to be brittle, remove the tube from the heat. Alternatively, heat the tube on a steam bath for 90 minutes with the wood stick in place. Allow the tube to cool; then cool it in ice. Often, the polymer will shrink enough so that it can be pulled from the tube; otherwise break the tube. If the polymer is sticky, the polymerization can be completed in an oven overnight at a temperature of about 85°C.

Photo: *Filtration Using a Pasteur Pipette*

Cleaning Up. Shake the alumina out of the Pasteur pipette and place it in the styrene/alumina hazardous waste container. Clean up spills of benzoyl peroxide immediately. It can be destroyed by reacting each gram with 1.4 g of sodium iodide in 28 mL of acetic acid. After 30 minutes, neutralize the brown solution with sodium carbonate, dilute with water, and flush down the drain.

PART 2: RECYCLING PLASTICS

2.2 lb = 1 kg

Garbage. Trash. Hardly the concern of the organic chemistry lab. Or is it? In the United States, more than 250 million tons of trash is generated each year. Eighty-five percent of this trash is dumped into landfills, which are filling up fast. Plastics make up 11% of this waste. In any given year, we produce about 55 billion pounds of plastic, of which 22 billion pounds are discarded. More than half of this plastic has been used in packaging. By volume, this waste plastic makes up 30% of municipal solid waste. Some 28% of milk bottles and soft drink containers are recycled. Some plastic can be recycled simply by melting it and remolding it; other polymers must be treated chemically.

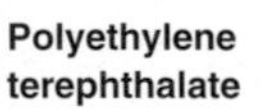

Polyethylene terephthalate

Plastics must be carefully sorted before recycling because some are incompatible with others. The Society of the Plastics Industry has adopted a coding system in which manufacturers place an alphanumeric identification code at or near the bottom of each container to identify the type of plastic resin being used. Poly(ethylene terephthalate) (PET) is the leading recycled plastic.

High-density polyethylene

In the experiments in this part of the chapter, a plastic soft-drink bottle made of PET is converted to fiberglass-reinforced polyester laminates and polyurethane foam.

Polyvinyl chloride

3 V

PET is the plastic from which clear plastic bottles for soft drinks are made by blow-molding. PET is made of repeating units of ethylene glycol and terephthalic acid (benzene 1,4-dicarboxylic acid). Each repeating unit has a molecular weight of 192; with the number of repeating units more than 100, the polymer has a molecular weight of about 20,000. When this polymer is formed into sheets, it is called *Mylar*, and when it is spun into fibers, it is called polyester (*Dacron*). In the United Kingdom, it is called *Terylene*. Millions of kilograms of PET are produced each year, and most of this ends up in waste dumps wherever it is not recycled. As we become increasingly aware of the environmental toll such waste entails, we are turning to recycling. This experiment demonstrates the commercial processes by which PET bottles are chemically recycled.

Low-density polyethylene

4 LDPE

Polypropylene

Polystyrene

Other, which includes polycarbonate

The process is fairly straightforward. The polymer is broken down (depolymerized) into small fragments that can then be used as the starting material for making a new polymer. In the case of PET, the new polymer could be fiberglass or polyurethane foam. It cannot be blow-molded again to make new beverage bottles because of possible contamination that decreases the strength of the polymer.

The chemical structure of PET is as follows:

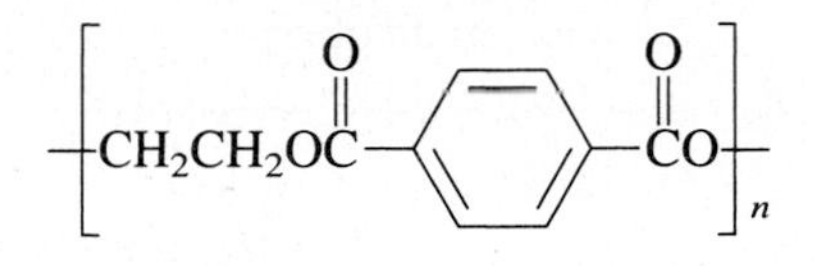

PET is an ester. Low-molecular-weight esters are prepared and hydrolyzed according to the following equilibrium reaction, which can be either acid- or base-catalyzed (see Chapter 40):

$$\mathrm{RC(=O)OH} + \mathrm{R'OH} \underset{}{\overset{\mathrm{H^+}}{\rightleftharpoons}} \mathrm{RC(=O)OR'} + \mathrm{H_2O}$$

If you put small pieces of PET in a flask with either aqueous acid or base and reflux the mixture, there will be no apparent reaction. It is remarkably resistant to hydrolysis. Try it for yourself. Esters undergo a related reaction called *transesterification* that can be used to depolymerize PET:

$$\mathrm{RC(=O)OR'} + \mathrm{R''OH} \overset{\mathrm{cat}}{\rightleftharpoons} \mathrm{RC(=O)OR''} + \mathrm{R'OH}$$

When applied to PET, the reaction produces a mixture containing three esters when the alcohol is diethylene glycol:

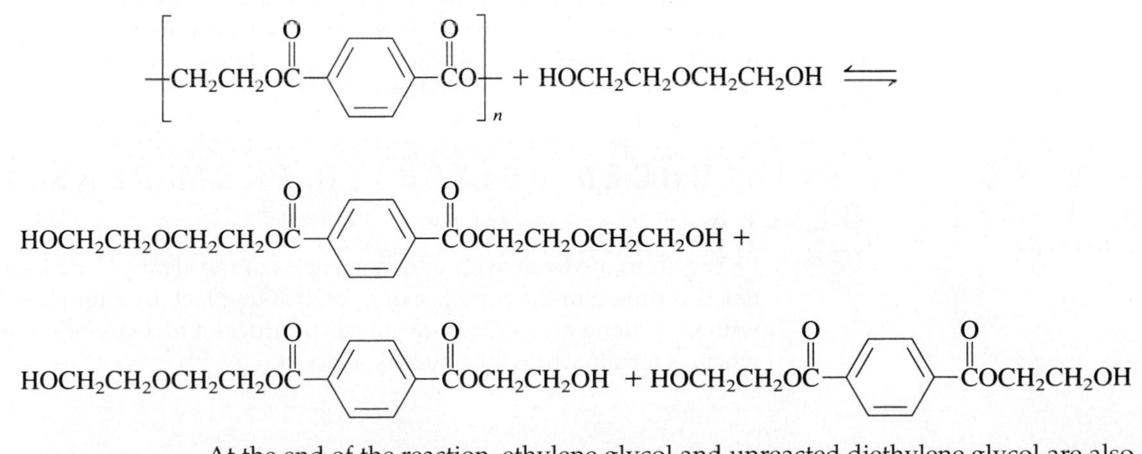

At the end of the reaction, ethylene glycol and unreacted diethylene glycol are also present. This reaction is catalyzed by a mixture of manganese acetate and manganese carbonate.

POLYURETHANE FOAM FROM PLASTIC BEVERAGE BOTTLES

The familiar foam-rubber cushions that you sit on are most likely made of polyurethane. It is also widely used in the building trades as a rigid insulating material in the form of lightweight boards.

A urethane (also called a carbamate) is a functional group that looks like an amide on one side and an ester on the other. It is made by reacting an isocyanate with an alcohol:

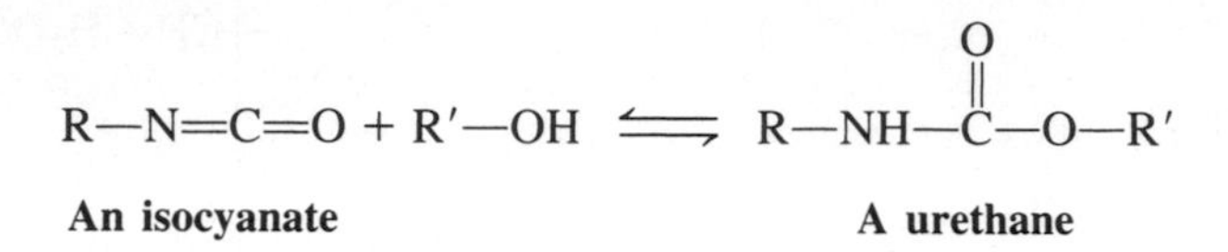

Polyurethanes, like many other polymers, are made from difunctional starting materials:

$$\mathrm{O{=}C{=}N{-}R{-}N{=}C{=}O + H{-}O{-}R'{-}O{-}H \rightleftharpoons O{=}C{=}N{-}\left[R{-}NH{-}\overset{\overset{\displaystyle O}{\|}}{C}{-}O{-}R'\right]_n{-}O{-}H}$$

A polyurethane

Isocyanates react with water to form an amine and carbon dioxide:

$$\mathrm{R{-}N{=}C{=}O + H_2O \longrightarrow R{-}\overset{\overset{\displaystyle H}{|}}{N}{-}\overset{\overset{\displaystyle O}{\|}}{C}{-}OH \longrightarrow R{-}\overset{\overset{\displaystyle H}{|}}{N}{-}H + CO_2}$$

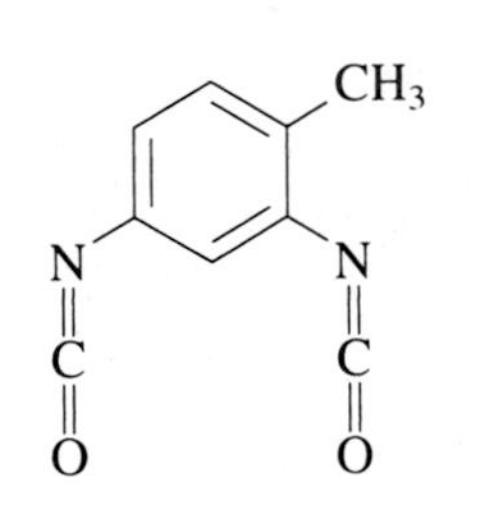

Tolylene 2,4-diisocyanate

For this reason, a small amount of water can be added during the polymerization step. The creation of carbon dioxide is desirable because it is one means of forming the bubbles in the foam.

As noted earlier, commercial polyurethane can be rigid or flexible, depending on the nature of the R and R′ group. In this experiment, the foamed polymer will be rigid. The most common diisocyanate is tolylene diisocyanate.

FIBERGLASS-REINFORCED POLYESTER FROM PLASTIC BEVERAGE BOTTLES

PET resin in the form of shredded particles of the clear plastic from beverage bottles is refluxed in the presence of a catalyst to effect an ester interchange reaction with propylene glycol. This glycolysis produces a mixture of three esters, each of which is terminated with a hydroxyl group.

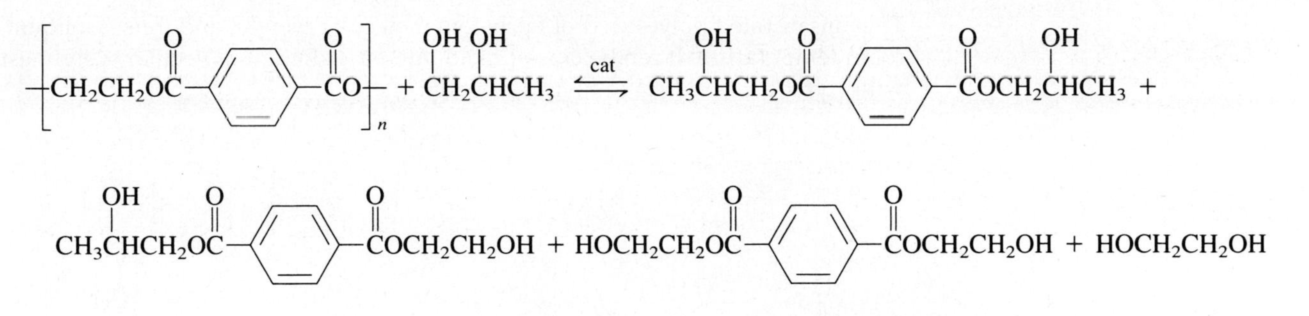

Alcohols react with anhydrides to give esters. In a very simple example, acetic anhydride reacts with ethyl alcohol to give the ester ethyl acetate and acetic acid:

$$CH_3\overset{O}{\overset{\|}{C}}O\overset{O}{\overset{\|}{C}}CH_3 + CH_3CH_2OH \longrightarrow CH_3\overset{O}{\overset{\|}{C}}OCH_2CH_3 + CH_3\overset{O}{\overset{\|}{C}}OH$$

If the anhydride is maleic anhydride, then the resulting ester terminates in a carboxyl group, which, in an equilibrium reaction, can react with another molecule of alcohol to give an unsaturated ester and water:

$$\text{(maleic anhydride)} + HOCH_2CH_3 \longrightarrow HO\overset{O}{\overset{\|}{C}}CH{=}CH\overset{O}{\overset{\|}{C}}OCH_2CH_3$$

$$CH_3CH_2OH + HO\overset{O}{\overset{\|}{C}}CH{=}CH\overset{O}{\overset{\|}{C}}OCH_2CH_3 \rightleftharpoons CH_3CH_2O\overset{O}{\overset{\|}{C}}CH{=}CH\overset{O}{\overset{\|}{C}}OCH_2CH_3 + H_2O$$

This reaction is employed with the mixture of diols produced in the glycolysis of PET. When this mixture reacts with maleic anhydride and water is allowed to distill from the reaction mixture, the result is an unsaturated polymeric ester:

$$CH_3\overset{OH}{\overset{|}{C}}HCH_2O\overset{O}{\overset{\|}{C}}-C_6H_4-\overset{O}{\overset{\|}{C}}OCH_2\overset{OH}{\overset{|}{C}}HCH_3 + \text{(maleic anhydride)} \longrightarrow$$

$$\left[CH_3\overset{O-}{\overset{|}{C}}HCH_2O\overset{O}{\overset{\|}{C}}-C_6H_4-\overset{O}{\overset{\|}{C}}OCH_2\overset{O\overset{O}{\overset{\|}{C}}CH=CH\overset{O}{\overset{\|}{C}}-}{\overset{|}{C}}HCH_3 \right]_n + H_2O$$

This unsaturated polyester is of fairly low molecular weight, and thus is a liquid at room temperature. It can be cross-linked into a rigid, high molecular weight ester with styrene in a radical-catalyzed reaction:

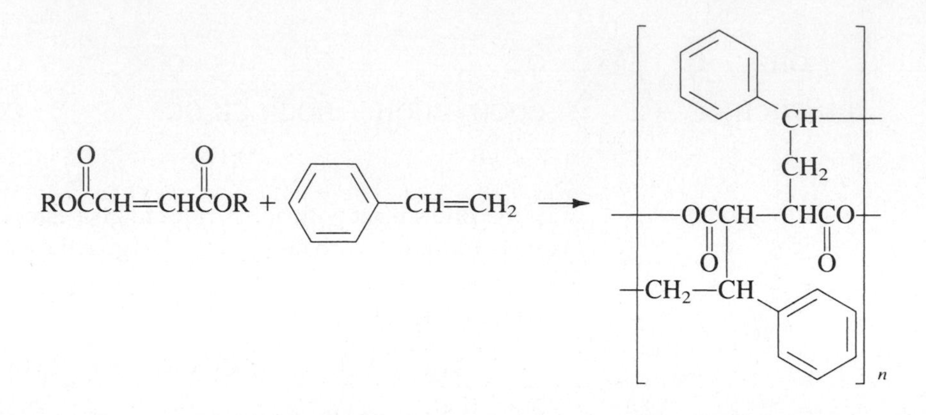

This polyester, when reinforced with fiberglass, can be molded and polymerized in place to form such things as automobile bodies and boats up to 100 ft in length.

EXPERIMENTS

1. DEPOLYMERIZATION OF PET WITH DIETHYLENE GLYCOL

IN THIS EXPERIMENT, shredded PET is heated with diethylene glycol, a 4-carbon dialcohol, and a catalyst. Depolymerization results in the formation of three esters in a clear solution. This material is used to prepare polyurethane foam.

Microscale Procedure

Into a 13 × 100-mm test tube, weigh 1.74 g of PET from a clear soft-drink bottle that has been cut into 6-mm ($\frac{1}{4}$-in.) squares or shredded material from a commercial shredder.[4] Add 3 mg of manganese acetate [$Mn(OAc)_2$] and 3 mg of manganese carbonate ($MnCO_3$), followed by 1.26 g of diethylene glycol. Add a boiling chip and reflux the mixture. Note the time at which all the chips of PET disappear. If the mixture is cooled at this point, it will have a cloudy white appearance because the material still has a high molecular weight and comes out of solution. Continue to reflux for a total of 45 minutes; at the end of this time, the mixture should remain clear after cooling. At this point, the glycolysis (trans-esterification) is finished, and the tube contains the mixture of three esters shown on page 773. This material will be used for making fiberglass and polyurethane foam in the next two experiments.

[4]In places where a deposit on bottles is required, there is often a machine that will shred the bottles and return the deposit. Obtain this shredded material from the store and treat it first with toluene to remove the label glue and then, after drying, separate it by flotation in water. Any paper and the black polyethylene base of the bottle will float, the aluminum cap will sink, and the PET (density about 1) will be suspended in the water.

Look up information on PET. How much is produced each year? Are Mylar and Dacron simply different forms of this polymer? See if you can discover anything about the use of manganese acetate and manganese carbonate to catalyze a reaction of this type.

Macroscale Procedure

Into a 20 × 150-mm test tube, weigh 7.5 g of PET from a clear soft-drink bottle that has been cut into 6-mm ($\frac{1}{4}$-in.) squares or shredded material from a commercial shredder.[5] Add 18 mg of manganese acetate [$Mn(OAc)_2$] or antimony acetate [$Sb(OAc)_3$] and 18 mg of manganese carbonate ($MnCO_3$), followed by 5.0 g of diethylene glycol. Add a boiling chip and reflux the mixture. Poke the shredded plastic down into the liquid as the reaction proceeds. Note the time at which all the chips of PET disappear. If the mixture is cooled at this point, it will have a cloudy white appearance because the material still has a high molecular weight and comes out of solution. Continue to reflux for a total of 45 minutes; at the end of this time, the mixture should remain clear after cooling. At this point, the glycolysis (transesterification) is finished, and the tube contains the mixture of three esters shown on page 773. This material will be used for making fiberglass and polyurethane foam in the next two experiments.

Look up information on PET. How much is produced each year? Are Mylar and Dacron, now known as polyester, simply different forms of this polymer? See if you can discover anything about the use of manganese acetate or antimony acetate and manganese carbonate to catalyze a reaction of this type.

2. PREPOLYMER FOR POLYURETHANE FOAM

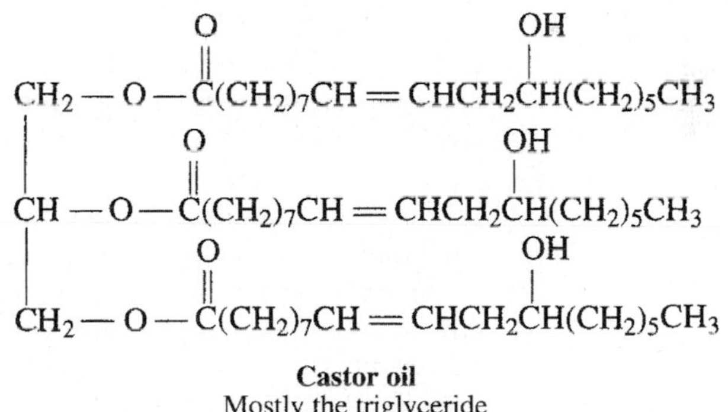

Castor oil
Mostly the triglyceride of ricinoleic acid

CAUTION: Tolylene diisocyanate is extremely toxic, a strong irritant to the respiratory tract and skin, and a suspected carcinogen. Neutralize spills with 5% aqueous ammonia solution. Wash from the skin with 30% aqueous isopropyl alcohol, and then with soap and water. Handle the prepolymer with as much care as the starting diisocyanate.

Into a 100-mL round-bottomed flask, place 21.5 g of castor oil and 7.5 g of transesterified PET from the preceding experiment. Add a magnetic stirring bar and thermometer and, with vigorous stirring in the hood, add 27.5 g of tolylene 2,4-diisocyanate all at once from an Erlenmeyer flask. Over about the next 15 minutes, the temperature of the mixture rises to 80°C. Some foaming takes place. Once the temperature starts to drop, place the flask in a Thermowell and heat the mixture at 120°C for 1 hour. Remove it from the heat and cool it in a water bath to room temperature. This prepolymer should contain approximately 15% free isocyanate. It will keep overnight or longer, if refrigerated.

[5]*See* footnote 4.

CAUTION: Do not carry out this experiment unless good hood facilities are available!

3. Polyurethane Foam

To prepare foamed polyurethane, in a small paper cup, mix 0.175 g of buffered catalyst (prepared from 7 g of diethylaminoethanol, 3.32 mL of concentrated hydrochloric acid, and 6.6 mL of water), 1 drop of Dow Corning silicone oil, and 0.55 mL of water. To this, add 5 g of the prepolymer mix (from the preceding experiment) and stir the mixture thoroughly with a metal spatula. Carry out this operation in the hood. Leave the cup with your name on it in the hood until the next lab period.

4. CONVERSION OF PET INTO FIBERGLASS-REINFORCED POLYESTER

Microscale Procedure

Into a 20 × 150-mm test tube, place 3 g of PET that is cut into 6-mm ($\frac{1}{4}$-in.) squares or shredded material from a commercial shredder,[6] add 1.4 g of propylene glycol, 5 mg of manganese acetate tetrahydrate, and 2 mg of manganese carbonate. Insert a thermometer into the mixture as far as possible, and heat the mixture on a sand bath until the propylene glycol boils (187°C). A rapid increase in the temperature is taken as an indication that the glycolysis reaction is over. When all the particles have disappeared and the temperature reaches 200°C, cool the mixture to about 150°C and add 1.56 g of maleic anhydride. Heat the mixture to boiling, and allow about 0.3 mL of water to escape (you probably will not see it leaving). Again, monitor the temperature. When it reaches 215°C, the reaction is probably over. Cool the mixture to 140°C and, with stirring, add to it 4.6 g of styrene in the hood.

CAUTION: Cobalt acetate is a suspected carcinogen and a mutagen. Handle with care.

The mixture of styrene and unsaturated polyester resin can now be polymerized. To do this, add 30 mg of cobalt acetate, stir well with a stirring rod, and then add 100 mg of methyl ethyl ketone peroxide (the radical catalyst) with thorough stirring. The mixture can be poured onto aluminum foil to polymerize; alternatively, the mixture can be poured onto 2.5 g of fiberglass on aluminum foil, the air squeezed out, and then allowed to polymerize in the hood. This fiberglass-reinforced polyester should be about 3-mm ($\frac{1}{8}$-in.) thick. The original procedure says that these castings are allowed to cure at room temperature for 16 hours, followed by a post-cure at 70°C for 1 hour and a 100°C cure for 1 hour. You could simply place the unpolymerized material in a well-ventilated oven at 80°C overnight.

Macroscale Procedure

Into a 20 × 150-mm test tube, place 6 g of PET cut into 6-mm ($\frac{1}{4}$-in.) squares or material from a commercial shredder.[7] Add 2.8 g of propylene glycol, 10 mg of manganese acetate tetrahydrate, and 4 mg of manganese carbonate. Insert a thermometer into the mixture as far as possible, and heat the mixture on a sand bath until the propylene glycol boils (187°C). A rapid increase in the temperature is taken as an indication that the glycolysis reaction is over. When all the particles have disappeared and the temperature reaches 200°C, cool the mixture to about 150°C and

[6] *See* footnote 4.
[7] *See* footnote 4.

add 3.12 g of maleic anhydride. Heat the mixture to boiling, and allow about 0.6 mL of water to escape (you probably will not see it leaving). Again, monitor the temperature. When it reaches 215°C, the reaction is probably over. Cool the mixture to 140°C and, with stirring, add to it 9.2 g of styrene in the hood.

CAUTION: Cobalt acetate is a suspected carcinogen and a mutagen. Handle with care.

The mixture of styrene and unsaturated polyester resin can now be polymerized. To do this, add 60 mg of cobalt acetate, stir well with a stirring rod, and then add 0.20 g of methyl ethyl ketone peroxide (the radical catalyst) with thorough stirring. This mixture can be poured onto aluminum foil to polymerize; alternatively, the mixture can be poured onto 5 g of fiberglass on aluminum foil, the air squeezed out, and then allowed to polymerize in the hood. This fiberglass-reinforced polyester should be about 3-mm ($\frac{1}{8}$-in.) thick. The original procedure says that these castings are allowed to cure at room temperature for 16 hour, followed by a post-cure at 70°C for 1 hour, and a 100°C cure for 1 hour. You could simply place the unpolymerized material in a well-ventilated oven at 80°C overnight.

QUESTIONS

1. What might the products be from an explosion of smokeless gunpowder? How many moles of carbon dioxide and water would come from 1 mol of trinitroglucose? Does the molecule contain enough oxygen for the production of these two substances?
2. Write a balanced equation for the reaction of sebacoyl chloride with water.
3. In the final step in the synthesis of Bakelite, the partially polymerized material is heated at 85°C for several hours. What other product is produced in this reaction?

See *Web Links*

4. Give the detailed mechanism of the first step of the ring-opening polymerization of caprolactam.

68 CHAPTER

Searching the Chemical Literature

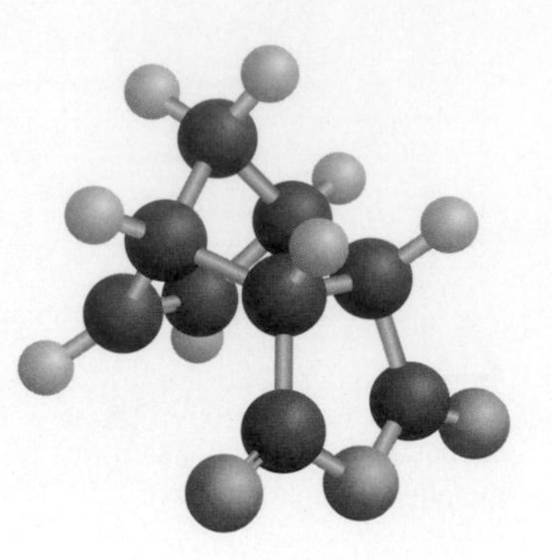

In planning a synthesis or investigating the properties of compounds, the organic chemist is faced with the problem of locating information on the chemical and physical properties of substances that have been previously prepared, as well as the best methods and reagents for carrying out a synthesis. With tens of millions of compounds known and new ones being reported at the rate of more than 800,000 per year, this task might seem formidable. However, the field of chemistry has developed one of the best information retrieval systems of all the sciences. In a university chemistry library or over the internet (at an academic institution that has subscribed to the necessary databases), it is possible to locate information on almost any known compound in a few minutes. Even a modest library will have information on several million different substances in printed form.

The ultimate source of information is the primary literature: articles written by individual chemists and published in journals. The secondary literature consists of compilations of information taken from these primary sources. For example, to find the melting point of benzoic acid, one would naturally turn to a secondary reference, a handbook or on the web, containing a table of physical constants of organic compounds. If, on the other hand, one wished to have information about the phase changes of benzoic acid under high pressure, one would need to consult the primary literature. If a piece of information, such as a melting point, is crucial to an investigation, the primary literature should be consulted because, for instance, transcription errors can occur in the preparation of secondary references.

In this chapter, we will first consider the resources available on the internet and then the traditional chemistry library, although a distinction between the real and virtual library is becoming less clear each day.

THE MODERN CHEMISTRY LIBRARY: ON THE WEB

SCIFINDER SCHOLAR

Although there will probably always be a chemistry library, more and more of the resources of the traditional library are now on the internet.

The American Chemical Society (ACS) owns the Chemical Abstracts Service (CAS) that publishes *Chemical Abstracts* and maintains access to over 200 databases through *SciFinder*.

The place to start searching this mass of information is *SciFinder Scholar* on the web, which will lead you into *SciFinder*, a graphical software interface for both

Windows and Mac computers that allows the search of the 200 *Chemical Abstracts* databases with minimal or no training. Just draw the structure of the molecule of interest using the included software and obtain a wealth of information about it—physical properties, references to its synthesis and uses, etc. *SciFinder Scholar* has a number of tutorials available to aid in the use of *SciFinder*.

It is possible to conduct substructure searches of databases that are impossible via the printed page. For example, one can ask for a list of all three-membered rings bearing a single hydroxyl group, even if the ring is buried in a much larger structure.

CrossFire Beilstein is a complementary database of chemical substances and reactions particularly useful to the synthetic organic chemist, and is not a part of *SciFinder*.

Chemical Abstracts was a telephone-directory-size book that was published once a week in order to keep up with abstracting the world's chemical literature, now coming in at the rate of 900,000 references per year. These 52 volumes became unwieldy and expensive to print and mail. The printed version ceased publication at the end of 2009 and now it is only available on the web, to subscribers, like the printed version.

All 2.5 million pages of all 34 research journals published by the American Chemical Society since it began in 1879 are now online and searchable. No longer will it be necessary to go to the library stacks and pull off, for example, volume 26 of the *Journal of Organic Chemistry*, turn to page 4563, and write down the details of the syntheses of steroidal α-acetoxyketones. If your institution subscribes to *SciFinder*, you can obtain the same information in a fraction of the time on your computer.

A subscription to the database, which costs thousands of dollars per year ($41/hr of connect time) might seem expensive until you consider how many cubic feet those volumes would occupy in a library, and what it costs to build and maintain that space. There are several other commercial vendors of databases in addition to the Chemical Abstracts Service, but it is the primary one, allowing you through *SciFinder* to carry out structural searches of the 50 million substances in their databases.

The Conventional Chemistry Library

HANDBOOKS

For rapid access to information such as melting point, boiling point, density, solubility, optical rotation, λ_{max}, and crystal form, one turns first to the *Handbook of Chemistry and Physics*, now in its 89th edition.[1] This book contains information on some 15,000 organic compounds, including the *Beilstein* reference to each compound. The compounds presented are well known and completely characterized. Most are commercially available. *The Merck Index*[2] contains information on more than 10,000 compounds, especially those of pharmaceutical interest. The usual information such as physical properties and literature references to synthesis and

[1]Lide DR, Ed. *CRC Handbook of Chemistry and Physics*, 89th ed; CRC Press: Boca Raton, FL (or go online).
[2]O'Neill MJ, Smith A, Heckelman PE, Koch CB, Roman KJ, Eds. *The Merck Index*, 14th ed; Merck and Co, Inc: Rahway, NJ.

isolation, as well as medicinal properties, such as toxicity data, can be found in this large volume. Approximately one-third of *The Merck Index* is devoted to a long cross index of names (which is very useful for looking up drugs), a table of organic name reactions, an excellent section on first aid for poisons, a list of chemical poisons, and a list of the locations of many poison control centers. *The Merck Index* is also available online for a fee.

In nearly every organic chemistry laboratory throughout the world can be found a copy of the *Aldrich Handbook*, a comprehensive catalog of chemicals and equipment related to organic chemistry.[3] Not only does this catalog list the prices of more than 48,000 chemicals (primarily organic) as well as their molecular weights, melting points, boiling points, and optical rotations, it also gives references to infrared (IR) and nuclear magnetic resonance (NMR) spectral data and to *Beilstein's Handbüch der organischen Chemie, The Merck Index*, and *Fiesers' Reagents for Organic Synthesis*. In addition, the catalog provides the reference to the *Registry of Toxic Effects of Chemical Substances* (RTECS #) for each chemical[4] and, if appropriate, the reference to *Sax's Dangerous Properties of Industrial Materials*.[5] Special hazards are noted (e.g., "severe poison," "lachrymator," and "corrosive"). The Aldrich catalog also provides pertinent information (uses, physiological effects, etc.) with literature references for many compounds. Material Safety Data Sheets (MSDSs) are also available from Aldrich.

A complete list of all chemicals sold commercially and the addresses of the companies making them is found in *Chem Sources*,[6] which is published yearly.

After consulting these three single-volume references, *The Handbook of Chemistry and Physics*, the *Aldrich Handbook*, and *The Merck Index*, one would turn to more comprehensive multivolume sources such as the *Dictionary of Organic Compounds*.[7] This dictionary, still known as "Heilbron," the name of its former editor, now comprises nine volumes of specific information, with primary literature references on the syntheses, reactions, and derivatives of more than 50,000 compounds.

The ultimate reference for chemistry as a whole, including articles on individual substances, environmental aspects, industrial processes, pharmaceuticals, polymers, food additives, and so on, is the 27-volume *Kirk-Othmer Encyclopedia of Chemical Technology*.[8] It is also available, for a fee, at Kirk-Othmer Online.[9]

Beilstein's Handbüch der organischen Chemie is certainly not a "handbook" in the traditional sense—it can occupy an entire alcove of a chemistry library! This reference covers every well-characterized organic compound that has been reported in the literature up to 1979—syntheses, properties, and reactions. Included with the main reference set are three supplements covering the periods 1910–1919, 1920–1929, and 1930–1949; later supplements are available as well. Although written in German, this reference can provide much information even to those who possess no knowledge of the language. Physical constants and primary literature references are easy to pick out. A German-English dictionary for *Beilstein* is available on the Internet, as well as a number of guides and leading references to its use.

[3]The *Aldrich Handbook* is free from the Aldrich Chemical Company. Obtain a copy at http://www.sigmaaldrich.com/technical-service-home/literature-request.html#catalogs.
[4]RTECS is a compendium of toxicity data extracted from the open scientific literature. Formerly maintained by the National Institute for Occupational Safety and Health, it is now managed by MDL Information Systems, Inc., a wholly owned subsidiary of Elsevier Science, Inc.
[5]Lewis RJ. *Sax's Dangerous Properties of Industrial Materials*; 11th ed, John Wiley & Sons: New York, 2004.
[6]*Chem Sources* is recompiled annually and published by Chemical Sources International, Inc.
[7]Buckingham J. *Dictionary of Organic Compounds*, 6th ed; Chapman & Hall/CRC Press: Boca Raton, FL, 1996.
[8]*Kirk-Othmer Encyclopedia of Chemical Technology*, 5th ed; John Wiley and Sons: New York, 2004.
[9]The online version is at http://www.MRW.interscience.wiley.com/kirk/

The *Handbüch* is organized around a complex classification scheme that is explained on the Internet and in *The Beilstein Guide*,[10] but the casual user can gain access through the *Handbook of Chemistry and Physics* and the *Aldrich Catalog*. In the *Handbook of Chemistry and Physics* the reference for 2-iodobenzoic acid is listed as **B9**^{2}, 239, which means the reference will be found on p. 239 of Vol. (Band) 9 of the second supplement (Zweites Ergänzungwerk, EII). In the *Aldrich Catalog* the reference is given as Beil **9**, 363, which indicates information can be found in the main series on p. 363, Vol. 9. At the top of that page will be a system number. The system number assigned to each compound can be traced through the supplements. The easiest access to the *Handbüch* itself is through the formula index (General Formel register) of the second supplement. A rudimentary knowledge of German will enable one to pick out, for example, iodo (Jod) benzoic acid (Saure).

A modern, up-to-date English version of the *Handbüch* is available as *CrossFire Beilstein* for a fee on the Internet.[11] Some 100 properties of 9 million compounds and 9.5 million reactions are available for searching.

CHEMICAL ABSTRACTS

Secondary references are incomplete; some suffer from transcription errors, and the best—*Beilstein's Handbüch der organischen Chemie*—is not completely up-to-date in its survey of the organic literature, so one must often turn to the primary chemical literature. However, there are more than 14,000 periodicals where chemical information might appear. Chemists are fortunate to have *Chemical Abstracts*, an index that covers this huge volume of literature very promptly. The information in each abstract is then compiled into author, chemical substance, general subject, formula, and patent indexes. It is now available exclusively on the internet.

Since publication of *Chemical Abstracts* began in 1907, chemical nomenclature has changed. Now, no trivial (common) names are used, so acetone appears as 2-propanone and *o*-cresol appears as benzene, 1-methyl, 2-hydroxy. It takes some experience to adapt to these changing names; even now, *Chemical Abstracts* does not follow exactly the rules of the International Union of Pure and Applied Chemistry (IUPAC). The formula index is useful because it provides not only reference to specific abstracts, but also a correct name that can be found in the chemical substances index.

Current Contents

As long as a year can elapse between the time a paper appears and the time an abstract is published. *Current Contents* helps fill the gap by providing much of the information found in an abstract: a list of the titles, authors, and keywords that appear in titles of papers recently published. (In some cases, these may include papers about to be published.)

Science Citation Index

The *Science Citation Index* is unique in that it allows for a type of search not possible with any other index—a search forward in time. For example, if one wanted to learn what recent applications have been made of the coupling reaction first

[10]Weissbach O. *The Beilstein Guide: A Manual for the Use of Beilstein's Handbuch Der Organischen Chemie;* Springer: New York, 1976.
[11]*CrossFire Beilstein* can be found online at http://www.info.crossfirebeilstein.com

reported by Stansbury and Proops in 1962, one would look up their paper [*J. Org. Chem.*, 27, 320 (1962)] in a current volume, say 1998, of *Science Citation Index* and find there a list of those articles published in 1998 in which an author cited the 1962 work of Stansbury and Proops in a footnote. This database is available by subscription on the Internet.

PLANNING A SYNTHESIS

In planning a synthesis, the organic chemist is faced with at least three overlapping considerations: the chemical reactions, the reagents, and the experimental procedure to be employed. For students, a good place to start might be an advanced textbook such as *Advanced Organic Chemistry*[12] or *March's Advanced Organic Chemistry*,[13] both containing literature references. An excellent one-volume reference is *Comprehensive Organic Transformations: A Guide to Functional Group Preparation*.[14] Often, a good lead is obtained through named reactions, for which *Name Reactions and Reagents in Organic Chemistry*[15] is a useful reference.

It is instructive for the beginning synthetic chemist to read about some elegant and classical syntheses that have been carried out in the past. For this purpose, see *Art in Organic Synthesis*[16] for some of the most elegant syntheses carried out. Similarly, *Creativity in Organic Synthesis*[17] and *Selected Organic Syntheses*[18] are compendia of elegant syntheses. At a more advanced level, see *The Logic of Chemical Synthesis*.[19] For natural product synthesis, see *The Total Synthesis of Natural Products*,[20] an excellent series in nine volumes, with the most recent published in 1999. *Modern Synthetic Reactions*[21] is a more comprehensive source of information with several thousand references to the original literature. It has been brought further up-to-date by the modern synthetic methods that appear in *Some Modern Methods of Organic Synthesis*.[22]

An extremely useful reference is the *Compendium of Organic Synthetic Methods*,[23] presently in twelve volumes, the most recent published in 2009. The material is organized by reaction type. In addition, *Organic Reactions*,[24] published annually, should be consulted. Over 100 preparative reactions, with examples of experimental details, are covered in great detail in 67 volumes. *Annual Reports in Organic Synthesis*[25] is an excellent review of new reactions in a given year, organized by reaction type. *Stereoselective Synthesis*[26] treats one of the newest and most important areas of organic chemistry.

[12]Carey FJ, Sundberg RJ. *Advanced Organic Chemistry: Structure and Mechanisms* (part A) and *Reactions and Synthesis* (part B), 5th ed; Plenum: New York, 2007.
[13]Smith MB, March J. *March's Advanced Organic Chemistry: Reactions, Mechanisms, and Structure*, 6th ed; Wiley-Interscience: New York, 2007.
[14]Larock RC. *Comprehensive Organic Transformations: A Guide to Functional Group Preparation*, John Wiley & Sons, New York 1999.
[15]Mundy BP, Ellerd MG, Favaloro FG. *Name Reactions and Reagents in Organic Chemistry*, 2nd ed; Wiley-Interscience: New York, 2005.
[16]Anand N, Bindra JS, Randanathan S. *Art in Organic Synthesis*, 2nd ed; Wiley-Interscience: New York, 1988.
[17]Bindra JS. *Creativity in Organic Synthesis*; Academic Press: San Diego, 1975.
[18]Fleming I. *Selected Organic Syntheses: A Guide for Organic Chemists*; University Microfilms International: Ann Arbor, MI, 1991.
[19]Corey EJ, Cheng XM. *The Logic of Chemical Synthesis*, 1st ed; Wiley-Interscience: New York, 1995.
[20]ApSimon J. *The Total Synthesis of Natural Products*; Wiley-Interscience: New York, 1973–1999.
[21]House HO. *Modern Synthetic Reactions*, 2nd ed; Benjamin-Cummings: San Francisco, 1972.
[22]Carruthers W, Coldham I. *Some Modern Methods of Organic Synthesis*, 4th ed; Cambridge University Press: Oxford, UK, 2004.
[23]Smith MB. *Compendium of Organic Synthetic Methods*; John Wiley and Sons: New York, 1995; e-book, 2004.
[24]Paquette LA. *Organic Reactions*, Vols 1–67; Wiley: New York.
[25]*Annual Reports in Organic Synthesis*; Academic Press: San Diego; Volume 100 was published in 2004.
[26]Atkinson RS. *Stereoselective Synthesis*, 1st ed; John Wiley and Sons: New York, 1995.

The late Mary Fieser was responsible for publishing the first 17 volumes of *Fiesers' Reagents for Organic Synthesis*.[27] The last three volumes have been edited by Tse-Lok Ho.[28] Begun in 1967, this series critically surveys the reagents employed to carry out organic synthesis. Included are references to the original literature and suppliers of reagents, and to *Organic Syntheses* (see below), as well as the critical reviews that have been written about various reagents. The index of reagents according to type of reaction is very useful when planning a synthesis. On a much larger scale is the comprehensive *Encyclopedia of Reagents for Organic Synthesis*,[29] which gives a critical assessment of most of the reagents one would use to carry out a synthesis.

To carry out a reaction on one functional group without affecting another functional group, it is often necessary to put on a protective group. This rather specialized subject is admirably covered in *Protective Groups in Organic Synthesis*.[30]

Before carrying out the synthesis itself, one should consult *Organic Syntheses*,[31] an annual series published since 1921 that is grouped in nine collective volumes with reaction and reagent indexes. *Organic Syntheses* gives detailed laboratory procedures for carrying out more than 1000 different reactions. Each synthesis is submitted for review, and then the procedure is sent to an independent laboratory for checking and verification. The reactions, unlike many that are reported in the primary literature, are carried out a number of times and can therefore be relied on to work. This valuable series is now available at no cost on the Internet.

Laboratory techniques are covered in *Techniques of Organic Chemistry*,[32] an uneven, multiauthored compendium that is outdated yet unique in many respects. An excellent advanced text on the detailed mechanisms of many representative reactions is *Mechanism and Theory in Organic Chemistry*.[33]

Methoden der Organischen Chemie (*Methods of Organic Chemistry*,[34] often called Houben-Weyl) is a multivolume work, published in English since 1990. Detailed experimental procedures are presented for the syntheses of thousands of compounds, making this a valuable resource. *Organic Reactions*,[35]as the name implies, has long chapters or even an entire volume devoted to one reaction. Each article of this 59-volume series is written by an authority.

Although the journals may disappear from the shelves, books, monographs, and treatises, including many of the secondary references, will undoubtedly remain in the chemistry libraries of the future.

[27]Fieser, M. *Fiesers' Reagents for Organic Synthesis*, Vols 1–17; Wiley-Interscience: New York.
[28]Ho, Tse-Lok. *Fiesers' Reagents for Organic Synthesis*, Vols 18–25 (Vol 25, 2009); Wiley-Interscience: New York.
[29]Paquette LA. *Encyclopedia of Reagents for Organic Synthesis*, Vols 1–8; John Wiley and Sons: New York, 1995.
[30]Greene TW, Wuts PGM. *Protective Groups in Organic Synthesis*, 3rd ed; Wiley-Interscience: New York, 1999.
[31]*Organic Syntheses* is a publication of the American Chemical Society. All past and present procedures are available online at http://www.orgsyn.org.
[32]Weissberger A, Ed. *Techniques of Organic Chemistry*, Vols 1–14; Wiley-Interscience: New York; 1955.
[33]Lowry TH, Richardson KS. *Mechanism and Theory in Organic Chemistry*, 3rd ed; Benjamin Cummings: San Francisco, 1987.
[34]Houben J, Weyl T. *Methoden der Organischen Chemie* (*Methods of Organic Chemistry*), 4th ed; Georg Thieme: Stuttgart, Germany, 1999.
[35]*See* footnote 24.

Index